Applied Calculus

For Business, Economics, and the Social and Life Sciences

EXPANDED
Eleventh Edition

Applied Calculus

For Business, Economics, and the Social and Life Sciences

Laurence Hoffmann
Morgan Stanley Smith Barney

Gerald Bradley
Claremont McKenna College

Dave Sobecki
Miami University of Ohio

Michael Price
University of Oregon

The McGraw-Hill Companies

Connect
Learn
Succeed™

APPLIED CALCULUS FOR BUSINESS, ECONOMICS, AND THE SOCIAL AND LIFE SCIENCES:
EXPANDED EDITION, ELEVENTH EDITION

Published by McGraw-Hill, a business unit of The McGraw-Hill Companies, Inc., 1221 Avenue of the Americas,
New York, NY 10020. Copyright © 2013 by The McGraw-Hill Companies, Inc. All rights reserved. Printed in
the United States of America. Previous editions © 2010, 2007, and 2005. No part of this publication may be
reproduced or distributed in any form or by any means, or stored in a database or retrieval system, without the
prior written consent of The McGraw-Hill Companies, Inc., including, but not limited to, in any network or
other electronic storage or transmission, or broadcast for distance learning.

Some ancillaries, including electronic and print components, may not be available to customers outside the
United States.

This book is printed on acid-free paper.

4 5 6 7 8 9 LWI 21 20 19 18 17

ISBN 978–0–07–353237–0
MHID 0–07–353237–1

Vice President, Editor-in-Chief: *Marty Lange*
Vice President, EDP: *Kimberly Meriwether David*
Senior Director of Development: *Kristine Tibbetts*
Editorial Director: *Michael Lange*
Developmental Editor: *Eve L. Lipton*
Marketing Manager: *Alexandra Coleman*
Senior Project Manager: *Vicki Krug*
Senior Buyer: *Kara Kudronowicz*
Lead Media Project Manager: *Judi David*
Senior Designer: *Laurie B. Janssen*
Cover Designer: *Ron Bissell*
Cover Image: *Jillis van Nes, Gettyimages*
Senior Photo Research Coordinator: *Lori Hancock*
Compositor: *Aptara®, Inc.*
Typeface: *10/12 Times*
Printer: *LSC Communications*

All credits appearing on page or at the end of the book are considered to be an extension of the copyright page.

CO 1, CO 2: © Corbis RF; p. 195(right): © Nigel Cattlin/Photo Researchers, Inc.; p. 195(left): Courtesy of Ricardo
Bessin; CO 3: © Getty RF; CO 4: © The McGraw-Hill Companies, Inc./Jill Braaten, photographer; p. 384: ©
Getty RF; CO 5: © Richard Klune/Corbis; p. 489: © Corbis RF; CO 6: BJORNBAKK JAN-MORTEN/AFP/Getty
Images; p. 523: © Alamy RF; CO 7(right): US Geological Survey; CO 7(left): Courtesy of Trails.com CO 7
B:(left): Courtesy of Trails.com; CO 8: © Getty RF; p. 691: © Goodshoot/Fotosearch RF; CO 9: © Mug
Shots/Corbis; p. 769: © Corbis RF; CO 10, p. 820: © Jack Hollingsworth/Getty Images; CO 11: © Corbis RF;
p. 894: Courtesy of Zimmer Inc.; p. 911: © Getty R.

Library of Congress Cataloging-in-Publication Data

Applied calculus for business, economics, and the social and life sciences / Laurence Hoffmann ... [et al.]. —
Expanded 11th ed.
    p. cm.
  Previously entered under: Hoffmann, Laurence.
  Includes index.
  ISBN 978–0–07–353237–0 — ISBN 0–07–353237–1 (hard copy : alk. paper) 1. Calculus—Textbooks.
I. Hoffmann, Laurence D., 1943-
    QA303.2.H64 2013
    515—dc23

                                 2011015843

www.mhhe.com

In memory of our parents
Doris and Banesh Hoffmann
and
Mildred and Gordon Bradley

CONTENTS

(handwritten: 11 as)

CHAPTER 4   Exponential and Logarithmic Functions

(handwritten: Fall 2019)

CHAPTER 5   *(handwritten: 9 cls)* Integration

(handwritten: Fall 2019 Final)

(handwritten: Area Between Two Curves (1)   Average Value (1))

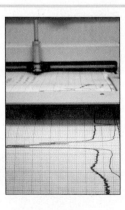

CHAPTER **9** Differential Equations

CHAPTER **10** Infinite Series and Taylor Series Approximations

CHAPTER **11** Probability and Calculus

APPENDIX **A** Algebra Review

PREFACE

Overview of the Eleventh Edition

Applied Calculus for Business, Economics, and the Social and Life Sciences, Expanded Edition, provides a sound, intuitive understanding of the basic concepts students need as they pursue careers in business, economics, and the life and social sciences. Students achieve success using this text as a result of the authors' applied and real-world orientation to concepts, problem-solving approach, straightforward and concise writing style, and comprehensive exercise sets. More than 100,000 students worldwide have studied from this text!

Applied Calculus for Business, Economics, and the Social and Life Sciences, Expanded Eleventh Edition, contains all of the material present in the Brief Edition of *Calculus for Business, Economics, and the Social and Life Sciences,* plus four additional chapters covering Trigonometric Functions, Differential Equations, Infinite Series and Taylor Series Approximations, and Probability and Calculus.

Improvements to This Edition

Revised Content

Every section in the text underwent careful analysis and extensive review to ensure the most beneficial and clear presentation. Additional steps and definition boxes were added when necessary for greater clarity and precision, graphs and figures were revised as necessary, and discussions and introductions were added or rewritten as needed to improve presentation.

Enhanced Topic Coverage

As technology advances, trigonometric functions are becoming more and more important for all applied topics. For this reason, the chapter on Trigonometric Functions has been moved from Chapter 11 to Chapter 8. This allows classes who cover this topic to do so earlier in the term, following the syllabi of most applied calculus classes and making it clear that trigonometric functions are important for future study. Trig functions are identified and integrated throughout the later chapters, allowing those professors who cover trig functions to assign items that relate to what they have covered while making it simple for students who have not seen trigonometric functions before to skip this material.

Material on the Extreme Value Property for functions of two variables and finding extreme values on closed, bounded regions has been added to Section 7.3. This completes the analogy with the one-variable case and better prepares students for future study of statistics and finite mathematics.

Improved Exercise Sets

Almost 250 new routine and application exercises have been added to the already extensive problem sets. A wealth of new applied problems has been added to help demonstrate the practicality of the material, and existing applications have been updated. Exercise sets have been rearranged so that applications are categorized by topic (business/economics, life and social sciences, and miscellaneous).

New Pedagogical Design Elements

Titles have been added to each example in the text, and learning objectives have been specified at the beginning of each section. Example titles allow both students and instructors to quickly find items of interest to them. These pedagogical improvements make the topics clear and comprehensible for all students, help to organize ideas, and aid both students and professors with review and evaluation.

Online Matrix Supplement

The authors have fully revised the matrix supplement. Problems and examples have been revised and updated to include more contemporary applications. The revised supplement in PDF format is posted online for instructors to download at www.mhhe.com/hoffmann.

Chapter-by-Chapter Changes

- Titles have been added to all worked examples throughout the book.
- A list of learning objectives has been added at the beginning of every section.
- End-of-section exercises have been grouped according to subject.

Chapter 1

- New applied exercises have been added to Sections 1.1 through 1.5.
- Material in Section 1.2 on the rectangular coordinate system, the distance formula, intercepts, and quadratic functions has been added and rewritten. New and revised examples support these changes.
- In Section 1.4, the coverage of modeling has been revised and includes both new and revised examples.
- New notes, modified language, and new and revised examples in Section 1.5 help to clarify the topics of limits and infinity.
- A new example on break-even analysis has been added to Section 1.6.
- New Just-In-Time Reviews have been added to Sections 1.2 and 1.5.

Chapter 2

- The boxes for the constant multiple rule and sum rule have been rewritten to include the prime notation versions of the rules.
- Many introductions have been rewritten with an eye toward achieving better focus on describing a concept with greater clarity.
- Ten new exercises have been added to Section 2.3 (Exercises 36 through 39) and 2.4 (Exercises 61 through 64, 89, and 90).
- A new example using the chain rule twice has been added to Section 2.4.
- Section 2.5 includes a new introduction to marginal cost with a new example illustrating marginal cost and revenue. New exercises on marginal cost and revenue have also been added.
- A new introduction to implicit differentiation has been added, and there is a new Just-In-Time Review on related rates.

Chapter 3

- A new introductory example for increasing and decreasing functions has been added.
- There is a new discussion of worker efficiency and point of diminishing returns.
- The discussion and definition of inflection points and the box summarizing curve sketching with the second derivative have been modified.
- New exercises have been added in Sections 3.2 and 3.4.
- The material on price elasticity of demand has been completely rewritten.
- The chapter summary has been modified.

Chapter 4

- Boxes on the present and future values of an investment have been updated.
- New exercises on investment have been added in Section 4.1 and on elasticity of demand in Section 4.3.
- A new example with a new demand function has been added to Section 4.3.

Chapter 5

- In Section 5.1, there is a new introduction to differential equations and a new example on continuous compounding.
- There are 20 new exercises in Sections 5.1 and 5.3 (a total of 35 new exercises have been added throughout Chapter 5).
- A new subsection on the price-adjustment model in economics has been moved from Section 6.2 to Section 5.2, and new examples on price adjustment and a separable differential equation using substitution have been added to Section 5.2.
- There is a new introduction to Section 5.5.
- The table of Gini indices for various countries has been updated.
- The subsections on Consumer Willingness to Spend and Consumers' Surplus have been completely rewritten.

Chapter 6

- Old examples have been deleted from Section 6.1 in favor of a new applied example using the integral table to solve a logistic equation.
- Twenty-seven new exercises have been added to Chapter 6.
- A new introduction to improper integrals, new discussion and summary boxes for improper integrals involving $-\infty$, and a new example are included in Section 6.3.

Chapter 7

- Twenty-six new exercises have been added to Chapter 7.
- Data in Section 7.4 have been updated.
- Section 7.3 has been substantially revised. There is a new introduction to practical optimization, and a subsection involving optimization on a closed, bounded region (the extreme value property) has been added. This material helps students see how one- and two-dimensional optimization problems are related.
- A new subsection on finding population from population density has been added to Section 7.6.

Chapter 8

- This is a revision of the old Chapter 11.
- Sections 8.2 and 8.3 involve a complete reorganization of old Sections 11.2 and 11.3 and the Chapter Summary has been reorganized.
- The chapter introduction is new.
- In Section 8.1, there is a revised table and description of modeling periodic phenomena.
- New applied examples have been added to Sections 8.2 and 8.3.
- Forty new exercises have been added to Chapter 8.

Chapter 9

- This is a revision of the old Chapter 8.
- New subsections on modeling with differential equations and population models, and new examples on learning models, inventory models, and logistic growth have been added.
- There is a new introduction to compartmental analysis.
- There are 40 new exercises in Chapter 9.

Chapter 10

- This is a revision of the old Chapter 9.
- There is a new Just-In-Time Review, a new summary statement regarding the Taylor series, and a new note on using the Taylor series to evaluate the integral of a normal probability density function.
- New exercises have been added to Section 10.3 and the Chapter 10 Review Exercises to reflect this new material.

Chapter 11

- This is a revision of the old Chapter 10.
- Section 11.1, the old Section 10.1, has been extensively revised with an emphasis on probability distributions.
- There is a new introduction to continuous random variables and probability distributions in Section 11.2.
- Several new routine and applied exercises have been added to Sections 11.1 and 11.2.
- There is a new interpretation of variance in Section 11.3.

KEY FEATURES OF THIS TEXT

Learning Objectives

Each section begins with a list of objectives for that section. In addition to preparing students for what they will learn, these help students organize information for study and review and make connections between topics.

Applications

Throughout the text great effort is made to ensure that topics are applied to practical problems soon after their introduction, providing methods for dealing with both routine computations and applied problems. These problem-solving methods and strategies are introduced in applied examples and practiced throughout in the exercise sets.

EXPLORE!

Refer to Example 3.5.4. Store
$$P(x) = \frac{75}{x} + \frac{300}{15 - x}$$
into Y1, and graph using the modified decimal window [0, 14]1 by [0, 350]10. Now use **TRACE** to move the cursor from X = 1 to 14 and confirm the location of minimal pollution. To view the behavior of the derivative P'(x), enter Y2 = nDeriv(Y1, X, X) and graph using the window [0, 14]1 by [−75, 300]10. What do you observe?

EXAMPLE 3.5.4 Finding a Location That Minimizes Pollution

Two industrial plants, A and B, are located 15 miles apart and emit 75 ppm (parts per million) and 300 ppm of particulate matter, respectively. Each plant is surrounded by a restricted area of radius 1 mile in which no housing is allowed, and the concentration of pollutant arriving at any other point Q from each plant decreases with the reciprocal of the distance between that plant and Q. Where should a house be located on a road joining the two plants to minimize the total pollution arriving from both plants?

Solution

Suppose a house H is located x miles from plant A and, hence, $15 - x$ miles from plant B, where x satisfies $1 \le x \le 14$ since there is a 1-mile restricted area around each plant (Figure 3.49). Since the concentration of particulate matter arriving at H from each plant decreases with the reciprocal of the distance from the plant to H, the concentration of pollutant from plant A is $\frac{75}{x}$ and from plant B is $\frac{300}{15 - x}$. Thus, the total concentration of particulate matter arriving at H is given by the function

$$P(x) = \underbrace{\frac{75}{x}}_{\substack{\text{pollution} \\ \text{from } A}} + \underbrace{\frac{300}{15 - x}}_{\substack{\text{pollution} \\ \text{from } B}}$$

"The example titles are a really excellent idea. From the student's perspective, they stimulate interest and get the students to read examples that interest them. From the instructor's perspective, they allow the instructor to make a decision about what to include without spending a lot of preparation time reading each example."

—*Jay Zimmerman, Towson University*

Procedural Examples and Boxes

Each new topic is approached with careful clarity by providing step-by-step problem-solving techniques through frequent procedural examples and summary boxes.

A General Procedure for Sketching the Graph of f(x)

Step 1. Find the domain of $f(x)$ [that is, where $f(x)$ is defined].

Step 2. Find and plot all intercepts. The y intercept (where $x = 0$) is usually easy to find, but the x intercepts [where $f(x) = 0$] may require a calculator.

Step 3. Determine all vertical and horizontal asymptotes of the graph. Draw the asymptotes in a coordinate plane.

Step 4. Find $f'(x)$, and use it to determine the critical numbers of $f(x)$ and intervals of increase and decrease.

Step 5. Determine all relative extrema (both coordinates). Plot each relative maximum with a cap ($\frown$) and each relative minimum with a cup ($\smile$).

Step 6. Find $f''(x)$, and use it to determine intervals of concavity and points of inflection. Plot each inflection point with a "twist" to suggest the shape of the graph near the point.

Step 7. You now have a preliminary graph, with asymptotes in place, intercepts plotted, arrows indicating the direction of the graph, and caps, cups, and twists suggesting the shape at key points. Plot additional points if needed, and complete the sketch by joining the plotted points in the directions indicated. Be sure to remember that the graph cannot cross a vertical asymptote.

Relative and Percentage Rates of Change ■ The **relative rate of change** of a quantity $Q(x)$ with respect to x is given by the ratio

$$\frac{\text{Relative rate of}}{\text{change of } Q(x)} = \frac{Q'(x)}{Q(x)}$$

The corresponding **percentage rate of change** of $Q(x)$ with respect to x is

$$\frac{\text{Percentage rate of}}{\text{change of } Q(x)} = \frac{100\,Q'(x)}{Q(x)}$$

Definitions

Definitions and key concepts are set off in shaded boxes to provide easy referencing for the student.

Just-In-Time Reviews

These references, located in the margins, are used to quickly remind students of important concepts from college algebra or precalculus as they are being used in examples and review.

Just-In-Time REVIEW

In Example 1.5.6, we perform the multiplication
$$(\sqrt{x} - 1)(\sqrt{x} + 1) = x - 1$$
using the identity
$$(a - b)(a + b) = a^2 - b^2$$
with $\sqrt{x}$ corresponding to a and 1 corresponding to b.

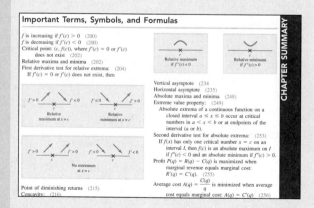

Exercise Sets

Almost 250 new problems have been added to increase the effectiveness of the highly praised exercise sets. Routine problems have been added where needed to ensure students have enough practice to master basic skills, and a variety of applied problems have been added to help demonstrate the practicality of the material.

Writing Exercises

These problems, designated by writing icons, challenge a student's critical thinking skills and invite students to research topics on their own.

Calculator Exercises

Calculator icons designate problems within each section that can only be completed with a graphing calculator.

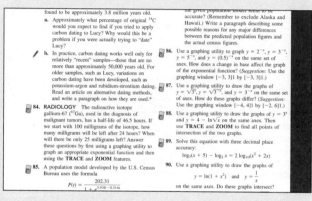

"[Hoffmann-Bradley] has excellent application problems in the social science, life science, economics, and finance fields."

—*Rebecca Leefers, Michigan State University–East Lansing*

Chapter Review

Chapter Review material aids the student in synthesizing the important concepts discussed within the chapter, including a master list of key technical terms and formulas introduced in the chapter.

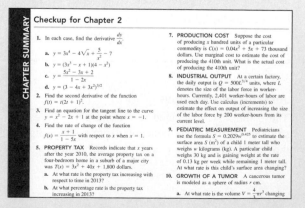

Chapter Checkup

Chapter Checkups provide a quick quiz for students to test their understanding of the concepts introduced in the chapter.

Review Exercises

In Exercises 1 through 12, find the general solution of the given differential equation.

1. $\dfrac{dy}{dx} = x^3 - 3x^2 + 5$

2. $\dfrac{dy}{dx} = 0.02y$

3. $\dfrac{dy}{dx} = k(80 - y)$

4. $\dfrac{dy}{dx} = y(1 - y)$

5. $\dfrac{dy}{dx} = e^{x+y}$ [*Hint:* $e^{x+y} = e^x e^y$]

6. $\dfrac{dy}{dx} = ye^{-2x}$

7. $\dfrac{dy}{dx} + \dfrac{4y}{x} = e^{-x}$

8. $\dfrac{dy}{dx} + \dfrac{y}{x} = \dfrac{2}{x+1}$

9. $\dfrac{dy}{dx} + \dfrac{2y}{x+1} = \dfrac{4}{x+2}$

10. $x^3 \dfrac{dy}{dx} + xy = 5$

11. $\dfrac{dy}{dx} = \dfrac{\ln x}{xy}$

18. $\dfrac{dy}{dx} + \dfrac{y}{x+1} = x;\ y = 0$ when $x = 3$

19. $\dfrac{dy}{dx} - xy = e^{x^2/2};\ y = 4$ when $x = 0$

20. $\dfrac{dy}{dx} + 3x^2y = x^2;\ y = 2$ when $x = 0$

21. $\dfrac{dy}{dx} = y \ln \sqrt{x};\ y = -4$ when $x = 1$

22. $\dfrac{dy}{dx} = \dfrac{\ln x}{y};\ y = 100$ when $x = 1$

23. $\dfrac{dy}{dx} = \dfrac{xy}{\sqrt{1 - x^2}};\ y = 2$ when $x = 0$

24. $\dfrac{d^2y}{dx^2} = 2;\ y = 5$ and $\dfrac{dy}{dx} = 3$ when $x = 0$

Review Problems

A wealth of additional routine and applied problems is provided within the end-of-chapter exercise sets, offering further opportunities for practice.

Explore! Technology

Utilizing the graphing calculator, Explore Boxes challenge a student's understanding of the topics presented with explorations tied to specific examples. Explore! Updates provide solutions and hints to selected boxes throughout the chapter.

EXPLORE! UPDATE

Complete solutions for all EXPLORE! boxes throughout the text can be accessed at the book-specific website, www.mhhe.com/hoffmann.

Solution for Explore! on Page 377

Store the constants $\{-4, -2, 2, 4\}$ into L1, and write Y1 = X^3 and Y2 = Y1 + L1. Graph Y1 in bold, using the modified decimal window $[-4.7, 4.7]1$ by $[-6, 6]1$. At $x = 1$ (where we have drawn a vertical line), the slopes for each curve appear equal.

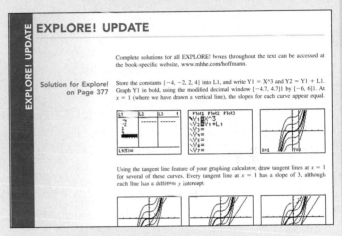

Using the tangent line feature of your graphing calculator, draw tangent lines at $x = 1$ for several of these curves. Every tangent line at $x = 1$ has a slope of 3, although each line has a different y intercept.

"The book as a whole is one of the best calculus books I have used.... I really like how calculators are included on every section and that at the end of the chapter there is opportunity for students to explore the calculators even more."

—*Joseph Oakes, Indiana University Southeast*

THINK ABOUT IT

ALLOMETRIC MODELS

When developing a mathematical model, the first task is to identify quantities of interest, and the next is to find equations that express relationships between these quantities. Such equations can be quite complicated, but there are many important relationships that can be expressed in the relatively simple form $y = Cx^k$, in which one quantity y is expressed as a constant multiple of a power function of another quantity x.

In biology, the study of the relative growth rates of various parts of an organism is called *allometry*, from the Greek words *allo* (other or different) and *metry* (measure). In allometric models, equations of the form $y = Cx^k$ are often used to describe the relationship between two biological measurements. For example, the size a of the antlers of an elk from tip to tip has been shown to be related to h, the shoulder height of the elk, by the allometric equation

Think About It Essays

The modeling-based Think About It essays show students how material introduced in the chapter can be used to construct useful mathematical models while explaining the modeling process and providing an excellent starting point for projects or group discussions.

Supplements

Student's Solutions Manual

The *Student's Solutions Manual* contains comprehensive, worked-out solutions for all odd-numbered problems in the text with the exception of the Checkup section for which solutions to all problems are provided. Detailed calculator instructions and keystrokes are also included for problems marked by the calculator icon. Written by an instructor with years of classroom experience, it guides professors to demonstrate solutions in a manner consistent with the methods used throughout the text.
ISBN–13: 978-0-07-742746-7 (ISBN–10: 0-07-742746-7)

Instructor's Solutions Manual

The *Instructor's Solutions Manual* contains comprehensive, worked-out solutions for all problems in the text and is available on the book's website, www.mhhe.com/hoffmann.

Computerized Test Bank

Brownstone Diploma testing software, available on the book's website, offers instructors a quick and easy way to create customized exams and view student results. The software utilizes an electronic test bank of short answer, multiple choice, and true/false questions tied directly to the text, with many new questions added for the Eleventh Edition. Sample chapter tests and final exams in Microsoft Word and PDF formats are also provided.

Connect www.mcgraw-hillconnect.com   connect

McGraw-Hill Connect® is a complete online system for mathematics. A variety of study tools are available at any time, including videos, applets, worked examples, algorithmic exercises, and lecture capture software, offering unlimited practice and accommodating many different learning styles. An answer palette is available for many exercises, allowing the entry of mathematical symbols without struggling with keyboard commands. Students can also benefit from immediate feedback like guided solutions, access to the textbook online, and checking an answer, all directly accessible right from a Connect assignment. Connect can be used on Macs and PCs as well as on several different browsers, making it flexible to meet a wide range of student needs. Connect offers full Blackboard integration, making electronic record keeping even easier.

For more information, visit the book's website (www.mhhe.com/hoffmann) or contact your local McGraw-Hill sales representative (www.mhhe.com/rep).

Electronic Textbooks
McGraw-Hill ConnectPlus®   connect (plus)

ConnectPlus provides students with all the advantages of **Connect,** plus 24/7 access to a media-rich, interactive eBook. Included is a powerful suite of built-in tools that allow detailed searching, highlighting, note taking, and student-to-student or instructor-to-student note sharing. In addition, ConnectPlus integrates relevant animations, interactives, and videos into the textbook content for a true multimedia learning experience.

McGraw-Hill Create™ create

Craft your teaching resources to match the way you teach. With Create, at www.mcgrawhillcreate.com, you can easily rearrange chapters, combine material from other content sources, and quickly upload content you have written like your course syllabus or teaching notes. Find the content you need in Create by searching through thousands of leading McGraw-Hill textbooks. Arrange your book to fit your teaching style. Create even allows you to personalize your book's appearance by selecting the cover and adding your name, school, and course information. Order a Create book and you'll receive a complimentary print review copy in 3 to 5 business days or a complimentary electronic review copy (eComp) via e-mail in about one hour. Go to www.mcgrawhillcreate.com today and register. Experience how Create empowers you to teach *your* students *your* way.

CourseSmart

CourseSmart is a new way for faculty to find and review e-textbooks. With the CourseSmart eTextbook version of *Applied Calculus for Business, Economics, and the Social and Life Sciences,* Eleventh Edition, students can save up to 50 percent off the cost of a print book, reduce their impact on the environment, and access powerful Web tools for learning, including full text search, notes, highlighting, and e-mail tools for sharing notes between classmates. Faculty can also review and compare the full text online without having to wait for a print desk copy. CourseSmart is an online e-textbook, which means users need to be connected to the Internet to access it. Students can also print sections of the book for maximum portability.

ALEKS®

ALEKS (**A**ssessment and **LE**arning in **K**nowledge **S**paces) is a unique, online program that dramatically raises student proficiency and success rates in mathematics, while reducing faculty workload and office-hour lines. ALEKS uses artificial intelligence and adaptive, open-response questioning to assess precisely a student's preparedness and provide personalized instruction on the exact topics the student is most ready to learn. As a result, ALEKS interacts with students much like a skilled human tutor to improve student confidence and performance. Additionally, ALEKS includes robust course management tools that allow instructors to spend less time on administrative tasks and more time directing student learning.

ALEKS Prep for Calculus

ALEKS Prep for Calculus focuses on prerequisite and introductory material for Calculus, and can be used during the first six weeks of the term to prepare students for success in the course. Through comprehensive explanations, practice, and feedback, ALEKS enables students to quickly fill gaps in prerequisite knowledge. As a result, instructors can focus on core course concepts instead of review material, and see fewer drops from the course.

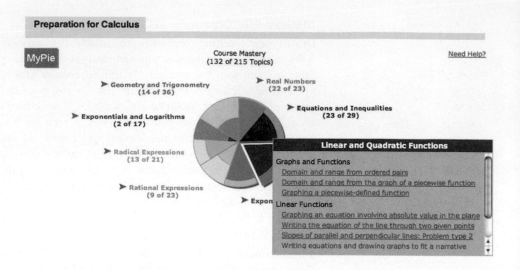

ALEKS Prep for Calculus Features:

- **Artificial Intelligence:** Targets gaps in student knowledge
- **Individualized Assessment and Learning:** Ensure student mastery
- **Dynamic, Automated Reports:** Monitor student and class progress
- **Adaptive, Open-Response Environment:** Avoids multiple-choice

For more information about ALEKS, please visit: www.aleks.com/highered/math.

Acknowledgments

As in past editions, we have enlisted the feedback of professors teaching from our text as well as those using other texts to point out possible areas for improvement. Our reviewers provided a wealth of detailed information on both our content and the changing needs of their course, and many changes we've made were a direct result of consensus among these review panels. This text owes its considerable success to their valuable contributions, and we thank every individual involved in this process.

James N. Adair, *Missouri Valley College*

Wendy Ahrendsen, *South Dakota State University*

Faiz Al-Rubaee, *University of North Florida*

George Anastassiou, *University of Memphis*

Dan Anderson, *University of Iowa*

Randy Anderson, *Craig School of Business*

John Avioli, *Christopher Newport University*

Christina Bacuta, *University of Delaware*

Ratan Barua, *Miami Dade College*

John Beachy, *Northern Illinois University*

Jay H. Beder, *University of Wisconsin—Milwaukee*

Denis Bell, *University of North Florida*

Don Bensy, *Suffolk County Community College*

Adel Boules, *University of North Florida*

Neal Brand, *University of North Texas*

Lori Braselton, *Georgia Southern University*

Randall Brian, *Vincennes University*

Paul W. Britt, *Louisiana State University—Baton Rouge*

Albert Bronstein, *Purdue University*

James F. Brooks, *Eastern Kentucky University*

Beverly Broomell, *State University of New York—Suffolk*

Roxanne Byrne, *University of Colorado at Denver*

Laura Cameron, *University of New Mexico*

Rick Carey, *University of Kentucky*

Debra Carney, *University of Denver*

Jamylle Carter, *Diablo Valley College*

Steven Castillo, *Los Angeles Valley College*

Rose Marie Castner, *Canisius College*

Deanna Caveny, *College of Charleston*

Gerald R. Chachere, *Howard University*

Terry Cheng, *Irvine Valley College*

William Chin, *DePaul University*

Lynn Cleaveland, *University of Arkansas*

Dominic Clemence, *North Carolina Agricultural and Technical State University*

Charles C. Clever, *South Dakota State University*

Allan Cochran, *University of Arkansas*

Flavia Colonna, *George Mason University*

Peter Colwell, *Iowa State University*

Cecil Coone, *Southwest Tennessee Community College*

Charles Brian Crane, *Emory University*

Daniel Curtin, *Northern Kentucky University*

Raul Curto, *University of Iowa*

Jean F. Davis, *Texas State University—San Marcos*

John Davis, *Baylor University*

Shirley Davis, *South Plains College*

Yulia Dementieva, *Emmanuel College*

Karahi Dints, *Northern Illinois University*

Ken Dodaro, *Florida State University*

Eugene Don, *Queens College*

Dora Douglas, *Wright State University*

Peter Dragnev, *Indiana University–Purdue University, Fort Wayne*

Bruce Edwards, *University of Florida*

Margaret Ehrlich, *Georgia State University*

Maurice Ekwo, *Texas Southern University*

George Evanovich, *St. Peter's College*

Haitao Fan, *Georgetown University*

Brad Feldser, *Kennesaw State University*

Klaus Fischer, *George Mason University*

Guy Forrest, *Louisiana State University—Baton Rouge*

Michael Freeze, *University of North Carolina—Wilmington*

Constantine Georgakis, *DePaul University*

Sudhir Goel, *Valdosta State University*

Hurlee Gonchigdanzan, *University of Wisconsin—Stevens Point*

Ronnie Goolsby, *Winthrop College*

Lauren Gordon, *Bucknell University*

Michael Grady, *Loyola Marymount University*

Angela Grant, *University of Memphis*

John Gresser, *Bowling Green State University*

Murli Gupta, *George Washington University*

James Hager, *Pennsylvania State University*

Doug Hardin, *Vanderbilt University*

Marc Harper, *University of Illinois at Urbana—Champaign*

Sheyleah V. Harris-Plant, *South Plains College*

Jonathan Hatch, *University of Delaware*
John B. Hawkins, *Georgia Southern University*
Damon Hay, *University of North Florida*
Celeste Hernandez, *Richland College*
William Hintzman, *San Diego State University*
Frederick Hoffman, *Florida Atlantic University*
Matthew Hudock, *St. Philips College*
Joel W. Irish, *University of Southern Maine*
Zonair Issac, *Vanderbilt University*
Erica Jen, *University of Southern California*
Jun Ji, *Kennesaw State University*
Shafiu Jibrin, *Northern Arizona University*
Victor Kaftal, *University of Cincinnati*
Sheldon Kamienny, *University of Southern California*
Georgia Katsis, *DePaul University*
Victoria Kauffman, *University of New Mexico*
Fritz Keinert, *Iowa State University*
Melvin Kiernan, *St. Peter's College*
Marko Kranjc, *Western Illinois University*
Donna Krichiver, *Johnson County Community College*
Harvey Lambert, *University of Nevada*
Kamila Larripa, *Humboldt State University*
Donald R. LaTorre, *Clemson University*
Melvin Lax, *California State University, Long Beach*
Rebecca Leefers, *Michigan State University*
Steffen Lempp, *University of Wisconsin—Madison*
Robert Lewis, *El Camino College*
Shlomo Libeskind, *University of Oregon*
W. Conway Link, *Louisiana State University—Shreveport*
James Liu, *James Madison University*
Yingjie Liu, *University of Illinois at Chicago*
Bin Lu, *California State University—Sacramento*
Jeanette Martin, *Washington State University*
James E. McClure, *University of Kentucky*
Mark McCombs, *University of North Carolina*
Ennis McCune, *Stephen F. Austin State University*
Ann B. Megaw, *University of Texas at Austin*
Fabio Milner, *Purdue University*
Kailash Misra, *North Carolina State University*
Mohammad Moazzam, *Salisbury State University*
Rebecca Muller, *Southeastern Louisiana University*
Sanjay Mundkur, *Kennesaw State University*
Kandasamy Muthuvel, *University of Wisconsin—Oshkosh*
Charlie Nazemian, *University of Nevada—Reno*
Karla Neal, *Louisiana State University*
Cornelius Nelan, *Quinnipiac University*
Said Ngobi, *Troy University eCampus*
Devi Nichols, *Purdue University—West Lafayette*

Joseph Oakes, *Indiana University Southeast*
Richard O'Beirne, *George Mason University*
Jaynes Osterberg, *University of Cincinnati*
Ray Otto, *Wright State University*
Hiram Paley, *University of Illinois*
Virginia Parks, *Georgia Perimeter College*
Shahla Peterman, *University of Missouri—St. Louis*
Murray Peterson, *College of Marin*
Lefkios Petevis, *Kirkwood Community College*
Boris Petracovici, *Western Illinois University*
Lia Petracovici, *Western Illinois University*
Cyril Petras, *Lord Fairfax Community College*
Robert E. Plant, II, *South Plains College*
Kimberley Polly, *Indiana University at Bloomington*
Natalie Priebe, *Rensselaer Polytechnic Institute*
Georgia Pyrros, *University of Delaware*
Richard Randell, *University of Iowa*
Mohsen Razzaghi, *Mississippi State University*
Nathan P. Ritchey, *Youngstown State University*
Arthur Rosenthal, *Salem State College*
Judith Ross, *San Diego State University*
Robert Sacker, *University of Southern California*
Katherine Safford, *St. Peter's College*
Mansour Samimi, *Winston-Salem State University*
Ronda Sanders, *University of South Carolina*
Subhash Saxena, *Coastal Carolina University*
Daniel Schaal, *South Dakota State University*
Dolores Schaffner, *University of South Dakota*
Thomas J. Sharp, *West Georgia College*
Robert E. Sharpton, *Miami-Dade Community College*
Anthony Shershin, *Florida International University*
Minna Shore, *University of Florida International University*
Ken Shores, *Arkansas Tech University*
Gordon Shumard, *Kennesaw State University*
Jane E. Sieberth, *Franklin University*
Marlene Sims, *Kennesaw State University*
Brian Smith, *Parkland College*
Nancy Smith, *Kent State University*
Jim Stein, *California State University, Long Beach*
Joseph F. Stokes, *Western Kentucky University*
Keith Stroyan, *University of Iowa*
Hugo Sun, *California State University—Fresno*
Martin Tangora, *University of Illinois at Chicago*
Tuong Ton-That, *University of Iowa*
Lee Topham, *North Harris Community College*
George Trowbridge, *University of New Mexico*
Boris Vainberg, *University of North Carolina at Charlotte*
Nader Vakil, *Western Illinois University*

Dinh Van Huynh, *Ohio University*
Mildred Vernia, *Indiana University Southwest*
Maria Elena Verona, *University of Southern California*
Tilaka N. Vijithakumara, *Illinois State University*
Kimberly Vincent, *Washington State University*
Karen Vorwerk, *Westfield State College*
Charles C. Votaw, *Fort Hays State University*
Hiroko Warshauer, *Southwest Texas State University*
Pam Warton, *Bowling Green State University*

Jonathan Weston-Dawkes, *University of North Carolina*
Donald Wilkin, *University at Albany, SUNY*
Dr. John Woods, *Southwestern Oklahoma State University*
Henry Wyzinski, *Indiana University—Northwest*
Yangbo Ye, *University of Iowa*
Paul Yun, *El Camino College*
Xiao-Dong Zhang, *Florida Atlantic University*
Jay Zimmerman, *Towson University*

Special thanks go to those instrumental in checking each problem and page for accuracy, including Devilyna Nichols, Lucy Mullins, Kurt Norlin, Hal Whipple, and Jaqui Bradley. Special thanks also go to Steffen Lempp and Amadou Gaye for providing specific, detailed suggestions for improvement that were particularly helpful in preparing this Eleventh Edition. Finally, we wish to thank our McGraw-Hill team, Michael Lange, John Osgood, Vicki Krug, Christina Lane, and Eve Lipton for their patience, dedication, and sustaining support.

CHAPTER **1**

Supply and demand determine the price of stock and other commodities.

Functions, Graphs, and Limits

SECTION 1.1   Functions

Learning Objectives

1. Identify the domain of a function, and evaluate a function from an equation.
2. Gain familiarity with piecewise-defined functions.
3. Introduce and illustrate functions used in economics.
4. Form and use composite functions in applied problems.

The word *function* is often used conversationally in connection with the act of playing a role, as seen in the following statements obtained in a Google search for the string "is a function of":

Just-In-Time **REVIEW**

Appendices A.1 and A.2
contain a brief review of
algebraic properties needed
in calculus.

"Intelligence is a function of experience."

"Human population is a function of food supply."

"Freedom is a function of economics."

What these statements have in common is that some quantity or characteristic (intelligence, population, freedom) depends on another (experience, food supply, economics). This is the essence of the mathematical concept of function.

Loosely speaking, a function consists of two sets and a rule that associates elements in one set with elements in the other. For instance, suppose you want to determine the effect of price on the number of iPods that can be sold at that price. To study this relationship, you need to know the acceptable prices, the set of possible sales levels, and a rule for associating each price with a particular sales level. Here is the definition we will use for function.

Function ■   A **function** is a rule that assigns to each object in a set A exactly one object in a set B. The set A is called the **domain** of the function, and the set of assigned objects in B is called the **range.**

For most functions we will consider, the domain and range will be collections of real numbers, and the function itself will be denoted by a letter like f. The value that the function f assigns to a number x in its domain is then denoted by $f(x)$. This is read as "f of x" (*never* as "f times x"). In many cases, we will use a formula like $f(x) = x^2 + 4$ to describe the value of a function.

You can also think of a function as a mapping from numbers in the domain set A to numbers in the range set B (Figure 1.1a) or as a machine that takes a given number from A and converts it into a specific number in B through a process prescribed

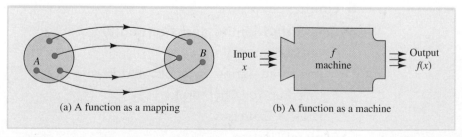

(a) A function as a mapping   (b) A function as a machine

FIGURE 1.1   Interpretations of the function f.

by the functional rule (Figure 1.1b). Thus, the function $f(x) = x^2 + 4$ can be regarded as an f machine that accepts an input x, squares it, and then adds 4 to produce an output $y = x^2 + 4$.

Regardless of how you choose to think of a functional relationship, it is important to remember that *a function assigns one and only one number in the range (output) to each number in the domain (input)*. Example 1.1.1 illustrates the convenience of functional notation.

EXAMPLE 1.1.1 Evaluating a Function

Find and simplify $f(-3)$ if $f(x) = x^2 + 4$.

Solution

We interpret $f(-3)$ to mean "replace all x values in the formula for the function whose name is f with the number -3." Thus, we write

$$f(-3) = (-3)^2 + 4 = 13$$

Note the efficiency of this notation. In Example 1.1.1 the compact formula $f(x) = x^2 + 4$ completely defines the function, and you can indicate that 13 is the unique number the function assigns to -3 by simply writing $f(-3) = 13$.

It is often convenient to represent a functional relationship by an equation $y = f(x)$, and in this context, x and y are called **variables.** In particular, since the numerical value of y is determined by that of x, we refer to y as the **dependent variable** and to x as the **independent variable.** There is nothing sacred about the symbols x and y. For example, the function $y = x^2 + 4$ can just as easily be represented by $s = t^2 + 4$ or by $w = u^2 + 4$. These formulas are equivalent because in each the independent variable is squared and the result is increased by 4 to produce the value for the dependent variable.

Functional notation can also be used to describe tabular data. For instance, Table 1.1 lists the average tuition and fees for private 4-year colleges at 5-year intervals from 1973 to 2008.

TABLE 1.1 Average Tuition and Fees for 4-Year Private Colleges

Academic Year Ending in	Period p	Tuition and Fees T
1973	1	$1,898
1978	2	$2,700
1983	3	$4,639
1988	4	$7,048
1993	5	$10,448
1998	6	$13,785
2003	7	$18,273
2008	8	$25,177

We can describe these data as a function T whose rule is "assign to each value of p the average tuition and fees in dollars, $T(p)$, at the beginning of the pth 5-year period." Thus, $T(1) = \$1{,}898$, $T(2) = \$2{,}700, \ldots , T(8) = \$25{,}177$. Note that in this example we departed from the traditional function named f and independent variable named x. Instead, we chose T to represent the function name because it is suggestive of "tuition" just as using p for the independent variable is suggestive of "period."

In the absence of additional directions or restrictions, we will assume that the domain of a function f is the set of all numbers x for which $f(x)$ is defined. Thus, the domain of the function in Example 1.1.1 is the set of all real numbers since any number x can be squared and added to 4. On the other hand, the college tuition function T illustrated in Table 1.1 has the set of numbers $\{1, 2, \ldots , 8\}$ as its domain since $T(p)$ is given (defined) only for inputs $p = 1, 2, 3, \ldots , 8$. Here is the definition we will follow for the domain convention.

> **Domain Convention** ■ Unless otherwise specified, we assume the domain of a function f to be all real numbers x for which $f(x)$ is defined as a real number. We refer to this as the **natural domain** of f.

Determining the natural domain of a function often amounts to excluding all inputs that result in dividing by 0 or in taking the square root of a negative number, as illustrated in Examples 1.1.2 and 1.1.3.

EXAMPLE 1.1.2 Finding the Domain of a Function

Find the domain of each of the functions.

a. $f(x) = \dfrac{1}{x - 3}$ **b.** $g(t) = \dfrac{\sqrt{3 - 2t}}{t^2 + 4}$

Solution

a. Because division by any number other than 0 is possible, the domain of f is the set of all numbers x such that $x - 3 \neq 0$; that is, $x \neq 3$.

b. The denominator $t^2 + 4$ of $g(t)$ is always positive, so we need not be concerned with dividing by 0. However, all numbers t such that $3 - 2t < 0$ must be excluded from the domain to prevent taking the square root of a negative number. Thus, the domain is the set of all numbers t such that $3 - 2t \geq 0$; that is, $t \leq \dfrac{3}{2}$.

EXAMPLE 1.1.3 Evaluating an Applied Function

A satellite TV company commissions a study that finds the number of customers who can be accommodated each hour by its customer service call center is given by the function $N(w) = 30(w - 1)^{1/2}$, where w is the number of workers at the center. Find $N(5)$, $N(17)$, $N(1)$, and $N(0)$, and interpret your results.

Solution

First, rewrite the function as $N(w) = 30\sqrt{w-1}$. (Fractional exponents are discussed in Appendix A.1 if you need a quick review.) Then

$$N(5) = 30\sqrt{5-1} = 30\sqrt{4} = 30(2) = 60$$
$$N(17) = 30\sqrt{17-1} = 30\sqrt{16} = 30(4) = 120$$
$$N(1) = 30\sqrt{1-1} = 30(0) = 0$$

but $N(0)$ is not defined since $30\sqrt{0-1} = 30\sqrt{-1}$ and negative numbers do not have real square roots.

This tells us that the call center can accommodate 60 callers per hour with 5 workers, 120 callers per hour with 17 workers, and no callers with only 1 worker. It also tells us that 0 workers is not an acceptable input for this function.

Functions are often defined using more than one formula, where each individual formula describes the function on a subset of the domain. A function defined in this way is sometimes called a **piecewise-defined function.** Such functions appear often in business, biology, and physics applications. In Example 1.1.4, we use a piecewise-defined function to describe sales.

EXPLORE!

Create a simple piecewise-defined function using the boolean algebra features of your graphing utility. Store Y1 = 2(X < 1) + (−1)(X ≥ 1) in the function editor. Examine the graph of this function, using the **ZOOM** decimal window. What values does Y1 assume at X = −2, 0, 1, and 3?

EXAMPLE 1.1.4   Evaluating a Piecewise-Defined Function

Suppose we use a function to model the stock price over time of Deckers Outdoor Corporation, the company that produces the popular Ugg boots. While Uggs have been on the market since 1979, during 2003 Ugg sales, and consequently stock values, increased dramatically. It makes sense to use one formula to model stock prices before 2003 and another to model it afterward. Let $S(t)$ represent the stock price of Deckers Outdoor Corporation t years after January 1, 2000. Then

$$S(t) = \begin{cases} 8.1 - 1.7t & \text{if } t < 3 \\ 6t^2 - 36t + 57 & \text{if } t \ge 3 \end{cases}$$

Find and interpret $S(2)$, $S(3)$, and $S(7.5)$.

Solution

Because $t = 2$ satisfies $t < 3$, we use the first formula to calculate the value of the function. Then $S(2) = 8.1 - 1.7(2) = 4.7$. In terms of the model, this means that on January 1, 2002, the share price of Deckers Outdoor Corporation was predicted to be $4.70.

Both $t = 3$ and $t = 7.5$ satisfy $t \ge 3$, so we use the second formula to evaluate $S(3)$ and $S(7.5)$. We find that

$$S(3) = 6(3)^2 - 36(3) + 57 = 3$$

and

$$S(7.5) = 6(7.5)^2 - 36(7.5) + 57 = 124.5$$

Therefore, share prices were predicted to be $3 per share on January 1, 2003, and $124.50 per share on July 1, 2007, the day 7.5 years after January 1, 2000.

Functions Used in Economics

We will study several functions associated with the marketing of a particular commodity.

The **demand function** $D(x)$ for the commodity is the price $p = D(x)$ that must be charged for each unit of the commodity if x units are to be sold (demanded).

The **supply function** $S(x)$ for the commodity is the unit price $p = S(x)$ at which producers are willing to supply x units to the market.

The **revenue** $R(x)$ obtained from selling x units of the commodity is given by the product

$$R(x) = (\text{number of items sold})(\text{price per item})$$
$$= xp(x)$$

The **cost function** $C(x)$ is the cost of producing x units of the commodity.

The **profit function** $P(x)$ is the profit obtained from selling x units of the commodity and is given by the difference

$$P(x) = \text{revenue} - \text{cost}$$
$$= R(x) - C(x) = xp(x) - C(x)$$

The **average cost function** is $AC(x) = \dfrac{C(x)}{x}$. Similarly, the average revenue function $AR(x)$ and average profit function $AP(x)$ are given by

$$AR(x) = \frac{R(x)}{x} \quad \text{and} \quad AP(x) = \frac{P(x)}{x}$$

Generally speaking, the higher the unit price, the fewer the number of units demanded, and vice versa. Similarly, an increase in unit price leads to an increase in the number of units supplied. Thus, demand functions are typically decreasing ("falling" from left to right), while supply functions are increasing ("rising"), as illustrated in the margin. Example 1.1.5 uses several of these special economic functions.

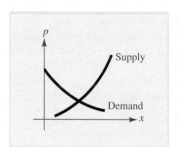

EXAMPLE 1.1.5   Studying a Production Process

Market research indicates that consumers will buy x thousand units of a particular kind of coffee maker when the unit price is

$$p(x) = -0.27x + 51$$

dollars. The cost of producing the x thousand units is

$$C(x) = 2.23x^2 + 3.5x + 85$$

thousand dollars.

  a. What is the average cost of producing 4,000 coffee makers?
  b. How much revenue $R(x)$ and profit $P(x)$ are obtained from producing x thousand units (coffee makers)?
  c. For what values of x is production of the coffee makers profitable?

Solution

a. A production level of 4,000 coffee makers corresponds to $x = 4$ (since x is in thousands of units), and the corresponding average cost is

$$AC(4) = \frac{C(4)}{4} = \frac{2.23(4)^2 + 3.5(4) + 85}{4}$$

$$= \frac{134.68}{4} = 33.67 \text{ thousand dollars per thousand units}$$

So the average cost is $33.67 per coffee maker produced.

b. The revenue is the price $p(x)$ times the number of units x:

$$R(x) = xp(x) = -0.27x^2 + 51x$$

thousand dollars. The profit is the revenue minus the cost:

$$P(x) = R(x) - C(x)$$
$$= -0.27x^2 + 51x - (2.23x^2 + 3.5x + 85)$$
$$= -2.5x^2 + 47.5x - 85$$

thousand dollars.

Just-In-Time **REVIEW**

The product of two numbers is positive if they have the same sign and negative if they have different signs. That is, $ab > 0$ if $a > 0$ and $b > 0$ and also if $a < 0$ and $b < 0$. On the other hand, $ab < 0$ if $a < 0$ and $b > 0$ or if $a > 0$ and $b < 0$.

c. Production is profitable when the profit function has a positive output, that is, when $P(x) > 0$. First, we factor the profit function:

$$P(x) = -2.5x^2 + 47.5x - 85$$
$$= -2.5(x^2 - 19x + 34)$$
$$= -2.5(x - 2)(x - 17)$$

Since -2.5 is negative, the profit $P(x) = -2.5(x - 2)(x - 17)$ is positive only when the product $(x - 2)(x - 17)$ is also negative. This happens when the separate factors $x - 2$ and $x - 17$ have opposite signs. Since there are no x values for which $x - 2 < 0$ and $x - 17 > 0$, we must have $x - 2 > 0$ and $x - 17 < 0$, that is, $2 < x < 17$. So production is profitable when the level of production is between 2,000 and 17,000 units.

Example 1.1.6 provides another illustration of functional notation in a practical situation. Once again, the letters assigned for the function and the independent variable are suggestive of the real quantities they represent.

EXPLORE!

Refer to Example 1.1.6 and store the cost function $C(q)$ into Y1 as

X³ − 30X² + 500X + 200

Construct a **TABLE** of values for $C(q)$ using your calculator, setting TblStart at X = 5 with an increment ΔTbl = 1 unit. On the table of values observe the cost of manufacturing the 10th unit.

EXAMPLE 1.1.6 Evaluating a Cost Function

Suppose the total cost in dollars of manufacturing m treadmills is given by the function $C(m) = m^3 - 30m^2 + 500m + 200$.

a. Find the cost of manufacturing 10 treadmills. What is the average cost of producing these treadmills?

b. Compute the cost of manufacturing the 10th treadmill.

Solution

a. The cost of manufacturing 10 treadmills is the value of the total cost function when $m = 10$; that is,

$$\text{Cost of 10 treadmills} = C(10)$$
$$= (10)^3 - 30(10)^2 + 500(10) + 200$$
$$= 3,200$$

The average cost of producing the 10 treadmills is

$$AC(10) = \frac{C(10)}{10} = \frac{3,200}{10} = 320$$

So the total cost of producing 10 treadmills is \$3,200, and the average cost is \$320 per treadmill.

b. The cost of manufacturing the 10th treadmill is the difference between the cost of manufacturing 10 treadmills and the cost of manufacturing 9 treadmills:

$$\text{Cost of 10th treadmill} = C(10) - C(9) = 3,200 - 2,999 = \$201$$

Composition of Functions

There are many situations in which a quantity is given as a function of one variable that, in turn, can be written as a function of a second variable. By combining the functions in an appropriate way, you can express the original quantity as a function of the second variable. This process is called **composition of functions** or **functional composition.**

For instance, consider a factory that produces GPS units. The number of units produced depends on the amount of material available which, in turn, depends on the amount of capital spent on material. So overall, the production level depends on the amount of capital spent on material. In this sense, production is a composite function of capital expenditure. Here is a definition of functional composition.

Composition of Functions ■ Given functions $f(u)$ and $g(x)$, the composition $f(g(x))$ is the function of x formed by substituting $u = g(x)$ for u in the formula for $f(u)$.

Note that the composite function $f(g(x))$ makes sense only if the domain of f contains the range of g. In Figure 1.2, the definition of composite function is illustrated as an assembly line in which raw input x is first converted into a transitional product $g(x)$ that acts as input the f machine uses to produce $f(g(x))$.

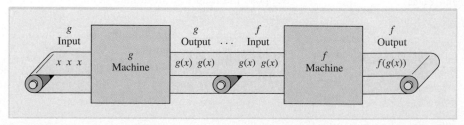

FIGURE 1.2   The composition $f(g(x))$ as an assembly line.

The construction of a composite function is illustrated in Example 1.1.7.

EXAMPLE 1.1.7 Finding a Composite Function

Find the composite function $f(g(x))$, where $f(u) = u^2 + 3u + 1$ and $g(x) = x + 1$.

Solution

Replace u by $x + 1$ in the formula for $f(u)$ to get

$$f(g(x)) = (x + 1)^2 + 3(x + 1) + 1$$
$$= (x^2 + 2x + 1) + (3x + 3) + 1$$
$$= x^2 + 5x + 5$$

EXPLORE!

Store the functions $f(x) = x^2$ and $g(x) = x + 3$ into Y1 and Y2, respectively, of the function editor. Deselect (turn off) Y1 and Y2. Set Y3 = Y1(Y2) and Y4 = Y2(Y1). Show graphically (using **ZOOM** Standard) and analytically (by table values) that $f(g(x))$ represented by Y3 and $g(f(x))$ represented by Y4 are not the same functions. What are the explicit equations for both of these composites?

NOTE   By reversing the roles of f and g in the definition of composite function, you can define the composition $g(f(x))$. In general, $f(g(x))$ and $g(f(x))$ will *not* be the same. For instance, with the functions in Example 1.1.7, you first write

$$g(w) = w + 1 \qquad \text{and} \qquad f(x) = x^2 + 3x + 1$$

and then replace w by $x^2 + 3x + 1$ to get

$$g(f(x)) = (x^2 + 3x + 1) + 1$$
$$= x^2 + 3x + 2$$

which is quite different from $f(g(x)) = x^2 + 5x + 5$ found in Example 1.1.7. In fact, $f(g(x)) = g(f(x))$ only when

$$x^2 + 5x + 5 = x^2 + 3x + 2$$
$$2x = -3$$
$$x = -\frac{3}{2} \qquad \blacksquare$$

Example 1.1.7 could have been worded more compactly as follows: Find the composite function $f(x + 1)$ where $f(x) = x^2 + 3x + 1$. The use of this compact notation is illustrated further in Example 1.1.8.

EXAMPLE 1.1.8 Expressing Cost as a Composite Function

Neal, the owner of a small furniture company, finds that if r recliners are produced per hour, the cost will be $C(r)$ dollars, where

$$C(r) = r^3 - 50r + \frac{1}{r + 1}$$

Suppose, in turn, the production level satisfies $r = 4 + 0.3w$, where w is the hourly wage of the workers.

a. Express the cost of production as a composite function of hourly wage.

b. How much should Neal expect to pay for production when workers earn $20 per hour?

Solution

a. To obtain the required composite function, we replace each r in the expression for $C(r)$ by $4 + 0.3w$. As a preliminary step, it may help to write C in more neutral terms, say

$$C(\square) = (\square)^3 - 50(\square) + \frac{1}{\square + 1}$$

where the box $\square$ is to be filled by $4 + 0.3w$ in each case. Thus, we have

$$C(r(w)) = C([4 + 0.3w]) = [4 + 0.3w]^3 - 50[4 + 0.3w] + \frac{1}{[(4 + 0.3w)] + 1}$$

b. An hourly wage of $20 means that $w = 20$, and the corresponding cost is

$$C(r(20)) = [4 + 0.3(20)]^3 - 50[4 + 0.3(20)] + \frac{1}{[4 + 0.3(20)] + 1}$$

$$= 500.091$$

so if workers earn $20 per hour, the production cost is roughly $500.09.

Occasionally, you will have to take apart a given composite function $g(h(x))$ and identify the outer function $g(u)$ and inner function $h(x)$ from which it was formed. The procedure is demonstrated in Example 1.1.9.

EXAMPLE 1.1.9 Finding Functions That Form a Given Composition

If $f(x) = \dfrac{5}{x - 2} + 4(x - 2)^3$, find functions $g(u)$ and $h(x)$ such that $f(x) = g(h(x))$.

Solution

The form of the given function is

$$f(x) = \frac{5}{\square} + 4(\square)^3$$

where each box contains the expression $x - 2$. Thus, $f(x) = g(h(x))$, where

$$g(u) = \underbrace{\frac{5}{u} + 4u^3}_{\text{outer function}} \quad \text{and} \quad h(x) = \underbrace{x - 2}_{\text{inner function}}$$

Actually, in Example 1.1.9 there are infinitely many pairs of functions $g(u)$ and $h(x)$ that combine to produce the given composite function $f(x) = g(h(x))$. For instance, another such pair is

$$g(u) = \frac{5}{u + 1} + 4(u + 1)^3 \quad \text{and} \quad h(x) = x - 3$$

The particular pair selected in the solution to Example 1.1.9 is the most natural one and reflects most clearly the structure of the original function.

In Example 1.1.10, we examine an application in which the level of air pollution in a community is expressed as a composite function of time.

EXAMPLE 1.1.10 Using a Composite Function to Study Air Pollution

An environmental study of a certain community suggests that the average daily level of carbon monoxide in the air will be $c(p) = 0.5p + 1$ parts per million when the population is p thousand. It is estimated that t years from now the population of the community will be $p(t) = 10 + 0.1t^2$ thousand.

a. Express the level of carbon monoxide in the air as a function of time.

b. When will the carbon monoxide level reach 6.8 parts per million?

Solution

a. Because the level of carbon monoxide is related to the variable p by the equation

$$c(p) = 0.5p + 1$$

and the variable p is related to the variable t by the equation

$$p(t) = 10 + 0.1t^2$$

it follows that the composite function

$$c(p(t)) = c(10 + 0.1t^2) = 0.5(10 + 0.1t^2) + 1 = 6 + 0.05t^2$$

expresses the level of carbon monoxide in the air as a function of the variable t.

b. Set $c(p(t))$ equal to 6.8 and solve for t to get

$$6 + 0.05t^2 = 6.8 \qquad \text{subtract 6 from both sides}$$
$$0.05t^2 = 0.8 \qquad \text{divide both sides by 0.05}$$
$$t^2 = \frac{0.8}{0.05} = 16 \qquad \text{take square roots of both sides}$$
$$t = \sqrt{16} = 4 \qquad \text{discard } t = -4$$

So the level of carbon monoxide will be 6.8 parts per million in 4 years.

A **difference quotient** for a function $f(x)$ is a composite function of the form

$$\frac{f(x + h) - f(x)}{h}$$

where h is a constant. Difference quotients are used in Chapter 2 to compute the average rate of change and the slope of a tangent line and then to define the **derivative,** a concept of central importance in calculus. Example 1.1.11 illustrates how to compute a difference quotient.

EXAMPLE 1.1.11 Finding a Difference Quotient

Find the difference quotient for $f(x) = x^2 - 3x$.

Solution

Using the given form, the difference quotient for f can be written as

$$\frac{f(x+h)-f(x)}{h} = \frac{[(x+h)^2-3(x+h)]-[x^2-3x]}{h} \qquad \text{expand the numerator}$$

$$= \frac{[x^2+2xh+h^2-3x-3h]-[x^2-3x]}{h} \qquad \begin{array}{l}\text{combine like terms}\\ \text{in the numerator}\end{array}$$

$$= \frac{2xh+h^2-3h}{h} \qquad \begin{array}{l}\text{divide the numerator}\\ \text{and denominator by } h\end{array}$$

$$= 2x+h-3$$

EXERCISES ■ 1.1

In Exercises 1 through 14, compute the indicated values of the given function.

1. $f(x) = 3x+5$; $f(0), f(-1), f(2)$

2. $f(x) = -7x+1$; $f(0), f(1), f(-2)$

3. $f(x) = 3x^2+5x-2$; $f(0), f(-2), f(1)$

4. $h(t) = (2t+1)^3$; $h(-1), h(0), h(1)$

5. $g(x) = x+\dfrac{1}{x}$; $g(-1), g(1), g(2)$

6. $f(x) = \dfrac{x}{x^2+1}$; $f(2), f(0), f(-1)$

7. $h(t) = \sqrt{t^2+2t+4}$; $h(2), h(0), h(-4)$

8. $g(u) = (u+1)^{3/2}$; $g(0), g(-1), g(8)$

9. $f(t) = (2t-1)^{-3/2}$; $f(1), f(5), f(13)$

10. $f(t) = \dfrac{1}{\sqrt{3-2t}}$; $f(1), f(-3), f(0)$

11. $f(x) = x-|x-2|$; $f(1), f(2), f(3)$

12. $g(x) = 4+|x|$; $g(-2), g(0), g(2)$

13. $h(x) = \begin{cases} -2x+4 & \text{if } x \le 1 \\ x^2+1 & \text{if } x > 1 \end{cases}$; $h(3), h(1), h(0), h(-3)$

14. $f(t) = \begin{cases} 3 & \text{if } t < -5 \\ t+1 & \text{if } -5 \le t \le 5 \\ \sqrt{t} & \text{if } t > 5 \end{cases}$; $f(-6), f(-5), f(16)$

In Exercises 15 through 18, determine whether or not the given function has the set of all real numbers as its domain.

15. $g(x) = \dfrac{x}{1+x^2}$

16. $f(x) = \dfrac{x+1}{x^2-1}$

17. $f(t) = \sqrt{1-t}$

18. $h(t) = \sqrt{t^2+1}$

In Exercises 19 through 24, determine the domain of the given function.

19. $g(x) = \dfrac{x^2+5}{x+2}$

20. $f(x) = x^3-3x^2+2x+5$

21. $f(x) = \sqrt{2x+6}$

22. $f(t) = \dfrac{t+1}{t^2-t-2}$

23. $f(t) = \dfrac{t+2}{\sqrt{9-t^2}}$

24. $h(s) = \sqrt{s^2-4}$

In Exercises 25 through 32, find the composite function $f(g(x))$.

25. $f(u) = 3u^2+2u-6$, $g(x) = x+2$

26. $f(u) = u^2+4$, $g(x) = x-1$

27. $f(u) = (u-1)^3+2u^2$, $g(x) = x+1$

28. $f(u) = (2u+10)^2$, $g(x) = x-5$

29. $f(u) = \dfrac{1}{u^2}$, $g(x) = x-1$

30. $f(u) = \dfrac{1}{u}$, $g(x) = x^2+x-2$

31. $f(u) = \sqrt{u+1}$, $g(x) = x^2-1$

32. $f(u) = u^2$, $g(x) = \dfrac{1}{x-1}$

In Exercises 33 through 38, find the difference quotient,
$$\frac{f(x + h) - f(x)}{h}.$$

33. $f(x) = 4 - 5x$

34. $f(x) = 2x + 3$

35. $f(x) = 4x - x^2$

36. $f(x) = x^2$

37. $f(x) = \dfrac{x}{x + 1}$

38. $f(x) = \dfrac{1}{x}$

In Exercises 39 through 42, first obtain the composite functions $f(g(x))$ and $g(f(x))$, and then find all numbers x (if any) such that $f(g(x)) = g(f(x))$.

39. $f(x) = \sqrt{x}, g(x) = 1 - 3x$

40. $f(x) = x^2 + 1, g(x) = 1 - x$

41. $f(x) = \dfrac{2x + 3}{x - 1}, g(x) = \dfrac{x + 3}{x - 2}$

42. $f(x) = \dfrac{1}{x}, g(x) = \dfrac{4 - x}{2 + x}$

In Exercises 43 through 50, find the indicated composite function.

43. $f(x - 2)$ where $f(x) = 2x^2 - 3x + 1$

44. $f(x + 1)$ where $f(x) = x^2 + 5$

45. $f(x - 1)$ where $f(x) = (x + 1)^5 - 3x^2$

46. $f(x + 3)$ where $f(x) = (2x - 6)^2$

47. $f(x^2 + 3x - 1)$ where $f(x) = \sqrt{x}$

48. $f\left(\dfrac{1}{x}\right)$ where $f(x) = 3x + \dfrac{2}{x}$

49. $f(x + 1)$ where $f(x) = \dfrac{x - 1}{x}$

50. $f(x^2 - 2x + 9)$ where $f(x) = 2x - 20$

In Exercises 51 through 56, find functions $h(x)$ and $g(u)$ such that $f(x) = g(h(x))$.

51. $f(x) = (x - 1)^2 + 2(x - 1) + 3$

52. $f(x) = (x^5 - 3x^2 + 12)^3$

53. $f(x) = \dfrac{1}{x^2 + 1}$

54. $f(x) = \sqrt{3x - 5}$

55. $f(x) = \sqrt[3]{2 - x} + \dfrac{4}{2 - x}$

56. $f(x) = \sqrt{x + 4} - \dfrac{1}{(x + 4)^3}$

57. PRODUCTION COST Suppose the total cost of manufacturing q units of a certain product is $C(q)$ thousand dollars, where
$$C(q) = 0.01q^2 + 0.9q + 2$$
   a. Find the total cost and the average cost of producing 10 units.
   b. Find the cost of producing the 10th unit.

58. PRODUCTION COST Answer the questions in Exercise 57 for the cost function
$$C(q) = q^3 - 30q^2 + 400q + 500$$

PROFITABILITY *In Exercises 59 through 62, the demand function $p = D(x)$ and the total cost function $C(x)$ for a particular commodity are given in terms of the level of production x. In each case, find:*
   (a) The revenue $R(x)$ and profit $P(x)$.
   (b) All values of x for which production of the commodity is profitable.

59. $D(x) = -0.02x + 29$
      $C(x) = 1.43x^2 + 18.3x + 15.6$

60. $D(x) = -0.37x + 47$
      $C(x) = 1.38x^2 + 15.15x + 115.5$

61. $D(x) = -0.5x + 39$
      $C(x) = 1.5x^2 + 9.2x + 67$

62. $D(x) = -0.09x + 51$
      $C(x) = 1.32x^2 + 11.7x + 101.4$

63. DISTRIBUTION COST Suppose that the number of worker-hours required to distribute new telephone books to $x\%$ of the households in a certain rural community is given by the function
$$W(x) = \frac{600x}{300 - x}$$
   a. What is the domain of the function W?
   b. For what values of x does $W(x)$ have a practical interpretation in this context?
   c. How many worker-hours were required to distribute new telephone books to the first 50% of the households?
   d. How many worker-hours were required to distribute new telephone books to the entire community?
   e. What percentage of the households in the community had received new telephone books by the time 150 worker-hours had been expended?

64. DATA TRANSFER In the year 2000, Digicorp, a data management firm, began transferring files from antiquated databases and storing them on more modern systems. Measured in years after 2010, the function $R(t) = 30\sqrt{6 - t}$ represents the number of databases remaining to be transferred.
  a. What is the domain of R?
  b. How many databases were present when Digicorp began the transfer?
  c. How many databases still needed to be transferred in 2007?
  d. Approximately how many databases had been transferred as of 2011?
  e. The data transfer was scheduled to be complete by 2015. Will the engineers accomplish this goal? Explain.

65. STOCK PRICES Apple Inc. (stock symbol AAPL) produces popular products such as the iPhone, iPad, and MacBook laptop computers. The company was not always as wildly successful, however. Taking into account stock splits, prices (in dollars) for AAPL shares can be represented by the following piecewise-defined function:

$$S(t) = \begin{cases} 14.7 + 0.6t & \text{if } t \le 4 \\ 14.2t^2 - 128t + 304 & \text{if } t > 4 \end{cases}$$

where t is the number of years after 2000.
  a. Using this function, what was the share price of AAPL in 1990 (when $t = -10$)? In 2006?
  b. In what year does the function predict that AAPL shares first reached the $200 level?
  c. What share value does the function predict for AAPL in the year 2012?

66. WORKER EFFICIENCY An efficiency study of the morning shift at a certain factory indicates that an average worker who arrives on the job at 8:00 A.M. will have assembled

$$f(x) = -x^3 + 6x^2 + 15x$$

television sets x hours later.
  a. How many sets will such a worker have assembled by 10:00 A.M.? [*Hint:* At 10:00 A.M., $x = 2$.]
  b. How many sets will such a worker assemble between 9:00 and 10:00 A.M.?

67. CONSUMER DEMAND An importer of Brazilian coffee estimates that local consumers will buy approximately $Q(p) = \dfrac{4{,}374}{p^2}$ kilograms of the coffee per week when the price is p dollars per kilogram. It is estimated that t weeks from now the price of this coffee will be

$$p(t) = 0.04t^2 + 0.2t + 12$$

dollars per kilogram.
  a. Express the weekly demand (kilograms sold) for the coffee as a function of t.
  b. How many kilograms of the coffee will consumers be buying from the importer 10 weeks from now?
  c. When will the demand for the coffee be 30.375 kilograms?

68. PRODUCTION COST Arthur, the manager of a furniture factory, finds that the cost of producing q bookcases during the morning production run is $C(q) = q^2 + q + 500$ dollars. On a typical workday, $q(t) = 25t$ bookcases are produced during the first t hours of a production run for $0 \le t \le 5$.
  a. Express the production cost C in terms of t.
  b. How much will have been spent on production by the end of the 3rd hour? What is the average cost of production during the first 3 hours?
  c. Arthur's budget allows no more than $11,000 for production during the morning production run. When will this limit be reached?

LIFE AND SOCIAL SCIENCE APPLIED PROBLEMS

69. IMMUNIZATION Suppose that during a nationwide program to immunize the population against a new strain of influenza, public health officials found that the cost of inoculating $x\%$ of the population was approximately $C(x) = \dfrac{150x}{200 - x}$ million dollars.
  a. What is the domain of the function C?
  b. For what values of x does $C(x)$ have a practical interpretation in this context?
  c. What was the cost of inoculating the first 50% of the population?
  d. What was the cost of inoculating the second 50% of the population?
  e. What percentage of the population had been inoculated by the time 37.5 million dollars had been spent?

70. TEMPERATURE CHANGE Suppose that t hours past midnight, the temperature in Miami was $C(t) = -\dfrac{1}{6}t^2 + 4t + 10$ degrees Celsius.

a. What was the temperature at 2:00 A.M.?

b. By how much did the temperature increase or decrease between 6:00 and 9:00 P.M.?

71. POPULATION GROWTH   It is estimated that t years from now, the population of a certain suburban community will be $P(t) = 20 - \dfrac{6}{t+1}$ thousand people.

a. What will the population of the community be 9 years from now?

b. By how much will the population increase during the 9th year?

c. What happens to $P(t)$ as t gets larger and larger? Interpret your result.

72. EXPERIMENTAL PSYCHOLOGY   To study the rate at which animals learn, Becky performed an experiment in which a rat was sent repeatedly through a laboratory maze. Suppose that the time required for the rat to traverse the maze on the nth trial was approximately

$$T(n) = 3 + \frac{12}{n}$$

minutes.

a. What is the domain of the function T?

b. For what values of n does $T(n)$ have meaning in the context of Becky's experiment?

c. How long did it take the rat to traverse the maze on the 3rd trial?

d. On which trial did the rat first traverse the maze in 4 minutes or less?

e. According to the function T, what will happen to the time required for the rat to traverse the maze as the number of trials increases? Will the rat ever be able to traverse the maze in less than 3 minutes?

73. BLOOD FLOW   Biologists have found that the speed of blood in an artery is a function of the distance of the blood from the artery's central axis. According to **Poiseuille's law,*** the speed (in centimeters per second) of blood that is r centimeters from the central axis of an artery is given by the function $S(r) = C(R^2 - r^2)$, where C is a constant and R is the radius of the artery. Suppose that for a certain artery, $C = 1.76 \times 10^5$ and $R = 1.2 \times 10^{-2}$ centimeters.

a. Compute the speed of the blood at the central axis of this artery.

b. Compute the speed of the blood midway between the artery's wall and central axis.

74. POPULATION DENSITY   Observations suggest that for herbivorous mammals, the number of animals N per square kilometer can be estimated by the formula $N = \dfrac{91.2}{m^{0.73}}$, where m is the average mass of the animal in kilograms.

a. Assuming that the average elk on a particular reserve has mass 300 kilograms, approximately how many elk would you expect to find per square kilometer in the reserve?

b. Using this formula, it is estimated that there is less than one animal of a certain species per square kilometer. How large can the average animal of this species be?

c. One species of large mammal has twice the average mass as a second species. If a particular reserve contains 100 animals of the larger species, how many animals of the smaller species would you expect to find there?

75. ISLAND ECOLOGY   Observations show that on an island of area A square miles, the average number of animal species is approximately equal to $s(A) = 2.9\sqrt[3]{A}$.

a. On average, how many animal species would you expect to find on an island of area 8 square miles?

b. If s_1 is the average number of species on an island of area A and s_2 is the average number of species on an island of area $2A$, what is the relationship between s_1 and s_2?

c. How big must an island be to have an average of 100 animal species?

76. AIR POLLUTION   An environmental study of a certain suburban community suggests that the average daily level of carbon monoxide in the air will be $c(p) = 0.4p + 1$ parts per million when the population is p thousand. It is estimated that t years from now the population of the community will be $p(t) = 8 + 0.2t^2$ thousand.

a. Express the level of carbon monoxide in the air as a function of time.

b. What will the carbon monoxide level be 2 years from now?

c. When will the carbon monoxide level reach 6.2 parts per million?

---

*Edward Batschelet, *Introduction to Mathematics for Life Scientists,* 3rd ed., New York: Springer-Verlag, 1979, pp. 101–103.

77. **POSITION OF A MOVING OBJECT**   A ball has been dropped from the top of a building. Its height (in feet) after t seconds is given by the function $H(t) = -16t^2 + 256$.
   a. What is the height of the ball after 2 seconds?
   b. How far will the ball travel during the third second?
   c. How tall is the building?
   d. When will the ball hit the ground?

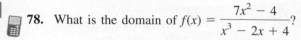

78. What is the domain of $f(x) = \dfrac{7x^2 - 4}{x^3 - 2x + 4}$?

79. What is the domain of $f(x) = \dfrac{4x^2 - 3}{2x^2 + x - 3}$?

80. For $f(x) = 2\sqrt{x - 1}$ and $g(x) = x^3 - 1.2$, find $g(f(4.8))$. Use two decimal places.

81. For $f(x) = 2\sqrt{x - 1}$ and $g(x) = x^3 - 1.2$, find $f(g(2.3))$. Use two decimal places.

SECTION 1.2   The Graph of a Function

Learning Objectives

1. Review the rectangular coordinate system.
2. Graph several functions.
3. Study intersections of graphs, the vertical line test, and intercepts.
4. Sketch and use graphs of quadratic functions in applications.

Graphs have visual impact. They also reveal information that may not be evident from verbal or algebraic descriptions. Two graphs depicting practical relationships are shown in Figure 1.3.

The graph in Figure 1.3a describes the variation in total industrial production in a certain country over a 4-year period of time. Notice that the highest point on the graph occurs near the end of the third year, indicating that production was greatest at that time. The graph in Figure 1.3b represents population growth when environmental factors impose an upper bound on the possible size of the population. It indicates that the *rate* of population growth increases at first and then decreases as the size of the population gets closer and closer to the upper bound.

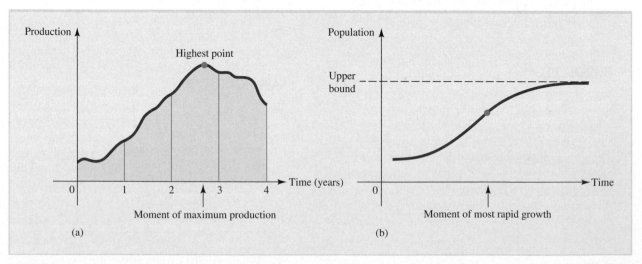

FIGURE 1.3   (a) A production function. (b) Bounded population growth.

The Rectangular Coordinate System

To represent graphs in the plane, we will use a **rectangular coordinate system,** also called a **Cartesian coordinate system** in honor of its inventor, the 17th century French mathematician-philosopher, Rene Descartes. To construct such a system, we begin by choosing two perpendicular number lines that intersect at the origin of each line. For convenience, one line is drawn horizontally and is called the **x axis,** with positive direction to the right. The other line, called the **y axis,** is vertical with positive direction upward. The scales on the two coordinate axes are often different, with the choice of scale depending on the quantities being represented by the two variables. The coordinate axes separate the plane into four parts called **quadrants,** which are numbered counterclockwise I through IV, as shown in Figure 1.4.

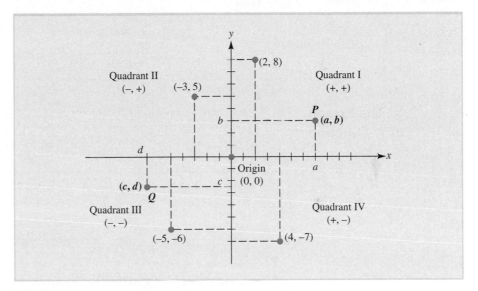

FIGURE 1.4 A rectangular coordinate system.

Any point P in the plane can be associated with a unique ordered pair of numbers (a, b) called the **coordinates** of P. Specifically, a is called the **x coordinate** (or **abscissa**) and b is called the **y coordinate** (or **ordinate**). To find a and b, draw the vertical and horizontal lines through P. The vertical line intersects the x axis at a, and the horizontal line intersects the y axis at b. Conversely, if c and d are given, the vertical line through c and horizontal line through d intersect at the unique point Q with coordinates (c, d).

Several points are plotted in Figure 1.4. In particular, note that the point $(2, 8)$ is 2 units to the right of the vertical axis and 8 units above the horizontal axis, while $(-3, 5)$ is 3 units to the left of the vertical axis and 5 units above the horizontal axis.

The Distance Formula

There is a simple formula for finding the distance D between two points P and Q with coordinates (x_1, y_1) and (x_2, y_2), respectively. In Figure 1.5, note that the differences $x_2 - x_1$ of the x coordinates and $y_2 - y_1$ of the y coordinates represent the lengths of the sides of a right triangle. Then the hypotenuse of the triangle has length equal to the required distance D between P and Q, and the Pythagorean theorem gives the distance formula $D = \sqrt{(x_2 - x_1)^2 + (y_2 - y_1)^2}$. Figure 1.5 shows only the special case where Q is above and to the right of P, but the distance formula holds for all points in the plane. To summarize:

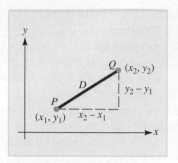

FIGURE 1.5   The distance formula
$$D = \sqrt{(x_2 - x_1)^2 + (y_2 - y_1)^2}$$

The Distance Formula ■ The distance between the points (x_1, y_1) and (x_2, y_2) is given by

$$D = \sqrt{(x_2 - x_1)^2 + (y_2 - y_1)^2}$$

EXAMPLE 1.2.1   Using the Distance Formula

Find the distance between the points $(-2, 5)$ and $(4, -1)$.

Solution

In the distance formula, we have $x_1 = -2$, $y_1 = 5$, $x_2 = 4$, and $y_2 = -1$, so the distance between the points is

$$D = \sqrt{[4 - (-2)]^2 + (-1 - 5)^2} = \sqrt{72} = 6\sqrt{2}$$

The Graph of a Function

To represent a function $y = f(x)$ geometrically as a graph, we plot values of the independent variable x on the (horizontal) x axis and values of the dependent variable y on the (vertical) y axis. The graph of the function is defined as follows.

The Graph of a Function ■ The graph of a function f consists of all points (x, y) where x is in the domain of f and $y = f(x)$, that is, all points of the form $(x, f(x))$.

In Chapter 3, you will study efficient techniques involving calculus that can be used to draw accurate graphs of functions. For many functions, however, you can make a fairly good sketch by plotting a few points, as illustrated in Example 1.2.2.

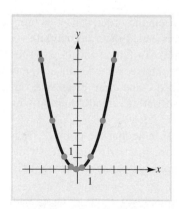

FIGURE 1.6   The graph of $y = x^2$.

EXAMPLE 1.2.2   Graphing by Plotting Points

Graph the function $f(x) = x^2$.

Solution

Begin by constructing the table

x	-3	-2	-1	$-\dfrac{1}{2}$	0	$\dfrac{1}{2}$	1	2	3
$y = x^2$	9	4	1	$\dfrac{1}{4}$	0	$\dfrac{1}{4}$	1	4	9

Then plot the points (x, y) and connect them with the smooth curve shown in Figure 1.6.

NOTE   Many different curves pass through the points in Example 1.2.2. Several of these curves are shown in Figure 1.7. There is no way to guarantee that the curve we pass through the plotted points is the actual graph of f. However, in general, the more points that are plotted, the more likely the graph is to be reasonably accurate. ■

EXPLORE!

Store $f(x) = x^2$ into Y1 of the equation editor, using a bold graphing style. Represent $g(x) = x^2 + 2$ by Y2 = Y1 + 2 and $h(x) = x^2 - 3$ by Y3 = Y1 - 3. Use **ZOOM** decimal graphing to show how the graphs of $g(x)$ and $h(x)$ relate to that of $f(x)$. Now deselect Y2 and Y3 and write Y4 = Y1(X + 2) and Y5 = Y1(X - 3). Explain how the graphs of Y1, Y4, and Y5 relate.

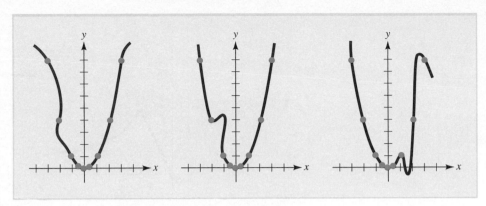

FIGURE 1.7   Other graphs through the points in Example 1.2.2.

Example 1.2.3 illustrates how to sketch the graph of a function defined by more than one formula.

EXAMPLE 1.2.3   Graphing a Piecewise-Defined Function

Graph the function

$$f(x) = \begin{cases} 2x & \text{if } 0 \le x < 1 \\ \dfrac{2}{x} & \text{if } 1 \le x < 4 \\ 3 & \text{if } x \ge 4 \end{cases}$$

EXPLORE!

Certain functions that are defined piecewise can be entered into a graphing calculator using indicator functions in sections. For example, the absolute value function,

$$f(x) = |x| = \begin{cases} x & \text{if } x \ge 0 \\ -x & \text{if } x < 0 \end{cases}$$

can be represented by Y1 = X(X ≥ 0) + (−X)(X < 0). Now represent the function in Example 1.2.3, using indicator functions and graph it with an appropriate viewing window. [*Hint:* You will need to represent the interval 0 < X < 1 by the boolean expression (0 < X)(X < 1).]

Solution

When making a table of values for this function, remember to use the formula that is appropriate for each particular value of x. Using the formula $f(x) = 2x$ when $0 \le x < 1$, the formula $f(x) = \dfrac{2}{x}$ when $1 \le x < 4$, and the formula $f(x) = 3$ when $x \ge 4$, you can compile this table:

x	0	$\dfrac{1}{2}$	1	2	3	4	5	6
$f(x)$	0	1	2	1	$\dfrac{2}{3}$	3	3	3

Now plot the corresponding points $(x, f(x))$ and draw the graph as in Figure 1.8. Notice that the pieces for $0 \le x < 1$ and $1 \le x < 4$ are connected to one another at $(1, 2)$ but that the piece for $x \ge 4$ is separated from the rest of the graph. [The "open dot" at $\left(4, \dfrac{1}{2}\right)$ indicates that the graph approaches this point but that the point is not actually on the graph.]

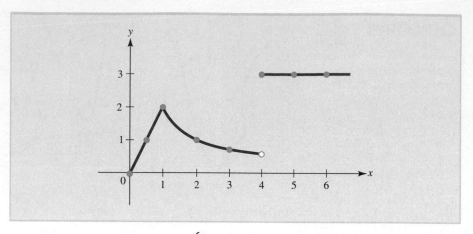

FIGURE 1.8   The graph of $f(x) = \begin{cases} 2x & 0 \le x < 1 \\ \dfrac{2}{x} & 1 \le x < 4 \\ 3 & x \ge 4 \end{cases}$

Intersections of Graphs

Sometimes it is necessary to determine when two functions are equal. For instance, an economist may wish to compute the market price at which the consumer demand for a commodity will be equal to supply. Or a political analyst may wish to predict how long it will take for the popularity of a certain challenger to reach that of the incumbent. We shall examine some of these applications in Section 1.4.

In geometric terms, the values of x for which two functions $f(x)$ and $g(x)$ are equal are the x coordinates of the points where their graphs intersect. In Figure 1.9, the graph of $y = f(x)$ intersects that of $y = g(x)$ at two points, labeled P and Q. To find the points of intersection algebraically, set $f(x)$ equal to $g(x)$ and solve for x. This procedure is illustrated in Example 1.2.4.

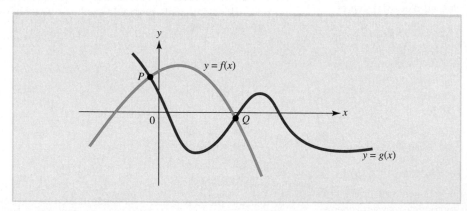

FIGURE 1.9   The graphs of $y = f(x)$ and $y = g(x)$ intersect at P and Q.

Just-In-Time REVIEW

The **quadratic formula** applies to equations of the form

$$Ax^2 + Bx + C = 0$$

like the one in Example 1.2.4. It provides solutions of the form

$$r_1 = \frac{-B + \sqrt{B^2 - 4AC}}{2A}$$

and

$$r_2 = \frac{-B - \sqrt{B^2 - 4AC}}{2A}$$

If $B^2 - 4AC$ is negative, there are no real solutions. You can find a more extensive review of the quadratic formula in Appendix A.2.

EXAMPLE 1.2.4 Finding Points of Intersection

Find all points of intersection of the graphs of $f(x) = 3x + 2$ and $g(x) = x^2$.

Solution

You must solve the equation $x^2 = 3x + 2$. Rewrite the equation as $x^2 - 3x - 2 = 0$ and apply the quadratic formula to obtain

$$x = \frac{-(-3) \pm \sqrt{(-3)^2 - 4(1)(-2)}}{2(1)} = \frac{3 \pm \sqrt{17}}{2}$$

The solutions are

$$x = \frac{3 + \sqrt{17}}{2} \approx 3.56 \quad \text{and} \quad x = \frac{3 - \sqrt{17}}{2} \approx -0.56$$

(The computations were done on a calculator, with results rounded off to two decimal places.)

Computing the corresponding y coordinates from the equation $y = x^2$, you find that the points of intersection are approximately $(3.56, 12.67)$ and $(-0.56, 0.31)$. (As a result of round-off errors, you will get slightly different values for the y coordinates if you substitute into the equation $y = 3x + 2$.) The graphs and the intersection points are shown in Figure 1.10.

EXPLORE!

Refer to Example 1.2.4. Use your graphing utility to find all points of intersection of the graphs of $f(x) = 3x + 2$ and $g(x) = x^2$. Also find the roots of $g(x) - f(x) = x^2 - 3x - 2$. What can you conclude?

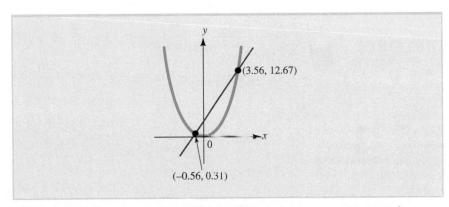

FIGURE 1.10 The intersection of the graphs of $f(x) = 3x + 2$ and $g(x) = x^2$.

The Vertical Line Test

It is important to realize that not every curve is the graph of a function (Figure 1.11). For instance, suppose the circle $x^2 + y^2 = 5$ were the graph of some function $y = f(x)$. Then, since the points $(1, 2)$ and $(1, -2)$ both lie on the circle, we would have $f(1) = 2$ and $f(1) = -2$, contrary to the requirement that a function assigns one and *only* one value to each number in its domain. Geometrically, this happens because the vertical line $x = 1$ intersects the graph of the circle more than once. The **vertical line test** is a geometric rule for determining whether a curve is the graph of a function.

> **The Vertical Line Test** ■ A curve is the graph of a function if and only if no vertical line intersects the curve more than once.

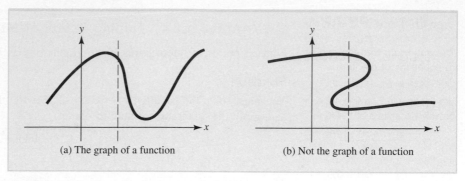

(a) The graph of a function   (b) Not the graph of a function

FIGURE 1.11   The vertical line test.

Intercepts   The points (if any) where a graph crosses the x axis are called **x intercepts,** and similarly, a **y intercept** is a point where the graph crosses the y axis. The graph of a function can have many x intercepts but at most one y intercept since the vertical line test says that the graph can cross the vertical line $x = 0$ (the y axis) no more than once. Intercepts are key features of a graph and can be found using algebra or technology in conjunction with these criteria.

> **How to Find the x and y Intercepts**   ■   To find any x intercept of a graph, set $y = 0$ and solve for x. To find any y intercept, set $x = 0$ and solve for y. For a function f, the only y intercept is $y_0 = f(0)$, but finding x intercepts may be difficult.

EXPLORE!

Using your graphing utility, locate the x intercepts of $f(x) = -x^2 + x + 2$. These intercepts can be located by first using the **ZOOM** button and then confirmed by using the root-finding feature of the graphing utility. Do the same for $g(x) = x^2 + x - 4$. What radical form do these roots have?

EXAMPLE 1.2.5   Finding the Intercepts of a Graph

Find the intercepts of the graph of each of the following functions:

a. $f(x) = -x^2 + x + 2$

b. $g(x) = x\sqrt{x^2 - 1}$

Solution

a. Since $f(0) = 2$, the y intercept of the graph of $f(x)$ is $(0, 2)$. To find the x intercepts, we solve the equation $f(x) = 0$. Factoring, we find that

$$-x^2 + x + 2 = 0 \quad \text{factor}$$
$$-(x + 1)(x - 2) = 0 \quad uv = 0 \text{ if and only if } u = 0 \text{ or } v = 0$$
$$x = -1, x = 2$$

so the x intercepts are $(-1, 0)$ and $(2, 0)$.

b. The function $g(x) = x\sqrt{x^2 - 1}$ is not defined for $-1 < x < 1$ since $x^2 - 1$, the quantity under the square root, is negative on this interval. Since 0 is in this excluded interval, $g(0)$ is not defined and there is no y intercept. To get the x intercepts, we set $g(x) = 0$ and find that the solutions are $x = 0, 1$, and -1. Again, the function is not defined at $x = 0$, so $(1, 0)$ and $(-1, 0)$ are the only x intercepts.

NOTE   The factoring in Example 1.2.5 is fairly straightforward, but in other problems, you may need to review the factoring procedure provided in Appendix A.2.   ■

**Power Functions,
Polynomials, and
Rational Functions**

A **power function** is a function of the form $f(x) = x^n$, where n is a real number. For example, $f(x) = x^2$, $f(x) = x^{-3}$, and $f(x) = x^{1/2}$ are all power functions. So are $f(x) = \dfrac{1}{x^2}$ and $f(x) = \sqrt[3]{x}$ since they can be rewritten as $f(x) = x^{-2}$ and $f(x) = x^{1/3}$, respectively.

A **polynomial** is a function of the form

$$p(x) = a_n x^n + a_{n-1} x^{n-1} + \cdots + a_1 x + a_0$$

where n is a nonnegative integer and $a_0, a_1, \ldots, a_n$ are constants. If $a_n \neq 0$, the integer n is called the **degree** of the polynomial. For example, $f(x) = 3x^5 - 6x^2 + 7$ is a polynomial of degree 5. It can be shown that the graph of a polynomial of degree n is an unbroken curve that crosses the x axis no more than n times. To illustrate some of the possibilities, the graphs of three polynomials of degree 3 are shown in Figure 1.12.

EXPLORE!

Use your calculator to graph the third-degree polynomial $f(x) = x^3 - x^2 - 6x + 3$. Conjecture the values of the x intercepts and confirm them using the root-finding feature of your calculator.

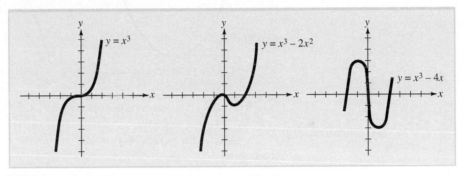

FIGURE 1.12   Three polynomials of degree 3.

A quotient $\dfrac{p(x)}{q(x)}$ of two polynomials $p(x)$ and $q(x)$ is called a **rational function.** Such functions appear throughout this text in examples and exercises. Graphs of three rational functions are shown in Figure 1.13. You will learn how to sketch such graphs in Section 3.3.

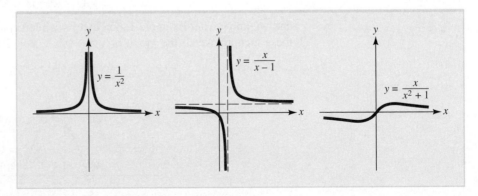

FIGURE 1.13   Graphs of three rational functions.

**Sketching and
Applying the Graph
of a Quadratic
Function**

A polynomial of degree 2 is called a **quadratic function.** It has the general form $f(x) = Ax^2 + Bx + C$ with $A \neq 0$, and its graph is a curve called a **parabola.** Quadratic functions occur frequently in applications that can be analyzed graphically, so it is useful to have a procedure for quickly sketching the graph of such a function.

All parabolas have a U shape like the graphs in Figure 1.14, but the U opens up if $A > 0$ (Figure 1.14a) and down if $A < 0$ (Figure 1.14b). The point at the "peak" or the "valley" of the parabola is called its **vertex** and occurs where $x = \dfrac{-B}{2A}$.

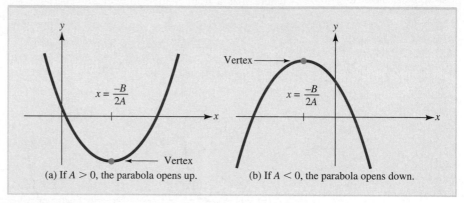

(a) If $A > 0$, the parabola opens up.

(b) If $A < 0$, the parabola opens down.

FIGURE 1.14   The graph of the parabola $y = Ax^2 + Bx + C$.

To sketch the parabola $y = Ax^2 + Bx + C$, you only need two pieces of information:

1. The location of the vertex (where $x = -B/2A$)
2. Any two additional points on the parabola (often these are intercepts)

For instance, the parabola $y = -x^2 + x + 2$ opens downward since $A = -1$ is negative; its vertex (high point) is at $\left(\dfrac{1}{2}, \dfrac{9}{4}\right)$, the point where

$$x = \frac{-B}{2A} = \frac{-1}{2(-1)} = \frac{1}{2}$$

and we know from Example 1.2.5a that its x intercepts are $(-1, 0)$ and $(2, 0)$. We use these facts to sketch the graph of $y = -x^2 + x + 2$ in Figure 1.15.

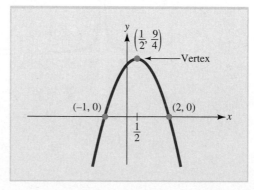

FIGURE 1.15   The graph of $y = -x^2 + x + 2$.

In Example 1.2.6, we use what we know about the graph of a parabola to optimize a quadratic revenue function.

EXAMPLE 1.2.6   Finding Maximum Revenue

Jorge determines that when x hundred units of his product are produced, they can all be sold for a unit price given by the demand function $p = 60 - x$ dollars. At what level of production is Jorge's revenue maximized? What is the maximum revenue?

Solution

The revenue derived from producing x hundred units and selling them all at $60 - x$ dollars apiece is $R(x) = x(60 - x)$ hundred dollars. Note that $R(x) \geq 0$ only for $0 \leq x \leq 60$. The graph of the revenue function

$$R(x) = x(60 - x) = -x^2 + 60x$$

is a parabola that opens downward (since $A = -1 < 0$) and has its high point (vertex) where

$$x = \frac{-B}{2A} = \frac{-60}{2(-1)} = 30$$

as shown in Figure 1.16. Thus, revenue is maximized when $x = 30$ hundred (3,000) units are produced, and the corresponding maximum revenue is

$$R(30) = 30(60 - 30) = 900$$

hundred (90,000) dollars. Jorge should produce 3,000 units and at that level of production should expect maximum revenue of $90,000.

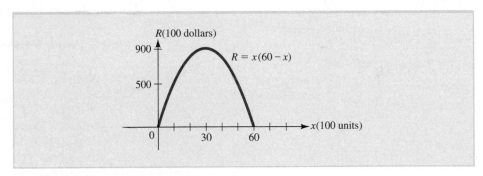

FIGURE 1.16   A revenue function.

Just-In-Time REVIEW

Completing the square is reviewed in Appendix A.2 and illustrated in Examples A.2.12 and A.2.13.

Note that we can also find the largest value of $R(x) = -x^2 + 60x$ by completing the square:

$$R(x) = -x^2 + 60x \qquad \text{factor out } -1, \text{ the coefficient of } x^2$$
$$= -(x^2 - 60x) \qquad \text{complete the square inside parentheses}$$
$$\text{by adding } \left(\frac{-60}{2}\right)^2 = 900$$
$$= -(x^2 - 60x + 900) + 900$$
$$\qquad\qquad \uparrow \qquad\quad \uparrow$$
$$\qquad\qquad -900 \; + \; 900$$
$$= (x - 30)^2 + 900$$

Thus, $R(30) = 0 + 900 = 900$ and if c is any number other than 30, then

$$R(c) = -(c - 30)^2 + 900 < 900 \quad \text{since } -(c - 30)^2 < 0$$

so the maximum revenue is \$90,000 when $x = 30$ (3,000 units).

EXERCISES ■ 1.2

In Exercises 1 through 6, plot the given points in a rectangular coordinate plane.

1. (4, 3)   **2.** $(-2, 7)$

3. $(5, -1)$ **4.** $(-1, -8)$

5. $(0, -2)$ **6.** (3, 0)

In Exercises 7 through 10, find the distance between the given points.

7. $(3, -1)$ and (7, 1)   **8.** (4, 5) and $(-2, -1)$

9. $(7, -3)$ and (5, 3)   **10.** $\left(0, \dfrac{1}{2}\right)$ and $\left(-\dfrac{1}{5}, \dfrac{3}{8}\right)$

In Exercises 11 and 12, classify each function as a polynomial, a power function, or a rational function. If the function is not one of these types, classify it as "different."

11. a. $f(x) = x^{1.4}$
    b. $f(x) = -2x^3 - 3x^2 + 8$
    c. $f(x) = (3x - 5)(4 - x)^2$
    d. $f(x) = \dfrac{3x^2 - x + 1}{4x + 7}$

12. a. $f(x) = -2 + 3x^2 + 5x^4$
    b. $f(x) = \sqrt{x} + 3x$
    c. $f(x) = \dfrac{(x - 3)(x + 7)}{-5x^3 - 2x^2 + 3}$
    d. $f(x) = \left(\dfrac{2x + 9}{x^2 - 3}\right)^3$

In Exercises 13 through 28, sketch the graph of the given function. Include all x and y intercepts.

13. $f(x) = x$ **14.** $f(x) = x^2$

15. $f(x) = \sqrt{x}$ **16.** $f(x) = \sqrt{1 - x}$

17. $f(x) = 2x - 1$ **18.** $f(x) = 2 - 3x$

19. $f(x) = x(2x + 5)$ **20.** $f(x) = (x - 1)(x + 2)$

21. $f(x) = -x^2 - 2x + 15$

22. $f(x) = x^2 + 2x - 8$

23. $f(x) = x^3$

24. $f(x) = -x^3 + 1$

25. $f(x) = \begin{cases} x - 1 & \text{if } x \le 0 \\ x + 1 & \text{if } x > 0 \end{cases}$

26. $f(x) = \begin{cases} 2x - 1 & \text{if } x < 2 \\ 3 & \text{if } x \ge 2 \end{cases}$

27. $f(x) = \begin{cases} x^2 + x - 3 & \text{if } x < 1 \\ 1 - 2x & \text{if } x \ge 1 \end{cases}$

28. $f(x) = \begin{cases} 9 - x & \text{if } x \le 2 \\ x^2 + x - 2 & \text{if } x > 2 \end{cases}$

In Exercises 29 through 34, find the points of intersection (if any) of the given pair of curves and draw the graphs.

29. $y = 3x + 5$ and $y = -x + 3$

30. $y = 3x + 8$ and $y = 3x - 2$

31. $y = x^2$ and $y = 3x - 2$

32. $y = x^2 - x$ and $y = x - 1$

33. $3y - 2x = 5$ and $y + 3x = 9$

34. $2x - 3y = -8$ and $3x - 5y = -13$

In Exercises 35 through 38, the graph of a function f(x) is given. In each case find:
    (a) The y intercept.
    (b) All x intercepts.
    (c) The largest value of f(x) and the value(s) of x for which it occurs.
    (d) The smallest value of f(x) and the value(s) of x for which it occurs.

35.

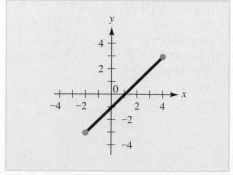

36.

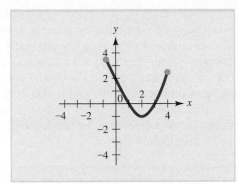

37.

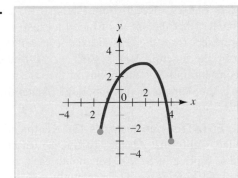

38.

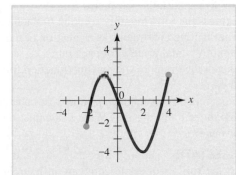

39. MANUFACTURING COST Vicki's company can produce digital recorders at a cost of $40 apiece. It is estimated that if the recorders are sold for p dollars apiece, consumers will buy $120 - p$ of them a month. Express Vicki's monthly profit as a function of price, graph this function, and use the graph to estimate the optimal selling price.

40. MANUFACTURING COST A manufacturer can produce tires at a cost of $20 each. It is estimated that if the tires are sold for p dollars apiece, consumers will buy $1{,}560 - 12p$ of them each month. Express the manufacturer's monthly

profit as a function of price, graph this function, and use the graph to determine the optimal selling price. How many tires will be sold each month at the optimal price?

41. RETAIL SALES The owner of a toy store can obtain a popular board game at a cost of $15 per set. She estimates that if each set sells for x dollars, then $5(27 - x)$ sets will be sold each week. Express the owner's weekly profit from the sales of this game as a function of price, graph this function, and estimate the optimal selling price. How many sets will be sold each week at the optimal price?

42. RETAIL SALES A bookstore can obtain an atlas from the publisher at a cost of $10 per copy and estimates that if it sells the atlas for x dollars per copy, approximately $20(22 - x)$ copies will be sold each month. Express the bookstore's monthly profit from the sale of the atlas as a function of price, graph this function, and use the graph to estimate the optimal selling price.

43. CONSUMER EXPENDITURE Suppose $x = -200p + 12{,}000$ units of a particular commodity are sold each month when the market price is p dollars per unit. The total monthly consumer expenditure E is the total amount of money spent by consumers during each month.
   a. Express the total monthly consumer expenditure E as a function of the unit price p, and sketch the graph of $E(p)$.
   b. Discuss the economic significance of the p intercepts of the expenditure function $E(p)$.
   c. Use the graph in part (a) to determine the market price that generates the greatest total monthly consumer expenditure.

44. CONSUMER EXPENDITURE Suppose that x thousand units of a particular commodity are sold each month when the price is p dollars per unit, where

$$p(x) = 5(24 - x)$$

The total monthly consumer expenditure E is the total amount of money consumers spend during each month.
   a. Express total monthly expenditure E as a function of the unit price p, and sketch the graph of $E(p)$.
   b. Discuss the economic significance of the p intercepts of the expenditure function $E(p)$.

c. Use the graph in part (a) to determine the market price that generates the greatest total monthly consumer expenditure. How many units will be sold during each month at the optimal price?

45. PROFIT Suppose that when the price of a certain commodity is p dollars per unit, then x hundred units will be purchased by consumers, where $p = -0.05x + 38$. The cost of producing x hundred units is $C(x) = 0.02x^2 + 3x + 574.77$ hundred dollars.

    a. Express the profit P obtained from the sale of x hundred units as a function of x. Sketch the graph of the profit function.

    b. Find the average profit AP when the price is $37 per unit.

    c. Use the profit curve found in part (a) to determine the level of production x that results in maximum profit. What unit price p corresponds to maximum profit?

46. EQUIPMENT RENTAL It costs $90 to rent a piece of equipment plus $21 for every day of use.

    a. Make a table showing the number of days the equipment is rented and the cost of renting for 2 days, 5 days, 7 days, and 10 days.

    b. Write an algebraic expression representing the cost y as a function of the number of days x.

    c. Graph the expression in part (b).

47. MANUFACTURING OUTPUT A company that manufactures lawnmowers has determined that a new employee can assemble N mowers per day after t days of training, where

$$N(t) = \frac{45t^2}{5t^2 + t + 8}$$

    a. Make a table showing the numbers of mowers assembled for training periods of lengths $t = 2$ days, 3 days, 5 days, 10 days, and 50 days.

    b. Based on the table in part (a), what do you think happens to $N(t)$ for very long training periods?

    c. Use your calculator to graph $N(t)$.

48. PROFIT Chuck owns several hot dog carts in a large downtown area. He finds that he can sell x hot dogs per day when the price of each hot dog is $p = 4.2 - 0.01x$ dollars. The cost of preparing x hot dogs per day is $C(x) = 0.002x^2 + 30$ dollars.

    a. Express Chuck's profit P as a function of sales x. Sketch the graph of the profit function.

    b. What price p should Chuck charge to maximize profit?

49. CELL PHONE COST A pay-as-you-go cell phone company offers a monthly plan for $19 that includes 200 minutes of calls. After that, calls are an additional 4 cents per minute up to a maximum of 1,000 minutes. Let $C(m)$ be the cost in dollars of making m minutes of calls with a phone on this plan, for $0 \leq m \leq 1,000$.

    a. Write $C(m)$ as a piecewise-defined function.

    b. Sketch the graph of C.

LIFE AND SOCIAL SCIENCE APPLIED PROBLEMS

50. BLOOD FLOW Recall from Exercise 73, Section 1.1, that the speed of blood located r centimeters from the central axis of an artery is given by the function $S(r) = C(R^2 - r^2)$, where C is a constant and R is the radius of the artery. What is the domain of this function? Sketch the graph of $S(r)$.

51. REAL ESTATE Yuri manages 150 apartments in Irvine, California. All the apartments can be rented at a price of $1,200 per month each, but for each $100 increase in the monthly rent, there will be five additional vacancies.

    a. Express the total monthly revenue R obtained from renting apartments as a function of the monthly rental price p for each unit.

    b. Sketch the graph of the revenue function found in part (a).

    c. What monthly rental price p should Yuri charge to maximize total revenue? What is the maximum revenue?

52. REAL ESTATE Suppose it costs $500 each month for Yuri's real estate company in Exercise 51 to maintain and advertise each unit that is unrented.

    a. Express the total monthly revenue R obtained from renting apartments as a function of the monthly rental price p for each unit.

    b. Sketch the graph of the revenue function found in part (a).

    c. What monthly rental price p should Yuri charge to maximize total revenue? What is the maximum revenue?

53. AIR POLLUTION Lead emissions are a major source of air pollution. Using data gathered by the U.S. Environmental Protection Agency in the 1990s, it can be shown that the formula

$$N(t) = -35t^2 + 299t + 3,347$$

estimates the total amount of lead emission N (in thousands of tons) occurring in the United States t years after the base year 1990.
a. Sketch the graph of the pollution function $N(t)$.
b. Approximately how much lead emission did the formula predict for the year 1995? (The actual amount was about 3,924 thousand tons.)
c. Based on this formula, when during the decade 1990–2000 would you expect the maximum lead emission to have occurred?
d. Can this formula be used to accurately predict the current level of lead emission? Explain.

54. **ARCHITECTURE** An arch over a road has a parabolic shape. It is 6 meters wide at the base and is just tall enough to allow a truck 5 meters high and 4 meters wide to pass.
a. Assuming that the arch has an equation of the form $y = ax^2 + b$, use the given information to find a and b. Explain why this assumption is reasonable.
b. Sketch the graph of the arch equation you found in part (a).

55. **ROAD SAFETY** When an automobile is being driven at v miles per hour, the average driver requires D feet of visibility to stop safely, where $D = 0.065v^2 + 0.148v$. Sketch the graph of $D(v)$.

56. **POSTAL RATES** Effective May 11, 2009, the cost of mailing a first-class letter was 44 cents for the first ounce and 17 cents for each additional ounce or fraction of an ounce. Let $P(w)$ be the postage required for mailing a letter weighing w ounces, for $0 \le w \le 3$.
a. Describe $P(w)$ as a piecewise-defined function.
b. Sketch the graph of P.

57. **MOTION OF A PROJECTILE** If an object is thrown vertically upward from the ground with an initial speed of 160 feet per second, its height (in feet) t seconds later is given by the function $H(t) = -16t^2 + 160t$.
a. Graph the function $H(t)$.
b. Use the graph in part (a) to determine when the object will hit the ground.
c. Use the graph in part (a) to estimate how high the object will rise.

58. **MOTION OF A PROJECTILE** A missile is projected vertically upward from an underground bunker in such a way that t seconds after launch, it is s feet above the ground, where
$$s(t) = -16t^2 + 800t - 15$$
a. How deep is the bunker?
b. Sketch the graph of $s(t)$.
c. Use the graph in part (b) to determine when the missile is at its highest point. What is its maximum height?

In Exercises 59 through 62, use the vertical line test to determine whether or not the given curve is the graph of a function.

59.

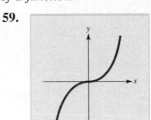

60.

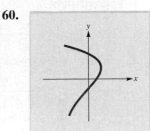

61.

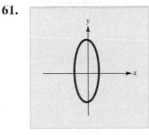

62.

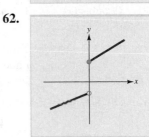

63. What viewing rectangle should be used to get an adequate graph for the quadratic function
$$f(x) = -9x^2 + 3,600x - 358,200?$$

64. What viewing rectangle should be used to get an adequate graph for the quadratic function
$$f(x) = 4x^2 - 2,400x + 355,000?$$

65. a. Graph the functions $y = x^2$ and $y = x^2 + 3$. How are the graphs related?
b. Without further computation, graph the function $y = x^2 - 5$.
c. Suppose $g(x) = f(x) + c$, where c is a constant. How are the graphs of f and g related? Explain.

66. a. Graph the functions $y = x^2$ and $y = -x^2$. How are the graphs related?
b. Suppose $g(x) = -f(x)$. How are the graphs of f and g related? Explain.

67. a. Graph the functions $y = x^2$ and $y = (x - 2)^2$. How are the graphs related?

b. Without further computation, graph the function $y = (x + 1)^2$.

c. Suppose $g(x) = f(x - c)$, where c is a constant. How are the graphs of f and g related? Explain.

68. Use your graphing utility to graph $y = x^4$, $y = x^4 - x$, $y = x^4 - 2x$, and $y = x^4 - 3x$ on the same coordinate axes, using $[-2, 2]1$ by $[-2, 5]1$. What effect does the added term involving x have on the shape of the graph? Repeat using $y = x^4$, $y = x^4 - x^3$, $y = x^4 - 2x^3$, and $y = x^4 - 3x^3$. Adjust the viewing rectangle appropriately.

69. Graph $f(x) = \dfrac{-9x^2 - 3x - 4}{4x^2 + x - 1}$. Determine the values of x for which the function is defined.

70. Graph $f(x) = \dfrac{8x^2 + 9x + 3}{x^2 + x - 1}$. Determine the values of x for which the function is defined.

71. Graph $g(x) = -3x^3 + 7x + 4$. Find the x intercepts.

72. Use the distance formula to show that the circle with center (a, b) and radius R has the equation
$$(x - a)^2 + (y - b)^2 = R^2$$

73. Use the result in Exercise 72 to solve the following problems.

a. Find an equation for the circle with center $(2, -3)$ and radius 4.

b. Find the center and radius of the circle with the equation
$$x^2 + y^2 - 4x + 6y = 11$$

c. Describe the set of all points (x, y) such that
$$x^2 + y^2 + 4y = 2x - 10$$

74. Show that the vertex of the parabola $y = Ax^2 + Bx + C$ $(A \neq 0)$ occurs at the point where $x = \dfrac{-B}{2A}$. $\{$*Hint:* First verify that
$$Ax^2 + Bx + C = A\left[\left(x + \frac{B}{2A}\right)^2 + \left(\frac{C}{A} - \frac{B^2}{4A^2}\right)\right].$$
Then note that the largest or smallest value of $f(x) = Ax^2 + Bx + C$ must occur where $x = \dfrac{-B}{2A}\}$.

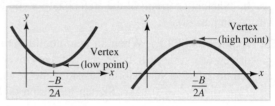

EXERCISE 74

SECTION 1.3 Lines and Linear Functions

Learning Objectives

1. Review properties of lines: slope, horizontal and vertical lines, and forms for the equation of a line.
2. Solve applied problems involving linear functions.
3. Study parallel and perpendicular lines.
4. Explore a least-squares linear approximation of data.

A **linear function** is one with the general form $f(x) = mx + b$, and the graph of such a function is a line. Linear functions play an important role in many practical applications, some of which we shall examine later in this section, but first we provide a brief review of the key properties of lines.

The Slope of a Line   A surveyor might say that a hill with a *rise* of 2 feet for every foot of *run* has a **slope** of

$$m = \frac{\text{rise}}{\text{run}} = \frac{2}{1} = 2$$

The steepness of a line can be measured by slope in much the same way. In particular, suppose (x_1, y_1) and (x_2, y_2) lie on a line as indicated in Figure 1.17. Between these points, x changes by the amount $x_2 - x_1$ and y by the amount $y_2 - y_1$. The slope is the ratio

$$\text{Slope} = \frac{\text{change in } y}{\text{change in } x} = \frac{y_2 - y_1}{x_2 - x_1}$$

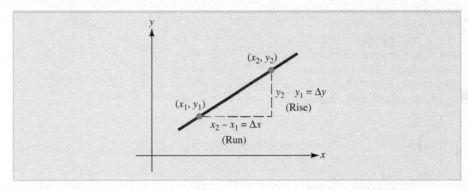

FIGURE 1.17   Slope $= \dfrac{y_2 - y_1}{x_2 - x_1} = \dfrac{\Delta y}{\Delta x}$.

It is sometimes convenient to use the symbol Δy instead of $y_2 - y_1$ to denote the change in y. The symbol Δy is read "delta y." Similarly, the symbol Δx is used to denote the change $x_2 - x_1$.

The Slope of a Line ■ The slope of the nonvertical line passing through the points (x_1, y_1) and (x_2, y_2) is given by the formula

$$\text{Slope} = \frac{\Delta y}{\Delta x} = \frac{y_2 - y_1}{x_2 - x_1}$$

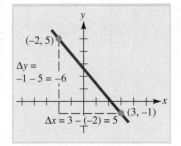

FIGURE 1.18   The line joining $(-2, 5)$ and $(3, -1)$.

EXAMPLE 1.3.1 Finding the Slope of a Line

Find the slope of the line joining the points $(-2, 5)$ and $(3, -1)$.

Solution

$$\text{Slope} = \frac{\Delta y}{\Delta x} = \frac{-1 - 5}{3 - (-2)} = \frac{-6}{5}$$

The line is shown in Figure 1.18.

The sign and magnitude of the slope of a line indicate the line's direction and steepness, respectively. The slope is positive if the height of the line increases as x increases and is negative if the height decreases as x increases. The absolute value of the slope is large if the slant of the line is severe and small if the slant of the line is gradual. The situation is illustrated in Figure 1.19.

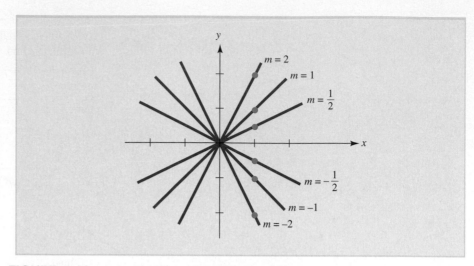

FIGURE 1.19 The direction and steepness of a line.

Horizontal and Vertical Lines

Horizontal and vertical lines (Figures 1.20a and 1.20b) have particularly simple equations. The y coordinates of all points on a horizontal line are the same. Hence, a horizontal line is the graph of a linear function of the form $y = b$, where b is a constant. The slope of a horizontal line is zero, since changes in x produce no changes in y.

The x coordinates of all points on a vertical line are equal. Hence, vertical lines are characterized by equations of the form $x = c$, where c is a constant. The slope of a vertical line is undefined because only the y coordinates of points on the line can change, so the denominator of the quotient $\dfrac{\text{change in } y}{\text{change in } x}$ is zero.

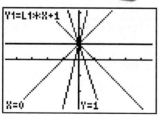
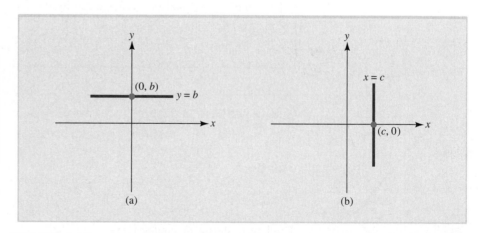

FIGURE 1.20 Horizontal and vertical lines.

Forms for the Equation of a Line

The constants m and b in the equation $y = mx + b$ of a nonvertical line have geometric interpretations. The coefficient m is the slope of the line. To see this, suppose that (x_1, y_1) and (x_2, y_2) are two points on the line $y = mx + b$. Then, $y_1 = mx_1 + b$

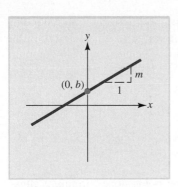

FIGURE 1.21 The slope and y intercept of the line $y = mx + b$.

and $y_2 = mx_2 + b$, so

$$\text{Slope} = \frac{y_2 - y_1}{x_2 - x_1} = \frac{(mx_2 + b) - (mx_1 + b)}{x_2 - x_1}$$

$$= \frac{mx_2 - mx_1}{x_2 - x_1} = \frac{m(x_2 - x_1)}{x_2 - x_1} = m$$

The constant b in the equation $y = mx + b$ is the value of y corresponding to $x = 0$; so $(0, b)$ is the point where the line crosses the y axis, that is, the y intercept of the line, as shown in Figure 1.21.

Because the constants m and b in the equation $y = mx + b$ correspond to the slope and y intercept, respectively, this form of the equation of a line is known as the **slope-intercept form.**

The Slope-Intercept Form of the Equation of a Line ■ The equation

$$y = mx + b$$

is the equation of the line whose slope is m and whose y intercept is $(0, b)$.

EXAMPLE 1.3.2 Using Intercepts to Graph a Line

Find the slope and y intercept of the line $3y + 2x = 6$, and draw the graph.

Solution

First put the equation $3y + 2x = 6$ in slope-intercept form $y = mx + b$. To do this, solve for y to get

$$3y = -2x + 6 \qquad \text{or} \qquad y = -\frac{2}{3}x + 2$$

It follows that the slope is $-\frac{2}{3}$ and the y intercept is $(0, 2)$.

The easiest way to graph any line is to find two of its points and pass a straight line through them. In the present case, you already have the y intercept $(0, 2)$, and by setting $y = 0$ in the equation $3y + 2x = 6$ to get $x = 3$, you find that the x intercept is $(3, 0)$. Passing a line through $(0, 2)$ and $(3, 0)$, you get the graph shown in Figure 1.22.

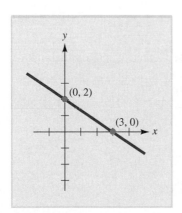

FIGURE 1.22 The line $3y + 2x = 6$.

The slope-intercept form for the equation of a line is especially useful if you have the equation of a particular line and want to know its slope and y intercept. In practical applications, you are more likely to be given information that allows you to find the slope of a line and one of its points. Here is a formula for finding the equation of the line in such circumstances.

The Point-Slope Form of the Equation of a Line ■ The equation

$$y - y_0 = m(x - x_0)$$

is an equation of the line that passes through the point (x_0, y_0) and that has slope equal to m.

EXPLORE!

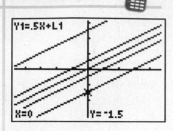

Find the y intercept values needed in List L1 so that the function Y1 = 0.5X + L1 creates the screen shown here.

The point-slope form of the equation of a line is simply the formula for slope in disguise. To see this, suppose the point (x, y) lies on the line that passes through a given point (x_0, y_0) and has slope m. Using the points (x, y) and (x_0, y_0) to compute the slope, you get

$$\frac{y - y_0}{x - x_0} = m$$

which you can put in point-slope form

$$y - y_0 = m(x - x_0)$$

by simply multiplying both sides by $x - x_0$. Use of the point-slope formula is illustrated in Example 1.3.3.

EXAMPLE 1.3.3 Finding the Equation of a Line

Find the equation of the line that passes through the point $(5, 1)$ with slope $\dfrac{1}{2}$.

Solution

Use the formula $y - y_0 = m(x - x_0)$ with $(x_0, y_0) = (5, 1)$ and $m = \dfrac{1}{2}$ to get

$$y - 1 = \frac{1}{2}(x - 5)$$

which you can rewrite as

$$y = \frac{1}{2}x - \frac{3}{2}$$

The graph is shown in Figure 1.23.

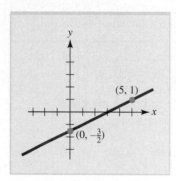

FIGURE 1.23 The line $y = \dfrac{1}{2}x - \dfrac{3}{2}$.

In Chapter 2, the point-slope formula will be used extensively for finding the equation of the tangent line to the graph of a function at a given point. Another useful application of the formula is in finding the equation of a line through two given points, as illustrated in Example 1.3.4.

EXAMPLE 1.3.4 Finding the Equation of a Line

Find the equation of the line that passes through the points $(3, -2)$ and $(1, 6)$.

Solution

First compute the slope

$$m = \frac{6 - (-2)}{1 - 3} = \frac{8}{-2} = -4$$

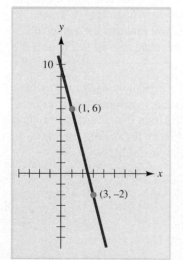

FIGURE 1.24 The line $y = -4x + 10$.

Then use the point-slope formula with $(1, 6)$ as the given point (x_0, y_0) to get

$$y - 6 = -4(x - 1) \qquad \text{or} \qquad y = -4x + 10$$

Note that if we had used the point $(3, -2)$ rather than $(1, 6)$, the resulting equation would have been the same. (Try it for practice.) The graph is shown in Figure 1.24.

NOTE   The general form for the equation of a line is $Ax + By + C = 0$, where A, B, C are constants, with A and B not both equal to 0. If $B = 0$, the line is vertical, and when $B \neq 0$, the equation $Ax + By + C = 0$ can be rewritten as

$$y = \left(\frac{-A}{B}\right)x + \left(\frac{-C}{B}\right)$$

Comparing this equation with the slope-intercept form $y = mx + b$, we see that the line has slope $m = -\dfrac{A}{B}$ and y intercept $\left(0, -\dfrac{C}{B}\right)$. The line is horizontal (slope 0) when $A = 0$. ∎

Applications Involving Linear Functions

A quantity y that varies at a constant rate m with respect to another quantity x can be described as a linear function $y = f(x) = mx + b$. This feature of linear functions is used in Examples 1.3.5 and 1.3.6.

EXAMPLE 1.3.5 Writing a Linear Cost Function

A manufacturer's total cost consists of a fixed overhead of $200 plus production costs of $50 per unit. Express the total cost as a function of the number of units produced, and draw the graph.

Solution

Let x denote the number of units produced and $C(x)$ the corresponding total cost. Then,

$$\text{Total cost} = (\text{cost per unit})(\text{number of units}) + \text{overhead}$$

where

$$\text{Cost per unit} = 50$$
$$\text{Number of units} = x$$
$$\text{Overhead} = 200$$

Hence,

$$C(x) = 50x + 200$$

The graph of this linear cost function is the line shown in Figure 1.25 with slope $m = 50$ equal to the constant increase in cost per unit of production and y intercept $(0, 200)$ corresponding to the overhead.

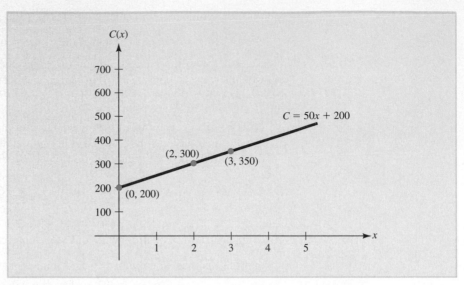

FIGURE 1.25 The cost function, $C(x) = 50x + 200$.

EXAMPLE 1.3.6 Writing a Linear Price Function

Since the beginning of the year, the price of a bottle of soda at a local discount super-market has been rising at a constant rate of 2 cents per month. By November 1, the price had reached $1.56 per bottle. Express the price of the soda as a function of time, and determine the price at the beginning of the year.

Solution

Let x denote the number of months that have elapsed since the first of the year and y the price of a bottle of soda (in cents). Since y changes at a constant rate with respect to x, the function relating y to x must be linear, and its graph is a straight line. Since the price y increases by 2 each time x increases by 1, the slope of the line must be 2. The fact that 10 months after the first of the year (November 1), the price was 156 cents ($1.56) tells us that the line passes through the point (10, 156). Now we can use the point-slope formula to express y as a function of x:

$$y - y_0 = m(x - x_0)$$

with $\quad\quad\quad\quad\quad m = 2, x_0 = 10, y_0 = 156$

to get $\quad\quad\quad y - 156 = 2(x - 10) \quad\quad$ or $\quad\quad y = 2x + 136$

The corresponding line is shown in Figure 1.26. Notice that the y intercept is (0, 136), which indicates that the price of soda at the beginning of the year was $1.36 per bottle.

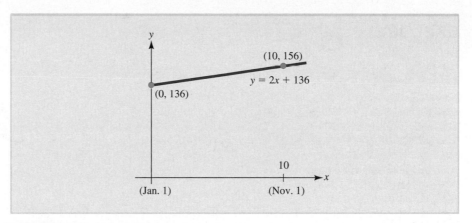

FIGURE 1.26   The rising price of soda: $y = 2x + 136$.

Parallel and Perpendicular Lines

In applications, it is sometimes necessary or useful to know whether two given lines are parallel or perpendicular. A vertical line is parallel only to other vertical lines and is perpendicular to any horizontal line. Cases involving nonvertical lines can be handled by the following slope criteria.

> **Parallel and Perpendicular Lines** ■ Let m_1 and m_2 be the slopes of the nonvertical lines L_1 and L_2. Then
>
> L_1 and L_2 are **parallel** if and only if $m_1 = m_2$.
>
> L_1 and L_2 are **perpendicular** if and only if $m_2 = \dfrac{-1}{m_1}$.

These criteria are demonstrated in Figure 1.27, and geometric proofs are outlined in Exercises 64 and 65. Example 1.3.7 illustrates one way the criteria can be used.

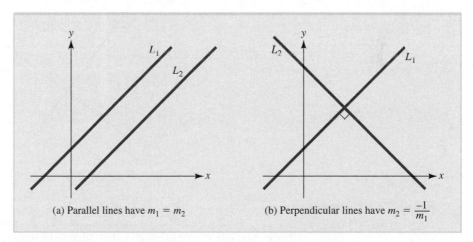

(a) Parallel lines have $m_1 = m_2$ (b) Perpendicular lines have $m_2 = \dfrac{-1}{m_1}$

FIGURE 1.27   Slope criteria for parallel and perpendicular lines.

EXAMPLE 1.3.7   Finding Parallel and Perpendicular Lines

Let L be the line $4x + 3y = 3$.
   a. Find the equation of a line L_1 parallel to L through $(-1, 4)$.
   b. Find the equation of a line L_2 perpendicular to L through $(2, -3)$.

EXPLORE!

Write Y1 = AX + 2 and Y2 = (−1/A)X + 5 in the equation editor of your graphing calculator. On the home screen, store different values into A and then graph both lines using a **ZOOM** Square Window. What do you notice for different values of A (A ≠ 0)? Can you solve for the point of intersection in terms of the value A?

Solution

By rewriting the equation $4x + 3y = 3$ in the slope-intercept form $y = -\dfrac{4}{3}x + 1$, we see that L has slope $m_L = -\dfrac{4}{3}$.

a. Any line parallel to L must also have slope $m = -\dfrac{4}{3}$. The required line L_1 contains $(-1, 4)$, so by using the point-slope formula, we get

$$y - 4 = -\frac{4}{3}(x + 1)$$

$$y = -\frac{4}{3}x + \frac{8}{3}$$

b. A line perpendicular to L must have slope $m = -\dfrac{1}{m_L} = \dfrac{3}{4}$. Since the required line L_2 contains $(2, -3)$, we have

$$y + 3 = \frac{3}{4}(x - 2)$$

$$y = \frac{3}{4}x - \frac{9}{2}$$

The given line L and the required lines L_1 and L_2 are shown in Figure 1.28.

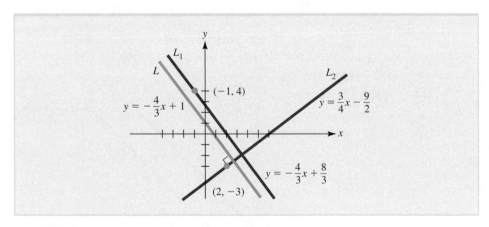

FIGURE 1.28 Lines parallel and perpendicular to a given line L.

Preview of Least-Squares Linear Approximation of Data

Suppose a researcher plots a set of data for two variables on a coordinate plane and observes that the data are roughly linear. It may be useful to the researcher to have a linear formula relating x and y, but such a formula cannot be obtained by the algebraic methods we have discussed in this section since the data are only *approximately* linear.

Least-squares linear approximation of data is a procedure for finding a line positioned so that the sum of squares of vertical distances from the data points to the line is minimized. The least-squares method is developed in Section 7.4. For now, you can use your calculator to find and apply the line that best fits a given set of data in the least-squares sense. This procedure is illustrated in Example 1.3.8.

TABLE 1.2 Percentage of Civilian Unemployment 1991–2000

Year	Number of Years from 1991	Percentage of Unemployed
1991	0	6.8
1992	1	7.5
1993	2	6.9
1994	3	6.1
1995	4	5.6
1996	5	5.4
1997	6	4.9
1998	7	4.5
1999	8	4.2
2000	9	4.0

SOURCE: U.S. Bureau of Labor Statistics, Bulletin 2307; and *Employment and Earnings,* monthly.

EXPLORE!

Place the data in Table 1.2 into L1 and L2 of the **STAT** data editor, where L1 is the number of years from 1991 and L2 is the percentage of unemployed. Following the Calculator Introduction for statistical graphing using the **STAT** and **STAT PLOT** keys, verify the scatterplot and best-fit line displayed in Figure 1.29.

EXAMPLE 1.3.8 Finding a Best-Fitting Line

Table 1.2 lists the percentage of the labor force that was unemployed during the decade 1991–2000. Plot a graph with time (years after 1991) on the x axis and percentage of unemployment on the y axis. Do the points follow a clear pattern? Based on these data, what would you expect the percentage of unemployment to be in the year 2005?

Solution

The unemployment data are plotted in the graph in Figure 1.29. Note that except for the initial point (0, 6.8) the distribution of data points is roughly linear. The Explore! box in the margin describes how to use your calculator to apply the least-squares procedure to these data, and you get $y = -0.389x + 7.338$ as the equation of the best-fitting line. By evaluating the corresponding linear function $f(x) = -0.389x + 7.338$ at $x = 14$ (the year 2005), you find that

$$f(14) = -0.389(14) + 7.338 = 1.892$$

so least-squares approximation of data predicts an approximate 1.9% unemployment rate in 2005.

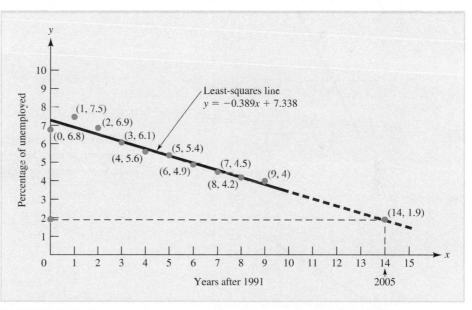

FIGURE 1.29 Percentage of unemployed in the United States for 1991–2000.

> **NOTE** Care must be taken when making predictions by extrapolating from known data, especially when the data set is as small as the one in Example 1.3.8. In particular, the economy began to weaken after the year 2000, but the least-squares line in Figure 1.29 predicts a steadily decreasing unemployment rate. Is this reasonable? In Exercise 48, you are asked to explore this question by using unemployment data from 2001 to 2009 and then comparing these new data with the values predicted by the least-squares line. ■

EXERCISES ▪ 1.3

In Exercises 1 through 8, find the slope (if defined) of the line that passes through the given pair of points.

1. $(2, -3)$ and $(0, 4)$ **2.** $(-1, 2)$ and $(2, 5)$

3. $(2, 0)$ and $(0, 2)$ **4.** $(5, -1)$ and $(-2, -1)$

5. $(2, 6)$ and $(2, -4)$

6. $\left(\dfrac{2}{3}, -\dfrac{1}{5}\right)$ and $\left(-\dfrac{1}{7}, \dfrac{1}{8}\right)$

7. $\left(\dfrac{1}{7}, 5\right)$ and $\left(-\dfrac{1}{11}, 5\right)$

8. $(-1.1, 3.5)$ and $(-1.1, -9)$

In Exercises 9 through 12, find the slope and intercepts of the line shown. Then find an equation for the line.

9.
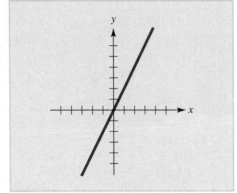

10.
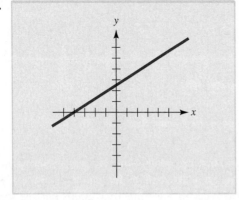

11.
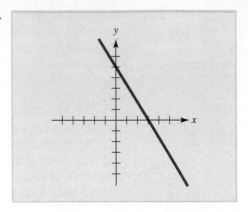

12.
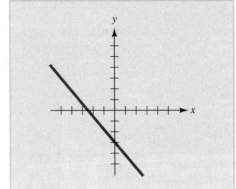

In Exercises 13 through 20, find the slope and intercepts of the line whose equation is given and sketch the graph of the line.

13. $x = 3$ **14.** $y = 5$

15. $y = 3x$ **16.** $y = 3x - 6$

17. $3x + 2y = 6$ **18.** $5y - 3x = 4$

19. $\dfrac{x}{2} + \dfrac{y}{5} = 1$ **20.** $\dfrac{x + 3}{-5} + \dfrac{y - 1}{2} = 1$

In Exercises 21 through 36, write an equation for the line with the given properties.

21. Through $(2, 0)$ with slope 1

22. Through $(-1, 2)$ with slope $\dfrac{2}{3}$

23. Through $(5, -2)$ with slope $-\dfrac{1}{2}$

24. Through $(0, 0)$ with slope 5

25. Through $(2, 5)$ and parallel to the x axis

26. Through (2, 5) and parallel to the y axis

27. Through (1, 0) and (0, 1)

28. Through (2, 5) and (1, −2)

29. Through $\left(-\dfrac{1}{5}, 1\right)$ and $\left(\dfrac{2}{3}, \dfrac{1}{4}\right)$

30. Through (−2, 3) and (0, 5)

31. Through (1, 5) and (3, 5)

32. Through (1, 5) and (1, −4)

33. Through (4, 1) and parallel to the line $2x + y = 3$

34. Through (−2, 3) and parallel to the line
$x + 3y = 5$

35. Through (3, 5) and perpendicular to the line
$x + y = 4$

36. Through $\left(-\dfrac{1}{2}, 1\right)$ and perpendicular to the line
$2x + 5y = 3$

BUSINESS AND ECONOMICS APPLIED PROBLEMS

37. **MANUFACTURING COST** A manufacturer's total cost consists of a fixed overhead of $5,000 plus production costs of $60 per unit.
    a. Express the total cost as a function of the number of units produced, and sketch its graph.
    b. Find the average cost function $AC(x)$. What is the average cost of producing 20 units?

38. **MANUFACTURING COST** A manufacturer estimates that it costs $75 to produce each unit of a particular commodity. The fixed overhead is $4,500.
    a. Express the total cost of production as a function of the number of units produced, and sketch the graph.
    b. Find the average cost function $AC(x)$. What is the average cost of producing 50 units?

39. **CREDIT CARD DEBT** A credit card company estimates that the average cardholder owed $7,853 in the year 2005 and $9,127 in 2010. Suppose average cardholder debt D grows at a constant rate.
    a. Express D as a linear function of time t, where t is the number of years after 2005. Draw the graph.
    b. Use the function in part (a) to predict the average cardholder debt in the year 2015.
    c. Approximately when will the average cardholder debt be double the amount in the year 2005?

40. **CAR RENTAL** A car rental agency charges $75 per day plus 70 cents per mile.
    a. Express the cost of renting a car from this agency for 1 day as a function of the number of miles driven, and draw the graph.
    b. How much does it cost to rent a car for a 1-day trip of 50 miles?
    c. The agency also offers a rental for a flat fee of $125 per day. How many miles must you drive on a 1-day trip for this to be the better deal?

41. **LINEAR DEPRECIATION** Dr. Adams owns $1,500 worth of medical books which, for tax purposes, are assumed to depreciate linearly to zero over a 10-year period. That is, the value of the books decreases at a constant rate so that it is equal to zero at the end of 10 years. Express the value of the doctor's books as a function of time, and draw the graph.

42. **LINEAR DEPRECIATION** A manufacturer buys $20,000 worth of machinery that depreciates linearly so that its trade-in value after 10 years will be $1,000.
    a. Express the value of the machinery as a function of its age, and draw the graph.
    b. Compute the value of the machinery after 4 years.
    c. When does the machinery become worthless?
    d. The manufacturer might not wait this long to dispose of the machinery. Discuss the issues the manufacturer may consider in deciding when to sell.

43. **ACCOUNTING** For tax purposes, the book value of certain assets is determined by depreciating the original value of the asset linearly over a fixed period of time. Suppose an asset originally worth V dollars is linearly depreciated over a period of N years, at the end of which it has a scrap value of S dollars.
    a. Express the book value B of the asset t years into the N-year depreciation period as a linear function of t. [*Hint:* Note that $B = V$ when $t = 0$ and $B = S$ when $t = N$.]
    b. Suppose a $50,000 piece of office equipment is depreciated linearly over a 5-year period, with a scrap value of $18,000. What is the book value of the equipment after 3 years?

44. **PRINTING COST** A publisher estimates that the cost of producing between 1,000 and 10,000 copies of a certain textbook is $50 per copy;

between 10,001 and 20,000, the cost is $40 per copy; and between 20,001 and 50,000, the cost is $35 per copy.

a. Find a function $F(N)$ that gives the total cost of producing N copies of the text for $1,000 \leq N \leq 50,000$.

b. Sketch the graph of the function $F(N)$ you found in part (a).

45. **STOCK PRICES**   A certain stock had an initial public offering (IPO) price of $10 per share and is traded 24 hours a day. Sketch the graph of the share price over a 2-year period for each of the following cases:

a. The price increases steadily to $50 a share over the first 18 months and then decreases steadily to $25 per share over the next 6 months.

b. The price takes just 2 months to rise at a constant rate to $15 a share and then slowly drops to $8 over the next 9 months before steadily rising to $20.

c. The price steadily rises to $60 a share during the first year, at which time an accounting scandal is uncovered. The price gaps down to $25 a share and then steadily decreases to $5 over the next 3 months before rising at a constant rate to close at $12 at the end of the 2-year period.

46. **EQUIPMENT RENTAL**   A rental company rents a piece of equipment for a $60.00 flat fee plus an hourly fee of $5.00 per hour.

a. Make a chart showing the number of hours the equipment is rented and the cost for renting the equipment for 2 hours, 5 hours, 10 hours, and t hours of time.

b. Write an algebraic expression representing the cost y as a function of the number of hours t. Assume t can be measured to any decimal portion of an hour. (In other words, assume t is any nonnegative real number.)

c. Graph the expression from part (b).

d. Use the graph to approximate, to two decimal places, the number of hours the equipment was rented if the bill is $216.25 (before taxes).

47. **APPRECIATION OF ASSETS**   Aria owns a rare book that doubles in value every 10 years. In 1900, the book was worth $100.

a. How much was the book worth in 1930? In 2000? How much should Aria expect her book to be worth in the year 2020?

b. Is the value of the book a linear function of its age? Answer the question by interpreting an appropriate graph.

48. **UNEMPLOYMENT RATE**   In the solution to Example 1.3.8, we noted that the line that best fits the unemployment data in the example in the sense of least-squares approximation has the equation $y = -0.389x + 7.338$. The data in the example stop at the year 2000. The accompanying table gives the unemployment data for the years 2001 through 2009.

Year	Number of Years from 2001	Percentage of Unemployed
2001	0	4.7
2002	1	5.8
2003	2	6.0
2004	3	5.5
2005	4	5.1
2006	5	4.6
2007	6	4.6
2008	7	5.8
2009	8	9.3

TABLE FOR EXERCISE 48   Percentage of Civilian Unemployment 2001–2009

a. Plot the data in the table in a coordinate plane. Then apply the calculator instructions in the Explore! box on page 39 to show that the line that best fits these data in the sense of least-squares approximation has the equation $y = 0.245x + 4.731$.

b. Interpret the slope of the line in part (a) in terms of the rate of unemployment.

c. Does the least-squares line do a good job of predicting these new data? Explain.

LIFE AND SOCIAL SCIENCE APPLIED PROBLEMS

49. **COURSE REGISTRATION**   Students at a state college may preregister for their fall classes by mail during the summer. Those who do not preregister must register in person in September. The registrar can process 35 students per hour during the September registration period. Suppose that after 4 hours in September, a total of 360 students (including those who preregistered) have been registered.

a. Express the number of students registered as a function of time and draw the graph.

b. How many students were registered after 3 hours?

c. How many students preregistered during the summer?

50. **ENTOMOLOGY** It has been observed that the number of chirps made by a cricket each minute depends on the temperature. Crickets won't chirp if the temperature is 38°F or less, and observations yield the following data:

Number of chirps (C)	0	5	10	20	60
Temperature T (°F)	38	39	40	42	50

a. Express T as a linear function of C.

b. How many chirps would you expect to hear if the temperature is 75°F? If you hear 37 chirps in a 30-second period of time, what is the approximate temperature?

51. **GROWTH OF A CHILD** The average height H in centimeters of a child of age A years can be estimated by the linear function $H = 6.5A + 50$. Use this formula to answer these questions.

a. What is the average height of a 7-year-old child?

b. How old must a child be to have an average height of 150 cm?

c. What is the average height of a newborn baby? Does this answer seem reasonable?

d. What is the average height of a 20-year-old? Does this answer seem reasonable?

52. **MEMBERSHIP FEES** Membership in a swimming club costs $250 for the 12-week summer season. If a member joins after the start of the season, the fee is prorated; that is, it is reduced linearly.

a. Express the membership fee as a function of the number of weeks that have elapsed by the time the membership is purchased and draw the graph.

b. Compute the cost of a membership that is purchased 5 weeks after the start of the season.

53. **WATER CONSUMPTION** Since the beginning of the month, a local reservoir has been losing water at a constant rate. On the 12th of the month the reservoir held 200 million gallons of water, and on the 21st it held only 164 million gallons.

a. Express the amount of water in the reservoir as a function of time, and draw the graph.

b. How much water was in the reservoir on the 8th of the month?

54. **CAR POOLING** To encourage motorists to form car pools, the transit authority in a certain metropolitan area has been offering a special reduced rate at toll bridges for vehicles containing four or more persons. When the program began 30 days ago, 157 vehicles qualified for the reduced rate during the morning rush hour. Since then, the number of vehicles qualifying has been increasing at a constant rate, and today 247 vehicles qualified.

a. Express the number of vehicles qualifying each morning for the reduced rate as a function of time, and draw the graph.

b. If the trend continues, how many vehicles will qualify during the morning rush hour 14 days from now?

55. **TEMPERATURE CONVERSION**

a. Temperature measured in degrees Fahrenheit is a linear function of temperature measured in degrees Celsius. Use the fact that 0° Celsius is equal to 32° Fahrenheit and 100° Celsius is equal to 212° Fahrenheit to write an equation for this linear function.

b. Use the function you obtained in part (a) to convert 15° Celsius to Fahrenheit.

c. Convert 68° Fahrenheit to Celsius.

d. What temperature is the same in both the Celsius and Fahrenheit scales?

56. **NUTRITION** Each ounce of Food I contains 3 g of carbohydrate and 2 g of protein, and each ounce of Food II contains 5 g of carbohydrate and 3 g of protein. Suppose x ounces of Food I are mixed with y ounces of Food II. The foods are combined to produce a blend that contains exactly 73 g of carbohydrate and 46 g of protein.

a. Explain why there are $3x + 5y$ g of carbohydrate in the blend and why we must have $3x + 5y = 73$. Find a similar equation for protein. Sketch the graphs of both equations.

b. Where do the two graphs in part (a) intersect? Interpret the significance of this point of intersection.

57. **COLLEGE ADMISSIONS** The average scores of incoming students at an eastern liberal arts college in the SAT mathematics examination have been declining at a constant rate in recent years. In 2005, the average SAT score was 575, while in 2010 it was 545.

a. Express the average SAT score as a function of time.

b. If the trend continues, what will the average SAT score of incoming students be in 2015?

c. If the trend continues, when will the average SAT score be 527?

58. ALCOHOL ABUSE CONTROL   Ethyl alcohol is metabolized by the human body at a constant rate (independent of concentration). Suppose the rate is 10 milliliters per hour.

a. How much time is required to eliminate the effects of a liter of beer containing 3% ethyl alcohol?

b. Express the time T required to metabolize the effects of drinking ethyl alcohol as a function of the amount A of ethyl alcohol consumed.

c. Discuss how the function in part (b) can be used to determine a reasonable "cutoff" value for the amount of ethyl alcohol A that each individual may be served at a party.

59. AIR POLLUTION   In certain parts of the world, the number of deaths N per week have been observed to be linearly related to the average concentration x of sulfur dioxide in the air. Suppose there are 97 deaths when $x = 100$ mg/m^3 and 110 deaths when $x = 500$ mg/m^3.

a. What is the functional relationship between N and x?

b. Use the function in part (a) to find the number of deaths per week when $x = 300$ mg/m^3. What concentration of sulfur dioxide corresponds to 100 deaths per week?

c. Research data on how air pollution affects the death rate in a population.* Summarize your results in a one-paragraph essay.

MISCELLANEOUS PROBLEMS

60. ASTRONOMY   The following table gives the length of year L (in earth years) of each planet in the solar system, along with the mean (average) distance D of the planet from the sun, in astronomical units (1 astronomical unit is the mean distance of the earth from the sun).

*You may find the following articles helpful: D. W. Dockery, J. Schwartz, and J. D. Spengler, "Air Pollution and Daily Mortality: Associations with Particulates and Acid Aerosols," *Environ. Res.,* Vol. 59, 1992, pp. 362–373; Y. S. Kim, "Air Pollution, Climate, Socioeconomics Status and Total Mortality in the United States," *Sci. Total Environ.,* Vol. 42, 1985, pp. 245–256.

Planet	Mean Distance from Sun, D	Length of Year, L
Mercury	0.388	0.241
Venus	0.722	0.615
Earth	1.000	1.000
Mars	1.523	1.881
Jupiter	5.203	11.862
Saturn	9.545	29.457
Uranus	19.189	84.013
Neptune	30.079	164.783

a. Plot the point (D, L) for each planet on a coordinate plane. Do these quantities appear to be linearly related?

b. For each planet, compute the ratio $\dfrac{L^2}{D^3}$. Interpret what you find by expressing L as a function of D.

c. What you have discovered in part (b) is one of Kepler's laws, named for the German astronomer Johannes Kepler (1571–1630). Read an article about Kepler, and describe his place in the history of science.

61. AN ANCIENT FABLE   In Aesop's fable about the race between the tortoise and the hare, the tortoise trudges along at a slow, constant rate from start to finish. The hare starts out running steadily at a much more rapid pace, but halfway to the finish line, stops to take a nap. Finally, the hare wakes up, sees the tortoise near the finish line, desperately resumes his old rapid pace, and is nosed out by a hair. On the same coordinate plane, graph the respective distances of the tortoise and the hare from the starting line, regarded as functions of time.

62. Graph $y = \dfrac{54}{270}x - \dfrac{63}{19}$ and $y = \dfrac{139}{695}x - \dfrac{346}{14}$ on the same set of coordinate axes using $[-10, 10]1$ by $[-10, 10]1$ for a starting range. Adjust the range settings until both lines are displayed. Are the two lines parallel?

63. Graph $y = \dfrac{25}{7}x + \dfrac{13}{2}$ and $y = \dfrac{144}{45}x + \dfrac{630}{229}$ on the same set of coordinate axes using $[-10, 10]1$ by $[-10, 10]1$. Are the two lines parallel?

64. PARALLEL LINES Show that two nonvertical lines are parallel if and only if they have the same slope.

65. PERPENDICULAR LINES Show that if a nonvertical line L_1 with slope m_1 is perpendicular to a line L_2 with slope m_2, then $m_2 = \dfrac{-1}{m_1}$.

[*Hint:* Find expressions for the slopes of the lines L_1 and L_2 in the accompanying figure. Then apply the Pythagorean theorem along with the distance formula from Section 1.2 to the right triangle OAB to obtain the required relationship between m_1 and m_2.]

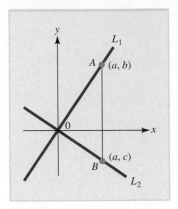

EXERCISE 65

SECTION 1.4 Functional Models

Learning Objectives
1. Study general modeling procedure.
2. Explore a variety of applied models.
3. Investigate market equilibrium and break-even analysis in economics.

Practical problems in business, economics, and the physical and life sciences are often too complicated to be precisely described by simple formulas, and one of our basic goals is to develop mathematical methods for dealing with such problems. Toward this end, we shall use a procedure called **mathematical modeling,** which may be described in terms of four stages displayed schematically in Figure 1.30:

Stage 1 (Formulation): Given a real-world situation (for example, the U.S. trade deficit, the AIDS epidemic, global weather patterns), we make enough simplifying assumptions to allow a mathematical formulation. This may require gathering and analyzing data and using knowledge from a variety of different areas to identify key variables and establish equations relating those variables. This formulation is called a **mathematical model.**

Stage 2 (Analysis of the Model): We use mathematical methods to analyze or "solve" the mathematical model. Calculus will be the primary tool of analysis in this text, but in practice, a variety of tools, such as algebra, statistics, numerical analysis, and computer methods may be brought to bear on a particular model.

Stage 3 (Interpretation): After the mathematical model has been analyzed, any conclusions that may be drawn from the analysis are applied to the original real-world problem, both to gauge the accuracy of the model and to make predictions. For instance, analysis of a model of a particular business may predict that profit will be maximized by producing 200 units of a certain commodity.

Stage 4 (Testing and Adjustment): In this final stage, the model is tested by gathering new data to check the accuracy of any predictions inferred from the analysis. If the predictions are not confirmed by the new evidence, the assumptions of the model are adjusted and the modeling process is repeated. Referring to the business example described in stage 3, it may be found that profit begins to wane at a production level significantly less than 200 units, which would indicate that the model requires modification.

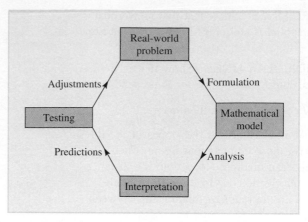

FIGURE 1.30   A diagram illustrating the mathematical modeling procedure.

In a good model, the real-world problem is idealized just enough to allow mathematical analysis but not so much that the essence of the underlying situation is compromised. For instance, if we assume that weather is strictly periodic, with rain occurring every 10 days, the resulting model would be relatively easy to analyze but would clearly be a poor representation of reality.

In preceding sections, you have seen models representing quantities such as manufacturing cost, price and demand, air pollution levels, and population size, and you will encounter many more throughout the rest of this text. Some of these models, especially those analyzed in the Think About It essays at the end of each chapter, are more detailed and illustrate how decisions are made about assumptions and predictions.

Constructing and analyzing mathematical models is one of the most important skills you will learn in calculus, and in this section you will encounter a number of techniques for setting up and solving various kinds of applied problems. In Example 1.4.1, we begin with an optimization problem in which the solution is presented in four parts corresponding to the stages of the modeling procedure.

EXAMPLE 1.4.1   Finding Maximum Profit

A manufacturer can produce printer paper at a cost of $2 per ream. The paper has been selling for $5 per ream, and at that price consumers have been buying 4,000 reams per month. The manufacturer is planning to raise the price of the paper and estimates that for each $1 increase in the price, 400 fewer reams will be sold each month. What price corresponds to the maximum profit, and what is the maximum profit?

Solution

Formulation: Begin by stating the desired relationship in words:

$$\text{Profit} = (\text{number of reams sold})(\text{profit per ream})$$

Since the goal is to express profit as a function of price, the independent variable is price and the dependent variable is profit. Let x denote the price at which each ream will be sold, and let $P(x)$ be the corresponding monthly profit. [Ordinarily, we would use p to represent price, but writing $P(p)$ for profit could be confusing.]

Next, express the number of reams sold in terms of the variable x. You know that 4,000 reams are sold each month when the price is $5 and that

400 fewer will be sold each month for each $1 increase in price. Since the number of $1 increases is the difference $x - 5$ between the new and old selling prices, you must have

$$\text{Number of reams sold} = 4{,}000 - 400(\text{number of \$1 increases})$$
$$= 4{,}000 - 400(x - 5)$$
$$= 6{,}000 - 400x$$

The profit per ream is the difference $x - 2$ between the selling price x and the cost $2, so the total profit is

$$P(x) = (\text{number of reams sold})(\text{profit per ream})$$
$$= (6{,}000 - 400x)(x - 2)$$
$$= -400x^2 + 6{,}800x - 12{,}000$$

Analysis: The graph of $P(x)$ is the downward-opening parabola shown in Figure 1.31. The maximum profit will occur at the value of x that corresponds to the highest point on the profit graph. This is the vertex of the parabola, which we know occurs where

$$x = \frac{-B}{2A} = \frac{-(6{,}800)}{2(-400)} = 8.5$$

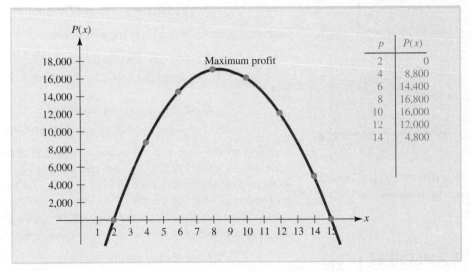

p	$P(x)$
2	0
4	8,800
6	14,400
8	16,800
10	16,000
12	12,000
14	4,800

FIGURE 1.31 The profit function $P(x) = (6{,}000 - 400x)(x - 2)$.

Interpretation: We conclude that the profit is maximized when the manufacturer charges $8.50 for each ream, and at that price the maximum monthly profit is

$$P_{\max} = P(8.5) = -400(8.5)^2 + 6{,}800(8.5) - 12{,}000$$
$$= \$16{,}900$$

Notice that if the manufacturer decides to charge less than $2 or more than $15 per ream, the profit function $P(x)$ becomes negative and the manufacturer experiences a loss. This fact is represented by the portion of the profit curve in Figure 1.31 that lies below the x axis.

Based on this analysis, the manufacturer might raise the price of paper to $8.50 per ream, expecting profit to be maximized at the $16,900 level. If sales actually begin to lag or are more robust than would be expected, the assumptions in the model would be adjusted and the analysis repeated.

In many applied problems, we have a quantity involving more than one variable and want to express it in terms of a single variable. The most common way to do this involves using a side condition, as illustrated in Example 1.4.2.

EXAMPLE 1.4.2 Modeling Construction Cost

Armando wants to build a closed cylindrical water tank for his farm with a capacity of 11,000 cubic feet. The cost of the material used for the top is $3 per square foot, while the material used for the bottom and the curved side is $5 per square foot. Express Armando's total cost of constructing the tank as a function of its radius.

Solution

Let r denote the radius of the circular top and bottom, and let h be the height of the tank. Then the total cost of construction is

$$C = \text{cost of top} + \text{cost of bottom} + \text{cost of curved side}$$

Since the area of the circular top and bottom of the tank is πr^2, we have

Cost of constructing the top
$$= (\text{cost per unit area of top})(\text{area of top}) = 3\pi r^2 \text{ dollars}$$

and

Cost of constructing the bottom
$$= (\text{cost per unit area of bottom})(\text{area of bottom}) = 5\pi r^2 \text{ dollars}$$

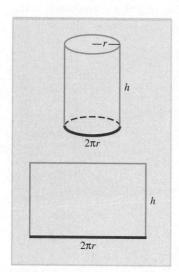

FIGURE 1.32 Cylindrical tank for Example 1.4.2.

To find the cost of constructing the curved side of the tank, we first need a formula for its area. Imagine the top and bottom of the tank removed and the side cut to form a rectangle, as shown in Figure 1.32. The height of the rectangle is the height h of the tank, and the length of the rectangle is the circumference $2\pi r$ of the tank's circular cross section. Thus, the area of the rectangle (and hence the curved side) is $2\pi rh$ square feet, and it follows that

Cost of constructing the sides
$$= (\text{cost per unit area of side})(\text{area of side}) = 5(2\pi rh) \text{ dollars}$$

Adding the cost of the top, bottom, and side of the tank, we get the total cost

$$C = 3\pi r^2 + 5\pi r^2 + 5(2\pi rh) = 8\pi r^2 + 10\pi rh$$

EXPLORE!

Use the table feature **(TBLSET)** of your graphing calculator to discover an appropriate viewing window for graphing

$$C(r) = 8\pi r^2 + \frac{110{,}000}{r}$$

Now employ the **TRACE, ZOOM** or **MINIMUM** commands of your calculator to determine the radius for which the cost is minimum.

Since the goal is to express the cost as a function of the radius only, we must eliminate the variable h from our cost expression, so we need an equation relating the variables r and h. To get such an equation, we recall that a cylinder with circular cross-sectional area πr^2 and height h has volume $V = \pi r^2 h$, and since the cylindrical tank is to have volume 11,000 cubic feet, it follows that

$$V = \pi r^2 h = 11{,}000$$

so

$$h = \frac{11{,}000}{\pi r^2}$$

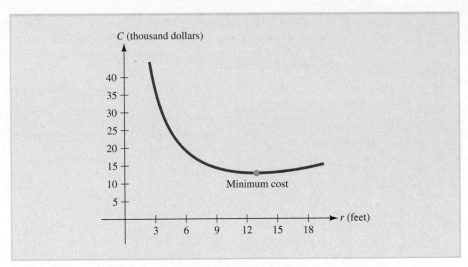

FIGURE 1.33 The cost function $C(r) = 8\pi r^2 + \dfrac{110,000}{r}$.

Substituting this expression for h in the formula for the total cost C, we get

$$C(r) = 8\pi r^2 + 10\pi rh = 8\pi r^2 + 10\pi r\left(\frac{11,000}{\pi r^2}\right)$$

$$= 8\pi r^2 + \frac{110,000}{r}$$

The graph of the cost function $C(r)$ is shown in Figure 1.33. Notice that the lowest point on the curve appears to be where $r \approx 13$, so Armando can minimize total cost by constructing a tank with a radius of about 13 feet. In Chapter 3, you will learn how to find this optimal radius more precisely using calculus.

Example 1.4.3 illustrates how a piecewise-defined function can be used in modeling an applied problem.

EXAMPLE 1.4.3   Modeling with a Piecewise-Defined Function

During a drought, residents of Marin County, California, were faced with a severe water shortage. To discourage excessive use of water, the county water district initiated drastic rate increases. The monthly rate for a family of four was $1.22 per 100 cubic feet of water for the first 1,200 cubic feet, $10 per 100 cubic feet for the next 1,200 cubic feet, and $50 per 100 cubic feet thereafter. Express the monthly water bill for a family of four as a function of the amount of water used.

Solution

Let x denote the number of hundred-cubic-feet units of water used by the family during the month and $C(x)$ the corresponding cost in dollars. If $0 \le x \le 12$, the cost is simply the cost per unit times the number of units used:

$$C(x) = 1.22x$$

If $12 < x \le 24$, each of the first 12 units costs \$1.22, and so the total cost of these 12 units is $1.22(12) = 14.64$ dollars. Each of the remaining $x - 12$ units costs \$10, and hence the total cost of these units is $10(x - 12)$ dollars. The cost of all x units is the sum

$$C(x) = 14.64 + 10(x - 12) = 10x - 105.36$$

If $x > 24$, the cost of the first 12 units is $1.22(12) = 14.64$ dollars, the cost of the next 12 units is $10(12) = 120$ dollars, and that of the remaining $x - 24$ units is $50(x - 24)$ dollars. The cost of all x units is the sum

$$C(x) = 14.64 + 120 + 50(x - 24) = 50x - 1,065.36$$

Combining these three formulas, we can express the total cost as the piecewise-defined function

$$C(x) = \begin{cases} 1.22x & \text{if } 0 \le x \le 12 \\ 10x - 105.36 & \text{if } 12 < x \le 24 \\ 50x - 1,065.36 & \text{if } x > 24 \end{cases}$$

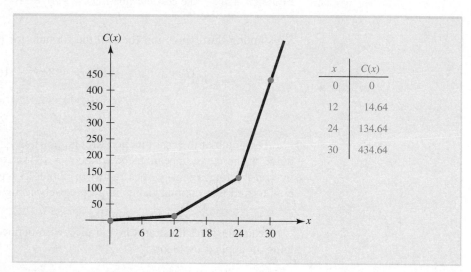

FIGURE 1.34 The cost of water in Marin County.

The graph of this function is shown in Figure 1.34. Notice that the graph consists of three line segments, each one steeper than the preceding one. What aspect of the practical situation is reflected by the increasing steepness of the lines?

Proportionality

In constructing mathematical models, it is often important to consider proportionality relationships. Three important kinds of proportionality are defined as follows:

Proportionality ■ The quantity Q is said to be:

directly proportional to x if $Q = kx$ for some constant k

inversely proportional to x if $Q = \dfrac{k}{x}$ for some constant k

jointly proportional to x and y if $Q = kxy$ for some constant k

Example 1.4.4 shows the use of proportionality with a biological model.

EXAMPLE 1.4.4 Modeling with Proportionality

When environmental factors impose an upper bound on its size, population grows at a rate that is jointly proportional to its current size and the difference between its current size and the upper bound. Express the rate of population growth as a function of the size of the population.

Solution

Let p denote the size of the population, $R(p)$ the corresponding rate of population growth, and b the upper bound placed on the population by the environment. Then

$$\text{Difference between population and bound} = b - p$$

and so
$$R(p) = kp(b - p)$$

where k is the constant of proportionality.

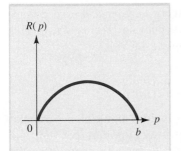

A graph of this factored polynomial is sketched in Figure 1.35. In Chapter 3, you will use calculus to compute the population size for which the rate of population growth is greatest.

FIGURE 1.35 The rate of bounded population growth: $R(p) = kp(b - p)$.

Market Equilibrium

Recall from Section 1.1 that the **demand function** $D(x)$ for a commodity relates the number of units x that are produced to the unit price $p = D(x)$ at which all x units are demanded (sold) in the marketplace. Similarly, the **supply function** $S(x)$ gives the corresponding price $p = S(x)$ at which producers are willing to supply x units to the marketplace. Usually, as the price of a commodity increases, more units of the commodity will be supplied and fewer will be demanded. Likewise, as the production level x increases, the supply price $p = S(x)$ also increases but the demand price $p = D(x)$ decreases. This means that a typical supply curve is rising, while a typical demand curve is falling, as indicated in Figure 1.36.

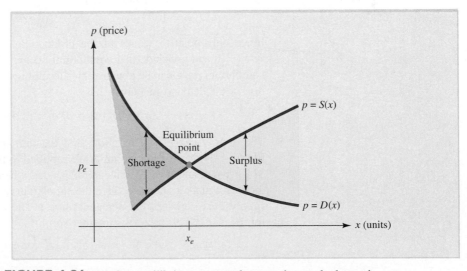

FIGURE 1.36 Market equilibrium occurs when supply equals demand.

The **law of supply and demand** says that in a competitive market environment, supply tends to equal demand, and when this occurs, the market is said to be in **equilibrium.** Thus, market equilibrium occurs precisely at the production level x_e, where $S(x_e) = D(x_e)$. The corresponding unit price p_e is called the **equilibrium price;** that is,

$$p_e = D(x_e) = S(x_e)$$

When the market is not in equilibrium, it has a **shortage** when demand exceeds supply $[D(x) > S(x)]$ and a **surplus** when supply exceeds demand $[S(x) > D(x)]$. This terminology is illustrated in Figure 1.36 and in Example 1.4.5.

EXPLORE!

Following Example 1.4.5 store $S(x) = x^2 + 14$ into Y1 and $D(x) = 174 - 6x$ into Y2. Use a viewing window [0, 35]5 by [0, 200]50 to observe the shortage and surplus sectors. Check if your calculator can shade these sectors by a command such as **SHADE (Y2, Y1).** What sector is this?

EXAMPLE 1.4.5 Modeling Market Equilibrium

Market research indicates that manufacturers will supply x units of a particular commodity to the marketplace when the price is $p = S(x)$ dollars per unit and that the same number of units will be demanded (bought) by consumers when the price is $p = D(x)$ dollars per unit, where the supply and demand functions are given by

$$S(x) = x^2 + 14 \qquad \text{and} \qquad D(x) = 174 - 6x$$

a. At what level of production x and unit price p is market equilibrium achieved?

b. Sketch the supply and demand curves, $p = S(x)$ and $p = D(x)$, on the same graph and interpret.

Solution

a. Market equilibrium occurs when

$$
\begin{aligned}
S(x) &= D(x) \\
x^2 + 14 &= 174 - 6x \qquad \text{subtract } 174 - 6x \text{ from both sides} \\
x^2 + 6x - 160 &= 0 \qquad \text{factor} \\
(x - 10)(x + 16) &= 0 \\
x = 10 \quad &\text{or} \quad x = -16
\end{aligned}
$$

Since only positive values of the production level x are meaningful, we reject $x = -16$ and conclude that equilibrium occurs when $x_e = 10$. The corresponding equilibrium price can be obtained by substituting $x = 10$ into either the supply function or the demand function. Thus,

$$p_e = D(10) = 174 - 6(10) = 114$$

b. The supply curve is a parabola and the demand curve is a line, as shown in Figure 1.37. Notice that no units are supplied to the market until the price reaches $14 per unit and that 29 units are demanded when the price is 0. For $0 \le x < 10$, there is a market shortage since the supply curve is below the demand curve. The supply curve crosses the demand curve at the equilibrium point (10, 114), and for $10 < x \le 29$, there is a market surplus.

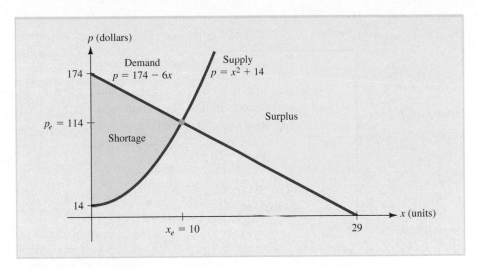

FIGURE 1.37   Supply, demand, and equilibrium point for Example 1.4.5.

Break-Even Analysis   Intersections of graphs arise in business in the context of **break-even analysis.** In a typical situation, a manufacturer wishes to determine how many units of a certain commodity have to be sold for total revenue to equal total cost. Suppose x denotes the number of units manufactured and sold, and let $C(x)$ and $R(x)$ be the corresponding total cost and total revenue, respectively. A pair of cost and revenue curves is sketched in Figure 1.38.

Because of fixed overhead costs, the total cost curve is initially higher than the total revenue curve. Hence, at low levels of production, the manufacturer suffers a loss. At higher levels of production, however, the total revenue curve is the higher one and the manufacturer realizes a profit. The point at which the two curves cross is called the **break-even point,** because when total revenue equals total cost, the manufacturer breaks even, experiencing neither a profit nor a loss. Example 1.4.6 illustrates break-even analysis.

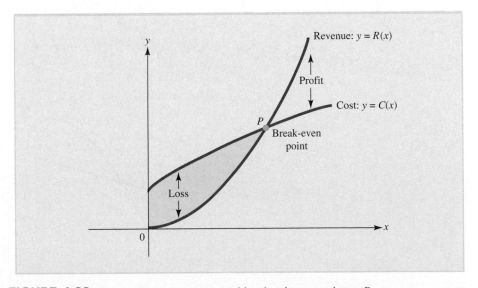

FIGURE 1.38   Cost and revenue curves, with a break-even point at P.

EXAMPLE 1.4.6 Break-Even Analysis

Emory's Furniture can sell a luxury reclining chair for $p = 1,500 - 3x$ dollars per unit when x units (chairs) are produced and sold, and the total cost of production consists of a fixed overhead of \$66,500 plus \$20 per unit. Plant capacity limits production to no more than 300 units.

a. How many chairs must Emory sell to break even? At what unit price should the chairs be sold for Emory to break even?

b. What is Emory's profit or loss if 35 chairs are sold?

c. How many chairs must be sold for Emory to realize a profit of \$120,000?

Solution

a. The demand function is $p = 1,500 - 3x$, and the revenue realized from selling x chairs is $R(x) = (1,500 - 3x)x$ dollars. The total cost is

$$C(x) = \underbrace{66,500}_{\text{overhead}} + \underbrace{20x}_{\text{variable cost}}$$

so Emory breaks even when

$$(1,500 - 3x)x = 66,500 + 20x$$

or, equivalently,

$$-3x^2 + 1,480x - 66,500 = 0$$

Applying the quadratic formula, we find that this equation is satisfied when

$$x = \frac{-1,480 \pm \sqrt{(1,480)^2 - 4(-3)(66,500)}}{2(-3)}$$
$$= 50 \text{ and } 443.33$$

So Emory breaks even when 50 units are sold and again when about 443 units are sold. Since plant capacity requires $0 \le x \le 300$, the break-even point occurs when 50 units are produced and sold (see the graph in Figure 1.39.) The corresponding unit price is

$$p(50) = 1,500 - 3(50) = \$1,350$$

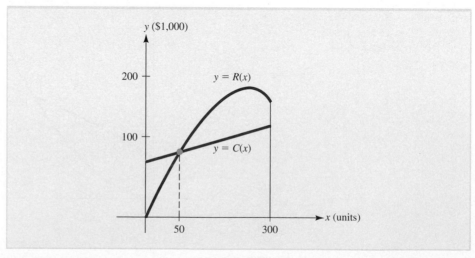

FIGURE 1.39 Revenue $R(x) = (1,500 - 3x)x$, and cost $C(x) = 66,500 + 20x$.

b. The profit $P(x)$ is revenue minus cost; that is,

$$P(x) = R(x) - C(x) = (1,500 - 3x)x - (66,500 + 20x)$$
$$= -3x^2 + 1,480x - 66,500$$

The profit when 35 units are sold is

$$P(35) = -3(35)^2 + 1,480(35) - 66,500 = -18,375$$

The minus sign indicates a negative profit, which should be expected since $x = 35$ is less than the break-even level of 50 units. It follows that if 35 units are sold, Emory will lose $18,375.

c. To determine the number of units that must be sold to generate a profit of $120,000, set the formula for the profit $P(x)$ equal to 120,000 and solve for x. You get

$$-3x^2 + 1,480x - 66,500 = 120,000$$

or

$$-3x^2 + 1,480x - 186,500 = 0$$

which has no real solution since the discriminant

$$(1,480)^2 - 4(-3)(-186,500) = -47,600$$

is negative. Thus, it is impossible for Emory to achieve a profit of $120,000.

Example 1.4.7 illustrates how break-even analysis can be used as a tool for decision making.

EXPLORE!

Refer to Example 1.4.7. Place $C_1(x) = 25 + 0.6x$ into Y1 and $C_2(x) = 30 + 0.5x$ into Y2 of the equation editor of your graphing calculator. Use the viewing window [−25, 250]25 by [−10, 125]50 to determine the range of mileage for which each agency gives the better deal. Would a person be better off using $C_1(x)$, $C_2(x)$, or $C_3(x) = 23 + 0.55x$ if more than 100 miles are to be driven?

EXAMPLE 1.4.7 Comparative Cost Analysis

A certain car rental agency charges $25 plus 60 cents per mile. A second agency charges $30 plus 50 cents per mile. Which agency offers the better deal?

Solution

The answer depends on the number of miles the car is driven. For short trips, the first agency charges less than the second, but for long trips, the second agency charges less. Break-even analysis can be used to find the number of miles for which the two agencies charge the same.

Suppose a car is to be driven x miles. Then the first agency will charge $C_1(x) = 25 + 0.60x$ dollars and the second will charge $C_2(x) = 30 + 0.50x$ dollars. If you set these expressions equal to one another and solve, you get

$$25 + 0.60x = 30 + 0.50x$$

so that

$$0.1x = 5 \quad \text{or} \quad x = 50$$

This shows that the two agencies charge the same amount if the car is driven 50 miles. For shorter distances, the first agency offers the better deal, and for longer distances, the second agency is better. The situation is illustrated in Figure 1.40.

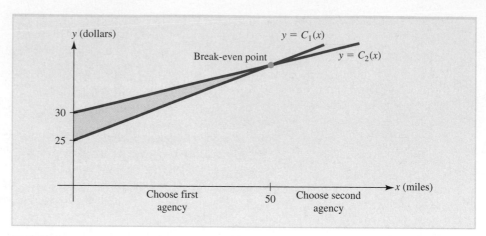

FIGURE 1.40 Car rental costs at competing agencies.

EXERCISES ▪ 1.4

1. The product of two numbers is 318. Express the sum of the numbers as a function of the smaller number.

2. The sum of two numbers is 18. Express the product of the numbers as a function of the smaller number.

3. **POPULATION GROWTH** In the absence of environmental constraints, population grows at a rate proportional to its size. Express the rate of population growth as a function of the size of the population.

4. **LANDSCAPING** A landscaper wishes to make a rectangular flower garden that is twice as long as it is wide. Express the area of the garden as a function of its width.

5. **FENCING** A farmer wishes to fence off a rectangular field with 1,000 feet of fencing. If the long side of the field is along a stream (and does not require fencing), express the area of the field as a function of its width.

6. **FENCING** A city recreation department plans to build a rectangular playground 3,600 square meters in area. The playground is to be surrounded by a fence. Express the length of the fencing as a function of the length of one of the sides of the playground, draw the graph, and estimate the dimensions of the playground requiring the least amount of fencing.

7. **AREA** Express the area of a rectangular field whose perimeter is 320 meters as a function of the length of one of its sides. Draw the graph, and estimate the dimensions of the field of maximum area.

8. **PACKAGING** A closed box with a square base is to have a volume of 1,500 cubic inches. Express its surface area as a function of the length of its base.

9. **PACKAGING** A closed box with a square base has a surface area of 4,000 square centimeters. Express its volume as a function of the length of its base.

10. **RADIOACTIVE DECAY** A sample of radium decays at a rate proportional to the amount of radium remaining. Express the rate of decay of the sample as a function of the amount remaining.

11. **TEMPERATURE CHANGE** The rate at which the temperature of an object changes is proportional to the difference between its own temperature and the temperature of the surrounding medium. Express this rate as a function of the temperature of the object.

12. **SPREAD OF AN EPIDEMIC** The rate at which an epidemic spreads through a community is jointly proportional to the number of people who have caught the disease and the number who have not. Express this rate as a function of the number of people who have caught the disease.

13. **POLITICAL CORRUPTION** The rate at which people are implicated in a government scandal is jointly proportional to the number of people already implicated and the number of people involved who have not yet been implicated.

Express this rate as a function of the number of people who have been implicated.

14. PRODUCTION COST At a certain factory, setup cost is directly proportional to the number of machines used and operating cost is inversely proportional to the number of machines used. Express the total cost as a function of the number of machines used.

15. TRANSPORTATION COST A truck is hired to transport goods from a factory to a warehouse. The driver's wages are figured by the hour and so are inversely proportional to the speed at which the truck is driven. The cost of gasoline is directly proportional to the speed. Express the total cost of operating the truck as a function of the speed at which it is driven.

BUSINESS AND ECONOMICS APPLIED PROBLEMS

16. MANUFACTURING PROFIT Rafael estimates that it costs $14 to produce each unit of a particular commodity that sells for $23 per unit. There is also a fixed cost of $1,200.
   a. Express the cost $C(x)$, the revenue $R(x)$, and the profit $P(x)$ as functions of the number of units x that are produced and sold.
   b. How much profit is generated when $x = 2,000$ units are produced? When $x = 100$ units? What is the smallest number of units that must be produced for Rafael's company to be profitable?
   c. What is the average profit function $AP(x)$? What is the average profit when 2,500 units are produced?

17. MANUFACTURING PROFIT A manufacturer estimates that each unit of a particular commodity can be sold for $3 more than it costs to produce. There is also a fixed cost of $17,000 associated with the production of the commodity.
   a. Express total profit $P(x)$ as a function of the level of production x.
   b. How much profit (or loss) is generated when $x = 5,000$ units are produced? When $x = 20,000$? What is the smallest number of units that must be sold for production to be profitable?
   c. Find the average profit function $AP(x)$. What is the average profit when 10,000 units are produced?

18. SALES REVENUE When x units of a certain luxury commodity are produced, they can all be sold at a price of p thousand dollars per unit, where $p = -6x + 100$.
   a. Express the revenue function $R(x)$ as a function of x. How much revenue is obtained when $x = 15$ units are produced?
   b. Find the average revenue function $AR(x)$. What is the average revenue when 10 units are produced?

19. RETAIL SALES Sally's company has been selling lamps at the price of $50 per lamp, and at this price consumers have been buying 3,000 lamps a month. Sally wishes to lower the price and estimates that for each $1 decrease in the price, 1,000 more lamps will be sold each month. She can produce the lamps at a cost of $29 per lamp. Express Sally's monthly profit as a function of the price that the lamps are sold, draw the graph, and estimate the optimal selling price.

20. RETAIL SALES A bookstore can obtain a certain book from the publisher at a cost of $3 per copy. The bookstore has been offering the book at the price of $15 per copy, and at this price, has been selling 200 copies a month. The bookstore is planning to lower its price to stimulate sales and estimates that for each $1 reduction in the price, 20 more books will be sold each month. Express the bookstore's monthly profit from the sale of this book as a function of the selling price, draw the graph, and estimate the optimal selling price.

21. INCOME TAX The accompanying table expresses the 2010 federal income tax rate schedule for single taxpayers.
   a. Express an individual's income tax as a function of the taxable income x for $0 \le x \le 171,850$, and draw the graph.
   b. The graph in part (a) should consist of four line segments. Compute the slope of each segment. What happens to these slopes as the taxable income increases? Interpret the slopes in practical terms.

If the Taxable Income Is		The Income Tax Is	
Over	But Not Over		Of the Amount Over
0	$8,375	10%	0
$8,375	$34,000	$837.50 + 15%	$8,375
$34,000	$82,400	$4,681.25 + 25%	$34,000
$82,400	$171,850	$16,781.25 + 28%	$82,400

22. MARKETING A company makes two products, A and B. The manager estimates that if $x\%$ of the total marketing budget is spent on marketing product A, then the total profit gained from both products will be P thousand dollars, where

$$P(x) = \begin{cases} 20 + 0.7x & \text{for } 0 \le x < 30 \\ 26 + 0.5x & \text{for } 30 \le x < 72 \\ 80 - 0.25x & \text{for } 72 \le x \le 100 \end{cases}$$

a. Sketch the graph of $P(x)$.
b. What is the company's profit when the marketing budget is split equally between the two products?
c. Express the total profit P in terms of the percentage y of the budget that is spent on marketing product B.

23. HARVESTING Farmers can get $8 per bushel for their potatoes on July 1, and after that, the price drops by 5 cents per bushel per day. On July 1, a farmer has 140 bushels of potatoes in the field and estimates that the crop is increasing at the rate of 1 bushel per day. Express the farmer's revenue from the sale of the potatoes as a function of the time at which the crop is harvested, draw the graph, and estimate when the farmer should harvest the potatoes to maximize revenue.

24. AUCTION BUYER'S PREMIUM Usually, when you purchase a lot in an auction, you pay not only your winning bid price but also a buyer's premium. At one auction house, the buyer's premium is 17.5% of the winning bid price for purchases up to $50,000. For larger purchases, the buyer's premium is 17.5% of the first $50,000 plus 10% of the purchase price above $50,000.

a. Find the total price a buyer pays (bid price plus buyer's premium) at this auction house for purchases of $1,000, $25,000, and $100,000.
b. Express the total purchase price of a lot at this auction house as a function of the final (winning) bid price. Sketch the graph of this function.

25. TRANSPORTATION COST A bus company has adopted the following pricing policy for groups that wish to charter its buses. Groups containing no more than 40 people will be charged a fixed amount of $2,400 (40 times $60). In groups containing between 40 and 80 people, everyone will pay $60 minus 50 cents for each person in excess of 40. The company's lowest fare of $40 per person will be offered to groups that

have 80 members or more. Express the bus company's revenue as a function of the size of the group. Draw the graph.

26. ADMISSION FEES A local natural history museum charges admission to groups according to the following policy. Groups of fewer than 50 people are charged a rate of $3.50 per person, while groups of 50 people or more are charged a reduced rate of $3 per person.

a. Express the amount a group will be charged for admission as a function of its size and draw the graph.
b. How much money will a group of 49 people save in admission costs if it can recruit 1 additional member?

27. PUBLISHING DECISION Ahmad must decide between two publishers who are vying for his new book. Publisher A offers royalties of 1% of net proceeds on the first 30,000 copies and 3.5% on all copies in excess of that figure, and expects to net $2 on each copy sold. Publisher B will pay no royalties on the first 4,000 copies sold but will pay 2% on the net proceeds of all copies sold in excess of 4,000 copies, and expects to net $3 on each copy sold. Suppose Ahmad expects to sell N copies. State a simple criterion based on N for deciding how to choose between the publishers.

28. CHECKING ACCOUNT The charge for maintaining a checking account at a certain bank is $12 per month plus 10 cents for each check that is written. A competing bank charges $10 per month plus 14 cents per check. Find a criterion for deciding which bank offers the better deal.

29. PRODUCTION COST A company has received an order from the city recreation department to manufacture 8,000 Styrofoam kickboards for its summer swimming program. The company owns several machines, each of which can produce 30 kickboards an hour. The cost of setting up the machines to produce these particular kickboards is $20 per machine. Once the machines have been set up, the operation is fully automated and can be overseen by a single production supervisor earning $19.20 per hour. Express the cost of producing the 8,000 kickboards as a function of the number of machines used, draw the graph, and estimate the number of machines the company should use to minimize cost.

MARKET EQUILIBRIUM *In Exercises 30 through 33, supply and demand functions, S(x) and D(x), are given for a particular commodity in terms of the level of production x. In each case:*

(a) *Find the value of x_e for which equilibrium occurs and the corresponding equilibrium price p_e.*

(b) *Sketch the graphs of the supply and demand curves, p = S(x) and p = D(x), on the same graph.*

(c) *For what values of x is there a market shortage? A market surplus?*

30. $S(x) = 4x + 200$ and $D(x) = -3x + 480$

31. $S(x) = 3x + 150$ and $D(x) = -2x + 275$

32. $S(x) = x^2 + x + 3$ and $D(x) = 21 - 3x^2$

33. $S(x) = 2x + 7.43$ and $D(x) = -0.21x^2 - 0.84x + 50$

34. **SUPPLY AND DEMAND** When electric blenders are sold for p dollars apiece, manufacturers will supply $\dfrac{p^2}{10}$ blenders to local retailers, while the local demand will be $60 - p$ blenders. At what market price will the manufacturers' supply of electric blenders be equal to the consumers' demand for the blenders? How many blenders will be sold at this price?

35. **SUPPLY AND DEMAND** Producers will supply x units of a certain commodity to the market when the price is $p = S(x)$ dollars per unit, and consumers will demand (buy) x units when the price is $p = D(x)$ dollars per unit, where

$$S(x) = 2x + 15 \quad \text{and} \quad D(x) = \frac{385}{x + 1}$$

a. Find the equilibrium production level x_e and the equilibrium price p_e.

b. Draw the supply and demand curves on the same graph.

c. Where does the supply curve cross the y axis? Describe the economic significance of this point.

36. **BREAK-EVEN ANALYSIS** A furniture manufacturer can sell dining room tables for $500 each. The manufacturer's total cost consists of a fixed overhead of $30,000 plus production costs of $350 per table.

a. How many tables must the manufacturer sell to break even?

b. How many tables must the manufacturer sell to make a profit of $6,000?

c. What will be the manufacturer's profit or loss if 150 tables are sold?

d. On the same set of axes, graph the manufacturer's total revenue and total cost functions. Explain how the overhead can be read from the graph.

37. **BREAK-EVEN ANALYSIS** Julia can sell a certain product for $110 per unit. Total cost consists of a fixed overhead of $7,500 plus production costs of $60 per unit.

a. Express Julia's total revenue, total cost, and total profit in terms of x, the number of units sold. Sketch the total revenue and total cost functions on the same set of axes.

b. How many units must be sold for Julia to break even?

c. What is Julia's profit or loss if 100 units are sold?

d. How many units must be sold for Julia to realize a profit of $1,250?

38. **BREAK-EVEN ANALYSIS** A greeting card company can sell cards for $2.75 each. The company's total cost consists of a fixed overhead of $12,000 plus variable costs of 35 cents per card.

a. Express the company's total revenue, total cost, and total profit in terms of x, the number of cards sold.

b. How many cards must the company sell to break even?

c. What will be the company's profit or loss if it sells 5,000 cards?

d. How many cards must be sold for the company to make a profit of $9,000?

e. On the same set of axes, graph the company's total revenue and total cost functions. What is the significance of the point where the cost function crosses the y axis?

39. **CONSTRUCTION COST** A company plans to construct a new building and parking lot on a rectangular plot of land 100 meters wide and 120 meters long. The building is to be 20 meters high and to have a rectangular footprint with perimeter 320 meters, as shown in the accompanying figure.

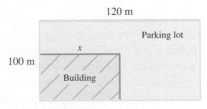

EXERCISE 39

a. Express the volume $V(x)$ of the building as a function of the length of its longer side x.

b. Graph the volume function in part (a), and determine the dimensions of the building of greatest volume that satisfies the stated requirements.

c. Suppose the company decides to construct the building of maximum volume. If it costs \$75 per cubic meter to construct the building and \$50 per square meter for the parking lot, what is the total cost of construction?

40. PUBLISHING PROFIT   It costs a publisher \$74,200 to prepare a book for publication (typesetting, illustrating, editing, and so on); printing and binding costs are \$5.50 per book. The book is sold to bookstores for \$19.50 per copy.

a. Make a table showing the cost of producing 2,000, 4,000, 6,000, and 8,000 books.

b. Make a table showing the revenue from selling 2,000, 4,000, 6,000, and 8,000 books.

c. Write an algebraic expression representing the cost y as a function of the number of books x that are produced.

d. Write an algebraic expression representing the revenue y as a function of the number of books x sold.

e. Graph both functions on the same coordinate axes.

f. Use **TRACE** and **ZOOM** to find where cost equals revenue.

g. Use the graph to determine how many books need to be made to produce revenue of at least \$85,000. How much profit is earned for this number of books?

41. SUPPLY AND DEMAND   Producers will supply q units of a certain commodity to the market when the price is $p = S(q)$ dollars per unit, and consumers will demand (buy) q units when the price is $p = D(q)$ dollars per unit, where

$$S(q) = aq + b \qquad \text{and} \qquad D(q) = cq + d$$

for constants a, b, c, and d.

a. What can you say about the signs of the constants a, b, c, and d if the supply and demand curves are as shown in the accompanying figure?

b. Express the equilibrium production level q_e and the equilibrium price p_e in terms of the coefficients a, b, c, and d.

c. Use your answer in part (b) to determine what happens to the equilibrium production level q_e as a increases. What happens to q_e as d increases?

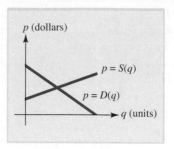

EXERCISE 41

42. AGRICULTURAL YIELD   Belinda, a Florida citrus grower, estimates that if 60 orange trees are planted, the average yield per tree will be 400 oranges. The average yield will decrease by 4 oranges per tree for each additional tree planted on the same acreage. Express Belinda's total yield as a function of the number of additional trees planted, draw the graph, and estimate the total number of trees Belinda should plant to maximize yield.

43. PHYSIOLOGY   The pupil of the human eye is roughly circular. If the intensity of light I entering the eye is proportional to the area of the pupil, express I as a function of the radius r of the pupil.

44. RECYCLING   To raise money, a service club has been collecting used bottles that it plans to deliver to a local glass company for recycling. Since the project began 80 days ago, the club has collected 24,000 pounds of glass for which the glass company currently offers 1 cent per pound. However, since bottles are accumulating faster than they can be recycled, the company plans to reduce by 1 cent each day the price it will pay for 100 pounds of used glass. Assuming that the club can continue to collect bottles at the same rate and that transportation costs make more than one trip to the glass company unfeasible, express the club's revenue from its recycling project as a function of the number of additional days the project runs. Draw the graph, and estimate when the club should conclude the project and deliver the bottles to maximize its revenue.

45. LIFE EXPECTANCY In 1900, the life expectancy of a newborn child was 46 years, and by 2000, it had grown to 77 years. Over the same century, the life expectancy of a person at age 65 grew from 76 years to 83 years. Both of these life expectancies increased linearly with time between 1900 and 2000.

a. Find the function $B(t)$ that represents the life expectancy of a newborn child t years after 1900 and the function $E(t)$ that represents the life expectancy of a 65-year-old.

b. Sketch the graphs of $E(t)$ and $B(t)$. Then determine the age A to which both a newborn child and a 65-year-old person can expect to live.

c. Some scientists and others regard the age A you found in part (b) as the ultimate age imposed by limitations in the human body. That is, a person older than age A is living on "borrowed time." Write a paragraph on the concept of "ultimate" life expectancy for a human being.

PEDIATRIC DRUG DOSAGE *Several different formulas have been proposed for determining the appropriate dose of a drug for a child in terms of the adult dosage. Suppose that A milligrams (mg) is the adult dose of a certain drug and C is the appropriate dosage for a child of age N years. Then* Cowling's rule *says that*

$$C = \left(\frac{N + 1}{24}\right)A$$

while Friend's rule *says that*

$$C = \frac{2NA}{25}$$

Exercises 46 through 48 require these formulas.

46. If an adult dose of ibuprofen is 300 mg, what dose does Cowling's rule suggest for an 11-year-old child? What dose does Friend's rule suggest for the same child?

47. Assume an adult dose of $A = 300$ mg, so that Cowling's rule and Friend's rule become functions of the child's age N. Sketch the graphs of these two functions.

48. For what child's age is the dosage suggested by Cowling's rule the same as that predicted by Friend's rule? For what ages does Cowling's rule suggest a larger dosage than Friend's rule? For what ages does Friend's rule suggest the larger dosage?

49. PEDIATRIC DRUG DOSAGE As an alternative to Friend's rule and Cowling's rule, pediatricians sometimes use the formula

$$C = \frac{SA}{1.7}$$

to estimate an appropriate drug dosage for a child whose surface area is S square meters, when the adult dosage of the drug is A milligrams (mg). In turn, the surface area of the child's body is estimated by the formula

$$S = 0.0072W^{0.425}H^{0.725}$$

where W and H are, respectively, the child's weight in kilograms (kg) and height in centimeters (cm).

a. The adult dosage for a certain drug is 250 mg. How much of the drug should be given to a child who is 91 cm tall and weighs 18 kg?

b. A drug is prescribed for two children, one of whom is twice as tall and twice as heavy as the other. Show that the larger child should receive approximately 2.22 times as much of the drug as the smaller child.

50. VOLUME OF A TUMOR The shape of a cancerous tumor is roughly spherical and has volume

$$V = \frac{4}{3}\pi r^3$$

where r is the radius in centimeters.

a. When first observed, the tumor has radius 0.73 cm, and 45 days later, the radius is 0.95 cm. By how much does the volume of the tumor increase during this period?

b. After being treated with chemotherapy, the radius of the tumor decreases by 23%. What is the corresponding percentage decrease in the volume of the tumor?

51. BIOCHEMISTRY In biochemistry, the rate R of an enzymatic reaction is found to be given by the equation

$$R = \frac{R_m[S]}{K_m + [S]}$$

where K_m is a constant (called the **Michaelis constant**), R_m is the maximum possible rate, and $[S]$ is the substrate concentration.* Rewrite this equation so that $y = \dfrac{1}{R}$ is expressed as a function of $x = \dfrac{1}{[S]}$, and sketch the graph of this function. (This graph is called the **Lineweaver-Burk double-reciprocal plot.**)

---

*Mary K. Campbell, *Biochemistry,* Philadelphia: Saunders College Publishing, 1991, pp. 221–226.

MISCELLANEOUS PROBLEMS

In Exercises 52 through 56, you need to know that a cylinder of radius r and height h has volume $V = \pi r^2 h$ and lateral (side) surface area $S = 2\pi rh$. A circular disk of radius r has area $A = \pi r^2$.

52. PACKAGING   A soda can holds 12 fluid ounces (approximately 6.89π cubic inches). Express the surface area of the can as a function of its radius.

53. PACKAGING   A closed cylindrical can has surface area 120π square inches. Express the volume of the can as a function of its radius.

54. PACKAGING   A closed cylindrical can has a radius r and height h.
  a. If the surface area S of the can is a constant, express the volume V of the can in terms of S and r.
  b. If the volume V of the can is a constant, express the surface area S in terms of V and r.

55. PACKAGING   A cylindrical can is to hold 4π cubic inches of frozen orange juice. The cost per square inch of constructing the metal top and bottom is twice the cost per square inch of constructing the cardboard side. Express the cost of constructing the can as a function of its radius if the cost of the side is 0.02 cent per square inch.

56. PACKAGING   A cylindrical can with no top has been made from 27π square inches of metal. Express the volume of the can as a function of its radius.

57. POSTER DESIGN   A rectangular poster contains 25 square centimeters of print surrounded by margins of 2 centimeters on each side and 4 centimeters on the top and bottom. Express the total area of the poster (printing plus margins) as a function of the width of the printed portion.

58. CONSTRUCTION COST   A closed box with a square base is to have a volume of 250 cubic meters. The material for the top and bottom of the box costs $2 per square meter, and the material for the sides costs $1 per square meter. Express the construction cost of the box as a function of the length of its base.

59. CONSTRUCTION COST   An open box with a square base is to be built for $48. The sides of the box will cost $3 per square meter, and the base will cost $4 per square meter. Express the volume of the box as a function of the length of its base.

60. CONSTRUCTION COST   An open box is to be made from a square piece of cardboard, 18 inches by 18 inches, by removing a small square from each corner and folding up the flaps to form the sides. Express the volume of the resulting box as a function of the length x of a side of the removed squares. Draw the graph, and estimate the value of x for which the volume of the resulting box is greatest.

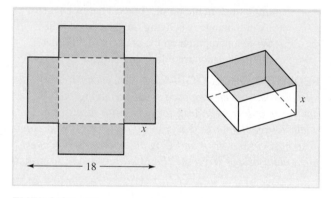

EXERCISE 60

61. AIR TRAVEL   Two jets bound for Los Angeles leave New York 30 minutes apart. The first travels at 550 miles per hour, while the second travels at 650 miles per hour. At what time will the second plane pass the first?

 62. SPY STORY   The hero of a popular spy story has escaped from the headquarters of an international diamond smuggling ring in the tiny Mediterranean country of Azusa. Our hero, driving a stolen milk truck at 72 kilometers per hour, has a 40-minute head start on his pursuers, who are chasing him in a Ferrari going 168 kilometers per hour. The distance from the smugglers' headquarters to the border, and freedom, is 83.8 kilometers. Will the spy make it?

SECTION 1.5 Limits

Learning Objectives

1. Examine the limit concept and general properties of limits.
2. Compute limits using a variety of techniques.
3. Investigate limits involving infinity.

As you will see in subsequent chapters, calculus is an enormously powerful branch of mathematics with a wide range of applications, including optimization of functions, analysis of rates of change, curve sketching, and computation of area and probability. What gives calculus its power and distinguishes it from algebra is the concept of limit, and the purpose of this section is to provide an introduction to this important concept. Our approach will be intuitive rather than formal. The ideas outlined here form the basis for a more rigorous development of the laws and procedures of calculus and lie at the heart of much of modern mathematics.

Intuitive Introduction to the Limit

Roughly speaking, the limit process involves examining the behavior of a function $f(x)$ as x approaches a number c that may or may not be in the domain of f. Limiting behavior occurs in a variety of practical situations. For instance, absolute zero, the temperature T_c at which all molecular activity ceases, can be approached but never actually attained in practice. Similarly, economists who speak of profit under ideal conditions or engineers profiling the ideal specifications of a new engine are really dealing with limiting behavior.

To illustrate the limit process, suppose the manager of a real estate firm determines that t years from now, roughly S units in a certain neighborhood will be sold, where

$$S(t) = \frac{-2t^3 + 19t^2 - 8t - 9}{-t^2 + 8t - 7}$$

How many sales should be expected 1 year from now?

Your first instinct may be to simply evaluate $S(t)$ at $t = 1$, but that computation results in the meaningless fraction $\frac{0}{0}$. However, it is still possible to make the required computation by evaluating $S(t)$ for values of t that are very close to the 1-year mark, both slightly before the year is up ($t < 1$) and just afterward ($t > 1$). A few such calculations are summarized in the following table:

t	0.95	0.98	0.99	0.999	1	1.001	1.01	1.1	1.2
$S(t)$	3.859	3.943	3.972	3.997	Undefined	4.003	4.028	4.285	4.572

The numbers on the bottom line of this table suggest that $S(t)$ approaches the number 4 as t gets closer and closer to 1. Thus, it is reasonable to expect 4 sales to be made in the target neighborhood 1 year from now.

The functional behavior in this example can be described by saying "$S(t)$ has the limiting value 4 as t approaches 1," or, equivalently, by writing

$$\lim_{t \to 1} S(t) = 4$$

More generally, the limit of $f(x)$ as x approaches the number c can be defined informally as follows:

> **The Limit of a Function**   ■   If $f(x)$ gets closer and closer to a number L as x gets closer and closer to c from both sides, then L is the **limit** of $f(x)$ as x approaches c. The behavior is expressed by writing
>
> $$\lim_{x \to c} f(x) = L$$

Geometrically, the limit statement $\lim\limits_{x \to c} f(x) = L$ means that the height of the graph $y = f(x)$ approaches L as x approaches c, as shown in Figure 1.41. This interpretation is illustrated along with the tabular approach to computing limits in Example 1.5.1.

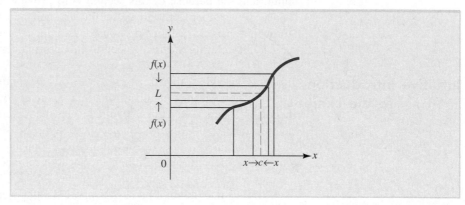

FIGURE 1.41   If $\lim\limits_{x \to c} f(x) = L$, the height of the graph of f approaches L as x approaches c.

EXPLORE!

Graph $f(x) = \dfrac{\sqrt{x} - 1}{x - 1}$, using the modified decimal viewing window

    [0, 4.7]1 by [−1.1, 2.1]1

Trace values near $x = 1$. Also construct a table of values, using an initial value of 0.97 for x with an incremental change of 0.01. Describe what you observe. Now use an initial value of 0.997 for x with an incremental change of 0.001. Specifically, what happens as x approaches 1 from either side? What would be the most appropriate value for $f(x)$ at $x = 1$ to fill the hole in the graph?

EXAMPLE 1.5.1   Estimating a Limit with a Table

Use a table to estimate the limit

$$\lim_{x \to 1} \frac{\sqrt{x} - 1}{x - 1}$$

Solution

Let

$$f(x) = \frac{\sqrt{x} - 1}{x - 1}$$

and compute $f(x)$ for a succession of values of x approaching 1 from the left and from the right:

				$x \to 1 \leftarrow x$			
x	0.99	0.999	0.9999	1	1.00001	1.0001	1.001
$f(x)$	0.50126	0.50013	0.50001	✕	0.499999	0.49999	0.49988

The numbers on the bottom line of the table suggest that $f(x)$ approaches 0.5 as x approaches 1; that is,

$$\lim_{x \to 1} \frac{\sqrt{x} - 1}{x - 1} = 0.5$$

The graph of $f(x)$ is shown in Figure 1.42. The limit computation says that the height of the graph of $y = f(x)$ approaches $L = 0.5$ as x approaches 1. This corresponds to the "hole" in the graph of $f(x)$ at $(1, 0.5)$. We will compute this same limit using an algebraic procedure in Example 1.5.6.

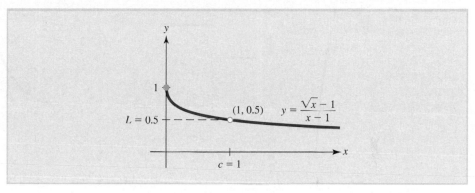

FIGURE 1.42   The function $f(x) = \dfrac{\sqrt{x} - 1}{x - 1}$ tends toward $L = 0.5$ as x approaches $c = 1$.

It is important to remember that limits describe the behavior of a function *near* a particular point, not necessarily *at* the point itself. This is illustrated in Figure 1.43. For all three functions graphed, the limit of $f(x)$ as x approaches 3 is equal to 4. Yet the functions behave quite differently at $x = 3$ itself. In Figure 1.43a, $f(3)$ is equal to the limit 4; in Figure 1.43b, $f(3)$ is different from 4; and in Figure 1.43c, $f(3)$ is not defined at all.

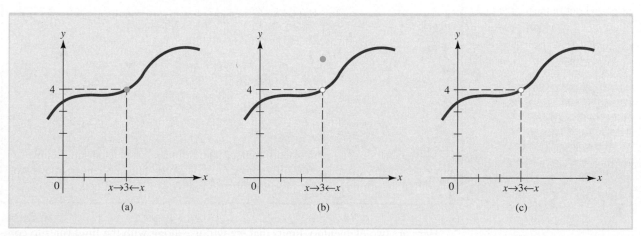

FIGURE 1.43   Three functions for which $\lim\limits_{x \to 3} f(x) = 4$.

Figure 1.44 shows the graph of two functions that do not have a limit as x approaches 2. The limit does not exist in Figure 1.44a because $f(x)$ tends toward 5 as x approaches 2 from the right and tends toward a different value, 3, as x approaches 2 from the left. The function in Figure 1.44b has no finite limit as x approaches 2 because the values of $f(x)$ increase without bound as x tends toward 2 and hence tend to no finite number L. Such so-called *infinite limits* will be discussed later in this section.

EXPLORE!

Graph $f(x) = \dfrac{2}{(x-2)^2}$ using the window [0, 4]1 by [−5, 40]5. Trace the graph on both sides of $x = 2$ to view the behavior of $f(x)$ about $x = 2$. Also display the table value of the function with the incremental change of x set to 0.01 and the initial value $x = 1.97$. What happens to the values of $f(x)$ as x approaches 2?

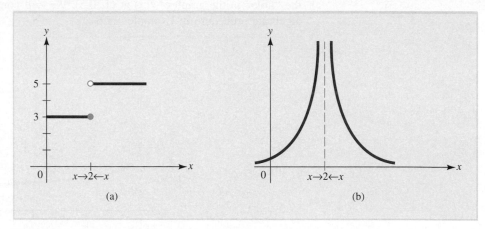

(a)   (b)

FIGURE 1.44   Two functions for which $\lim\limits_{x \to 2} f(x)$ does not exist.

Properties of Limits

Limits obey certain algebraic rules that can be used to simplify computations. These rules, which should seem plausible on the basis of our informal definition of limit, are proved formally in more theoretical courses.

EXPLORE!

Graph the function

$$f(x) = \begin{cases} 3 & x \le 2 \\ 5 & x > 2 \end{cases}$$

using the dot graphing style and writing

Y1 = 3(X ≤ 2) + 5(X > 2)

in the equation editor of your graphing calculator. Use your **TRACE** key to determine the values of y when x is near 2. Does it make a difference from which side $x = 2$ is approached? Also evaluate $f(2)$.

> **Algebraic Properties of Limits** ■ If $\lim\limits_{x \to c} f(x)$ and $\lim\limits_{x \to c} g(x)$ exist, then
>
> $$\lim_{x \to c} [f(x) + g(x)] = \lim_{x \to c} f(x) + \lim_{x \to c} g(x)$$
>
> $$\lim_{x \to c} [f(x) - g(x)] = \lim_{x \to c} f(x) - \lim_{x \to c} g(x)$$
>
> $$\lim_{x \to c} [kf(x)] = k \lim_{x \to c} f(x) \quad \text{for any constant } k$$
>
> $$\lim_{x \to c} [f(x)g(x)] = [\lim_{x \to c} f(x)][\lim_{x \to c} g(x)]$$
>
> $$\lim_{x \to c} \frac{f(x)}{g(x)} = \frac{\lim\limits_{x \to c} f(x)}{\lim\limits_{x \to c} g(x)} \quad \text{if } \lim_{x \to c} g(x) \ne 0$$
>
> $$\lim_{x \to c} [f(x)]^p = [\lim_{x \to c} f(x)]^p \quad \text{if } [\lim_{x \to c} f(x)]^p \text{ exists}$$
>
> That is, the limit of a sum, a difference, a multiple, a product, a quotient, or a power exists and is the sum, difference, multiple, product, quotient, or power of the individual limits, as long as all expressions involved are defined.

Here are two elementary limits that we will use along with the limit rules to compute limits involving more complex expressions.

> **Limits of Two Linear Functions** ■ For any constant k,
>
> $$\lim_{x \to c} k = k \quad \text{and} \quad \lim_{x \to c} x = c$$
>
> That is, the limit of a constant is the constant itself, and the limit of $f(x) = x$ as x approaches c is c.

These statements are illustrated in Figure 1.45. Note that in geometric terms, the limit statement $\lim_{x \to c} x = c$ says that the height of the linear function $f(x) = x$ approaches c as x approaches c.

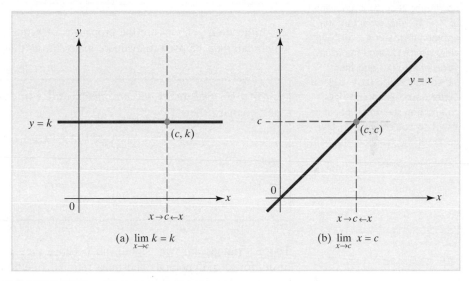

FIGURE 1.45 Limits of two linear functions.

Computation of Limits

Examples 1.5.2 through 1.5.6 illustrate how the properties of limits can be used to calculate limits of algebraic functions. In Example 1.5.2, you will see how to find the limit of a polynomial.

EXAMPLE 1.5.2 Finding the Limit of a Polynomial

Find $\lim_{x \to -1} (3x^3 - 4x + 8)$.

Solution

Apply the properties of limits to obtain

$$\lim_{x \to -1} (3x^3 - 4x + 8) = 3(\lim_{x \to -1} x)^3 - 4(\lim_{x \to -1} x) + \lim_{x \to -1} 8$$
$$= 3(-1)^3 - 4(-1) + 8 = 9$$

In Example 1.5.3, you will see how to find the limit of a rational function whose denominator does not approach zero.

EXPLORE!

Graph $f(x) = \dfrac{x^2 + x - 2}{x - 1}$ using the viewing window [0, 2]0.5 by [0, 5]0.5. **TRACE** to $x = 1$, and notice there is no corresponding y value. Create a table with an initial value of 0.5 for x, increasing in increments of 0.1. Notice that an error is displayed for $x = 1$, confirming that $f(x)$ is undefined at $x = 1$. What would be the appropriate y value if this gap were filled? Change the initial value of x to 0.9 and the increment size to 0.01 to get a better approximation. Finally, zoom in on the graph about $x = 1$ to conjecture a limiting value for the function at $x = 1$.

EXAMPLE 1.5.3 Finding the Limit of a Rational Function

Find $\displaystyle\lim_{x \to 1} \dfrac{3x^3 - 8}{x - 2}$.

Solution

Since $\displaystyle\lim_{x \to 1} (x - 2) \neq 0$, you can use the quotient rule for limits to get

$$\lim_{x \to 1} \frac{3x^3 - 8}{x - 2} = \frac{\displaystyle\lim_{x \to 1}(3x^3 - 8)}{\displaystyle\lim_{x \to 1}(x - 2)} = \frac{3 \displaystyle\lim_{x \to 1} x^3 - \displaystyle\lim_{x \to 1} 8}{\displaystyle\lim_{x \to 1} x - \displaystyle\lim_{x \to 1} 2} = \frac{3 - 8}{1 - 2} = 5$$

In general, you can use the properties of limits to obtain the following formulas, which can then be used to evaluate many limits that occur in practical problems.

Limits of Polynomials and Rational Functions ■ If $p(x)$ and $q(x)$ are polynomials, then

$$\lim_{x \to c} p(x) = p(c)$$

and

$$\lim_{x \to c} \frac{p(x)}{q(x)} = \frac{p(c)}{q(c)} \quad \text{if } q(c) \neq 0$$

These formulas are very significant because they give us a simple way to compute limits for all polynomials and most rational functions: just evaluate the function for the value that the variable is approaching. If the result is a real number, that number is the limit.

In Example 1.5.4, the denominator of the given rational function approaches zero, while the numerator does not. The absolute value of such a quotient increases without bound and hence does not approach any finite number. When this happens, you can conclude that the limit does not exist.

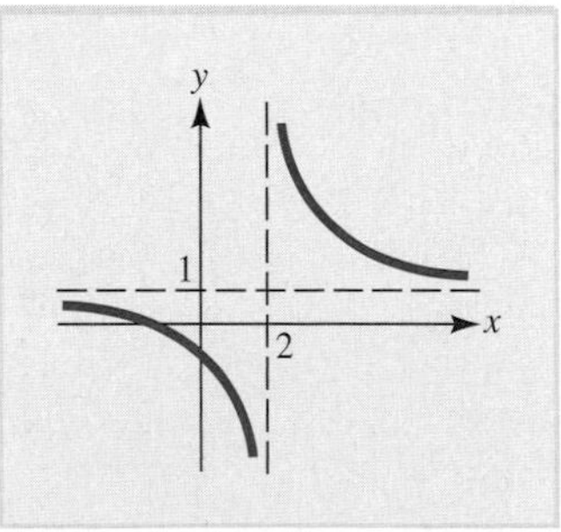

FIGURE 1.46 The graph of $f(x) = \dfrac{x + 1}{x - 2}$.

EXAMPLE 1.5.4 Showing That a Limit Does Not Exist

Find $\displaystyle\lim_{x \to 2} \dfrac{x + 1}{x - 2}$.

Solution

The quotient rule for limits does not apply in this case since the limit of the denominator is

$$\lim_{x \to 2} (x - 2) = 0$$

Since the limit of the numerator is $\displaystyle\lim_{x \to 2} (x + 1) = 3$, which is not equal to zero, you can conclude that the limit of the quotient does not exist.

Graph $y = \dfrac{x+1}{x-2}$ using an enlarged decimal window $[-9.4, 9.4]1$ by $[-6.2, 6.2]1$. Use the **TRACE** key to approach $x = 2$ from the left side and the right side. Also create a table of values, using an initial value of 1.97 for x and increasing in increments of 0.01. Describe what you observe.

The graph of the function $f(x) = \dfrac{x+1}{x-2}$ in Figure 1.46 gives you a better idea of what is actually happening in this example. Note that $f(x)$ increases without bound as x approaches 2 from the right and decreases without bound as x approaches 2 from the left.

In Example 1.5.5, the numerator and denominator of the given rational function both approach zero. When this happens, you should try to simplify the function algebraically and then use the fact that if $f(x) = g(x)$ for $x \neq c$, then $\lim\limits_{x \to c} f(x) = \lim\limits_{x \to c} g(x)$. This is another way of saying that the limit as x approaches c is about what happens close to c but not at c.

EXAMPLE 1.5.5 Finding a Limit Using Algebra

Find $\lim\limits_{x \to 1} \dfrac{x^2 - 1}{x^2 - 3x + 2}$.

Solution

As x approaches 1, both the numerator and the denominator approach zero, and you can draw no conclusion about the size of the quotient. To proceed, observe that the given function is not defined when $x = 1$ but that for all other values of x, you can divide out the common factor $x - 1$ to obtain

$$\frac{x^2 - 1}{x^2 - 3x + 2} = \frac{(x-1)(x+1)}{(x-1)(x-2)} = \frac{x+1}{x-2} \qquad x \neq 1$$

(Since $x \neq 1$, you are not dividing by zero.) Now take the limit as x approaches (but is not equal to) 1 to get

$$\lim_{x \to 1} \frac{x^2 - 1}{x^2 - 3x + 2} = \frac{\lim\limits_{x \to 1}(x+1)}{\lim\limits_{x \to 1}(x-2)} = \frac{2}{-1} = -2$$

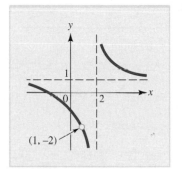

FIGURE 1.47 The graph of $f(x) = \dfrac{x^2 - 1}{x^2 - 3x + 2}$.

The graph of the function $f(x) = \dfrac{x^2 - 1}{x^2 - 3x + 2}$ is shown in Figure 1.47. Note that it is like the graph in Figure 1.46 with a hole at the point $(1, -2)$.

In general, when both the numerator and denominator of a quotient approach zero as x approaches c, your strategy will be to simplify the quotient algebraically (as in Example 1.5.5 by dividing out $x - 1$). In most cases, the simplified form of the quotient will give the same values as the original form except for $x = c$. Since you are interested in the behavior of the quotient *near* $x = c$ and not *at* $x = c$, you may use the simplified form of the quotient to evaluate the limit. In Example 1.5.6, we use this strategy to obtain the limit we estimated using a table in Example 1.5.1.

EXAMPLE 1.5.6 Finding a Limit Using Algebra

Find $\lim\limits_{x \to 1} \dfrac{\sqrt{x} - 1}{x - 1}$.

Just-In-Time **REVIEW**

In Example 1.5.6, we perform
the multiplication
   $(\sqrt{x} - 1)(\sqrt{x} + 1) = x - 1$
using the identity
   $(a - b)(a + b) = a^2 - b^2$
with $\sqrt{x}$ corresponding to a
and 1 corresponding to b.

Solution

Both the numerator and denominator approach 0 as x approaches 1. To simplify the quotient, we rationalize the numerator (that is, multiply the numerator and denominator by $\sqrt{x} + 1$) to get

$$\frac{\sqrt{x} - 1}{x - 1} = \frac{(\sqrt{x} - 1)(\sqrt{x} + 1)}{(x - 1)(\sqrt{x} + 1)} = \frac{x - 1}{(x - 1)(\sqrt{x} + 1)} = \frac{1}{\sqrt{x} + 1} \qquad x \neq 1$$

and then take the limit to obtain

$$\lim_{x \to 1} \frac{\sqrt{x} - 1}{x - 1} = \lim_{x \to 1} \frac{1}{\sqrt{x} + 1} = \frac{1}{2}$$

Limits Involving Infinity

"Long-term" behavior is often a matter of interest in business and economics or the physical and life sciences. For example, a biologist may wish to know the population of a bacterial colony or a population of fruit flies after an indefinite period of time, or a business manager may wish to know how the average cost of producing a particular commodity is affected as the level of production increases indefinitely.

In mathematics, the infinity symbol ∞ is used to represent either unbounded growth or the result of such growth. Here are definitions of limits involving infinity that we will use to study long-term behavior.

Limits at Infinity ■ If the values of the function $f(x)$ approach the number L as x increases without bound, we write

$$\lim_{x \to +\infty} f(x) = L$$

Similarly, we write

$$\lim_{x \to -\infty} f(x) = M$$

when the functional values $f(x)$ approach the number M as x decreases without bound.

NOTE   The statement "x decreases without bound" means that the input x of the function becomes larger and larger in absolute value while remaining negative. ■

Geometrically, the limit statement $\lim_{x \to +\infty} f(x) = L$ means that as x increases without bound, the graph of $f(x)$ approaches the horizontal line $y = L$, while $\lim_{x \to -\infty} f(x) = M$ means that the graph of $f(x)$ approaches the line $y = M$ as x decreases without bound. The lines $y = L$ and $y = M$ that appear in this context are called **horizontal asymptotes** of the graph of $f(x)$. There are many different ways for a graph to have horizontal asymptotes, one of which is shown in Figure 1.48. We will have more to say about asymptotes in Chapter 3 as part of a general discussion of graphing with calculus.

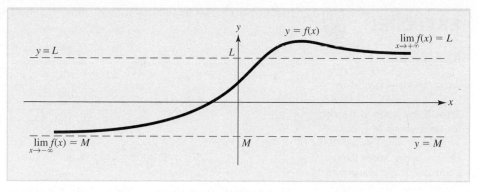

FIGURE 1.48 A graph illustrating limits at infinity and horizontal asymptotes.

The algebraic properties of limits listed earlier in this section also apply to limits at infinity. In addition, since any reciprocal power $\dfrac{1}{x^k}$ for $k > 0$ becomes smaller and smaller in absolute value as x either increases or decreases without bound, we have these useful rules:

Reciprocal Limit Rules ■ If A and k are constants with $k > 0$ and x^k is defined for all x, then

$$\lim_{x \to +\infty} \frac{A}{x^k} = 0 \quad \text{and} \quad \lim_{x \to -\infty} \frac{A}{x^k} = 0$$

The use of these rules is illustrated in Example 1.5.7.

EXAMPLE 1.5.7 Finding a Limit at Infinity

Find $\displaystyle\lim_{x \to +\infty} \frac{x^2}{1 + x + 2x^2}$.

Solution

To get a feeling for what happens with this limit, we evaluate the function

$$f(x) = \frac{x^2}{1 + x + 2x^2}$$

at $x = 100, 1{,}000, 10{,}000$, and $100{,}000$ and display the results in the table:

				$x \to +\infty$
x	100	1,000	10,000	100,000
$f(x)$	0.49749	0.49975	0.49997	0.49999

The functional values on the bottom line in the table suggest that $f(x)$ tends toward 0.5 as x grows larger and larger. To confirm this observation analytically, we divide each term in $f(x)$ by the highest power that appears in the denominator $1 + x + 2x^2$,

EXPLORE!

Graph $f(x) = \dfrac{x^2}{1 + x + 2x^2}$ using the viewing window $[-20, 20]5$ by $[0, 1]1$. Now **TRACE** the graph to the right for large values of x, past $x = 30$, 40, and beyond. What do you notice about the corresponding y values and the behavior of the graph? What would you conjecture as the value of $f(x)$ as $x \to \infty$?

that is, by x^2. This enables us to find $\lim\limits_{x \to +\infty} f(x)$ by applying reciprocal power rules as follows:

$$\lim_{x \to +\infty} \frac{x^2}{1 + x + 2x^2} = \lim_{x \to +\infty} \frac{x^2/x^2}{1/x^2 + x/x^2 + 2x^2/x^2}$$

several algebraic properties of limits

$$= \frac{\lim\limits_{x \to +\infty} 1}{\lim\limits_{x \to +\infty} 1/x^2 + \lim\limits_{x \to +\infty} 1/x + \lim\limits_{x \to +\infty} 2}$$

reciprocal limit rule

$$= \frac{1}{0 + 0 + 2} = \frac{1}{2}$$

The graph of $f(x)$ is shown in Figure 1.49. For practice, verify that $\lim\limits_{x \to -\infty} f(x) = \dfrac{1}{2}$ also.

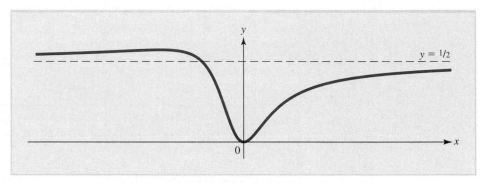

FIGURE 1.49 The graph of $f(x) = \dfrac{x^2}{1 + x + 2x^2}$.

Here is a general description of the procedure for evaluating a limit of a rational function at infinity.

Procedure for Evaluating a Limit at Infinity of $f(x) = p(x)/q(x)$

Step 1. Divide each term in $f(x)$ by the highest power x^k that appears in the denominator polynomial $q(x)$.

Step 2. Compute $\lim\limits_{x \to +\infty} f(x)$ or $\lim\limits_{x \to -\infty} f(x)$ using algebraic properties of limits and the reciprocal power rules.

Examples 1.5.8 and 1.5.9 illustrate how to compute and use a limit at infinity.

EXAMPLE 1.5.8 Finding a Limit at Infinity

Find $\lim\limits_{x \to +\infty} \dfrac{2x^2 + 3x + 1}{3x^2 - 5x + 2}$.

Solution

The highest power in the denominator is x^2. Divide the numerator and denominator by x^2 to get

$$\lim_{x \to +\infty} \frac{2x^2 + 3x + 1}{3x^2 - 5x + 2} = \lim_{x \to +\infty} \frac{2 + 3/x + 1/x^2}{3 - 5/x + 2/x^2} = \frac{2 + 0 + 0}{3 - 0 + 0} = \frac{2}{3}$$

EXAMPLE 1.5.9 Applying a Limit at Infinity

If a crop is planted in soil where the nitrogen level is N, then the crop yield Y can be modeled by the *Michaelis-Menten* function

$$Y(N) = \frac{AN}{B + N} \qquad N \geq 0$$

where A and B are positive constants. What happens to crop yield as the nitrogen level is increased indefinitely?

Solution

We wish to compute

$$\lim_{N \to +\infty} Y(N) = \lim_{N \to +\infty} \frac{AN}{B + N} \qquad \text{divide numerator and denominator by } N$$

$$= \lim_{N \to +\infty} \frac{AN/N}{B/N + N/N}$$

$$= \lim_{N \to +\infty} \frac{A}{B/N + 1} = \frac{A}{0 + 1}$$

$$= A$$

Thus, the crop yield tends toward the constant value A as the nitrogen level N increases indefinitely. For this reason, A is called the *maximum attainable yield*.

If the functional values $f(x)$ increase or decrease without bound as x approaches c, then technically $\lim_{x \to c} f(x)$ does not exist. However, the behavior of the function in such a case may be more precisely described by using the following notation, which is illustrated in Example 1.5.10.

Infinite Limits ■ We say that $\lim_{x \to c} f(x)$ is an **infinite limit** if $f(x)$ increases or decreases without bound as $x \to c$. We write

$$\lim_{x \to c} f(x) = +\infty$$

if $f(x)$ increases without bound as $x \to c$, or

$$\lim_{x \to c} f(x) = -\infty$$

if $f(x)$ decreases without bound as $x \to c$.

NOTE Saying "$f(x)$ decreases without bound as $x \to c$" means that the output $f(x)$ of the function becomes negative and grows larger and larger in absolute value as x approaches the number c. ■

EXAMPLE 1.5.10 Using an Infinite Limit to Study Average Profit

A manufacturer determines that when x hundred units of a particular product are produced and sold, the profit will be $P(x) = 4x - \sqrt{x}$ thousand dollars. What happens to the average profit when the production level is very small?

Solution

The average profit is

$$AP(x) = \frac{4x - \sqrt{x}}{x} = 4 - \frac{1}{\sqrt{x}}$$

thousand dollars per hundred units. To find what happens at a very low level of production, we examine the limit of $AP(x)$ as $x \to 0$:

$$\lim_{x \to 0} AP(x) = \lim_{x \to 0} \frac{4x - \sqrt{x}}{x} = \lim_{x \to 0} 4 - \frac{1}{\sqrt{x}}$$

$4 - 1/\sqrt{x}$ becomes negative and grows large in absolute value

$$= -\infty$$

We interpret this limit as saying that as fewer and fewer units are produced, the average profit derived from producing each unit is actually a huge loss. This makes sense because when only a few units are produced, fixed start-up costs dominate any revenue that may be derived from sales.

EXERCISES ■ 1.5

In Exercises 1 through 6, find $\lim_{x \to a} f(x)$ *if it exists.*

1.   **2.**   **3.**

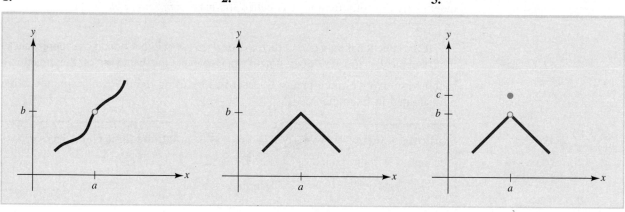

4.   **5.**   **6.**

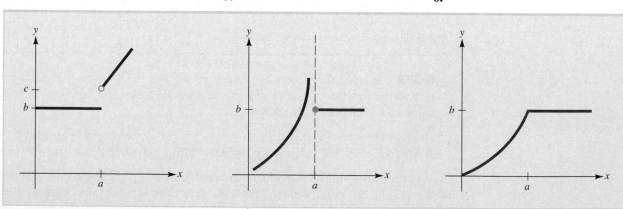

In Exercises 7 through 26, find the indicated limit if it exists.

7. $\lim_{x \to 2} (3x^2 - 5x + 2)$

8. $\lim_{x \to -1} (x^3 - 2x^2 + x - 3)$

9. $\lim_{x \to 0} (x^5 - 6x^4 + 7)$

10. $\lim_{x \to -1/2} (1 - 5x^3)$

11. $\lim_{x \to 3} (x - 1)^2(x + 1)$

12. $\lim_{x \to -1} (x^2 + 1)(1 - 2x)^2$

13. $\lim_{x \to 1/3} \dfrac{x + 1}{x + 2}$

14. $\lim_{x \to 1} \dfrac{2x + 3}{x + 1}$

15. $\lim_{x \to 5} \dfrac{x + 3}{5 - x}$

16. $\lim_{x \to 3} \dfrac{2x + 3}{x - 3}$

17. $\lim_{x \to 1} \dfrac{x^2 - 1}{x - 1}$

18. $\lim_{x \to 3} \dfrac{9 - x^2}{x - 3}$

19. $\lim_{x \to 5} \dfrac{x^2 - 3x - 10}{x - 5}$

20. $\lim_{x \to 2} \dfrac{x^2 + x - 6}{x - 2}$

21. $\lim_{x \to 4} \dfrac{(x + 1)(x - 4)}{(x - 1)(x - 4)}$

22. $\lim_{x \to 0} \dfrac{x(x^2 - 1)}{x^2}$

23. $\lim_{x \to -2} \dfrac{x^2 - x - 6}{x^2 + 3x + 2}$

24. $\lim_{x \to 1} \dfrac{x^2 + 4x - 5}{x^2 - 1}$

25. $\lim_{x \to 4} \dfrac{\sqrt{x} - 2}{x - 4}$

26. $\lim_{x \to 9} \dfrac{\sqrt{x} - 3}{x - 9}$

For Exercises 27 through 36, find $\lim_{x \to +\infty} f(x)$ and $\lim_{x \to -\infty} f(x)$. If the limiting value is infinite, indicate whether it is $+\infty$ or $-\infty$.

27. $f(x) = x^3 - 4x^2 - 4$

28. $f(x) = 1 - x + 2x^2 - 3x^3$

29. $f(x) = (1 - 2x)(x + 5)$

30. $f(x) = (1 + x^2)^3$

31. $f(x) = \dfrac{x^2 - 2x + 3}{2x^2 + 5x + 1}$

32. $f(x) = \dfrac{1 - 3x^3}{2x^3 - 6x + 2}$

33. $f(x) = \dfrac{2x + 1}{3x^2 + 2x - 7}$

34. $f(x) = \dfrac{x^2 + x - 5}{1 - 2x - x^3}$

35. $f(x) = \dfrac{3x^2 - 6x + 2}{2x - 9}$

36. $f(x) = \dfrac{1 - 2x^3}{x + 1}$

In Exercises 37 and 38, the graph of a function $f(x)$ is given. Use the graph to determine $\lim_{x \to +\infty} f(x)$ and $\lim_{x \to -\infty} f(x)$.

37.

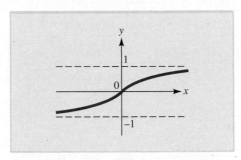

38.

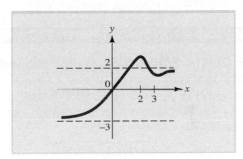

In Exercises 39 through 42, complete the table by evaluating $f(x)$ at the specified values of x. Then use the table to estimate the indicated limit or show it does not exist.

39. $f(x) = x^2 - x$; $\lim_{x \to 2} f(x)$

x	1.9	1.99	1.999	2	2.001	2.01	2.1
$f(x)$							

40. $f(x) = x - \dfrac{1}{x}$; $\lim_{x \to 0} f(x)$

x	−0.09	−0.009	0	0.0009	0.009	0.09
$f(x)$						

41. $f(x) = \dfrac{x^3 + 1}{x - 1}$; $\lim_{x \to 1} f(x)$

x	0.9	0.99	0.999	1	1.001	1.01	1.1
$f(x)$							

42. $f(x) = \dfrac{x^3 + 1}{x + 1}$; $\lim_{x \to -1} f(x)$

x	−1.1	−1.01	−1.001	−1	−0.999	−0.99	−0.9
$f(x)$							

In Exercises 43 through 50, find the indicated limit or show that it does not exist using the following facts about limits involving the functions f(x) and g(x):

$$\lim_{x \to c} f(x) = 5 \quad \text{and} \quad \lim_{x \to \infty} f(x) = -3$$

$$\lim_{x \to c} g(x) = -2 \quad \text{and} \quad \lim_{x \to \infty} g(x) = 4$$

43. $\lim_{x \to c} [2f(x) - 3g(x)]$ **44.** $\lim_{x \to c} f(x)\, g(x)$

45. $\lim_{x \to c} \sqrt{f(x) + g(x)}$ **46.** $\lim_{x \to c} f(x)[g(x) - 3]$

47. $\lim_{x \to c} \dfrac{f(x)}{g(x)}$ **48.** $\lim_{x \to c} \dfrac{2f(x) - g(x)}{5g(x) + 2f(x)}$

49. $\lim_{x \to \infty} \dfrac{2f(x) + g(x)}{x + f(x)}$ **50.** $\lim_{x \to \infty} \sqrt{g(x)}$

BUSINESS AND ECONOMICS APPLIED PROBLEMS

51. PER CAPITA EARNINGS Studies indicate that t years from now, the population of a certain country will be $p = 0.2t + 1{,}500$ thousand people, and that the gross earnings of the country will be E million dollars, where

$$E(t) = \sqrt{9t^2 + 0.5t + 179}$$

a. Express the per capita earnings of the country $P = \dfrac{E}{p}$ as a function of time t. (Take care with the units.)

b. What happens to the per capita earnings in the long run (as $t \to \infty$)?

52. PRODUCTION A business manager determines that t months after production begins on a new product, the number of units produced will be P thousand, where

$$P(t) = \dfrac{6t^2 + 5t}{(t + 1)^2}$$

What happens to production in the long run (as $t \to \infty$)?

53. AVERAGE COST A business manager determines that the total cost of producing x units of a particular commodity may be modeled by the function

$$C(x) = 7.5x + 120{,}000$$

(dollars). The average cost is $A(x) = \dfrac{C(x)}{x}$. Find $\lim\limits_{x \to +\infty} A(x)$, and interpret your result.

54. REVENUE Tomas, the organizer of a sports event, estimates that if the event is announced x days in advance, the revenue obtained will be $R(x)$ thousand dollars, where

$$R(x) = 400 + 120x - x^2$$

The cost of advertising the event for x days is $C(x)$ thousand dollars, where

$$C(x) = 2x^2 + 300$$

a. Find the profit function $P(x) = R(x) - C(x)$, and sketch its graph.

b. How many days in advance should Tomas announce the event to maximize profit? What is the maximum profit?

c. What is the ratio of revenue to cost

$$Q(x) = \dfrac{R(x)}{C(x)}$$

at the optimal announcement time found in part (b)? What happens to this ratio as $x \to 0$? Interpret these results.

55. PLANT MANAGEMENT Alicia, the manager of a plant, determines that when $x\%$ of the plant's capacity is being used, the total cost of operation is C hundred dollars, where

$$C(x) = \dfrac{8x^2 - 636x - 320}{x^2 - 68x - 960}$$

The company has a policy of rotating maintenance in an attempt to ensure that approximately 80% of capacity is always in use. What cost should Alicia expect when the plant is operating at this ideal capacity?

56. PRODUCTIVITY When starting a new job at a production facility, employees can be expected to assemble n items per hour after t weeks of work experience, where

$$n(t) = 70 - \dfrac{150}{t + 4}$$

Employees are paid 20 cents for each item they assemble.

a. Find an expression for the amount of money $A(t)$ earned per hour by an employee with t weeks of experience.

b. How much money per hour should an employee expect to earn in the long run (as $t \to \infty$)?

57. CONTINUOUS COMPOUNDING If $1{,}000 is invested at 5% compounded n times per year, the balance after 1 year will be $1{,}000(1 + 0.05x)^{1/x}$, where $x = 1/n$ is the length of the compounding period. For example, if $n = 4$, the compounding period is 1/4 year long. When the length of the compounding period approaches zero, we say that interest is *compounded continuously*, and the balance after 1 year is given by the limit.

$$B = \lim_{x \to 0} 1{,}000(1 + 0.05x)^{1/x}$$

Estimate the value of this limit by filling in the second line of the following table:

x	1	0.1	0.01	0.001	0.0001
$1{,}000(1 + 0.05x)^{1/x}$					

LIFE AND SOCIAL SCIENCE APPLIED PROBLEMS

58. POPULATION Scott, an urban planner, models the population $P(t)$ (in thousands) of his community t years from now by the function

$$P(t) = \frac{40t}{t^2 + 10} - \frac{50}{t + 1} + 70$$

  a. What is the current population of the community?
  b. By how much does the population change during the 3rd year? Is the population increasing or decreasing over this time period?
  c. What population should Scott plan for in the long run (as $t \to \infty$)?

59. CONCENTRATION OF DRUG The concentration of drug in a patient's bloodstream t hours after an injection is $C(t)$ milligrams per milliliter where

$$C(t) = \frac{0.4}{t^{1.2} + 1} + 0.013$$

  a. What is the concentration of drug immediately after the injection (when $t = 0$)?
  b. By how much does the concentration change during the 5th hour? Does it increase or decrease over this time period?
  c. What is the residual concentration of drug, that is, the concentration that remains in the long run (as $t \to \infty$)?

60. EXPERIMENTAL PSYCHOLOGY To study the rate at which animals learn, a psychology student performed an experiment in which a rat was sent repeatedly through a laboratory maze. Suppose the time required for the rat to traverse the maze on the nth trial was approximately

$$T(n) = \frac{5n + 17}{n}$$

minutes. What happens to the time of traverse as the number of trials n increases indefinitely? Interpret your result.

61. EXPLOSION AND EXTINCTION Two species coexist in the same ecosystem. Species I has population $P(t)$ in t years, while Species II has population $Q(t)$, both in thousands, where P and Q are modeled by the functions

$$P(t) = \frac{30}{3 + t} \qquad \text{and} \qquad Q(t) = \frac{64}{4 - t}$$

for all times $t \geq 0$ for which the respective populations are nonnegative.

  a. What is the initial population of each species?
  b. What happens to $P(t)$ as t increases? What happens to $Q(t)$?
  c. Sketch the graphs of $P(t)$ and $Q(t)$.
  d. Species I is said to face **extinction** in the long run, while the population of Species II **explodes.** Write a paragraph on what kind of circumstances might result in either explosion or extinction of a species.

62. BACTERIAL GROWTH The accompanying graph shows how the growth rate $R(T)$ of a bacterial colony changes with temperature T.*

  a. Over what range of values of T does the growth rate $R(T)$ double?
  b. What can be said about the growth rate for $25 < T < 45$?
  c. What happens when the temperature reaches roughly 45°C? Does it make sense to compute $\lim_{T \to 50} R(T)$?
  d. Write a paragraph describing how temperature affects the growth rate of a species.

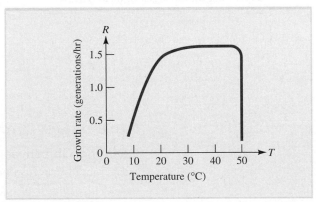

EXERCISE 62

Source: Michael D. La Grega, Phillip L. Buckingham, and Jeffrey C. Evans, *Hazardous Waste Management.* New York: McGraw-Hill, 1994, pp. 565–566. Reprinted by permission.

63. **ANIMAL BEHAVIOR**   In some animal species, the intake of food is affected by the amount of vigilance maintained by the animal while feeding. In essence, it is hard to eat heartily while watching for predators that may eat you. In one model,[†] if the animal is foraging on plants that offer a bite of size S, the intake rate of food, $I(S)$, is given by a function of the form

$$I(S) = \frac{aS}{S + c}$$

where a and c are positive constants.
   a. What happens to the intake $I(S)$ as bite size S increases indefinitely? Interpret your result.
   b. Read an article on various ways that the food intake rate may be affected by scanning for predators. Then write a paragraph on how mathematical models may be used to study such behavior in zoology. The reference cited in this problem offers a good starting point.

MISCELLANEOUS PROBLEMS

64. Solve Exercises 17 through 26 by using the **TRACE** feature of your calculator to make a table of x and $f(x)$ values near the number x is approaching.

65. Evaluate the limit

$$\lim_{x \to +\infty} \frac{a_n x^n + a_{n-1}x^{n-1} + \cdots + a_1 x + a_0}{b_m x^m + b_{m-1}x^{m-1} + \cdots + b_1 x + b_0}$$

for constants $a_0, a_1, \ldots, a_n$ and $b_0, b_1, \ldots, b_m$ in each of the following cases:
   a. $n < m$
   b. $n = m$
   c. $n > m$ [*Note:* There are two possible answers, depending on the signs of a_n and b_m.]

[†]A. W. Willius and C. Fitzgibbon, "Costs of Vigilance in Foraging Ungulates," *Animal Behavior*, Vol. 47, Pt. 2 (Feb. 1994).

66. The accompanying graph represents a function $f(x)$ that oscillates between 1 and -1 more and more frequently as x approaches 0 from either the right or the left. Does $\lim_{x \to 0} f(x)$ exist? If so, what is its value? [*Note:* For students with experience in trigonometry, the function $f(x) = \sin\left(\dfrac{1}{x}\right)$ behaves in this way.]

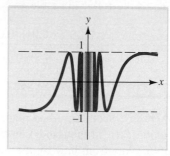

EXERCISE 66

67. A wire is stretched horizontally, as shown in the accompanying figure. An experiment is conducted in which different weights are attached at the center and the corresponding vertical displacements are measured. When too much weight is added, the wire snaps. Based on the data in the following table, what do you think is the maximum possible displacement for this kind of wire?

Weight W (lb)	15	16	17	18	17.5	17.9	17.99
Displacement y (in.)	1.7	1.75	1.78	Snaps	1.79	1.795	Snaps

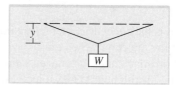

EXERCISE 67

SECTION 1.6   One-Sided Limits and Continuity

Learning Objectives

1. Compute and use one-sided limits.
2. Explore the concept of continuity and examine the continuity of several functions.
3. Investigate the intermediate value property.

The dictionary defines continuity as an "unbroken or uninterrupted succession." Continuous behavior is certainly an important part of our lives. For instance, the growth of a tree is continuous, as are the motion of a rocket and the volume of water flowing into a bathtub. In this section, we shall discuss what it means for a function to be continuous and shall examine a few important properties of such functions.

One-Sided Limits   Informally, a continuous function is one whose graph can be drawn without the pen leaving the paper (Figure 1.50a). Not all functions have this property, but those that do play a special role in calculus. A function is *not* continuous where its graph has a hole or gap (Figure 1.50b), but what do we really mean by holes and gaps in a graph? To describe such features mathematically, we require the concept of a **one-sided limit** of a function; that is, a limit in which the approach is either from the right or from the left, rather than from both sides as required for the two-sided limit introduced in Section 1.5.

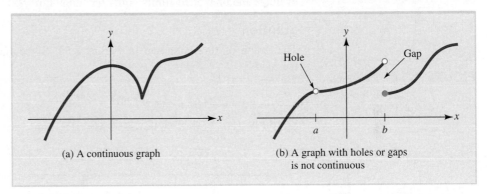

(a) A continuous graph   (b) A graph with holes or gaps is not continuous

FIGURE 1.50   Continuity and discontinuity.

For instance, Figure 1.51 shows the graph of inventory I as a function of time t for a company that immediately restocks to level L_1 whenever the inventory falls to a certain minimum level L_2 (this is called *just-in-time inventory*). Suppose the first restocking time occurs at $t = t_1$. Then as t tends toward t_1 from the left, the limiting value of $I(t)$ is L_2, while if the approach is from the right, the limiting value is L_1.

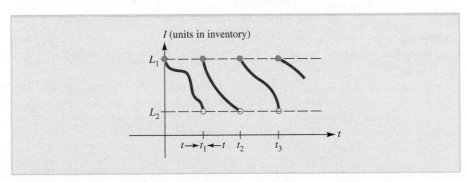

FIGURE 1.51   One-sided limits in a *just-in-time* inventory example.

Here is the notation we will use to describe one-sided limiting behavior.

One-Sided Limits   ▪   If $f(x)$ approaches L as x tends toward c from the left ($x < c$), we write $\lim\limits_{x \to c^-} f(x) = L$. Likewise, if $f(x)$ approaches M as x tends toward c from the right ($c < x$), then $\lim\limits_{x \to c^+} f(x) = M$.

If this notation is used in our inventory example, we would write

$$\lim_{t \to t_1^-} I(t) = L_2 \quad \text{and} \quad \lim_{t \to t_1^+} I(t) = L_1$$

Examples 1.6.1 and 1.6.2 show how to evaluate one-sided limits.

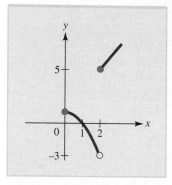

FIGURE 1.52   The graph of $f(x) = \begin{cases} 1 - x^2 & \text{if } 0 \le x < 2 \\ 2x + 1 & \text{if } x \ge 2 \end{cases}$

EXAMPLE 1.6.1   Evaluating One-Sided Limits

For the function

$$f(x) = \begin{cases} 1 - x^2 & \text{if } 0 \le x < 2 \\ 2x + 1 & \text{if } x \ge 2 \end{cases}$$

evaluate the one-sided limits $\lim_{x \to 2^-} f(x)$ and $\lim_{x \to 2^+} f(x)$.

Solution

The graph of $f(x)$ is shown in Figure 1.52. Since $f(x) = 1 - x^2$ for $0 \le x < 2$, we have

$$\lim_{x \to 2^-} f(x) = \lim_{x \to 2^-} (1 - x^2) = -3$$

Similarly, $f(x) = 2x + 1$ if $x \ge 2$, so

$$\lim_{x \to 2^+} f(x) = \lim_{x \to 2^+} (2x + 1) = 5$$

EXPLORE!

Refer to Example 1.6.2. Graph $f(x) = \dfrac{x - 2}{x - 4}$ using the window [0, 9.4]1 by [−4, 4]1 to verify the limit results as x approaches 4 from the left and the right. Now **TRACE** $f(x)$ for large positive or negative values of x. What do you observe?

EXAMPLE 1.6.2   Evaluating Infinite One-Sided Limits

Find the limit of $\dfrac{x - 2}{x - 4}$ as x approaches 4 from the left and from the right.

Solution

First, note that for $2 < x < 4$ the quantity

$$f(x) = \frac{x - 2}{x - 4}$$

is negative, so as x approaches 4 from the left, the denominator approaches zero, and $f(x)$ *decreases* without bound. We denote this fact by writing

$$\lim_{x \to 4^-} \frac{x - 2}{x - 4} = -\infty$$

Likewise, as x approaches 4 from the right (with $x > 4$), $f(x)$ increases without bound and we write

$$\lim_{x \to 4^+} \frac{x - 2}{x - 4} = +\infty$$

The graph of f is shown in Figure 1.53.

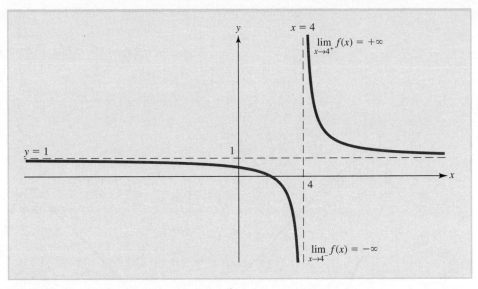

FIGURE 1.53 The graph of $f(x) = \dfrac{x-2}{x-4}$.

EXPLORE!

Re-create the piecewise linear function $f(x)$ defined in Example 1.6.1. Verify graphically that

$\lim\limits_{x\to 2^-} f(x) = -3$ and $\lim\limits_{x\to 2^+} f(x) = 5$.

Notice that the two-sided limit $\lim\limits_{x\to 2} f(x)$ does *not* exist for the function in Example 1.6.1 since the functional values $f(x)$ do not approach a single value L as x tends toward 2 from each side. In general, we have the following useful criterion for the existence of a limit.

Existence of a Limit ■ The two-sided limit $\lim\limits_{x\to c} f(x)$ exists if and only if the two one-sided limits $\lim\limits_{x\to c^-} f(x)$ and $\lim\limits_{x\to c^+} f(x)$ both exist and are equal, and then

$$\lim\limits_{x\to c} f(x) = \lim\limits_{x\to c^-} f(x) = \lim\limits_{x\to c^+} f(x)$$

EXAMPLE 1.6.3 **Using One-Sided Limits to Find a Two-Sided Limit**

Determine whether $\lim\limits_{x\to 1} f(x)$ exists, where

$$f(x) = \begin{cases} x + 1 & \text{if } x < 1 \\ -x^2 + 4x - 1 & \text{if } x \geq 1 \end{cases}$$

Solution

Computing the one-sided limits at $x = 1$, we find

$$\lim\limits_{x\to 1^-} f(x) = \lim\limits_{x\to 1^-} (x + 1) = (1) + 1 = 2 \quad \text{since } f(x) = x + 1 \text{ when } x < 1$$

and

$$\lim\limits_{x\to 1^+} f(x) = \lim\limits_{x\to 1^+} (-x^2 + 4x - 1) \quad \text{since } f(x) = -x^2 + 4x - 1 \text{ when } x \geq 1$$
$$= -(1)^2 + 4(1) - 1 = 2$$

Since the two one-sided limits are equal, it follows that the two-sided limit of $f(x)$ at $x = 1$ exists, and we have

$$\lim_{x \to 1} f(x) = \lim_{x \to 1^-} f(x) = \lim_{x \to 1^+} f(x) = 2$$

The graph of $f(x)$ is shown in Figure 1.54.

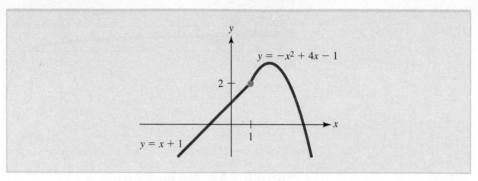

FIGURE 1.54 The graph of $f(x) = \begin{cases} x + 1 & \text{if } x < 1 \\ -x^2 + 4x - 1 & \text{if } x \geq 1 \end{cases}$.

Continuity

At the beginning of this section, we observed that a continuous function is one whose graph has no holes or gaps. A hole at $x = c$ can arise in several ways, three of which are shown in Figure 1.55.

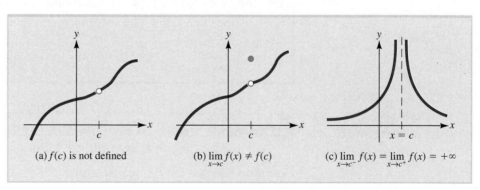

(a) $f(c)$ is not defined   (b) $\lim_{x \to c} f(x) \neq f(c)$ (c) $\lim_{x \to c^-} f(x) = \lim_{x \to c^+} f(x) = +\infty$

FIGURE 1.55 Three ways the graph of a function can have a hole at $x = c$.

The graph of $f(x)$ will have a gap at $x = c$ if the one-sided limits $\lim_{x \to c^-} f(x)$ and $\lim_{x \to c^+} f(x)$ are not equal. Three ways this can happen are shown in Figure 1.56.

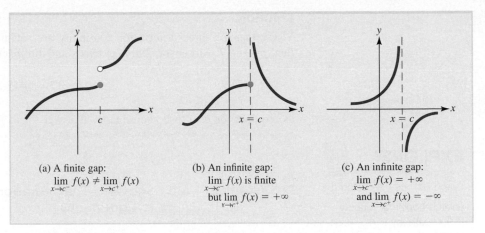

FIGURE 1.56   Three ways for the graph of a function to have a gap at $x = c$.

So what properties will guarantee that $f(x)$ does not have a hole or gap at $x = c$? The answer is surprisingly simple. The function must be defined at $x = c$, it must have a finite, two-sided limit at $x = c$; and $\lim\limits_{x \to c} f(x)$ must equal $f(c)$. To summarize:

Continuity ■ A function f is **continuous** at $x = c$ if all three of these conditions are satisfied:

a. $f(c)$ is defined.

b. $\lim\limits_{x \to c} f(x)$ exists.

c. $\lim\limits_{x \to c} f(x) = f(c)$.

If $f(x)$ is not continuous at $x = c$, it is said to have a **discontinuity** there.

Continuity of Polynomials and Rational Functions

Recall that if $p(x)$ and $q(x)$ are polynomials, then

$$\lim_{x \to c} p(x) = p(c)$$

and

$$\lim_{x \to c} \frac{p(x)}{q(x)} = \frac{p(c)}{q(c)} \quad \text{if } q(c) \neq 0$$

These limit formulas can be interpreted as saying that **a polynomial or a rational function is continuous wherever it is defined.** This is illustrated in Examples 1.6.4 through 1.6.7.

EXAMPLE 1.6.4   Showing That a Polynomial Is Continuous

Show that the polynomial $p(x) = 3x^3 - x + 5$ is continuous at $x = 1$.

Solution

Verify that the three criteria for continuity are satisfied. Clearly $p(1)$ is defined; in fact, $p(1) = 7$. Moreover, $\lim\limits_{x \to 1} p(x)$ exists and $\lim\limits_{x \to 1} p(x) = 7$. Thus,

$$\lim_{x \to 1} p(x) = 7 = p(1)$$

as required for $p(x)$ to be continuous at $x = 1$.

EXPLORE!

Graph $f(x) = \dfrac{x + 1}{x - 2}$ using the enlarged decimal window $[-9.4, 9.4]1$ by $[-6.2, 6.2]1$. Is the function continuous? Is it continuous at $x = 2$? How about at $x = 3$? Also examine this function using a table with an initial value of x at 1.8, increasing in increments of 0.2.

EXAMPLE 1.6.5 Showing That a Rational Function Is Continuous

Show that the rational function $f(x) = \dfrac{x + 1}{x - 2}$ is continuous at $x = 3$.

Solution

Note that $f(3) = \dfrac{3 + 1}{3 - 2} = 4$. Since $\lim\limits_{x \to 3} (x - 2) \neq 0$, you find that

$$\lim_{x \to 3} f(x) = \lim_{x \to 3} \frac{x + 1}{x - 2} = \frac{\lim\limits_{x \to 3}(x + 1)}{\lim\limits_{x \to 3}(x - 2)} = \frac{4}{1} = 4 = f(3)$$

as required for $f(x)$ to be continuous at $x = 3$.

EXPLORE!

Store $h(x)$ of Example 1.6.6(c) into the equation editor as $Y1 = (X + 1)(X < 1) + (2 - X)(X \geq 1)$. Use a decimal window with a dot graphing style. Is this function continuous at $x = 1$? Use the **TRACE** key to display the value of the function at $x = 1$ and to find the limiting values of y as x approaches 1 from the left side and from the right side.

EXAMPLE 1.6.6 Deciding if a Function Is Continuous

Discuss the continuity of each of the following functions:

a. $f(x) = \dfrac{1}{x}$ **b.** $g(x) = \dfrac{x^2 - 1}{x + 1}$ **c.** $h(x) = \begin{cases} x + 1 & \text{if } x < 1 \\ 2 - x & \text{if } x \geq 1 \end{cases}$

Solution

The functions in parts (a) and (b) are rational and are therefore continuous wherever they are defined (that is, wherever their denominators are not zero).

a. $f(x) = \dfrac{1}{x}$ is defined everywhere except $x = 0$, so it is continuous for all $x \neq 0$ (Figure 1.57a).

b. Since $x = -1$ is the only value of x for which $g(x)$ is undefined, $g(x)$ is continuous except at $x = -1$ (Figure 1.57b).

c. This function is defined in two pieces. First check for continuity at $x = 1$, the value of x that separates the two pieces. You find that $\lim\limits_{x \to 1} h(x)$ does not exist, since $h(x)$ approaches 2 from the left and 1 from the right. Thus, $h(x)$ is not continuous at 1 (Figure 1.57c). However, since the polynomials $x + 1$ and $2 - x$ are each continuous for every value of x, it follows that $h(x)$ is continuous at every number x other than 1.

EXPLORE!

Graph $f(x) = \dfrac{x^3 - 8}{x - 2}$, using a standard window. Does this graph appear continuous? Now use a modified decimal window $[-4.7, 4.7]1$ by $[0, 14.4]1$ and describe what you observe. Which case in Example 1.6.6 does this resemble?

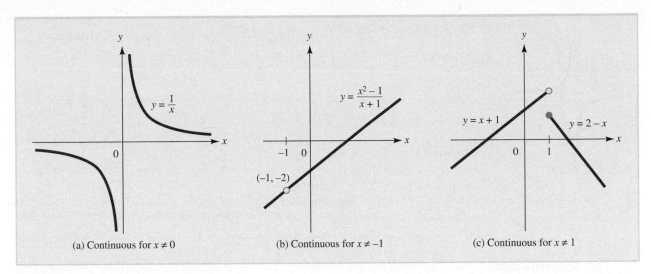

FIGURE 1.57 Functions for Example 1.6.6.

EXAMPLE 1.6.7 Making a Piecewise-Defined Function Continuous

For what value of the constant A is the following function continuous for all real x?

$$f(x) = \begin{cases} Ax + 5 & \text{if } x < 1 \\ x^2 - 3x + 4 & \text{if } x \geq 1 \end{cases}$$

Solution

Since $Ax + 5$ and $x^2 - 3x + 4$ are both polynomials, it follows that $f(x)$ will be continuous everywhere except possibly at $x = 1$. Moreover, $f(x)$ approaches $A + 5$ as x approaches 1 from the left and approaches 2 as x approaches 1 from the right. Thus, for $\lim_{x \to 1} f(x)$ to exist, we must have $A + 5 = 2$ or $A = -3$, in which case

$$\lim_{x \to 1} f(x) = 2 = f(1)$$

This means that f is continuous for all x only when $A = -3$.

Continuity on an Interval

For many applications of calculus, it is useful to have definitions of continuity on open and closed intervals.

> **Continuity on an Interval** ■ A function $f(x)$ is said to be continuous on an open interval $a < x < b$ if it is continuous at each point $x = c$ in that interval.
> Moreover, f is continuous on the closed interval $a \leq x \leq b$ if it is continuous on the open interval $a < x < b$ and
>
> $$\lim_{x \to a^+} f(x) = f(a) \qquad \text{and} \qquad \lim_{x \to b^-} f(x) = f(b)$$

In other words, continuity on an interval means that the graph of f is "one piece" throughout the interval. Example 1.6.8 illustrates how to determine the continuity of a function on an open interval.

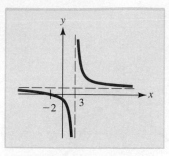

FIGURE 1.58   The graph of $f(x) = \dfrac{x+2}{x-3}$.

EXAMPLE 1.6.8   Deciding Where a Function Is Continuous

Discuss the continuity of the function

$$f(x) = \frac{x+2}{x-3}$$

on the open interval $-2 < x < 3$ and on the closed interval $-2 \le x \le 3$.

Solution

The rational function $f(x)$ is continuous for all x except $x = 3$. Therefore, it is continuous on the open interval $-2 < x < 3$ but not on the closed interval $-2 \le x \le 3$, since it is discontinuous at the endpoint 3 (where its denominator is zero). The graph of f is shown in Figure 1.58.

The Intermediate Value Property

An important feature of continuous functions is the **intermediate value property,** which says that if $f(x)$ is continuous on the interval $a \le x \le b$ and L is a number between $f(a)$ and $f(b)$, then $f(c) = L$ for some number c between a and b (see Figure 1.59). In other words, **a continuous function attains all values between any two of its values.** For instance, a girl who weighs 5 pounds at birth and 100 pounds at age 12 must have weighed exactly 50 pounds at some time during her 12 years of life, since her weight is a continuous function of time.

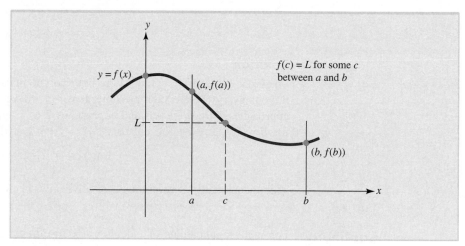

FIGURE 1.59   The intermediate value property.

The intermediate value property has a variety of applications. In Example 1.6.9, we show how it may be used to estimate a break-even point in a production process.

EXAMPLE 1.6.9   Break-Even Analysis Using an Intermediate Value

Martina can sell x hundred units of a suction pool-sweep device for $p = 400 - 3x^2$ dollars per unit and finds that total cost of production consists of a fixed overhead of $120,000 plus $7 per unit. Show that Martina breaks even for some level of production less than 500 units.

Solution

The revenue obtained from producing x hundred units is

$$R(x) = 100xp(x) = 100x(400 - 3x^2)$$

and the total cost is

$$C(x) = 120{,}000 + 7(100x)$$

For the manufacturing process to break even, we want revenue to equal cost; that is,

$$\underbrace{100x(400 - 3x^2)}_{R(x)} = \underbrace{120{,}000 + 700x}_{C(x)}$$

or, equivalently, in terms of the profit function $P(x) = R(x) - C(x)$,

$$
\begin{aligned}
P(x) &= 100x(400 - 3x^2) - (120{,}000 + 700x) \quad \text{combine like terms and factor out 100}\\
&= 100(-3x^3 + 393x - 1{,}200)\\
&= 0
\end{aligned}
$$

Note that $P(x)$ is a polynomial and hence is continuous. When $x = 0$ units are produced, we have $P(0) = -120{,}000 < 0$, and when $x = 5$, the profit is $P(5) = 39{,}000 > 0$. Since $P(x)$ changes sign from negative to positive as x varies between $x = 0$ and $x = 5$, it follows from the intermediate value property that $P(x) = 0$ for some x in the interval $0 < x < 5$. This is equivalent to saying that Martina's business breaks even for some positive level of production less than 500 units ($x = 5$).

EXERCISES ▪ 1.6

In Exercises 1 through 4, find the one-sided limits
$\lim\limits_{x \to 2^-} f(x)$ *and* $\lim\limits_{x \to 2^+} f(x)$ *from the given graph of f and*
determine whether $\lim\limits_{x \to 2} f(x)$ *exists.*

1.

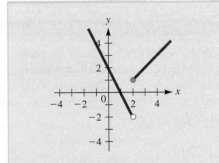

2.

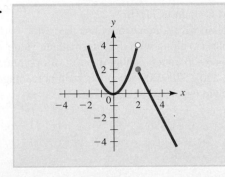

3.

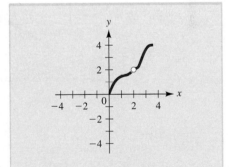

4.

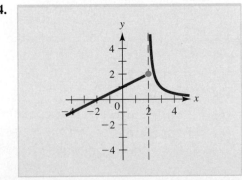

In Exercises 5 through 16, find the indicated one-sided limit. If the limiting value is infinite, indicate whether it is $+\infty$ or $-\infty$.

5. $\displaystyle\lim_{x\to 4^+}(3x^2 - 9)$

6. $\displaystyle\lim_{x\to 1^-} x(2 - x)$

7. $\displaystyle\lim_{x\to 3^+}\sqrt{3x - 9}$

8. $\displaystyle\lim_{x\to 2^-}\sqrt{4 - 2x}$

9. $\displaystyle\lim_{x\to 2^-}\frac{x + 3}{x + 2}$

10. $\displaystyle\lim_{x\to 2^-}\frac{x^2 + 4}{x - 2}$

11. $\displaystyle\lim_{x\to 0^+}(x - \sqrt{x})$

12. $\displaystyle\lim_{x\to 1^-}\frac{x - \sqrt{x}}{x - 1}$

13. $\displaystyle\lim_{x\to 3^+}\frac{\sqrt{x + 1} - 2}{x - 3}$

14. $\displaystyle\lim_{x\to 5^+}\frac{\sqrt{2x - 1} - 3}{x - 5}$

15. $\displaystyle\lim_{x\to 3^-} f(x)$ and $\displaystyle\lim_{x\to 3^+} f(x)$,

   where $f(x) = \begin{cases} 2x^2 - x & \text{if } x < 3 \\ 3 - x & \text{if } x \geq 3 \end{cases}$

16. $\displaystyle\lim_{x\to -1^-} f(x)$ and $\displaystyle\lim_{x\to -1^+} f(x)$

   where $f(x) = \begin{cases} \dfrac{1}{x - 1} & \text{if } x < -1 \\ x^2 + 2x & \text{if } x \geq -1 \end{cases}$

In Exercises 17 through 28, decide if the given function is continuous at the specified value of x.

17. $f(x) = 5x^2 - 6x + 1$ at $x = 2$

18. $f(x) = x^3 - 2x^2 + x - 5$ at $x = 0$

19. $f(x) = \dfrac{x + 2}{x + 1}$ at $x = 1$

20. $f(x) = \dfrac{2x - 4}{3x - 2}$ at $x = 2$

21. $f(x) = \dfrac{x + 1}{x - 1}$ at $x = 1$

22. $f(x) = \dfrac{2x + 1}{3x - 6}$ at $x = 2$

23. $f(x) = \dfrac{\sqrt{x} - 2}{x - 4}$ at $x = 4$

24. $f(x) = \dfrac{\sqrt{x} - 2}{x - 4}$ at $x = 2$

25. $f(x) = \begin{cases} x + 1 & \text{if } x \leq 2 \\ 2 & \text{if } x > 2 \end{cases}$ at $x = 2$

26. $f(x) = \begin{cases} x + 1 & \text{if } x < 0 \\ x - 1 & \text{if } x \geq 0 \end{cases}$ at $x = 0$

27. $f(x) = \begin{cases} x^2 + 1 & \text{if } x \leq 3 \\ 2x + 4 & \text{if } x > 3 \end{cases}$ at $x = 3$

28. $f(x) = \begin{cases} \dfrac{x^2 - 1}{x + 1} & \text{if } x < -1 \\ x^2 - 3 & \text{if } x \geq -1 \end{cases}$ at $x = -1$

In Exercises 29 through 42, list all the values of x for which the given function is not continuous.

29. $f(x) = 3x^2 - 6x + 9$

30. $f(x) = x^5 - x^3$

31. $f(x) = \dfrac{x + 1}{x - 2}$

32. $f(x) = \dfrac{3x - 1}{2x - 6}$

33. $f(x) = \dfrac{3x + 3}{x + 1}$

34. $f(x) = \dfrac{x^2 - 1}{x + 1}$

35. $f(x) = \dfrac{3x - 2}{(x + 3)(x - 6)}$

36. $f(x) = \dfrac{x}{(x + 5)(x - 1)}$

37. $f(x) = \dfrac{x}{x^2 - x}$

38. $f(x) = \dfrac{x^2 - 2x + 1}{x^2 - x - 2}$

39. $f(x) = \begin{cases} 2x + 3 & \text{if } x \leq 1 \\ 6x - 1 & \text{if } x > 1 \end{cases}$

40. $f(x) = \begin{cases} x^2 & \text{if } x \leq 2 \\ 9 & \text{if } x > 2 \end{cases}$

41. $f(x) = \begin{cases} 3x - 2 & \text{if } x < 0 \\ x^2 + x & \text{if } x \geq 0 \end{cases}$

42. $f(x) = \begin{cases} 2 - 3x & \text{if } x \leq -1 \\ x^2 - x + 3 & \text{if } x > -1 \end{cases}$

BUSINESS AND ECONOMICS APPLIED PROBLEMS

43. **COST MANAGEMENT** Recall Excercise 55, Section 1.5. Alicia, the manager of a plant, determines that when $x\%$ of her plant's capacity is being used, the total cost of operation is C hundred thousand dollars, where

$$C(x) = \frac{8x^2 - 636x - 320}{x^2 - 68x - 960}$$

  a. Find $C(0)$ and $C(100)$.
  b. Explain why Alicia cannot use the result of part (a) along with the intermediate value property to conclude that the cost of operation is exactly \$700,000 when a certain percentage of plant capacity is being used.

44. EARNINGS On January 1, 2012, Sam started working for Acme Corporation with an annual salary of $48,000, paid each month on the last day of that month. On July 1, he received a commission of $2,000 for his work, and on September 1, his base salary was raised to $54,000 per year. Finally, on December 21, he received a Christmas bonus of 1% of his base salary.

  a. Sketch the graph of Sam's cumulative earnings E as a function of time t (days) during the year 2012.

  b. For what values of t is the graph of $E(t)$ discontinuous?

45. COST-BENEFIT ANALYSIS In certain situations, it is necessary to weigh the benefit of pursuing a certain goal against the cost of achieving that goal. For instance, suppose that to remove x% of the pollution from an oil spill, it costs C thousands of dollars, where

$$C(x) = \frac{12x}{100 - x}$$

  a. How much does it cost to remove 25% of the pollution? 50%?

  b. Sketch the graph of the cost function.

  c. What happens as $x \to 100^-$? Is it possible to remove all the pollution?

46. INVENTORY The accompanying graph shows the number of units in inventory at a certain business over a 2-year period. When is the graph discontinuous? What do you think is happening at those times?

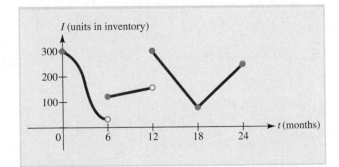

EXERCISE 46

LIFE AND SOCIAL SCIENCE APPLIED PROBLEMS

47. ENERGY CONSUMPTION The accompanying graph shows the amount of gasoline in the tank of a car over a 30-day period. When is the graph discontinuous? What do you think happens at these times?

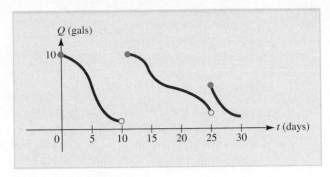

EXERCISE 47

48. WATER POLLUTION A ruptured pipe in a North Sea oil rig produces a circular oil slick that is y meters thick at a distance x meters from the rupture. Turbulence makes it difficult to directly measure the thickness of the slick at the source (where $x = 0$), but for $x > 0$, it is found that

$$y = \frac{0.5(x^2 + 3x)}{x^3 + x^2 + 4x}$$

Assuming the oil slick is continuously distributed, how thick would you expect it to be at the source?

49. AIR POLLUTION It is estimated that t years from now the population of a certain suburban community will be p thousand people, where

$$p(t) = 20 - \frac{7}{t + 2}$$

An environmental study indicates that the average level of carbon monoxide in the air will be c parts per million when the population is p thousand, where

$$c(p) = 0.4\sqrt{p^2 + p + 21}$$

What happens to the level of pollution c in the long run (as $t \to \infty$)?

50. WEATHER Suppose air temperature on a certain day is 30°F. Then the equivalent windchill temperature (in °F) produced by a wind with speed v miles per hour (mph) is given by*

$$W(v) = \begin{cases} 30 & \text{if } 0 \le v \le 4 \\ 1.25v - 18.67\sqrt{v} + 62.3 & \text{if } 4 < v < 45 \\ -7 & \text{if } v \ge 45 \end{cases}$$

---

*Adapted from *UMAP Module No. 658,* "Windchill," by W. Bosch and L. G. Cobb, 1984, pp. 244–247.

a. What is the windchill temperature when $v = 20$ mph? When $v = 50$ mph?

b. What wind speed produces a windchill temperature of 0°F?

c. Is the windchill function $W(v)$ continuous at $v = 4$? What about at $v = 45$?

51. WEATHER If the air temperature on a given day is 80°F, the heat index $I(h)$ (also in °F) can be approximated by the following function, where h is the relative humidity as a percentage:

$$I(h) = \begin{cases} 80 & \text{if } 0 \le h \le 40 \\ 80 + 0.1(h - 40) & \text{if } 40 < h \le 80 \\ 0.005h^2 - 0.65h + 104 & \text{if } 80 < h \le 100 \end{cases}$$

a. What is the heat index if the relative humidity is 30%? What if it is 90%?

b. What relative humidity produces a heat index of 83°F?

c. Is the heat index function $I(h)$ continuous at $h = 40$? What about at $h = 80$?

MISCELLANEOUS PROBLEMS

52. ELECTRIC FIELD INTENSITY If a hollow sphere of radius R is charged with one unit of static electricity, then the field intensity $E(x)$ at a point P located x units from the center of the sphere satisfies:

$$E(x) = \begin{cases} 0 & \text{if } 0 < x < R \\ \dfrac{1}{2x^2} & \text{if } x = R \\ \dfrac{1}{x^2} & \text{if } x > R \end{cases}$$

Sketch the graph of $E(x)$. Is $E(x)$ continuous for $x > 0$?

53. POSTAGE In 2010, the cost $p(x)$ in cents of mailing a letter weighing x ounces was

$$p(x) = \begin{cases} 44 & \text{if } 0 < x \le 1 \\ 61 & \text{if } 1 < x \le 2 \\ 78 & \text{if } 2 < x \le 3.5 \end{cases}$$

Sketch the graph of $p(x)$ for $0 < x \le 3.5$. For what values of x is $p(x)$ discontinuous for $0 < x \le 3.5$?

54. Discuss the continuity of the function

$$f(x) = \begin{cases} x^2 - 3x & \text{if } x < 2 \\ 4 + 2x & \text{if } x \ge 2 \end{cases}$$

on the open interval $0 < x < 2$ and on the closed interval $0 \le x \le 2$.

55. Discuss the continuity of the function

$$f(x) = x\left(1 + \frac{1}{x}\right)$$ on the open interval $0 < x < 1$

and on the closed interval $0 \le x \le 1$.

In Exercises 56 and 57, find the values of the constant A so that the function $f(x)$ will be continuous for all x.

56. $f(x) = \begin{cases} 1 - 3x & \text{if } x < 4 \\ Ax^2 + 2x - 3 & \text{if } x \ge 4 \end{cases}$

57. $f(x) = \begin{cases} Ax - 3 & \text{if } x < 2 \\ 3 - x + 2x^2 & \text{if } x \ge 2 \end{cases}$

58. Consider the equation

$$x^2 - x - 1 = \frac{1}{x + 1}$$

a. Use the intermediate value property to show that the equation has a solution between $x = 1$ and $x = 2$.

b. Is there a solution between $x = 1.5$ and $x = 1.6$? What about between $x = 1.7$ and $x = 1.8$? Explain.

59. Show that the equation $\sqrt[3]{x - 8} + 9x^{2/3} = 29$ has at least one solution for the interval $0 \le x \le 8$.

60. Show that the equation $\sqrt[3]{x} = x^2 + 2x - 1$ must have at least one solution on the interval $0 \le x \le 1$.

61. Investigate the behavior of $f(x) = \dfrac{2x^2 - 5x + 2}{x^2 - 4}$ when x is near to (a) 2 and (b) -2. Does the limit exist at these values of x? Is the function continuous at these values of x?

62. Explain why there must have been some time in your life when your weight in pounds was the same as your height in inches.

63. Explain why there is a time every hour when the hour hand and minute hand of a clock coincide.

64. At age 15, Michaela is twice as tall as her 5-year-old brother Juan, but on Juan's 21st birthday, they find that he is 6 inches taller. Explain why there must have been a time when Michaela and Juan were exactly the same height.

Important Terms, Symbols, and Formulas

Function   (2)
Functional notation: $f(x)$ (2)
Domain and range of a function   (2)
Independent and dependent variables   (3)
Domain convention   (4)
Piecewise-defined functions   (5)
Functions used in economics:
   Demand   (6)
   Supply   (6)
   Revenue   (6)
   Cost   (6)
   Profit   (6)
Average cost, revenue, and profit   (6)
Composition of functions: $f(g(x))$ (8)
Rectangular coordinate system   (17)
x and y axes   (17)
Quadrants   (17)
Coordinates: (17)
   x (abscissa)
   y (ordinate)
Distance formula: The distance between
   points (x_1, y_1) and (x_2, y_2) is
$$D = \sqrt{(x_2 - x_1)^2 + (y_2 - y_1)^2} \quad (18)$$
Graph of a function: the points $(x, f(x))$ (18)
Vertical line test   (21)
x and y intercepts   (22)
Power function   (23)
Polynomial   (23)
Rational function   (23)
Linear function; constant rate of change   (30)
Slope: $m = \dfrac{\Delta y}{\Delta x} = \dfrac{y_2 - y_1}{x_2 - x_1}$ (31)
Slope-intercept formula: $y = mx + b$ (33)
Point-slope formula: $y - y_0 = m(x - x_0)$ (33)
Criteria for lines to be parallel or perpendicular   (37)
Mathematical modeling   (45)

Direct proportionality: $Q = kx$ (50)
Inverse proportionality: $Q = \dfrac{k}{x}$ (50)
Joint proportionality: $Q = kxy$ (50)
Market equilibrium; law of supply and demand   (52)
Shortage and surplus   (52)
Break-even analysis   (53)
Limit of a function: $\lim\limits_{x \to c} f(x)$ (64)
Algebraic properties of limits   (66)
Limits at infinity:
$$\lim_{x \to +\infty} f(x) \quad \text{and} \quad \lim_{x \to -\infty} f(x) \quad (70)$$
Horizontal asymptote   (70)
Reciprocal limit rules:
$$\lim_{x \to +\infty} \frac{A}{x^k} = 0 \quad \text{and} \quad \lim_{x \to -\infty} \frac{A}{x^k} = 0 \quad k > 0 \quad (71)$$
Limits at infinity of a rational function $f(x) = \dfrac{p(x)}{q(x)}$:
   Divide all terms in $f(x)$ by the highest power x^k
   in the denominator $q(x)$ and use the reciprocal
   power rules.   (72)
Infinite limit:
$$\lim_{x \to c} f(x) = +\infty \quad \text{or} \quad \lim_{x \to c} f(x) = -\infty \quad (73)$$
One-sided limits:
$$\lim_{x \to c^-} f(x) \quad \text{and} \quad \lim_{x \to c^+} f(x) \quad (79)$$
Existence of a limit: $\lim\limits_{x \to c} f(x)$ exists if and only if
$$\lim_{x \to c^-} f(x) \text{ and } \lim_{x \to c^+} f(x) \text{ exist and are equal.} \quad (81)$$
Continuity of $f(x)$ at $x = c$:
$$\lim_{x \to c} f(x) = f(c) \quad (83)$$
Discontinuity   (83)
Continuity of polynomials and rational functions   (83)
Continuity on an interval   (85)
Intermediate value property   (86)

Checkup for Chapter 1

1. Specify the domain of the function
$$f(x) = \frac{2x - 1}{\sqrt{4 - x^2}}$$

2. Find the composite function $g(h(x))$, where
$$g(u) = \frac{1}{2u + 1} \quad \text{and} \quad h(x) = \frac{x + 2}{2x + 1}$$

3. Find an equation for each of these lines:
   a. Through the point $(-1, 2)$ with slope $-\dfrac{1}{2}$
   b. With slope 2 and y intercept -3

4. Sketch the graph of each of these functions. Be sure to show all intercepts and any high or low points.
   a. $f(x) = 3x - 5$ b. $f(x) = -x^2 + 3x + 4$

5. Find each of these limits. If the limit is infinite, indicate whether it is $+\infty$ or $-\infty$.

a. $\lim\limits_{x \to -1} \dfrac{x^2 + 2x - 3}{x - 1}$ b. $\lim\limits_{x \to 1} \dfrac{x^2 + 2x - 3}{x - 1}$

c. $\lim\limits_{x \to 1} \dfrac{x^2 - x - 1}{x - 2}$ d. $\lim\limits_{x \to +\infty} \dfrac{2x^3 + 3x - 5}{-x^2 + 2x + 7}$

6. Determine whether this function $f(x)$ is continuous at $x = 1$:

$$f(x) = \begin{cases} 2x + 1 & \text{if } x \le 1 \\ \dfrac{x^2 + 2x - 3}{x - 1} & \text{if } x > 1 \end{cases}$$

7. **PRICE OF GASOLINE** Since the beginning of the year, the price of unleaded gasoline has been increasing at a constant rate of 2 cents per gallon per month. By June first, the price had reached $3.80 per gallon.

a. Express the price of unleaded gasoline as a function of time, and draw the graph.

b. What was the price at the beginning of the year?

c. What will be the price on October 1?

8. **DISTANCE** A truck is 300 miles due east of a car and is traveling west at the constant speed of 30 miles per hour. Meanwhile, the car is going north at the constant speed of 60 miles per hour. Express the distance between the car and truck as a function of time.

9. **SUPPLY AND DEMAND** Suppose it is known that producers will supply x units of a certain commodity to the market when the price is $p = S(x)$ dollars per unit and that the same number of units will be demanded (bought) by consumers when the price is $p = D(x)$ dollars per unit, where

$$S(x) = x^2 + A \quad \text{and} \quad D(x) = Bx + 59$$

for constants A and B. It is also known that no units will be supplied until the unit price is $3 and that market equilibrium occurs when $x = 7$ units.

a. Use this information to find A and B and the equilibrium unit price.

b. Sketch the supply and demand curves on the same graph.

c. What is the difference between the supply price and the demand price when 5 units are produced? When 10 units are produced?

10. **BACTERIAL POPULATION** The population (in thousands) of a colony of bacteria t minutes after the introduction of a toxin is given by the function

$$f(t) = \begin{cases} t^2 + 7 & \text{if } 0 \le t < 5 \\ -8t + 72 & \text{if } t \ge 5 \end{cases}$$

a. When does the colony die out?

b. Explain why the population must be 10,000 some time between $t = 1$ and $t = 7$.

11. **MUTATION** In a study of mutation in fruit flies, researchers radiate flies with X-rays and determine that the mutation percentage M increases linearly with the X-ray dosage D, measured in kilo-Roentgens (kR). When a dose of $D = 3$ kR is used, the percentage of mutations is 7.7%, while a dose of 5 kR results in a 12.7% mutation percentage. Express M as a function of D. What percentage of the flies will mutate even if no radiation is used?

Review Exercises

1. Specify the domain of each of these functions:

a. $f(x) = x^2 - 2x + 6$

b. $f(x) = \dfrac{x - 3}{x^2 + x - 2}$

c. $f(x) = \sqrt{x^2 - 9}$

2. Specify the domain of each of these functions.

a. $f(x) = 4 - (3 - x)^2$

b. $f(x) = \dfrac{x - 1}{x^2 - 2x + 1}$

c. $f(x) = \dfrac{1}{\sqrt{4 - 3x}}$

3. Find the composite function $g(h(x))$.

a. $g(u) = u^2 + 2u + 1, h(x) = 1 - x$

b. $g(u) = \dfrac{1}{2u + 1}, h(x) = x + 2$

4. Find the composite function $g(h(x))$.

a. $g(u) = (1 - 2x)^2; h(x) = \sqrt{x + 1}$

b. $g(u) = \sqrt{1 - u}, h(x) = 2x + 4$

5. a. Find $f(3 - x)$ if $f(x) = 4 - x - x^2$.

b. Find $f(x^2 - 3)$ if $f(x) = x - 1$.

c. Find $f(x + 1) - f(x)$ if $f(x) = \dfrac{1}{x - 1}$.

6. **a.** Find $f(x - 2)$ if $f(x) = x^2 - x + 4$.

  b. Find $f(x^2 + 1)$ if $f(x) = \sqrt{x} + \dfrac{2}{x - 1}$.

  c. Find $f(x + 1) - f(x)$ if $f(x) = x^2$.

7. Find functions $h(x)$ and $g(u)$ such that $f(x) = g(h(x))$.

  a. $f(x) = (x^2 + 3x + 4)^5$

  b. $f(x) = (3x + 1)^2 + \dfrac{5}{2(3x + 2)^3}$

8. Find functions $h(x)$ and $g(u)$ so $f(x) = g(h(x))$.

  a. $f(x) = (x - 1)^2 - 3(x - 1) + 1$

  b. $f(x) = \dfrac{2(x + 4)}{2x - 3}$

9. Graph the quadratic function
$$f(x) = x^2 + 2x - 8$$

10. Graph the quadratic function
$$f(x) = 3 + 4x - 2x^2$$

11. Find the slope and y intercept of the given line and draw the graph.

  a. $y = 3x + 2$

  b. $5x - 4y = 20$

12. Find the slope and y intercept of the given line and draw the graph.

  a. $2y + 3x = 0$ **b.** $\dfrac{x}{3} + \dfrac{y}{2} = 4$

13. Find equations for these lines:

  a. Slope 5 and y intercept $(0, -4)$

  b. Slope -2 and contains $(1, 3)$

  c. Contains $(5, 4)$ and is parallel to $2x + y = 3$

14. Find equations for these lines:

  a. Passes through the points $(-1, 3)$ and $(4, 1)$.

  b. x intercept $(3, 0)$ and y intercept $\left(0, -\dfrac{2}{3}\right)$

  c. Contains $(-1, 3)$ and is perpendicular to $5x - 3y = 7$

15. Find the points of intersection (if any) of the given pair of curves, and draw the graphs.

  a. $y = -3x + 5$ and $y = 2x - 10$

  b. $y = x + 7$ and $y = -2 + x$

16. Find the points of intersection (if any) of the given pair of curves, and draw the graphs.

  a. $y = x^2 - 1$ and $y = 1 - x^2$

  b. $y = x^2$ and $y = 15 - 2x$

17. Find c so that the curve $y = 3x^2 - 2x + c$ passes through the point $(2, 4)$.

18. Find c so that the curve $y = 4 - x - cx^2$ passes through the point $(-2, 1)$.

In Exercises 19 through 32, either find the given limit or show it does not exist. If the limit is infinite, indicate whether it is $+\infty$ or $-\infty$.

19. $\lim\limits_{x \to 1} \dfrac{x^2 + x - 2}{x^2 - 1}$ **20.** $\lim\limits_{x \to 2} \dfrac{x^2 - 3x}{x + 1}$

21. $\lim\limits_{x \to 2} \dfrac{x^3 - 8}{2 - x}$ **22.** $\lim\limits_{x \to 1} \left(\dfrac{1}{x^2} - \dfrac{1}{x}\right)$

23. $\lim\limits_{x \to 0} \left(2 - \dfrac{1}{x^3}\right)$ **24.** $\lim\limits_{x \to -\infty} \left(2 + \dfrac{1}{x^2}\right)$

25. $\lim\limits_{x \to -\infty} \dfrac{x}{x^2 + 5}$ **26.** $\lim\limits_{x \to 0^-} \left(x^3 - \dfrac{1}{x^2}\right)$

27. $\lim\limits_{x \to -\infty} \dfrac{x^4 + 3x^2 - 2x + 7}{x^3 + x + 1}$ **28.** $\lim\limits_{x \to -\infty} \dfrac{x^3 - 3x + 5}{2x + 3}$

29. $\lim\limits_{x \to -\infty} \dfrac{1 + \dfrac{1}{x} + \dfrac{1}{x^2}}{x^2 + 3x - 1}$ **30.** $\lim\limits_{x \to -\infty} \dfrac{x(x - 3)}{7 - x^2}$

31. $\lim\limits_{x \to 0^-} x\sqrt{1 - \dfrac{1}{x}}$ **32.** $\lim\limits_{x \to 0^+} \sqrt{x\left(1 + \dfrac{1}{x^2}\right)}$

In Exercises 33 through 36, list all values of x for which the given function is not continuous.

33. $f(x) = \dfrac{x^2 - 1}{x + 3}$

34. $f(x) = 5x^3 - 3x + \sqrt{x}$

35. $h(x) = \begin{cases} x^3 + 2x - 33 & \text{if } x \le 3 \\ \dfrac{x^2 - 6x + 9}{x - 3} & \text{if } x > 3 \end{cases}$

36. $g(x) = \dfrac{x^3 + 5x}{(x - 2)(2x + 3)}$

37. **PRICE** As advances in technology result in the production of increasingly powerful, compact calculators, the price of calculators currently on the market drops. Suppose that x months from now, the price of a certain model will be P dollars per unit, where
$$P(x) = 40 + \dfrac{30}{x + 1}$$

  a. What will be the price 5 months from now?

  b. By how much will the price drop during the fifth month?

c. When will the price be $43?

d. What happens to the price in the long run (as $x \to \infty$)?

38. ENVIRONMENTAL ANALYSIS An environmental study of a certain community suggests that the average daily level of smog in the air will be $Q(p) = \sqrt{0.5p + 19.4}$ units when the population is p thousand. It is estimated that t years from now, the population will be $p(t) = 8 + 0.2t^2$ thousand.

a. Express the level of smog in the air as a function of time.

b. What will the smog level be 3 years from now?

c. When will the smog level reach 5 units?

39. EDUCATIONAL FUNDING A private college in the southwest has launched a fund-raising campaign. Suppose that college officials estimate that it will take $f(x) = \dfrac{10x}{150 - x}$ weeks to reach $x\%$ of their goal.

a. Sketch the relevant portion of the graph of this function.

b. How long will it take to reach 50% of the campaign's goal?

c. How long will it take to reach 100% of the goal?

40. CONSUMER EXPENDITURE The demand for a certain commodity is $D(x) = -50x + 800$; that is, x units of the commodity will be demanded by consumers when the price is $p = D(x)$ dollars per unit. Total consumer expenditure $E(x)$ is the amount of money consumers pay to buy x units of the commodity.

a. Express consumer expenditure as a function of x, and sketch the graph of $E(x)$.

b. Use the graph in part (a) to determine the level of production x at which consumer expenditure is largest. What price p corresponds to maximum consumer expenditure?

41. MICROBIOLOGY A spherical cell of radius r has volume $V = \dfrac{4}{3}\pi r^3$ and surface area $S = 4\pi r^2$. Express V as a function of S. If S is doubled, what happens to V?

42. NEWSPAPER CIRCULATION The circulation of a newspaper is increasing at a constant rate. Three months ago the circulation was 3,200. Today it is 4,400.

a. Express the circulation as a function of time, and draw the graph.

b. What will be the circulation 2 months from now?

43. MANUFACTURING EFFICIENCY A manufacturing firm has received an order to make 400,000 souvenir silver medals commemorating the anniversary of the landing of Apollo 11 on the moon. The firm owns several machines, each of which can produce 200 medals per hour. The cost of setting up the machines to produce the medals is $80 per machine, and the total operating cost is $5.76 per hour. Express the cost of producing the 400,000 medals as a function of the number of machines used. Draw the graph and estimate the number of machines the firm should use to minimize cost.

44. OPTIMAL SELLING PRICE Chloe's company can produce bookcases at a cost of $80 apiece. Sales figures indicate that if the bookcases are sold for x dollars apiece, approximately $150 - x$ will be sold each month. Express Chloe's monthly profit as a function of the selling price x, draw the graph, and estimate the optimal selling price.

45. OPTIMAL SELLING PRICE A retailer can obtain digital cameras from the manufacturer at a cost of $150 apiece. The retailer has been selling the cameras at the price of $340 apiece, and at this price, consumers have been buying 40 cameras a month. The retailer is planning to lower the price to stimulate sales and estimates that for each $5 reduction in the price, 10 more cameras will be sold each month. Express the retailer's monthly profit from the sale of the cameras as a function of the selling price. Draw the graph, and estimate the optimal selling price.

46. STRUCTURAL DESIGN A cylindrical can with no top is to be constructed for 80 cents. The cost of the material used for the bottom is 3 cents per square inch, and the cost of the material used for the curved side is 2 cents per square inch. Express the volume of the can as a function of its radius.

47. PROPERTY TAX Khalil is trying to decide between two competing property tax propositions. With Proposition A, he will pay $100 plus 8% of the assessed value of his home, while Proposition B requires a payment of $1,900 plus 2% of the assessed value. Assuming Khalil's only consideration is to minimize his tax payment, develop a criterion based on the assessed value V of his home for deciding between the propositions.

48. **INVENTORY ANALYSIS** A businessman maintains inventory over a particular 30-day month as follows:

days 1–9   30 units

days 10–15   17 units

days 16–23   12 units

days 24–30   steadily decreasing from 12 units to 0 units

Sketch the graph of the inventory as a function of time t (days). At what times is the graph discontinuous?

49. **BREAK-EVEN ANALYSIS** A manufacturer can sell a certain product for $80 per unit. Total cost consists of a fixed overhead of $4,500 plus production costs of $50 per unit.

   a. How many units must the manufacturer sell to break even?

   b. What is the manufacturer's profit or loss if 200 units are sold?

   c. How many units must the manufacturer sell to realize a profit of $900?

50. **PRODUCTION MANAGEMENT** During the summer, a group of students builds kayaks in a converted garage. The rental for the garage is $1,500 for the summer, and the materials needed to build a kayak cost $125. The kayaks can be sold for $275 apiece.

   a. How many kayaks must the students sell to break even?

   b. How many kayaks must the students sell to make a profit of at least $1,000?

51. **LEARNING** Some psychologists believe that when a person is asked to recall a set of facts, the rate at which the facts are recalled is proportional to the number of relevant facts in the subject's memory that have not yet been recalled. Express the recall rate as a function of the number of facts that have been recalled.

52. **COST-EFFICIENT DESIGN** A cable is to be run from a power plant on one side of a river 900 meters wide to a factory on the other side, 3,000 meters downstream. The cable will be run

in a straight line from the power plant to some point P on the opposite bank and then along the bank to the factory. The cost of running the cable across the water is $5 per meter, while the cost over land is $4 per meter. Let x be the distance from P to the point directly across the river from the power plant. Express the cost of installing the cable as a function of x.

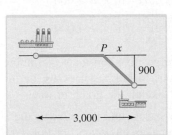 

EXERCISE 52   **EXERCISE 53**

53. **CONSTRUCTION COST** A window with a 20-foot perimeter (frame) is to be comprised of a semicircular stained glass pane above a rectangular clear pane, as shown in the accompanying figure. Clear glass costs $3 per square foot and stained glass costs $10 per square foot. Express the cost of the window as a function of the radius of the stained glass pane.

54. **MANUFACTURING OVERHEAD** A furniture manufacturer can sell end tables for $125 apiece. It costs the manufacturer $85 to produce each table, and it is estimated that revenue will equal cost when 200 tables are sold. What is the overhead associated with the production of the tables? [*Note:* Overhead is the cost when 0 units are produced.]

55. **MANUFACTURING COST** A manufacturer is capable of producing 5,000 units per day. There is a fixed (overhead) cost of $1,500 per day and a variable cost of $2 per unit produced.

   a. Express the daily cost $C(x)$ as a function of the number of units produced, and sketch the graph of $C(x)$.

   b. Find the average daily cost $AC(x)$. What is the average daily cost of producing 3,000 units per day?

   c. Is $C(x)$ continuous? If not, where do its discontinuities occur?

56. At what time between 3 P.M. and 4 P.M. will the minute hand coincide with the hour hand?

[*Hint:* The hour hand moves $\dfrac{1}{12}$ as fast as the minute hand.]

57. The radius of the earth is roughly 4,000 miles, and an object located x miles from the center of the earth weighs $w(x)$ lb, where

$$w(x) = \begin{cases} Ax & \text{if } x \le 4{,}000 \\ \dfrac{B}{x^2} & \text{if } x > 4{,}000 \end{cases}$$

and A and B are positive constants. Assuming that $w(x)$ is continuous for all x, what must be true about A and B? Sketch the graph of $w(x)$.

58. In each of these cases, find the value of the constant A that makes the given function $f(x)$ continuous for all x.

a. $f(x) = \begin{cases} 2x + 3 & \text{if } x < 1 \\ Ax - 1 & \text{if } x \ge 1 \end{cases}$

b. $f(x) = \begin{cases} \dfrac{x^2 - 1}{x + 1} & \text{if } x < -1 \\ Ax^2 + x - 3 & \text{if } x \ge -1 \end{cases}$

59. The accompanying graph represents a function $g(x)$ that oscillates more and more frequently as x approaches 0 from either the right or the left but with decreasing magnitude. Does $\lim\limits_{x \to 0} g(x)$ exist? If so, what is its value? [*Note:* For students with experience in trigonometry, the function $g(x) = |x| \sin(1/x)$ behaves in this way.]

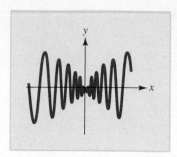

EXERCISE 59

60. Graph $f(x) = \dfrac{3x^2 - 6x + 9}{x^2 + x - 2}$. Determine the values of x where the function is undefined.

61. Graph $y = \dfrac{21}{9}x - \dfrac{84}{35}$ and $y = \dfrac{654}{279}x - \dfrac{54}{10}$ on the same set of coordinate axes using $[-10, 10]1$ by $[-10, 10]1$. Are the two lines parallel?

62. For $f(x) = \sqrt{x + 3}$ and $g(x) = 5x^2 + 4$, find:

a. $f(g(-1.28))$

b. $g(f(\sqrt{2}))$

Use three decimal place accuracy.

63. Graph $y = \begin{cases} x^2 + 1 & \text{if } x \le 1 \\ x^2 - 1 & \text{if } x > 1 \end{cases}$. Find the points of discontinuity.

64. Graph the function $f(x) = \dfrac{x^2 - 3x - 10}{1 - x} - 2$.

Find the x and y intercepts. For what values of x is this function defined?

EXPLORE! UPDATE

Explore! Updates are included at the end of each chapter of this textbook. These updates provide additional instruction and exercises for exploring calculus using a graphing calculator or giving hints and solutions for selected Explore! boxes found in each chapter. An attempt has been made to use function keys that are standard on most handheld graphing utilities. The exact names of the function keys on your particular calculator may vary, depending on the brand. Please consult your calculator user manual for more specific details.

Complete solutions for all Explore! boxes throughout the text can be accessed via the book-specific website www.mhhe.com/hoffmann.

Solution for Explore! Exercise on Page 3

Write $f(x) = x^2 + 4$ in the function editor (**Y=** key) of your graphing calculator. While on a cleared homescreen (**2nd MODE CLEAR**), locate the symbol for Y1 through the **VARS** key, arrowing right to **Y-VARS** and selecting **1:Function** and **1:Y1.** (Also see the Calculator Introduction found in the website www.mhhe.com/hoffmann). Y1({-3, -1, 0, 1, 3}) yields the desired functional values, all at once. Or you can do each value individually, such as Y1(-3).

NOTE An Explore! tip is that it is easier to view a table of values, especially for several functions. Set up the desired table parameters through **TBLSET (2nd WINDOW).** Now enter $g(x) = x^2 - 1$ into **Y2** of the equation editor (**Y=**). Observing the values of Y1 and Y2, we notice that they differ by a fixed constant of -5, since the two functions are simply vertical translations of $f(x) = x^2$. ∎

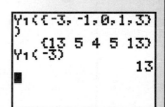

Solution for Explore! on Page 4

The function $f(x) = 1/(x - 3)$ appears to have a break at $x = 3$ and not be defined there. So the domain consists of all real numbers $x \neq 3$.

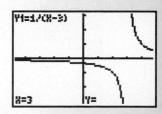

EXPLORE! UPDATE

Solution for Explore! Exercise on Page 5

The graph of the piecewise-defined function Y1 = 2(X < 1) + (−1)(X ≥ 1) is shown in the following figure, and the table to its right shows the functional values at X = −2, 0, 1, and 3. Recall that the inequality symbols can be accessed through the **TEST (2nd MATH)** key.

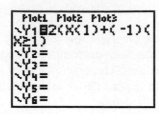 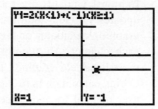 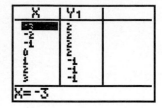

Solution for Explore! on Page 48

Use **TBLSET** with **TblStart = 1** and **ΔTbl = 1** and scroll down to obtain the table on the left which indicates a minimal cost in the twelve thousand range occurring within the interval [0, 25]. Then, graph using **WINDOW** dimensions of [0, 25]5 by [0, 30000]5000 to obtain the second screen, where we have used the **MINIMUM** on the **CALC** menu (see the Calculator Introduction found in the website www.mhhe.com/hoffmann) to locate an apparent minimal value of about $12,709 at a radius of 13 feet.

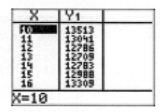 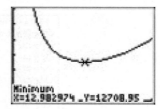

Solution for Explore! on Page 84

The graph of $f(x) = (x^3 - 8)/(x - 2)$ appears continuous based on the window, [−6, 6]1 by [−2, 10]1. However, examination of this graph using a modified decimal window, [−4.7, 4.7]1 by [0, 14.4]1, shows an exaggerated hole at $x = 2$.

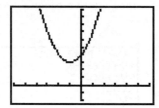 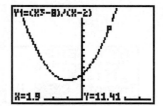 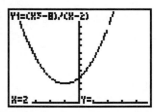

The function $f(x)$ is not continuous, specifically at $x = 2$, where it is not defined. The situation is similar to Figure 1.57(b) on page 85. What value of y would fill up the hole in the graph here? Answer: $\lim_{x \to 2} f(x) = 12$.

THINK ABOUT IT

ALLOMETRIC MODELS

When developing a mathematical model, the first task is to identify quantities of interest, and the next is to find equations that express relationships between these quantities. Such equations can be quite complicated, but there are many important relationships that can be expressed in the relatively simple form $y = Cx^k$, in which one quantity y is expressed as a constant multiple of a power function of another quantity x.

In biology, the study of the relative growth rates of various parts of an organism is called *allometry,* from the Greek words *allo* (other or different) and *metry* (measure). In allometric models, equations of the form $y = Cx^k$ are often used to describe the relationship between two biological measurements. For example, the size a of the antlers of an elk from tip to tip has been shown to be related to h, the shoulder height of the elk, by the allometric equation

$$a = 0.026h^{1.7}$$

where a and h are both measured in centimeters (cm).* This relationship is shown in the accompanying figure.

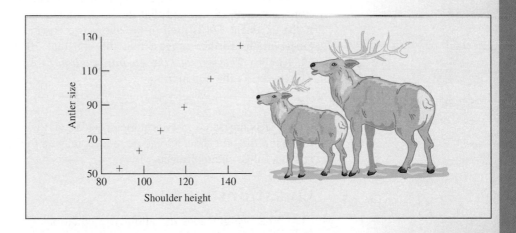

Whenever possible, allometric models are developed using basic assumptions from biological (or other) principles. For example, it is reasonable to assume that the body volume and hence the weight of most animals is proportional to the cube of the linear dimension of the body, such as height for animals that stand up or length for four-legged animals. Thus, it is reasonable to expect the weight of a snake to be proportional to the cube of its length, and indeed, observations of the hognose snake of

---

*Frederick R. Adler, *Modeling the Dynamics of Life,* Pacific Grove, CA: Brooks-Cole Publishing, 1998, p. 61.

Kansas indicate that the weight w (grams) and length L (meters) of a typical such snake are related by the equation*

$$w = 440L^3$$

At times, observed data may be quite different from the results predicted by the model. In such a case, we seek a better model. For the western hognose snake, it turns out that the equations $w = 446L^{2.99}$ and $w = 429L^{2.90}$ provide better approximations to the weight of male and female western hognose snakes, respectively. However, there is no underlying biological reason why we should use exponents less than 3. It just turns out these equations are slightly better fits.

The *basal metabolic rate M* of an animal is the rate of heat produced by its body when it is resting and fasting. Basal metabolic rates have been studied since the 1830s. Allometric equations of the form $M = cw^r$, for constants c and r, have long been used to build models relating the basal metabolic rate to the body weight w of an animal. The development of such a model is based on the assumption that basal metabolic rate M is proportional to S, the surface area of the body, so that $M = aS$ where a is a constant. To set up an equation relating M and w, we need to relate the weight w of an animal to its surface area S. Assuming that all animals are shaped like spheres or cubes, and that the weight of an animal is proportional to its volume, we can show (see Questions 1 and 2) that the surface area is proportional to $w^{2/3}$, so that $S = bw^{2/3}$ where b is a constant. Putting the equations $M = aS$ and $S = bw^{2/3}$ together, we obtain the allometric equation

$$M = abw^{2/3} = kw^{2/3}$$

where $k = ab$.

However, this is not the end of the story. When more refined modeling assumptions are used, it is found that the basal metabolic rate M is better approximated if the exponent 3/4 is used in the allometric equation rather than the exponent 2/3. Observations further suggest that the constant 70 be used in this equation (see M. Kleiber, *The Fire of Life, An Introduction to Animal Energetics,* Wiley, 1961). This gives us the equation

$$M = 70w^{3/4}$$

where M is measured in kilocalories per day and w is measured in kilograms. Additional information about allometric models may be found in our Web Resources Guide at www.mhhe.com/hoffmann.

Questions

1. What weight does the allometric equation $w = 440L^3$ predict for a western hognose snake 0.7 meters long? If this snake is male, what does the equation $w = 446L^{2.99}$ predict for its weight? If it is female, what does the equation $w = 429L^{2.90}$ predict for its weight?

2. What basal metabolic rates are predicted by the equation $M = 70w^{3/4}$ for animals with weights of 50 kg, 100 kg, and 350 kg?

---

*Edward Batschelet, *Introduction to Mathematics for Life Scientists,* 3rd ed., New York: Springer-Verlag, 1979, p. 178.

3. Observations show that the brain weight b, measured in grams, of adult female primates is given by the allometric equation $b = 0.064w^{0.822}$, where w is the weight in grams (g) of the primate. What are the predicted brain weights of female primates weighing 5 kg, 10 kg, 25 kg, 50 kg, and 100 kg? (Recall that 1 kg = 1,000 g.)

4. If $y = Cx^k$ where C and k are constants, then y and x are said to be related by a positive allometry if $k > 1$ and a negative allometry if $k < 1$. The *weight-specific metabolic rate* of an animal is defined to be the basal metabolic rate M of the animal divided by its weight w; that is, M/w. Show that if the basal metabolic rate is the positive allometry of the weight in Question 2, then the weight-specific metabolic rate is a negative allometry equation of the weight.

5. Show that if we assume all animal bodies are shaped like cubes, then the surface area S of an animal body is proportional to $V^{2/3}$, where V is the volume of the body. By combining this fact with the assumption that the weight w of an animal is proportional to its volume, show that S is proportional to $w^{2/3}$.

6. Show that if we assume all animal bodies are shaped like spheres, then the surface area S of an animal body is proportional to $V^{2/3}$, where V is the volume of the body. By combining this fact with the assumption that the weight w of an animal is proportional to its volume, show that S is proportional to $w^{2/3}$.

    [*Hint:* Recall that a sphere of radius r has surface area $4\pi r^2$ and volume $\frac{4}{3}\pi r^3$.]

CHAPTER **2**

The acceleration of a moving object is found by differentiating its velocity.

Differentiation: Basic Concepts

SECTION 2.1 The Derivative

Learning Objectives

1. Examine slopes of tangent lines and rates of change.
2. Define the derivative, and study its basic properties.
3. Compute and interpret a variety of derivatives using the definition.
4. Study the relationship between differentiability and continuity.

Calculus is the mathematics of change, and the primary tool for studying change is a procedure called **differentiation.** In this section, we shall introduce this procedure and examine some of its uses, especially in computing rates of change. Here, and later in this chapter, we shall encounter rates such as velocity, acceleration, production rates with respect to labor level or capital expenditure, the rate of growth of a population, the infection rate of a susceptible population during an epidemic, and many others.

Calculus was developed in the 17th century by Isaac Newton (1642–1727) and G. W. Leibniz (1646–1716) and others at least in part in an attempt to deal with two geometric problems:

Tangent problem: Find a tangent line at a particular point on a given curve.
Area problem: Find the area of the region under a given curve.

The area problem involves a procedure called **integration** in which quantities such as area, average value, present value of an income stream, and blood flow rate are computed as a special kind of limit of a sum. We shall study this procedure in Chapters 5 and 6. The tangent problem is closely related to the study of rates of change, and we will begin our study of calculus by examining this connection.

Slope and Rates of Change

Recall from Section 1.3 that a linear function $L(x) = mx + b$ changes at the constant rate m with respect to the independent variable x. That is, the rate of change of $L(x)$ is given by the slope or steepness of its graph, the line $y = mx + b$ (Figure 2.1a). For a function $f(x)$ that is not linear, the rate of change is not constant but varies with x. In particular, when $x = c$, the rate is given by the steepness of the graph of $f(x)$ at the point P with coordinates $(c, f(c))$, which can be measured by the slope of the tangent line to the graph at P (Figure 2.1b). The relationship between rate of change and slope is illustrated in Example 2.1.1.

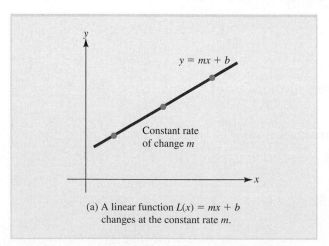

(a) A linear function $L(x) = mx + b$
changes at the constant rate m.

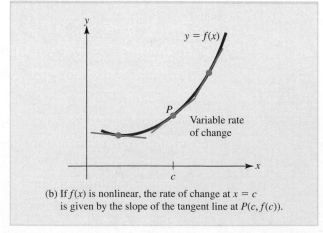

(b) If $f(x)$ is nonlinear, the rate of change at $x = c$
is given by the slope of the tangent line at $P(c, f(c))$.

FIGURE 2.1   Rate of change is measured by slope.

EXAMPLE 2.1.1 Estimating Rates of Change

The graph shown in Figure 2.2 gives the relationship between the percentage of unemployment U and the corresponding percentage of inflation I. Use the graph to estimate the rate at which I changes with respect to U when the level of unemployment is 3% and again when it is 10%.

Solution

From the figure, we estimate the slope of the tangent line at the point (3, 14), corresponding to $U = 3$, to be approximately -14. That is, when unemployment is 3%, inflation I is *decreasing* at the rate of 14 percentage points for each percentage point increase in unemployment U.

At the point (10, -5), the slope of the tangent line is approximately -0.4, which means that when there is 10% unemployment, inflation is decreasing at the rate of only 0.4 percentage point for each percentage point increase in unemployment.

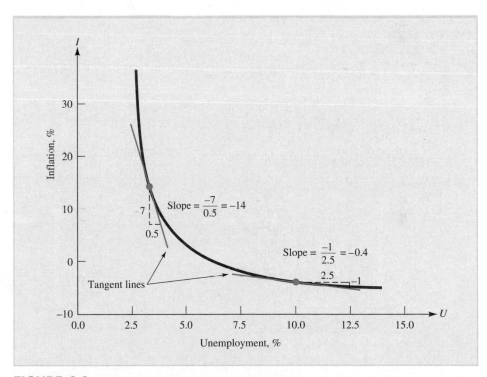

FIGURE 2.2 Inflation as a function of unemployment.

Source: Adapted from Robert Eisner, *The Misunderstood Economy: What Counts and How to Count It,* Boston, MA: Harvard Business School Press, 1994, p. 173.

In Example 2.1.2, we show how slope and a rate of change can be computed analytically using a limit.

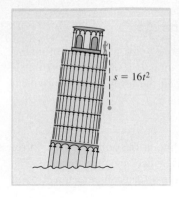

EXAMPLE 2.1.2   Computing Instantaneous Velocity

If air resistance is neglected, an object dropped from a great height will fall $s(t) = 16t^2$ feet in t seconds.

a. What is the object's velocity after $t = 2$ seconds?

b. How is the velocity found in part (a) related to the graph of $s(t)$?

Solution

a. The velocity after 2 seconds is the *instantaneous* rate of change of $s(t)$ at $t = 2$. Unless the falling object has a speedometer, it is hard to simply "read" its velocity. However, we can measure the distance it falls as time t changes by a small amount h from $t = 2$ to $t = 2 + h$ and then compute the *average* rate of change of $s(t)$ over the time period $[2, 2 + h]$ by the ratio

$$v_{ave} = \frac{\text{distance traveled}}{\text{elapsed time}} = \frac{s(2 + h) - s(2)}{(2 + h) - 2}$$

$$= \frac{16(2 + h)^2 - 16(2)^2}{h} = \frac{16(4 + 4h + h^2) - 16(4)}{h}$$

$$= \frac{64h + 16h^2}{h} = 64 + 16h$$

Just-In-Time **REVIEW**

To simplify certain discussions, we sometimes use the notation $P(a, b)$ to represent the point P with coordinates (a, b). Take care not to confuse this with function notation.

Since the elapsed time h is small, we would expect the average velocity v_{ave} to closely approximate the instantaneous ("speedometer") velocity v_{ins} at time $t = 2$. Thus, it is reasonable to compute the instantaneous velocity by the limit

$$v_{ins} = \lim_{h \to 0} v_{ave} = \lim_{h \to 0} (64 + 16h) = 64$$

That is, after 2 seconds, the object is traveling at the rate of 64 feet per second.

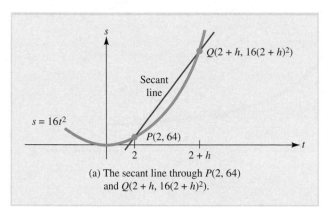

(a) The secant line through $P(2, 64)$
and $Q(2 + h, 16(2 + h)^2)$.

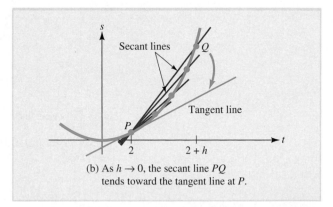

(b) As $h \to 0$, the secant line PQ
tends toward the tangent line at P.

FIGURE 2.3   Computing the slope of the tangent line to $s = 16t^2$ at $P(2, 64)$.

b. The procedure described in part (a) is represented geometrically in Figure 2.3. Figure 2.3a shows the graph of $s(t) = 16t^2$, along with the points $P(2, 64)$ and $Q(2 + h, 16(2 + h)^2)$. The line joining P and Q is called a *secant line* of the graph of $s(t)$ and has slope

$$m_{sec} = \frac{16(2 + h)^2 - 64}{(2 + h) - 2} = 64 + 16h$$

As indicated in Figure 2.3b, when we take smaller and smaller values of h, the corresponding secant lines PQ tend toward the position of what we intuitively think of as the tangent line to the graph of $s(t)$ at P. This suggests that we can compute the slope m_{tan} of the tangent line by finding the limiting value of m_{sec} as h tends toward 0; that is,

$$m_{tan} = \lim_{h \to 0} m_{sec} = \lim_{h \to 0} (64 + 16h) = 64$$

Thus, the slope of the tangent line to the graph of $s(t) = 16t^2$ at the point where $t = 2$ is the same as the instantaneous rate of change of $s(t)$ with respect to t when $t = 2$.

The procedure used in Example 2.1.2 to find the velocity of a falling body can be used to find other rates of change. Suppose we wish to find the rate at which the function $f(x)$ is changing with respect to x when $x = c$. We begin by finding the **average rate of change** of $f(x)$ as x varies from $x = c$ to $x = c + h$, which is given by the ratio

$$\text{Rate}_{ave} = \frac{\text{change in } f(x)}{\text{change in } x} = \frac{f(c + h) - f(c)}{(c + h) - c}$$
$$= \frac{f(c + h) - f(c)}{h}$$

This ratio can be interpreted geometrically as the slope of the secant line from the point $P(c, f(c))$ to the point $Q(c + h, f(c + h))$, as shown in Figure 2.4a.

EXPLORE!

A graphing calculator can simulate secant lines approaching a tangent line. Store $f(x) = (x - 2)^2 + 1$ into Y1 of the equation editor, selecting a **BOLD** graphing style. Store the values $\{-2.0, -1.6, -1.1, -0.6, -0.2\}$ into L1, list 1, using the **STAT** edit menu. Store

$$f(x) = L1 \times (x - 2) + 1$$

into Y2 of the equation editor. Graph using a modified decimal window [0, 4.7]1 by [0, 3.1]1, and describe what you observe. What is the equation for the limiting tangent line?

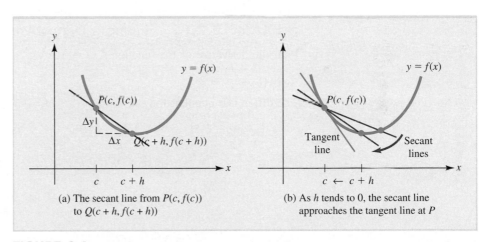

(a) The secant line from $P(c, f(c))$ to $Q(c + h, f(c + h))$

(b) As h tends to 0, the secant line approaches the tangent line at P

FIGURE 2.4   Secant lines approximating a tangent line.

We then compute the instantaneous rate of change of $f(x)$ at $x = c$ by finding the limiting value of the average rate as h tends to 0; that is,

$$\text{Rate}_{ins} = \lim_{h \to 0} \text{Rate}_{ave} = \lim_{h \to 0} \frac{f(c + h) - f(c)}{h}$$

This limit also gives the slope of the tangent line to the curve $y = f(x)$ at the point $P(c, f(c))$, as indicated in Figure 2.4b.

The Derivative

The expression

$$\frac{f(x + h) - f(x)}{h}$$

is called a **difference quotient** for the function $f(x)$, and we have just seen that rates of change and slope can both be computed by finding the limit of an appropriate difference quotient as h tends to 0. To unify the study of these and other applications that involve taking the limit of a difference quotient, we introduce the following terminology and notation.

The Derivative of a Function ■ The **derivative** of the function $f(x)$ with respect to x is the function $f'(x)$ given by

$$f'(x) = \lim_{h \to 0} \frac{f(x + h) - f(x)}{h}$$

[read $f'(x)$ as "f prime of x"]. The process of computing the derivative is called **differentiation,** and we say that $f(x)$ is **differentiable** at $x = c$ if $f'(c)$ exists; that is, if the limit that defines $f'(x)$ exists when $x = c$.

NOTE We use "h" to increment the independent variable in difference quotients to simplify algebraic computations. However, when it is important to emphasize that, say, the variable x is being incremented, we will denote the increment by Δx (read as "delta x"). Similarly, Δt and Δs denote small (incremental) changes in the variables t and s, respectively. This notation is used extensively in Section 2.5. ■

EXAMPLE 2.1.3   Finding a Derivative

Find the derivative of the function $f(x) = 16x^2$.

Solution

The difference quotient for $f(x)$ is

$$\frac{f(x + h) - f(x)}{h} = \frac{16(x + h)^2 - 16x^2}{h} \qquad \text{square } (x + h)$$

$$= \frac{16(x^2 + 2hx + h^2) - 16x^2}{h} \qquad \text{distribute and combine like terms}$$

$$= \frac{32hx + 16h^2}{h} \qquad \text{divide out common } h \text{ factors}$$

$$= 32x + 16h$$

Thus, the derivative of $f(x) = 16x^2$ is the function

$$f'(x) = \lim_{h \to 0} \frac{f(x + h) - f(x)}{h} = \lim_{h \to 0} (32x + 16h)$$

$$= 32x$$

Once we have computed the derivative of a function $f(x)$, it can be used to streamline any computation involving the limit of the difference quotient of $f(x)$ as h tends to 0. For instance, notice that the function in Example 2.1.3 is essentially the same as the distance function $s = 16t^2$ that appears in the falling body problem in Example 2.1.2. Using the result of Example 2.1.3, the velocity of the falling body at time $t = 2$ can now be found by simply substituting $t = 2$ into the formula for the derivative $s'(t)$:

$$\text{Velocity} = s'(2) = 32(2) = 64$$

Likewise, the slope of the tangent line to the graph of $s(t)$ at the point $(2, 64)$ is given by

$$\text{Slope} = s'(2) = 64$$

For future reference, our observations about rates of change and slope may be summarized as follows in terms of the derivative notation.

> **Slope as a Derivative** ■ The slope of the tangent line to the curve $y = f(x)$ at the point $(c, f(c))$ is $m_{\tan} = f'(c)$.

> **Instantaneous Rate of Change as a Derivative** ■ The rate of change of $f(x)$ with respect to x when $x = c$ is given by $f'(c)$.

In Example 2.1.4, we find the equation of a tangent line. Then in Example 2.1.5, we consider a business application involving a rate of change.

EXAMPLE 2.1.4   Using a Derivative to Find Slope

First compute the derivative of $f(x) = x^3$ and then use it to find the slope of the tangent line to the curve $y = x^3$ at the point where $x = -1$. What is the equation of the tangent line at this point?

Solution
According to the definition of the derivative

$$f'(x) = \lim_{h \to 0} \frac{f(x + h) - f(x)}{h} = \lim_{h \to 0} \frac{(x + h)^3 - x^3}{h}$$

$$= \lim_{h \to 0} \frac{(x^3 + 3x^2 h + 3xh^2 + h^3) - x^3}{h} = \lim_{h \to 0} (3x^2 + 3xh + h^2)$$

$$= 3x^2$$

Thus, the slope of the tangent line to the curve $y = x^3$ at the point where $x = -1$ is $f'(-1) = 3(-1)^2 = 3$ (Figure 2.5). To find an equation for the tangent line, we also need the y coordinate of the point of tangency, $y = (-1)^3 = -1$. Therefore, the tangent line passes through the point $(-1, -1)$ with slope 3. By applying the point-slope formula, we obtain the equation

$$y - (-1) = 3[x - (-1)]$$

or

$$y = 3x + 2$$

Just-In-Time REVIEW

Recall that $(a + b)^3 = a^3 + 3a^2 b + 3ab^2 + b^3$. This is a special case of the binomial theorem for the exponent 3 and is used to expand the numerator of the difference quotient in Example 2.1.4.

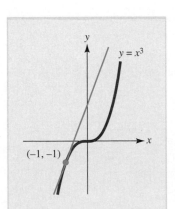

FIGURE 2.5 The graph of $y = x^3$.

EXAMPLE 2.1.5   Studying a Rate of Change in Profit

Gordon owns a small manufacturing firm. He determines that when x thousand units of one of his products are produced and sold, the profit generated will be

$$P(x) = -400x^2 + 6,800x - 12,000$$

dollars. At what rate should Gordon expect profit to be changing with respect to the level of production x when 9,000 units are produced? Is the profit increasing or decreasing at this level of production?

Solution

We find that

$$
\begin{aligned}
P'(x) &= \lim_{h \to 0} \frac{P(x+h) - P(x)}{h} \\
&= \lim_{h \to 0} \frac{[-400(x+h)^2 + 6,800(x+h) - 12,000] - (-400x^2 + 6,800x - 12,000)}{h} \\
&= \lim_{h \to 0} \frac{-400h^2 - 800hx + 6,800h}{h} \\
&= \lim_{h \to 0} (-400h - 800x + 6,800) \\
&= -800x + 6,800
\end{aligned}
$$

Thus, when the level of production is $x = 9$ (9,000 units), the profit is changing at the rate of

$$P'(9) = -800(9) + 6,800 = -400$$

dollars per thousand units.

Since $P'(9) = -400$ is negative, the tangent line to the profit curve $y = P(x)$ has a negative slope at the point Q where $x = 9$, so the tangent line is sloped downward at Q, as shown in Figure 2.6. This means that the profit curve itself is "falling" (tending downward) at Q, and the profit must be *decreasing* when 9,000 units are produced.

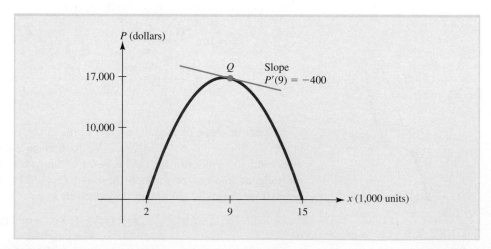

FIGURE 2.6   The graph of the profit function $P(x) = -400x^2 + 6,800x - 12,000$.

The significance of the sign of the derivative is summarized in the following box and in Figure 2.7. We will have much more to say about the relationship between the shape of a curve and the sign of derivatives in Chapter 3, where we develop a general procedure for curve sketching.

Significance of the Sign of the Derivative $f'(x)$. ■ If the function f is differentiable at $x = c$, then

$$f \text{ is } \textbf{increasing} \text{ at } x = c \text{ if } f'(c) > 0$$

and

$$f \text{ is } \textbf{decreasing} \text{ at } x = c \text{ if } f'(c) < 0$$

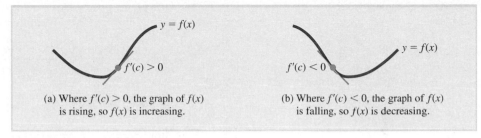

(a) Where $f'(c) > 0$, the graph of $f(x)$ is rising, so $f(x)$ is increasing.

(b) Where $f'(c) < 0$, the graph of $f(x)$ is falling, so $f(x)$ is decreasing.

FIGURE 2.7   The significance of the sign of the derivative $f'(c)$.

Derivative Notation

The derivative $f'(x)$ of $y = f(x)$ is sometimes written as $\dfrac{dy}{dx}$ or $\dfrac{df}{dx}$ (read as "dee y, dee x" or "dee f, dee x"). In this notation, the value of the derivative at $x = c$ [that is, $f'(c)$] is written as

$$\left.\frac{dy}{dx}\right|_{x=c} \qquad \text{or} \qquad \left.\frac{df}{dx}\right|_{x=c}$$

For example, if $y = x^2$, then

$$\frac{dy}{dx} = 2x$$

and the value of this derivative at $x = -3$ is

$$\left.\frac{dy}{dx}\right|_{x=-3} = 2x\Big|_{x=-3} = 2(-3) = -6$$

The $\dfrac{dy}{dx}$ notation for derivative suggests slope, $\dfrac{\Delta y}{\Delta x}$, and can also be thought of as "the rate of change of y with respect to x." Sometimes it is convenient to condense a statement such as

$$\text{"when } y = x^2, \text{ then } \frac{dy}{dx} = 2x\text{"}$$

by writing simply

$$\frac{d}{dx}(x^2) = 2x$$

which reads, "the derivative of x^2 with respect to x is $2x$."

Example 2.1.6 illustrates how the different notational forms for the derivative can be used.

EXAMPLE 2.1.6   Finding Slope and Rate of Change

First compute the derivative of $f(x) = \sqrt{x}$ and then use it to:

a. Find the equation of the tangent line to the curve $y = \sqrt{x}$ at the point where $x = 4$.

b. Find the rate at which $y = \sqrt{x}$ is changing with respect to x when $x = 1$.

Solution

The derivative of $y = \sqrt{x}$ with respect to x is given by

$$
\begin{aligned}
\frac{d}{dx}(\sqrt{x}) &= \lim_{h \to 0} \frac{f(x+h) - f(x)}{h} = \lim_{h \to 0} \frac{\sqrt{x+h} - \sqrt{x}}{h} \\
&= \lim_{h \to 0} \frac{(\sqrt{x+h} - \sqrt{x})(\sqrt{x+h} + \sqrt{x})}{h(\sqrt{x+h} + \sqrt{x})} \\
&= \lim_{h \to 0} \frac{x+h-x}{h(\sqrt{x+h} + \sqrt{x})} = \lim_{h \to 0} \frac{h}{h(\sqrt{x+h} + \sqrt{x})} \\
&= \lim_{h \to 0} \frac{1}{\sqrt{x+h} + \sqrt{x}} = \frac{1}{2\sqrt{x}} \qquad \text{if } x > 0
\end{aligned}
$$

a. When $x = 4$, the corresponding y coordinate on the graph of $f(x) = \sqrt{x}$ is $y = \sqrt{4} = 2$, so the point of tangency is (4, 2). Since $f'(x) = \dfrac{1}{2\sqrt{x}}$, the slope of the tangent line to the graph of $f(x)$ at the point (4, 2) is given by

$$
f'(4) = \frac{1}{2\sqrt{4}} = \frac{1}{4}
$$

and by substituting into the point-slope formula, we find that the equation of the tangent line at the point (4, 2) is

$$
y - 2 = \frac{1}{4}(x - 4)
$$

or

$$
y = \frac{1}{4}x + 1
$$

b. The rate of change of $y = \sqrt{x}$ when $x = 1$ is

$$
\left.\frac{dy}{dx}\right|_{x=1} = \frac{1}{2\sqrt{1}} = \frac{1}{2}
$$

NOTE   Notice that the function $f(x) = \sqrt{x}$ in Example 2.1.6 is defined at $x = 0$ but its derivative $f'(x) = \dfrac{1}{2\sqrt{x}}$ is not. This example shows that a function and its derivative do not always have the same domain. ∎

Differentiability and Continuity

If the function $f(x)$ is differentiable where $x = c$, then the graph of $y = f(x)$ has a nonvertical tangent line at the point P with coordinates $(c, f(c))$ and at all points on the graph that are near P. We would expect such a function to be continuous at $x = c$ since a graph with a tangent line at the point P certainly cannot have a hole or gap at P. To summarize:

> **Continuity of a Differentiable Function** ▪ If the function $f(x)$ is differentiable at $x = c$, then it is also continuous at $x = c$.

Verification of this observation is outlined in Exercise 64. Notice that we are *not* claiming that a continuous function must be differentiable. Indeed it can be shown that a continuous function $f(x)$ will not be differentiable at $x = c$ if $f'(x)$ becomes infinite at $x = c$ or if the graph of $f(x)$ has a "sharp" point at $(c, f(c))$, that is, a point where the curve makes an abrupt change in direction. If $f(x)$ is continuous at $x = c$ but $f'(c)$ is infinite, the graph of f may have a *vertical tangent* at the point $(c, f(c))$ (Figure 2.8a) or a *cusp* (Figure 2.8b). The absolute value function $f(x) = |x|$ is continuous for all x but has a *sharp point* at the origin $(0, 0)$ (see Figure 2.8c and Exercise 63). Another graph with a sharp point is shown in Figure 2.8d.

(a) Vertical tangent: the graph of $f(x) = x^{1/3}$

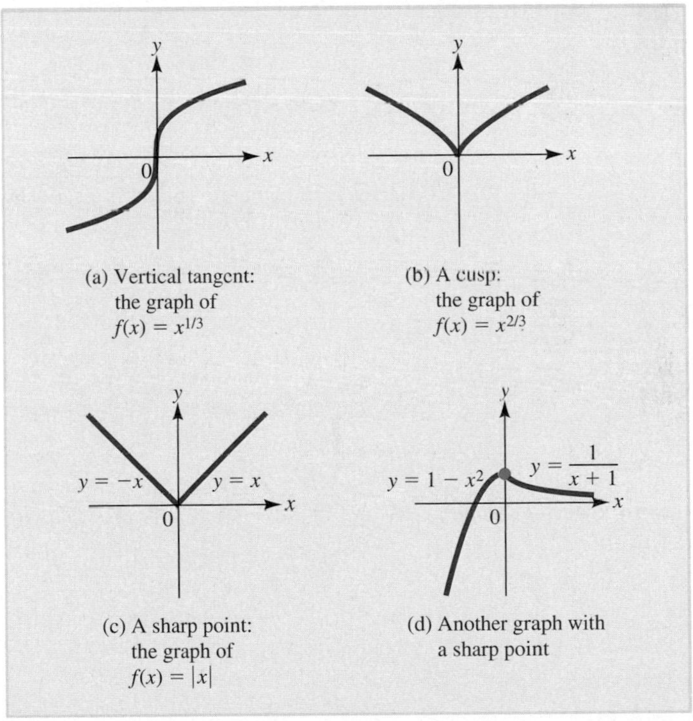

(b) A cusp: the graph of $f(x) = x^{2/3}$

(c) A sharp point: the graph of $f(x) = |x|$

$y = -x$ $y = x$

(d) Another graph with a sharp point

$y = 1 - x^2$ $y = \dfrac{1}{x + 1}$

FIGURE 2.8 Graphs of four continuous functions that are not differentiable at $x = 0$.

In general, the functions you encounter in this text will be differentiable at almost all points. In particular, polynomials are everywhere differentiable and rational functions are differentiable wherever they are defined.

An example of a practical situation involving a continuous function that is not always differentiable is provided by the circulation of blood in the human body.* It is tempting to think of blood coursing smoothly through the veins and arteries in a constant flow, but in actuality, blood is ejected by the heart into the arteries in discrete bursts. This results in an *arterial pulse*, which can be used to measure the heart rate by applying pressure to an accessible artery, as at the wrist. A blood pressure curve showing a pulse is sketched in Figure 2.9. Notice how the curve quickly changes direction at the minimum (*diastolic*) pressure, as a burst of blood is injected by the heart into the arteries, causing the blood pressure to rise rapidly until the maximum (*systolic*) pressure is reached, after which the pressure gradually declines as blood is distributed to tissue by the arteries. The blood pressure function is continuous but is not differentiable at $t = 0.75$ seconds where the infusion of blood occurs.

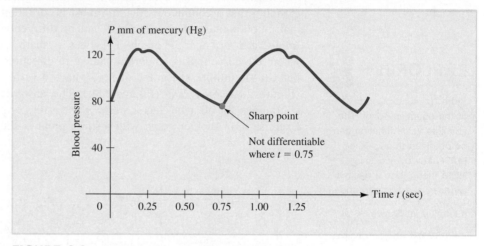

FIGURE 2.9   Graph of blood pressure showing an arterial pulse.

*This example is adapted from F. C. Hoppensteadt and C. S. Peskin, *Mathematics in Medicine and the Life Sciences*, New York: Springer-Verlag, 1992, p. 131.

EXERCISES ▪ 2.1

In Exercises 1 through 12, compute the derivative of the given function and find the slope of the line that is tangent to its graph for the specified value of the independent variable.

1. $f(x) = 4; x = 0$

2. $f(x) = -3; x = 1$

3. $f(x) = 5x - 3; x = 2$

4. $f(x) = 2 - 7x; x = -1$

5. $f(x) = 2x^2 - 3x - 5; x = 0$

6. $f(x) = x^2 - 1; x = -1$

7. $f(x) = x^3 - 1; x = 2$ **8.** $f(x) = -x^3; x = 1$

9. $g(t) = \dfrac{2}{t}; t = \dfrac{1}{2}$ **10.** $f(x) = \dfrac{1}{x^2}; x = 2$

11. $H(u) = \dfrac{1}{\sqrt{u}}; u = 4$ **12.** $f(x) = \sqrt{x}; x = 9$

In Exercises 13 through 24, compute the derivative of the given function and find the equation of the line that is tangent to its graph for the specified value $x = c$.

13. $f(x) = 2; c = 13$ **14.** $f(x) = 3; c = -4$

15. $f(x) = 7 - 2x; c = 5$ **16.** $f(x) = 3x; c = 1$

17. $f(x) = x^2; c = 1$ **18.** $f(x) = 2 - 3x^2; c = 1$

19. $f(x) = \dfrac{-2}{x}; c = -1$ **20.** $f(x) = \dfrac{3}{x^2}; c = \dfrac{1}{2}$

21. $f(x) = 2\sqrt{x}; c = 4$ **22.** $f(x) = \dfrac{1}{\sqrt{x}}; c = 1$

23. $f(x) = \dfrac{1}{x^3}; c = 1$

24. $f(x) = x^3 - 1; c = 1$

In Exercises 25 through 32, find the rate of change $\dfrac{dy}{dx}$ where $x = x_0$.

25. $y = 3;\ x_0 = 2$ **26.** $y = -17;\ x_0 = 14$

27. $y = 3x + 5;\ x_0 = -1$ **28.** $y = 6 - 2x;\ x_0 = 3$

29. $y = x(1 - x);\ x_0 = -1$

30. $y = x^2 - 2x;\ x_0 = 1$

31. $y = x - \dfrac{1}{x};\ x_0 = 1$ **32.** $y = \dfrac{1}{2 - x};\ x_0 = -3$

33. Let $f(x) = x^2$.
   a. Compute the slope of the secant line joining the points on the graph of f whose x coordinates are $x = -2$ and $x = -1.9$.
   b. Use calculus to compute the slope of the line that is tangent to the graph when $x = -2$, and compare with the slope found in part (a).

34. Let $f(x) = 2x - x^2$.
   a. Compute the slope of the secant line joining the points where $x = 0$ and $x = \dfrac{1}{2}$.
   b. Use calculus to compute the slope of the tangent line to the graph of $f(x)$ at $x = 0$, and compare with the slope found in part (a).

35. Let $f(x) = x^3$.
   a. Compute the slope of the secant line joining the points on the graph of f whose x coordinates are $x = 1$ and $x = 1.1$.
   b. Use calculus to compute the slope of the line that is tangent to the graph when $x = 1$, and compare with the slope found in part (a).

36. Let $f(x) = \dfrac{x}{x - 1}$.
   a. Compute the slope of the secant line joining the points where $x = -1$ and $x = -\dfrac{1}{2}$.

b. Use calculus to compute the slope of the tangent line to the graph of $f(x)$ at $x = -1$, and compare with the slope found in part (a).

37. Let $f(x) = 3x^2 - x$.
   a. Find the average rate of change of $f(x)$ with respect to x as x changes from $x = 0$ to $x = \dfrac{1}{16}$.
   b. Use calculus to find the instantaneous rate of change of $f(x)$ at $x = 0$, and compare with the average rate found in part (a).

38. Let $f(x) = x(1 - 2x)$.
   a. Find the average rate of change of $f(x)$ with respect to x as x changes from $x = 0$ to $x = \dfrac{1}{2}$.
   b. Use calculus to find the instantaneous rate of change of $f(x)$ at $x = 0$, and compare with the average rate found in part (a).

39. Let $s(t) = \dfrac{t - 1}{t + 1}$.
   a. Find the average rate of change of $s(t)$ with respect to t as t changes from $t = -\dfrac{1}{2}$ to $t = 0$.
   b. Use calculus to find the instantaneous rate of change of $s(t)$ at $t = -\dfrac{1}{2}$, and compare with the average rate found in part (a).

40. Let $s(t) = \sqrt{t}$.
   a. Find the average rate of change of $s(t)$ with respect to t as t changes from $t = 1$ to $t = \dfrac{1}{4}$.
   b. Use calculus to find the instantaneous rate of change of $s(t)$ at $t = 1$, and compare with the average rate found in part (a).

41. Fill in the missing interpretations in the table using the example as a guide.

	If $f(t)$ represents ...	then $\dfrac{f(t_0 + h) - f(t_0)}{h}$ represents ... and	$\displaystyle\lim_{h \to 0} \dfrac{f(t_0 + h) - f(t_0)}{h}$ represents ...
Example	The number of bacteria in a colony at time t	The average rate of change of the bacteria population during the time interval $[t_0, t_0 + h]$	The instantaneous rate of change of the bacteria population at time $t = t_0$
	a. The temperature in San Francisco t hours after midnight on a certain day		
	b. The blood alcohol level t hours after a person consumes an ounce of alcohol		
	c. The 30-year fixed mortgage rate t years after 2005		

42. Fill in the missing interpretations in the table using the example as a guide.

If $f(x)$ represents ...	then $\dfrac{f(x_0 + h) - f(x_0)}{h}$ represents ... and	$\lim\limits_{h \to 0} \dfrac{f(x_0 + h) - f(x_0)}{h}$ represents ...
Example The cost of producing x units of a particular commodity	The average rate of change of cost as production changes from x_0 to $x_0 + h$ units	The instantaneous rate of change of cost with respect to production level when x_0 units are produced
a. The revenue obtained when x units of a particular commodity are produced		
b. The amount of unexpended fuel (lbs) left in a rocket when it is x feet above ground		
c. The volume (cm^3) of a cancerous growth 6 months after injection of x mg of an experimental drug		

BUSINESS AND ECONOMICS APPLIED PROBLEMS

43. PROFIT   A manufacturer determines that when x hundred units of a particular commodity are produced, the profit will be

$$P(x) = 4{,}000(15 - x)(x - 2) \text{ dollars}$$

a. Find $P'(x)$.

b. Determine where $P'(x) = 0$. What is the significance of the level of production x_m where this occurs?

44. PROFIT   A manufacturer can produce digital recorders at a cost of $50 apiece. It is estimated that if the recorders are sold for p dollars apiece, consumers will buy $q = 120 - p$ recorders each month.

a. Express the manufacturer's profit P as a function of q.

b. What is the average rate of change in profit obtained as the level of production increases from $q = 0$ to $q = 20$?

c. At what rate is profit changing when $q = 20$ recorders are produced? Is the profit increasing or decreasing at this level of production?

45. COST OF PRODUCTION   Malik, the business manager of a company that produces in-ground outdoor spas, determines that the cost of producing x spas is C thousand dollars, where

$$C(x) = 0.04x^2 + 2.1x + 60$$

a. If Malik decides to increase the level of production from $x = 10$ to $x = 11$ spas, what is the corresponding average rate of change of cost?

b. Next, Malik computes the instantaneous rate of change of cost with respect to the level of production x. What is this rate when $x = 10$, and how does the instantaneous rate compare with

the average rate found in part (a)? By raising the level of production from $x = 10$, should Malik expect the cost to increase or decrease?

46. MANUFACTURING OUTPUT   At a certain factory, it is determined that an output of Q units is to be expected when L worker-hours of labor are employed, where

$$Q(L) = 3{,}100\sqrt{L}$$

a. Find the average rate of change of output as the labor employment changes from $L = 3{,}025$ worker-hours to 3,100 worker-hours.

b. Use calculus to find the instantaneous rate of change of output with respect to labor level when $L = 3{,}025$.

47. UNEMPLOYMENT   In economics, the inflation graph in Figure 2.2 is called the **Phillips curve,** after A. W. Phillips, a New Zealander associated with the London School of Economics. Until Phillips published his ideas in the 1950s, many economists believed that unemployment and inflation were linearly related. Read an article on the Phillips curve (the source cited with Figure 2.2 would be a good place to start), and write a paragraph on the nature of unemployment in the U.S. economy.

48. CONSUMER EXPENDITURE   The demand for a particular commodity is given by $D(x) = -35x + 200$; that is, x units will be sold (demanded) at a price of $p = D(x)$ dollars per unit.

a. Consumer expenditure $E(x)$ is the total amount of money consumers pay to buy x units. Express consumer expenditure E as a function of x.

b. Find the average rate of change in consumer expenditure as x changes from $x = 4$ to $x = 5$.

c. Use calculus to find the instantaneous rate of change of expenditure with respect to x when $x = 4$. Is expenditure increasing or decreasing when $x = 4$?

LIFE AND SOCIAL SCIENCE APPLIED PROBLEMS
---

49. RENEWABLE RESOURCES The accompanying graph shows how the volume of lumber V in a tree varies with time t (the age of the tree). Use the graph to estimate the rate at which V is changing with respect to time when $t = 30$ years. What seems to be happening to the rate of change of V as t increases without bound (that is, in the long run)?

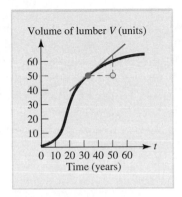

EXERCISE 49 Graph showing how the volume of lumber V in a tree varies with time t.

Source: Adapted from Robert H. Frank, *Microeconomics and Behavior,* 2nd ed., New York: McGraw-Hill, 1994, p. 623.

50. POPULATION GROWTH The accompanying graph shows how a population P of fruit flies (*Drosophila*) changes with time t (days) during an experiment. Use the graph to estimate the rate at which the population is growing after 20 days and also after 36 days. At what time is the population growing at the greatest rate?

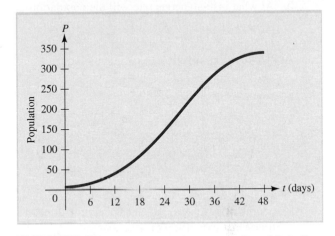

EXERCISE 50 Growth curve for a population of fruit flies.

Source: Adapted from E. Batschelet, *Introduction to Mathematics for Life Scientists,* 3rd ed., New York: Springer-Verlag, 1979, p. 355.

51. THERMAL INVERSION Air temperature usually decreases with increasing altitude. However, during the winter, thanks to a phenomenon called *thermal inversion,* the temperature of air warmed by the sun in mountains above a fog may rise above the freezing point, while the temperature at lower elevations remains near or below 0°C. Use the accompanying graph to estimate the rate at which temperature T is changing with respect to altitude h at an altitude of 1,000 meters and also at 2,000 meters.

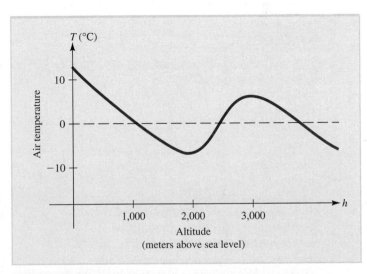

EXERCISE 51

Source: E. Batschelet, *Introduction to Mathematics for Life Scientists,* 3rd ed., New York: Springer-Verlag, 1979, p. 150.

52. POPULATION GROWTH   A 5-year projection of population trends suggests that t years after 2010, the population of a certain community will be P thousand, where
$$P(t) = -6t^2 + 12t + 151$$
a. At what average rate will the population be growing between 2010 and 2012?
b. At what instantaneous rate will the population be growing in 2012 ($t = 2$)? Is the population increasing or decreasing at this time?

53. ANIMAL BEHAVIOR   Experiments indicate that when a flea jumps, its height (in meters) after t seconds is given by the function
$$H(t) = 4.4t - 4.9t^2$$
a. Find $H'(t)$. At what rate is $H(t)$ changing after 1 second? Is it increasing or decreasing?
b. At what time is $H'(t) = 0$? What is the significance of this time?
c. When does the flea land (return to its initial height)? At what rate is $H(t)$ changing at this time? Is it increasing or decreasing?

54. BLOOD PRESSURE   Refer to the graph of blood pressure as a function of time shown in Figure 2.9.
a. Estimate the average rate of change in blood pressure over the time periods [0.7, 0.75] and [0.75, 0.8]. Interpret your results.
b. Write a paragraph on the dynamics of the arterial pulse. Pages 131–136 in the reference given with Figure 2.9 is a good place to start, and there is an excellent list of annotated references to related topics on pp. 137–138.

55. CARDIOLOGY   A study conducted on a patient undergoing cardiac catheterization indicated that the diameter of the aorta was approximately D millimeters (mm) when the aortic pressure was p (mm of mercury), where
$$D(p) = -0.0009p^2 + 0.13p + 17.81$$
for $50 \le p \le 120$.
a. Find the average rate of change of the aortic diameter D as p changes from $p = 60$ to $p = 61$.
b. Use calculus to find the instantaneous rate of change of diameter D with respect to aortic pressure p when $p = 60$. Is the pressure increasing or decreasing when $p = 60$?
c. For what value of p is the instantaneous rate of change of D with respect to p equal to 0? What is the significance of this pressure?

MISCELLANEOUS PROBLEMS

56. VELOCITY   A toy rocket rises vertically in such a way that t seconds after liftoff, it is $h(t) = -16t^2 + 200t$ feet above ground.
a. How high is the rocket after 6 seconds?
b. What is the average velocity of the rocket over the first 6 seconds of flight (between $t = 0$ and $t = 6$)?
c. What is the (instantaneous) velocity of the rocket at liftoff ($t = 0$)? What is its velocity after 40 seconds?

57. VELOCITY   An object moves along a straight line in such a way that t seconds after it begins to move, it is $s(t) = 4\sqrt{t + 1} - 4$ meters from its starting point.
a. What is the object's (instantaneous) velocity at time t?
b. What is the object's initial velocity (at time $t = 0$)?
c. How far from the starting point is the object at time $t = 3$ seconds? What is its velocity at this time?

58. a. Find the derivative of the linear function $f(x) = 3x - 2$.
b. Find the equation of the tangent line to the graph of this function at the point where $x = -1$.
c. Explain how the answers to parts (a) and (b) could have been obtained from geometric considerations with no calculation whatsoever.

59. a. Find the derivatives of the functions $y = x^2$ and $y = x^2 - 3$, and account geometrically for their similarity.
b. Without further computation, find the derivative of the function $y = x^2 + 5$.

60. a. Find the derivative of the function $y = x^2 + 3x$.
b. Find the derivatives of the functions $y = x^2$ and $y = 3x$ separately.
c. How is the derivative in part (a) related to those in part (b)?
d. In general, if $f(x) = g(x) + h(x)$, what would you guess is the relationship between the derivative of f and those of g and h?

61. a. Compute the derivatives of the functions $y = x^2$ and $y = x^3$.
b. Examine your answers in part (a). Can you detect a pattern? What do you think is the derivative of $y = x^4$? How about the derivative of $y = x^{27}$?

62. Use calculus to prove that if $y = mx + b$, the rate of change of y with respect to x is constant.

63. Let f be the absolute value function; that is,

$$f(x) = \begin{cases} x & \text{if } x \geq 0 \\ -x & \text{if } x < 0 \end{cases}$$

Show that

$$f'(x) = \begin{cases} 1 & \text{if } x > 0 \\ -1 & \text{if } x < 0 \end{cases}$$

and explain why f is not differentiable at $x = 0$.

64. Let f be a function that is differentiable at $x = c$.

a. Explain why

$$f'(c) = \lim_{x \to c} \frac{f(x) - f(c)}{x - c}$$

b. Use the result of part (a) together with the fact that

$$f(x) - f(c) = \left[\frac{f(x) - f(c)}{x - c} \right] (x - c)$$

to show that

$$\lim_{x \to c} [f(x) - f(c)] = 0$$

c. Explain why the result obtained in part (b) shows that f is continuous at $x = c$.

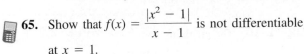

65. Show that $f(x) = \dfrac{|x^2 - 1|}{x - 1}$ is not differentiable at $x = 1$.

66. Find the x values at which the peaks and valleys of the graph of $y = 2x^3 - 0.8x^2 + 4$ occur. Use four decimal places.

67. Find the slope of the line that is tangent to the graph of the function $f(x) = \sqrt{x^2 + 2x} - \sqrt{3x}$ at the point where $x = 3.85$ by filling in the following table.

h	−0.02	−0.01	−0.001	0	0.001	0.01	0.02
$x + h$							
$f(x)$							
$f(x + h)$							
$\dfrac{f(x + h) - f(x)}{h}$							

SECTION 2.2 Techniques of Differentiation

Learning Objectives

1. Use the constant multiple rule, sum rule, and power rule to find derivatives.
2. Find relative and percentage rates of change.
3. Study rectilinear motion and the motion of a projectile.

If we had to use the limit definition every time we wanted to compute a derivative, it would be both tedious and difficult to use calculus in applications. Fortunately, this is not necessary, and in this section and Section 2.3, we develop techniques that greatly simplify the process of differentiation. We begin with a rule for the derivative of a constant.

The Constant Rule ■ For any constant c,

$$\frac{d}{dx}[c] = 0$$

That is, the derivative of a constant is zero.

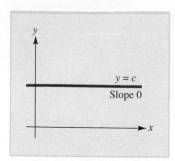

FIGURE 2.10   The graph of $f(x) = c$.

You can see this by considering the graph of a constant function $f(x) = c$, which is a horizontal line (see Figure 2.10). Since the slope of such a line is 0 at all its points, it follows that $f'(x) = 0$. Here is a proof using the limit definition:

$$f'(x) = \lim_{h \to 0} \frac{f(x + h) - f(x)}{h} \quad \text{since } f(x + h) = c \text{ for all } x \text{ and any } h$$

$$= \lim_{h \to 0} \frac{c - c}{h} = 0$$

EXAMPLE 2.2.1 Finding the Derivative of a Constant Function

Differentiate the constant function $f(x) = -15$.

Solution

$$\frac{d}{dx}[-15] = 0$$

The next rule is one of the most useful because it tells you how to take the derivative of any power function $f(x) = x^n$. Note that the rule applies not only to functions like $f(x) = x^5$ but also to those such as $g(x) = \sqrt[5]{x^4} = x^{4/5}$ and $h(x) = \frac{1}{x^3} = x^{-3}$.

The Power Rule ■ For any real number n,

$$\frac{d}{dx}[x^n] = nx^{n-1}$$

In words, to find the derivative of x^n, multiply by the original exponent n then subtract 1 from the exponent of x.

According to the power rule, the derivative of $y = x^3$ is $\frac{d}{dx}(x^3) = 3x^2$, which agrees with the result found directly in Example 2.1.4 of Section 2.1. You can use the power rule to differentiate radicals and reciprocals by first converting them to power functions with fractional and negative exponents, respectively. (You can find a review of exponential notation in Appendix A.1 at the back of the book.) For example, recall that $\sqrt{x} = x^{1/2}$, so the derivative of $y = \sqrt{x}$ is

$$\frac{d}{dx}(\sqrt{x}) = \frac{d}{dx}(x^{1/2}) = \frac{1}{2}x^{-1/2} = \frac{1}{2\sqrt{x}}$$

which agrees with the result of Example 2.1.6 of Section 2.1. In Example 2.2.2, we verify the power rule for a reciprocal power function.

EXAMPLE 2.2.2 Verifying the Power Rule
for Negative Exponents

Verify the power rule for the function $F(x) = \frac{1}{x^2} = x^{-2}$ by showing that its derivative is $F'(x) = -2x^{-3}$.

Just-In-Time REVIEW

Here is the rule for simplifying
a complex fraction:

$$\frac{A/B}{C/D} = \frac{A}{B} \cdot \frac{D}{C} = \frac{AD}{BC}$$

Just-In-Time REVIEW

Recall that $x^{-n} = \dfrac{1}{x^n}$ when n
is a positive integer, and that
$x^{a/b} = \sqrt[b]{x^a}$ whenever a and b
are positive integers.

Solution

The derivative of $F(x)$ is given by

$$F'(x) = \lim_{h\to 0} \frac{F(x+h) - F(x)}{h} = \lim_{h\to 0} \frac{\dfrac{1}{(x+h)^2} - \dfrac{1}{x^2}}{h} \quad \text{put the numerator over a common denominator}$$

$$= \lim_{h\to 0} \frac{\dfrac{x^2 - (x+h)^2}{x^2(x+h)^2}}{h} \quad \text{simplify complex fraction}$$

$$= \lim_{h\to 0} \frac{x^2 - (x^2 + 2hx + h^2)}{x^2 h(x+h)^2} \quad \text{combine like terms in the numerator}$$

$$= \lim_{h\to 0} \frac{-2xh - h^2}{x^2 h(x+h)^2} \quad \text{divide out common } h \text{ factors}$$

$$= \lim_{h\to 0} \frac{-2x - h}{x^2(x+h)^2}$$

$$= \frac{-2x}{x^2(x^2)}$$

$$= \frac{-2}{x^3} = -2x^{-3}$$

as claimed by the power rule.

Here are several additional applications of the power rule:

$$\frac{d}{dx}(x^7) = 7x^{7-1} = 7x^6$$

$$\frac{d}{dx}(\sqrt[3]{x^2}) = \frac{d}{dx}(x^{2/3}) = \frac{2}{3}x^{2/3-1} = \frac{2}{3}x^{-1/3}$$

$$\frac{d}{dx}\left(\frac{1}{x^5}\right) = \frac{d}{dx}(x^{-5}) = -5x^{-5-1} = -5x^{-6}$$

$$\frac{d}{dx}(x^{1.3}) = 1.3x^{1.3-1} = 1.3x^{0.3}$$

A general proof for the power rule in the case where n is a positive integer is outlined in Exercise 78. The case where n is a negative integer and the case where n is a rational number ($n = \dfrac{r}{s}$ for integers r and s with $s \neq 0$) are outlined in exercises in Section 2.3 and Section 2.6, respectively.

The constant rule and power rule provide simple formulas for finding derivatives of a class of important functions, but to differentiate more complicated expressions, we need to know how to manipulate derivatives algebraically. The next two rules tell us that derivatives of multiples and sums of functions are multiples and sums of the corresponding derivatives.

The Constant Multiple Rule ▪ If c is a constant and $f(x)$ is differentiable, then so is $cf(x)$ and

$$\frac{d}{dx}[cf(x)] = c\frac{d}{dx}[f(x)]$$

or equivalently,

$$(cf)' = cf'$$

That is, *the derivative of a multiple is the multiple of the derivative.*

EXAMPLE 2.2.3 Using the Constant Multiple Rule

Find the derivatives of $f(x) = 3x^4$ and $g(x) = \dfrac{-7}{\sqrt{x}}$.

Solution

$$\frac{d}{dx}(3x^4) = 3\frac{d}{dx}(x^4) = 3(4x^3) = 12x^3$$

$$\frac{d}{dx}\left(\frac{-7}{\sqrt{x}}\right) = \frac{d}{dx}(-7x^{-1/2}) = -7\left(\frac{-1}{2}x^{-3/2}\right) = \frac{7}{2}x^{-3/2}$$

The Sum Rule ▪ If $f(x)$ and $g(x)$ are differentiable, then so is the sum $f(x) + g(x)$ and its derivative is given by

$$\frac{d}{dx}[f(x) + g(x)] = \frac{d}{dx}[f(x)] + \frac{d}{dx}[g(x)]$$

or, equivalently,

$$(f + g)' = f' + g'$$

That is, *the derivative of a sum is the sum of the separate derivatives.*

A proof of the sum rule is outlined in Exercise 79.

EXAMPLE 2.2.4 Using the Sum Rule

Differentiate these functions:
  a. $f(x) = x^{-2} + 7$
  b. $g(x) = 2x^5 - 3x^{-7}$

Solution

a. $\dfrac{d}{dx}[x^{-2} + 7] = \dfrac{d}{dx}[x^{-2}] + \dfrac{d}{dx}[7] = -2x^{-3} + 0 = -2x^{-3}$

b. $\dfrac{d}{dx}[2x^5 - 3x^{-7}] = 2\dfrac{d}{dx}(x^5) + (-3)\dfrac{d}{dx}(x^{-7}) = 2(5x^4) + (-3)(-7x^{-8})$

$$= 10x^4 + 21x^{-8}$$

EXPLORE!

Use your graphing calculator to store $f(x) = x^3 - 3x + 1$ into Y1, using a bold graphing style. Follow the setup in the Explore! box on page 120 to place nDeriv(Y1, X, X) into Y2 and graph both of these functions, using the modified decimal window $[-4.7, 4.7]1$ by $[-5, 5]1$. Trace the Y2 = $f'(x)$ function, and find the x values for which $f'(x) = 0$. What do you notice about the graph of $f(x)$ at those points? What is the equation for $f'(x)$?

By combining the power rule, the constant multiple rule, and the sum rule, you can differentiate any polynomial. This is demonstrated in Examples 2.2.5 and 2.2.6.

EXAMPLE 2.2.5 Differentiating a Polynomial

Differentiate the polynomial $y = 5x^3 - 4x^2 + 12x - 8$.

Solution

Differentiate this sum term by term to get

$$\frac{dy}{dx} = \frac{d}{dx}[5x^3] + \frac{d}{dx}[-4x^2] + \frac{d}{dx}[12x] + \frac{d}{dx}[-8]$$
$$= 5[3x^2] - 4[2x] + 12[1] - 8[0]$$
$$= 15x^2 - 8x + 12$$

EXAMPLE 2.2.6 Using a Derivative to Study Population Change

It is estimated that x months from now, the population of a certain community will be $P(x) = x^2 + 20x + 8,000$.

a. At what rate will the population be changing with respect to time 15 months from now?

b. By how much will the population actually change during the 16th month?

Solution

a. The rate of change of the population with respect to time is the derivative of the population function. That is,

$$\text{Rate of change} = P'(x) = 2x + 20$$

The rate of change of the population 15 months from now will be

$$P'(15) = 2(15) + 20 = 50 \text{ people per month}$$

b. The actual change in the population during the 16th month is the difference between the population at the end of 16 months and the population at the end of 15 months. That is,

$$\text{Change in population} = P(16) - P(15) = 8,576 - 8,525$$
$$= 51 \text{ people}$$

NOTE   In Example 2.2.6, the actual change in population during the 16th month in part (b) differs from the monthly rate of change at the beginning of the month in part (a) because the rate varies during the month. The instantaneous rate in part (a) can be thought of as the change in population that would occur during the 16th month if the rate of change of population were to remain constant.   ■

Relative and Percentage Rates of Change

In many practical situations, the rate of change of a quantity Q is not as significant as its *relative* rate of change, which is defined as the ratio

$$\text{Relative change} = \frac{\text{change in } Q}{\text{size of } Q}$$

For example, a yearly rate of change of 500 individuals in a city whose total population is 5 million yields a negligible relative (per capita) rate of change of

$$\frac{500}{5,000,000} = 0.0001$$

or 0.01%, while the same rate of change in a town of 2,000 would yield a relative rate of change of

$$\frac{500}{2,000} = 0.25$$

or 25%, which would have enormous impact on the town.

Since the rate of change of a quantity $Q(x)$ is measured by the derivative $Q'(x)$, we can express the relative rate of change and the associated percentage rate of change in the following forms.

Relative and Percentage Rates of Change ■ The **relative rate of change** of a quantity $Q(x)$ with respect to x is given by the ratio

$$\frac{\text{Relative rate of}}{\text{change of } Q(x)} = \frac{Q'(x)}{Q(x)}$$

The corresponding **percentage rate of change** of $Q(x)$ with respect to x is

$$\frac{\text{Percentage rate of}}{\text{change of } Q(x)} = \frac{100\, Q'(x)}{Q(x)}$$

EXPLORE!

Refer to Example 2.2.7. Compare the rate of change of GDP at time $t = 10$ for $N(t)$ with the rate of a new GDP given by

$$N_1(t) = 2t^2 + 2t + 100$$

Graph both functions using x as the independent variable and the graphing window

[3, 12.4]1 by [90, 350]0

where a 0 scale corresponds to a display of no markings for the y axis. How do the percentage rates of change of GDP compare for 2010?

EXAMPLE 2.2.7   Finding a Percentage Rate of Change

The gross domestic product (GDP) of a certain country was $N(t) = t^2 + 5t + 106$ billion dollars t years after 2000.

a. At what rate was the GDP changing with respect to time in 2010?

b. At what percentage rate was the GDP changing with respect to time in 2010?

Solution

a. The rate of change of the GDP is the derivative $N'(t) = 2t + 5$. The rate of change in 2010 was $N'(10) = 2(10) + 5 = 25$ billion dollars per year.

b. The percentage rate of change of the GDP in 2010 was

$$100\,\frac{N'(10)}{N(10)} = 100\,\frac{25}{256} \approx 9.77 \qquad (9.77\% \text{ per year})$$

EXAMPLE 2.2.8   Finding a Percentage Rate of Growth

Experiments indicate that the biomass $Q(t)$ of a fish species in a given area of the ocean changes at the rate

$$\frac{dQ}{dt} = rQ\left(1 - \frac{Q}{a}\right)$$

where r is the natural growth rate of the species and a is a constant.* Find the percentage rate of growth of the species. If the biomass satisfies $Q(t) > a$, what can be said about its growth rate?

Solution

The percentage rate of change of $Q(t)$ is

$$\frac{100\, Q'(t)}{Q(t)} = \frac{100\, rQ\left(1 - \dfrac{Q}{a}\right)}{Q} = 100r\left(1 - \frac{Q}{a}\right)$$

Notice that the percentage rate decreases as Q increases and becomes zero when $Q = a$. If $Q > a$, the percentage rate is negative, which means the biomass is actually decreasing.

Rectilinear Motion

The motion of an object along a line is called *rectilinear motion*. For example, the motion of a rocket early in its flight can be regarded as rectilinear. When studying rectilinear motion, we will assume that the object involved is moving along a coordinate axis. If the function $s(t)$ gives the *position* of the object at time t, then the rate of change of $s(t)$ with respect to t is its *velocity* $v(t)$ and the derivative of velocity with respect to t is its *acceleration* $a(t)$. That is, $v(t) = s'(t)$ and $a(t) = v'(t)$.

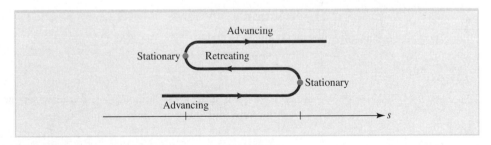

FIGURE 2.11   A diagram for rectilinear motion.

The object is said to be *advancing* (moving forward) when $v(t) > 0$, and *retreating* (moving backward) when $v(t) < 0$. When $v(t) = 0$, the object is neither advancing nor retreating and is said to be *stationary*. See Figure 2.11. Finally, the object is *accelerating* when $a(t) > 0$ and *decelerating* when $a(t) < 0$. To summarize:

---

*Adapted from W. R. Derrick and S. I. Grossman, *Introduction to Differential Equations,* 3rd ed., St. Paul, MN: West Publishing, 1987, p. 52, problem 20. The authors note that the problem was originally one of many models described in the text *Mathematical Bioeconomics,* by C. W. Clark (Wiley-Interscience, 1976).

> **Rectilinear Motion** ▪ If the **position** at time t of an object moving along a straight line is given by $s(t)$, then the object has
>
> $$\text{velocity} \quad v(t) = s'(t) = \frac{ds}{dt}$$
>
> and
>
> $$\text{acceleration} \quad a(t) = v'(t) = \frac{dv}{dt}$$
>
> The object is **advancing** when $v(t) > 0$, **retreating** when $v(t) < 0$, and **stationary** when $v(t) = 0$. It is **accelerating** when $a(t) > 0$ and **decelerating** when $a(t) < 0$.

If position is measured in meters and time in seconds, velocity is measured in meters per second (m/sec) and acceleration in meters per second per second (written as m/sec^2). Similarly, if position is measured in feet, velocity is measured in feet per second (ft/sec) and acceleration in feet per second per second (ft/sec^2).

EXPLORE!

Set your graphing utility in parametric, dotted, and simultaneous modes.
Store X1T = T, Y1T = 0.5, X2T = T, and
Y2T = T^3 − 6T^2 + 9T + 5.

Graph using a viewing rectangle of [0, 5]1 by [0, 10]1, and $0 \le t \le 4$ with a step of 0.2. Describe what you observe. What does the vertical axis represent? What does the horizontal line of dots mean?

EXAMPLE 2.2.9 Studying the Motion of an Object

The position at time t of an object moving along a line is given by $s(t) = t^3 - 6t^2 + 9t + 5$.

a. Find the velocity of the object and discuss its motion between times $t = 0$ and $t = 4$.

b. Find the total distance traveled by the object between times $t = 0$ and $t = 4$.

c. Find the acceleration of the object and determine when the object is accelerating and decelerating between times $t = 0$ and $t = 4$.

Solution

a. The velocity is $v(t) = \dfrac{ds}{dt} = 3t^2 - 12t + 9$. The object will be stationary when

$$v(t) = 3t^2 - 12t + 9 = 3(t - 1)(t - 3) = 0$$

that is, at times $t = 1$ and $t = 3$. Otherwise, the object is either advancing or retreating, as described in the following table.

Interval	Sign of $v(t)$	Description of the Motion
$0 < t < 1$	+	Advances from $s(0) = 5$ to $s(1) = 9$
$1 < t < 3$	−	Retreats from $s(1) = 9$ to $s(3) = 5$
$3 < t < 4$	+	Advances from $s(3) = 5$ to $s(4) = 9$

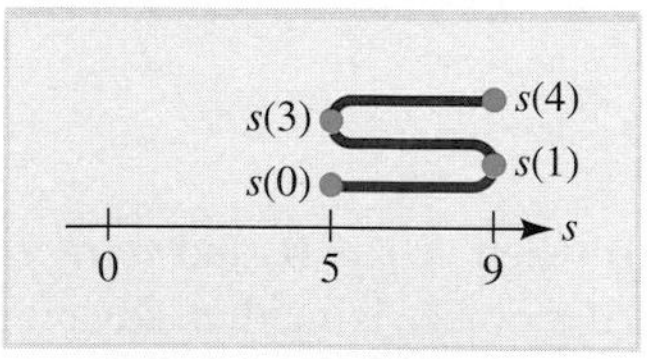

FIGURE 2.12 The motion of an object:

$$s(t) = t^3 - 6t^2 + 9t + 5$$

The motion of the object is summarized in the diagram in Figure 2.12.

b. The object travels from $s(0) = 5$ to $s(1) = 9$, then back to $s(3) = 5$, and finally to $s(4) = 9$. Thus, the total distance traveled by the object is

$$D = \underbrace{|9 - 5|}_{0 < t < 1} + \underbrace{|5 - 9|}_{1 < t < 3} + \underbrace{|9 - 5|}_{3 < t < 4} = 12$$

c. The acceleration of the object is

$$a(t) = \frac{dv}{dt} = 6t - 12 = 6(t - 2)$$

The object will be accelerating $[a(t) > 0]$ when $2 < t < 4$ and decelerating $[a(t) < 0]$ when $0 < t < 2$.

NOTE   An accelerating object is not necessarily "speeding up," nor is a decelerating object always "slowing down." For instance, the object in Example 2.2.9 has negative velocity and is accelerating for $2 < t < 3$. This means that the velocity is increasing over this time period, that is, becoming *less negative*. In other words, the object is actually slowing down. ▪

The Motion of a Projectile   An important example of rectilinear motion is the motion of a projectile. Suppose an object is projected (e.g., thrown, fired, or dropped) vertically in such a way that the only acceleration acting on the object is the constant downward acceleration g due to gravity. Near sea level, g is approximately 32 ft/sec^2 (or 9.8 m/sec^2). It can be shown that at time t, the height of the object is given by the formula

$$H(t) = -\frac{1}{2}gt^2 + V_0 t + H_0$$

where H_0 and V_0 are the initial height and velocity of the object, respectively. Example 2.2.10 illustrates how to use this formula.

EXAMPLE 2.2.10   Studying Projectile Motion

Suppose a person standing at the top of a building 112 feet high throws a ball vertically upward with an initial velocity of 96 ft/sec (see Figure 2.13).

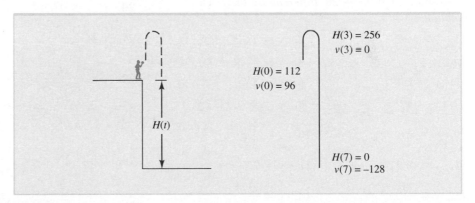

FIGURE 2.13   The motion of a ball thrown upward from the top of a building.

a. Find the ball's height and velocity at time t.
b. When does the ball hit the ground, and what is its impact velocity?
c. When is the velocity 0? What is the significance of this time?
d. How far does the ball travel during its flight?

Solution

a. Since $g = 32$, $V_0 = 96$, and $H_0 = 112$, the height of the ball above the ground at time t is

$$H(t) = -16t^2 + 96t + 112 \text{ feet}$$

The velocity at time t is

$$v(t) = \frac{dH}{dt} = -32t + 96 \text{ ft/sec}$$

b. The ball hits the ground when $H = 0$. Solve the equation $-16t^2 + 96t + 112 = 0$ to find that this occurs when $t = 7$ and $t = -1$ (verify). Disregarding the negative time $t = -1$, which is not meaningful in this context, we can conclude that impact occurs when $t = 7$ seconds and that the impact velocity is

$$v(7) = -32(7) + 96 = -128 \text{ ft/sec}$$

(The negative sign means the ball is coming down at the moment of impact.)

c. The velocity is zero when $v(t) = -32t + 96 = 0$, which occurs when $t = 3$ seconds. For $t < 3$, the velocity is positive and the ball is rising, and for $t > 3$, the ball is falling (see Figure 2.13). Thus, the ball is at its highest point when $t = 3$ seconds.

d. The ball starts at $H(0) = 112$ feet and rises to a maximum height of $H(3) = 256$ feet before falling to the ground. Thus,

$$\text{Total distance traveled} = \underbrace{(256 - 112)}_{\text{up}} + \underbrace{256}_{\text{down}} = 400 \text{ feet}$$

EXERCISES ▪ 2.2

In Exercises 1 through 28, differentiate the given function.

1. $y = -2$
2. $y = 3$
3. $y = 5x - 3$
4. $y = -2x + 7$
5. $y = x^{-4}$
6. $y = x^{7/3}$
7. $y = x^{3.7}$
8. $y = 4 - x^{-1.2}$
9. $y = \pi r^2$
10. $y = \frac{4}{3}\pi r^3$
11. $y = \sqrt{2x}$
12. $y = 2\sqrt[4]{x^3}$
13. $y = \frac{9}{\sqrt{t}}$
14. $y = \frac{3}{2t^2}$
15. $y = x^2 + 2x + 3$
16. $y = 3x^5 - 4x^3 + 9x - 6$
17. $f(x) = x^9 - 5x^8 + x + 12$
18. $f(x) = \frac{1}{4}x^8 - \frac{1}{2}x^6 - x + 2$
19. $f(x) = -0.02x^3 + 0.3x$
20. $f(u) = 0.07u^4 - 1.21u^3 + 3u - 5.2$
21. $y = \frac{1}{t} + \frac{1}{t^2} - \frac{1}{\sqrt{t}}$
22. $y = \frac{3}{x} - \frac{2}{x^2} + \frac{2}{3x^3}$
23. $f(x) = \sqrt{x^3} + \frac{1}{\sqrt{x^3}}$
24. $f(t) = 2\sqrt{t^3} + \frac{4}{\sqrt{t}} - \sqrt{2}$
25. $y = -\frac{x^2}{16} + \frac{2}{x} - x^{3/2} + \frac{1}{3x^2} + \frac{x}{3}$
26. $y = -\frac{7}{x^{1.2}} + \frac{5}{x^{-2.1}}$
27. $y = \frac{x^5 - 4x^2}{x^3}$ [*Hint:* Divide first.]
28. $y = x^2(x^3 - 6x + 7)$ [*Hint:* Multiply first.]

In Exercises 29 through 34, find the equation of the line that is tangent to the graph of the given function at the specified point.

29. $y = -x^3 - 5x^2 + 3x - 1; (-1, -8)$

30. $y = x^5 - 3x^3 - 5x + 2; (1, -5)$

31. $y = 1 - \dfrac{1}{x} + \dfrac{2}{\sqrt{x}}; \left(4, \dfrac{7}{4}\right)$

32. $y = \sqrt{x^3} - x^2 + \dfrac{16}{x^2}; (4, -7)$

33. $y = (x^2 - x)(3 + 2x); (-1, 2)$

34. $y = 2x^4 - \sqrt{x} + \dfrac{3}{x}; (1, 4)$

In Exercises 35 through 40, find the equation of the line that is tangent to the graph of the given function at the point $(c, f(c))$ for the specified value of $x = c$.

35. $f(x) = -2x^3 + \dfrac{1}{x^2}; x = -1$

36. $f(x) = x^4 - 3x^3 + 2x^2 - 6; x = 2$

37. $f(x) = x - \dfrac{1}{x^2}; x = 1$

38. $f(x) = x^3 + \sqrt{x}; x = 4$

39. $f(x) = -\dfrac{1}{3}x^3 + \sqrt{8x}; x = 2$

40. $f(x) = x(\sqrt{x} - 1); x = 4$

In Exercises 41 through 46, find the rate of change of the given function $f(x)$ with respect to x for the prescribed value $x = c$.

41. $f(x) = 2x^4 + 3x + 1; x = -1$

42. $f(x) = x^3 - 3x + 5; x = 2$

43. $f(x) = x - \sqrt{x} + \dfrac{1}{x^2}; x = 1$

44. $f(x) = \sqrt{x} + 5x; x = 4$

45. $f(x) = \dfrac{x + \sqrt{x}}{\sqrt{x}}; x = 1$

46. $f(x) = \dfrac{2}{x} - x\sqrt{x}; x = 1$

In Exercises 47 through 50, find the relative rate of change of $f(x)$ with respect to x for the prescribed value $x = c$.

47. $f(x) = 2x^3 - 5x^2 + 4; c = 1$

48. $f(x) = x + \dfrac{1}{x}; c = 1$

49. $f(x) = x\sqrt{x} + x^2; c = 4$

50. $f(x) = (4 - x)x^{-1}; c = 3$

BUSINESS AND ECONOMIC APPLIED PROBLEMS

51. ANNUAL EARNINGS The gross annual earnings of a certain company were $A(t) = 0.1t^2 + 10t + 20$ thousand dollars t years after its formation in 2008.
  a. At what rate were the gross annual earnings of the company growing with respect to time in 2012?
  b. At what percentage rate were the gross annual earnings growing with respect to time in 2012?

52. WORKER EFFICIENCY Lupe Garcia is the efficiency expert for a large manufacturing company. She finds that the average worker who arrives on the job when the morning shift begins at 8:00 A.M. will have assembled
$$f(x) = -x^3 + 6x^2 + 15x$$
units x hours later.
  a. Lupe wants to find the rate at which the average worker is assembling units x hours after the morning shift begins. What is this rate?
  b. At what rate is the average worker assembling units at 9:00 A.M.? How many units does the average worker actually assemble between 9:00 A.M. and 10:00 A.M.?
  c. Lupe expects the assembly rate to be less at noon when the morning shift ends than at 9:00 A.M. Is this true?

53. PROPERTY TAX Records indicate that x years after 2008, the average property tax on a three-bedroom home in a certain community was $T(x) = 20x^2 + 40x + 600$ dollars.
  a. At what rate was the property tax increasing with respect to time in 2008?
  b. By how much did the tax change between the years 2008 and 2012?

54. ADVERTISING A manufacturer of motorcycles estimates that if x thousand dollars is spent on advertising, then
$$M(x) = 2{,}300 + \dfrac{125}{x} - \dfrac{517}{x^2} \qquad 3 \le x \le 18$$
cycles will be sold. At what rate will sales be changing when \$9,000 is spent on advertising? Are sales increasing or decreasing for this level of advertising expenditure?

55. COST MANAGEMENT   A company uses a truck to deliver its products. To estimate costs, the manager models gas consumption by the function

$$G(x) = \frac{1}{250}\left(\frac{1,200}{x} + x\right)$$

gal/mile, assuming that the truck is driven at a constant speed of x miles per hour, for $x \geq 5$. The driver is paid \$20 per hour to drive the truck 250 miles, and gasoline costs \$4 per gallon.
a. Find an expression for the total cost $C(x)$ of the trip.
b. At what rate is the cost $C(x)$ changing with respect to x when the truck is driven at 40 miles per hour? Is the cost increasing or decreasing at that speed?

56. NEWSPAPER CIRCULATION   It is estimated that t years from now, the circulation of a local newspaper will be $C(t) = 100t^2 + 400t + 5,000$.
a. Derive an expression for the rate at which the circulation will be changing with respect to time t years from now.
b. At what rate will the circulation be changing with respect to time 5 years from now? Will the circulation be increasing or decreasing at that time?
c. By how much will the circulation actually change during the sixth year?

57. SALARY INCREASES   Gary's starting salary is \$45,000, and he gets a raise of \$2,000 each year.
a. Express the percentage rate of change of Gary's salary as a function of time, and draw the graph.
b. At what percentage rate will Gary's salary be increasing after 1 year?
c. What will happen to the percentage rate of change of Gary's salary in the long run?

58. GROSS DOMESTIC PRODUCT   The gross domestic product of a certain country is growing at a constant rate. In 2000 the GDP was 125 billion dollars, and in 2008 it was 155 billion dollars. If this trend continues, at what percentage rate will the GDP be growing in 2015?

LIFE AND SOCIAL SCIENCE APPLIED PROBLEMS

59. EDUCATIONAL TESTING   It is estimated that x years from now, the average SAT mathematics score of the incoming students at an eastern liberal arts college will be $f(x) = -6x + 582$.
a. Derive an expression for the rate at which the average SAT score will be changing with respect to time x years from now.

b. What is the significance of the fact that the expression in part (a) is a constant? What is the significance of the fact that the constant in part (a) is negative?

60. PUBLIC TRANSPORTATION   After x weeks, the number of people using a new rapid transit system was approximately $N(x) = 6x^3 + 500x + 8,000$.
a. At what rate was the use of the system changing with respect to time after 8 weeks?
b. By how much did the use of the system change during the eighth week?

61. POPULATION GROWTH   It is projected that x months from now, the population of a certain town will be $P(x) = 2x + 4x^{3/2} + 5,000$.
a. At what rate will the population be changing with respect to time 9 months from now?
b. At what percentage rate will the population be changing with respect to time 9 months from now?

62. POPULATION GROWTH   It is estimated that t years from now, the population of a certain town will be $P(t) = t^2 + 200t + 10,000$.
a. Express the percentage rate of change of the population as a function of t, simplify this function algebraically, and draw its graph.
b. What will happen to the percentage rate of change of the population in the long run (that is, as $t \rightarrow \infty$)?

63. SPREAD OF AN EPIDEMIC   A medical research team determines that t days after an epidemic begins, $N(t) = 10t^3 + 5t + \sqrt{t}$ people will be infected, for $0 \leq t \leq 20$. At what rate is the infected population increasing on the ninth day?

64. SPREAD OF AN EPIDEMIC   A disease is spreading in such a way that after t weeks, the number of people infected is
$$N(t) = 5,175 - t^3(t - 8) \qquad 0 \leq t \leq 8$$
a. At what rate is the epidemic spreading after 3 weeks?
b. Suppose health officials declare the disease to have reached epidemic proportions when the percentage rate of change of N is at least 25%. Over what time period is this epidemic criterion satisfied?
c. Read an article on epidemiology, and write a paragraph on how public health policy is related to the spread of an epidemic.

65. ORNITHOLOGY   An ornithologist determines that the body temperature of a certain bird species fluctuates over roughly a 17-hour period according to the cubic formula
$$T(t) = -68.07t^3 + 30.98t^2 + 12.52t + 37.1$$

for $0 \le t \le 0.713$, where T is the temperature in degrees Celsius measured t days from the beginning of a period.

a. Compute and interpret the derivative $T'(t)$.

b. At what rate is the temperature changing at the beginning of the period ($t = 0$) and at the end of the period ($t = 0.713$)? Is the temperature increasing or decreasing at each of these times?

c. At what time is the temperature not changing (neither increasing nor decreasing)? What is the bird's temperature at this time? Interpret your result.

66. **POLLUTION CONTROL** It has been suggested that one way to reduce worldwide carbon dioxide (CO_2) emissions is to impose a single tax that would apply to all nations. The accompanying graph shows the relationship between different levels of the carbon tax and the percentage of reduction in CO_2 emissions.

a. What tax rate would have to be imposed to achieve a worldwide reduction of 50% in CO_2 emissions?

b. Use the graph to estimate the rate of change of the percentage reduction in CO_2 emissions when the tax rate is $200 per ton.

c. Read an article on CO_2 emissions, and write a paragraph on how public policy can be used to control air pollution.*

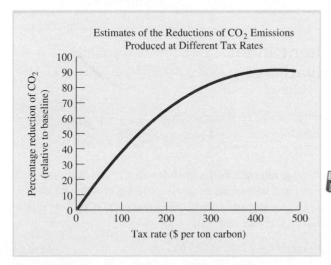

Estimates of the Reductions of CO_2 Emissions Produced at Different Tax Rates

EXERCISE 66 Sources: Barry C. Field, *Environmental Economics: An Introduction*. New York: McGraw-Hill, 1994, p. 441.

*You may wish to begin your research by reading Chapter 12, "Incentive-Based Strategies: Emission Taxes and Subsidies," and Chapter 15, "Federal Air Pollution-Control Policy," of Barry C. Field, *Environmental Economics: An Introduction*, New York: McGraw-Hill, 1994.

67. **AIR POLLUTION** An environmental study of a certain suburban community suggests that t years from now, the average level of carbon monoxide in the air will be $Q(t) = 0.05t^2 + 0.1t + 3.4$ parts per million.

a. At what rate will the carbon monoxide level be changing with respect to time 1 year from now?

b. By how much will the carbon monoxide level change this year?

c. By how much will the carbon monoxide level change over the next 2 years?

MISCELLANEOUS PROBLEMS

68. **PHYSICAL CHEMISTRY** According to Debye's formula in physical chemistry, the orientation polarization P of a gas satisfies

$$P = \frac{4}{3}\pi N\left(\frac{\mu^2}{3kT}\right)$$

where μ, k, and N are positive constants, and T is the temperature of the gas. Find the rate of change of P with respect to T.

69. **SPY STORY** Our friend, the spy who escaped from the diamond smugglers in Chapter 1 (Problem 62 of Section 1.4), is on a secret mission in space. An encounter with an enemy agent leaves him with a mild concussion and temporary amnesia. Fortunately, he has a book that gives the formula for the motion of a projectile and the values of g for various heavenly bodies (32 ft/sec^2 on earth, 5.5 ft/sec^2 on the moon, 12 ft/sec^2 on Mars, and 28 ft/sec^2 on Venus). To deduce his whereabouts, he throws a rock vertically upward (from ground level) and notes that it reaches a maximum height of 37.5 ft and hits the ground 5 seconds after it leaves his hand. Where is he?

RECTILINEAR MOTION *In Exercises 70 through 73, $s(t)$ is the position of a particle moving along a straight line at time t.*

(a) Find the velocity and acceleration of the particle.

(b) Find all times in the given interval when the particle is stationary.

70. $s(t) = t^2 - 2t + 6$ for $0 \le t \le 2$

71. $s(t) = 3t^2 + 2t - 5$ for $0 \le t \le 1$

72. $s(t) = t^3 - 9t^2 + 15t + 25$ for $0 \le t \le 6$

73. $s(t) = t^4 - 4t^3 + 8t$ for $0 \le t \le 4$

74. MOTION OF A PROJECTILE A stone is dropped from a height of 144 feet.
   a. When will the stone hit the ground?
   b. With what velocity does it hit the ground?

75. MOTION OF A PROJECTILE You are standing on the top of a building and throw a ball vertically upward. After 2 seconds, the ball passes you on the way down, and 2 seconds after that, it hits the ground below.
   a. What is the initial velocity of the ball?
   b. How high is the building?
   c. What is the velocity of the ball when it passes you on the way down?

   d. What is the velocity of the ball as it hits the ground?

76. Find the equations of all the tangents to the graph of the function

$$f(x) = x^2 - 4x + 25$$

that pass through the origin $(0, 0)$.

77. Find numbers a, b, and c such that the graph of the function $f(x) = ax^2 + bx + c$ will have x intercepts at $(0, 0)$ and $(5, 0)$, and a tangent with slope 1 when $x = 2$.

78. a. If $f(x) = x^4$, show that $\dfrac{f(x + h) - f(x)}{h} = 4x^3 + 6x^2 h + 4xh^2 + h^3$.

   b. If $f(x) = x^n$ for positive integer n, show that

$$\frac{f(x + h) - f(x)}{h} = nx^{n-1} + \frac{n(n - 1)}{2}x^{n-2}h + \cdots + nxh^{n-2} + h^{n-1}$$

   c. Use the result in part (b) in the definition of the derivative to prove the power rule:

$$\frac{d}{dx}[x^n] = nx^{n-1}$$

79. Prove the sum rule for derivatives. *Hint:* Note that the difference quotient for $f + g$ can be written as

$$\frac{(f + g)(x + h) - (f + g)(x)}{h} = \frac{[f(x + h) + g(x + h)] - [f(x) + g(x)]}{h}$$

SECTION 2.3 Product and Quotient Rules; Higher-Order Derivatives

Learning Objectives

1. Use the product and quotient rules to find derivatives.
2. Define and study the second derivative and higher-order derivatives.

Based on your experience with the multiple and sum rules in Section 2.2, you may think that the derivative of a product of functions is the product of separate derivatives, but it is easy to see that this conjecture is false. For instance, if $f(x) = x^2$ and $g(x) = x^3$, then $f'(x) = 2x$ and $g'(x) = 3x^2$, so

$$f'(x)g'(x) = (2x)(3x^2) = 6x^3$$

while $f(x)g(x) = x^2 x^3 = x^5$ and

$$[f(x)g(x)]' = (x^5)' = 5x^4$$

so $(fg)' \neq f'g'$. But if the derivative of a product is not generally equal to the product of separate derivatives, what rule can you count on for differentiating a product? Here is the answer.

> **The Product Rule** ▪ If $f(x)$ and $g(x)$ are differentiable at x, then so is their product $f(x)g(x)$ and
>
> $$\frac{d}{dx}[f(x)g(x)] = f(x)\frac{d}{dx}[g(x)] + g(x)\frac{d}{dx}[f(x)]$$
>
> or equivalently,
>
> $$(fg)' = fg' + gf'$$
>
> In words, the derivative of the product fg is f times the derivative of g plus g times the derivative of f.

Applying the product rule to our introductory example, we find that

$$(x^2x^3)' = x^2(x^3)' + (x^3)(x^2)'$$
$$= (x^2)(3x^2) + (x^3)(2x) = 3x^4 + 2x^4 = 5x^4$$

which is the same as the result obtained by direct computation:

$$(x^2x^3)' = (x^5)' = 5x^4$$

Examples 2.3.1 and 2.3.2 show two additional ways to use the product rule.

EXPLORE!

Use a graphing utility to graph $f(x) = (x - 1)(3x - 2)$ using a viewing rectangle of $[0, 2]0.1$ by $[-1, 1]0.1$. Find $f'(x)$, and graph it on the same coordinate axes. Explain why the graph of $f'(x)$ is a straight line. Explain what feature $f(x)$ has when $f'(x) = 0$.

EXAMPLE 2.3.1 Using the Product Rule

Differentiate the product $P(x) = (x - 1)(3x - 2)$ by

  a. Expanding $P(x)$.

  b. The product rule.

Solution

  a. We have $P(x) = 3x^2 - 5x + 2$, so $P'(x) = 6x - 5$.

  b. By the product rule

$$P'(x) = (x - 1)\frac{d}{dx}[3x - 2] + (3x - 2)\frac{d}{dx}[x - 1]$$
$$= (x - 1)(3) + (3x - 2)(1) = 6x - 5$$

EXPLORE!

Store the function in Example 2.3.2, $y = (2x + 1)(2x^2 - x - 1)$, into Y1 of the equation editor. Graph using the modified decimal window $[-2.35, 2.35]1$ by $[-3.1, 3.1]1$, and trace the cursor to X = 1. Construct the tangent line to the curve at this point using the Tangent feature of the **DRAW (2nd PRGM)** key. Does the equation of the tangent agree with what is calculated in the example?

EXAMPLE 2.3.2 Studying Tangent Lines to a Curve

For the curve $y = (2x + 1)(2x^2 - x - 1)$:

  a. Find y'.

  b. Find an equation for the tangent line to the curve at the point where $x = 1$.

  c. Find all points on the curve where the tangent line is horizontal.

Solution

  a. Using the product rule, we get

$$y' = (2x + 1)\frac{d}{dx}[2x^2 - x - 1] + (2x^2 - x - 1)\frac{d}{dx}[2x + 1]$$
$$= (2x + 1)(4x - 1) + (2x^2 - x - 1)(2)$$

b. When $x = 1$, the corresponding y value is

$$y(1) = [2(1) + 1][2(1)^2 - 1 - 1] = 0$$

so the point of tangency is $(1, 0)$. The slope at $x = 1$ is

$$y'(1) = [2(1) + 1][4(1) - 1] + [2(1)^2 - 1 - 1](2) = 9$$

Substituting into the point-slope formula, we find that the tangent line at $(1, 0)$ has the equation

$$y - 0 = 9(x - 1)$$

or $$y = 9x - 9$$

c. Horizontal tangents occur where the slope is zero, that is, where $y' = 0$. Expanding the expression for the derivative and combining like terms, we obtain

$$y' = (2x + 1)(4x - 1) + (2x^2 - x - 1)(2) = 12x^2 - 3$$

Solving $y' = 0$, we find that

$$y' = 12x^2 - 3 = 0 \qquad \text{add 3 to each side and divide by 12}$$

$$x^2 = \frac{3}{12} = \frac{1}{4} \qquad \text{take square roots on each side}$$

$$x = \frac{1}{2} \quad \text{and} \quad x = -\frac{1}{2}$$

Substituting $x = \dfrac{1}{2}$ and $x = -\dfrac{1}{2}$ into the formula for y, we get $y\left(\dfrac{1}{2}\right) = -2$ and $y\left(-\dfrac{1}{2}\right) = 0$, so horizontal tangents occur at the points $\left(\dfrac{1}{2}, -2\right)$ and $\left(-\dfrac{1}{2}, 0\right)$ on the curve.

In Example 2.3.3, the product rule is used in an economics application.

EXAMPLE 2.3.3   Finding the Rate of Change of Revenue

A manufacturer determines that t months after a new product is introduced to the market, $x(t) = t^2 + 3t$ hundred units can be produced and then sold at a price of $p(t) = -2t^{3/2} + 30$ dollars per unit.

a. Express the revenue $R(t)$ for this product as a function of time.

b. At what rate is revenue changing with respect to time after 4 months? Is revenue increasing or decreasing at this time?

Solution

a. The revenue is given by

$$R(t) = x(t)p(t) = (t^2 + 3t)(-2t^{3/2} + 30)$$

hundred dollars.

b. The rate of change of revenue $R(t)$ with respect to time is given by the derivative $R'(t)$, which we find using the product rule:

$$R'(t) = (t^2 + 3t)\frac{d}{dt}[-2t^{3/2} + 30] + (-2t^{3/2} + 30)\frac{d}{dt}[t^2 + 3t]$$

$$= (t^2 + 3t)\left[-2\left(\frac{3}{2}t^{1/2}\right)\right] + (-2t^{3/2} + 30)[2t + 3]$$

At time $t = 4$, the revenue is changing at the rate

$$R'(4) = [(4)^2 + 3(4)][-3(4)^{1/2}] + [-2(4)^{3/2} + 30][2(4) + 3]$$
$$= -14$$

Thus, after 4 months, the revenue is changing at the rate of 14 hundred dollars ($1,400) per month. It is *decreasing* at that time since $R'(4)$ is negative.

A proof of the product rule is given at the end of this section. It is also important to be able to differentiate quotients of functions, and for this purpose, we have the following rule, a proof of which is outlined in Exercise 73.

CAUTION: A common error is to assume that $\left(\dfrac{f}{g}\right)' = \dfrac{f'}{g'}$.

The Quotient Rule ■ If $f(x)$ and $g(x)$ are differentiable functions, then so is the quotient $Q(x) = \dfrac{f(x)}{g(x)}$ and

$$\frac{d}{dx}\left[\frac{f(x)}{g(x)}\right] = \frac{g(x)\dfrac{d}{dx}[f(x)] - f(x)\dfrac{d}{dx}[g(x)]}{[g(x)]^2} \quad \text{if } g(x) \neq 0$$

or equivalently,

$$\left(\frac{f}{g}\right)' = \frac{gf' - fg'}{g^2}$$

NOTE The quotient rule is probably the most complicated formula you have had to learn so far in this book. It may help to note that the quotient rule resembles the product rule except it contains a minus sign, which makes the order of terms in the numerator very important. Begin by squaring the denominator g, and then, while still thinking of g, copy it into the numerator. This gets you started with the proper order in the numerator, and you can easily write down the rest while thinking of the product rule. Don't forget to insert the minus sign, without which the quotient rule would not have been so hard to remember in the first place. This whimsical version of the quotient rule may help you remember its form:

$$d\left[\frac{\text{hi}}{\text{ho}}\right] = \frac{\text{ho } d(\text{hi}) - \text{hi } d(\text{ho})}{\text{ho ho}} \quad ■$$

EXAMPLE 2.3.4 Using the Quotient Rule

Differentiate the quotient $Q(x) = \dfrac{x^2 - 5x + 7}{2x}$ by

a. Dividing through first.   **b.** Using the quotient rule.

Just-In-Time **REVIEW**

Recall that
$$\frac{A + B}{C} = \frac{A}{C} + \frac{B}{C}$$
but that
$$\frac{A}{B + C} \neq \frac{A}{B} + \frac{A}{C}$$

Solution

a. Dividing by the denominator $2x$, we get

$$Q(x) = \frac{1}{2}x - \frac{5}{2} + \frac{7}{2}x^{-1}$$

so

$$Q'(x) = \frac{1}{2} - 0 + \frac{7}{2}(-x^{-2}) = \frac{1}{2} - \frac{7}{2x^2}$$

b. By the quotient rule

$$Q'(x) = \frac{(2x)\dfrac{d}{dx}[x^2 - 5x + 7] - (x^2 - 5x + 7)\dfrac{d}{dx}[2x]}{(2x)^2}$$

$$= \frac{(2x)(2x - 5) - (x^2 - 5x + 7)(2)}{4x^2} = \frac{2x^2 - 14}{4x^2} = \frac{1}{2} - \frac{7}{2x^2}$$

EXAMPLE 2.3.5   Studying How a Population Changes

Linda Grant, a biologist studying the effects of a toxin on a bacterial culture, determines that t hours after the toxin has been introduced, the population of bacteria in the culture is P million where

$$P(t) = \frac{t + 1}{t^2 + t + 4}$$

a. At what rate is the population of the culture changing with respect to time when Linda introduces the toxin (at time $t = 0$)? Is the population increasing or decreasing at this time?

b. Linda is especially interested in knowing when the population peaks out and begins to decline. At what time does this occur, and by how much does the population increase before it begins to decline?

Solution

a. The rate of change of the population with respect to time is given by the derivative $P'(t)$, which we compute using the quotient rule:

$$P'(t) = \frac{(t^2 + t + 4)\dfrac{d}{dt}[t + 1] - (t + 1)\dfrac{d}{dt}[t^2 + t + 4]}{(t^2 + t + 4)^2}$$

$$= \frac{(t^2 + t + 4)(1) - (t + 1)(2t + 1)}{(t^2 + t + 4)^2}$$

$$= \frac{-t^2 - 2t + 3}{(t^2 + t + 4)^2}$$

The toxin is introduced when $t = 0$, and at that time the population is changing at the rate

$$P'(0) = \frac{0 + 0 + 3}{(0 + 0 + 4)^2} = \frac{3}{16} = 0.1875$$

That is, the population is initially changing at the rate of 0.1875 million (187,500) bacteria per hour, and it is increasing since $P'(0) > 0$.

b. The population is decreasing when $P'(t) < 0$. Since the numerator of $P'(t)$ can be factored as

$$-t^2 - 2t + 3 = -(t^2 + 2t - 3) = -(t - 1)(t + 3)$$

we can write

$$P'(t) = \frac{-(t - 1)(t + 3)}{(t^2 + t + 4)^2}$$

The denominator $(t^2 + t + 4)^2$ and the factor $t + 3$ are both positive for all $t \geq 0$, which means that

$$\text{for } 0 \leq t < 1 \quad P'(t) > 0 \text{ and } P(t) \text{ is increasing}$$
$$\text{for } t > 1 \quad P'(t) < 0 \text{ and } P(t) \text{ is decreasing}$$

Thus, the population begins to decline after 1 hour.

The initial population of the colony is

$$P(0) = \frac{0 + 1}{0 + 0 + 4} = \frac{1}{4}$$

million, and after 1 hour, the population is

$$P(1) = \frac{1 + 1}{1 + 1 + 4} = \frac{1}{3}$$

million. Therefore, before the population begins to decline, it increases by

$$P(1) - P(0) = \frac{1}{3} - \frac{1}{4} = \frac{1}{12}$$

million, that is, by approximately 83,333 bacteria.

The quotient rule is somewhat cumbersome, so don't use it unnecessarily. Consider Example 2.3.6.

EXAMPLE 2.3.6 Using the Power Rule with Quotients

Differentiate the function $y = \dfrac{2}{3x^2} - \dfrac{x}{3} + \dfrac{4}{5} + \dfrac{x + 1}{x}$.

Solution

You can use the quotient rule here, but it is quicker and simpler to rewrite the function as

$$y = \frac{2}{3}x^{-2} - \frac{1}{3}x + \frac{4}{5} + 1 + x^{-1}$$

and then apply the power rule term by term to get

$$\frac{dy}{dx} = \frac{2}{3}(-2x^{-3}) - \frac{1}{3} + 0 + 0 + (-1)x^{-2}$$

$$= -\frac{4}{3}x^{-3} - \frac{1}{3} - x^{-2}$$

$$= -\frac{4}{3x^3} - \frac{1}{3} - \frac{1}{x^2}$$

The Second Derivative

In applications, it may be necessary to compute the rate of change of a function that is itself a rate of change. For example, the acceleration of a car is the rate of change with respect to time of its velocity, which in turn is the rate of change with respect to time of its position. If the position is measured in miles and time in hours, the velocity (rate of change of position) is measured in miles per hour, and the acceleration (rate of change of velocity) is measured in miles per hour, per hour.

Statements about the rate of change of a rate of change are used frequently in economics. In inflationary times, for example, you may hear a government economist assure the nation that although prices are inflating, they are doing so at a decreasing rate. That is, prices are still going up, but not as rapidly as they were before.

The rate of change of the function $f(x)$ with respect to x is the derivative $f'(x)$, and likewise, the rate of change of the function $f'(x)$ with respect to x is *its* derivative $(f'(x))'$. This notation is awkward, so we write the derivative of the derivative of $f(x)$ as $f''(x)$ and refer to it as the *second derivative* of $f(x)$ (read $f''(x)$ as "f double prime of x".) If $y = f(x)$, then the second derivative of y with respect to x is written as y'' or as $\frac{d^2y}{dx^2}$. Here is a summary of the terminology and notation used for second derivatives.

The Second Derivative ■ The second derivative of a function is the derivative of its derivative. If $y = f(x)$, the second derivative is denoted by

$$f''(x) \qquad \text{or} \qquad \frac{d^2y}{dx^2}$$

The second derivative gives the rate of change of the rate of change of the original function.

NOTE   The derivative $f'(x)$ is sometimes called the **first derivative** to distinguish it from the **second derivative** $f''(x)$. ■

You don't have to use any new rules to find the second derivative of a function. Just find the first derivative and then differentiate again.

EXAMPLE 2.3.7   Finding a Second Derivative

Find the second derivative of the function $f(x) = 5x^4 - 3x^2 - 3x + 7$.

Solution

Compute the first derivative

$$f'(x) = 20x^3 - 6x - 3$$

and then differentiate again to get

$$f''(x) = 60x^2 - 6$$

EXAMPLE 2.3.8 Finding a Second Derivative

Find the second derivative of $y = x^2(3x + 1)$.

Solution

According to the product rule,

$$\frac{d}{dx}[x^2(3x + 1)] = x^2\frac{d}{dx}[3x + 1] + (3x + 1)\frac{d}{dx}[x^2]$$

$$= x^2(3) + (3x + 1)(2x)$$

$$= 9x^2 + 2x$$

Therefore, the second derivative is

$$\frac{d^2y}{dx^2} = \frac{d}{dx}[9x^2 + 2x]$$

$$= 18x + 2$$

NOTE Before computing the second derivative of a function, always take time to simplify the first derivative as much as possible. The more complicated the form of the first derivative, the more tedious the computation of the second derivative. ▪

The second derivative will be used in Section 3.2 to obtain information about the shape of a graph, and in Sections 3.4 and 3.5 in optimization problems. Example 2.3.9 features the interpretation of the second derivative as the rate of change of a rate of change.

EXAMPLE 2.3.9 Finding the Rate of Change of a Production Rate

An efficiency study of the morning shift at a certain factory indicates that an average worker who arrives on the job at 8:00 A.M. will have produced

$$Q(t) = -t^3 + 6t^2 + 24t$$

units t hours later.

a. Compute the worker's rate of production at 11:00 A.M.

b. At what rate is the worker's rate of production changing with respect to time at 11:00 A.M.?

Solution

a. The worker's rate of production is the first derivative

$$R(t) = Q'(t) = -3t^2 + 12t + 24$$

of the output $Q(t)$. The rate of production at 11:00 A.M. ($t = 3$) is

$$R(3) = Q'(3) = -3(3)^2 + 12(3) + 24 = 33$$
$$= 33 \text{ units per hour}$$

b. The rate of change of the rate of production is the second derivative

$$R'(t) = Q''(t) = -6t + 12$$

of the output function. At 11:00 A.M., this rate is

$$R'(3) = Q''(3) = -6(3) + 12$$
$$= -6 \text{ units per hour per hour}$$

The minus sign indicates that the worker's rate of production is decreasing; that is, the worker is slowing down. The rate of this decrease in efficiency at 11:00 A.M. is 6 units per hour per hour.

Recall from Section 2.2 that the **acceleration** $a(t)$ of an object moving along a straight line is the derivative of the velocity $v(t)$, which in turn is the derivative of the position function $s(t)$. Thus, the acceleration may be thought of as the second derivative of position; that is,

$$a(t) = \frac{d^2s}{dt^2}$$

This notation is used in Example 2.3.10.

EXPLORE!

Change t to x in $s(t)$, $v(t)$, and $a(t)$ in Example 2.3.10. Use a graphing utility to graph $v(x)$ and $a(x)$ on the same coordinate axes using a viewing rectangle of [0, 2]0.1 by [−5, 5]0.5. Explain what is happening to $v(x)$ when $a(x)$ is zero. Then use your calculator to see what effect changing $s(t)$ to $s_1(t) = 2t^3 - 3t^2 + 4t$ has on $v(t)$ and $a(t)$.

EXAMPLE 2.3.10   Finding Velocity and Acceleration

If the position of an object moving along a straight line is given by $s(t) = t^3 - 3t^2 + 4t$ at time t, find its velocity and acceleration.

Solution

The velocity of the object is

$$v(t) = \frac{ds}{dt} = 3t^2 - 6t + 4$$

and its acceleration is

$$a(t) = \frac{dv}{dt} = \frac{d^2s}{dt^2} = 6t - 6$$

Higher-Order Derivatives   If you differentiate the second derivative $f''(x)$ of a function $f(x)$ one more time, you get the third derivative $f'''(x)$. Differentiate again and you get the fourth derivative, which is denoted by $f^{(4)}(x)$ since the prime notation $f''''(x)$ is cumbersome. In general, the derivative obtained from $f(x)$ after n successive differentiations is called the **nth derivative** or **derivative of order n** and is denoted by $f^{(n)}(x)$.

> **The nth Derivative** ▪ For any positive integer n, the nth derivative of a function is obtained from the function by differentiating successively n times. If the original function is $y = f(x)$, the nth derivative is denoted by
>
> $$f^{(n)}(x) \quad \text{or} \quad \frac{d^n y}{dx^n}$$

EXAMPLE 2.3.11   Finding Higher-Order Derivatives

Find the fifth derivative of each of these functions:

a. $f(x) = 4x^3 + 5x^2 + 6x - 1$ **b.** $y = \dfrac{1}{x}$

Solution

a. $f'(x) = 12x^2 + 10x + 6$

$f''(x) = 24x + 10$

$f'''(x) = 24$

$f^{(4)}(x) = 0$

$f^{(5)}(x) = 0$

b. $\dfrac{dy}{dx} = \dfrac{d}{dx}(x^{-1}) = -x^{-2} = -\dfrac{1}{x^2}$

$\dfrac{d^2 y}{dx^2} = \dfrac{d}{dx}(-x^{-2}) = 2x^{-3} = \dfrac{2}{x^3}$

$\dfrac{d^3 y}{dx^3} = \dfrac{d}{dx}(2x^{-3}) = -6x^{-4} = -\dfrac{6}{x^4}$

$\dfrac{d^4 y}{dx^4} = \dfrac{d}{dx}(-6x^{-4}) = 24x^{-5} = \dfrac{24}{x^5}$

$\dfrac{d^5 y}{dx^5} = \dfrac{d}{dx}(24x^{-5}) = -120x^{-6} - -\dfrac{120}{x^6}$

Derivation of the Product Rule

The product and quotient rules are not easy to prove. In both cases, the key is to express the difference quotient of the given expression $\left(\text{the product } fg \text{ or quotient } \dfrac{f}{g}\right)$ in terms of difference quotients of f and g. Here is a proof of the product rule. The proof of the quotient rule is outlined in Exercise 73.

To show that $\dfrac{d}{dx}(fg) = f\dfrac{dg}{dx} + g\dfrac{df}{dx}$, begin with the appropriate difference quotient for $f(x)g(x)$ and then rewrite the numerator by subtracting and adding the term $f(x + h)g(x)$ to produce difference quotients for $f(x)$ and $g(x)$ as follows:

$$\frac{d}{dx}(fg) = \lim_{h \to 0} \frac{f(x + h)g(x + h) - f(x)g(x)}{h}$$

$$= \lim_{h \to 0} \left[\frac{f(x + h)g(x + h) - f(x + h)g(x)}{h} + \frac{f(x + h)g(x) - f(x)g(x)}{h} \right]$$

$$= \lim_{h \to 0} \left(f(x + h)\left[\frac{g(x + h) - g(x)}{h} \right] + g(x)\left[\frac{f(x + h) - f(x)}{h} \right] \right)$$

Now let h approach zero. Since

$$\lim_{h \to 0} \frac{f(x + h) - f(x)}{h} = \frac{df}{dx}$$

$$\lim_{h \to 0} \frac{g(x + h) - g(x)}{h} = \frac{dg}{dx}$$

and $$\lim_{h \to 0} f(x + h) = f(x) \qquad \text{continuity of } f(x)$$

it follows that $$\frac{d}{dx}(fg) = f\frac{dg}{dx} + g\frac{df}{dx}$$

EXERCISES ■ 2.3

In Exercises 1 through 18, differentiate the given function.

1. $f(x) = (2x + 1)(3x - 2)$

2. $f(x) = (x - 5)(1 - 2x)$

3. $y = 10(3u + 1)(1 - 5u)$

4. $y = 400(15 - x^2)(3x - 2)$

5. $f(x) = \frac{1}{3}(x^5 - 2x^3 + 1)\left(x - \frac{1}{x}\right)$

6. $f(x) = -3(5x^3 - 2x + 5)(\sqrt{x} + 2x)$

7. $y = \dfrac{x + 1}{x - 2}$ **8.** $y = \dfrac{2x - 3}{5x + 4}$

9. $f(t) = \dfrac{t}{t^2 - 2}$ **10.** $f(x) = \dfrac{1}{x - 2}$

11. $y = \dfrac{3}{x + 5}$ **12.** $y = \dfrac{t^2 + 1}{1 - t^2}$

13. $f(x) = \dfrac{x^2 - 3x + 2}{2x^2 + 5x - 1}$

14. $g(x) = \dfrac{(x^2 + x + 1)(4 - x)}{2x - 1}$

15. $f(x) = (2 + 5x)^2$

16. $f(x) = \left(x + \dfrac{1}{x}\right)^2$ **17.** $g(t) = \dfrac{t^2 + \sqrt{t}}{2t + 5}$

18. $h(x) = \dfrac{x}{x^2 - 1} + \dfrac{4 - x}{x^2 + 1}$

In Exercises 19 through 23, find an equation for the tangent line to the given curve at the point where $x = x_0$.

19. $y = (5x - 1)(4 + 3x)$; $x_0 = 0$

20. $y = (x^2 + 3x - 1)(2 - x)$; $x_0 = 1$

21. $y = \dfrac{x}{2x + 3}$; $x_0 = -1$

22. $y = \dfrac{x + 7}{5 - 2x}$; $x_0 = 0$

23. $y = (3\sqrt{x} + x)(2 - x^2)$; $x_0 = 1$

In Exercises 24 through 27, find all points on the graph of the given function where the tangent line is horizontal.

24. $f(x) = (x - 1)(x^2 - 8x + 7)$

25. $f(x) = (x + 1)(x^2 - x - 2)$

26. $f(x) = \dfrac{x^2 + x - 1}{x^2 - x + 1}$ **27.** $f(x) = \dfrac{x + 1}{x^2 + x + 1}$

In Exercises 28 through 31, find the rate of change $\dfrac{dy}{dx}$ for the prescribed value of x_0.

28. $y = (x^2 + 2)(x + \sqrt{x})$; $x_0 = 4$

29. $y = (x^2 + 3)(5 - 2x^3)$; $x_0 = 1$

30. $y = \dfrac{2x - 1}{3x + 5}$; $x_0 = 1$

31. $y = x + \dfrac{3}{2 - 4x}$; $x_0 = 0$

The normal line to the curve $y = f(x)$ at the point P with coordinates $(x_0, f(x_0))$ is the line perpendicular to the tangent line at P. In Exercises 32 through 35, find an equation for the normal line to the given curve at the prescribed point.

32. $y = x^2 + 3x - 5$; $(0, -5)$

33. $y = \dfrac{2}{x} - \sqrt{x}$; $(1, 1)$

34. $y = (x + 3)(1 - \sqrt{x})$; $(1, 0)$

35. $y = \dfrac{5x + 7}{2 - 3x}$; $(1, -12)$

36. Find $h'(2)$ if $h(x) = (x^2 + 3)g(x)$ where $g(2) = 3$ and $g'(2) = -2$.

37. Find $h'(-3)$ if $h(x) = [3x^2 - 2g(x)][g(x) + 5x]$ where $g(-3) = 1$ and $g'(-3) = 2$.

38. Find $h'(0)$ if $h(x) = \dfrac{3x^2 - 5g(x)}{g(x) + 4}$ where $g(0) = 2$ and $g'(0) = -3$.

39. Find $h'(-1)$ if $h(x) = \dfrac{x^3 + xg(x)}{3x - 5}$ where $g(-1) = 0$ and $g'(-1) = 1$.

40. a. Differentiate the function $y = 2x^2 - 5x - 3$.
   b. Now factor the function in part (a) as $y = (2x + 1)(x - 3)$ and differentiate using the product rule. Show that the two answers are the same.

41. a. Use the quotient rule to differentiate the function $y = \dfrac{2x - 3}{x^3}$.
   b. Rewrite the function as $y = x^{-3}(2x - 3)$, and differentiate using the product rule.
   c. Rewrite the function as $y = 2x^{-2} - 3x^{-3}$ and differentiate.
   d. Show that your answers to parts (a), (b), and (c) are the same.

In Exercises 42 through 47, find the second derivative of the given function. In each case, use the appropriate notation for the second derivative and simplify your answer. (Don't forget to simplify the first derivative as much as possible before computing the second derivative.)

42. $f(x) = 5x^{10} - 6x^5 - 27x + 4$

43. $f(x) = \dfrac{2}{5}x^5 - 4x^3 + 9x^2 - 6x - 2$

44. $y = 5\sqrt{x} + \dfrac{3}{x^2} + \dfrac{1}{3\sqrt{x}} + \dfrac{1}{2}$

45. $y = \dfrac{2}{3x} - \sqrt{2x} + \sqrt{2}x - \dfrac{1}{6\sqrt{x}}$

46. $y = (x^2 - x)\left(2x - \dfrac{1}{x}\right)$

47. $y = (x^3 + 2x - 1)(3x + 5)$

48. DEMAND AND REVENUE   The manager of a company that produces graphing calculators determines that when x thousand calculators are produced, they will all be sold when the price is

$$p(x) = \dfrac{1,000}{0.3x^2 + 8}$$

dollars per calculator.
   a. At what rate is demand $p(x)$ changing with respect to the level of production x when 3,000 ($x = 3$) calculators are produced?
   b. The revenue derived from the sale of x thousand calculators is $R(x) = xp(x)$ thousand dollars. At what rate is revenue changing when 3,000 calculators are produced? Is revenue increasing or decreasing at this level of production?

49. SALES   The manager of the Many Facets jewelry store models total sales by the function

$$S(t) = \dfrac{2,000t}{4 + 0.3t}$$

where t is the time (years) since the year 2010 and S is measured in thousands of dollars.
   a. At what rate were sales changing in the year 2012?
   b. What happens to sales in the long run (that is, as $t \to +\infty$)?

50. PROFIT   Bea Johnson, the owner of the Bea Nice boutique, estimates that when a particular kind of perfume is priced at p dollars per bottle, she will sell

$$B(p) = \dfrac{500}{p + 3} \qquad p \geq 5$$

bottles per month at a total cost of

$$C(p) = 0.2p^2 + 3p + 200 \text{ dollars}$$

   a. Express Bea's profit $P(p)$ as a function of the price p per bottle.
   b. At what rate is the profit changing with respect to p when the price is \$12 per bottle? Is profit increasing or decreasing at that price?

51. ADVERTISING   A company manufactures a "thin" DVD burner kit that can be plugged into personal computers. The marketing manager determines that t weeks after an advertising campaign begins, $P(t)$ percent of the potential market is aware of the burners, where

$$P(t) = 100\left(\dfrac{t^2 + 5t + 5}{t^2 + 10t + 30}\right)$$

a. At what rate is the market percentage $P(t)$ changing with respect to time after 5 weeks? Is the percentage increasing or decreasing at this time?

b. What happens to the percentage $P(t)$ in the long run, that is, as $t \to +\infty$? What happens to the rate of change of $P(t)$ as $t \to +\infty$?

52. WORKER EFFICIENCY An efficiency study of the morning shift at a certain factory indicates that an average worker arriving on the job at 8:00 A.M. will have produced $Q(t) = -t^3 + 8t^2 + 15t$ units t hours later.

a. Compute the worker's rate of production $R(t) = Q'(t)$.

b. At what rate is the worker's rate of production changing with respect to time at 9:00 A.M.?

53. REVENUE Currently, a company is selling 1,000 units of a certain commodity at a price of $5 per unit. The manager of the company estimates that the price is currently increasing at the rate of 5 cents per week, while demand is decreasing at the rate of 4 units per week.

a. If x is the level of production at time t, what is $R(x)$? Find the rate at which $R(x)$ is currently changing with respect to x. Is the revenue currently increasing or decreasing?

b. At what rate is the average revenue $\dfrac{R(x)}{x}$ currently changing? Is the average revenue currently increasing or decreasing?

54. PROFIT The manager of a company estimates that it will cost $10,000 to produce 400 units of her product 1 year from now and that all those units can then be sold at a price of $30 per unit. She also estimates that in 1 year, the price will be increasing at the rate of 75 cents per unit per month, while the level of production will be decreasing at the rate of 2 units per month and the cost will stay constant.

a. If x is the level of production at time t, where $t = 0$ is 1 year from now, what is the profit $P(x)$? Find the rate at which the profit will be changing 1 year from now with respect to x. Will the profit be increasing or decreasing at that time?

b. At what rate will the average profit $\dfrac{P(x)}{x}$ be changing 1 year from now? Will the average profit be increasing or decreasing at that time?

55. POLLUTION CONTROL A study indicates that spending money on pollution control is effective up to a point but eventually becomes wasteful. Suppose it is known that when x million dollars is spent on controlling pollution, the percentage of pollution removed is given by

$$P(x) = \frac{100\sqrt{x}}{0.03x^2 + 9}$$

a. At what rate is the percentage of pollution removal $P(x)$ changing when 16 million dollars is spent? Is the percentage increasing or decreasing at this level of expenditure?

b. For what values of x is $P(x)$ increasing? For what values of x is $P(x)$ decreasing?

56. BACTERIAL POPULATION A bacterial colony is estimated to have a population of

$$P(t) = \frac{24t + 10}{t^2 + 1}$$

million t hours after the introduction of a toxin.

a. At what rate is the population changing 1 hour after the toxin is introduced ($t = 1$)? Is the population increasing or decreasing at this time?

b. At what time does the population begin to decline?

57. DRUG DOSAGE The human body's reaction to a dose of medicine can be modeled* by a function of the form

$$F = \frac{1}{3}(KM^2 - M^3)$$

where K is a positive constant and M is the amount of medicine absorbed in the blood. The derivative $S = \dfrac{dF}{dM}$ can be thought of as a measure of the sensitivity of the body to the medicine.

a. Find the sensitivity S.

b. Find $\dfrac{dS}{dM} = \dfrac{d^2F}{dM^2}$, and give an interpretation of the second derivative.

58. PHARMACOLOGY An oral painkiller is administered to a patient, and t hours later, the concentration of drug in the patient's bloodstream is given by

$$C(t) = \frac{2t}{3t^2 + 16}$$

---

*Thrall et al., *Some Mathematical Models in Biology,* U.S. Dept. of Commerce, 1967.

a. At what rate $R(t)$ is the concentration of drug in the patient's bloodstream changing t hours after being administered? At what rate is $R(t)$ changing at time t?

b. At what rate is the concentration of drug changing after 1 hour? Is the concentration changing at an increasing or decreasing rate at this time?

c. When does the concentration of the drug begin to decline?

d. Over what time period is the concentration changing at a declining rate?

59. POPULATION GROWTH It is estimated that t years from now, the population of a certain suburban community will be $P(t) = 20 - \dfrac{6}{t+1}$ thousand.

a. Derive a formula for the rate at which the population will be changing with respect to time t years from now.

b. At what rate will the population be growing 1 year from now?

c. By how much will the population actually increase during the second year?

d. At what rate will the population be growing 9 years from now?

e. What will happen to the rate of population growth in the long run?

60. BLOOD CELL PRODUCTION A biological model[†] measures the production of a certain type of white blood cell (*granulocytes*) by the function

$$p(x) = \frac{Ax}{B + x^m}$$

where A and B are positive constants, the exponent m is positive, and x is the number of cells present.

a. Find the rate of production $p'(x)$.

b. Find $p''(x)$, and determine all values of x for which $p''(x) = 0$ (your answer will involve m).

c. Read an article on blood cell production, and write a paragraph on how mathematical methods can be used to model such production. A good place to start is with the article, "Blood Cell Population Model, Dynamical Diseases, and Chaos" by W. B. Gearhart and M. Martelli, UMAP Module 1990, Arlington, MA: Consortium for Mathematics and Its Applications, Inc., 1991.

†M. C. Mackey and L. Glass, "Oscillations and Chaos in Physiological Control Systems," *Science*, Vol. 197, pp. 287–289.

MISCELLANEOUS PROBLEMS

In Exercises 61 through 64, the position s(t) of an object moving along a straight line is given. In each case:

(a) Find the object's velocity v(t) and acceleration a(t).

(b) Find all times t when the acceleration is 0.

61. $s(t) = 3t^5 - 5t^3 - 7$

62. $s(t) = 2t^4 - 5t^3 + t - 3$

63. $s(t) = -t^3 + 7t^2 + t + 2$

64. $s(t) = 4t^{5/2} - 15t^2 + t - 3$

65. VELOCITY An object moves along a straight line so that after t minutes, its distance from its starting point is $D(t) = 10t + \dfrac{5}{t+1} - 5$ meters.

a. At what velocity is the object moving at the end of 4 minutes?

b. How far does the object actually travel during the fifth minute?

66. ACCELERATION After t hours of an 8-hour trip, a car has gone $D(t) = 64t + \dfrac{10}{3}t^2 - \dfrac{2}{9}t^3$ kilometers.

a. Derive a formula expressing the acceleration of the car as a function of time.

b. At what rate is the velocity of the car changing with respect to time at the end of 6 hours? Is the velocity increasing or decreasing at this time?

c. By how much does the velocity of the car actually change during the seventh hour?

67. ACCELERATION If an object is dropped or thrown vertically, its height (in feet) after t seconds is $H(t) = -16t^2 + V_0t + H_0$, where V_0 is the initial speed of the object and H_0 is its initial height.

a. Derive an expression for the acceleration of the object.

b. How does the acceleration vary with time?

c. What is the significance of the fact that the answer to part (a) is negative?

68. Find $f^{(4)}(x)$ if $f(x) = x^5 - 2x^4 + x^3 - 3x^2 + 5x - 6$.

69. Find $\dfrac{d^3y}{dx^3}$ if $y = \sqrt{x} - \dfrac{1}{2x} + \dfrac{x}{\sqrt{2}}$.

70. a. Show that

$$\frac{d}{dx}[fgh] = fg\frac{dh}{dx} + fh\frac{dg}{dx} + gh\frac{df}{dx}$$

[*Hint:* Apply the product rule twice.]

b. Find $\frac{dy}{dx}$ where $y = (2x + 1)(x - 3)(1 - 4x)$.

71. a. By combining the product rule and the quotient rule, find an expression for $\frac{d}{dx}\left[\frac{fg}{h}\right]$.

b. Find $\frac{dy}{dx}$, where $y = \frac{(2x + 7)(x^2 + 3)}{3x + 5}$.

72. The product rule tells you how to differentiate the product of any two functions, while the constant multiple rule tells you how to differentiate products in which one of the factors is constant. Show that the two rules are consistent. In particular, use the product rule to show that

$$\frac{d}{dx}[cf] = c\frac{df}{dx} \text{ if } c \text{ is a constant.}$$

73. Derive the quotient rule. [*Hint:* Show that the difference quotient for $\frac{f}{g}$ is

$$\frac{1}{h}\left[\frac{f(x + h)}{g(x + h)} - \frac{f(x)}{g(x)}\right] = \frac{g(x)f(x + h) - f(x)g(x + h)}{g(x + h)g(x)h}$$

Before letting h approach zero, rewrite this quotient using the trick of subtracting and adding $g(x)f(x)$ in the numerator.]

74. Prove the power rule $\frac{d}{dx}[x^n] = nx^{n-1}$ for the case where $n = -p$ is a negative integer. $\left[\textit{Hint:} \text{ Apply the quotient rule to } y = x^{-p} = \frac{1}{x^p}.\right]$

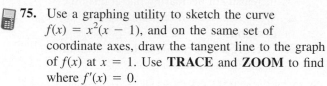

75. Use a graphing utility to sketch the curve $f(x) = x^2(x - 1)$, and on the same set of coordinate axes, draw the tangent line to the graph of $f(x)$ at $x = 1$. Use **TRACE** and **ZOOM** to find where $f'(x) = 0$.

76. Use a graphing utility to sketch the curve $f(x) = \frac{3x^2 - 4x + 1}{x + 1}$, and on the same set of coordinate axes, draw the tangent lines to the graph of $f(x)$ at $x = -2$ and at $x = 0$. Use **TRACE** and **ZOOM** to find where $f'(x) = 0$.

77. Graph $f(x) = x^4 + 2x^3 - x + 1$ using a graphing utility with a viewing rectangle of $[-5, 5]1$ by $[0, 2]0.5$. Use **TRACE** and **ZOOM,** or the maximum and minimum commands, to find the minima and maxima of this function. Find the derivative function $f'(x)$ algebraically, and graph $f(x)$ and $f'(x)$ on the same axes using a viewing rectangle of $[-5, 5]1$ by $[-2, 2]0.5$. Use **TRACE** and **ZOOM** or the zero command to find the x intercepts of $f'(x)$. Explain why the maxima or minima of $f(x)$ occurs at the x intercepts of $f'(x)$.

78. Repeat Exercise 77 for the product function $f(x) = x^3(x - 3)^2$.

SECTION 2.4   The Chain Rule

Learning Objectives

1. Discuss the chain rule.
2. Use the chain rule to find and apply derivatives.

In many practical situations, the rate at which one quantity is changing can be expressed as the product of other rates. For example, suppose a car is traveling at 50 miles/hour at a particular time when gasoline is being consumed at the rate of 0.1 gal/mile. Then, to find out how much gasoline is being used each hour, you would multiply the rates:

$$(0.1 \text{ gal/mile})(50 \text{ miles/hour}) = 5 \text{ gal/hour}$$

Or, suppose the total manufacturing cost at a certain factory is a function of the number of units produced, which in turn is a function of the number of hours the factory has been operating. If C, q, and t denote the cost, units produced, and time, respectively, then

$$\begin{bmatrix} \text{rate of change of cost} \\ \text{with respect to output} \end{bmatrix} = \frac{dC}{dq} \quad \text{dollars per unit}$$

and

$$\begin{bmatrix} \text{rate of change of output} \\ \text{with respect to time} \end{bmatrix} = \frac{dq}{dt} \quad \text{units per hour}$$

The product of these two rates is the rate of change of cost with respect to time; that is,

$$\frac{dC}{dt} = \frac{dC}{dq}\frac{dq}{dt} \quad \text{(dollars per hour)}$$

This formula is a special case of an important result in calculus called the **chain rule.**

The Chain Rule ■ If $y = f(u)$ is a differentiable function of u and $u = g(x)$ is in turn a differentiable function of x, then the composite function $y = f(g(x))$ is a differentiable function of x whose derivative is given by the product

$$\frac{dy}{dx} = \frac{dy}{du}\frac{du}{dx}$$

or, equivalently, by

$$\frac{dy}{dx} = f'(g(x))g'(x)$$

NOTE   One way to remember the chain rule is to pretend the derivatives $\dfrac{dy}{du}$ and $\dfrac{du}{dx}$ are fractions and to divide out du; that is,

$$\frac{dy}{dx} = \frac{dy}{d\!\!\!/u}\frac{d\!\!\!/u}{dx} \quad ■$$

To illustrate the use of the chain rule, suppose you wish to differentiate the function $y = (3x + 1)^2$. Your first instinct may be to guess that the derivative is

$$\frac{dy}{dx} = \frac{d}{dx}[(3x + 1)^2] = 2(3x + 1)$$
$$= 6x + 2$$

But this guess cannot be correct since expanding $(3x + 1)^2$ and differentiating each term yields

$$\frac{dy}{dx} = \frac{d}{dx}[(3x + 1)^2] = \frac{d}{dx}[9x^2 + 6x + 1] = 18x + 6$$

which is 3 times our guess of $6x + 2$. However, if you write $y = (3x + 1)^2$ as $y = u^2$ where $u = 3x + 1$, then

$$\frac{dy}{du} = \frac{d}{du}[u^2] = 2u \qquad \text{and} \qquad \frac{du}{dx} = \frac{d}{dx}[3x + 1] = 3$$

and the chain rule tells you that

$$\frac{dy}{dx} = \frac{dy}{du}\frac{du}{dx} = (2u)(3)$$
$$= 6(3x + 1) = 18x + 6$$

which coincides with the correct answer found earlier by expanding $(3x + 1)^2$. Examples 2.4.1 and 2.4.2 illustrate various ways of using the chain rule.

EXAMPLE 2.4.1 Using the Chain Rule

Find $\dfrac{dy}{dx}$ if $y = (x^2 + 2)^3 - 3(x^2 + 2)^2 + 1$.

Solution

Note that $y = u^3 - 3u^2 + 1$, where $u = x^2 + 2$. Thus,

$$\frac{dy}{du} = 3u^2 - 6u \qquad \text{and} \qquad \frac{du}{dx} = 2x$$

and according to the chain rule,

$$\frac{dy}{dx} = \frac{dy}{du}\frac{du}{dx} = (3u^2 - 6u)(2x) \qquad \text{substitute } x^2 + 2 \text{ for } u$$
$$= [3(x^2 + 2)^2 - 6(x^2 + 2)](2x)$$
$$= 6x^5 + 12x^3$$

In Sections 2.1 through 2.3, you have seen several applications (e.g. slope, rates of change) that require the evaluation of a derivative at a particular value of the independent variable. There are two basic ways of doing this when the derivative is computed using the chain rule.

For instance, suppose in Example 2.4.1 we wish to evaluate $\dfrac{dy}{dx}$ when $x = -1$. One way to proceed would be to simply substitute $x = -1$ in the expression for the derivative:

$$\frac{dy}{dx}\bigg|_{x=-1} = 6x^5 + 12x^3 \bigg|_{x=-1}$$
$$= 6(-1)^5 + 12(-1)^3 = -18$$

Alternatively, we could compute $u(-1) = (-1)^2 + 2 = 3$ and then substitute directly into the formula $\dfrac{dy}{dx} = (3u^2 - 6u)(2x)$ to obtain

$$\frac{dy}{dx}\bigg|_{x=-1} = (3u^2 - 6u)(2x)\bigg|_{\substack{x=-1 \\ u=3}}$$
$$= [3(3)^2 - 6(3)][2(-1)] = (9)(-2) = -18$$

Both methods yield the correct result, but since it is easier to substitute numbers than algebraic expressions, the second (numerical) method is often preferable, unless for some reason you need to have the derivative function $\dfrac{dy}{dx}$ expressed in terms of x alone. In Example 2.4.2, the numerical method for evaluating a derivative computed with the chain rule is used to find the slope of a tangent line.

EXAMPLE 2.4.2 Finding a Tangent Line Using the Chain Rule

Consider the function $y = \dfrac{u}{u+1}$, where $u = 3x^2 - 1$.

a. Use the chain rule to find $\dfrac{dy}{dx}$ in terms of x and u.

b. Find an equation for the tangent line to the graph of $y(x)$ at the point where $x = 1$.

Solution

a. We find that

$$\frac{dy}{du} = \frac{(u+1)(1) - u(1)}{(u+1)^2} = \frac{1}{(u+1)^2} \quad \text{quotient rule}$$

and

$$\frac{du}{dx} = 6x$$

According to the chain rule, it follows that

$$\frac{dy}{dx} = \frac{dy}{du}\frac{du}{dx} = \left[\frac{1}{(u+1)^2}\right](6x) = \frac{6x}{(u+1)^2}$$

b. To find an equation for the tangent line to the graph of $y(x)$ at $x = 1$, we need to know the value of y and the slope at the point of tangency. Since

$$u(1) = 3(1)^2 - 1 = 2$$

the value of y when $x = 1$ is

$$y(1) = \frac{(2)}{(2)+1} = \frac{2}{3} \quad \text{substitute } u(1) = 2 \text{ for } u$$

and the slope is

$$\left.\frac{dy}{dx}\right|_{\substack{x=1\\u=2}} = \frac{6(1)}{(2+1)^2} = \frac{6}{9} = \frac{2}{3} \quad \begin{array}{l}\text{substitute } x = 1\\ \text{and } u(1) = 2 \text{ for } u\end{array}$$

Therefore, by applying the point-slope formula for the equation of a line, we find that the tangent line to the graph of $y(x)$ at the point where $x = 1$ has the equation

$$y - \frac{2}{3} = \frac{2}{3}(x - 1)$$

or, equivalently, $y = \dfrac{2}{3}x$.

In many practical problems, a quantity is given as a function of one variable, which, in turn, can be written as a function of a second variable, and the goal is to find the rate of change of the original quantity with respect to the second variable. Such problems can be addressed by means of the chain rule, as demonstrated in Example 2.4.3.

EXAMPLE 2.4.3   Finding a Rate of Change of Cost

The cost of producing x units of a particular commodity is $C(x) = \frac{1}{3}x^2 + 4x + 53$ dollars, and the production level t hours into a particular production run is $x(t) = 0.2t^2 + 0.03t$ units. At what rate is cost changing with respect to time after 4 hours?

Solution

We find that

$$\frac{dC}{dx} = \frac{2}{3}x + 4 \qquad \text{and} \qquad \frac{dx}{dt} = 0.4t + 0.03$$

so according to the chain rule,

$$\frac{dC}{dt} = \frac{dC}{dx}\frac{dx}{dt} = \left(\frac{2}{3}x + 4\right)(0.4t + 0.03)$$

When $t = 4$, the level of production is

$$x(4) = 0.2(4)^2 + 0.03(4) = 3.32 \text{ units}$$

and by substituting $t = 4$ and $x = 3.32$ into the formula for $\frac{dC}{dt}$, we get

$$\frac{dC}{dt}\bigg|_{\substack{t=4\\x=3.32}} = \left[\frac{2}{3}(3.32) + 4\right][0.4(4) + 0.03] = 10.1277$$

Thus, after 4 hours, cost is increasing at the rate of approximately \$10.13 per hour.

Sometimes when dealing with a composite function $y = f(g(x))$ it may help to think of f as the *outer* function and g as the *inner* function, as indicated here:

outer function
$$y = f(g(x))$$
inner function

Then the chain rule

$$\frac{dy}{dx} = f'(g(x))g'(x)$$

says that *the derivative of $y = f(g(x))$ with respect to x is given by the derivative of the outer function evaluated at the inner function times the derivative of the inner function.* In Example 2.4.4, we emphasize this interpretation by using a box ($\Box$) to indicate the location and role of the inner function in computing a derivative with the chain rule.

EXAMPLE 2.4.4 Using the Chain Rule with a Root Function

Differentiate the function $f(x) = \sqrt{x^2 + 3x + 2}$.

Solution

The form of the function is

$$f(x) = (\Box)^{1/2}$$

where the box $\Box$ contains the expression $x^2 + 3x + 2$. Then

$$(\Box)' = (x^2 + 3x + 2)' = 2x + 3$$

and according to the chain rule, the derivative of the composite function $f(x)$ is

$$f'(x) = \frac{1}{2}(\Box)^{-1/2}(\Box)'$$

$$= \frac{1}{2}(\Box)^{-1/2}(2x + 3)$$

$$= \frac{1}{2}(x^2 + 3x + 2)^{-1/2}(2x + 3) = \frac{2x + 3}{2\sqrt{x^2 + 3x + 2}}$$

The General Power Rule

In Section 2.2, you learned the rule

$$\frac{d}{dx}[x^n] = nx^{n-1}$$

for differentiating power functions. By combining this rule with the chain rule, you obtain the following rule for differentiating functions of the general form $[h(x)]^n$.

> **The General Power Rule** ■ For any real number n and differentiable function h,
>
> $$\frac{d}{dx}[h(x)]^n = n[h(x)]^{n-1}\frac{d}{dx}[h(x)]$$

To derive the general power rule, think of $[h(x)]^n$ as the composite function

$$[h(x)]^n = g[h(x)] \quad \text{where } g(u) = u^n$$

Then, $g'(u) = nu^{n-1}$ and $h'(x) = \frac{d}{dx}[h(x)]$

and, by the chain rule,

$$\frac{d}{dx}[h(x)]^n = \frac{d}{dx}g[h(x)] = g'[h(x)]h'(x) = n[h(x)]^{n-1}\frac{d}{dx}[h(x)]$$

The use of the general power rule is illustrated in Examples 2.4.5 and 2.4.6.

EXAMPLE 2.4.5 Using the General Power Rule

Differentiate the function $f(x) = (2x^4 - x)^3$.

Solution

One way to solve this problem is to expand the function and rewrite it as

$$f(x) = 8x^{12} - 12x^9 + 6x^6 - x^3$$

and then differentiate this polynomial term by term to get

$$f'(x) = 96x^{11} - 108x^8 + 36x^5 - 3x^2$$

The algebra involved in this approach is fairly complicated, but using the general power rule is much simpler:

$$f'(x) = [3(2x^4 - x)^2]\frac{d}{dx}[2x^4 - x] = 3(2x^4 - x)^2(8x^3 - 1)$$

Not only is this method easier, but the answer even comes out in factored form!

EXAMPLE 2.4.6 Using the General Power Rule
                  with a Reciprocal Power

Differentiate the function $f(x) = \dfrac{1}{(2x + 3)^5}$.

Solution

Although you can use the quotient rule, it is easier to rewrite the function as

$$f(x) = (2x + 3)^{-5}$$

and apply the general power rule to get

$$f'(x) = [-5(2x + 3)^{-6}]\frac{d}{dx}[2x + 3] = -5(2x + 3)^{-6}(2) = -\frac{10}{(2x + 3)^6}$$

The chain rule is often used in combination with the other rules you learned in Sections 2.2 and 2.3. Example 2.4.7 involves the product rule.

EXAMPLE 2.4.7 Finding Where a Tangent Line Is Horizontal

Differentiate the function $f(x) = (3x + 1)^4(2x - 1)^5$, and simplify your answer. Then find all values of $x = c$ for which the tangent line to the graph of $f(x)$ at $(c, f(c))$ is horizontal.

Solution

First apply the product rule to get

$$f'(x) = (3x + 1)^4\frac{d}{dx}[(2x - 1)^5] + (2x - 1)^5\frac{d}{dx}[(3x + 1)^4]$$

Continue by applying the general power rule to each term:

$$f'(x) = (3x + 1)^4[5(2x - 1)^4(2)] + (2x - 1)^5[4(3x + 1)^3(3)]$$
$$= 10(3x + 1)^4(2x - 1)^4 + 12(2x - 1)^5(3x + 1)^3$$

Finally, simplify your answer by factoring:

$$f'(x) = 2(3x + 1)^3(2x - 1)^4[5(3x + 1) + 6(2x - 1)]$$
$$= 2(3x + 1)^3(2x - 1)^4[15x + 5 + 12x - 6]$$
$$= 2(3x + 1)^3(2x - 1)^4(27x - 1)$$

The tangent line to the graph of $f(x)$ is horizontal at points $(c, f(c))$ where $f'(c) = 0$. By solving

$$f'(x) = 2(3x + 1)^3(2x - 1)^4(27x - 1) = 0$$

we see that $f'(c) = 0$ where

$$3c + 1 = 0 \quad \text{or} \quad 2c - 1 = 0 \quad \text{or} \quad 27c - 1 = 0$$

that is, at $c = -\dfrac{1}{3}$, $c = \dfrac{1}{2}$, and $c = \dfrac{1}{27}$.

Sometimes it is necessary to apply the chain rule more than once. In Example 2.4.8, we apply the general power rule twice to find the derivative of a root function.

EXAMPLE 2.4.8 Using the General Power Rule Twice

Find the derivative of $f(x) = \sqrt{(x^2 - 4)^5 + 2x}$.

Solution

First, write the function in the form

$$f(x) = [(x^2 - 4)^5 + 2x]^{1/2}$$

and then apply the general power rule twice to find the derivative:

$$f(x) = [(x^2 - 4)^5 + 2x]^{1/2} \qquad \text{general power rule}$$

$$f'(x) = \frac{1}{2}[(x^2 - 4)^5 + 2x]^{-1/2} \frac{d}{dx}[(x^2 - 4)^5 + 2x] \qquad \text{general power rule again}$$

$$= \frac{1}{2}[(x^2 - 4)^5 + 2x]^{-1/2}\left[5(x^2 - 4)^4 \frac{d}{dx}(x^2 - 4) + 2\right] \qquad \text{power rule}$$

$$= \frac{1}{2}[(x^2 - 4)^5 + 2x]^{-1/2}[5(x^2 - 4)^4(2x) + 2] \qquad \begin{array}{l}\text{combine like}\\\text{terms and simplify}\end{array}$$

$$= \frac{5x(x^2 - 4)^4 + 1}{\sqrt{(x^2 - 4)^5 + 2x}}$$

The chain rule is often used in computing higher-order derivatives. In Example 2.4.9, we illustrate the general procedure by using the general power rule to find a second derivative of a rational function.

> **EXAMPLE 2.4.9** Using the General Power Rule to Find a Second Derivative

Find the second derivative of the function $f(x) = \dfrac{3x - 2}{(x - 1)^2}$.

Solution

Using the quotient rule, along with the general power rule [applied to $(x - 1)^2$], we get

$$f'(x) = \frac{(x - 1)^2(3) - (3x - 2)[2(x - 1)(1)]}{(x - 1)^4} \quad \text{factor out } (x - 1)$$

$$= \frac{(x - 1)[3(x - 1) - 2(3x - 2)]}{(x - 1)^4} \quad \text{divide out } (x - 1) \text{ and distribute}$$

$$= \frac{3x - 3 - 6x + 4}{(x - 1)^3}$$

$$= \frac{1 - 3x}{(x - 1)^3}$$

Using the quotient rule again, this time applying the general power rule to $(x - 1)^3$, we find that

$$f''(x) = \frac{(x - 1)^3(-3) - (1 - 3x)[3(x - 1)^2(1)]}{(x - 1)^6}$$

$$= \frac{-3(x - 1)^2[(x - 1) + (1 - 3x)]}{(x - 1)^6}$$

$$= \frac{-3(-2x)}{(x - 1)^4} = \frac{6x}{(x - 1)^4}$$

Examples 2.4.10 and 2.4.11 illustrate how the chain rule can be used in applications.

> **EXAMPLE 2.4.10** Finding a Rate of Change of Demand

Jarvis manages an appliance manufacturing firm. He determines that when blenders are priced at p dollars apiece, the number sold each month will be

$$D(p) = \frac{8,000}{p}$$

Furthermore, he estimates that t months from now, blenders will be selling at a price of $p(t) = 0.06t^{3/2} + 22.5$ dollars apiece. At what rate should Jarvis expect the monthly demand $D(p)$ to be changing with respect to time 25 months from now? Will the demand be increasing or decreasing at that time?

Solution

We want to find $\dfrac{dD}{dt}$ when $t = 25$. We have

$$\frac{dD}{dp} = \frac{d}{dp}\left[\frac{8{,}000}{p}\right] = -\frac{8{,}000}{p^2}$$

and

$$\frac{dp}{dt} = \frac{d}{dt}[0.06t^{3/2} + 22.5] = 0.06\left(\frac{3}{2}t^{1/2}\right) = 0.09t^{1/2}$$

so it follows from the chain rule that

$$\frac{dD}{dt} = \frac{dD}{dp}\frac{dp}{dt} = \left[-\frac{8{,}000}{p^2}\right](0.09t^{1/2})$$

When $t = 25$, the unit price is

$$p(25) = 0.06(25)^{3/2} + 22.5 = 30 \text{ dollars}$$

and we have

$$\frac{dD}{dt}\bigg|_{\substack{t=25\\p=30}} = \left[-\frac{8{,}000}{30^2}\right][0.09(25)^{1/2}] = -4$$

That is, 25 months from now, the demand for blenders will be changing at the rate of 4 units per month and will be decreasing since $\dfrac{dD}{dt}$ is negative.

EXAMPLE 2.4.11 Finding a Rate of Change of an Air Pollutant

An environmental study of a certain suburban community suggests that the average daily level of carbon monoxide in the air will be $c(p) = \sqrt{0.5p^2 + 17}$ parts per million when the population is p thousand. It is estimated that t years from now, the population of the community will be $p(t) = 3.1 + 0.1t^2$ thousand. At what rate will the carbon monoxide level be changing with respect to time 3 years from now?

Solution

The goal is to find $\dfrac{dc}{dt}$ when $t = 3$. Since

$$\frac{dc}{dp} = \frac{1}{2}(0.5p^2 + 17)^{-1/2}[0.5(2p)] = \frac{1}{2}p(0.5p^2 + 17)^{-1/2}$$

and

$$\frac{dp}{dt} = 0.2t$$

it follows from the chain rule that

$$\frac{dc}{dt} = \frac{dc}{dp}\frac{dp}{dt} = \frac{1}{2}p(0.5p^2 + 17)^{-1/2}(0.2t) = \frac{0.1pt}{\sqrt{0.5p^2 + 17}}$$

When $t = 3$,

$$p(3) = 3.1 + 0.1(3)^2 = 4$$

and by substituting $t = 3$ and $p = 4$ into the formula for $\dfrac{dc}{dt}$, we get

$$\frac{dc}{dt} = \frac{0.1(4)(3)}{\sqrt{0.5(4)^2 + 17}}$$

$$= \frac{1.2}{\sqrt{25}} = \frac{1.2}{5} = 0.24 \text{ parts per million per year}$$

EXERCISES ▪ 2.4

In Exercises 1 through 12, use the chain rule to compute the derivative $\dfrac{dy}{dx}$ and simplify your answer.

1. $y = u^2 + 1;\ u = 3x - 2$

2. $y = 1 - 3u^2;\ u = 3 - 2x$

3. $y = \sqrt{u};\ u = x^2 + 2x - 3$

4. $y = 2u^2 - u + 5;\ u = 1 - x^2$

5. $y = \dfrac{1}{u^2};\ u = x^2 + 1$ 6. $y = \dfrac{1}{u};\ u = 3x^2 + 5$

7. $y = \dfrac{1}{u - 1};\ u = x^2$ 8. $y = \dfrac{1}{\sqrt{u}};\ u = x^2 - 9$

9. $y = u^2 + 2u - 3;\ u = \sqrt{x}$

10. $y = u^3 + u;\ u = \dfrac{1}{\sqrt{x}}$

11. $y = u^2 + u - 2;\ u = \dfrac{1}{x}$

12. $y = u^2;\ u = \dfrac{1}{x - 1}$

In Exercises 13 through 20, use the chain rule to compute the derivative $\dfrac{dy}{dx}$ for the given value of x.

13. $y = u^2 - u;\ u = 4x + 3$ for $x = 0$

14. $y = u + \dfrac{1}{u};\ u = 5 - 2x$ for $x = 0$

15. $y = 3u^4 - 4u + 5;\ u = x^3 - 2x - 5$ for $x = 2$

16. $y = u^5 - 3u^2 + 6u - 5;\ u = x^2 - 1$ for $x = 1$

17. $y = \sqrt{u};\ u = x^2 - 2x + 6$ for $x = 3$

18. $y = 3u^2 - 6u + 2;\ u = \dfrac{1}{x^2}$ for $x = \dfrac{1}{3}$

19. $y = \dfrac{1}{u};\ u = 3 - \dfrac{1}{x^2}$ for $x = \dfrac{1}{2}$

20. $y = \dfrac{1}{u + 1};\ u = x^3 - 2x + 5$ for $x = 0$

In Exercises 21 through 38, differentiate the given function and simplify your answer.

21. $f(x) = (2x + 3)^{1.4}$ 22. $f(x) = \dfrac{1}{\sqrt{5 - 3x}}$

23. $f(x) = (2x + 1)^4$ 24. $f(x) = \sqrt{5x^6 - 12}$

25. $f(x) = (x^5 - 4x^3 - 7)^8$

26. $f(t) = (3t^4 - 7t^2 + 9)^5$

27. $f(t) = \dfrac{1}{5t^2 - 6t + 2}$ 28. $f(x) = \dfrac{2}{(6x^2 + 5x + 1)^2}$

29. $g(x) = \dfrac{1}{\sqrt{4x^2 + 1}}$ 30. $f(s) = \dfrac{1}{\sqrt{5s^3 + 2}}$

31. $f(x) = \dfrac{3}{(1 - x^2)^4}$ 32. $f(x) = \dfrac{2}{3(5x^4 + 1)^2}$

33. $h(s) = (1 + \sqrt{3s})^5$ 34. $g(x) = \sqrt{1 + \dfrac{1}{3x}}$

35. $f(x) = (x + 2)^3(2x - 1)^5$

36. $f(x) = 2(3x + 1)^4(5x - 3)^2$

37. $f(x) = \dfrac{(x + 1)^5}{(1 - x)^4}$

38. $f(x) = \dfrac{1 - 5x^2}{\sqrt[3]{3 + 2x}}$

In Exercises 39 through 46, find an equation of the line that is tangent to the graph of f for the given value of x.

39. $f(x) = \sqrt{3x + 4}; x = 0$

40. $f(x) = (9x - 1)^{-1/3}; x = 1$

41. $f(x) = (3x^2 + 1)^2; x = -1$

42. $f(x) = (x^2 - 3)^5(2x - 1)^3; x = 2$

43. $f(x) = \dfrac{1}{(2x - 1)^6}; x = 1$

44. $f(x) = \left(\dfrac{x + 1}{x - 1}\right)^3; x = 3$

45. $f(x) = \sqrt[3]{\dfrac{x}{x + 2}}; x = -1$

46. $f(x) = x^2\sqrt{2x + 3}; x = -1$

In Exercises 47 through 52, find all values of x = c so that the tangent line to the graph of f(x) at (c, f(c)) will be horizontal.

47. $f(x) = (x^2 + x)^2$

48. $f(x) = x^3(2x^2 + x - 3)^2$

49. $f(x) = \dfrac{x}{(3x - 2)^2}$

50. $f(x) = \dfrac{2x + 5}{(1 - 2x)^3}$

51. $f(x) = \sqrt{x^2 - 4x + 5}$

52. $f(x) = (x - 1)^2(2x + 3)^3$

In Exercises 53 and 54, differentiate the given function f(x) by two different methods, first by using the general power rule and then by using the product rule. Show that the two answers are the same.

53. $f(x) = (3x + 5)^2$

54. $f(x) = (7 - 4x)^2$

In Exercises 55 through 60, find the second derivative of the given function.

55. $f(x) = (3x + 1)^5$

56. $f(t) = \dfrac{2}{5t + 1}$

57. $h(t) = (t^2 + 5)^8$

58. $y = (1 - 2x^3)^4$

59. $f(x) = \sqrt{1 + x^2}$

60. $f(u) = \dfrac{1}{(3u^2 - 1)^2}$

61. Find $h'(0)$ if $h(x) = \sqrt{5x^2 + g(x)}$, where $g(0) = 4$ and $g'(0) = 2$.

62. Find $h'(-1)$ if $g(-1) = -1$ and $g'(-1) = 1$, where $h(x) = [3g^2(x) + 4g(x) + 2]^5[g(x) + x]$.

63. Find $h'(1)$ if $h(x) = \left[3x + \dfrac{1}{g(x)}\right]^{3/2}$, where $g(1) = g'(1) = 1$.

64. Find $h'(0)$ if $h(x) = \left[\dfrac{g(x) - x}{3 + g(x)}\right]^2$, where $g(0) = 3$ and $g'(0) = -2$.

BUSINESS AND ECONOMICS APPLIED PROBLEMS

65. ANNUAL EARNINGS The gross annual earnings of a certain company are $f(t) = \sqrt{10t^2 + t + 229}$ thousand dollars t years after its formation in January 2010.
  a. At what rate will the gross annual earnings of the company be growing in January 2015?
  b. At what percentage rate will the gross annual earnings be growing in January 2015?

66. MANUFACTURING COST At a certain factory, the total cost of manufacturing q units is $C(q) = 0.2q^2 + q + 900$ dollars. It has been determined that approximately $q(t) = t^2 + 100t$ units are manufactured during the first t hours of a production run. Compute the rate at which the total manufacturing cost is changing with respect to time 1 hour after production begins.

67. CONSUMER DEMAND An importer of Brazilian coffee estimates that local consumers will buy approximately $D(p) = \dfrac{4{,}374}{p^2}$ pounds of the coffee per week when the price is p dollars per pound. It is also estimated that t weeks from now, the price of Brazilian coffee will be $p(t) = 0.02t^2 + 0.1t + 6$ dollars per pound.
  a. At what rate will the demand for coffee be changing with respect to price when the price is \$9?
  b. At what rate will the demand for coffee be changing with respect to time 10 weeks from now? Will the demand be increasing or decreasing at this time?

68. CONSUMER DEMAND When a certain commodity is sold for p dollars per unit, consumers will buy $D(p) = \dfrac{40{,}000}{p}$ units per month. It is estimated that t months from now, the price of the commodity will be $p(t) = 0.4t^{3/2} + 6.8$ dollars per unit. At what percentage rate will the monthly demand for the commodity be changing with respect to time 4 months from now?

69. **PRODUCTION**   After t months, the production output at a certain factory is $N(t)$ thousand units, where

$$N(t) = \sqrt{t^2 + 3t + 6}$$

At what rate is the production level changing after 2 months? Is production increasing or decreasing at this time?

70. **PRODUCTION**   After t weeks, a factory is producing $N(t)$ thousand DVD players, where

$$N(t) = \frac{2t}{t^2 + 3t + 12}$$

At what rate is the production level changing after 4 weeks? Is production increasing or decreasing at this time?

71. **PRODUCTION**   The number of units Q of a particular commodity that will be produced with K thousand dollars of capital expenditure is modeled by

$$Q(K) = 500K^{2/3}$$

Suppose that capital expenditure varies with time in such a way that t months from now there will be $K(t)$ thousand dollars of capital expenditure, where

$$K(t) = \frac{2t^4 + 3t + 149}{t + 2}$$

a. What will be the capital expenditure 3 months from now? How many units will be produced at this time?
b. At what rate will production be changing with respect to time 5 months from now? Will production be increasing or decreasing at this time?

72. **PRODUCTION**   The number of units Q of a particular commodity that will be produced when L worker-hours of labor are employed is modeled by

$$Q(L) = 300L^{1/3}$$

Suppose that the labor level varies with time in such a way that t months from now $L(t)$ worker-hours will be employed, where

$$L(t) = \sqrt{739 + 3t - t^2}$$

for $0 \le t \le 12$.
a. How many worker-hours will be employed in producing the commodity 5 months from now? How many units will be produced at this time?
b. At what rate will production be changing with respect to time 5 months from now? Will production be increasing or decreasing at this time?

73. **COMPOUND INTEREST**   If \$10,000 is invested at an annual rate of $r\%$ compounded monthly, then the total amount (principal and interest) accumulated after 10 years is given by the formula

$$A = 10,000\left(1 + \frac{0.01r}{12}\right)^{120}$$

a. Find the instantaneous rate of change of A with respect to r. What is $A'(5)$? Include units.
b. How much does the value of the account actually change if the interest rate is raised from 5% to 6%?

74. **DEPRECIATION**   The value V (in thousands of dollars) of an industrial machine is modeled by

$$V(N) = \left(\frac{3N + 430}{N + 1}\right)^{2/3}$$

where N is the number of hours the machine is used each day. Suppose further that usage varies with time in such a way that

$$N(t) = \sqrt{t^2 - 10t + 45}$$

where t is the number of months the machine has been in operation.
a. How many hours per day will the machine be used 9 months from now? What will be the value of the machine at this time?
b. At what rate is the value of the machine changing with respect to time 9 months from now? Will the value be increasing or decreasing at this time?

LIFE AND SOCIAL SCIENCE APPLIED PROBLEMS

75. **AIR POLLUTION**   It is estimated that t years from now, the population of a certain suburban community will be $p(t) = 20 - \dfrac{6}{t + 1}$ thousand. An environmental study indicates that the average daily level of carbon monoxide in the air will be $c(p) = 0.5\sqrt{p^2 + p + 58}$ parts per million when the population is p thousand.
a. At what rate will the level of carbon monoxide be changing with respect to population when the population is 18 thousand people?
b. At what rate will the carbon monoxide level be changing with respect to time 2 years from now? Will the level be increasing or decreasing at this time?

76. **ANIMAL BEHAVIOR** In a research paper,* V. A. Tucker and K. Schmidt-Koenig demonstrated that a species of Australian parakeet (the Budgerigar) expends energy (calories per gram of mass per kilometer) according to the formula

$$E = \frac{1}{v}[0.074(v - 35)^2 + 22]$$

where v is the bird's velocity (in km/hr). Find a formula for the rate of change of E with respect to velocity v.

77. **MAMMALIAN GROWTH** Observations show that the length L in millimeters (mm) from nose to tip of tail of a Siberian tiger can be estimated using the function $L = 0.25w^{2.6}$, where w is the weight of the tiger in kilograms (kg). Furthermore, when a tiger is less than 6 months old, its weight (kg) can be estimated in terms of its age A in days by the function $w = 3 + 0.21A$.
   a. At what rate is the length of a Siberian tiger increasing with respect to its weight when it weighs 60 kg?
   b. How long is a Siberian tiger when it is 100 days old? At what rate is its length increasing with respect to time at this age?

78. **QUALITY OF LIFE** A demographic study models the population p (in thousands) of a community by the function

$$p(Q) = 3Q^2 + 4Q + 200$$

where Q is a quality-of-life index that ranges from $Q = 0$ (extremely poor quality) to $Q = 10$ (excellent quality). Suppose the index varies with time in such a way that t years from now,

$$Q(t) = \frac{t^2 + 2t + 3}{2t + 1}$$

for $0 \le t \le 10$.
   a. What value of the quality-of-life index should be expected 4 years from now? What will be the corresponding population at this time?
   b. At what rate is the population changing with respect to time 4 years from now? Is the population increasing or decreasing at this time?

79. **WATER POLLUTION** When organic matter is introduced into a body of water, the oxygen content of the water is temporarily reduced by oxidation. Suppose that t days after untreated sewage is dumped into a particular lake, the proportion of the usual oxygen content in the water of the lake that remains is given by the function

$$P(t) = 1 - \frac{12}{t + 12} + \frac{144}{(t + 12)^2}$$

   a. At what rate is the oxygen proportion $P(t)$ changing after 10 days? Is the proportion increasing or decreasing at this time?
   b. Is the oxygen proportion increasing or decreasing after 15 days?
   c. If there is no new dumping, what would you expect to eventually happen to the proportion of oxygen? Use a limit to verify your conjecture.

80. **LEARNING** When you first begin to study a topic or practice a skill, you may not be very good at it, but in time, you will approach the limits of your ability. One model for describing this behavior involves the function

$$T = aL\sqrt{L - b}$$

where T is the time required for a particular person to learn the items on a list of L items and a and b are positive constants.
   a. Find the derivative $\dfrac{dT}{dL}$, and interpret it in terms of the learning model.
   b. Read and discuss in one paragraph an article on how learning curves can be used to study worker productivity.[†]

81. **INSECT GROWTH** The growth of certain insects varies with temperature. Suppose a particular species of insect grows in such a way that the volume of an individual is

$$V(T) = 0.41(-0.01T^2 + 0.4T + 3.52) \text{ cm}^3$$

when the temperature is $T°C$, and that its mass is m grams, where

$$m(V) = \frac{0.39V}{1 + 0.09V}$$

   a. Find the rate of change of the insect's volume with respect to temperature.
   b. Find the rate of change of the insect's mass with respect to volume.
   c. When $T = 10°C$, what is the insect's volume? At what rate is the insect's mass changing with respect to temperature when $T = 10°C$?

*V. A. Tucker and K. Schmidt-Koenig, "Flight Speeds of Birds in Relation to Energetics and Wind Directions," *The Auk*, Vol. 88, 1971, pp. 97–107.

[†]You may wish to begin your research by consulting Philip E. Hicks, *Industrial Engineering and Management: A New Perspective*, 2nd ed., Chapter 6, New York: McGraw-Hill, 1994, pp. 267–293.

MISCELLANEOUS PROBLEMS

82. An object moves along a straight line with velocity
$$v(t) = (2t + 9)^2(8 - t)^3 \qquad \text{for } 0 \le t \le 5$$

  a. Find the acceleration $a(t)$ of the object at time t.

  b. When is the object stationary for $0 \le t \le 5$? Find the acceleration at each such time.

  c. When is the acceleration zero for $0 \le t \le 5$? Find the velocity at each such time.

  d. Use the graphing utility of your calculator to draw the graphs of the velocity $v(t)$ and acceleration $a(t)$ on the same screen.

  e. The object is said to be *speeding up* when $v(t)$ and $a(t)$ have the same sign (both positive or both negative). Use your calculator to determine when (if ever) this occurs for $0 \le t \le 5$.

83. An object moves along a straight line in such a way that its position at time t is given by
$$s(t) = (3 + t - t^2)^{3/2} \qquad \text{for } 0 \le t \le 2$$

  a. What are the object's velocity $v(t)$ and acceleration $a(t)$ at time t?

  b. When is the object stationary for $0 \le t \le 2$? Where is the object and what is its acceleration at each such time?

  c. When is the acceleration zero for $0 \le t \le 2$? What are the object's position and velocity at each such time?

  d. Use the graphing utility of your calculator to draw the graphs of the object's position $s(t)$, velocity $v(t)$, and acceleration $a(t)$ on the same screen for $0 \le t \le 2$.

  e. The object is said to be *slowing down* when $v(t)$ and $a(t)$ have opposite signs (one positive, the other negative). Use your calculator to determine when (if ever) this occurs for $0 \le t \le 2$.

84. Suppose $L(x)$ is a function with the property that $L'(x) = \dfrac{1}{x}$. Use the chain rule to find the derivatives of the following functions and simplify your answers.

  a. $f(x) = L(x^2)$
  b. $f(x) = L\left(\dfrac{1}{x}\right)$

  c. $f(x) = L\left(\dfrac{2}{3\sqrt{x}}\right)$
  d. $f(x) = L\left(\dfrac{2x + 1}{1 - x}\right)$

85. Prove the general power rule for $n = 2$ by using the product rule to compute $\dfrac{dy}{dx}$ if $y = [h(x)]^2$.

86. Prove the general power rule for $n = 3$ by using the product rule and the result of Exercise 85 to compute $\dfrac{dy}{dx}$ if $y = [h(x)]^3$. [*Hint:* Begin by writing y as $h(x)[h(x)]^2$.]

87. Store the function $f(x) = \sqrt[3]{3.1x^2 + 19.4}$ in your graphing utility. Use the numeric differentiation feature of your utility to calculate $f'(1)$ and $f'(-3)$. Explore the graph of $f(x)$. How many horizontal tangents does the graph have?

88. Store the function $f(x) = (2.7x^3 - 3\sqrt{x} + 5)^{2/3}$ in your graphing utility. Use the numeric differentiation feature of the utility to calculate $f'(0)$ and $f'(4.3)$. Explore the graph of $f(x)$. How many horizontal tangents does it have?

89. Let f be a function such that $f'(x) = \dfrac{1}{1 + x^2}$. If $g(x) = f(2x + 1)$, what is $g'(x)$?

90. Let f be a function such that $f(3) = -1$ and $f'(x) = \sqrt{x^2 + 3}$. If $g(x) = x^3 f\left(\dfrac{x}{x - 2}\right)$, what is $g'(3)$?

SECTION 2.5   Marginal Analysis and Approximations Using Increments

Learning Objectives

1. Study marginal analysis in economics.
2. Approximate derivatives using increments and the differential.

Calculus is an important tool in economics. We briefly discussed sales and production in Chapter 1, where we introduced economic concepts such as cost, revenue, profit and supply, demand, and market equilibrium. In this section, we will use the derivative to explore rates of change involving economic quantities.

Marginal Analysis In economics,* the use of the derivative to approximate the change in a quantity that results from a 1-unit increase in production is called **marginal analysis.** For instance, suppose $C(x)$ is the total cost of producing x units of a particular commodity. If x_0 units are currently being produced, then the derivative

$$C'(x_0) = \lim_{h \to 0} \frac{C(x_0 + h) - C(x_0)}{h}$$

is called the **marginal cost** of producing x_0 units. Notice that if we take $h = 1$, the difference quotient for $C(x_0)$ becomes

$$\frac{C(x_0 + 1) - C(x_0)}{1} = C(x_0 + 1) - C(x_0)$$

which is the cost of producing the $(x_0 + 1)^{st}$ unit. For x_0 large in relation $h = 1$, this difference quotient is approximately equal to the derivative $C'(x_0)$; that is,

$$C'(x_0) \approx \text{cost of producing the } (x_0 + 1)^{st} \text{ unit}$$

To summarize:

Marginal Cost ■ If $C(x)$ is the total cost of producing x units of a commodity, then the **marginal cost** of producing x_0 units is the derivative $C'(x_0)$.

For x_0 sufficiently large, the marginal cost $C'(x_0)$ can be used to estimate the additional cost $C(x_0 + 1) - C(x_0)$ incurred when the level of production is increased from x_0 to $x_0 + 1$.

The geometric relationship between the marginal cost $C'(x_0)$ and the additional cost $C(x_0 + 1) - C(x_0)$ is shown in Figure 2.14.

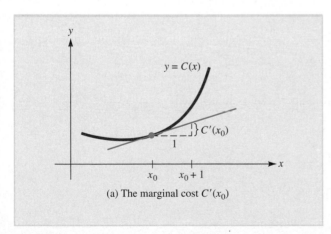

(a) The marginal cost $C'(x_0)$

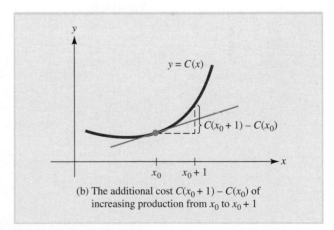

(b) The additional cost $C(x_0 + 1) - C(x_0)$ of increasing production from x_0 to $x_0 + 1$

FIGURE 2.14 Marginal cost $C'(x_0)$ approximates $C(x_0 + 1) - C(x_0)$.

---

*Economists and business managers view the topics we are about to discuss from a slightly different perspective. A good source for the economist's view is the text by J. M. Henderson and R. E. Quandt, *Microeconomic Theory,* New York: McGraw-Hill, 1986. The viewpoint of business management may be found in D. Salvatore, *Management Economics,* New York: McGraw-Hill, 1989, which is an excellent source of practical applications and case studies.

The preceding discussion applies not only to cost, but also to other economic quantities. Here is a summary of what is meant by marginal revenue and marginal profit and how these marginal quantities can be used to estimate 1-unit changes in revenue and profit.

Marginal Revenue and Marginal Profit ■ Suppose $R(x)$ is the revenue generated when x units of a particular commodity are produced, and $P(x)$ is the corresponding profit. When $x = x_0$ units are being produced, then:

The **marginal revenue** is $R'(x_0)$. It approximates $R(x_0 + 1) - R(x_0)$, the additional revenue generated by producing one more unit.

The **marginal profit** is $P'(x_0)$. It approximates $P(x_0 + 1) - P(x_0)$, the additional profit obtained by producing one more unit.

EXPLORE!

Refer to Example 2.5.1. Graph the cost function $C(x)$ and revenue function $R(x)$ together using the window [0, 50]10 by [0, 1000]100. Graph the tangent line to $C(x)$ at $x = 36$. Notice why the marginal cost, represented by the slope of the tangent line, is a good approximation to the amount that $C(x)$ changes at $x = 36$ for a one-unit change in x. Repeat for the tangent line to $R(x)$ at $x = 36$.

EXAMPLE 2.5.1   Studying Marginal Cost and Marginal Revenue

A manufacturer estimates that when x units of a particular commodity are produced, the total cost will be $C(x) = \frac{1}{8}x^2 + 3x + 98$ dollars, and furthermore, that all x units will be sold when the price is $p(x) = \frac{1}{3}(75 - x)$ dollars per unit.

a. Find the marginal cost and the marginal revenue.

b. Use marginal cost to estimate the cost of producing the 37th unit. What is the actual cost of producing the 37th unit?

c. Use marginal revenue to estimate the revenue derived from the sale of the 37th unit. What is the actual revenue derived from the sale of the 37th unit?

Solution

a. The marginal cost is $C'(x) = \frac{1}{4}x + 3$. Since x units of the commodity are sold at a price of $p(x) = \frac{1}{3}(75 - x)$ dollars per unit, the total revenue is

$$R(x) = (\text{number of units sold})(\text{price per unit})$$
$$= xp(x) = x\left[\frac{1}{3}(75 - x)\right] = 25x - \frac{1}{3}x^2$$

The marginal revenue is

$$R'(x) = 25 - \frac{2}{3}x$$

b. The cost of producing the 37th unit is the change in cost as x increases from 36 to 37 and can be estimated by the marginal cost when producing 36 units:

$$C'(36) = \frac{1}{4}(36) + 3 = \$12$$

The actual cost of producing the 37th unit is

$$C(37) - C(36) = 380.125 - 368 \approx \$12.13$$

which is reasonably well approximated by the marginal cost $C'(36) = \$12$.

c. The revenue obtained from the sale of the 37th unit is approximated by the marginal revenue at $x = 36$:

$$R'(36) = 25 - \frac{2}{3}(36) = \$1$$

The actual revenue obtained from the sale of the 37th unit is

$$R(37) - R(36) = 468.67 - 468 = 0.67$$

that is, 67 cents, which is approximately the same as the marginal revenue $R'(36) = \$1$.

In Example 2.5.2, a marginal economic quantity is used to analyze a production process.

EXAMPLE 2.5.2   Using Marginal Analysis to Make a Business Decision

Quentin is the business manager for a company that manufactures digital cameras. He determines that when x hundred cameras are produced, the total profit will be P thousand dollars where

$$P(x) = -0.0035x^3 + 0.07x^2 + 25x - 200$$

Quentin plans to use marginal profit to make decisions regarding future production.

a. What is the marginal profit function?
b. The current level of production is $x = 10$ (1,000 cameras). Based on the marginal profit at this level of production, should Quentin recommend increasing or decreasing production to increase profit?
c. What decision should Quentin make if the current level of production is $x = 50$ (5,000 cameras)? What if $x = 80$ (8,000 cameras are being produced)?

Solution

a. The marginal profit is given by the derivative

$$P'(x) = -0.0035(3x^2) + 0.07(2x) + 25$$
$$= -0.0105x^2 + 0.14x + 25$$

b. We find that $P'(10) = 25.35$. This means that a 1-unit increase in production from 10 to 11 hundred cameras increases profit by approximately 25.35 thousand dollars ($25,350), so Quentin may be inclined to increase production at this level.

c. The marginal profit when $x = 50$ is $P'(50) = 5.75$, so increasing the level of production from 50 units to 51 (5,000 cameras to 5,100) increases the profit by only about $5,750. This provides little incentive for Quentin to change the level of production.

At a level of production of 80 units, the marginal profit $P'(80) = -31$ is negative, so raising the level of production by 1 unit (from 8,000 units to 8,100) actually *decreases* the profit by approximately $31,000. Quentin may wish to decrease production in this case.

Approximation by Increments

Marginal analysis is an important example of a general approximation procedure based on the fact that since

$$f'(x_0) = \lim_{h \to 0} \frac{f(x_0 + h) - f(x_0)}{h}$$

then for small h, the derivative $f'(x_0)$ is approximately equal to the difference quotient

$$\frac{f(x_0 + h) - f(x_0)}{h}$$

We indicate this approximation by writing

$$f'(x_0) \approx \frac{f(x_0 + h) - f(x_0)}{h}$$

or, equivalently,

$$f(x_0 + h) - f(x_0) \approx f'(x_0)h$$

To emphasize that the incremental change is in the variable x, we write $h = \Delta x$ (read Δx as "delta x") and summarize the incremental approximation formula as follows.

Approximation by Increments ■ If $f(x)$ is differentiable at $x = x_0$ and Δx is a small change in x, then

$$f(x_0 + \Delta x) \approx f(x_0) + f'(x_0)\Delta x$$

or, equivalently, if $\Delta f = f(x_0 + \Delta x) - f(x_0)$, then

$$\Delta f \approx f'(x_0)\Delta x$$

Example 2.5.3 illustrates how this approximation formula can be used in economics.

EXAMPLE 2.5.3 **Estimating Change in Cost Using a Derivative**

Suppose the total cost of manufacturing q hundred units of a certain commodity is C thousand dollars where $C(q) = 3q^2 + 5q + 10$. If the current level of production is 4,000 units, estimate how the total cost will change if 4,050 units are produced.

Solution

In this problem, the current level of production is $q = 40$ (4,000 units) and the change in production is $\Delta q = 0.5$ (an additional 50 units). By the approximation formula, the corresponding change in cost is

$$\Delta C = C(40.5) - C(40) \approx C'(40)\Delta q = C'(40)(0.5)$$

Since

$$C'(q) = 6q + 5 \quad \text{and} \quad C'(40) = 6(40) + 5 = 245$$

it follows that

$$\Delta C \approx [C'(40)](0.5) = 245(0.5) = 122.50$$

That is, the difference is 122.5 thousand dollars ($122,500).

For practice, compute the actual change in cost caused by the increase in the level of production from 40 to 40.5 and compare your answer with the approximation. Is the approximation a good one?

Suppose you wish to compute a quantity Q using a formula $Q(x)$. If the value of x used in the computation is not precise, then its inaccuracy is passed on or *propagated* to the computed value of Q. Example 2.5.4 shows how such **propagated error** can be estimated.

EXPLORE!

Refer to Example 2.5.4. Store the volume $V = \dfrac{1}{6}\pi x^3$ into Y1, where **x** measures the diameter of the spherical tumor. Write

Y2 = Y1(X + 0.05) − Y1(X)

to compute the incremental change in volume, and

Y3 = nDeriv(Y1, X, X)∗(0.05)

for the differential change in volume. Set **TblStart** = 2.4 and ΔTbl = 0.05 in **TBLSET (2nd WINDOW)**. Now examine the **TABLE** of values, particularly for Y2 and Y3. Observe the results for X = 2.5. Finally, perform similar calculations for the accuracy of the volume V if the diameter **x** is measured accurately within 1%.

EXAMPLE 2.5.4 Estimating Error in Measurement

During a medical procedure, the size of a roughly spherical tumor is estimated by measuring its diameter and using the formula $V = \dfrac{4}{3}\pi R^3$ to compute its volume. If the diameter is measured as 2.5 cm with a maximum error of 2%, how accurate is the volume measurement?

Solution

A sphere of radius R and diameter $x = 2R$ has volume

$$V = \frac{4}{3}\pi R^3 = \frac{4}{3}\pi \left(\frac{x}{2}\right)^3 = \frac{1}{6}\pi x^3$$

so the volume using the estimated diameter $x = 2.5$ cm is

$$V = \frac{1}{6}\pi(2.5)^3 \approx 8.181 \text{ cm}^3$$

The error made in computing this volume using the diameter 2.5 when the actual diameter is $2.5 + \Delta x$ is

$$V = V(2.5 + \Delta x) - V(2.5) \approx V'(2.5)\Delta x$$

The measurement of the diameter can be off by as much as 2%, that is, by as much as $0.02(2.5) = 0.05$ cm in either direction. Hence, the maximum error in the measurement of the diameter is $\Delta x = \pm 0.05$, and the corresponding maximum error in the calculation of volume is

$$\text{Maximum error in volume} = \Delta V \approx [V'(2.5)](\pm 0.05)$$

Since

$$V'(x) = \frac{1}{6}\pi(3x^2) = \frac{1}{2}\pi x^2 \qquad \text{and} \qquad V'(2.5) = \frac{1}{2}\pi(2.5)^2 \approx 9.817$$

it follows that

$$\text{Maximum error in volume} = (9.817)(\pm 0.05) \approx \pm 0.491$$

Thus, at worst, the calculation of the volume as 8.181 cm³ is off by 0.491 cm³, so the actual volume V must satisfy

$$7.690 \le V \le 8.672$$

In Example 2.5.5, the desired change in the function is given, and the goal is to estimate the necessary corresponding change in the variable.

EXAMPLE 2.5.5 Estimating a Change in Required Labor

The daily output at a certain factory is $Q(L) = 900L^{1/3}$ units, where L denotes the size of the labor force measured in worker-hours. Currently, 1,000 worker-hours of labor are used each day. Use calculus to estimate the number of additional worker-hours of labor that will be needed to increase daily output by 15 units.

Solution

Solve for ΔL using the approximation formula

$$\Delta Q \approx Q'(L)\Delta L$$

with $\Delta Q = 15$ $L = 1{,}000$ and $Q'(L) = 300L^{-2/3}$

to get $15 \approx 300(1{,}000)^{-2/3}\Delta L$

so $\Delta L \approx \dfrac{15}{300}(1{,}000)^{2/3} = \dfrac{15}{300}(10)^2 = 5$ worker-hours

In certain applications, we are less interested in the actual change $\Delta Q = Q(x_0 + \Delta x) - Q(x_0)$ in a quantity $Q(x)$ than the **relative change**

$$\frac{\Delta Q}{Q} = \frac{Q'(x)\Delta x}{Q(x)}$$

or the **percentage change**

$$100\frac{\Delta Q}{Q} = 100\frac{Q'(x)\Delta x}{Q(x)}$$

The same expressions can also be thought of as measuring the **relative error** and **percentage error** incurred by computing a quantity $Q(x)$ with an inaccurate value $x = x_0$ rather than the exact value $x_0 + \Delta x$. Example 2.5.6 illustrates the use of percentage change in an economics application.

EXAMPLE 2.5.6 Estimating Percentage Change in GDP

The GDP of a certain country was $N(t) = t^2 + 5t + 200$ billion dollars t years after 2005. Use calculus to estimate the percentage change in the GDP during the first quarter of 2013.

Solution

Use the formula

$$\text{Percentage change in } N \approx 100\frac{N'(t)\Delta t}{N(t)}$$

with $t = 8$ $\Delta t = 0.25$ and $N'(t) = 2t + 5$

to get

$$\text{Percentage change in } N \approx 100\frac{N'(8)0.25}{N(8)}$$

$$= 100\frac{[2(8) + 5](0.25)}{(8)^2 + 5(8) + 200}$$

$$\approx 1.73\%$$

Differentials Sometimes the increment Δx is referred to as the *differential of x* and is denoted by dx, and then our approximation formula can be written as $df \approx f'(x)dx$. If $y = f(x)$, the *differential of y* is defined to be $dy = f'(x)dx$. To summarize:

> **Differentials** ■ The **differential of x** is $dx = \Delta x$, and if $y = f(x)$ is a differentiable function of x, then $dy = f'(x)dx$ is the **differential of y.**

EXAMPLE 2.5.7 Finding Differentials

In each case, find the differential of $y = f(x)$.
a. $f(x) = x^3 - 7x^2 + 2$
b. $f(x) = (x^2 + 5)(3 - x - 2x^2)$

Solution
a. $dy = f'(x)dx = [3x^2 - 7(2x)]\ dx = (3x^2 - 14x)\ dx$
b. By the product rule,

$$dy = f'(x)dx = [(x^2 + 5)(-1 - 4x) + (2x)(3 - x - 2x^2)]dx$$

A geometric interpretation of the approximation of Δy by the differential dy is shown in Figure 2.15. Note that since the slope of the tangent line at $(x, f(x))$ is $f'(x)$, the differential $dy = f'(x)dx$ is the change in the height of the tangent that corresponds to a change from x to $x+\Delta x$. On the other hand, Δy is the change in the height of the curve corresponding to this change in x. Hence, approximating Δy by the differential dy is the same as approximating the change in the height of a curve by the change in height of the tangent line. If Δx is small, it is reasonable to expect this to be a good approximation.

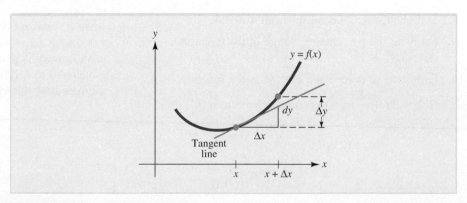

FIGURE 2.15 Approximation of Δy by the differential dy.

EXERCISES ■ 2.5

In Exercises 1 through 6, C(x) is the total cost of producing x units of a particular commodity and p(x) is the unit price at which all x units will be sold. Assume p(x) and C(x) are in dollars.

    (a) Find the marginal cost and the marginal revenue.

    (b) Use marginal cost to estimate the cost of producing the 21st unit. What is the actual cost of producing the 21st unit?

    (c) Use marginal revenue to estimate the revenue derived from the sale of the 21st unit. What is the actual revenue obtained from the sale of the 21st unit?

1. $C(x) = \dfrac{1}{5}x^2 + 4x + 57; p(x) = \dfrac{1}{4}(48 - x)$

2. $C(x) = \dfrac{1}{4}x^2 + 3x + 67; p(x) = \dfrac{1}{5}(45 - x)$

3. $C(x) = \dfrac{1}{3}x^2 + 2x + 39; p(x) = -x^2 - 10x + 4{,}000$

4. $C(x) = \dfrac{5}{9}x^2 + 5x + 73; p(x) = -2x^2 - 15x + 6{,}000$

5. $C(x) = \dfrac{1}{4}x^2 + 43; p(x) = \dfrac{3 + 2x}{1 + x}$

6. $C(x) = \dfrac{2}{7}x^2 + 65; p(x) = \dfrac{12 + 2x}{3 + x}$

In Exercises 7 through 10, use increments to make the required estimate.

7. Estimate how much the function $f(x) = x^2 - 3x + 5$ will change as x increases from 5 to 5.3.

8. Estimate how much the function $f(x) = \dfrac{x}{x + 1} - 3$ will change as x decreases from 4 to 3.8.

9. Estimate the percentage change in the function $f(x) = x^2 + 2x - 9$ as x increases from 4 to 4.3.

10. Estimate the percentage change in the function $f(x) = 3x + \dfrac{2}{x}$ as x decreases from 5 to 4.6.

BUSINESS AND ECONOMICS APPLIED PROBLEMS

11. BUSINESS MANAGEMENT Leticia manages a company whose total weekly revenue is
$$R(q) = 240q - 0.05q^2$$
dollars when q units are produced and sold. Currently, the company produces and sells 80 units a week.

    a. Using marginal analysis, Leticia estimates the additional revenue that will be generated by the production and sale of the 81st unit. What does she discover? Based on this result, should she recommend increasing the level of production?

    b. To check her results, Leticia uses the revenue function to compute the actual revenue generated by the production and sale of the 81st unit. How accurate was her result found by marginal analysis?

12. MARGINAL ANALYSIS A manufacturer's total cost is $C(q) = 0.001q^3 - 0.05q^2 + 40q + 4{,}000$ dollars, where q is the number of units produced.

    a. Use marginal analysis to estimate the cost of producing the 251st unit.

    b. Compute the actual cost of producing the 251st unit.

13. MARGINAL ANALYSIS Suppose the total cost in dollars of manufacturing q units is $C(q) = 3q^2 + q + 500$.

    a. Use marginal analysis to estimate the cost of manufacturing the 41st unit.

    b. Compute the actual cost of manufacturing the 41st unit.

14. MANUFACTURING A manufacturer's total cost is $C(q) = 0.1q^3 - 0.5q^2 + 500q + 200$ dollars when q thousand units are produced. Currently, 4,000 units $(q = 4)$ are being produced and the manufacturer is planning to increase the level of production to 4,100. Use marginal analysis to estimate how this change will affect total cost.

15. **MANUFACTURING** A manufacturer's total monthly revenue is $R(q) = 240q - 0.05q^2$ dollars when q hundred units are produced during the month. Currently, the manufacturer is producing 8,000 units a month and is planning to decrease the monthly output by 65 units. Estimate how the total monthly revenue will change as a result.

16. **EFFICIENCY** An efficiency study of the morning shift at a certain factory indicates that an average worker arriving on the job at 8:00 A.M. will have assembled $f(x) = -x^3 + 6x^2 + 15x$ units x hours later. Approximately how many units will the worker assemble between 9:00 and 9:15 A.M.?

17. **PRODUCTION** At a certain factory, the daily output is $Q(K) = 600K^{1/2}$ units, where K denotes the capital investment measured in units of $1,000. The current capital investment is $900,000. Estimate the effect that an additional capital investment of $800 will have on the daily output.

18. **PRODUCTION** At a certain factory, the daily output is $Q(L) = 60,000L^{1/3}$ units, where L denotes the size of the labor force measured in worker-hours. Currently 1,000 worker-hours of labor are used each day. Estimate the effect on output that will be produced if the labor force is cut to 940 worker-hours.

19. **BUSINESS MANAGEMENT** Matthew manages a company that produces
$$Q = 3,000K^{1/2}L^{1/3}$$
units per day, where K is the capital investment in thousands of dollars and L is the size of the labor force measured in worker-hours. Currently, the capital investment is $400,000 ($K = 400$) and the labor force is 1,331 worker-hours per day.

Matthew is trying to decide whether increasing capital expenditure by $10,000 or increasing the labor force by 10 worker-hours will produce the greatest increase in daily output. Use marginal analysis to help him make his decision.

20. **BUSINESS MANAGEMENT** Aurelia determines that her company produces
$$Q(L) = 300L^{2/3}$$
units per day, where L is the size of her labor force, measured in worker-hours.

Currently, she employs 512 worker-hours each day. Use marginal analysis to help her estimate the number of additional worker-hours required to increase daily output by 12.5 units.

21. **MANUFACTURING** A manufacturer's total cost is $C(q) = \dfrac{1}{6}q^3 + 642q + 400$ dollars when q units are produced. The current level of production is 4 units. Estimate the amount by which the manufacturer should decrease production to reduce the total cost by $130.

22. **PROPERTY TAX** A projection made in January of 2005 determined that x years later, the average property tax on a three-bedroom home in a certain community will be $T(x) = 60x^{3/2} + 40x + 1,200$ dollars. Estimate the percentage change by which the property tax will increase during the first half of the year 2013.

LIFE AND SOCIAL SCIENCE APPLIED PROBLEMS

23. **NEWSPAPER CIRCULATION** It is projected that t years from now, the circulation of a local newspaper will be $C(t) = 100t^2 + 400t + 5,000$. Estimate the amount by which the circulation will increase during the next 6 months.

24. **AIR POLLUTION** An environmental study of a certain community suggests that t years from now, the average level of carbon monoxide in the air will be $Q(t) = 0.05t^2 + 0.1t + 3.4$ parts per million. By approximately how much will the carbon monoxide level change during the coming 6 months?

25. **POPULATION GROWTH** A 5-year projection of population trends suggests that t years from now, the population of a certain community will be $P(t) = -t^3 + 9t^2 + 48t + 200$ thousand.
   a. Find the rate of change of population $R(t) = P'(t)$ with respect to time t.
   b. At what rate does the population growth rate $R(t)$ change with respect to time?
   c. Use increments to estimate how much $R(t)$ changes during the first month of the fourth year. What is the actual change in $R(t)$ during this time period?

26. **GROWTH OF A CELL** A certain cell has the shape of a sphere. The formulas $S = 4\pi r^2$ and $V = \dfrac{4}{3}\pi r^3$ are used to compute the surface area and volume of the cell, respectively. Estimate the effect on S and V produced by a 1% increase in the radius r.

27. **CARDIAC OUTPUT**   *Cardiac output* is the volume (cubic centimeters) of blood pumped by a person's heart each minute. One way of measuring cardiac output C is by Fick's formula

$$C = \frac{a}{x - b}$$

where x is the concentration of carbon dioxide in the blood entering the lungs from the right side of the heart and a and b are positive constants. If x is measured as $x = c$ with a maximum error of 3%, what is the maximum percentage error that can be incurred by measuring cardiac output with Fick's formula? (Your answer will be in terms of a, b, and c.)

28. **MEDICINE**   A tiny spherical balloon is inserted into a clogged artery. If the balloon has an inner diameter of 0.01 millimeter (mm) and is made from material 0.0005 mm thick, approximately how much material is inserted into the artery?

$$\left[\text{\textit{Hint}: Think of the amount of material as a change in volume } \Delta V, \text{ where } V = \frac{4}{3}\pi r^3.\right]$$

29. **ARTERIOSCLEROSIS**   In *arteriosclerosis,* fatty material called plaque gradually builds up on the walls of arteries, impeding the flow of blood, which, in turn, can lead to stroke and heart attacks. Consider a model in which the carotid artery is represented as a circular cylinder with cross-sectional radius $R = 0.3$ cm and length L. Suppose it is discovered that plaque 0.07 cm thick is distributed uniformly over the inner wall of the carotid artery of a particular patient. Use increments to estimate the percentage of the total volume of the artery that is blocked by plaque. [*Hint:* The volume of a cylinder of radius R and length L is $V = \pi R^2 L$. Does it matter that we have not specified the length L of the artery?]

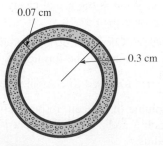

EXERCISE 29

30. **BLOOD CIRCULATION**   In Exercise 73, Section 1.1, we introduced an important law attributed to the French physician, Jean Poiseuille. Another law discovered by Poiseuille says that the volume of the fluid flowing through a small tube in unit time under fixed pressure is given by the formula $V = kR^4$, where k is a positive constant and R is the radius of the tube. This formula is used in medicine to determine how wide a clogged artery must be opened to restore a healthy flow of blood.

   a. Suppose the radius of a certain artery is increased by 5%. Approximately what effect does this have on the volume of blood flowing through the artery?

   b. Read an article on the cardiovascular system, and write a paragraph on the flow of blood.*

MISCELLANEOUS PROBLEMS

31. **EXPANSION OF MATERIAL**   The (linear) **thermal expansion coefficient** of an object is defined to be

$$\sigma = \frac{L'(T)}{L(T)}$$

where $L(T)$ is the length of the object when the temperature is T. Suppose a 50-meter span of a bridge is built of steel with $\sigma = 1.4 \times 10^{-5}$ per degree Celsius. Approximately how much will the length change during a year when the temperature varies from $-20°C$ (winter) to $35°C$ (summer)?

32. **RADIATION**   Stefan's law in physics states that a body emits radiant energy according to the formula $R(T) = kT^4$, where R is the amount of energy emitted from a surface whose temperature is T (in degrees kelvin) and k is a positive constant. Estimate the percentage change in R that results from a 2% increase in T.

---

*You may wish to begin your research by consulting such textbooks as Elaine N. Marieb, *Human Anatomy and Physiology,* 2nd ed., Redwood City, CA: The Benjamin/Cummings Publishing Co., 1992, and Kent M. Van De Graaf and Stuart Ira Fox, *Concepts of Human Anatomy and Physiology,* 3rd ed., Dubuque, IA: Wm. C. Brown Publishers, 1992.

EXPLORE!

Store into Y1 the function $f(x) = x^3 - x^2 - 1$ and graph using a decimal window to observe that there is a root between X = 1 and 2. Now type Y2 = nDeriv(Y1, X, X). On a clear homescreen, store the starting value 1 into X, 1 → X, and type X − Y1(X)/Y2(X) → X. Next press **ENTER** successively and observe the sequence of resulting values. Note how many iterations were needed so that two consecutive approximations agree to four decimal places. Repeat this process using X = −2. Note whether a stable result occurs and how many iterations are required.

Newton's Method ■ Tangent line approximations can be used in a variety of ways. *Newton's method* for approximating the roots of an equation $f(x) = 0$ is based on the idea that if we start with a guess x_0 that is close to an actual root c, we can often obtain an improved estimate by finding the x intercept x_1 of the tangent line to the curve $y = f(x)$ at $x = x_0$ (see the figure). The process can then be repeated until a desired degree of accuracy is attained. In practice, it is usually easier and faster to use the graphing utility **ZOOM** and **TRACE** features of your graphing utility to find roots, but the ideas behind Newton's method are still important. Exercises 33 through 37 involve Newton's method.

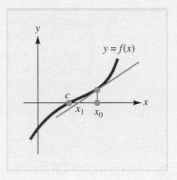

33. Show that when Newton's method is applied repeatedly, the nth approximation is obtained from the $(n − 1)$st approximation by the formula

$$x_n = x_{n-1} - \frac{f(x_{n-1})}{f'(x_{n-1})} \quad n = 1, 2, 3, \ldots$$

[*Hint:* First find x_1 as the x intercept of the tangent line to $y = f(x)$ at $x = x_0$. Then use x_1 to find x_2 in the same way.]

 34. Let $f(x) = x^3 - x^2 - 1$.
   a. Use your graphing utility to graph $f(x)$. Note that there is only one root located between 1 and 2. Use **ZOOM** and **TRACE** or other utility features to find the root.
   b. Using $x_0 = 1$, estimate the root by applying Newton's method until two consecutive approximations agree to four decimal places.
   c. Take the root you found graphically in part (a) and the root you found by Newton's method in part (b) and substitute each into the equation $f(x) = 0$. Which is more accurate?

35. Let $f(x) = x^4 - 4x^3 + 10$. Use your graphing utility to graph $f(x)$. Note that there are two real roots of the equation $f(x) = 0$. Estimate each root using Newton's method, and then check your results using **ZOOM** and **TRACE** or other utility features.

36. The ancient Babylonians (circa 1700 B.C.) approximated $\sqrt{N}$ by applying the formula

$$x_{n+1} = \frac{1}{2}\left(x_n + \frac{N}{x_n}\right) \quad \text{for } n = 1, 2, 3, \ldots$$

a. Show that this formula can be derived from the formula for Newton's method in Exercise 33, and then use it to estimate $\sqrt{1{,}265}$. Repeat the formula until two consecutive approximations agree to four decimal places. Use your calculator to check your result.

 b. The spy (Exercise 69, Section 2.2) wakes up one morning in Babylonia and finds that his calculator has been stolen. Create a spy story problem based on using the ancient formula to compute a square root.

37. Sometimes Newton's method fails no matter what initial value x_0 is chosen (unless you are lucky enough to choose the root itself). Let $f(x) = \sqrt[3]{x}$ and choose x_0 arbitrarily ($x_0 \neq 0$).

a. Show that $x_{n+1} = -2x_n$ for $n = 0, 1, 2, \ldots$ so that the successive guesses generated by Newton's method are $x_0, -2x_0, 4x_0, \ldots$

b. Use your graphing utility to graph $f(x)$, and use an appropriate utility to draw the tangent lines to the graph of $y = f(x)$ at the points that correspond to $x_0, -2x_0, 4x_0, \ldots$ Why do these numbers fail to estimate a root of $f(x) = 0$?

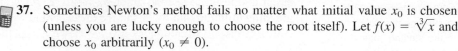

SECTION 2.6   Implicit Differentiation and Related Rates

Learning Objectives

1. Use implicit differentiation to find slope and rates of change.
2. Examine applied problems involving related rates of change.

The functions you have worked with so far have all been given by equations of the form $y = f(x)$ in which the dependent variable y on the left is given explicitly by an expression on the right involving the independent variable x. A function in this form is said to be in **explicit form.** For example, the functions

$$y = x^2 + 3x + 1 \qquad y = \frac{x^3 + 1}{2x - 3} \qquad \text{and} \qquad y = \sqrt{1 - x^2}$$

are all functions in explicit form.

Unfortunately, certain equations in x and y, such as

$$x^2 y^3 - 6 = 5y^2 + x \qquad \text{or} \qquad x^2 y + 2y^3 = 3x + 2y$$

either cannot be solved explicitly for y in terms of x or can be done so only with great effort. Sometimes, but not always, such an equation can be used to define one or more functions implicitly by placing appropriate restrictions on the variables. For instance, the equation $x^2 + y^2 = 16$ produces two implicitly defined functions

$$y = \sqrt{16 - x^2} \quad \text{for } y \geq 0 \qquad \text{and} \qquad y = -\sqrt{16 - x^2} \quad \text{for } y < 0$$

Note, however, that no real-valued function $y = f(x)$ satisfies the equation $x^2 + y^2 = -16$ (do you see why?).

Suppose you have an equation that defines y implicitly in terms of x and you want to find the derivative $\dfrac{dy}{dx}$. For instance, if the equation relates the cost C of a commodity to the level of production x, you may want to find the marginal cost $C'(x)$. If the equation cannot be easily solved to express y explicitly in terms of x, how should you proceed? The answer is provided by a method called **implicit differentiation,** which consists of differentiating both sides of the defining equation with respect to x and then solving algebraically for $\dfrac{dy}{dx}$. The procedure is illustrated in Example 2.6.1.

EXAMPLE 2.6.1   Finding the Derivative of an Implicitly Defined Function

Suppose $y = f(x)$ is a differentiable function of x that satisfies the equation $x^2y + y^2 = x^3$. Find the derivative $\dfrac{dy}{dx}$.

Solution

You are going to differentiate both sides of the given equation with respect to x. So that you won't forget that y is actually a function of x, temporarily replace y by $f(x)$ and begin by rewriting the equation as

$$x^2 f(x) + (f(x))^2 = x^3$$

Now differentiate both sides of this equation term by term with respect to x:

$$\frac{d}{dx}[x^2 f(x) + (f(x))^2] = \frac{d}{dx}[x^3] \qquad \begin{array}{l}\text{product rule and} \\ \text{general power rule}\end{array}$$

$$\underbrace{\left[x^2\frac{df}{dx} + f(x)\frac{d}{dx}(x^2)\right]}_{\frac{d}{dx}[x^2 f(x)]} + \underbrace{2f(x)\frac{df}{dx}}_{\frac{d}{dx}[(f(x))^2]} = \underbrace{3x^2}_{\frac{d}{dx}(x^3)}$$

Thus, we have

$$x^2\frac{df}{dx} + f(x)(2x) + 2f(x)\frac{df}{dx} = 3x^2 \qquad \begin{array}{l}\text{gather all } \frac{df}{dx} \text{ terms on} \\ \text{one side of the equation}\end{array}$$

$$x^2\frac{df}{dx} + 2f(x)\frac{df}{dx} = 3x^2 - 2xf(x) \qquad \text{factor out } \frac{df}{dx}$$

$$[x^2 + 2f(x)]\frac{df}{dx} = 3x^2 - 2xf(x) \qquad \text{solve for } \frac{df}{dx}$$

$$\frac{df}{dx} = \frac{3x^2 - 2xf(x)}{x^2 + 2f(x)}$$

Finally, replace $f(x)$ by y to get

$$\frac{dy}{dx} = \frac{3x^2 - 2xy}{x^2 + 2y}$$

NOTE From now on, in all examples and exercises involving implicit differentiation, you may assume that the given equation defines y implicitly as a differentiable function of x.

Temporarily replacing y by $f(x)$ as in Example 2.6.1 is a useful device for illustrating the implicit differentiation process, but as soon as you feel comfortable with the technique, try to leave out this unnecessary step and differentiate across the equation directly. Just keep in mind that y is really a function of x, and remember to use the chain rule when it is appropriate. ▪

Here is an outline of the implicit differentiation procedure.

Implicit Differentiation ▪ Suppose an equation defines y implicitly as a differentiable function of x. To find $\dfrac{dy}{dx}$:

1. Differentiate both sides of the equation with respect to x. Remember that y is really a function of x, and use the chain rule when differentiating terms containing y.

2. Solve the differentiated equation algebraically for $\dfrac{dy}{dx}$ in terms of x and y.

Computing the Slope of a Tangent Line by Implicit Differentiation

In Examples 2.6.2 and 2.6.3, you will see how to use implicit differentiation to find the slope of a tangent line.

EXAMPLE 2.6.2 Finding Slope for an Implicit Function

Find the slope of the tangent line to the circle $x^2 + y^2 = 25$ at the point $(3, 4)$. What is the slope at the point $(3, -4)$?

Solution

Differentiating both sides of the equation $x^2 + y^2 = 25$ with respect to x, you get

$$2x + 2y\frac{dy}{dx} = 0$$

$$\frac{dy}{dx} = -\frac{x}{y}$$

The slope at $(3, 4)$ is the value of the derivative when $x = 3$ and $y = 4$:

$$\left.\frac{dy}{dx}\right|_{(3,\,4)} = \left.-\frac{x}{y}\right|_{\substack{x=3 \\ y=4}} = -\frac{3}{4}$$

Similarly, the slope at $(3, -4)$ is the value of $\dfrac{dy}{dx}$ when $x = 3$ and $y = -4$:

$$\left.\frac{dy}{dx}\right|_{(3,\,-4)} = \left.-\frac{x}{y}\right|_{\substack{x=3 \\ y=-4}} = -\frac{3}{-4} = \frac{3}{4}$$

The graph of the circle is shown in Figure 2.16 together with the tangent lines at $(3, 4)$ and $(3, -4)$.

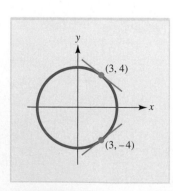

FIGURE 2.16 The graph of the circle $x^2 + y^2 = 25$.

EXAMPLE 2.6.3 Finding Horizontal Tangent Lines Using Implicit Differentiation

Find all points on the graph of the equation $x^2 - y^2 = 2x + 4y$ where the tangent line is horizontal. Does the graph have any vertical tangents?

Solution

Differentiate both sides of the given equation with respect to x to get

$$2x - 2y\frac{dy}{dx} = 2 + 4\frac{dy}{dx}$$

$$\frac{dy}{dx} = \frac{2x - 2}{4 + 2y}$$

There will be a horizontal tangent at each point on the graph where the slope is zero, that is, where the *numerator* $2x - 2$ of $\dfrac{dy}{dx}$ is zero:

$$2x - 2 = 0$$

$$x = 1$$

To find the corresponding value of y, substitute $x = 1$ into the given equation and solve using the quadratic formula (or your calculator):

$$1 - y^2 = 2(1) + 4y$$

$$y^2 + 4y + 1 = 0$$

$$y = -0.27, -3.73$$

Thus, the given graph has horizontal tangents at the points $(1, -0.27)$ and $(1, -3.73)$.

Since the slope of a vertical line is undefined, the given graph can have a vertical tangent only where the *denominator* $4 + 2y$ of $\dfrac{dy}{dx}$ is zero:

$$4 + 2y = 0$$

$$y = -2$$

To find the corresponding value of x, substitute $y = -2$ into the given equation:

$$x^2 - (-2)^2 = 2x + 4(-2)$$

$$x^2 - 2x + 4 = 0$$

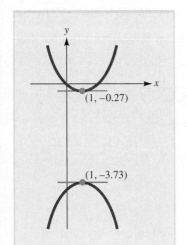

FIGURE 2.17 The graph of the equation $x^2 - y^2 = 2x + 4y$.

But this quadratic equation has no real solutions, which, in turn, means that the given graph has no vertical tangents. The graph is shown in Figure 2.17.

Application to Economics

Implicit differentiation is used in economics in both practical and theoretical work. In Section 4.3, it will be used to derive certain theoretical relationships. A more practical application of implicit differentiation is given in Example 2.6.4, which is a preview of the discussion of level curves of a function of two variables given in Section 7.1.

EXAMPLE 2.6.4 Using Marginal Analysis in Labor Management

Daiyu manages a factory whose daily output is $Q = 2x^3 + x^2y + y^3$ units, where x is the number of hours of skilled labor used and y is the number of hours of unskilled labor. The current labor force consists of 30 hours of skilled labor and 20 hours of unskilled labor. Daiyu wants to increase the skilled labor level by 1 hour without affecting the daily output. Show how she can use calculus to estimate the change in unskilled labor required to carry out this policy.

Solution

The current level of output is the value of Q when $x = 30$ and $y = 20$. That is,

$$Q = 2(30)^3 + (30)^2(20) + (20)^3 = 80,000 \text{ units}$$

If output is to be maintained at this level, the relationship between skilled labor x and unskilled labor y is given by the equation

$$80,000 = 2x^3 + x^2y + y^3$$

which defines y implicitly as a function of x.

The goal is to estimate the change in y that corresponds to a 1-unit increase in x when x and y are related by this equation. As you saw in Section 2.5, the change in y caused by a 1-unit increase in x can be approximated by the derivative $\dfrac{dy}{dx}$. To find this derivative, use implicit differentiation. (Remember that the derivative of the constant 80,000 on the left-hand side is zero.)

$$0 = 6x^2 + x^2\frac{dy}{dx} + y\frac{d}{dx}(x^2) + 3y^2\frac{dy}{dx}$$

$$0 = 6x^2 + x^2\frac{dy}{dx} + 2xy + 3y^2\frac{dy}{dx}$$

$$-(x^2 + 3y^2)\frac{dy}{dx} = 6x^2 + 2xy$$

$$\frac{dy}{dx} = -\frac{6x^2 + 2xy}{x^2 + 3y^2}$$

Now evaluate this derivative when $x = 30$ and $y = 20$ to conclude that

$$\text{Change in } y \approx \frac{dy}{dx}\bigg|_{\substack{x=30 \\ y=20}} = -\frac{6(30)^2 + 2(30)(20)}{(30)^2 + 3(20)^2} \approx -3.14 \text{ hours}$$

That is, to maintain the current level of output, unskilled labor should be decreased by approximately 3.14 hours to offset a 1-hour increase in skilled labor.

NOTE In general, if $Q(x, y)$ gives the production corresponding to x units of one input and y units of another, then an equation of the form $Q(x, y) = C$ for constant C is called an **isoquant.** Such equations are used by economists to explore the different combinations of inputs x and y that result in the same level of production. In this context, the rate $\dfrac{dy}{dx}$, often found by implicit differentiation as in Example 2.6.4, is called the **marginal rate of technical substitution** (MRTS). ■

Related Rates　In certain practical problems, x and y are related by an equation and can be regarded as functions of a third variable t, which often represents time. Then implicit differentiation can be used to relate $\dfrac{dx}{dt}$ to $\dfrac{dy}{dt}$. This kind of problem is said to involve **related rates.** Here is a general procedure for analyzing related rates problems.

A Procedure for Solving Related Rates Problems

1. Draw a figure (if appropriate) and assign variables.
2. Find a formula relating the variables.
3. Use implicit differentiation to find how the rates are related.
4. Substitute any given numerical information into the equation in step 3 to find the desired rate of change.

Examples 2.6.5 through 2.6.8 illustrate various ways that related rates appear in applications.

EXAMPLE 2.6.5　Finding a Related Rate of Cost

The manager of a company determines that when q hundred units of a particular commodity are produced, the total cost of production is C thousand dollars, where $C^2 - 3q^3 = 4{,}275$. When 1,500 units are being produced, the level of production is increasing at the rate of 20 units per week. What is the total cost at this time, and at what rate is it changing?

Solution

We want to find $\dfrac{dC}{dt}$ when $q = 15$ (1,500 units) and $\dfrac{dq}{dt} = 0.2$ (20 units per week with q measured in hundreds of units). Differentiating the equation $C^2 - 3q^3 = 4{,}275$ implicitly with respect to time, we get

$$2C\frac{dC}{dt} - 3\left[3q^2\frac{dq}{dt}\right] = 0$$

so that

$$2C\frac{dC}{dt} = 9q^2\frac{dq}{dt}$$

and

$$\frac{dC}{dt} = \frac{9q^2}{2C}\frac{dq}{dt}$$

When $q = 15$, the cost C satisfies

$$C^2 - 3(15)^3 = 4{,}275$$
$$C^2 = 4{,}275 + 3(15)^3 = 14{,}400$$
$$C = 120$$

and by substituting $q = 15$, $C = 120$, and $\dfrac{dq}{dt} = 0.2$ into the formula for $\dfrac{dC}{dt}$, we obtain

$$\frac{dC}{dt} = \left[\frac{9(15)^2}{2(120)}\right](0.2) = 1.6875$$

thousand dollars ($1,687.50) per week. To summarize, the cost of producing 1,500 units is $120,000 ($C = 120$) and at this level of production, total cost is increasing at the rate of $1,687.50 per week.

EXAMPLE 2.6.6 Using Related Rates to Study an Oil Spill

A storm at sea has damaged an oil rig. Oil spills from the rupture at the constant rate of 60 ft³/min, forming a slick that is roughly circular in shape and 3 inches thick.

  a. How fast is the radius of the slick increasing when the radius is 70 feet?

  b. Suppose the rupture is repaired in such a way that the flow is shut off instantaneously. If the radius of the slick is increasing at the rate of 0.2 ft/min when the flow stops, what is the total volume of oil that spilled onto the sea?

Solution

We can think of the slick as a cylinder of oil of radius r feet and thickness $h = \dfrac{3}{12} = 0.25$ feet. Such a cylinder will have volume

$$V = \pi r^2 h = 0.25\pi r^2 \text{ ft}^3$$

Differentiating implicitly in this equation with respect to time t, we get

$$\frac{dV}{dt} = 0.25\pi\left(2r\frac{dr}{dt}\right) = 0.5\pi r\frac{dr}{dt}$$

and since $\dfrac{dV}{dt} = 60$ at all times, we obtain the rate relationship

$$60 = 0.5\pi r\frac{dr}{dt}$$

  a. We want to find $\dfrac{dr}{dt}$ when $r = 70$. Substituting into the rate relationship we have just obtained, we find that

$$60 = 0.5\pi(70)\frac{dr}{dt}$$

so that

$$\frac{dr}{dt} = \frac{60}{(0.5)\pi(70)} \approx 0.55$$

Thus, when the radius is 70 feet, it is increasing at about 0.55 ft/min.

b. We can compute the total volume of oil in the spill if we know the radius of the slick at the instant the flow stops. Since $\dfrac{dr}{dt} = 0.2$ at that instant, we have

$$60 = 0.5\pi r(0.2)$$

and the radius is

$$r = \frac{60}{0.5\pi(0.2)} \approx 191 \text{ feet}$$

Therefore, the total amount of oil spilled is

$$V = 0.25\pi(191)^2 \approx 28{,}652 \text{ ft}^3$$

(about 214,332 gallons).

EXAMPLE 2.6.7 Using Related Rates to Study a Fish Population

A lake is polluted by waste from a plant located on its shore. Ecologists determine that when the level of pollutant is x parts per million (ppm), there will be F fish of a certain species in the lake, where

$$F = \frac{32{,}000}{3 + \sqrt{x}}$$

When there are 4,000 fish left in the lake, the pollution is increasing at the rate of 1.4 ppm/year. At what rate is the fish population changing at this time?

Solution

We want to find $\dfrac{dF}{dt}$ when $F = 4{,}000$ and $\dfrac{dx}{dt} = 1.4$. When there are 4,000 fish in the lake, the level of pollution x satisfies

$$4{,}000 = \frac{32{,}000}{3 + \sqrt{x}}$$
$$4{,}000(3 + \sqrt{x}) = 32{,}000$$
$$3 + \sqrt{x} = 8$$
$$\sqrt{x} = 5$$
$$x = 25$$

We find that

$$\frac{dF}{dx} = \frac{32{,}000(-1)}{(3 + \sqrt{x})^2}\left(\frac{1}{2}\frac{1}{\sqrt{x}}\right) = \frac{-16{,}000}{\sqrt{x}(3 + \sqrt{x})^2}$$

and according to the chain rule

$$\frac{dF}{dt} = \frac{dF}{dx}\frac{dx}{dt} = \left[\frac{-16{,}000}{\sqrt{x}(3 + \sqrt{x})^2}\right]\frac{dx}{dt}$$

Substituting $x = 25$ and $\dfrac{dx}{dt} = 1.4$, we find that

$$\frac{dF}{dt} = \left[\frac{-16{,}000}{\sqrt{25}(3 + \sqrt{25})^2} \right](1.4) = -70$$

so the fish population is decreasing by 70 fish per year.

EXAMPLE 2.6.8 Using Related Rates to Study Supply

When the price of a certain commodity is p dollars per unit, the manufacturer is willing to supply x thousand units, where

$$x^2 - 2x\sqrt{p} - p^2 = 31$$

How fast is the supply changing when the price is \$9 per unit and is increasing at the rate of 20 cents per week?

Solution

We know that when $p = 9$, $\dfrac{dp}{dt} = 0.20$. We are asked to find $\dfrac{dx}{dt}$ at this time. First, note that when $p = 9$, we have

$$x^2 - 2x\sqrt{9} - 9^2 = 31$$
$$x^2 - 6x - 112 = 0$$
$$(x + 8)(x - 14) = 0$$
$$x = 14 \quad (x = -8 \text{ has no practical value})$$

Next, we differentiate both sides of the supply equation implicitly with respect to time to obtain

$$2x\frac{dx}{dt} - 2\left[\left(\frac{dx}{dt}\right)\sqrt{p} + x\left(\frac{1}{2}\frac{1}{\sqrt{p}}\frac{dp}{dt}\right) \right] - 2p\frac{dp}{dt} = 0$$

Finally, by substituting $x = 14$, $p = 9$, and $\dfrac{dp}{dt} = 0.20$ into this rate equation and then solving for the required rate $\dfrac{dx}{dt}$, we get

$$2(14)\frac{dx}{dt} - 2\left[\sqrt{9}\frac{dx}{dt} + 14\left(\frac{1}{2}\frac{1}{\sqrt{9}}\right)(0.20) \right] - 2(9)(0.20) = 0$$

$$[28 - 2(3)]\frac{dx}{dt} = 2(14)\left(\frac{1}{2\sqrt{9}}\right)(0.20) + 2(9)(0.20)$$

$$\frac{dx}{dt} = \frac{14\left(\dfrac{1}{3}\right)(0.20) + 18(0.20)}{22}$$

$$\approx 0.206$$

Since the supply is given in terms of thousands of units, it follows that the supply is increasing at the rate of $0.206(1{,}000) = 206$ units per week.

EXERCISES ▪ 2.6

In Exercises 1 through 8, find $\dfrac{dy}{dx}$ in two ways:

　　(a) by implicit differentiation
　　(b) by differentiating an explicit formula for y.
In each case, show that the two answers are the same.

1. $2x + 3y = 7$

2. $5x - 7y = 3$

3. $x^3 - y^2 = 5$

4. $x^2 + y^3 = 12$

5. $xy = 4$

6. $x + \dfrac{1}{y} = 5$

7. $xy + 2y = 3$

8. $xy + 2y = x^2$

In Exercises 9 through 22, find $\dfrac{dy}{dx}$ by implicit differentiation.

9. $x^2 + y^2 = 25$

10. $x^2 + y = x^3 + y^2$

11. $x^3 + y^3 = xy$

12. $5x - x^2y^3 = 2y$

13. $y^2 + 2xy^2 - 3x + 1 = 0$

14. $\dfrac{1}{x} + \dfrac{1}{y} = 1$

15. $\sqrt{x} + \sqrt{y} = 1$

16. $\sqrt{2x} + y^2 = 4$

17. $xy - x = y + 2$

18. $y^2 + 3xy - 4x^2 = 9$

19. $(2x + y)^3 = x$

20. $(x - 2y)^2 = y$

21. $(x^2 + 3y^2)^5 = 2xy$

22. $(3xy^2 + 1)^4 = 2x - 3y$

In Exercises 23 through 30, find the equation of the tangent line to the given curve at the specified point.

23. $x^2 = y^3$; $(8, 4)$

24. $x^2 - y^3 = 2x$; $(1, -1)$

25. $xy = 2$; $(2, 1)$

26. $\dfrac{1}{x} - \dfrac{1}{y} = 2$; $\left(\dfrac{1}{4}, \dfrac{1}{2}\right)$

27. $xy^2 - x^2y = 6$; $(2, -1)$

28. $x^2y^3 - 2xy = 6x + y + 1$; $(0, -1)$

29. $(1 - x + y)^3 = x + 7$; $(1, 2)$

30. $(x^2 + 2y)^3 = 2xy^2 + 64$; $(0, 2)$

In Exercises 31 through 36, find all points (both coordinates) on the given curve where the tangent line is (a) horizontal and (b) vertical.

31. $x + y^2 = 9$

32. $x^2 + xy + y = 3$

33. $xy = 16y^2 + x$

34. $\dfrac{y}{x} - \dfrac{x}{y} = 5$

35. $x^2 + xy + y^2 = 3$

36. $x^2 - xy + y^2 = 3$

In Exercises 37 and 38, use implicit differentiation to find the second derivative $\dfrac{d^2y}{dx^2}$.

37. $x^2 + 3y^2 = 5$

38. $xy + y^2 - 1$

BUSINESS AND ECONOMICS APPLIED PROBLEMS

39. **MANUFACTURING** The output at a certain plant is $Q = 0.08x^2 + 0.12xy + 0.03y^2$ units per day, where x is the number of hours of skilled labor used and y is the number of hours of unskilled labor used. Currently, 80 hours of skilled labor and 200 hours of unskilled labor are used each day. Use calculus to estimate the change in unskilled labor that should be made to offset a 1-hour increase in skilled labor so that output will be maintained at its current level.

40. **MANUFACTURING** The output of a certain plant is $Q = 0.06x^2 + 0.14xy + 0.05y^2$ units per day, where x is the number of hours of skilled labor used and y is the number of hours of unskilled labor used. Currently, 60 hours of skilled

labor and 300 hours of unskilled labor are used each day. Use calculus to estimate the change in unskilled labor that should be made to offset a 1-hour increase in skilled labor so that output will be maintained at its current level.

41. SUPPLY RATE   When the price of a certain commodity is p dollars per unit, the manufacturer is willing to supply x hundred units, where

$$3p^2 - x^2 = 12$$

How fast is the supply changing when the price is \$4 per unit and is increasing at the rate of 87 cents per month?

42. DEMAND RATE   When the price of a certain commodity is p dollars per unit, customers demand x hundred units of the commodity, where

$$x^2 + 3px + p^2 = 79$$

How fast is the demand x changing with respect to time when the price is \$5 per unit and is decreasing at the rate of 30 cents per month?

43. CONSUMER DEMAND   When electric toasters are sold for p dollars apiece, local consumers will buy $D(p) = \dfrac{32,670}{2p + 1}$ toasters. It is estimated that t months from now, the unit price of the toasters will be $p(t) = 0.04t^{3/2} + 44$ dollars. Compute the rate of change of the monthly demand for toasters with respect to time 25 months from now. Will the demand be increasing or decreasing?

44. MANUFACTURING   At a certain factory, output Q is related to inputs u and v by the equation

$$Q = 3u^2 + \frac{2u + 3v}{(u + v)^2}$$

If the current levels of input are $u = 10$ and $v = 25$, use calculus to estimate the change in input v that should be made to offset a decrease of 0.7 unit in input u so that output will be maintained at its current level.

45. PRODUCTION   At a certain factory, output is given by $Q = 60K^{1/3}L^{2/3}$ units, where K is the capital investment (in thousands of dollars) and L is the size of the labor force, measured in worker-hours. If output is kept constant, at what rate is capital investment changing at a time when $K = 8$, $L = 1,000$, and L is increasing at the rate of 25 worker-hours per week?

[*Note:* Output functions of the general form $Q = AK^{\alpha}L^{1-\alpha}$, where A and α are constants with $0 \le \alpha \le 1$, are called **Cobb-Douglas production functions.** Such functions appear in examples and exercises throughout this text, especially in Chapter 7.]

46. MANUFACTURING   At a certain factory, output Q is related to inputs x and y by the equation

$$Q = 2x^3 + 3x^2y^2 + (1 + y)^3$$

If the current levels of input are $x = 30$ and $y = 20$, use calculus to estimate the change in input y that should be made to offset a decrease of 0.8 unit in input x so that output will be maintained at its current level.

LIFE AND SOCIAL SCIENCE APPLIED PROBLEMS

47. MEDICINE   A tiny spherical balloon is inserted into a clogged artery and is inflated at the rate of 0.002π mm^3/min. How fast is the radius of the balloon growing when the radius is $R = 0.005$ mm?

$$\left[Note\text{: A sphere of radius } R \text{ has volume } V = \frac{4}{3}\pi R^3. \right]$$

48. POLLUTION CONTROL   An environmental study for a certain community indicates that there will be $Q(p) = p^2 + 4p + 900$ units of a harmful pollutant in the air when the population is p thousand people. If the population is currently 50,000 and is increasing at the rate of 1,500 per year, at what rate is the level of pollution increasing?

49. GROWTH OF A TUMOR   A tumor is modeled as being roughly spherical, with radius R. If the radius of the tumor is currently $R = 0.54$ cm and is increasing at the rate of 0.13 cm per month, what is the corresponding rate of change of the volume $V = \dfrac{4}{3}\pi R^3$?

50. WATER POLLUTION   A circular oil slick spreads in such a way that its radius is increasing at the rate of 20 ft/hr. How fast is the area of the slick changing when the radius is 200 feet?

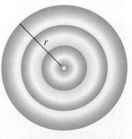

EXERCISE 50

51. METABOLIC RATE The *basal metabolic rate* is the rate of heat produced by an animal per unit time. Observations indicate that the basal metabolic rate of a warm-blooded animal of mass w kilograms (kg) is given by

$$M = 70w^{3/4} \text{ kilocalories per day}$$

  a. Find the rate of change of the metabolic rate of an 80-kg cougar that is gaining mass at the rate of 0.8 kg per day.

  b. Find the rate of change of the metabolic rate of a 50-kg ostrich that is losing mass at the rate of 0.5 kg per day.

52. SPEED OF A LIZARD Herpetologists have proposed using the formula $s = 1.1w^{0.2}$ to estimate the maximum sprinting speed s (meters per second) of a lizard of mass w (grams). At what rate is the maximum sprinting speed of an 11-gram lizard increasing if the lizard is growing at the rate of 0.02 grams per day?

53. BLOOD FLOW One of Poiseuille's laws (see Exercise 73, Section 1.1) says that the speed of blood flowing under constant pressure in a blood vessel at a distance r from the center of the vessel is given by

$$v = \frac{K}{L}(R^2 - r^2)$$

where K is a positive constant, R is the radius of the vessel, and L is the length of the vessel.* Suppose the radius R and length L of the vessel change with time in such a way that the speed of blood flowing at the center is unaffected; that is, v does not change with time. Show that in this case, the relative rate of change of L with respect to time must be twice the relative rate of change of R.

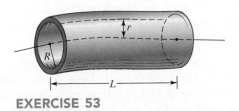

EXERCISE 53

*E. Batschelet, *Introduction to Mathematics for Life Scientists*, 3rd ed., New York: Springer-Verlag, 1979, pp. 102–103.

54. POLLUTION CONTROL It is estimated that t years from now, the population of a certain suburban community will be $p(t) = 10 - \dfrac{20}{(t + 1)^2}$ thousand. An environmental study indicates that the average daily level of carbon monoxide in the air will be $c(p) = 0.8\sqrt{p^2 + p + 139}$ units when the population is p thousand. At what percentage rate will the level of carbon monoxide be changing with respect to time 1 year from now?

55. BLOOD FLOW Physiologists have observed that the flow of blood from an artery into a small capillary is given by the formula

$$F = kD^2\sqrt{A - C} \quad (\text{cm}^3/\text{sec})$$

where D is the diameter of the capillary, A is the pressure in the artery, C is the pressure in the capillary, and k is a positive constant.

  a. By how much is the flow of blood F changing with respect to pressure C in the capillary if A and D are kept constant? Does the flow increase or decrease with increasing C?

  b. What is the percentage rate of change of flow F with respect to A if C and D are kept constant?

MISCELLANEOUS PROBLEMS

56. LUMBER PRODUCTION To estimate the amount of wood in the trunk of a tree, it is reasonable to assume that the trunk is a cutoff cone (see the figure).

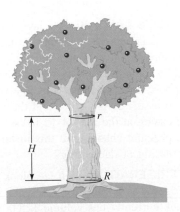

EXERCISE 56

If the upper radius of the trunk is r, the lower radius is R, and the height is H, the volume of the wood is given by

$$V = \frac{\pi}{3}H(R^2 + rR + r^2)$$

Suppose r, R, and H are increasing at the respective rates of 4 in/yr, 5 in/yr, and 9 in/yr. At what rate is V increasing at the time when $r = 2$ feet, $R = 3$ feet, and $H = 15$ feet?

57. A 6-foot-tall man walks at the rate of 4 ft/sec away from the base of a street light 12 feet above the ground. At what rate is the length of his shadow changing when he is 20 feet away from the base of the light?

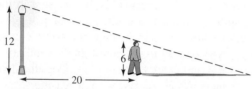

EXERCISE 57

58. CHEMISTRY   In an *adiabatic* chemical process, there is no net change (gain or loss) of heat. Suppose a container of oxygen is subjected to such a process. Then if the pressure on the oxygen is P and its volume is V, it can be shown that $PV^{1.4} = C$, where C is a constant. At a certain time, $V = 5$ m^3, $P = 0.6$ kg/m^2, and P is increasing at 0.23 kg/m^2 per sec. What is the rate of change of V? Is V increasing or decreasing?

59. BOYLE'S LAW   Boyle's law states that when gas is compressed at a constant temperature, the pressure P and volume V of a given sample satisfy the equation $PV = C$, where C is constant. Suppose that at a certain time the volume is 40 in^3, the pressure is 70 lb/in^2, and the volume is increasing at the rate of 12 in^3/sec. How fast is the pressure changing at this instant? Is it increasing or decreasing?

60. REFRIGERATION   An ice block used for refrigeration is modeled as a cube of side s. The block currently has volume 125,000 cm^3 and is melting at the rate of 1,000 cm^3 per hour.

a. What is the current length s of each side of the cube? At what rate is s currently changing with respect to time t?

b. What is the current rate of change of the surface area S of the block with respect to time? [*Note:* A cube of side s has volume $V = s^3$ and surface area $S = 6s^2$.]

61. Show that the tangent line to the curve

$$\frac{x^2}{a^2} + \frac{y^2}{b^2} = 1$$

at the point (x_0, y_0) is

$$\frac{x_0 x}{a^2} + \frac{y_0 y}{b^2} = 1$$

62. Consider the equation $x^2 + y^2 = 6y - 10$.

a. Show that there are no points (x, y) that satisfy this equation. [*Hint:* Complete the square.]

b. Show that by applying implicit differentiation to the given equation, you obtain

$$\frac{dy}{dx} = \frac{x}{3 - y}$$

The point of this exercise is to show that one must be careful when applying implicit differentiation. Just because it is possible to find a derivative formally by implicit differentiation does not mean that the derivative has any meaning.

63. Prove the power rule $\frac{d}{dx}[x^n] = nx^{n-1}$ for the case where $n = \frac{r}{s}$ is a rational number. [*Hint:* Note that if $y = x^{r/s}$, then $y^s = x^r$ and use implicit differentiation.]

64. Use a graphing utility to graph the curve $5x^2 - 2xy + 5y^2 = 8$. Draw the tangent line to the curve at $(1, 1)$. How many horizontal tangent lines does the curve have? Find the equation of each horizontal tangent.

65. Use a graphing utility to graph the curve $11x^2 + 4xy + 14y^2 = 21$. Draw the tangent line to the curve at $(-1, 1)$. How many horizontal tangent lines does the curve have? Find the equation of each horizontal tangent.

66. Answer the following questions about the curve $x^3 + y^3 = 3xy$ (called the **folium of Descartes**).
   a. Find the equation of each horizontal tangent to the curve.
   b. The curve intersects the line $y = x$ at exactly one point other than the origin. What is the equation of the tangent line at this point?

   c. Try to figure out a way to sketch the graph of the curve using your calculator.

 67. Use a graphing utility to graph the curve $x^2 + y^2 = \sqrt{x^2 + y^2} + x$. Find the equation of each horizontal tangent to the curve. (The curve is called a **cardioid**.)

Important Terms, Symbols, and Formulas

Secant line and tangent line   (105 and 106)
Average and instantaneous rates of change   (107)
Difference quotient

$$\frac{f(x + h) - f(x)}{h} \quad (108)$$

Derivative of $f(x)$: $f'(x) = \lim_{h \to 0} \dfrac{f(x + h) - f(x)}{h}$ (108)

Differentiation   (108)
Differentiable function   (108)
The derivative $f'(x_0)$ as slope of the tangent line to $y = f(x)$ at $(x_0, f(x_0))$ (109)
The derivative $f'(x_0)$ as the rate of change of $f(x)$ with respect to x when $x = x_0$ (109)

Derivative notation for $f(x)$: $f'(x)$ and $\dfrac{df}{dx}$ (111)

A differentiable function is continuous   (113)

The constant rule: $\dfrac{d}{dx}[c] = 0$ (119)

The power rule: $\dfrac{d}{dx}[x^n] = nx^{n-1}$ (120)

The constant multiple rule: $\dfrac{d}{dx}[cf(x)] = c\dfrac{df}{dx}$ (122)

The sum rule: $\dfrac{d}{dx}[f(x) + g(x)] = \dfrac{df}{dx} + \dfrac{dg}{dx}$ (122)

Relative rate of change of $Q(x)$: $\dfrac{Q'(x)}{Q(x)}$ (124)

Percentage rate of change of $Q(x)$: $\dfrac{100Q'(x)}{Q(x)}$ (124)

Rectilinear motion: position $s(t)$
   velocity $v(t) = s'(t)$
   acceleration $a(t) = v'(t)$ (126)
The motion of a projectile:

$$\text{height } H(t) = -\frac{1}{2}gt^2 + V_0t + H_0 \quad (127)$$

The product rule:

$$\frac{d}{dx}[f(x)g(x)] = f(x)\frac{dg}{dx} + g(x)\frac{df}{dx} \quad (133)$$

The quotient rule: For $g(x) \neq 0$

$$\frac{d}{dx}\left[\frac{f(x)}{g(x)}\right] = \frac{g(x)\dfrac{df}{dx} - f(x)\dfrac{dg}{dx}}{g^2(x)} \quad (135)$$

The second derivative of $f(x)$: $f''(x) = [f'(x)]' = \dfrac{d^2f}{dx^2}$
   gives the rate of change of $f'(x)$ (138)
Notation for the nth derivative of $f(x)$:

$$f^{(n)}(x) = \frac{d^nf}{dx^n} \quad (141)$$

The chain rule for $y = f(u(x))$: $\dfrac{dy}{dx} = \dfrac{dy}{du}\dfrac{du}{dx}$
   or $[f(u(x))]' = f'(u)\,u'(x)$ (147)
The general power rule:

$$\frac{d}{dx}[h(x)]^n = n[h(x)]^{n-1}\frac{dh}{dx} \quad (151)$$

Marginal cost $C'(x_0)$ estimates $C(x_0 + 1) - C(x_0)$, the extra cost of producing the $(x_0 + 1)^{st}$ unit   (161)
Marginal revenue $R'(x_0)$ estimates $R(x_0 + 1) - R(x_0)$, the extra revenue from producing the $(x_0 + 1)^{st}$ unit   (162)
Marginal profit: $P'(x_0)$ estimates $P(x_0 + 1) - P(x_0)$, the extra profit from producing the $(x_0 + 1)^{st}$ unit   (162)
Approximation by increments:
   $f(x_0 + \Delta x) \approx f(x_0) + f'(x_0)\Delta x$ (164)
Propagated error   (165)
Differential of $y = f(x)$ is $dy = f'(x)\,dx$ (167)
Implicit differentiation   (173–174)
Related rates   (177)

Checkup for Chapter 2

1. In each case, find the derivative $\dfrac{dy}{dx}$.

  a. $y = 3x^4 - 4\sqrt{x} + \dfrac{5}{x^2} - 7$

  b. $y = (3x^3 - x + 1)(4 - x^2)$

  c. $y = \dfrac{5x^2 - 3x + 2}{1 - 2x}$

  d. $y = (3 - 4x + 3x^2)^{3/2}$

2. Find the second derivative of the function $f(t) = t(2t + 1)^2$.

3. Find an equation for the tangent line to the curve $y = x^2 - 2x + 1$ at the point where $x = -1$.

4. Find the rate of change of the function $f(x) = \dfrac{x + 1}{1 - 5x}$ with respect to x when $x = 1$.

5. PROPERTY TAX Records indicate that x years after the year 2010, the average property tax on a four-bedroom home in a suburb of a major city was $T(x) = 3x^2 + 40x + 1{,}800$ dollars.

  a. At what rate is the property tax increasing with respect to time in 2013?

  b. At what percentage rate is the property tax increasing in 2013?

6. MOTION ON A LINE An object moves along a line in such a way that its position at time t is given by $s(t) = 2t^3 - 3t^2 + 2$ for $t \geq 0$.

  a. Find the velocity $v(t)$ and acceleration $a(t)$ of the object.

  b. When is the object stationary? When is it advancing? Retreating?

  c. What is the total distance traveled by the object for $0 \leq t \leq 2$?

7. PRODUCTION COST Suppose the cost of producing x hundred units of a particular commodity is $C(x) = 0.04x^2 + 5x + 73$ thousand dollars. Use marginal cost to estimate the cost of producing the 410th unit. What is the actual cost of producing the 410th unit?

8. INDUSTRIAL OUTPUT At a certain factory, the daily output is $Q = 500L^{3/4}$ units, where L denotes the size of the labor force in worker-hours. Currently, 2,401 worker-hours of labor are used each day. Use calculus (increments) to estimate the effect on output of increasing the size of the labor force by 200 worker-hours from its current level.

9. PEDIATRIC MEASUREMENT Pediatricians use the formula $S = 0.2029w^{0.425}$ to estimate the surface area S (m^2) of a child 1 meter tall who weighs w kilograms (kg). A particular child weighs 30 kg and is gaining weight at the rate of 0.13 kg per week while remaining 1 meter tall. At what rate is this child's surface area changing?

10. GROWTH OF A TUMOR A cancerous tumor is modeled as a sphere of radius r cm.

  a. At what rate is the volume $V = \dfrac{4}{3}\pi r^3$ changing with respect to r when $r = 0.75$ cm?

  b. Estimate the percentage error that can be allowed in the measurement of the radius r to ensure that there will be no more than an 8% error in the calculation of volume.

Review Exercises

In Exercises 1 and 2, use the definition of the derivative to find $f'(x)$.

1. $f(x) = x^2 - 3x + 1$

2. $f(x) = \dfrac{1}{x - 2}$

In Exercises 3 through 13, find the derivative of the given function.

3. $f(x) = 6x^4 - 7x^3 + 2x + \sqrt{2}$

4. $f(x) = x^3 - \dfrac{1}{3x^5} + 2\sqrt{x} - \dfrac{3}{x} + \dfrac{1 - 2x}{x^3}$

5. $y = \dfrac{2 - x^2}{3x^2 + 1}$

6. $y = (x^3 + 2x - 7)(3 + x - x^2)$

7. $f(x) = (5x^4 - 3x^2 + 2x + 1)^{10}$

8. $f(x) = \sqrt{x^2 + 1}$

9. $y = \left(x + \dfrac{1}{x}\right)^2 - \dfrac{5}{\sqrt{3x}}$

10. $y = \left(\dfrac{x + 1}{1 - x}\right)^2$

11. $f(x) = (3x + 1)\sqrt{6x + 5}$

12. $f(x) = \dfrac{(3x + 1)^3}{(1 - 3x)^4}$

13. $y = \sqrt{\dfrac{1 - 2x}{3x + 2}}$

In Exercises 14 through 17, find an equation for the tangent line to the graph of the given function at the specified point.

14. $f(x) = x^2 - 3x + 2$; $x = 1$

15. $f(x) = \dfrac{4}{x - 3}$; $x = 1$

16. $f(x) = \dfrac{x}{x^2 + 1}$; $x = 0$

17. $f(x) = \sqrt{x^2 + 5}$; $x = -2$

18. In each of these cases, find the rate of change of $f(t)$ with respect to t at the given value of t.
   a. $f(t) = t^3 - 4t^2 + 5t\sqrt{t} - 5$ at $t = 4$
   b. $f(t) = \dfrac{2t^2 - 5}{1 - 3t}$ at $t = -1$

19. In each of these cases, find the rate of change of $f(t)$ with respect to t at the given value of t.
   a. $f(t) = t^3(t^2 - 1)$ at $t = 0$
   b. $f(t) = (t^2 - 3t + 6)^{1/2}$ at $t = 1$

20. In each of these cases, find the percentage rate of change of the function $f(t)$ with respect to t at the given value of t.
   a. $f(t) = t^2 - 3t + \sqrt{t}$ at $t = 4$
   b. $f(t) = \dfrac{t}{t - 3}$ at $t = 4$

21. In each of these cases, find the percentage rate of change of the function $f(t)$ with respect to t at the given value of t.
   a. $f(t) = t^2(3 - 2t)^3$ at $t = 1$
   b. $f(t) = \dfrac{1}{t + 1}$ at $t = 0$

22. Use the chain rule to find $\dfrac{dy}{dx}$.
   a. $y = 5u^2 + u - 1$; $u = 3x + 1$
   b. $y = \dfrac{1}{u^2}$; $u = 2x + 3$

23. Use the chain rule to find $\dfrac{dy}{dx}$.
   a. $y = (u + 1)^2$; $u = 1 - x$
   b. $y = \dfrac{1}{\sqrt{u}}$; $u = 2x + 1$

24. Use the chain rule to find $\dfrac{dy}{dx}$ for the given value of x.
   a. $y = u - u^2$; $u = x - 3$; for $x = 0$
   b. $y = \left(\dfrac{u - 1}{u + 1}\right)^{1/2}$, $u = \sqrt{x} - 1$; for $x = \dfrac{34}{9}$

25. Use the chain rule to find $\dfrac{dy}{dx}$ for the given value of x.
   a. $y = u^3 - 4u^2 + 5u + 2$; $u = x^2 + 1$; for $x = 1$
   b. $y = \sqrt{u}$, $u = x^2 + 2x - 4$; for $x = 2$

26. Find the second derivative:
   a. $f(x) = 6x^5 - 4x^3 + 5x^2 - 2x + \dfrac{1}{x}$
   b. $z = \dfrac{2}{1 + x^2}$
   c. $y = (3x^2 + 2)^4$

27. Find the second derivative:
   a. $f(x) = 4x^3 - 3x$
   b. $f(x) = 2x(x + 4)^3$
   c. $f(x) = \dfrac{x - 1}{(x + 1)^2}$

28. Find $\dfrac{dy}{dx}$ by implicit differentiation.
   a. $5x + 3y = 12$
   b. $(2x + 3y)^5 = x + 1$

29. Find $\dfrac{dy}{dx}$ by implicit differentiation.
   a. $x^2y = 1$
   b. $(1 - 2xy^3)^5 = x + 4y$

30. Use implicit differentiation to find the slope of the line that is tangent to the given curve at the specified point.

  a. $xy^3 = 8$; $(1, 2)$

  b. $x^2y - 2xy^3 + 6 = 2x + 2y$; $(0, 3)$

31. Use implicit differentiation to find the slope of the line that is tangent to the given curve at the specified point.

  a. $x^2 + 2y^3 = \dfrac{3}{xy}$; $(1, 1)$

  b. $y = \dfrac{x + y}{x - y}$; $(6, 2)$

32. Use implicit differentiation to find $\dfrac{d^2y}{dx^2}$ if $4x^2 + y^2 = 1$.

33. Use implicit differentiation to find $\dfrac{d^2y}{dx^2}$ if $3x^2 - 2y^2 = 6$.

34. A projectile is launched vertically upward from ground level with an initial velocity of 160 ft/sec.

  a. When will the projectile hit the ground?

  b. What is the impact velocity?

  c. When will the projectile reach its maximum height? What is the maximum height?

35. POPULATION GROWTH   Suppose that a 5-year projection of population trends suggests that t years from now, the population of a certain community will be P thousand, where

$$P(t) = -2t^3 + 9t^2 + 8t + 200$$

  a. At what rate will the population be growing 3 years from now?

  b. At what rate will the rate of population growth be changing with respect to time 3 years from now?

In Exercises 36 and 37, s(t) denotes the position of an object moving along a line.

  (a) Find the velocity and acceleration of the object, and describe its motion during the indicated time interval.

  (b) Compute the total distance traveled by the object during the indicated time interval.

36. $s(t) = 2t^3 - 21t^2 + 60t - 25$; $1 \le t \le 6$

37. $s(t) = \dfrac{2t + 1}{t^2 + 12}$; $0 \le t \le 4$

38. RAPID TRANSIT   After x weeks, the number of people using a new rapid transit system was approximately $N(x) = 6x^3 + 500x + 8{,}000$.

  a. At what rate was the use of the system changing with respect to time after 8 weeks?

  b. By how much did the use of the system change during the eighth week?

39. PRODUCTION   It is estimated that the weekly output at a certain plant is $Q(x) = 50x^2 + 9{,}000x$ units, where x is the number of workers employed at the plant. Currently there are 30 workers employed at the plant.

  a. Use calculus to estimate the change in the weekly output that will result from the addition of 1 worker to the force.

  b. Compute the actual change in output that will result from the addition of 1 worker.

40. POPULATION   It is projected that t months from now, the population of a certain town will be $P(t) = 3t + 5t^{3/2} + 6{,}000$. At what percentage rate will the population be changing with respect to time 4 months from now?

41. PRODUCTION   At a certain factory, the daily output is $Q(L) = 20{,}000L^{1/2}$ units, where L denotes the size of the labor force measured in worker-hours. Currently 900 worker-hours of labor are used each day. Use calculus to estimate the effect on output that will be produced if the labor force is cut to 885 worker-hours.

42. GROSS DOMESTIC PRODUCT   The gross domestic product of a certain country was $N(t) = t^2 + 6t + 300$ billion dollars t years after 2004. Use calculus to predict the percentage change in the GDP during the second quarter of 2012.

43. POLLUTION   The level of air pollution in a certain city is proportional to the square of the population. Use calculus to estimate the percentage by which the air pollution level will increase if the population increases by 5%.

44. AIDS EPIDEMIC   In its early phase, specifically the period 1984–1990, the AIDS epidemic could be modeled* by the cubic function

$$C(t) = -170.36t^3 + 1{,}707.5t^2 + 1{,}998.4t + 4{,}404.8$$

*Mortality and Morbidity Weekly Report, U.S. Centers for Disease Control, Vol. 40, No. 53, October 2, 1992.

for $0 \le t \le 6$, where C is the number of reported cases t years after the base year 1984.

a. Compute and interpret the derivative $C'(t)$.

b. At what rate was the epidemic spreading in the year 1984?

c. At what percentage rate was the epidemic spreading in 1984? In 1990?

45. POPULATION DENSITY   The formula $D = 36\,m^{-1.14}$ is sometimes used to determine the ideal population density D (individuals per square kilometer) for a large animal of mass m kilograms (kg).

a. What is the ideal population density for humans, assuming that a typical human weighs about 70 kg?

b. The area of the United States is about 9.2 million square kilometers. What would the population of the United States have to be for the population density to be ideal?

c. Consider an island of area 3,000 km². Two hundred animals of mass $m = 30$ kg are brought to the island, and t years later, the population is given by
$$P(t) = 0.43t^2 + 13.37t + 200$$
How long does it take for the ideal population density to be reached? At what rate is the population changing when the ideal density is attained?

46. BACTERIAL GROWTH   The population P of a bacterial colony t days after observation begins is modeled by the cubic function
$$P(t) = 1.035t^3 + 103.5t^2 + 6,900t + 230,000$$

a. Compute and interpret the derivative $P'(t)$.

b. At what rate is the population changing after 1 day? After 10 days?

c. What is the initial population of the colony? How long does it take for the population to double? At what rate is the population growing at the time it doubles?

47. PRODUCTION   The output at a certain factory is $Q(L) = 600L^{2/3}$ units, where L is the size of the labor force. The manufacturer wishes to increase output by 1%. Use calculus to estimate the percentage increase in labor that will be required.

48. PRODUCTION   The output Q at a certain factory is related to inputs x and y by the equation
$$Q = x^3 + 2xy^2 + 2y^3$$
If the current levels of input are $x = 10$ and $y = 20$, use calculus to estimate the change in input y that should be made to offset an increase of 0.5 in input x so that output will be maintained at its current level.

49. You measure the radius r of a circle to be 12 cm with an error no greater than 3%. Use calculus to estimate the error incurred by using this approximate value of r in the formula $A = \pi r^2$ to compute the area of the circle.

50. Estimate what will happen to the volume of a cube if the length of each side is decreased by 2%. Express your answer as a percentage.

51. PRODUCTION   The output at a certain factory is $Q = 600K^{1/2}L^{1/3}$ units, where K denotes the capital investment and L is the size of the labor force. Estimate the percentage increase in output that will result from a 2% increase in the size of the labor force if capital investment is not changed.

52. BLOOD FLOW   The speed of blood flowing along the central axis of a certain artery is $S(R) = 1.8 \times 10^5 R^2$ centimeters per second, where R is the radius of the artery. A medical researcher measures the radius of the artery to be 1.2×10^{-2} centimeter and makes an error of 5×10^{-4} centimeter. Estimate the amount by which the calculated value of the speed of the blood will differ from the true speed if the incorrect value of the radius is used in the formula.

53. SURFACE AREA OF A TUMOR   You measure the radius r of a spherical tumor to be 1.2 cm with an error no greater than 3%. Use calculus to estimate the error incurred by using this approximate value of r in the formula $S = 4\pi r^2$ to compute the surface area of the tumor.

54. CARDIOVASCULAR SYSTEM   One model of the cardiovascular system relates $V(t)$, the stroke volume of blood in the aorta at a time t during systole (the contraction phase), to the pressure $P(t)$ in the aorta during systole by the equation
$$V(t) = [C_1 + C_2 P(t)]\left(\frac{3t^2}{T^2} - \frac{2t^3}{T^3}\right)$$

CHAPTER SUMMARY

where C_1 and C_2 are positive constants and T is the (fixed) time length of the systole phase.* Find a relationship between the rates $\dfrac{dV}{dt}$ and $\dfrac{dP}{dt}$.

55. **DEMAND RATE**   When the price of a certain commodity is p dollars per unit, consumers demand x hundred units of the commodity, where
$$75x^2 + 17p^2 = 5{,}300$$
How fast is the demand x changing with respect to time when the price is $7 and is decreasing at the rate of 75 cents per months? $\left(\text{That is, } \dfrac{dp}{dt} = -0.75.\right)$

56. At noon, a truck is at the intersection of two roads and is moving north at 70 km/hr. An hour later, a car passes through the same intersection, traveling east at 105 km/hr. How fast is the distance between the car and truck changing at 2 P.M.?

57. **POPULATION GROWTH**   It is projected that t years from now, the population of a certain suburban community will be $P(t) = 20 - \dfrac{6}{t+1}$ thousand. By approximately what percentage will the population grow during the next quarter year?

58. **WORKER EFFICIENCY**   An efficiency study of the morning shift at a certain factory indicates that an average worker arriving on the job at 8:00 A.M. will have produced $Q(t) = -t^3 + 9t^2 + 12t$ units t hours later.
   a. Compute the worker's rate of production $R(t) = Q'(t)$.
   b. At what rate is the worker's rate of production changing with respect to time at 9:00 A.M.?
   c. Use calculus to estimate the change in the worker's rate of production between 9:00 and 9:06 A.M.
   d. Compute the actual change in the worker's rate of production between 9:00 and 9:06 A.M.

59. **TRAFFIC SAFETY**   A car is traveling at 88 ft/sec when the driver applies the brakes to avoid hitting a child. After t seconds, the car is $s = 88t - 8t^2$ feet from the point where the brakes were applied. How long does it take for the car to come to a stop, and how far does it travel before stopping?

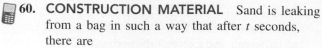

60. **CONSTRUCTION MATERIAL**   Sand is leaking from a bag in such a way that after t seconds, there are
$$S(t) = 50\left(1 - \frac{t^2}{15}\right)^3$$
pounds of sand left in the bag.
   a. How much sand was originally in the bag?
   b. At what rate is sand leaking from the bag after 1 second?
   c. How long does it take for all the sand to leak from the bag? At what rate is the sand leaking from the bag at the time it empties?

61. **PRICE INFLATION**   It is projected that t months from now, the average price per unit for goods in a certain sector of the economy will be P dollars, where
$$P(t) = -t^3 + 7t^2 + 200t + 300$$
   a. At what rate will the price per unit be increasing with respect to time 5 months from now?
   b. At what rate will the rate of price increase be changing with respect to time 5 months from now?
   c. Use calculus to estimate the change in the rate of price increase during the first half of the sixth month.
   d. Compute the actual change in the rate of price increase during the first half of the sixth month.

62. **PRODUCTION COST**   At a certain factory, approximately $q(t) = t^2 + 50t$ units are manufactured during the first t hours of a production run, and the total cost of manufacturing q units is $C(q) = 0.1q^2 + 10q + 400$ dollars. Find the rate at which the manufacturing cost is changing with respect to time 2 hours after production commences.

63. **PRODUCTION COST**   It is estimated that the monthly cost of producing x units of a particular commodity is $C(x) = 0.06x + 3x^{1/2} + 20$ hundred dollars. Suppose production is decreasing at the rate of 11 units per month when the monthly production is 2,500 units. At what rate is the cost changing at this level of production?

64. Estimate the largest percentage error you can allow in the measurement of the radius of a sphere if you want the error in the calculation of its surface area using the formula $S = 4\pi r^2$ to be no greater than 8%.

65. A soccer ball made of leather 1/8 inch thick has an inner diameter of 8.5 inches. Estimate the volume of its leather shell. [*Hint:* Think of the volume of the shell as a certain change ΔV in volume.]

66. A car traveling north at 60 mph and a truck traveling east at 45 mph leave an intersection at the same time. At what rate is the distance between them changing 2 hours later?

67. A child is flying a kite at a height of 80 feet above her hand. If the kite moves horizontally at a constant speed of 5 ft/sec, at what rate is string being paid out when the string is 100 feet long?

68. A person stands at the end of a pier 8 feet above the water and pulls in a rope attached to a buoy. If the rope is hauled in at the rate of 2 ft/min, how fast is the buoy moving in the water when it is 6 feet from the pier?

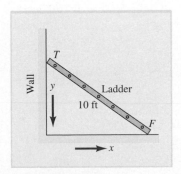

EXERCISE 68

69. A 10-foot-long ladder leans against the side of a wall. The top of the ladder is sliding down the wall at the rate of 3 ft/sec. How fast is the foot of the ladder moving away from the building when the top is 6 feet above the ground?

EXERCISE 69

70. A lantern falls from the top of a building in such a way that after t seconds, it is $h(t) = 150 - 16t^2$ feet above ground. A woman 5 feet tall originally standing directly under the lantern sees it start to fall and walks away at the constant rate of 5 ft/sec. How fast is the length of the woman's shadow changing when the lantern is 10 feet above the ground?

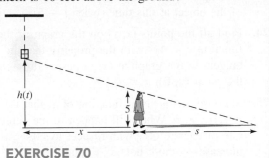

EXERCISE 70

71. A baseball diamond is a square, 90 feet on a side. A runner runs from second base to third at 20 ft/sec. How fast is the distance s between the runner and home plate changing when he is 15 feet from third base?

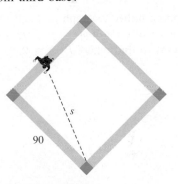

EXERCISE 71

72. MANUFACTURING COST Suppose the total manufacturing cost C at a certain factory is a function of the number q of units produced, which in turn is a function of the number t of hours during which the factory has been operating.

  a. What quantity is represented by the derivative $\dfrac{dC}{dq}$? In what units is this quantity measured?

  b. What quantity is represented by the derivative $\dfrac{dq}{dt}$? In what units is this quantity measured?

  c. What quantity is represented by the product $\dfrac{dC}{dq}\dfrac{dq}{dt}$? In what units is this quantity measured?

73. An object projected from a point P moves along a straight line. It is known that the velocity of the object is directly proportional to the product of the time the object has been moving and the distance it has moved from P. It is also known that at the end of 5 seconds, the object is 20 feet from P and is moving at the rate of 4 ft/sec. Find the acceleration of the object at this time (when $t = 5$).

74. Find all the points (x, y) on the graph of the function $y = 4x^2$ with the property that the tangent to the graph at (x, y) passes through the point $(2, 0)$.

75. Suppose y is a linear function of x; that is, $y = mx + b$. What will happen to the percentage rate of change of y with respect to x as x increases without bound? Explain.

76. Find an equation for the tangent line to the curve

$$\frac{x^2}{a^2} - \frac{y^2}{b^2} = 1$$

at the point (x_0, y_0).

77. Let $f(x) = (3x + 5)(2x^3 - 5x + 4)$. Use a graphing utility to graph $f(x)$ and $f'(x)$ on the same set of coordinate axes. Use **TRACE** and **ZOOM** to find where $f'(x) = 0$.

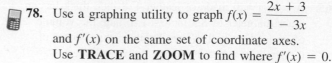

78. Use a graphing utility to graph $f(x) = \dfrac{2x + 3}{1 - 3x}$ and $f'(x)$ on the same set of coordinate axes. Use **TRACE** and **ZOOM** to find where $f'(x) = 0$.

79. The curve $y^2(2 - x) = x^3$ is called a **cissoid.**
    a. Use a graphing utility to sketch the curve.
    b. Find an equation for the tangent line to the curve at all points where $x = 1$.
    c. What happens to the curve as x approaches 2 from the left?
    d. Does the curve have a tangent line at the origin? If so, what is its equation?

80. An object moves along a straight line in such a way that its position at time t is given by
    $$s(t) = t^{5/2}(0.73t^2 - 3.1t + 2.7) \quad \text{for } 0 \le t \le 2$$
    a. Find the velocity $v(t)$ and the acceleration $a(t)$, and then use a graphing utility to graph $s(t)$, $v(t)$, and $a(t)$ on the same axes for $0 \le t \le 2$.
    b. Use your calculator to find a time when $v(t) = 0$ for $0 \le t \le 2$. What is the object's position at this time?
    c. When does the smallest value of $a(t)$ occur? Where is the object at this time and what is its velocity?

EXPLORE! UPDATE

Finding Roots

1. The x intercepts of a function, also called **roots,** are an important feature of a function. Later we will see what characteristics of a function the roots of its derivative can tell us. For now, we explore the various ways a graphing calculator can find roots.

Check that your calculator is in function mode (Func) by accessing the **MODE** menu. Store $f(x) = x^2 - 2x - 10$ into Y1 of the equation editor **(Y=).** Graph using a standard **ZOOM** window. Find x intercepts by tracing the graph or using the Zoom In option of **ZOOM.** Or press **CALC (2nd TRACE)** and select 2:zero. You will need to specify a left bound, right bound, and an estimate in the process. Verify algebraically that the negative root of this function is $x = 1 - \sqrt{11}$.

 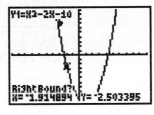 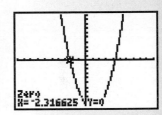

Another method for finding roots is to use the equation solver feature of your graphing calculator. Access the last item of your **MATH** key (0:Solver). Write the equation to be solved, in this case, Y1, into the equation solver, eqn: 0 = Y1, and then press **ENTER.** Suppose we are trying to locate the positive root for $f(x)$ shown in the previous graph. That root looks close to $x = 4$, so we use this value as a starting point. Next press the green **ALPHA** key and then the **ENTER** key **(SOLVE).** The resulting value $x = 4.317$ can be verified to be $x = 1 + \sqrt{11}$.

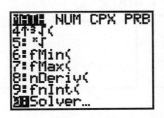 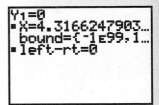

2. Another way to locate a root is Newton's method, an iterative root-finding algorithm that uses tangent line approximations to zero in on a root. See page 171.

Let us apply this method to $f(x) = x^2 - 2x - 10$. Place $f(x)$ into Y1 and write Y2 = nDeriv(Y1, X, X), which is the numerical derivative of Y1, obtained from item 8 of the **MATH** key. Newton's method requires a starting value. From the graph of $f(x)$ it appears that the positive root is smaller than $x = 5$. On a cleared home-screen **(QUIT = 2nd MODE),** store the value 5 into X by writing $5 \rightarrow X$, and then write Newton's algorithm, $X - Y1(X)/Y2(X) \rightarrow X$. Now press **ENTER** successively and observe the sequence of resulting values. Note how many iterations were needed so that two consecutive approximations agree to four (or more) decimal places.

Try a starting value of $x = 4$. Do you construct a sequence of approximations to the same root? How about if you started at $x = -2$?

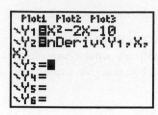 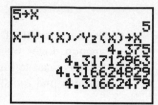 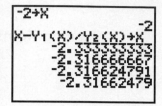

3. We can also use the equation solver feature to locate the roots of the derivative of a function. As an example, store $f(x) = x^2(3x - 1)$ into Y1 of the equation editor and write Y2 = nDeriv(Y1, X, X). Use a bold graphing style for Y2 and a small size **WINDOW,** $[-1, 1]0.2$ by $[-1, 1]0.2$. From the following graph, it is evident that one of the roots of the derivative is $x = 0$. The other root appears to be close to $x = 0.2$.

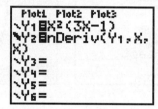 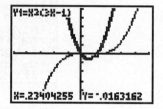

To find this other root, place Y2 into the equation solver position, eqn: 0= (use the up arrow to return to the equation solver screen). Use $x = 0.2$ as a starting point, and then press **ALPHA ENTER (SOLVE)** to obtain the second root of the derivative. What happens if you set the starting x value as -0.2?

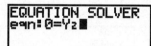  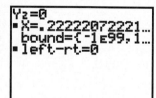

THINK ABOUT IT

MODELING BEHAVIOR OF THE CODLING MOTH

The **codling moth** is an insect that causes serious damage to apples. Adult codling moths emerge from their cocoons in the spring. They soon mate, and the female lays as many as 130 tiny eggs on the leaves of apple trees. After the codling moth larva, also known as the common apple worm, hatches from its egg, it goes looking for an apple. The time between hatching and finding an apple is called the **searching period.** Once a codling moth finds an apple, it squirms inside and eats the fruit and seeds of the apple, thereby ruining it. After about 4 weeks, the codling moth backs out of the apple and crawls under the bark of the tree or into the soil where it forms a cocoon.

Observations regarding the behavior of codling moths indicate that the length of the searching period, $S(T)$, and the percentage of larvae that survive the searching period, $N(T)$, depend on the air temperature, T. Methods of data analysis (polynomial regression) applied to data recorded from observations suggest that if T is measured in degrees Celsius with $20 \leq T \leq 30$, then $S(T)$ and $N(T)$ may be modeled* by

$$S(T) = (-0.03T^2 + 1.6T - 13.65)^{-1} \text{ days}$$

and

$$N(T) = -0.85T^2 + 45.4T - 547$$

Use these formulas to answer the following questions.

Questions

1. What do these formulas for $S(T)$ and $N(T)$ predict for the length of searching period and percentage of larvae surviving the searching period when the air temperature is 25 degrees Celsius?

---

*P. L. Shaffer and H. J. Gold, "A Simulation Model of Population Dynamics of the Codling Moth *Cydia Pomonella*," *Ecological Modeling,* Vol. 30, 1985, pp. 247–274.

2. Sketch the graph of $N(T)$, and determine the temperature at which the largest percentage of codling moth larvae survive. Then determine the temperature at which the smallest percentage of larvae survive. (Remember, $20 \leq T \leq 30$.)

3. Find $\dfrac{dS}{dT}$, the rate of change of the searching period with respect to temperature T.

   When does this rate equal zero? What (if anything) occurs when $\dfrac{dS}{dT} = 0$?

4. Find $\dfrac{dN}{dS}$, the rate of change of the percentage of larvae surviving the searching period with respect to the length of the searching period using the chain rule,

$$\frac{dN}{dT} = \frac{dN}{dS} \frac{dS}{dT}$$

   What (if any) information does this rate provide?

Determining when an assembly line is working at peak efficiency is one application of the derivative.

Additional Applications of the Derivative

SECTION 3.1   Increasing and Decreasing Functions; Relative Extrema

Learning Objectives

1. Discuss increasing and decreasing functions.
2. Define critical points and relative extrema.
3. Use the first derivative test to study relative extrema and sketch graphs.

The graph in Figure 3.1 shows the U.S. federal budget surplus or deficit for the years 2000 to 2009, with positive values representing a surplus and negative values a deficit. Notice that the values decreased steadily from 2000 to 2004 and then started to increase before again decreasing dramatically after 2007. Borrowing terminology from this description, we will say that a function $f(x)$ is *increasing* if its graph is rising from left to right and *decreasing* if its graph is falling.

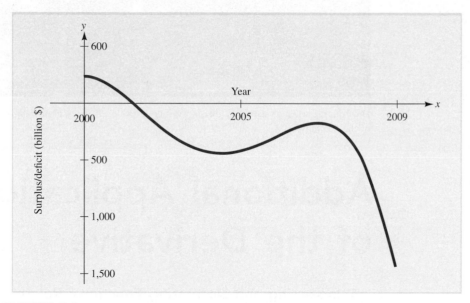

FIGURE 3.1   U.S. federal budget surplus and deficit.

Here is a more formal definition of increasing and decreasing functions, which is illustrated graphically in Figure 3.2.

Increasing and Decreasing Functions   ■   Let $f(x)$ be a function defined on the interval $a < x < b$, and let x_1 and x_2 be two numbers in the interval. Then

$f(x)$ is **increasing** on the interval if $f(x_2) > f(x_1)$ whenever $x_2 > x_1$.
$f(x)$ is **decreasing** on the interval if $f(x_2) < f(x_1)$ whenever $x_2 > x_1$.

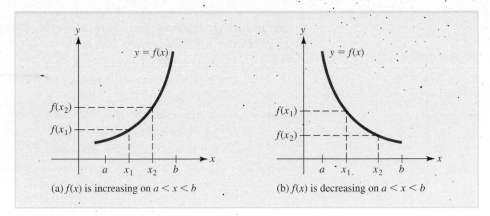

FIGURE 3.2 Intervals of increase and decrease.

As demonstrated in Figure 3.3a, if the graph of a function $f(x)$ has tangent lines with only positive slopes on the interval $a < x < b$, then the graph will be rising and $f(x)$ will be increasing on the interval. Since the slope of each such tangent line is given by the derivative $f'(x)$, it follows that $f(x)$ is increasing (graph rising) on intervals where $f'(x) > 0$. Similarly, $f(x)$ is decreasing (graph falling) on intervals where $f'(x) < 0$ (Figure 3.3b).

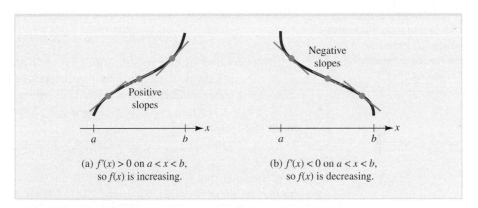

FIGURE 3.3 Derivative criteria for increasing and decreasing functions.

Thanks to the intermediate value property (Section 1.6), we know that a continuous function cannot change sign without becoming 0. This means that if we mark on a number line all numbers x where $f'(x)$ is discontinuous or $f'(x) = 0$, the line will be divided into intervals where $f'(x)$ does not change sign. Therefore, if we pick a test number c from each such interval and find that $f'(c) > 0$, we know that $f'(x) > 0$ for *all* numbers x in the interval and $f(x)$ must be increasing (graph rising) throughout the interval. Similarly, if $f'(c) < 0$, it follows that $f(x)$ is decreasing (graph falling) throughout the interval. These observations may be summarized as follows.

> **Procedure for Using the Derivative to Determine Intervals of Increase and Decrease for a Function *f*.**
>
> **Step 1.** Find all values of x for which $f'(x) = 0$ or $f'(x)$ is not continuous, and mark these numbers on a number line. This divides the line into a number of intervals.
>
> **Step 2.** Choose a test number c from each interval $a < x < b$ determined in step 1 and evaluate $f'(c)$. Then,
>
> If $f'(c) > 0$, the function $f(x)$ is increasing (graph rising) on $a < x < b$.
> If $f'(c) < 0$, the function $f(x)$ is decreasing (graph falling) on $a < x < b$.

This procedure is illustrated in Examples 3.1.1 and 3.1.2.

EXAMPLE 3.1.1 Finding Intervals of Increase and Decrease

Find the intervals of increase and decrease for the function

$$f(x) = 2x^3 + 3x^2 - 12x - 7$$

Solution

The derivative of $f(x)$ is

$$f'(x) = 6x^2 + 6x - 12 = 6(x + 2)(x - 1)$$

which is continuous everywhere, with $f'(x) = 0$ when $x = 1$ and $x = -2$. The numbers -2 and 1 divide the x axis into three intervals: $x < -2$, $-2 < x < 1$, and $x > 1$. Choose a test number c from each of these intervals; say, $c = -3$ from $x < -2$, $c = 0$ from $-2 < x < 1$, and $c = 2$ from $x > 1$. Then evaluate $f'(c)$ for each test number:

$$f'(-3) = 24 > 0 \qquad f'(0) = -12 < 0 \qquad f'(2) = 24 > 0$$

We conclude that $f'(x) > 0$ for $x < -2$ and for $x > 1$, so $f(x)$ is increasing (graph rising) on these intervals. Similarly, $f'(x) < 0$ on $-2 < x < 1$, so $f(x)$ is decreasing (graph falling) on this interval. These results are summarized in Table 3.1. The graph of $f(x)$ is shown in Figure 3.4.

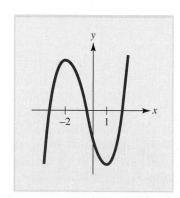

FIGURE 3.4 The graph of $f(x) = 2x^3 + 3x^2 - 12x - 7$.

TABLE 3.1   Intervals of Increase and Decrease for
$f(x) = 2x^3 + 3x^2 - 12x - 7$

Interval	Test Number c	$f'(c)$	Conclusion	Direction of Graph
$x < -2$	-3	$f'(-3) > 0$	f is increasing	Rising
$-2 < x < 1$	0	$f'(0) < 0$	f is decreasing	Falling
$x > 1$	2	$f'(2) > 0$	f is increasing	Rising

NOTATION Henceforth, we shall indicate an interval where $f(x)$ is increasing by an "up arrow" (⟋) and an interval where $f(x)$ is decreasing by a "down arrow" (⟍). Thus, the results in Example 3.1.1 can be represented by this arrow diagram:

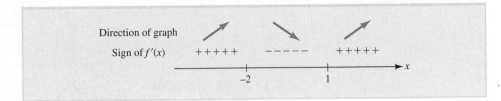

EXPLORE!

Graph

$$f(x) = \frac{x^2}{x - 2}$$

in regular style and

$$g(x) = \frac{x^2}{x - 4}$$

in bold, using the window [−9.4, 9.4]1 by [−20, 30]5. What effect does the change in the denominator have on the high and low points of the graph? Over what interval is $g(x)$ decreasing?

EXAMPLE 3.1.2 Finding Intervals of Increase and Decrease

Find the intervals of increase and decrease for the function

$$f(x) = \frac{x^2}{x - 2}$$

Solution

The function is defined for $x \neq 2$, and its derivative is

$$f'(x) = \frac{(x - 2)(2x) - x^2(1)}{(x - 2)^2} = \frac{x(x - 4)}{(x - 2)^2}$$

which is discontinuous at $x = 2$ and has $f'(x) = 0$ at $x = 0$ and $x = 4$. Thus, there are four intervals on which the sign of $f'(x)$ does not change: $x < 0$, $0 < x < 2$, $2 < x < 4$, and $x > 4$. Choosing test numbers in these intervals (say, -2, 1, 3, and 5, respectively), we find that

$$f'(-2) = \frac{3}{4} > 0 \qquad f'(1) = -3 < 0 \qquad f'(3) = -3 < 0 \qquad f'(5) = \frac{5}{9} > 0$$

We conclude that $f(x)$ is increasing (graph rising) for $x < 0$ and for $x > 4$ and that it is decreasing (graph falling) for $0 < x < 2$ and for $2 < x < 4$. These results are summarized in the following arrow diagram [the dashed vertical line indicates that $f(x)$ is not defined at $x = 2$].

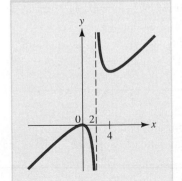

FIGURE 3.5 The graph of $f(x) = \dfrac{x^2}{x - 2}$.

Intervals of increase and decrease for $f(x) = \dfrac{x^2}{x - 2}$.

The graph of $f(x)$ is shown in Figure 3.5. Notice how the graph approaches the vertical line $x = 2$ as x tends toward 2. This behavior identifies $x = 2$ as a *vertical asymptote* of the graph of $f(x)$. We will discuss asymptotes in Section 3.3.

Relative Extrema   The simplicity of the graphs in Figures 3.4 and 3.5 may be misleading. A more general graph is shown in Figure 3.6. Note that *peaks* occur at C and E and *valleys* at B, D, and G, and while there are horizontal tangents at B, C, D, and G, no tangent can be drawn at the "sharp" point E. Moreover, there is a horizontal tangent at F that is neither a peak nor a valley. In this section and Section 3.2, you will see how the methods of calculus can be used to locate and identify the peaks and valleys of a graph. This, in turn, provides the basis for a curve sketching procedure and for methods of optimization.

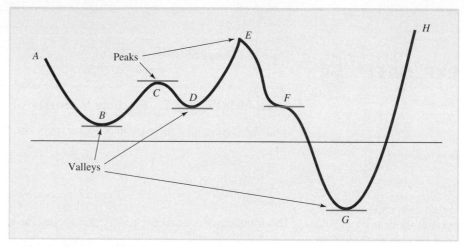

FIGURE 3.6 A graph with various kinds of peaks and valleys.

Less informally, a peak on the graph of a function f is known as a **relative maximum** of f, and a valley is a **relative minimum.** Thus, a relative maximum is a point on the graph of f that is at least as high as any nearby point on the graph, while a relative minimum is at least as low as any nearby point. Collectively, relative maxima and minima are called **relative extrema.** In Figure 3.6, the relative maxima are located at C and E, and the relative minima are at B, D, and G. Note that a relative extremum does not have to be the highest or lowest point on the entire graph. For instance, in Figure 3.6, the lowest point on the graph is at the relative minimum G, but the highest point occurs at the right endpoint H. Here is a summary of this terminology.

> **Relative Extrema** ■ The graph of the function $f(x)$ is said to have a *relative maximum* at $x = c$ if $f(c) \geq f(x)$ for all x in some interval $a < x < b$ containing c. Similarly, the graph has a *relative minimum* at $x = c$ if $f(c) \leq f(x)$ on such an interval. Collectively, the relative maxima and minima of f are called its *relative extrema.*

Since a function $f(x)$ is increasing when $f'(x) > 0$ and decreasing when $f'(x) < 0$, the only points where $f(x)$ can have a relative extremum are where $f'(x) = 0$ or $f'(x)$ does not exist. Such points are so important that we give them a special name.

> **Critical Numbers and Critical Points** ■ A number c in the domain of $f(x)$ is called a **critical number** if either $f'(c) = 0$ or $f'(c)$ does not exist. The corresponding point $(c, f(c))$ on the graph of $f(x)$ is called a **critical point** for $f(x)$. *Relative extrema can only occur at critical points.*

It is important to note that while relative extrema occur at critical points, *not all critical points correspond to relative extrema*. For example, Figure 3.7 shows three different situations where $f'(c) = 0$, so a horizontal tangent line occurs at the critical point $(c, f(c))$. The critical point corresponds to a relative maximum in Figure 3.7a and to a relative minimum in Figure 3.7b, but neither kind of relative extremum occurs at the critical point in Figure 3.7c.

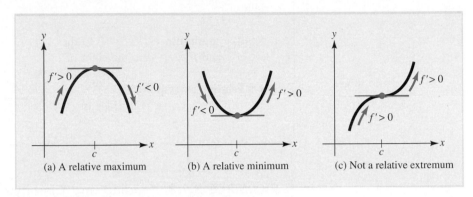

FIGURE 3.7   Three critical points $(c, f(c))$ where $f'(c) = 0$.

Three functions with critical points at which the derivative is undefined are shown in Figure 3.8. In Figure 3.8c, the tangent line is vertical at $(c, f(c))$, so the slope $f'(c)$ is undefined. In Figure 3.8a and b, no (unique) tangent line can be drawn at the sharp point $(c, f(c))$.

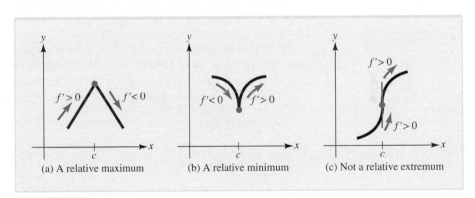

FIGURE 3.8   Three critical points $(c, f(c))$ where $f'(c)$ is undefined.

The First Derivative Test for Relative Extrema

Figures 3.7 and 3.8 also suggest a method for using the sign of the derivative to classify critical points as relative maxima, relative minima, or neither. Suppose c is a critical number of f and that $f'(x) > 0$ to the left of c, while $f'(x) < 0$ to the right. Geometrically, this means the graph of f goes up before the critical point P with coordinates $(c, f(c))$ and then comes down, which implies that P is a relative maximum. Similarly, if $f'(x) < 0$ to the left of c and $f'(x) > 0$ to the right, the graph goes down before P and up afterward, so there must be a relative minimum at P. On the other hand, if the derivative has the same sign on both sides of c, then the graph either rises through P or falls through P, and no relative extremum occurs there. These observations can be summarized as follows.

The First Derivative Test for Relative Extrema ■ Let c be a critical number for $f(x)$ [that is, $f(c)$ is defined and either $f'(c) = 0$ or $f'(c)$ does not exist]. Then the critical point $(c, f(c))$ is

a **relative maximum** if $f'(x) > 0$ to the left of c and $f'(x) < 0$ to the right of c

a **relative minimum** if $f'(x) < 0$ to the left of c and $f'(x) > 0$ to the right of c

not a relative extremum if $f'(x)$ has the same sign on both sides of c

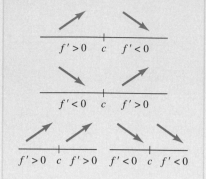

EXAMPLE 3.1.3   Finding and Classifying Critical Numbers

Find all critical numbers of the function

$$f(x) = 2x^4 - 4x^2 + 3$$

and classify each critical point as a relative maximum, a relative minimum, or neither.

Solution

The polynomial $f(x)$ is defined for all x, and its derivative is

$$f'(x) = 8x^3 - 8x = 8x(x^2 - 1) = 8x(x - 1)(x + 1)$$

Since the derivative exists for all x, the only critical numbers are where $f'(x) = 0$; that is, $x = 0$, $x = 1$, and $x = -1$. These numbers divide the x axis into four intervals, on each of which the sign of the derivative does not change: $x < -1$, $-1 < x < 0$, $0 < x < 1$, and $x > 1$. Choose a test number c in each of these intervals (say, -5, $-\dfrac{1}{2}$, $\dfrac{1}{4}$, and 2, respectively) and evaluate $f'(c)$ in each case:

$$f'(-5) = -960 < 0 \quad f'\left(-\frac{1}{2}\right) = 3 > 0 \quad f'\left(\frac{1}{4}\right) = -\frac{15}{8} < 0 \quad f'(2) = 48 > 0$$

Thus, the graph of f falls for $x < -1$ and for $0 < x < 1$, and rises for $-1 < x < 0$ and for $x > 1$, so there must be a relative maximum at $x = 0$ and relative minima at $x = -1$ and $x = 1$, as indicated in the following arrow diagram.

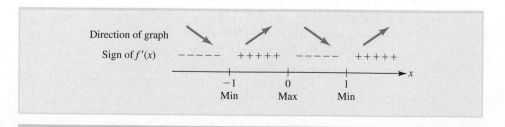

Applications   Once you determine the intervals of increase and decrease of a function f and find its relative extrema, you can obtain a rough sketch of the graph of the function. Here is a step-by-step description of the procedure for sketching the graph of a continuous function $f(x)$ using the derivative $f'(x)$. In Section 3.3, we will extend this procedure to cover the situation where $f(x)$ is discontinuous.

A Procedure for Sketching the Graph of a Function $f(x)$ Continuous on Its Domain Using the Derivative $f'(x)$

Step 1. Determine the domain of $f(x)$. Set up a number line restricted to include only those numbers in the domain of $f(x)$.

Step 2. Find $f'(x)$, and mark each critical number on the restricted number line obtained in step 1. Then analyze the sign of the derivative to determine intervals of increase and decrease for $f(x)$ on the restricted number line.

Step 3. For each critical number c, find $f(c)$ and plot the critical point $(c, f(c))$ on a coordinate plane, with a "cap" ⌒ if it is a relative maximum (↗↘), or a "cup" ⌣ if it is a relative minimum (↘↗). Plot intercepts and other key points that can be easily found.

Step 4. Sketch the graph of f as a smooth curve joining the critical points in such a way that it rises where $f'(x) > 0$, falls where $f'(x) < 0$, and has a horizontal tangent where $f'(x) = 0$.

EXAMPLE 3.1.4 Graphing Using a Derivative

Sketch the graph of the function $f(x) = x^4 + 8x^3 + 18x^2 - 8$.

Solution

Since $f(x)$ is a polynomial, it is defined for all x. Its derivative is

$$f'(x) = 4x^3 + 24x^2 + 36x = 4x(x^2 + 6x + 9) = 4x(x + 3)^2$$

Since the derivative exists for all x, the only critical numbers are where $f'(x) = 0$: at $x = 0$ and $x = -3$. These numbers divide the x axis into three intervals, on each of which the sign of the derivative $f'(x)$ does not change: $x < -3$, $-3 < x < 0$, and $x > 0$. Choose a test number c in each interval (say, -5, -1, and 1, respectively), and determine the sign of $f'(c)$:

$$f'(-5) = -80 < 0 \qquad f'(-1) = -16 < 0 \qquad f'(1) = 64 > 0$$

Thus, the graph of f has horizontal tangents where x is -3 and 0, and the graph is falling (f decreasing) on the intervals $x < -3$ and $-3 < x < 0$ and is rising (f increasing) for $x > 0$, as indicated in this arrow diagram:

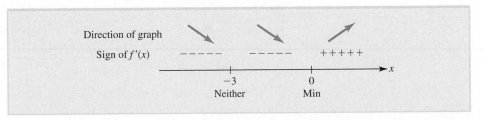

Interpreting the diagram, we see that the graph falls to a horizontal tangent at $x = -3$, then continues falling to the relative minimum at $x = 0$, after which it rises indefinitely. We find that $f(-3) = 19$ and $f(0) = -8$. To begin your sketch, plot a cup ⌣ at the critical point $(0, -8)$ to indicate that a relative minimum occurs there (if it had been a relative maximum, you would have used a cap ⌢), and a twist ↷ at $(-3, 19)$ to indicate a falling graph with a horizontal tangent at this point. This is shown in the preliminary graph in Figure 3.9a. Finally, complete the sketch by passing a smooth curve through the critical points in the directions indicated by the arrows, as shown in Figure 3.9b.

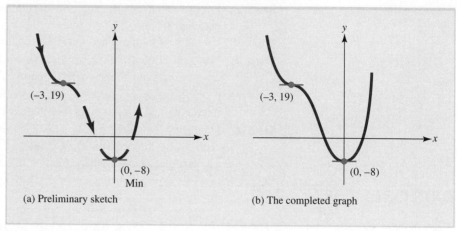

(a) Preliminary sketch   (b) The completed graph

FIGURE 3.9   The graph of $f(x) = x^4 + 8x^3 + 18x^2 - 8$.

EXAMPLE 3.1.5   Graphing Using a Derivative

Find the intervals of increase and decrease and the relative extrema of the function $g(t) = \sqrt{3 - 2t - t^2}$. Sketch the graph.

Just-In-Time **REVIEW**

The product ab satisfies $ab \geq 0$ *only* when both $a \geq 0$ and $b \geq 0$ or when $a \leq 0$ and $b \leq 0$. If a and b have opposite signs, then their product is negative.

Solution

Since $\sqrt{u}$ is defined only for $u \geq 0$, the domain of g will be the set of all t such that $3 - 2t - t^2 \geq 0$. Factoring the expression $3 - 2t - t^2$, we get

$$3 - 2t - t^2 = (3 + t)(1 - t)$$

Note that $3 + t \geq 0$ when $t \geq -3$ and that $1 - t \geq 0$ when $t \leq 1$. We have $(3 + t)(1 - t) \geq 0$ when *both* terms are nonnegative, that is, when $t \geq -3$ *and* $t \leq 1$, or equivalently, when $-3 \leq t \leq 1$. We would also have $(3 + t)(1 - t) \geq 0$ if $3 + t \leq 0$ and $1 - t \leq 0$, but this is not possible (do you see why?). Thus, $g(t)$ is defined only for $-3 \leq t \leq 1$.

Next, using the chain rule, we compute the derivative of $g(t)$:

$$g'(t) = \frac{1}{2} \frac{1}{\sqrt{3 - 2t - t^2}}(-2 - 2t)$$

$$= \frac{-1 - t}{\sqrt{3 - 2t - t^2}}$$

Note that $g'(t)$ does not exist at the endpoints $t = -3$ and $t = 1$ of the domain of $g(t)$, and that $g'(t) = 0$ only when $t = -1$. Next, mark these three critical numbers on a number line restricted to the domain of g, $-3 \leq t \leq 1$, and determine the sign of the derivative $g'(t)$ on the subintervals $-3 < t < -1$ and $-1 < t < 1$ to obtain the arrow diagram shown in Figure 3.10a. Finally, compute $g(-3) = g(1) = 0$ and $g(-1) = 2$ and observe that the arrow diagram suggests there is a relative maximum at $(-1, 2)$. The completed graph is shown in Figure 3.10b.

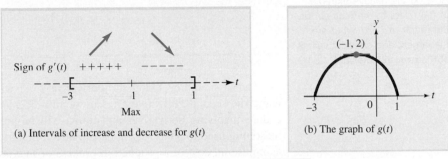

(a) Intervals of increase and decrease for $g(t)$ (b) The graph of $g(t)$

FIGURE 3.10   Sketching the graph of $g(t) = \sqrt{3 - 2t - t^2}$.

Sometimes, the graph of $f(x)$ is known and the relationship between the sign of the derivative $f'(x)$ and intervals of increase and decrease can be used to determine the general shape of the graph of $f'(x)$. The procedure is illustrated in Example 3.1.6.

EXAMPLE 3.1.6   Using the Graph of f to Graph the Derivative f'

The graph of a function $f(x)$ is shown. Sketch a possible graph for the derivative $f'(x)$.

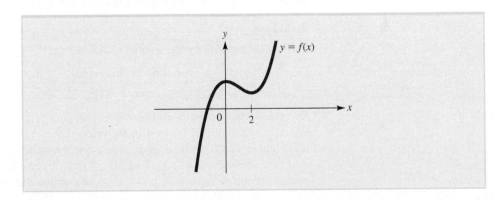

Solution

Since the graph of $f(x)$ is falling for $0 < x < 2$, we have $f'(x) < 0$ and the graph of $f'(x)$ is below the x axis on this interval. Similarly, for $x < 0$ and for $x > 2$, the graph of $f(x)$ is rising, so $f'(x) > 0$ and the graph of $f'(x)$ is above the x axis on both these intervals. The graph of $f(x)$ is "flat" (horizontal tangent line) at $x = 0$ and $x = 2$, so $f'(0) = f'(2) = 0$, and $x = 0$ and $x = 2$ are the x intercepts of the graph of $f'(x)$. Here is one possible graph that satisfies these conditions:

EXPLORE!

Store $f(x) = x^3 - x^2 - 4x + 4$ into Y1 and $f'(x)$ (via the numerical derivative) into Y2 using the bold graphing style and a window size of $[-4.7, 4.7]1$ by $[-10, 10]2$. How do the relative extrema of $f(x)$ relate to the features of the graph of $f'(x)$? What are the largest and smallest values of $f(x)$ over the interval $[-2, 1]$?

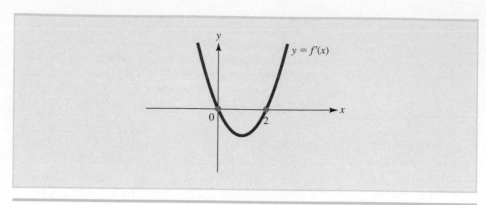

Curve sketching using the derivative $f'(x)$ will be refined by adding features using the second derivative $f''(x)$ in Section 3.2, and a general curve sketching procedure involving derivatives and limits will be presented in Section 3.3. The same reasoning used to analyze graphs can be applied to determining optimal values, such as minimum cost of a production process or maximal sustainable yield for a salmon fishery. Optimization is illustrated in Example 3.1.7 and is discussed in more detail in Sections 3.4 and 3.5.

EXAMPLE 3.1.7 Finding Maximum Revenue

The revenue derived from the sale of a new kind of motorized skateboard t weeks after its introduction is given by

$$R(t) = \frac{63t - t^2}{t^2 + 63} \qquad 0 \le t \le 63$$

million dollars. When does maximum revenue occur? What is the maximum revenue?

Solution

Differentiating $R(t)$ by the quotient rule, we get

$$R'(t) = \frac{(t^2 + 63)(63 - 2t) - (63t - t^2)(2t)}{(t^2 + 63)^2} = \frac{-63(t - 7)(t + 9)}{(t^2 + 63)^2}$$

By setting the numerator in this expression for $R'(t)$ equal to 0, we find that $t = 7$ is the only solution of $R'(t) = 0$ in the interval $0 \le t \le 63$, and hence is the only critical number of $R(t)$ in its domain. The critical number divides the domain $0 \le t \le 63$ into two intervals: $0 \le t < 7$ and $7 < t \le 63$. Evaluating $R'(t)$ at test numbers in each interval (say, at $t = 1$ and $t = 9$), we obtain the arrow diagram shown here.

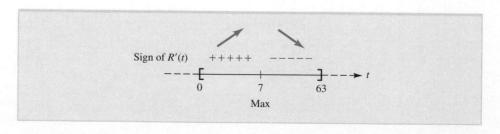

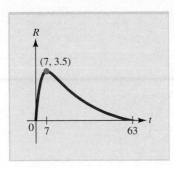

FIGURE 3.11   The graph of $R(t) = \dfrac{63t - t^2}{t^2 + 63}$ for $0 \le t \le 63$.

The arrow pattern indicates that the revenue increases to a maximum at $t = 7$, after which it decreases. At the optimal time $t = 7$, the revenue produced is

$$R(7) = \frac{63(7) - (7)^2}{(7)^2 + 63} = 3.5 \text{ million dollars}$$

The graph of the revenue function $R(t)$ is shown in Figure 3.11. It suggests that immediately after its introduction, the motorized skateboard is very popular, producing peak revenue of 3.5 million dollars after only 7 weeks. However, its popularity, as measured by revenue, then begins to wane. After 63 weeks, revenue ceases altogether as presumably the skateboards are taken off the shelves and replaced by something new. A product that exhibits this kind of revenue pattern, steep increase followed by a steady decline toward 0, is sometimes referred to as a *fad*.

EXERCISES ■ 3.1

In Exercises 1 through 4, specify the intervals on which the derivative of the given function is positive and those on which it is negative.

1.

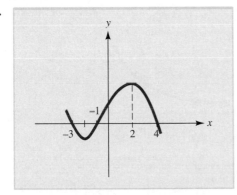

2.

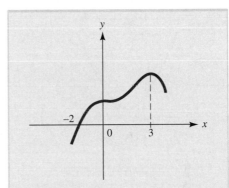

3.

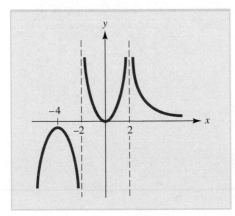

4.

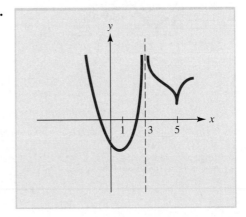

In each of Exercises 5 through 8, match the graph of the derivative of the given function with one of the graphs A, B, C, D shown in the right column.

5.

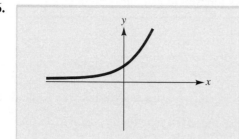

A

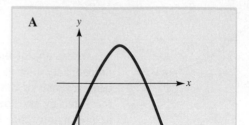

6.

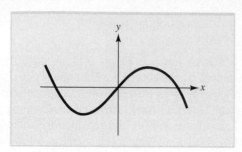

B

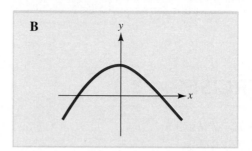

7.

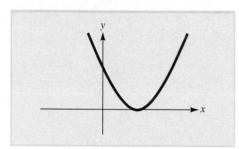

C

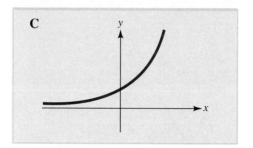

8.

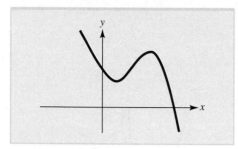

D

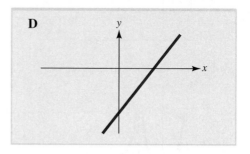

In Exercises 9 through 22, find the intervals of increase and decrease for the given function.

9. $f(x) = x^2 - 4x + 5$

10. $f(t) = t^3 + 3t^2 + 1$

11. $f(x) = x^3 - 3x - 4$

12. $f(x) = \frac{1}{3}x^3 - 9x + 2$

13. $g(t) = t^5 - 5t^4 + 100$

14. $f(x) = 3x^5 - 5x^3$

15. $f(t) = \frac{1}{4 - t^2}$

16. $g(t) = \frac{1}{t^2 + 1} - \frac{1}{(t^2 + 1)^2}$

17. $h(u) = \sqrt{9 - u^2}$

18. $f(x) = \sqrt{6 - x - x^2}$

19. $F(x) = x + \dfrac{9}{x}$

20. $f(t) = \dfrac{t}{(t + 3)^2}$

21. $f(x) = \sqrt{x} + \dfrac{1}{\sqrt{x}}$

22. $G(x) = x^2 - \dfrac{1}{x^2}$

In Exercises 23 through 34, determine the critical numbers of the given function and classify each critical point as a relative maximum, a relative minimum, or neither.

23. $f(x) = 3x^4 - 8x^3 + 6x^2 + 2$

24. $f(x) = 324x - 72x^2 + 4x^3$

25. $f(t) = 2t^3 + 6t^2 + 6t + 5$

26. $f(t) = 10t^6 + 24t^5 + 15t^4 + 3$

27. $g(x) = (x - 1)^5$

28. $F(x) = 3 - (x + 1)^3$

29. $f(t) = \dfrac{t}{t^2 + 3}$

30. $f(t) = t\sqrt{9 - t}$

31. $h(t) = \dfrac{t^2}{t^2 + t - 2}$

32. $g(x) = 4 - \dfrac{2}{x} + \dfrac{3}{x^2}$

33. $S(t) = (t^2 - 1)^4$

34. $F(x) = \dfrac{x^2}{x - 1}$

In Exercises 35 through 44, use calculus to sketch the graph of the given function.

35. $f(x) = x^3 - 3x^2$

36. $f(x) = 3x^4 - 4x^3$

37. $f(x) = 3x^4 - 8x^3 + 6x^2 + 2$

38. $g(x) = 3 - (x + 1)^3$

39. $f(t) = 2t^3 + 6t^2 + 6t + 5$

40. $f(x) = x^3(x + 5)^2$

41. $g(t) = \dfrac{t}{t^2 + 3}$

42. $f(x) = \dfrac{x + 1}{x^2 + x + 1}$

43. $f(x) = 3x^5 - 5x^3 + 4$

44. $H(x) = \dfrac{1}{50}(3x^4 - 8x^3 - 90x^2 + 70)$

In Exercises 45 through 48, the derivative of a function $f(x)$ is given. In each case, find the critical numbers of $f(x)$ and classify each as corresponding to a relative maximum, a relative minimum, or neither.

45. $f'(x) = x^2(4 - x^2)$

46. $f'(x) = \dfrac{x(2 - x)}{x^2 + x + 1}$

47. $f'(x) = \dfrac{(x + 1)^2(4 - 3x)^3}{(x^2 + 1)^2}$

48. $f'(x) = x^3(2x - 7)^2(x + 5)$

In Exercises 49 through 52, the graph of a function f is given. In each case, sketch a possible graph for f'.

49.

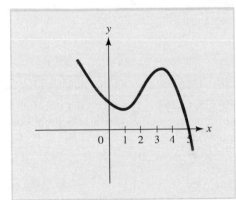

50.

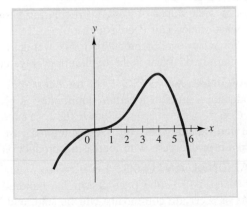

51.

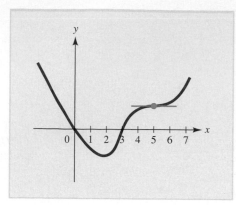

52.

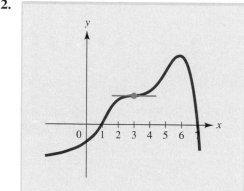

BUSINESS AND ECONOMICS APPLIED PROBLEMS

53. AVERAGE COST   The total cost of producing x units of a certain commodity is $C(x)$ thousand dollars, where

$$C(x) = x^3 - 20x^2 + 179x + 242$$

  a. Find $A'(x)$, where $A(x) = \dfrac{C(x)}{x}$ is the average cost function.
  b. For what values of x is $A(x)$ increasing? For what values is it decreasing?
  c. For what level of production x is average cost minimized? What is the minimum average cost?

54. MARGINAL ANALYSIS   The total cost of producing x units of a certain commodity is given by $C(x) = \sqrt{5x + 2} + 3$. Sketch the cost curve and find the marginal cost. Does marginal cost increase or decrease with increasing production?

55. MARGINAL ANALYSIS   Let $p = (10 - 3x)^2$ for $0 \le x \le 3$ be the price at which x hundred units of a certain commodity will be sold, and let

$R(x) = xp(x)$ be the revenue obtained from the sale of the x units. Find the marginal revenue $R'(x)$, and sketch the revenue and marginal revenue curves on the same graph. For what level of production is revenue maximized?

56. PROFIT UNDER A MONOPOLY   To produce x units of a particular commodity, a monopolist has a total cost of

$$C(x) = 2x^2 + 3x + 5$$

and total revenue of $R(x) = xp(x)$, where $p(x) = 5 - 2x$ is the price at which the x units will be sold. Find the profit function $P(x) = R(x) - C(x)$, and sketch its graph. For what level of production is profit maximized?

57. ADVERTISING   A company determines that if x thousand dollars are spent on advertising a certain product, then $S(x)$ units of the product will be sold, where

$$S(x) = -2x^3 + 27x^2 + 132x + 207$$
$$\text{for } 0 \le x \le 17$$

  a. Sketch the graph of $S(x)$.
  b. How many units will be sold if nothing is spent on advertising?
  c. How much should be spent on advertising to maximize sales? What is the maximum sales level?

58. ADVERTISING   Answer the questions in Exercise 57 for the sales function

$$S(x) = \frac{200x + 1{,}500}{0.02x^2 + 5} \quad \text{for } x \ge 0$$

59. MORTGAGE REFINANCING   When interest rates are low, many homeowners take the opportunity to refinance their mortgages. As rates start to rise, there is often a flurry of activity as latecomers rush in to refinance while they still can do so profitably. Eventually, however, rates reach a level where refinancing begins to wane.

Suppose in a certain community, there will be $M(r)$ thousand refinanced mortgages when the 30-year fixed mortgage rate is $r\%$, where

$$M(r) = \frac{1 + 0.05r}{1 + 0.004r^2} \quad \text{for } 1 \le r \le 8$$

  a. For what values of r is $M(r)$ increasing?
  b. For what interest rate r is the number of refinanced mortgages maximized? What is this maximum number?

60. DEPRECIATION The value V (in thousands of dollars) of an industrial machine is modeled by

$$V(N) = \left(\frac{3N + 430}{N + 1}\right)^{2/3}$$

where N is the number of hours the machine is used each day. Suppose further that usage varies with time in such a way that

$$N(t) = \sqrt{t^2 - 10t + 61}$$

where t is the number of months the machine has been in operation.

a. Over what time interval is the value of the machine increasing? When is it decreasing?

b. At what time t is the value of the machine the largest? What is this maximum value?

61. GROSS DOMESTIC PRODUCT The accompanying graph shows the consumption of the baby boom generation, measured as a percentage of total GDP (gross domestic product) during the time period 1970–1997.

a. At what years do relative maxima occur?

b. At what years do relative minima occur?

c. At roughly what rate was consumption increasing in 1987?

d. At roughly what rate was consumption decreasing in 1972?

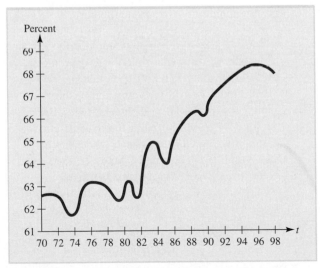

EXERCISE 61 Baby Boom Consumption as Percentage of GDP.

Source: Bureau of Economic Analysis.

62. POPULATION DISTRIBUTION A demographic study of a certain city indicates that $P(r)$ hundred people live r miles from the civic center, where

$$P(r) = \frac{5(3r + 1)}{r^2 + r + 2}$$

a. What is the population at the city center?

b. For what values of r is $P(r)$ increasing? For what values is it decreasing?

c. At what distance from the civic center is the population largest? What is this largest population?

63. MEDICINE The concentration of a drug t hours after being injected into the arm of a patient is given by

$$C(t) = \frac{0.15t}{t^2 + 0.81}$$

Sketch the graph of the concentration function. When does the maximum concentration occur?

64. POLLUTION CONTROL Commissioners of a certain city determine that when x million dollars are spent on controlling pollution, the percentage of pollution removed is given by

$$P(x) = \frac{100\sqrt{x}}{0.04x^2 + 12}$$

a. Sketch the graph of $P(x)$.

b. What expenditure results in the largest percentage of pollution removal?

65. FISHERY MANAGEMENT Dom, the manager of a fishery, determines that t weeks after 300 fish of a particular species are released into a pond, the average weight of an individual fish (in pounds) for the first 10 weeks will be

$$w(t) = 3 + t - 0.05t^2$$

Dom further determines that the proportion of the fish that are still alive after t weeks is given by

$$p(t) = \frac{31}{31 + t}$$

a. The expected yield $Y(t)$ of the fish after t weeks is the total weight of the fish that are still alive. Express $Y(t)$ in terms of $w(t)$ and $p(t)$, and sketch the graph of $Y(t)$ for $0 \le t \le 10$.

b. When should Dom expect the yield $Y(t)$ to be maximized? What is the maximum yield?

66. FISHERY MANAGEMENT Suppose for the situation described in Exercise 65, it costs the fishery $C(t) = 50 + 1.2t$ hundred dollars to maintain and monitor the pond for t weeks after the fish are released, and that each fish harvested after t weeks can be sold for $2.75 per pound.
   a. If all fish that remain alive in the pond after t weeks are harvested, express the profit obtained by the fishery as a function of t.
   b. When should the fish be harvested to maximize profit? What is the maximum profit?

67. GROWTH OF A SPECIES The percentage of codling moth eggs* that hatch at a given temperature T (degrees Celsius) is given by
$$H(T) = -0.53T^2 + 25T - 209 \quad \text{for } 15 \le T \le 30$$
Sketch the graph of the hatching function $H(T)$. At what temperature T ($15 \le T \le 30$) does the maximum percentage of eggs hatch? What is the maximum percentage?

MISCELLANEOUS PROBLEMS

68. Sketch a graph of a function that has all the following properties:
   a. $f'(0) = f'(1) = f'(2) = 0$
   b. $f'(x) < 0$ when $x < 0$ and $x > 2$
   c. $f'(x) > 0$ when $0 < x < 1$ and $1 < x < 2$

69. Sketch a graph of a function that has all the following properties:
   a. $f'(0) = f'(1) = f'(2) = 0$
   b. $f'(x) < 0$ when $0 < x < 1$
   c. $f'(x) > 0$ when $x < 0$, $1 < x < 2$, and $x > 2$

70. Sketch a graph of a function that has all the following properties:
   a. $f'(x) > 0$ when $x < -5$ and when $x > 1$
   b. $f'(x) < 0$ when $-5 < x < 1$
   c. $f(-5) = 4$ and $f(1) = -1$

71. Sketch a graph of a function that has all the following properties:
   a. $f'(x) < 0$ when $x < -1$
   b. $f'(x) > 0$ when $-1 < x < 3$ and when $x > 3$
   c. $f'(-1) = 0$ and $f'(3) = 0$

72. Find constants a, b, and c so that the graph of the function $f(x) = ax^2 + bx + c$ has a relative maximum at $(5, 12)$ and crosses the y axis at $(0, 3)$.

73. Find constants a, b, c, and d so that the graph of the function $f(x) = ax^3 + bx^2 + cx + d$ will have a relative maximum at $(-2, 8)$ and a relative minimum at $(1, -19)$.

74. Sketch the graph of $f(x) = (x - 1)^{2/5}$. Explain why $f'(x)$ is not defined at $x = 1$.

75. Sketch the graph of $f(x) = 1 - x^{3/5}$.

76. Use calculus to prove that the relative extremum of the quadratic function
$$f(x) = ax^2 + bx + c$$
occurs when $x = -\dfrac{b}{2a}$. Where is $f(x)$ increasing and decreasing?

77. Use calculus to prove that the relative extremum of the quadratic function $y = (x - p)(x - q)$ occurs midway between its x intercepts.

*In Exercises 78 through 81, use a graphing utility to sketch the graph of $f(x)$. Then find $f'(x)$ and graph it on the same axes (screen) as $f(x)$. Finally, use **TRACE**, **ZOOM**, or the **ZERO** command to find the values of x where $f'(x) = 0$.*

78. $f(x) = x^4 + 3x^3 - 9x^2 + 4$

79. $f(x) = (x^2 + x - 1)^3(x + 3)^2$

80. $f(x) = x^5 - 7.6x^3 + 2.1x^2 - 5$

81. $f(x) = (1 - x^{1/2})^{1/2}$

82. Use a graphing utility to sketch the graph of $f(x) = x^3 + 3x^2 - 5x + 11$. Then sketch the graph of $g(x) = (x + 1)^3 + 3(x + 1)^2 - 5(x + 1) + 11$ on the same axes. What function $h(x)$ would have a graph that is the same as that of $f(x)$, only shifted upward by 2 units and to the left by 3 units?

83. Let $f(x) = 4 + \sqrt{9 - 2x - x^2}$. Before actually graphing this function, what do you think the graph looks like? Now use a graphing utility to sketch the graph. Were you right?

84. Let $f(x) = x^3 - 6x^2 + 5x - 11$. Use a graphing utility to sketch the graph of $f(x)$. Then, on the same axes, sketch the graph of
$$g(x) = f(2x) = (2x)^3 - 6(2x)^2 + 5(2x) - 11$$
What (if any) relation exists between the two graphs?

*P. L. Shaffer and H. J. Gold, "A Simulation Model of Population Dynamics of the Codling Moth *Cydia Pomonella*," *Ecological Modeling*, Vol. 30, 1985, pp. 247–274.

SECTION 3.2   Concavity and Points of Inflection

Learning Objectives

1. Discuss concavity.
2. Use the sign of the second derivative to find intervals of concavity.
3. Locate and examine inflection points.
4. Apply the second derivative test for relative extrema.

In Section 3.1, you saw how to use the sign of the derivative $f'(x)$ to determine where $f(x)$ is increasing and decreasing and where its graph has relative extrema. In this section, you will see that the second derivative $f''(x)$ also provides useful information about the graph of $f(x)$. By way of introduction, here is a brief description of a situation from industry that can be analyzed using the second derivative.

The number of units that a factory worker can produce in t hours after arriving at work is often given by a function $Q(t)$ like the one whose graph is shown in Figure 3.12. Notice that at first the graph is not very steep. The steepness increases, however, until the graph reaches a point of maximum steepness, after which the steepness begins to decrease. This reflects the fact that at first the worker's efficiency is low. The efficiency increases, however, as the worker settles into a routine and continues to increase until the worker is performing at maximum efficiency. Then fatigue sets in and the efficiency begins to decline. The point P on the output curve where maximum efficiency occurs is known in economics as the **point of diminishing returns.**

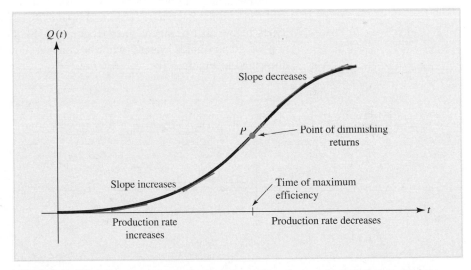

FIGURE 3.12   The output $Q(t)$ of a factory worker t hours after coming to work.

In this example, the worker's efficiency is the rate of production and is measured by the first derivative of the production function, so the *rate of efficiency* is measured by the *second derivative* of the production function. The efficiency rate increases before the point of diminishing returns P and decreases after P. Geometrically, this means that with increasing t, the slope of the tangent line increases to the left of P and decreases to the right, as indicated in Figure 3.12. Later in this section (in Example 3.2.6) we will return to the analysis of worker efficiency, but first we examine this geometric interpretation of the second derivative and show how it can be used in curve sketching and optimization.

Concavity   The increase and decrease of the slope of a tangent line will be described in terms of a graphical feature called **concavity.** Here is the definition we will use.

> **Concavity** ■ If the function $f(x)$ is differentiable on the interval $a < x < b$, then the graph of f is
>
> **concave upward** on $a < x < b$ if $f'(x)$ is increasing on the interval
> **concave downward** on $a < x < b$ if $f'(x)$ is decreasing on the interval

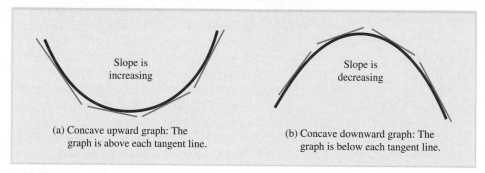

(a) Concave upward graph: The graph is above each tangent line.

(b) Concave downward graph: The graph is below each tangent line.

FIGURE 3.13   Concavity.

Equivalently, a graph is concave upward on an interval if it lies above all its tangent lines on the interval (Figure 3.13a), and concave downward on an interval where it lies below all its tangent lines (Figure 3.13b). Informally, a concave upward portion of graph "holds water," while a concave downward portion "spills water," as illustrated in Figure 3.14.

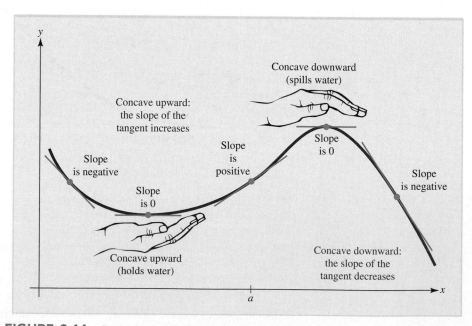

FIGURE 3.14   Concavity and the slope of the tangent.

Determining Intervals of Concavity Using the Sign of f''

There is a simple characterization of the concavity of the graph of a function $f(x)$ in terms of the second derivative $f''(x)$. In particular, in Section 3.1, we observed that a function $f(x)$ is increasing where its derivative is positive. Thus, the derivative function $f'(x)$ must be increasing where *its* derivative $f''(x)$ is positive. Suppose $f''(x) > 0$ on an interval $a < x < b$. Then $f'(x)$ is increasing, which in turn means that the graph of $f(x)$ is concave upward on this interval. Similarly, on an interval $a < x < b$, where $f''(x) < 0$, the derivative $f'(x)$ will be decreasing and the graph of $f(x)$ will be concave downward. Using these observations, we can modify the procedure for determining intervals of increase and decrease developed in Section 3.1 to obtain this procedure for determining intervals of concavity.

Second Derivative Procedure for Determining Intervals of Concavity for a Function f

Step 1. Find all values of x for which $f''(x) = 0$ or $f''(x)$ does not exist, and mark these numbers on a number line. This divides the line into a number of open intervals.

Step 2. Choose a test number c from each interval $a < x < b$ determined in step 1, and evaluate $f''(c)$. Then:

If $f''(c) > 0$, the graph of $f(x)$ is concave upward on $a < x < b$.

If $f''(c) < 0$, the graph of $f(x)$ is concave downward on $a < x < b$.

EXAMPLE 3.2.1 Finding Intervals of Concavity

Determine intervals of concavity for the function

$$f(x) = 2x^6 - 5x^4 + 7x - 3$$

Solution

We find that

$$f'(x) = 12x^5 - 20x^3 + 7$$

and

$$f''(x) = 60x^4 - 60x^2 = 60x^2(x^2 - 1) = 60x^2(x - 1)(x + 1)$$

The second derivative $f''(x)$ is continuous for all x and $f''(x) = 0$ for $x = 0$, $x = 1$, and $x = -1$. These numbers divide the x axis into four intervals on which $f''(x)$ does not change sign: $x < -1$, $-1 < x < 0$, $0 < x < 1$, and $x > 1$. Evaluating $f''(x)$ at test numbers in each of these intervals (say, at $x = -2$, $x = -\dfrac{1}{2}$, $x = \dfrac{1}{2}$, and $x = 5$, respectively), we find that

$$f''(-2) = 720 > 0 \qquad f''\left(\frac{-1}{2}\right) = -\frac{45}{4} < 0$$

$$f''\left(\frac{1}{2}\right) = -\frac{45}{4} < 0 \qquad f''(5) = 36{,}000 > 0$$

Thus, the graph of $f(x)$ is concave up for $x < -1$ and for $x > 1$ and concave down for $-1 < x < 0$ and for $0 < x < 1$, as indicated in this concavity diagram.

EXPLORE!

Following Example 3.2.1, store $f(x) = 2x^6 - 5x^4 + 7x - 3$ into Y1 using a bold graphing style. Write

Y2 = nDeriv(Y1, X, X)

but deselect it by moving the cursor to the equal sign and pressing **ENTER,** which will define but not graph Y2. Write Y3 = nDeriv(Y2, X, X). Graph Y1 and Y3 using the window [−2.35, 2.35]1 by [−18, 8]2. How does the behavior of Y3, representing $f''(x)$, relate to the concavity of $f(x)$, specifically at X = −1, 0, and 1?

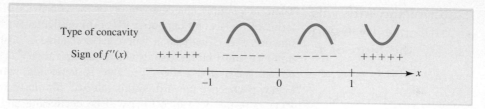

Intervals of concavity for $f(x) = 2x^6 - 5x^4 + 7x - 3$

The graph of $f(x)$ is shown in Figure 3.15.

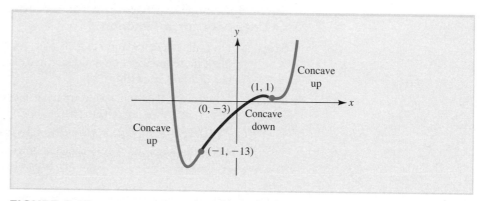

FIGURE 3.15   The graph of $f(x) = 2x^6 - 5x^4 + 7x - 3$.

NOTE Do not confuse the concavity of a graph with its "direction" (rising or falling). A function f may be increasing or decreasing on an interval regardless of whether its graph is concave upward or concave downward on that interval. The four possibilities are illustrated in Figure 3.16. ■

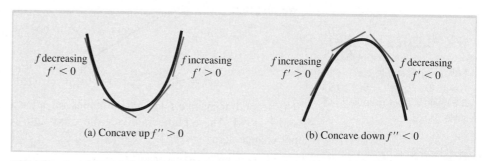

FIGURE 3.16   Possible combinations of increase, decrease, and concavity.

Inflection Points A point P on the graph of a function f is called an **inflection point** of f if the concavity of the graph changes at P; that is, the graph of f is concave upward on one side of P and concave downward on the other side. Such transition points provide useful information about the graph of f. For instance, the concavity diagram for the function $f(x) = 2x^6 - 5x^4 + 7x - 3$ analyzed in Example 3.2.1 shows that the concavity of the graph of f changes from upward to downward at $x = -1$ and from downward to upward at $x = 1$, so the corresponding points $(-1, -13)$ and $(1, 1)$ on the graph are inflection points of f. Inflection points may also be of practical interest when interpreting a mathematical model based on f. For example, the point of diminishing returns on the production curve shown in Figure 3.12 is an inflection point.

At an inflection point $(c, f(c))$, the graph of f can be neither concave upward $(f''(c) > 0)$ nor concave downward $(f''(c) < 0)$. Therefore, if $f''(c)$ exists, we must have $f''(c) = 0$. To summarize:

Inflection Points ■ An **inflection point** of the function f is a point $(c, f(c))$ on the graph of f where f is continuous and the concavity changes.

Note that $f(c)$ is defined at an inflection point and either $f''(c) = 0$ or $f''(c)$ does not exist. However, the derivative $f'(c)$ may or may not exist at such a point.

Procedure for Finding the Inflection Points for a Function f

Step 1. Compute $f''(x)$ and determine all points in the domain of f where either $f''(c) = 0$ or $f''(c)$ does not exist.

Step 2. For each number c found in step 1, determine the sign of $f''(x)$ to the left and to the right of $x = c$, that is, for $x < c$ and for $x > c$. If $f''(x) > 0$ on one side of $x = c$ and $f''(x) < 0$ on the other side, then $(c, f(c))$ is an inflection point for f.

EXAMPLE 3.2.2 Finding Inflection Points

In each case, find all inflection points of the given function.

a. $f(x) = 3x^5 - 5x^4 - 1$ **b.** $g(x) = x^{1/3}$

Solution

a. Note that $f(x)$ exists for all x and that

$$f'(x) = 15x^4 - 20x^3$$
$$f''(x) = 60x^3 - 60x^2 = 60x^2(x - 1)$$

Thus, $f''(x)$ is continuous for all x and $f''(x) = 0$ when $x = 0$ and $x = 1$. Testing the sign of $f''(x)$ on each side of $x = 0$ and $x = 1$ (say, at $x = -1, \frac{1}{2}$, and 2), we get

$$f''(-1) = -120 < 0 \qquad f''\left(\frac{1}{2}\right) = -\frac{15}{2} < 0 \qquad f''(2) = 240 > 0$$

which leads to the concavity pattern shown in this diagram:

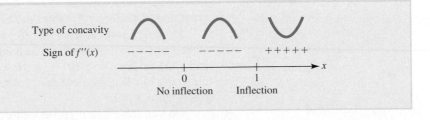

We see that the concavity does not change at $x = 0$ but changes from downward to upward at $x = 1$. Since $f(1) = -3$, it follows that $(1, -3)$ is an inflection point of f. The graph of f is shown in Figure 3.17a.

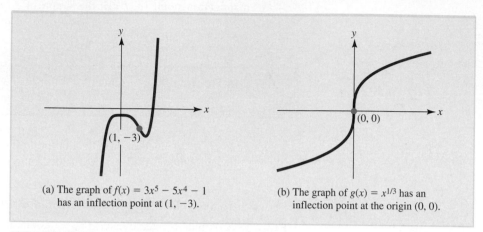

(a) The graph of $f(x) = 3x^5 - 5x^4 - 1$ has an inflection point at $(1, -3)$.

(b) The graph of $g(x) = x^{1/3}$ has an inflection point at the origin $(0, 0)$.

FIGURE 3.17 Two graphs with inflection points.

b. The function $g(x)$ is continuous for all x, and since

$$g'(x) = \frac{1}{3}x^{-2/3} \quad \text{and} \quad g''(x) = -\frac{2}{9}x^{-5/3}$$

it follows that $g''(x)$ is never 0 but does not exist when $x = 0$. Testing the sign of $g''(x)$ on each side of $x = 0$, we obtain the results displayed in this concavity diagram:

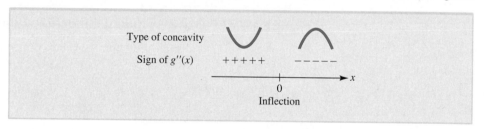

Since the concavity of the graph changes at $x = 0$ and $g(0) = 0$, there is an inflection point at the origin, $(0, 0)$. The graph of g is shown in Figure 3.17b.

NOTE A function can have an inflection point only where it is continuous. In particular, if $f(c)$ is not defined, there cannot be an inflection point corresponding to $x = c$ even if $f''(x)$ changes sign at $x = c$. For example, if $f(x) = \frac{1}{x}$, then $f''(x) = \frac{2}{x^3}$, so $f''(x) < 0$ if $x < 0$ and $f''(x) > 0$ if $x > 0$. The concavity changes from downward to upward at $x = 0$ (see Figure 3.18a), but there is no inflection point at $x = 0$ since $f(0)$ is not defined.

Moreover, just knowing that $f(c)$ is defined and that $f''(c) = 0$ does not guarantee that $(c, f(c))$ is an inflection point. For instance, if $f(x) = x^4$, then $f(0) = 0$ and $f''(x) = 12x^2$, so $f''(0) = 0$. However, $f''(x) > 0$ for any number $x \neq 0$, so the graph of f is always concave upward, and there is no inflection point at $(0, 0)$ (see Figure 3.18b).

Do you think that if $f(c)$ is defined and $f''(c) = 0$, then you can at least conclude that either an inflection point or a relative extremum occurs at $x = c$? For the answer, see Exercise 69. ∎

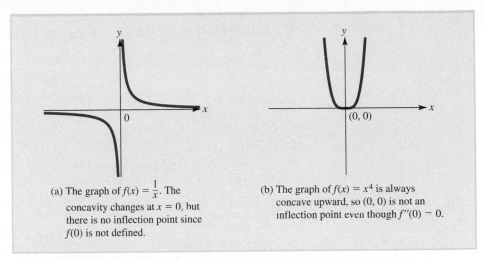

(a) The graph of $f(x) = \frac{1}{x}$. The concavity changes at $x = 0$, but there is no inflection point since $f(0)$ is not defined.

(b) The graph of $f(x) = x^4$ is always concave upward, so $(0, 0)$ is not an inflection point even though $f''(0) = 0$.

FIGURE 3.18   A graph need not have an inflection point where $f'' = 0$ or f'' does not exist.

Curve Sketching with the Second Derivative

Geometrically, inflection points occur at "twists" on a graph. Here is a summary of the graphical possibilities.

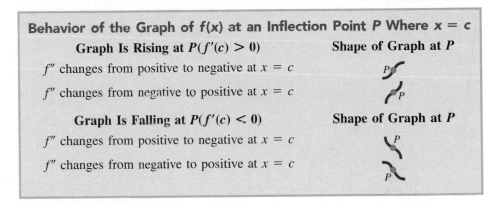

Behavior of the Graph of $f(x)$ at an Inflection Point P Where $x = c$	
Graph Is Rising at $P(f'(c) > 0)$	**Shape of Graph at P**
f'' changes from positive to negative at $x = c$	
f'' changes from negative to positive at $x = c$	
Graph Is Falling at $P(f'(c) < 0)$	**Shape of Graph at P**
f'' changes from positive to negative at $x = c$	
f'' changes from negative to positive at $x = c$	

Figure 3.19 shows several ways that inflection points can occur on a graph.

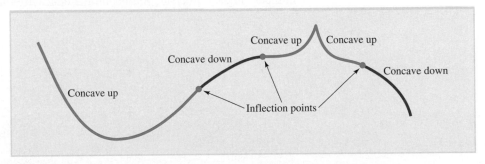

FIGURE 3.19   A graph showing concavity and inflection points.

By adding the criteria for concavity and inflection points to the first derivative methods discussed in Section 3.1, you can sketch a variety of graphs with considerable detail. This is illustrated in Example 3.2.3.

> **EXAMPLE 3.2.3** Using Concavity in Graphing

Determine where the function

$$f(x) = 3x^4 - 2x^3 - 12x^2 + 18x + 15$$

is increasing and decreasing and where its graph is concave up and concave down. Find all relative extrema and points of inflection and sketch the graph.

Solution

First, note that since $f(x)$ is a polynomial, it is continuous for all x, as are the derivatives $f'(x)$ and $f''(x)$. The first derivative of $f(x)$ is

$$f'(x) = 12x^3 - 6x^2 - 24x + 18 = 6(x - 1)^2(2x + 3)$$

and $f'(x) = 0$ only when $x = 1$ and $x = -1.5$. Evaluating $f'(x)$ at test numbers (say, at -2, 0, and 3), you obtain the arrow diagram shown. Note that there is a relative minimum at $x = -1.5$ but no extremum at $x = 1$.

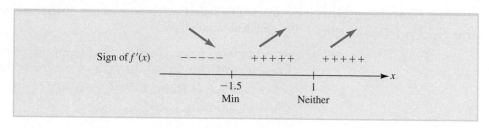

The second derivative is

$$f''(x) = 36x^2 - 12x - 24 = 12(x - 1)(3x + 2)$$

and $f''(x) = 0$ only when $x = 1$ and $x = -\frac{2}{3}$. The sign of $f''(x)$ does not change on each of the intervals $x < -\frac{2}{3}$, $-\frac{2}{3} < x < 1$, and $x > 1$. Evaluating $f''(x)$ at test numbers in each interval, you obtain the concavity diagram shown.

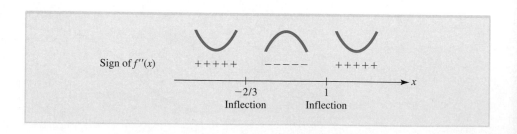

The patterns in these two diagrams suggest that there is a relative minimum at $x = -1.5$ and inflection points at $x = -\frac{2}{3}$ and $x = 1$ (since the concavity changes at both points).

To sketch the graph, you find that

$$f(-1.5) = -17.06 \qquad f\left(-\frac{2}{3}\right) = -1.15 \qquad f(1) = 22$$

and plot a cup ⌣ at $(-1.5, -17.06)$ to mark the relative minimum located there. Likewise, plot twists ∫ at $\left(-\frac{2}{3}, -1.15\right)$ and ⌐ at $(1, 22)$ to mark the shape of the graph near the inflection points. Using the arrow and concavity diagrams, you get the preliminary diagram shown in Figure 3.20a. Finally, complete the sketch as shown in Figure 3.20b by passing a smooth curve through the minimum point, the inflection points, and the y intercept $(0, 15)$.

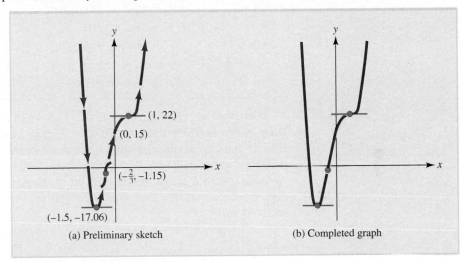

(a) Preliminary sketch   (b) Completed graph

FIGURE 3.20 The graph of $f(x) = 3x^4 - 2x^3 - 12x^2 + 18x + 15$.

Sometimes you are given the graph of a derivative function $f'(x)$ and asked to analyze the graph of $f(x)$ itself. For instance, it would be quite reasonable for a manufacturer who knows the marginal cost $C'(x)$ associated with producing x units of a particular commodity to want to know as much as possible about the total cost $C(x)$. Example 3.2.4 illustrates a procedure for carrying out this kind of analysis.

EXAMPLE 3.2.4 Using the Graph of a Derivative f' to Graph f

The graph of the derivative $f'(x)$ of a function $f(x)$ is shown in the figure. Find intervals of increase and decrease and concavity for $f(x)$, and locate all relative extrema and inflection points. Then sketch a curve that has all these features.

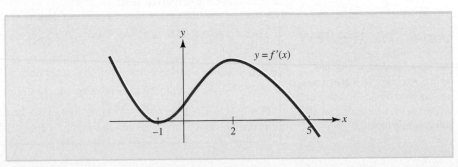

Solution

First, note that for $x < -1$, the graph of $f'(x)$ is above the x axis, so $f'(x) > 0$ and the graph of $f(x)$ is rising. The graph of $f'(x)$ is also falling for $x < -1$, which means that $f''(x) < 0$ and the graph of $f(x)$ is concave down. Other intervals can be analyzed in a similar fashion, and the results are summarized in the table.

x	Feature of $y = f'(x)$	Feature of $y = f(x)$
$x < -1$	f' is positive; decreasing	f is increasing; concave down
$x = -1$	x intercept; horizontal tangent	Horizontal tangent; possible inflection point ($f'' = 0$)
$-1 < x < 2$	f' is positive; increasing	f is increasing; concave up
$x = 2$	Horizontal tangent	Possible inflection point
$2 < x < 5$	f' is positive; decreasing	f is increasing; concave down
$x = 5$	x intercept	Horizontal tangent
$x > 5$	f' is negative; decreasing	f is decreasing; concave down

Since the concavity changes at $x = -1$ (down to up), an inflection point occurs there, along with a horizontal tangent. At $x = 2$, there is also an inflection point (concavity changes from up to down), but no horizontal tangent. The graph of $f(x)$ is rising to the left of $x = 5$ and falling to the right, so there must be a relative maximum at $x = 5$.

One possible graph that has all the features required for $y = f(x)$ is shown in Figure 3.21. Note, however, that since you are not given the values of $f(-1)$, $f(2)$, and $f(5)$, many other graphs will also satisfy the requirements.

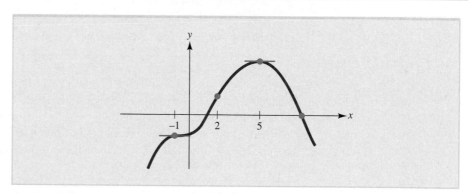

FIGURE 3.21 A possible graph for $y = f(x)$ in Example 3.2.4.

The Second Derivative Test

The second derivative can also be used to classify a critical point of a function as a relative maximum or minimum. Here is a statement of the procedure, which is known as the **second derivative test.**

Just-In-Time **REVIEW**

An open interval is one like $a < x < b$ or $x > a$ that does not include its endpoints. See Appendix A.1 for a review of interval terminology.

The Second Derivative Test ■ Suppose $f''(x)$ exists on an open interval containing $x = c$ and that $f'(c) = 0$.

If $f''(c) > 0$, then f has a relative minimum at $x = c$.

If $f''(c) < 0$, then f has a relative maximum at $x = c$.

However, if $f''(c) = 0$ or if $f''(c)$ does not exist, the test is inconclusive and f may have a relative maximum, a relative minimum, or no relative extremum at all at $x = c$.

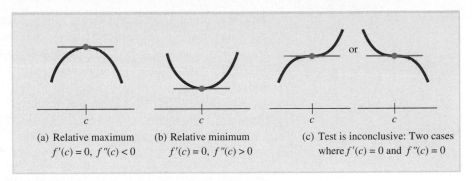

FIGURE 3.22 The second derivative test.

To see why the second derivative test works, look at Figure 3.22, which shows four possibilities that can occur when $f'(c) = 0$. Figure 3.22a suggests that at a relative maximum, the graph of f must be concave downward, so $f''(c) < 0$. Likewise, at a relative minimum (Figure 3.22b), the graph of f must be concave upward, so $f''(c) > 0$. On the other hand, if $f'(c) = 0$ and $f''(c)$ is neither positive nor negative, the test is inconclusive. Figure 3.22c suggests that if $f'(c) = 0$ and $f''(c) = 0$, there may be an inflection point at $x = c$. There may also be a relative extremum. For example, $f(x) = x^4$ has a relative minimum at $x = 0$ and $g(x) = -x^4$ has a relative maximum (see Figure 3.24).

The second derivative test is illustrated in Example 3.2.5.

EXAMPLE 3.2.5   Applying the Second Derivative Test

Find the critical points of $f(x) = 2x^3 + 3x^2 - 12x - 7$, and use the second derivative test to classify each critical point as a relative maximum or minimum.

Solution

Since the first derivative

$$f'(x) = 6x^2 + 6x - 12 = 6(x + 2)(x - 1)$$

is zero when $x = -2$ and $x = 1$, the corresponding points $(-2, 13)$ and $(1, -14)$ are the critical points of f. To test these points, compute the second derivative

$$f''(x) = 12x + 6$$

and evaluate it at $x = -2$ and $x = 1$. Since

$$f''(-2) = -18 < 0$$

it follows that the critical point $(-2, 13)$ is a relative maximum, and since

$$f''(1) = 18 > 0$$

it follows that the critical point $(1, -14)$ is a relative minimum. For reference, the graph of f is sketched in Figure 3.23.

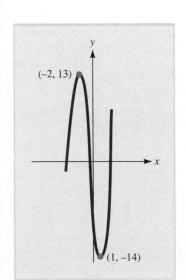

FIGURE 3.23 The graph of $f(x) = 2x^3 + 3x^2 - 12x - 7$.

NOTE   Although it was easy to use the second derivative test to classify critical points in Example 3.2.5, the test does have some limitations. For some functions, the work involved in computing the second derivative is time-consuming, which may diminish the appeal of the test. Moreover, the test applies only to critical points at which the derivative is zero and not to those where the derivative is undefined. Finally, if $f'(c)$ and $f''(c)$ are both zero, the second derivative test tells you nothing about the nature of the critical point. This is illustrated in Figure 3.24, which shows the graphs of three functions whose first and second derivatives are both zero when $x = 0$. When it is inconvenient or impossible to apply the second derivative test, you may still be able to use the first derivative test described in Section 3.1 to classify critical points.   ■

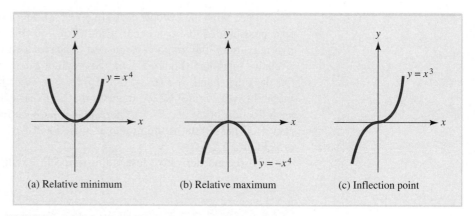

(a) Relative minimum   (b) Relative maximum   (c) Inflection point

FIGURE 3.24   Three functions whose first and second derivatives are zero at $x = 0$.

In Example 3.2.6, we return to the questions of worker efficiency and diminishing returns examined in the illustration at the beginning of this section. Our goal is to find a maximum for a worker's *rate* of production, that is, the derivative of the worker's output. Hence, we will set the *second* derivative of output equal to zero and find an inflection point of the output function, which we interpret as the point of diminishing returns for production.

EXAMPLE 3.2.6   Finding a Point of Diminishing Returns for Production

Anish is an efficiency expert for an electronics firm. He conducts a study of the morning shift (8:00 A.M. to 12:00 noon) at one of the firm's factories which indicates that the average worker who arrives on the job at 8:00 A.M. will have produced Q units t hours later, where $Q(t) = -t^3 + 9t^2 + 12t$. At what time during the morning shift should Anish expect the average worker's production to reach the point of diminishing returns?

Solution

The worker's efficiency is the rate of production, given by the derivative of the output $Q(t)$; that is,

$$R(t) = Q'(t) = -3t^2 + 18t + 12$$

Since the morning shift runs from 8:00 A.M. until noon, Anish's goal is to find the largest rate $R(t)$ for $0 \leq t \leq 4$. The derivative of the rate function is

$$R'(t) = Q''(t) = -6t + 18$$

which is zero when $t = 3$, positive for $0 < t < 3$, and negative for $3 < t < 4$, as indicated in the arrow diagram shown.

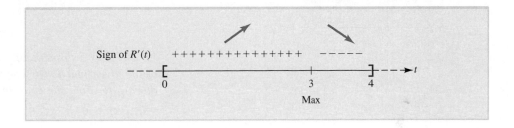

Anish finds that the rate of worker production (efficiency) $R(t)$ increases for $0 < t < 3$ and decreases for $3 < t < 4$, so its maximum value occurs when $t = 3$ (at 11:00 A.M.), as shown in Figure 3.25a. This means that the point of maximum worker efficiency (point of diminishing returns) on the graph of the production function $Q(t)$ is an inflection point of the graph, since $Q''(t) = R'(t)$ changes sign there (Figure 3.25b).

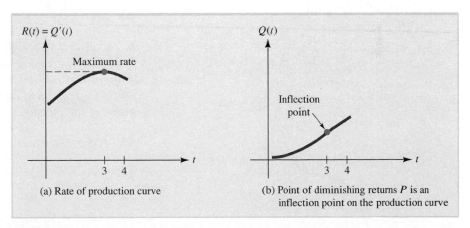

(a) Rate of production curve

(b) Point of diminishing returns P is an inflection point on the production curve

FIGURE 3.25 The production of an average worker.

EXERCISES ■ 3.2

In Exercises 1 through 4, determine where the second derivative of the function is positive and where it is negative.

1.

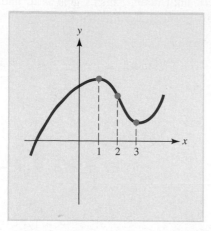

2.

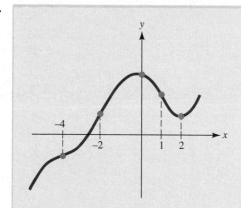

3.

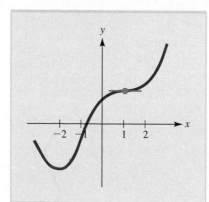

4.

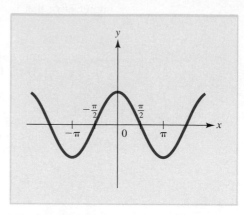

In Exercises 5 through 12, determine where the graph of the given function is concave upward and concave downward. Find the coordinates of all inflection points.

5. $f(x) = x^3 + 3x^2 + x + 1$

6. $f(x) = x^4 - 4x^3 + 10x - 9$

7. $f(x) = x(2x + 1)^2$

8. $f(s) = s(s + 3)^2$

9. $g(t) = t^2 - \dfrac{1}{t}$

10. $F(x) = (x - 4)^{7/3}$

11. $f(x) = x^4 - 6x^3 + 7x - 5$

12. $g(x) = 3x^5 - 25x^4 + 11x - 17$

In Exercises 13 through 26, determine where the given function is increasing and decreasing, and where its graph is concave up and concave down. Find the relative extrema and inflection points, and sketch the graph of the function.

13. $f(x) = \dfrac{1}{3}x^3 - 9x + 2$

14. $f(x) = x^3 + 3x^2 + 1$

15. $f(x) = x^4 - 4x^3 + 10$

16. $f(x) = x^3 - 3x^2 + 3x + 1$

17. $f(x) = (x - 2)^3$

18. $f(x) = x^5 - 5x$

19. $f(x) = (x^2 - 5)^3$

20. $f(x) = (x - 2)^4$

21. $f(s) = 2s(s + 4)^3$

22. $f(x) = (x^2 - 3)^3$

23. $g(x) = \sqrt{x^2 + 1}$

24. $f(x) = \dfrac{x^2}{x^2 + 3}$

25. $f(x) = \dfrac{1}{x^2 + x + 1}$

26. $f(x) = x^4 + 6x^3 - 24x^2 + 24$

In Exercises 27 through 38, use the second derivative test to find the relative maxima and minima of the given function.

27. $f(x) = x^3 + 3x^2 + 1$

28. $f(x) = x^4 - 2x^2 + 3$

29. $f(x) = (x^2 - 9)^2$

30. $f(x) = x + \dfrac{1}{x}$

31. $f(x) = 2x + 1 + \dfrac{18}{x}$

32. $f(x) = \dfrac{x^2}{x - 2}$

33. $f(x) = x^2(x - 5)^2$

34. $f(x) = \left(\dfrac{x}{x + 1}\right)^2$

35. $h(t) = \dfrac{2}{1 + t^2}$

36. $f(s) = \dfrac{s + 1}{(s - 1)^2}$

37. $f(x) = \dfrac{(x - 2)^3}{x^2}$

38. $h(t) = \dfrac{(t + 3)^3}{(t - 1)^2}$

In Exercises 39 through 42, the second derivative $f''(x)$ of a function is given. In each case, use this information to determine where the graph of $f(x)$ is concave upward and concave downward and find all values of x for which an inflection point occurs. [You are not required to find $f(x)$ or the y coordinates of the inflection points.]

39. $f''(x) = x^2(x - 3)(x - 1)$

40. $f''(x) = x^3(x^2 + 2x - 3)$

41. $f''(x) = (x - 1)^{1/3}$

42. $f''(x) = \dfrac{x^2 + x - 2}{x^2 + 4}$

In Exercises 43 through 46, the first derivative $f'(x)$ of a certain function $f(x)$ is given. In each case,

    (a) Find intervals on which f is increasing and decreasing.

    (b) Find intervals on which the graph of f is concave up and concave down.

    (c) Find the x coordinates of the relative extrema and inflection points of f.

    (d) Sketch a possible graph for $f(x)$.

43. $f'(x) = x^2 - 4x$

44. $f'(x) = x^2 - 2x - 8$

45. $f'(x) = 5 - x^2$

46. $f'(x) = x(1 - x)$

47. Sketch the graph of a function that has all the following properties:

    a. $f'(x) > 0$ when $x < -1$ and when $x > 3$

    b. $f'(x) < 0$ when $-1 < x < 3$

    c. $f''(x) < 0$ when $x < 2$

    d. $f''(x) > 0$ when $x > 2$

48. Sketch the graph of a function f that has all the following properties:

    a. The graph has discontinuities at $x = -1$ and at $x = 3$

    b. $f'(x) > 0$ for $x < 1$, $x \neq -1$

    c. $f'(x) < 0$ for $x > 1$, $x \neq 3$

    d. $f''(x) > 0$ for $x < -1$ and $x > 3$ and $f''(x) < 0$ for $-1 < x < 3$

    e. $f(0) = 0 = f(2)$, $f(1) = 3$

In Exercises 49 through 52, the graph of a derivative function $y = f'(x)$ is given. Describe the function $f(x)$, and sketch a possible graph of $y = f(x)$.

49.

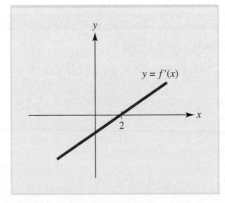

50.

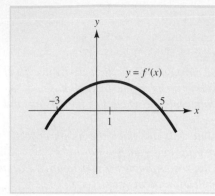

51.

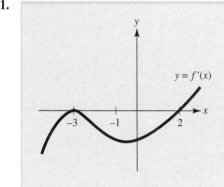

52.

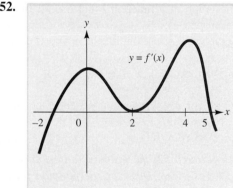

BUSINESS AND ECONOMICS APPLIED PROBLEMS

53. MARGINAL ANALYSIS   The cost of producing
x units of a commodity per week is
$$C(x) = 0.3x^3 - 5x^2 + 28x + 200$$
a. Find the marginal cost $C'(x)$, and sketch its
graph along with the graph of $C(x)$ on the
same coordinate plane.
b. Find all values of x where $C''(x) = 0$. How are
these levels of production related to the graph
of the marginal cost?

54. MARGINAL ANALYSIS   The profit obtained
from producing x thousand units of a particular
commodity each year is $P(x)$ dollars, where
$$P(x) = -x^{9/2} + 90x^{7/2} - 5,000$$
a. Find the marginal profit $P'(x)$, and determine
all values of x such that $P'(x) = 0$.
b. Sketch the graph of marginal profit along with
the graph of $P(x)$ on the same coordinate plane.
c. Find $P''(x)$, and determine all values of x such
that $P''(x) = 0$. How are these levels of
production related to the graph of marginal
profit?

55. SALES   A company estimates that if x thousand
dollars are spent on marketing a certain product,
then $S(x)$ units of the product will be sold each
month, where
$$S(x) = -x^3 + 33x^2 + 60x + 1,000$$
a. How many units will be sold if no money is
spent on marketing?
b. Sketch the graph of $S(x)$. For what value of x
does the graph have an inflection point? What
is the significance of this marketing
expenditure?

56. SALES   A company estimates that when x
thousand dollars are spent on the marketing of a
certain product, $Q(x)$ units of the product will be
sold, where
$$Q(x) = -4x^3 + 252x^2 - 3,200x + 17,000$$
for $10 \le x \le 40$. Sketch the graph of $Q(x)$.
Where does the graph have an inflection point?
What is the significance of the marketing
expenditure that corresponds to this point?

57. DIMINISHING RETURNS OF PRODUCTION
An efficiency study of the morning shift at a
factory (7:00 A.M. to 12:00 noon) indicates that
an average worker who arrives on the job at
7:00 A.M. will have produced Q units t hours
later, where
$$Q(t) = -t^3 + \frac{9}{2}t^2 + 15t$$
a. When during the morning shift does the
worker's production reach the point of
diminishing returns?
b. When during the morning shift is the worker
performing least efficiently?

58. DIMINISHING RETURNS OF PRODUCTION An efficiency study of the afternoon shift at a factory (noon to 5 P.M.) indicates that an average worker who arrives on the job at noon will have produced

$$Q(t) = -t^3 + 6t^2 + 15t$$

units t hours later.

a. When during the afternoon shift does the worker's production reach the point of diminishing returns?

b. At what time during the afternoon shift is the worker performing least efficiently?

59. HOUSING STARTS Suppose that in a certain community, there will be $M(r)$ thousand new houses built when the 30-year fixed mortgage rate is $r\%$, where

$$M(r) = \frac{1 + 0.02r}{1 + 0.009r^2}$$

a. Find $M'(r)$ and $M''(r)$.

b. Sketch the graph of the construction function $M(r)$.

c. At what rate of interest r is the rate of construction of new houses minimized?

60. GOVERNMENT SPENDING During a recession, Congress decides to stimulate the economy by providing funds to hire unemployed workers for government projects. Suppose that t months after the stimulus program begins, there are $N(t)$ thousand people unemployed, where

$$N(t) = -t^3 + 45t^2 + 408t + 3{,}078$$

a. What is the maximum number of unemployed workers? When does the maximum level of unemployment occur?

b. To avoid overstimulating the economy (and inducing inflation), a decision is made to terminate the stimulus program as soon as the rate of unemployment begins to decline. When does this occur? At this time, how many people are unemployed?

61. ADVERTISING Madge, the manager of the Footloose sandal company, determines that t months after initiating an advertising campaign, $S(t)$ hundred pairs of sandals will be sold, where

$$S(t) = \frac{3}{t + 2} - \frac{12}{(t + 2)^2} + 5$$

a. Find $S'(t)$ and $S''(t)$.

b. At what time will sales be maximized? What is the maximum level of sales?

c. Madge plans to terminate the advertising campaign when the sales rate is minimized. When does this occur? What are the sales level and sales rate at this time?

LIFE AND SOCIAL SCIENCE APPLIED PROBLEMS

62. SPREAD OF AN EPIDEMIC Let $Q(t)$ denote the number of people in a city of population N_0 who have been infected with a certain disease t days after the beginning of an epidemic. Studies indicate that the rate $R(Q)$ at which an epidemic spreads is jointly proportional to the number of people who have contracted the disease and the number who have not, so $R(Q) = kQ(N_0 - Q)$. Sketch the graph of the rate function, and interpret your graph. In particular, what is the significance of the highest point on the graph of $R(Q)$?

63. SPREAD OF AN EPIDEMIC An epidemiologist determines that a particular epidemic spreads in such a way that t weeks after the outbreak, N hundred new cases will be reported, where

$$N(t) = \frac{5t}{12 + t^2}$$

a. Find $N'(t)$ and $N''(t)$.

b. At what time is the epidemic at its worst? What is the maximum number of reported new cases?

c. Health officials declare the epidemic to be under control when the rate of reported new cases is minimized. When does this occur? What number of new cases will be reported at that time?

64. POPULATION GROWTH Studies show that when environmental factors impose an upper bound on the possible size of a population $P(t)$, the population often tends to grow in such a way that the percentage rate of change of $P(t)$ satisfies

$$\frac{100\, P'(t)}{P(t)} = A - BP(t)$$

where A and B are positive constants. Where does the graph of $P(t)$ have an inflection point? What is the significance of this point? (Your answer will be in terms of A and B.)

65. POPULATION GROWTH   A 5-year projection of population trends suggests that t years from now, the population of a certain community will be $P(t) = -t^3 + 9t^2 + 48t + 50$ thousand.
   a. At what time during the 5-year period will the population be growing most rapidly?
   b. At what time during the 5-year period will the population be growing least rapidly?
   c. At what time is the rate of population growth changing most rapidly?

66. SPREAD OF A RUMOR   The rate at which a rumor spreads through a community of P people is jointly proportional to the number of people N who have heard the rumor and the number who have not. Show that the rumor is spreading most rapidly when half the people have heard it.

67. TISSUE GROWTH   Suppose a particular tissue culture has area $A(t)$ at time t and a potential maximum area M. Based on properties of cell division, it is reasonable to assume that the area A grows at a rate jointly proportional to $\sqrt{A(t)}$ and $M - A(t)$; that is,

$$\frac{dA}{dt} = k\sqrt{A(t)}[M - A(t)]$$

where k is a positive constant.
   a. Let $R(t) = A'(t)$ be the rate of tissue growth. Show that $R'(t) = 0$ when $A(t) = M/3$.
   b. Is the rate of tissue growth greatest or least when $A(t) = M/3$? [*Hint:* Use the first derivative test or the second derivative test.]
   c. Based on the given information and what you discovered in part (a), what can you say about the graph of $A(t)$?

MISCELLANEOUS PROBLEMS

68. Water is poured at a constant rate into the vase shown in the accompanying figure. Let $h(t)$ be the height of the water in the vase at time t (assume the vase is empty when $t = 0$). Sketch a rough graph of the function $h(t)$. In particular, what happens when the water level reaches the neck of the vase?

EXERCISE 68

69. Let $f(x) = x^4 + x$. Show that even though $f''(0) = 0$, the graph of f has neither a relative extremum nor an inflection point where $x = 0$. Sketch the graph of $f(x)$.

70. Use calculus to show that the graph of the quadratic function $y = ax^2 + bx + c$ is concave upward if a is positive and concave downward if a is negative.

71. If $f(x)$ and $g(x)$ are continuous functions that both have an inflection point at $x = c$, is it true that the sum $h(x) = f(x) + g(x)$ must also have an inflection point at $x = c$? Either explain why this must always be true or find functions $f(x)$ and $g(x)$ for which it is false.

72. Suppose $f(x)$ and $g(x)$ are continuous functions with $f'(c) = 0$. If both f and g have an inflection point at $x = c$, does $P(x) = f(x)g(x)$ have an inflection point at $x = c$? Either explain why this must always be true or find functions $f(x)$ and $g(x)$ for which it is false.

73. Given the function $f(x) = 2x^3 + 3x^2 - 12x - 7$, complete these steps:

   a. Graph using $[-10, 10]1$ by $[-10, 10]1$ and $[-10, 10]1$ by $[-20, 20]2$.
   b. Fill in this table:

x	-4	-2	-1	0	1	2
$f(x)$						
$f'(x)$						
$f''(x)$						

   c. Approximate the x intercepts and y intercept to two decimal places.
   d. Find the relative maximum and relative minimum points.
   e. Find the intervals over which $f(x)$ is increasing.
   f. Find the intervals over which $f(x)$ is decreasing.
   g. Find any inflection points.
   h. Find the intervals over which the graph of $f(x)$ is concave upward.
   i. Find the intervals over which the graph of $f(x)$ is concave downward.
   j. Show that the concavity changes from upward to downward, or vice versa, when x moves from a little less than the point of inflection to a little greater than the point of inflection.
   k. Find the largest and smallest values for this function for $-4 \le x \le 2$.

74. Repeat Exercise 73 for the function
$$f(x) = 3.7x^4 - 5.03x^3 + 2x^2 - 0.7$$

SECTION 3.3   Curve Sketching

Learning Objectives
1. Determine horizontal and vertical asymptotes of a graph.
2. Discuss and apply a general procedure for sketching graphs.

Just-In-Time **REVIEW**

It is important to remember that ∞ is *not* a number. It is used only to represent a process of unbounded growth or the result of such growth.

So far in this chapter, you have seen how to use the derivative $f'(x)$ to determine where the graph of $f(x)$ is rising and falling and to use the second derivative $f''(x)$ to determine the concavity of the graph. While these tools are adequate for locating the high and low points of a graph and for sculpting its twists and turns, there are other graphical features that are best described using limits.

Recall from Section 1.5 that a limit of the form $\lim\limits_{x \to +\infty} f(x)$ or $\lim\limits_{x \to -\infty} f(x)$, in which the independent variable x either increases or decreases without bound, is called a *limit at infinity*. On the other hand, if the functional values $f(x)$ themselves grow without bound as x approaches a number c, we say that $f(x)$ has an *infinite limit* at $x = c$ and write $\lim\limits_{x \to c} f(x) = +\infty$ if $f(x)$ increases indefinitely as x approaches c, or $\lim\limits_{x \to c} f(x) = -\infty$ if $f(x)$ decreases indefinitely. Collectively, limits at infinity and infinite limits are referred to as **limits involving infinity.** Our first goal in this section is to see how limits involving infinity may be interpreted as graphical features. This information will then be combined with the derivative methods of Sections 3.1 and 3.2 to form a general procedure for sketching graphs.

Vertical Asymptotes

Limits involving infinity can be used to describe graphical features called *asymptotes*. In particular, the graph of a function $f(x)$ is said to have a **vertical asymptote** at $x = c$ if $f(x)$ increases or decreases without bound as x tends toward c, from either the right or the left.

For instance, consider the rational function

$$f(x) = \frac{x + 1}{x - 2}$$

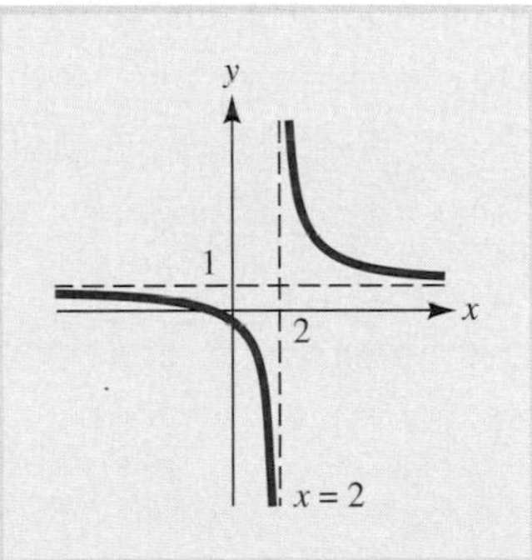

As x approaches 2 from the left ($x < 2$), the functional values decrease without bound, but they increase without bound if the approach is from the right ($x > 2$). This behavior is illustrated in the table and demonstrated graphically in Figure 3.26.

FIGURE 3.26   The graph of $f(x) = \dfrac{x + 1}{x - 2}$.

x	1.95	1.97	1.99	1.999	2	2.001	2.005	2.01
$f(x) = \dfrac{x + 1}{x - 2}$	−59	−99	−299	−2,999	Undefined	3,001	601	301

The behavior in this example can be summarized as follows using the one-sided limit notation introduced in Section 1.6:

$$\lim_{x \to 2^-} \frac{x + 1}{x - 2} = -\infty \qquad \text{and} \qquad \lim_{x \to 2^+} \frac{x + 1}{x - 2} = +\infty$$

In a similar fashion, we use the limit notation to define the concept of vertical asymptote.

Vertical Asymptotes ■ The line $x = c$ is a *vertical asymptote* of the graph of $f(x)$ if one or both of these conditions holds:

$$\lim_{x \to c^-} f(x) = +\infty \quad (\text{or } -\infty)$$

$$\lim_{x \to c^+} f(x) = +\infty \quad (\text{or } -\infty)$$

In general, a rational function $R(x) = \dfrac{p(x)}{q(x)}$ has a vertical asymptote $x = c$ whenever the denominator $q(c)$ is zero but the numerator $p(c)$ is not. Example 3.3.1 features a function with a vertical asymptote.

EXAMPLE 3.3.1   Finding Vertical Asymptotes

Determine all vertical asymptotes of the graph of

$$f(x) = \frac{x^2 - 9}{x^2 + 3x}$$

Solution

Let $p(x) = x^2 - 9$ and $q(x) = x^2 + 3x$ be the numerator and denominator of $f(x)$, respectively. Then $q(x) = 0$ when $x = -3$ and when $x = 0$. However, for $x = -3$, we also have $p(-3) = 0$ and

$$\lim_{x \to -3} \frac{x^2 - 9}{x^2 + 3x} = \lim_{x \to -3} \frac{x - 3}{x} = 2$$

This means that the graph of $f(x)$ has a hole at the point $(-3, 2)$ and $x = -3$ is *not* a vertical asymptote of the graph.

On the other hand, for $x = 0$ we have $q(0) = 0$ but $p(0) \neq 0$, which suggests that the y axis (the vertical line $x = 0$) is a vertical asymptote for the graph of $f(x)$. This asymptotic behavior is verified by noting that

$$\lim_{x \to 0^-} \frac{x^2 - 9}{x^2 + 3x} = +\infty \quad \text{and} \quad \lim_{x \to 0^+} \frac{x^2 - 9}{x^2 + 3x} = -\infty$$

The graph of $f(x)$ is shown in Figure 3.27.

EXPLORE!

Referring to Example 3.3.1, store $f(x) = \dfrac{x^2 - 9}{x^2 + 3x}$ into Y1 and graph using the expanded decimal window $[-4.7, 4.7]1$ by $[-6.2, 6.2]1$. Use the **TRACE** feature to confirm that $f(x)$ is not defined at X = −3 and X = 0. How does your calculator indicate these undefined points?

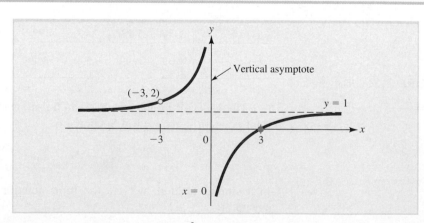

FIGURE 3.27   The graph of $f(x) = \dfrac{x^2 - 9}{x^2 + 3x}$.

Horizontal Asymptotes

In Figure 3.27, note that the graph approaches the horizontal line $y = 1$ as x increases or decreases without bound; that is,

$$\lim_{x \to -\infty} \frac{x^2 - 9}{x^2 + 3x} = 1 \qquad \text{and} \qquad \lim_{x \to +\infty} \frac{x^2 - 9}{x^2 + 3x} = 1$$

In general, when a function $f(x)$ tends toward a finite value b as x either increases or decreases without bound (or both as in our example), then the horizontal line $y = b$ is called a **horizontal asymptote** of the graph of $f(x)$. Here is a definition.

Horizontal Asymptotes ■ The horizontal line $y = b$ is called a *horizontal asymptote* of the graph of $y = f(x)$ if

$$\lim_{x \to -\infty} f(x) = b \qquad \text{or} \qquad \lim_{x \to +\infty} f(x) = b$$

EXAMPLE 3.3.2 Finding Horizontal Asymptotes

Determine all horizontal asymptotes of the graph of

$$f(x) = \frac{x^2}{x^2 + x + 1}$$

Solution

Dividing each term in the rational function $f(x)$ by x^2 (the highest power of x in the denominator), we find that

$$\lim_{x \to +\infty} f(x) = \lim_{x \to +\infty} \frac{x^2}{x^2 + x + 1} = \lim_{x \to +\infty} \frac{x^2/x^2}{x^2/x^2 + x/x^2 + 1/x^2}$$

$$= \lim_{x \to +\infty} \frac{1}{1 + 1/x + 1/x^2} = \frac{1}{1 + 0 + 0} = 1 \qquad \begin{matrix}\text{reciprocal power}\\\text{rule}\end{matrix}$$

and similarly,

$$\lim_{x \to -\infty} f(x) = \lim_{x \to -\infty} \frac{x^2}{x^2 + x + 1} = 1$$

Thus, the graph of $f(x)$ has $y = 1$ as a horizontal asymptote. The graph is shown in Figure 3.28.

Just-In-Time **REVIEW**

Recall the reciprocal power rules for limits (Section 1.5):

$$\lim_{x \to +\infty} \frac{A}{x^k} = 0$$

and

$$\lim_{x \to -\infty} \frac{A}{x^k} = 0$$

for constants A and k, with $k > 0$ and x^k defined for all x.

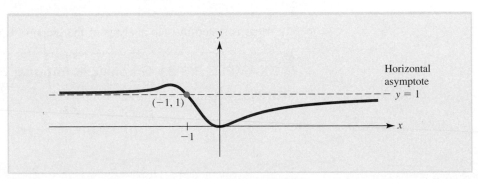

FIGURE 3.28 The graph of $f(x) = \dfrac{x^2}{x^2 + x + 1}$.

NOTE   The graph of a function $f(x)$ can never cross a vertical asymptote $x = c$ because at least one of the one-sided limits $\lim_{x \to c^-} f(x)$ and $\lim_{x \to c^+} f(x)$ must be infinite. However, it is quite possible for a graph to cross its horizontal asymptotes. For instance, in Example 3.3.2, the graph of $y = \dfrac{x^2}{x^2 + x + 1}$ crosses the horizontal asymptote $y = 1$ at the point where

$$\frac{x^2}{x^2 + x + 1} = 1$$
$$x^2 = x^2 + x + 1$$
$$x = -1$$

that is, at the point $(-1, 1)$.   ∎

A General Graphing Procedure

We now have the tools we need to describe a general procedure for sketching a variety of graphs.

A General Procedure for Sketching the Graph of $f(x)$

Step 1. Find the domain of $f(x)$ [that is, where $f(x)$ is defined].

Step 2. Find and plot all intercepts. The y intercept (where $x = 0$) is usually easy to find, but the x intercepts [where $f(x) = 0$] may require a calculator.

Step 3. Determine all vertical and horizontal asymptotes of the graph. Draw the asymptotes in a coordinate plane.

Step 4. Find $f'(x)$, and use it to determine the critical numbers of $f(x)$ and intervals of increase and decrease.

Step 5. Determine all relative extrema (both coordinates). Plot each relative maximum with a cap ($\frown$) and each relative minimum with a cup ($\smile$).

Step 6. Find $f''(x)$, and use it to determine intervals of concavity and points of inflection. Plot each inflection point with a "twist" to suggest the shape of the graph near the point.

Step 7. You now have a preliminary graph, with asymptotes in place, intercepts plotted, arrows indicating the direction of the graph, and caps, cups, and twists suggesting the shape at key points. Plot additional points if needed, and complete the sketch by joining the plotted points in the directions indicated. Be sure to remember that the graph cannot cross a vertical asymptote.

Here is a step-by-step analysis of the graph of a rational function.

EXAMPLE 3.3.3   Graphing a Rational Function

Sketch the graph of the function

$$f(x) = \frac{x}{(x + 1)^2}$$

Solution

Steps 1 and 2. The function is defined for all x except $x = -1$, and the only intercept is the origin $(0, 0)$.

Step 3. The line $x = -1$ is a vertical asymptote of the graph of $f(x)$ since $f(x)$ decreases indefinitely as x approaches -1 from either side; that is,

$$\lim_{x \to -1^-} \frac{x}{(x+1)^2} = \lim_{x \to -1^+} \frac{x}{(x+1)^2} = -\infty$$

Moreover, since

$$\lim_{x \to -\infty} \frac{x}{(x+1)^2} = \lim_{x \to +\infty} \frac{x}{(x+1)^2} = 0$$

the line $y = 0$ (the x axis) is a horizontal asymptote. Draw dashed lines $x = -1$ and $y = 0$ on a coordinate plane. (In this case, drawing the dashed vertical line $y = 0$ is not really necessary since it is a coordinate axis.)

Step 4. Applying the quotient rule, compute the derivative of $f(x)$:

$$f'(x) = \frac{(x+1)^2(1) - x[2(x+1)(1)]}{(x+1)^4} = \frac{1-x}{(x+1)^3}$$

Since $f'(1) = 0$, it follows that $x = 1$ is a critical number. Note that even though $f'(-1)$ does not exist, $x = -1$ is not a critical number since it is not in the domain of $f(x)$. Place $x = 1$ and $x = -1$ on a number line with a dashed vertical line at $x = -1$ to indicate the vertical asymptote there. Then evaluate $f'(x)$ at appropriate test numbers (say, at -2, 0, and 3) to obtain the arrow diagram shown.

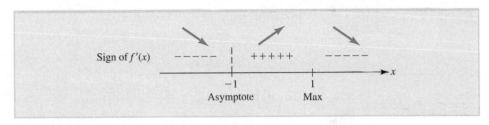

Step 5. The arrow pattern in the diagram obtained in step 4 indicates there is a relative maximum at $x = 1$. Since $f(1) = \dfrac{1}{4}$, we plot a cap at $\left(1, \dfrac{1}{4}\right)$.

Step 6. Apply the quotient rule again to get

$$f''(x) = \frac{2(x-2)}{(x+1)^4}$$

Since $f''(x) = 0$ at $x = 2$ and $f''(x)$ does not exist at $x = -1$, plot -1 and 2 on a number line and check the sign of $f''(x)$ on the intervals $x < -1$, $-1 < x < 2$, and $x > 2$ to obtain the concavity diagram shown.

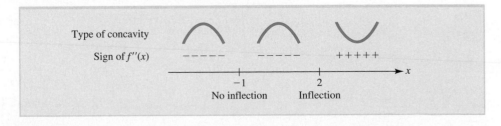

Note that the concavity changes at $x = 2$. Since $f(2) = \dfrac{2}{9}$, plot a twist ($\smallsmile$) at $\left(2, \dfrac{2}{9}\right)$ to indicate the inflection point there.

Step 7. The preliminary graph is shown in Figure 3.29a. Note that the vertical asymptote (dashed line) breaks the graph into two parts. Join the features in each separate part by a smooth curve to obtain the completed graph shown in Figure 3.29b.

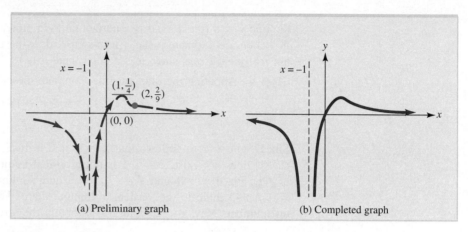

(a) Preliminary graph   (b) Completed graph

FIGURE 3.29 The graph of $f(x) = \dfrac{x}{(x + 1)^2}$.

In Example 3.3.4, we sketch the graph of a more complicated rational function using a condensed form of the step-by-step solution featured in Example 3.3.3.

EXAMPLE 3.3.4 Graphing a Rational Function

Sketch the graph of

$$f(x) = \frac{3x^2}{x^2 + 2x - 15}$$

Solution

Since $x^2 + 2x - 15 = (x + 5)(x - 3)$, the function $f(x)$ is defined for all x except $x = -5$ and $x = 3$. The only intercept is the origin $(0, 0)$.

We see that $x = 3$ and $x = -5$ are vertical asymptotes because if we write $f(x) = \dfrac{p(x)}{q(x)}$, where $p(x) = 3x^2$ and $q(x) = x^2 + 2x - 15$, then $q(3) = 0$ and $q(-5) = 0$, while $p(3) \neq 0$ and $p(-5) \neq 0$. Moreover, $y = 3$ is a horizontal asymptote since

$$\lim_{x \to +\infty} f(x) = \lim_{x \to +\infty} \frac{3x^2}{x^2 + 2x - 15} = \lim_{x \to +\infty} \frac{3}{1 + 2/x - 15/x^2} = \frac{3}{1 + 0 - 0} = 3$$

and similarly, $\lim\limits_{x \to -\infty} f(x) = 3$. Begin the preliminary sketch by drawing the asymptotes $x = 3$, $x = -5$, and $y = 3$ as dashed lines on a coordinate plane.

Next, use the quotient rule to obtain

$$f'(x) = \frac{(x^2 + 2x - 15)(6x) - (3x^2)(2x + 2)}{(x^2 + 2x - 15)^2} = \frac{6x(x - 15)}{(x^2 + 2x - 15)^2}$$

We see that $f'(x) = 0$ when $x = 0$ and $x = 15$ and that $f'(x)$ does not exist when $x = -5$ and $x = 3$. Put $x = -5, 0, 3,$ and 15 on a number line, and obtain the arrow diagram shown in Figure 3.30a by determining the sign of $f'(x)$ at appropriate test numbers (say, at $-7, -1, 2, 5,$ and 20). Interpreting the arrow diagram, we see that there is a relative maximum at $x = 0$ and a relative minimum at $x = 15$. Since $f(0) = 0$ and $f(15) \approx 2.81$, we plot a cap at $(0, 0)$ on our preliminary graph and a cup at $(15, 2.81)$.

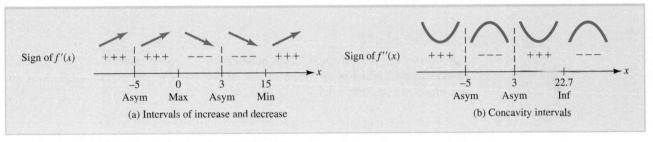

FIGURE 3.30   Arrow and concavity diagrams for $f(x) = \dfrac{3x^2}{x^2 + 2x - 15}$.

Apply the quotient rule again to get

$$f''(x) = \frac{-6(2x^3 - 45x^2 - 225)}{(x^2 + 2x - 15)^3}$$

We see that $f''(x)$ does not exist when $x = -5$ and $x = 3$, and that $f''(x) = 0$ when

$$2x^3 - 45x^2 - 225 = 0$$

$$x \approx 22.7 \quad \text{found by calculator}$$

Put $x = -5, 3,$ and 22.7 on a number line, and obtain the concavity diagram shown in Figure 3.30b by determining the sign of $f''(x)$ at appropriate test numbers (say, at $-6, 0, 4,$ and 25). The concavity changes at $x = -5, 3,$ and 22.7, but only $x = 22.7$ corresponds to an inflection point since $x = -5$ and $x = 3$ are not in the domain of $f(x)$. We find that $f(22.7) \approx 2.83$ and plot a twist ($\smile$) at $(22.7, 2.83)$.

The preliminary graph is shown in Figure 3.31a. Note that the two vertical asymptotes break the graph into three parts. Join the features in each separate part by a smooth curve to obtain the completed graph shown in Figure 3.31b.

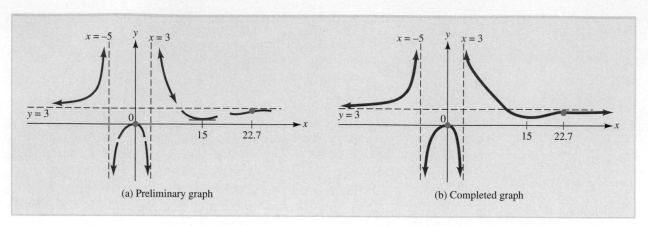

(a) Preliminary graph

(b) Completed graph

FIGURE 3.31 The graph of $f(x) = \dfrac{3x^2}{x^2 + 2x - 15}$.

When $f'(c)$ does not exist at a number $x = c$ in the domain of $f(x)$, there are several possibilities for the graph of $f(x)$ at the point $(c, f(c))$. Two such cases are examined in Example 3.3.5.

EXAMPLE 3.3.5   Graphing with Cusps and Vertical Tangents

Sketch the graphs of $f(x) = x^{2/3}$ and $g(x) = (x - 1)^{1/3}$.

Solution

Both functions are defined for all x. For $f(x) = x^{2/3}$, we have

$$f'(x) = \frac{2}{3}x^{-1/3} \quad \text{and} \quad f''(x) = -\frac{2}{9}x^{-4/3} \quad x \neq 0$$

The only critical point is $(0, 0)$, and the intervals of increase and decrease and concavity are as shown:

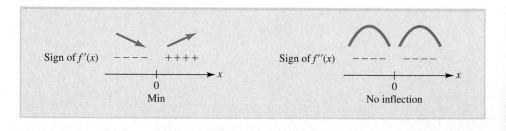

Interpreting these diagrams, we conclude that the graph of $f(x)$ is concave downward for all $x \neq 0$ but is falling for $x < 0$ while rising for $x > 0$. Thus, the graph of $f(x)$ has a relative minimum at the origin $(0, 0)$ and its shape there is ∨, called a *cusp*. The graph of $f(x)$ is shown in Figure 3.32a.

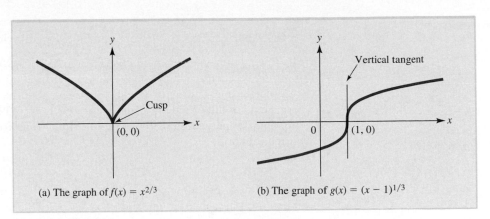

(a) The graph of $f(x) = x^{2/3}$ (b) The graph of $g(x) = (x - 1)^{1/3}$

FIGURE 3.32 A graph with a cusp and another with a vertical tangent.

The derivatives of $g(x) = (x - 1)^{1/3}$ are given by

$$g'(x) = \frac{1}{3}(x - 1)^{-2/3} \quad \text{and} \quad g''(x) = -\frac{2}{9}(x - 1)^{-5/3} \quad x \neq 1$$

The only critical point is (1, 0), and we obtain the intervals of increase and decrease and concavity shown:

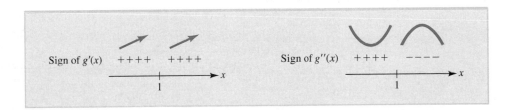

Thus, the graph of $g(x)$ is rising for all $x \neq 1$ but is concave upward for $x < 1$ and concave downward for $x > 1$. This means that (1, 0) is an inflection point, but in addition, note that

$$\lim_{x \to 1^-} g'(x) = \lim_{x \to 1^+} g'(x) = +\infty$$

Geometrically, this means that as x approaches 0 from either side, the tangent line at $(x, g(x))$ becomes steeper and steeper (with positive slope). This can be interpreted as saying that the graph of $g(x)$ has a tangent line at (1, 0) with "infinite" slope, that is, a **vertical tangent** line. The graph of $g(x)$ is shown in Figure 3.32b.

Sometimes, it is useful to represent observations about a quantity in graphical form. This procedure is illustrated in Example 3.3.6.

EXAMPLE 3.3.6 Representing Population Graphically

The population of a community is 230,000 in 1995 and grows at an increasing rate for 5 years, reaching the 300,000 level in 2000. It then continues to rise, but at a decreasing rate until it peaks out at 350,000 in 2007. After that, the population starts to fall at a decreasing rate for 3 years to 320,000 and then at an increasing rate, approaching 280,000 in the long run. Represent this information in graphical form.

Solution

Let $P(t)$ denote the population of the community t years after the base year 1995, where P is measured in units of 10,000 people. Since the population increases at an increasing rate for 5 years from 230,000 to 300,000, the graph of $P(t)$ rises from $(0, 23)$ to $(5, 30)$ and is concave upward for $0 < t < 5$. The population then continues to increase until 2007, but at a decreasing rate, until it reaches a maximum value of 350,000. That is, the graph continues to rise from $(5, 30)$ to a high point at $(12, 35)$ but is now concave downward. Since the graph changes concavity at $x = 5$ (from up to down), it has an inflection point at $(5, 30)$.

For the next 3 years, the population decreases at a decreasing rate, so the graph of $P(t)$ is falling and concave downward for $12 < t < 15$. Since the population continues to decrease from the 320,000 level reached in 2010 but at an increasing rate, the graph falls from the point $(15, 32)$ for $t > 15$ but is concave upward. The change in concavity at $x = 15$ (from down to up) means that $(15, 32)$ is another inflection point.

The statement that "the population decreases at an increasing rate" for $t > 15$ means that the population is changing at a negative rate, which is becoming less negative with increasing time. In other words, the decline in population "slows down" after 2010. This, coupled with the statement that the population "approaches 280,000 in the long run," suggests that the population curve $y = P(t)$ "flattens out" and approaches $y = 28$ asymptotically as $t \to +\infty$.

These observations are summarized in Table 3.2 and are represented in graphical form in Figure 3.33.

TABLE 3.2 Behavior of a Population $P(t)$

Time Period	The function $P(t)$ is . . .	and the graph of $P(t)$ is . . .
$t = 0$	$P(0) = 23$	at the point $(0, 23)$
$0 < t < 5$	increasing at an increasing rate	rising and concave upward
$t = 5$	$P(5) = 30$	at the inflection point $(5, 30)$
$5 < t < 12$	increasing at a decreasing rate	rising and concave downward
$t = 12$	$P(12) = 35$	at the high point $(12, 35)$
$12 < t < 15$	decreasing at a decreasing rate	falling and concave downward
$t = 15$	$P(15) = 32$	at the inflection point $(15, 32)$
$t > 15$	decreasing at an increasing rate and gradually tending toward 28	falling and concave upward asymptotically approaching $y = 28$

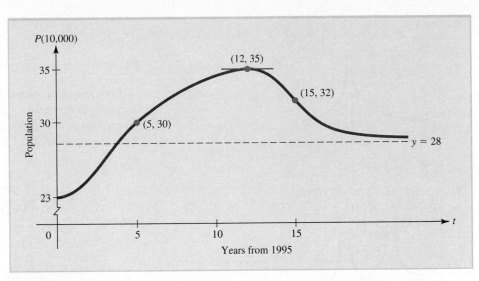

FIGURE 3.33   Graph of a population function.

EXERCISES ■ 3.3

In Exercises 1 through 8, determine the vertical and horizontal asymptotes of the given graph.

1.

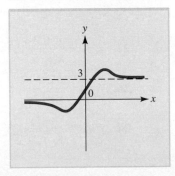

2.

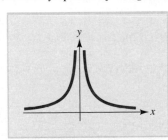

3.

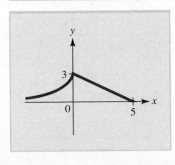

4.

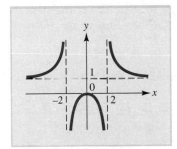

5.

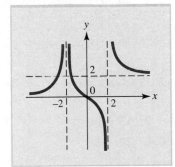

6.

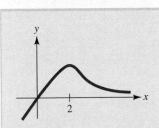

7.

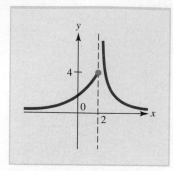

8.

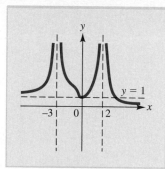

In Exercises 9 through 16, find all vertical and horizontal asymptotes of the graph of the given function.

9. $f(x) = \dfrac{3x - 1}{x + 2}$

10. $f(x) = \dfrac{x}{2 - x}$

11. $f(x) = \dfrac{x^2 + 2}{x^2 + 1}$

12. $f(t) = \dfrac{t + 2}{t^2}$

13. $f(t) = \dfrac{t^2 + 3t - 5}{t^2 - 5t + 6}$

14. $g(x) = \dfrac{5x^2}{x^2 - 3x - 4}$

15. $h(x) = \dfrac{1}{x} - \dfrac{1}{x - 1}$

16. $g(t) = \dfrac{t}{\sqrt{t^2 - 4}}$

In Exercises 17 through 32, sketch the graph of the given function.

17. $f(x) = x^3 + 3x^2 - 2$ **18.** $f(x) = x^5 - 5x^4 + 93$

19. $f(x) = x^4 + 4x^3 + 4x^2$ **20.** $f(x) = 3x^4 - 4x^2 + 3$

21. $f(x) = (2x - 1)^2(x^2 - 9)$

22. $f(x) = x^3 - 3x^4$

23. $f(x) = \dfrac{1}{2x + 3}$

24. $f(x) = \dfrac{x + 3}{x - 5}$

25. $f(x) = x - \dfrac{1}{x}$

26. $f(x) = \dfrac{x^2}{x + 2}$

27. $f(x) = \dfrac{1}{x^2 - 9}$

28. $f(x) = \dfrac{1}{\sqrt{1 - x^2}}$

29. $f(x) = \dfrac{x^2 - 9}{x^2 + 1}$

30. $f(x) = \dfrac{1}{\sqrt{x}} - \dfrac{1}{x}$

31. $f(x) = x^{3/2}$

32. $f(x) = x^{4/3}$

In Exercises 33 through 38, diagrams indicating intervals of increase or decrease and concavity are given. Sketch a possible graph for a function with these characteristics.

33.

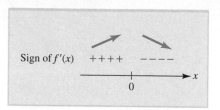

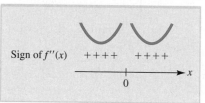

34.

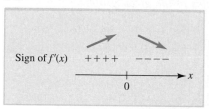

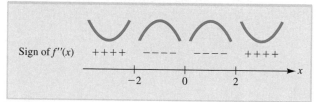

35.

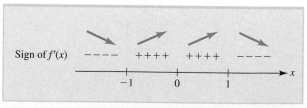

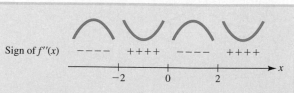

36.

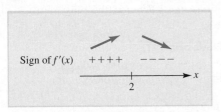

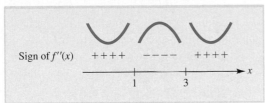

37.

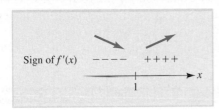

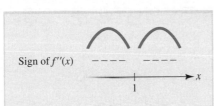

38.

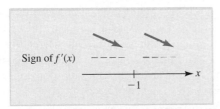

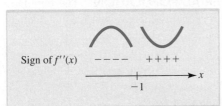

In Exercises 39 through 42, the derivative $f'(x)$ of a differentiable function $f(x)$ is given. In each case,

(a) Find intervals of increase and decrease for $f(x)$.

(b) Determine values of x for which relative maxima and minima occur on the graph of $f(x)$.

(c) Find $f''(x)$, and determine intervals of concavity for the graph of $f(x)$.

(d) For what values of x do inflection points occur on the graph of $f(x)$?

39. $f'(x) = x^3(x - 2)^2$ **40.** $f'(x) = x^2(x + 1)^3$

41. $f'(x) = \dfrac{x + 3}{(x - 2)^2}$ **42.** $f'(x) = \dfrac{x + 2}{(x - 1)^2}$

43. Find constants A and B so that the graph of the function

$$f(x) = \frac{Ax - 3}{5 + Bx}$$

will have $x = 2$ as a vertical asymptote and $y = 4$ as a horizontal asymptote. Once you find A and B, sketch the graph of $f(x)$.

44. Find constants A and B so that the graph of the function

$$f(x) = \frac{Ax + 2}{8 - Bx}$$

will have $x = 4$ as a vertical asymptote and $y = -1$ as a horizontal asymptote. Once you find A and B, sketch the graph of $f(x)$.

BUSINESS AND ECONOMICS APPLIED PROBLEMS

45. AVERAGE COST The total cost of producing x units of a particular commodity is C thousand dollars, where $C(x) = 3x^2 + x + 48$, and the average cost is

$$A(x) = \frac{C(x)}{x} = 3x + 1 + \frac{48}{x}$$

a. Find all vertical and horizontal asymptotes of the graph of $A(x)$.

b. Note that as x gets larger and larger, the term $\dfrac{48}{x}$ in $A(x)$ gets smaller and smaller. What does this say about the relationship between the average cost curve $y = A(x)$ and the line $y = 3x + 1$?

c. Sketch the graph of $A(x)$, incorporating the result of part (b) in your sketch. [*Note:* The line $y = 3x + 1$ is called an *oblique* (or *slant*) *asymptote* of the graph.]

46. INVENTORY COST A manufacturer estimates that if each shipment of raw materials contains x units, the total cost in dollars of obtaining and storing the year's supply of raw materials will be

$$C(x) = 2x + \frac{80,000}{x}.$$

a. Find all vertical and horizontal asymptotes of the graph of $C(x)$.

b. Note that as x gets larger and larger, the term $\dfrac{80{,}000}{x}$ in $C(x)$ gets smaller and smaller. What does this say about the relationship between the cost curve $y = C(x)$ and the line $y = 2x$?

c. Sketch the graph of $C(x)$, incorporating the result of part (b) in your sketch. [*Note:* The line $y = 2x$ is called an *oblique* (or *slant*) *asymptote* of the graph.]

47. **DISTRIBUTION** The number of worker-hours W required to distribute new telephone books to $x\%$ of the households in a certain community is modeled by the function

$$W(x) = \frac{200x}{100 - x}$$

a. Sketch the graph of $W(x)$.

b. Suppose only 1,500 worker-hours are available for distributing telephone books. What percentage of households do not receive new books?

48. **PRODUCTION** A business manager determines that t months after production begins on a new product, the number of units produced will be P million per month, where

$$P(t) = \frac{t}{(t + 1)^2}$$

a. Find $P'(t)$ and $P''(t)$.

b. Sketch the graph of $P(t)$.

c. What happens to production in the long run (as $t \to \infty$)?

49. **SALES** A company estimates that if x thousand dollars are spent on the marketing of a certain product, then $Q(x)$ thousand units of the product will be sold, where

$$Q(x) = \frac{7x}{27 + x^2}$$

a. Sketch the graph of the sales function $Q(x)$.

b. For what marketing expenditure x are sales maximized? What is the maximum sales level?

c. For what value of x is the sales rate minimized?

50. **ADVERTISING** A manufacturer of motorcycles estimates that if x thousand dollars are spent on advertising, then for $x > 0$,

$$M(x) = 2{,}300 + \frac{125}{x} - \frac{500}{x^2}$$

cycles will be sold.

a. Sketch the graph of the sales function $M(x)$.

b. What level of advertising expenditure results in maximum sales? What is the maximum sales level?

51. **COST MANAGEMENT** A company uses a truck to deliver its products. To estimate costs, the manager models gas consumption by the function

$$G(x) = \frac{1}{2{,}000}\left(\frac{800}{x} + 5x\right)$$

gal/mile, assuming that the truck is driven at a constant speed of x miles per hour for $x \geq 5$. The driver is paid \$18 per hour to drive the truck 400 miles, and gasoline costs \$4.25 per gallon. Highway regulations require $30 \leq x \leq 65$.

a. Find an expression for the total cost $C(x)$ of the trip. Sketch the graph of $C(x)$ for the legal speed interval $30 \leq x \leq 65$.

b. What legal speed will minimize the total cost of the trip? What is the minimal total cost?

LIFE AND SOCIAL SCIENCE APPLIED PROBLEMS

52. **CONCENTRATION OF DRUG** Hal Cooper is given an injection of a particular drug at noon, and samples of blood are taken at regular intervals to determine the concentration of drug in his system. It is found that the concentration increases at an increasing rate with respect to time until 1 P.M., and for the next 3 hours, continues to increase but at a decreasing rate until the peak concentration is reached at 4 P.M. The concentration then decreases at a decreasing rate until 5 P.M., after which it decreases at an increasing rate toward zero. Sketch a possible graph for the concentration of drug $C(t)$ as a function of time.

53. **BACTERIAL POPULATION** The population of a bacterial colony increases at an increasing rate for 1 hour, after which it continues to increase but at a rate that gradually decreases toward zero. Sketch a possible graph for the population $P(t)$ as a function of time t.

54. **EPIDEMIOLOGY** Epidemiologists studying a contagious disease observe that the number of newly infected people increases at an increasing rate during the first 3 years of the epidemic. At that time, a new drug is introduced, and the number of infected people declines at a decreasing rate. Two years after its introduction, the drug begins to lose effectiveness. The number of new cases continues to decline for 1 more year but at an increasing rate before rising again at an

increasing rate. Draw a possible graph for the number of new cases $N(t)$ as a function of time.

55. **EXPERIMENTAL PSYCHOLOGY** To study the rate at which animals learn, a psychology student performed an experiment in which a rat was sent repeatedly through a laboratory maze. Suppose the time required for the rat to traverse the maze on the nth trial was approximately

$$f(n) = 3 + \frac{12}{n} \text{ minutes.}$$

    a. Graph the function $f(n)$.
    b. What portion of the graph is relevant to the practical situation under consideration?
    c. What happens to the graph as n increases without bound? Interpret your answer in practical terms.

56. **IMMUNIZATION** During a nationwide program to immunize the population against a new strain of influenza, public health officials determined that the cost of inoculating $x\%$ of the susceptible population would be approximately

$$C(x) = \frac{1.7x}{100 - x}$$

million dollars.

    a. Sketch the graph of the cost function $C(x)$.
    b. Suppose 40 million dollars are available for providing immunization. What percentage of the susceptible population will not be inoculated?

57. **POLITICAL POLLING** Congresswoman Celesta Gomez has just come out in favor of a controversial bill. A poll suggests that t days after she declares her position on the bill, the percentage of her constituency (those who favor her at the time she declares her position) that still supports her is

$$S(t) = \frac{100(t^2 - 3t + 25)}{t^2 + 7t + 25}$$

The vote is to be taken 10 days after she announces her position.

    a. Sketch the graph of $S(t)$ for $0 \le t \le 10$.
    b. When is the congresswoman's support at its lowest point? What is the minimum support level?
    c. The derivative $S'(t)$ may be thought of as an approval rate. Is the congresswoman's approval rate positive or negative when the vote is taken? Is the approval rate increasing or decreasing at this time? Interpret your results.

MISCELLANEOUS PROBLEMS

58. **ADOPTION OF TECHNOLOGY** Draw a possible graph for the percentage of households adopting a new type of consumer electronic technology if the percentage grows at an increasing rate for the first 2 years, after which the rate of increase declines, with the market penetration of the technology eventually approaching 90%.

59. **AVERAGE TEMPERATURE** A researcher models the temperature T (in degrees Celsius) during the time period from 6 A.M. to 6 P.M. in a certain city by the function

$$T(t) = \frac{-1}{36}t^3 + \frac{1}{8}t^2 + \frac{7}{3}t - 2 \qquad \text{for } 0 \le t \le 12$$

where t is the number of hours after 6 A.M.

    a. Sketch the graph of $T(t)$.
    b. At what time is the temperature the greatest? What is the highest temperature of the day?

60. Let $f(x) = x^{1/3}(x - 4)$.
    a. Find $f'(x)$, and determine intervals of increase and decrease for $f(x)$. Locate all relative extrema on the graph of $f(x)$.
    b. Find $f''(x)$, and determine intervals of concavity for $f(x)$. Find all inflection points on the graph of $f(x)$.
    c. Find all intercepts for the graph of $f(x)$. Does the graph have any asymptotes?
    d. Sketch the graph of $f(x)$.

61. Repeat Exercise 60 for the function
$$f(x) = x^{2/3}(2x - 5)$$

62. Repeat Exercise 60 for the function
$$f(x) = \frac{x + 9.4}{25 - 1.1x - x^2}$$

63. Let $f(x) = \frac{x - 1}{x^2 - 1}$ and let $g(x) = \frac{x - 1.01}{x^2 - 1}$.
    a. Use a graphing utility to sketch the graph of $f(x)$. What happens at $x = 1$?
    b. Sketch the graph of $g(x)$. Now what happens at $x = 1$?

64. Find constants A, B, and C so that the function $f(x) = Ax^3 + Bx^2 + C$ will have a relative extremum at $(2, 11)$ and an inflection point at $(1, 5)$. Sketch the graph of f.

SECTION 3.4 Optimization; Elasticity of Demand

Learning Objectives

1. Use the extreme value property to find absolute extrema.
2. Compute absolute extrema in applied problems.
3. Study optimization principles in economics.
4. Define and examine price elasticity of demand.

You have already seen several situations where the methods of calculus were used to determine the largest or smallest value of a function of interest (for example, maximum profit or minimum cost). In most such optimization problems, the goal is to find the absolute maximum or absolute minimum of a particular function on some relevant interval. The absolute maximum of a function on an interval is the largest value of the function on that interval, and the absolute minimum is the smallest value. Here is a definition of absolute extrema.

> **Absolute Maxima and Minima of a Function** ■ Let f be a function defined on an open interval I that contains the number c. Then
>
> $$f(c) \text{ is the } absolute \ maximum \text{ of } f \text{ on } I \text{ if } f(c) \geq f(x) \text{ for all } x \text{ in } I$$
> $$f(c) \text{ is the } absolute \ minimum \text{ of } f \text{ on } I \text{ if } f(c) \leq f(x) \text{ for all } x \text{ in } I$$
>
> Collectively, absolute maxima and minima are called *absolute extrema*.

Absolute extrema often coincide with relative extrema but not always. For example, in Figure 3.34 the absolute maximum and relative maximum on the interval $a \leq x \leq b$ are the same, but the absolute minimum occurs at the left endpoint, $x = a$.

EXPLORE!

Use a graphing calculator to graph $f(x) = \dfrac{x^3}{\sqrt{x+2}}$ with the modified decimal window [0, 4.7]1 by [0, 60]5. Trace or use other utility methods to find the absolute maximum and absolute minimum of $f(x)$ over the interval [1, 3].

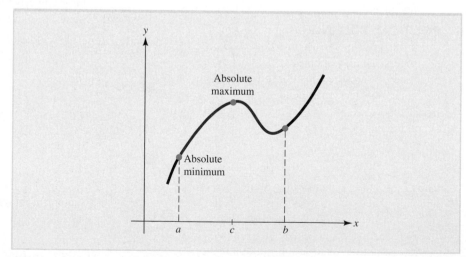

FIGURE 3.34 Absolute extrema.

In this section, you will learn how to find absolute extrema of functions on intervals. We begin by considering intervals $a \leq x \leq b$ that are "closed" in the sense they include both endpoints, a and b. It can be shown that a function continuous on such an interval has both an absolute maximum and an absolute minimum on the interval. Moreover, each absolute extremum must occur either at an endpoint of the interval (at a or b) or at a critical number c between a and b (Figure 3.35). To summarize:

The Extreme Value Property ■ A function $f(x)$ that is continuous on the closed interval $a \leq x \leq b$ attains its absolute extrema on the interval either at an endpoint of the interval (a or b) or at a critical number c such that $a < c < b$.

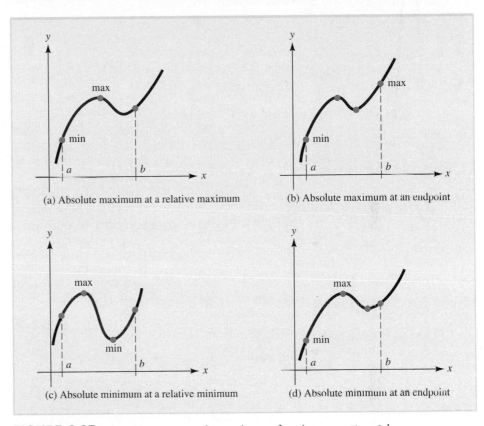

(a) Absolute maximum at a relative maximum

(b) Absolute maximum at an endpoint

(c) Absolute minimum at a relative minimum

(d) Absolute minimum at an endpoint

FIGURE 3.35   Absolute extrema of a continuous function on $a \leq x \leq b$.

Thanks to the extreme value property, you can find the absolute extrema of a continuous function on a closed interval $a \leq x \leq b$ by using this straightforward procedure.

How to Find the Absolute Extrema of a Continuous Function f on a Closed Interval $a \leq x \leq b$

Step 1. Find all critical numbers of f in the open interval $a < x < b$.

Step 2. Compute $f(x)$ at the critical numbers found in step 1 and at the endpoints $x = a$ and $x = b$.

Step 3. Interpretation: The largest and smallest values found in step 2 are, respectively, the absolute maximum and absolute minimum values of $f(x)$ on the closed interval $a \leq x \leq b$.

The procedure is illustrated in Examples 3.4.1 through 3.4.3.

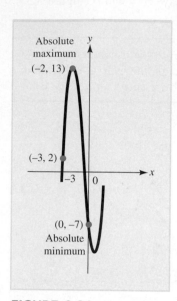

FIGURE 3.36 The absolute extrema on $-3 \leq x \leq 0$ for $y = 2x^3 + 3x^2 - 12x - 7.$

EXAMPLE 3.4.1 Finding Absolute Extrema

Find the absolute maximum and absolute minimum of the function

$$f(x) = 2x^3 + 3x^2 - 12x - 7$$

on the interval $-3 \leq x \leq 0.$

Solution

From the derivative

$$f'(x) = 6x^2 + 6x - 12 = 6(x + 2)(x - 1)$$

we see that the critical numbers are $x = -2$ and $x = 1$. Of these, only $x = -2$ lies in the interval $-3 \leq x \leq 0$. Compute $f(x)$ at $x = -2$ and at the endpoints $x = -3$ and $x = 0$.

$$f(-2) = 13 \qquad f(-3) = 2 \qquad f(0) = -7$$

Compare these values to conclude that the absolute maximum of f on the interval $-3 \leq x \leq 0$ is $f(-2) = 13$ and the absolute minimum is $f(0) = -7$.

Notice that we did not have to classify the critical points or draw the graph to locate the absolute extrema. The sketch in Figure 3.36 is presented only for the sake of illustration.

EXAMPLE 3.4.2 Studying Maximum and Minimum Traffic Flow

For several weeks, the highway department has been recording the speed of freeway traffic flowing past a certain downtown exit. The data suggest that between 1:00 and 6:00 P.M. on a normal weekday, the speed of the traffic at the exit is approximately $S(t) = t^3 - 10.5t^2 + 30t + 20$ miles per hour, where t is the number of hours past noon. At what time between 1:00 and 6:00 P.M. is the traffic moving the fastest, and at what time is it moving the slowest?

Solution

The goal is to find the absolute maximum and absolute minimum of the function $S(t)$ on the interval $1 \leq t \leq 6$. From the derivative

$$S'(t) = 3t^2 - 21t + 30 = 3(t^2 - 7t + 10) = 3(t - 2)(t - 5)$$

we get the critical numbers $t = 2$ and $t = 5$, both of which lie in the interval $1 \leq t \leq 6$.

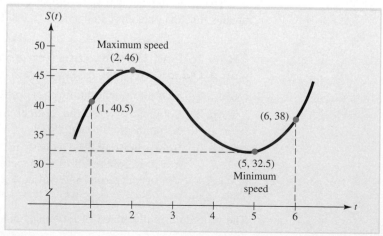

FIGURE 3.37 Traffic speed $S(t) = t^3 - 10.5t^2 + 30t + 20.$

Refer to Example 3.4.2.

EXPLORE!

Because of an increase in the speed limit, the speed past the exit is now

$$S_1(t) = t^3 - 10.5t^2 + 30t + 25$$

Graph $S(t)$ and $S_1(t)$ using the window [0, 6]1 by [20, 60]5. At what time between 1 P.M. and 6 P.M. is the maximum speed achieved using $S_1(t)$? At what time is the minimum speed achieved?

Compute $S(t)$ for these values of t and at the endpoints $t = 1$ and $t = 6$ to get

$$S(1) = 40.5 \qquad S(2) = 46 \qquad S(5) = 32.5 \qquad S(6) = 38$$

Since the largest of these values is $S(2) = 46$ and the smallest is $S(5) = 32.5$, we can conclude that the traffic is moving fastest at 2:00 P.M., when its speed is 46 miles per hour, and slowest at 5:00 P.M., when its speed is 32.5 miles per hour. For reference, the graph of S is sketched in Figure 3.37.

EXAMPLE 3.4.3 Finding Maximum Air Speed During a Cough

When you cough, the radius of your trachea (windpipe) decreases, affecting the speed of the air in the trachea. If r_0 is the normal radius of the trachea, the relationship between the speed S of the air and the radius r of the trachea during a cough is given by a function of the form $S(r) = ar^2(r_0 - r)$, where a is a positive constant.* Find the radius r for which the speed of the air is greatest.

Solution

The radius r of the contracted trachea cannot be greater than the normal radius r_0 or less than zero. Hence, the goal is to find the absolute maximum of $S(r)$ on the interval $0 \leq r \leq r_0$.

First differentiate $S(r)$ with respect to r using the product rule and factor the derivative as follows (note that a and r_0 are constants):

$$S'(r) = -ar^2 + (r_0 - r)(2ar) = ar[-r + 2(r_0 - r)] = ar(2r_0 - 3r)$$

Then set the factored derivative equal to zero and solve to get the critical numbers:

$$ar(2r_0 - 3r) = 0$$
$$r = 0 \quad \text{or} \quad r = \frac{2}{3}r_0$$

Both of these values of r lie in the interval $0 \leq r \leq r_0$, and one is actually an endpoint of the interval. Compute $S(r)$ for these two values of r and for the other endpoint $r = r_0$ to get

$$S(0) = 0 \qquad S\left(\frac{2}{3}r_0\right) = \frac{4a}{27}r_0^3 \qquad S(r_0) = 0$$

Compare these values and conclude that the speed of the air is greatest when the radius of the contracted trachea is $\frac{2}{3}r_0$, that is, when it is two-thirds the radius of the uncontracted trachea.

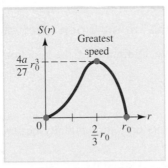

FIGURE 3.38 The speed of air during a cough
$$S(r) = ar^2(r_0 - r).$$

A graph of the function $S(r)$ is given in Figure 3.38. Note that the r intercepts of the graph are obvious from the factored function $S(r) = ar^2(r_0 - r)$. Notice also that the graph has a horizontal tangent when $r = 0$, reflecting the fact that $S'(0) = 0$.

---

*Philip M. Tuchinsky, "The Human Cough," *UMAP Modules 1976: Tools for Teaching*, Lexington, MA: Consortium for Mathematics and Its Application, Inc., 1977.

More General Optimization

When the interval on which you wish to maximize or minimize a continuous function is not of the form $a \leq x \leq b$, the procedure illustrated in Examples 3.4.1 through 3.4.3 no longer applies. This is because there is no longer any guarantee that the function actually has an absolute maximum or minimum on the interval in question. On the other hand, if an absolute extremum does exist and the function is continuous on the interval, the absolute extremum will still occur at a relative extremum or endpoint contained in the interval. Several possibilities for functions on unbounded intervals are illustrated in Figure 3.39.

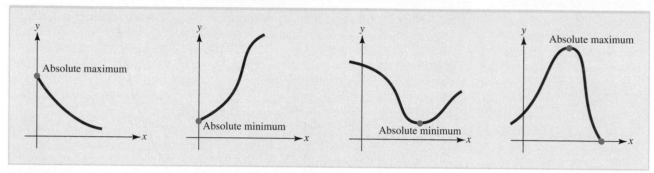

FIGURE 3.39   Extrema for functions defined on unbounded intervals.

To find the absolute extrema of a continuous function on an interval that is not of the form $a \leq x \leq b$, you still evaluate the function at all the critical points and endpoints that are contained in the interval. However, before you can draw any final conclusions, you must find out if the function actually has relative extrema on the interval. One way to do this is to use the first derivative to determine where the function is increasing and where it is decreasing and then to sketch the graph. This technique is illustrated in Example 3.4.4.

EXAMPLE 3.4.4   Finding Absolute Extrema on an Open Interval

If they exist, find the absolute maximum and absolute minimum of the function $f(x) = x^2 + \dfrac{16}{x}$ on the interval $x > 0$.

Solution

The function is continuous on the interval $x > 0$ since its only discontinuity occurs at $x = 0$. The derivative is

$$f'(x) = 2x - \frac{16}{x^2} = \frac{2x^3 - 16}{x^2} = \frac{2(x^3 - 8)}{x^2}$$

which is zero when

$$x^3 - 8 = 0 \qquad x^3 = 8 \qquad \text{or} \qquad x = 2$$

Since $f'(x) < 0$ for $0 < x < 2$ and $f'(x) > 0$ for $x > 2$, the graph of f is decreasing for $0 < x < 2$ and increasing for $x > 2$, as shown in Figure 3.40. It follows that

$$f(2) = 2^2 + \frac{16}{2} = 12$$

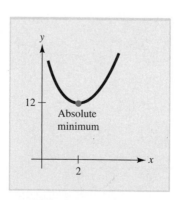

FIGURE 3.40   The function $f(x) = x^2 + \dfrac{16}{x}$ on the interval $x > 0$.

is the absolute minimum of f on the interval $x > 0$ and that there is no absolute maximum.

The procedure illustrated in Example 3.4.4 can be used whenever we wish to find the largest or smallest value of a function f that is continuous on an interval I on which it has *exactly one* critical number c. In particular, if this condition is satisfied and $f(x)$ has a *relative* maximum (minimum) at $x = c$, it also has an *absolute* maximum (minimum) there. To see why, suppose the graph has a relative minimum at $x = c$. Then the graph is always falling before c and always rising after c, since to change direction would require the presence of a second critical point (Figure 3.41). Thus, the relative minimum is also the absolute minimum. These observations suggest that any test for relative extrema becomes a test for absolute extrema in this special case. Here is a statement of the second derivative test for absolute extrema.

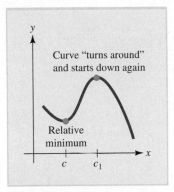

FIGURE 3.41 The relative minimum is not the absolute minimum.

> **The Second Derivative Test for Absolute Extrema** ■ Suppose that $f(x)$ is continuous on an interval I where $x = c$ is the only critical number and that $f'(c) = 0$. Then,
>
> if $f''(c) > 0$, the absolute minimum of $f(x)$ on I is $f(c)$
> if $f''(c) < 0$, the absolute maximum of $f(x)$ on I is $f(c)$

Example 3.4.5 illustrates how the second derivative test for absolute extrema can be used in practice.

EXAMPLE 3.4.5 Maximizing Profit and Minimizing Average Cost

Adam Goodman determines that when q thousand units of his product are produced each month, they will all be sold at a price of $p(q) = 22.2 - 1.2q$ dollars per unit. The total cost of producing the q units will be $C(q) = 0.4q^2 + 3q + 40$ thousand dollars.

a. How many units should Adam produce to maximize profit? What is the maximum profit he can expect?

b. How many units should Adam produce to minimize the average cost per unit of production $A(q) = \dfrac{C(q)}{q}$? What is the minimal average cost?

c. How many units should Adam produce to guarantee that average cost per unit equals the marginal cost of production $C'(q)$?

Solution

a. The revenue is

$$R(q) = qp(q) = q(22.2 - 1.2q) = -1.2q^2 + 22.2q$$

thousand dollars, so the profit is

$$P(q) = R(q) - C(q) = -1.2q^2 + 22.2q - (0.4q^2 + 3q + 40)$$
$$= -1.6q^2 + 19.2q - 40$$

thousand dollars. We have

$$P'(q) = -1.6(2q) + 19.2 = -3.2q + 19.2$$
$$= 0$$

when

$$-3.2q + 19.2 = 0$$

$$q = \frac{19.2}{3.2} = 6$$

Since $q = 6$ is the only critical number for $P(q)$, the second derivative test for absolute extrema applies. We find that $P''(q) = -3.2$, so $P''(6) < 0$, and the second derivative test tells us that the maximum profit occurs when $q = 6$, that is, when 6,000 units are produced. The maximum profit is

$$P(6) = -1.6(6)^2 + 19.2(6) - 40 = 17.6$$

thousand dollars ($17,600). The graph of the profit function is shown in Figure 3.42a.

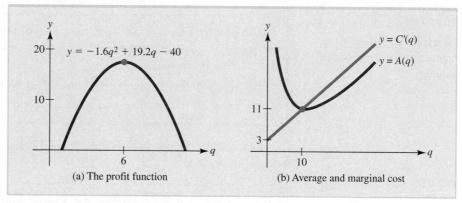

FIGURE 3.42   Graphs of profit, average cost, and marginal cost for Example 3.4.5.

b. The average cost is

$$A(q) = \frac{C(q)}{q} = \frac{0.4q^2 + 3q + 40}{q} \qquad \frac{\text{thousand dollars}}{\text{thousand units}}$$

$$= 0.4q + 3 + \frac{40}{q} \qquad \frac{\text{dollars}}{\text{unit}}$$

for $q > 0$ (the level of production cannot be negative or zero). We find

$$A'(q) = 0.4 - \frac{40}{q^2} = \frac{0.4q^2 - 40}{q^2}$$

which is 0 for $q > 0$ only when $q = 10$. Using the quotient rule, we find that

$$A''(q) = \frac{80}{q^3} > 0 \quad \text{when } q > 0$$

so the second derivative test for absolute extrema tells us that average cost $A(q)$ is minimized when $q = 10$ (thousand) units. The minimal average cost is

$$A(10) = 0.4(10) + 3 + \frac{40}{10} = 11 \qquad \frac{\text{dollars}}{\text{unit}}$$

c. The marginal cost is $C'(q) = 0.8q + 3$, and it equals average cost when

$$0.8q + 3 = 0.4q + 3 + \frac{40}{q}$$

$$0.4q = \frac{40}{q}$$

$$0.4q^2 = 40$$

$$q = 10 \text{ (thousand) units}$$

which equals the optimal level of production in part (b). The graphs of the marginal cost $C'(q)$ and average cost $A(q) = \dfrac{C(q)}{q}$ are shown in Figure 3.42b.

Two General Principles of Marginal Analysis

If the revenue derived from the sale of q units is $R(q)$ and the cost of producing those units is $C(q)$, then the profit is $P(q) = R(q) - C(q)$. Since

$$P'(q) = [R(q) - C(q)]' = R'(q) - C'(q)$$

it follows that $P'(q) = 0$ when $R'(q) = C'(q)$. If it is also true that $P''(q) < 0$, or equivalently, that $R''(q) < C''(q)$, then the profit will be maximized.

> **Marginal Analysis Criterion for Maximum Profit** ■ The profit $P(q) = R(q) - C(q)$ is maximized at a level of production q where marginal revenue equals marginal cost and the rate of change of marginal cost exceeds the rate of change of marginal revenue, that is, where
>
> $$R'(q) = C'(q) \qquad \text{and} \qquad R''(q) < C''(q)$$

For instance, in Example 3.4.5, the revenue is $R(q) = -1.2q^2 + 22.2q$, and the cost is $C(q) = 0.4q^2 + 3q + 40$, so the marginal revenue is $R'(q) = -2.4q + 22.2$ and the marginal cost is $C'(q) = 0.8q + 3$. Thus, marginal revenue equals marginal cost when

$$R'(q) = C'(q)$$

$$-2.4q + 22.2 = 0.8q + 3$$

$$3.2q = 19.2$$

$$q = 6$$

which is the level of production for maximum profit found in part (a) of Example 3.4.5. Note that $R'' < C''$ is also satisfied since $R'' = -2.4$ and $C'' = 0.8$.

In part (c) of Example 3.4.5, you found that marginal cost equals average cost at the level of production where average cost is minimized. This, too, is no accident. To see why, let $C(q)$ be the cost of producing q units of a commodity. Then the average cost per unit is $A(q) = \dfrac{C(q)}{q}$, and by applying the quotient rule, you get

$$A'(q) = \frac{qC'(q) - C(q)}{q^2}$$

Thus, $A'(q) = 0$ when the numerator on the right is zero. That is, when

$$qC'(q) = C(q)$$

or equivalently, when

$$\underbrace{C'(q)}_{\substack{\text{marginal} \\ \text{cost}}} = \underbrace{\frac{C(q)}{q}}_{\substack{\text{average} \\ \text{cost}}} = A(q)$$

To show that average cost is actually minimized where average cost equals marginal cost, it is necessary to make a few reasonable assumptions about total cost (see Exercise 40).

Marginal Analysis Criterion for Minimal Average Cost ■ The average cost is minimized at the level of production where average cost equals marginal cost, that is, when $A(q) = C'(q)$.

Here is an informal explanation of the relationship between average and marginal cost that is often given in economics texts. The marginal cost (MC) is approximately the same as the cost of producing one additional unit. If the additional unit costs less to produce than the average cost (AC) of the existing units (if MC < AC), then this less expensive unit will cause the average cost per unit to decrease. On the other hand, if the additional unit costs more than the average cost of the existing units (if MC > AC), then this more expensive unit will cause the average cost per unit to increase. However, if the cost of the additional unit is equal to the average cost of the existing units (if MC = AC), then the average cost will neither increase nor decrease, which means (AC)′ = 0.

The relation between average cost and marginal cost can be generalized to apply to any pair of average and marginal quantities. The only possible modification involves the nature of the critical point that occurs when the average quantity equals the marginal quantity. For example, average revenue usually has a relative *maximum* (instead of a minimum) when average revenue equals marginal revenue.

Price Elasticity of Demand

Consumers tend to respond to an increase in the price of a commodity by decreasing demand, but the degree of response varies greatly for different products. For instance, an increase in the price of milk, soap, or batteries will not greatly affect the demand for those products, but an increase in the cost of home loans may cause demand to decline sharply as people opt to rent instead of buying.

Economists use a function $E(p)$, called **price elasticity of demand,** to measure the responsiveness of demand q for a particular commodity to a change in the unit price p. The elasticity function is defined as the negative of the ratio of the percentage rate of change of quantity demanded to the corresponding percentage rate of change of price. *This is approximately the same as the percentage decrease in demand produced by a 1% increase in unit price.*

Recall from Section 2.2 that the percentage rate of change of a differentiable function $f(x)$ with respect to x is given by the ratio $100\frac{f'(x)}{f(x)}$. Therefore, we have

$$\begin{bmatrix} \text{Percentage rate of} \\ \text{change of quantity } q \end{bmatrix} = \frac{100\dfrac{dq}{dp}}{q}$$

and

$$\begin{bmatrix} \text{Percentage rate of} \\ \text{change of price } p \end{bmatrix} = \frac{100\dfrac{dp}{dp}}{p} = \frac{100}{p}$$

so the elasticity function is given by

$$E(p) = -\left[\frac{\text{Percentage rate of change of quantity } q}{\text{Percentage rate of change of price } p}\right] = -\frac{100\dfrac{dq}{dp}}{\dfrac{100}{p}}$$

$$= -\frac{p}{q}\frac{dq}{dp}$$

To summarize:

Price Elasticity of Demand ■ If $q = D(p)$ units of a commodity are demanded by the market at a unit price p, where D is a differentiable function, then the price elasticity of demand for the commodity is given by

$$E(p) = -\frac{p}{q}\frac{dq}{dp}$$

and has the interpretation

$$E(p) \approx \left[\begin{array}{c}\text{percentage rate of decrease in demand } q\\\text{produced by a 1\% increase in price } p\end{array}\right]$$

NOTE You may wonder why the negative sign is part of the definition of $E(p)$. Since demand generally decreases with increasing price, the percentage rate of change of quantity demanded with respect to price will be negative. By introducing a minus sign in the formula for $E(p)$, we guarantee that elasticity of demand is a positive quantity, which is more convenient to deal with, especially when making comparisons. ■

EXAMPLE 3.4.6 Finding Elasticity of Demand

Suppose the demand q and price p for a certain commodity are related by the linear equation $q = 240 - 2p$ (for $0 \le p \le 120$).

  a. Express the elasticity of demand as a function of p.

  b. Calculate the elasticity of demand when the price is $p = 100$. Interpret your answer.

  c. Calculate the elasticity of demand when the price is $p = 50$. Interpret your answer.

  d. At what price is the elasticity of demand equal to 1? What is the economic significance of this price?

Solution

  a. Since $q = 240 - 2p$, the derivative of q with respect to p is $\dfrac{dq}{dp} = -2$ and the elasticity of demand is

$$E(p) = -\frac{p}{q}\frac{dq}{dp} = -\frac{p}{q}(-2) = -\left[\frac{-2p}{240 - 2p}\right] = \frac{p}{120 - p}$$

b. When $p = 100$, the elasticity of demand is

$$E(100) = \frac{100}{120 - 100} = 5$$

That is, when the price is $p = 100$, a 1% increase in price will produce a decrease in demand of approximately 5%.

c. When $p = 50$, the elasticity of demand is

$$E(50) = \frac{50}{120 - 50} \approx 0.71$$

That is, when the price is $p = 50$, a 1% increase in price will produce a decrease in demand of approximately 0.71%.

d. The elasticity of demand will be equal to 1 when

$$1 = \frac{p}{120 - p} \qquad 120 - p = p \qquad 2p = 120 \qquad \text{or} \qquad p = 60$$

At this price, a 1% increase in price will result in a decrease in demand of approximately the same percentage.

There are three levels of elasticity, depending on whether $E(p)$ is greater than, less than, or equal to 1. Here is a description and economic interpretation of each level.

Levels of Elasticity

$E(p) > 1$ **Elastic demand.** The percentage decrease in demand is greater than the percentage increase in price that caused it. Thus, demand is relatively sensitive to changes in price.

$E(p) < 1$ **Inelastic demand.** The percentage decrease in demand is less than the percentage increase in price that caused it. When this occurs, demand is relatively insensitive to changes in price.

$E(p) = 1$ **Demand is of unit elasticity (or unitary).** The percentage changes in price and demand are (approximately) equal.

For instance, in Example 3.4.6b, we found that $E(100) = 5 > 1$, so the demand is elastic with respect to price when $p = 100$. In part (c) of the same example, we found that $E(50) = 0.7 < 1$, so the demand is inelastic when $p = 50$. Finally, in part (d), we found $E(60) = 1$, which means that the demand is of unit elasticity when $p = 60$.

The level of elasticity of demand for a commodity gives useful information about the total revenue R obtained from the sale of q units of the commodity at p dollars per unit. Assuming that the demand q is a differentiable function of unit price p, the revenue is $R(p) = p \cdot q(p)$ and by differentiating implicitly with respect to p, we find that

$$\frac{dR}{dp} = p\frac{dq}{dp} + q \quad \text{by the product rule}$$

To get the elasticity $E(p) = -\dfrac{p}{q}\dfrac{dq}{dp}$ into the picture, simply multiply the expression on the right-hand side by $\dfrac{q}{q}$ as follows:

$$\frac{dR}{dp} = \frac{q}{q}\left(p\frac{dq}{dp} + q\right) = q\left(\frac{p}{q}\frac{dq}{dp} + 1\right) = q[-E(p) + 1]$$

In Example 3.4.7, we use this formula to study the relationship between the revenue obtained in a production process and the level of elasticity.

EXAMPLE 3.4.7 Relating Change in Revenue to Levels of Elasticity

The manager of a bookstore determines that when a certain new paperback novel is priced at p dollars per copy, the daily demand will be $q = 300 - p^2$ copies, where $0 \le p \le \sqrt{300}$.

  a. Determine where the demand is elastic, inelastic, and of unit elasticity with respect to price.

  b. Interpret the results of part (a) in terms of the behavior of total revenue as a function of price.

Solution

  a. The elasticity of demand is

$$E(p) = -\frac{p}{q}\frac{dq}{dp} = \frac{-p}{300 - p^2}(-2p) = \frac{2p^2}{300 - p^2}$$

and since $0 \le p \le \sqrt{300}$,

$$E(p) = \frac{2p^2}{300 - p^2}$$

The demand is of unit elasticity when $E = 1$, that is, when

$$\frac{2p^2}{300 - p^2} = 1$$
$$2p^2 = 300 - p^2$$
$$3p^2 = 300$$
$$p = \pm 10$$

of which only $p = 10$ is in the relevant interval $0 \le p \le \sqrt{300}$. If $0 \le p < 10$, then

$$E = \frac{2p^2}{300 - p^2} < \frac{2(10)^2}{300 - (10)^2} = 1$$

so the demand is inelastic. Likewise, if $10 < p < \sqrt{300}$, then

$$E = \frac{2p^2}{300 - p^2} > \frac{2(10)^2}{300 - (10)^2} = 1$$

and the demand is elastic.

b. Recall from the discussion preceding this example that the derivative of the revenue function $R = pq$ with respect to p satisfies

$$R'(p) = q(p)[-E(p) + 1]$$

For $0 \le p < 10$, the demand is inelastic, so $E(p) < 1$ and the term $-E(p) + 1$ is positive, so in this case, $R'(q) > 0$ and the revenue is increasing. For this range of prices, a specified percentage increase in price results in a smaller percentage decrease in demand, so the bookstore will take in more money for each increase in price up to $10 per copy.

For the price range $10 < p \le \sqrt{300}$, the demand is elastic. This means $E(p) > 1$ and $-E(p) + 1 < 0$, so $R'(p) < 0$ and the revenue is decreasing. If the book is priced in this range, a specified percentage increase in price results in a larger percentage decrease in demand. This means that if the bookstore increases the price beyond the $10 per copy, it will lose revenue.

Finally, the revenue is optimized when $R'(p) = 0$. This occurs when $E(p) = 1$, that is, when $p = 10$ (unit elasticity). The graphs of the demand and revenue functions are shown in Figure 3.43.

EXPLORE!

Suppose the demand/price equation in Example 3.4.7 is $q = 300 - ap^2$. Describe how the price is affected when the demand is of unit elasticity for $a = 0, 1, 3,$ or 5 by examining the graph of $x(300 - ax^2)$ for these values of a. Use a viewing rectangle of [0, 20]1 by [0, 3,000]500.

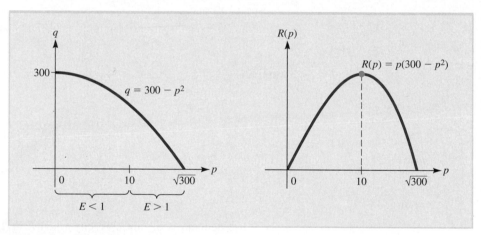

FIGURE 3.43 Demand and revenue curves for Example 3.4.7.

By generalizing the approach illustrated in the solution of Example 3.4.7b (see Exercise 61), we can summarize the relationship between revenue and the level of elasticity as follows

---

Levels of Elasticity and the Effect on Revenue

If demand is **elastic** $[E(p) > 1]$, revenue R decreases as price p increases.

If demand is **inelastic** $[E(p) < 1]$, revenue R increases as price p increases.

If demand is of **unit elasticity** $[E(p) = 1]$, revenue is unaffected by a small increase in price.

The relationship between revenue and price is shown in Figure 3.44. Note that the revenue curve is rising where demand is inelastic, falling where demand is elastic, and has a horizontal tangent line where the demand is of unit elasticity.

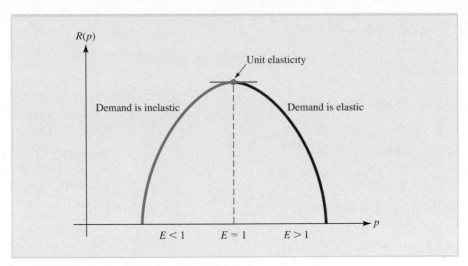

FIGURE 3.44 Revenue as a function of price.

EXERCISES ■ 3.4

In Exercises 1 through 16, find the absolute maximum and absolute minimum (if any) of the given function on the specified interval.

1. $f(x) = x^2 + 4x + 5; \; -3 \leq x \leq 1$

2. $f(x) = x^3 + 3x^2 + 1; \; -3 \leq x \leq 2$

3. $f(x) = \frac{1}{3}x^3 - 9x + 2; \; 0 \leq x \leq 2$

4. $f(x) = x^5 - 5x^4 + 1; \; 0 \leq x \leq 5$

5. $f(t) = 3t^5 - 5t^3; \; -2 \leq t \leq 0$

6. $f(x) = 10x^6 + 24x^5 + 15x^4 + 3; \; -1 \leq x \leq 1$

7. $f(x) = (x^2 - 4)^5; \; -3 \leq x \leq 2$

8. $f(t) = \frac{t^2}{t - 1}; \; -2 \leq t \leq -\frac{1}{2}$

9. $g(x) = x + \frac{1}{x}; \; \frac{1}{2} \leq x \leq 3$

10. $g(x) = \frac{1}{x^2 - 9}; \; 0 \leq x \leq 2$

11. $f(u) = u + \frac{1}{u}; \; u > 0$

12. $f(u) = 2u + \frac{32}{u}; \; u > 0$

13. $f(x) = \frac{1}{x}; \; x > 0$

14. $f(x) = \frac{1}{x^2}; \; x > 0$

15. $f(x) = \frac{1}{x + 1}; \; x \geq 0$

16. $f(x) = \frac{1}{(x + 1)^2}; \; x \geq 0$

MAXIMUM PROFIT AND MINIMUM AVERAGE COST
In Exercises 17 through 22, you are given the price p(q) at which q units of a particular commodity can be sold and the total cost C(q) of producing the q units. In each case:

(a) *Find the revenue function R(q), the profit function P(q), the marginal revenue R'(q), and marginal cost C'(q). Sketch the graphs of P(q), R'(q), and C'(q) on the same coordinate axes and determine the level of production q where P(q) is maximized.*

(b) *Find the average cost $A(q) = \frac{C(q)}{q}$, and sketch the graphs of A(q), and the marginal cost C'(q) on the same axes. Determine the level of production q at which A(q) is minimized.*

17. $p(q) = 49 - q; \; C(q) = \frac{1}{8}q^2 + 4q + 200$

18. $p(q) = 37 - 2q; \; C(q) = 3q^2 + 5q + 75$

19. $p(q) = 180 - 2q; \; C(q) = q^3 + 5q + 162$

20. $p(q) = 710 - 1.1q^2;$
      $C(q) = 2q^3 - 23q^2 + 90.7q + 151$

21. $p(q) = 1.0625 - 0.0025q$; $C(q) = \dfrac{q^2 + 1}{q + 3}$

22. $p(q) = 81 - 3q$; $C(q) = \dfrac{q + 1}{q + 3}$

ELASTICITY OF DEMAND   *In Exercises 23 through 28, compute the elasticity of demand for the given demand function $D(p)$ and determine whether the demand is elastic, inelastic, or of unit elasticity at the indicated price p.*

23. $D(p) = -1.3p + 10$; $p = 4$

24. $D(p) = -1.5p + 25$; $p = 12$

25. $D(p) = 200 - p^2$; $p = 10$

26. $D(p) = \sqrt{400 - 0.01p^2}$; $p = 120$

27. $D(p) = \dfrac{3{,}000}{p} - 100$; $p = 10$

28. $D(p) = \dfrac{2{,}000}{p^2}$; $p = 5$

29. For what value of x in the interval $-1 \le x \le 4$ is the graph of the function

$$f(x) = 2x^2 - \frac{1}{3}x^3$$

steepest? What is the slope of the tangent at this point?

30. At what point does the tangent to the curve $y = 2x^3 - 3x^2 + 6x$ have the smallest slope? What is the slope of the tangent at this point?

BUSINESS AND ECONOMICS APPLIED PROBLEMS

31. AVERAGE PROFIT   A manufacturer estimates that when q units of a certain commodity are produced, the profit obtained is $P(q)$ thousand dollars, where

$$P(q) = -2q^2 + 68q - 128$$

a. Find the average profit and the marginal profit functions.
b. At what level of production $\bar{q}$ is average profit equal to marginal profit?
c. Show that average profit is maximized at the level of production $\bar{q}$ found in part (b).
d. On the same set of axes, graph the relevant portions of the average and marginal profit functions.

32. MARGINAL ANALYSIS   A manufacturer estimates that if x units of a particular commodity

are produced, the total cost will be $C(x)$ dollars, where

$$C(x) = x^3 - 24x^2 + 350x + 338$$

a. At what level of production will the marginal cost $C'(x)$ be minimized?
b. At what level of production will the average cost $A(x) = \dfrac{C(x)}{x}$ be minimized?

33. ELASTICITY OF DEMAND   When a particular commodity is priced at p dollars per unit, consumers demand q units, where p and q are related by the equation $q^2 + 3pq = 22$.
a. Find the elasticity of demand for this commodity.
b. For a unit price of \$3, is the demand elastic, inelastic, or of unit elasticity?

34. ELASTICITY OF DEMAND   When an electronics store prices a certain brand of stereo at p hundred dollars per set, it is found that q sets will be sold each month, where $q^2 + 2p^2 = 41$.
a. Find the elasticity of demand for the stereos.
b. For a unit price of $p = 4$ (\$400), is the demand elastic, inelastic, or of unit elasticity?

35. DEMAND FOR ART   An art gallery offers 50 prints by a famous artist. If each print in the limited edition is priced at p dollars, it is expected that $q = 500 - 2p$ prints will be sold.
a. What limitations are there on the possible range of the price p?
b. Find the elasticity of demand. Determine the values of p for which the demand is elastic, inelastic, and of unit elasticity.
c. Interpret the results of part (b) in terms of the behavior of the total revenue as a function of unit price p.
d. If you were the owner of the gallery, what price would you charge for each print? Explain the reasoning behind your decision.

36. DEMAND FOR AIRLINE TICKETS   An airline determines that when a round-trip ticket between Los Angeles and San Francisco costs p dollars ($0 \le p \le 160$), the daily demand for tickets is $q = 256 - 0.01p^2$.
a. Find the elasticity of demand. Determine the values of p for which the demand is elastic, inelastic, and of unit elasticity.
b. Interpret the results of part (a) in terms of the behavior of the total revenue as a function of unit price p.

c. What price would you advise the airline to charge for each ticket? Explain your reasoning.

37. **PRODUCTION CONTROL** Gina manages a company that makes toys. Her firm produces an inexpensive doll (Floppsy) and an expensive doll (Moppsy) in units of x hundreds and y hundreds, respectively. Gina finds that it is possible to produce the dolls in such a way that

$$y = \frac{82 - 10x}{10 - x}$$

for $0 \le x \le 8$ and that the company receives twice as much for selling a Moppsy doll as for a Floppsy doll.

How many units of each kind of doll (both x and y) should Gina recommend producing to maximize the total revenue? (You may assume that the company is able to sell every doll it produces.)

38. **NATIONAL CONSUMPTION** Assume that total national consumption is given by a function $C(x)$, where x is the total national income. The derivative $C'(x)$ is called the **marginal propensity to consume.** Then $S = x - C$ represents total national savings, and $S'(x)$ is called **marginal propensity to save.** Suppose the consumption function is $C(x) = 8 - 0.8x - 0.8\sqrt{x}$. Find the marginal propensity to consume, and determine the value of x that results in the smallest total savings.

39. **WORKER EFFICIENCY** An efficiency study of the morning shift at a certain factory indicates that an average worker who is on the job at 8.00 A.M. will have assembled $f(x) = -x^3 + 6x^2 + 15x$ units x hours later. The study indicates further that after a 15-minute coffee break the worker can assemble $g(x) = -\frac{1}{3}x^3 + x^2 + 23x$ units in x hours.

Determine the time between 8:00 A.M. and noon at which a 15-minute coffee break should be scheduled so that the worker will assemble the maximum number of units by lunchtime at 12:15 P.M. [*Hint:* If the coffee break begins x hours after 8:00 A.M., $4 - x$ hours will remain after the break.]

40. **MARGINAL ANALYSIS** Suppose $q > 0$ units of a commodity are produced at a total cost of $C(q)$ dollars and an average cost of $A(q) = \frac{C(q)}{q}$.

In this section, we showed that $q = q_c$ satisfies $A'(q_c) = 0$ if and only if $C'(q_c) = A(q_c)$; that is, when marginal cost equals average cost. The purpose of this problem is to show that $A(q)$ is *minimized* when $q = q_c$.

a. Generally speaking, the cost of producing a commodity increases at an increasing rate as more and more goods are produced. Using this economic principle, what can be said about the sign of $C''(q)$ as q increases?

b. Show that $A''(q_c) > 0$ if and only if $C''(q_c) > 0$. Then use part (a) to argue that average cost $A(q)$ is minimized when $q = q_c$.

41. **ELASTICITY AND REVENUE** Suppose the demand for a certain commodity is given by $q = b - ap$, where a and b are positive constants, and $0 \le p \le \frac{b}{a}$.

a. Express elasticity of demand as a function of p.

b. Show that the demand is of unit elasticity at the midpoint $p = \frac{b}{2a}$ of the interval $0 \le p \le \frac{b}{a}$.

c. For what values of p is the demand elastic? Inelastic?

42. **INCOME ELASTICITY OF DEMAND** **Income elasticity of demand** is defined to be the percentage change in quantity purchased divided by the percentage change in real income.

a. Write a formula for income elasticity of demand E in terms of real income I and quantity purchased Q.

b. In the United States, which would you expect to be greater, the income elasticity of demand for cars or for food? Explain your reasoning.

c. What do you think is meant by a *negative* income elasticity of demand? Which of the following goods would you expect to have $E < 0$: used clothing, personal computers, bus tickets, refrigerators, used cars? Explain your reasoning.

d. Read an article on income elasticity of demand, and write a paragraph on why the income elasticity of demand for food is much larger in a developing country than in a country such as the United States or Japan.*

---

*You may find the following text a helpful starting point for your research: Campbell R. McConnell and Stanley L. Brue, *Microeconomics,* 12th ed., New York: McGraw-Hill, 1993.

43. **ELASTICITY** Suppose that the demand equation for a certain commodity is $q = \dfrac{a}{p^m}$, where a and m are positive constants. Show that the elasticity of demand is equal to m for all values of p. Interpret this result.

44. **MARGINAL ANALYSIS** Let $R(x)$ be the revenue obtained from the production and sale of x units of a commodity, and let $C(x)$ be the total cost of producing the x units. Show that the ratio $Q(x) = \dfrac{R(x)}{C(x)}$ is optimized when the relative rate of change of revenue equals the relative rate of change of cost. Would you expect this optimum to be a maximum or a minimum?

LIFE AND SOCIAL SCIENCE APPLIED PROBLEMS

45. **GROUP MEMBERSHIP** A national consumers' association determines that x years after its founding in 1998, it will have $P(x)$ members, where
$$P(x) = 100(2x^3 - 45x^2 + 264x)$$
   a. At what time between 2000 and 2013 was the membership largest? Smallest?
   b. What were the largest and smallest membership levels between 2008 and 2013?

46. **GROWTH OF A SPECIES** The percentage of codling moths* that survive the pupa stage at a given temperature T (degrees Celsius) is modeled by the formula
$$P(T) = -1.42T^2 + 68T - 746$$
$$\text{for } 20 \le T \le 30$$
Find the temperatures at which the greatest and smallest percentage of moths survive.

47. **BLOOD CIRCULATION** Poiseuille's law asserts that the speed of blood that is r centimeters from the central axis of an artery of radius R is $S(r) = c(R^2 - r^2)$, where c is a positive constant. Where is the speed of the blood greatest?

48. **POLITICS** A poll indicates that x months after a particular candidate for public office declares her candidacy, she will have the support of $S(x)$ percent of the voters, where
$$S(x) = \frac{1}{29}(-x^3 + 6x^2 + 63x + 1{,}080)$$
$$\text{for } 0 \le x \le 12$$
If the election is held in November, when should the politician announce her candidacy? Should she expect to win if she needs at least 50% of the vote?

49. **LEARNING** In a learning model, two responses (A and B) are possible for each of the series of observations. If there is a probability p of getting response A in any individual observation, the probability of getting response A exactly n times in a series of m observations is $F(p) = p^n(1 - p)^{m-n}$. The **maximum likelihood estimate** is the value of p that maximizes $F(p)$ for $0 \le p \le 1$. For what value of p does this occur?

50. **VOTING PATTERN** After a presidential election, the proportion $h(p)$ of seats in the House of Representatives won by the party of the winning presidential candidate may be modeled by the *cube rule*
$$h(p) = \frac{p^3}{p^3 + (1 - p)^3} \quad \text{for } 0 \le p \le 1$$
where p is the proportion of the popular vote received by the winning presidential candidate.
   a. Find $h'(p)$ and $h''(p)$.
   b. Sketch the graph of $h(p)$.
   c. In 1964, the Democrat Lyndon Johnson received 61% of the popular vote. What percentage of seats in the House does the cube rule predict should have gone to Democrats? (They actually won 72%.)
   d. For the most part, the cube rule has been extremely accurate for presidential elections since 1900. Use the Internet to research the actual proportions that occurred during these elections, and write a paragraph on your findings.

51. **ORNITHOLOGY** According to the results[†] of Tucker and Schmidt-Koenig, the energy expended by a certain species of parakeet is given by
$$E(v) = \frac{1}{v}[0.074(v - 35)^2 + 22]$$

---

*P. L. Shaffer and H. J. Gold, "A Simulation Model of Population Dynamics of the Codling Moth *Cydia Pomonella*," *Ecological Modeling*, Vol. 30, 1985, pp. 247–274.

[†]V. A. Tucker and K. Schmidt-Koenig, "Flight Speeds of Birds in Relation to Energetics and Wind Directions," *The Auk*, Vol. 88, 1971, pp. 97–107.

where v is the bird's velocity (in km/hr).

a. What velocity minimizes energy expenditure?

b. Read an article on how mathematical methods can be used to study animal behavior, and write a paragraph on whether you think such methods are valid. You may wish to begin with the reference cited in this problem.

52. **SPEED OF FLIGHT** In a model[‡] developed by C. J. Pennycuick, the power P required by a bird to maintain flight is given by the formula

$$P = \frac{w^2}{2\rho S v} + \frac{1}{2}\rho A v^3$$

where v is the relative speed of the bird, w is the weight, ρ is the density of air, and S and A are positive constants associated with the bird's size and shape. What relative speed v will minimize the power required by the bird?

53. **SENSITIVITY TO DRUGS** Body reaction to drugs is often modeled[§] by an equation of the form

$$R(D) = D^2\left(\frac{C}{2} - \frac{D}{3}\right)$$

where D is the dosage and C (a constant) is the maximum dosage that can be given. The rate of change of $R(D)$ with respect to D is called the **sensitivity.**

a. Find the value of D for which the sensitivity is the greatest. What is the greatest sensitivity? (Express your answer in terms of C.)

b. What is the reaction (in terms of C) when the dosage resulting in greatest sensitivity is used?

54. **SURVIVAL OF AQUATIC LIFE** It is known that a quantity of water that occupies 1 liter at 0°C will occupy

$$V(T) = \left(\frac{-6.8}{10^8}\right)T^3 + \left(\frac{8.5}{10^6}\right)T^2 - \left(\frac{6.4}{10^5}\right)T + 1$$

liters when the temperature is $T°C$, for $0 \le T \le 30$.

a. Use a graphing utility to graph $V(T)$ for $0 \le T \le 10$. The density of water is maximized when $V(T)$ is minimized. At what temperature does this occur?

b. Does the answer to part (a) surprise you? It should. Water is the only common liquid whose maximum density occurs *above* its freezing point (0°C for water). Read an article on the survival of aquatic life during the winter, and then write a paragraph on how the property of water examined in this problem is related to such survival.

55. **BLOOD PRODUCTION** A useful model for the production $p(x)$ of blood cells involves a function of the form

$$p(x) = \frac{Ax}{B + x^m}$$

where x is the number of cells present, and A, B, and m are positive constants.[†]

a. Find the rate of blood production $R(x) = p'(x)$, and determine where $R(x) = 0$.

b. Find the rate at which $R(x)$ is changing with respect to x, and determine where $R'(x) = 0$.

c. If $m > 1$, does the nonzero critical number you found in part (b) correspond to a relative maximum or a relative minimum? Explain.

56. **RESPIRATION** Physiologists define the flow F of air in the trachea by the formula $F = SA$, where S is the speed of the air and A is the area of a cross section of the trachea.

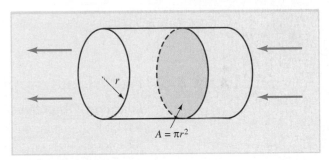

$A = \pi r^2$

EXERCISE 56

a. Assume the trachea has a circular cross section with radius r. Use the formula for the speed of air in the trachea during a cough given in Example 3.4.3 to express air flow F in terms of r.

b. Find the radius r for which the flow is greatest.

[‡]C. J. Pennycuick, "The Mechanics of Bird Migration," *Ibis* III, 1969, pp. 525–556.

[§]R. M. Thrall et al., *Some Mathematical Models in Biology*, U. of Michigan, 1967.

[†]M. C. Mackey and L. Glass, "Oscillations and Chaos in Physiological Control Systems," *Science*, Vol. 197, pp. 287–289.

57. ELECTRICITY   When a resistor of R ohms is connected across a battery with electromotive force E volts and internal resistance r ohms, a current of I amperes will flow, generating P watts of power, where

$$I = \frac{E}{r + R} \quad \text{and} \quad P = I^2 R$$

Assuming r is constant, what choice of R results in maximum power?

58. BROADCASTING   An all-news radio station has made a survey of the listening habits of local residents between the hours of 5:00 P.M. and midnight. The survey indicates that the percentage of the local adult population that is tuned in to the station x hours after 5:00 P.M. is

$$f(x) = \frac{1}{8}(-2x^3 + 27x^2 - 108x + 240)$$

a. At what time between 5:00 P.M. and midnight are the most people listening to the station? What percentage of the population is listening at this time?

b. At what time between 5:00 P.M. and midnight are the fewest people listening? What percentage of the population is listening at this time?

59. AERODYNAMICS   In designing airplanes, an important feature is the so-called drag factor; that is, the retarding force exerted on the plane by the air. One model measures drag by a function of the form

$$F(v) = Av^2 + \frac{B}{v^2}$$

where v is the velocity of the plane and A and B are constants. Find the velocity (in terms of A and B) that minimizes $F(v)$. Show that you have found the minimum rather than a maximum.

60. AMPLITUDE OF OSCILLATION   In physics, it can be shown that a particle forced to oscillate in a resisting medium has amplitude $A(r)$ given by

$$A(r) = \frac{1}{(1 - r^2)^2 + kr^2}$$

where r is the ratio of the forcing frequency to the natural frequency of oscillation and k is a positive constant that measures the damping effect of the resisting medium. Show that $A(r)$ has exactly one positive critical number. Does it correspond to a relative maximum or a relative minimum? Can anything be said about the *absolute* extrema of $A(r)$?

61.   Verify the relationships between revenue and levels of elasticity given in the summary box on page 260.

SECTION 3.5 Additional Applied Optimization

Learning Objectives

1. List and explore guidelines for solving optimization problems.
2. Model and analyze a variety of optimization problems.
3. Examine inventory control.

In Section 3.4, you saw a number of applications in which a formula was given and it was required to determine either a maximum or a minimum value. In practice, things are often not that simple, and it is necessary to first gather information about a quantity of interest and then formulate and analyze an appropriate mathematical model.

In this section, you will learn how to combine the techniques of model-building from Section 1.4 with the optimization techniques of Section 3.4. Here is a procedure for dealing with such problems.

Guidelines for Solving Optimization Problems

Step 1. Begin by deciding precisely what you want to maximize or minimize. Once this has been done, assign names to all variables of interest. It may help to pick letters that suggest the nature or role of the variable, such as "R" for "revenue."

Step 2. After assigning variables, express the relationships between the variables in terms of equations or inequalities. A figure may help.

Step 3. Express the quantity to be optimized (maximized or minimized) in terms of just one variable (the independent variable). To do this, you may need to use one or more of the available equations from step 2 to eliminate other variables. Also determine any necessary restrictions on the independent variable.

Step 4. If $f(x)$ is the quantity to be optimized, find $f'(x)$ and determine all critical numbers of f. Then find the required maximum or minimum value using the methods of Section 3.4 (the extreme value theorem or the second derivative test for absolute extrema). Remember, you may have to check the value of $f(x)$ at endpoints of an interval.

Step 5. Interpret your results in terms of appropriate physical, geometric, or economic quantities.

This procedure is illustrated in Example 3.5.1.

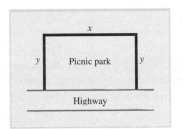

FIGURE 3.45 Rectangular picnic park.

EXAMPLE 3.5.1   Minimizing Amount of Fence

The highway department is planning to build a picnic park for motorists along a major highway. The park is to be rectangular with an area of 5,000 square yards and is to be fenced off on the three sides not adjacent to the highway. What is the least amount of fencing required for this job? How long and wide should the park be for the fencing to be minimized?

Solution

Step 1. Draw the picnic area, as in Figure 3.45. Let x (yards) be the length of the park (the side parallel to the highway) and let y (yards) be the width.

Step 2. Since the park is to have area 5,000 square yards, we must have $xy = 5,000$.

Step 3. The length of the fencing is $F = x + 2y$, where x and y are both positive, since otherwise there would be no park. Since

$$xy = 5,000 \qquad \text{or} \qquad y = \frac{5,000}{x}$$

we can eliminate y from the formula for F to obtain a formula in terms of x alone:

$$F(x) = x + 2y = x + 2\left(\frac{5,000}{x}\right) = x + \frac{10,000}{x} \qquad \text{for } x > 0$$

Step 4. The derivative of $F(x)$ is

$$F'(x) = 1 - \frac{10,000}{x^2}$$

and we get the critical numbers for $F(x)$ by setting $F'(x) = 0$ and solving for x:

$$F'(x) = 1 - \frac{10,000}{x^2} = 0 \qquad \text{put the fraction over a common denominator}$$

$$\frac{x^2 - 10,000}{x^2} = 0 \qquad \text{equate the numerator to 0}$$

$$x^2 = 10,000 \qquad \text{reject } -100 \text{ since } x > 0$$
$$x = 100$$

Since $x = 100$ is the only critical number in the interval $x > 0$, we can apply the second derivative test for absolute extrema. The second derivative of $F(x)$ is

$$F''(x) = \frac{20,000}{x^3}$$

so $F''(100) > 0$ and an absolute minimum of $F(x)$ occurs where $x = 100$. For reference, the graph of the fencing function $F(x)$ is shown in Figure 3.46.

Step 5. We have shown the minimal amount of fencing is

$$F(100) = 100 + \frac{10,000}{100} = 200 \text{ yards}$$

which is achieved when the park is $x = 100$ yards long and

$$y = \frac{5,000}{100} = 50 \text{ yards}$$

wide.

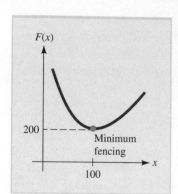

FIGURE 3.46 The graph of $F(x) = x + \dfrac{10,000}{x}$ for $x > 0$.

In Section 1.4 (Example 1.4.1), we illustrated the modeling procedure by examining a problem in which profit is maximized using geometric properties of parabolas. In Example 3.5.2, we analyze a similar problem, this time using calculus as an optimization tool.

EXAMPLE 3.5.2 Maximizing Profit

Mateo owns a small company that makes souvenir T-shirts. He can produce the shirts at a cost of $2 apiece. The shirts have been selling for $5 apiece, and at this price, tourists have been buying 4,000 shirts a month. Mateo plans to raise the price of shirts and expects that for each $1 increase in price, 400 fewer shirts will be sold each month. What price should Mateo charge per shirt to maximize profit?

Solution

Let x denote the new price at which the shirts will be sold and $P(x)$ the corresponding profit. The goal is to maximize the profit. Begin by stating the formula for profit in words:

$$\text{Profit} = (\text{number of shirts sold})(\text{profit per shirt})$$

Since 4,000 shirts are sold each month when the price is $5 and 400 fewer will be sold each month for each $1 increase in the price, it follows that

$$\text{Number of shirts sold} = 4,000 - 400(\text{number of \$1 increases})$$

The number of $1 increases in the price is the difference $x - 5$ between the new and old selling prices. Hence,

$$\text{Number of shirts sold} = 4{,}000 - 400(x - 5)$$
$$= 400[10 - (x - 5)]$$
$$= 400(15 - x)$$

The profit per shirt is simply the difference between the selling price x and the cost $2. That is,

$$\text{Profit per shirt} = x - 2$$

Putting it all together,

$$P(x) = 400(15 - x)(x - 2)$$

The goal is to find the absolute maximum of the profit function $P(x)$. To determine the relevant interval for this problem, note that since the new price x is to be at least as high as the old price $5, we must have $x \geq 5$. On the other hand, the number of shirts sold is $400(15 - x)$, which will be negative if $x > 15$. If you assume that Mateo will not price the shirts so high that no one buys them, you can restrict the optimization problem to the closed interval $5 \leq x \leq 15$.

To find the critical numbers, compute the derivative using the product and constant multiple rules to get

$$P'(x) = 400[(15 - x)(1) + (x - 2)(-1)]$$
$$= 400(15 - x - x + 2) = 400(17 - 2x)$$

which is zero when

$$17 - 2x = 0 \quad \text{or} \quad x = 8.5$$

Comparing the values of the profit function

$$P(5) = 12{,}000 \quad P(8.5) = 16{,}900 \quad \text{and} \quad P(15) = 0$$

at the critical number and at the endpoints of the interval, you can conclude that the maximum possible profit is $16,900, which will be generated if the shirts are sold for $8.50 apiece. For reference, the graph of the profit function is shown in Figure 3.47.

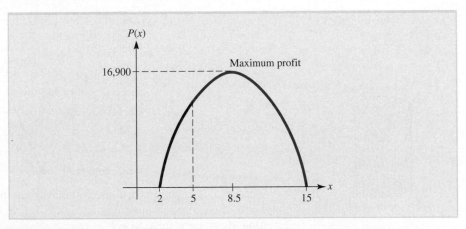

FIGURE 3.47 The profit function $P(x) = 400(15 - x)(x - 2)$.

NOTE   An Alternative Solution to the Profit Maximization Problem in Example 3.5.2

Since the number of shirts sold in Example 3.5.2 is described in terms of the number N of \$1 increases in price, you may wish to use N as the independent variable in your solution rather than the new price itself. With this choice, you find that

$$\text{Number of shirts} = 4{,}000 - 400N$$
$$\text{Profit per shirt} = (N + 5) - 2 = N + 3$$

Thus, the total profit is

$$P(N) = (4{,}000 - 400N)(N + 3) = 400(10 - N)(N + 3)$$

and the relevant interval is $0 \le N \le 10$. (Do you see why?) The absolute maximum in this case will occur when $N = 3.5$ (provide the details), that is, when the old price is increased from 5 to $5 + 3.5 = 8.5$ dollars (\$8.50). As you would expect, this is the same as the result obtained using price as the independent variable in Example 3.5.2. ∎

In Example 1.4.2, we used a graph to estimate the radius resulting in the minimum cost of constructing a water well required to have a certain volume. In Example 3.5.3, we consider a similar problem, this time using calculus to algebraically determine the radius that minimizes the cost of constructing a cylindrical can with specified volume.

EXAMPLE 3.5.3   Minimizing Cost of Construction

A cylindrical can is to be constructed to hold a fixed volume of liquid. The cost of the material used for the top and bottom of the can is 3 cents per square inch, and the cost of the material used for the curved side is 2 cents per square inch. Use calculus to derive a simple relationship between the radius and height of the can that is the least costly to construct.

Solution

Let r denote the radius, h the height, C the cost (in cents), and V the (fixed) volume. The goal is to minimize total cost, which comes from three sources:

$$\text{Cost} = \text{cost of top} + \text{cost of bottom} + \text{cost of side}$$

where, for each component of the cost,

$$\text{Cost} = (\text{cost per square inch})(\text{area})$$

Hence, $\text{Cost of top} = \text{cost of bottom} = 3(\pi r^2)$

and $\text{Cost of side} = 2(2\pi rh) = 4\pi rh$

so the total cost is

$$C = \underbrace{3\pi r^2}_{\text{top}} + \underbrace{3\pi r^2}_{\text{bottom}} + \underbrace{4\pi rh}_{\text{sides}} = 6\pi r^2 + 4\pi rh$$

Before you can apply calculus, you must write the cost in terms of just one variable. To do this, use the fact that the can is to have a fixed volume V_0 and solve the equation $V_0 = \pi r^2 h$ for h to get

$$h = \frac{V_0}{\pi r^2}$$

Just-In-Time **REVIEW**

A cylinder of radius r and height h has lateral (curved) area $A = 2\pi rh$ and volume $V = \pi r^2 h$.

Circumference
$2\pi r$

Then substitute this expression for h into the formula for C to express the cost in terms of r alone:

$$C(r) = 6\pi r^2 + 4\pi r\left(\frac{V_0}{\pi r^2}\right) = 6\pi r^2 + \frac{4V_0}{r}$$

The radius r can be any positive number, so the goal is to find the absolute minimum of $C(r)$ for $r > 0$. Differentiating $C(r)$, you get

$$C'(r) = 12\pi r - \frac{4V_0}{r^2} \qquad \text{remember } V_0 \text{ is a constant}$$

Since $C'(r)$ exists for all $r > 0$, any critical number $r = R$ must satisfy $C'(R) = 0$; that is,

$$C'(R) = 12\pi R - \frac{4V_0}{R^2} = 0$$

$$12\pi R = \frac{4V_0}{R^2}$$

$$R^3 = \frac{4V_0}{12\pi}$$

$$R = \sqrt[3]{\frac{V_0}{3\pi}}$$

If H is the height of the can that corresponds to the radius R, then $V_0 = \pi r^2 H$. Since R must satisfy

$$12\pi R = \frac{4V_0}{R^2}$$

you find that

$$12\pi R = \frac{4(\pi R^2 H)}{R^2} = 4\pi H$$

or

$$H = \frac{12\pi R}{4\pi} = 3R$$

Finally, note that the second derivative of $C(r)$ satisfies

$$C''(r) = 12\pi + \frac{8V_0}{r^3} > 0 \qquad \text{for all } r > 0$$

Therefore, since $r = R$ is the only critical number for $C(r)$ and since $C''(R) > 0$, the second derivative test for absolute extrema assures you that when the total cost of constructing the can is minimized, the height of the can is three times its radius. The graph of the cost function $C(r)$ is shown in Figure 3.48.

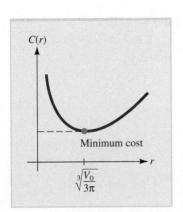

FIGURE 3.48 The cost function $C(r) = 6\pi r^2 + \dfrac{4V}{r}$ for $r > 0$.

In modern cities, where residential areas are often developed in the vicinity of industrial plants, it is important to carefully monitor and control the emission of pollutants. In Example 3.5.4, we examine a modeling problem in which calculus is used to determine the location in a community where pollution is minimized.

EXPLORE!

Refer to Example 3.5.4. Store
$$P(x) = \frac{75}{x} + \frac{300}{15 - x}$$
into Y1, and graph using the modified decimal window [0, 14]1 by [0, 350]10. Now use **TRACE** to move the cursor from X = 1 to 14 and confirm the location of minimal pollution. To view the behavior of the derivative $P'(x)$, enter Y2 = nDeriv(Y1, X, X) and graph using the window [0, 14]1 by [−75, 300]10. What do you observe?

EXAMPLE 3.5.4   Finding a Location That Minimizes Pollution

Two industrial plants, A and B, are located 15 miles apart and emit 75 ppm (parts per million) and 300 ppm of particulate matter, respectively. Each plant is surrounded by a restricted area of radius 1 mile in which no housing is allowed, and the concentration of pollutant arriving at any other point Q from each plant decreases with the reciprocal of the distance between that plant and Q. Where should a house be located on a road joining the two plants to minimize the total pollution arriving from both plants?

Solution

Suppose a house H is located x miles from plant A and, hence, $15 - x$ miles from plant B, where x satisfies $1 \leq x \leq 14$ since there is a 1-mile restricted area around each plant (Figure 3.49). Since the concentration of particulate matter arriving at H from each plant decreases with the reciprocal of the distance from the plant to H, the concentration of pollutant from plant A is $\frac{75}{x}$ and from plant B is $\frac{300}{15 - x}$. Thus, the total concentration of particulate matter arriving at H is given by the function

$$P(x) = \underbrace{\frac{75}{x}}_{\substack{\text{pollution} \\ \text{from } A}} + \underbrace{\frac{300}{15 - x}}_{\substack{\text{pollution} \\ \text{from } B}}$$

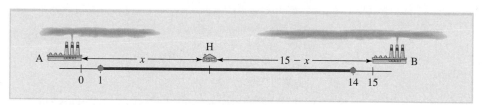

FIGURE 3.49   Pollution at a house located between two industrial plants.

To minimize the total pollution $P(x)$, we first find the derivative $P'(x)$ and solve $P'(x) = 0$. Applying the quotient rule and chain rule, we obtain

$$P'(x) = \frac{-75}{x^2} + \frac{-300(-1)}{(15 - x)^2} = \frac{-75}{x^2} + \frac{300}{(15 - x)^2}$$

Solving $P'(x) = 0$, we find that

Just-In-Time REVIEW

Cross-multiplying means that if
$$\frac{A}{B} = \frac{C}{D}$$
then
$$AD = CB$$

$$\frac{-75}{x^2} + \frac{300}{(15 - x)^2} = 0$$

$$\frac{75}{x^2} = \frac{300}{(15 - x)^2} \qquad \text{cross multiply}$$

$$75(15 - x)^2 = 300x^2 \qquad \text{divide by 75 and expand}$$

$$x^2 - 30x + 225 = 4x^2 \qquad \text{combine like terms}$$

$$3x^2 + 30x - 225 = 0 \qquad \text{factor}$$

$$3(x - 5)(x + 15) = 0$$

$$x = 5, x = -15$$

The symbol $\approx$ means "approximately equal to," so $a \approx b$ means "*a* is approximately equal to *b*."

The only critical number in the allowable interval $1 \le x \le 14$ is $x = 5$. Evaluating $P(x)$ at $x = 5$ and at the two endpoints of the interval, $x = 1$ and $x = 14$, we get

$$P(1) \approx 96.43 \text{ ppm}$$
$$P(5) = 45 \text{ ppm}$$
$$P(14) \approx 305.36 \text{ ppm}$$

Thus, total pollution is minimized when the house is located 5 miles from plant A.

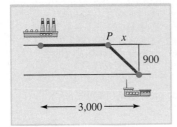

FIGURE 3.50 Relative positions of factory, river, and power plant.

EXAMPLE 3.5.5   Minimizing Cost of Construction

A cable is to be run from a power plant on one side of a river 900 meters wide to a factory on the other side, 3,000 meters downstream. The cost of running the cable under the water is \$5 per meter, while the cost over land is \$4 per meter. What is the most economical route over which to run the cable?

Solution

To help you visualize the situation, begin by drawing a diagram as shown in Figure 3.50. (Notice that in drawing the diagram, we have already assumed that the cable should be run in a *straight line* from the power plant to some point P on the opposite bank. Do you see why this assumption is justified?)

The goal is to minimize the cost of installing the cable. Let C denote this cost and represent C as follows:

$$C = 5(\text{number of meters of cable under water})$$
$$+ 4(\text{number of meters of cable over land})$$

Since you wish to describe the optimal route over which to run the cable, it will be convenient to choose a variable in terms of which you can easily locate the point P. Two reasonable choices for the variable x are illustrated in Figure 3.51.

The Pythagorean theorem says that the square of the hypotenuse of a right triangle equals the sum of the squares of the other two sides.

$$a^2 + b^2 = c^2$$

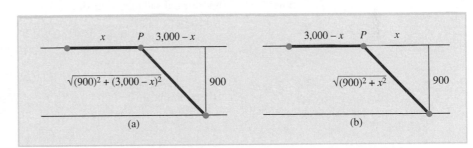

FIGURE 3.51 Two choices for the variable x.

Before plunging into the calculations, take a minute to decide which choice of variables is more advantageous. In Figure 3.51a, the distance across the water from the power plant to the point P is (by the Pythagorean theorem) $\sqrt{(900)^2 + (3,000 - x)^2}$, and the corresponding total cost function is

$$C(x) = 5\sqrt{(900)^2 + (3,000 - x)^2} + 4x$$

In Figure 3.51b, the distance across the water is $\sqrt{(900)^2 + x^2}$, and the total cost function is

$$C(x) = 5\sqrt{(900)^2 + x^2} + 4(3,000 - x)$$

The second function is easier to work with since the term $3,000 - x$ is only multiplied by 4, while in the first function it is squared and appears under the radical. Hence, you should choose x as in Figure 3.51b and work with the total cost function

$$C(x) = 5\sqrt{(900)^2 + x^2} + 4(3,000 - x)$$

Since the distances x and $3,000 - x$ cannot be negative, the relevant interval is $0 \le x \le 3,000$, and your goal is to find the absolute minimum of the function $C(x)$ on this closed interval. To find the critical values, compute the derivative

$$C'(x) = \frac{5}{2}[(900)^2 + x^2]^{-1/2}(2x) - 4 = \frac{5x}{\sqrt{(900)^2 + x^2}} - 4$$

and set it equal to zero to get

$$\frac{5x}{\sqrt{(900)^2 + x^2}} - 4 = 0 \qquad \text{or} \qquad \sqrt{(900)^2 + x^2} = \frac{5}{4}x$$

Square both sides of the equation and solve for x to get

$$(900)^2 + x^2 = \frac{25}{16}x^2 \qquad\qquad \text{subtract } \frac{25x^2}{16} \text{ and} \\ (900)^2 \text{ from each side}$$

$$x^2 - \frac{25}{16}x^2 = -(900)^2 \qquad\qquad \text{combine like terms on the left}$$

$$-\frac{9}{16}x^2 = -(900)^2 \qquad\qquad \text{multiply by } -\frac{16}{9}$$

$$x^2 = \frac{16}{9}(900)^2 \qquad\qquad \text{take square roots on} \\ \text{each side}$$

$$x = \pm\frac{4}{3}(900) = \pm 1,200$$

Since only the positive value $x = 1,200$ is in the interval $0 \le x \le 3,000$, compute $C(x)$ at this critical value and at the endpoints $x = 0$ and $x = 3,000$. Since

$$C(0) = 5\sqrt{(900)^2 + 0} + 4(3,000 - 0) = 16,500$$
$$C(1,200) = 5\sqrt{(900)^2 + (1,200)^2} + 4(3,000 - 1,200) = 14,700$$
$$C(3,000) = 5\sqrt{(900)^2 + (3,000)^2} + 4(3,000 - 3,000) = 15,660$$

it follows that the minimal installation cost is $14,700, which will occur if the cable reaches the opposite bank 1,200 meters downstream from the power plant.

In Example 3.5.6, the function to be maximized has practical meaning only when its independent variable is a whole number. However, the optimization procedure leads to a fractional value of this variable, and additional analysis is needed to obtain a meaningful solution.

EXAMPLE 3.5.6   Maximizing a Revenue Function with Integer Inputs

A bus company will charter a bus that holds 50 people to groups of 35 or more. If a group contains exactly 35 people, each person pays $60. In large groups, everybody's fare is reduced by $1 for each person in excess of 35. Determine the size of the group for which the bus company's revenue will be greatest.

Solution

Let R denote the bus company's revenue. Then,

$$R = \text{(number of people in the group)(fare per person)}$$

You could let x denote the total number of people in the group, but it is slightly more convenient to let x denote the number of people in excess of 35. Then,

$$\text{Number of people in the group} = 35 + x$$

and $$\text{Fare per person} = 60 - x$$

so the revenue function is

$$R(x) = (35 + x)(60 - x)$$

Since x represents the number of people in excess of 35 but no more than 50, you want to maximize $R(x)$ for a positive integer x in the interval $0 \leq x \leq 15$ (Figure 3.52a). However, to use the methods of calculus, consider the *continuous* function $R(x) = (35 + x)(60 - x)$ defined on the entire interval $0 \leq x \leq 15$ (Figure 3.52b).

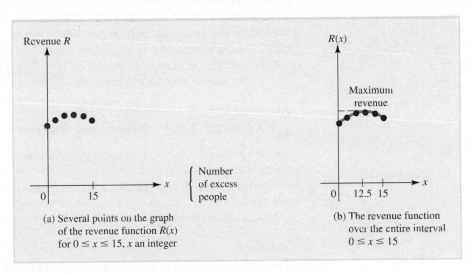

(a) Several points on the graph of the revenue function $R(x)$ for $0 \leq x \leq 15$, x an integer

(b) The revenue function over the entire interval $0 \leq x \leq 15$

FIGURE 3.52   The revenue function $R(x) = (35 + x)(60 - x)$.

The derivative is

$$R'(x) = (35 + x)(-1) + (60 - x)(1) = 25 - 2x$$

which is zero when $x = 12.5$. Since

$$R(0) = 2,100 \qquad R(12.5) = 2,256.25 \qquad R(15) = 2,250$$

it follows that the absolute maximum of $R(x)$ on the interval $0 \leq x \leq 15$ occurs when $x = 12.5$.

But x represents a certain number of people and must be a whole number. Hence, $x = 12.5$ cannot be the solution to this practical optimization problem. To find the optimal *integer* value of x, observe that R is increasing for $0 < x < 12.5$ and decreasing for $x > 12.5$, as shown in this diagram (see also Figure 3.52b).

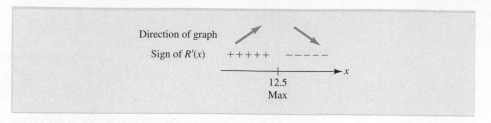

It follows that the optimal integer value of x is either $x = 12$ or $x = 13$. Since

$$R(12) = 2{,}256 \quad \text{and} \quad R(13) = 2{,}256$$

you can conclude that the bus company's revenue will be greatest when the group contains either 12 or 13 people in excess of 35, that is, for groups of 47 or 48. The revenue in either case will be $2,256.

Inventory Control

Inventory control is an important consideration in business. In particular, for each shipment of raw materials, a manufacturer must pay an ordering fee to cover handling and transportation. When the raw materials arrive, they must be stored until needed, and storage costs result. If each shipment of raw materials is large, few shipments will be needed, so ordering costs will be low, while storage costs will be high. On the other hand, if each shipment is small, ordering costs will be high because many shipments will be needed, but storage costs will be low. Example 3.5.7 shows how the methods of calculus can be used to determine the shipment size that minimizes total cost.

EXAMPLE 3.5.7 Minimizing Inventory Cost

Sheldon Evans, the manager of a company that manufactures bicycles, buys 6,000 tires a year from a distributor. Each tire costs $21, the ordering fee is $20 per shipment, and the storage cost is 96 cents per tire per year. Suppose that the tires are used at a constant rate throughout the year and that each shipment arrives just as the preceding shipment is being used up. How many tires should Sheldon order each time to minimize total cost?

Solution

The goal is to minimize the total cost, which can be written as

$$\text{Total cost} = \text{storage cost} + \text{ordering cost} + \text{purchase cost}$$

Let x denote the number of tires in each shipment and $C(x)$ the corresponding total cost in dollars. Then,

$$\text{Ordering cost} = (\text{ordering cost per shipment})(\text{number of shipments})$$

Since 6,000 tires are ordered during the year and each shipment contains x tires, the number of shipments is $\dfrac{6{,}000}{x}$ and so

$$\text{Ordering cost} = 20\left(\frac{6{,}000}{x}\right) = \frac{120{,}000}{x}$$

Moreover,

$$\text{Purchase cost} = (\text{total number of tires ordered})(\text{cost per tire})$$
$$= 6{,}000(21) = 126{,}000$$

The storage cost is slightly more complicated. When a shipment arrives, all x tires are placed in storage and then withdrawn for use at a constant rate. The inventory decreases linearly until there are no tires left, at which time the next shipment arrives. The situation is illustrated in Figure 3.53a. This is a basic example of what is called **just-in-time inventory management.**

EXPLORE!

Refer to Example 3.5.7 and the cost function $C(x)$, but vary the ordering fee $q = \$20$. Graph the cost functions for $q = 10, 15, 20,$ and 25 using a viewing window of [0, 6,000]500 by [126,000, 130,000]5,000 followed by a window of [0, 1,000]100 by [126,000, 127,000]1,000. Describe the difference in the graphs for these values of q. Find the minimum cost in each case. Describe how the minimum changes with the changing values of q.

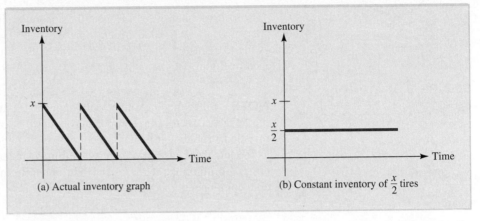

(a) Actual inventory graph   (b) Constant inventory of $\frac{x}{2}$ tires

FIGURE 3.53 Inventory graphs.

The average number of tires in storage during the year is $\frac{x}{2}$, and the total yearly storage cost is the same as if $\frac{x}{2}$ tires were kept in storage for the entire year (Figure 3.53b).

This assertion, although reasonable, is not really obvious, and you have every right to be unconvinced. In Chapter 5, you will learn how to prove this fact mathematically using integral calculus. It follows that

Storage cost = (average number of tires stored)(storage cost per tire)

$$= \frac{x}{2}(0.96) = 0.48x$$

Putting it all together, the total cost is

$$C(x) = \underbrace{0.48x}_{\substack{\text{storage} \\ \text{cost}}} + \underbrace{\frac{120,000}{x}}_{\substack{\text{ordering} \\ \text{cost}}} + \underbrace{126,000}_{\substack{\text{cost of} \\ \text{purchase}}}$$

and the goal is to find the absolute minimum of $C(x)$ on the interval

$$0 < x \le 6,000$$

The derivative of $C(x)$ is

$$C'(x) = 0.48 - \frac{120,000}{x^2}$$

which is zero when

$$x^2 = \frac{120,000}{0.48} = 250,000 \qquad \text{or} \qquad x = \pm 500$$

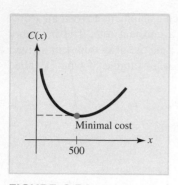

FIGURE 3.54 Total cost
$C(x) = 0.48x + \dfrac{120,000}{x} + 126,000$.

Since $x = 500$ is the only critical number in the relevant interval $0 < x \le 6,000$, you can apply the second derivative test for absolute extrema. You find that the second derivative of the cost function is

$$C''(x) = \frac{240,000}{x^3}$$

which is positive when $x > 0$. Hence, the absolute minimum of the total cost $C(x)$ on the interval $0 < x \le 6,000$ occurs when $x = 500$, that is, when the manufacturer orders the tires in lots of 500. For reference, the graph of the total cost function is sketched in Figure 3.54.

NOTE Since the derivative of the (constant) purchase price $126,000 in Example 3.5.7 was zero, this component of the total cost had no bearing on the optimization problem. In general, economists distinguish between **fixed costs** (such as the total purchase price) and **variable costs** (such as the storage and ordering costs). To minimize total cost, it is sufficient to minimize the sum of all the variable components of cost. ∎

EXERCISES ▪ 3.5

1. What number exceeds its square by the largest amount? [*Hint:* Find the number x that maximizes $f(x) = x - x^2$.]

2. What number is exceeded by its square root by the largest amount?

3. Find two positive numbers whose sum is 50 and whose product is as large as possible.

4. Find two positive numbers x and y whose sum is 30 and are such that xy^2 is as large as possible.

5. A city recreation department plans to build a rectangular playground having an area of 3,600 square meters and surround it by a fence. How can this be done using the least amount of fencing?

6. There are 320 yards of fencing available to enclose a rectangular field. How should this fencing be used so that the enclosed area is as large as possible?

7. Prove that of all rectangles with a given perimeter, the square has the largest area.

8. Prove that of all rectangles with a given area, the square has the smallest perimeter.

9. A rectangle is inscribed in a right triangle, as shown in the accompanying figure. If the triangle has sides of length 5, 12, and 13, what are the dimensions of the inscribed rectangle of greatest area?

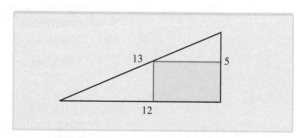

EXERCISE 9

10. A triangle is positioned with its hypotenuse on a diameter of a circle, as shown in the accompanying figure. If the circle has radius 4, what are the dimensions of the triangle of greatest area?

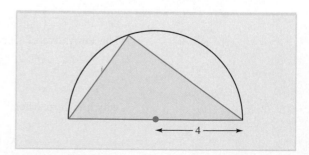

EXERCISE 10

BUSINESS AND ECONOMICS APPLIED PROBLEMS

11. **RETAIL SALES** A store has been selling a popular computer game at the price of $40 per unit, and at this price, players have been buying 50 units per month. The owner of the store wishes to raise the price of the game and estimates that for each $1 increase in price, three fewer units will be sold each month. If each unit costs the store $25, at what price should the game be sold to maximize profit?

12. **RETAIL SALES** A bookstore can obtain a certain novel from the publisher at a cost of $3 per book. The bookstore has been offering the novel at a price of $15 per copy and, at this price, has been selling 200 copies a month. The bookstore is planning to lower its price to stimulate sales and estimates that for each $1 reduction in the price, 20 more books will be sold each month. At what price should the bookstore sell the novel to generate the greatest possible profit?

13. **AGRICULTURAL YIELD** Granville Thomas is a citrus grower in Florida. He estimates that if 60 orange trees are planted, the average yield will be 400 oranges per tree. The average yield will decrease by 4 oranges per tree for each additional tree planted on the same acreage. How many trees should Granville plant to maximize the total yield?

14. **HARVESTING** Farmers can get $8 per bushel for their potatoes on July 1, and after that, the price drops by 8 cents per bushel per day. On July 1, a farmer has 80 bushels of potatoes in the field and estimates that the crop is increasing at the rate of 1 bushel per day. When should the farmer harvest the potatoes to maximize revenue?

15. **PROFIT** A baseball card store can obtain Mel Schlabotnic rookie cards at a cost of $5 per card. The store has been offering the cards at $10 apiece and, at this price, has been selling 25 cards per month. The store is planning to lower the price to stimulate sales and estimates that for each 25-cent reduction in the price, 5 more cards will be sold each month. At what price should the cards be sold to maximize total monthly profit?

16. **PROFIT** A manufacturer has been selling flashlights at $6 apiece, and at this price, consumers have been buying 3,000 flashlights per month. The manufacturer wishes to raise the price and estimates that for each $1 increase in the price, 1,000 fewer flashlights will be sold each month. The manufacturer can produce the flashlights at a cost of $4 per flashlight. At what price should the manufacturer sell the flashlights to generate the greatest possible profit?

17. **INSTALLATION COST** A cable is to be run from a power plant on one side of a river 1,200 meters wide to a factory on the other side, 1,500 meters downstream. The cost of running the cable under the water is $25 per meter, while the cost over land is $20 per meter. What is the most economical route over which to run the cable?

18. **INSTALLATION COST** Find the most economical route in Exercise 17 if the power plant is 2,000 meters downstream from the factory.

19. **RETAIL SALES** A retailer has bought several cases of a certain imported wine. As the wine ages, its value initially increases, but eventually the wine will pass its prime and its value will decrease. Suppose that x years from now, the value of a case will be changing at the rate of $53 - 10x$ dollars per year. Suppose, in addition, that storage rates will remain fixed at $3 per case per year. When should the retailer sell the wine to obtain the greatest possible profit?

20. **PRODUCTION COST** Each machine at a certain factory can produce 50 units per hour. The setup cost is $80 per machine, and the operating cost is $5 per hour. How many machines should be used to produce 8,000 units at the least possible cost? (Remember that the answer should be a whole number.)

21. **COST ANALYSIS** It is estimated that the cost of constructing an office building that is n floors high is $C(n) = 2n^2 + 500n + 600$ thousand dollars. How many floors should the building have to minimize the average cost per floor? (Remember that your answer should be a whole number.)

22. **INVENTORY** A manufacturer of medical monitoring devices uses 36,000 cases of components per year. The ordering cost is $54 per shipment, and the annual cost of storage is $1.20 per case. The components are used at a constant rate throughout the year, and each shipment arrives just as the preceding shipment is being used up. How many cases should be ordered in each shipment to minimize total cost?

23. INVENTORY MANAGEMENT Paula Perkins, the owner of Paula's Perfume Shoppe, expects to sell 800 bottles of a certain brand of perfume this year. The perfume costs $20 per bottle, the ordering fee is $10 per shipment, and the cost of storing the perfume is 40 cents per bottle per year. The perfume is sold at a constant rate throughout the year, and each shipment arrives just as the preceding shipment is being used up.
   a. How many bottles should Paula order in each shipment to minimize total cost?
   b. How often should Paula order the perfume?

24. TRANSPORTATION COST A truck is hired to transport goods from a factory to a warehouse. The driver's wages are figured by the hour and so are inversely proportional to the speed at which the truck is driven. The amount of gasoline used is directly proportional to the speed at which the truck is driven, and the price of gasoline remains constant during the trip. Show that the total cost is smallest at the speed for which the driver's wages are equal to the cost of the gasoline used.

25. PRODUCTION COST A plastics firm has received an order from the city recreation department to manufacture 8,000 special Styrofoam kickboards for its summer swimming program. The firm owns 10 machines, each of which can produce 30 kickboards an hour. The cost of setting up the machines to produce the kickboards is $20 per machine. Once the machines have been set up, the operation is fully automated and can be overseen by a single production supervisor earning $15 per hour.
   a. How many of the machines should be used to minimize the cost of production?
   b. How much will the supervisor earn during the production run if the optimal number of machines is used?
   c. How much will it cost to set up the optimal number of machines?

26. TRANSPORTATION COST For speeds between 40 and 65 miles per hour, a truck gets $\frac{480}{x}$ miles per gallon when driven at a constant speed of x miles per hour. Diesel gasoline costs $3.90 per gallon, and the driver is paid $19.50 per hour. What is the most economical constant speed between 40 and 65 miles per hour at which to drive the truck?

27. OPTIMAL SETUP COST Suppose that at a certain factory, setup cost is directly proportional to the number N of machines used and operating cost is inversely proportional to N. Show that when the total cost is minimal, the setup cost is equal to the operating cost.

28. AVERAGE PRODUCTIVITY The output Q at a certain factory is a function of the number L of worker-hours of labor that are used. Use calculus to prove that when the average output per worker-hour is greatest, the average output is equal to the marginal output per worker-hour. You may assume without proof that the critical point of the average output function corresponds to the desired absolute maximum. [*Hint:* The marginal output per worker-hour is the derivative of output Q with respect to labor L.]

29. EFFECT OF TAXATION ON A MONOPOLY A **monopolist** is a manufacturer who can manipulate the price of a commodity and usually does so with an eye toward maximizing profit. When the government taxes output, the tax effectively becomes an additional cost item, and the monopolist is forced to decide how much of the tax to absorb and how much to pass on to the consumer.
   Suppose a particular monopolist estimates that when x units are produced, the total cost will be $C(x) = \frac{7}{8}x^2 + 5x + 100$ dollars and the market price of the commodity will be $p(x) = 15 - \frac{3}{8}x$ dollars per unit. Further assume that the government imposes a tax of t dollars on each unit produced.
   a. Show that profit is maximized when $x = \frac{2}{5}(10 - t)$.
   b. Suppose the government assumes that the monopolist will always act so as to maximize total profit. What value of t should be chosen to guarantee maximum total tax revenue?
   c. If the government chooses the optimum rate of taxation found in part (b), how much of this tax will be absorbed by the monopolist and how much will be passed on to the consumer?
   d. Read an article on taxation and write a paragraph on how it affects consumer spending.*

*You may wish to begin your research by consulting the following references: Robert Eisner, *The Misunderstood Economics: What Counts and How to Count It*, Boston, MA: Harvard Business School Press, 1994, pp. 196–199; and Robert H. Frank, *Microeconomics and Behavior*, 2nd ed., New York: McGraw-Hill, 1994, pp. 656–657.

30. **INVENTORY** The inventory model analyzed in Example 3.5.7 is not the only such model possible. Suppose a company must supply N units per time period at a uniform rate. Assume that the storage cost per unit is D_1 dollars per time period and that the setup (ordering) cost is D_2 dollars. If production is at a uniform rate of m units per time period (with no items in inventory at the end of each period), it can be shown that the total storage cost is

$$C_1 = \frac{D_1 x}{2}\left(1 - \frac{N}{m}\right)$$

where x is the number of items produced in each run.

  a. Show that the total average cost per period is

$$C = \frac{D_1 x}{2}\left(1 - \frac{N}{m}\right) + \frac{D_2 N}{x}$$

  b. Find an expression for the number of items that should be produced in each run to minimize the total average cost per time period.

  c. The optimum quantity found in the inventory problem in Example 3.5.7 is sometimes called the **economic order quantity (EOQ),** while the optimum found in part (b) of this exercise is called the **economic production quantity (EPQ).** Modern inventory management goes far beyond the simple conditions in the EOQ and EPQ models, but elements of these models are still very important. For instance, just-in-time inventory management fits well with the production philosophy of many Japanese companies. Read an article on Japanese production methods, and write a paragraph on why the Japanese regard using space for the storage of materials as undesirable.*

LIFE AND SOCIAL SCIENCE APPLIED PROBLEMS

31. **URBAN PLANNING** Two industrial plants, A and B, are located 18 miles apart, and each day, respectively, emit 80 ppm (parts per million) and 720 ppm of particulate matter. Plant A is surrounded by a restricted area of radius 1 mile,

while the restricted area around plant B has a radius of 2 miles. The concentration of particulate matter arriving at any other point Q from each plant decreases with the reciprocal of the distance between that plant and Q. Where should a house be located on a road joining the two plants to minimize the total concentration of particulate matter arriving from both plants?

32. **URBAN PLANNING** In Exercise 31, suppose the concentration of particulate matter arriving at a point Q from each plant decreases with the reciprocal of the *square* of the distance between that plant and Q. With this alteration, now where should a house be located to minimize the total concentration of particulate matter arriving from both plants?

33. **AVIAN BEHAVIOR** Homing pigeons will rarely fly over large bodies of water unless forced to do so, presumably because it requires more energy to maintain altitude in flight in the heavy air over cool water.[†] Suppose a pigeon is released from a boat B floating on a lake 5 miles from a point A on the shore and 13 miles from the pigeon's loft L, as shown in the accompanying figure. Assuming the pigeon requires twice as much energy to fly over water as over land, what path should it follow to minimize the total energy expended in flying from the boat to its loft? Assume the shoreline is straight, and describe your path as a line from B to a point P on the shore followed by a line from P to L.

EXERCISE 33

34. **RECYCLING** To raise money, a service club has been collecting used bottles that it plans to deliver to a local glass company for recycling. Since the project began 80 days ago, the club has collected 24,000 pounds of glass for which the glass

---

*You may wish to begin your search with the text by Philip E. Hicks, *Industrial Engineering and Management: A New Perspective,* New York: McGraw-Hill, 1994, pp. 144–170.

†E. Batschelet, *Introduction to Mathematics for Life Sciences,* 3rd ed., New York: Springer-Verlag, 1979, pp. 276–277.

company currently offers 1 cent per pound. However, because bottles are accumulating faster than they can be recycled, the company plans to reduce by 1 cent each day the price it will pay for 100 pounds of used glass. Assume that the club can continue to collect bottles at the same rate and that transportation costs make more than one trip to the glass company unfeasible. What is the most advantageous time for the club to conclude its project and deliver the bottles?

35. **ELIMINATION OF HAZARDOUS WASTE** Certain hazardous waste products have the property that as the concentration of substrate (the substance undergoing change by enzymatic action) increases, there is a toxic inhibition effect. A mathematical model for this behavior is the **Haldane equation***

$$R(S) = \frac{cS}{a + S + bS^2}$$

where R is the specific growth rate of the substance (the rate at which cells divide); S is the substrate concentration; and a, b, and c are positive constants.
   a. Sketch the graph of $R(S)$. Does the graph appear to have a highest point? A lowest point? A point of inflection? What happens to the growth rate as S grows larger and larger? Interpret your observations.
   b. Read an article on hazardous waste management, and write a paragraph on how mathematical models are used to develop methods for eliminating waste. A good place to start is the reference cited here.

36. **BOTANY** An experimental garden plot contains N annual plants, each of which produces S seeds that are dropped within the same plot. A botanical model measures the number of offspring plants $A(N)$ that survive until the next year by the function

$$A(N) = \frac{NS}{1 + (cN)^p}$$

where c and p are positive constants.
   a. For what value of N is $A(N)$ maximized? What is the maximum value? Express your answer in terms of S, c, and p.

   b. For what value of N is the offspring survival rate $A'(N)$ minimized?
   c. The function $F(N) = \dfrac{A(N)}{N}$ is called the *net reproductive rate*. It measures the number of surviving offspring per plant. Find $F'(N)$, and use it to show that the greater the number of plants, the lower the number of surviving offspring per plant. This is called the principle of *density-dependent mortality*.

37. **MARINE BIOLOGY** When a fish swims upstream at a speed v against a constant current v_w, the energy it expends in traveling to a point upstream is given by a function of the form

$$E(v) = \frac{Cv^k}{v - v_w}$$

where $C > 0$ and $k > 2$ is a number that depends on the species of fish involved.[†]
   a. Show that $E(v)$ has exactly one critical number. Does it correspond to a relative maximum or a relative minimum?
   b. Note that the critical number in part (a) depends on k. Let $F(k)$ be the critical number. Sketch the graph of $F(k)$. What can be said about $F(k)$ if k is very large?

38. **PHYSICAL CHEMISTRY** In physical chemistry, it is shown that the pressure P of a gas is related to the volume V and temperature T by *van der Waals' equation*:

$$\left(P + \frac{a}{V^2}\right)(V - b) = nRT$$

where a, b, n, and R are constants. The *critical temperature* T_c of the gas is the highest temperature at which the gaseous and liquid phases can exist as separate phases.
   a. When $T = T_c$, the pressure P is given as a function $P(V)$ of volume alone. Sketch the graph of $P(V)$.
   b. The critical volume V_c is the volume for which $P'(V_c) = 0$ and $P''(V_c) = 0$. Show that $V_c = 3b$.
   c. Find the critical pressure $P_c = P(V_c)$ and then express T_c in terms of a, b, n, and R.

---

*Michael D. La Grega, Philip L. Buckingham, and Jeffrey C. Evans, *Hazardous Waste Management*, New York: McGraw-Hill, 1994, p. 578.

[†]E. Batschelet, *Introduction to Mathematics for Life Scientists*, 2nd ed., New York: Springer-Verlag, 1976, p. 280.

MISCELLANEOUS PROBLEMS

39. POSTER DESIGN   Kamal, the owner of a print shop, receives an order to produce a rectangular poster containing 648 square centimeters of print surrounded by margins of 2 centimeters on each side and 4 centimeters on the top and bottom. What dimensions should Kamal choose to minimize the total area of the poster while satisfying these requirements? [*Hint:* An unwise choice of variables will make the calculations unnecessarily complicated.]

40. PACKAGING   A cylindrical can is to hold 4π cubic inches of frozen orange juice. The cost per square inch of constructing the metal top and bottom is twice the cost per square inch of constructing the cardboard side. What are the dimensions of the least expensive can?

41. PACKAGING   Use the fact that 12 fluid ounces is approximately 6.89π cubic inches to find the dimensions of the 12-ounce soda can that can be constructed using the least amount of metal. Compare these dimensions with those of one of the soda cans in your refrigerator. What do you think accounts for the difference?

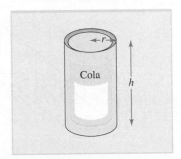

EXERCISE 41

42. PACKAGING   A cylindrical can (with top) is to be constructed using a fixed amount of metal. Use calculus to derive a simple relationship between the radius and height of the can having the greatest volume.

43. CONSTRUCTION COST   A cylindrical container with no top is to be constructed to hold a fixed volume of liquid. The cost of the material used for the bottom is 3 cents per square inch, and that for the curved side is 2 cents per square inch. Use calculus to derive a simple relationship between the radius and height of the least expensive container.

44. DISTANCE BETWEEN MOVING OBJECTS   A truck is 300 miles due east of a car and is traveling west at the constant speed of 30 miles per hour. Meanwhile, the car is going north at the constant speed of 60 miles per hour. At what time will the car and truck be closest to each other? [*Hint:* You will simplify the calculation if you minimize the *square* of the distance between the car and truck rather than the distance itself. Can you explain why this simplification is justified?]

45. CONSTRUCTION COST   A carpenter has been asked to build an open box with a square base. The sides of the box will cost $3 per square meter, and the base will cost $4 per square meter. What are the dimensions of the box of greatest volume that can be constructed for $48?

46. CONSTRUCTION COST   Carla is a carpenter who has been hired to make a closed box with a square base and volume of 250 cubic meters. The material for the top and bottom of the box costs $2 per square meter, and the material for the sides costs $1 per square meter. Can Carla construct the box for less than $300?

47. INSTALLATION COST   For the summer, the company installing the cable in Example 3.5.5 has hired Frank Kornercutter as a consultant. Frank, recalling a problem from first-year calculus, asserts that no matter how far downstream the factory is located (beyond 1,200 meters), it would be most economical to have the cable reach the opposite bank 1,200 meters downstream from the power plant. The supervisor, amused by Frank's naivete, replies, "Any fool can see that if the factory is farther away, the cable should reach the opposite bank farther downstream. It's just common sense!" Of course, Frank is no common fool, but is he right? Why?

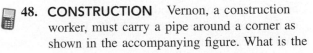 **48. CONSTRUCTION**   Vernon, a construction worker, must carry a pipe around a corner as shown in the accompanying figure. What is the

length of the longest pipe Vernon can carry horizontally around the corner?

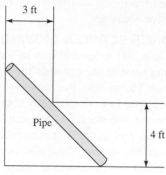

EXERCISE 48

49. SPY STORY   It is noon, and the spy is back from space (see Exercise 69 in Section 2.2) and driving a jeep through the sandy desert in the tiny principality of Alta Loma. He is 32 kilometers from the nearest point on a straight paved road. Down the road 16 kilometers is an abandoned power plant where a group of rival spies are holding captive his superior, code name "N." If the spy doesn't arrive with a ransom by 12:50 P.M., the bad guys have threatened to do N in. The jeep can travel at 48 km/hr in the sand and at 80 km/hr on the paved road. Can the spy make it in time, or is this the end of N? [*Hint:* The goal is to minimize time, which is distance divided by speed.]

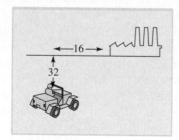

EXERCISE 49

50. CONSTRUCTION   The strength of a rectangular beam is proportional to the product of its width and the square of its depth. Find the dimensions of the strongest beam that can be cut from a wooden log of diameter 15 inches. (Note the figure accompanying Exercise 51.)

51. CONSTRUCTION   The stiffness of a rectangular beam is proportional to the product of its width w and the cube of its depth h. Find the dimensions of the stiffest beam that can be cut from a wooden log of diameter 15 inches. (Note the accompanying figure.)

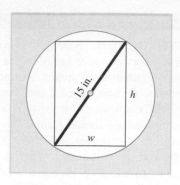

EXERCISES 50 AND 51

52. CONSTRUCTION   An open box is to be made from a square piece of cardboard, 18 inches by 18 inches, by removing a small square from each corner and folding up the flaps to form the sides. What are the dimensions of the box of greatest volume that can be constructed in this way?

53. POSTAL REGULATIONS   According to postal regulations, the girth plus length of parcels sent by fourth-class mail may not exceed 108 inches. What is the largest possible volume of a rectangular parcel with two square sides that can be sent by fourth-class mail?

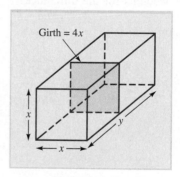

EXERCISE 53

54. POSTAL REGULATIONS   Refer to Exercise 53. What is the largest volume of a cylindrical parcel that can be sent by fourth-class mail?

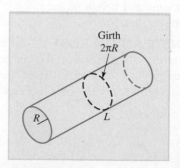

EXERCISE 54

Important Terms, Symbols, and Formulas

f is increasing if $f'(c) > 0$ (200)
f is decreasing if $f'(c) < 0$ (200)
Critical point: $(c, f(c))$, where $f'(c) = 0$ or $f'(c)$
   does not exist   (202)
Relative maxima and minima   (202)
First derivative test for relative extrema: (204)
   If $f'(c) = 0$ or $f'(c)$ does not exist, then

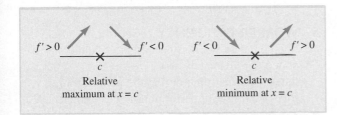

Relative
maximum at $x = c$

Relative
minimum at $x = c$

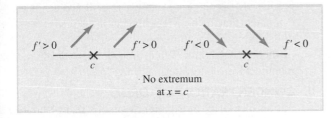

No extremum
at $x = c$

Point of diminishing returns   (215)
Concavity: (216)
   Upward if $f'(x)$ is increasing; that is, $f''(x) > 0$
   Downward if $f'(x)$ is decreasing; that is, $f''(x) < 0$
Inflection point: A point on the graph of a function f
   where f is continuous and the concavity
   changes   (219)
Second derivative test for relative extrema: (224)
   If $f'(c) = 0$, then

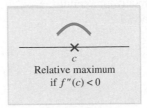

Relative maximum
if $f''(c) < 0$

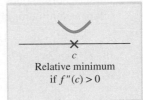

Relative minimum
if $f''(c) > 0$

Vertical asymptote   (234
Horizontal asymptote   (235)
Absolute maxima and minima   (248)
Extreme value property: (249)
   Absolute extrema of a continuous function on a
      closed interval $a \le x \le b$ occur at critical
      numbers in $a < x < b$ or at endpoints of the
      interval (a or b).
Second derivative test for absolute extrema: (253)
   If $f(x)$ has only one critical number $x = c$ on an
      interval I, then $f(c)$ is an absolute maximum on I
      if $f''(c) < 0$ and an absolute minimum if $f''(c) > 0$.
Profit $P(q) = R(q) - C(q)$ is maximized when
      marginal revenue equals marginal cost:
      $R'(q) = C'(q)$.   (255)
Average cost $A(q) = \dfrac{C(q)}{q}$ is minimized when average
      cost equals marginal cost: $A(q) = C'(q)$ (256)
Elasticity of demand $q = D(p)$: $E(p) = -\dfrac{p}{q}\dfrac{dq}{dp}$ (257)
Levels of elasticity of demand: (258)
                inelastic if $E(p) < 1$
                  elastic if $E(p) > 1$
          unit elasticity if $E(p) = 1$
Inventory models   (276)

Checkup for Chapter 3

1. Of the two curves shown here, one is the graph of a function $f(x)$ and the other is the graph of its derivative
   $f'(x)$. Determine which graph is the derivative, and give reasons for your decision.

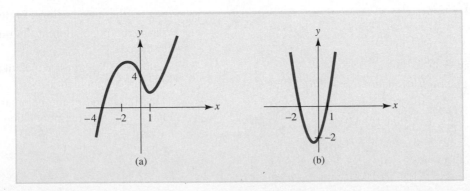

(a)   (b)

CHAPTER SUMMARY

2. Find intervals of increase and decrease for each of these functions, and determine whether each critical number corresponds to a relative maximum, a relative minimum, or neither.

   a. $f(x) = -x^4 + 4x^3 + 5$

   b. $f(t) = 2t^3 - 9t^2 + 12t + 5$

   c. $g(t) = \dfrac{t}{t^2 + 9}$

   d. $g(x) = \dfrac{4 - x}{x^2 + 9}$

3. Determine where the graph of each of these functions is concave upward and concave downward. Find the x (or t) coordinate of each point of inflection.

   a. $f(x) = 3x^5 - 10x^4 + 2x - 5$

   b. $f(x) = 3x^5 + 20x^4 - 50x^3$

   c. $f(t) = \dfrac{t^2}{t - 1}$

   d. $g(t) = \dfrac{3t^2 + 5}{t^2 + 3}$

4. Determine all vertical and horizontal asymptotes for the graph of each of these functions.

   a. $f(x) = \dfrac{2x - 1}{x + 3}$

   b. $f(x) = \dfrac{x}{x^2 - 1}$

   c. $f(x) = \dfrac{x^2 + x - 1}{2x^2 + x - 3}$

   d. $f(x) = \dfrac{1}{x} - \dfrac{1}{\sqrt{x}}$

5. Sketch the graph of each of the following functions. Be sure to show all key features such as intercepts, asymptotes, high and low points, points of inflection, cusps, and vertical tangents.

   a. $f(x) = 3x^4 - 4x^3$

   b. $f(x) = x^4 - 3x^3 + 3x^2 + 1$

   c. $f(x) = \dfrac{x^2 + 2x + 1}{x^2}$

   d. $f(x) = \dfrac{1 - 2x}{(x - 1)^2}$

6. Sketch the graph of a function $f(x)$ with all these properties:

   a. $f'(x) > 0$ for $x < 0$ and for $0 < x < 2$

   b. $f'(x) < 0$ for $x > 2$

   c. $f'(0) = f'(2) = 0$

   d. $f''(x) < 0$ for $x < 0$ and for $x > 1$

   e. $f''(x) > 0$ for $0 < x < 1$

   f. $f(-1) = f(4) = 0; f(0) = 1, f(1) = 2, f(2) = 3$

7. In each of the following cases, find the largest and smallest values of the given function on the specified interval.

   a. $f(x) = x^3 - 3x^2 - 9x + 1$ on $-2 \le x \le 4$

   b. $g(t) = -4t^3 + 9t^2 + 12t - 5$ on $-1 \le t \le 4$

   c. $h(u) = 8\sqrt{u} - u + 3$ on $0 \le u \le 25$

8. **WORKER EFFICIENCY**   Rita is a postal clerk. She comes to work at 6 A.M. and t hours later has sorted approximately $f(t) = -t^3 + 7t^2 + 200t$ letters. At what time during the period from 6 A.M. to 10 A.M. is Rita performing at peak efficiency?

9. **MAXIMIZING PROFIT**   A manufacturer can produce MP3 players at a cost of $90 apiece. It is estimated that if the MP3 players are sold for x dollars apiece, consumers will buy $20(180 - x)$ of them each month. What unit price should the manufacturer charge to maximize profit?

10. **CONCENTRATION OF DRUG**   The concentration of a drug in a patient's bloodstream t hours after it is injected is given by

$$C(t) = \frac{0.05t}{t^2 + 27}$$

milligrams per cubic centimeter.

   a. Sketch the graph of the concentration function.

   b. At what time is the concentration decreasing most rapidly?

   c. What happens to the concentration in the long run (as $t \to +\infty$)?

11. **BACTERIAL POPULATION**   A bacterial colony is estimated to have a population of

$$P(t) = \frac{15t^2 + 10}{t^3 + 6} \text{ million}$$

t hours after the introduction of a toxin.

   a. What is the population at the time the toxin is introduced?

   b. When does the largest population occur? What is the largest population?

   c. Sketch the graph of the population curve. What happens to the population in the long run (as $t \to +\infty$)?

Review Exercises

In Exercises 1 through 10, determine intervals of increase and decrease and intervals of concavity for the given function. Then sketch the graph of the function. Be sure to show all key features such as intercepts, asymptotes, high and low points, points of inflection, cusps, and vertical tangents.

1. $f(x) = -2x^3 + 3x^2 + 12x - 5$

2. $f(x) = x^2 - 6x + 1$

3. $f(x) = 3x^3 - 4x^2 - 12x + 17$

4. $f(x) = x^3 - 3x^2 + 2$

5. $f(t) = 3t^5 - 20t^3$

6. $f(x) = \dfrac{x^2 + 3}{x - 1}$

7. $g(t) = \dfrac{t^2}{t + 1}$

8. $G(x) = (2x - 1)^2(x - 3)^3$

9. $F(x) = 2x + \dfrac{8}{x} + 2$

10. $f(x) = \dfrac{1}{x^3} + \dfrac{2}{x^2} + \dfrac{1}{x}$

In Exercises 11 and 12, one of the two curves shown is the graph of a certain function $f(x)$ and the other is the graph of its derivative $f'(x)$. Determine which curve is the graph of the derivative, and give reasons for your decision.

11.

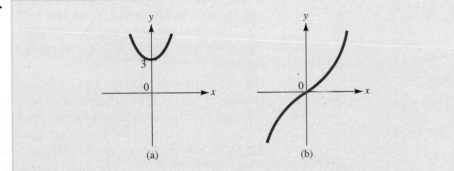

(a)   (b)

12.

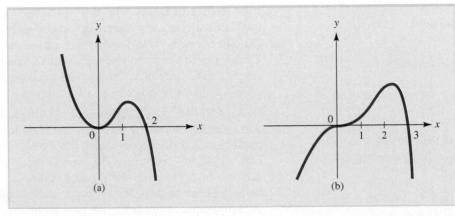

(a)   (b)

In Exercises 13 through 16, the derivative $f'(x)$ of a function is given. Use this information to classify each critical number of $f(x)$ as a relative maximum, a relative minimum, or neither.

13. $f'(x) = x^3(2x - 3)^2(x + 1)^5(x - 7)$

14. $f'(x) = \sqrt[3]{x}(3 - x)(x + 1)^2$

15. $f'(x) = \dfrac{x(x - 2)^2}{x^4 + 1}$

16. $f'(x) = \dfrac{x^2 + 2x - 3}{x^2(x^2 + 1)}$

In Exercises 17 through 20, sketch the graph of a function f that has all the given properties.

17. a. $f'(x) > 0$ when $x < 0$ and when $x > 5$
    b. $f'(x) < 0$ when $0 < x < 5$
    c. $f''(x) > 0$ when $-6 < x < -3$ and when $x > 2$
    d. $f''(x) < 0$ when $x < -6$ and when $-3 < x < 2$

18. a. $f'(x) > 0$ when $x < -2$ and when $-2 < x < 3$
    b. $f'(x) < 0$ when $x > 3$
    c. $f'(-2) = 0$ and $f'(3) = 0$

19. a. $f'(x) > 0$ when $1 < x < 2$
    b. $f'(x) < 0$ when $x < 1$ and when $x > 2$
    c. $f''(x) > 0$ for $x < 2$ and for $x > 2$
    d. $f'(1) = 0$ and $f'(2)$ is undefined.

20. a. $f'(x) > 0$ when $x < 1$
    b. $f'(x) < 0$ when $x > 1$
    c. $f''(x) > 0$ when $x < 1$ and when $x > 1$
    d. $f'(1)$ is undefined.

In Exercises 21 through 24, find all critical numbers for the given function $f(x)$ and use the second derivative test to determine which (if any) critical points are relative maxima or relative minima.

21. $f(x) = -2x^3 + 3x^2 + 12x - 5$

22. $f(x) = x(2x - 3)^2$

23. $f(x) = \dfrac{x^2}{x + 1}$

24. $f(x) = \dfrac{1}{x} - \dfrac{1}{x + 3}$

In Exercises 25 through 28, find the absolute maximum and the absolute minimum values (if any) of the given function on the specified interval.

25. $f(x) = -2x^3 + 3x^2 + 12x - 5; \ -3 \le x \le 3$

26. $f(t) = -3t^4 + 8t^3 - 10; \ 0 \le t \le 3$

27. $g(s) = \dfrac{s^2}{s + 1}; \ -\dfrac{1}{2} \le s \le 1$

28. $f(x) = 2x + \dfrac{8}{x} + 2; \ x > 0$

29. The first derivative of a certain function is $f'(x) = x(x - 1)^2$.
    a. On what intervals is f increasing? Decreasing?
    b. On what intervals is the graph of f concave up? Concave down?
    c. Find the x coordinates of the relative extrema and inflection points of f.
    d. Sketch a possible graph of $f(x)$.

30. The first derivative of a certain function is $f'(x) = x^2(5 - x)$.
    a. On what intervals is f increasing? Decreasing?
    b. On what intervals is the graph of f concave up? Concave down?
    c. Find the x coordinates of the relative extrema and inflection points of f.
    d. Sketch a possible graph of $f(x)$.

31. **PROFIT** A manufacturer can produce sunglasses at a cost of $5 apiece and estimates that if they are sold for x dollars apiece, consumers will buy $100(20 - x)$ sunglasses a day. At what price should the manufacturer sell the sunglasses to maximize profit?

32. **CONSTRUCTION COST** A box with a rectangular base is to be constructed of material costing $2/in.^{2} for the sides and bottom and $3/in.^{2} for the top. If the box is to have volume 1,215 in.^{3} and the length of its base is to be twice its width, what dimensions of the box will minimize its cost of construction? What is the minimal cost?

33. **CONSTRUCTION COST** A cylindrical container with no top is to be constructed for a fixed amount of money. The cost of the material used for the bottom is 3 cents per square inch, and the cost of the material used for the curved side is 2 cents per square inch. Use calculus to derive a simple relationship between the radius and height of the container having the greatest volume.

34. **REAL ESTATE DEVELOPMENT** Bernardo is a real estate developer. He estimates that if 60 luxury houses are built in a certain area, the average profit will be $47,500 per house. The average profit will decrease by $500 per house for each additional house built in the area. How many houses should Bernardo build to maximize the total profit? (Remember, the answer must be an integer.)

35. **OPTIMAL DESIGN**   A farmer wishes to enclose a rectangular pasture with 320 feet of fence. What dimensions give the maximum area if

   a. the fence is on all four sides of the pasture?

   b. the fence is on three sides of the pasture and the fourth side is bounded by a wall?

36. **TRAFFIC CONTROL**   It is estimated that between the hours of noon and 7:00 P.M., the speed of highway traffic flowing past a certain downtown exit is approximately

$$S(t) = t^3 - 9t^2 + 15t + 45$$

   miles per hour, where t is the number of hours past noon. At what time between noon and 7:00 P.M. is the traffic moving the fastest, and at what time between noon and 7:00 P.M. is it moving the slowest?

37. **MINIMIZING TRAVEL TIME**   Loni is standing on the bank of a river that is 1 mile wide and wants to get to a town on the opposite bank, 1 mile upstream. She plans to row on a straight line to some point P on the opposite bank and then walk the remaining distance along the bank. To what point P should Loni row to reach the town in the shortest possible time if she can row at 4 miles per hour and walk at 5 miles per hour?

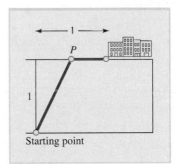

EXERCISE 37

38. **CONSTRUCTION**   Dan wants to use 300 meters of fencing to surround two identical adjacent rectangular plots, as shown in the accompanying figure. How should he do this to make the combined area of the plots as large as possible?

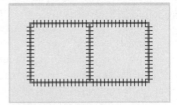

EXERCISE 38

39. **PRODUCTION**   A manufacturing firm has received an order to make 400,000 souvenir medals. The firm owns 20 machines, each of which can produce 200 medals per hour. The cost of setting up the machines to produce the medals is $80 per machine, and the total operating cost is $5.76 per hour. How many machines should be used to minimize the cost of producing the 400,000 medals? (Remember, the answer must be an integer.)

40. **ELASTICITY OF DEMAND**   Suppose that the demand equation for a certain commodity is $q = 60 - 0.1p$ (for $0 \le p \le 600$).

   a. Express the elasticity of demand as a function of p.

   b. Calculate the elasticity of demand when the price is $p = 200$. Interpret your answer.

   c. At what price is the elasticity of demand equal to 1?

41. **ELASTICITY OF DEMAND**   Suppose that the demand equation for a certain commodity is $q = 200 - 2p^2$ (for $0 \le p \le 10$).

   a. Express the elasticity of demand as a function of p.

   b. Calculate the elasticity of demand when the price is $p = 6$. Interpret your answer.

   c. At what price is the elasticity of demand equal to 1?

42. **ELASTICITY AND REVENUE**   Suppose that $q = 500 - 2p$ units of a certain commodity are demanded when p dollars per unit are charged, for $0 \le p \le 250$.

   a. Determine where the demand is elastic, inelastic, and of unit elasticity with respect to price.

   b. Use the results of part (a) to determine the intervals of increase and decrease of the revenue function and the price at which revenue is maximized.

   c. Find the total revenue function explicitly, and use its first derivative to determine its intervals of increase and decrease and the price at which revenue is maximized.

   d. Graph the demand and revenue functions.

43. **DEMAND FOR CRUISE TICKETS**   A cruise line estimates that when each deluxe balcony stateroom on a particular cruise is priced at p thousand dollars, then q tickets for staterooms will be demanded by travelers, where $q = 300 - 0.7p^2$.

CHAPTER SUMMARY

a. Find the elasticity of demand for the stateroom tickets.

b. When the price is $8,000 ($p = 8$) per stateroom, should the cruise line raise or lower the price to increase total revenue?

44. **PACKAGING**   What is the maximum possible volume of a cylindrical can with no top that can be made from 27π square inches of metal?

45. **ARCHITECTURE**   Lamar is an artist who has been commissioned to construct an ornate window. The window is to be in the form of an equilateral triangle surmounted on a rectangle, and the entire window is to have a perimeter of 20 feet. The rectangular part will be made of clear glass and will admit twice as much light as the stained glass triangular part. What dimensions should Lamar choose for the window if he wants it to admit the maximum amount of light?

46. **CONSTRUCTION COST**   Oil from an offshore rig located 3 miles from the shore is to be pumped to a location on the edge of the shore that is 8 miles east of the rig. The cost of constructing a pipe in the ocean from the rig to the shore is 1.5 times more expensive than the cost of construction on land. How should the pipe be laid to minimize cost?

47. **INVENTORY**   Through its franchised stations, an oil company gives out 16,000 road maps per year. The cost of setting up a press to print the maps is $100 for each production run. In addition, production costs are 6 cents per map and storage costs are 20 cents per map per year. The maps are distributed at a uniform rate throughout the year and are printed in equal batches timed so that each arrives just as the preceding batch has been used up. How many maps should the oil company print in each batch to minimize cost?

48. **INVENTORY**   An electronics firm uses 600 cases of components each year. Each case costs $1,000. The cost of storing one case for a year is 90 cents, and the ordering fee is $30 per shipment. How many cases should the firm order each time to keep total cost at a minimum? (Assume that the components are used at a constant rate throughout the year and that each shipment arrives just as the preceding shipment is being used up.)

49. **INVENTORY**   A manufacturing firm receives raw materials in equal shipments arriving at regular intervals throughout the year. The cost of storing the raw materials is directly proportional to the size of each shipment, while the total yearly ordering cost is inversely proportional to the shipment size. Show that the total cost is lowest when the total storage cost and total ordering cost are equal.

50. **ACCOUNT MANAGEMENT**   Austin needs $10,000 spending money each year, which he takes from his savings account by making N equal withdrawals. Each withdrawal incurs a transaction fee of $8, and money in his account earns interest at the simple interest rate of 4%.

a. The total cost C of managing the account is the transaction cost plus the loss of interest due to withdrawn funds. Express C as a function of N. $\left[Hint:\ \text{You may need the fact that } 1 + 2 + \cdots + N = \dfrac{N(N + 1)}{2}. \right]$

b. How many withdrawals should Austin make each year to minimize the total transaction cost $C(N)$?

51. **PRODUCTION COST**   A manufacturing firm receives an order for q units of a certain commodity. Each of the firm's machines can produce n units per hour. The setup cost is s dollars per machine, and the operating cost is p dollars per hour.

a. Derive a formula for the number of machines that should be used to keep total cost as low as possible.

b. Prove that when the total cost is minimal, the cost of setting up the machines is equal to the cost of operating the machines.

52. **MINIMAL COST**   A manufacturer finds that in producing x units per day (for $0 < x < 100$), three different kinds of cost are involved:

a. A fixed cost of $1,200 per day in wages

b. A production cost of $1.20 per day for each unit produced

c. An ordering cost of $\dfrac{100}{x^2}$ dollars per day

Express the total cost as a function of x and determine the level of production that results in minimal total cost.

53. **CRYSTALLOGRAPHY** A fundamental problem in crystallography is the determination of the **packing fraction** of a crystal lattice, which is the fraction of space occupied by the atoms in the lattice, assuming that the atoms are hard spheres. When the lattice contains exactly two different kinds of atoms, it can be shown that the packing fraction is given by the formula*

$$f(x) = \frac{K(1 + c^2 x^3)}{(1 + x)^3}$$

where $x = \dfrac{r}{R}$ is the ratio of the radii, r and R, of the two kinds of atoms in the lattice, and c and K are positive constants.

  a. The function $f(x)$ has exactly one critical number. Find it, and use the second derivative test to classify it as a relative maximum or a relative minimum.

  b. The numbers c and K and the domain of $f(x)$ depend on the cell structure in the lattice. For ordinary rock salt: $c = 1$, $K = \dfrac{2\pi}{3}$, and the domain is the interval $(\sqrt{2} - 1) \le x \le 1$. Find the largest and smallest values of $f(x)$.

  c. Repeat part (b) for β-cristobalite, for which $c = \sqrt{2}$, $K = \dfrac{\sqrt{3}\pi}{16}$, and the domain is $0 \le x \le 1$.

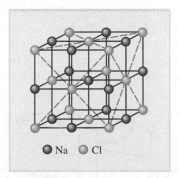

EXERCISE 53

  d. What can be said about the packing fraction $f(x)$ if r is much larger than R? Answer this question by computing $\lim\limits_{x \to \infty} f(x)$.

  e. Read the article on which this problem is based, and write a paragraph on how packing factors are computed and used in crystallography.

54. **FIREFIGHTING** If air resistance is neglected, it can be shown that the stream of water emitted by a fire hose will have height

$$y = -16(1 + m^2)\left(\frac{x}{v}\right)^2 + mx$$

feet above a point located x feet from the nozzle, where m is the slope of the nozzle and v is the velocity of the stream of water as it leaves the nozzle. Assume v is constant.

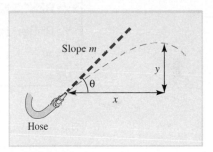

EXERCISE 54

  a. Suppose m is also constant. What is the maximum height reached by the stream of water? How far away from the nozzle does the stream reach (that is, what is x when $y = 0$)?

  b. If m is allowed to vary, find the slope that allows a firefighter to spray water on a fire from the greatest distance.

  c. Suppose the firefighter is $x = x_0$ feet from the base of a building. If m is allowed to vary, what is the highest point on the building that the firefighter can reach with the water from her hose?

---

*John C. Lewis and Peter P. Gillis, "Packing Factors in Diatomic Crystals," *American Journal of Physics*, Vol. 61, No. 5, 1993, pp. 434–438.

EXPLORE! UPDATE

Complete solutions for all EXPLORE! boxes throughout the text can be accessed at the book-specific website, www.mhhe.com/hoffmann.

Solution for Explore! on Page 200

Place the function $f(x) = 2x^3 + 3x^2 - 12x - 7$ into Y1, using a bold graphing style, and store $f'(x)$ into Y2, representing the numerical derivative. Using a modified decimal window $[-4.7, 4.7]1$ by $[-20, 20]2$, we obtain the following graphs. Using the tracing feature of your graphing calculator, you can locate where the derivative $f'(x)$ intersects the x axis (the y value displayed in scientific notation $2E^-6 = 0.000002$ can be considered negligible).

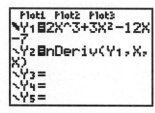
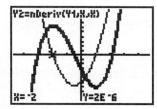
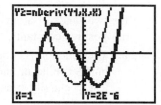

You will observe that the graph of $f(x)$ in bold is decreasing between $x = -2$ and $x = 1$, the same interval over which $f'(x)$ is negative. Note that the graph of $f'(x)$ is below the x axis in this interval. Where $f(x)$ is increasing, $f'(x)$ is positive, specifically, for $x < -2$ and $x > 1$. It appears that where the derivative is zero, $x = -2$ and 1, the graph of $f(x)$ has turns or bends in the curve, indicating possible high or low points.

Solution for Explore! on Page 201

Store $f(x) = \dfrac{x^2}{x - 2}$ into Y1 and $g(x) = \dfrac{x^2}{x - 4}$ into Y2 with a bold graphing style. Use the graphing window $[-9.4, 9.4]1$ by $[-20, 30]5$ to obtain the graphs shown here. How do the two graphs differ? The relative maximum (on the left branch of the graph before the asymptote) occurs at the origin for both graphs. But the relative minimum (on the right branch of the graph) appears to shift from $x = 4$ to $x = 8$. Note that in both graphs the relative maximum is actually smaller than any point in the right branch of the respective graph. The intervals over which $g(x)$ decreases are $(0, 4)$ and $(4, 8)$.

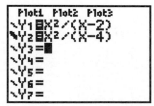
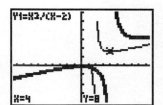
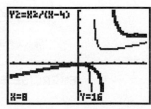

**Solution for Explore!
on Page 222**

Place $f(x) = 3x^4 - 2x^3 - 12x^2 + 18x + 15$ into Y1 of your equation editor using a bold graphing style, and let Y2 and Y3 be the first and second derivatives of Y1, respectively, as shown here. Now deselect Y3 and only consider the graphs of Y1 and Y2. Use the modified decimal window given here. The critical values of $f(x)$ are located at the zeros of $f'(x)$, namely, $x = -1.5$ and 1, the latter a double root.

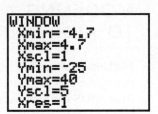
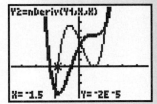

Next select (turn on) Y3 and deselect (turn off) Y2, as shown in the following window. The graph of Y3, the second derivative of $f(x)$, shows that there are two inflection points, at $x = -\dfrac{2}{3}$ and $x = 1$. The root-finding feature of your graphing calculator can verify that these are the roots of $f''(x)$, coinciding with the locations at which $f(x)$ changes concavity.

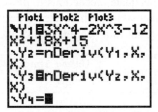
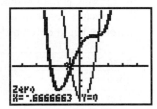
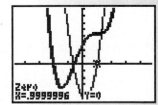

**Solution for Explore!
on Page 238**

Store $f(x) = \dfrac{3x^2}{x^2 + 2x - 15}$ into Y1 of your equation editor. As you trace the graph for large x values, $f(x)$ hovers below but appears to approach the value 3, visually demonstrating that $y = 3$ is a horizontal asymptote. Do you think that there is a relative minimum on the right branch of the graph of $f(x)$? Also how far out on the x axis do you need to trace to obtain a y value above 2.9? The table on the far right should help.

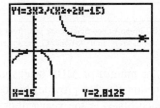
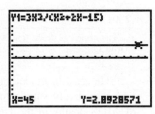
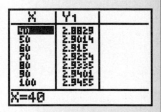

THINK ABOUT IT

MODELING AN EPIDEMIC: DEATHS DUE TO AIDS IN THE UNITED STATES

Mathematical models are used extensively to study infectious disease. In particular, during the past 20 years, many different mathematical models for the number of deaths from AIDS per year have been developed. In this essay, we will describe some of the simplest such models and discuss their accuracy. Before doing so, however, we will provide some background.

From 1980 until 1995, the number of people dying from AIDS in the United States increased rapidly from close to zero to more than 51,000. In the mid-1990s it appeared that the number of these deaths would continue to increase each year. However, with the advent of new treatments and drugs, the number of annual deaths from AIDS in the United States has steadily decreased, with less than 16,000 such deaths reported in 2001. That said, the future trend for the total number of annual AIDS deaths in the United States is not clear. The virus that causes AIDS has developed resistance to some of the drugs that have been effective in treating it, and in addition, there are indications that many susceptible people are not taking adequate precautions to prevent the spread of the disease. Will these factors result in an increase in AIDS-related deaths, or will the many new drugs currently being developed result in a continuation of the downward trend?

Year	United States AIDS Deaths
1989	27,408
1990	31,120
1991	36,175
1992	40,587
1993	45,850
1994	50,842
1995	51,670
1996	38,296
1997	22,245
1998	18,823
1999	18,249
2000	16,672
2001	15,603

The total number of deaths from AIDS in the United States each year from 1989 to 2001 is shown in the table. (SOURCE: Centers for Disease Control & Prevention, National Center for HIV, STD, and TB Prevention.)

The goal in modeling the number of AIDS deaths in the United States per year is to find a relatively simple function $f(t)$ that provides a close approximation to the number of AIDS deaths in the United States in the year t after the year 1989. One of the simplest approaches to construct such functions is to use *polynomial regression*, a technique that produces best-fit polynomials of specified degree that are good approximations to observed data. When using polynomial regression, we first need to specify the degree of the polynomial we want to use to approximate the number of AIDS deaths per year in the United States. The higher the degree, the better the approximation, but the more complicated the function. Suppose we decide to use a polynomial of degree 3, so that $f(t)$ has the general form $f(t) = at^3 + bt^2 + ct + d$. The coefficients a, b, c, and d of the best-fitting cubic curve we seek are then determined by requiring the sum of squares of the vertical distances between data points in the table and corresponding points on the cubic polynomial $y = f(t)$ to be as small as possible. Calculus methods for carrying out this optimization procedure are developed in Section 7.4. In practice, however, regression polynomials are almost always found using a computer or graphing calculator. There is a detailed description of how to use your calculator for finding regression polynomials in the Calculator Introduction on the textbook's website, www.mhhe.com/hoffmann.

It is important to realize that there are no underlying biological reasons why a polynomial should provide a good approximation to the number of AIDS deaths in the United States during a year. Nevertheless, we begin with such models because they are easy to construct. More sophisticated models, including some that are based on biological reasoning, can be constructed using exponential functions studied in Chapter 4.

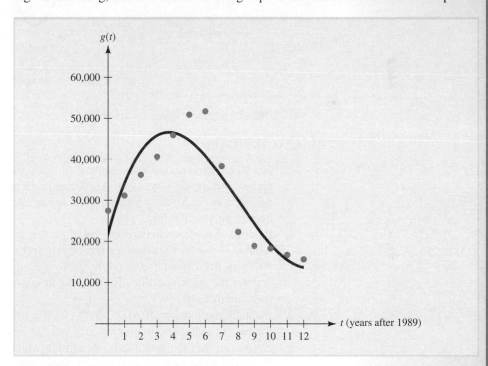

FIGURE 1   The best-fitting cubic curve.

We will take a look at both the degree three (cubic) and the degree four (quartic) polynomials produced with polynomial regression using our 13 points (t_j, d_j), where d_j is the number of AIDS deaths in the United States that occurred in the year j after 1989. (Our data corresponds to $t_j = 0, 1, \ldots, 12$.) Computations show that the best-fit cubic approximation to our data is

$$g(t) = 107.0023t^3 - 2,565.0889t^2 + 14,682.6031t + 21,892.5055$$

This cubic polynomial is graphed together with the data points in Figure 1. Similar computations show that the best-fit quartic approximation to our data is

$$h(t) = 29.6957t^4 - 605.6941t^3 + 2,801.3456t^2 + 1,599.5330t + 26,932.2873$$

which is graphed together with the given data points in Figure 2. These best-fitting graphs can then be used to analyze the given data and to make predictions. A few examples of this kind of analysis are provided in the accompanying questions.

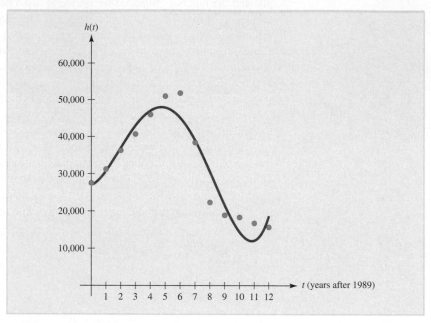

FIGURE 2   The best-fitting quartic curve.

Questions

1. For which time period does the best-fit cubic polynomial $g(t)$ do the best job of approximating the number of AIDS deaths in the United States? For what time period does it do the poorest job? Answer the same questions for the best-fit quartic polynomial $h(t)$.

2. Find all critical numbers of the best-fit cubic polynomial $g(t)$ for the given time interval $0 \le t \le 12$. What are the largest and smallest number of deaths due to AIDS in the United States predicted by this model for the period 1989–2001? Answer the same questions for the best-fit quartic polynomial $h(t)$. Comment on these predictions.

3. Estimate the number of AIDS deaths in the United States in 2002 projected by the best-fit cubic approximation $g(t)$ and the best-fit quartic approximation $h(t)$. If you can find the actual data, determine which produces the better prediction. What do these two models predict about the number of AIDS deaths in the United States in the years beyond 2002?

4. Why would a linear equation, such as that produced by linear regression, not do a good job of approximating the number of AIDS deaths in the United States from 1989 until 2001?

5. Use your graphing calculator to generate a best-fit quadratic (degree 2) polynomial $Q(t)$ for the given data. Using your $Q(t)$, answer questions 2 and 3. Comment on how well your polynomial fits the given data.

6. Could we ever find a polynomial that exactly agrees with the number of U.S. AIDS deaths for every year from 1989 until 2001? If there is such a polynomial, how large must we take the degree to be sure that we have found it?

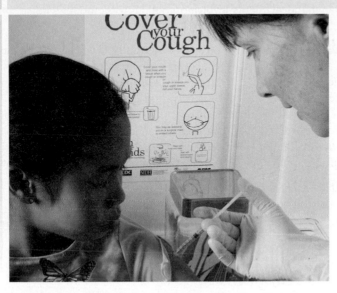

CHAPTER 4

Exponential functions can be used for describing the effect of medication.

Exponential and Logarithmic Functions

SECTION 4.1 Exponential Functions; Continuous Compounding

Learning Objectives

1. Define exponential functions.
2. Explore properties of the natural exponential function.
3. Examine investments involving continuous compounding of interest.

A population $Q(t)$ is said to grow **exponentially** if whenever it is measured at equally spaced time intervals, the population at the end of any particular interval is a fixed multiple (greater than 1) of the population at the end of the previous interval. For instance, according to the United Nations, in the year 2000, the population of the world was 6.1 billion people and was growing at an annual rate of about 1.4%. If this pattern were to continue, then every year, the population would be 1.014 times the population of the previous year. Thus, if $P(t)$ is the world population (in billions) t years after the base year 2000, the population would grow as follows:

2000	$P(0) = 6.1$
2001	$P(1) = 6.1(1.014) = 6.185$
2002	$P(2) = 6.185(1.014) = [6.1(1.014)](1.014) = 6.1(1.014)^2 = 6.272$
2003	$P(3) = 6.272(1.014) = [6.1(1.014)^2](1.014) = 6.1(1.014)^3 = 6.360$
$\vdots$	$\vdots$
$2000 + t$	$P(t) = 6.1(1.014)^t$

The graph of $P(t)$ is shown in Figure 4.1a. Notice that according to this model, the world population grows gradually at first but doubles after about 50 years (to 12.22 billion in 2050).

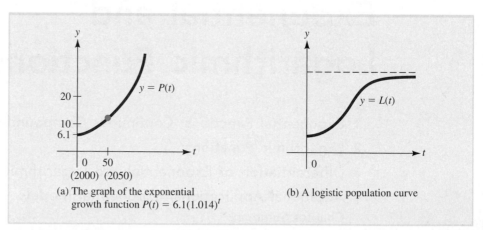

(a) The graph of the exponential growth function $P(t) = 6.1(1.014)^t$

(b) A logistic population curve

FIGURE 4.1 Two models for population growth.

Exponential population models are sometimes referred to as *Malthusian,* after Thomas Malthus (1766–1834), a British economist who predicted mass starvation would result if a population grows exponentially while the food supply grows at a constant rate (linearly). Fortunately, world population does not continue to grow exponentially as predicted by Malthus's model, and models that take into account various restrictions on the growth rate actually provide more accurate predictions. The population curve that results from one such model, the so-called *logistic* model, is shown in Figure 4.1b. Note how the logistic growth curve rises steeply at first, like an exponential curve, but then eventually turns over and flattens out as environmental factors act to brake the growth rate. We will examine logistic curves in Section 4.4 (Example 4.4.6) and again in Chapter 6 as part of a more detailed study of population models.

A function of the general form $f(x) = b^x$, where b is a positive number, is called an **exponential function.** Such functions can be used to describe exponential and logistic growth and a variety of other important quantities. For instance, exponential functions are used in demography to forecast population size, in finance to calculate the value of investments, in archaeology to date ancient artifacts, in psychology to study learning patterns, and in industry to estimate the reliability of products.

In this section, we will explore the properties of exponential functions and introduce a few basic models in which such functions play a prominent role. Additional applications such as the logistic model are examined in subsequent sections.

Working with exponential functions requires the use of exponential notation and the algebraic laws of exponents. Solved examples and practice problems involving this notation can be found in Appendix A.1. Here is a brief summary of the notation.

Definition of b^n for Rational Values of n (and $b > 0$) ■ Integer powers: If n is a positive integer,

$$b^n = \underbrace{b \cdot b \cdots b}_{n \text{ factors}}$$

Fractional powers: If n and m are positive integers,

$$b^{n/m} = (\sqrt[m]{b})^n = \sqrt[m]{b^n}$$

where $\sqrt[m]{b}$ denotes the positive mth root.

Negative powers: $b^{-n} = \dfrac{1}{b^n}$

Zero power: $b^0 = 1$

For example,

$$3^4 = 3 \cdot 3 \cdot 3 \cdot 3 = 81 \qquad\qquad 3^{-4} = \frac{1}{3^4} = \frac{1}{81}$$

$$4^{1/2} = \sqrt{4} = 2 \qquad\qquad 4^{3/2} = (\sqrt{4})^3 = 2^3 = 8$$

$$4^{-3/2} = \frac{1}{4^{3/2}} = \frac{1}{8} \qquad\qquad 27^{-2/3} = \frac{1}{(\sqrt[3]{27})^2} = \frac{1}{3^2} = \frac{1}{9}$$

EXPLORE!

Graph $y = (-1)^x$ using a modified decimal window $[-4.7, 4.7]1$ by $[-1, 7]1$. Why does the graph appear with dotted points? Next set b to decimal values, $b = 0.5, 0.25,$ and 0.1, and graph $y = b^x$ for each case. Explain the behavior of these graphs.

We know what is meant by b^r for any rational number r, but if we try to graph $y = b^x$ there will be a hole in the graph for each value of x that is not rational, such as $x = \sqrt{2}$. However, using methods beyond the scope of this book, it can be shown that irrational numbers can be approximated by rational numbers, which in turn means that there is only one unbroken curve passing through all points (r, b^r) for r rational. In other words, there exists a unique continuous function $f(x)$ that is defined for all real numbers x and is equal to b^r when r is rational. It is this function we define as $f(x) = b^r$.

Exponential Functions ■ If b is a positive number other than 1 ($b > 0$, $b \neq 1$), there is a unique function called the *exponential function* with base b that is defined by

$$f(x) = b^x \qquad \text{for every real number } x$$

To get a feeling for the appearance of the graph of an exponential function, consider Example 4.1.1.

EXAMPLE 4.1.1 Graphing Exponential Functions

Sketch the graphs of $y = 2^x$ and $y = \left(\dfrac{1}{2}\right)^x$.

Solution

Begin by constructing a table of values for $y = 2^x$ and $y = \left(\dfrac{1}{2}\right)^x$:

x	-15	-10	-1	0	1	3	5	10	15
$y = 2^x$	0.00003	0.001	0.5	1	2	8	32	1,024	32,768
$y = \left(\dfrac{1}{2}\right)^x$	32,768	1,024	2	1	0.5	0.125	0.313	0.001	0.00003

The pattern of values in this table suggests that the functions $y = 2^x$ and $y = \left(\dfrac{1}{2}\right)^x$ have the following features:

The function $y = 2^x$	**The function $y = \left(\dfrac{1}{2}\right)^x$**
always increasing	always decreasing
$\displaystyle\lim_{x \to -\infty} 2^x = 0$	$\displaystyle\lim_{x \to -\infty} \left(\dfrac{1}{2}\right)^x = +\infty$
$\displaystyle\lim_{x \to +\infty} 2^x = +\infty$	$\displaystyle\lim_{x \to +\infty} \left(\dfrac{1}{2}\right)^x = 0$

Using this information, we sketch the graphs shown in Figure 4.2. Notice that each graph has (0, 1) as its y intercept, has the x axis as a horizontal asymptote, and appears to be concave upward for all x. The graphs also appear to be reflections of one another in the y axis. You are asked to verify this observation in Exercise 74.

EXPLORE!

Graph $y = b^x$, for $b = 1, 2, 3$, and 4, using the modified decimal window $[-4.7, 4.7]1$ by $[-1, 7]1$. Explain what you observe. Conjecture and then check where the graph of $y = 4^x$ will lie relative to that of $y = 2^x$ and $y = 6^x$. Where does $y = e^x$ lie, assuming e is a value between 2 and 3?

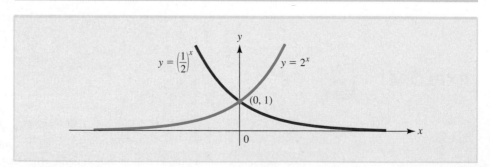

FIGURE 4.2 The graphs of $y = 2^x$ and $y = \left(\dfrac{1}{2}\right)^x$.

Figure 4.3 shows graphs of various members of the family of exponential functions $y = b^x$. Notice that the graph of any function of the form $y = b^x$ resembles that of $y = 2^x$ if $b > 1$ or $y = \left(\dfrac{1}{2}\right)^x$ if $0 < b < 1$. In the special case where $b = 1$, the function $y = b^x$ becomes the constant function $y = 1$.

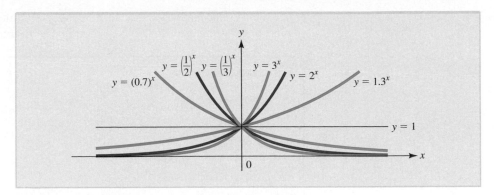

FIGURE 4.3   Graphs of the exponential form $y = b^x$.

Important graphical and analytical properties of exponential functions are summarized in the following box.

Properties of an Exponential Function ■ The exponential function $f(x) = b^x$ for $b > 0$, $b \neq 1$ has these properties:

1. It is defined, continuous, and positive ($b^x > 0$) for all x.
2. The x axis is a horizontal asymptote of the graph of f.
3. The y intercept of the graph is $(0, 1)$; there is no x intercept.
4. If $b > 1$, $\lim\limits_{x \to -\infty} b^x = 0$ and $\lim\limits_{x \to +\infty} b^x = +\infty$.

   If $0 < b < 1$, $\lim\limits_{x \to -\infty} b^x = +\infty$ and $\lim\limits_{x \to +\infty} b^x = 0$.
5. For all x, the function is increasing (graph rising) if $b > 1$ and decreasing (graph falling) if $0 < b < 1$.

NOTE   Students often confuse the *power* function $p(x) = x^b$ with the *exponential* function $f(x) = b^x$. Remember that in x^b, the variable x is the base and the exponent b is constant, while in b^x, the base b is constant and the variable x is the exponent. The graphs of $y = x^2$ and $y = 2^x$ are shown in Figure 4.4. Notice that after the crossover point $(4, 16)$, the exponential curve $y = 2^x$ rises much more steeply than the power curve $y = x^2$. For instance, when $x = 10$, the y value on the power curve is $y = 10^2 = 100$, while the corresponding y value on the exponential curve is $y = 2^{10} = 1{,}024$. ■

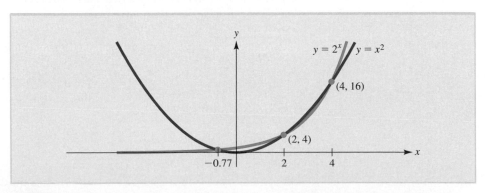

FIGURE 4.4   Comparing the power curve $y = x^2$ with the exponential curve $y = 2^x$.

Exponential functions obey the same algebraic rules as the rules for exponential numbers reviewed in Appendix A.1. These rules are summarized in the following box.

Exponential Rules ▪ For bases a, b ($a > 0$, $b > 0$) and any real numbers x, y, we have

The **equality rule:** For $b \neq 1$, $b^x = b^y$ if and only if $x = y$

The **product rule:** $b^x b^y = b^{x+y}$

The **quotient rule:** $\dfrac{b^x}{b^y} = b^{x-y}$

The **power rule:** $(b^x)^y = b^{xy}$

The **multiplication rule:** $(ab)^x = a^x b^x$

The **division rule:** $\left(\dfrac{a}{b}\right)^x = \dfrac{a^x}{b^x}$

EXAMPLE 4.1.2 Evaluating Exponential Expressions

Evaluate each of these exponential expressions:

a. $(3)^2(3)^3$ **b.** $(2^3)^2$ **c.** $(5^{1/3})(2^{1/3})$

d. $\dfrac{2^3}{2^5}$ **e.** $\left(\dfrac{4}{7}\right)^3$

Solution

a. $(3)^2(3)^3 = 3^{2+3} = 3^5 = 243$

b. $(2^3)^2 = 2^{(3)(2)} = 2^6 = 64$

c. $(5^{1/3})(2^{1/3}) = [(5)(2)]^{1/3} = 10^{1/3} = \sqrt[3]{10}$

d. $\dfrac{2^3}{2^5} = 2^{3-5} = 2^{-2} = \dfrac{1}{4}$

e. $\left(\dfrac{4}{7}\right)^3 = \dfrac{4^3}{7^3} = \dfrac{64}{343}$

EXAMPLE 4.1.3 Solving an Exponential Equation

If $f(x) = 5^{x^2+2x}$, find all values of x such that $f(x) = 125$.

Solution

The equation $f(x) = 125 = 5^3$ is satisfied if and only if

$$5^{x^2+2x} = 5^3 \quad b^x = b^y \text{ only when } x = y$$
$$x^2 + 2x = 3$$
$$x^2 + 2x - 3 = 0 \quad \text{factor}$$
$$(x - 1)(x + 3) = 0$$
$$x = 1, \; x = -3$$

Thus, $f(x) = 125$ if and only if $x = 1$ or $x = -3$.

The Natural Exponential Base e

In algebra, it is common practice to use the base $b = 10$ for exponential functions or, in some cases, $b = 2$, but in calculus, it turns out to be more convenient to use a number denoted by e and defined by the limit

$$e = \lim_{n \to +\infty} \left(1 + \frac{1}{n}\right)^n$$

"Hold on!" you say, "That limit has to be 1, since $1 + \frac{1}{n}$ certainly tends to 1 as n increases without bound, and $1^n = 1$ for any n." Not so. The limit process does not work this way, as you can see from this table:

n	10	100	1,000	10,000	100,000	1,000,000
$\left(1 + \frac{1}{n}\right)^n$	2.59374	2.70481	2.71692	2.71815	2.71827	2.71828

The number e is one of the most important numbers in all mathematics, and its value has been computed with great precision. To twelve decimal places, its value is

$$e = \lim_{n \to +\infty} \left(1 + \frac{1}{n}\right)^n = 2.718281828459 \ldots$$

The function $y = e^x$ is called the **natural exponential function.** To compute e^N for a particular number N, you can use a table of exponential values or, more likely, the "e^x" key on your calculator. For example, to find $e^{1.217}$ press the e^x key and then enter the number 1.217 to obtain $e^{1.217} \approx 3.37704$.

EXPLORE!

Store $\left(1 + \frac{1}{x}\right)^x$ into Y1 of the function editor, and examine its graph. Trace the graph to the right for large values of x. What number is y approaching as x gets larger and larger? Try using the table feature of the graphing calculator, setting both the initial value and the incremental change first to 10 and then successively to 1,000 and 100,000. Estimate the limit to five decimal places. Now do the same as x approaches $-\infty$, and observe this limit.

EXAMPLE 4.1.4   Evaluating an Exponential Demand Function

Lydia is the manager of a company that manufactures small appliances. She determines that x units of a particular appliance will be produced and sold at a price of p dollars per unit, where $p(x) = 200e^{-0.01x}$. How much revenue should she expect when 100 units of the appliance are produced?

Solution

The revenue is given by the product (price/unit)(number of units sold); that is,

$$R(x) = p(x)x = (200e^{-0.01x})x = 200xe^{-0.01x}$$

Using a calculator, we find that the revenue obtained by producing $x = 100$ units is

$$R(100) = 200(100)e^{-0.01(100)} \approx 7{,}357.59$$

or approximately \$7,357.59.

EXAMPLE 4.1.5   Evaluating an Exponential Population Function

Biologists have determined that the number of bacteria in a culture is given by

$$P(t) = 5{,}000e^{0.015t}$$

where t is the number of minutes after observation begins. What is the average rate of change of the bacterial population during the second hour?

Solution

During the second hour (from time $t = 60$ to $t = 120$), the population changes by $P(120) - P(60)$, so the average rate of change during this time period is given by

$$A = \frac{P(120) - P(60)}{120 - 60}$$

$$= \frac{[5{,}000e^{0.015(120)}] - [5{,}000e^{0.015(60)}]}{60}$$

$$= \frac{30{,}248 - 12{,}298}{60}$$

$$\approx 299$$

Thus, the population increases at the average rate of roughly 299 bacteria per minute during the second hour.

Continuous Compounding of Interest

The number e is called the "natural exponential base," but it may seem anything but "natural" to you. As an illustration of how this number appears in practical situations, we use it to describe the accounting practice known as continuous compounding of interest.

First, let us review the basic ideas behind compound interest. Suppose a sum of money is invested and the interest is compounded only once. If P is the initial investment (the *principal*) and r is the interest rate (expressed as a decimal), the balance B after the interest is added will be

$$B = P + Pr = P(1 + r) \quad \text{dollars}$$

That is, to compute the balance at the end of an interest period, you multiply the balance at the beginning of the period by the expression $1 + r$.

At most banks, interest is compounded more than once a year. The interest that is added to the account during one period will itself earn interest during the subsequent periods. If the annual interest rate is r and interest is compounded k times per year, then the year is divided into k equal compounding periods and the interest rate in each period is $\frac{r}{k}$. Hence, the balance at the end of the first period is

$$P_1 = P + P\left(\frac{r}{k}\right) = P\left(1 + \frac{r}{k}\right)$$

$$\underbrace{\hspace{1cm}}_{\text{principal}} \underbrace{\hspace{1cm}}_{\text{interest}}$$

At the end of the second period, the balance is

$$P_2 = P_1 + P_1\left(\frac{r}{k}\right) = P_1\left(1 + \frac{r}{k}\right)$$

$$= \left[P\left(1 + \frac{r}{k}\right)\right]\left(1 + \frac{r}{k}\right) = P\left(1 + \frac{r}{k}\right)^2$$

EXPLORE!

Suppose you have $1,000 to invest. Which is the better investment, 5% compounded monthly for 10 years or 6% compounded quarterly for 10 years? Write the expression 1,000(1 + R/K)^(K*T) on the Home Screen, and evaluate after storing appropriate values for R, K, and T.

and, in general, the balance at the end of the mth period is

$$P_m = P\left(1 + \frac{r}{k}\right)^m$$

Since there are k periods in a year, the balance after 1 year is

$$P\left(1 + \frac{r}{k}\right)^k$$

At the end of t years, interest has been compounded kt times and the balance, called the **future value** of the investment, is given by

$$B(t) = P\left(1 + \frac{r}{k}\right)^{kt}$$

As the frequency with which interest is compounded increases, the corresponding balance $B(t)$ also increases. Hence, a bank that compounds interest frequently may attract more customers than one that offers the same interest rate but compounds interest less often. But what happens to the balance at the end of t years as the compounding frequency increases without bound? More specifically, what will the balance be at the end of t years if interest is compounded not quarterly, not monthly, not daily, but continuously? In mathematical terms, this question is equivalent to asking what happens to the expression $P\left(1 + \frac{r}{k}\right)^{kt}$ as k increases without bound. The answer turns out to involve the number e. Here is the argument.

To simplify the calculation, let $n = \dfrac{k}{r}$. Then, $k = nr$ and so

$$P\left(1 + \frac{r}{k}\right)^{kt} = P\left(1 + \frac{1}{n}\right)^{nrt} = P\left[\left(1 + \frac{1}{n}\right)^n\right]^{rt}$$

Since n increases without bound as k does, and since $\left(1 + \dfrac{1}{n}\right)^n$ approaches e as n increases without bound, it follows that the balance after t years is

$$B(t) = \lim_{k \to +\infty} P\left(1 + \frac{r}{k}\right)^{kt} = P\left[\lim_{n \to +\infty}\left(1 + \frac{1}{n}\right)^n\right]^{rt} = Pe^{rt}$$

To summarize:

Future Value of an Investment ■ Suppose a principal P is invested at an annual interest rate r for t years to accumulate a **future value** $B(t)$. If interest is compounded k times per year, then

$$B(t) = P\left(1 + \frac{r}{k}\right)^{kt}$$

and if interest is compounded continuously

$$B(t) = Pe^{rt}$$

EXPLORE!

Enter the expressions

P*(1 + R/K)^(K*T)

and

P*e^(R*T)

on the Home Screen, and compare these two expressions for the P, R, T, and K values in Example 4.1.6. Repeat using the same values for P, R, and K, but with $T = 15$ years.

EXAMPLE 4.1.6 Computing Future Value

Suppose $1,000 is invested at an annual interest rate of 6% for a period of 10 years. Compute the future value of the investment if interest is compounded:

a. Quarterly   **b.** Monthly   **c.** Daily   **d.** Continuously

Solution

a. To compute the balance after 10 years if the interest is compounded quarterly, use the formula $B(t) = P\left(1 + \dfrac{r}{k}\right)^{kt}$, with $t = 10$, $P = 1,000$, $r = 0.06$, and $k = 4$:

$$B(10) = 1,000\left(1 + \frac{0.06}{4}\right)^{40} \approx \$1,814.02$$

b. This time, take $t = 10$, $P = 1,000$, $r = 0.06$, and $k = 12$ to get

$$B(10) = 1,000\left(1 + \frac{0.06}{12}\right)^{120} \approx \$1,819.40$$

c. Take $t = 10$, $P = 1,000$, $r = 0.06$, and $k = 365$ to obtain

$$B(10) = 1,000\left(1 + \frac{0.06}{365}\right)^{3,650} \approx \$1,822.03$$

d. For continuously compounded interest, use the formula $B(t) = Pe^{rt}$, with $t = 10$, $P = 1,000$, and $r = 0.06$:

$$B(10) = 1,000e^{0.6} \approx \$1,822.12$$

This value, $1,822.12, is an upper bound for the possible balance. No matter how often interest is compounded, $1,000 invested at an annual interest rate of 6% cannot grow to more than $1,822.12 in 10 years.

Present Value

In many situations, it is useful to know how much money P must be invested at a fixed compound interest rate to obtain a desired accumulated (future) value B over a given period of time T. This investment P is called the **present value of the amount B to be received in T years.** Present value may be regarded as a measure of the current worth of an investment and is used by economists to compare different investment possibilities.

To derive a formula for present value, we need only solve an appropriate future value formula for P. In particular, if the investment is compounded k times per year at an annual rate r for the term of T years, then

$$B = P\left(1 + \frac{r}{k}\right)^{kT}$$

and the present value of B dollars in T years is obtained by multiplying both sides of the equation by $\left(1 + \dfrac{r}{k}\right)^{-kT}$ to get

$$P = B\left(1 + \frac{r}{k}\right)^{-kT}$$

Just-In-Time REVIEW

If $C \neq 0$, you can solve the equation

$A = PC$

for P by multiplying both sides by

$$\frac{1}{C} = C^{-1}$$

to get

$P = AC^{-1}$

Likewise, if the compounding is continuous, then

$$B = Pe^{rT}$$

and the present value is given by

$$P = Be^{-rT}$$

To summarize:

Present Value of an Investment ■ The **present value** $P(T)$ of an investment B accumulated at the annual interest rate r compounded k times per year over a term of T years is given by

$$P(T) = B\left(1 + \frac{r}{k}\right)^{-kT}$$

If interest is compounded continuously at the same annual rate r over the same term of T years, the present value is

$$P(T) = Be^{-rT}$$

EXPLORE!

Find the present value so that the balance 25 years from now will be $40,000 if the annual interest rate of 6% is compounded continuously. To do this, place the equation $F - P*e^{\wedge}(R*T) = 0$ into the equation solver of your graphing calculator (using the **SOLVER** option), with $F = 40,000$, $R = 0.06$, and $T = 25$. Then solve for P.

EXAMPLE 4.1.7 Finding Present Value

Fay is about to enter college. When she graduates 4 years from now, she wants to take a trip to Europe that she estimates will cost $5,000. How much should she invest now at 7% to have enough for the trip if interest is compounded:

a. Quarterly   **b.** Continuously

Solution

The required future value is $B = \$5,000$ in $T = 4$ years with $r = 0.07$.

a. If the compounding is quarterly, then $k = 4$ and the present value is

$$P = 5,000\left(1 + \frac{0.07}{4}\right)^{-4(4)} = \$3,788.08$$

b. For continuous compounding, the present value is

$$P = 5,000e^{-0.07(4)} = \$3,778.92$$

Thus, Fay would have to invest about $9 more if interest is compounded quarterly than if the compounding is continuous.

Effective Interest Rate

Which is better, an investment that earns 10% compounded quarterly, one that earns 9.95% compounded monthly, or one that earns 9.9% compounded continuously? One way to answer this question is to determine the simple annual interest rate that is equivalent to each investment. This is known as the **effective interest rate,** and it can be easily obtained from the compound interest formulas.

Suppose interest is compounded k times per year at the annual rate r. This is called the **nominal** rate of interest. Then the balance at the end of 1 year is

$$A = P(1 + i)^k \qquad \text{where } i = \frac{r}{k}$$

On the other hand, if x is the effective interest rate, the corresponding balance at the end of 1 year is $A = P(1 + x)$. Equating the two expressions for A, we get

$$P(1 + i)^k = P(1 + x) \qquad \text{or} \qquad x = (1 + i)^k - 1$$

For continuous compounding, we have

$$Pe^r = P(1 + x) \qquad \text{so} \qquad x = e^r - 1$$

To summarize:

Effective Interest Rate Formulas ■ If interest is compounded at the nominal rate r, the effective interest rate is the simple annual interest rate r_e that yields the same interest after 1 year. If the compounding is k times per year, the effective rate is given by the formula

$$r_e = (1 + i)^k - 1 \qquad \text{where } i = \frac{r}{k}$$

while continuous compounding yields

$$r_e = e^r - 1$$

Example 4.1.8 answers the question raised in the introduction to this subsection.

EXAMPLE 4.1.8   Comparing Two Investments

Which is better, an investment that earns 10% compounded quarterly, one that earns 9.95% compounded monthly, or one that earns 9.9% compounded continuously?

Solution

We answer the question by comparing the effective interest rates of the three investments. For the first, the nominal rate is 10% and compounding is quarterly, so we have $r = 0.10$, $k = 4$, and

$$i = \frac{r}{k} = \frac{0.10}{4} = 0.025$$

Substituting into the formula for effective rate, we get

First effective rate $= (1 + 0.025)^4 - 1 = 0.10381$

For the second investment, the nominal rate is 9.95% and compounding is monthly, so $r = 0.0995$, $k = 12$, and

$$i = \frac{r}{k} = \frac{0.0995}{12} = 0.008292$$

We find that

Second effective rate $= (1 + 0.008292)^{12} - 1 = 0.10417$

Finally, if compounding is continuous with nominal rate 9.9%, we have $r = 0.099$ and the effective rate is

Third effective rate $= e^{0.099} - 1 = 0.10407$

The effective rates are, respectively, 10.38%, 10.42%, and 10.41%, so the second investment is best.

EXERCISES ▪ 4.1

In Exercises 1 and 2, use your calculator to find the indicated power of e. (Round your answers to three decimal places.)

1. e^2, e^{-2}, $e^{0.05}$, $e^{-0.05}$, e^0, e, $\sqrt{e}$, and $\dfrac{1}{\sqrt{e}}$

2. e^3, e^{-1}, $e^{0.01}$, $e^{-0.1}$, e^2, $e^{-1/2}$, $e^{1/3}$, and $\dfrac{1}{\sqrt[3]{e}}$

3. Sketch the curves $y = 3^x$ and $y = 4^x$ on the same set of axes.

4. Sketch the curves $y = \left(\dfrac{1}{3}\right)^x$ and $y = \left(\dfrac{1}{4}\right)^x$ on the same set of axes.

In Exercises 5 through 12, evaluate the given expressions.

5. a. $27^{2/3}$
   b. $\left(\dfrac{1}{9}\right)^{3/2}$

6. a. $(-128)^{3/7}$
   b. $\left(\dfrac{27}{64}\right)^{2/3}\left(\dfrac{64}{25}\right)^{3/2}$

7. a. $8^{2/3} + 16^{3/4}$
   b. $\left(\dfrac{27 + 36}{121}\right)^{3/2}$

8. a. $(2^3 - 3^2)^{11/7}$
   b. $(27^{2/3} + 8^{4/3})^{-3/2}$

9. a. $(3^3)(3^{-2})$
   b. $(4^{2/3})(2^{2/3})$

10. a. $\dfrac{5^2}{5^3}$
    b. $\left(\dfrac{\pi^2}{\sqrt{\pi}}\right)^{4/3}$

11. a. $(3^2)^{5/2}$
    b. $(e^2 e^{3/2})^{4/3}$

12. a. $\dfrac{(3^{1.2})(3^{2.7})}{3^{4.1}}$
    b. $\left(\dfrac{16}{81}\right)^{1/4}\left(\dfrac{125}{8}\right)^{-2/3}$

In Exercises 13 through 18, use the properties of exponents to simplify the given expressions.

13. a. $(27x^6)^{2/3}$
    b. $(8x^2y^3)^{1/3}$

14. a. $(x^{1/3})^{3/2}$
    b. $(x^{2/3})^{-3/4}$

15. a. $\dfrac{(x + y)^0}{(x^2y^3)^{1/6}}$
    b. $(x^{1.1}y^2)(x^2 + y^3)^0$

16. a. $(-2t^{-3})(3t^{2/3})$
    b. $(t^{-2/3})(t^{3/4})$

17. a. $(t^{5/6})^{-6/5}$
    b. $(t^{-3/2})^{-2/3}$

18. a. $(x^2y^{-3}z)^3$
    b. $\left(\dfrac{x^3y^{-2}}{z^4}\right)^{1/6}$

In Exercises 19 through 28, find all real numbers x that satisfy the given equation.

19. $4^{2x-1} = 16$

20. $3^x 2^{2x} = 144$

21. $2^{3-x} = 4^x$

22. $4^x\left(\dfrac{1}{2}\right)^{3x} = 8$

23. $(2.14)^{x-1} = (2.14)^{1-x}$

24. $(3.2)^{2x-3} = (3.2)^{2-x}$

25. $10^{x^2-1} = 10^3$

26. $\left(\dfrac{1}{10}\right)^{1-x^2} = 1{,}000$

27. $\left(\dfrac{1}{8}\right)^{x-1} = 2^{3-2x^2}$

28. $\left(\dfrac{1}{9}\right)^{1-3x^2} = 3^{4x}$

In Exercises 29 through 32, use a graphing calculator to sketch the graph of the given exponential function.

29. $y = 3^{1-x}$

30. $y = e^{x+2}$

31. $y = 4 - e^{-x}$

32. $y = 2^{x/2}$

In Exercises 33 and 34, find the values of the constants C and b so that the curve $y = Cb^x$ contains the indicated points.

33. (2, 12) and (3, 24)
34. (2, 3) and (3, 9)

35. Suppose $1,000 is invested at an annual interest rate of 7%. Compute the future value of the investment after 10 years if the interest is compounded:
   a. Annually
   b. Quarterly
   c. Monthly
   d. Continuously

36. Suppose $5,000 is invested at an annual interest rate of 10%. Compute the future value of the investment after 10 years if the interest is compounded:
   a. Annually
   b. Semiannually
   c. Daily (using 365 days per year)
   d. Continuously

37. Find the present value of $10,000 over a term of 5 years at an annual interest rate of 7% if interest is compounded:
   a. Annually
   b. Quarterly
   c. Daily (use 365 days)
   d. Continuously

38. Find the present value of $25,000 over a term of 10 years at an annual interest rate of 5% if interest is compounded:
    a. Semiannually
    b. Monthly
    c. Continuously

In Exercises 39 through 42, find the effective interest rate r_e for the given investment.

39. Annual interest rate 6%, compounded quarterly

40. Annual interest rate 8%, compounded daily (use $k = 365$)

41. Nominal annual rate of 5%, compounded continuously

42. Nominal annual rate of 7.3%, compounded continuously

BUSINESS AND ECONOMICS APPLIED PROBLEMS

43. **INVESTMENT**   Esmeralda needs $5,000 for a trip to Peru when she graduates from college in 4 years. How much must she invest now at an annual interest rate of 5% compounded continuously to achieve her goal?

44. **INVESTMENT**   Bob and Alice want to remodel their bathroom in 3 years. They estimate the job will cost $25,000. How much must they invest now at an annual interest rate of 7% compounded quarterly to achieve their goal? What if the compounding were continuous?

45. **DEMAND**   A manufacturer estimates that when x units of a particular commodity are produced, the market price p (dollars per unit) is given by the demand function

    $$p = 300e^{-0.02x}$$

    a. What market price corresponds to the production of $x = 100$ units?
    b. How much revenue is obtained when 100 units of the commodity are produced?
    c. How much more (or less) revenue is obtained when $x = 100$ units are produced than when $x = 50$ are produced?

46. **DEMAND**   A manufacturer estimates that when x units of a particular commodity are produced, the market price p (dollars per unit) is given by the demand function

    $$p = 7 + 50e^{-x/200}$$

    a. What market price corresponds to the production of $x = 0$ units?
    b. How much revenue is obtained when 200 units of the commodity are produced?
    c. How much more (or less) revenue is obtained when $x = 100$ units are produced than when $x = 50$ are produced?

47. **RANKING INVESTMENTS**   In terms of effective interest rate, order the following nominal rate investments from lowest to highest:
    a. 7.9% compounded semiannually
    b. 7.8% compounded quarterly
    c. 7.7% compounded monthly
    d. 7.65% compounded continuously

48. **RANKING INVESTMENTS**   In terms of effective interest rate, order the following nominal rate investments from lowest to highest:
    a. 4.87% compounded quarterly
    b. 4.85% compounded monthly
    c. 4.81% compounded daily (365 days)
    d. 4.79% compounded continuously

49. **EFFECT OF INFLATION**   Lyle buys a rare stamp for $500. If the annual rate of inflation is 4%, how much should he ask when he sells it in 5 years to break even?

50. **EFFECT OF INFLATION**   Suppose during a 10-year period of rapid inflation, it is estimated that prices inflate at an annual rate of 5% per year. If an item costs $3 at the beginning of the period, what would you expect to pay for the same item 10 years later?

51. **PRODUCT RELIABILITY**   A statistical study indicates that the fraction of the electric toasters manufactured by a certain company that are still in working condition after t years of use is approximately $f(t) = e^{-0.2t}$.
    a. What fraction of the toasters can be expected to work for at least 3 years?
    b. What fraction of the toasters can be expected to fail before 1 year of use?
    c. What fraction of the toasters can be expected to fail during the third year of use?

52. **ADVERTISING**   Jingfei is the marketing manager for a major firm. She estimates that t days after termination of an advertising campaign for a new product, $S(t)$ units will be sold, where

    $$S(t) = 4,000e^{-0.015t}$$

a. How many units are being sold at the time the advertising ends?

b. How many units should Jingfei expect her company to have sold 30 days after the advertising ends? After 60 days?

c. At what average rate do sales of the product change over the first 3 months (90 days) after the advertising ends?

53. **REAL ESTATE INVESTMENT** In 1626, Peter Minuit traded trinkets worth $24 to a tribe of Native Americans for land on Manhattan Island. Assume that in 1990 the same land was worth $25.2 billion. If the sellers in this transaction had invested their $24 at 7% annual interest compounded continuously during the entire 364-year period, who would have gotten the better end of the deal? By how much?

54. **SUPPLY** A manufacturer will supply $S(x) = 300e^{0.03x} - 310$ units of a particular commodity when the price is x dollars per unit.

a. How many units will be supplied when the unit price is $10?

b. How many more units will be supplied when the unit price is $100 than when it is $80?

55. **GROWTH OF GDP** The gross domestic product (GDP) of a certain country was $500 billion at the beginning of the year 2005 and increases at the rate of 2.7% per year.

a. Express the GDP of this country as a function of the number of years t after 2005. (*Hint:* Think of this as a compounding problem.)

b. What does this formula predict the GDP of the country will be at the beginning of the year 2015?

Amortization of Debt ■ If a loan of A dollars is amortized over n years at an annual interest rate r (expressed as a decimal) compounded monthly, the monthly payments are given by

$$M = \frac{Ai}{1 - (1 + i)^{-12n}}$$

where $i = \dfrac{r}{12}$ is the monthly interest rate. Use this formula in Exercises 56 and 57. [*Note:* To *amortize* a debt means to gradually pay it off with a sequence of regular payments.]

---

56. **FINANCE PAYMENTS** Determine the monthly car payment for a new car costing $15,675, if there is a down payment of $4,000 and the car is financed over a 5-year period at an annual rate of 6% compounded monthly.

57. **MORTGAGE PAYMENTS** A home loan is made for $150,000 at 9% annual interest, compounded monthly, for 30 years. What is the monthly mortgage payment on this loan?

58. **MORTGAGE PAYMENTS** Suppose a family figures it can handle monthly mortgage payments of no more than $1,200. What is the largest amount of money they can borrow, assuming the lender is willing to amortize over 30 years at 8% annual interest compounded monthly?

59. **TRUTH IN LENDING** You are selling your car for $6,000. A potential buyer says, "I will pay you $1,000 now for the car and pay off the rest at 12% interest with monthly payments for 3 years. Let's see . . . 12% of the $5,000 is $600 and $5,600 divided by 36 months is $155.56, but I'll pay you $160 per month for the trouble of carrying the loan. Is it a deal?"

a. If this deal sounds fair to you, I have a perfectly lovely bridge I think you should consider as your next purchase. If not, explain why the deal is fishy and compute a fair monthly payment (assuming you still plan to amortize the debt of $5,000 over 3 years at 12%).

b. Read an article on truth in lending, and think up some examples of plausible yet shady deals, such as the proposed used-car transaction in this exercise.

LIFE AND SOCIAL SCIENCE APPLIED PROBLEMS

60. **LEARNING** According to the Ebbinghaus model, the fraction $F(t)$ of subject matter you will remember from this course t months after the final exam can be estimated by the formula

$$F(t) = B + (1 - B)e^{-kt}$$

where B is the fraction of the material you will never forget and k is a constant that depends on the quality of your memory. Suppose you are tested and it is found that $B = 0.3$ and $k = 0.2$. What fraction of the material will you remember 1 month after the class ends? What fraction will you remember after 1 year?

61. POPULATION DENSITY The population density x miles from the center of a certain city is $D(x) = 12e^{-0.07x}$ thousand people per square mile.
  a. What is the population density at the center of the city?
  b. What is the population density 10 miles from the center of the city?

62. POPULATION GROWTH The size of a bacterial population $P(t)$ grows at the rate of 3.1% per day. If the initial population is 10,000, what is the population after 10 days? (*Hint:* Think of this as a compounding problem.)

63. POPULATION GROWTH It is projected that t years from now, the population of a certain country will be $P(t) = 50e^{0.02t}$ million.
  a. What is the current population?
  b. What will the population be 30 years from now?

64. POPULATION GROWTH It is estimated that t years after 2005, the population of a certain country will be $P(t)$ million people where
$$P(t) = 2 \cdot 5^{0.018t}$$
  a. What was the population in 2005?
  b. What will the population be in 2015?

65. DRUG CONCENTRATION The concentration of drug in a patient's bloodstream t hours after an injection is given by $C(t) = 3 \cdot 2^{-0.75t}$ milligrams per milliliter (mg/ml).
  a. What is the concentration when $t = 0$? After 1 hour?
  b. What is the average rate of change of concentration during the second hour?

66. DRUG CONCENTRATION The concentration of drug in a patient's bloodstream t hours after an injection is given by $C(t) = Ae^{-0.87t}$ milligrams per milliliter (mg/ml) for constant A. The concentration is 4 mg/ml after 1 hour.
  a. What is A?
  b. What is the initial concentration ($t = 0$)? The concentration after 2 hours?
  c. What is the average rate of change of concentration during the first 2 hours?

67. BACTERIAL GROWTH The size of a bacterial culture grows in such a way that after t minutes, there are $P(t) = A \cdot 2^{0.001t}$ bacteria present, for some constant A. After 10 minutes, there are 10,000 bacteria.
  a. What is A?

  b. How many bacteria are initially present ($t = 0$)? After 20 minutes? After 1 hour?
  c. At what average rate does the bacterial population change over the second hour?

68. DRUG CONCENTRATION The concentration of a certain drug in an organ t minutes after an injection is given by
$$C(t) = 0.05 - 0.04(1 - e^{-0.03t})$$
grams per cubic centimeter (g/cm³).
  a. What is the initial concentration of the drug (when $t = 0$)?
  b. What is the concentration 10 minutes after an injection? After 1 hour?
  c. What is the average rate of change of concentration during the first hour?
  d. What happens to the concentration of the drug in the long run (as $t \to \infty$)?
   e. Sketch the graph of $C(t)$.

69. DRUG CONCENTRATION The concentration of a certain drug in an organ t minutes after an injection is given by
$$C(t) = 0.065(1 + e^{-0.025t})$$
grams per cubic centimeter (g/cm³).
  a. What is the initial concentration of the drug (when $t = 0$)?
  b. What is the concentration 20 minutes after an injection? After 1 hour?
  c. What is the average rate of change of concentration during the first minute?
  d. What happens to the concentration of the drug in the long run (as $t \to \infty$)?
  e. Sketch the graph of $C(t)$.

70. POPULATION GROWTH It is estimated that t years after 1995, the population of a certain country will be $P(t)$ million people where
$$P(t) = Ae^{0.03t} - Be^{0.005t}$$
for certain constants A and B. The population was 100 million in 1997 and 200 million in 2010.
  a. Use the given information to find A and B.
  b. What was the population in 1995?
  c. What will the population be in 2015?

71. AQUATIC PLANT LIFE Plant life exists only in the top 10 meters of a lake or sea, primarily because the intensity of sunlight decreases exponentially with depth. Specifically, the **Bouguer-Lambert law** says that a beam of light that strikes the surface of a body of water with intensity I_0 will have intensity I at a depth of x meters, where $I = I_0 e^{-kx}$ with $k > 0$. The constant k, called the **absorption coefficient,** depends on the wavelength of the light and the density of the water. Suppose a beam of sunlight is only 10% as intense at a depth of 3 meters as at the surface. How intense is the beam at a depth of 1 meter? (Express your answer in terms of I_0.)

72. **LINGUISTICS** **Glottochronology** is the methodology used by linguists to determine how many years have passed since two modern languages "branched" from a common ancestor. Experiments suggest that if N words are in common use at a base time $t = 0$, then the number $N(t)$ of them still in use with essentially the same meaning t thousand years later is given by the so-called **fundamental glottochronology equation***
$$N(t) = N_0 e^{-0.217t}$$

 a. Out of a set of 500 basic words used in classical Latin in 200 B.C., how many would you expect to be still in use in modern Italian in the year 2010?

 b. The research of C. W. Feng and M. Swadesh indicated that out of a set of 210 words commonly used in classical Chinese in 950 A.D., 167 were still in use in modern Mandarin in 1950. Is this the same number that the fundamental glottochronology equation would predict? How do you account for the difference?

 c. Read an article on glottochronology, and write a paragraph on its methodology. You may wish to begin your research by reading the article cited in this exercise.

MISCELLANEOUS PROBLEMS

73. **RADIOACTIVE DECAY** The amount of a sample of a radioactive substance remaining after t years is given by a function of the form $Q(t) = Q_0 e^{-0.0001t}$. At the end of 5,000 years, 200 grams of the substance remain. How many grams were present initially?

74. Two graphs $y = f(x)$ and $y = g(x)$ are reflections of one another in the y axis if whenever (a, b) is a point on one of the graphs, then $(-a, b)$ is a point

on the other, as indicated in the accompanying figure. Use this criterion to show that the graphs of
$$y = b^x \text{ and } y = \left(\frac{1}{b}\right)^x \text{ for } b > 0, b \neq 1 \text{ are}$$
reflections of one another in the y axis.

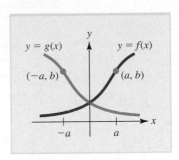

EXERCISE 74

75. Complete the following table for $f(x) = \frac{1}{2}\left(\frac{1}{4}\right)^x$.

x	-2.2	-1.5	0	1.5	2.3
$f(x)$					

76. Program a computer or use a calculator to evaluate $\left(1 + \frac{1}{n}\right)^n$ for $n = 1{,}000, 2{,}000, \ldots, 50{,}000$.

77. Program a computer or use a calculator to evaluate $\left(1 + \frac{1}{n}\right)^n$ for $n = -1{,}000, -2{,}000, \ldots, -50{,}000$. On the basis of these calculations, what can you conjecture about the behavior of $\left(1 + \frac{1}{n}\right)^n$ as n *decreases* without bound?

78. Program a computer or use a calculator to estimate
$$\lim_{n \to +\infty} \left(1 + \frac{3}{n}\right)^{2n}.$$

79. Program a computer or use a calculator to estimate
$$\lim_{n \to +\infty} \left(2 - \frac{5}{2n}\right)^{n/3}.$$

---

*Source: Anthony LoBello and Maurice D. Weir, "Glottochronology: An Application of Calculus to Linguistics," *UMAP Modules 1982: Tools for Teaching*, Lexington, MA: Consortium for Mathematics and Its Applications, Inc., 1983.

SECTION 4.2 Logarithmic Functions

Learning Objectives

1. Define and explore logarithmic functions and their properties.
2. Use logarithms to solve exponential equations.
3. Examine applications involving logarithms.

Suppose you invest \$1,000 at 8% compounded continuously and want to know how much time must pass for your investment to double in value to \$2,000. According to the formula derived in Section 4.1, the value of your account after t years will be $1,000e^{0.08t}$, so to find the doubling time for your account, you must solve for t in the equation

$$1,000e^{0.08t} = 2,000$$

or, by dividing both sides by 1,000,

$$e^{0.08t} = 2$$

We will answer the question about doubling time in Example 4.2.10. Solving an exponential equation such as this involves using *logarithms,* which reverse the process of exponentiation. Logarithms play an important role in a variety of applications, such as measuring the capacity of a transmission channel and in the famous Richter scale for measuring earthquake intensity. In this section, we examine the basic properties of logarithmic functions and a few applications. We begin with a definition.

> **Logarithmic Functions** ■ If x is a positive number, then the **logarithm** of x to the base b ($b > 0$, $b \neq 1$), denoted $\log_b x$, is the number y such that $b^y = x$; that is,
>
> $$y = \log_b x \quad \text{if and only if} \quad b^y = x \quad \text{for } x > 0$$

EXAMPLE 4.2.1 Evaluating Logarithms

Evaluate

a. $\log_{10} 1,000$ **b.** $\log_2 32$ **c.** $\log_5\left(\dfrac{1}{125}\right)$

Solution

a. $\log_{10} 1,000 = 3$ since $10^3 = 1,000$.

b. $\log_2 32 = 5$ since $2^5 = 32$.

c. $\log_5 \dfrac{1}{125} = -3$ since $5^{-3} = \dfrac{1}{125}$.

EXAMPLE 4.2.2 Solving Equations Involving Logarithms

Solve each of the following equations for x:

a. $\log_4 x = \dfrac{1}{2}$ **b.** $\log_{64} 16 = x$ **c.** $\log_x 27 = 3$

Solution

a. By definition, $\log_4 x = \dfrac{1}{2}$ is equivalent to $4^{1/2} = x$, so $x = 4^{1/2} = 2$.

b. $\log_{64} 16 = x$ means

$$64^x = 16$$
$$(2^6)^x = 2^4$$
$$2^{6x} = 2^4 \qquad\qquad \text{if } b^m = b^n, \text{ then } m = n$$
$$6x = 4 \quad\text{so}\quad x = \frac{2}{3}$$

c. $\log_x 27 = 3$ means

$$x^3 = 27$$
$$x = (27)^{1/3} = 3$$

Logarithms were introduced in the 17th century as a computational device, primarily because they can be used to convert expressions involving products and quotients into much simpler expressions involving sums and differences. Here are the rules for logarithms that facilitate such simplification.

> **Logarithmic Rules** ■ Let b be any logarithmic base $(b > 0, b \neq 1)$. Then
> $$\log_b 1 = 0 \qquad \text{and} \qquad \log_b b = 1$$
> and if u and v are any positive numbers, we have
>
> The **equality rule** $\qquad \log_b u = \log_b v$ if and only if $u = v$
> The **product rule** $\qquad \log_b(uv) = \log_b u + \log_b v$
> The **power rule** $\qquad \log_b u^r - r \log_b u$ for any real number r
> The **quotient rule** $\qquad \log_b\left(\dfrac{u}{v}\right) = \log_b u - \log_b v$
> The **inversion rule** $\qquad \log_b b^u = u$

All these logarithmic rules follow from corresponding exponential rules. For example,

$$\log_b 1 = 0 \qquad \text{since} \qquad b^0 = 1$$
$$\log_b b = 1 \qquad \text{since} \qquad b^1 = b$$

To prove the equality rule, let

$$m = \log_b u \quad\text{and}\quad n = \log_b v$$

so that by definition,

$$b^m = u \quad\text{and}\quad b^n = v$$

Therefore, if

$$\log_b u = \log_b v$$

then $m = n$, so

$$b^m = b^n \quad \text{equality rule for exponentials}$$

or, equivalently,

$$u = v$$

as stated in the equality rule for logarithms. Similarly, to prove the product rule for logarithms, note that

$$\log_b u + \log_b v = m + n \qquad \text{inversion rule}$$
$$= \log_b(b^{m+n}) \qquad \text{product rule for exponentials}$$
$$= \log_b(b^m b^n) \qquad \text{since } b^m = u \text{ and } b^n = v$$
$$= \log_b(uv)$$

Proofs of the power rule and the quotient rule are left as exercises (see Exercise 78). Table 4.1 displays the correspondence between basic properties of exponential and logarithmic functions.

TABLE 4.1 Comparison of Exponential and Logarithmic Rules

Exponential Rule	Logarithmic Rule
$b^x b^y = b^{x+y}$	$\log_b(xy) = \log_b x + \log_b y$
$\dfrac{b^x}{b^y} = b^{x-y}$	$\log_b\left(\dfrac{x}{y}\right) = \log_b x - \log_b y$
$b^{xp} = (b^x)^p$	$\log_b x^p = p \log_b x$

EXAMPLE 4.2.3 Using Logarithmic Rules

Use logarithm rules to rewrite each of the following expressions in terms of $\log_5 2$ and $\log_5 3$.

a. $\log_5\left(\dfrac{5}{3}\right)$ **b.** $\log_5 8$ **c.** $\log_5 36$

Solution

a. $\log_5\left(\dfrac{5}{3}\right) = \log_5 5 - \log_5 3$ use quotient rule
$$= 1 - \log_5 3 \qquad \text{since } \log_5 5 = 1$$
b. $\log_5 8 = \log_5 2^3 = 3\log_5 2$ use power rule
c. $\log_5 36 = \log_5(2^2 3^2)$ use product rule
$$= \log_5 2^2 + \log_5 3^2 \qquad \text{use power rule}$$
$$= 2\log_5 2 + 2\log_5 3$$

EXAMPLE 4.2.4 Using Logarithmic Rules

Use logarithmic rules to expand each of the following expressions.

a. $\log_3(x^3 y^{-4})$ **b.** $\log_2\left(\dfrac{y^5}{x^2}\right)$ **c.** $\log_7(x^3\sqrt{1-y^2})$

Solution

a. $\log_3(x^3 y^{-4}) = \log_3 x^3 + \log_3 y^{-4}$ product rule
$$= 3\log_3 x + (-4)\log_3 y \qquad \text{power rule}$$
$$= 3\log_3 x - 4\log_3 y$$

b. $\log_2\!\left(\dfrac{y^5}{x^2}\right) = \log_2 y^5 - \log_2 x^2$ quotient rule

$\qquad\qquad\quad = 5 \log_2 y - 2 \log_2 x$ power rule

c. $\log_7(x^3\sqrt{1 - y^2}) = \log_7[x^3(1 - y^2)^{1/2}]$ use product rule

$\qquad\qquad\qquad = \log_7 x^3 + \log_7(1 - y^2)^{1/2}$ use power rule

$\qquad\qquad\qquad = 3 \log_7 x + \dfrac{1}{2}\log_7(1 - y^2)$ factor $1 - y^2$

$\qquad\qquad\qquad = 3 \log_7 x + \dfrac{1}{2}\log_7[(1 - y)(1 + y)]$ use product rule

$\qquad\qquad\qquad = 3 \log_7 x + \dfrac{1}{2}[\log_7(1 - y) + \log_7(1 + y)]$ distribute

$\qquad\qquad\qquad = 3 \log_7 x + \dfrac{1}{2}\log_7(1 - y) + \dfrac{1}{2}\log_7(1 + y)$

Graphs of Logarithmic Functions

There is an easy way to obtain the graph of the logarithmic function $y = \log_b x$ from the graph of the exponential function $y = b^x$. The idea is that since $y = \log_b x$ is equivalent to $x = b^y$, the graph of $y = \log_b x$ is the same as the graph of $y = b^x$ with the roles of x and y reversed. That is, if (u, v) is a point on the curve $y = \log_b x$, then $v = \log_b u$, or equivalently, $u = b^v$, which means that (v, u) is on the graph of $y - b^x$. As illustrated in Figure 4.5a, the points (u, v) and (v, u) are mirror images of one another in the line $y = x$ (see Exercise 79). Thus, the graph of $y = \log_b x$ can be obtained by simply *reflecting* the graph of $y = b^x$ in the line $y = x$, as shown in Figure 4.5b for the case where $b > 1$. To summarize:

> **Relationship Between the Graphs of $y = \log_b x$ and $y = b^x$** ■ The graphs of $y = \log_b x$ and $y = b^x$ are mirror images of one another in the line $y = x$. Therefore, the graph of $y = \log_b x$ can be obtained by reflecting the graph of $y = b^x$ in the line $y = x$.

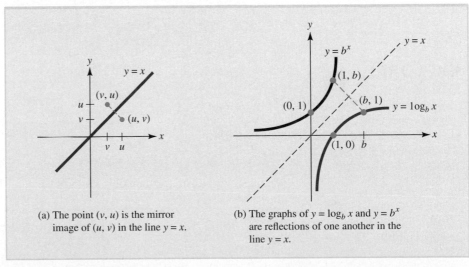

(a) The point (v, u) is the mirror image of (u, v) in the line $y = x$.

(b) The graphs of $y = \log_b x$ and $y = b^x$ are reflections of one another in the line $y = x$.

FIGURE 4.5 The graph of $y = \log_b x$ for $b > 1$ is obtained by reflecting the graph of $y = b^x$ in the line $y = x$.

Figure 4.5b reveals important properties of the logarithmic function $f(x) = \log_b x$ for the case where $b > 1$. The following box lists these properties along with similar properties for the case where $0 < b < 1$.

Properties of a Logarithmic Function ■ The logarithmic function $f(x) = \log_b x (b > 0, b \neq 1)$ has these properties:

1. It is defined and continuous for all $x > 0$.
2. The y axis is a vertical asymptote.
3. The x intercept is $(1, 0)$; there is no y intercept.
4. If $b > 1$, then $\lim\limits_{x \to 0^+} \log_b x = -\infty$ and $\lim\limits_{x \to +\infty} \log_b x = +\infty$.

   If $0 < b < 1$, then $\lim\limits_{x \to 0^+} \log_b x = +\infty$ and $\lim\limits_{x \to +\infty} \log_b x = -\infty$.
5. For all $x > 0$, the function is increasing (graph rising) if $b > 1$ and decreasing (graph falling) if $0 < b < 1$.

The Natural Logarithm

In calculus, the most frequently used logarithmic base is e. In this case, the logarithm $\log_e x$ is called the **natural logarithm** of x and is denoted by $\ln x$ (read as "el en x"); that is, for $x > 0$

$$y = \ln x \qquad \text{if and only if} \qquad e^y = x$$

The graph of the natural logarithm is shown in Figure 4.6.

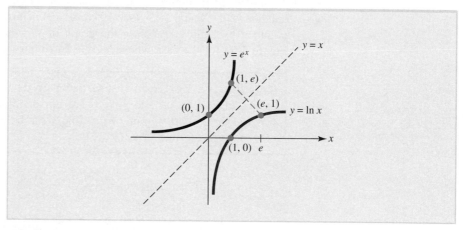

FIGURE 4.6 The graph of $y = \ln x$.

To evaluate $\ln a$ for a particular number $a > 0$, use the **LN** key on your calculator. For example, to find $\ln(2.714)$, you would press the **LN** key, and then enter the number 2.714 to get

$$\ln(2.714) = 0.9984 \qquad \text{(to four decimal places)}$$

Example 4.2.5 illustrates the computation of natural logarithms.

EXAMPLE 4.2.5 Evaluating Natural Logarithms

Find

a. $\ln e$ **b.** $\ln 1$ **c.** $\ln \sqrt{e}$ **d.** $\ln 2$

Solution

a. According to the definition, $\ln e$ is the unique number c such that $e = e^c$. Clearly this number is $c = 1$. Hence, $\ln e = 1$.

b. $\ln 1$ is the unique number c such that $1 = e^c$. Since $e^0 = 1$, it follows that $\ln 1 = 0$.

Just-In-Time **REVIEW**

Remember that a logarithm with any base is defined only for inputs that are positive numbers, so a quantity like $\ln(-3)$ makes no sense.

c. $\ln\sqrt{e} = \ln e^{1/2}$ is the unique number c such that $e^{1/2} = e^c$; that is, $c = \dfrac{1}{2}$. Hence, $\ln\sqrt{e} = \dfrac{1}{2}$.

d. $\ln 2$ is the unique number c such that $2 = e^c$. The value of this number is not obvious, and you will have to use your calculator to find that $\ln 2 \approx 0.69315$.

EXAMPLE 4.2.6 Using Properties of Natural Logarithms

a. Find $\ln\sqrt{ab}$ if $\ln a = 3$ and $\ln b = 7$.

b. Show that $\ln\dfrac{1}{x} = -\ln x$.

c. Find x if $2^x = e^3$.

Solution

a. $\ln\sqrt{ab} = \ln(ab)^{1/2} = \dfrac{1}{2}\ln ab = \dfrac{1}{2}(\ln a + \ln b) = \dfrac{1}{2}(3 + 7) = 5$

b. $\ln\dfrac{1}{x} = \ln 1 - \ln x = 0 - \ln x = -\ln x$

c. Take the natural logarithm of each side of the equation $2^x = e^3$, and solve for x to get

$$\ln 2^x = \ln e^3 \qquad \text{use power rule}$$
$$x \ln 2 = 3 \ln e = 3 \qquad \text{since } \ln e = 1$$

so

$$x = \dfrac{3}{\ln 2} \approx 4.33$$

Two functions f and g with the property that $f(g(x)) = x$ and $g(f(x)) = x$, whenever both composite functions are defined, are said to be **inverses** of one another. Such an inverse relationship exists between exponential and logarithmic functions with base b. For instance, we have

$$\ln e^x = x \ln e = x(1) = x \qquad \text{for all } x$$

Similarly, if $y = e^{\ln x}$ for $x > 0$, then by definition, $\ln y = \ln x$, so $y = x$; that is,

$$e^{\ln x} = y = x$$

This inverse relationship between the natural exponential and logarithmic functions is especially useful. It is summarized in the following box and used in Example 4.2.7.

> **The Inverse Relationship Between e^x and $\ln x$** ∎
> $$e^{\ln x} = x \quad \text{for } x > 0 \qquad \text{and} \qquad \ln e^x = x \quad \text{for all } x$$

EXPLORE!

Solve the equation

$$3 - e^x = \ln(x^2 + 1)$$

by placing the left side of the equation into Y1 and the right side into Y2. Use Standard Window (**ZOOM** 6) to find the x values of the intersection points.

EXPLORE!

Put the function $y = 10^x$ into Y1, and graph using a bold graphing style. Then put $y = x$ into Y2, $y = \ln x$ into Y3, and $y = \log_{10} x$ into Y4. What can you conclude from this series of graphs?

EXPLORE!

Refer to Example 4.2.8. Place $0 = Q - A*e^\wedge(-K*X)$ into the equation solver of your graphing calculator. Find the distance from the center of the city if $Q = 13{,}500$ people per square mile. Recall from the example that the density at the city center is 15,000 and the density 10 miles from the city center is 9,000 people per square mile.

EXAMPLE 4.2.7   Solving Exponential and Logarithmic Equations

Solve each of the following equations for x:

a. $3 = e^{20x}$ **b.** $2 \ln x = 1$

Solution

a. Take the natural logarithm of each side of the equation to get

$$\ln 3 = \ln e^{20x} \qquad \text{or} \qquad \ln 3 = 20x$$

Solve for x using a calculator to find ln 3:

$$x = \frac{\ln 3}{20} \approx \frac{1.0986}{20} \approx 0.0549$$

b. First isolate ln x on the left side of the equation by dividing both sides by 2:

$$\ln x = \frac{1}{2}$$

Then apply the exponential function to both sides of the equation to get

$$e^{\ln x} = e^{1/2} \qquad \text{or} \qquad x = e^{1/2} = \sqrt{e} \approx 1.6487$$

Example 4.2.8 illustrates how to use logarithms to find an exponential function that fits certain specified information.

EXAMPLE 4.2.8   Writing an Exponential Population Function

The population density x miles from the center of a city is given by a function of the form $Q(x) = Ae^{-kx}$. Find this function if it is known that the population density at the center of the city is 15,000 people per square mile and the density 10 miles from the center is 9,000 people per square mile.

Solution

For simplicity, express the density in units of 1,000 people per square mile. The fact that $Q(0) = 15$ tells you that $A = 15$. The fact that $Q(10) = 9$ means

$$9 = 15e^{-10k} \qquad \text{so} \qquad \frac{3}{5} = e^{-10k}$$

Taking the natural logarithm of each side of this equation, you get

$$\ln\frac{3}{5} = -10k \qquad \text{so} \qquad k = -\frac{\ln 3/5}{10} \approx 0.051$$

Hence the exponential function for the population density is $Q(x) = 15e^{-0.051x}$.

You have already seen how to use the **LN** key to compute natural logarithms, and most calculators have a **LOG** key for computing logarithms to base 10, but what about logarithms to bases other than e or 10? To be specific, suppose you need to calculate the logarithmic number $c = \log_b a$. You have

$$c = \log_b a \qquad \text{definition of the logarithm}$$
$$b^c = a \qquad \text{take natural logarithms on both sides}$$
$$\ln b^c = \ln a \qquad \text{use the power rule}$$
$$c \ln b = \ln a$$
$$c = \frac{\ln a}{\ln b}$$

EXPLORE!

Store $f(x) = B^x$ into Y1 and then $g(x) = \log_B x$ into Y2 as ln(x)/ln(B). Experiment with different values of $1 < B < 2$, using the **STO▸** key, to determine for which B values these two functions intersect, touch at a point, or are separated.

Thus, the logarithm $\log_b a$ can be computed by finding the ratio of two natural logarithms, $\ln a$ and $\ln b$. To summarize:

Conversion Formula for Logarithms ■ If a and b are positive numbers with $b \neq 1$, then

$$\log_b a = \frac{\ln a}{\ln b}$$

EXAMPLE 4.2.9 Using the Logarithm Conversion Formula

Find $\log_5 3.172$.

Solution

Using the conversion formula, you find

$$\log_5 3.172 = \frac{\ln 3.172}{\ln 5} \approx \frac{1.1544}{1.6094} \approx 0.7172$$

Compounding Applications

In the introductory paragraph at the beginning of this section, you were asked how long it would take for a particular investment to double in value. This question is answered in Example 4.2.10.

EXAMPLE 4.2.10 Finding Doubling Time for an Investment

If Hubert invests $1,000 at 8% annual interest, compounded continuously, how long does it take for his investment to double? Would the doubling time of Hubert's investment change if the principal were something other than $1,000?

Solution

With a principal of $1,000, the balance after t years is $B(t) = 1,000e^{0.08t}$, so the investment doubles when $B(t) = \$2,000$, that is, when

$$2,000 = 1,000e^{0.08t}$$

Dividing by 1,000 and taking the natural logarithm on each side of the equation, we get

$$2 = e^{0.08t}$$
$$\ln 2 = 0.08t$$
$$t = \frac{\ln 2}{0.08} \approx 8.66 \text{ years}$$

If the principal had been P_0 dollars instead of \$1,000, the doubling time would satisfy

$$2P_0 = P_0 e^{0.08t}$$
$$2 = e^{0.08t}$$

which is exactly the same equation we had with $P_0 = \$1,000$, so once again, the doubling time is 8.66 years.

EXPLORE!

Use the equation solver of your graphing calculator with the equation F − P∗e^(R∗T) = 0 to determine how long it will take for \$2,500 to double at 8.5% compounded continuously.

The situation illustrated in Example 4.2.10 applies to any quantity $Q(t) = Q_0 e^{kt}$ with $k > 0$. In particular, since at time $t = 0$, we have $Q(0) = Q_0 e^0 = Q_0$, the quantity doubles when

$$2Q_0 = Q_0 e^{kt}$$
$$2 = e^{kt}$$
$$\ln 2 = kt$$
$$t = \frac{\ln 2}{k}$$

To summarize:

> **Doubling Time** ■ A quantity $Q(t) = Q_0 e^{kt}$ $(k > 0)$ doubles when $t = d$, where
>
> $$d = \frac{\ln 2}{k}$$

Determining the time it takes for an investment to double is just one of several issues an investor may address when comparing various investment opportunities. Two additional issues are illustrated in Examples 4.2.11 and 4.2.12.

EXAMPLE 4.2.11 Reaching an Investment Goal

How long will it take \$5,000 to grow to \$7,000 in an investment earning interest at an annual rate of 6% if the compounding is

a. Quarterly   **b.** Continuous

Solution

a. We use the future value formula $B = P\left(1 + \dfrac{r}{k}\right)$. We have $B = 7,000$, $P = 5,000$, and $\dfrac{r}{k} = \dfrac{0.06}{4} = 0.015$, since $r = 0.06$ and there are $k = 4$ compounding periods per year. Substituting, we find that

$$7,000 = 5,000(1.015)^{4t}$$
$$(1.015)^{4t} = \frac{7,000}{5,000} = 1.4$$

Taking the natural logarithm on each side of this equation, we get

$$\ln(1.015)^{4t} = \ln 1.4 \qquad \text{use the power rule}$$
$$4t \ln 1.015 = \ln 1.4$$
$$t = \frac{\ln 1.4}{4 \, (\ln 1.015)} \approx 5.65$$

This means that since compounding is done quarterly, the investment will first be worth $7,000 in the third quarter of the sixth year.

b. With continuous compounding, we use the formula $B = Pe^{rt}$:

$$7,000 = 5,000e^{0.06t}$$

$$e^{0.06t} = \frac{7,000}{5,000} = 1.4$$

Taking logarithms, we get

$$\ln e^{0.06t} = \ln 1.4$$

$$0.06t = \ln 1.4$$

$$t = \frac{\ln 1.4}{0.06} = 5.61$$

So, with continuous compounding, it takes only 5.61 years to reach the investment objective.

EXAMPLE 4.2.12 Computing an Interest Rate

Millicent has $1,500 to invest and wants it to grow to $2,000 in 5 years. At what annual rate r compounded continuously must she invest her money to achieve her goal?

Solution

If the interest rate is r, the future value of $1,500 in 5 years is given by $1,500e^{r(5)}$. For this to equal $2,000, we must have

$$1,500e^{r(5)} = 2,000$$

$$e^{5r} = \frac{2,000}{1,500} = \frac{4}{3}$$

Taking natural logarithms on both sides of this equation, we get

$$\ln e^{5r} = \ln \frac{4}{3}$$

$$5r = \ln \frac{4}{3}$$

so

$$r = \frac{1}{5} \ln \frac{4}{3} \approx 0.575$$

The annual interest rate is approximately 5.75%.

Radioactive Decay and Carbon Dating

It has been experimentally determined that a radioactive sample of initial size Q_0 grams will decay to $Q(t) = Q_0 e^{-kt}$ grams in t years. The positive constant k in this formula measures the rate of decay, but this rate is usually given by specifying the amount of time $t = h$ required for half a given sample to decay. This time h is called the **half-life** of the radioactive substance. Example 4.2.13 shows how half-life is related to k.

> **EXAMPLE 4.2.13**   Calculating Half-Life
>
> Show that a radioactive substance that decays according to the formula $Q(t) = Q_0 e^{-kt}$ has half-life $h = \dfrac{\ln 2}{k}$.

Solution

The goal is to find the value of t for which $Q(h) = \dfrac{1}{2} Q_0$; that is,

$$\frac{1}{2} Q_0 = Q_0 e^{-kh}$$

Divide by Q_0 and take the natural logarithm of each side to get

$$\ln \frac{1}{2} = -kh$$

Thus, the half-life is

$$h = \frac{\ln \dfrac{1}{2}}{-k} \qquad\qquad \ln \frac{1}{2} = \ln 2^{-1} = -\ln 2$$

$$= \frac{-\ln 2}{-k} = \frac{\ln 2}{k}$$

as required.

In 1960, W. F. Libby won a Nobel prize for his discovery of **carbon dating,** a technique for determining the age of certain fossils and artifacts. Here is an outline of the technique.*

The carbon dioxide in the air contains the radioactive isotope ^{14}C (carbon-14) as well as the stable isotope ^{12}C (carbon-12). Living plants absorb carbon dioxide from the air, which means that the ratio of ^{14}C to ^{12}C in a living plant (or in an animal that eats plants) is the same as that in the air itself. When a plant or an animal dies, the absorption of carbon dioxide ceases. The ^{12}C already in the plant or animal remains the same as at the time of death while the ^{14}C decays, and the ratio of ^{14}C to ^{12}C decreases exponentially. It is reasonable to assume that the ratio R_0 of ^{14}C to ^{12}C in the atmosphere is the same today as it was in the past, so that the ratio of ^{14}C to ^{12}C in a sample (e.g., a fossil or an artifact) is given by a function of the form $R(t) = R_0 e^{-kt}$. The half-life of ^{14}C is 5,730 years. By comparing $R(t)$ to R_0, archaeologists can estimate the age of the sample. Example 4.2.14 illustrates the dating procedure.

---

*For instance, see Raymond J. Cannon, "Exponential Growth and Decay," *UMAP Modules 1977: Tools for Teaching,* Lexington, MA: Consortium for Mathematics and Its Applications, Inc., 1978. More advanced dating procedures are discussed in Paul J. Campbell, "How Old Is the Earth?" *UMAP Modules 1992: Tools for Teaching,* Lexington, MA: Consortium for Mathematics and Its Applications, Inc., 1993.

EXAMPLE 4.2.14 Using Carbon Dating

An archaeologist has found a fossil in which the ratio of ^{14}C to ^{12}C is $\dfrac{1}{5}$ the ratio found in the atmosphere. Approximately how old is the fossil?

Solution

The age of the fossil is the value of t for which $R(t) = \dfrac{1}{5} R_0$, that is, for which

$$\frac{1}{5} R_0 = R_0 e^{-kt}$$

Dividing by R_0 and taking logarithms, you find that

$$\frac{1}{5} = e^{-kt}$$

$$\ln \frac{1}{5} = -kt$$

and

$$t = \frac{-\ln \dfrac{1}{5}}{k} = \frac{\ln 5}{k}$$

In Example 4.2.13, you found that the half-life h satisfies $h = \dfrac{\ln 2}{k}$, and since ^{14}C has half-life $h = 5{,}730$ years, you have

$$k = \frac{\ln 2}{h} = \frac{\ln 2}{5{,}730} \approx 0.000121$$

Therefore, the age of the fossil is

$$t = \frac{\ln 5}{k} = \frac{\ln 5}{0.000121} \approx 13{,}300$$

That is, the fossil is approximately 13,300 years old.

EXERCISE ■ 4.2

In Exercises 1 and 2, use your calculator to find the indicated natural logarithms.

1. Find $\ln 1$, $\ln 2$, $\ln e$, $\ln 5$, $\ln \dfrac{1}{5}$, and $\ln e^2$. What happens if you try to find $\ln 0$ or $\ln(-2)$? Why?

2. Find $\ln 7$, $\ln \dfrac{1}{3}$, $\ln e^{-3}$, $\ln \dfrac{1}{e^{2.1}}$, and $\ln \sqrt[5]{e}$. What happens if you try to find $\ln(-7)$ or $\ln(-e)$?

In Exercises 3 through 8, evaluate the given expression using properties of the natural logarithm.

3. $\ln e^3$

4. $\ln \sqrt{e}$

5. $e^{\ln 5}$

6. $e^{2 \ln 3}$

7. $e^{3 \ln 2 - 2 \ln 5}$

8. $\ln \dfrac{e^3 \sqrt{e}}{e^{1/3}}$

In Exercises 9 through 12, use logarithmic rules to rewrite each expression in terms of $\log_3 2$ and $\log_3 5$.

9. $\log_3 270$

10. $\log_3(2.5)$

11. $\log_3 100$

12. $\log_3\left(\dfrac{64}{125}\right)$

In Exercises 13 through 20, use logarithmic rules to simplify each expression.

13. $\log_2(x^4 y^3)$

14. $\log_3(x^5 y^{-2})$

15. $\ln \sqrt[3]{x^2 - x}$

16. $\ln(x^2 \sqrt{4 - x^2})$

17. $\ln\left[\dfrac{x^2(3 - x)^{2/3}}{\sqrt{x^2 + x + 1}}\right]$

18. $\ln\left[\dfrac{1}{x} + \dfrac{1}{x^2}\right]$

19. $\ln(x^3 e^{-x^2})$

20. $\ln\left[\dfrac{\sqrt[4]{x}}{x^3 \sqrt{1 - x^2}}\right]$

In Exercises 21 through 36, solve the given equation for x.

21. $4^x = 53$

22. $\log_2 x = 4$

23. $\log_3(2x - 1) = 2$

24. $3^{2x-1} = 17$

25. $2 = e^{0.06x}$

26. $\dfrac{1}{2} Q_0 = Q_0 e^{-1.2x}$

27. $3 = 2 + 5e^{-4x}$

28. $-2 \ln x = b$

29. $-\ln x = \dfrac{t}{50} + C$

30. $5 = 3 \ln x - \dfrac{1}{2} \ln x$

31. $\ln x = \dfrac{1}{3}(\ln 16 + 2 \ln 2)$

32. $\ln x = 2(\ln 3 - \ln 5)$

33. $3^x = e^2$

34. $a^k = e^{kx}$

35. $\dfrac{25 e^{0.1x}}{e^{0.1x} + 3} = 10$

36. $\dfrac{5}{1 + 2e^{-x}} = 3$

37. If $\log_2 x = 5$, what is $\ln x$?

38. If $\log_{10} x = -3$, what is $\ln x$?

39. If $\log_5(2x) = 7$, what is $\ln x$?

40. If $\log_3(x - 5) = 2$, what is $\ln x$?

41. Find $\ln \dfrac{1}{\sqrt{ab^3}}$ if $\ln a = 2$ and $\ln b = 3$.

42. Find $\dfrac{1}{a} \ln\left(\dfrac{\sqrt{b}}{c}\right)^a$ if $\ln b = 6$ and $\ln c = -2$.

BUSINESS AND ECONOMICS APPLIED PROBLEMS

43. COMPOUND INTEREST How quickly will money double if it is invested at an annual interest rate of 6% compounded continuously?

44. COMPOUND INTEREST How quickly will money double if it is invested at an annual interest rate of 7% compounded continuously?

45. COMPOUND INTEREST Money deposited in a certain bank doubles every 13 years. The bank compounds interest continuously. What annual interest rate does the bank offer?

46. TRIPLING TIME How long will it take for a quantity of money A_0 to triple in value if it is invested at an annual interest rate r compounded continuously?

47. TRIPLING TIME If an account that earns interest compounded continuously takes 12 years to double in value, how long will it take to triple in value?

48. INVESTMENT The Morenos invest $10,000 in an account that grows to $12,000 in 5 years. What is the annual interest rate r if interest is compounded
a. Quarterly   **b.** Continuously

49. COMPOUND INTEREST A certain bank offers an interest rate of 6% per year compounded annually. A competing bank compounds its interest continuously. What (nominal) interest rate should the competing bank offer so that the effective interest rates of the two banks will be equal?

50. ADVERTISING Ravi Patel, the marketing editor at a major publishing house, estimates that if x thousand complimentary copies of a new text are distributed to instructors, the first-year sales will be approximately

$$N(x) = 20 - 12e^{-0.03x}$$

thousand copies. How many complimentary copies should Ravi send out to instructors if his goal is to generate first-year sales of 12,000 copies?

51. WORKER EFFICIENCY An efficiency expert hired by a manufacturing firm has compiled these data relating workers' output to their experience:

Experience t (months)	0	6
Output Q (units per hour)	300	410

Suppose output Q is related to experience t by a function of the form $Q(t) = 500 - Ae^{-kt}$. Find the function of this form that fits the data. What output is expected from a worker with 1 year's experience?

52. SUPPLY AND DEMAND A manufacturer determines that the supply function for x units of a particular commodity is $S(x) = e^{0.02x}$ and the corresponding demand function is $D(x) = 3e^{-0.03x}$.
a. Find the demand price $p = D(x)$ when the level of production is $x = 10$ units.
b. Find the supply price $p = S(x)$ when $x = 12$ units.
c. Find the level of production and unit price that correspond to market equilibrium (where supply = demand).

53. SUPPLY AND DEMAND A manufacturer determines that the supply function for x units of a particular commodity is $S(x) = \ln(x + 2)$ and the corresponding demand function is $D(x) = 10 - \ln(x + 1)$.

a. Find the demand price $p = D(x)$ when the level of production is $x = 10$ units.
b. Find the supply price $p = S(x)$ when $x = 100$ units.
c. Find the level of production and unit price that correspond to market equilibrium (where supply = demand).

54. **GROSS DOMESTIC PRODUCT** An economist has compiled these data on the gross domestic product (GDP) of a certain country:

Year	1995	2005
GDP (in billions)	100	180

Use these data to predict the GDP in the year 2015 if the GDP is growing:
a. Linearly, so that GDP $= at + b$.
b. Exponentially, so that GDP $= Ae^{kt}$.

55. **INVESTMENT** An investment firm estimates that the value of its portfolio after t years is A million dollars, where

$$A(t) = 300 \ln(t + 3)$$

a. What is the value of the account when $t = 0$?
b. How long does it take for the account to double its initial value?
c. How long does it take before the account is worth a billion dollars?

LIFE AND SOCIAL SCIENCE APPLIED PROBLEMS

56. **CONCENTRATION OF DRUG** A drug is injected into a patient's bloodstream and t seconds later, the concentration of the drug is C grams per cubic centimeter (g/cm³), where

$$C(t) = 0.1(1 + 3e^{-0.03t})$$

a. What is the drug concentration after 10 seconds?
b. How long does it take for the drug concentration to reach 0.12 g/cm³?

57. **CONCENTRATION OF DRUG** The concentration of a drug in a patient's kidneys at time t (seconds) is C grams per cubic centimeter (g/cm³), where

$$C(t) = 0.4(2 - 0.13e^{-0.02t})$$

a. What is the drug concentration after 20 seconds? After 60 seconds?
b. How long does it take for the drug concentration to reach 0.75 g/cm³?

58. **DEMOGRAPHICS** The world's population grows at the rate of approximately 2% per year. If it is assumed that the population growth is exponential, then the population t years from now will be given by a function of the form $P(t) = P_0 e^{0.02t}$, where P_0 is the current population. Assuming that this model of population growth is correct, how long will it take for the world's population to double?

59. **ALLOMETRY** Suppose that for the first 6 years of a moose's life, its shoulder height $H(t)$ and tip-to-tip antler length $A(t)$ increase with time t (years) according to the formulas $H(t) = 125e^{0.08t}$ and $A(t) = 50e^{0.16t}$, where H and A are both measured in centimeters (cm).
a. On the same graph, sketch $y = H(t)$ and $y = A(t)$ for the applicable period $0 \le t \le 6$.
b. Express antler length A as a function of height H. [*Hint:* First take logarithms on both sides of the equation $H = 125e^{0.08t}$ to express time t in terms of H and then substitute into the formula for $A(t)$.]

60. **ARCHAEOLOGY** An archaeologist has found a fossil in which the ratio of ^{14}C to ^{12}C is $\frac{1}{3}$ the ratio found in the atmosphere. Approximately how old is the fossil?

61. **ARCHAEOLOGY** Tests of an artifact discovered at the Debert site in Nova Scotia show that 28% of the original ^{14}C is still present. Approximately how old is the artifact?

62. **ARCHAEOLOGY** The Dead Sea Scrolls were written on parchment in about 100 B.C. What percentage of the original ^{14}C in the parchment remained when the scrolls were discovered in 1947?

63. **ART FORGERY** A forged painting allegedly painted by Rembrandt in 1640 is found to have 99.7% of its original ^{14}C. When was it actually painted? What percentage of the original ^{14}C should remain if it were legitimate?

64. **ARCHAEOLOGY** In 1389, Pierre d'Arcis, the bishop of Troyes, wrote a memo to the pope, accusing a colleague of passing off "a certain cloth, cunningly painted," as the burial shroud of Jesus Christ. Despite this early testimony of forgery, the image on the cloth is so compelling that many people regard it as a sacred relic. Known as the Shroud of Turin, the cloth was subjected to carbon dating in 1988. If authentic, the cloth would have been approximately 1,960 years old at that time.
a. If the Shroud were actually 1,960 years old, what percentage of the ^{14}C would have remained?
b. Scientists determined that 92.3% of the Shroud's original ^{14}C remained. Based on this information alone, what was the likely age of the Shroud in 1988?

65. POPULATION GROWTH   A community grows in such a way that t years from now, its population is $P(t)$ thousand, where

$$P(t) = 51 + 100 \ln(t + 3)$$

a. What is the population when $t = 0$?
b. How long does it take for the population to double its initial value?
c. What is the average rate of growth of the population over the first 10 years?

66. GROWTH OF BACTERIA   A medical student studying the growth of bacteria in a certain culture has compiled these data:

Number of minutes	0	20
Number of bacteria	6,000	9,000

Use these data to find an exponential function of the form $Q(t) = Q_0 e^{kt}$ expressing the number of bacteria in the culture as a function of time. How many bacteria are present after 1 hour?

67. RADIOLOGY   Radioactive iodine ^{133}I has a half-life of 20.9 hours. If injected into the bloodstream, the iodine accumulates in the thyroid gland.

a. After 24 hours, a medical technician scans a patient's thyroid gland to determine whether thyroid function is normal. If the thyroid has absorbed all the iodine, what percentage of the original amount should be detected?
b. A patient returns to the medical clinic 25 hours after having received an injection of ^{133}I. The medical technician scans the patient's thyroid gland and detects the presence of 41.3% of the original iodine. How much of the original ^{133}I remains in the rest of the patient's body?

68. ENTOMOLOGY   It is determined that the volume of the yolk of a house fly egg shrinks according to the formula $V(t) = 5e^{-1.3t}$ mm^3 (cubic millimeters), where t is the number of days from the time the egg is produced. The egg hatches after 4 days.

a. What is the volume of the yolk when the egg hatches?
b. Sketch the graph of the volume of the yolk over the time period $0 \le t \le 4$.
c. Find the half-life of the volume of the yolk, that is, the time it takes for the volume of the yolk to shrink to half its original size.

69. LEARNING   In an experiment designed to test short-term memory,* L. R. Peterson and M. J.

*L. R. Peterson and M. J. Peterson, "Short-Term Retention of Individual Verbal Items," *Journal of Experimental Psychology,* Vol. 58, 1959, pp. 193–198.

Peterson found that the probability $p(t)$ of a subject recalling a pattern of numbers and letters t seconds after being given the pattern is

$$p(t) = 0.89[0.01 + 0.99(0.85)^t]$$

a. What is the probability that the subject can recall the pattern immediately ($t = 0$)?
b. How much time passes before $p(t)$ drops to 0.5?
c. Sketch the graph of $p(t)$.

MISCELLANEOUS PROBLEMS

70. RADIOACTIVE DECAY   The amount of a certain radioactive substance remaining after t years is given by a function of the form $Q(t) = Q_0 e^{-0.003t}$. Find the half-life of the substance.

71. RADIOACTIVE DECAY   The half-life of radium is 1,690 years. How long will it take for a 50-gram sample of radium to be reduced to 5 grams?

72. AIR PRESSURE   The air pressure $f(s)$ at a height of s meters above sea level is given by

$$f(s) = e^{-0.000125s} \text{ atmospheres}$$

a. The atmospheric pressure outside an airplane is 0.25 atmosphere. How high is the plane?
b. A mountain climber decides she will wear an oxygen mask once she has reached an altitude of 7,000 meters. What is the atmospheric pressure at this altitude?

73. COOLING   Instant coffee is made by adding boiling water (212°F) to coffee mix. If the air temperature is 70°F, Newton's law of cooling says that after t minutes, the temperature of the coffee will be given by a function of the form $f(t) = 70 + Ae^{-kt}$. After cooling for 2 minutes, the coffee is still 15°F too hot to drink, but 2 minutes later it is just right. What is this "ideal" temperature for drinking?

74. SOUND LEVELS   A **decibel,** named for Alexander Graham Bell, is the smallest increase of the loudness of sound that is detectable by the human ear. In physics, it is shown that when two sounds of intensity I_1 and I_2 (watts/cm^3) occur, the difference in loudness is D decibels, where

$$D = 10 \log_{10}\left(\frac{I_1}{I_2}\right)$$

When sound is rated in relation to the threshold of human hearing ($I_0 = 10^{-12}$), the level of normal conversation is about 60 decibels, while a rock concert may be as loud as 110 decibels.

a. How much more intense is the rock concert than normal conversation?

b. The threshold of pain is reached at a sound level roughly 10 times as loud as a rock concert. What is the decibel level of the threshold of pain?

 75. SPY STORY Having ransomed his superior in Exercise 49 of Section 3.5, the spy returns home, only to learn that his best friend, Sigmund ("Siggy") Leiter, has been murdered. The police say that Siggy's body was discovered at 1 P.M. on Thursday, stuffed in a freezer where the temperature was 10°F. He is also told that the temperature of the corpse at the time of discovery was 40°F, and he remembers that t hours after death a body has temperature

$$T = T_a + (98.6 - T_a)(0.97)^t$$

where T_a is the air temperature adjacent to the body. The spy knows the dark deed was done by either Ernst Stavro Blohardt or André Scélérat. If Blohardt was in jail until noon on Wednesday and Scélérat was seen in Las Vegas from noon Wednesday until Friday, who "iced" Siggy, and when?

76. SEISMOLOGY On the Richter scale, the magnitude R of an earthquake of intensity I is given by

$$R = \frac{\ln I}{\ln 10}$$

a. Find the intensity of the 1906 San Francisco earthquake, which measured $R = 8.3$ on the Richter scale.

b. How much more intense was the San Francisco earthquake of 1906 than the devastating 2010 earthquake in Haiti, which measured $R = 7.0$?

77. SEISMOLOGY The magnitude formula for the Richter scale is

$$R = \frac{2}{3}\log_{10}\left(\frac{E}{E_0}\right)$$

where E is the energy released by the earthquake (in joules), and $E_0 = 10^{4.4}$ joules is the energy released by a small reference earthquake used as a standard of measurement.

a. The 1906 San Francisco earthquake released approximately 5.96×10^{16} joules of energy. What was its magnitude on the Richter scale?

b. How much energy was released by the Fukushima, Japan, earthquake of 2011, which measured 9.0 on the Richter scale?

78. In each case, use one of the laws of exponents to prove the indicated law of logarithms.

a. The quotient rule: $\ln \dfrac{u}{v} = \ln u - \ln v$

b. The power rule: $\ln u^r = r \ln u$

79. Show that the reflection of the point (a, b) in the line $y = x$ is (b, a). [*Hint:* Show that the line joining (a, b) and (b, a) is perpendicular to $y = x$ and that the distance from (a, b) to $y = x$ is the same as the distance from $y = x$ to (b, a).]

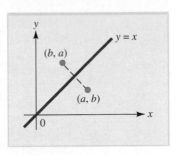

EXERCISE 79

80. Sketch the graph of $y = \log_b x$ for $0 < b < 1$ by reflecting the graph of $y = b^x$ in the line $y = x$. Then answer these questions:

a. Is the graph of $y = \log_b x$ rising or falling for $x > 0$?

b. Is the graph concave upward or concave downward for $x > 0$?

c. What are the intercepts of the graph? Does the graph have any horizontal or vertical asymptotes?

d. What can be said about

$$\lim_{x \to +\infty} \log_b x \quad \text{and} \quad \lim_{x \to 0^+} \log_b x?$$

81. Show that if y is a power function of x, so that $y = Cx^k$ where C and k are constants, then $\ln y$ is a linear function of $\ln x$. (*Hint:* Take the logarithm on both sides of the equation $y = Cx^k$.)

82. Use the graphing utility of your calculator to graph $y = 10^x$, $y = x$, and $y = \log_{10} x$ on the same coordinate axes (use $[-5, 5]1$ by $[-5, 5]1$). How are these graphs related?

 In Exercises 83 through 86 solve for x.

83. $x = \ln(3.42 \times 10^{-8.1})$

84. $3{,}500e^{0.31x} = \dfrac{e^{-3.5x}}{1 + 257e^{-1.1x}}$

85. $e^{0.113x} + 4.72 = 7.031 - x$

86. $\ln(x + 3) - \ln x = 5 \ln(x^2 - 4)$

87. Let a and b be any positive numbers other than 1.

a. Show that $(\log_a b)(\log_b a) = 1$.

b. Show that $\log_a x = \dfrac{\log_b x}{\log_b a}$ for any $x > 0$.

SECTION 4.3 Differentiation of Exponential and Logarithmic Functions

Learning Objectives

1. Differentiate exponential and logarithmic functions.
2. Examine applications involving exponential and logarithmic derivatives.
3. Explore logarithmic differentiation.

In the examples and exercises examined so far in this chapter we have seen how exponential functions can be used to model a variety of situations, ranging from compound interest to population growth and radioactive decay. To discuss rates of change and to determine extreme values in such situations, we need derivative formulas for exponential functions and their logarithmic counterparts. We obtain the formulas in this section and examine a few basic applications. Additional exponential and logarithmic models will be explored in Section 4.4. We begin by showing that the natural exponential function $f(x) = e^x$ has the remarkable property of being its own derivative.

The Derivative of e^x ■ For every real number x

$$\frac{d}{dx}(e^x) = e^x$$

To obtain this formula, let $f(x) = e^x$ and note that

$$f'(x) = \lim_{h \to 0} \frac{f(x + h) - f(x)}{h}$$

$$= \lim_{h \to 0} \frac{e^{x+h} - e^x}{h} \qquad e^{A+B} = e^A e^B$$

$$= \lim_{h \to 0} \frac{e^x e^h - e^x}{h} \qquad \text{factor } e^x \text{ out of the limit}$$

$$= e^x \lim_{h \to 0} \frac{e^h - 1}{h}$$

It can be shown that

$$\lim_{h \to 0} \frac{e^h - 1}{h} = 1$$

TABLE 4.2

h	$\dfrac{e^h - 1}{h}$
0.01	1.005017
0.001	1.000500
0.0001	1.000050
−0.00001	0.999995
−0.0001	0.999950

as indicated in Table 4.2. (A formal verification of this limit formula requires methods beyond the scope of this text.) Thus, we have

$$f'(x) = e^x \lim_{h \to 0} \frac{e^h - 1}{h}$$

$$= e^x(1)$$

$$= e^x$$

as claimed. This derivative formula is used in Example 4.3.1.

EXAMPLE 4.3.1 Differentiating Exponential Functions

Differentiate the following functions:

a. $f(x) = x^2 e^x$ **b.** $g(x) = \dfrac{x^3}{e^x + 2}$

Solution

a. Using the product rule, we find

$$f'(x) = x^2 \frac{d}{dx}(e^x) + \frac{d}{dx}(x^2)e^x \qquad \text{power rule and exponential rule}$$

$$= x^2 e^x + (2x)e^x \qquad \text{factor out } x \text{ and } e^x$$

$$= xe^x(x + 2)$$

b. To differentiate this function, we use the quotient rule:

$$g'(x) = \frac{(e^x + 2)\dfrac{d}{dx}(x^3) - x^3\dfrac{d}{dx}(e^x + 2)}{(e^x + 2)^2} \qquad \text{power rule and exponential rule}$$

$$= \frac{(e' + 2)[3x^2] - x^3[e^x + 0]}{(e^x + 2)^2} \qquad \text{factor } x^2 \text{ from the numerator and combine like terms}$$

$$= \frac{x^2(3e^x - xe^x + 6)}{(e^x + 2)^2}$$

The fact that e^x is its own derivative means that at each point (c, e^c) on the curve $y = e^x$, the slope is equal to e^c, the y coordinate of that point (Figure 4.7). This is one of the most important reasons for using e as the base for exponential functions in calculus.

EXPLORE!

Graph $y = e^x$ using a modified decimal window, $[-0.7, 8.7]1$ by $[-0.1, 6.1]1$. Trace the curve to any value of x, and determine the value of the derivative at this point. Observe how close the derivative value is to the y coordinate of the graph. Repeat this for several values of x.

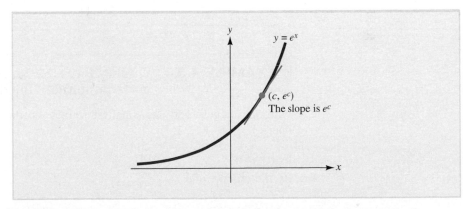

FIGURE 4.7   At each point $P(c, e^c)$ on the graph of $y = e^x$, the slope equals e^c.

By using the chain rule in conjunction with the differentiation formula

$$\frac{d}{dx}(e^x) = e^x$$

we obtain this formula for differentiating general exponential functions.

> **The Chain Rule for e^u** ■ If $u(x)$ is a differentiable function of x, then
>
> $$\frac{d}{dx}(e^{u(x)}) = e^{u(x)}\frac{du}{dx}$$

EXAMPLE 4.3.2   Using the Chain Rule with e^u

Differentiate the function $f(x) = e^{x^2+1}$.

Solution

Using the chain rule with $u = x^2 + 1$, we find

$$f'(x) = e^{x^2+1}\left[\frac{d}{dx}(x^2 + 1)\right] = 2xe^{x^2+1}$$

EXAMPLE 4.3.3   Combining the Chain and Quotient Rules

Differentiate the function

$$f(x) = \frac{e^{-3x}}{x^2 + 1}$$

Solution

Using the chain rule together with the quotient rule, you get

$$f'(x) = \frac{(x^2 + 1)(-3e^{-3x}) - (2x)e^{-3x}}{(x^2 + 1)^2}$$

$$= e^{-3x}\left[\frac{-3(x^2 + 1) - 2x}{(x^2 + 1)^2}\right] = e^{-3x}\left[\frac{-3x^2 - 2x - 3}{(x^2 + 1)^2}\right]$$

EXAMPLE 4.3.4   Finding Extreme Values of an Exponential Function

Find the largest and the smallest values of the function $f(x) = xe^{2x}$ on the interval $-1 \le x \le 1$.

Solution

By the product rule

$$f'(x) = x\frac{d}{dx}(e^{2x}) + e^{2x}\frac{d}{dx}(x) = x(2e^{2x}) + e^{2x}(1) = (2x + 1)e^{2x}$$

so $f'(x) = 0$ when

$$(2x + 1)e^{2x} = 0 \qquad e^{2x} > 0 \text{ for all } x$$
$$2x + 1 = 0$$
$$x = -\frac{1}{2}$$

Evaluating $f(x)$ at the critical number $x = -\dfrac{1}{2}$ and at the endpoints of the interval, $x = -1$ and $x = 1$, we find that

$$f(-1) = (-1)e^{-2} \approx -0.135$$

$$f\left(-\frac{1}{2}\right) = \left(-\frac{1}{2}\right)e^{-1} \approx -0.184 \quad \text{minimum}$$

$$f(1) = (1)e^{2} \approx 7.389 \qquad \text{maximum}$$

Thus, $f(x)$ has its largest value 7.389 at $x = 1$ and its smallest value -0.184 at $x = -\dfrac{1}{2}$.

Derivatives of Logarithmic Functions

Here is the derivative formula for the natural logarithmic function.

The Derivative of ln x ■ For all $x > 0$

$$\frac{d}{dx}(\ln x) = \frac{1}{x}$$

A proof using the definition of the derivative is outlined in Exercise 94. The formula can also be obtained as follows using implicit differentiation. Consider the equation

$$e^{\ln x} = x$$

Differentiating both sides with respect to x, we find that

$$\frac{d}{dx}[e^{\ln x}] = \frac{d}{dx}[x] \quad \text{chain rule for } e^{u}$$

$$e^{\ln x}\frac{d}{dx}[\ln x] = 1 \qquad e^{\ln x} = x$$

$$x\frac{d}{dx}[\ln x] = 1 \qquad \text{divide both sides by } x$$

so

$$\frac{d}{dx}[\ln x] = \frac{1}{x}$$

as claimed. The derivative formula for the natural logarithmic function is used in Examples 4.3.5 through 4.3.7.

EXAMPLE 4.3.5 Differentiating a Logarithmic Function

Differentiate the function $f(x) = x \ln x$.

Solution

Combine the product rule with the formula for the derivative of ln x to get

$$f'(x) = x\left(\frac{1}{x}\right) + \ln x = 1 + \ln x$$

EXAMPLE 4.3.6 Differentiating a Logarithmic Function

Differentiate $g(t) = (t + \ln t)^{3/2}$.

Solution

The function has the form $g(t) = u^{3/2}$, where $u = t + \ln t$, and by applying the general power rule, we find

$$g'(t) = \frac{d}{dt}(u^{3/2}) = \frac{3}{2}u^{1/2}\frac{du}{dt}$$

$$= \frac{3}{2}(t + \ln t)^{1/2}\frac{d}{dt}(t + \ln t)$$

$$= \frac{3}{2}(t + \ln t)^{1/2}\left(1 + \frac{1}{t}\right)$$

Using the rules for logarithms can simplify the differentiation of complicated expressions. In Example 4.3.7, we use the power rule for logarithms before differentiating.

EXAMPLE 4.3.7 Differentiating a Logarithmic Function

Differentiate $f(x) = \dfrac{\ln \sqrt[3]{x^2}}{x^4}$.

Solution

First, since $\sqrt[3]{x^2} = x^{2/3}$, the power rule for logarithms allows us to write

$$f(x) = \frac{\ln \sqrt[3]{x^2}}{x^4} = \frac{\ln x^{2/3}}{x^4} = \frac{\frac{2}{3}\ln x}{x^4}$$

Then, by the quotient rule, we find

$$f'(x) = \frac{2}{3}\left[\frac{x^4\dfrac{d}{dx}(\ln x) - \dfrac{d}{dx}(x^4)\ln x}{(x^4)^2}\right]$$

$$= \frac{2}{3}\left[\frac{x^4\left(\dfrac{1}{x}\right) - 4x^3 \ln x}{x^8}\right] \qquad \text{divide all terms by } x^3$$

$$= \frac{2}{3}\left[\frac{1 - 4\ln x}{x^5}\right]$$

If $f(x) = \ln u(x)$, where $u(x)$ is a differentiable function of x, then the chain rule yields the following formula for $f'(x)$.

The Chain Rule for ln u ■ If $u(x)$ is a differentiable function of x, then

$$\frac{d}{dx}[\ln u(x)] = \frac{1}{u(x)}\frac{du}{dx} \qquad \text{for } u(x) > 0$$

EXAMPLE 4.3.8   Using the Chain Rule with $\ln u$

Differentiate the function $f(x) = \ln(2x^3 + 1)$.

Solution

Here, we have $f(x) = \ln u$, where $u(x) = 2x^3 + 1$. Thus,

$$f'(x) = \frac{1}{u}\frac{du}{dx} = \frac{1}{2x^3 + 1}\frac{d}{dx}(2x^3 + 1)$$

$$= \frac{2(3x^2)}{2x^3 + 1} = \frac{6x^2}{2x^3 + 1}$$

EXAMPLE 4.3.9   Finding the Equation of a Tangent Line

Find an equation for the tangent line to the graph of $f(x) = x - \ln\sqrt{x}$ at the point where $x = 1$.

Solution

When $x = 1$, we have

$$y = f(1) = 1 - \ln(\sqrt{1}) = 1 - 0 = 1$$

so the point of tangency is $(1, 1)$. To find the slope of the tangent line at this point, we first write

$$f(x) = x - \ln\sqrt{x} = x - \ln x^{1/2} = x - \frac{1}{2}\ln x \qquad \text{power rule for logarithms}$$

and compute the derivative

$$f'(x) = 1 - \frac{1}{2}\left(\frac{1}{x}\right) = 1 - \frac{1}{2x}$$

Thus, the tangent line at the point $(1, 1)$ has slope

$$f'(1) = 1 - \frac{1}{2(1)} = \frac{1}{2}$$

so it has the equation

$$y - 1 = \frac{1}{2}(x - 1) \qquad \text{point-slope formula}$$

or equivalently,

$$y = \frac{1}{2}x + \frac{1}{2}$$

Formulas for differentiating exponential and logarithmic functions with bases other than e are similar to those obtained for $y = e^x$ and $y = \ln x$. These formulas are given in the following box.

Derivatives of b^x and $\log_b x$ for Base $b > 0$, $b \neq 1$

$$\frac{d}{dx}(b^x) = (\ln b)b^x \qquad \text{for all } x$$

and

$$\frac{d}{dx}(\log_b x) = \frac{1}{x \ln b} \qquad \text{for all } x > 0$$

For instance, to obtain the derivative formula for $y = \log_b x$, recall that

$$\log_b x = \frac{\ln x}{\ln b}$$

so we have

$$\frac{d}{dx}(\log_b x) = \frac{d}{dx}\left[\frac{\ln x}{\ln b}\right] = \frac{1}{\ln b}\frac{d}{dx}(\ln x)$$

$$= \frac{1}{x \ln b}$$

You are asked to obtain the derivative formula for $y = b^x$ in Exercise 89.

EXAMPLE 4.3.10   Using Derivative Rules for b^x and $\log_b x$

Differentiate each of the following functions:

   a. $f(x) = 5^{2x-3}$ **b.** $g(x) = (x^2 + \log_7 x)^4$

Solution

Using the chain rule, we find:

   a. $f'(x) = [(\ln 5)5^{2x-3}]\dfrac{d}{dx}[2x - 3] = (\ln 5)5^{2x-3}(2) = (2 \ln 5)5^{2x-3}$

   b. $g'(x) = 4(x^2 + \log_7 x)^3 \dfrac{d}{dx}[x^2 + \log_7 x]$ general power rule

$$= 4(x^2 + \log_7 x)^3\left[2x + \frac{1}{x \ln 7}\right]$$

Applications   Next, we will examine several applications of calculus involving exponential and logarithmic functions. In Example 4.3.11, we compute the marginal revenue for a commodity with logarithmic demand.

EXAMPLE 4.3.11   Estimating Additional Revenue Using Marginal Revenue

A manufacturer determines that x units of a particular luxury item will be sold when the price is $p(x) = 112 - x \ln(\sqrt{x})$ hundred dollars per unit.

   a. Find the revenue and marginal revenue as functions of the number of units produced and sold.

   b. Use marginal analysis to estimate the revenue obtained from producing the 12th unit. What is the actual revenue obtained from producing the 12th unit?

Solution

a. The revenue is

$$R(x) = xp(x) = x(112 - x\ln(\sqrt{x})) = 112x - x^2\left(\frac{1}{2}\ln x\right) \quad \text{power rule for logarithms}$$

hundred dollars, and the marginal revenue is

$$R'(x) = 112 - \frac{1}{2}\left[x^2\left(\frac{1}{x}\right) + (2x)\ln x\right] = 112 - \frac{1}{2}x - x\ln x$$

b. The additional revenue obtained from producing the 12th unit is estimated by the marginal revenue generated by the 11th unit, that is, by

$$R'(11) = 112 - \frac{1}{2}(11) - (11)\ln(11) \approx 80.12$$

So marginal analysis suggests that the manufacturer will receive approximately 80.12 hundred dollars ($8,012) in additional revenue by producing the 12th unit. To check the accuracy of this estimate, the actual additional revenue is

$$R(12) - R(11) = \left[112(12) - (12)^2\left(\frac{1}{2}\ln 12\right)\right] - \left[112(11) - (11)^2\left(\frac{1}{2}\ln 11\right)\right]$$

$$\approx 1{,}165.09 - 1{,}086.93 = 78.16$$

hundred dollars ($7,816).

In Example 4.3.12, we examine exponential demand and use marginal analysis to determine the price at which the revenue associated with such demand is maximized. Part of the example deals with the concept of elasticity of demand, which was introduced in Section 3.4.

EXAMPLE 4.3.12 Studying Elasticity of Demand

Edgar Morgan is the manager of a manufacturing firm. He determines that $D(p) = 5{,}000e^{-0.02p}$ units of the company's product will be demanded (sold) when the price is p dollars per unit.

a. Find the elasticity of demand for this commodity. For what values of p is the demand elastic, inelastic, and of unit elasticity?

b. If Edgar decides to increase the price 3% from the current level of $40, what change in demand should he expect?

c. Find the revenue $R(p)$ obtained by selling $q = D(p)$ units at p dollars per unit. For what value of p is the revenue maximized?

Solution

a. According to the formula derived in Section 3.4, the elasticity of demand is given by

$$E(p) = -\frac{p}{q}\frac{dq}{dp}$$

$$= -\left(\frac{p}{5{,}000e^{-0.02p}}\right)[5{,}000e^{-0.02p}(-0.02)]$$

$$= -\frac{p[5{,}000(-0.02)e^{-0.02p}]}{5{,}000e^{-0.02p}} = 0.02p$$

so

demand is of unit elasticity when $E(p) = 0.02p = 1$, that is, when $p = 50$

demand is elastic when $E(p) = 0.02p > 1$, or $p > 50$

demand is inelastic when $E(p) = 0.02p < 1$, or $p < 50$

The graph of the demand function, showing levels of elasticity, is displayed in Figure 4.8a.

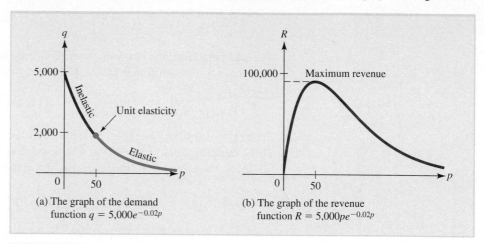

(a) The graph of the demand
function $q = 5,000e^{-0.02p}$

(b) The graph of the revenue
function $R = 5,000pe^{-0.02p}$

FIGURE 4.8   Demand and revenue curves for the commodity in Example 4.3.12.

b. When $p = 40$, the demand is

$$q(40) = 5,000e^{-0.02(40)} \approx 2,247 \text{ units}$$

and the elasticity of demand is

$$E(40) = 0.02(40) = 0.8$$

Thus, an increase of 1% in price from $p = \$40$ will result in a decrease in the quantity demanded by approximately 0.8%. Consequently, an increase of 3% in price, from \$40 to \$41.20, results in a decrease in demand of approximately $2,247[3(0.008)] = 54$ units, from 2,247 to 2,193 units.

c. The revenue function is

$$R(p) = pq = 5,000pe^{-0.02p}$$

for $p \geq 0$ (only nonnegative prices have economic meaning), with derivative

$$R'(p) = 5,000(-0.02pe^{-0.02p} + e^{-0.02p})$$
$$= 5,000(1 - 0.02p)e^{-0.02p}$$

Since $e^{-0.02p}$ is always positive, $R'(p) = 0$ if and only if

$$1 - 0.02p = 0 \qquad \text{or} \qquad p = \frac{1}{0.02} = 50$$

To verify that $p = 50$ actually gives the absolute maximum, note that

$$R''(p) = 5,000(0.0004p - 0.04)e^{-0.02p}$$

so

$$R''(50) = 5,000[0.0004(50) - 0.04]e^{-0.02(50)} \approx -37 < 0$$

Thus, the second derivative test tells you that the absolute maximum of $R(p)$ does indeed occur when $p = 50$ (Figure 4.8b).

Logarithmic Differentiation

Differentiating a function that involves products, quotients, or powers can often be simplified by first taking the logarithm of the function. This technique, called **logarithmic differentiation,** is illustrated in Example 4.3.13.

EXAMPLE 4.3.13 Using Logarithmic Differentiation

Differentiate the function $f(x) = \dfrac{\sqrt[3]{x+1}}{(1-3x)^4}$.

Solution

You could find the derivative using the quotient rule and the chain rule, but the resulting computation would be somewhat tedious. (Try it!)

A more efficient approach is to take logarithms of both sides of the expression for f and then use the quotient and power rules for logarithms:

$$\ln f(x) = \ln\left[\frac{\sqrt[3]{x+1}}{(1-3x)^4}\right] = \ln\sqrt[3]{(x+1)} - \ln(1-3x)^4$$

$$= \frac{1}{3}\ln(x+1) - 4\ln(1-3x)$$

Notice that by introducing the logarithm, you eliminate the quotient, the cube root, and the fourth power.

Now use the chain rule for logarithms to differentiate both sides of this equation to get

$$\frac{f'(x)}{f(x)} = \frac{1}{3}\frac{1}{x+1} - 4\left(\frac{-3}{1-3x}\right) = \frac{1}{3}\frac{1}{x+1} + \frac{12}{1-3x} \qquad \frac{d}{dx}\ln f(x) = \frac{f'(x)}{f(x)}$$

so that

$$f'(x) = f(x)\left[\frac{1}{3}\frac{1}{x+1} + \frac{12}{1-3x}\right]$$

$$= \left[\frac{\sqrt[3]{x+1}}{(1-3x)^4}\right]\left[\frac{1}{3}\frac{1}{x+1} + \frac{12}{1-3x}\right]$$

If $Q(x)$ is a differentiable function of x, note that

$$\frac{d}{dx}(\ln Q) = \frac{Q'(x)}{Q(x)}$$

where the ratio on the right is the relative rate of change of $Q(x)$. That is, *the relative rate of change of a quantity $Q(x)$ can be computed by finding the derivative of* $\ln Q$. This special kind of logarithmic differentiation can be used to simplify the computation of various growth rates, as illustrated in Example 4.3.14.

EXAMPLE 4.3.14 Finding a Relative Growth Rate

A country exports three goods, wheat W, steel S, and oil O. Suppose at a particular time $t = t_0$, the revenue (in billions of dollars) derived from each of these goods is

$$W(t_0) = 4 \qquad S(t_0) = 7 \qquad O(t_0) = 10$$

and that S is growing at 8%, O is growing at 15%, while W is declining at 3%. At what relative rate is total export revenue growing at this time?

Solution

Let $R = W + S + O$. At time $t = t_0$, we know that

$$R(t_0) = W(t_0) + S(t_0) + O(t_0) = 4 + 7 + 10 = 21$$

The percentage growth rates can be expressed as

$$\frac{W'(t_0)}{W(t_0)} = -0.03 \qquad \frac{S'(t_0)}{S(t_0)} = 0.08 \qquad \frac{O'(t_0)}{O(t_0)} = 0.15$$

so that

$$W'(t_0) = -0.03W(t_0) \qquad S'(t_0) = 0.08S(t_0) \qquad O'(t_0) = 0.15O(t_0)$$

Thus, at $t = t_0$, the relative rate of growth of R is

$$
\begin{aligned}
\frac{R'(t_0)}{R(t_0)} &= \frac{d(\ln R)}{dt} = \frac{d}{dt}\left[\ln(W + S + O)\right]\Big|_{t=t_0} \\
&= \frac{[W'(t_0) + S'(t_0) + O'(t_0)]}{[W(t_0) + S(t_0) + O(t_0)]} \\
&= \frac{-0.03W(t_0) + 0.08S(t_0) + 0.15O(t_0)}{W(t_0) + S(t_0) + O(t_0)} \\
&= \frac{-0.03W(t_0) + 0.08S(t_0) + 0.15O(t_0)}{R(t_0)} \\
&= \frac{-0.03W(t_0)}{R(t_0)} + \frac{0.08S(t_0)}{R(t_0)} + \frac{0.15O(t_0)}{R(t_0)} \\
&= \frac{-0.03(4)}{21} + \frac{0.08(7)}{21} + \frac{0.15(10)}{21} \\
&\approx 0.0924
\end{aligned}
$$

That is, at time $t = t_0$, the total revenue obtained from the three exported goods is increasing at the rate of 9.24%.

EXERCISES ■ 4.3

In Exercises 1 through 38, differentiate the given function.

1. $f(x) = e^{5x}$

2. $f(x) = 3e^{4x+1}$

3. $f(x) = xe^x$

4. $f(x) = \dfrac{e^x}{x}$

5. $f(x) = 30 + 10e^{-0.05x}$

6. $f(x) = e^{x^2+2x-1}$

7. $f(x) = (x^2 + 3x + 5)e^{6x}$

8. $f(x) = xe^{-x^2}$

9. $f(x) = (1 - 3e^x)^2$

10. $f(x) = \sqrt{1 + e^x}$

11. $f(x) = e^{\sqrt{3x}}$

12. $f(x) = e^{1/x}$

13. $f(x) = \ln x^3$

14. $f(x) = \ln 2x$

15. $f(x) = x^2 \ln x$

16. $f(x) = x \ln \sqrt{x}$

17. $f(x) = \sqrt[3]{e^{2x}}$

18. $f(x) = \dfrac{\ln x}{x}$

19. $f(x) = \ln\left(\dfrac{x+1}{x-1}\right)$

20. $f(x) = e^x \ln x$

21. $f(x) = e^{-2x} + x^3$

22. $f(t) = t^2 \ln \sqrt[3]{t}$

23. $g(s) = (e^s + s + 1)(2e^{-s} + s)$

24. $F(x) = \ln(2x^3 - 5x + 1)$

25. $h(t) = \dfrac{e^t + t}{\ln t}$

26. $g(u) = \ln(u^2 - 1)^3$

27. $f(x) = \dfrac{e^x + e^{-x}}{2}$

28. $h(x) = \dfrac{e^{-x}}{x^2}$

29. $f(t) = \sqrt{\ln t + t}$

30. $f(x) = \dfrac{e^x + e^{-x}}{e^x - e^{-x}}$

31. $f(x) = \ln(e^{-x} + x)$

32. $f(s) = e^{s + \ln s}$

33. $g(u) = \ln(u + \sqrt{u^2 + 1})$

34. $L(x) = \ln\left[\dfrac{x^2 + 2x - 3}{x^2 + 2x + 1}\right]$

35. $f(x) = \dfrac{2^x}{x}$

36. $f(x) = x^2 3^{x^2}$

37. $f(x) = x \log_{10} x$

38. $f(x) = \dfrac{\log_2 x}{\sqrt{x}}$

In Exercises 39 through 46, find the largest and smallest values of the given function over the prescribed closed, bounded interval.

39. $f(x) = e^{1-x}$ for $0 \le x \le 1$

40. $F(x) = e^{x^2 - 2x}$ for $0 \le x \le 2$

41. $f(x) = (3x - 1)e^{-x}$ for $0 \le x \le 2$

42. $g(x) = \dfrac{e^x}{2x + 1}$ for $0 \le x \le 1$

43. $g(t) = t^{3/2} e^{-2t}$ for $0 \le t \le 1$

44. $f(x) = e^{-2x} - e^{-4x}$ for $0 \le x \le 1$

45. $f(x) = \dfrac{\ln(x + 1)}{x + 1}$ for $0 \le x \le 2$

46. $h(s) = 2s \ln s - s^2$ for $0.5 \le s \le 2$

In Exercises 47 through 52, find an equation for the tangent line to $y = f(x)$ at the specified point.

47. $f(x) = xe^{-x}$; where $x = 0$

48. $f(x) = (x + 1)e^{-2x}$; where $x = 0$

49. $f(x) = \dfrac{e^{2x}}{x^2}$; where $x = 1$

50. $f(x) = \dfrac{\ln x}{x}$; where $x = 1$

51. $f(x) = x^2 \ln \sqrt{x}$; where $x = 1$

52. $f(x) = x - \ln x$; where $x = e$

In Exercises 53 through 56, find the second derivative of the given function.

53. $f(x) = e^{2x} + 2e^{-x}$

54. $f(x) = \ln(2x) + x^2$

55. $f(t) = t^2 \ln t$

56. $g(t) = t^2 e^{-t}$

In Exercises 57 through 64, use logarithmic differentiation to find the derivative $f'(x)$.

57. $f(x) = (2x + 3)^2 (x - 5x^2)^{1/2}$

58. $f(x) = x^2 e^{-x}(3x + 5)^3$

59. $f(x) = \dfrac{(x + 2)^5}{\sqrt[6]{3x - 5}}$

60. $f(x) = \sqrt[4]{\dfrac{2x + 1}{1 - 3x}}$

61. $f(x) = (x + 1)^3 (6 - x)^2 \sqrt[3]{2x + 1}$

62. $f(x) = \dfrac{e^{-3x}\sqrt{2x - 5}}{(6 - 5x)^4}$

63. $f(x) = 5^{x^2}$

64. $f(x) = \log_2(\sqrt{x})$

In Exercises 65 through 68, the demand function
$q = D(p)$ *for a particular commodity is given in terms*
of a price p per unit at which all q units can be sold.
In each case:

(a) *Find the elasticity of demand and determine*
the values of p for which the demand is
elastic, inelastic, and of unit elasticity.

(b) *If the price is increased by 2% from \$15,*
what is the approximate effect on demand?

(c) *Find the revenue R(p) obtained by selling q*
units at the unit price p. For what value of p
is revenue maximized?

65. $D(p) = 3{,}000e^{-0.04p}$

66. $D(p) = 10{,}000e^{-0.025p}$

67. $D(p) = 5{,}000(p + 11)e^{-0.1p}$

68. $D(p) = \dfrac{10{,}000e^{-p/10}}{p + 1}$

In Exercises 69 through 72, the cost C(x) of producing
x units of a particular commodity is given. In each
case:

(a) *Find the marginal cost C′(x).*

(b) *Determine the level of production x for which*
the average cost $A(x) = \dfrac{C(x)}{x}$ *is minimized.*

69. $C(x) = e^{0.2x}$

70. $C(x) = 100e^{0.01x}$

71. $C(x) = 12\sqrt{x}\, e^{x/10}$

72. $C(x) = x^2 + 10xe^{-x}$

BUSINESS AND ECONOMICS APPLIED PROBLEMS

73. DEPRECIATION   A certain industrial machine
depreciates so that its value after t years becomes
$Q(t) = 20{,}000e^{-0.4t}$ dollars.

a.   At what rate is the value of the machine changing
with respect to time after 5 years?

b.   At what percentage rate is the value of the
machine changing with respect to time after
t years? Does this percentage rate depend on
t or is it constant?

74. COMPOUND INTEREST   Money is deposited
into a bank offering interest at an annual rate of
6% compounded continuously. Find the percentage
rate of change of the balance with respect to time.

75. MARGINAL ANALYSIS   Jane McGee, the
mathematics editor at a major publishing house,
estimates that if x thousand complimentary copies
are distributed to professors, the first-year sales of
a certain new text will be $f(x) = 20 - 15e^{-0.2x}$
thousand copies. Currently, Jane is planning to
distribute 10,000 complimentary copies.

a.   Use marginal analysis to estimate the increase in
first-year sales that Jane should expect if 1,000
additional complimentary copies are distributed.

b.   Calculate the actual increase in first-year sales
that will result from the distribution of the
additional 1,000 complimentary copies. Is the
estimate in part (a) a good one?

76. PER CAPITA GROWTH   The national income
$I(t)$ of a particular country is increasing by 2.3%
per year, while the population $P(t)$ of the country
is decreasing at the annual rate of 1.75%. The per
capita income C is defined to be

$$C(t) = \frac{I(t)}{P(t)}$$

a.   Find the derivative of $\ln C(t)$.

b.   Use the result of part (a) to determine the
percentage rate of growth of per capita
income.

77. REVENUE GROWTH   A country exports
electronic components E and textiles T. Suppose
at a particular time $t = t_0$, the revenue (in billions
of dollars) derived from each of these goods is

$$E(t_0) = 11 \quad \text{and} \quad T(t_0) = 8$$

and that E is growing at 9%, while T is declining
at 2%. At what relative rate is total export revenue
$R = E + T$ changing at this time?

78. CONSUMER EXPENDITURE   The demand for
a certain commodity is $D(p) = 3{,}000e^{-0.01p}$ units
per month when the market price is p dollars
per unit.

a.   At what rate is the consumer expenditure
$E(p) = pD(p)$ changing with respect to
price p?

b.   At what price does consumer expenditure stop
increasing and begin to decrease?

c.   At what price does the *rate* of consumer
expenditure begin to increase? Interpret this
result.

79. ELASTICITY OF DEMAND Show that if $q(p)$ units of a quantity are demanded when the price is p, the price elasticity of demand is given by the ratio of derivatives

$$E(p) = \frac{-(\ln q)'}{(\ln p)'}$$

LIFE AND SOCIAL SCIENCE APPLIED PROBLEMS

80. ECOLOGY In a model developed by John Helms,* the water evaporation $E(T)$ for a ponderosa pine is given by

$$E(T) = 4.6e^{17.3T/(T+237)}$$

where T (degrees Celsius) is the surrounding air temperature.
  a. What is the rate of evaporation when $T = 30°C$?
  b. What is the percentage rate of evaporation? At what temperature does the percentage rate of evaporation first drop below 0.5?

81. LEARNING According to the Ebbinghaus model (recall Exercise 60, Section 4.1), the fraction $F(t)$ of subject matter you will remember from this course t months after the final exam can be estimated by the formula $F(t) = B + (1 - B)e^{-kt}$, where B is the fraction of the material you will never forget and k is a constant that depends on the quality of your memory.
  a. Find $F'(t)$, and explain what this derivative represents.
  b. Show that $F'(t)$ is proportional to $F - B$, and interpret this result. [*Hint:* What does $F - B$ represent in terms of what you remember?]
  c. Sketch the graph of $F(t)$ for the case where $B = 0.3$ and $k = 0.2$.

82. ALCOHOL ABUSE CONTROL Suppose the percentage of alcohol in the blood t hours after consumption is given by

$$C(t) = 0.12te^{-t/2}$$

  a. At what rate is the blood alcohol level changing at time t?
  b. How much time passes before the blood alcohol level begins to decrease?
  c. Suppose the legal limit for blood alcohol is 0.04%. How much time must pass before the blood alcohol reaches this level? At what rate is the blood alcohol level decreasing when it reaches the legal limit?

83. POPULATION GROWTH It is projected that t years from now, the population of a certain country will be $P(t) = 50e^{0.02t}$ million.
  a. At what rate will the population be changing with respect to time 10 years from now?
  b. At what percentage rate will the population be changing with respect to time t years from now? Does this percentage rate depend on t or is it constant?

84. POPULATION GROWTH It is projected that t years from now, the population of a certain town will be approximately $P(t)$ thousand people, where

$$P(t) = \frac{100}{1 + e^{-0.2t}}$$

At what rate will the population be changing 10 years from now? At what percentage rate will the population be changing at that time?

85. ENDANGERED SPECIES An international agency determines that the number of individuals of an endangered species that remain in the wild t years after a protection policy is instituted may be modeled by

$$N(t) = \frac{600}{1 + 3e^{-0.02t}}$$

  a. At what rate is the population changing at time t? When is the population increasing? When is it decreasing?
  b. When is the rate of change of the population increasing? When is it decreasing? Interpret your results.
  c. What happens to the population in the long run (as $t \to +\infty$)?

86. ENDANGERED SPECIES The agency in Exercise 85 studies a second endangered species but fails to receive funding to develop a policy of protection. The population of the species is modeled by

$$N(t) = \frac{30 + 500e^{-0.3t}}{1 + 5e^{-0.3t}}$$

  a. At what rate is the population changing at time t? When is the population increasing? When is it decreasing?
  b. When is the rate of change of the population increasing? When is it decreasing? Interpret your results.
  c. What happens to the population in the long run (as $t \to +\infty$)?

---

*John A. Helms, "Environmental Control of Net Photosynthesis in Naturally Growing Pinus Ponderosa Nets," *Ecology,* Winter, 1972, p. 92.

87. PLANT GROWTH   Two plants grow in such a way that t days after planting, they are $P_1(t)$ and $P_2(t)$ centimeters tall, respectively, where

$$P_1(t) = \frac{21}{1 + 25e^{-0.3t}} \quad \text{and} \quad P_2(t) = \frac{20}{1 + 17e^{-0.6t}}$$

   a. At what rate is the first plant growing at time $t = 10$ days? Is the rate of growth of the second plant increasing or decreasing at this time?

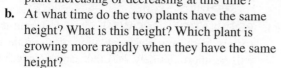

   b. At what time do the two plants have the same height? What is this height? Which plant is growing more rapidly when they have the same height?

88. LEARNING   In an experiment to test learning, a subject is confronted by a series of tasks, and it is found that t minutes after the experiment begins, the number of tasks successfully completed is

$$R(t) = \frac{15(1 - e^{-0.01t})}{1 + 1.5e^{-0.01t}}$$

   a. For what values of t is the learning function $R(t)$ increasing? For what values is it decreasing?

   b. When is the rate of change of the learning function $R(t)$ increasing? When is it decreasing? Interpret your results.

MISCELLANEOUS PROBLEMS

89. For base $b > 0$, $b \neq 1$, show that

$$\frac{d}{dx}(b^x) = (\ln b)b^x$$

   a. By using the fact that $b^x = e^{x \ln b}$.

   b. By using logarithmic differentiation.

90. COOLING   A cool drink is removed from a refrigerator on a hot summer day and placed in a room whose temperature is $30°$ Celsius. According to Newton's law of cooling, the temperature of the drink t minutes later is given by a function of the form $f(t) = 30 - Ae^{-kt}$. Show that the rate of change of the temperature of the drink with respect to time is proportional to the difference between the temperature of the room and that of the drink.

91. Use a numerical differentiation utility to find $f'(c)$, where $c = 0.65$ and

$$f(x) = \ln\left[\frac{\sqrt[3]{x + 1}}{(1 + 3x)^4}\right]$$

Then use a graphing utility to sketch the graph of $f(x)$ and to draw the tangent line at the point where $x = c$.

92. Repeat Exercise 91 with the function

$$f(x) = (3.7x^2 - 2x + 1)e^{-3x+2}$$

and $c = -2.17$.

93. A quantity grows so that $Q(t) = Q_0 \frac{e^{kt}}{t}$. Find the percentage rate of change of Q with respect to t.

94. Prove that the derivative of $f(x) = \ln x$ is $f'(x) = \dfrac{1}{x}$ by completing these steps.

   a. Show that the difference quotient of $f(x)$ can be expressed as

$$\frac{f(x + h) - f(x)}{h} = \ln\left(1 + \frac{h}{x}\right)^{1/h}$$

   b. Let $n = \dfrac{x}{h}$ so that $x = nh$. Show that the difference quotient in part (a) can be rewritten as

$$\ln\left[\left(1 + \frac{1}{n}\right)^n\right]^{1/x}$$

   c. Show that the limit of the expression in part (b) as $n \to \infty$ is $\ln e^{1/x} = \dfrac{1}{x}$. $\left[\text{\textit{Hint:} What is } \lim_{n \to \infty}\left(1 + \frac{1}{n}\right)^n ?\right]$

   d. Complete the proof by finding the limit of the difference quotient in part (a) as $h \to \infty$. [*Hint:* How is this related to the limit you found in part (c)?]

SECTION 4.4   Additional Applications; Exponential Models

Learning Objectives
1. Use exponential and logarithmic derivatives in curve sketching.
2. Examine applications involving exponential models.

Earlier in this chapter we saw how continuous compounding and radioactive decay can be modeled using exponential functions. In this section, we introduce several additional exponential models from a variety of areas such as business and economics, biology, psychology, demography, and sociology. We will begin with two examples illustrating the particular issues that arise when sketching exponential and logarithmic graphs.

Curve Sketching   As when graphing polynomials or rational functions, the key to graphing a function $f(x)$ involving e^x or $\ln x$ is to use the derivative $f'(x)$ to find intervals of increase and decrease and then use the second derivative $f''(x)$ to determine concavity.

EXAMPLE 4.4.1   Sketching a Logarithmic Curve

Sketch the graph of $f(x) = x^2 - 8 \ln x$.

Solution
The function $f(x)$ is defined only for $x > 0$. Its derivative is

$$f'(x) = 2x - \frac{8}{x} = \frac{2x^2 - 8}{x}$$

and $f'(x) = 0$ if and only if $2x^2 = 8$ or $x = 2$ (since $x > 0$). Testing the sign of $f'(x)$ for $0 < x < 2$ and for $x > 2$, you obtain the intervals of increase and decrease shown in the figure.

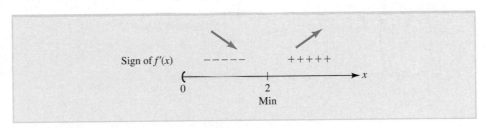

Notice that the arrow pattern indicates there is a relative minimum at $x = 2$, and since $f(2) = 2^2 - 8 \ln 2 \approx -1.5$, the minimum point is $(2, -1.5)$.

The second derivative

$$f''(x) = 2 + \frac{8}{x^2}$$

satisfies $f''(x) > 0$ for all $x > 0$, so the graph of $f(x)$ is always concave up and there are no inflection points.

EXPLORE!

Refer to Example 4.4.1. Graph $f(x)$ in regular style together with $f'(x)$ in bold, using the modified decimal window $[0, 4.7]1$ by $[-3.1, 3.1]1$. Locate the exact minimum point of the graph either using the minimum finding feature of the graphing calculator applied to $f(x)$ or the root-finding feature applied to $f'(x)$.

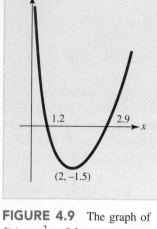

FIGURE 4.9 The graph of $f(x) = x^2 - 8 \ln x$.

Checking for asymptotes, it can be shown that

$$\lim_{x \to 0} (x^2 - 8 \ln x) = +\infty \qquad \text{and} \qquad \lim_{x \to +\infty} (x^2 - 8 \ln x) = +\infty$$

so the y axis ($x = 0$) is a vertical asymptote, but there is no horizontal asymptote. The x intercepts are found by using your calculator to solve the equation

$$x^2 - 8 \ln x = 0$$
$$x \approx 1.2 \qquad \text{and} \qquad x \approx 2.9$$

To summarize, the graph falls (down from the vertical asymptote) to the minimum at $(2, -1.5)$, after which it rises indefinitely, while maintaining a concave up shape. It passes through the x intercept $(1.2, 0)$ on the way down and through $(2.9, 0)$ as it rises. The graph is shown in Figure 4.9.

EXAMPLE 4.4.2 Sketching an Exponential Curve

Determine where the function

$$f(x) = \frac{1}{\sqrt{2\pi}} e^{-x^2/2}$$

is increasing and decreasing, and where its graph is concave up and concave down. Find the relative extrema and inflection points, and draw the graph.

Solution

The first derivative is

$$f'(x) = \frac{-x}{\sqrt{2\pi}} e^{-x^2/2}$$

Since $e^{-x^2/2}$ is always positive, $f'(x)$ is zero if and only if $x = 0$. Since $f(0) = \frac{1}{\sqrt{2\pi}} \approx 0.4$, the only critical point is $(0, 0.4)$. By the product rule, the second derivative is

$$f''(x) = \frac{x^2}{\sqrt{2\pi}} e^{-x^2/2} - \frac{1}{\sqrt{2\pi}} e^{-x^2/2} = \frac{1}{\sqrt{2\pi}} (x^2 - 1) e^{-x^2/2}$$

which is zero if $x = \pm 1$. Since

$$f(1) = \frac{e^{-1/2}}{\sqrt{2\pi}} \approx 0.24 \qquad \text{and} \qquad f(-1) = \frac{e^{-1/2}}{\sqrt{2\pi}} \approx 0.24$$

the potential inflection points are $(1, 0.24)$ and $(-1, 0.24)$.

Plot the critical points and then check the signs of the first and second derivatives on each of the intervals defined by the x coordinates of these points:

EXPLORE!

Refer to Example 4.4.2. Graph $f(x)$ in regular style along with $f''(x)$ in bold, using the modified decimal window $[-4.7, 4.7]1$ by $[-0.5, 0.5]0.1$. Find the x intercepts of $f''(x)$, and explain why they are the x coordinates of the inflection points of $f(x)$.

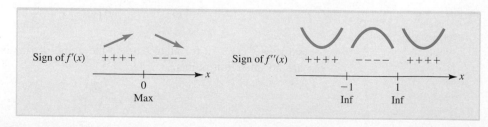

The arrow pattern indicates there is a relative maximum at $(0, 0.4)$, and since the concavity changes at $x = -1$ (from up to down) and at $x = 1$ (from down to up), both $(-1, 0.24)$ and $(1, 0.24)$ are inflection points.

Complete the graph, as shown in Figure 4.10, by connecting the key points with a curve of appropriate shape on each interval. Notice that the graph has no x intercepts since $e^{-x^2/2}$ is always positive and that the graph approaches the x axis as a horizontal asymptote since $e^{-x^2/2}$ approaches zero as $|x|$ increases without bound.

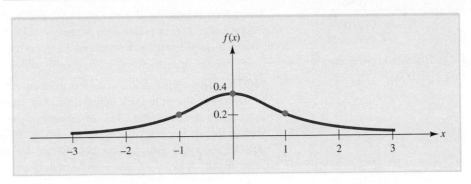

FIGURE 4.10   The standard normal density function: $f(x) = \dfrac{1}{\sqrt{2\pi}}e^{-x^2/2}$.

NOTE   The function $f(x) = \dfrac{1}{\sqrt{2\pi}}e^{-x^2/2}$ whose graph was sketched in Example 4.4.2 is known as the **standard normal probability density function** and plays a vital role in probability and statistics. The famous bell shape of the graph is used by physicists and social scientists to describe the distributions of IQ scores, measurements on large populations of living organisms, the velocity of a molecule in a gas, and numerous other important phenomena. ■

Optimal Holding Time

Suppose you own an asset whose value increases with time. The longer you hold the asset, the more it will be worth, but there may come a time when you could do better by selling the asset and reinvesting the proceeds. Economists determine the optimal time for selling by maximizing the present value of the asset in relation to the prevailing rate of interest, compounded continuously. The application of this criterion is illustrated in Example 4.4.3.

EXPLORE!

Store the function

$P(x) = 20{,}000*e^{\wedge}(\sqrt{x} - 0.07x)$

from Example 4.4.3 into Y1 of the equation editor of your graphing calculator. Find an appropriate window to view the graph, and determine its maximum value.

EXAMPLE 4.4.3   Finding Optimal Holding Time

Bartolo owns a parcel of land whose market value t years from now is estimated to be $V(t) = 20{,}000e^{\sqrt{t}}$ dollars. If the prevailing interest rate remains constant at 7% compounded continuously, when should Bartolo sell if his goal is to maximize the present value of the investment?

Solution

In t years, the market price of the land will be $V(t) = 20{,}000e^{\sqrt{t}}$. Using the present value formula obtained in Section 4.1, we find that the present value of this investment is

$$P(t) = V(t)e^{-0.07t} = 20{,}000e^{\sqrt{t}}e^{-0.07t} = 20{,}000e^{\sqrt{t} - 0.07t}$$

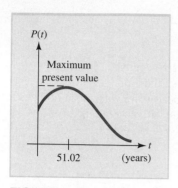

FIGURE 4.11 Present value $P(t) = 20{,}000e^{\sqrt{t}-0.07t}$.

The goal is to maximize $P(t)$ for $t \geq 0$. The derivative of P is

$$P'(t) = 20{,}000e^{\sqrt{t}-0.07t}\left(\frac{1}{2\sqrt{t}} - 0.07\right)$$

Thus, $P'(t)$ is undefined when $t = 0$ and $P'(t) = 0$ when

$$\frac{1}{2\sqrt{t}} - 0.07 = 0 \qquad \text{or} \qquad t = \left[\frac{1}{2(0.07)}\right]^2 \approx 51.02$$

Since $P'(t)$ is positive if $0 < t < 51.02$ and negative if $t > 51.02$, it follows that the present value $P(t)$ is maximized when $t = 51.02$ as shown in Figure 4.11. Therefore, Bartolo should sell his land about 51 years from now.

NOTE   The optimization criterion used in Example 4.4.3 is not the only way to determine the optimal holding time. For instance, you may decide to sell the asset when the percentage rate of growth of the asset's value just equals the prevailing rate of interest (7% in the example). Which criterion seems more reasonable to you? Actually, it does not matter, for the two criteria yield exactly the same result! A proof of this equivalence is outlined in Exercise 38. ■

Exponential Growth and Decay

A quantity $Q(t)$ is said to experience **exponential growth** if $Q(t) = Q_0 e^{kt}$ for $k > 0$ and **exponential decay** if $Q(t) = Q_0 e^{-kt}$. Many important quantities in business and economics, and the physical, social, and biological sciences can be modeled in terms of exponential growth and decay. The future value of a continuously compounded investment grows exponentially, as does population in the absence of restrictions. We discussed the decay of radioactive substances in Section 4.2. Other examples of exponential decay include present value of a continuously compounded investment, sales of certain commodities once advertising is discontinued, and the concentration of drug in a patient's bloodstream.

Exponential Growth and Decay ■ A quantity $Q(t)$ grows *exponentially* if $Q(t) = Q_0 e^{kt}$ for $k > 0$ and *decays exponentially* if $Q(t) = Q_0 e^{-kt}$ for $k > 0$.

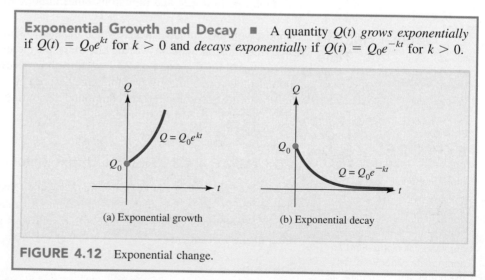

(a) Exponential growth

(b) Exponential decay

FIGURE 4.12 Exponential change.

Typical exponential growth and decay graphs are shown in Figure 4.12. It is customary to display such graphs only for $t \geq 0$ since the variable t typically represents time in the corresponding exponential model. Note that the graph of $Q(t) = Q_0 e^{kt}$ begins at Q_0 on the vertical axis since

$$Q(0) = Q_0 e^{k(0)} = Q_0$$

Note also that the graph of $Q(t) = Q_0 e^{kt}$ rises sharply since

$$Q'(t) = Q_0 k e^{kt} = kQ(t)$$

which means that $Q(t)$ always increases at a rate proportional to its current value, so the larger the value of Q, the larger the slope. The graph of $Q(t) = Q_0 e^{-kt}$ also begins at Q_0, but falls sharply, approaching the t axis asymptotically. Example 4.4.4 shows a business model involving exponential decay.

EXAMPLE 4.4.4 Finding Exponential Sales Rates

Lester Cummings is a marketing manager. He determines that sales of a particular commodity produced by his firm will decline exponentially once an advertising campaign for the commodity is terminated. Lester finds that the sales level is 21,000 units at the time the advertising ends and 19,000 units 5 weeks later.

a. Find S_0 and k so that $S(t) = S_0 e^{-kt}$ gives the sales level t weeks after the advertising campaign ends. What sales level should Lester expect 8 weeks after the advertising ends?

b. At what rate should Lester expect sales to be changing t weeks after the advertising ends? What is the percentage rate of change?

Solution

For simplicity, express sales S in terms of thousands of units.

a. We know that $S = 21$ when $t = 0$ and $S = 19$ when $t = 5$. Substituting $t = 0$ into the formula $S(t) = S_0 e^{-kt}$, we get

$$S(0) = 21 = S_0 e^{-k(0)} = S_0(1) = S_0$$

so $S_0 = 21$ and $S(t) = 21 e^{-kt}$ for all t. Substituting $S = 19$ and $t = 5$, we find that $19 = 21 e^{-k(5)}$ or

$$e^{-5k} = \frac{19}{21}$$

Taking the natural logarithm on each side of this equation, we find that

$$\ln(e^{-5k}) = \ln\frac{19}{21}$$

$$-5k = \ln\frac{19}{21}$$

$$k = -\frac{1}{5}\ln\left(\frac{19}{21}\right) \approx 0.02$$

Thus, for all $t > 0$, we have

$$S(t) = 21 e^{-0.02t}$$

For $t = 8$,

$$S(8) = 21 e^{-0.02(8)} \approx 17.9$$

so the model predicts sales of about 17,900 units 8 weeks after the advertising campaign ends.

b. The rate of change of sales is given by the derivative

$$S'(t) = 21[e^{-0.02t}(-0.02)] = -0.42e^{-0.02t}$$

and the percentage rate of change PR is

$$PR = \frac{100\,S'(t)}{S(t)} = \frac{100[-0.02(21)e^{-0.02t}]}{21e^{-0.02t}}$$

$$= -2$$

That is, sales are declining at the rate of 2% per week.

Notice that the percentage rate of change obtained in Example 4.4.4(b) is the same as k, expressed as the percentage. This is no accident, since for any function of the form $Q(t) = Q_0e^{rt}$, the percentage rate of change is

$$PR = \frac{100Q'(t)}{Q(t)} = \frac{100[Q_0e^{rt}(r)]}{Q_0e^{rt}} = 100r$$

For instance, an investment in an account that earns interest at an annual rate of 5% compounded continuously has a future value $B = Pe^{0.05t}$. Thus, the percentage rate of change of the future value is $100(0.05) = 5\%$, which is exactly what we would expect.

Learning Curves   The graph of a function of the form $Q(t) = B - Ae^{-kt}$, where A, B, and k are positive constants, is sometimes called a **learning curve**. The name arose when psychologists discovered that for $t \geq 0$, functions of this form often realistically model the relationship between the efficiency with which an individual performs a task and the amount of training time or experience the "learner" has had.

To sketch the graph of $Q(t) = B - Ae^{-kt}$ for $t \geq 0$, note that

$$Q'(t) = -Ae^{-kt}(-k) = Ake^{-kt}$$

and

$$Q''(t) = Ake^{-kt}(-k) = -Ak^2e^{-kt}$$

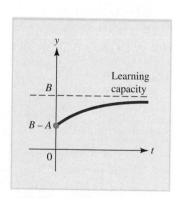

FIGURE 4.13   A learning curve $y = B - Ae^{-kt}$.

Since A and k are positive, it follows that $Q'(t) > 0$ and $Q''(t) < 0$ for all t, so the graph of $Q(t)$ is always rising and is always concave down. Furthermore, the vertical (Q axis) intercept is $Q(0) = B - A$, and $Q = B$ is a horizontal asymptote since

$$\lim_{t \to +\infty} Q(t) = \lim_{t \to +\infty} (B - Ae^{-kt}) = B - 0 = B$$

A graph with these features is sketched in Figure 4.13. The behavior of the learning curve as $t \to +\infty$ reflects the fact that in the long run, an individual approaches his or her learning capacity, and additional training time will result in only marginal improvement in performance efficiency. A typical learning model is examined in Example 4.4.5.

EXAMPLE 4.4.5   Examining a Learning Model

The rate at which a postal clerk can sort mail is a function of the clerk's experience. Suppose the postmaster of a large city estimates that after t months on the job, the average clerk can sort $Q(t) = 700 - 400e^{-0.5t}$ letters per hour.

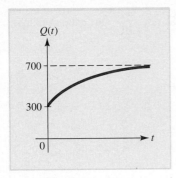

FIGURE 4.14 Worker efficiency $Q(t) = 700 - 400e^{-0.5t}$

a. How many letters can a new employee sort per hour?

b. How many letters can a clerk with 6 months' experience sort per hour?

c. Approximately how many letters will the average clerk ultimately be able to sort per hour?

Solution

a. The number of letters a new employee can sort per hour is

$$Q(0) = 700 - 400e^0 = 300$$

b. After 6 months, the average clerk can sort

$$Q(6) = 700 - 400e^{-0.5(6)} = 700 - 400e^{-3} \approx 680 \quad \text{letters per hour}$$

c. As t increases without bound, $Q(t)$ approaches 700. Hence, the average clerk will ultimately be able to sort approximately 700 letters per hour. The graph of the function $Q(t)$ is sketched in Figure 4.14.

Logistic Curves

The graph of a function of the form $Q(t) = \dfrac{B}{1 + Ae^{-Bkt}}$, where A, B, and k are positive constants, is called a **logistic curve**. A typical logistic curve is shown in Figure 4.15. Notice that it rises steeply like an exponential curve at first, and then turns over and flattens out, approaching a horizontal asymptote in much the same way as a learning curve. The asymptotic line represents a "saturation level" for the quantity represented by the logistic curve and is called the **carrying capacity** of the quantity. For instance, in population models, the carrying capacity represents the maximum number of individuals the environment can support, while in a logistic model for the spread of an epidemic, the carrying capacity is the total number of individuals susceptible to the disease, say, those who are unvaccinated or, at worst, the entire community.

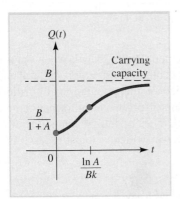

FIGURE 4.15 A logistic curve $Q(t) = \dfrac{B}{1 + Ae^{-Bkt}}$.

To sketch the graph of $Q(t) = \dfrac{B}{1 + Ae^{-Bkt}}$ for $t \geq 0$, note that

$$Q'(t) = \frac{AB^2 ke^{-Bkt}}{(1 + Ae^{-Bkt})^2}$$

and

$$Q''(t) = \frac{AB^3 k^2 e^{-Bkt}(-1 + Ae^{-Bkt})}{(1 + Ae^{-Bkt})^3}$$

Verify these formulas and also verify the fact that $Q'(t) > 0$ for all t, which means the graph of $Q(t)$ is always rising. The equation $Q''(t) = 0$ has exactly one solution, when

$$-1 + Ae^{-Bkt} = 0 \qquad \text{add 1 to both sides and divide by } A$$

$$e^{-Bkt} = \frac{1}{A} \qquad \text{take logarithms on both sides}$$

$$-Bkt = \ln\left(\frac{1}{A}\right) = -\ln A \qquad \text{divide both sides by } -Bk$$

$$t = \frac{\ln A}{Bk}$$

As shown in this diagram, there is an inflection point at $t = \dfrac{\ln A}{Bk}$ since the concavity changes there (from up to down).

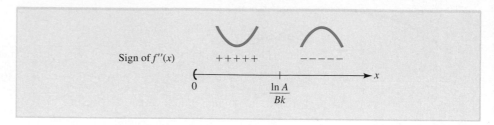

The vertical intercept of the logistic curve is

$$Q(0) = \frac{B}{1 + Ae^0} = \frac{B}{1 + A}$$

Since $Q(t)$ is defined for all $t \geq 0$, the logistic curve has no vertical asymptotes, but $y = B$ is a horizontal asymptote since

$$\lim_{t \to +\infty} Q(t) = \lim_{t \to +\infty} \frac{B}{1 + Ae^{-Bkt}} = \frac{B}{1 + A(0)} = B$$

To summarize, as shown in Figure 4.15, the logistic curve begins at $Q(0) = \dfrac{B}{1 + A}$, rises sharply (concave up) until it reaches the inflection point at $t = \dfrac{\ln A}{Bk}$, and then flattens out as it continues to rise (concave down) toward the horizontal asymptote $y = B$. Thus, B is the carrying capacity of the quantity $Q(t)$ represented by the logistic curve, and the inflection point at $t = \dfrac{\ln A}{Bk}$ can be interpreted as a point of *diminishing growth.*

Logistic curves often provide accurate models of population growth when environmental factors such as restricted living space, inadequate food supply, or urban pollution impose an upper bound on the possible size of the population. Logistic curves are also often used to describe the dissemination of privileged information or rumors in a community, where the restriction is the number of individuals susceptible to receiving such information. Here is an example in which a logistic curve is used to describe the spread of an infectious disease.

EXPLORE!

Graph the function in Example 4.4.6 using the window [0, 10]1 by [0, 25]5. Trace out this function for large values of *x*. What do you observe? Determine a graphic way to find out when 90% of the population has caught the disease.

EXAMPLE 4.4.6 Studying the Progress of an Epidemic

Public health records indicate that t weeks after the outbreak of a certain strain of influenza, approximately $Q(t) = \dfrac{20}{1 + 19e^{-1.2t}}$ thousand people had caught the disease.

a. How many people had the disease when it broke out? How many had it 2 weeks later?

b. At what time does the rate of infection begin to decline?

c. If the trend continues, approximately how many people will eventually contract the disease?

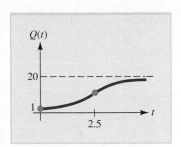

FIGURE 4.16   The spread of an epidemic

$$Q(t) = \frac{20}{1 + 19e^{-1.2t}}$$

Solution

a. Since $Q(0) = \dfrac{20}{1 + 19} = 1$, it follows that 1,000 people initially had the disease. When $t = 2$,

$$Q(2) = \frac{20}{1 + 19e^{-1.2(2)}} \approx 7.343$$

so about 7,343 had contracted the disease by the second week.

b. The rate of infection begins to decline at the inflection point on the graph of $Q(t)$. By comparing the given formula with the logistic formula $Q(t) = \dfrac{B}{1 + Ae^{-Bkt}}$, you find that $B = 20$, $A = 19$, and $Bk = 1.2$. Thus, the inflection point occurs when

$$t = \frac{\ln A}{Bk} = \frac{\ln 19}{1.2} \approx 2.454$$

so the epidemic begins to fade about 2.5 weeks after it starts.

c. Since $Q(t)$ approaches 20 as t increases without bound, it follows that approximately 20,000 people will eventually contract the disease. For reference, the graph is sketched in Figure 4.16.

Optimal Age for Reproduction

An organism such as Pacific salmon or bamboo that breeds only once during its lifetime is said to be *semelparous*. Biologists measure the per capita rate of reproduction of such an organism by the function*

$$R(x) = \frac{\ln[p(x)f(x)]}{x}$$

where $p(x)$ is the likelihood of an individual organism surviving to age x and $f(x)$ is the number of female births to an individual reproducing at age x. The larger the value of $R(x)$, the more offspring will be produced. Hence, the age at which $R(x)$ is maximized is regarded as the optimal age for reproduction.

EXAMPLE 4.4.7   Finding Optimal Age for Reproduction

Suppose that for a particular semelparous organism, the likelihood of an individual surviving to age x (years) is given by $p(x) = e^{-0.15x}$ and that the number of female births at age x is $f(x) = 3x^{0.85}$. What is the optimal age for reproduction?

Solution

The per capita rate of increase function for this model is

$$
\begin{aligned}
R(x) &= \frac{\ln[e^{-0.15x}(3x^{0.85})]}{x} && \text{product rule for logarithms}\\
&= x^{-1}(\ln e^{-0.15x} + \ln 3 + \ln x^{0.85}) && \text{power rule for logarithms}\\
&= x^{-1}(-0.15x + \ln 3 + 0.85 \ln x) && \text{distribute } x^{-1}\\
&= -0.15 + (\ln 3 + 0.85 \ln x)x^{-1}
\end{aligned}
$$

---

*Adapted from Claudia Neuhauser, *Calculus for Biology and Medicine*, Upper Saddle River, NJ: Prentice-Hall, 2000, p. 199 (Problem 22).

Differentiating $R(x)$ by the product rule, we find that

$$R'(x) = 0 + (\ln 3 + 0.85 \ln x)(-x^{-2}) + \left[0.85\left(\frac{1}{x}\right)\right]x^{-1}$$

$$= \frac{-\ln 3 - 0.85 \ln x + 0.85}{x^2}$$

Therefore, $R'(x) = 0$ when

$$-\ln 3 - 0.85 \ln x + 0.85 = 0$$

$$\ln x = \frac{0.85 - \ln 3}{0.85} \approx -0.2925$$

$$x = e^{-0.2925} \approx 0.7464$$

To show that this critical number corresponds to a maximum, we apply the second derivative test. We find that

$$R''(x) = \frac{1.7 \ln x + 2 \ln 3 - 2.55}{x^3}$$

(supply the details), and since

$$R''(0.7464) \approx -2.0441 < 0$$

it follows that the largest value of $R(x)$ occurs when $x \approx 0.7464$. Thus, the optimal age for an individual organism to reproduce is when it is 0.7464 years old (approximately 9 months).

EXERCISES ▪ 4.4

Each of the curves shown in Exercises 1 through 4 is the graph of one of the six functions listed here. In each case, match the given curve to the proper function.

$$f_1(x) = 2 - e^{-2x} \qquad f_2(x) = x \ln x^5$$

$$f_3(x) = \frac{2}{1 - e^{-x}} \qquad f_4(x) = \frac{2}{1 + e^{-x}}$$

$$f_5(x) = \frac{\ln x^5}{x} \qquad f_6(x) = (x - 1)e^{-2x}$$

2.

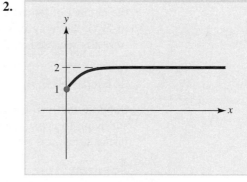

1.

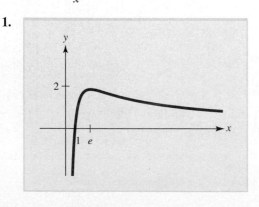

3.

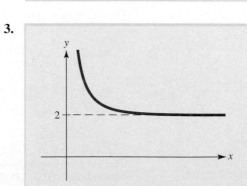

4.

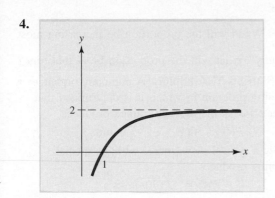

In Exercises 5 through 20, determine where the given function is increasing and decreasing and where its graph is concave upward and concave downward. Sketch the graph of the function. Show as many key features as possible (high and low points, points of inflection, vertical and horizontal asymptotes, intercepts, cusps, vertical tangents).

5. $f(t) = 2 + e^t$

6. $g(x) = 3 + e^{-x}$

7. $g(x) = 2 - 3e^x$

8. $f(t) = 3 - 2e^t$

9. $f(x) = \dfrac{2}{1 + 3e^{-2x}}$

10. $h(t) = \dfrac{2}{1 + 3e^{2t}}$

11. $f(x) = xe^x$

12. $f(x) = xe^{-x}$

13. $f(x) = xe^{2-x}$

14. $f(x) = e^{-x^2}$

15. $f(x) = x^2 e^{-x}$

16. $f(x) = e^x + e^{-x}$

17. $f(x) = \dfrac{6}{1 + e^{-x}}$

18. $f(x) = x - \ln x$ (for $x > 0$)

19. $f(x) = (\ln x)^2$ (for $x > 0$)

20. $f(x) = \dfrac{\ln x}{x}$ (for $x > 0$)

BUSINESS AND ECONOMICS APPLIED PROBLEMS

21. RETAIL SALES The total number of hamburgers sold by a national fast-food chain is growing exponentially. If 4 billion had been sold by 2005 and 12 billion had been sold by 2010, how many will have been sold by 2015?

22. SALES Once the initial publicity surrounding the release of a new book is over, sales of the hardcover edition tend to decrease exponentially. At the time publicity was discontinued, a certain book was experiencing sales of 25,000 copies per month. One month later, sales of the book had dropped to 10,000 copies per month. What will the sales be after one more month?

23. PRODUCT RELIABILITY A manufacturer of toys has found that the fraction of its plastic battery-operated toy oil tankers that sink in fewer than t days is approximately $f(t) = 1 - e^{-0.03t}$.
 a. Sketch this reliability function. What happens to the graph as t increases without bound?
 b. What fraction of the tankers can be expected to float for at least 10 days?
 c. What fraction of the tankers can be expected to sink between the 15th and 20th days?

24. DEPRECIATION When a certain industrial machine has become t years old, its resale value will be $V(t) = 4,800e^{-t/5} + 400$ dollars.
 a. Sketch the graph of $V(t)$. What happens to the value of the machine as t increases without bound?
 b. How much is the machine worth when it is new?
 c. How much will the machine be worth after 10 years?

25. INDUSTRIAL EFFICIENCY Devi Singh is an efficiency expert working for a major industrial firm. She determines that the daily output of a worker who has been on the job for t weeks is given by a function of the form $Q(t) = 40 - Ae^{-kt}$. Devi observes that a typical worker produces 20 units a day initially and after 1 week, produces 30 units a day. How many units per day should Devi expect a typical worker to produce after 3 weeks?

26. ADVERTISING When professors select texts for their courses, they usually choose from among the books already on their shelves. For this reason, most publishers send complimentary copies of new texts to professors teaching related courses. The mathematics editor at a major publishing house estimates that if x thousand complimentary copies are distributed, the first-year sales of a certain new mathematics text will be approximately $f(x) = 20 - 15e^{-0.2x}$ thousand copies.
 a. Sketch this sales function.

b. How many copies can the editor expect to sell in the first year if no complimentary copies are sent out?

c. How many copies can the editor expect to sell in the first year if 10,000 complimentary copies are sent out?

d. If the editor's estimate is correct, what is the most optimistic projection for the first-year sales of the text?

27. **MARGINAL ANALYSIS** The economics editor at a major publishing house estimates that if x thousand complimentary copies are distributed to professors, the first-year sales of a certain new text will be $f(x) = 15 - 20e^{-0.3x}$ thousand copies. Currently, the editor is planning to distribute 9,000 complimentary copies.

a. Use marginal analysis to estimate the increase in first-year sales that will result if 1,000 additional complimentary copies are distributed.

b. Calculate the actual increase in first-year sales that will result from the distribution of the additional 1,000 complimentary copies. Is the estimate in part (a) a good one?

28. **LABOR MANAGEMENT** Paul Edwards owns an electronics firm. He determines that when he employs x thousand people, the profit will be P million dollars, where

$$P(x) = \ln(4x + 1) + 3x - x^2$$

How many workers should Paul employ to maximize profit? What is the maximum profit?

29. **STOCK SPECULATION** In a classic paper on the theory of conflict,* L. F. Richardson claimed that the proportion p of a population advocating war or other aggressive action at a time t satisfies

$$p(t) = \frac{Ce^{kt}}{1 + Ce^{kt}}$$

where k and C are positive constants. Speculative day-trading in the stock market can be regarded as aggressive action. Suppose that initially, $\frac{1}{200}$ of total daily market volume is attributed to day-trading and that 4 weeks later, the proportion is

$\frac{1}{100}$. When will the proportion be increasing most rapidly? What will the proportion be at that time?

30. **BUSINESS TRAINING** A company organizes a training program in which it is determined that after t weeks the average trainee produces

$$P(t) = 50(1 - e^{-0.15t}) \text{ units}$$

while a typical new worker without special training produces

$$W(t) = \sqrt{150t} \text{ units}$$

a. How many units does the average trainee produce during the third week of the training period?

b. Explain how the function $F(t) = P(t) - W(t)$ can be used to evaluate the effectiveness of the training program. Is the program effective if it lasts just 5 weeks? What if it lasts at least 7 weeks? Explain your reasoning.

31. **OPTIMAL HOLDING TIME** Pilar owns a parcel of land whose value t years from now will be $V(t) = 8{,}000e^{\sqrt{t}}$ dollars. If the prevailing interest rate remains constant at 6% per year compounded continuously, when should Pilar sell the land to maximize its present value?

32. **OPTIMAL HOLDING TIME** Suppose your family owns a rare book whose value t years from now will be $V(t) = 200e^{\sqrt{2t}}$ dollars. If the prevailing interest rate remains constant at 6% per year compounded continuously, when will it be most advantageous for your family to sell the book and invest the proceeds?

33. **OPTIMAL HOLDING TIME** Aaron owns a stamp collection that is currently worth $1,200 and whose value increases linearly at the rate of $200 per year. If the prevailing interest rate remains constant at 8% per year compounded continuously, when will it be most advantageous for Aaron to sell the collection and invest the proceeds?

34. **OPTIMAL HOLDING TIME** Suppose you win a parcel of land whose market value t years from now is estimated to be $V(t) = 20{,}000te^{\sqrt{0.4t}}$ dollars. If the prevailing interest rate remains constant at 7% compounded continuously, when will it be most advantageous to sell the land? (Use a graphing utility and **ZOOM** and **TRACE** to make the required determination.)

---

*Richardson's original work appeared in *Generalized Foreign Politics*, Monograph Supplement 23 of the *British Journal of Psychology* (1939). His work was also featured in the text *Mathematical Models of Arms Control and Disarmament* by T. L. Saaty, New York: John Wiley & Sons, 1968.

35. ACCOUNTING The **double declining balance** formula in accounting is

$$V(t) = V_0\left(1 - \frac{2}{L}\right)^t$$

where $V(t)$ is the value after t years of an article that originally cost V_0 dollars and L is a constant, called the "useful life" of the article.

a. A refrigerator costs $875 and has a useful life of 8 years. What is its value after 5 years? What is its annual rate of depreciation?
b. In general, what is the percentage rate of change of $V(t)$?

36. MARGINAL ANALYSIS A manufacturer can produce digital recorders at a cost of $125 apiece and estimates that if they are sold for x dollars apiece, consumers will buy approximately $1,000e^{-0.02x}$ each week.

a. Express the profit P as a function of x. Sketch the graph of $P(x)$.
b. At what price should the manufacturer sell the recorders to maximize profit?

37. MARKET RESEARCH A company is trying to use television advertising to introduce as many people as possible to a new product in a large metropolitan area with 2 million possible viewers. A model for the number of people N (in millions) who are aware of the product after t days is found to be

$$N = 2(1 - e^{-0.037t})$$

Use a graphing utility to graph this function. What happens as $t \to +\infty$? (*Suggestion:* Set the range on your viewing screen to [0, 200]10 by [0, 3]1.)

38. OPTIMAL HOLDING TIME Let $V(t)$ be the value of an asset t years from now, and assume that the prevailing annual interest rate remains fixed at r (expressed as a decimal) compounded continuously.

a. Show that the present value of the asset $P(t) = V(t)e^{-rt}$ has a critical number where $V'(t) = V(t)r$. (Using economic arguments, it can be shown that the critical number corresponds to a maximum.)
b. Explain why the present value of $V(t)$ is maximized at a value of t where the percentage rate of change (expressed in decimal form) equals r.

39. POPULATION GROWTH It is estimated that the population of a certain country grows exponentially. If the population was 60 million in 1997 and 90 million in 2002, what will the population be in 2012?

40. POPULATION GROWTH It is estimated that t years from now, the population of a certain country will be $P(t) = \dfrac{20}{2 + 3e^{-0.06t}}$ million.

a. Sketch the graph of $P(t)$.
b. What is the current population?
c. What will the population be 50 years from now?
d. What will happen to the population in the long run?

41. CHILDHOOD LEARNING A psychologist measures a child's capability to learn and remember by the function

$$L(t) = \frac{\ln(t + 1)}{t + 1}$$

where t is the child's age in years, for $0 \le t \le 5$. Answer these questions about this model.

a. At what age does a child have the greatest learning capability?
b. At what age is a child's learning capability increasing most rapidly?

42. AEROBIC RATE The aerobic rating of a person x years old is modeled by the function

$$A(x) = \frac{110(\ln x - 2)}{x} \quad \text{for } x \ge 10$$

a. At what age is a person's aerobic rating largest?
b. At what age is a person's aerobic rating decreasing most rapidly?

43. OZONE DEPLETION It is known that fluorocarbons have the effect of depleting ozone in the upper atmosphere. Suppose it is found that the amount of original ozone Q_0 that remains after t years is given by

$$Q = Q_0e^{-0.0015t}$$

a. At what percentage rate is the ozone level decreasing at time t?
b. How many years will it take for 10% of the ozone to be depleted? At what percentage rate is the ozone level decreasing at this time?

44. SPREAD OF DISEASE In the Think About It essay at the end of Chapter 3, we examined several models associated with the AIDS epidemic. Using a data analysis program, we obtain the function

$$C(t) = 456 + 1{,}234te^{-0.137t}$$

as a model for the number of cases of AIDS reported t years after the base year of 1990.
   a. According to this model, in what year will the largest number of cases be reported? What will the maximum number of reported cases be?
   b. When will the number of reported cases be the same as the number reported in 1990?

45. SPREAD OF A RUMOR A traffic accident was witnessed by 10% of the residents of a small town, and 25% of the residents had heard about the accident 2 hours later. Suppose the number $N(t)$ of residents who had heard about the accident t hours after it occurred is given by a function of the form

$$N(t) = \frac{B}{1 + Ce^{-kt}}$$

where B is the population of the town and C and k are constants.
   a. Use this information to find C and k.
   b. How long does it take for half the residents of the town to know about the accident?
   c. When is the news about the accident spreading most rapidly? [*Hint:* See Example 4.4.6(b).]

46. SPREAD OF AN EPIDEMIC An epidemic spreads through a community so that t weeks after its outbreak, the number of people who have been infected is given by a function of the form $f(t) = \dfrac{B}{1 + Ce^{-kt}}$, where B is the number of residents in the community who are susceptible to the disease. If $\dfrac{1}{5}$ of the susceptible residents were infected initially and $\dfrac{1}{2}$ had been infected by the end of the fourth week, what fraction of the susceptible residents will have been infected by the end of the eighth week?

47. SPREAD OF AN EPIDEMIC Public health records indicate that t weeks after the outbreak of a certain strain of influenza, approximately $f(t) = \dfrac{2}{1 + 3e^{-0.8t}}$ thousand people had contracted the disease.

   a. Sketch the graph of $f(t)$.
   b. How many people had the disease initially?
   c. How many had caught the disease by the end of 3 weeks?
   d. If the trend continues, approximately how many people in all will contract the disease?

48. EFFECT OF A TOXIN A medical researcher determines that t hours from the time a toxin is introduced to a bacterial colony, the population will be

$$P(t) = 10{,}000(7 + 15e^{-0.05t} + te^{-0.05t})$$

   a. What is the population at the time the toxin is introduced?
   b. When does the maximum bacterial population occur? What is the maximum population?
   c. What eventually happens to the bacterial population as $t \rightarrow +\infty$?

49. OPTIMAL AGE FOR REPRODUCTION Suppose that for a particular semelparous organism, the likelihood of an individual surviving to age x years is $p(x) = e^{-0.2x}$ and that the number of female births to an individual at age x is $f(x) = 5x^{0.9}$. What is the ideal age for reproduction for an individual organism of this species? (See Example 4.4.7.)

50. OPTIMAL AGE FOR REPRODUCTION Suppose an environmental change affects the organism in Exercise 49 in such a way that an individual is only half as likely as before to survive to age x years. If the number of female births $f(x)$ remains the same, how is the ideal age for reproduction affected by this change?

51. RESPONSE TO STIMULATION According to **Hoorweg's law,** when a nerve is stimulated by discharges from an electrical condenser of capacitance C, the electric energy required to elicit a minimal response (a muscle contraction) is given by

$$E(C) = C\left(aR + \frac{b}{C}\right)^2$$

where a, b, and R are positive constants.
   a. For what value of C is the energy $E(C)$ minimized? How do you know you have found the minimum value? (Your answer will be in terms of a, b, and R.)
   b. Another model for E has the form

$$E(C) = mCe^{k/C}$$

where m and k are constants. What must m

and k be for the two models to have the same minimum value for the same value of C?

52. **WORLD POPULATION** According to a certain logistic model, the world's population (in billions) t years after 1960 will be approximately

$$P(t) = \frac{40}{1 + 12e^{-0.08t}}$$

   a. If this model is correct, at what rate was the world's population increasing with respect to time in the year 2000? At what *percentage* rate was the population increasing at this time?
   b. When will the population be growing most rapidly?
   c. Sketch the graph of $P(t)$. What feature occurs on the graph at the time found in part (b)? What happens to $P(t)$ in the long run?
   d. Do you think such a population model is reasonable? Why or why not?

53. **BUREAUCRATIC GROWTH** **Parkinson's law*** states that in any administrative department not actively at war, the staff will grow by about 6% per year, regardless of need.
   a. Parkinson applied his law to the size of the British Colonial Office. He noted that the Colonial Office had 1,139 staff members in 1947. How many staff members did the law predict for the year 1954? (The actual number was 1,661.)
   b. Based on Parkinson's law, how long should it take for a staff to double in size?
   c. Read about Parkinson's law, and write a paragraph about how it applies to business practices such as project scheduling.

54. **POPULATION GROWTH** Based on the estimate that there are 10 billion acres of arable land on the earth and that each acre can produce enough food to feed 4 people, some demographers believe that the earth can support a population of no more than 40 billion people. The population of the earth was approximately 3 billion in 1960 and 4 billion in 1975. If the population of the earth were growing exponentially, when would it reach the theoretical limit of 40 billion?

55. **FISHERY MANAGEMENT** Corwin Lewis, the manager of a fishery, determines that t days after 1,000 fish of a particular species are released into a pond, the average weight of an individual fish will be $w(t)$ pounds and the proportion of the fish still alive after t days will be $p(t)$, where

$$w(t) = \frac{10}{1 + 15e^{-0.05t}} \quad \text{and} \quad p(t) = e^{-0.01t}$$

   a. The expected yield $E(t)$ from harvesting after t days is the total weight of the fish that are still alive. Express $E(t)$ in terms of $w(t)$ and $p(t)$.

   b. When should Corwin harvest the fish to maximize the expected yield $E(t)$? What is the maximum expected yield?
   c. Sketch the yield curve $y = E(t)$.

56. **FISHERY MANAGEMENT** The manager of a fishery determines that t weeks after 3,000 fish of a particular species are hatched, the average weight of an individual fish will be $w(t) = 0.8te^{-0.05t}$ pounds, for $0 \leq t \leq 20$. Moreover, the proportion of the fish that will still be alive after t weeks is estimated to be

$$p(t) = \frac{10}{10 + t}$$

   a. The expected yield $E(t)$ from harvesting after t weeks is the total weight of the fish that are still alive. Express $E(t)$ in terms of $w(t)$ and $p(t)$.
   b. For what value of t is the expected yield $E(t)$ the largest? What is the maximum expected yield?
   c. Sketch the yield curve $y = E(t)$ for $0 \leq t \leq 20$.

57. **RECALL FROM MEMORY** Psychologists believe that when a person is asked to recall a set of facts, the number of facts recalled after t minutes is given by a function of the form $Q(t) = A(1 - e^{-kt})$, where k is a positive constant and A is the total number of relevant facts in the person's memory.
   a. Sketch the graph of $Q(t)$.
   b. What happens to the graph as t increases without bound? Explain this behavior in practical terms.

58. **LEARNING THEORY** In a learning model proposed by C. L. Hull, the habit strength H of an individual is related to the number r of reinforcements by the equation

$$H(r) = M(1 - e^{-kr})$$

---

*C. N. Parkinson, *Parkinson's Law,* Boston, MA: Houghton-Mifflin, 1957.

a. Sketch the graph of $H(r)$. What happens to $H(r)$ as $r \rightarrow +\infty$?

b. Show that if the number of reinforcements is doubled from r to $2r$, the habit strength is multiplied by $1 + e^{-kr}$.

59. SPREAD OF AN EPIDEMIC An epidemic spreads throughout a community so that t weeks after its outbreak, the number of residents who have been infected is given by a function of the form $f(t) = \dfrac{A}{1 + Ce^{-kt}}$, where A is the total number of susceptible residents. Show that the epidemic is spreading most rapidly when half of the susceptible residents have been infected.

60. CONCENTRATION OF DRUG The concentration of a certain drug in a patient's bloodstream t hours after being administered orally is assumed to be given by the function $C(t) = Ate^{-kt}$, where C is measured in micrograms of drug per milliliter of blood. Monitoring devices indicate that a maximum concentration of 5 occurs 20 minutes after the drug is administered.

a. Use this information to find A and k.

b. What is the drug concentration in the patient's blood after 1 hour?

c. At what time after the maximum concentration occurs will the concentration be half the maximum?

d. If you double the time in part (c), will the resulting concentration of drug be $\dfrac{1}{4}$ the maximum? Explain.

61. DRUG CONCENTRATION In a classic paper,* E. Heinz modeled the concentration $y(t)$ of a drug injected into the body intramuscularly by the function

$$y(t) = \frac{c}{b - a}(e^{-at} - e^{-bt}) \qquad t \geq 0$$

where t is the number of hours after the injection and a, b, and c are positive constants, with $b > a$.

a. When does the maximum concentration occur? What happens to the concentration in the long run?

b. Sketch the graph of $y(t)$.

c. Write a paragraph on the reliability of the Heinz model. In particular, is it more reliable when t is small or large? You may wish to begin your research with the article cited in this exercise.

62. CONCENTRATION OF DRUG A function of the form $C(t) = Ate^{-kt}$, where A and k are positive constants, is called a **surge function** and is sometimes used to model the concentration of drug in a patient's bloodstream t hours after the drug is administered. Assume $t \geq 0$.

a. Find $C'(t)$, and determine the time interval where the drug concentration is increasing and where it is decreasing. For what value of t is the concentration maximized? What is the maximum concentration? (Your answers will be in terms of A and k.)

b. Find $C''(t)$, and determine intervals of concavity for the graph of $C(t)$. Find all points of inflection, and explain what is happening to the rate of change of drug concentration at the times that correspond to the inflection points.

c. Sketch the graph of $C(t) = te^{-kt}$ for $k = 0.2$, $k = 0.5$, $k = 1.0$, and $k = 2.0$. Describe how the shape of the graph changes as k increases.

63. CORPORATE ORGANIZATION A **Gompertz curve** is the graph of a function of the general form

$$N(t) = CA^{B^t}$$

where A, B, and C are constants. Such curves are used by psychologists and others to describe such things as learning and growth within an organization.[†]

a. Suppose the personnel director of a large corporation determines that after t years, the corporation will have

$$N(t) = 500(0.03)^{(0.4)^t}$$

employees. How many employees are there originally (at time $t = 0$)? How many are there after 5 years? When will there be 300 employees? How many employees will there be in the long run?

---

*E. Heinz, "Probleme bei der Diffusion kleiner Substanzmengen innerhalb des menschlichen Körpers," *Biochem. Z.,* Vol. 319, 1949, pp. 482–492.

[†]A discussion of models of Gompertz curves and other models of differential growth can be found in an article by Roger V. Jean titled, "Differential Growth, Huxley's Allometric Formula, and Sigmoid Growth," *UMAP Modules 1983: Tools for Teaching,* Lexington, MA: Consortium for Mathematics and Its Applications, Inc., 1984.

b. Sketch the graph of $N(t)$. Then, on the same graph sketch the graph of the Gompertz function

$$F(t) = 500(0.03)^{-(0.4)^{-t}}$$

How would you describe the relationship between the two graphs?

64. MORTALITY RATES An actuary measures the probability that a person in a certain population will die at age x by the formula

$$P(x) = \lambda^2 x e^{-\lambda x}$$

where λ is a constant such that $0 < \lambda < e$.

a. Find the maximum value of $P(x)$ in terms of λ.

b. Sketch the graph of $P(x)$.

c. Read an article about actuarial formulas of this kind. Write a paragraph on what is represented by the number λ.

65. CANCER RESEARCH In Exercise 60, Section 2.3, you were given a function to model the production of blood cells. Such models are useful in the study of leukemia and other so-called *dynamical diseases* in which certain physiological systems begin to behave erratically. An alternative model* for blood cell production developed by A. Lasota involves the exponential production function

$$p(x) = A x^s e^{-sx/r}$$

where A, s, and r are positive constants and x is the number of granulocytes (a type of white blood cell) present.

a. Find the blood cell level x that maximizes the production function $p(x)$. How do you know the optimum level is a maximum?

b. If $s > 1$, show that the graph of $p(x)$ has two inflection points. Sketch the graph. Give a physical interpretation of the inflection points.

c. Sketch the graph of $p(x)$ in the case where $0 \le s \le 1$. What is different in this case?

d. Read an article on mathematical methods in the study of dynamical diseases, and write a paragraph on these methods. A good place to start is the article referenced in this exercise.

MISCELLANEOUS PROBLEMS

66. STRUCTURAL DESIGN When a chain, a telephone line, or a TV cable is strung between supports, the curve it forms is called a **catenary.** A typical catenary curve is

$$y = 0.125(e^{4x} + e^{-4x})$$

a. Sketch this catenary curve.

b. Catenary curves are important in architecture. Read an article on the Gateway Arch to the West in St. Louis, Missouri, and write a paragraph on the use of the catenary shape in its design.[†]

67. COOLING A hot drink is taken outside on a cold winter day when the air temperature is $-5°C$. According to a principle of physics called Newton's law of cooling, the temperature T (in degrees Celsius) of the drink t minutes after being taken outside is given by a function of the form

$$T(t) = -5 + A e^{-kt}$$

where A and k are constants. Suppose the temperature of the drink is $80°C$ when it is taken outside and 20 minutes later is $25°C$.

a. Use this information to determine A and k.

b. Sketch the graph of the temperature function $T(t)$. What happens to the temperature as t increases indefinitely ($t \to +\infty$)?

c. What will the temperature be after 30 minutes?

d. When will the temperature reach $0°C$?

68. Use the graphing utility of your calculator to sketch the graph of $f(x) = x(e^{-x} + e^{-2x})$. Use **ZOOM** and **TRACE** to find the largest value of $f(x)$. What happens to $f(x)$ as $x \to +\infty$?

69. PROBABILITY DENSITY FUNCTION The general probability density function has the form

$$f(x) = \frac{1}{\sigma \sqrt{2\pi}} e^{-(x-\mu)^2/2\sigma^2}$$

where μ and σ are constants, with $\sigma > 0$.

a. Show that $f(x)$ has an absolute maximum at $x = \mu$ and inflection points at $x = \mu + \sigma$ and $x = \mu - \sigma$.

b. Show that $f(\mu + c) = f(\mu - c)$ for every number c. What does this tell you about the graph of $f(x)$?

---

*W. B. Gearhart and M. Martelli, "A Blood Cell Population Model, Dynamical Diseases, and Chaos," *UMAP Modules 1990: Tools for Teaching,* Arlington, MA: Consortium for Mathematics and Its Applications, Inc., 1991.

[†]A good place to start is the article by William V. Thayer, "The St. Louis Arch Problem," *UMAP Modules 1983: Tools for Teaching,* Lexington, MA: Consortium for Mathematics and Its Applications, Inc., 1984.

CHAPTER SUMMARY

Important Terms, Symbols, and Formulas

Exponential function $y = b^x$ (299)

Properties of $y = b^x$ ($b > 0$, $b \neq 1$):   (301)

 It is defined and continuous for all x.

 The x axis is a horizontal asymptote.

 The y intercept is $(0, 1)$.

 If $b > 1$, $\lim\limits_{x \to -\infty} b^x = 0$ and $\lim\limits_{x \to +\infty} b^x = +\infty$.

 If $0 < b < 1$, $\lim\limits_{x \to +\infty} b^x = 0$ and $\lim\limits_{x \to -\infty} b^x = +\infty$.

 For all x, it is increasing if $b > 1$ and decreasing if $0 < b < 1$.

Exponential rules: (302)

$$b^x = b^y \quad \text{if and only if } x = y$$
$$b^x b^y = b^{x+y}$$
$$\frac{b^x}{b^y} = b^{x-y}$$
$$(b^x)^y = b^{xy}$$
$$b^0 = 1$$

The natural exponential base e: (303)

$$e = \lim_{n \to +\infty}\left(1 + \frac{1}{n}\right)^n = 2.71828\ldots$$

Logarithmic function $y = \log_b x$ (314)

Logarithmic rules: (315)

$$\log_b u = \log_b v \quad \text{if and only if } u = v$$
$$\log_b uv = \log_b u + \log_b v$$
$$\log_b\left(\frac{u}{v}\right) = \log_b u - \log_b v$$
$$\log_b u^r = r \log_b u$$
$$\log_b 1 = 0 \text{ and } \log_b b = 1$$
$$\log_b b^u = u$$

Properties of $y = \log_b x$ ($b > 0$, $b \neq 1$):   (318)

 It is defined and continuous for all $x > 0$.

 The y axis is a vertical asymptote.

 The x intercept is $(1, 0)$.

 If $b > 1$, $\lim\limits_{x \to +\infty} \log_b x = +\infty$ and $\lim\limits_{x \to 0^+} \log_b x = -\infty$.

 If $0 < b < 1$, $\lim\limits_{x \to +\infty} \log_b x = -\infty$ and

 $\lim\limits_{x \to 0^+} \log_b x = +\infty$.

 For all $x > 0$, it is increasing if $b > 1$ and decreasing if $0 < b < 1$.

Graphs of $y = b^x$ and $y = \log_b x$ ($b > 1$):   (317)

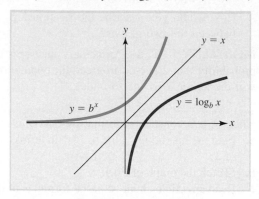

Natural exponential and logarithmic functions:

 $y = e^x$ (303)

 $y = \ln x$ (318)

Inversion formulas: (319)

 $e^{\ln x} = x, \quad \text{for } x > 0$

 $\ln e^x = x \quad \text{for all } x$

Conversion formula for logarithms: (321)

$$\log_b a = \frac{\ln a}{\ln b}$$

Derivatives of exponential functions: (330 and 332)

$$\frac{d}{dx}(e^x) = e^x \quad \text{and} \quad \frac{d}{dx}[e^{u(x)}] = e^{u(x)}\frac{du}{dx}$$

Derivatives of logarithmic functions: (333 and 334)

$$\frac{d}{dx}(\ln x) = \frac{1}{x} \quad \text{and} \quad \frac{d}{dx}[\ln u(x)] = \frac{1}{u(x)}\frac{du}{dx}$$

Logarithmic differentiation   (339)

Applications

Compounding interest k times per year at an annual interest rate r for t years:

 Future value of P dollars is $B = P\left(1 + \dfrac{r}{k}\right)^{kt}$.   (305)

 Present value of B dollars is $P = B\left(1 + \dfrac{r}{k}\right)^{-kt}$.   (307)

 Effective interest rate is $r_e = \left(1 + \dfrac{r}{k}\right)^k - 1$.   (308)

Continuous compounding at an annual interest rate r for t years:

 Future value of P dollars is $B = Pe^{rt}$.   (305)

 Present value of B dollars is $P = Be^{-rt}$.   (307)

 Effective interest rate is $r_e = e^r - 1$.   (308)

Optimal holding time   (347)
Exponential growth   (348)

Carbon dating   (324)
Learning curve $y = B - Ae^{-kt}$ (350)

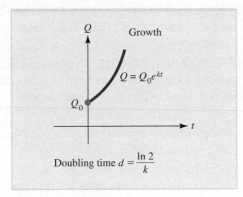

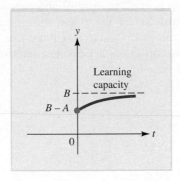

Exponential decay   (348)

Logistic curve $y = \dfrac{B}{1 + Ae^{-Bkt}}$ (351)

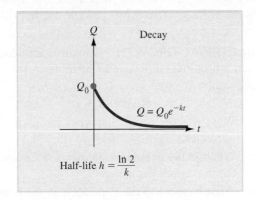

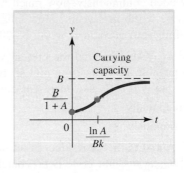

Checkup for Chapter 4

1. Evaluate these expressions:

   a. $\dfrac{(3^{-2})(9^2)}{(27)^{2/3}}$

   b. $\sqrt[3]{(25)^{1.5}\left(\dfrac{8}{27}\right)}$

   c. $\log_2 4 + \log_4 16^{-1}$

   d. $\left(\dfrac{8}{27}\right)^{-2/3}\left(\dfrac{16}{81}\right)^{3/2}$

2. Simplify these expressions:

   a. $(9x^4y^2)^{3/2}$

   b. $(3x^2y^{4/3})^{-1/2}$

   c. $\left(\dfrac{y}{x}\right)^{3/2}\left(\dfrac{x^{2/3}}{y^{1/6}}\right)^2$

   d. $\left(\dfrac{x^{0.2}y^{-1.2}}{x^{1.5}y^{0.4}}\right)^5$

3. Find all real numbers x that satisfy these equations.

   a. $4^{2x-x^2} = \dfrac{1}{64}$

   b. $e^{1/x} = 4$

   c. $\log_4 x^2 = 2$

   d. $\dfrac{25}{1 + 2e^{-0.5t}} = 3$

4. In each case, find the derivative $\dfrac{dy}{dx}$. (In some cases, it may help to use logarithmic differentiation.)

   a. $y = \dfrac{e^x}{x^2 - 3x}$

   b. $y = \ln(x^3 + 2x^2 - 3x)$

   c. $y = x^3 \ln x$

   d. $y = \dfrac{e^{-2x}(2x - 1)^3}{1 - x^2}$

5. In each of these cases, determine where the given function is increasing and decreasing and where its graph is concave upward and concave downward. Sketch the graph, showing as many key features as possible (high and low points, points of inflection, asymptotes, intercepts, cusps, vertical tangents).

   a. $y = x^2 e^{-x}$

   b. $y = \dfrac{\ln \sqrt{x}}{x^2}$

   c. $y = \ln(\sqrt{x} - x)^2$

   d. $y = \dfrac{4}{1 + e^{-x}}$

6. **FUTURE VALUE** If you invest $2,000 at 5% compounded continuously, how much will your account be worth in 3 years? How long does it take before your account is worth $3,000?

7. **PRESENT VALUE** Find the present value of $8,000 payable 10 years from now if the annual interest rate is 6.25% and interest is compounded:

   a. Semiannually

   b. Continuously

8. **PRICE ANALYSIS** A product is introduced and t months later, its unit price is $p(t)$ hundred dollars, where

$$p = \frac{\ln(t + 1)}{t + 1} + 5$$

a. For what values of t is the price increasing? When is it decreasing?

b. When is the price decreasing most rapidly?

c. What happens to the price in the long run (as $t \to +\infty$)?

9. **MAXIMIZING REVENUE** It is determined that q units of a commodity can be sold when the price is p hundred dollars per unit, where

$$q(p) = 1,000(p + 2)e^{-p}$$

a. Verify that the demand function $q(p)$ decreases as p increases for $p \geq 0$.

b. For what price p is revenue $R = pq$ maximized? What is the maximum revenue?

10. **CARBON DATING** An archaeological artifact is found to have 45% of its original ^{14}C. How old is the artifact? (Use 5,730 years as the half-life of ^{14}C.)

11. **BACTERIAL GROWTH** A toxin is introduced into a bacterial colony, and t hours later, the population is given by

$$N(t) = 10,000(8 + t)e^{-0.1t}$$

a. What was the population when the toxin was introduced?

b. When is the population maximized? What is the maximum population?

c. What happens to the population in the long run (as $t \to +\infty$)?

Review Exercises

In Exercises 1 through 4, sketch the graph of the given exponential or logarithmic function without using calculus.

1. $f(x) = 5^x$

2. $f(x) = -2e^{-x}$

3. $f(x) = \ln x^2$

4. $f(x) = \log_3 x$

5. a. Find $f(4)$ if $f(x) = Ae^{-kx}$ and $f(0) = 10$, $f(1) = 25$.

   b. Find $f(3)$ if $f(x) = Ae^{kx}$ and $f(1) = 3$, $f(2) = 10$.

   c. Find $f(9)$ if $f(x) = 30 + Ae^{-kx}$ and $f(0) = 50$, $f(3) = 40$.

   d. Find $f(10)$ if $f(t) = \dfrac{6}{1 + Ae^{-kt}}$ and $f(0) = 3$, $f(5) = 2$.

6. Evaluate the following expressions without using tables or a calculator.

   a. $\ln e^5$

   b. $e^{\ln 2}$

   c. $e^{3 \ln 4 - \ln 2}$

   d. $\ln 9e^2 + \ln 3e^{-2}$

In Exercises 7 through 14, find all real numbers x that satisfy the given equation.

7. $8 = 2e^{0.04x}$

8. $5 = 1 + 4e^{-6x}$

9. $4 \ln x = 8$

10. $5^x = e^3$

11. $\log_9(4x - 1) = 2$

12. $\ln(x - 2) + 3 = \ln(x + 1)$

13. $e^{2x} + e^x - 2 = 0$ [*Hint:* Let $u = e^x$.]

14. $e^{2x} + 2e^x - 3 = 0$ [*Hint:* Let $u = e^x$.]

In Exercises 15 through 30, find the derivative $\dfrac{dy}{dx}$. In some of these problems, you may need to use implicit differentiation or logarithmic differentiation.

15. $y = x^2 e^{-x}$

16. $y = 2e^{3x+5}$

17. $y = x \ln x^2$

18. $y = \ln\sqrt{x^2 + 4x + 1}$

19. $y = \log_3(x^2)$

20. $y = \dfrac{x}{\ln 2x}$

21. $y = \dfrac{e^{-x} + e^x}{1 + e^{-2x}}$

22. $y = \dfrac{e^{3x}}{e^{3x} + 2}$

23. $y = \ln(e^{-2x} + e^{-x})$

24. $y = (1 + e^{-x})^{4/5}$

25. $y = \dfrac{e^{-x}}{x + \ln x}$

26. $y = \ln\left(\dfrac{e^{3x}}{1 + x}\right)$

27. $ye^{x-x^2} = x + y$

28. $xe^{-y} + ye^{-x} = 3$

29. $y = \dfrac{(x^2 + e^{2x})^3 e^{-2x}}{(1 + x - x^2)^{2/3}}$

30. $y = \dfrac{e^{-2x}(2 - x^3)^{3/2}}{\sqrt{1 + x^2}}$

In Exercises 31 through 38, determine where the given function is increasing and decreasing and where its graph is concave upward and concave downward. Sketch the graph, showing as many key features as possible (high and low points, points of inflection, asymptotes, intercepts, cusps, vertical tangents).

31. $f(x) = e^x - e^{-x}$

32. $f(x) = xe^{-2x}$

33. $f(t) = t + e^{-t}$

34. $f(x) = \dfrac{4}{1 + e^{-x}}$

35. $F(u) = u^2 + 2\ln(u + 2)$

36. $g(t) = \dfrac{\ln(t + 1)}{t + 1}$

37. $G(x) = \ln(e^{-2x} + e^{-x})$

38. $f(u) = e^{2u} + e^{-u}$

In Exercises 39 through 42, find the largest and smallest values of the given function over the prescribed closed, bounded interval.

39. $f(x) = \ln(4x - x^2)$ for $1 \le x \le 3$

40. $f(x) = \dfrac{e^{-x/2}}{x^2}$ for $-5 \le x \le -1$

41. $h(t) = (e^{-t} + e^t)^5$ for $-1 \le t \le 1$

42. $g(t) = \dfrac{\ln(\sqrt{t})}{t^2}$ for $1 \le t \le 2$

In Exercises 43 through 46, find an equation for the tangent line to the given curve at the specified point.

43. $y = x \ln x^2$ where $x = 1$

44. $y = (x^2 - x)e^{-x}$ where $x = 0$

45. $y = x^3 e^{2-x}$ where $x = 2$

46. $y = (x + \ln x)^3$ where $x = 1$

47. Find $f(9)$ if $f(x) = e^{kx}$ and $f(3) = 2$.

48. Find $f(8)$ if $f(x) = A2^{kx}$, $f(0) = 20$, and $f(2) = 40$.

49. COMPOUND INTEREST A sum of money is invested at a certain fixed interest rate, and the interest is compounded quarterly. After 15 years, the money has doubled. How will the balance at the end of 30 years compare with the initial investment?

50. COMPOUND INTEREST A bank pays 5% interest compounded quarterly, and a savings institution pays 4.9% interest compounded continuously. Over a 1-year period, which account pays more interest? What about a 5-year period?

51. RADIOACTIVE DECAY A radioactive substance decays exponentially. If 500 grams of the substance were present initially and 400 grams are present 50 years later, how many grams will be present after 200 years?

CHAPTER SUMMARY

52. COMPOUND INTEREST A sum of money is invested at a certain fixed interest rate, and the interest is compounded continuously. After 10 years, the money has doubled. How will the balance at the end of 20 years compare with the initial investment?

53. GROWTH OF BACTERIA The following data were compiled by a researcher during the first 10 minutes of an experiment designed to study the growth of bacteria:

Number of minutes	0	10
Number of bacteria	5,000	8,000

Assuming that the number of bacteria grows exponentially, how many bacteria will be present after 30 minutes?

54. RADIOACTIVE DECAY The following data were compiled by a researcher during an experiment designed to study the decay of a radioactive substance:

Number of hours	0	5
Grams of substance	1,000	700

Assuming that the sample of radioactive substance decays exponentially, how much is left after 20 hours?

55. SALES FROM ADVERTISING It is estimated that if x thousand dollars are spent on advertising, approximately $Q(x) = 50 - 40e^{-0.1x}$ thousand units of a certain commodity will be sold.
a. Sketch the sales curve for $x \geq 0$.
b. How many units will be sold if no money is spent on advertising?
c. How many units will be sold if $8,000 is spent on advertising?
d. How much should be spent on advertising to generate sales of 35,000 units?
e. According to this model, what is the most optimistic sales projection?

56. WORKER PRODUCTION An employer determines that the daily output of a worker on the job for t weeks is $Q(t) = 120 - Ae^{-kt}$ units. Initially, the worker can produce 30 units per day, and after 8 weeks, can produce 80 units per day. How many units can the worker produce per day after 4 weeks?

57. COMPOUND INTEREST How quickly will $2,000 grow to $5,000 when invested at an annual interest rate of 8% if interest is compounded:
a. Quarterly
b. Continuously

58. COMPOUND INTEREST How much should you invest now at an annual interest rate of 6.25% so that your balance 10 years from now will be $2,000 if interest is compounded:
a. Monthly
b. Continuously

59. DEBT REPAYMENT You have a debt of $10,000, which is scheduled to be repaid at the end of 10 years. If you want to repay your debt now, how much should your creditor demand if the prevailing annual interest rate is:
a. 7% compounded monthly
b. 6% compounded continuously

60. COMPOUND INTEREST A bank compounds interest continuously. What annual interest rate does it offer if $1,000 grows to $2,054.44 in 12 years?

61. EFFECTIVE RATE OF INTEREST Which investment has the greater effective interest rate: 8.25% per year compounded quarterly or 8.20% per year compounded continuously?

62. DEPRECIATION The value of a certain industrial machine decreases exponentially. If the machine was originally worth $50,000 and was worth $20,000 five years later, how much will it be worth when it is 10 years old?

63. POPULATION GROWTH It is estimated that t years from now the population of a certain country will be P million people, where

$$P(t) = \frac{30}{1 + 2e^{-0.05t}}$$

a. Sketch the graph of $P(t)$.
b. What is the current population?
c. What will be the population in 20 years?
d. What happens to the population in the long run?

64. BACTERIAL GROWTH The number of bacteria in a certain culture grows exponentially. If 5,000 bacteria were initially present and 8,000 were present 10 minutes later, how long will it take for the number of bacteria to double?

65. **AIR POLLUTION** An environmental study of a certain suburban community suggests that t years from now, the average level of carbon monoxide in the air will be $Q(t) = 4e^{0.03t}$ parts per million.

   a. At what rate will the carbon monoxide level be changing with respect to time 2 years from now?

   b. At what percentage rate will the carbon monoxide level be changing with respect to time t years from now? Does this percentage rate of change depend on t or is it constant?

66. **PROFIT** A manufacturer of digital cameras estimates that when cameras are sold for x dollars apiece, consumers will buy $8,000e^{-0.02x}$ cameras each week. He also determines that profit is maximized when the selling price x is 1.4 times the cost of producing each unit. What price maximizes weekly profit? How many units are sold each week at this optimal price?

67. **OPTIMAL HOLDING TIME** Beth owns an asset whose value t years from now will be $V(t) = 2,000e^{\sqrt{2t}}$ dollars. If the prevailing interest rate remains constant at 5% per year compounded continuously, when will it be most advantageous to sell the collection and invest the proceeds?

68. **RULE OF 70** Investors are often interested in knowing how long it takes for a particular investment to double. A simple means for making this determination is the "rule of 70," which says: The doubling time of an investment with an annual interest rate $r\%$ compounded continuously is given by $d = \dfrac{70}{r}$.

   a. For interest rate r, use the formula $B = Pe^{rt}$ to find the doubling time for $r = 4, 6, 9, 10,$ and 12. In each case, compare the value with the value obtained from the rule of 70.

   b. Some people prefer a "rule of 72" and others use a "rule of 69." Test these alternative rules as in part (a), and write a paragraph on which rule you would prefer to use.

69. **RADIOACTIVE DECAY** A radioactive substance decays exponentially with half-life λ. Suppose the amount of the substance initially present (when $t = 0$) is Q_0.

   a. Show that the amount of the substance that remains after t years will be $Q(t) = Q_0 e^{-(\ln 2/\lambda)t}$.

   b. Find a number k so that the amount in part (a) can be expressed as $Q(t) = Q_0(0.5)^{kt}$.

70. **ANIMAL DEMOGRAPHY** A naturalist at an animal sanctuary has determined that the function
$$f(x) = \frac{4e^{-(\ln x)^2}}{\sqrt{\pi}\, x}$$
provides a good measure of the number of animals in the sanctuary that are x years old. Sketch the graph of $f(x)$ for $x > 0$, and find the most likely age among the animals, that is, the age for which $f(x)$ is largest.

71. **CARBON DATING** "Ötzi the Iceman" is the name given a neolithic corpse found frozen in an Alpine glacier in 1991. He was originally thought to be from the Bronze Age because of the hatchet he was carrying. However, the hatchet proved to be made of copper rather than bronze. Read an article on the Bronze Age, and determine the least age of the Ice Man assuming that he dates before the Bronze Age. What is the *largest* percentage of ^{14}C that can remain in a sample taken from his body?

72. **FICK'S LAW** According to Fick's law* $f(t) = C(1 - e^{-kt})$, where $f(t)$ is the concentration of solute inside a cell at time t, C is the (constant) concentration of solute surrounding the cell, and k is a positive constant. Suppose that for a particular cell, the concentration on the inside of the cell after 2 hours is 0.8% of the concentration outside the cell.

   a. What is k?

   b. What is the percentage rate of change of $f(t)$ at time t?

   c. Write a paragraph on the role played by Fick's law in ecology.

73. **COOLING** Jayla falls into a lake where the water temperature is $-3°C$. Her body temperature after t minutes in the water is $T(t) = 35e^{-0.32t}$. She will lose consciousness when her body temperature reaches $27°C$. How long do rescuers have to save her? How fast is Jayla's body temperature dropping at the time it reaches $27°C$?

---

*Fick's law plays an important role in ecology. For instance, see M. D. LaGrega, P. L. Buckingham, and J. C. Evans, *Hazardous Waste Management,* New York: McGraw-Hill, 1994, pp. 95, 464, and especially p. 813, where the authors discuss contaminant transport through landfill.

74. FORENSIC SCIENCE The temperature T (in degrees Celsius) of the body of a murder victim found in a room where the air temperature is 20°C is given by

$$T(t) = 20 + 17e^{-0.07t}$$

where t is the number of hours after the victim's death.

a. Graph the body temperature $T(t)$ for $t \geq 0$. What is the horizontal asymptote of this graph, and what does it represent?

b. What is the temperature of the victim's body after 10 hours? How long does it take for the body's temperature to reach 25°C?

c. Abel Baker is a clerk in the firm of Dewey, Cheatum, and Howe. He comes to work early one morning and finds the corpse of his boss, Will Cheatum, draped across his desk. He calls the police, and at 8 A.M., they determine that the temperature of the corpse is 33°C. Since the last item entered on the victim's notepad was, "Fire that idiot, Baker," Abel is considered the prime suspect. Actually, Abel is bright enough to have been reading this text in his spare time. He glances at the thermostat to confirm that the room temperature is 20°C. For what time will he need an alibi to establish his innocence?

75. CONCENTRATION OF DRUG Suppose that t hours after an antibiotic is administered orally, its concentration in the patient's bloodstream is given by a surge function of the form $C(t) = Ate^{-kt}$, where A and k are positive constants and C is measured in micrograms per milliliter of blood. Blood samples are taken periodically, and it is determined that the maximum concentration of drug occurs 2 hours after it is administered and is 10 micrograms per milliliter.

a. Use this information to determine A and k.

b. A new dose will be administered when the concentration falls to 1 microgram per milliliter. When does this occur?

76. CHEMICAL REACTION RATE The effect of temperature on the reaction rate of a chemical reaction is given by the **Arrhenius equation**

$$k = Ae^{-E_0/RT}$$

where k is the rate constant, T (in Kelvin) is the temperature, and R is the gas constant. The quantities A and E_0 are fixed once the reaction is specified. Let k_1 and k_2 be the reaction rate

constants associated with temperatures T_1 and T_2. Find an expression for $\ln\left(\dfrac{k_1}{k_2}\right)$ in terms of E_0, R, T_1, and T_2.

77. POPULATION GROWTH According to a logistic model based on the assumption that the earth can support no more than 40 billion people, the world's population (in billions) t years after 1960 is given by a function of the form $P(t) = \dfrac{40}{1 + Ce^{-kt}}$ where C and k are positive constants. Find the function of this form that is consistent with the fact that the world's population was approximately 3 billion in 1960 and 4 billion in 1975. What does your model predict for the population in the year 2010? Check the accuracy of the model by consulting the Internet.

78. THE SPREAD OF AN EPIDEMIC Public health records indicate that t weeks after the outbreak of a certain strain of influenza, approximately

$$Q(t) = \dfrac{80}{4 + 76e^{-1.2t}}$$

thousand people had caught the disease. At what rate was the disease spreading at the end of the second week? At what time is the disease spreading most rapidly?

79. ACIDITY (pH) OF A SOLUTION The acidity of a solution is measured by its pH value, which is defined by pH $= -\log_{10} [H_3O^+]$, where $[H_3O^+]$ is the hydronium ion concentration (moles/liter) of the solution. On average, milk has a pH value that is three times the pH value of a lime, which in turn has half the pH value of an orange. If the average pH of an orange is 3.2, what is the average hydronium ion concentration of a lime?

80. CARBON DATING A Cro-Magnon cave painting at Lascaux, France, is approximately 15,000 years old. Approximately what ratio of ^{14}C to ^{12}C would you expect to find in a fossil from the same period as the painting?

81. MORTALITY RATES It is sometimes useful for actuaries to be able to project mortality rates within a given population. A formula sometimes used for computing the mortality rate $D(t)$ for women in the age group 25–29 is

$$D(t) = (D_0 - 0.00046)e^{-0.162t} + 0.00046$$

where t is the number of years after a fixed base year and D_0 is the mortality rate when $t = 0$.

a. Suppose the initial mortality rate of a particular group is 0.008 (8 deaths per 1,000 women). What is the mortality rate of this group 10 years later? What is the rate 25 years later?

b. Sketch the graph of the mortality function $D(t)$ for the group in part (a) for $0 \leq t \leq 25$.

82. GROSS DOMESTIC PRODUCT   The gross domestic product (GDP) of a certain country was 100 billion dollars in 1995 and 165 billion dollars in 2005. Assuming that the GDP is growing exponentially, what will it be in the year 2015?

83. ARCHAEOLOGY   "Lucy," the famous prehuman whose skeleton was discovered in Africa, has been found to be approximately 3.8 million years old.

a. Approximately what percentage of original ^{14}C would you expect to find if you tried to apply carbon dating to Lucy? Why would this be a problem if you were actually trying to "date" Lucy?

b. In practice, carbon dating works well only for relatively "recent" samples—those that are no more than approximately 50,000 years old. For older samples, such as Lucy, variations on carbon dating have been developed, such as potassium-argon and rubidium-strontium dating. Read an article on alternative dating methods, and write a paragraph on how they are used.*

84. RADIOLOGY   The radioactive isotope gallium-67 (^{67}Ga), used in the diagnosis of malignant tumors, has a half-life of 46.5 hours. If we start with 100 milligrams of the isotope, how many milligrams will be left after 24 hours? When will there be only 25 milligrams left? Answer these questions by first using a graphing utility to graph an appropriate exponential function and then using the **TRACE** and **ZOOM** features.

85. A population model developed by the U.S. Census Bureau uses the formula

$$P(t) = \frac{202.31}{1 + e^{3.938 - 0.314t}}$$

to estimate the population of the United States (in millions) for every 10th year from the base year 1790. Thus, for instance, $t = 0$ corresponds to 1790, $t = 1$ to 1800, $t = 10$ to 1890, and so on. The model excludes Alaska and Hawaii.

a. Use this formula to compute the population of the contiguous United States for the years 1790, 1800, 1830, 1860, 1880, 1900, 1920, 1940, 1960, 1980, 1990, and 2000.

b. Sketch the graph of $P(t)$. When does this model predict that the population of the contiguous United States will be increasing most rapidly?

c. Use the Internet to find the actual population figures for the years listed in part (a). Does the given population model seem to be accurate? (Remember to exclude Alaska and Hawaii.) Write a paragraph describing some possible reasons for any major differences between the predicted population figures and the actual census figures.

86. Use a graphing utility to graph $y = 2^{-x}$, $y = 3^{-x}$, $y = 5^{-x}$, and $y = (0.5)^{-x}$ on the same set of axes. How does a change in base affect the graph of the exponential function? (*Suggestion:* Use the graphing window $[-3, 3]1$ by $[-3, 3]1$.)

87. Use a graphing utility to draw the graphs of $y = \sqrt{3^x}$, $y = \sqrt{3^{-x}}$, and $y = 3^{-x}$ on the same set of axes. How do these graphs differ? (*Suggestion:* Use the graphing window $[-4, 4]1$ by $[-2, 6]1$.)

88. Use a graphing utility to draw the graphs of $y = 3^x$ and $y = 4 - \ln \sqrt{x}$ on the same axes. Then use **TRACE** and **ZOOM** to find all points of intersection of the two graphs.

89. Solve this equation with three decimal place accuracy:

$$\log_5(x + 5) - \log_2 x = 2 \log_{10}(x^2 + 2x)$$

90. Use a graphing utility to draw the graphs of

$$y = \ln(1 + x^2) \quad \text{and} \quad y = \frac{1}{x}$$

on the same axes. Do these graphs intersect?

91. Make a table for the quantities $(\sqrt{n})^{\sqrt{n+1}}$ and $(\sqrt{n + 1})^{\sqrt{n}}$, with $n = 8, 9, 12, 20, 25, 31, 37, 38, 43, 50, 100,$ and $1,000$. Which of the two quantities seems to be larger? Do you think this inequality holds for all $n \geq 8$?

---

*A good place to start your research is the article by Paul J. Campbell, "How Old Is the Earth?", *UMAP Modules 1992: Tools for Teaching,* Arlington, MA: Consortium for Mathematics and Its Applications, 1993.

EXPLORE! UPDATE

Complete solutions for all EXPLORE! boxes throughout the text can be accessed at the book specific website, www.mhhe.com/hoffmann.

Solution for Explore! on Page 300

One method to display all the desired graphs is to list the desired bases in the function form. First, write Y1 = {1, 2, 3, 4}^X into the equation editor of your graphing calculator. Observe that for $b > 1$, the functions increase exponentially, with steeper growth for larger bases. Also all the curves pass through the point (0, 1). Why? Now try Y1 = {2, 4, 6}^X. Note that $y = 4^x$ lies between $y = 2^x$ and $y = 6^x$. Likewise the graph of $y = e^x$ would lie between $y = 2^x$ and $y = 3^x$.

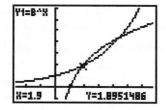
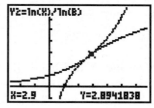
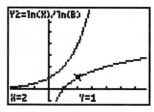

Solution for Explore! on Page 321

Store $f(x) = B^x$ into Y1 and $g(x) = \log_B x$ into Y2 as $\dfrac{\ln(x)}{\ln(B)}$. Experimenting with different values of B, we find this: For $B < e^{1/e} \approx 1.444668$, the two curves intersect in two places (where, in terms of B?), for $B = e^{1/e}$ they touch only at one place, and for $B > e^{1/e}$ there is no intersection. (See Classroom Capsules, "An Overlooked Calculus Question," *The College Mathematics Journal,* Vol. 33, No. 5, November 2002.)

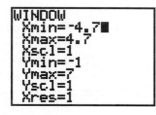
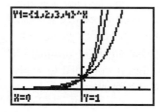
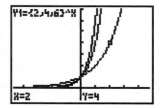

Solution for Explore! on Page 338

Following Example 4.3.12, store the function $f(x) = Axe^{-Bx}$ into Y1 of the equation editor. We attempt to find the maximum of $f(x)$ in terms of A and B. We can set $A = 1$ and vary the value of B (say, 1, 0.5, and 0.01). Then we can fix B to be 0.1 and let A vary (say, 1, 10, 100).

For instance, when $A = 1$ and $B = 1$, the maximal y value occurs at $x = 1$ (see the figure on the left). When $A = 1$ and $B = 0.1$, it occurs at $x = 10$ (middle figure).

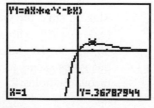
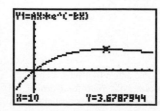
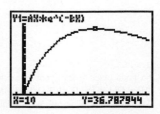

When $A = 10$ and $B = 0.1$, this maximum remains at $x = 10$ (figure on the right). In this case, the y coordinate of the maximum increases by the A factor. In general, it can be shown that the x value of the maximal point is just $\dfrac{1}{B}$. The A factor does not change the location of the x value of the maximum, but it does affect the y value as a multiplier. To confirm this analytically, set the derivative of $y = Axe^{-Bx}$ equal to zero and solve for the location of the maximal point.

Solution for Explore!
on Page 346

Following Example 4.4.2, store $f(x)$ into Y1 and $f'(x)$ into Y2 (but deselected), and $f''(x)$ into Y3 in bold, using the window $[-4.7, 4.7]1$ by $[-0.5, 0.5]0.1$. Using the **TRACE** or the root-finding feature of the calculator, you can determine that the two x intercepts of $f''(x)$ are located at $x = -1$ and $x = 1$. Since the second derivative $f''(x)$ represents the concavity function of $f(x)$, we know that at these values $f(x)$ changes concavity. At the inflection point $(-1, 0.242)$, $f(x)$ changes concavity from positive (concave upward) to negative (concave downward). At $x = 1$, $y = 0.242$, concavity changes from downward to upward.

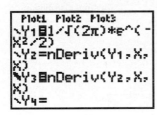
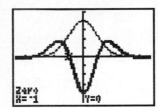
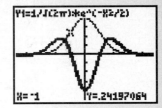

Solution for Explore!
on Page 352

Following Example 4.4.6, store $Q(t) = \dfrac{20}{1 + 19e^{-1.2t}}$ into Y1 and graph using the window $[0, 10]1$ by $[0, 25]5$. We can trace the function for large values of the independent variable, time. As x approaches 10 (weeks), the function attains a value close to 20,000 people infected ($Y > 19.996$ thousand). Since 90% of the population is 18,000, by setting Y2 = 18 and using the intersection feature of the calculator, you can determine that 90% of the population becomes infected after $x = 4.28$ weeks (about 30 days).

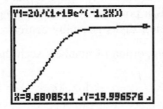
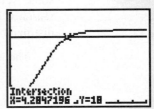

THINK ABOUT IT

FORENSIC ACCOUNTING: BENFORD'S LAW

You might guess that the first digit of each number in a collection of numbers has an equal chance of being any of the digits 1 through 9, but it was discovered in 1938 by physicist Frank Benford that the chance of the digit being a 1 is more than 30%! Naturally occurring numbers exhibit a curious pattern in the proportions of the first digit: smaller digits such as 1, 2, and 3 occur much more often than larger digits, as seen in the following table:

First Digit	Proportion
1	30.1%
2	17.6
3	12.5
4	9.7
5	7.9
6	6.7
7	5.8
8	5.1
9	4.6

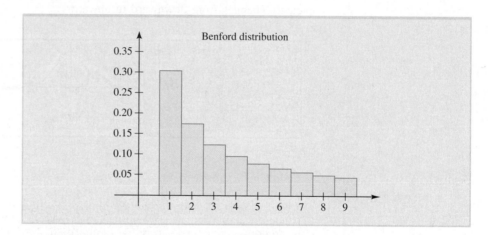

Naturally occurring, in this case, means that the numbers arise without explicit bound and describe similar quantities, such as the populations of cities or the amounts paid out on checks. This pattern also holds for exponentially growing numbers and some types of randomly sampled (but not randomly generated) numbers, and for this reason it is a powerful tool for determining if these types of data are genuine. The distribution of digits generally conforms closely to the following rule:

The proportion of numbers such that the first digit is n is given by

$$\log_{10}(n + 1) - \log_{10} n = \log_{10}\frac{n + 1}{n}$$

This rule, known as Benford's law, is used to detect fraud in accounting and is one of several techniques in a field called *forensic accounting*. Often people who write fraudulent checks, such as an embezzler at a corporation, try to make the first digits (or even all the digits) occur *equally often* so as to appear to be random. Benford's law predicts that the first digits of such a set of accounting data should not be evenly proportioned, but rather show a higher occurrence of smaller digits. If an employee is writing a large number of checks or committing many monetary transfers to a suspicious target, the check values can be analyzed to see if they follow Benford's law, indicating possible fraud if they do not.

This technique can be applied to various types of accounting data (such as for taxes and expenses) and has been used to analyze socioeconomic data such as the gross domestic products of many countries. Benford's law is also used by the Internal Revenue Service to detect fraud and has been applied to locate errors in data entry and analysis.

Questions

1. Verify that the formula given for the proportions of digits,

$$P(n) = \log_{10} \frac{n+1}{n}$$

produces the values in the given table. Use calculus to show that the proportion is a decreasing function of n.

2. The proportions of first digits depend on the base of the number system used. Computers generally use number systems that are powers of 2. Benford's law for base b is

$$P(n) = \log_b \frac{n+1}{n}$$

where n ranges from 1 to b. Compute a table like the one given for base $b = 10$ for the bases 4 and 8. Using these computed tables and the given table, do the proportions seem to be evening out or becoming more uneven as the size of the base increases?

Use calculus to justify your assertion by viewing $P(n)$ as a function of b, for particular values of n. For instance, for $n = 1$:

$$f(b) = \log_b 2 = \frac{\ln 2}{\ln b}$$

Use this new function to determine if the proportion of numbers with leading digit 1 are increasing or decreasing as the size of base b increases. What happens for the other digits?

THINK ABOUT IT

3. In the course of a potential fraud investigation, it is found that an employee wrote checks with the following values to a suspicious source: $234, $444, $513, $1,120, $2,201, $3,614, $4,311, $5,557, $5,342, $6,710, $8,712, and $8,998. Compute the proportions corresponding to each of the first digits. Do you think that fraud may be present? (In actual investigations, statistical tests are used to determine if the deviation is statistically significant.)

4. Select a collection of numbers arbitrarily from a newspaper or magazine and record the first digit (the first nonzero digit if it is a decimal less than 1). Do the numbers appear to follow Benford's law?

5. The following list of numbers is a sample of heights of prominent mountain peaks in California, measured in feet. Do they appear to follow Benford's law?

10,076	1,677	7,196	2,894	9,822
373	1,129	1,558	1,198	343
331	1,119	932	2,563	1,936
1,016	364	1,003	833	765
755	545	1,891	2,027	512
675	2,648	2,601	1,480	719
525	570	884	560	1,362
571	1,992	745	541	385
971	1,220	984	879	1,135
604	2,339	1,588	594	587

Source: http://en.wikipedia.org/wiki/Mountain_peaks_of_California.

References

T. P. Hill, "The First Digit Phenomenon," *American Scientist,* Vol. 86, 1998, p. 358.
Steven W. Smith, "The Scientist's and Engineer's Guide to Signal Processing," Chapter 34. Available online at http://www.dspguide.com/ch34/1.htm.
C. Durtschi et al. "The Effective Use of Benford's Law in Detecting Fraud in Accounting Data." Available online at http://www.auditnet.org/articles/JFA-V-1-17-34.pdf.

Computing area under a curve, like the area of the region spanned by the scaffolding under the roller coaster track, is an application of integration.

Integration

SECTION 5.1 Indefinite Integration and Differential Equations

Learning Objectives

1. Study and compute indefinite integrals.
2. Explore differential equations and initial value problems.
3. Set up and solve separable differential equations.

How can a known rate of inflation be used to determine future prices? What is the velocity of an object moving along a straight line with known acceleration? How can knowing the rate at which a population is changing be used to predict future population levels? In all these situations, the derivative (rate of change) of a quantity is known and the quantity itself is required. Here is the terminology we will use in connection with obtaining a function from its derivative.

Antidifferentiation ■ A function $F(x)$ is said to be an *antiderivative* of $f(x)$ if

$$F'(x) = f(x)$$

for every x in the domain of $f(x)$. The process of finding antiderivatives is called *antidifferentiation* or *indefinite integration*.

NOTE   Sometimes we write the equation

$$F'(x) = f(x)$$

as

$$\frac{dF}{dx} = f(x)$$ ■

Later in this section, you will learn techniques you can use to find antiderivatives. Once you have found what you believe to be an antiderivative of a function, you can always check your answer by differentiating. You should get the original function back. Example 5.1.1 shows how to verify an antiderivative.

EXAMPLE 5.1.1 Verifying an Antiderivative

Verify that $F(x) = \dfrac{1}{3}x^3 + 5x + 2$ is an antiderivative of $f(x) = x^2 + 5$.

Solution

$F(x)$ is an antiderivative of $f(x)$ if and only if $F'(x) = f(x)$. Differentiate F and you will find that

$$F'(x) = \frac{1}{3}(3x^2) + 5$$
$$= x^2 + 5 = f(x)$$

as required.

**The General
Antiderivative
of a Function**

A function has more than one antiderivative. For example, one antiderivative of the function $f(x) = 3x^2$ is $F(x) = x^3$, since

$$F'(x) = 3x^2 = f(x)$$

but so are $x^3 + 12$ and $x^3 - 5$ and $x^3 + \pi$, since

$$\frac{d}{dx}(x^3 + 12) = 3x^2 \qquad \frac{d}{dx}(x^3 - 5) = 3x^2 \qquad \frac{d}{dx}(x^3 + \pi) = 3x^2$$

In general, if F is one antiderivative of f, then so is any function of the form $G(x) = F(x) + C$, for constant C since

$$
\begin{aligned}
G'(x) &= [F(x) + C]' && \text{sum rule for derivatives} \\
&= F'(x) + C' && \text{derivative of a constant is 0} \\
&= F'(x) + 0 && \text{since } F \text{ is an antiderivative of } f \\
&= f(x)
\end{aligned}
$$

Conversely, it can be shown that if F and G are both antiderivatives of f, then $G(x) = F(x) + C$, for some constant C (Exercise 74). To summarize:

> **Fundamental Property of Antiderivatives** ■ If $F(x)$ is an antiderivative of the continuous function $f(x)$, then any other antiderivative of $f(x)$ has the form $G(x) = F(x) + C$ for some constant C.

There is a simple geometric interpretation for the fundamental property of antiderivatives. If F and G are both antiderivatives of f, then

$$G'(x) = F'(x) = f(x)$$

Just-In-Time **REVIEW**

Recall that lines are parallel whenever their slopes are equal.

This means that the slope $F'(x)$ of the tangent line to $y = F(x)$ at the point $(x, F(x))$ is the same as the slope $G'(x)$ of the tangent line to $y = G(x)$ at $(x, G(x))$. Since the slopes are equal, it follows that the tangent lines at $(x, F(x))$ and $(x, G(x))$ are parallel, as shown in Figure 5.1a. Since this is true for all x, the entire curve $y = G(x)$ must be parallel to the curve $y = F(x)$, so that

$$y = G(x) = F(x) + C$$

EXPLORE!

Store the function $F(x) = x^3$ into Y1 of the equation editor in a bold graphing style. Generate a family of vertical transformations Y2 = Y1 + L1, where L1 is a list of constants, $\{-4, -2, 2, 4\}$. Use the graphing window $[-4.7, 4.7]1$ by $[-6, 6]1$. What do you observe about the slopes of all these curves at $x = 1$?

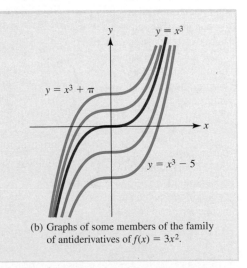

(a) If $F'(x) = G'(x)$, the tangent lines at $(x, F(x))$ and $(x, G(x))$ are parallel.

(b) Graphs of some members of the family of antiderivatives of $f(x) = 3x^2$.

FIGURE 5.1   Graphs of antiderivatives of a function f form a family of parallel curves.

In general, the collection of graphs of all antiderivatives of a given function f is a family of parallel curves that are vertical translations of one another. This is illustrated in Figure 5.1b for the family of antiderivatives of $f(x) = 3x^2$.

The Indefinite Integral

You have just seen that if $F(x)$ is one antiderivative of the continuous function $f(x)$, then all such antiderivatives may be characterized by $F(x) + C$ for constant C. The family of all antiderivatives of $f(x)$ is written

$$\int f(x)\, dx = F(x) + C$$

and is called the **indefinite integral** of $f(x)$. The integral is "indefinite" because it involves a constant C that can take on any value. In Section 5.3, we introduce a **definite integral** that has a specific numerical value and is used to represent a variety of quantities, such as area, average value, present value of an income flow, and cardiac output, to name a few. The connection between definite and indefinite integrals is made in Section 5.3 through a result so important that it is referred to as the **fundamental theorem of calculus.**

In the context of the indefinite integral $\int f(x)\, dx = F(x) + C$, the **integral symbol** is $\int$, the function $f(x)$ is called the **integrand,** C is the **constant of integration,** and dx is a differential that specifies x as the **variable of integration.** These features are displayed in this diagram for the indefinite integral of $f(x) = 3x^2$:

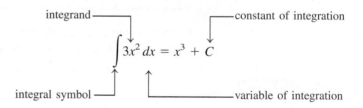

For any differentiable function F, we have

$$\int F'(x)\, dx = F(x) + C$$

since by definition, $F(x)$ is an antiderivative of $F'(x)$. Equivalently,

$$\int \frac{dF}{dx}\, dx = F(x) + C$$

which can be interpreted as saying that *the integral of the derivative of a function is the function itself.* This property of integrals is especially useful in applied problems where a rate of change $F'(x)$ is given and we wish to find $F(x)$. Several such problems are examined later in this section.

It is useful to remember that if you have performed an indefinite integration calculation that leads you to believe that $\int f(x)\, dx = G(x) + C$, then you can check your calculation by differentiating $G(x)$:

> *If $G'(x) = f(x)$, then the integration $\int f(x)\, dx = G(x) + C$ is correct, but if $G'(x)$ is anything other than $f(x)$, you've made a mistake.*

This relationship between differentiation and antidifferentiation enables us to establish the following integration rules by "reversing" analogous differentiation rules.

Rules for Integrating Common Functions

The **constant rule:** $\int k\,dx = kx + C$ for constant k

The **power rule:** $\int x^n\,dx = \dfrac{x^{n+1}}{n+1} + C$ for all $n \neq -1$

The **logarithmic rule:** $\int \dfrac{1}{x}\,dx = \ln|x| + C$ for all $x \neq 0$

The **exponential rule:** $\int e^{kx}\,dx = \dfrac{1}{k}e^{kx} + C$ for constant $k \neq 0$

To verify the power rule, it is enough to show that the derivative of $\dfrac{x^{n+1}}{n+1}$ is x^n:

$$\frac{d}{dx}\left(\frac{x^{n+1}}{n+1}\right) = \frac{1}{n+1}[(n+1)x^n] = x^n \quad \text{power rule for differentiation}$$

For the logarithmic rule, if $x > 0$, then $|x| = x$ and

$$\frac{d}{dx}(\ln|x|) = \frac{d}{dx}(\ln x) = \frac{1}{x}$$

If $x < 0$, then $-x > 0$ and $\ln|x| = \ln(-x)$, and it follows from the chain rule that

$$\frac{d}{dx}(\ln|x|) = \frac{d}{dx}[\ln(-x)] = \frac{1}{(-x)}(-1) = \frac{1}{x}$$

Thus, for all $x \neq 0$,

$$\frac{d}{dx}(\ln|x|) = \frac{1}{x}$$

so

$$\int \frac{1}{x}\,dx = \ln|x| + C$$

You are asked to verify the constant rule and exponential rule in Exercise 76.

> **NOTE**   Notice that the logarithm rule "fills the gap" in the power rule, namely, the case where $n = -1$. You may wish to blend the two rules into this combined form:
>
> $$\int x^n\,dx = \begin{cases} \dfrac{x^{n+1}}{n+1} + C & \text{if } n \neq -1 \\[2ex] \ln|x| + C & \text{if } n = -1 \end{cases}$$ ∎

EXPLORE!

Graph $y = F(x)$, where

$F(x) = \ln|x| = \ln(\text{abs}(x))$

in bold and $f(x) = \dfrac{1}{x}$ in the regular graphing style using a decimal graphing window. At any point $x \neq 0$, show that the derivative of $F(x)$ is equal to the value of $f(x)$ at that particular point, confirming that $F(x)$ is the antiderivative of $f(x)$.

EXAMPLE 5.1.2   Computing Indefinite Integrals

Find these integrals:

a. $\displaystyle\int 3\,dx$ **b.** $\displaystyle\int x^{17}\,dx$ **c.** $\displaystyle\int \frac{1}{\sqrt{x}}\,dx$ **d.** $\displaystyle\int e^{-3x}\,dx$

Solution

a. Use the constant rule with $k = 3$: $\displaystyle\int 3\,dx = 3x + C$

b. Use the power rule with $n = 17$: $\displaystyle\int x^{17}\,dx = \frac{1}{18}x^{18} + C$

c. Use the power rule with $n = -\dfrac{1}{2}$: Since $n + 1 = \dfrac{1}{2}$,

$$\int \frac{dx}{\sqrt{x}} = \int x^{-1/2}\,dx = \frac{1}{1/2}x^{1/2} + C = 2\sqrt{x} + C$$

d. Use the exponential rule with $k = -3$:

$$\int e^{-3x}\,dx = \frac{1}{-3}e^{-3x} + C$$

Example 5.1.2 illustrates how certain basic functions can be integrated, but what about combinations of functions, such as the polynomial $x^5 + 2x^3 + 7$ or an expression like $5e^{-x} + \sqrt{x}$? Here are algebraic rules that will enable you to handle such expressions in a natural fashion.

Algebraic Rules for Indefinite Integration

The **constant multiple rule:** $\displaystyle\int kf(x)\,dx = k\int f(x)\,dx$ for constant k

The **sum rule:** $\displaystyle\int [f(x) + g(x)]\,dx = \int f(x)\,dx + \int g(x)\,dx$

The **difference rule:** $\displaystyle\int [f(x) - g(x)]\,dx = \int f(x)\,dx - \int g(x)\,dx$

To prove the constant multiple rule, note that if $\dfrac{dF}{dx} = f(x)$, then

$$\frac{d}{dx}[kF(x)] = k\frac{dF}{dx} = kf(x)$$

which means that

$$\int kf(x)\,dx = k\int f(x)\,dx$$

The sum and difference rules can be established in a similar fashion.

EXAMPLE 5.1.3 Using Algebraic Rules for Integration

Find the following integrals:

a. $\displaystyle\int (2x^5 + 8x^3 - 3x^2 + 5)\,dx$

b. $\displaystyle\int \left(\frac{x^3 + 2x - 7}{x}\right)dx$

c. $\displaystyle\int (3e^{-5t} + \sqrt{t})\,dt$

Solution

a. By using the power rule in conjunction with the sum and difference rules and the constant multiple rule, you get

$$\int (2x^5 + 8x^3 - 3x^2 + 5)\,dx = 2\int x^5\,dx + 8\int x^3\,dx - 3\int x^2\,dx + \int 5\,dx$$

$$= 2\left(\frac{x^6}{6}\right) + 8\left(\frac{x^4}{4}\right) - 3\left(\frac{x^3}{3}\right) + 5x + C$$

$$= \frac{1}{3}x^6 + 2x^4 - x^3 + 5x + C$$

b. There is no "quotient rule" for integration, but at least in this case, you can still divide the denominator into the numerator and then integrate using the method in part (a):

$$\int \left(\frac{x^3 + 2x - 7}{x}\right)dx = \int \left(x^2 + 2 - \frac{7}{x}\right)dx$$

$$= \frac{1}{3}x^3 + 2x - 7\ln|x| + C$$

c. $\displaystyle\int (3e^{-5t} + \sqrt{t})\,dt = \int (3e^{-5t} + t^{1/2})\,dt$

$$= 3\left(\frac{1}{-5}e^{-5t}\right) + \frac{1}{3/2}t^{3/2} + C = -\frac{3}{5}e^{-5t} + \frac{2}{3}t^{3/2} + C$$

 EXPLORE!

Refer to Example 5.1.4. Store the function $f(x) = 3x^2 + 1$ into Y1. Graph using a bold graphing style and the window [0, 2.35]0.5 by [−2, 12]1. Place into Y2 the family of antiderivatives

$F(x) = x^3 + x + \mathbf{L1}$

where L1 is the list of integer values −5 to 5. Which of these antiderivatives passes through the point (2, 6)? Repeat this exercise for $f(x) = 3x^2 - 2$.

EXAMPLE 5.1.4 Finding a Function from Its Slope Function

Find the function $f(x)$ whose tangent has slope $3x^2 + 1$ for each value of x and whose graph passes through the point (2, 6).

Solution

The slope of the tangent at each point $(x, f(x))$ is the derivative $f'(x)$. Thus,

$$f'(x) = 3x^2 + 1$$

and so $f(x)$ is the antiderivative

$$f(x) = \int f'(x)\,dx = \int (3x^2 + 1)\,dx = x^3 + x + C$$

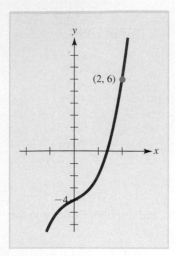

FIGURE 5.2   The graph of $y = x^3 + x - 4$

To find C, use the fact that the graph of f passes through $(2, 6)$. That is, substitute $x = 2$ and $f(2) = 6$ into the equation for $f(x)$ and solve for C to get

$$6 = (2)^3 + 2 + C \qquad \text{or} \qquad C = -4$$

Thus, the desired function is $f(x) = x^3 + x - 4$. The graph of this function is shown in Figure 5.2.

Example 5.1.5 illustrates how integration can be used to find the total cost of production when marginal cost is known.

EXAMPLE 5.1.5   Finding Total Cost from Marginal Cost

A manufacturer has found that the marginal cost of a certain product is $3q^2 - 60q + 400$ dollars per unit when q units have been produced. The total cost of producing the first 2 units is \$900. What is the total cost of producing the first 5 units?

Solution

Recall that the marginal cost is the derivative of the total cost function $C(q)$. Thus,

$$\frac{dC}{dq} = 3q^2 - 60q + 400$$

and so $C(q)$ must be the antiderivative

$$C(q) = \int \frac{dC}{dq}\, dq = \int (3q^2 - 60q + 400)\, dq = q^3 - 30q^2 + 400q + K$$

for some constant K. (The letter K was used for the constant to avoid confusion with the cost function C.)

The value of K is determined by the fact that $C(2) = 900$. In particular,

$$900 = (2)^3 - 30(2)^2 + 400(2) + K \qquad \text{or} \qquad K = 212$$

Hence, $\qquad\qquad\qquad\qquad C(q) = q^3 - 30q^2 + 400q + 212$

and the cost of producing the first 5 units is

$$C(5) = (5)^3 - 30(5)^2 + 400(5) + 212 = \$1,587$$

Recall from Section 2.2 that if an object moving along a straight line is at the position $s(t)$ at time t, then its velocity is given by $v = \dfrac{ds}{dt}$ and its acceleration by $a = \dfrac{dv}{dt}$. Turning things around, if the acceleration of the object is known, then its velocity and position can be found by integration. This is illustrated in Example 5.1.6.

EXAMPLE 5.1.6   Finding Velocity and Position from Acceleration

A car is traveling along a straight, level road at 45 miles per hour (66 feet per second) when the driver is forced to apply the brakes to avoid an accident. If the brakes supply a constant deceleration of 22 ft/sec² (feet per second, per second), how far does the car travel before coming to a complete stop?

Solution

Let $s(t)$ denote the distance traveled by the car in t seconds after the brakes are applied. Since the car decelerates at 22 ft/sec^2, it follows that $a(t) = -22$; that is,

$$\frac{dv}{dt} = a(t) = -22$$

Integrating, you find that the velocity at time t is given by

$$v(t) = \int \frac{dv}{dt}\, dt = \int -22 \, dt = -22t + C_1$$

EXPLORE!

Refer to Example 5.1.6. Graph the position function $s(t)$ in the equation editor of your calculator as Y1 $= -11x^2 + 66x$, using the window [0, 9.4]1 by [0, 200]10. Locate the stopping time and the corresponding position on the graph. Work the problem again for a car traveling 60 mph (88 ft/sec). In this case, what is happening at the 3-second mark?

To evaluate C_1, note that $v = 66$ when $t = 0$ so that

$$66 = v(0) = -22(0) + C_1$$

and $C_1 = 66$. Thus, the velocity at time t is $v(t) = -22t + 66$.

Next, to find the distance $s(t)$, begin with the fact that

$$\frac{ds}{dt} = v(t) = -22t + 66$$

and use integration to show that

$$s(t) = \int \frac{ds}{dt}\, dt = \int (-22t + 66)\, dt = -11t^2 + 66t + C_2$$

Since $s(0) = 0$ (Do you see why?), it follows that $C_2 = 0$ and

$$s(t) = -11t^2 + 66t$$

Finally, to find the stopping distance, note that the car stops when $v(t) = 0$, and this occurs when

$$v(t) = -22t + 66 = 0$$

Solving this equation, you find that the car stops after 3 seconds of deceleration, and in that time it has traveled

$$s(3) = -11(3)^2 + 66(3) = 99 \text{ feet}$$

Introduction to Differential Equations

A **differential equation** is an equation that involves differentials or derivatives, and a function that satisfies such an equation is called a **solution.** For instance,

$$\frac{dy}{dx} = 3x^2 + 5 \qquad \frac{dP}{dt} = kP \qquad \text{and} \qquad \left(\frac{dy}{dx}\right)^2 + 3\frac{dy}{dx} + 2y = e^x$$

are all differential equations. Differential equations are used extensively in modeling applications, especially those that involve one or more rates of change.

The simplest kind of differential equation has the form

$$\frac{dQ}{dx} = g(x)$$

in which the derivative of the quantity $Q(x)$ is given explicitly as a function of the independent variable x. Such a differential equation is solved by simply finding the indefinite integral of $g(x)$; that is,

$$Q(x) = \int g(x)\, dx$$

For example, the differential equation

$$\frac{dy}{dx} = x^2 + 3x$$

has the solution

$$y = \int (x^2 + 3x)\, dx = \frac{1}{3}x^3 + \frac{3}{2}x^2 + C$$

This is called a **general solution** of the equation because it characterizes an entire family of solutions, one for each choice of the arbitrary constant C.

A more general type of differential equation, called a **separable equation,** has the form

$$\frac{dy}{dx} = \frac{h(x)}{g(y)}$$

and can be solved by algebraically separating the variables and integrating both sides, like this:

$$g(y)\, dy = h(x)\, dx$$

so

$$\int g(y)\, dy = \int h(x)\, dx$$

A brief justification of this procedure is given at the end of this section. Example 5.1.7 illustrates the method for solving a separable differential equation.

EXPLORE!

Refer to Example 5.1.7. Store the family of solution curves $y = (3x^2 + L1)^{\wedge}(1/3)$ into Y1 of the equation editor, with the list $L1 = \{-16, -12, -8, -4, 0, 4, 8, 12, 16\}$. Use the modified decimal window $[-4.7, 4.7]1$ by $[-3, 4]1$ to graph this family of curves, and describe what you observe. Which of these curves passes through the point $(0, 2)$?

EXAMPLE 5.1.7   Solving a Separable Differential Equation

Find the general solution of the differential equation $\dfrac{dy}{dx} = \dfrac{2x}{y^2}$.

Solution

To separate the variables, pretend that the derivative $\dfrac{dy}{dx}$ is actually a quotient and cross multiply to obtain.

$$y^2\, dy = 2x\, dx$$

Now integrate both sides of this equation to get

$$\int y^2\, dy = \int 2x\, dx$$

$$\frac{1}{3}y^3 + C_1 = x^2 + C_2$$

where C_1 and C_2 are arbitrary constants. Solving for y, we get

$$\frac{1}{3}y^3 = x^2 + (C_2 - C_1) = x^2 + C_3 \quad \text{where } C_3 = C_2 - C_1$$

$$y^3 = 3x^2 + 3C_3 = 3x^2 + C \qquad \text{where } C = 3C_3$$

$$y = (3x^2 + C)^{1/3}$$

An **initial value problem** is a differential equation coupled with a side condition that can be used to evaluate the arbitrary constant C in the general solution and so obtain a specific solution to the equation. Examples 5.1.4 through 5.1.6 can be viewed as initial value problems. In Example 5.1.8, we show how a business manager can use an initial value problem to model the flow of revenue.

EXAMPLE 5.1.8   Solving a Differential Equation to Find Revenue

Willis Jenkins owns an oil well that is expected to yield 200 barrels of crude oil per month and, at that rate, is expected to run dry in 3 years. It is estimated that t months from now, the price of crude oil will be $p(t) = 140 + 2.4\sqrt{t}$ dollars per barrel. If Willis is able to sell the oil as soon as it is extracted from the ground, what total revenue should he expect to receive during the well's time of operation?

Solution

Let $R(t)$ be the revenue generated during the first t months after the well is opened, so that $R(0) = 0$. To construct the relevant differential equation, use the rate relationship

$$\text{Rate of change of revenue with respect to time (dollars/month)} = \left(\frac{\text{dollars per}}{\text{barrel}}\right)\left(\frac{\text{barrels per}}{\text{month}}\right)$$

so the flow of revenue from the well can be modeled by the initial value problem

$$\underbrace{\frac{dR}{dt}}_{\substack{\text{rate of change} \\ \text{of revenue}}} = \underbrace{(140 + 2.4\sqrt{t})}_{\substack{\text{dollars per} \\ \text{barrel}}}\underbrace{(200)}_{\substack{\text{barrels} \\ \text{per month}}} \qquad \text{with } R(0) = 0$$

Solving the differential equation

$$\frac{dR}{dt} = (140 + 2.4\sqrt{t})(200)$$

$$= 28{,}000 + 480t^{1/2}$$

we integrate to obtain the general solution

$$R(t) = \int (28{,}000 + 480t^{1/2})\, dt = 28{,}000t + 480\left(\frac{t^{3/2}}{3/2}\right) + C$$

$$= 28{,}000t + 320t^{3/2} + C$$

Since $R(0) = 0$, we find that

$$R(0) = 0 = 28{,}000(0) + 320(0)^{3/2} + C$$

so $C = 0$, and the appropriate particular solution is

$$R(t) = 28{,}000t + 320t^{3/2}$$

Finally, since the well will run dry in 36 months, the total revenue generated by the well during its operation is

$$R(36) = 28{,}000(36) + 320(36)^{3/2}$$

$$= \$1{,}077{,}120$$

Continuous Compounding Using a Differential Equation

A savings account with value $B(t)$ at time t is said to *compound continuously* if the percentage growth rate of the account is equal to the prevailing rate of interest. For example, if the interest rate is 5%, we have

$$\underbrace{\frac{100 \dfrac{dB}{dt}}{B}}_{\substack{\text{percentage} \\ \text{growth rate of } B}} = \underbrace{5}_{\substack{\text{interest} \\ \text{rate}}}$$

It is customary to express the interest rate in decimal form, so 5% becomes $r = 0.05$ and the continuous compounding formula can be expressed in terms of the relative growth rate of $B(t)$:

Just-In-Time **REVIEW**

Recall from Section 2.2 that the *percentage rate of change* of a quantity $Q(t)$ is

$$\frac{100\dfrac{dQ}{dt}}{Q}$$

and the *relative rate of change* of $Q(t)$ is

$$\frac{\dfrac{dQ}{dt}}{Q}$$

$$\frac{100\dfrac{dB}{dt}}{B} = 5 = 100(0.05) \quad \text{divide both sides by 100}$$

so

$$\underbrace{\frac{\dfrac{dB}{dt}}{B}}_{\substack{\text{relative} \\ \text{growth rate of } B}} = \underbrace{0.05}_{\substack{\text{interest rate} \\ \textit{in decimal form}}}$$

In Section 4.1, we used a limit to show that if P dollars (the principal) are invested in an account that pays interest at an annual rate r compounded continuously, then the account will be worth $B(t) = Pe^{rt}$ dollars after t years. In Example 5.1.9, we show how this formula can be obtained by solving a separable initial value problem.

EXAMPLE 5.1.9 Deriving the Continuous Compounding Formula

Solve an initial value problem to show that if P dollars are invested at an annual interest rate r compounded continuously, then the (future) value of the account after t years is $B(t) = Pe^{rt}$ dollars, where r is expressed in decimal form.

Solution

According to the preceding discussion, the relative growth rate of the account value $B(t)$ must equal the interest rate r, so we have the initial value problem

$$\underbrace{\frac{\dfrac{dB}{dt}}{B}}_{\substack{\text{relative} \\ \text{growth rate of } B}} = \underbrace{r}_{\substack{\text{interest rate} \\ \textit{in decimal form}}} \quad \text{with } B(0) = P$$

or, equivalently,

$$\frac{dB}{dt} = rB$$

Separating the variables in this differential equation and integrating, we get

$$\int \frac{dB}{B} = \int r \, dt$$

$$\ln |B| = \ln B = rt + C_1 \quad |B(t)| = B(t) \text{ since } B(t) > 0 \text{ for all } t$$

Taking exponentials on both sides, we find that

$$B(t) = e^{\ln B} = e^{rt + C_1} = e^{rt} e^{C_1} \quad \text{product rule for exponentials}$$
$$= C e^{rt}$$

where $C = e^{C_1}$. Since $B(0) = P$, it follows that

$$P = B(0) = C e^0 = C \quad e^0 = 1$$

so $B(t) = Pe^{rt}$ as claimed.

Any quantity $Q(t)$ that changes at a constant relative rate like a continuously compounding savings account is said to **grow exponentially** if the growth rate is positive or **decay exponentially** if the rate is negative. Another way of saying this is that $Q(t)$ grows or decays exponentially if its derivative is proportional to Q itself; that is,

$$\frac{dQ}{dt} = kQ \quad \text{exponential growth if } k > 0 \text{ and decay if } k < 0$$

For any such quantity, the argument in Example 5.1.9 can be used to show that

$$Q(t) = Q_0 e^{kt} \quad \text{for all } t$$

where $Q_0 = Q(0)$ is the amount of the quantity at $t = 0$. Exponential growth and decay models are extremely important and occur in such diverse phenomena as the decay of radioactive substances, the growth of certain populations, and the elimination of a drug from a person's bloodstream.

Why the Method of Separation of Variables Works

To see why separation of variables works, consider the separable differential equation

$$\frac{dy}{dx} = \frac{h(x)}{g(y)}$$

or, equivalently,

$$g(y) \frac{dy}{dx} - h(x) = 0$$

The left-hand side of this equation can be rewritten in terms of the antiderivatives of g and h. In particular, if G is an antiderivative of g and H an antiderivative of h, it follows from the chain rule that

$$\frac{d}{dx}[G(y) - H(x)] = G'(y)\frac{dy}{dx} - H'(x) = g(y)\frac{dy}{dx} - h(x)$$

Hence, the differential equation $g(y)\dfrac{dy}{dx} - h(x) = 0$ says that

$$\frac{d}{dx}[G(y) - H(x)] = 0$$

But constants are the only functions whose derivatives are identically zero, and so

$$G(y) - H(x) = C$$

for some constant C. That is,

$$G(y) = H(x) + C$$

or, equivalently,

$$\int g(y)\, dy = \int h(x)\, dx + C$$

and the proof is complete.

EXERCISES ▪ 5.1

In Exercises 1 through 30, find the indicated integral. Check your answers by differentiation.

1. $\displaystyle\int -3\, dx$

2. $\displaystyle\int dx$

3. $\displaystyle\int x^5\, dx$

4. $\displaystyle\int \sqrt{t}\, dt$

5. $\displaystyle\int \frac{1}{x^2}\, dx$

6. $\displaystyle\int 3e^x\, dx$

7. $\displaystyle\int \frac{2}{\sqrt{t}}\, dt$

8. $\displaystyle\int x^{-0.3}\, dx$

9. $\displaystyle\int u^{-2/5}\, du$

10. $\displaystyle\int \left(\frac{1}{x^2} - \frac{1}{x^3}\right) dx$

11. $\displaystyle\int (3t^2 - \sqrt{5}\,t + 2)\, dt$

12. $\displaystyle\int (x^{1/3} - 3x^{-2/3} + 6)\, dx$

13. $\displaystyle\int (3\sqrt{y} - 2y^{-3})\, dy$

14. $\displaystyle\int \left(\frac{1}{2y} - \frac{2}{y^2} + \frac{3}{\sqrt{y}}\right) dy$

15. $\displaystyle\int \left(\frac{e^x}{2} + x\sqrt{x}\right) dx$

16. $\displaystyle\int \left(\sqrt{x^3} - \frac{1}{2\sqrt{x}} + \sqrt{2}\right) dx$

17. $\displaystyle\int u^{1.1}\left(\frac{1}{3u} - 1\right) du$

18. $\displaystyle\int \left(2e^u + \frac{6}{u} + \ln 2\right) du$

19. $\displaystyle\int \left(\frac{x^2 + 2x + 1}{x^2}\right) dx$

20. $\displaystyle\int \frac{x^2 + 3x - 2}{\sqrt{x}}\, dx$

21. $\displaystyle\int (x^3 - 2x^2)\left(\frac{1}{x} - 5\right) dx$

22. $\displaystyle\int y^3\left(2y + \frac{1}{y}\right) dy$

23. $\displaystyle\int \sqrt{t}(t^2 - 1)\, dt$

24. $\displaystyle\int x(2x + 1)^2\, dx$

25. $\displaystyle\int (e^t + 1)^2\, dt$

26. $\displaystyle\int e^{-0.02t}(e^{-0.13t} + 4)\, dt$

27. $\displaystyle\int \left(\frac{1}{3y} - \frac{5}{\sqrt{y}} + e^{-y/2}\right) dy$

28. $\displaystyle\int \frac{1}{x}(x + 1)^2\, dx$

29. $\displaystyle\int t^{-1/2}(t^2 - t + 2)\, dt$

30. $\displaystyle\int \ln(e^{-x^2})\, dx$

In Exercises 31 through 34, solve the given initial value problem for $y = f(x)$.

31. $\dfrac{dy}{dx} = 3x - 2$ where $y = 2$ when $x = -1$

32. $\dfrac{dy}{dx} = e^{-x}$ where $y = 3$ when $x = 0$

33. $\dfrac{dy}{dx} = \dfrac{2}{x} - \dfrac{1}{x^2}$ where $y = -1$ when $x = 1$

34. $\dfrac{dy}{dx} = \dfrac{x + 1}{\sqrt{x}}$ where $y = 5$ when $x = 4$

In Exercises 35 through 42, the slope $f'(x)$ at each point (x, y) on a curve $y = f(x)$ is given along with a particular point (a, b) on the curve. Use this information to find $f(x)$.

35. $f'(x) = 4x + 1$; $(1, 2)$

36. $f'(x) = 3 - 2x$; $(0, -1)$

37. $f'(x) = -x(x + 1)$; $(-1, 5)$

38. $f'(x) = 3x^2 + 6x - 2$; $(0, 6)$

39. $f'(x) = x^3 - \dfrac{2}{x^2} + 2$; $(1, 3)$

40. $f'(x) = x^{-1/2} + x$; $(1, 2)$

41. $f'(x) = e^{-x} + x^2$; $(0, 4)$

42. $f'(x) = \dfrac{3}{x} - 4$; $(1, 0)$

In Exercises 43 through 46, solve the given separable initial value problem.

43. $\dfrac{dy}{dx} = 2y$; $y = 3$ when $x = 0$

44. $\dfrac{dy}{dx} = xy$; $y = 1$ when $x = 0$

45. $\dfrac{dy}{dx} = e^{x+y}$; $y = 0$ when $x = 0$

46. $\dfrac{dy}{dx} = \sqrt{\dfrac{y}{x}}$; $y = 1$ when $x = 1$

BUSINESS AND ECONOMICS APPLIED PROBLEMS

47. MARGINAL COST A manufacturer estimates that the marginal cost of producing q units of a certain commodity is $C'(q) = 3q^2 - 24q + 48$ dollars per unit. If the cost of producing 10 units is $5,000, what is the cost of producing 30 units?

48. MARGINAL REVENUE The marginal revenue derived from producing q units of a certain commodity is $R'(q) = 4q - 1.2q^2$ dollars per unit. If the revenue derived from producing 20 units is $30,000, how much revenue should be expected from producing 40 units?

49. MARGINAL PROFIT A manufacturer estimates marginal revenue to be $R'(q) = 100q^{-1/2}$ dollars per unit when the level of production is q units. The corresponding marginal cost has been found to be $0.4q$ dollars per unit. Suppose the manufacturer's profit is $520 when the level of production is

16 units. What is the manufacturer's profit when the level of production is 25 units?

50. SALES The monthly sales at an import store are currently $10,000 but are expected to be declining at the rate of

$$S'(t) = -10t^{2/5} \quad \text{dollars per month}$$

t months from now. The store is profitable as long as the sales level is above $8,000 per month.

  a. Find a formula for the expected sales in t months.

  b. What sales figure should be expected 2 years from now?

  c. For how many months will the store remain profitable?

51. ADVERTISING After initiating an advertising campaign, Lazaro Jimenez, the manager of a satellite dish provider in an urban area, estimates that t months from now, the number of new subscribers $N(t)$ will be growing at the rate of $N'(t) = 154t^{2/3} + 37$ subscribers per month. How many new subscribers should Lazaro expect 8 months from now?

52. MARGINAL REVENUE Suppose it has been determined that the marginal revenue associated with the production of x units of a particular commodity is $R'(x) = 240 - 4x$ dollars per unit. What is the revenue function $R(x)$? You may assume $R(0) = 0$. What price will be paid for each unit when the level of production is $x = 5$ units?

53. MARGINAL PROFIT The marginal profit of a certain commodity is $P'(q) = 100 - 2q$ when q units are produced. When 10 units are produced, the profit is $700.

  a. Find the profit function $P(q)$.

  b. What production level q results in maximum profit? What is the maximum profit?

54. PRODUCTION At a certain factory, when K thousand dollars is invested in the plant, the production Q is changing at a rate given by

$$Q'(K) = 200K^{-2/3}$$

units per thousand dollars invested. When $8,000 is invested, the level of production is 5,500 units.

  a. Find a formula for the level of production Q to be expected when K thousand dollars is invested.

  b. How many units will be produced when $27,000 is invested?

  c. What capital investment K is required to produce 7,000 units?

55. MARGINAL PROPENSITY TO CONSUME
Suppose the consumption function for a particular country is $c(x)$, where x is national disposable income. Then the **marginal propensity to consume** is $c'(x)$. Suppose x and c are both measured in billions of dollars and
$$c'(x) = 0.9 + 0.3\sqrt{x}$$
If consumption is 10 billion dollars when $x = 0$, find $c(x)$.

56. MARGINAL ANALYSIS   A manufacturer estimates marginal revenue to be $200q^{-1/2}$ dollars per unit when the level of production is q units. The corresponding marginal cost has been found to be $0.4q$ dollars per unit. If the manufacturer's profit is \$2,000 when the level of production is 25 units, what is the profit when the level of production is 36 units?

57. OIL PRODUCTION   A certain oil well that yields 400 barrels of crude oil per month will run dry in 2 years. The price of crude oil is currently \$98 per barrel and is expected to rise at the constant rate of 40 cents per barrel per month. If the oil is sold as soon as it is extracted from the ground, how much total revenue will be obtained from the well?

58. OIL PRODUCTION   Suppose the owner of the oil well in Exercise 57 decides to step up production so that 600 barrels per month are extracted but everything else remains the same.
a. How many months pass before the well runs dry?
b. How much total revenue will be obtained from the well?

LIFE AND SOCIAL SCIENCE APPLIED PROBLEMS

59. POPULATION GROWTH   It is estimated that t months from now the population of a certain town will be increasing at the rate of $4 + 5t^{2/3}$ people per month. If the current population is 10,000, what will the population be 8 months from now?

60. NET CHANGE IN A BIOMASS   A biomass is growing at the rate of $M'(t) = 0.5e^{0.2t}$ g/hr. By how much does the mass change during the second hour?

61. BACTERIAL POPULATION   After t hours of observation, the population $P(t)$ of a bacterial colony is found to be changing at the rate
$$\frac{dP}{dt} = 200e^{0.1t} + 150e^{-0.03t}$$
If the population was 200,000 bacteria when observation began, what will it be 12 hours later?

62. TREE GROWTH   An environmentalist finds that a certain type of tree grows in such a way that its height $h(t)$ after t years is changing at the rate of
$$h'(t) = 0.2t^{2/3} + \sqrt{t} \quad \text{ft/yr}$$
If the tree was 2 feet tall when it was planted, how tall will it be in 27 years?

63. LEARNING   Let $f(x)$ represent the total number of items a subject has memorized x minutes after being presented with a long list of items to learn. Psychologists refer to the graph of $y = f(x)$ as a **learning curve** and to $f'(x)$ as the **learning rate.** The time of **peak efficiency** is the time when the learning rate is maximized. Suppose the learning rate is
$$f'(x) = 0.1(10 + 12x - 0.6x^2) \quad \text{for } 0 \le x \le 25$$
a. When does peak efficiency occur? What is the learning rate at peak efficiency?
b. What is $f(x)$? You may assume that $f(0) = 0$.
c. What is the largest number of items memorized by the subject?

64. CORRECTIONAL FACILITY MANAGEMENT
Statistics compiled by the local department of corrections indicate that x years from now the number of inmates in county prisons will be increasing at the rate of $280e^{0.2x}$ per year. Currently, 2,000 inmates are housed in county prisons. How many inmates should the county expect 10 years from now?

65. LEARNING   Rob is taking a learning test in which the time he takes to memorize items from a given list is recorded. Let $M(t)$ be the number of items he can memorize in t minutes. His learning rate is found to be
$$M'(t) = 0.4t - 0.005t^2$$
a. How many items can Rob memorize during the first 10 minutes?
b. How many additional items can he memorize during the next 10 minutes (from time $t = 10$ to $t = 20$)?

66. ENDANGERED SPECIES   A conservationist finds that the population $P(t)$ of a certain endangered species is growing at a rate given by $P'(t) = 0.51e^{-0.03t}$, where t is the number of years after records began to be kept.
a. If the population is $P_0 = 500$ now (at time $t = 0$), what will it be in 10 years?
b. Read an article on endangered species, and write a paragraph on the use of mathematical models in studying populations of such species.

67. **CANCER THERAPY** A new medical procedure is applied to a cancerous tumor with volume 30 cm³, and t days later the volume is found to be changing at the rate

$$V'(t) = 0.15 - 0.09e^{0.006t} \quad \text{cm}^3/\text{day}$$

   a. Find a formula for the volume of the tumor after t days.
   b. What is the volume after 60 days? After 120 days?
   c. For the procedure to be successful, it should take no longer than 90 days for the tumor to begin to shrink. Based on this criterion, does the procedure succeed?

68. **RESPONSE TO STIMULUS** The Weber-Fechner law in experimental psychology specifies that the rate of change of a response R with respect to the level of stimulus S is inversely proportional to the stimulus; that is,

$$\frac{dR}{dS} = \frac{k}{S}$$

   Let S_0 be the threshold stimulus, that is, the highest level of stimulus for which there is no response, so that $R = 0$ when $S = S_0$.
   a. Solve this separable differential equation for $R(S)$. Your answer will involve k and S_0.
   b. Sketch the graph of the response function $R(S)$ found in part (a).

69. **FLOW OF BLOOD** One of Poiseuille's laws for the flow of blood in an artery says that if $v(r)$ is the velocity of flow r cm from the central axis of the artery, then the velocity decreases at a rate proportional to r. That is,

$$v'(r) = -ar$$

   where a is a positive constant.* Find an expression for $v(r)$. Assume $v(R) = 0$, where R is the radius of the artery.

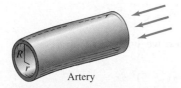

Artery

EXERCISE 69

70. **ALLOMETRY** The different members or organs of an individual often grow at different rates, and an important part of **allometry** involves the study of relationships among these growth rates (recall

---

*E. Batschelet, *Introduction to Mathematics for Life Scientists*, 2nd ed., New York: Springer-Verlag, 1979, pp. 101–103.

the Think About It essay at the end of Chapter 1). Suppose $x(t)$ is the size (length, volume, or weight) at time t of one organ or member of an individual organism and $y(t)$ is the size of another organ or member of the same individual. Then the **law of allometry** states that the relative growth rates of x and y are proportional; that is,

$$\frac{y'(t)}{y(t)} = k\frac{x'(t)}{x(t)} \quad \text{for some constant } k > 0$$

First show that the allometry law can be written as

$$\frac{dy}{dx} = k\frac{y}{x}$$

then solve this separable equation for y in terms of x.

MISCELLANEOUS PROBLEMS

71. **DEFROSTING** A roast is removed from the freezer of a refrigerator and left on the counter to defrost. The temperature of the roast was $-4°C$ when it was removed from the freezer and t hours later, was increasing at the rate of

$$T'(t) = 7e^{-0.35t} \quad °C/\text{hr}$$

   a. Find a formula for the temperature of the roast after t hours.
   b. What is the temperature after 2 hours?
   c. Assume the roast is defrosted when its temperature reaches 10°C. How long does it take for the roast to defrost?

72. **DISTANCE AND VELOCITY** An object is moving so that its velocity after t minutes is $v(t) = 3 + 2t + 6t^2$ meters per minute. How far does the object travel during the second minute?

73. **SPY STORY** Our spy, intent on avenging the death of Siggy Leiter (Exercise 75 in Section 4.2), is driving a sports car toward the lair of the fiend who killed his friend. To remain as inconspicuous as possible, he is traveling at the legal speed of 60 mph (88 feet per second) when suddenly, he sees a camel in the road, 199 feet in front of him. It takes him 00.7 seconds to react to the crisis. Then he hits the brakes, and the car decelerates at the constant rate of 28 ft/sec² (28 feet per second, per second). Does he stop before hitting the camel?

74. If $H'(x) = 0$ for all real numbers x, what must be true about the graph of $H(x)$? Explain how your observation can be used to show that if $G'(x) = F'(x)$ for all x, then $G(x) = F(x) + C$ for constant C.

75. A car traveling at 67 ft/sec decelerates at the constant rate of 23 ft/sec² when the brakes are applied.
  a. Find the velocity $v(t)$ of the car t seconds after the brakes are applied. Then find its distance $s(t)$ from the point where the brakes are applied.
  b. Use the graphing utility of your calculator to sketch the graphs of $v(t)$ and $s(t)$ on the same screen (use [0, 5]1 by [0, 200]10).
  c. Use **TRACE** and **ZOOM** to determine when the car comes to a complete stop and how far it travels in that time. How fast is the car traveling when it has traveled 45 feet?

76. a. Prove the constant rule: $\int k \, dx = kx + C$.

  b. Prove the exponential rule: $\int e^{kx} \, dx = \frac{1}{k} e^{kx} + C$.

77. What is $\int b^x \, dx$ for base b ($b > 0, b \neq 1$)?
  [*Hint:* Recall that $b^x = e^{x \ln b}$.]

 78. It is estimated that x months from now, the population of a certain town will be changing at the rate of $P'(x) = 2 + 1.5\sqrt{x}$ people per month. The current population is 5,000.
  a. Find a function $P(x)$ that satisfies these conditions. Use the graphing utility of your calculator to graph this function.
  b. Use **TRACE** and **ZOOM** to determine the level of population 9 months from now. When will the population be 7,590?
  c. Suppose the current population were 2,000 (instead of 5,000). Sketch the graph of $P(x)$ with this assumption. Then sketch the graph of $P(x)$ assuming current populations of 4,000 and 6,000. What is the difference between the graphs?

SECTION 5.2   Integration by Substitution

Learning Objectives
1. Use the method of substitution to find indefinite integrals.
2. Solve initial value problems using substitution.
3. Explore a price-adjustment model in economics.

The majority of functions that occur in practical situations can be differentiated by applying rules and formulas such as those you learned in Chapter 2. Integration, however, is at least as much an art as a science, and many integrals that appear deceptively simple may actually require a special technique or clever insight.

For example, we easily find that

$$\int x^7 \, dx = \frac{1}{8} x^8 + C$$

by applying the power rule, but suppose we wish to compute

$$\int (3x + 5)^7 \, dx$$

We could proceed by expanding the integrand $(3x + 5)^7$ and then integrating term by term, but the algebra involved in this approach is daunting. Instead, we make the change of variable

Just-In-Time **REVIEW**

Recall that the differential of $y = f(x)$ is $dy = f'(x) \, dx$.

$$u = 3x + 5 \quad \text{so that} \quad du = 3 \, dx \quad \text{or} \quad dx = \frac{1}{3} \, du$$

Then, by substituting these quantities into the given integral, we get

$$\int (3x + 5)^7 \, dx = \int u^7 \left(\frac{1}{3} \, du \right) \qquad \text{power rule}$$

$$= \frac{1}{3} \left(\frac{1}{8} u^8 \right) + C = \frac{1}{24} u^8 + C \qquad \begin{array}{l} \text{substitute} \\ u = 3x + 5 \end{array}$$

$$= \frac{1}{24} (3x + 5)^8 + C$$

We can check this computation by differentiating using the chain rule (Section 2.4):

$$\frac{d}{dx} \left[\frac{1}{24} (3x + 5)^8 \right] = \frac{1}{24} [8(3x + 5)^7 (3)] = (3x + 5)^7$$

which verifies that $\dfrac{1}{24}(3x + 5)^8$ is indeed an antiderivative of $(3x + 5)^7$.

The change of variable procedure we have just demonstrated is called **integration by substitution,** and it amounts to reversing the chain rule for differentiation. To see why, consider an integral that can be written as

$$\int f(x) \, dx = \int g(u(x)) u'(x) \, dx$$

for certain functions g and u. Suppose G is an antiderivative of g, so that $G' = g$. Then, according to the chain rule

$$\frac{d}{dx} [G(u(x))] = G'(u(x)) u'(x)$$

$$= g(u(x)) u'(x)$$

and by integrating both sides of this equation with respect to x, we find that

$$\int f(x) \, dx = \int g(u(x)) u'(x) \, dx$$

$$= \int \left(\frac{d}{dx} [G(u(x))] \right) dx \qquad \int G' = G$$

$$= G(u(x)) + C$$

In other words, once we have an antiderivative for $g(u)$, we also have one for $f(x)$.

A useful device for remembering the substitution procedure is to think of $u = u(x)$ as a change of variable whose differential $du = u'(x) \, dx$ can be manipulated algebraically. Then

$$\int f(x) \, dx = \int g(u(x)) u'(x) \, dx \qquad \text{substitute } du \text{ for } u'(x) \, dx$$

$$= \int g(u) \, du \qquad\qquad\qquad G \text{ is an antiderivative of } g$$

$$= G(u) + C \qquad\qquad\qquad \text{substitute } u(x) \text{ for } u$$

$$= G(u(x)) + C$$

Here is a step-by-step procedure for integrating by substitution.

Using Substitution to Integrate $\int f(x)\, dx$

Step 1. Choose a substitution $u = u(x)$ that "simplifies" the integrand $f(x)$.

Step 2. Express the entire integral in terms of u and $du = u'(x)\, dx$. This means that *all* terms involving x and dx must be transformed into terms involving u and du.

Step 3. When step 2 is complete, the given integral should have the form

$$\int f(x)\, dx = \int g(u)\, du$$

If possible, evaluate this transformed integral by finding an antiderivative $G(u)$ for $g(u)$.

Step 4. Replace u by $u(x)$ in $G(u)$ to obtain an antiderivative $G(u(x))$ for $f(x)$, so that

$$\int f(x)\, dx = G(u(x)) + C$$

An old saying goes, "The first step in making rabbit stew is to catch a rabbit." Likewise, the first step in integrating by substitution is to find a suitable change of variable $u = u(x)$ that simplifies the integrand of the given integral $\int f(x)\, dx$ without adding undesired complexity when dx is replaced by $du = u'(x)\, dx$. Here are a few guidelines for choosing $u(x)$:

1. If possible, try to choose u so that $u'(x)$ is part of the integrand $f(x)$.
2. Consider choosing u as the part of the integrand that makes $f(x)$ difficult to integrate directly, such as the quantity inside a radical, the denominator of a fraction, or the exponent of an exponential function.
3. Don't "oversubstitute." For instance, in our introductory example $\int (3x + 5)^7\, dx$, a common mistake is to use $u = (3x + 5)^7$. This certainly simplifies the integrand, but then $du = 7(3x + 5)^6(3)\, dx$, and you are left with a transformed integral that is more complicated than the original.
4. Persevere. If you try a substitution that does not result in a transformed integral you can evaluate, try a different substitution.

Examples 5.2.1 through 5.2.7 illustrate how substitutions are chosen and used in various kinds of integrals.

EXAMPLE 5.2.1   Integration by a Linear Substitution

Find $\int \sqrt{2x + 7}\, dx$.

Solution

We choose $u = 2x + 7$ and obtain

$$du = 2\ dx \qquad \text{so that} \qquad dx = \frac{1}{2}\ du$$

Then the integral becomes

$$\int \sqrt{2x + 7}\ dx = \int \sqrt{u}\left(\frac{1}{2}\ du\right) \qquad \text{rewrite } \sqrt{u} \text{ as } u^{1/2}$$

$$= \frac{1}{2}\int u^{1/2}\ du \qquad \text{use the power rule}$$

$$= \frac{1}{2}\frac{u^{3/2}}{3/2} + C = \frac{1}{3}u^{3/2} + C \qquad \begin{array}{l}\text{substitute } 2x + 7 \\ \text{for } u\end{array}$$

$$= \frac{1}{3}(2x + 7)^{3/2} + C$$

EXAMPLE 5.2.2 **Integration by a Quadratic Substitution**

Find $\displaystyle\int 8x(4x^2 - 3)^5\ dx$.

Solution

First, note that the integrand $8x(4x^2 - 3)^5$ is a product in which one of the factors, $8x$, is the derivative of an expression, $4x^2 - 3$, that appears in the other factor. This suggests that you make the substitution

$$u = 4x^2 - 3 \qquad \text{with} \qquad du = 4(2x\ dx) = 8x\ dx$$

to obtain

$$\int 8x(4x^2 - 3)^5\ dx = \int (4x^2 - 3)^5(8x\ dx)$$

$$= \int u^5\ du$$

$$= \frac{1}{6}u^6 + C \qquad \text{substitute } 4x^2 - 3 \text{ for } u$$

$$= \frac{1}{6}(4x^2 - 3)^6 + C$$

EXAMPLE 5.2.3 **Using Substitution on an Exponential Function**

Find $\displaystyle\int x^3 e^{x^4 + 2}\ dx$.

Solution

If the integrand of an integral contains an exponential function, it is often useful to substitute for the exponent. In this case, we choose

$$u = x^4 + 2 \qquad \text{so that} \qquad du = 4x^3\, dx$$

and

$$\int x^3 e^{x^4+2}\, dx = \int e^{x^4+2}(x^3\, dx) \qquad du = 4x^3\, dx \text{ so } x^3\, dx = \frac{1}{4}\, du$$

$$= \int e^u \left(\frac{1}{4}\, du\right) \qquad \text{use exponential rule}$$

$$= \frac{1}{4}e^u + C \qquad \text{substitute } x^4 + 2 \text{ for } u$$

$$= \frac{1}{4}e^{x^4+2} + C$$

EXAMPLE 5.2.4   Using Substitution on a Rational Function

Find $\displaystyle\int \frac{x}{x-1}\, dx$.

Solution

Following our guidelines, we substitute for the denominator of the integrand, so that $u = x - 1$ and $du = dx$. Since $u = x - 1$, we also have $x = u + 1$. Thus,

$$\int \frac{x}{x-1}\, dx = \int \frac{u+1}{u}\, du \qquad \text{divide}$$

$$= \int \left[1 + \frac{1}{u}\right] du \qquad \begin{array}{l}\text{use the constant and}\\ \text{logarithmic rules}\end{array}$$

$$= u + \ln|u| + C \qquad \text{substitute } x - 1 \text{ for } u$$

$$= x - 1 + \ln|x-1| + C$$

EXAMPLE 5.2.5   Using Substitution on a Radical Function

Find $\displaystyle\int \frac{3x+6}{\sqrt{2x^2+8x+3}}\, dx$.

Solution

This time, our guidelines suggest substituting for the quantity inside the radical in the denominator; that is,

$$u = 2x^2 + 8x + 3 \qquad du = (4x+8)\, dx$$

At first glance, it may seem that this substitution fails, since $du = (4x + 8)\, dx$ appears quite different from the term $(3x + 6)\, dx$ in the integral. However, note that

$$(3x+6)\, dx = 3(x+2)\, dx = \frac{3}{4}(4)[(x+2)\, dx]$$

$$= \frac{3}{4}[(4x+8)\, dx] = \frac{3}{4}\, du$$

Substituting, we find that

$$\int \frac{3x + 6}{\sqrt{2x^2 + 8x + 3}} \, dx = \int \frac{1}{\sqrt{2x^2 + 8x + 3}} [(3x + 6) \, dx]$$

$$= \int \frac{1}{\sqrt{u}} \left(\frac{3}{4} \, du \right) = \frac{3}{4} \int u^{-1/2} \, du \qquad \text{rewrite } \frac{1}{\sqrt{u}} \text{ as } u^{-1/2}$$

$$= \frac{3}{4} \left(\frac{u^{1/2}}{1/2} \right) + C = \frac{3}{2} \sqrt{u} + C \qquad \begin{array}{l}\text{substitute}\\ u = 2x^2 + 8x + 3\end{array}$$

$$= \frac{3}{2} \sqrt{2x^2 + 8x + 3} + C$$

EXAMPLE 5.2.6 Using Substitution on a Logarithmic Function

Find $\int \frac{(\ln x)^2}{x} \, dx$.

Solution

Because

$$\frac{d}{dx} (\ln x) = \frac{1}{x}$$

the integrand

$$\frac{(\ln x)^2}{x} = (\ln x)^2 \left(\frac{1}{x} \right)$$

is a product in which one factor $\frac{1}{x}$ is the derivative of an expression $\ln x$ that appears in the other factor. This suggests substituting $u - \ln x$ with $du = \frac{1}{x} \, dx$ so that

$$\int \frac{(\ln x)^2}{x} \, dx = \int (\ln x)^2 \left(\frac{1}{x} \, dx \right)$$

$$= \int u^2 \, du = \frac{1}{3} u^3 + C \quad \text{substitute } \ln x \text{ for } u$$

$$= \frac{1}{3} (\ln x)^3 + C$$

EXAMPLE 5.2.7 Using Algebra Before Substituting

Find $\int \frac{x^2 + 3x + 5}{x + 1} \, dx$.

Solution

There is no easy way to approach this integral as it stands (remember, there is no quotient rule for integration). However, suppose we simply divide the denominator into the numerator:

$$
\begin{array}{r}
x + 2 \\
x + 1 \overline{\smash{\big)}\ x^2 + 3x + 5} \\
\underline{-x(x + 1)} \\
2x + 5 \\
\underline{-2(x + 1)} \\
3
\end{array}
$$

that is,

$$
\frac{x^2 + 3x + 5}{x + 1} = x + 2 + \frac{3}{x + 1}
$$

We can integrate $x + 2$ directly, and for the term $\dfrac{3}{x + 1}$, we use the substitution $u = x + 1$ with $du = dx$:

$$
\int \frac{x^2 + 3x + 5}{x + 1}\, dx = \int \left[x + 2 + \frac{3}{x + 1} \right] dx \qquad \begin{aligned} u &= x + 1 \\ du &= dx \end{aligned}
$$

$$
= \int x\, dx + \int 2\, dx + \int \frac{3}{u}\, du
$$

$$
= \frac{1}{2}x^2 + 2x + 3 \ln |u| + C \qquad \text{substitute } x + 1 \text{ for } u
$$

$$
= \frac{1}{2}x^2 + 2x + 3 \ln |x + 1| + C
$$

The method of substitution does not always succeed. In Example 5.2.8, we consider an integral very similar to the one in Example 5.2.3 but just different enough so no substitution will work.

EXAMPLE 5.2.8   An Integral Where Substitution Fails

Evaluate $\displaystyle\int x^4 e^{x^4 + 2}\, dx.$

Solution

The natural substitution is $u = x^4 + 2$, as in Example 5.2.3. Again, you find $du = 4x^3\, dx$, so $x^3\, dx = \dfrac{1}{4}\, du$, but this integrand involves x^4, not x^3. The "extra" factor of x satisfies $x = \sqrt[4]{u - 2}$, so when the substitution is made, you have

$$
\int x^4 e^{x^4 + 2}\, dx = \int x e^{x^4 + 2}(x^3\, dx) = \int \sqrt[4]{u - 2}\ e^u \left(\frac{1}{4}\, du \right)
$$

which is hardly an improvement on the original integral! Try a few other possible substitutions (say, $u = x^2$ or $u = x^3$) to convince yourself that nothing works. This does not mean that the integrand has no antiderivative, only that substitution fails.

Differential Equations Involving Substitution

The method of substitution is frequently used in solving differential equations. The procedure is illustrated in Examples 5.2.9 and 5.2.10.

EXAMPLE 5.2.9   Using Substitution to Solve a Separable Differential Equation

Find the general solution of the separable differential equation

$$\frac{dy}{dx} = \frac{\sqrt{4 - y^2}}{xy}$$

Solution

First, separate the variables to obtain

$$\int \frac{y\,dy}{\sqrt{4 - y^2}} = \int \frac{dx}{x}$$

The integration on the right side is straightforward. On the left, we use the substitution

$$u = 4 - y^2 \qquad du = -2y\,dy \qquad y\,dy = -\frac{1}{2}\,du$$

Integrating, we obtain

$$\int \frac{y\,dy}{\sqrt{4 - y^2}} = \int \frac{dx}{x}$$

$$\int \frac{(-1/2)\,du}{\sqrt{u}} = \int \frac{dx}{x} \qquad \text{rewrite } \frac{1}{\sqrt{u}} \text{ as } u^{-1/2}$$

$$\int -\frac{1}{2}u^{-1/2}\,du = \int x^{-1}\,dx \qquad \text{power and logarithmic rules}$$

$$\frac{(-1/2)\,u^{1/2}}{1/2} = \ln|x| + C$$

Finally, substituting $u = 4 - y^2$ and simplifying, we get

$$-\sqrt{4 - y^2} = \ln|x| + C$$

NOTE   In Example 5.2.9, we left the solution to the differential equation in implicit form (an equation involving x and y). This is often necessary since solving to obtain an explicit solution for y in terms of x may be impossible or extremely time consuming. ∎

EXAMPLE 5.2.10   Obtaining Price from Its Rate of Change

The price p (dollars) of each unit of a particular commodity is estimated to be changing at the rate

$$\frac{dp}{dx} = \frac{-135x}{\sqrt{9 + x^2}}$$

where x (hundred) units is the consumer demand (the number of units purchased at that price). Suppose 400 units ($x = 4$) are demanded when the price is $30 per unit.

a. Find the demand function $p(x)$.

b. At what price will 300 units be demanded? At what price will no units be demanded?

c. How many units are demanded at a price of $20 per unit?

Solution

a. The price per unit $p(x)$ is found by integrating $\dfrac{dp}{dx}$ with respect to x. To perform this integration, use the substitution

$$u = 9 + x^2 \qquad du = 2x\,dx \qquad x\,dx = \frac{1}{2}\,du$$

to get

$$p(x) = \int \frac{-135x}{\sqrt{9 + x^2}}\,dx = \int \frac{-135}{u^{1/2}}\left(\frac{1}{2}\right)du$$

$$= -\frac{135}{2}\int u^{-1/2}\,du$$

$$= -\frac{135}{2}\left(\frac{u^{1/2}}{1/2}\right) + C \qquad \text{substitute } 9 + x^2 \text{ for } u$$

$$= -135\sqrt{9 + x^2} + C$$

Since $p = 30$ when $x = 4$, you find that

$$30 = -135\sqrt{9 + 4^2} + C$$
$$C = 30 + 135\sqrt{25} = 705$$

so

$$p(x) = -135\sqrt{9 + x^2} + 705$$

b. When 300 units are demanded, $x = 3$ and the corresponding price is

$$p(3) = -135\sqrt{9 + 3^2} + 705 = \$132.24 \text{ per unit}$$

No units are demanded when $x = 0$, and the corresponding price is

$$p(0) = -135\sqrt{9 + 0} + 705 = \$300 \text{ per unit}$$

c. To determine the number of units demanded at a unit price of $20 per unit, you need to solve the equation

$$-135\sqrt{9 + x^2} + 705 = 20$$
$$135\sqrt{9 + x^2} = 685$$
$$\sqrt{9 + x^2} = \frac{685}{135} \qquad \text{square both sides}$$
$$9 + x^2 \approx 25.75$$
$$x^2 \approx 16.75$$
$$x \approx 4.09$$

That is, roughly 4.09 hundred units (409 units) will be demanded when the price is $20 per unit.

A Price Adjustment Model

Let $S(p)$ denote the number of units of a particular commodity supplied to the market at a price of p dollars per unit, and let $D(p)$ denote the corresponding number of units demanded by the market at the same price. In static circumstances, market equilibrium occurs at the price where demand equals supply (recall the discussion in Section 1.4). However, certain economic models consider a more dynamic kind of economy in which price, supply, and demand are assumed to vary with time. One of these, the *Evans price adjustment model*,* assumes that the rate of change of price with respect to time t is proportional to the shortage $D - S$, so that

$$\frac{dp}{dt} = k(D - S)$$

where k is a positive constant. Example 5.2.11 involves this model.

EXAMPLE 5.2.11 Studying a Price-Adjustment Model

Byung-Soon is the sales manager for a product that is introduced with an initial price of \$5 per unit. She determines that t months later, the price $p(t)$ will be changing at a rate equal to 2% of the shortage $D - S$, where the supply $S(p)$ and demand $D(p)$ of the product (each in thousands of units) are given by

$$D = 50 - p \quad \text{and} \quad S = 23 + 2p \quad \text{for } 0 \le p < 9$$

a. Set up and solve an initial value problem for the unit price $p(t)$.

b. How much revenue should Byung-Soon expect from the sale of the product after 6 months?

c. What is the equilibrium price p_e (where supply equals demand)? Show that $p(t)$ approaches p_e in the long run (as $t \to \infty$).

Solution

a. According to the given information, the unit price $p(t)$ satisfies the initial value problem

$$\frac{dp}{dt} = k(D - S) = \underbrace{0.02[(50 - p) - (23 + 2p)]}_{\text{2\% of the shortage } D - S} \quad \text{with } \underbrace{p(0) = 5}_{\text{initial price}}$$

$$\frac{dp}{dt} = 0.02(27 - 3p) = 0.06(9 - p)$$

Separating the variables and integrating, we find that

$$\int \frac{dp}{9 - p} = \int 0.06 \, dt \qquad\qquad \begin{array}{l}\text{substitute } u = 9 - p, \, du = -dp \\ |9 - p| = 9 - p \text{ since } p < 9\end{array}$$

$$-\ln(9 - p) = 0.06t + C_1 \qquad\qquad \begin{array}{l}\text{multiply by } -1 \text{ and take} \\ \text{exponentials on both sides}\end{array}$$

$$9 - p = e^{-0.06t}e^{-C_1} = Ce^{-0.06t} \qquad\qquad \text{where } C = e^{-C_1}$$

$$p = 9 - Ce^{-0.06t}$$

---

*The Evans price adjustment model and several other dynamic economic models are examined in the text by J. E. Draper and J. S. Klingman, *Mathematical Analysis with Business and Economic Applications*, New York: Harper and Row, 1967, pp. 430–434.

Since $p = 5$ when $t = 0$, we have

$$5 = 9 - Ce^0 = 9 - C$$
$$C = 9 - 5 = 4$$

so for all $t \geq 0$,

$$p(t) = 9 - 4e^{-0.06t}$$

(See Figure 5.3.)

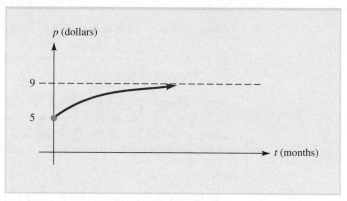

FIGURE 5.3   The graph of the unit price function $p(t) = 9 - 4e^{-0.06t}$.

b. When $t = 6$, the price of the product is

$$p(6) = 9 - 4e^{-0.06(6)} \approx 6.21$$

Since demand is given in thousands of units, the revenue $1,000D(p)p(t)$ generated by a price of $p = 6.21$ dollars per unit is

$$R(6) = 1,000D[p(6)]p(6) = 1,000[50 - p(6)]p(6)$$
$$= 1,000(50 - 6.21)(6.21) = 1,000(271.936) = 271,936$$

so the revenue generated after 6 months is roughly \$271,936.

c. The equilibrium price p_e satisfies

$$\underset{\text{supply}}{23 + 2p_e} = \underset{\text{demand}}{50 - p_e}$$

and solving, we find that

$$p_e = 9$$

As $t \to \infty$, we find that

$$\lim_{t \to \infty} p(t) = \lim_{t \to \infty} (9 - 6e^{-0.3465t}) = 9 \qquad e^{-mt} \to 0 \text{ as } t \to +\infty$$
$$\text{for } m > 0$$
$$= p_e$$

as claimed.

EXERCISES ▪ 5.2

In Exercises 1 and 2, fill in the table by specifying the substitution you would choose to find each of the four given integrals.

1.

Integral	Substitution *u*
a. $\int (3x + 4)^{5/2}\, dx$	
b. $\int \dfrac{4}{3 - x}\, dx$	
c. $\int t e^{2 - t^2}\, dt$	
d. $\int t(2 + t^2)^3\, dt$	

2.

Integral	Substitution *u*
a. $\int \dfrac{3}{(2x - 5)^4}\, dx$	
b. $\int x^2 e^{-x^3}\, dx$	
c. $\int \dfrac{e^t}{e^t + 1}\, dt$	
d. $\int \dfrac{t + 3}{\sqrt[3]{t^2 + 6t + 5}}\, dt$	

In Exercises 3 through 36, find the indicated integral and check your answer by differentiation.

3. $\int (2x + 6)^5\, dx$

4. $\int e^{5x + 3}\, dx$

5. $\int \sqrt{4x - 1}\, dx$

6. $\int \dfrac{1}{3x + 5}\, dx$

7. $\int e^{1 - x}\, dx$

8. $\int [(x - 1)^5 + 3(x - 1)^2 + 5]\, dx$

9. $\int x e^{x^2}\, dx$

10. $\int 2x e^{x^2 - 1}\, dx$

11. $\int t(t^2 + 1)^5\, dt$

12. $\int 3t \sqrt{t^2 + 8}\, dt$

13. $\int x^2 (x^3 + 1)^{3/4}\, dx$

14. $\int x^5 e^{1 - x^6}\, dx$

15. $\int \dfrac{2y^4}{y^5 + 1}\, dy$

16. $\int \dfrac{y^2}{(y^3 + 5)^2}\, dy$

17. $\int (x + 1)(x^2 + 2x + 5)^{12}\, dx$

18. $\int (3x^2 - 1)e^{x^3 - x}\, dx$

19. $\int \dfrac{3x^4 + 12x^3 + 6}{x^5 + 5x^4 + 10x + 12}\, dx$

20. $\int \dfrac{10x^3 - 5x}{\sqrt{x^4 - x^2 + 6}}\, dx$

21. $\int \dfrac{3u - 3}{(u^2 - 2u + 6)^2}\, du$

22. $\int \dfrac{6u - 3}{4u^2 - 4u + 1}\, du$

23. $\int \dfrac{\ln 5x}{x}\, dx$

24. $\int \dfrac{1}{x \ln x}\, dx$

25. $\int \dfrac{1}{x(\ln x)^2}\, dx$

26. $\int \dfrac{\ln x^2}{x}\, dx$

27. $\int \dfrac{2x \ln(x^2 + 1)}{x^2 + 1}\, dx$

28. $\int \dfrac{e^{\sqrt{x}}}{\sqrt{x}}\, dx$

29. $\int \dfrac{e^x + e^{-x}}{e^x - e^{-x}}\, dx$

30. $\int e^{-x}(1 + e^{2x})\, dx$

31. $\int \dfrac{x}{2x + 1}\, dx$

32. $\int \dfrac{t - 1}{t + 1}\, dt$

33. $\int x \sqrt{2x + 1}\, dx$

34. $\int \dfrac{x}{\sqrt[3]{4 - 3x}}\, dx$

35. $\int \dfrac{1}{\sqrt{x}(\sqrt{x} + 1)}\, dx$
[*Hint:* Let $u = \sqrt{x} + 1$.]

36. $\int \dfrac{1}{x^2}\left(\dfrac{1}{x} - 1\right)^{2/3}\, dx$

$\left[\text{*Hint:* Let } u = \dfrac{1}{x} - 1. \right]$

In Exercises 37 through 42, solve the given initial value problem for $y = f(x)$.

37. $\dfrac{dy}{dx} = (3 - 2x)^2$ where $y = 0$ when $x = 0$

38. $\dfrac{dy}{dx} = \sqrt{4x + 5}$ where $y = 3$ when $x = 1$

39. $\dfrac{dy}{dx} = \dfrac{1}{x + 1}$ where $y = 1$ when $x = 0$

40. $\dfrac{dy}{dx} = e^{2 - x}$ where $y = 0$ when $x = 2$

41. $\dfrac{dy}{dx} = \dfrac{x + 2}{x^2 + 4x + 5}$ where $y = 3$ when $x = -1$

42. $\dfrac{dy}{dx} = \dfrac{\ln\sqrt{x}}{x}$ where $y = 2$ when $x = 1$

In Exercises 43 through 46, the slope $f'(x)$ at each point (x, y) on a curve $y = f(x)$ is given, along with a particular point (a, b) on the curve. Use this information to find $f(x)$.

43. $f'(x) = (1 - 2x)^{3/2}$; $(0, 0)$

44. $f'(x) = x\sqrt{x^2 + 5}$; $(2, 10)$

45. $f'(x) = xe^{4 - x^2}$; $(-2, 1)$

46. $f'(x) = \dfrac{2x}{1 + 3x^2}$; $(0, 5)$

In Exercises 47 through 50, use an appropriate substitution to solve the given separable differential equation.

47. $\dfrac{dy}{dx} = \dfrac{2 - y}{(x + 1)^2}$ **48.** $\dfrac{dy}{dx} = \dfrac{xy}{2x - 1}$

49. $\dfrac{dy}{dx} = \dfrac{2 - y^2}{xy}$ **50.** $\dfrac{dx}{dt} = \dfrac{\ln t}{xt}$

BUSINESS AND ECONOMICS APPLIED PROBLEMS

51. MARGINAL COST   At a certain factory, the marginal cost is $3(q - 4)^2$ dollars per unit when the level of production is q units.
   a. Express the total production cost in terms of the overhead (the cost of producing 0 units) and the number of units produced.
   b. What is the cost of producing 14 units if the overhead is $436?

52. DEPRECIATION   The resale value of a certain industrial machine decreases at a rate that depends on its age. When the machine is t years old, the rate at which its value is changing is $-960e^{-t/5}$ dollars per year.
   a. Express the value of the machine in terms of its age and initial value.
   b. If the machine was originally worth $5,200, how much will it be worth when it is 10 years old?

53. REVENUE   The marginal revenue from the sale of x units of a particular commodity is estimated to be
$$R'(x) = 50 + 3.5xe^{-0.01x^2} \quad \text{dollars per unit}$$
where $R(x)$ is revenue in dollars.
   a. Find $R(x)$, assuming that $R(0) = 0$.
   b. What revenue should be expected from the sale of 1,000 units?

54. RETAIL PRICES   In a certain section of the country, it is estimated that t weeks from now, the price of chicken will be increasing at the rate of $p'(t) = 3\sqrt{t + 1}$ cents per kilogram per week. If chicken currently costs $2.30 per kilogram, what will it cost 8 weeks from now?

55. MARGINAL PROFIT   A company determines that the marginal revenue from the production of x units of its product is $R'(x) = x(5 - x)^3$ hundred dollars per unit, while the corresponding marginal cost is $C'(x) = 5 + 2x$ hundred dollars per unit. By how much does the profit change when the level of production is raised from 1 to 5 units?

56. MARGINAL PROFIT   Repeat Exercise 55 for marginal revenue $R'(x) = \dfrac{11 - x}{\sqrt{14 - x}}$ and for the marginal cost $C'(x) = 2 + x + x^2$.

57. DEMAND   The manager of a shoe store determines that the price p (dollars) for each pair of a popular brand of sports sneakers is changing at the rate of
$$p'(x) = \dfrac{-300x}{(x^2 + 9)^{3/2}}$$
when x (hundred) pairs are demanded by consumers. When the price is $75 per pair, 400 pairs ($x = 4$) are demanded by consumers.
   a. Find the demand (price) function $p(x)$.
   b. At what price will 500 pairs of sneakers be demanded? At what price will no sneakers be demanded?
   c. How many pairs will be demanded at a price of $90 per pair?

58. SUPPLY   The price p (dollars per unit) of a particular commodity is increasing at the rate
$$p'(x) = \dfrac{20x}{(7 - x)^2}$$
when x hundred units of the commodity are supplied to the market. The manufacturer supplies 200 units ($x = 2$) when the price is $2 per unit.
   a. Find the supply function $p(x)$.
   b. What price corresponds to a supply of 500 units?

59. SUPPLY   The owner of a fast-food chain determines that if x thousand units of a new meal item are supplied, then the marginal price at that level of supply is given by
$$p'(x) = \dfrac{x}{(x + 3)^2} \quad \text{dollars per meal}$$

where $p(x)$ is the price (in dollars) per unit at which all x thousand meal units will be sold. Currently, 5,000 units are being supplied at a price of $2.20 per unit.

a. Find the supply (price) function $p(x)$.

b. If 10,000 meal units are supplied to restaurants in the chain, what unit price should be charged so that all the units will be sold?

60. **INVESTMENT PLAN** An investor makes regular deposits totaling D dollars each year into an account that earns interest at the annual rate r compounded continuously.

a. Explain why the account grows at the rate

$$\frac{dV}{dt} = rV + D$$

where $V(t)$ is the value of the account t years after the initial deposit. Solve this separable differential equation to find $V(t)$. Your answer will involve r and D.

b. Amanda wants to retire in 20 years. To build up a retirement fund, she makes regular annual deposits of $8,000. If the prevailing interest rate stays constant at 4% compounded continuously, how much will she have in her account at the end of the 20-year period?

c. Anibal estimates that he will need $800,000 to retire. If the prevailing annual rate of interest is 5% compounded continuously, how large should his regular annual deposits be so that he can retire in 30 years?

61. **RETIREMENT INCOME** A retiree deposits S dollars into an account that earns interest at an annual rate r compounded continuously, and annually withdraws W dollars.

a. Explain why the account changes at the rate

$$\frac{dV}{dt} = rV - W$$

where $V(t)$ is the value of the account t years after the account is started. Solve this separable differential equation to find $V(t)$. Your answer will involve r, W, and S.

b. Frank and Jessie Jones deposit $500,000 in an account that pays 5% interest compounded continuously. If they withdraw $50,000 annually, what is their account worth at the end of 10 years?

c. What annual amount W can the couple in part (b) withdraw if their goal is to keep their account unchanged at $500,000?

d. If the couple in part (b) decide to withdraw $80,000 annually, how long does it take to exhaust their account?

62. **LAND VALUE** It is estimated that x years from now, the value $V(x)$ of an acre of Carter Hightower's farmland will be increasing at the rate of

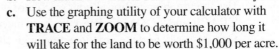

$$V'(x) = \frac{0.4x^3}{\sqrt{0.2x^4 + 8,000}}$$

dollars per year. The land is currently worth $500 per acre.

a. Find $V(x)$.

b. How much will Carter's land be worth in 10 years?

c. Use the graphing utility of your calculator with **TRACE** and **ZOOM** to determine how long it will take for the land to be worth $1,000 per acre.

PRICE ADJUSTMENTS FOR SUPPLY/DEMAND

In Exercises 63 through 66, the supply $S(t)$ and demand $D(t)$ functions for a commodity are given in terms of the unit price $p(t)$ at time t. Assume that price changes at a rate proportional to the shortage $D(t) - S(t)$, with the indicated constant of proportionality k and initial price p_0. In each exercise:

(a) Set up and solve a differential equation for $p(t)$.

(b) Find the unit price of the commodity when $t = 4$.

(c) Determine what happens to the price as $t \to \infty$.

63. $S(t) = 3 + t;\ D(t) = 9 - t;\ k = 0.01;\ p_0 = 1$

64. $S(t) = 5 + 2t;\ D(t) = 17 - t;\ k = 0.03;\ p_0 = 2$

65. $S(t) = 2 + 3p(t);\ D(t) = 10 - p(t);\ k = 0.02;\ p_0 = 1$

66. $S(t) = 1 + 4p(t);\ D(t) = 15 - 3p(t);\ k = 0.015;$
$p_0 = 3$

LIFE AND SOCIAL SCIENCE APPLIED PROBLEMS

67. **TREE GROWTH** A tree has been transplanted and after x years is growing at the rate of

$$1 + \frac{1}{(x + 1)^2}$$ meters per year. After 2 years, it

has reached a height of 5 meters. How tall was it when it was transplanted?

68. **WATER POLLUTION** An oil spill in the ocean is roughly circular in shape, with radius $R(t)$ feet t minutes after the spill begins. The radius is increasing at the rate

$$R'(t) = \frac{21}{0.07t + 5}\ \text{ft/min}$$

a. Find an expression for the radius $R(t)$, assuming that $R = 0$ when $t = 0$.

b. What is the area $A = \pi R^2$ of the spill after 1 hour?

69. DRUG CONCENTRATION The concentration $C(t)$ in milligrams per cubic centimeter (mg/cm^3) of a drug in a patient's bloodstream is 0.5 mg/cm^3 immediately after an injection and t minutes later is decreasing at the rate

$$C'(t) = \frac{-0.01e^{0.01t}}{(e^{0.01t} + 1)^2} \text{ mg/cm}^3 \text{ per minute}$$

A new injection is given when the concentration drops below 0.05 mg/cm^3.
a. Find an expression for $C(t)$.
b. What is the concentration after 1 hour? After 3 hours?
 c. Use the graphing utility of your calculator with **TRACE** and **ZOOM** to determine how much time passes before the next injection is given.

70. AGRICULTURAL PRODUCTION The **Mitscherlich model** for agricultural production specifies that the size $Q(t)$ of a crop changes at a rate proportional to the difference between the maximum possible crop size B and Q; that is,

$$\frac{dQ}{dt} = k(B - Q)$$

a. Solve this separable equation for $Q(t)$. Express your answer in terms of k and the initial crop size $Q_0 = Q(0)$.
b. A particular crop has a maximum size of 200 bushels per acre. At the start of the growing season ($t = 0$), the crop size is 50 bushels and 1 month later, it is 60 bushels. How large is the crop 3 months later ($t = 3$)?
c. Note that this model is similar to the learning model discussed in Example 5.2.11. Is this just a coincidence or is there some meaningful analogy linking the two situations? Explain.

71. AIR POLLUTION In a certain suburb of Los Angeles, the level of ozone $L(t)$ at 7:00 A.M. is 0.25 parts per million (ppm). A 12-hour weather forecast predicts that the ozone level t hours later will be changing at the rate of

$$L'(t) = \frac{0.24 - 0.03t}{\sqrt{36 + 16t - t^2}}$$

parts per million per hour (ppm/hr).
a. Express the ozone level $L(t)$ as a function of t. When does the peak ozone level occur? What is the peak level?

 b. Use the graphing utility of your calculator to sketch the graph of $L(t)$, and use **TRACE** and **ZOOM** to answer the questions in part (a). Then determine at what other time the ozone level will be the same as it is at 11:00 A.M.

MISCELLANEOUS PROBLEMS

In Exercises 72 through 75, the velocity $v(t) = x'(t)$ at time t of an object moving along the x axis is given, along with the initial position $x(0)$ of the object. In each case, find:
 (a) The position $x(t)$ at time t.
 (b) The position of the object at time $t = 4$.
 (c) The time when the object is at $x = 3$.

72. $x'(t) = \dfrac{-1}{1 + 0.5t}$; $x(0) = 5$

73. $x'(t) = -2(3t + 1)^{1/2}$; $x(0) = 4$

74. $x'(t) = \dfrac{-2t}{(1 + t^2)^{3/2}}$; $x(0) = 4$

75. $x'(t) = \dfrac{1}{\sqrt{2t + 1}}$; $x(0) = 0$

76. Find $\displaystyle\int x^3(4 - x^2)^{-1/2}\,dx$. [*Hint:* Substitute $u = 4 - x^2$ and use the fact that $x^2 = 4 - u$.]

77. Find $\displaystyle\int \frac{e^{2x}}{1 + e^x}\,dx$. [*Hint:* Let $u = 1 + e^x$.]

78. Find $\displaystyle\int e^{-x}(1 + e^x)^2\,dx$. [*Hint:* Is it better to set $u = 1 + e^x$ or $u = e^x$? Or is it better to not even use the method of substitution?]

79. Find $\displaystyle\int x^{1/3}(x^{2/3} + 1)^{3/2}\,dx$. [*Hint:* Substitute $u = x^{2/3} + 1$ and use $x^{2/3} = u - 1$.]

80. Find $\displaystyle\int \frac{1}{1 + e^{-x}}\,dx$. [*Hint:* Write e^{-x} as $\dfrac{1}{e^x}$, and simplify the complex fraction before making an appropriate substitution.]

81. Find $\displaystyle\int \frac{dx}{1 + e^x}$. [*Hint:* Write e^x as $\dfrac{1}{e^{-x}}$, and simplify the complex fraction before making an appropriate substitution.]

SECTION 5.3 The Definite Integral and the Fundamental Theorem of Calculus

Learning Objectives

1. Show how area under a curve can be expressed as the limit of a sum.
2. Define the definite integral and explore its properties.
3. State the fundamental theorem of calculus, and use it to compute definite integrals.
4. Use the fundamental theorem to solve applied problems involving net change.
5. Provide a geometric justification of the fundamental theorem.

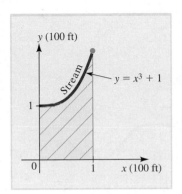

FIGURE 5.4 Determining land value by finding the area under a curve.

Suppose a real estate agent wants to evaluate an unimproved parcel of land that is 100 feet wide and is bounded by streets on three sides and by a stream on the fourth side. The agent determines that if a coordinate system is set up as shown in Figure 5.4, the stream can be described by the curve $y = x^3 + 1$, where x and y are measured in hundreds of feet. If the area of the parcel is A square feet and the agent estimates the land is worth \$12 per square foot, then the total value of the parcel is $12A$ dollars. If the parcel were rectangular in shape or triangular or even trapezoidal, its area A could be found by substituting into a well-known formula, but the upper boundary of the parcel is curved, so how can the agent find the area and hence the total value of the parcel?

Our goal in this section is to show how area under a curve, such as the area A in our real estate example, can be expressed as a limit of a sum of terms called a **definite integral.** We will then introduce a result called the **fundamental theorem of calculus** that allows us to compute *definite* integrals and thus find area and other quantities by using the *indefinite* integration (antidifferentiation) methods of Sections 5.1 and 5.2. In Example 5.3.3, we will illustrate this procedure by expressing the area A in our real estate example as a definite integral and evaluating it using the fundamental theorem of calculus.

Area as the Limit of a Sum

Consider the area of the region under the curve $y = f(x)$ over an interval $a \le x \le b$, where $f(x) \ge 0$ and f is continuous, as illustrated in Figure 5.5a. To find this area, we will follow a useful general policy:

> *When faced with something you don't know how to handle, try to relate it to something you do know how to handle.*

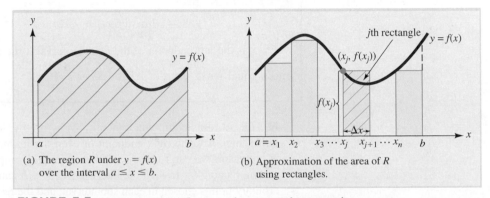

(a) The region R under $y = f(x)$ over the interval $a \le x \le b$.

(b) Approximation of the area of R using rectangles.

FIGURE 5.5 Approximation of area under a curve by rectangles.

In this particular case, we may not know the area under the given curve, but we do know how to find the area of a rectangle. Thus, we proceed by dividing the region into a number of rectangular regions and then approximating the area A under the curve $y = f(x)$ by adding the areas of the approximating rectangles.

To be specific, begin the approximation by dividing the interval $a \leq x \leq b$ into n equal subintervals, each of length $\Delta x = \dfrac{b - a}{n}$, and let x_j denote the left endpoint of the jth subinterval, for $j = 1, 2, \ldots, n$. Then draw n rectangles such that the jth rectangle has height $f(x_j)$ with the jth subinterval as its base, so its width is Δx. The approximation scheme is illustrated in Figure 5.5b.

The area of the jth rectangle is $f(x_j)\Delta x$ and approximates the area under the curve above the subinterval $x_j \leq x \leq x_{j+1}$. The sum of the areas of all n rectangles is

$$S_n = f(x_1)\Delta x + f(x_2)\Delta x + \cdots + f(x_n)\Delta x$$
$$= [f(x_1) + f(x_2) + \cdots + f(x_n)]\Delta x$$

which approximates the total area A under the curve.

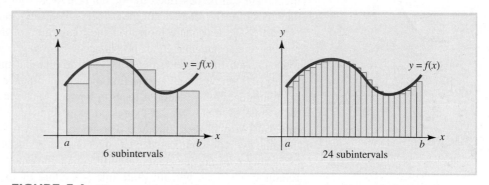

FIGURE 5.6   The approximation improves as the number of subintervals increases.

As the number of subintervals n increases, the approximating sum S_n gets closer and closer to what we intuitively think of as the area under the curve, as illustrated in Figure 5.6. Therefore, it is reasonable to define the actual area A under the curve as the limit of the sums. To summarize:

Area Under a Curve ■ Let $f(x)$ be continuous and satisfy $f(x) \geq 0$ on the interval $a \leq x \leq b$. Then the region under the curve $y = f(x)$ over the interval $a \leq x \leq b$ has area

$$A = \lim_{n \to +\infty} [f(x_1) + f(x_2) + \cdots + f(x_n)]\Delta x$$

where x_j is the left endpoint of the jth subinterval if the interval $a \leq x \leq b$ is divided into n equal parts, each of length $\Delta x = \dfrac{b - a}{n}$.

NOTE   At this point, you may ask, "Why use the left endpoint of the subintervals rather than, say, the right endpoint or even the midpoint?" The answer is that there is no reason we can't use those other points to compute the height of our approximating rectangles. In fact, the interval $a \leq x \leq b$ can be subdivided arbitrarily with arbitrary points chosen in each subinterval, and the result will still be the same. However, proving this equivalence is beyond the scope of this text. ■

Example 5.3.1 shows how to compute area as the limit of a sum and then how to check it using a geometric formula.

EXAMPLE 5.3.1 Finding Area Using the Limit of a Sum

Let R be the region under the graph of $f(x) = 2x + 1$ over the interval $1 \le x \le 3$, as shown in Figure 5.7a. Compute the area of R as the limit of a sum.

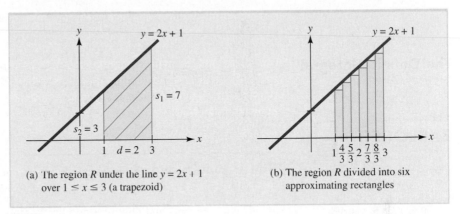

(a) The region R under the line $y = 2x + 1$ over $1 \le x \le 3$ (a trapezoid)

(b) The region R divided into six approximating rectangles

FIGURE 5.7 Approximating the area under a line with rectangles.

Solution

The region R is shown in Figure 5.7b with six approximating rectangles, each of width $\Delta x = \dfrac{3-1}{6} = \dfrac{1}{3}$. The left endpoints in the partition of $1 \le x \le 3$ are $x_1 = 1$, $x_2 = 1 + \dfrac{1}{3} = \dfrac{4}{3}$, and similarly, $x_3 = \dfrac{5}{3}$, $x_4 = 2$, $x_5 = \dfrac{7}{3}$, and $x_6 = \dfrac{8}{3}$. The corresponding values of $f(x) = 2x + 1$ are given in the following table:

x_j	1	$\frac{4}{3}$	$\frac{5}{3}$	2	$\frac{7}{3}$	$\frac{8}{3}$
$f(x_j) = 2x_j + 1$	3	$\frac{11}{3}$	$\frac{13}{3}$	5	$\frac{17}{3}$	$\frac{19}{3}$

Thus, the area A of the region R is approximated by the sum

$$S = \left(3 + \frac{11}{3} + \frac{13}{3} + 5 + \frac{17}{3} + \frac{19}{3}\right)\left(\frac{1}{3}\right) = \frac{28}{3} \approx 9.333$$

If you continue to subdivide the region R using more and more rectangles, the corresponding approximating sums S_n approach the actual area A of the region. The sum we have already computed for $n = 6$ is listed in the following table, along with those for $n = 10, 20, 50, 100,$ and 500. (If you have access to a computer or a programmable calculator, see if you can write a program for generating any such sum for given n.)

Number of rectangles n	6	10	20	50	100	500
Approximating sum S_n	9.333	9.600	9.800	9.920	9.960	9.992

The numbers on the bottom line of this table seem to be approaching 10 as n gets larger and larger. Thus, it is reasonable to conjecture that the region R has area

$$A = \lim_{n \to +\infty} S_n = 10$$

Just-In-Time REVIEW

A trapezoid is a four-sided polygon with at least two parallel sides. Its area is

$$A = \frac{1}{2}(s_1 + s_2)h$$

where s_1 and s_2 are the lengths of the two parallel sides and h is the distance between them.

Notice in Figure 5.7a that the region R is a trapezoid of width $d = 3 - 1 = 2$ with parallel sides of lengths

$$s_1 = 2(3) + 1 = 7 \qquad \text{and} \qquad s_2 = 2(1) + 1 = 3$$

Such a trapezoid has area

$$A = \frac{1}{2}(s_1 + s_2)d = \frac{1}{2}(7 + 3)(2) = 10$$

the same result we just obtained using the limit of a sum procedure.

The Definite Integral

Area is just one of many quantities that can be expressed as the limit of a sum. To handle all such cases, including those for which $f(x) \geq 0$ is *not* required and left endpoints are not used, we require the terminology and notation introduced in the following definition.

The Definite Integral ■ Let $f(x)$ be a function that is continuous on the interval $a \leq x \leq b$. Divide the interval $a \leq x \leq b$ into n equal parts, each of width $\Delta x = \dfrac{b - a}{n}$, and choose a number x_k from the kth subinterval for $k = 1$, $2, \ldots, n$. Form the sum

$$[f(x_1) + f(x_2) + \cdots + f(x_n)]\Delta x$$

called a **Riemann sum.**

Then the **definite integral** of f on the interval $a \leq x \leq b$, denoted by $\int_a^b f(x)\, dx$, is the limit of the Riemann sum as $n \to +\infty$; that is,

$$\int_a^b f(x)\, dx = \lim_{n \to +\infty} [f(x_1) + f(x_2) + \cdots + f(x_n)]\Delta x$$

The function $f(x)$ is called the **integrand,** and the numbers a and b are called the **lower and upper limits of integration,** respectively. The process of finding a definite integral is called **definite integration.**

Surprisingly, the fact that $f(x)$ is continuous on $a \leq x \leq b$ turns out to be enough to guarantee that the limit used to define the definite integral $\int_a^b f(x)\, dx$ exists and is the same regardless of how the subinterval representatives x_k are chosen.

The symbol $\int_a^b f(x)\, dx$ used for the definite integral is essentially the same as the symbol $\int f(x)\, dx$ for the indefinite integral, even though the definite integral is a specific number while the indefinite integral is a family of functions, the antiderivatives of f. You will soon see that these two apparently very different concepts are intimately related. Here is a compact form for the definition of area using integral notation.

> **Area as a Definite Integral** ■ If $f(x)$ is continuous and $f(x) \geq 0$ on the interval $a \leq x \leq b$, then the region R under the curve $y = f(x)$ over the interval $a \leq x \leq b$ has area A given by the definite integral $A = \int_a^b f(x)\, dx$.
>
>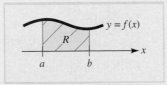

The Fundamental Theorem of Calculus

If computing the limit of a sum were the only way of evaluating a definite integral, the integration process probably would be little more than a mathematical novelty. Fortunately, there is an easier way of performing this computation, thanks to this remarkable result connecting the definite integral to antidifferentiation.

> **The Fundamental Theorem of Calculus** ■ If the function $f(x)$ is continuous on the interval $a \leq x \leq b$, then
>
> $$\int_a^b f(x)\, dx = F(b) - F(a)$$
>
> where $F(x)$ is any antiderivative of $f(x)$ on $a \leq x \leq b$.

A special case of the fundamental theorem of calculus is verified at the end of this section. When applying the fundamental theorem, we use the notation

$$F(x)\Big|_a^b = F(b) - F(a)$$

Thus,

$$\int_a^b f(x)\, dx - F(x)\Big|_a^b = F(b) - F(a)$$

NOTE You may wonder how the fundamental theorem of calculus can promise that if $F(x)$ is *any* antiderivative of $f(x)$, then

$$\int_a^b f(x)\, dx = F(b) - F(a)$$

To see why this is true, suppose $G(x)$ is another such antiderivative. Then $G(x) = F(x) + C$ for some constant C, so $F(x) = G(x) - C$ and

$$\int_a^b f(x)\, dx = F(b) - F(a)$$
$$= [G(b) - C] - [G(a) - C]$$
$$= G(b) - G(a)$$

since the C's cancel. Thus, the valuation is the same regardless of which antiderivative is used. ■

In Example 5.3.2, we demonstrate the computational value of the fundamental theorem of calculus by using it to compute the same area we estimated as the limit of a sum in Example 5.3.1.

EXAMPLE 5.3.2   Finding Area Using the Fundamental Theorem

Use the fundamental theorem of calculus to find the area of the region under the line $y = 2x + 1$ over the interval $1 \le x \le 3$.

Solution

Since $f(x) = 2x + 1$ satisfies $f(x) \ge 0$ on the interval $1 \le x \le 3$, the area is given by the definite integral $A = \int_1^3 (2x + 1)\, dx$. Since an antiderivative of $f(x) = 2x + 1$ is $F(x) = x^2 + x$, the fundamental theorem of calculus tells us that

$$A = \int_1^3 (2x + 1)\, dx = x^2 + x \Big|_1^3$$
$$= [(3)^2 + (3)] - [(1)^2 + (1)] = 10$$

as estimated in Example 5.3.1.

In Example 5.3.3, we show how definite integration can be used to find the area of the parcel of land described in the introduction to this section.

EXPLORE!

Refer to Example 5.3.3. Use the numerical integration feature of your calculator to confirm numerically that

$$\int_0^1 (x^3 + 1)\, dx = 1.25$$

EXAMPLE 5.3.3   Using Integration to Find the Area of a Plot of Land

A parcel of land is 100 feet wide and is bounded by streets on three sides and by a stream on the fourth side. A real estate agent determines that a coordinate system can be set up in which the streets are represented by the lines $y = 0$, $x = 0$, and $x = 1$, and the stream is represented by the curve $y = x^3 + 1$, where x and y are measured in hundreds of feet (see Figure 5.4). If the land is appraised at $12 per square foot, what is the total value of the parcel?

Solution

The area of the parcel is given by the definite integral

$$A = \int_0^1 (x^3 + 1)\, dx$$

Since an antiderivative of $f(x) = x^3 + 1$ is $F(x) = \frac{1}{4}x^4 + x$, the fundamental theorem of calculus tells us that

$$A = \int_0^1 (x^3 + 1)\, dx = \frac{1}{4}x^4 + x \Big|_0^1$$
$$= \left[\frac{1}{4}(1)^4 + 1\right] - \left[\frac{1}{4}(0)^4 + 0\right] = \frac{5}{4}$$

Because x and y are measured in hundreds of feet, the total area is

$$\frac{5}{4} \times 100 \times 100 = 12{,}500 \text{ ft}^2$$

and since the land in the parcel is worth \$12 per square foot, the total value of the parcel is

$$V = (\$12/\text{ft}^3)(12{,}500 \text{ ft}^2) = \$150{,}000$$

Our definition of the definite integral was motivated by computing area, which is a nonnegative quantity, but the definition itself does not require $f(x) \geq 0$. Two definite integrals are evaluated using the fundamental theorem of calculus in Examples 5.3.4 and 5.3.5. Note that the value of the integral in Example 5.3.5 is negative, which would not be possible if it represented the area under a curve.

Just-In-Time **REVIEW**

When the fundamental theorem of calculus

$$\int_a^b f(x)\, dx = F(b) - F(a)$$

is used to evaluate a definite integral, remember to compute **both** $F(b)$ and $F(a)$, even when $a = 0$.

EXAMPLE 5.3.4   Evaluating a Definite Integral

Evaluate the definite integral $\displaystyle\int_0^1 (e^{-x} + \sqrt{x})\, dx$.

Solution

An antiderivative of $f(x) = e^{-x} + \sqrt{x}$ is $F(x) = -e^{-x} + \dfrac{2}{3}x^{3/2}$, so the definite integral is

$$\int_0^1 (e^{-x} + \sqrt{x})\, dx = \left(-e^{-x} + \frac{2}{3}x^{3/2} \right)\Bigg|_0^1$$

$$= \left[-e^{-1} + \frac{2}{3}(1)^{3/2} \right] - \left[-e^0 + \frac{2}{3}(0) \right]$$

$$= \frac{5}{3} - \frac{1}{e} \approx 1.299$$

EXAMPLE 5.3.5   Evaluating a Definite Integral

Evaluate $\displaystyle\int_1^4 \left(\frac{1}{x} - x^2 \right) dx$.

Solution

An antiderivative of $f(x) = \dfrac{1}{x} - x^2$ is $F(x) = \ln |x| - \dfrac{1}{3}x^3$, so we have

$$\int_1^4 \left(\frac{1}{x} - x^2 \right) dx = \left(\ln |x| - \frac{1}{3}x^3 \right)\Bigg|_1^4$$

$$= \left[\ln 4 - \frac{1}{3}(4)^3 \right] - \left[\ln 1 - \frac{1}{3}(1)^3 \right]$$

$$= \ln 4 - 21 \approx -19.6137$$

Integration Rules   This list of rules can be used to simplify the computation of definite integrals.

Rules for Definite Integrals

Let f and g be any functions continuous on $a \le x \le b$. Then,

1. **Constant multiple rule:** $\displaystyle\int_a^b k f(x)\, dx = k \int_a^b f(x)\, dx$ for constant k

2. **Sum rule:** $\displaystyle\int_a^b [f(x) + g(x)]\, dx = \int_a^b f(x)\, dx + \int_a^b g(x)\, dx$

3. **Difference rule:** $\displaystyle\int_a^b [f(x) - g(x)]\, dx = \int_a^b f(x)\, dx - \int_a^b g(x)\, dx$

4. $\displaystyle\int_a^a f(x)\, dx = 0$

5. $\displaystyle\int_b^a f(x)\, dx = -\int_a^b f(x)\, dx$

6. **Subdivision rule:** $\displaystyle\int_a^b f(x)\, dx = \int_a^c f(x)\, dx + \int_c^b f(x)\, dx$

Rules 4 and 5 are really special cases of the definition of the definite integral. The first three rules can be proved by using the fundamental theorem of calculus along with an analogous rule for indefinite integrals. For instance, to verify the constant multiple rule, suppose $F(x)$ is an antiderivative of $f(x)$. Then, according to the constant multiple rule for indefinite integrals, $kF(x)$ is an antiderivative of $kf(x)$ and the fundamental theorem of calculus tells us that

$$\int_a^b k f(x)\, dx = kF(x)\Big|_a^b$$

$$= kF(b) - kF(a) = k[F(b) - F(a)]$$

$$= k\int_a^b f(x)\, dx$$

You are asked to verify the sum rule using similar reasoning in Exercise 82.

In the case where $f(x) \ge 0$ on the interval $a \le x \le b$, the subdivision rule is a geometric reflection of the fact that the area under the curve $y = f(x)$ over the interval $a \le x \le b$ is the sum of the areas under $y = f(x)$ over the subintervals $a \le x \le c$ and $c \le x \le b$, as illustrated in Figure 5.8. However, it is important to remember that the subdivision rule is true even if $f(x)$ does *not* satisfy $f(x) \ge 0$ on $a \le x \le b$.

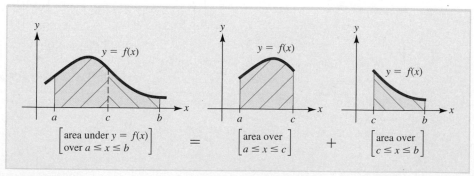

FIGURE 5.8   The subdivision rule for definite integrals [case where $f(x) \ge 0$].

EXAMPLE 5.3.6   Using Rules for Definite Integrals

Let $f(x)$ and $g(x)$ be functions that are continuous on the interval $-2 \le x \le 5$ and that satisfy

$$\int_{-2}^{5} f(x)\,dx = 3 \qquad \int_{-2}^{5} g(x)\,dx = -4 \qquad \int_{3}^{5} f(x)\,dx = 7$$

Use this information to evaluate each of these definite integrals:

a. $\displaystyle\int_{-2}^{5} [2\,f(x) - 3g(x)]\,dx$ **b.** $\displaystyle\int_{-2}^{3} f(x)\,dx$

Solution

a. By combining the difference rule and constant multiple rule and substituting the given information, we find that

$$\int_{-2}^{5} [2\,f(x) - 3g(x)]\,dx = \int_{-2}^{5} 2\,f(x)\,dx - \int_{-2}^{5} 3g(x)\,dx \quad \text{difference rule}$$

$$= 2 \int_{-2}^{5} f(x)\,dx - 3 \int_{-2}^{5} g(x)\,dx \quad \begin{array}{l}\text{constant}\\ \text{multiple rule}\end{array}$$

$$= 2(3) - 3(-4) = 18 \quad \begin{array}{l}\text{substitute given}\\ \text{information}\end{array}$$

b. According to the subdivision rule

$$\int_{-2}^{5} f(x)\,dx = \int_{-2}^{3} f(x)\,dx + \int_{3}^{5} f(x)\,dx$$

Solving this equation for the required integral $\int_{-2}^{3} f(x)\,dx$ and substituting the given information, we obtain

$$\int_{-2}^{3} f(x)\,dx = \int_{-2}^{5} f(x)\,dx - \int_{3}^{5} f(x)\,dx$$

$$= 3 - 7 = -4$$

Substituting in a Definite Integral

When using a substitution $u = g(x)$ to evaluate a definite integral $\int_{a}^{b} f(x)\,dx$, you can proceed in either of these two ways:

1. Use the substitution to find an antiderivative $F(x)$ for $f(x)$, and then evaluate the definite integral using the fundamental theorem of calculus.
2. Use the substitution to express the integrand and dx in terms of u and du and to replace the original limits of integration, a and b, with transformed limits $c = g(a)$ and $d = g(b)$. The original integral can then be evaluated by applying the fundamental theorem of calculus to the transformed definite integral.

These procedures are illustrated in Examples 5.3.7 and 5.3.8.

EXAMPLE 5.3.7 Definite Integration Using Substitution

Evaluate $\int_0^1 8x(x^2 + 1)^3\,dx$.

Solution

The integrand is a product in which one factor $8x$ is a constant multiple of the derivative of an expression $x^2 + 1$ that appears in the other factor. This suggests that you let $u = x^2 + 1$. Then $du = 2x\,dx$, and so

$$\int 8x(x^2 + 1)^3\,dx = \int 4u^3\,du = u^4$$

The limits of integration, 0 and 1, refer to the variable x and not to u. You can, therefore, proceed in one of two ways. Either you can rewrite the antiderivative in terms of x, or you can find the values of u that correspond to $x = 0$ and $x = 1$.

If you choose the first alternative, you find that

$$\int 8x(x^2 + 1)^3\,dx = u^4 = (x^2 + 1)^4$$

and so

$$\int_0^1 8x(x^2 + 1)^3\,dx = (x^2 + 1)^4\Big|_0^1 = 16 - 1 = 15$$

If you choose the second alternative, use the fact that $u = x^2 + 1$ to conclude that $u = 1$ when $x = 0$ and $u = 2$ when $x = 1$. Hence,

$$\int_0^1 8x(x^2 + 1)^3\,dx = \int_1^2 4u^3\,du = u^4\Big|_1^2 = 16 - 1 = 15$$

EXPLORE!

Refer to Example 5.3.8. Use a graphing utility with the window [0, 3]1 by [−4, 1]1 to graph $f(x) = \dfrac{\ln x}{x}$. Explain in terms of area why the integral of $f(x)$ over $\dfrac{1}{4} \le x \le 2$ is negative.

EXAMPLE 5.3.8 Definite Integration Using Substitution

Evaluate $\int_{1/4}^2 \left(\dfrac{\ln x}{x}\right) dx$.

Solution

Let $u = \ln x$, so $du = \dfrac{1}{x}\,dx$. Then

$$\int \frac{\ln x}{x}\,dx = \int \ln x\left(\frac{1}{x}\,dx\right) = \int u\,du$$

$$= \frac{1}{2}u^2 = \frac{1}{2}(\ln x)^2$$

Thus,

$$\int_{1/4}^2 \frac{\ln x}{x}\,dx = \left[\frac{1}{2}(\ln x)^2\right]\Big|_{1/4}^2 = \frac{1}{2}(\ln 2)^2 - \frac{1}{2}\left(\ln \frac{1}{4}\right)^2$$

$$= -0.721$$

Alternatively, use the substitution $u = \ln x$ to transform the limits of integration:

$$\text{when } x = \frac{1}{4}, \text{ then } u = \ln \frac{1}{4}$$

$$\text{when } x = 2, \text{ then } u = \ln 2$$

Substituting, we find

$$\int_{1/4}^{2} \frac{\ln x}{x}\, dx = \int_{\ln 1/4}^{\ln 2} u\, du = \frac{1}{2} u^2 \Big|_{\ln 1/4}^{\ln 2}$$

$$= \frac{1}{2}(\ln 2)^2 - \frac{1}{2}\left(\ln \frac{1}{4}\right)^2 \approx -0.721$$

Net Change

In certain applications, we are given the rate of change $Q'(x)$ of a quantity $Q(x)$ and required to compute the **net change** $Q(b) - Q(a)$ in $Q(x)$ as x varies from $x = a$ to $x = b$. We did this in Section 5.1 by solving initial value problems (recall Examples 5.1.5 through 5.1.7). However, since $Q(x)$ is an antiderivative of $Q'(x)$, the fundamental theorem of calculus allows us to compute net change by the following definite integration formula.

> **Net Change** ■ If $Q'(x)$ is continuous on the interval $a \le x \le b$, then the **net change** in $Q(x)$ as x varies from $x = a$ to $x = b$ is given by
>
> $$Q(b) - Q(a) = \int_a^b Q'(x)\, dx$$

EXAMPLE 5.3.9 Finding Net Change in Cost

At a certain factory, the marginal cost is $3(q - 4)^2$ dollars per unit when the level of production is q units. By how much will the total manufacturing cost increase if the level of production is raised from 6 units to 10 units?

Solution

Let $C(q)$ denote the total cost of producing q units. Then the marginal cost is the derivative $\dfrac{dC}{dq} = 3(q - 4)^2$, and the increase in cost if production is raised from 6 units to 10 units is given by the definite integral

$$C(10) - C(6) = \int_6^{10} \frac{dC}{dq}\, dq$$

$$= \int_6^{10} 3(q - 4)^2\, dq \qquad\qquad u = q - 4,\; du = dq$$

$$= (q - 4)^3 \Big|_6^{10} = (10 - 4)^3 - (6 - 4)^3$$

$$= \$208$$

EXAMPLE 5.3.10 Finding Net Change in Protein Mass

A protein with mass m (grams) disintegrates into amino acids at a rate given by

$$\frac{dm}{dt} = \frac{-30}{(t + 3)^2} \quad \text{g/hr}$$

What is the net change in mass of the protein during the first 2 hours?

Solution

The net change is given by the definite integral

$$m(2) - m(0) = \int_0^2 \frac{dm}{dt}\, dt = \int_0^2 \frac{-30}{(t + 3)^2}\, dt$$

If we substitute $u = t + 3$, $du = dt$, and change the limits of integration accordingly ($t = 0$ becomes $u = 3$ and $t = 2$ becomes $u = 5$), we find

$$m(2) - m(0) = \int_0^2 \frac{-30}{(t + 3)^2}\, dt = \int_3^5 -30u^{-2}\, du$$

$$= -30\left(\frac{u^{-1}}{-1}\right)\Bigg|_3^5 = 30\left[\frac{1}{5} - \frac{1}{3}\right]$$

$$= -4$$

Thus, the mass of the protein has a net decrease of 4 g over the first 2 hours.

Area Justification of the Fundamental Theorem of Calculus

We close this section with a justification of the fundamental theorem of calculus for the case where $f(x) \geq 0$. In this case, the definite integral $\int_a^b f(x)\, dx$ represents the area under the curve $y = f(x)$ over the interval $[a, b]$. For fixed x between a and b, let $A(x)$ denote the area under $y = f(x)$ over the interval $[a, x]$. Then the difference quotient of $A(x)$ is

$$\frac{A(x + h) - A(x)}{h}$$

and the expression $A(x + h) - A(x)$ in the numerator is just the area under the curve $y = f(x)$ between x and $x + h$. If h is small, this area is approximately the same as the area of the rectangle with height $f(x)$ and width h as indicated in Figure 5.9. That is,

$$A(x + h) - A(x) \approx f(x)h$$

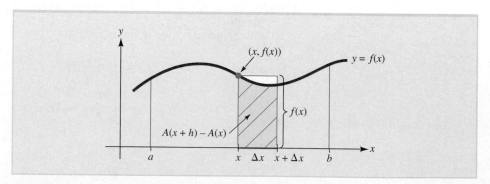

FIGURE 5.9 The area $A(x + h) - A(x)$.

or, equivalently,

$$\frac{A(x + h) - A(x)}{h} \approx f(x)$$

As h approaches 0, the error in the approximation approaches 0, and it follows that

$$\lim_{h \to 0} \frac{A(x + h) - A(x)}{h} = f(x)$$

But by the definition of the derivative,

$$\lim_{h \to 0} \frac{A(x + h) - A(x)}{h} = A'(x)$$

so that

$$A'(x) = f(x)$$

In other words, $A(x)$ is an antiderivative of $f(x)$.

Suppose $F(x)$ is any other antiderivative of $f(x)$. Then, according to the fundamental property of antiderivatives (Section 5.1), we have

$$A(x) = F(x) + C$$

for some constant C and all x in the interval $a \le x \le b$. Since $A(x)$ represents the area under $y = f(x)$ between a and x, it is certainly true that $A(a)$, the area between a and a, is 0, so that

$$A(a) = 0 = F(a) + C$$

and $C = -F(a)$. The area under $y = f(x)$ between $x = a$ and $x = b$ is $A(b)$, which satisfies

$$A(b) = F(b) + C = F(b) - F(a)$$

Finally, since the area under $y = f(x)$ above $a \le x \le b$ is also given by the definite integral $\int_a^b f(x)\, dx$, it follows that

$$\int_a^b f(x)\, dx = A(b) = F(b) - F(a)$$

as claimed in the fundamental theorem of calculus.

EXERCISES ▪ 5.3

In Exercises 1 and 2, the table gives the coordinates $(x, f(x))$ of points on the graph of a function f over the interval $a \le x \le b$. In each case, estimate the value of the indicated definite integral $\int_a^b f(x)\, dx$ by forming a Riemann sum using left endpoints.

1. $\int_0^2 f(x)\, dx$

x	0	0.4	0.8	1.2	1.6	2.0
$f(x)$	1.1	1.7	2.3	2.5	2.4	2.1

2. $\int_1^2 f(x)\, dx$

x	1	1.2	1.4	1.6	1.8	2.0
$f(x)$	1.1	1.4	0.8	-0.3	-1.4	-1.1

In Exercises 3 through 8, estimate the value of the definite integral $\int_a^b f(x)\,dx$ by computing the Riemann sum of f on the interval $a \le x \le b$ for $n = 8$ subintervals, using left endpoints. Then find the actual value of the integral using the fundamental theorem of calculus.

3. $f(x) = 4 - x$ over $0 \le x \le 4$

4. $f(x) = 3 - 2x$ over $-1 \le x \le 2$

5. $f(x) = x^2$ over $1 \le x \le 2$

6. $f(x) = 1 - x^2$ over $1 \le x \le 4$

7. $f(x) = \dfrac{1}{x}$ over $1 \le x \le 2$

8. $f(x) = \sqrt{x}$ over $0 \le x \le 4$

In Exercises 9 through 14, estimate the area under the graph of $y = f(x)$ over the interval $0 \le x \le 4$ by computing a Riemann sum of f over the interval with 8 subintervals, using left endpoints. Then find the actual area using the fundamental theorem of calculus.

9. $f(x) = x$

10. $f(x) = 4 - x$

11. $f(x) = x^2 + 1$

12. $f(x) = \dfrac{x}{x + 1}$

13. $f(x) = \dfrac{1}{5 - x}$

14. $f(x) = x(4 - x)$

In Exercises 15 through 44, evaluate the given definite integral using the fundamental theorem of calculus.

15. $\displaystyle\int_{-1}^{2} 5\,dx$

16. $\displaystyle\int_{-2}^{1} \pi\,dx$

17. $\displaystyle\int_{0}^{5} (3x + 2)\,dx$

18. $\displaystyle\int_{1}^{4} (5 - 2t)\,dt$

19. $\displaystyle\int_{-1}^{1} 3t^4\,dt$

20. $\displaystyle\int_{1}^{4} 2\sqrt{u}\,du$

21. $\displaystyle\int_{-1}^{1} (2u^{1/3} - u^{2/3})\,du$

22. $\displaystyle\int_{4}^{9} x^{-3/2}\,dx$

23. $\displaystyle\int_{0}^{1} e^{-x}(4 - e^x)\,dx$

24. $\displaystyle\int_{-1}^{1} \left(\dfrac{1}{e^x} - \dfrac{1}{e^{-x}}\right)dx$

25. $\displaystyle\int_{0}^{1} (x^4 + 3x^3 + 1)\,dx$

26. $\displaystyle\int_{-1}^{0} (-3x^5 - 3x^2 + 2x + 5)\,dx$

27. $\displaystyle\int_{2}^{5} (2 + 2t + 3t^2)\,dt$

28. $\displaystyle\int_{1}^{9} \left(\sqrt{t} - \dfrac{4}{\sqrt{t}}\right)dt$

29. $\displaystyle\int_{1}^{3} \left(1 + \dfrac{1}{x} + \dfrac{1}{x^2}\right)dx$

30. $\displaystyle\int_{0}^{\ln 2} (e^t - e^{-t})\,dt$

31. $\displaystyle\int_{-3}^{-1} \dfrac{t + 1}{t^3}\,dt$

32. $\displaystyle\int_{1}^{6} x^2(x - 1)\,dx$

33. $\displaystyle\int_{1}^{2} (2x - 4)^4\,dx$

34. $\displaystyle\int_{-3}^{0} (2x + 6)^4\,dx$

35. $\displaystyle\int_{0}^{4} \dfrac{1}{\sqrt{6t + 1}}\,dt$

36. $\displaystyle\int_{1}^{2} \dfrac{x^2}{(x^3 + 1)^2}\,dx$

37. $\displaystyle\int_{0}^{1} (x^3 + x)\sqrt{x^4 + 2x^2 + 1}\,dx$

38. $\displaystyle\int_{0}^{1} \dfrac{6t}{t^2 + 1}\,dt$

39. $\displaystyle\int_{2}^{e+1} \dfrac{x}{x - 1}\,dx$

40. $\displaystyle\int_{1}^{2} (t + 1)(t - 2)^6\,dt$

41. $\displaystyle\int_{1}^{e^2} \dfrac{(\ln x)^2}{x}\,dx$

42. $\displaystyle\int_{e}^{e^2} \dfrac{1}{x \ln x}\,dx$

43. $\displaystyle\int_{1/3}^{1/2} \dfrac{e^{1/x}}{x^2}\,dx$

44. $\displaystyle\int_{1}^{4} \dfrac{(\sqrt{x} - 1)^{3/2}}{\sqrt{x}}\,dx$

In Exercises 45 through 50, $f(x)$ and $g(x)$ are functions that are continuous on the interval $-3 \le x \le 2$ and satisfy

$$\int_{-3}^{2} f(x)\,dx = 5 \qquad \int_{-3}^{2} g(x)\,dx = -2 \qquad \int_{-3}^{1} f(x)\,dx = 0 \qquad \int_{-3}^{1} g(x)\,dx = 4$$

In each case, use this information along with rules for definite integrals to evaluate the indicated integral.

45. $\displaystyle\int_{-3}^{2} [-2f(x) + 5g(x)]\,dx$

46. $\displaystyle\int_{-3}^{1} [4f(x) - 3g(x)]\,dx$

47. $\displaystyle\int_{4}^{4} g(x)\,dx$

48. $\displaystyle\int_{2}^{-3} f(x)\,dx$

49. $\displaystyle\int_{1}^{2} [3f(x) + 2g(x)]\,dx$

50. $\displaystyle\int_{-3}^{1} [2f(x) + 3g(x)]\,dx$

In Exercises 51 through 58, find the area of the region R that lies under the given curve y = f(x) over the indicated interval a ≤ x ≤ b.

51. Under $y = x^4$, over $-1 \le x \le 2$

52. Under $y = \sqrt{x}(x + 1)$, over $0 \le x \le 4$

53. Under $y = (3x + 4)^{1/2}$, over $0 \le x \le 4$

54. Under $y = \dfrac{3}{\sqrt{9 - 2x}}$, over $-8 \le x \le 0$

55. Under $y = e^{2x}$, over $0 \le x \le \ln 3$

56. Under $y = xe^{-x^2}$, over $0 \le x \le 3$

57. Under $y = \dfrac{3}{5 - 2x}$, over $-2 \le x \le 1$

58. Under $y = \dfrac{3}{x}$, over $1 \le x \le e^2$

BUSINESS AND ECONOMICS APPLIED PROBLEMS

59. LAND VALUES   It is estimated that t years from now the value of a certain parcel of land will be increasing at the rate of $V'(t)$ dollars per year. Find an expression for the amount by which the value of the land will increase during the next 5 years.

60. ADMISSION TO EVENTS   The promoters of a county fair estimate that t hours after the gates open at 9:00 A.M., visitors will be entering the fair at the rate of $N'(t)$ people per hour. Find an expression for the number of people who will enter the fair between 11:00 A.M. and 1:00 P.M.

61. STORAGE COST   A retailer receives a shipment of 12,000 pounds of soybeans that will be used at a constant rate of 300 pounds per week. If the cost of storing the soybeans is 0.2 cent per pound per week, how much will the retailer have to pay in storage costs over the next 40 weeks?

62. STORAGE COST   Anita Bellman is a retailer who specializes in grains. She receives a shipment of 10,000 kilograms of rice that will be used up over a 5-month period at the constant rate of 2,000 kilograms per month. If storage costs are 80 cents per kilogram per month, how much will Anita pay in storage costs over the next 5 months?

63. FARMING   It is estimated that t days from now a farmer's crop will be increasing at the rate of $0.3t^2 + 0.6t + 1$ bushels per day. By how much

will the value of the crop increase during the next 5 days if the market price remains fixed at $3 per bushel?

64. SALES REVENUE   It is estimated that the demand for a manufacturer's product is increasing exponentially at the rate of 2% per year. If the current demand is 5,000 units per year and if the price remains fixed at $400 per unit, how much revenue will the manufacturer receive from the sale of the product over the next 2 years?

65. DEPRECIATION   The resale value of a certain industrial machine decreases over a 10-year period at a rate that changes with time. When the machine is x years old, the rate at which its value is changing is $220(x - 10)$ dollars per year. By how much does the machine depreciate during the second year?

66. MARGINAL COST   The marginal cost of producing a certain commodity is $C'(q) = 6q + 1$ dollars per unit when q units are being produced.
   a. What is the total cost of producing the first 10 units?
   b. What is the cost of producing the *next* 10 units?

67. PRODUCTION   Bejax Corporation has set up a production line to manufacture a new type of cellular telephone. The rate of production of the telephones is

$$\frac{dP}{dt} = 1,500\left(2 - \frac{t}{2t + 5}\right) \quad \text{units/month}$$

How many telephones are produced during the third month?

68. PRODUCTION   The output of a factory is changing at the rate

$$Q'(t) = 2t^3 - 3t^2 + 10t + 3 \quad \text{units/hour}$$

where t is the number of hours after the morning shift begins at 8 A.M. How many units are produced between 10 A.M. and noon?

69. INVESTMENT   Ashok's investment portfolio changes value at the rate

$$V'(t) = 12e^{-0.05t}(e^{0.3t} - 3)$$

where V is in thousands of dollars and t is the number of years after 2006. By how much does the value of Ashok's portfolio change between the years:
   a. 2006 and 2010
   b. 2010 and 2012

70. **ADVERTISING** An advertising agency begins a campaign to promote a new product and determines that t days later, the number of people $N(t)$ who have heard about the product is changing at a rate given by

$$N'(t) = 5t^2 - \frac{0.04t}{t^2 + 3} \quad \text{people per day}$$

How many people learn about the product during the first week? During the second week?

LIFE AND SOCIAL SCIENCE APPLIED PROBLEMS

71. **AIR POLLUTION** An environmental study of a certain community suggests that t years from now the level $L(t)$ of carbon monoxide in the air will be changing at the rate of $L'(t) = 0.1t + 0.1$ parts per million (ppm) per year. By how much will the pollution level change during the next 3 years?

72. **WATER POLLUTION** It is estimated that t years from now the population of a certain lakeside community will be changing at the rate of $0.6t^2 + 0.2t + 0.5$ thousand people per year. Environmentalists have found that the level of pollution in the lake increases at the rate of approximately 5 units per 1,000 people. By how much will the pollution in the lake increase during the next 2 years?

73. **NET GROWTH OF POPULATION** A study indicates that t months from now the population of a certain town will be growing at the rate of $P'(t) = 5 + 3t^{2/3}$ people per month. By how much will the population of the town increase over the next 8 months?

74. **WATER CONSUMPTION** The city manager of Playa Linda estimates that water is being consumed by his community at the rate of $C'(t) = 10 + 0.3e^{0.03t}$ billion gallons per year, where $C(t)$ is the water consumption t years after the year 2005. How much water will be consumed by the community during the decade 2005–2015?

75. **CHANGE IN BIOMASS** A protein with mass m (grams) disintegrates into amino acids at a rate given by

$$\frac{dm}{dt} = \frac{-2}{t + 1} \quad \text{g/hr}$$

How much more protein is there after 2 hours than after 5 hours?

76. **CHANGE IN BIOMASS** Answer the question in Exercise 75 if the rate of disintegration is given by

$$\frac{dm}{dt} = -(0.1t + e^{0.1t})$$

77. **RATE OF LEARNING** In a learning experiment, subjects are given a series of facts to memorize, and it is determined that t minutes after the experiment begins, the average subject is learning at the rate

$$L'(t) = \frac{4}{\sqrt{t + 1}} \quad \text{facts per minute}$$

where $L(t)$ is the total number of facts memorized by time t. About how many facts does the typical subject learn during the second 5 minutes (between $t = 5$ and $t = 10$)?

78. **ENDANGERED SPECIES** A study conducted by an environmental group in the year 2005 determined that t years later, the population of a certain endangered bird species will be decreasing at the rate of $P'(t) = -0.75t\sqrt{10 - 0.2t}$ individuals per year. By how much is the population expected to change during the decade 2005–2015?

79. **CONCENTRATION OF DRUG** The concentration of a drug in a patient's bloodstream t hours after an injection is decreasing at the rate

$$C'(t) = \frac{-0.33t}{\sqrt{0.02t^2 + 10}} \quad \text{mg/cm}^3 \text{ per hour}$$

By how much does the concentration change over the first 4 hours after the injection?

MISCELLANEOUS PROBLEMS

80. **DISTANCE AND VELOCITY** Paloma, while driving at a constant speed of 45 mph, begins to speed up in such a way that her velocity t hours later is $v(t) = 45 + 12t$ mph. How far does she travel in the first 2 hours?

81. **PROJECTILE MOTION** A ball is thrown upward from the top of a building, and t seconds later has velocity $v(t) = -32t + 80$ ft/sec. What is the change in the ball's position over the first 3 seconds?

82. Verify the sum rule for definite integrals; that is, if $f(x)$ and $g(x)$ are continuous on the interval $a \leq x \leq b$, then

$$\int_a^b [f(x) + g(x)] \, dx = \int_a^b f(x) \, dx + \int_a^b g(x) \, dx$$

83. You have seen that the definite integral can be used to compute the area under a curve, but the "area as an integral" formula works both ways.

  a. Compute $\int_0^1 \sqrt{1 - x^2} \, dx$. [*Hint:* Note that the integral is part of the area under the circle $x^2 + y^2 = 1$.]

  b. Compute $\int_1^2 \sqrt{2x - x^2} \, dx$. [*Hint:* Describe the graph of $y = \sqrt{2x - x^2}$, and look for a geometric solution as in part (a).]

84. Given the function of $f(x) = 2\sqrt{x} + \dfrac{1}{x + 1}$, approximate the value of the integral $\int_0^2 f(x) \, dx$ by completing these steps:

  a. Find the numbers x_1, x_2, x_3, x_4, and x_5 that divide the interval $0 \leq x \leq 2$ into four equal subintervals. Use these numbers to form four rectangles that approximate the area under the curve $y = f(x)$ over $0 \leq x \leq 2$.

  b. Estimate the value of the given integral by computing the sum of the areas of the four approximating rectangles in part (a).

  c. Repeat steps (a) and (b) with eight subintervals instead of four.

SECTION 5.4 Applying Definite Integration: Distribution of Wealth and Average Value

Learning Objectives

1. Explore a general procedure for using definite integration in applications.
2. Find area between two curves, and use it to compute net excess profit and distribution of wealth (Lorenz curves).
3. Derive and apply a formula for the average value of a function.
4. Interpret average value in terms of rate and area.

We introduced the definite integration process in Section 5.3 by expressing area as a special kind of limit of sums called a definite integral and then computing the integral by using the fundamental theorem of calculus. Area was used in the introduction of definite integration because it is easy to visualize, but there are many other applications in which the definite integration process plays an important role.

In this section, we extend the ideas introduced in Section 5.3 to find the area between two curves and the average value of a function. As part of our study of area between curves, we will examine an important socioeconomic device called a Lorenz curve, which is used to measure relative wealth within a society.

Applying the Definite Integral

Intuitively, definite integration can be thought of as a process that "accumulates" an infinite number of small pieces of a quantity to obtain the total quantity. Here is a step-by-step description of how to use this process in applications.

A Procedure for Using Definite Integration in Applications

To use definite integration to accumulate a quantity Q over an interval $a \le x \le b$, proceed as follows:

Step 1. Divide the interval $a \le x \le b$ into n equal subintervals, each of length $\Delta x = \dfrac{b - a}{n}$. Choose a number x_j from the jth subinterval, for $j = 1, 2, \ldots, n$.

Step 2. Approximate small parts of the quantity Q by products of the form $f(x_j)\Delta x$, where $f(x)$ is an appropriate function that is continuous on $a \le x \le b$.

Step 3. Add the individual approximating products to estimate the total quantity Q by the Riemann sum

$$[f(x_1) + f(x_2) + \cdots + f(x_n)]\Delta x$$

Step 4. Make the approximation in step 3 exact by taking the limit of the Riemann sum as $n \to +\infty$ to express Q as a definite integral; that is,

$$Q = \lim_{n \to +\infty} [f(x_1) + f(x_2) + \cdots + f(x_n)]\Delta x = \int_a^b f(x)\, dx$$

Then use the fundamental theorem of calculus to compute $\int_a^b f(x)\, dx$ and thus to obtain the required quantity Q.

Just-In-Time **REVIEW**

Summation notation is reviewed in Appendix A.4 including examples. Note that there is nothing special about using "j" for the index variable in the notation. The indices most frequently used are i, j, and k.

NOTATION   We can use *summation notation* to represent the Riemann sums that occur when quantities are modeled using definite integration. Specifically, to describe the sum

$$a_1 + a_2 + \cdots + a_n$$

it suffices to specify the general term a_j in the sum and to indicate that n terms of this form are to be added, starting with the term where $j = 1$ and ending with $j = n$. For this purpose, it is customary to use the uppercase Greek letter sigma (Σ) and to write the sum as $\displaystyle\sum_{j=1}^{n} a_j$, that is,

$$\sum_{j=1}^{n} a_j = a_1 + a_2 + \cdots + a_n$$

In particular, the Riemann sum

$$[f(x_1) + f(x_2) + \cdots + f(x_n)]\Delta x$$

can be written in the compact form

$$\sum_{j=1}^{n} f(x_j)\Delta x$$

Thus, the limit statement

$$\lim_{n \to +\infty} [f(x_1) + f(x_2) + \cdots + f(x_n)]\Delta x = \int_a^b f(x)\, dx$$

used to define the definite integral can be expressed as

$$\lim_{n \to +\infty} \sum_{j=1}^{n} f(x_j)\Delta x = \int_a^b f(x)\, dx \quad \blacksquare$$

Area Between Two Curves

In certain practical applications, you may find it useful to represent a quantity of interest in terms of area between two curves. First, suppose that f and g are continuous, nonnegative [that is, $f(x) \geq 0$ and $g(x) \geq 0$], and satisfy $f(x) \geq g(x)$ on the interval $a \leq x \leq b$, as shown in Figure 5.10a.

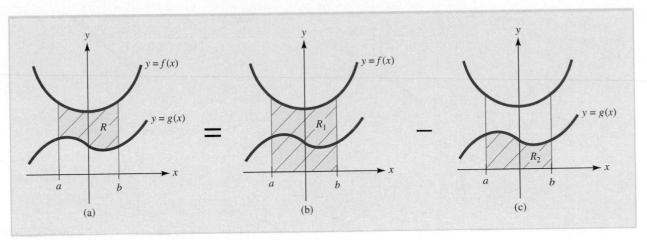

FIGURE 5.10   Area of R = area of R_1 − area of R_2.

Then, to find the area of the region R between the curves $y = f(x)$ and $y = g(x)$ over the interval $a \leq x \leq b$, we simply subtract the area under the lower curve $y = g(x)$ (Figure 5.10c) from the area under the upper curve $y = f(x)$ (Figure 5.10b), so that

$$\text{Area of } R = [\text{area under } y = f(x)] - [\text{area under } y = g(x)]$$

$$= \int_a^b f(x)\, dx - \int_a^b g(x)\, dx = \int_a^b [f(x) - g(x)]\, dx$$

This formula still applies whenever $f(x) \geq g(x)$ on the interval $a \leq x \leq b$, even when the curves $y = f(x)$ and $y = g(x)$ are not always both above the x axis. We will show that this is true by using the procedure for applying definite integration described on page 424.

Step 1. Divide the interval $a \leq x \leq b$ into n equal subintervals, each of width $\Delta x = \dfrac{b - a}{n}$. For $j = 1, 2, \ldots, n$, let x_j be the left endpoint of the jth subinterval.

Step 2. Construct approximating rectangles of width Δx and height $f(x_j) - g(x_j)$. This is possible since $f(x) \geq g(x)$ on $a \leq x \leq b$, which guarantees that the height is nonnegative; that is, $f(x_j) - g(x_j) \geq 0$. For $j = 1, 2, \ldots, n$, the area $[f(x_j) - g(x_j)]\Delta x$ of the jth rectangle you have just constructed is approximately the same as the area between the two curves over the jth subinterval, as shown in Figure 5.11a.

Step 3. Add the individual approximating areas $[f(x_j) - g(x_j)]\Delta x$ to estimate the total area A between the two curves over the interval $a \leq x \leq b$ by the Riemann sum

$$A \approx [f(x_1) - g(x_1)]\Delta x + [f(x_2) - g(x_2)]\Delta x + \cdots + [f(x_n) - g(x_n)]\Delta x$$

$$= \sum_{j=1}^{n} [f(x_j) - g(x_j)]\Delta x$$

(See Figure 5.11b.)

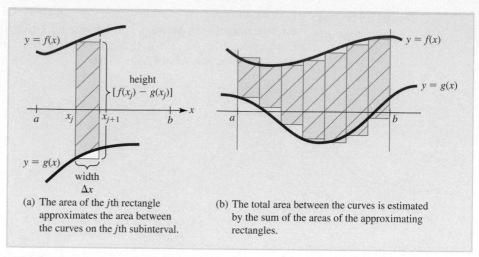

(a) The area of the jth rectangle
   approximates the area between
   the curves on the jth subinterval.

(b) The total area between the curves is estimated
   by the sum of the areas of the approximating
   rectangles.

FIGURE 5.11   Computing area between curves by definite integration.

Step 4. Make the approximation exact by taking the limit of the Riemann sum in step 3 as $n \to +\infty$ to express the total area A between the curves as a definite integral; that is,

$$A = \lim_{n \to +\infty} \sum_{j=1}^{n} [f(x_j) - g(x_j)]\Delta x = \int_a^b [f(x) - g(x)]\, dx$$

To summarize:

The Area Between Two Curves   ■   If $f(x)$ and $g(x)$ are continuous with $f(x) \geq g(x)$ on the interval $a \leq x \leq b$, then the area A between the curves $y = f(x)$ and $y = g(x)$ over the interval is given by

$$A = \int_a^b [f(x) - g(x)]\, dx$$

Just-In-Time REVIEW

Note that $x^2 \geq x^3$ for $0 \leq x \leq 1$. For example,

$$\left(\frac{1}{3}\right)^2 > \left(\frac{1}{3}\right)^3$$

This tells us that the graph of $y = x^2$ is above that of $y = x^3$ between $x = 0$ and $x = 1$.

EXAMPLE 5.4.1   Finding the Area Between Two Curves

Find the area of the region R enclosed by the curves $y = x^3$ and $y = x^2$.

Solution

To find the points where the curves intersect, solve the equations simultaneously as follows:

$$
\begin{aligned}
x^3 &= x^2 &&\text{subtract } x^2 \text{ from both sides}\\
x^3 - x^2 &= 0 &&\text{factor out } x^2\\
x^2(x - 1) &= 0 &&uv = 0 \text{ if and only if } u = 0 \text{ or } v = 0\\
x &= 0, 1
\end{aligned}
$$

The corresponding points $(0, 0)$ and $(1, 1)$ are the only points of intersection.

The region R enclosed by the two curves is bounded above by $y = x^2$ and below by $y = x^3$, over the interval $0 \leq x \leq 1$ (Figure 5.12). The area of this region is given by the integral

$$A = \int_0^1 (x^2 - x^3)\, dx = \frac{1}{3}x^3 - \frac{1}{4}x^4 \bigg|_0^1$$

$$= \left[\frac{1}{3}(1)^3 - \frac{1}{4}(1)^4\right] - \left[\frac{1}{3}(0)^3 - \frac{1}{4}(0)^4\right] = \frac{1}{12}$$

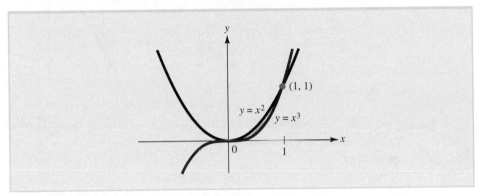

FIGURE 5.12   The region enclosed by the curves $y = x^2$ and $y = x^3$.

In certain applications, you may need to find the area A between the two curves $y = f(x)$ and $y = g(x)$ over an interval $a < x \leq b$, where $f(x) \geq g(x)$ for $a \leq x \leq c$ but $g(x) \geq f(x)$ for $c \leq x \leq b$. In this case, we have

$$A = \underbrace{\int_a^c [f(x) - g(x)]\, dx}_{f(x) \,\geq\, g(x)\text{ on }a\,<\,x\,\leq\,c} + \underbrace{\int_c^b [g(x) - f(x)]\, dx}_{g(x) \,\geq\, f(x)\text{ on }c\,\leq\,x\,\leq\,b}$$

Consider Example 5.4.2.

EXPLORE!

Refer to Example 5.4.2. Set Y1 = 4X and Y2 = X³ + 3X² in the equation editor of your graphing calculator. Graph using the window [−6, 2]1 by [−25, 10]5. Determine the points of intersection of the two curves. Another view of the area between the two curves is to set Y3 = Y2 − Y1, deselect (turn off) Y1 and Y2, and graph using [−4.5, 1.5]0.5 by [−5, 15]5. Numerical integration can be applied to this difference function.

EXAMPLE 5.4.2   Finding the Area Between Two Curves

Find the area of the region enclosed by the line $y = 4x$ and the curve $y = x^3 + 3x^2$.

Solution

To find where the line and curve intersect, solve the equations simultaneously as follows:

$$x^3 + 3x^2 = 4x \qquad \text{subtract } 4x \text{ from both sides}$$
$$x^3 + 3x^2 - 4x = 0 \qquad \text{factor out } x$$
$$x(x^2 + 3x - 4) = 0 \qquad \text{factor } x^2 + 3x - 4$$
$$x(x - 1)(x + 4) = 0 \qquad uv = 0 \text{ if and only if } u = 0 \text{ or } v = 0$$
$$x = 0, 1, -4$$

The corresponding points of intersection are $(0, 0)$, $(1, 4)$, and $(-4, -16)$. The curve and the line are sketched in Figure 5.13.

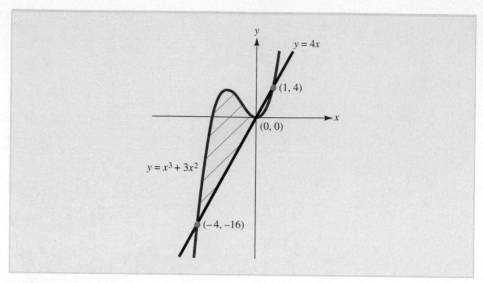

FIGURE 5.13   The region enclosed by the line $y = 4x$ and the curve $y = x^3 + 3x^2$.

Over the interval $-4 \leq x \leq 0$, the curve is above the line, so $x^3 + 3x^2 \geq 4x$, and the region enclosed by the curve and line has area

$$A_1 = \int_{-4}^{0} [(x^3 + 3x^2) - 4x]\,dx = \frac{1}{4}x^4 + x^3 - 2x^2 \Big|_{-4}^{0}$$

$$= \left[\frac{1}{4}(0)^4 + (0)^3 - 2(0)^2 \right] - \left[\frac{1}{4}(-4)^4 + (-4)^3 - 2(-4)^2 \right] = 32$$

Over the interval $0 \leq x \leq 1$, the line is above the curve and the enclosed region has area

$$A_2 = \int_{0}^{1} [4x - (x^3 + 3x^2)]\,dx = 2x^2 - \frac{1}{4}x^4 - x^3 \Big|_{0}^{1}$$

$$= \left[2(1)^2 - \frac{1}{4}(1)^4 - (1)^3 \right] - \left[2(0)^2 - \frac{1}{4}(0)^4 - (0)^3 \right] = \frac{3}{4}$$

Therefore, the total area enclosed by the line and the curve is given by the sum

$$A = A_1 + A_2 = 32 + \frac{3}{4} = 32.75$$

Net Excess Profit   The area between curves can sometimes be used as a way of measuring the amount of a quantity that has been accumulated during a particular procedure. For instance, suppose that t years from now, two investment plans will be generating profit $P_1(t)$ and $P_2(t)$, respectively, and that their respective rates of profitability, $P_1'(t)$ and $P_2'(t)$, are expected to satisfy $P_2'(t) \geq P_1'(t)$ for the next N years, that is, over the time interval $0 \leq t \leq N$. Then $E(t) = P_2(t) - P_1(t)$ represents the **excess profit** of plan 2 over plan 1 at time t, and the **net excess profit** $NE = E(N) - E(0)$ over the time interval $0 \le t \le N$ is given by the definite integral

$$NE = E(N) - E(0) = \int_0^N E'(t)\, dt \quad \begin{aligned} E'(t) &= [P_2(t) - P_1(t)]' \\ &= P_2'(t) - P_1'(t) \end{aligned}$$

$$= \int_0^N [P_2'(t) - P_1'(t)]\, dt$$

This integral can be interpreted geometrically as the area between the rate of profitability curves $y = P_1'(t)$ and $y = P_2'(t)$ as shown in Figure 5.14. Example 5.4.3 illustrates the computation of net excess profit.

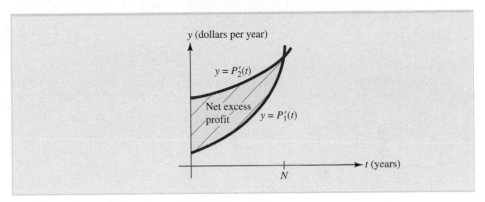

FIGURE 5.14   Net excess profit as the area between rate of profitability curves.

EXAMPLE 5.4.3   Finding Net Excess Profit

Suppose that t years from now, one investment will be generating profit at the rate of $P_1'(t) = 50 + t^2$ hundred dollars per year, while a second investment will be generating profit at the rate of $P_2'(t) = 200 + 5t$ hundred dollars per year.

a. For how many years does the rate of profitability of the second investment exceed that of the first?

b. Compute the net excess profit for the time period determined in part (a). Interpret the net excess profit as an area.

Solution

a. The rate of profitability of the second investment exceeds that of the first until

$$\begin{aligned} P_1'(t) &= P_2'(t) \\ 50 + t^2 &= 200 + 5t \quad \text{subtract } 200 + 5t \text{ from both sides} \\ t^2 - 5t - 150 &= 0 \quad \text{factor} \\ (t - 15)(t + 10) &= 0 \quad uv = 0 \text{ if and only if } u = 0 \text{ or } v = 0 \\ t &= 15, -10 \quad \text{reject the negative time } t = -10 \\ t &= 15 \text{ years} \end{aligned}$$

b. The excess profit of plan 2 over plan 1 is $E(t) = P_2(t) - P_1(t)$, and the net excess profit NE over the time period $0 \le t \le 15$ determined in part (a) is given by the definite integral

$$\text{NE} = E(15) - E(0) = \int_0^{15} E'(t)\, dt \qquad \text{fundamental theorem of calculus}$$

$$= \int_0^{15} [P_2'(t) - P_1'(t)]\, dt \qquad \text{since } E(t) = P_2(t) - P_1(t)$$

$$= \int_0^{15} [(200 + 5t) - (50 + t^2)]\, dt \qquad \text{combine like terms}$$

$$= \int_0^{15} [150 + 5t - t^2]\, dt$$

$$= \left[150t + 5\left(\frac{1}{2}t^2\right) - \left(\frac{1}{3}t^3\right) \right]\Big|_0^{15}$$

$$= \left[150(15) + \frac{5}{2}(15)^2 - \frac{1}{3}(15)^3 \right] - \left[150(0) + \frac{5}{2}(0)^2 - \frac{1}{3}(0)^3 \right]$$

$$= 1{,}687.50 \text{ hundred dollars}$$

Thus, the net excess profit is $168,750.

The graphs of the rate of profitability functions $P_1'(t)$ and $P_2'(t)$ are shown in Figure 5.15. The net excess profit

$$\text{NE} = \int_0^{15} [P_2'(t) - P_1'(t)]\, dt$$

can be interpreted as the area of the (shaded) region between the rate of profitability curves over the interval $0 \le t \le 15$.

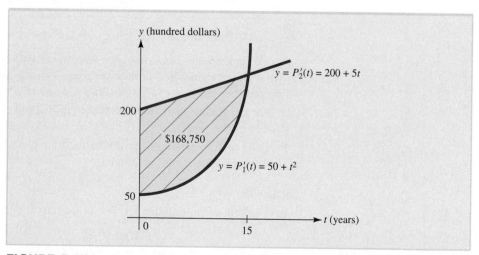

FIGURE 5.15   Net excess profit for one investment plan over another.

Lorenz Curves   Area also plays an important role in the study of **Lorenz curves,** a device used by both economists and sociologists to measure the percentage of a society's wealth that is possessed by a given percentage of its people. To be more specific, the Lorenz curve for a particular society's economy is the graph of the function $L(x)$, which denotes the fraction of total annual national income earned by the lowest-paid $100x\%$ of the wage-earners in the society, for $0 \le x \le 1$. For instance, if the lowest-paid 30% of all wage-earners receive 23% of the society's total income, then $L(0.3) = 0.23$.

Note that $L(x)$ is an increasing function on the interval $0 \le x \le 1$ and has these properties:

1. $0 \le L(x) \le 1$ because $L(x)$ is a percentage
2. $L(0) = 0$ because no wages are earned when no wage-earners are employed
3. $L(1) = 1$ because 100% of wages are earned by 100% of the wage-earners
4. $L(x) \le x$ because the lowest-paid $100x\%$ of wage-earners cannot receive more than $100x\%$ of total income

A typical Lorenz curve is shown in Figure 5.16a.

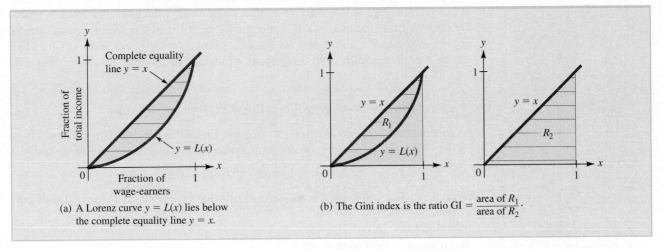

(a) A Lorenz curve $y = L(x)$ lies below the complete equality line $y = x$.

(b) The Gini index is the ratio $\mathrm{GI} = \dfrac{\text{area of } R_1}{\text{area of } R_2}$.

FIGURE 5.16 A Lorenz curve $y = L(x)$ and its Gini index.

The line $y = x$ represents the case corresponding to complete equality in the distribution of income (wage-earners with the lowest $100x\%$ of income receive $100x\%$ of the society's wealth). The closer a particular Lorenz curve is to this line, the more equitable the distribution of wealth in the corresponding society. We represent the total deviation of the actual distribution of wealth in the society from complete equality by the area of the region R_1 between the Lorenz curve $y = L(x)$ and the line $y = x$. The ratio of this area to the area of the region R_2 under the complete equality line $y = x$ over $0 \le x \le 1$ is used as a measure of the inequality in the distribution of wealth in the society. This ratio, called the **Gini index,** denoted GI (also called the **index of income inequality**), may be computed by the formula

$$\mathrm{GI} = \frac{\text{area of } R_1}{\text{area of } R_2} = \frac{\text{area between } y = L(x) \text{ and } y = x}{\text{area under } y = x \text{ over } 0 \le x \le 1}$$

$$= \frac{\displaystyle\int_0^1 [x - L(x)]\, dx}{\displaystyle\int_0^1 x\, dx} = \frac{\displaystyle\int_0^1 [x - L(x)]\, dx}{1/2}$$

$$= 2\int_0^1 [x - L(x)]\, dx$$

(see Figure 5.16b). To summarize:

> **Gini Index** ■ If $y = L(x)$ is the equation of a Lorenz curve, then the inequality in the corresponding distribution of wealth is measured by the *Gini index*, which is given by the formula
>
> $$\text{Gini index} = 2 \int_0^1 [x - L(x)]\, dx$$

The Gini index always lies between 0 and 1. An index of 0 corresponds to total equity in the distribution of income, while an index of 1 corresponds to total inequity (all income belongs to 0% of the population). The smaller the index, the more equitable the distribution of income, and the larger the index, the more the wealth is concentrated in only a few hands. Example 5.4.4 illustrates how Lorenz curves and the Gini index can be used to compare the relative equity of income distribution for two professions.

EXAMPLE 5.4.4 Studying Distribution of Income

A governmental agency determines that the Lorenz curves for the distribution of income for dentists and contractors in a certain state are given by the functions

$$L_1(x) = x^{1.7} \qquad \text{and} \qquad L_2(x) = 0.8x^2 + 0.2x$$

respectively. For which profession is the distribution of income more fairly distributed?

Solution

The respective Gini indices are

$$G_1 = 2 \int_0^1 (x - x^{1.7})\, dx = 2\left(\frac{x^2}{2} - \frac{x^{2.7}}{2.7} \right)\Bigg|_0^1 = 0.2593$$

and

$$G_2 = 2 \int_0^1 [x - (0.8x^2 + 0.2x)]\, dx$$

$$= 2\left[-0.8\left(\frac{x^3}{3}\right) + 0.8\left(\frac{x^2}{2}\right) \right]\Bigg|_0^1 = 0.2667$$

Since the Gini index for dentists is smaller, it follows that in this state, the incomes of dentists are more evenly distributed than those of contractors.

Using the Gini index, we can see how the distribution of income in the United States compares to that in other countries. Table 5.1 lists the Gini index for selected industrial and developing nations. Note that with an index of 0.450, the distribution of income in the United States is about the same as that of Thailand, is less equitable than that of the United Kingdom, Germany, or Denmark, but much more equitable than that of Brazil or Panama. (Is there anything you know about the sociopolitical nature of these countries that would explain the difference in income equity?)

TABLE 5.1   Gini Indices for Selected Countries

Country	Gini Index
United States	0.450
Brazil	0.567
Canada	0.321
Denmark	0.290
Germany	0.270
Japan	0.381
South Africa	0.650
Panama	0.561
Thailand	0.430
United Kingdom	0.340

Source: CIA World Factbook.

Average Value of a Function

As a second illustration of how definite integration can be used in applications, we will compute the **average value of a function,** which is of interest in a variety of situations. First, let us take a moment to clarify our thinking about what we mean by *average value.* A teacher who wants to compute the average score on an examination simply adds all the individual scores and then divides by the number of students taking the exam, but how should one go about finding, say, the average pollution level in a city during the daytime hours? The difficulty is that since time is continuous, there are too many pollution levels to add up in the usual way, so how should we proceed?

Consider the general case in which we wish to find the average value of the function $f(x)$ over an interval $a \leq x \leq b$ on which f is continuous. We begin by dividing the interval $a \leq x \leq b$ into n equal parts, each of length $\Delta x = \dfrac{b-a}{n}$. If x_j is a number taken from the jth subinterval for $j = 1, 2, \ldots, n$, then the average of the corresponding functional values $f(x_1), f(x_2), \ldots, f(x_n)$ is

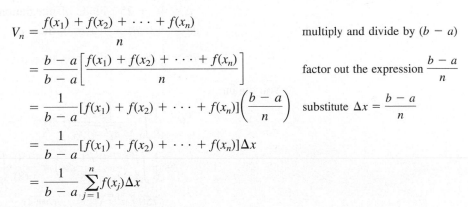

$$V_n = \frac{f(x_1) + f(x_2) + \cdots + f(x_n)}{n} \qquad \text{multiply and divide by } (b-a)$$

$$= \frac{b-a}{b-a}\left[\frac{f(x_1) + f(x_2) + \cdots + f(x_n)}{n}\right] \qquad \text{factor out the expression } \frac{b-a}{n}$$

$$= \frac{1}{b-a}[f(x_1) + f(x_2) + \cdots + f(x_n)]\left(\frac{b-a}{n}\right) \qquad \text{substitute } \Delta x = \frac{b-a}{n}$$

$$= \frac{1}{b-a}[f(x_1) + f(x_2) + \cdots + f(x_n)]\Delta x$$

$$= \frac{1}{b-a}\sum_{j=1}^{n} f(x_j)\Delta x$$

which we recognize as a Riemann sum.

If we refine the partition of the interval $a \leq x \leq b$ by taking more and more subdivision points, then V_n becomes more and more like what we may intuitively think of as the average value V of $f(x)$ over the entire interval $a \leq x \leq b$. Thus, it is

EXPLORE!

Suppose you wish to calculate the average value of $f(x) = x^3 - 6x^2 + 10x - 1$ over the interval [1, 4]. Store $f(x)$ in Y1, and obtain its graph using the window [0, 4.7]1 by [−2, 8]1. Shade the region under the curve over the interval [1, 4], and compute its area A. Set Y2 equal to the constant function $\dfrac{A}{b-a} = \dfrac{A}{3}$. This is the average value. Plot Y2 and Y1 on the same screen. At what number(s) between 1 and 4 does $f(x)$ equal the average value?

reasonable to *define* the average value V as the limit of the Riemann sum V_n as $n \to +\infty$, that is, as the definite integral

$$V = \lim_{n \to +\infty} V_n = \lim_{n \to +\infty} \frac{1}{b-a} \sum_{j=1}^{n} f(x_j) \Delta x$$

$$= \frac{1}{b-a} \int_a^b f(x)\, dx$$

To summarize:

> **The Average Value of a Function** ■ Let $f(x)$ be a function that is continuous on the interval $a \le x \le b$. Then the *average value* V of $f(x)$ over $a \le x \le b$ is given by the definite integral
>
> $$V = \frac{1}{b-a} \int_a^b f(x)\, dx$$

EXAMPLE 5.4.5   Finding Average Monthly Sales

A manufacturer determines that t months after introducing a new product, the company's sales will be $S(t)$ thousand dollars, where

$$S(t) = \frac{750t}{\sqrt{4t^2 + 25}}$$

What are the average monthly sales of the company over the first 6 months after the introduction of the new product?

Solution

The average monthly sales V over the time period $0 \le t \le 6$ is given by the integral

$$V = \frac{1}{6-0} \int_0^6 \frac{750t}{\sqrt{4t^2 + 25}}\, dt$$

To evaluate this integral, make the substitution

$$u = 4t^2 + 25 \quad \text{limits of integration:}$$
$$du = 4(2t\, dt) \quad \text{if } t = 0, \text{ then } u = 4(0)^2 + 25 = 25$$
$$t\, dt = \frac{1}{8}\, du \quad \text{if } t = 6, \text{ then } u = 4(6)^2 + 25 = 169$$

You obtain

$$V = \frac{1}{6} \int_0^6 \frac{750}{\sqrt{4t^2 + 25}} (t\, dt)$$

$$= \frac{1}{6} \int_{25}^{169} \frac{750}{\sqrt{u}} \left(\frac{1}{8}\, du \right) = \frac{750}{6(8)} \int_{25}^{169} u^{-1/2}\, du$$

$$= \frac{750}{6(8)} \left(\frac{u^{1/2}}{1/2} \right) \Bigg|_{25}^{169} = \frac{750(2)}{6(8)} [(169)^{1/2} - (25)^{1/2}]$$

$$= 250$$

Thus, for the 6-month period immediately after the introduction of the new product, the company's sales average $250,000 per month.

EXAMPLE 5.4.6 Finding Average Temperature

As part of her research, Prof. Ellen McGuire models the temperature T (in °C) in a certain northern city during the time period from 6 A.M. to 6 P.M. by the function

$$T(t) = 3 - \frac{1}{3}(t - 4)^2 \qquad \text{for } 0 \le t \le 12$$

where t is the number of hours after 6 A.M.

a. What is the average temperature in the city during the workday, from 8 A.M. to 5 P.M.?

b. At what time (or times) during the workday should Prof. McGuire expect the temperature to be the same as the average temperature found in part (a)?

Solution

a. Since 8 A.M. and 5 P.M. are, respectively, $t = 2$ and $t = 11$ hours after 6 A.M., we want to compute the average of the temperature $T(t)$ for $2 \le t \le 11$, which is given by the definite integral

$$T_{\text{ave}} = \frac{1}{11 - 2} \int_2^{11} \left[3 - \frac{1}{3}(t - 4)^2 \right] dt$$

$$-\frac{1}{9}\left[3t - \frac{1}{3}\frac{1}{3}(t - 4)^3 \right]\Bigg|_2^{11}$$

$$= \frac{1}{9}\left[3(11) - \frac{1}{9}(11 - 4)^3 \right] - \frac{1}{9}\left[3(2) - \frac{1}{9}(2 - 4)^3 \right]$$

$$= -\frac{4}{3} \approx -1.33$$

Thus, the average temperature during the workday is approximately -1.33°C (or 29.6°F).

b. We want to find a time $t = t_a$ with $2 \le t_a \le 11$ such that $T(t_a) = -\frac{4}{3}$. Solving this equation, we find that

$$3 - \frac{1}{3}(t_a - 4)^2 = -\frac{4}{3} \qquad \text{subtract 3 from both sides}$$

$$-\frac{1}{3}(t_a - 4)^2 = -\frac{4}{3} - 3 = -\frac{13}{3} \qquad \text{multiply both sides by } -3$$

$$(t_a - 4)^2 = (-3)\left(-\frac{13}{3}\right) = 13 \qquad \text{take square roots on both sides}$$

$$t_a - 4 = \pm\sqrt{13} \qquad \text{add 4 to both sides}$$

$$t_a = 4 \pm \sqrt{13}$$

$$\approx 0.39 \quad \text{or} \quad 7.61$$

Since $t = 0.39$ is outside the time interval $2 \le t_a \le 11$ (8 A.M. to 5 P.M.), it follows that the temperature in the city is the same as the average temperature only when $t = 7.61$, that is, at approximately 1:37 P.M.

Just-In-Time **REVIEW**

Fahrenheit temperature F is related to Celsius temperature C by the formula

$$F = \frac{9}{5}C + 32$$

Just-In-Time **REVIEW**

Since there are 60 minutes in an hour, 0.61 hour is the same as 0.61(60) ≈ 37 minutes. So, 7.61 hours after 6 A.M. is 37 minutes past 1 P.M., or 1:37 P.M.

Two Interpretations of Average Value

The average value of a function has several useful interpretations. First, note that if $f(x)$ is continuous on the interval $a \leq x \leq b$ and $F(x)$ is any antiderivative of $f(x)$ over the same interval, then the average value V of $f(x)$ over the interval satisfies

$$V = \frac{1}{b-a} \int_a^b f(x)\, dx$$

$$= \frac{1}{b-a}[F(b) - F(a)] \quad \text{fundamental theorem of calculus}$$

$$= \frac{F(b) - F(a)}{b-a}$$

We recognize this difference quotient as the average rate of change of $F(x)$ over $a \leq x \leq b$ (see Section 2.1). Thus, we have this interpretation:

> **Rate Interpretation of Average Value** ■ The average value of a function $f(x)$ over an interval $a \leq x \leq b$ where $f(x)$ is continuous is the same as the average rate of change of any antiderivative $F(x)$ of $f(x)$ over the same interval.

For instance, since the total cost $C(x)$ of producing x units of a commodity is an antiderivative of marginal cost $C'(x)$, it follows that the *average rate of change of cost over a range of production $a \leq x \leq b$ equals the average value of the marginal cost over the same range.*

The average value of a function $f(x)$ on an interval $a \leq x \leq b$ where $f(x) \geq 0$ can also be interpreted geometrically by rewriting the integral formula for average value

$$V = \frac{1}{b-a} \int_a^b f(x)\, dx$$

in the form

$$(b-a)V = \int_a^b f(x)\, dx$$

In the case where $f(x) \geq 0$ on the interval $a \leq x \leq b$, the integral on the right can be interpreted as the area under the curve $y = f(x)$ over $a \leq x \leq b$, and the product on the left as the area of a rectangle of height V and width $b - a$ equal to the length of the interval. In other words:

> **Geometric Interpretation of Average Value** ■ Let V be the average value of $f(x)$ over an interval $a \leq x \leq b$, where $f(x)$ is continuous and $f(x) \geq 0$. Then the rectangle with height V and base $a \leq x \leq b$ has the same area as the region under the curve $y = f(x)$ over $a \leq x \leq b$.

This geometric interpretation is illustrated in Figure 5.17.

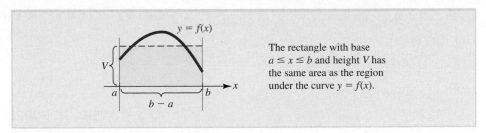

FIGURE 5.17   Geometric interpretation of average value V.

EXERCISES ▪ 5.4

In Exercises 1 through 4, find the area of the shaded region.

1.

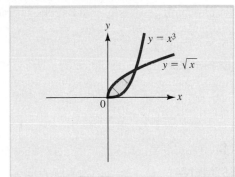

2.

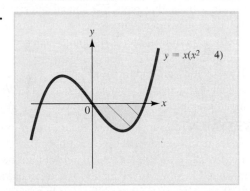

3.

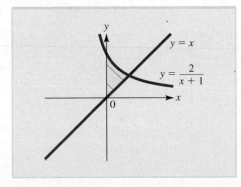

4.

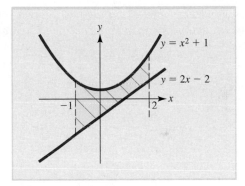

In Exercises 5 through 18, sketch the given region R and then find its area.

5. R is the region bounded by the lines $y = x$, $y = -x$, and $x = 1$.

6. R is the region bounded by the curves $y = x^2$, $y = -x^2$, and the line $x = 1$.

7. R is the region bounded by the x axis and the curve $y = -x^2 + 4x - 3$.

8. R is the region bounded by the curves $y = e^x$, $y = e^{-x}$, and the line $x = \ln 2$.

9. R is the region bounded by the curve $y = x^2 - 2x$ and the x axis. [*Hint:* Note that the region is below the x axis.]

10. R is the region bounded by the curve $y = \dfrac{1}{x^2}$ and the lines $y = x$ and $y = \dfrac{x}{8}$.

11. R is the region bounded by the curves $y = x^2 - 2x$ and $y = -x^2 + 4$.

12. R is the region between the curve $y = x^3$ and the line $y = 9x$, for $x \geq 0$.

CHAPTER 5 Integration

5-64

13. R is the region between the curves $y = x^3 - 3x^2$ and $y = x^2 + 5x$.

14. R is the triangle bounded by the line $y = 4 - 3x$ and the coordinate axes.

15. R is the triangle with vertices $(-4, 0)$, $(2, 0)$, and $(2, 6)$.

16. R is the rectangle with vertices $(1, 0)$, $(-2, 0)$, $(-2, 5)$, and $(1, 5)$.

17. R is the trapezoid bounded by the lines $y = x + 6$ and $x = 2$ and the coordinate axes.

18. R is the trapezoid bounded by the lines $y = x + 2$, $y = 8 - x$, $x = 2$, and the y axis.

In Exercises 19 through 24, find the average value of the given function $f(x)$ over the specified interval $a \leq x \leq b$.

19. $f(x) = 1 - x^2$ over $-3 \leq x \leq 3$

20. $f(x) = x^2 - 3x + 5$ over $-1 \leq x \leq 2$

21. $f(x) = e^{-x}(4 - e^{2x})$ over $-1 \leq x \leq 1$

22. $f(x) = e^{2x} + e^{-x}$ over $0 \leq x \leq \ln 2$

23. $f(x) = \dfrac{e^x - e^{-x}}{e^x + e^{-x}}$ over $0 \leq x \leq \ln 3$

24. $f(x) = \dfrac{x + 1}{x^2 + 2x + 6}$ over $-1 \leq x \leq 1$

In Exercises 25 through 28, find the average value V of the given function over the specified interval. In each case, sketch the graph of the function along with the rectangle whose base is the given interval and whose height is the average value V.

25. $f(x) = 2x - x^2$ over $0 \leq x \leq 2$

26. $f(x) = x$ over $0 \leq x \leq 4$

27. $h(u) = \dfrac{1}{u}$ over $2 \leq u \leq 4$

28. $g(t) = e^{-2t}$ over $-1 \leq t \leq 2$

BUSINESS AND ECONOMICS APPLIED PROBLEMS

LORENZ CURVES *In Exercises 29 through 34, find the Gini index for the given Lorenz curve.*

29. $L(x) = x^3$

30. $L(x) = x^2$

31. $L(x) = 0.55x^2 + 0.45x$

32. $L(x) = 0.7x^2 + 0.3x$

33. $L(x) = \dfrac{2}{3}x^{3.7} + \dfrac{1}{3}x$

34. $L(x) = \dfrac{e^x - 1}{e - 1}$

35. AVERAGE SUPPLY A manufacturer supplies $S(p) = 0.5p^2 + 3p + 7$ hundred units of a certain commodity to the market when the price is p dollars per unit. Find the average supply as the price varies from $p = \$2$ to $p = \$5$.

36. EFFICIENCY After t months on the job, a postal clerk can sort $Q(t) = 700 - 400e^{-0.5t}$ letters per hour. What is the average rate at which the clerk sorts mail during the first 3 months on the job?

37. INVENTORY An inventory of 60,000 kilograms of a certain commodity is used at a constant rate and is exhausted after 1 year. What is the average inventory for the year?

38. FOOD PRICES Records indicate that t months after the beginning of the year, the price of ground beef in local supermarkets was
$$P(t) = 0.09t^2 - 0.2t + 4$$
dollars per pound. What was the average price of ground beef during the first 3 months of the year?

39. INVESTMENT Marya invests \$10,000 for 5 years in a bank that pays 5% annual interest.
   a. What is the average value of her account over this time period if interest is compounded continuously?
   b. How would you find the average value of the account if interest is compounded quarterly? Write a paragraph to explain how you would proceed.

40. INVESTMENT Suppose that t years from now, one investment plan will be generating profit at the rate of $P_1'(t) = 100 + t^2$ hundred dollars per year, while a second investment will be generating profit at the rate of $P_2'(t) = 220 + 2t$ hundred dollars per year.
   a. For how many years does the rate of profitability of the second investment exceed that of the first?
   b. Compute the net excess profit assuming that you invest in the second plan for the time period determined in part (a).
   c. Sketch the rate of profitability curves $y = P_1'(t)$ and $y = P_2'(t)$, and shade the region whose area represents the net excess profit computed in part (b).

41. **INVESTMENT**   Answer the questions in Exercise 40 for two investments with respective rates of profitability $P_1'(t) = 130 + t^2$ hundred dollars per year and $P_2'(t) = 306 + 5t$ hundred dollars per year.

42. **INVESTMENT**   Answer the questions in Exercise 40 for two investments with respective rates of profitability $P_1'(t) = 60e^{0.12t}$ thousand dollars per year and $P_2'(t) = 160e^{0.08t}$ thousand dollars per year.

43. **INVESTMENT**   Answer the questions in Exercise 40 for two investments with respective rates of profitability $P_1'(t) = 90e^{0.1t}$ thousand dollars per year and $P_2'(t) = 140e^{0.07t}$ thousand dollars per year.

44. **ADVERTISING**   Everett Dunn's advertising firm has been hired to promote a new television series for 3 weeks before its debut and 2 weeks afterward. After t weeks of the advertising campaign, Everett estimates that $P(t)$ percent of the viewing public is aware of the series, where

    $$P(t) = \frac{59t}{0.7t^2 + 16} + 6$$

    a. What is the average percentage of the viewing public that is aware of the show during the 5 weeks of the advertising campaign?
    b. At what time during the 5 weeks of the advertising campaign should Everett expect the percentage of viewers to be the same as the average percentage found in part (a)?

45. **AVERAGE PRODUCTION**   A company determines that if L worker-hours of labor are employed, then Q units of a particular commodity will be produced, where

    $$Q(L) = 500L^{2/3}$$

    a. What is the average production as labor varies from 1,000 to 2,000 worker-hours?
    b. What labor level between 1,000 and 2,000 worker-hours results in the average production found in part (a)?

46. **EFFICIENCY**   After t hours on the job, one factory worker is producing $Q_1'(t) = 60 - 2(t - 1)^2$ units per hour, while a second worker is producing $Q_2'(t) = 50 - 5t$ units per hour.
    a. If both arrive on the job at 8:00 A.M., how many more units will the first worker have produced by noon than the second worker?
    b. Interpret the answer in part (a) as the area between two curves.

47. **AVERAGE COST**   The cost of producing x units of a new product is $C(x) = 3x\sqrt{x} + 10$ hundred dollars. What is the average cost of producing the first 81 units?

48. **REAL ESTATE**   The territory occupied by a certain community is bounded on one side by a river and on all other sides by mountains, forming the shaded region shown in the accompanying figure. If a coordinate system is set up as indicated, the mountainous boundary is given roughly by the curve $y = 4 - x^2$, where x and y are measured in miles. What is the total area occupied by the community?

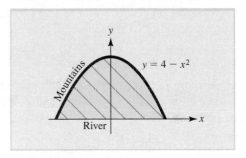

EXERCISE 48

49. **REAL ESTATE EVALUATION**   A square cabin with a plot of land adjacent to a lake is shown in the accompanying figure. If a coordinate system is set up as indicated, with distances given in yards, the lakefront boundary of the property is part of the curve $y = 10e^{0.04x}$. Assuming that the cabin costs $2,000 per square yard and the land in the plot outside the cabin (the shaded region in the figure) costs $800 per square yard, what is the total cost of this vacation property?

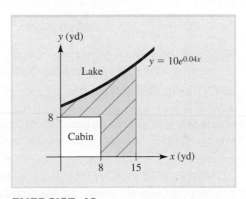

EXERCISE 49

50. DISTRIBUTION OF INCOME   In a certain state, it is found that the distribution of income for lawyers is given by the Lorenz curve $y = L_1(x)$, where

$$L_1(x) = \frac{4}{5}x^2 + \frac{1}{5}x$$

while that of surgeons is given by $y = L_2(x)$, where

$$L_2(x) = \frac{5}{8}x^4 + \frac{3}{8}x$$

Compute the Gini index for each Lorenz curve. Which profession has the more equitable income distribution?

51. DISTRIBUTION OF INCOME   Suppose a study indicates that the distribution of income for professional baseball players is given by the Lorenz curve $y = L_1(x)$, where

$$L_1(x) = \frac{2}{3}x^3 + \frac{1}{3}x$$

while those of professional football players and basketball players are $y = L_2(x)$ and $y = L_3(x)$, respectively, where

$$L_2(x) = \frac{5}{6}x^2 + \frac{1}{6}x$$

and

$$L_3(x) = \frac{3}{5}x^4 + \frac{2}{5}x$$

Find the Gini index for each professional sport, and determine which has the most equitable income distribution. Which has the least equitable distribution?

LIFE AND SOCIAL SCIENCE APPLIED PROBLEMS

52. AVERAGE POPULATION   The population of a certain community t years after the year 2000 is given by

$$P(t) = \frac{e^{0.2t}}{4 + e^{0.2t}} \quad \text{million people}$$

What was the average population of the community during the decade from 2000 to 2010?

53. BACTERIAL GROWTH   The number of bacteria present in a certain culture after t minutes of an experiment was $Q(t) = 2,000e^{0.05t}$. What was the average number of bacteria present during the first 5 minutes of the experiment?

54. COMPARATIVE GROWTH   The population of a third-world country grows exponentially at the unsustainable rate of

$$P_1'(t) = 10e^{0.02t} \text{ thousand people per year}$$

where t is the number of years after 2005. A study indicates that if certain socioeconomic changes are instituted in this country, then the population will instead grow at the restricted rate

$$P_2'(t) = 10 + 0.02t + 0.002t^2$$

thousand people per year. How much smaller will the population of this country be in the year 2015 if the changes are made than if they are not?

55. COMPARATIVE GROWTH   A second study of the country in Exercise 54 indicates that natural restrictive forces are at work that make the actual rate of growth

$$P_3'(t) = \frac{20e^{0.02t}}{1 + e^{0.02t}}$$

instead of the exponential rate $P_1'(t) = 10e^{0.02t}$. If this rate is correct, how much smaller will the population be in the year 2015 than if the exponential rate were correct?

56. AVERAGE AEROBIC RATING   The aerobic rating of a person x years old is given by

$$A(x) = \frac{110(\ln x - 2)}{x} \quad \text{for } x \geq 10$$

What is a person's average aerobic rating from age 15 to age 25? From age 60 to age 70?

57. THERMAL EFFECT OF FOOD   Normally, the metabolism of an organism functions at an essentially constant rate, called the *basal metabolic rate* of the organism. However, the metabolic rate may increase or decrease depending on the activity of the organism. In particular, after ingesting nutrients, the organism often experiences a surge in its metabolic rate, which then gradually returns to the basal level.

Michelle has just finished her Thanksgiving dinner, and her metabolic rate has surged from its basal level M_0. She then works off the meal over the next 12 hours. Suppose that t hours after the meal, her metabolic rate is given by

$$M(t) = M_0 + 50te^{-0.1t^2} \quad 0 \leq t \leq 12$$

kilojoules per hour (kJ/hr).
a. What is Michelle's average metabolic rate over the 12-hour period?
b. Sketch the graph of $M(t)$. What is the peak metabolic rate and when does it occur? [*Note:* Both the graph and the peak rate will involve M_0.]

58. **TRAFFIC MANAGEMENT** For several weeks, the highway department has been recording the speed of freeway traffic flowing past a certain downtown exit. The data suggest that between the hours of 1:00 and 6:00 P.M. on a normal weekday, the speed of traffic at the exit is approximately $S(t) = t^3 - 10.5t^2 + 30t + 20$ miles per hour, where t is the number of hours past noon.
   a. Compute the average speed of the traffic between the hours of 1:00 and 6:00 P.M.
   b. At what time between 1:00 and 6:00 P.M. is the traffic speed at the exit the same as the average speed found in part (a)?

59. **AVERAGE DRUG CONCENTRATION** A patient is injected with a drug, and t hours later, the concentration of the drug remaining in the patient's bloodstream is given by

   $$C(t) = \frac{3t}{(t^2 + 36)^{3/2}} \text{ mg/cm}^3$$

   What is the average concentration of drug during the first 8 hours after the injection?

60. **VOLUME OF BLOOD DURING SYSTOLE** A model* of the cardiovascular system relates the stroke volume $V(t)$ of blood in the aorta at time t during systole (the contraction phase) to the pressure $P(t)$ in the aorta at the same time by the equation

   $$V(t) = [C_1 + C_2 P(t)]\left(\frac{3t^2}{T^2} - \frac{2t^3}{T^3}\right)$$

   where C_1 and C_2 are positive constants and T is the period of the systolic phase (a fixed time). Assume that aortic pressure $P(t)$ rises at a constant rate from P_0 when $t = 0$ to P_1 when $t = T$.
   a. Show that

   $$P(t) = \left(\frac{P_1 - P_0}{T}\right)t + P_0$$

   b. Find the average volume of blood in the aorta during the systolic phase $(0 \le t \le T)$. [*Note:* Your answer will be in terms of C_1, C_2, P_0, P_1, and T.]

61. **REACTION TO MEDICATION** In certain biological models, the human body's reaction to a particular kind of medication is measured by a function of the form

   $$F(M) = \frac{1}{3}(kM^2 - M^3) \qquad 0 \le M \le k$$

   where k is a positive constant and M is the amount of medication absorbed in the blood.

The sensitivity of the body to the medication is measured by the derivative $S = F'(M)$.
   a. Show that the body is most sensitive to the medication when $M = \dfrac{k}{3}$.
   b. What is the average reaction to the medication for $0 \le M \le \dfrac{k}{3}$?

MISCELLANEOUS PROBLEMS

62. **TEMPERATURE** Records indicate that t hours past midnight, the temperature at the local airport was $f(t) = -0.3t^2 + 4t + 10$ degrees Celsius. What was the average temperature at the airport between 9:00 A.M. and noon?

63. **TEMPERATURE** A researcher models the temperature T (in °C) during the time period from 6 A.M. to 6 P.M. in a certain northern city by the function

   $$T(t) = 3 - \frac{1}{3}(t - 5)^2 \quad \text{for } 0 \le t \le 12$$

   where t is the number of hours after 6 A.M.
   a. What is the average temperature in the city during the workday, from 8 A.M. to 5 P.M.?
   b. At what time (or times) during the workday is the temperature in the city the same as the average temperature found in part (a)?

64. Use the graphing utility of your calculator to draw the graphs of the curves $y = x^2 e^{-x}$ and $y = \dfrac{1}{x}$ on the same screen. Use **ZOOM** and **TRACE** or some other feature of your calculator to find where the curves intersect. Then compute the area of the region bounded by the curves using the numeric integration feature.

65. Repeat Exercise 64 for the curves

   $$\frac{x^2}{5} - \frac{y^2}{2} = 1 \quad \text{and} \quad y = x^3 - 8.9x^2 + 26.7x - 27$$

66. Show that the average value V of a continuous function $f(x)$ over the interval $a \le x \le b$ may be computed as the slope of the line joining the points $(a, F(a))$ and $(b, F(b))$ on the curve $y = F(x)$, where $F(x)$ is any antiderivative of $f(x)$ over $a \le x \le b$.

67. Consider an object moving along a straight line. Explain why the object's average velocity over any time interval equals the average value of its velocity over that interval.

---

*J. G. Defares, J. J. Osborn, and H. H. Hura, *Acta Physiol. Pharm. Neerl.*, Vol. 12, 1963, pp. 189–265.

SECTION 5.5   Additional Applications of Integration to Business and Economics

Learning Objectives

1. Use integration to compute the future and present value of an income flow.
2. Define consumer willingness to spend as a definite integral, and use it to explore consumers' surplus and producers' surplus.

In this section, we examine several applications of definite integration to business and economics, specifically to finding the future and present value of an income flow, consumer willingness to spend, and consumers' surplus and producers' surplus.

Future Value and Present Value of an Income Flow

The revenue generated by a business operation can often be regarded as a continuous income stream which may then be invested to generate even more income. The future value of the income stream over a specified term is the total amount (money transferred into the account plus interest) that is accumulated during the term.

An **annuity** is a special kind of income flow in which payments are made (or received) at regular time intervals over a specified term. Home mortgage payments are one kind of annuity as are the payout arrangements for certain kinds of retirement plans. Annuity payments are often constant amounts (like a monthly car loan payment). In Example 5.5.1, we illustrate how the future value of any income stream can be computed by finding the future value of an annuity.

EXAMPLE 5.5.1   Finding the Future Value of an Annuity

Yuanxi has an annuity that pays $1,200 per year and earns interest at the annual rate of 8% compounded continuously. How much will Yuanxi's account be worth at the end of 2 years? Assume that the annuity is deposited continuously into the account.

Solution

Recall from Section 4.1 that P dollars invested at 8% compounded continuously will be worth $Pe^{0.08t}$ dollars t years later.

To approximate the future value of the income stream, divide the 2-year time interval $0 \le t \le 2$ into n equal subintervals of length Δt years and let t_j denote the beginning of the jth subinterval. Then, during the jth subinterval (of length Δt years),

$$\text{Money deposited} = (\text{dollars per year})(\text{number of years}) = 1{,}200\Delta t$$

If all this money were deposited at the beginning of the subinterval (at time t_j), it would remain in the account for $2 - t_j$ years and therefore would grow to $(1{,}200\Delta t)e^{0.08(2 - t_j)}$ dollars. Thus,

$$\begin{array}{c}\text{Future value of money deposited} \\ \text{during } j\text{th subinterval}\end{array} \approx 1{,}200e^{0.08(2 - t_j)}\Delta t$$

The situation is illustrated in Figure 5.18.

The future value of the entire income stream is the sum of the future values of the money deposited during each of the n subintervals. Hence,

$$\text{Future value of income stream} \approx \sum_{j=1}^{n} 1{,}200e^{0.08(2 - t_j)}\Delta t$$

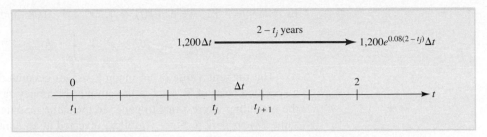

FIGURE 5.18   The (approximate) future value of the money deposited during the jth subinterval.

(Note that this is only an approximation because it is based on the assumption that all $1{,}200\Delta t_n$ dollars are deposited at time t_j rather than continuously throughout the jth subinterval.)

As n increases without bound, the length of each subinterval approaches zero and the approximation approaches the true future value of the income stream. Hence,

$$\begin{aligned}
\text{Future value of income stream} &= \lim_{n \to +\infty} \sum_{j=1}^{n} 1{,}200 e^{0.08(2 - t_j)} \Delta t \\
&= \int_0^2 1{,}200 e^{0.08(2 - t)}\, dt = 1{,}200 e^{0.16} \int_0^2 e^{-0.08t}\, dt \\
&= -\frac{1{,}200}{0.08} e^{0.16} (e^{-0.08t}) \Big|_0^2 = -15{,}000 e^{0.16}(e^{-0.16} - 1) \\
&= -15{,}000 + 15{,}000 e^{0.16} \approx 2{,}602.66
\end{aligned}$$

so the annuity account is worth roughly \$2,602.66 at the end of the 2-year term.

By generalizing the reasoning illustrated in Example 5.5.1, we are led to this integration formula for the future value of an income stream with rate of flow given by $f(t)$ for a term of T years with interest rate r:

$$\begin{aligned}
\text{FV} &= \int_0^T f(t)\, e^{r(T - t)}\, dt \qquad e^{r(T - t)} = e^{(rT - rt)} = e^{rT} e^{-rt} \\
&= \int_0^T f(t)\, e^{rT} e^{-rt}\, dt \quad \text{factor constant } e^{rT} \text{ outside integral} \\
&= e^{rT} \int_0^T f(t)\, e^{-rt}\, dt
\end{aligned}$$

The first and last forms of the formula for future value are both listed next for future reference.

Future Value of an Income Stream ■   Suppose money is being transferred continuously into an account over a time period $0 \le t \le T$ at a rate given by the function $f(t)$ and that the account earns interest at an annual rate r compounded continuously. Then the **future value FV of the income** stream over the term T is given by the definite integral

$$\text{FV} = \int_0^T f(t)\, e^{r(T - t)}\, dt = e^{rT} \int_0^T f(t)\, e^{-rt}\, dt$$

In Example 5.5.1, we had $f(t) = 1{,}200$, $r = 0.08$, and $T = 2$, so that

$$\text{FV} = e^{0.08(2)} \int_0^2 1{,}200 e^{-0.08t}\, dt$$

The **present value** of an income stream generated at a continuous rate $f(t)$ over a specified term of T years is the amount of money A that must be deposited now at the prevailing interest rate to generate the same income as the income stream over the same T-year period. Since A dollars invested at an annual interest rate r compounded continuously will be worth Ae^{rT} dollars in T years, we must have

$$Ae^{rT} = e^{rT} \int_0^T f(t)\, e^{-rt}\, dt \quad \text{divide both sides by } e^{rT}$$

$$A = \int_0^T f(t)\, e^{-rt}\, dt$$

To summarize:

Present Value of an Income Stream ■ The present value **PV of an income stream** that is deposited continuously at the rate $f(t)$ into an account that earns interest at an annual rate r compounded continuously for a term of T years is given by

$$\text{PV} = \int_0^T f(t)\, e^{-rt}\, dt$$

Example 5.5.2 illustrates how present value can be used in making certain financial decisions.

EXAMPLE 5.5.2   Using Present Value to Compare Two Income Streams

June is trying to decide between two investments. The first costs \$9,000 and is expected to generate a continuous income stream at the rate of $f_1(t) = 3{,}000 e^{0.03t}$ dollars per year. The second investment is an annuity that costs \$12,000 to purchase and generates income at the constant rate of $f_2(t) = 4{,}000$ per year. If the prevailing annual interest rate remains fixed at 5% compounded continuously, which account is better over a 5-year term?

Solution

The net value of each investment over the 5-year time period is the present value of the investment less its initial cost. For each investment, we have $r = 0.05$ and $T = 5$.

For the first investment, the net value is

$$\text{PV} - \text{cost} = \int_0^5 (3{,}000 e^{0.03t}) e^{-0.05t}\, dt - 9{,}000$$

$$= 3{,}000 \int_0^5 e^{0.03t - 0.05t}\, dt - 9{,}000$$

$$= 3{,}000 \int_0^5 e^{-0.02t}\, dt - 9{,}000$$

$$= 3{,}000 \left(\frac{e^{-0.02t}}{-0.02} \right) \Big|_0^5 - 9{,}000$$

$$= -150{,}000 [e^{-0.02(5)} - e^0] - 9{,}000$$

$$= 5{,}274.39$$

The net value of the annuity is

$$PV - \text{cost} = \int_0^5 (4,000)e^{-0.05t}\, dt - 12,000$$

$$= 4,000\left(\frac{e^{-0.05t}}{-0.05}\right)\Bigg|_0^5 - 12,000$$

$$= -80,000[e^{-0.05(5)} - e^0] - 12,000$$

$$= 5,695.94$$

Thus, the net income generated by the first investment is $5,274.39, while the annuity generates net income of $5,695.94. The annuity is a slightly better investment.

A general formula for the present value of an annuity is given in Exercise 45.

Consumer Willingness to Spend and Consumers' Surplus

Suppose a young couple is willing to spend up to $500 for a television set. For the convenience of having two sets (say, to settle disputes about which show to watch), they are willing to spend an additional $300 for an additional set, but since there would be relatively little advantage in having more than two sets, they might be willing to spend no more than $50 for a third set. Thus, the couple's demand function $p = D(q)$ for television sets would satisfy

$$500 = D(1) \qquad 300 = D(2) \qquad 50 = D(3)$$

and their total willingness to spend for as many as three television sets would be

$$\$500 + \$300 + \$50 = \$850$$

Now consider a commodity like grain that can be sold in any quantity q up to q_0 units (so $0 \le q \le q_0$), and let $p = D(q)$ be the demand function for the commodity. To find the total consumer willingness to buy as many as q_0 units, we cannot simply add up potential payments (demand values) as we did for the television set example because there are too many available levels of production q between 0 and q_0, so instead we use a definite integral.

Specifically, as shown in Figure 5.19, we divide the interval $0 \le q \le q_0$ into n evenly spaced subintervals and assume the demand is $D(q_{k-1})$ for all values of q in the kth subinterval, where q_{k-1} is the left endpoint of that subinterval, for $k = 1, 2, \ldots, n$. Then the consumer willingness to buy between q_{k-1} and q_k units is approximately

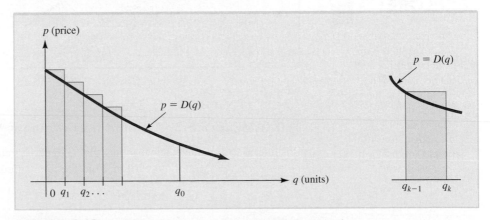

FIGURE 5.19   Computing total consumer willingness to spend.

$D(q_{k-1})\Delta q$ where $\Delta q = \dfrac{q_0 - 0}{n}$ is the width of each subinterval, so the *total* consumer willingness to spend for as many as q_0 units is estimated by the sum $\sum\limits_{k=1}^{n} D(q_{k-1})\Delta q$. This suggests that we *define* the total willingness to spend by the limit

$$\text{WS} = \lim_{n \to \infty} \sum_{k=1}^{n} D(q_{k-1})\Delta q$$

which we recognize as the definite integral of the demand function $p = D(q)$ over the interval $0 \le q \le q_0$. Note that since the demand function is always above the q axis, this integral can be interpreted geometrically as the area under the demand curve, as indicated in the following summary box.

Consumer Willingness to Spend ■ **Total consumer willingness to spend** for up to q_0 units of a commodity is given by

$$\text{WS} = \int_0^{q_0} D(q)\, dq$$

where $p = D(q)$ is the demand function for the commodity. Geometrically, this is the area under the demand function over the range of sales $0 \le q \le q_0$.

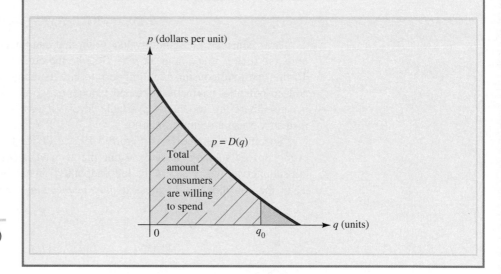

EXPLORE!

In Example 5.5.3, change $D(q)$ to $D_{\text{new}}(q) = 4(23 - q^2)$. Will the amount of money consumers are willing to spend to obtain 3 units of the commodity increase or decrease? Graph $D_{\text{new}}(q)$ in bold to compare with the graph of $D(q)$, using the viewing window [0, 5]1 by [0, 150]10.

EXAMPLE 5.5.3 Computing Consumer Willingness to Spend

Rashid, a farm manager, determines that q tons of grain will be sold when the price is $p = 10(25 - q^2)$ dollars per ton. Find the total amount buyers are willing to spend for as many as 3 tons of grain.

Solution

Since the demand function is $p = 10(25 - q^2)$, the total consumer willingness to pay for $q_0 = 3$ tons is

$$\text{WS} = \int_0^{q_0} D(q)\, dq = \int_0^3 10(25 - q^2)\, dq$$

$$= 250q - 10\left(\frac{1}{3}q^3\right)\bigg|_0^3 = 660$$

So consumers are willing to pay Rashid $660 for as many as 3 tons of grain.

Consumers' surplus is a quantity in economics that is closely related to the total consumer willingness to spend. In a competitive economy, consumers often expect to pay more for a commodity than they actually do. For instance, you may expect to pay $60 for a new video game and be pleasantly surprised to find that it only costs $40. The perceived savings $60 − $40 = $20 is your consumers' surplus in this case.

More generally, suppose consumers are willing to purchase as many as q_0 units of a commodity with demand function $p = D(q)$. Consumers (as a group) are *willing* to pay $\text{WS} = \int_0^{q_0} D(q)\, dq$ dollars for the q_0 units but actually pay only $p_0 q_0$ dollars, where $p_0 = D(q_0)$, and the consumers' surplus is the difference.

$$\text{CS} = \int_0^{q_0} D(q)\, dq - p_0 q_0$$

To summarize:

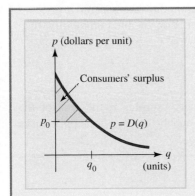

Consumers' Surplus ■ If q_0 units of a commodity are sold at a price of p_0 per unit and if $p = D(q)$ is the consumers' demand function for the commodity, then

$$\left[\begin{array}{c}\text{Consumers'}\\ \text{surplus}\end{array}\right] = \left[\begin{array}{c}\text{total amount consumers}\\ \text{are willing to spend}\\ \text{for } q_0 \text{ units}\end{array}\right] - \left[\begin{array}{c}\text{actual consumer}\\ \text{expenditure}\\ \text{for } q_0 \text{ units}\end{array}\right]$$

$$\text{CS} = \int_0^{q_0} D(q)\, dq - p_0 q_0$$

Consumers' surplus has the geometric interpretation shown in Figure 5.20. Figure 5.20a shows the area under the demand curve over the interval $0 \le q \le q_0$, which represents the consumers' willingness to buy q_0 units of the product. The rectangular area in Figure 5.20b represents the actual consumer expenditure for q_0 units at p_0 dollars per unit, so the consumers' surplus is given by the difference of the two areas, as shown in Figure 5.20c. Note that the consumers' surplus is the area of the region bounded above by the demand curve and below by the horizontal line $p = p_0$.

Producers' surplus is the suppliers' version of consumers' surplus. Recall that the supply function $p = S(q)$ gives the price per unit that producers are willing to accept to supply q units of a commodity to the market. If the established market price is $p_0 = S(q_0)$ dollars per unit, then any producer who would be willing to supply the commodity at a lower price enjoys a perceived gain. The producers' surplus is

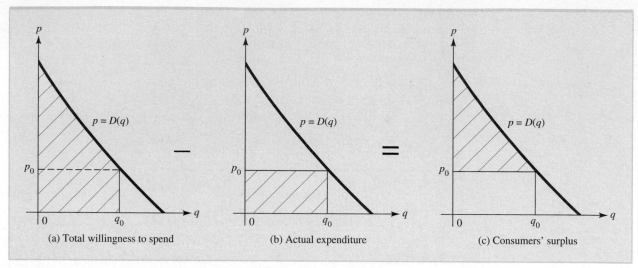

FIGURE 5.20   Geometric interpretation of consumers' surplus.

the difference between what producers would be willing to accept and the amount they actually receive. It may be computed by integration involving the supply function, just as consumers' surplus was computed by integrating the demand function. Here is the formula we will use to compute producers' surplus.

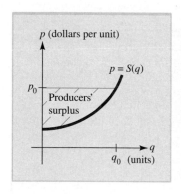

Producers' Surplus ■ If q_0 units of a commodity are sold at a price of p_0 dollars per unit and $p = S(q)$ is the supply function for the commodity, then the producers' surplus PS is given by

$$\text{PS} = p_0 q_0 - \int_0^{q_0} S(q)\, dq$$

As indicated in the accompanying figure, producers' surplus is the area of the region bounded above by the price line $p = p_0$ and below by the supply curve $p = S(q)$, over the production interval $0 \le q \le q_0$.

EXAMPLE 5.5.4   Studying Consumers' and Producers' Surplus

A tire manufacturer estimates that q (thousand) radial tires will be purchased (demanded) by wholesalers when the price is

$$p = D(q) = -0.1q^2 + 90$$

dollars per tire, and the same number of tires will be supplied when the price is

$$p = S(q) = 0.2q^2 + q + 50$$

dollars per tire.

a. Find the equilibrium price (where supply equals demand) and the quantity supplied and demanded at that price.

b. Determine the consumers' and producers' surplus at the equilibrium price.

Solution

a. The supply and demand curves are shown in Figure 5.21. Supply equals demand when

$$-0.1q^2 + 90 = 0.2q^2 + q + 50$$
$$0.3q^2 + q - 40 = 0$$
$$q = 10 \quad (\text{reject } q \approx -13.33)$$

and $p = -0.1(10)^2 + 90 = 80$ dollars per tire. Thus, equilibrium occurs at a price of \$80 per tire, and then 10,000 tires are supplied and demanded.

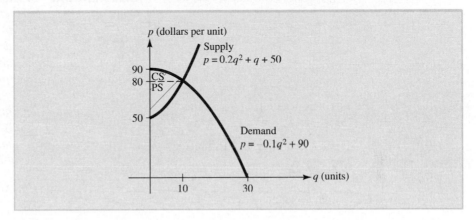

FIGURE 5.21 Consumers' surplus and producers' surplus for the demand and supply functions in Example 5.5.4.

b. Using $p_0 = 80$ and $q_0 = 10$, we find that the consumers' surplus is

$$CS = \int_0^{10} (-0.1q^2 + 90)\, dq - (80)(10)$$

$$= \left[-0.1\left(\frac{q^3}{3}\right) + 90q \right]\Bigg|_0^{10} - (80)(10)$$

$$\approx 866.67 - 800 = 66.67$$

or \$66,670 (since $q_0 = 10$ is really 10,000). The consumers' surplus is the area of the shaded region labeled CS in Figure 5.21.

The producers' surplus is

$$PS = (80)(10) - \int_0^{10} (0.2q^2 + q + 50)\, dq$$

$$= (80)(10) - \left[0.2\left(\frac{q^3}{3}\right) + \left(\frac{q^2}{2}\right) + 50q \right]\Bigg|_0^{10}$$

$$\approx 800 - 616.67 = 183.33$$

or \$183,330. The producers' surplus is the area of the shaded region labeled PS in Figure 5.21.

EXERCISES ▪ 5.5

CONSUMERS' WILLINGNESS TO SPEND   *For the consumers' demand functions $D(q)$ in Exercises 1 through 6:*

(a) *Find the total amount of money consumers are willing to spend to obtain q_0 units of the commodity.*

(b) *Sketch the demand curve and interpret the consumer willingness to spend in part (a) as an area.*

1. $D(q) = 2(64 - q^2)$ dollars per unit; $q_0 = 6$ units

2. $D(q) = \dfrac{300}{(0.1q + 1)^2}$ dollars per unit; $q_0 = 5$ units

3. $D(q) = \dfrac{400}{0.5q + 2}$ dollars per unit; $q_0 = 12$ units

4. $D(q) = \dfrac{300}{4q + 3}$ dollars per unit; $q_0 = 10$ units

5. $D(q) = 40e^{-0.05q}$ dollars per unit; $q_0 = 10$ units

6. $D(q) = 50e^{-0.04q}$ dollars per unit; $q_0 = 15$ units

CONSUMERS' SURPLUS   *In Exercises 7 through 10, $p = D(q)$ is the price (dollars per unit) at which q units of a particular commodity will be demanded by the market (that is, all q units will be sold at this price), and q_0 is a specified level of production. In each case, find the price $p_0 = D(q_0)$ at which q_0 units will be demanded and compute the corresponding consumers' surplus CS. Sketch the demand curve $y = D(q)$, and shade the region whose area represents the consumers' surplus.*

7. $D(q) = 2(64 - q^2)$; $q_0 = 3$ units

8. $D(q) = 150 - 2q - 3q^2$; $q_0 = 6$ units

9. $D(q) = 40e^{-0.05q}$; $q_0 = 5$ units

10. $D(q) = 75e^{-0.04q}$; $q_0 = 3$ units

PRODUCERS' SURPLUS   *In Exercises 11 through 14, $p = S(q)$ is the price (dollars per unit) at which q units of a particular commodity will be supplied to the market by producers, and q_0 is a specified level of production. In each case, find the price $p_0 = S(q_0)$ at which q_0 units will be supplied and compute the corresponding producers' surplus PS. Sketch the supply curve $y = S(q)$ and shade the region whose area represents the producers' surplus.*

11. $S(q) = 0.3q^2 + 30$; $q_0 = 4$ units

12. $S(q) = 0.5q + 15$; $q_0 = 5$ units

13. $S(q) = 10 + 15e^{0.03q}$; $q_0 = 3$ units

14. $S(q) = 17 + 11e^{0.01q}$; $q_0 = 7$ units

CONSUMERS' AND PRODUCERS' SURPLUS AT EQUILIBRIUM   *In Exercises 15 through 19, the demand and supply functions, $D(q)$ and $S(q)$, for a particular commodity are given. Specifically, q units of the commodity will be demanded (sold) at a price of $p = D(q)$ dollars per unit, while q units will be supplied by producers when the price is $p = S(q)$ dollars per unit. In each case:*

(a) *Find the equilibrium price p_e (where supply equals demand).*

(b) *Find the consumers' surplus and the producers' surplus at equilibrium.*

15. $D(q) = 131 - \dfrac{1}{3}q^2$; $S(q) = 50 + \dfrac{2}{3}q^2$

16. $D(q) = 65 - q^2$; $S(q) = \dfrac{1}{3}q^2 + 2q + 5$

17. $D(q) = -0.3q^2 + 70$; $S(q) = 0.1q^2 + q + 20$

18. $D(q) = \sqrt{245 - 2q}$; $S(q) = 5 + q$

19. $D(q) = \dfrac{16}{q + 2} - 3$; $S(q) = \dfrac{1}{3}(q + 1)$

20. **PROFIT OVER THE USEFUL LIFE OF A MACHINE**   Suppose that when it is t years old, a particular industrial machine generates revenue at the rate $R'(t) = 6{,}025 - 8t^2$ dollars per year, while operating and maintenance costs accumulate at the rate $C'(t) = 4{,}681 + 13t^2$ dollars per year.

a. The **useful life** of a machine is the number of years T before the profit it generates starts to decline. What is the useful life of this machine?

b. Compute the net profit generated by the machine over its useful lifetime.

c. Sketch the revenue rate curve $y = R'(t)$ and the cost rate curve $y = C'(t)$, and shade the region whose area represents the net profit computed in part (b).

21. PROFIT OVER THE USEFUL LIFE OF A MACHINE   Answer the questions in Exercise 20 for a machine that generates revenue at the rate $R'(t) = 7,250 - 18t^2$ dollars per year and for which costs accumulate at the rate $C'(t) = 3,620 + 12t^2$ dollars per year.

22. FUND-RAISING   It is estimated that t weeks from now, contributions in response to a fund-raising campaign will be coming in at the rate of $R'(t) = 5,000e^{-0.2t}$ dollars per week, while campaign expenses are expected to accumulate at the constant rate of $676 per week.
   a. For how many weeks does the rate of revenue exceed the rate of cost?
   b. What net earnings will be generated by the campaign during the period of time determined in part (a)?
   c. Interpret the net earnings in part (b) as an area between two curves.

23. FUND-RAISING   Answer the questions in Exercise 22 for a charity campaign in which contributions are made at the rate of $R'(t) = 6,537e^{-0.3t}$ dollars per week and expenses accumulate at the constant rate of $593 per week.

24. AMOUNT OF AN INCOME STREAM   Money is transferred continuously into an account at the constant rate of $2,400 per year. The account earns interest at the annual rate of 6% compounded continuously. How much will be in the account at the end of 5 years?

25. AMOUNT OF AN INCOME STREAM   Money is transferred continuously into an account at the constant rate of $1,000 per year. The account earns interest at the annual rate of 10% compounded continuously. How much will be in the account at the end of 10 years?

26. CONSTRUCTION DECISION   Magda wants to expand and renovate her import store and is presented with two plans for making the improvements. The first plan will cost her $40,000, and the second will cost only $25,000. However, she expects the improvements resulting from the first plan to provide income at the continuous rate of $10,000 per year, while the income flow from the second plan provides $8,000 per year. Which plan will result in more net income over the next 3 years if the prevailing rate of interest is 5% per year compounded continuously?

27. RETIREMENT ANNUITY   At age 25, Tim starts making annual deposits of $2,500 into an IRA account that pays interest at an annual rate of 5% compounded continuously. Assuming that his payments are made as a continuous income flow, how much money will be in his account if he retires at age 60? At age 65?

28. RETIREMENT ANNUITY   When she is 30, Sue starts making annual deposits of $2,000 into a bond fund that pays 8% annual interest compounded continuously. Assuming that her deposits are made as a continuous income flow, how much money will be in her account if she retires at age 55?

29. PRESENT VALUE OF AN INVESTMENT   An investment will generate income continuously at the constant rate of $1,200 per year for 5 years. If the prevailing annual interest rate remains fixed at 5% compounded continuously, what is the present value of the investment?

30. PRESENT VALUE OF A FRANCHISE   The management of a national chain of fast-food outlets is selling a 10-year franchise in Cleveland, Ohio. Past experience in similar localities suggests that t years from now the franchise will be generating profit at the rate of $f(t) = 100,000$ dollars per year. If the prevailing annual interest rate remains fixed at 4% compounded continuously, what is the present value of the franchise?

31. INVESTMENT ANALYSIS   Felipe is trying to choose between two investment opportunities. The first will cost $50,000 and is expected to produce income at the continuous rate of $15,000 per year. The second will cost $30,000 and is expected to produce income at the rate of $9,000 per year. If the prevailing rate of interest stays constant at 6% per year compounded continuously, which investment is better for Felipe over the next 5 years?

32. INVESTMENT ANALYSIS   Kevin spends $4,000 for an investment that generates a continuous income stream at the rate of $f_1(t) = 3,000$ dollars per year. His friend, Molly, makes a separate investment that also generates income continuously, but at a rate of $f_2(t) = 2,000e^{0.04t}$ dollars per year. The couple discovers that their investments have exactly the same net value over a 4-year period. If the prevailing annual interest rate stays fixed at 5% compounded continuously, how much did Molly pay for her investment?

33. **CONSUMERS' SURPLUS** A manufacturer of machinery parts determines that q units of a particular piece will be sold when the price is $p = 110 - q$ dollars per unit. The total cost of producing those q units is $C(q)$ dollars, where
$$C(q) = q^3 - 25q^2 + 2q + 3,000$$
   a. How much profit is derived from the sale of the q units at p dollars per unit? [*Hint:* First find the revenue $R = pq$; then find profit = revenue − cost.]
   b. For what value of q is profit maximized?
   c. Find the consumers' surplus for the level of production q_0 that corresponds to maximum profit.

34. **CONSUMERS' SURPLUS** Repeat Exercise 33 for $C(q) = 2q^3 - 59q^2 + 4q + 7,600$ and $p = 124 - 2q$.

35. **DEPLETION OF ENERGY RESOURCES** Oil is being pumped from an oil field t years after its opening at the rate of $P'(t) = 1.3e^{0.04t}$ billion barrels per year. The field has a reserve of 20 billion barrels, and the price of oil holds steady at $112 per barrel.
   a. Find $P(t)$, the amount of oil pumped from the field at time t. How much oil is pumped from the field during the first 3 years of operation? The next 3 years?
   b. For how many years T does the field operate before it runs dry?
   c. If the prevailing annual interest rate stays fixed at 5% compounded continuously, what is the present value of the continuous income stream $V = 112P'(t)$ over the period of operation of the field $0 \le t \le T$?
   d. If the owner of the oil field decides to sell on the first day of operation, do you think the present value determined in part (c) would be a fair asking price? Explain your reasoning.

36. **DEPLETION OF ENERGY RESOURCES** Answer the questions in Exercise 35 for another oil field with a pumping rate of $P'(t) = 1.5e^{0.03t}$ and with a reserve of 16 billion barrels. You may assume that the price of oil is still $112 per barrel and that the prevailing annual interest rate is 5%.

37. **DEPLETION OF ENERGY RESOURCES** Answer the questions in Exercise 35 for an oil field with a pumping rate of $P'(t) = 1.2e^{0.02t}$ and with a reserve of 12 billion barrels. Assume that the prevailing interest rate is 5% as before, but that the price of oil after t years is given by $A(t) = 112e^{0.015t}$.

38. **LOTTERY PAYOUT** Luisa, a $2 million state lottery winner, is given a $250,000 check now and a continuous income flow at the rate of $200,000 per year for 10 years. If the prevailing rate of interest is 5% per year compounded continuously, is this a good deal for Luisa or not? Explain.

39. **LOTTERY PAYOUT** The winner of a state lottery is offered a choice of either receiving $10 million now as a lump sum or of receiving A dollars a year for the next 6 years as a continuous income stream. If the prevailing annual interest rate is 5% compounded continuously and the two payouts are worth the same, what is A?

40. **SPORTS CONTRACTS** A star baseball free agent is the object of a bidding war between two rival teams. The first team offers a 3 million dollar signing bonus and a 5-year contract guaranteeing him 8 million dollars this year and an increase of 3% per year for the remainder of the contract. The second team offers $9 million per year for 5 years with no incentives. If the prevailing annual interest rate stays fixed at 4% compounded continuously, which offer is worth more? [*Hint:* Assume that with both offers, the salary is paid as a continuous income stream.]

41. **PRESENT VALUE OF AN INVESTMENT** An investment produces a continuous income stream at the rate of $A(t)$ thousand dollars per year at time t, where
$$A(t) = 10e^{1 - 0.05t}$$
The prevailing rate of interest is 5% per year compounded continuously.
   a. What is the future value of the investment over a term of 5 years ($0 \le t \le 5$)?
   b. What is the present value of the investment over the time period $1 \le t \le 3$?

42. **PROFIT FROM AN INVENTION** A marketing survey indicates that t months after a new type of computerized air purifier is introduced to the market, sales will be generating profit at the rate of $P'(t)$ thousand dollars per month, where
$$P'(t) = \frac{500[1.4 - \ln(0.5t + 1)]}{t + 2}$$
   a. When is the rate of profitability positive and when is it negative? When is the rate increasing and when is it decreasing?
   b. At what time $t = t_m$ is monthly profit maximized? Find the net change in profit over the time period $0 \le t \le t_m$.

c. It costs the manufacturer $100,000 to develop the purifier product, so $P(0) = -100$. Use this information along with integration to find $P(t)$.

d. Sketch the graph of $P(t)$ for $t \geq 0$. A *fad* is a product that gains rapid success in the market, then just as quickly fades from popularity. Based on the graph $P(t)$, would you call the purifiers a fad? Explain.

43. **TOTAL REVENUE**   Consider the following problem: A certain oil well that yields 300 barrels of crude oil a month will run dry in 3 years. It is estimated that t months from now the price of crude oil will be $P(t) = 118 + 0.3\sqrt{t}$ dollars per barrel. If the oil is sold as soon as it is extracted from the ground, what will be the total future revenue from the well?

a. Solve the problem using definite integration. [*Hint:* Divide the 3-year (36-month) time interval $0 \leq t \leq 36$ into n equal subintervals of length Δt, and let t_j denote the beginning of the jth subinterval. Find an expression that estimates the revenue $R(t_j)$ obtained during the jth subinterval. Then express the total revenue as the limit of a sum.]

b. Read an article on the petroleum industry, and write a paragraph on mathematical methods of modeling oil production.*

_____
*A good place to start is the article by J. A. Weyland and D. W. Ballew, "A Relevant Calculus Problem: Estimation of U.S. Oil Reserves," *The Mathematics Teacher,* Vol. 69, 1976, pp. 125–126.

44. **INVENTORY STORAGE COSTS**   A manufacturer receives N units of a certain raw material that are initially placed in storage and then withdrawn and used at a constant rate until the supply is exhausted 1 year later. Suppose storage costs remain fixed at p dollars per unit per year. Use definite integration to find an expression for the total storage cost the manufacturer will pay during the year. [*Hint:* Let $Q(t)$ denote the number of units in storage after t years, and find an expression for $Q(t)$. Then divide the interval $0 \leq t \leq 1$ into n equal subintervals and express the total storage cost as the limit of a sum.]

45. **FUTURE AND PRESENT VALUE OF AN ANNUITY**   An annuity pays a continuous income stream of M dollars per year into an account that pays interest at an annual rate r compounded continuously for a term of T years.

a. Show that the future value of the annuity is

$$FV = \frac{M}{r}(e^{rT} - 1)$$

b. Show that the present value of the annuity is

$$PV = \frac{M}{r}(1 - e^{-rT})$$

SECTION 5.6 Additional Applications of Integration to the Life and Social Sciences

Learning Objectives

1. Examine survival and renewal functions.
2. Use definite integration to compute population from population density and to explore the flow of blood through an artery.
3. Derive an integration formula for the volume of a solid of revolution, and use it to estimate the size of a tumor.

We have already seen how definite integration can be used to compute quantities of interest in the social and life sciences, such as net change, average value, and the Gini index of a Lorenz curve. In this section, we examine several additional applications, including survival and renewal within a group, blood flow through an artery, and estimating the volume of a tumor.

Survival and Renewal   In Example 5.6.1, a **survival function** gives the fraction of individuals in a group or population that can be expected to remain in the group for any specified period of time. A **renewal function** giving the rate at which new members arrive is also known, and the goal is to predict the size of the group at some future time. Problems of this type arise in many fields, including sociology, ecology, demography, and even finance, where the *population* is the number of dollars in an investment account and *survival and renewal* refer to features of an investment strategy.

EXAMPLE 5.6.1   Studying Survival and Renewal

A new county mental health clinic has just opened. Statistics from similar facilities suggest that the fraction of patients who will still be receiving treatment at the clinic t months after their initial visit is given by the function $f(t) = e^{-t/20}$. The clinic initially accepts 300 people for treatment and plans to accept new patients at the constant rate of $g(t) = 10$ patients per month. Approximately how many people will be receiving treatment at the clinic 15 months from now?

Solution

Since $f(15)$ is the fraction of patients whose treatment continues at least 15 months, it follows that of the current 300 patients, only $300f(15)$ will still be receiving treatment 15 months from now.

To approximate the number of *new* patients who will be receiving treatment 15 months from now, divide the 15-month time interval $0 \leq t \leq 15$ into n equal subintervals of length Δt months and let t_j denote the beginning of the jth subinterval. Since new patients are accepted at the rate of 10 per month, the number of new patients accepted during the jth subinterval is $10\Delta t$. Fifteen months from now, approximately $15 - t_j$ months will have elapsed since these $10\Delta t$ new patients had their initial visits, and so approximately $(10\Delta t)f(15 - t_j)$ of them will still be receiving treatment at that time (Figure 5.22). It follows that the total number of new patients still receiving treatment 15 months from now can be approximated by the sum

$$\sum_{j=1}^{n} 10f(15 - t_j)\Delta t$$

Adding this to the number of current patients who will still be receiving treatment in 15 months, you get

$$P \approx 300f(15) + \sum_{j=1}^{n} 10f(15 - t_j)\Delta t$$

where P is the total number of *all* patients (current and new) who will be receiving treatment 15 months from now.

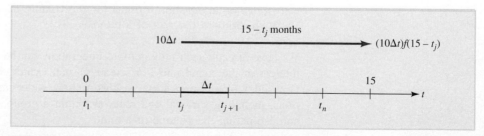

FIGURE 5.22   New members arriving during the jth subinterval.

As n increases without bound, the approximation improves and approaches the true value of P. It follows that

$$P = 300f(15) + \lim_{n \to +\infty} \sum_{j=1}^{n} 10f(15 - t_j)\Delta t$$

$$= 300f(15) + \int_{0}^{15} 10f(15 - t)\,dt$$

Since $f(t) = e^{-t/20}$, we have $f(15) = e^{-3/4}$ and $f(15 - t) = e^{-(15-t)/20} = e^{-3/4}e^{t/20}$. Hence,

$$P = 300e^{-3/4} + 10e^{-3/4}\int_{0}^{15} e^{t/20}\,dt$$

$$= 300e^{-3/4} + 10e^{-3/4}\left(\frac{e^{t/20}}{1/20}\right)\Big|_{0}^{15}$$

$$= 300e^{-3/4} + 200(1 - e^{-3/4})$$

$$\approx 247.24$$

That is, 15 months from now, the clinic will be treating approximately 247 patients.

In Example 5.6.1, we considered a variable survival function $f(t)$ and a constant renewal rate function $g(t)$. Essentially the same analysis applies when the renewal function also varies with time. Here is the result. Note that for definiteness, time is given in years, but the same basic formula would also apply for other units of time, like minutes, weeks, or months, as in Example 5.6.1.

Survival and Renewal ■ Suppose a population initially has P_0 members and that new members are added at the (renewal) rate of $R(t)$ individuals per year. Further suppose that the fraction of the population that remain for at least t years after arriving is given by the (survival) function $S(t)$. Then, at the end of a term of T years, the population will be

$$P(T) = P_0 S(T) + \int_{0}^{T} R(t)\, S(T - t)\,dt$$

In Example 5.6.1, each time period is 1 month, the initial "population" (membership) is $P_0 = 300$, the renewal function is $R(t) = 10$, the survival function is $S(t) = f(t) = e^{-t/20}$, and the term is $T = 15$ months. Example 5.6.2 demonstrates survival/renewal again, this time in the field of biology.

EXAMPLE 5.6.2 Finding Population Using Survival and Renewal

A mild toxin is introduced to a bacterial colony whose current population is 600,000. Observations indicate that $R(t) = 200e^{0.01t}$ bacteria per hour are born in the colony at time t and that the fraction of the population that survives for t hours after birth is $S(t) = e^{-0.015t}$. What is the population of the colony after 10 hours?

Solution

Substituting $P_0 = 600,000$, $R(t) = 200e^{0.01t}$, and $S(t) = e^{-0.015t}$ into the formula for survival and renewal, we find that the population at the end of the term of $T = 10$ hours is

$$P(10) = \underbrace{600,000e^{-0.015(10)}}_{\substack{P_0 \quad S(10)}} + \int_0^{10} \underbrace{200e^{0.01t}}_{R(t)}\underbrace{e^{-0.015(10-t)}}_{S(T-t)}\,dt \qquad e^{a-b} = e^a e^{-b}$$

$$\approx 516,425 + \int_0^{10} 200e^{0.01t}[e^{-0.015(10)}e^{0.015t}]\,dt \qquad \text{factor } 200e^{-0.015(10)} \text{ outside the integral}$$

$$\approx 516,425 + 200e^{-0.015(10)} \int_0^{10} [e^{0.01t}e^{0.015t}]\,dt \qquad \begin{array}{l} e^{a+b} = e^a e^b \text{ and} \\ 200e^{-0.015(10)} \approx 172.14 \end{array}$$

$$\approx 516,425 + 172.14 \int_0^{10} e^{0.025t}\,dt \qquad \begin{array}{l} \text{exponential rule for} \\ \text{integration} \end{array}$$

$$\approx 516,425 + 172.14 \left[\frac{e^{0.025t}}{0.025}\right]\Big|_0^{10}$$

$$\approx 516,425 + \frac{172.14}{0.025}[e^{0.025(10)} - e^0]$$

$$\approx 518,381$$

Thus, the population of the colony declines from 600,000 to about 518,381 during the first 10 hours after the toxin is introduced.

Population Density

The **population density** of an urban area is the number of people $p(r)$ per square mile that are located a distance r miles from the city center. We can determine the total population P of the portion of the city that lies within R miles of the city center by using integration.

Divide the interval $0 \le r \le R$ into n subintervals each of width $\Delta r = \dfrac{R}{n}$, and let r_k denote the left endpoint of the kth subinterval, for $k = 1, 2, \ldots, n$. These subintervals determine n concentric rings, centered on the city center as shown in Figure 5.23.

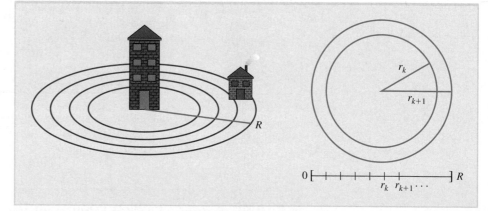

FIGURE 5.23   Dividing an urban area into concentric rings.

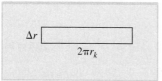

FIGURE 5.24   The area of the kth ring is approximately the area of a rectangle of length $2\pi r_k$ (the circumference) and width Δr.

As indicated in Figure 5.24, if the kth ring is cut and flattened out, the resulting figure is very close to a rectangle whose length is the circumference of the inner boundary of the ring. (It's not exactly a rectangle because the two long sides come from the circumferences of the inner and outer boundaries of the ring, which are just a bit different.) The width of that ring is Δr, so

$$\text{Area of } k\text{th ring} \approx 2\pi r_k \Delta r$$

and since the population density is $p(r)$ people per square mile, it follows that

$$\text{Population within } k\text{th ring} \approx \underbrace{p(r_k)}_{\substack{\text{population} \\ \text{per unit area}}} \cdot \underbrace{[2\pi r_k \Delta r]}_{\substack{\text{area} \\ \text{of ring}}} = 2\pi r_k p(r_k) \Delta r$$

We can estimate the total area with the bounding radius R by adding up the populations within the approximating rings; that is, by the Riemann sum

$$\begin{bmatrix} \text{Total population} \\ \text{within radius } R \end{bmatrix} = P(R) \approx \sum_{k=1}^{n} 2\pi r_k p(r_k) \Delta r$$

By taking the limit as $n \to \infty$, the estimate approaches the true value of the total population P, and since the limit of a Riemann sum is a definite integral, we have

$$P(R) = \lim_{n \to \infty} \sum_{k=1}^{n} 2\pi r_k p(r_k) \Delta r = \int_0^R 2\pi r p(r)\, dr$$

To summarize:

Total Population from Population Density ■ If a concentration of individuals has population density $p(r)$ individuals per square unit at a distance r from the center of concentration, then the total population $P(R)$ located within distance R from the center is given by

$$P(R) = \int_0^R 2\pi r p(r)\, dr$$

NOTE We found it convenient to derive the population density formula by considering the population of a city. However, the formula also applies to more general population concentrations, such as bacterial colonies or even the population of water drops from a sprinkler system. ■

EXAMPLE 5.6.3 Finding Population from Population Density

A city has population density $p(r) = 3e^{-0.01r^2}$, where $p(r)$ is the number of people (in thousands) per square mile at a distance of r miles from the city center.

a. What population lives within a 5-mile radius of the city center?

b. The city limits are set at a radius R where the population density is 1,000 people per square mile. What is the total population within the city limits?

Solution

a. The population within a 5-mile radius is

$$P(5) = \int_0^5 2\pi r (3e^{-0.01r^2})\, dr = 6\pi \int_0^5 e^{-0.01r^2} r\, dr$$

Using the substitution $u = -0.01r^2$, we find that

$$du = -0.01(2r\, dr) \quad \text{or} \quad r\, dr = \frac{du}{-0.02} = -50\, du$$

In addition, the limits of integration are transformed as follows:

If $r = 5$, then $u = -0.01(5)^2 = -0.25$.

If $r = 0$, then $u = -0.01(0)^2 = 0$.

Therefore, we have

$$P(5) = 6\pi \int_0^5 e^{-0.01r^2} r \, dr \qquad \text{substitute } u = -0.01r^2$$
$$-50 \, du = r \, dr$$

$$= 6\pi \int_0^{-0.25} e^u (-50 \, du)$$

$$= 6\pi(-50)[e^u] \Big|_{u=0}^{u=-0.25}$$

$$= -300\pi[e^{-0.25} - e^0]$$

$$\approx 208.5$$

So roughly 208,500 people live within a 5-mile radius of the city center.

b. To find the radius R that corresponds to the city limits, we want the population density to be 1 (one thousand), so we solve

$$3e^{-0.01R^2} = 1$$

$$e^{-0.01R^2} = \frac{1}{3} \qquad \text{take natural logarithms on both sides}$$

$$-0.01R^2 = \ln\left(\frac{1}{3}\right)$$

$$R^2 = \frac{\ln\left(\frac{1}{3}\right)}{-0.01} = 109.86$$

$$R \approx 10.48$$

Finally, using the substitution $u = -0.01r^2$ from part (a), we find that the population within the city limits is

$$P(10.48) = 6\pi \int_0^{10.48} e^{-0.01r^2} r \, dr \qquad \begin{array}{l} \text{Limits of integration:} \\ \text{when } r = 10.48, \text{ then } u = -0.01(10.48)^2 \approx -1.1 \\ \text{when } r = 0, \text{ then } u = 0 \end{array}$$

$$= 6\pi \int_0^{-1.1} e^u (-50 \, du)$$

$$\approx -300\pi[e^u] \Big|_{u=0}^{u=-1.1}$$

$$\approx -300\pi[e^{-1.1} - e^0]$$

$$\approx 628.75$$

Thus, approximately 628,750 people live within the city limits.

The Flow of Blood Through an Artery

Physiologists have found that the speed of blood in an artery is a function of the distance of the blood from the artery's central axis. According to Poiseuille's law, the speed (in centimeters per second) of blood that is r centimeters from the central axis of the artery is $S(r) = k(R^2 - r^2)$, where R is the radius of the artery and k is a constant. In Example 5.6.4, you will see how to use this information to compute the rate at which blood flows through the artery.

> **EXAMPLE 5.6.4** Finding a Formula for the Flow of Blood
> Through an Artery

Find an expression for the rate (in cubic centimeters per second) at which blood flows through an artery of radius R if the speed of the blood r centimeters from the central axis is $S(r) = k(R^2 - r^2)$, where k is a constant.

Solution

Our approach to computing blood flow is similar to the method just used for finding population from population density in Example 5.6.3.

To approximate the volume of blood flowing though a cross section of the artery each second, divide the interval $0 \leq r \leq R$ into n equal subintervals of width Δr centimeters and let r_j denote the left endpoint of the jth subinterval. These subintervals determine n concentric rings as illustrated in Figure 5.25.

If Δr is small, then by cutting the jth ring and flattening it out (recall Figure 5.24), you can see that the area of the jth ring is approximately the same as the area of a rectangle whose length is the circumference of the inner boundary of the ring and whose width is Δr. That is,

$$\text{Area of } j\text{th ring} \approx 2\pi r_j \, \Delta r$$

If you multiply the area of the jth ring (square centimeters) by the speed (centimeters per second) of the blood flowing through this ring, you get the volume rate (cubic centimeters per second) at which blood flows through the jth ring. Since the speed of blood flowing through the jth ring is approximately $S(r_j)$ centimeters per second, it follows that

$$\begin{pmatrix} \text{Volume rate of flow} \\ \text{through } j\text{th ring} \end{pmatrix} \approx \begin{pmatrix} \text{area of} \\ j\text{th ring} \end{pmatrix} \begin{pmatrix} \text{speed of blood} \\ \text{through } j\text{th ring} \end{pmatrix}$$

$$\approx (2\pi r_j \, \Delta r) S(r_j)$$
$$\approx (2\pi r_j \, \Delta r)[k(R^2 - r_j^2)]$$
$$\approx 2\pi k(R^2 r_j - r_j^3)\Delta r$$

The volume rate of flow of blood through the entire cross section is the sum of n such terms, one for each of the n concentric rings. That is,

$$\text{Volume rate of flow} \approx \sum_{j=1}^{n} 2\pi k(R^2 r_j - r_j^3)\Delta r$$

As n increases without bound, this approximation approaches the true value of the rate of flow. In other words,

$$\text{Volume rate of flow} = \lim_{n \to +\infty} \sum_{j=1}^{n} 2\pi k(R^2 r_j - r_j^3)\Delta r$$

$$= \int_0^R 2\pi k(R^2 r - r^3) \, dr$$

$$= 2\pi k \left(R^2 \frac{r^2}{2} - \frac{1}{4}r^4 \right)\Big|_0^R$$

$$= \frac{\pi k R^4}{2}$$

Thus, the blood is flowing at the rate of $\dfrac{\pi k R^4}{2}$ cubic centimeters per second.

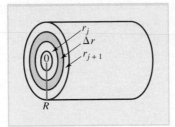

FIGURE 5.25 Dividing a cross section of an artery into concentric rings.

Volume of a Solid: The Size of a Tumor

Definite integration can also be used to compute the volume of certain objects. In this subsection, we demonstrate the general procedure by finding the volume of a solid formed by rotating a region R in the xy plane about the x axis. We then show how integration can be used to estimate the volume of a tumor.

The technique is to express the volume of the solid as the limit of a sum of the volumes of approximating disks. In particular, suppose that S is the solid formed by rotating the region R under the curve $y = f(x)$ between $x = a$ and $x = b$ about the x axis, as shown in Figure 5.26a. Divide the interval $a \leq x \leq b$ into n equal subintervals of length Δx. Then approximate the region R by n rectangles and the solid S by the corresponding n cylindrical disks formed by rotating these rectangles about the x axis. The general approximation procedure is illustrated in Figure 5.26b for the case where $n = 3$.

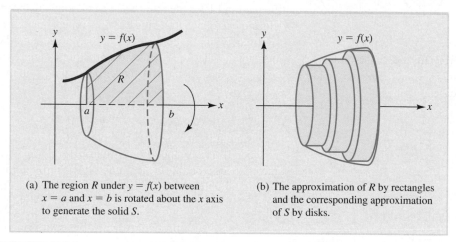

(a) The region R under $y = f(x)$ between $x = a$ and $x = b$ is rotated about the x axis to generate the solid S.

(b) The approximation of R by rectangles and the corresponding approximation of S by disks.

FIGURE 5.26   A solid S formed by rotating the region R about the x axis.

If x_j denotes the beginning (left endpoint) of the jth subinterval, then the jth rectangle has height $f(x_j)$ and width Δx as shown in Figure 5.27a. The jth approximating disk formed by rotating this rectangle about the x axis is shown in Figure 5.27b.

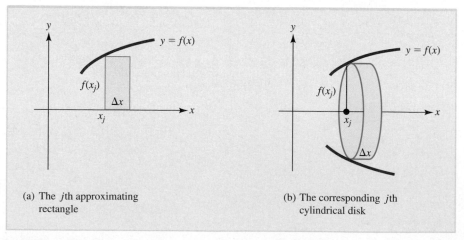

(a) The jth approximating rectangle

(b) The corresponding jth cylindrical disk

FIGURE 5.27   The volume of the solid S is approximated by adding volumes of approximating disks.

Since the jth approximating cylindrical disk has radius $r_j = f(x_j)$ and thickness Δx, its volume is

$$\text{Volume of } j\text{th disk} = (\text{area of circular cross section})(\text{width})$$
$$= \pi r_j^2(\text{width}) = \pi[f(x_j)]^2 \Delta x$$

The total volume of S is approximately the sum of the volumes of the n disks; that is,

$$\text{Volume of } S \approx \sum_{j=1}^{n} \pi[f(x_j)]^2 \Delta x$$

The approximation improves as n increases and

$$\text{Volume of } S = \lim_{n \to \infty} \sum_{j=1}^{n} \pi[f(x_j)]^2 \Delta x = \pi \int_a^b [f(x)]^2\, dx$$

To summarize:

Volume Formula

Suppose $f(x)$ is continuous and $f(x) \geq 0$ on $a \leq x \leq b$, and let R be the region under the curve $y = f(x)$ between $x = a$ and $x = b$. Then the volume of the solid S formed by rotating R about the x axis is

$$\text{Volume of } S = \pi \int_a^b [f(x)]^2\, dx$$

Examples 5.6.5 and 5.6.6 demonstrate the use of this formula.

EXAMPLE 5.6.5 Finding the Volume of a Solid

Find the volume of the solid S formed by rotating the region under the curve $y = x^2 + 1$ from $x = 0$ to $x = 2$ about the x axis.

Solution

The region, the solid, and the jth disk are shown in Figure 5.28.

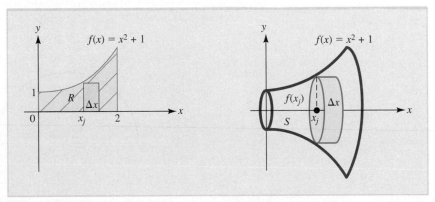

FIGURE 5.28 The solid formed by rotating the region under the curve $y = x^2 + 1$ between $x = 0$ and $x = 2$ about the x axis.

The radius of the jth disk is $f(x_j) = x_j^2 + 1$. Hence,

$$\text{Volume of } j\text{th disk} = \pi[f(x_j)]^2 \Delta x = \pi(x_j^2 + 1)^2 \Delta x$$

and

$$\text{Volume of } S = \lim_{n \to \infty} \sum_{j=1}^{n} \pi(x_j^2 + 1)^2 \Delta x$$

$$= \pi \int_0^2 (x^2 + 1)^2 \, dx$$

$$= \pi \int_0^2 (x^4 + 2x^2 + 1) \, dx$$

$$= \pi \left(\frac{1}{5}x^5 + \frac{2}{3}x^3 + x \right) \Bigg|_0^2 = \frac{206}{15}\pi \approx 43.14$$

EXAMPLE 5.6.6   Estimating the Volume of a Tumor

A tumor has approximately the same shape as the solid formed by rotating the region under the curve $y = \frac{1}{3}\sqrt{16 - 4x^2}$ about the x axis, where x and y are measured in centimeters. Find the volume of the tumor.

Solution
The curve intersects the x axis where $y = 0$, that is, where

$$\frac{1}{3}\sqrt{16 - 4x^2} = 0 \qquad \sqrt{a - b} = 0 \text{ only if } a = b$$

$$16 = 4x^2$$

$$x^2 = 4$$

$$x = \pm 2$$

The curve (called an *ellipse*) is shown in Figure 5.29.

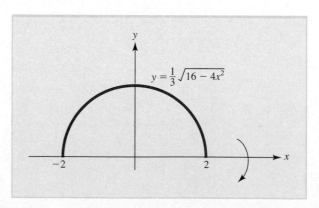

FIGURE 5.29   Tumor with the approximate shape of the solid formed by rotating the curve $y = \frac{1}{3}\sqrt{16 - 4x^2}$ about the x axis.

Let $f(x) = \dfrac{1}{3}\sqrt{16 - 4x^2}$. Then the volume of the solid is given by

$$V = \int_{-2}^{2} \pi[f(x)]^2 \, dx = \int_{-2}^{2} \pi\left[\frac{1}{3}\sqrt{16 - 4x^2}\right]^2 dx$$

$$= \int_{-2}^{2} \frac{\pi}{9}(16 - 4x^2) \, dx$$

$$= \frac{\pi}{9}\left[16x - \frac{4}{3}x^3\right]\Bigg|_{-2}^{2}$$

$$= \frac{\pi}{9}\left[16(2) - \frac{4}{3}(2)^3\right] - \frac{\pi}{9}\left[16(-2) - \frac{4}{3}(-2)^3\right] = \frac{128}{27}\pi$$

$$\approx 14.89$$

Thus, the volume of the tumor is approximately 15 cm³.

EXERCISES ■ 5.6

SURVIVAL AND RENEWAL   *In Exercises 1 through 6, an initial population P_0 is given along with a renewal rate R, and a survival function S(t). In each case, use the given information to find the population at the end of the indicated term T.*

1. $P_0 = 50{,}000$; $R(t) = 40$; $S(t) = e^{-0.1t}$, t in months; term $T = 5$ months

2. $P_0 = 100{,}000$; $R(t) = 300$; $S(t) = e^{0.02t}$, t in days; term $T = 10$ days

3. $P_0 = 500{,}000$; $R(t) = 800$; $S(t) = e^{-0.011t}$, t in years; term $T = 3$ years

4. $P_0 = 800{,}000$; $R(t) = 500$; $S(t) = e^{-0.005t}$, t in months; term $T = 5$ months

5. $P_0 = 500{,}000$; $R(t) = 100e^{0.01t}$; $S(t) = e^{-0.013t}$, t in years; term $T = 8$ years

6. $P_0 = 300{,}000$; $R(t) = 150e^{0.012t}$; $S(t) = e^{-0.02t}$, t in months; term $T = 20$ months

VOLUME OF A SOLID   *In Exercises 7 through 14, find the volume of the solid formed by rotating the region R about the x axis.*

7. R is the region under the line $y = 3x + 1$ from $x = 0$ to $x = 1$.

8. R is the region under the curve $y = \sqrt{x}$ from $x = 1$ to $x = 4$.

9. R is the region under the curve $y = x^2 + 2$ from $x = -1$ to $x = 3$.

10. R is the region under the curve $y = 4 - x^2$ from $x = -2$ to $x = 2$.

11. R is the region under the curve $y = \sqrt{4 - x^2}$ from $x = -2$ to $x = 2$.

12. R is the region under the curve $y = \dfrac{1}{x}$ from $x = 1$ to $x = 10$.

13. R is the region under the curve $y = \dfrac{1}{\sqrt{x}}$ from $x = 1$ to $x = e^2$.

14. R is the region under the curve $y = e^{-0.1x}$ from $x = 0$ to $x = 10$.

15. **NET POPULATION GROWTH**   It is projected that t years from now the population of a certain country will be changing at the rate of $e^{0.02t}$ million per year. If the current population is 50 million, what will be the population 10 years from now?

16. **NET POPULATION GROWTH**   A study indicates that x months from now, the population of a certain town will be increasing at the rate of $10 + 2\sqrt{x}$ people per month. By how much will the population increase over the next 9 months?

17. **GROUP MEMBERSHIP**   A national consumers' association has compiled statistics suggesting that the fraction of its members who are still active t months after joining is given by $f(t) = e^{-0.2t}$. A new local chapter has 200 charter members and expects to attract new members at the rate of 10 per month. How many members can the chapter expect to have at the end of 8 months?

18. **POLITICAL TRENDS**   Sarah Greene is running for mayor. Polls indicate that the fraction of those

who support her t weeks after first learning of her candidacy is given by $f(t) = e^{-0.03t}$. At the time she declared her candidacy, 25,000 people supported her, and new converts are being added at the constant rate of 100 people per week. Approximately how many people are likely to vote for her if the election is held 20 weeks from the day she entered the race?

19. **SPREAD OF DISEASE**   A new strain of influenza has just been declared an epidemic by health officials. Currently, 5,000 people have the disease and 60 more victims are added each day. If the fraction of infected people who still have the disease t days after contracting it is given by $f(t) = e^{-0.02t}$, how many people will have the flu 30 days from now?

20. **NUCLEAR WASTE**   A certain nuclear power plant produces radioactive waste in the form of strontium-90 at the constant rate of 500 pounds per year. The waste decays exponentially with a half-life of 28 years. How much of the radioactive waste from the nuclear plant will be present after 140 years? [*Hint:* Think of this as a survival and renewal problem.]

21. **ENERGY CONSUMPTION**   The government of a small country estimates that the demand for oil is increasing exponentially at the rate of 10% per year. If the demand is currently 30 billion barrels per year, how much oil will be consumed in this country during the next 10 years?

22. **POPULATION GROWTH**   The administrators of a town estimate that the fraction of people who will still be residing in the town t years after they arrive is given by the function $f(t) = e^{-0.04t}$. If the current population is 20,000 people and new townspeople arrive at the rate of 500 per year, what will be the population 10 years from now?

23. **COMPUTER DATING**   The operators of a new computer dating service estimate that the fraction of people who will retain their membership in the service for at least t months is given by the function $f(t) = e^{-t/10}$. There are 8,000 charter members, and the operators expect to attract 200 new members per month. How many members will the service have 10 months from now?

24. **FLOW OF BLOOD**   Calculate the rate (in cubic centimeters per second) at which blood flows through an artery of radius 0.1 centimeter if the speed of the blood r centimeters from the central axis is $8 - 800r^2$ centimeters per second.

25. **POISEUILLE'S LAW**   Blood flows through an artery of radius R. At a distance r centimeters from the central axis of the artery, the speed of the blood is given by $S(r) = k(R^2 - r^2)$. Show that the average velocity of the blood is one-half the maximum speed.

26. **POPULATION DENSITY**   The population density r miles from the center of a certain city is $D(r) = 5,000(1 + 0.5r^2)^{-1}$ people per square mile.
   a. How many people live within 5 miles of the city center?
   b. The city limits are set at a radius L where the population density is 1,000 people per square mile. What is L, and what is the total population within the city limits?

27. **POPULATION DENSITY**   The population density r miles from the center of a certain city is $D(r) = 25,000e^{-0.05r^2}$ people per square mile. How many people live between 1 and 2 miles from the city center?

28. **CHOLESTEROL REGULATION**   During his annual medical checkup, Byron is advised by his doctor to adopt a regimen of exercise, diet, and medication to lower his blood cholesterol level to 220 milligrams per deciliter (mg/dL). Suppose Byron finds that his cholesterol level t days after beginning the regimen is
   $$L(t) = 190 + 65e^{-0.003t}$$
   a. What is Byron's cholesterol level when he begins the regimen?
   b. How many days N must Byron remain on the regimen to lower his cholesterol level to 220 mg/dL?
   c. What was Byron's average cholesterol level during the first 30 days of the regimen? What was the average level over the entire period $0 \leq t \leq N$ of the regimen?

29. **CHOLESTEROL REGULATION**   Fat travels through the bloodstream attached to protein in a combination called a *lipoprotein*. Low-density lipoprotein (LDL) picks up cholesterol from the liver and delivers it to the cells, dropping off any excess cholesterol on the artery walls. Too much LDL in the bloodstream increases the risk of heart disease and stroke. A patient with a high level of LDL receives a drug that is found to reduce the level at a rate given by
   $$L'(t) = -0.3t(49 - t^2)^{0.4} \quad \text{units/day}$$
   where t is the number of days after the drug is administered, for $0 \leq t \leq 7$.

a. By how much does the patient's LDL level change during the first 3 days after the drug is administered?

b. Suppose the patient's LDL level is 120 at the time the drug is administered. Find $L(t)$.

c. The recommended "safe" LDL level is 100. How many days does it take for the patient's LDL level to be "safe"?

30. GROUP MEMBERSHIP A group has just been formed with an initial membership of 10,000. Suppose that the fraction of the membership of the group that remain members for at least t years after joining is $S(t) = e^{-0.03t}$, and that at time t, new members are being added at the rate of $R(t) = 10e^{0.017t}$ members per year. How many members will the group have 5 years from now?

31. GROWTH OF AN ENDANGERED SPECIES Environmentalists estimate that the population of a certain endangered species is currently 3,000. The population is expected to be growing at the rate of $R(t) = 10e^{0.01t}$ individuals per year t years from now, and the fraction that survive t years is given by $S(t) = e^{-0.07t}$. What will the population of the species be in 10 years?

32. POPULATION TRENDS The population of a small town is currently 85,000. A study commissioned by the mayor's office finds that people are settling in the town at the rate of $R(t) = 1,200e^{0.01t}$ per year and that the fraction of the population who continue to live in the town t years after arriving is given by $S(t) = e^{-0.02t}$. How many people will live in the town in 10 years?

33. POPULATION TRENDS Answer the question in Exercise 32 for a constant renewal rate $R = 1,000$ and the survival function

$$S(t) = \frac{1}{t+1}$$

34. EVALUATING DRUG EFFECTIVENESS A pharmaceutical firm has been granted permission by the FDA to test the effectiveness of a new drug in combating a virus. The firm administers the drug to a test group of uninfected but susceptible individuals, and using statistical methods, determines that t months after the test begins, people in the group are becoming infected at the rate of $D'(t)$ hundred individuals per month, where

$$D'(t) = 0.2 - 0.04t^{1/4}$$

Government figures indicate that without the drug, the infection rate would have been $W'(t)$ hundred

individuals per month, where

$$W'(t) = \frac{0.8e^{0.13t}}{(1 + e^{0.13t})^2}$$

If the test is evaluated 1 year after it begins, how many people does the drug protect from infection? What percentage of the people who would have been infected if the drug had not been used were protected from infection by the drug?

35. EVALUATING DRUG EFFECTIVENESS Repeat Exercise 34 for another drug for which the infection rate is

$$D'(t) = 0.12 + \frac{0.08}{t+1}$$

Assume the government comparison rate stays the same; that is,

$$W'(t) = \frac{0.8e^{0.13t}}{(1 + e^{0.13t})^2}$$

36. LIFE EXPECTANCY In a certain undeveloped country, the life expectancy of a person t years old is $L(t)$ years, where

$$L(t) = 41.6(1 + 1.07t)^{0.13}$$

a. What is the life expectancy of a person in this country at birth? At age 50?

b. What is the average life expectancy of all people in this country between the ages of 10 and 70?

c. Find the age T such that $L(T) = T$. Call T the *life limit* for this country. What can be said about the life expectancy of a person older than T years?

d. Find the average life expectancy L_e over the age interval $0 \le t \le T$. Why is it reasonable to call L_e the *expected length of life* for people in this country?

37. LIFE EXPECTANCY Answer the questions in Exercise 36 for a country whose life expectancy function is

$$L(t) = \frac{110e^{0.015t}}{1 + e^{0.015t}}$$

38. ENERGY EXPENDED IN FLIGHT In an investigation by V. A. Tucker and K. Schmidt-Koenig,* it was determined that the energy E expended by a bird in flight varies with the speed v (km/hr) of the bird. For a particular kind of parakeet, the energy expenditure changes at a rate given by

$$\frac{dE}{dv} = \frac{0.31v^2 - 471.75}{v^2} \quad \text{for } v > 0$$

*E. Batschelet, *Introduction to Mathematics for Life Scientists*, 3rd ed., New York, Springer-Verlag, 1979, p. 299.

where E is given in joules per gram mass per kilometer. Observations indicate that the parakeet tends to fly at the speed v_{min} that minimizes E.
a. What is the most economical speed v_{min}?
b. Suppose that when the parakeet flies at the most economical speed v_{min} its energy expenditure is E_{min}. Use this information to find $E(v)$ for $v > 0$ in terms of E_{min}.

39. **MEASURING RESPIRATION**   A pneumotacho-graph is a device used by physicians to graph the rate of air flow into and out of the lungs as a patient breathes. The graph in the accompanying figure shows the rate of inspiration (breathing in) for a particular patient. The area under the graph measures the total volume of air inhaled by the patient during the inspiration phase of one breathing cycle. Assume the inspiration rate is given by
$$R(t) = -0.41t^2 + 0.97t \quad \text{liters/sec}$$
a. How long is the inspiration phase?
b. Find the volume of air taken into the patient's lungs during the inspiration phase.
c. What is the average flow rate of air into the lungs during the inspiration phase?

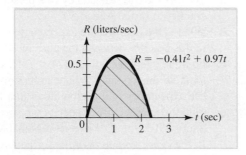

EXERCISE 39

40. **MEASURING RESPIRATION**   Repeat Exercise 39 with the inspiration rate function
$$R(t) = -1.2t^3 + 5.72t \quad \text{liters/sec}$$
and sketch the graph of $R(t)$.

41. **WATER POLLUTION**   A ruptured pipe in an off-shore oil rig produces a circular oil slick that is T feet thick at a distance r feet from the rupture, where
$$T(r) = \frac{3}{2 + r} \quad r \geq 0$$
At the time the spill is contained, the radius of the slick is 7 feet. We wish to find the volume of oil that has been spilled.
a. Sketch the graph of $T(r)$. Notice that the volume we want is obtained by rotating the curve $T(r)$ about the T axis (vertical axis) rather than the r axis (horizontal axis).

b. Solve the equation $T = \dfrac{3}{2 + r}$ for r in terms of T. Sketch the graph of $r(T)$, with T on the horizontal axis.
c. Find the required volume by rotating the graph of $r(T)$ found in part (b) about the T axis.

42. **WATER POLLUTION**   Rework Exercise 41 for a situation with spill thickness
$$T(r) = \frac{2}{1 + r^2}$$
(T and r in feet) and radius of containment 9 feet.

43. **AIR POLLUTION**   Particulate matter emitted from a smokestack is distributed in such a way that r miles from the stack, the pollution density is $p(r)$ units per square mile, where
$$p(r) = \frac{200}{5 + 2r^2}$$
a. What is the total amount of pollution within a 3-mile radius of the smokestack?
b. Suppose a health agency determines that it is unsafe to live within a radius L of the smoke-stack where the pollution density is at least four units per square mile. What is L, and what is the total amount of pollution in the unsafe zone?

44. **SIZE OF A TUMOR**   A tumor has approximately the same shape as the solid formed by rotating the region under the curve $y = x(4 - x)$ about the x axis, where x and y are measured in centimeters. Use integration to estimate the volume of the tumor.

45. **SIZE OF A TUMOR**   A tumor has approximately the same shape as the solid formed by rotating the region under the curve $y = \sqrt{x}(3 - x)$ about the x axis, where x and y are measured in centimeters. Use integration to estimate the volume of the tumor.

46. **SIZE OF A TUMOR**   A tumor has approximately the same shape as the solid formed by rotating the region under the curve $y = \sqrt{x}(A - x)$ about the x axis, where x and y are measured in centimeters, for some constant A. If the tumor is found to have volume 67 cm^3, what is A?

47. **BACTERIAL GROWTH**   An experiment is conducted with two bacterial colonies, each of which initially has a population of 100,000. In the first colony, a mild toxin is introduced that restricts growth so that only 50 new individuals are added per day and the fraction of individuals that survive at least t days is given by $f(t) = e^{-0.011t}$. The growth of the second colony is restricted indirectly,

by limiting food supply and space for expansion, and after t days, it is found that this colony contains

$$P(t) = \frac{5,000}{1 + 49e^{0.009t}}$$

thousand individuals. Which colony is larger after 50 days? After 100 days? After 300 days?

48. VOLUME OF A SPHERE Use integration to show that a sphere of radius r has volume

$$V = \frac{4}{3}\pi r^3$$

[*Hint:* Think of the sphere as the solid formed by rotating the region under the semicircle shown in the accompanying figure about the x axis.]

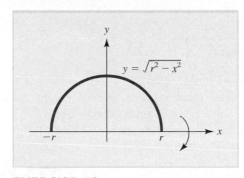

EXERCISE 48

49. VOLUME OF A CONE Use integration to show that a right circular cone of height h and top radius r has volume

$$V = \frac{1}{3}\pi r^2 h$$

[*Hint:* Think of the cone as a solid formed by rotating the triangle shown in the accompanying figure about the x axis.]

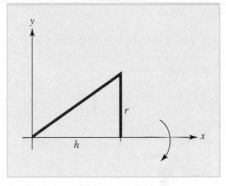

EXERCISE 49

Important Terms, Symbols, and Formulas

Antiderivative; indefinite integral: (376, 378)

$$\int f(x)\, dx = F(x) + C \text{ if and only if } F'(x) = f(x)$$

Power rule: (379)

$$\int x^n\, dx = \frac{x^{n+1}}{n+1} + C \quad \text{for } n \neq -1$$

Logarithmic rule: $\int \frac{1}{x}\, dx = \ln|x| + C$ (379)

Exponential rule: $\int e^{kx}\, dx = \frac{1}{k}e^{kx} + C$ (379)

Constant multiple rule: (380)

$$\int kf(x)\, dx = k\int f(x)\, dx$$

Sum rule: (380)

$$\int [f(x) + g(x)]\, dx = \int f(x)\, dx + \int g(x)\, dx$$

Differential equation (383)
Solution of a differential equation (384)
Initial value problem (385)

Separable differential equation (384)
Integration by substitution: (393)

$$\int g(u(x))u'(x)\, dx = \int g(u)\, du \quad \text{where } u = u(x) \atop du = u'(x)\, dx$$

Price-adjustment model (401)
Definite integral: (410)

$$\int_a^b f(x)\, dx = \lim_{n \to +\infty} [f(x_1) + \cdots + f(x_n)]\Delta x$$

Area under a curve: (408, 411)

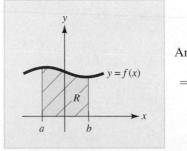

Area of R

$$= \int_a^b f(x)\, dx$$

Special rules for definite integrals: (414)

$$\int_a^a f(x)\, dx = 0$$

$$\int_b^a f(x)\, dx = -\int_a^b f(x)\, dx$$

Constant multiple rule: (414)

$$\int_a^b k f(x)\, dx = k \int_a^b f(x)\, dx \quad \text{for constant } k$$

Sum rule: (414)

$$\int_a^b [f(x) + g(x)]\, dx = \int_a^b f(x)\, dx + \int_a^b g(x)\, dx$$

Difference rule: (414)

$$\int_a^b [f(x) - g(x)]\, dx = \int_a^b f(x)\, dx - \int_a^b g(x)\, dx$$

Subdivision rule: (414)

$$\int_a^b f(x)\, dx = \int_a^c f(x)\, dx + \int_c^b f(x)\, dx$$

Fundamental theorem of calculus: (411)

$$\int_a^b f(x)\, dx = F(b) - F(a) \quad \text{where } F'(x) = f(x)$$

Net change of $Q(x)$ over the interval $a \le x \le b$:   (417)

$$Q(b) - Q(a) = \int_a^b Q'(x)\, dx$$

Area between two curves: (426)

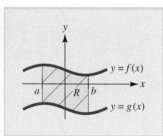

Area of R

$$= \int_a^b [f(x) - g(x)]\, dx$$

Average value of a function $f(x)$ over an interval $a \le x \le b$:   (434)

$$V = \frac{1}{b-a}\int_a^b f(x)\, dx$$

Lorenz curve   (430)

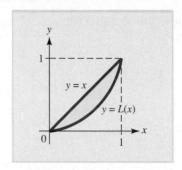

Gini index $= 2 \int_0^1 [x - L(x)]\, dx$ (432)

Net excess profit   (429)
Future value (amount) of an income stream   (443)
Present value of an income stream   (444)
Consumer willingness to spend   (446)
Consumers' surplus: (447)

$$CS = \int_0^{q_0} D(q)\, dq - p_0 q_0, \text{ where } p = D(q) \text{ is demand}$$

Producers' surplus: (448)

$$PS = p_0 q_0 - \int_0^{q_0} S(q)\, dq, \text{ where } p = S(q) \text{ is supply}$$

Survival and renewal   (455)
Flow of blood through an artery   (459)
Population from population density   (457)
Volume of a solid   (461)

Checkup for Chapter 5

1. Find these indefinite integrals (antiderivatives).

a. $\displaystyle\int (x^3 - \sqrt{3x} + 5e^{-2x})\, dx$

b. $\displaystyle\int \frac{x^2 - 2x + 4}{x}\, dx$

c. $\displaystyle\int \sqrt{x}\left(x^2 - \frac{1}{x}\right) dx$

d. $\displaystyle\int \frac{x}{(3 + 2x^2)^{3/2}}\, dx$

e. $\displaystyle\int \frac{\ln \sqrt{x}}{x}\, dx$

f. $\displaystyle\int x e^{1+x^2}\, dx$

2. Evaluate each of these definite integrals.

a. $\displaystyle\int_1^4 \left(x^{3/2} + \frac{2}{x}\right) dx$

b. $\displaystyle\int_0^3 e^{3-x}\, dx$

c. $\displaystyle\int_0^1 \frac{x}{x+1}\, dx$

d. $\displaystyle\int_0^3 \frac{x+3}{\sqrt{x^2+6x+4}}\, dx$

3. In each case, find the area of the specified region.

   a. The region bounded by the curve $y = x + \sqrt{x}$, the x axis, and the lines $x = 1$ and $x = 4$.

   b. The region bounded by the curve $y = x^2 - 3x$ and the line $y = x + 5$.

4. Find the average value of the function $f(x) = \dfrac{x-2}{x}$ over the interval $1 \le x \le 2$.

5. **NET CHANGE IN REVENUE** The marginal revenue of producing q units of a certain commodity is $R'(q) = q(10 - q)$ hundred dollars per unit. How much additional revenue is generated as the level of production is increased from 4 to 9 units?

6. **BALANCE OF TRADE** The government of a certain country estimates that t years from now, imports will be increasing at the rate $I'(t)$ and exports at the rate $E'(t)$, both in billions of dollars per year, where

$$I'(t) = 12.5e^{0.2t} \quad \text{and} \quad E'(t) = 1.7t + 3$$

The trade deficit is $D(t) = I(t) - E(t)$. By how much will the trade deficit for this country change over the next 5 years? Will it increase or decrease during this time period?

7. **CONSUMERS' SURPLUS** Suppose q hundred units of a certain commodity are demanded by consumers when the price is $p = 25 - q^2$ dollars per unit. What is the consumers' surplus for the commodity when the level of production is $q_0 = 4$ (400 units)?

8. **FUTURE VALUE OF AN ANNUITY** Money is transferred continuously into Rosa's account at the constant rate of $5,000 per year. The account earns interest at the annual rate of 5% compounded continuously. How much will be in Rosa's account at the end of 3 years?

9. **POPULATION GROWTH** Demographers estimate that the fraction of people who will still be residing in a particular town t years after they arrive is given by the function $f(t) = e^{-0.02t}$. If the current population is 50,000 and new townspeople arrive at the rate of 700 per year, what will the population be 20 years from now?

10. **AVERAGE DRUG CONCENTRATION** A patient is injected with a drug, and t hours later, the concentration of the drug remaining in the patient's bloodstream is given by

$$C(t) = \frac{0.3t}{(t^2 + 16)^{1/2}} \quad \text{mg/cm}^3$$

What is the average concentration of the drug during the first 3 hours after the injection?

Review Exercises

In Exercises 1 through 20, find the indicated indefinite integral.

1. $\displaystyle\int (x^3 + \sqrt{x} - 9)\, dx$

2. $\displaystyle\int \left(x^{2/3} - \frac{1}{x} + 5 + \sqrt{x} \right) dx$

3. $\displaystyle\int (x^4 - 5e^{-2x})\, dx$

4. $\displaystyle\int \left(2\sqrt[3]{s} + \frac{5}{s} \right) ds$

5. $\displaystyle\int \left(\frac{5x^3 - 3}{x} \right) dx$

6. $\displaystyle\int \left(\frac{3e^{-x} + 2e^{3x}}{e^{2x}} \right) dx$

7. $\displaystyle\int \left(t^5 - 3t^2 + \frac{1}{t^2} \right) dt$

8. $\displaystyle\int (x + 1)(2x^2 + \sqrt{x})\, dx$

9. $\displaystyle\int \sqrt{3x + 1}\, dx$

10. $\displaystyle\int (3x + 1)\sqrt{3x^2 + 2x + 5}\, dx$

11. $\displaystyle\int (x + 2)(x^2 + 4x + 2)^5\, dx$

12. $\displaystyle\int \frac{x + 2}{x^2 + 4x + 2}\, dx$

13. $\displaystyle\int \frac{3x + 6}{(2x^2 + 8x + 3)^2}\, dx$

14. $\int (t - 5)^{12}\, dt$

15. $\int v(v - 5)^{12}\, dv$

16. $\int \dfrac{\ln(3x)}{x}\, dx$

17. $\int 5xe^{-x^2}\, dx$

18. $\int \left(\dfrac{x}{x - 4} \right) dx$

19. $\int \left(\dfrac{\sqrt{\ln x}}{x} \right) dx$

20. $\int \left(\dfrac{e^x}{e^x + 5} \right) dx$

In Exercises 21 through 30, evaluate the indicated definite integral.

21. $\int_0^1 (5x^4 - 8x^3 + 1)\, dx$

22. $\int_1^4 (\sqrt{t} + t^{-3/2})\, dt$

23. $\int_0^1 (e^{2x} + 4\sqrt[3]{x})\, dx$

24. $\int_1^9 \dfrac{x^2 + \sqrt{x} - 5}{x}\, dx$

25. $\int_{-1}^2 30(5x - 2)^2\, dx$

26. $\int_{-1}^1 \dfrac{(3x + 6)}{(x^2 + 4x + 5)^2}\, dx$

27. $\int_0^1 2te^{t^2 - 1}\, dt$

28. $\int_0^1 e^{-x}(e^{-x} + 1)^{1/2}\, dx$

29. $\int_0^{e-1} \left(\dfrac{x}{x + 1} \right) dx$

30. $\int_e^{e^2} \dfrac{1}{x(\ln x)^2}\, dx$

AREA BETWEEN CURVES   In Exercises 31 through 38, sketch the indicated region R and find its area by integration.

31. R is the region under the curve $y = x + 2\sqrt{x}$ over the interval $1 \le x \le 4$.

32. R is the region under the curve $y = e^x + e^{-x}$ over the interval $-1 \le x \le 1$.

33. R is the region under the curve $y = \dfrac{1}{x} + x^2$ over the interval $1 \le x \le 2$.

34. R is the region under the curve $y = \sqrt{9 - 5x^2}$ over the interval $0 \le x \le 1$.

35. R is the region bounded by the curve $y = \dfrac{4}{x}$ and the line $x + y = 5$.

36. R is the region bounded by the curves $y = \dfrac{8}{x}$ and $y = \sqrt{x}$ and the line $x = 8$.

37. R is the region bounded by the curve $y = 2 + x - x^2$ and the x axis.

38. R is the triangular region with vertices $(0, 0)$, $(2, 4)$, and $(0, 6)$.

AVERAGE VALUE OF A FUNCTION   In Exercises 39 through 42, find the average value of the given function over the indicated interval.

39. $f(x) = x^3 - 3x + \sqrt{2x}$; over $1 \le x \le 8$

40. $f(t) = t\sqrt[3]{8 - 7t^2}$; over $0 \le t \le 1$

41. $g(v) = ve^{-v^2}$; over $0 \le v \le 2$

42. $h(x) = \dfrac{e^x}{1 + 2e^x}$; over $0 \le x \le 1$

CONSUMERS' SURPLUS   In Exercises 43 through 46, $p = D(q)$ is the demand curve for a particular commodity; that is, q units of the commodity will be demanded when the price is $p = D(q)$ dollars per unit. In each case, for the given level of production q_0, find $p_0 = D(q_0)$ and compute the corresponding consumers' surplus.

43. $D(q) = 4(36 - q^2)$; $q_0 = 2$ units

44. $D(q) = 100 - 4q - 3q^2$; $q_0 = 5$ units

45. $D(q) = 10e^{-0.1q}$; $q_0 = 4$ units

46. $D(q) = 5 + 3e^{-0.2q}$; $q_0 = 10$ units

LORENZ CURVES   In Exercises 47 through 50, sketch the Lorenz curve $y = L(x)$ and find the corresponding Gini index.

47. $L(x) = x^{3/2}$

48. $L(x) = x^{1.2}$

49. $L(x) = 0.3x^2 + 0.7x$

50. $L(x) = 0.75x^2 + 0.25x$

SURVIVAL AND RENEWAL   *In Exercises 51 through 54, an initial population P_0 is given along with a renewal rate $R(t)$ and a survival function $S(t)$. In each case, use the given information to find the population at the end of the indicated term T.*

51. $P_0 = 75,000$; $R(t) = 60$; $S(t) = e^{-0.09t}$; t in months; term $T = 6$ months

52. $P_0 = 125,000$; $R(t) = 250$; $S(t) = e^{-0.015t}$; t in years; term $T = 5$ years

53. $P_0 = 100,000$; $R(t) = 90e^{0.1t}$; $S(t) = e^{-0.2t}$; t in years; term $T = 10$ years

54. $P_0 = 200,000$; $R(t) = 50e^{0.12t}$; $S(t) = e^{-0.017t}$; t in hours; term $T = 20$ hours

VOLUME OF SOLID OF REVOLUTION   *In Exercises 55 through 58, find the volume of the solid of revolution formed by rotating the specified region R about the x axis.*

55. R is the region under the curve $y = x^2 + 1$ from $x = -1$ to $x = 2$.

56. R is the region under the curve $y = e^{-x/10}$ from $x = 0$ to $x = 10$.

57. R is the region under the curve $y = \dfrac{1}{\sqrt{x}}$ from $x = 1$ to $x = 3$.

58. R is the region under the curve $y = \dfrac{x + 1}{\sqrt{x}}$ from $x = 1$ to $x = 4$.

In Exercises 59 through 62, solve the given initial value problem.

59. $\dfrac{dy}{dx} = 2$, where $y = 4$ when $x = -3$

60. $\dfrac{dy}{dx} = x(x - 1)$, where $y = 1$ when $x = 1$

61. $\dfrac{dx}{dt} = e^{-2t}$, where $x = 4$ when $t = 0$

62. $\dfrac{dy}{dt} = \dfrac{t + 1}{t}$, where $y = 3$ when $t = 1$

63. Find the function whose tangent line has slope $x(x^2 + 1)^{-1}$ for each x and whose graph passes through the point $(1, 5)$.

64. Find the function whose tangent line has slope xe^{-2x^2} for each x and whose graph passes through the point $(0, -3)$.

65. **NET ASSET VALUE**   It is estimated that t days from now a farmer's crop will be increasing at the rate of $0.5t^2 + 4(t + 1)^{-1}$ bushels per day. By how much will the value of the crop increase during the next 6 days if the market price remains fixed at $2 per bushel?

66. **DEPRECIATION**   The resale value of a certain industrial machine decreases at a rate that changes with time. When the machine is t years old, the rate at which its value is changing is $200(t - 6)$ dollars per year. If the machine was bought new for $12,000, how much will it be worth 10 years later?

67. **TICKET SALES**   The promoters of a county fair estimate that t hours after the gates open at 9:00 A.M. visitors will be entering the fair at the rate of $-4(t + 2)^3 + 54(t + 2)^2$ people per hour. How many people will enter the fair between 10:00 A.M. and noon?

68. **MARGINAL COST**   At a certain factory, the marginal cost is $6(q - 5)^2$ dollars per unit when the level of production is q units. By how much will the total manufacturing cost increase if the level of production is raised from 10 to 13 units?

69. **PUBLIC TRANSPORTATION**   It is estimated that x weeks from now, the number of commuters using a new subway line will be increasing at the rate of $18x^2 + 500$ per week. Currently, 8,000 commuters use the subway. How many will be using it 5 weeks from now?

70. **NET CHANGE IN BIOMASS**   A protein with mass m (grams) disintegrates into amino acids at a rate given by

$$\frac{dm}{dt} = \frac{-15t}{t^2 + 5}$$

What is the net change in mass of the protein during the first 4 hours?

71. **CONSUMPTION OF OIL**   It is estimated that t years from the beginning of the year 2012, the demand for oil in a certain country will be changing at the rate of $D'(t) = (1 + 2t)^{-1}$ billion barrels per year. Will more oil be consumed (demanded) during 2013 or during 2014? How much more?

72. **FUTURE VALUE OF AN INCOME STREAM**   Money is transferred continuously into an account at the rate of $5,000e^{0.015t}$ dollars per year at time t (years). The account earns interest at the annual rate of 5% compounded continuously. How much will be in the account at the end of 3 years?

73. FUTURE VALUE OF AN INCOME STREAM Money is transferred continuously into an account at the constant rate of $1,200 per year. The account earns interest at the annual rate of 8% compounded continuously. How much will be in the account at the end of 5 years?

74. PRESENT VALUE OF AN INCOME STREAM What is the present value of an investment that will generate income continuously at a constant rate of $1,000 per year for 10 years if the prevailing annual interest rate remains fixed at 7% compounded continuously?

75. REAL ESTATE INVENTORY In a certain community the fraction of the homes placed on the market that remain unsold for at least t weeks is approximately $f(t) = e^{-0.2t}$. If 200 homes are currently on the market and if additional homes are placed on the market at the rate of 8 per week, approximately how many homes will be on the market 10 weeks from now?

76. AVERAGE REVENUE A bicycle manufacturer expects that x months from now consumers will be buying 5,000 bicycles per month at the price of $P(x) = 200 + 3\sqrt{x}$ dollars per bicycle. What is the average revenue the manufacturer can expect from the sale of the bicycles over the next 16 months?

77. NUCLEAR WASTE A nuclear power plant produces radioactive waste at a constant rate of 300 pounds per year. The waste decays exponentially with a half-life of 35 years. How much of the radioactive waste from the plant will remain after 200 years?

78. GROWTH OF A TREE A tree has been transplanted and after x years is growing at the rate of

$$h'(x) = 0.5 + \frac{1}{(x + 1)^2}$$

meters per year. By how much does the tree grow during the second year?

79. FUTURE REVENUE A certain oil well that yields 900 barrels of crude oil per month will run dry in 3 years. The price of crude oil is currently $92 per barrel and is expected to rise at the constant rate of 80 cents per barrel per month. If the oil is sold as soon as it is extracted from the ground, how much total future revenue will be obtained from the well?

80. CONSUMERS' SURPLUS Suppose that the consumers' demand function for a certain commodity is $D(q) = 50 - 3q - q^2$ dollars per unit.

a. Find the number of units that will be bought if the market price is $32 per unit.

b. Compute the consumer willingness to spend to get the number of units in part (a).

c. Compute the consumers' surplus when the market price is $32 per unit.

d. Use the graphing utility of your calculator to graph the demand curve. Interpret the consumer willingness to spend and the consumers' surplus as areas in relation to this curve.

81. AVERAGE PRICE Records indicate that t months after the beginning of the year, the price of bacon in local supermarkets was $P(t) = 0.06t^2 - 0.2t + 6.2$ dollars per pound. What was the average price of bacon during the first 6 months of the year?

82. SURFACE AREA OF A HUMAN BODY The surface area S of the body of an average person 4 feet tall who weighs w lb changes at the rate

$$S'(w) = 110w^{-0.575} \quad \text{in}^2/\text{lb}$$

The body of a particular child who is 4 feet tall and weighs 50 lb has surface area 1,365 in^2. If the child gains 3 lb while remaining the same height, by how much will the surface area of the child's body increase?

83. TEMPERATURE CHANGE At t hours past midnight, the temperature T (°C) in a certain northern city is found to be changing at a rate given by

$$T'(t) = -0.02(t - 7)(t - 14) \quad \text{°C/hour}$$

By how much does the temperature change between 8 A.M. and 8 P.M.?

84. EFFECT OF A TOXIN A toxin is introduced to a bacterial colony, and t hours later, the population $P(t)$ of the colony is changing at the rate

$$\frac{dP}{dt} = -(\ln 3)3^{4-t}$$

If there were 1 million bacteria in the colony when the toxin was introduced, what is $P(t)$? [*Hint:* Note that $3^x = e^{x \ln 3}$.]

85. MARGINAL ANALYSIS   In a certain section of the country, the price of large Grade A eggs is currently $2.50 per dozen. Studies indicate that x weeks from now, the price $p(x)$ will be changing at the rate of $p'(x) = 0.2 + 0.003x^2$ cents per week.

   a. Use integration to find $p(x)$, and then use the graphing utility of your calculator to sketch the graph of $p(x)$. How much will the eggs cost 10 weeks from now?

   b. Suppose the rate of change of the price was $p'(x) = 0.3 + 0.003x^2$. How does this affect $p(x)$? Check your conjecture by sketching the new price function on the same screen as the original. Now how much will the eggs cost in 10 weeks?

86. INVESTING IN A DOWN MARKET PERIOD Jan opens a stock account with $5,000 at the beginning of January and subsequently deposits $200 a month. Unfortunately, the market is depressed, and she finds that t months after depositing a dollar, only $100f(t)$ cents remain, where $f(t) = e^{-0.01t}$. If this pattern continues, what will her account be worth after 2 years? [*Hint:* Think of this as a survival and renewal problem.]

87. DISTANCE AND VELOCITY   After t minutes, an object moving along a line has velocity $v(t) = 1 + 4t + 3t^2$ meters per minute. How far does the object travel during the third minute?

88. AVERAGE POPULATION   The population (in thousands) of a certain city t years after January 1, 2005, is given by the function

$$P(t) = \frac{150e^{0.03t}}{1 + e^{0.03t}}$$

What is the average population of the city during the decade 2005–2015?

89. DISTRIBUTION OF INCOME   A study suggests that the distribution of incomes for social workers and physical therapists may be represented by the Lorenz curves $y = L_1(x)$ and $y = L_2(x)$, respectively, where

$$L_1(x) = x^{1.6} \quad \text{and} \quad L_2(x) = 0.65x^2 + 0.35x$$

For which profession is the distribution of income more equitable?

90. DISTRIBUTION OF INCOME   A study conducted by a certain state determines that the Lorenz curves for high school teachers and real estate brokers are given by the functions

$$L_1(x) = 0.67x^4 + 0.33x^3$$
$$L_2(x) = 0.72x^2 + 0.28x$$

respectively. For which profession is the distribution of income more equitable?

91. CONSERVATION   A lake has roughly the same shape as the bottom half of the solid formed by rotating the curve $2x^2 + 3y^2 = 6$ about the x axis, for x and y measured in miles. Conservationists want the lake to contain 1,000 trout per cubic mile. If the lake currently contains 5,000 trout, how many more must be added to meet this requirement?

92. HORTICULTURE   A sprinkler system sprays water onto a garden in such a way that $11e^{-r^2/10}$ inches of water per hour are delivered at a distance of r feet from the sprinkler. What is the total amount of water laid down by the sprinkler within a 5-foot radius during a 20-minute watering period?

93. SPEED AND DISTANCE   A car is driven so that after t hours its speed is $S(t)$ miles per hour.

   a. Write down a definite integral that gives the average speed of the car during the first N hours.

   b. Write down a definite integral that gives the total distance the car travels during the first N hours.

   c. Discuss the relationship between the integrals in parts (a) and (b).

94. Use the graphing utility of your calculator to draw the graphs of the curves $y = -x^3 - 2x^2 + 5x - 2$ and $y = x \ln x$ on the same screen. Use **ZOOM** and **TRACE** or some other feature of your calculator to find where the curves intersect, and then compute the area of the region bounded by the curves.

95. Repeat Exercise 94 for the curves

$$y = \frac{x - 2}{x + 1} \quad \text{and} \quad y = \sqrt{25 - x^2}$$

EXPLORE! UPDATE

Complete solutions for all EXPLORE! boxes throughout the text can be accessed at the book-specific website, www.mhhe.com/hoffmann.

Solution for Explore! on Page 377

Store the constants $\{-4, -2, 2, 4\}$ into L1, and write Y1 = X^3 and Y2 = Y1 + L1. Graph Y1 in bold, using the modified decimal window $[-4.7, 4.7]1$ by $[-6, 6]1$. At $x = 1$ (where we have drawn a vertical line), the slopes for each curve appear equal.

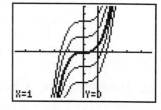

Using the tangent line feature of your graphing calculator, draw tangent lines at $x = 1$ for several of these curves. Every tangent line at $x = 1$ has a slope of 3, although each line has a different y intercept.

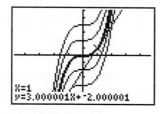
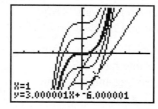
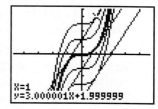

Solution for Explore! on Page 378

The numerical integral, **fnInt**(expression, variable, lower limit, upper limit) can be found via the **MATH** key, **9:fnInt(,** which we use to write Y1 below. We obtain a family of graphs that appear to be parabolas with vertices on the y axis at $y = 0, -1$, and -4. The antiderivative of $f(x) = 2x$ is $F(x) = x^2 + C$, where $C = 0, -1$, and -4, in our case.

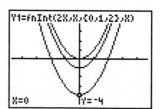

Solution for Explore! on Page 379

Place $y = F(x) = \ln |x|$ into Y1 as ln(abs(x)), using a bold graphing style, and store $f(x) = \dfrac{1}{x}$ into Y2; then graph using a decimal window. Choose $x = 1$, and compare the derivative $F'(1)$, which is displayed in the graph on the left as $\dfrac{dy}{dx} = 1.0000003$, with the value $y = 1$ of $f(1)$ displayed on the right. The negligible difference in value in this case can be attributed to the use of numerical differentiation. In general, choosing any other nonzero x value, we can verify that $F'(x) = f(x)$. For instance, when $x = -2$, we have $F'(-2) = -0.5 = f(-2)$.

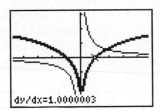
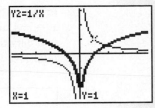

Solution for Explore! on Page 381

Place the integers from -5 to 5 into L1 (**STAT EDIT 1**). Set up the functions in the equation editor as shown here. Now graph with the designated window and notice that the antiderivative curves are generated sequentially from the lower to the upper levels. **TRACE** to the point $(2, 6)$, and observe that the antiderivative that passes through this point is second in the listing of L1. This curve is $F(x) = x^3 + x - 4$, which can also be calculated analytically, as in Example 5.1.4. For $f(x) = 3x^2 - 2$, the family of antiderivatives is $F(x) = x^3 - 2x + L1$ and the same window dimensions can be used to produce the screen on the right. The desired antiderivative is the eighth in L1, corresponding to $F(x) = x^3 - 2x + 2$, whose constant term can also be confirmed algebraically.

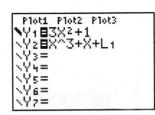
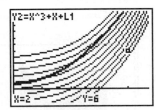
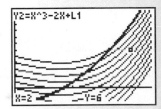

Solution for Explore! on Page 412

Following Example 5.3.3, set $Y1 = x^3 + 1$ and graph using a window $[-1, 3]1$ by $[-1, 2]1$. Access the numerical integration feature through **CALC, 7:∫f(x) dx,** specifying the lower limit as $X = 0$ and the upper limit as $X = 1$ to obtain $\int_0^1 (x^3 + 1)\, dx = 1.25$. Numerical integration can also be performed from the home screen via **MATH, 9: fnInt(,** as shown in the screen on the right.

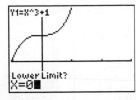
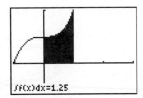
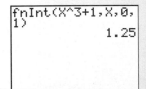

Solution for Explore! on Page 427

Following Example 5.4.2, set Y1 = 4X and Y2 = X^3 + 3X^2 in the equation editor of your graphing calculator. Graph using the window [−6, 2]1 by [−25, 10]5. The points of intersection are at $x = -4$, 0, and 1. Considering $y = 4x$ as a horizontal baseline, the area between Y1 and Y2 can be viewed as that of the difference curve Y3 = Y2 − Y1. Deselect (turn off) Y1 and Y2, and graph Y3 using the window [−4.5, 1.5]0.5 by [−5, 15]5. Numerical integration applied to this curve between $x = -4$ and 0 yields an area of 32 square units for the first sector enclosed by the two curves. The area of the second sector, between $x = 0$ and 1, has area −0.75. The total area enclosed by the two curves is $32 + |-0.75| = 32.75$.

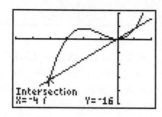 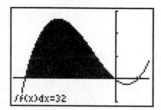 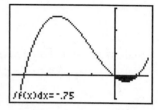

Solution for Explore! on Page 433

Set Y1 = $x^3 - 6x^2 + 10x - 1$, and use the **CALC, 7:∫(x) dx** feature to determine that the area under the curve from $x = 1$ to $x = 4$ is 9.75 square units, which equals the rectangular portion under Y2 = 9.75/(4 − 1) = 3.25 of length 3. It is as though the area under $f(x)$ over [1, 4] turned to water and became a level surface of height 3.25, the average $f(x)$ value. This value is attained at $x \approx 1.874$ (shown on the right) and also at $x = 3.473$. Note that you must clear the previous shading, using **DRAW, 1:ClrDraw,** before constructing the next drawing.

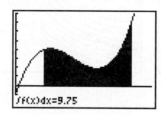 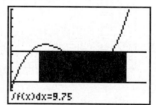 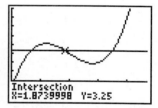

Solution for Explore! on Page 446

We graph $D_{new}(q)$ in bold as Y2 = 4(23 − X^2) with $D(q)$ in Y1 = 4(25 − X^2), using the viewing window [0, 5]1 by [0, 150]10. Visually, $D_{new}(q)$ is less than $D(q)$ for the observable range of values, supporting the conjecture that the area under the curve of $D_{new}(q)$ will be less than that of $D(q)$ over the range of values [0, 3]. This area is calculated to be $240, less than the $264 shown for $D(q)$ in Figure 5.21.

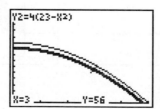 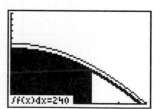

THINK ABOUT IT

JUST NOTICEABLE DIFFERENCES IN PERCEPTION

Calculus can help us answer questions about human perception, including questions relating to the number of different frequencies of sound or the number of different hues of light people can distinguish (see the accompanying figure). Our present goal is to show how integral calculus can be used to estimate the number of steps a person can distinguish as the frequency of sound increases from the lowest audible frequency of 15 hertz (Hz) to the highest audible frequency of 18,000 Hz. (Here hertz, abbreviated Hz, equals cycles per second.)

A mathematical model* for human auditory perception uses the formula $y = 0.767x^{0.439}$, where y Hz is the smallest change in frequency that is detectable at frequency x Hz. Thus, at the low end of the range of human hearing, 15 Hz, the smallest change of frequency a person can detect is $y = 0.767 \times 15^{0.439} \approx 2.5$ Hz, while at the upper end of human hearing, near 18,000 Hz, the least noticeable difference is approximately $y = 0.767 \times 18,000^{0.439} \approx 57$ Hz. If the smallest noticeable change of frequency were the same for all frequencies that people can hear, we could find the number of noticeable steps in human hearing by simply dividing the total frequency range by the size of this smallest noticeable change. Unfortunately, we have just seen that the smallest noticeable change of frequency increases as frequency increases, so the simple approach will not work. However, we can estimate the number of distinguishable steps using integration.

Toward this end, let $y = f(x)$ represent the just noticeable difference of frequency people can distinguish at frequency x. Next, choose numbers $x_0, x_1, \ldots, x_n$ beginning at $x_0 = 15$ Hz and working up through higher frequencies to $x_n = 18,000$ Hz in such a way that for $j = 0, 2, \ldots, n - 1$,

$$x_j + f(x_j) = x_{j+1}$$

In other words, x_{j+1} is the number we get by adding the just noticeable difference at x_j to x_j itself. Thus, the jth step has length

$$\Delta x_j = x_{j+1} - x_j = f(x_j)$$

Dividing by $f(x_j)$, we get

$$\frac{\Delta x_j}{f(x_j)} = \frac{x_{j+1} - x_j}{f(x_j)} = 1$$

and it follows that

$$\sum_{j=0}^{n-1} \frac{\Delta x_j}{f(x_j)} = \sum_{j=0}^{n-1} \frac{x_{j+1} - x_j}{f(x_j)} = \frac{x_1 - x_0}{f(x_0)} + \frac{x_2 - x_1}{f(x_1)} + \cdots + \frac{x_n - x_{n-1}}{f(x_n)}$$

$$= \underbrace{1 + 1 + \cdots + 1}_{n \text{ terms}} = n$$

*Part of this essay is based on *Applications of Calculus: Selected Topics from the Environmental and Life Sciences,* by Anthony Barcellos, New York: McGraw-Hill, 1994, pp. 21–24.

The sum on the left side of this equation is a Riemann sum, and since the step sizes $\Delta x_j = x_{j+1} - x_j$ are very small, the sum is approximately equal to a definite integral. Specifically, we have

$$\int_{x_0}^{x_n} \frac{dx}{f(x)} \approx \sum_{j=0}^{n-1} \frac{\Delta x_j}{f(x_j)} = n$$

Finally, using the modeling formula $f(x) = 0.767x^{0.439}$ along with $x_0 = 15$ and $x_n = 18,000$, we find that

$$\begin{aligned}
\int_{x_0}^{x_n} \frac{dx}{f(x)} &= \int_{15}^{18,000} \frac{dx}{0.767x^{0.439}} \\
&= \frac{1}{0.767} \left(\frac{x^{0.561}}{0.561} \right) \Big|_{15}^{18,000} \\
&= 2.324(18,000^{0.561} - 15^{0.561}) \\
&= 556.2
\end{aligned}$$

Thus, there are approximately 556 just noticeable steps in the audible range from 15 Hz to 18,000 Hz.

Here are some questions in which you are asked to apply these principles to issues involving both auditory and visual perception.

Questions

1. The 88 keys of a piano range from 15 Hz to 4,186 Hz. If the number of keys were based on the number of just noticeable differences, how many keys would a piano have?

2. An 8-bit gray-scale monitor can display 256 shades of gray. Let x represent the darkness of a shade of gray, where $x = 0$ for white and $x = 1$ for totally black. One model for gray-scale perception uses the formula $y = Ax^{0.3}$, where A is a positive constant and y is the smallest change detectable by the human eye at gray-level x. Experiments show that the human eye is incapable of distinguishing as many as 256 different shades of gray, so the number n of just noticeable shading differences from $x = 0$ to $x = 1$ must be less than 256. Using the assumption that $n < 256$, find a lower bound for the constant A in the modeling formula $y = Ax^{0.3}$.

3. One model of the ability of human vision to distinguish colors of different hue uses the formula $y = 2.9 \times 10^{-24}x^{8.52}$, where y is the just noticeable difference for a color of wavelength x, with both x and y measured in nanometers (nm).
   a. Blue-green light has a wavelength of 580 nm. What is the least noticeable difference at this wavelength?
   b. Red light has a wavelength of 760 nm. What is the least noticeable difference at this wavelength?
   c. How many just noticeable steps are there in hue from blue-green light to red light?

4. Find a model of the form $y = ax^k$ for just noticeable differences in hue for the color spectrum from blue-green light at 580 nm to violet light at 400 nm. Use the fact that the minimum noticeable difference at the wavelength of blue-green light is 1 nm, while at the wavelength of violet light, the minimum noticeable difference is 0.043 nm.

CHAPTER 6

This is a nuclear dump site in the Kola Peninsula of Russia. The long-term effects of accumulating nuclear waste can be studied using improper integration.

Additional Topics in Integration

Section 6.1   Integration by Parts; Integral Tables

Learning Objectives

1. Use integration by parts to find integrals and solve applied problems.
2. Examine and use a table of integrals.

Integration by parts is a technique of integration based on the product rule for differentiation. In particular, if $u(x)$ and $v(x)$ are both differentiable functions of x, then

$$\frac{d}{dx}[u(x)v(x)] = u(x)\frac{dv}{dx} + v(x)\frac{du}{dx}$$

so that

$$u(x)\frac{dv}{dx} = \frac{d}{dx}[u(x)v(x)] - v(x)\frac{du}{dx}$$

Integrating both sides of this equation with respect to x, we obtain

$$\int \left[u(x)\frac{dv}{dx}\right] dx = \int \frac{d}{dx}[u(x)v(x)] \, dx - \int \left[v(x)\frac{du}{dx}\right] dx$$

$$= u(x)v(x) - \int \left[v(x)\frac{du}{dx}\right] dx$$

since $u(x)v(x)$ is an antiderivative of $\dfrac{d}{dx}[u(x)v(x)]$. Moreover, we can write this integral formula in the more compact form

$$\int u \, dv = uv - \int v \, du$$

since

$$dv = \frac{dv}{dx}dx \qquad \text{and} \qquad du = \frac{du}{dx}dx$$

The equation $\displaystyle\int u \, dv = uv - \int v \, du$ is called the **integration by parts formula.**
The great value of this formula is that if we can find functions u and v so that a given integral $\displaystyle\int f(x) \, dx$ can be expressed in the form $\displaystyle\int f(x) \, dx = \int u \, dv$, then we have

$$\int f(x) \, dx = \int u \, dv = uv - \int v \, du$$

and the given integral is effectively exchanged for the integral $\displaystyle\int v \, du$. If the integral $\displaystyle\int v \, du$ is easier to compute than $\displaystyle\int u \, dv$, the exchange facilitates finding $\displaystyle\int f(x) \, dx$. Example 6.1.1 demonstrates how to use integration by parts.

EXAMPLE 6.1.1   Using Integration by Parts

Find $\displaystyle\int x^2 \ln x \, dx$.

Solution

Our strategy is to express $\int x^2 \ln x \, dx$ as $\int u \, dv$ by choosing u and v so that $\int v \, du$ is easier to evaluate than $\int u \, dv$. This strategy suggests that we choose

$$u = \ln x \qquad \text{and} \qquad dv = x^2 \, dx$$

since

$$du = \frac{1}{x} \, dx$$

is a simpler expression than $\ln x$, while v can be obtained by the relatively easy integration

$$v = \int x^2 \, dx = \frac{1}{3} x^3$$

(For simplicity, we leave the "+ C" out of the calculation until the final step.) Substituting this choice for u and v into the integration by parts formula, we obtain

$$\int x^2 \ln x \, dx = \int \underbrace{(\ln x)}_{u} \underbrace{(x^2 \, dx)}_{dv} = \underbrace{(\ln x)}_{u} \underbrace{\left(\frac{1}{3} x^3\right)}_{v} - \int \underbrace{\left(\frac{1}{3} x^3\right)}_{v} \underbrace{\left(\frac{1}{x} \, dx\right)}_{du}$$

$$= \frac{1}{3} x^3 \ln x - \frac{1}{3} \int x^2 \, dx = \frac{1}{3} x^3 \ln x - \frac{1}{3}\left(\frac{1}{3} x^3\right) + C$$

$$= \frac{1}{3} x^3 \ln x - \frac{1}{9} x^3 + C$$

Here is a summary of the procedure we have just illustrated.

Integration by Parts

To find an integral $\int f(x) \, dx$ using the integration by parts formula:

Step 1. Choose functions u and v so that $f(x) \, dx = u \, dv$. Try to pick u so that du is simpler than u and a dv that is easy to integrate.

Step 2. Organize the computation of du and v as

$$u \qquad\qquad dv$$

$$du \qquad v = \int dv$$

Step 3. Complete the integration by finding $\int v \, du$. Then

$$\int f(x) \, dx = \int u \, dv = uv - \int v \, du$$

Add "+ C" only at the end of the computation.

Choosing a suitable u and dv for integration by parts requires insight and experience. For instance, in Example 6.1.1, things would not have gone so smoothly if we had chosen $u = x^2$ and $dv = \ln x \, dx$. Certainly $du = 2x \, dx$ is simpler than $u = x^2$, but what is $v = \int \ln x \, dx$? In fact, finding this integral is just as hard as finding the original integral $\int x^2 \ln x \, dx$ (see Example 6.1.4). Examples 6.1.2 through 6.1.4 illustrate several ways of choosing u and dv in integrals that can be handled using integration by parts.

EXAMPLE 6.1.2   Using Integration by Parts

Find $\int x e^{2x} \, dx$.

Solution

Although both factors x and e^{2x} are easy to integrate, only x becomes simpler when differentiated. Therefore, we choose $u = x$ and $dv = e^{2x} \, dx$ and find

$$u = x \qquad dv = e^{2x} \, dx$$
$$du = dx \qquad v = \frac{1}{2} e^{2x}$$

Just-In-Time REVIEW

Recall from Section 5.1 that
$$\int e^{kx} \, dx = \frac{1}{k} e^{kx} + C$$

Substituting into the integration by parts formula, we obtain

$$\int \underbrace{x}_{u} \underbrace{(e^{2x} \, dx)}_{dv} = \underbrace{x}_{u} \underbrace{\left(\frac{1}{2} e^{2x}\right)}_{v} - \int \underbrace{\left(\frac{1}{2} e^{2x}\right)}_{v} \underbrace{dx}_{du}$$

$$= \frac{1}{2} x e^{2x} - \frac{1}{2}\left(\frac{1}{2} e^{2x}\right) + C \quad \text{factor out } \frac{1}{2} e^{2x}$$

$$= \frac{1}{2}\left(x - \frac{1}{2}\right) e^{2x} + C$$

EXAMPLE 6.1.3   Using Integration by Parts

Find $\int x \sqrt{x + 5} \, dx$.

Solution

Again, both factors x and $\sqrt{x + 5}$ are easy to differentiate and to integrate, but x is simplified by differentiation, while the derivative of $\sqrt{x + 5}$ is even more complicated than $\sqrt{x + 5}$ itself. This observation suggests that you choose

$$u = x \qquad dv = \sqrt{x + 5} \, dx = (x + 5)^{1/2} \, dx$$

so that

$$du = dx \qquad v = \frac{2}{3}(x+5)^{3/2}$$

Substituting into the integration by parts formula, you obtain

$$\int x(\sqrt{x+5}\,dx) = \underbrace{x}_{u}\underbrace{\left[\frac{2}{3}(x+5)^{3/2}\right]}_{dv} - \int \underbrace{\left[\frac{2}{3}(x+5)^{3/2}\right]}_{v}\underbrace{dx}_{du}$$

$$= \frac{2}{3}x\,(x+5)^{3/2} - \frac{2}{3}\left[\frac{2}{5}(x+5)^{5/2}\right] + C$$

$$= \frac{2}{3}x\,(x+5)^{3/2} - \frac{4}{15}(x+5)^{5/2} + C$$

NOTE Some integrals can be evaluated by either substitution or integration by parts. For instance, the integral in Example 6.1.3 can be found by substituting as follows:

Let $u = x + 5$. Then $du = dx$ and $x = u - 5$, and

$$\int x\sqrt{x+5}\,dx = \int (u-5)\sqrt{u}\,du = \int (u^{3/2} - 5u^{1/2})\,du$$

$$= \frac{u^{5/2}}{5/2} - \frac{5u^{3/2}}{3/2} + C$$

$$= \frac{2}{5}(x+5)^{5/2} - \frac{10}{3}(x+5)^{3/2} + C$$

This form of the integral is not the same as that found in Example 6.1.3. To show that the two forms are equivalent, note that the antiderivative in Example 6.1.3 can be expressed as

$$\frac{2x}{3}(x+5)^{3/2} - \frac{4}{15}(x+5)^{5/2} = (x+5)^{3/2}\left[\frac{2x}{3} - \frac{4}{15}(x+5)\right]$$

$$= (x+5)^{3/2}\left(\frac{2x}{5} - \frac{4}{3}\right) = (x+5)^{3/2}\left[\frac{2}{5}(x+5) - \frac{10}{3}\right]$$

$$= \frac{2}{5}(x+5)^{5/2} - \frac{10}{3}(x+5)^{3/2}$$

which is the form of the antiderivative obtained by substitution. This example shows that it is quite possible for you to do everything right and still not get the answer given at the back of the book. ∎

Definite Integration by Parts

The integration by parts formula can be applied to definite integrals by noting that

$$\int_a^b u\,dv = uv\Big|_a^b - \int_a^b v\,du$$

Definite integration by parts is used in Example 6.1.4 to find an area.

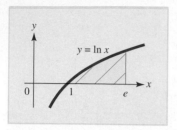

FIGURE 6.1   The region under $y = \ln x$ over $1 \leq x \leq e$.

EXAMPLE 6.1.4   Finding Area Using Integration by Parts

Find the area of the region bounded by the curve $y = \ln x$, the x axis, and the line $x = e$.

Solution

The region is shown in Figure 6.1. Since $\ln x \geq 0$ for $1 \leq x \leq e$, the area is given by the definite integral

$$A = \int_1^e \ln x \, dx$$

To evaluate this integral using integration by parts, think of $\ln x \, dx$ as $(\ln x)(1 \, dx)$ and use

$$u = \ln x \qquad dv = 1 \, dx$$
$$du = \frac{1}{x} \, dx \qquad v = \int 1 \, dx = x$$

Thus, the required area is

$$A = \int_1^e \ln x \, dx = x \ln x \Big|_1^e - \int_1^e x\left(\frac{1}{x} dx\right)$$
$$= x \ln x \Big|_1^e - \int_1^e 1 \, dx = (x \ln x - x)\Big|_1^e$$
$$= [e \ln e - e] - [1 \ln 1 - 1] \qquad \ln e = 1 \text{ and } \ln 1 = 0$$
$$= [e(1) - e] - [1(0) - 1]$$
$$= 1$$

As another illustration of the role played by integration by parts in applications, we use it in Example 6.1.5 to compute the future value of a continuous income stream (see Section 5.5).

EXAMPLE 6.1.5   Calculating Future Value of an Investment

Raza is considering a 5-year investment, and estimates that t years from now it will be generating a continuous income stream of $3,000 + 50t$ dollars per year. If the prevailing annual interest rate remains fixed at 4% compounded continuously during the entire 5-year term, what should her investment be worth in 5 years?

Solution

We measure the "worth" of Raza's investment by the future value of the income flow over the 5-year term. Recall (from Section 5.5) that an income stream deposited continuously at the rate $f(t)$ into an account that earns interest at an annual rate r compounded continuously for a term of T years has future value FV given by the integral

$$FV = \int_0^T f(t)e^{r(T-t)} \, dt$$

For this investment, we have $f(t) = 3{,}000 + 50t$, $r = 0.04$, and $T = 5$, so the future value is given by the integral

$$\text{FV} = \int_0^5 (3{,}000 + 50t)e^{0.04(5-t)}\,dt$$

Integrating by parts with

$$u = 3{,}000 + 50t \qquad dv = e^{0.04(5-t)}\,dt$$

$$du = 50\,dt \qquad\qquad v = \frac{e^{0.04(5-t)}}{-0.04} = -25e^{0.04(5-t)}$$

we get

$$\text{FV} = \left[(3{,}000 + 50t)(-25)e^{0.04(5-t)}\right]\Big|_0^5 - \int_0^5 (50)(-25)e^{0.04(5-t)}\,dt$$

$$= \left[(-75{,}000 - 1{,}250t)e^{0.04(5-t)}\right]\Big|_0^5 + 1{,}250\left[\frac{e^{0.04(5-t)}}{-0.04}\right]\Big|_0^5 \quad \text{combine terms}$$

$$= \left[(\ 106{,}250 - 1{,}250t)e^{0.04(5-t)}\right]\Big|_0^5$$

$$= [-106{,}250 - 1{,}250(5)]e^0 - [-106{,}250 - 1{,}250(0)]e^{0.04(5)}$$

$$\approx 17{,}274.04$$

Thus, in 5 years, Raza's investment will be worth roughly \$17,274.

Repeated Application of Integration by Parts

Sometimes integration by parts leads to a new integral that also must be integrated by parts. This situation is illustrated in Example 6.1.6.

EXAMPLE 6.1.6   Integrating by Parts Twice

Find $\displaystyle\int x^2 e^{2x}\,dx$.

Solution

Since the factor e^{2x} is easy to integrate and x^2 is simplified by differentiation, we choose

$$u = x^2 \qquad dv = e^{2x}\,dx$$

so that

$$du = 2x\,dx \qquad v = \int e^{2x}\,dx = \frac{1}{2}e^{2x}$$

Integrating by parts, we get

$$\int x^2 e^{2x}\,dx = x^2\left(\frac{1}{2}e^{2x}\right) - \int\left(\frac{1}{2}e^{2x}\right)(2x\,dx)$$

$$= \frac{1}{2}x^2 e^{2x} - \int x e^{2x}\,dx$$

The integral $\int xe^{2x}\,dx$ that remains can also be obtained using integration by parts. Indeed, in Example 6.1.2, we found that

$$\int xe^{2x}\,dx = \frac{1}{2}\left(x - \frac{1}{2}\right)e^{2x} + C$$

Thus,

$$\int x^2 e^{2x}\,dx = \frac{1}{2}x^2 e^{2x} - \int xe^{2x}\,dx$$

$$= \frac{1}{2}x^2 e^{2x} - \left[\frac{1}{2}\left(x - \frac{1}{2}\right)e^{2x}\right] + C$$

$$= \frac{1}{4}(2x^2 - 2x + 1)e^{2x} + C$$

Using Integral Tables

Most integrals you will encounter in the social, managerial, and life sciences can be evaluated using the basic formulas given in Section 5.1 along with substitution and integration by parts. However, occasionally you may encounter an integral that can not be handled by these methods. Some integrals, such as $\int \dfrac{e^x}{x}\,dx$, cannot be evaluated by any method, but others can be found by using a **table of integrals.**

Table 6.1, a short table of integrals,* is listed on pages 488 and 489. Note that the table is divided into sections such as "forms involving $u^2 - a^2$," and that formulas are given in terms of constants denoted a, b, and n. The use of this table is demonstrated in Examples 6.1.7 through 6.1.10.

EXAMPLE 6.1.7   Using a Table of Integrals

Find $\int \dfrac{1}{6 - 3x^2}\,dx$.

Solution

If the coefficient of x^2 were 1 instead of 3, you could use Formula 16. This suggests that you first rewrite the integrand as

$$\frac{1}{6 - 3x^2} = \frac{1}{3}\left(\frac{1}{2 - x^2}\right)$$

and then apply Formula 16 with $a = \sqrt{2}$:

$$\int \frac{1}{6 - 3x^2}\,dx = \frac{1}{3}\int \frac{1}{2 - x^2}\,dx$$

$$= \frac{1}{3}\left(\frac{1}{2\sqrt{2}}\right)\ln\left|\frac{\sqrt{2} + x}{\sqrt{2} - x}\right| + C$$

---

*A longer table of integrals can be found online, or in reference books, such as Murray R. Spiegel, *Mathematical Handbook of Formulas and Tables,* Schaum Outline Series, New York: McGraw-Hill.

EXAMPLE 6.1.8   **An Integral Not in the Table**

Find $\displaystyle\int \frac{1}{3x^2 + 6}\, dx$.

Solution

It is natural to try to match this integral to the one in Formula 16 by writing

$$\int \frac{1}{3x^2 + 6}\, dx = -\frac{1}{3}\int \frac{1}{-2 - x^2}\, dx$$

However, since -2 is negative, it cannot be written as the square a^2 of any real number a, so the formula does not apply.

EXAMPLE 6.1.9   **Using a Table of Integrals**

Find $\displaystyle\int \frac{1}{\sqrt{4x^2 - 9}}\, dx$.

Solution

To put this integral in the form of Formula 20, rewrite the integrand as

$$\frac{1}{\sqrt{4x^2 - 9}} = \frac{1}{\sqrt{4(x^2 - 9/4)}} = \frac{1}{2\sqrt{x^2 - 9/4}}$$

Then apply the formula with $a^2 = \dfrac{9}{4}$ to get

$$\int \frac{1}{\sqrt{4x^2 - 9}}\, dx = \frac{1}{2}\int \frac{1}{\sqrt{x^2 - 9/4}}\, dx = \frac{1}{2}\ln\left|x + \sqrt{x^2 - 9/4}\right| + C$$

A differential equation of the general form

$$\frac{dQ}{dt} = kQ(M - Q)$$

with constants k and M is called a **logistic equation,** and the graph of its solution $y = Q(t)$ is called a **logistic curve.** Logistic equations are used to model the spread of epidemics, populations whose growth is restricted, and various other quantities of interest. In Example 6.1.10, we use a formula from the integral table to solve a logistic equation involving the spread of a rumor.

EXAMPLE 6.1.10   **Using the Integral Table to Study the Spread of a Rumor**

At 6 A.M., two junior account executives at a brokerage firm hear a rumor that a new stock offering is to be presented at noon. The rumor spreads throughout the 26 junior executives in the firm at the rate

$$\frac{dN}{dt} = 0.025\, N(26 - N)$$

where $N(t)$ is the number of executives who have heard the rumor t hours after 6 A.M.

a. Find $N(t)$.

b. How many junior executives have not heard the rumor by noon?

TABLE 6.1 A Short Table of Integrals

Forms Involving $a + bu$

1. $\displaystyle\int \frac{u\,du}{a + bu} = \frac{1}{b^2}[a + bu - a\ln|a + bu|] + C$

2. $\displaystyle\int \frac{u^2\,du}{a + bu} = \frac{1}{2b^3}[(a + bu)^2 - 4a(a + bu) + 2a^2\ln|a + bu|] + C$

3. $\displaystyle\int \frac{u\,du}{(a + bu)^2} = \frac{1}{b^2}\left[\frac{a}{a + bu} + \ln|a + bu|\right] + C$

4. $\displaystyle\int \frac{u\,du}{\sqrt{a + bu}} = \frac{2}{3b^2}(bu - 2a)\sqrt{a + bu} + C$

5. $\displaystyle\int \frac{du}{u\sqrt{a + bu}} = \frac{1}{\sqrt{a}}\ln\left|\frac{\sqrt{a + bu} - \sqrt{a}}{\sqrt{a + bu} + \sqrt{a}}\right| + C \qquad a > 0$

6. $\displaystyle\int \frac{du}{u(a + bu)} = \frac{1}{a}\ln\left|\frac{u}{a + bu}\right| + C$

7. $\displaystyle\int \frac{du}{u^2(a + bu)} = \frac{-1}{a}\left[\frac{1}{u} + \frac{b}{a}\ln\left|\frac{u}{a + bu}\right|\right] + C$

8. $\displaystyle\int \frac{du}{u^2(a + bu)^2} = \frac{-1}{a^2}\left[\frac{a + 2bu}{u(a + bu)} + \frac{2b}{a}\ln\left|\frac{u}{a + bu}\right|\right] + C$

Forms Involving $a^2 + u^2$

9. $\displaystyle\int \sqrt{a^2 + u^2}\,du = \frac{u}{2}\sqrt{a^2 + u^2} + \frac{a^2}{2}\ln|u + \sqrt{a^2 + u^2}| + C$

10. $\displaystyle\int \frac{du}{\sqrt{a^2 + u^2}} = \ln|u + \sqrt{a^2 + u^2}| + C$

11. $\displaystyle\int \frac{du}{u\sqrt{a^2 + u^2}} = \frac{-1}{a}\ln\left|\frac{\sqrt{a^2 + u^2} + a}{u}\right| + C$

12. $\displaystyle\int \frac{du}{(a^2 + u^2)^{3/2}} = \frac{u}{a^2\sqrt{a^2 + u^2}} + C$

13. $\displaystyle\int u^2\sqrt{a^2 + u^2}\,du = \frac{u}{8}(a^2 + 2u^2)\sqrt{a^2 + u^2} - \frac{a^4}{8}\ln|u + \sqrt{a^2 + u^2}| + C$

Forms Involving $a^2 - u^2$

14. $\displaystyle \int \frac{du}{u\sqrt{a^2 - u^2}} = \frac{-1}{a}\ln\left|\frac{a + \sqrt{a^2 - u^2}}{u}\right| + C$

15. $\displaystyle \int \frac{du}{u^2\sqrt{a^2 - u^2}} = -\frac{\sqrt{a^2 - u^2}}{a^2 u} + C$

16. $\displaystyle \int \frac{du}{a^2 - u^2} = \frac{1}{2a}\ln\left|\frac{a + u}{a - u}\right| + C$

17. $\displaystyle \int \frac{\sqrt{a^2 - u^2}}{u}\,du = \sqrt{a^2 - u^2} - a\ln\left|\frac{a + \sqrt{a^2 - u^2}}{u}\right| + C$

Forms Involving $u^2 - a^2$

18. $\displaystyle \int \sqrt{u^2 - a^2}\,du = \frac{u}{2}\sqrt{u^2 - a^2} - \frac{a^2}{2}\ln\left|u + \sqrt{u^2 - a^2}\right| + C$

19. $\displaystyle \int \frac{\sqrt{u^2 - a^2}}{u^2}\,du = \frac{-\sqrt{u^2 - a^2}}{u} + \ln\left|u + \sqrt{u^2 - a^2}\right| + C$

20. $\displaystyle \int \frac{du}{\sqrt{u^2 - a^2}} - \ln\left|u + \sqrt{u^2 - a^2}\right| + C$

21. $\displaystyle \int \frac{du}{u^2\sqrt{u^2 - a^2}} = \frac{\sqrt{u^2 - a^2}}{a^2 u} + C$

Forms Involving e^{au} and $\ln u$

22. $\displaystyle \int u e^{au}\,du = \frac{1}{a^2}(au - 1)e^{au} + C$

23. $\displaystyle \int \ln u\,du = u\ln u - u + C$

24. $\displaystyle \int \frac{du}{u\ln u} = \ln|\ln u| + C$

25. $\displaystyle \int u^m \ln u\,du = \frac{u^{m+1}}{m + 1}\left(\ln u - \frac{1}{m + 1}\right) \qquad m \neq -1$

Reduction Formulas

26. $\displaystyle \int u^n e^{au}\,du = \frac{1}{a}u^n e^{au} - \frac{n}{a}\int u^{n-1}e^{au}\,du$

27. $\displaystyle \int (\ln u)^n\,du = u(\ln u)^n - n\int (\ln u)^{n-1}\,du$

28. $\displaystyle \int u^n\sqrt{a + bu}\,du = \frac{2}{b(2n + 3)}\left[u^n(a + bu)^{3/2} - na\int u^{n-1}\sqrt{a + bu}\,du\right] \qquad n \neq -\frac{3}{2}$

Solution

a. Separate the variables in the differential equation, and integrate to get

$$\int \frac{dN}{N(26 - N)} = \int 0.025 \, dt$$
integral table formula 6
with $u = N$, $a = 26$, $b = -1$

$$\frac{1}{26} \ln \left| \frac{N}{26 - N} \right| = 0.025t + C_1$$
multiply both sides by 26
$|26 - N| = 26 - N$ since $N \leq 26$

$$\ln \left(\frac{N}{26 - N} \right) = 0.65t + 26C_1$$
take exponentials on both sides;
$e^{26C_1} = C$

$$\frac{N}{26 - N} = e^{0.65t} e^{26C_1} = Ce^{0.65t}$$
multiply both sides by $26 - N$

$$N = (26 - N)Ce^{0.65t}$$
solve for N

$$(1 + Ce^{0.65t})N = 26Ce^{0.65t}$$

so that

$$N(t) = \frac{26Ce^{0.65t}}{1 + Ce^{0.65t}}$$

To evaluate the constant C, we use the fact that $N(0) = 2$ junior executives initially know the rumor, so

$$N(0) = 2 = \frac{26Ce^0}{1 + Ce^0}$$
multiply both sides by $1 + C$

$$2 + 2C = 26C$$
solve for C

$$C = \frac{1}{12}$$
substitute into the formula
for $N(t)$ and simplify

and

$$N(t) = \frac{26\left(\frac{1}{12}\right)e^{0.65t}}{1 + \left(\frac{1}{12}\right)e^{0.65t}} = \frac{26e^{0.65t}}{12 + e^{0.65t}}$$

b. At noon ($t = 6$),

$$N(6) = \frac{26e^{0.65(6)}}{12 + e^{0.65(6)}} \approx 21$$

junior executives have heard the rumor, so $26 - 21 = 5$ have not.

EXERCISES ▪ 6.1

In Exercises 1 through 26, use integration by parts to find the given integral.

1. $\displaystyle\int xe^{-x} \, dx$

2. $\displaystyle\int xe^{x/2} \, dx$

3. $\displaystyle\int (1 - x)e^x \, dx$

4. $\displaystyle\int (3 - 2x)e^{-x} \, dx$

5. $\displaystyle\int t \ln 2t \, dt$

6. $\displaystyle\int t \ln t^2 \, dt$

7. $\displaystyle\int ve^{-v/5} \, dv$

8. $\displaystyle\int we^{0.1w} \, dw$

9. $\displaystyle\int x\sqrt{x - 6} \, dx$

10. $\displaystyle\int x\sqrt{1 - x} \, dx$

11. $\displaystyle\int x(x + 1)^8 \, dx$

12. $\displaystyle\int (x + 1)(x + 2)^6 \, dx$

13. $\displaystyle\int \frac{x}{\sqrt{x + 2}} \, dx$

14. $\displaystyle\int \frac{x}{\sqrt{2x + 1}} \, dx$

15. $\displaystyle\int_{-1}^{4} \frac{x}{\sqrt{x + 5}} \, dx$

16. $\displaystyle\int_{0}^{2} \frac{x}{\sqrt{4x + 1}} \, dx$

17. $\displaystyle\int_{0}^{1} \frac{x}{e^{2x}} \, dx$

18. $\displaystyle\int_{1}^{e} \frac{\ln x}{x^2} \, dx$

19. $\int_{1}^{e^2} x \ln \sqrt[3]{x}\, dx$ **20.** $\int_{0}^{1} x(e^{-2x} + e^{-x})\, dx$

21. $\int_{1/2}^{e/2} t \ln 2t\, dt$ **22.** $\int_{1}^{2} (t - 1)e^{1 - t}\, dt$

23. $\int \dfrac{\ln x}{x^2}\, dx$ **24.** $\int x(\ln x)^2\, dx$

25. $\int x^3 e^{x^2}\, dx$ [*Hint:* Use $dv = xe^{x^2}\, dx$.]

26. $\int \dfrac{x^3}{\sqrt{x^2 + 1}}\, dx$ $\left[\textit{Hint:}\ \text{Use}\ dv = \dfrac{x}{\sqrt{x^2 + 1}}\, dx. \right]$

Use the table of integrals (Table 6.1) to find the integrals in Exercises 27 through 38.

27. $\int \dfrac{x\, dx}{3 - 5x}$ **28.** $\int \sqrt{x^2 - 9}\, dx$

29. $\int \dfrac{\sqrt{4x^2 - 9}}{x^2}\, dx$ **30.** $\int \dfrac{dx}{(9 + 2x^2)^{3/2}}$

31. $\int \dfrac{dx}{x(2 + 3x)}$ **32.** $\int \dfrac{t\, dt}{\sqrt{4 - 5t}}$

33. $\int \dfrac{du}{16 - 3u^2}$ **34.** $\int we^{-3w}\, dw$

35. $\int (\ln x)^3\, dx$ **36.** $\int x^2\sqrt{2 + 5x}\, dx$

37. $\int \dfrac{dx}{x^2(5 + 2x)^2}$ **38.** $\int \dfrac{\sqrt{9 - x^2}}{x}\, dx$

In Exercises 39 through 42, solve the given initial value problem for $y = f(x)$. Note that Exercises 41 and 42 involve separable differential equations.

39. $\dfrac{dy}{dx} = xe^{-2x}$, where $y = 0$ when $x = 0$

40. $\dfrac{dy}{dx} = x^2 \ln x$, where $y = 0$ when $x = 1$

41. $\dfrac{dy}{dx} = \dfrac{xy}{\sqrt{x + 1}}$, where $y = 1$ when $x = 0$

42. $\dfrac{dy}{dx} = xye^{x/2}$, where $y = 1$ when $x = 0$

43. Find the function whose tangent line has slope $(x + 1)e^{-x}$ for each value of x and whose graph passes through the point $(1, 5)$.

44. Find the function whose tangent line has slope $x \ln\sqrt{x}$ for each value of $x > 0$ and whose graph passes through the point $(2, -3)$.

BUSINESS AND ECONOMICS APPLIED PROBLEMS

45. EFFICIENCY   After t hours on the job, a factory worker can produce $100te^{-0.5t}$ units per hour. How many units does the worker produce during the first 3 hours?

46. MARGINAL COST   A manufacturer has found that marginal cost is $(0.1q + 1)e^{0.03q}$ dollars per unit when q units have been produced. The total cost of producing 10 units is $200. What is the total cost of producing the first 20 units?

Exercises 47 through 55 involve applications developed in Sections 5.4 and 5.5.

47. FUTURE VALUE OF AN INVESTMENT
Money is transferred into an account at the rate of $R(t) = 3,000 + 5t$ dollars per year for 10 years, where t is the number of years after 2000. If the account pays 5% interest compounded continuously, how much will be in the account at the end of the 10-year investment period (in 2010)?

48. FUTURE VALUE OF AN INVESTMENT
Money is transferred into an account at the rate of $R(t) = 1,000te^{-0.3t}$ dollars per year for 5 years. If the account pays 4% interest compounded continuously, how much will accumulate in the account over a 5-year period?

49. PRESENT VALUE OF AN INVESTMENT   An investment will generate income continuously at the rate of $R(t) = 20 + 3t$ hundred dollars per year for 5 years. If the prevailing interest rate is 7% compounded continuously, what is the present value of the investment?

50. INVESTMENT EVALUATION   Joel Evans, the manager of a national chain of pizza parlors, is selling a 6-year franchise to operate its newest outlet in Orlando, Florida. Experience in similar localities suggests that t years from now, the franchise will be generating profit continuously at the rate of $R(t) = 300 + 5t$ thousand dollars per year. If the prevailing rate of interest remains fixed during the next 6 years at an annual rate of 6% compounded continuously, what would be a fair price for Joel to charge for the franchise? [*Hint:* Use present value to measure what the franchise is worth.]

51. CONSUMERS' SURPLUS   A manufacturer has determined that when q thousand units of a particular commodity are produced, the price at which all the units can be sold is $p = D(q)$ dollars per unit, where D is the demand function

$$D(q) = 10 - qe^{0.02q}$$

a. At what price are 5,000 ($q_0 = 5$) units demanded?
b. Find the consumers' surplus when 5,000 units are demanded.

52. **CONSUMERS' SURPLUS** Answer the questions in Exercise 51 for the demand function

$$D(q) = \ln\left(\frac{52}{q + 1}\right)$$

when 12,000 ($q_0 = 12$) units are demanded.

53. **LORENZ CURVE** Find the Gini index for an income distribution whose Lorenz curve is the graph of the function $L(x) = xe^{x-1}$ for $0 \le x \le 1$.

54. **COMPARING INCOME DISTRIBUTIONS** In a certain state, the Lorenz curves for the distributions of income for lawyers and engineers are $y = L_1(x)$ and $y = L_2(x)$, respectively, where
$L_1(x) = 0.6x^2 + 0.4x$ and $L_2(x) = x^2e^{x-1}$
Find the Gini index for each curve. Which profession has the more equitable distribution of income?

55. **AVERAGE PROFIT** A manufacturer determines that when x hundred units of a particular commodity are produced, the profit generated is $P(x)$ thousand dollars, where

$$P(x) = \frac{500 \ln(x + 1)}{(x + 1)^2}$$

What is the average profit over the production range $0 \le x \le 10$?

LIFE AND SOCIAL SCIENCE APPLIED PROBLEMS

56. **POPULATION GROWTH** The population $P(t)$ (thousands) of a bacterial colony t hours after the introduction of a toxin is changing at the rate $P'(t) = (1 - 0.5t)e^{0.5t}$ thousand bacteria per hour. By how much does the population change during the fourth hour?

57. **POPULATION GROWTH** It is projected that t years from now the population $P(t)$ of a certain city will be changing at the rate of $P'(t) = t \ln \sqrt{t + 1}$ thousand people per year. If the current population is 2 million, what will the population be 5 years from now?

58. **FUND-RAISING** After t weeks, contributions in response to a local fund-raising campaign were coming in at the rate of $2{,}000te^{-0.2t}$ dollars per week. How much money was raised during the first 5 weeks?

Exercises 59 through 61 involve applications developed in Sections 5.4 and 5.6.

59. **AVERAGE DRUG CONCENTRATION** The concentration of a drug t hours after injection into a patient's bloodstream is $C(t) = 4te^{(2-0.3t)}$ mg/mL. What is the average concentration of drug in the patient's bloodstream over the first 6 hours after the injection?

60. **SPREAD OF DISEASE** A new virus has just been declared an epidemic by health officials. Currently, 10,000 people have the disease and it is estimated that t days from now new cases will be reported at the rate of $R(t) = 10te^{-0.1t}$ people per day. If the fraction of victims who still have the virus t days after first contracting it is given by $S(t) = e^{-0.015t}$, how many people will be infected by the virus 90 days from now? One year (365 days) from now? [*Hint:* Think of this as a survival/renewal problem.]

61. **GROUP MEMBERSHIP** In her role as editor of a new national book club, Denise Briggs has compiled statistics suggesting that the fraction of its members who are still active t months after joining is given by the function $S(t) = e^{-0.02t}$. The club currently has 5,000 members, and Denise expects new members to join at the rate of $R(t) = 5t$ per month. How many members should Denise expect to be active in the club 9 months from now? [*Hint:* Think of this as a survival/renewal problem.]

62. **CORRUPTION IN GOVERNMENT** The number of people implicated in a major government scandal increases at a rate jointly proportional to the number of people $G(t)$ already implicated and the number involved who have not yet been implicated, so

$$\frac{dG}{dt} = kG(M - G)$$

where M is the total number of people involved in the scandal. Suppose that 7 people are implicated when a Washington newspaper makes the scandal public, 9 more are implicated over the next 3 months, and a total of 28 people are implicated at the end of 6 months.
a. Use a formula from the integral table to find $G(t)$.
b. Approximately how many people are involved in the scandal?

MISCELLANEOUS PROBLEMS

63. SPY STORY After escaping from several angry camel owners (Exercise 73, Section 5.1), the spy learns that his enemy, Scelerat, is holed up in a chateau in the Alps. Disguised as an old duck plucker, the spy enters a nearby village to gather information. On the day he arrives, his identity is known only to the Redselig twins, Hans and Franz, but the next day, Hans' girlfriend, Blabba, finds out. Soon, word about the spy's true identity begins to spread among the 60 villagers at a rate jointly proportional to the number of people who know, $N(t)$, and the number who do not, so that

$$\frac{dN}{dt} = kN(60 - N)$$

He needs a week to gather enough information before attacking the chateau, but figures that as soon as 20 or more villagers know his identity, Scelerat will find out, too. Does our hero complete his mission or is he a dead duck plucker?

64. DISTANCE After t seconds, an object is moving with velocity $te^{-t/2}$ meters per second. Express the position of the object as a function of time.

CENTER OF A REGION *Let $f(x)$ be a continuous function with $f(x) \geq 0$ for $a \leq x \leq b$. Let R be the region with area A that is bounded by the curve $y = f(x)$, the x axis, and the lines $x = a$ and $x = b$. Then the* **centroid** *(or center) of R is the point $(\bar{x}, \bar{y})$, where*

$$\bar{x} = \frac{1}{A} \int_a^b xf(x)\, dx \qquad \text{and} \qquad \bar{y} = \frac{1}{2A} \int_a^b [f(x)]^2\, dx$$

In Exercises 65 and 66, find the centroid of the region shown. You may need to use one or more formulas from Table 6.1.

65.

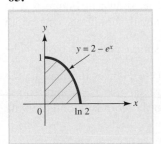

EXERCISE 65

66.

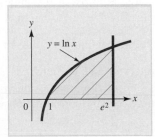

EXERCISE 66

67. SHOPPING MALL SECURITY The shaded region shown in the accompanying figure is a parking lot for a shopping mall. The dimensions are in hundreds of feet. To improve parking

security, the mall manager plans to place a surveillance kiosk in the center of the lot (see the definition preceding Exercises 65 and 66).

a. Where would you place the kiosk? Describe the location you choose in terms of the coordinate system indicated in the figure. [*Hint:* You may require a formula from Table 6.1.]

b. Write a paragraph on the security of shopping mall parking lots. In particular, do you think geometric centrality would be the prime consideration in locating a security kiosk or are there other, more important issues?

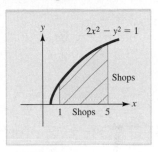

EXERCISE 67

68. Use integration by parts to verify reduction Formula 27:

$$\int (\ln u)^n\, du = u(\ln u)^n - n \int (\ln u)^{n-1}\, du$$

69. Use integration by parts to verify reduction Formula 26:

$$\int u^n e^{au}\, du = \frac{1}{a} u^n e^{au} - \frac{n}{a} \int u^{n-1} e^{au}\, du$$

70. Find a reduction formula for

$$\int u^n (\ln u)^m\, du$$

71. Find a reduction formula for

$$\int \frac{e^{kx}}{x^n}\, dx$$

72. Use the graphing utility of your calculator to draw the graphs of the curves $y = x^2 e^{-x}$ and

$$y = \frac{1}{x}$$ on the same screen. Use **ZOOM** and

TRACE or some other feature of your calculator to find where the curves intersect, and then compute the area of the region bounded by the curves.

73. Repeat Exercise 72 for the curves

$$\frac{x^2}{5} - \frac{y^2}{2} = 1 \qquad \text{and} \qquad y = x^3 - 3.5x^2 + 2x$$

74. Repeat Exercise 72 for the curves
$$y = \ln x \quad \text{and} \quad y = x^2 - 5x + 4$$

75. Repeat Exercise 72 for the curves
$$y = e^{2x} + 4 \quad \text{and} \quad y = 5e^x$$

If your calculator has a numeric integration feature, use it to evaluate the integrals in Exercises 76 through 79. In each case, verify your result by applying an appropriate integration formula from Table 6.1.

76. $\int_1^2 x^2 \ln \sqrt{x} \, dx$

77. $\int_2^3 \sqrt{4x^2 - 7} \, dx$

78. $\int_0^1 x^3 \sqrt{4 + 5x} \, dx$

79. $\int_0^1 \dfrac{\sqrt{x^2 + 2x}}{(x + 1)^2} \, dx$

80. Use the integral table to find $\int \dfrac{1}{x(3x - 6)} \, dx$.
Then store
$$f(x) = \frac{1}{x(3x - 6)}$$
into Y1 and
$$F(x) = -\frac{1}{6} \ln\left(\text{abs}\left(\frac{x}{3x - 6} \right) \right)$$
into Y2 using the bold graphing style. Graph both functions using the modified window $[-3.7, 5.7]1$ by $[-2, 2]1$. Verify that $F'(x) = f(x)$ for $x = (-2, 1)$.

SECTION 6.2   Numerical Integration

Learning Objectives

1. Explore the trapezoidal rule and Simpson's rule for numerical integration.
2. Use error bounds for numerical integration.
3. Interpret data using numerical integration.

In this section, you will see some techniques you can use to approximate definite integrals. Numerical methods such as these are needed when the function to be integrated does not have an elementary antiderivative. For instance, neither $\sqrt{x^3 + 1}$ nor $\dfrac{e^x}{x}$ has an elementary antiderivative.

Approximation by Rectangles

If $f(x)$ is positive on the interval $a \le x \le b$, the definite integral $\int_a^b f(x) \, dx$ is equal to the area under the graph of f between $x = a$ and $x = b$. As you saw in Section 5.3, one way to approximate this area is to use n rectangles, as shown in Figure 6.2. In particular, you divide the interval $a \le x \le b$ into n equal subintervals of width $\Delta x = \dfrac{b - a}{n}$ and let x_j denote the beginning of the jth subinterval. The base of the jth rectangle is the jth subinterval, and its height is $f(x_j)$. Hence, the area of the jth rectangle is $f(x_j)\Delta x$. The sum of the areas of all n rectangles is an approximation to the area under the curve, so an approximation to the corresponding definite integral is

$$\int_a^b f(x) \, dx \approx f(x_1)\Delta x + \cdots + f(x_n)\Delta x$$

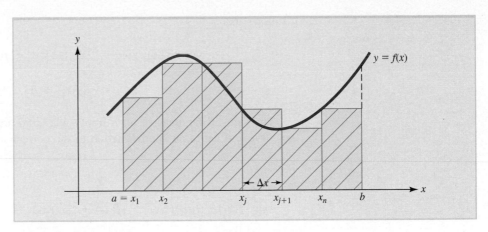

FIGURE 6.2 Approximation by rectangles.

This approximation improves as the number of rectangles increases, and you can estimate the integral to any desired degree of accuracy by making n large enough. However, since fairly large values of n are usually required to achieve reasonable accuracy, approximation by rectangles is rarely used in practice.

Approximation by Trapezoids

The accuracy of the approximation improves significantly if trapezoids are used instead of rectangles. Figure 6.3a shows the area from Figure 6.2 approximated by n trapezoids. Notice how much better the approximation is in this case.

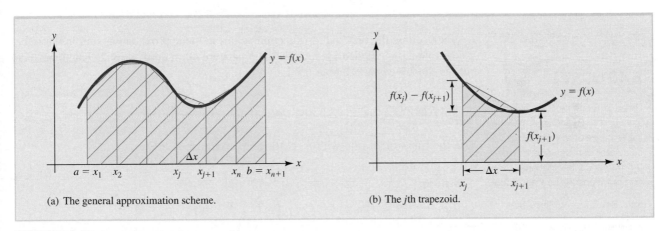

(a) The general approximation scheme.   (b) The jth trapezoid.

FIGURE 6.3 Approximation by trapezoids.

The jth trapezoid is shown in greater detail in Figure 6.3b. Notice that it consists of a rectangle with a right triangle on top of it. Since

$$\text{Area of rectangle} = f(x_{j+1})\Delta x$$

and

$$\text{Area of triangle} = \frac{1}{2}[f(x_j) - f(x_{j+1})]\Delta x$$

it follows that

$$\text{Area of } j\text{th trapezoid} = f(x_{j+1})\Delta x + \frac{1}{2}[f(x_j) - f(x_{j+1})]\Delta x$$

$$= \frac{1}{2}[f(x_j) + f(x_{j+1})]\Delta x$$

The sum of the areas of all n trapezoids is an approximation to the area under the curve and hence an approximation to the corresponding definite integral. Thus,

$$\int_a^b f(x)\,dx$$

$$\approx \frac{1}{2}[f(x_1) + f(x_2)]\Delta x + \frac{1}{2}[f(x_2) + f(x_3)]\Delta x + \cdots + \frac{1}{2}[f(x_n) + f(x_{n+1})]\Delta x$$

$$= \frac{\Delta x}{2}[f(x_1) + 2f(x_2) + \cdots + 2f(x_n) + f(x_{n+1})]$$

This approximation formula is known as the **trapezoidal rule** and applies even if the function f is not positive.

The Trapezoidal Rule

$$\int_a^b f(x)\,dx \approx \frac{\Delta x}{2}[f(x_1) + 2f(x_2) + \cdots + 2f(x_n) + f(x_{n+1})]$$

Notice that the first and last function values in the approximating sum in the trapezoidal rule are multiplied by 1, while the others are all multiplied by 2. The trapezoidal rule is illustrated in Example 6.2.1.

EXPLORE!

Refer to Example 6.2.1, where $a = 1$, $b = 2$, and $n = 10$. The list features of the graphing calculator can be used to aid the numerical integration computations in the trapezoidal rule. Set $Y1 = \frac{1}{x}$. Place the x values 1.0, 1.1, . . . , 1.9, 2.0 into L1, and the trapezoidal coefficients 1, 2, . . . , 2, 1 into L2. Write L3 = Y1(L1)∗L2∗H/2 where $H = \frac{b-a}{n}$. Confirm the result obtained in Example 6.2.1. See the Explore! Update at the end of this chapter for more details.

EXAMPLE 6.2.1 Using the Trapezoidal Rule

Use the trapezoidal rule with $n = 10$ to approximate $\int_1^2 \frac{1}{x}\,dx$.

Solution

Since
$$\Delta x = \frac{2-1}{10} = 0.1$$

the interval $1 \leq x \leq 2$ is divided into 10 subintervals by the points

$$x_1 = 1,\, x_2 = 1.1,\, x_3 = 1.2, \ldots,\, x_{10} = 1.9,\, x_{11} = 2$$

as shown in Figure 6.4.

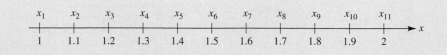

FIGURE 6.4 Division of the interval $1 \leq x \leq 2$ into 10 subintervals.

Then, by applying the trapezoidal rule for $f(x) = \dfrac{1}{x}$ and x_1 through x_{11}, we get

$$\int_1^2 \frac{1}{x}\,dx \approx \frac{0.1}{2}\left(\frac{1}{1} + \frac{2}{1.1} + \frac{2}{1.2} + \frac{2}{1.3} + \frac{2}{1.4} + \frac{2}{1.5} + \frac{2}{1.6} + \frac{2}{1.7} + \frac{2}{1.8} + \frac{2}{1.9} + \frac{1}{2}\right)$$

$$\approx 0.693771$$

The definite integral in Example 6.2.1 can be evaluated directly:

$$\int_1^2 \frac{1}{x}\,dx = \ln|x|\,\Big|_1^2 = \ln 2 \approx 0.693147$$

Thus, the approximation of this particular integral by the trapezoidal rule with $n = 10$ is accurate (after round-off) to two decimal places.

Accuracy of the Trapezoidal Rule

The difference between the true value of the integral $\int_a^b f(x)\,dx$ and the approximation generated by the trapezoidal rule when n subintervals are used is denoted by E_n. Here is an estimate for the absolute value of E_n that is proved in more advanced courses.

Error Estimate for the Trapezoidal Rule ■ If K is the maximum value of $|f''(x)|$ on the interval $a \le x \le b$, then

$$|E_n| \le \frac{K(b-a)^3}{12n^2}$$

The use of this formula is illustrated in Example 6.2.2.

EXAMPLE 6.2.2 Using Error Estimates for the Trapezoidal Rule

Estimate the accuracy of the approximation of $\int_1^2 \frac{1}{x}\,dx$ by the trapezoidal rule with $n = 10$.

Solution

Starting with $f(x) = \dfrac{1}{x}$, compute the derivatives

$$f'(x) = -\frac{1}{x^2} \qquad \text{and} \qquad f''(x) = \frac{2}{x^3}$$

and observe that the largest value of $|f''(x)|$ for $1 \le x \le 2$ is $|f''(1)| = 2$.
Apply the error formula with

$$K = 2 \qquad a = 1 \qquad b = 2 \qquad \text{and} \qquad n = 10$$

to get

$$|E_{10}| \le \frac{2(2-1)^3}{12(10)^2} \approx 0.00167$$

That is, the error in the approximation in Example 6.2.1 is guaranteed to be no greater than 0.00167. (In fact, to 5 decimal places, the error is 0.00062, as you can see by comparing the approximation obtained in Example 6.2.1 with the decimal representation of $\ln 2$.)

Just-In-Time REVIEW

Recall that $\dfrac{1}{x}$ can be differentiated using the power rule:

$$\frac{d}{dx}\left(\frac{1}{x^n}\right) = \frac{d}{dx}(x^{-n})$$

$$= -nx^{-n-1} = \frac{-n}{x^{n+1}}$$

It is important to keep the approximation error small when using the trapezoidal rule to estimate a definite integral. If you wish to achieve a certain degree of accuracy, you can determine how many intervals are required to attain your goal, as illustrated in Example 6.2.3.

EXAMPLE 6.2.3   Ensuring a Certain Level of Accuracy

How many subintervals are required to guarantee that the error will be less than 0.00005 in the approximation of $\int_1^2 \frac{1}{x}\,dx$ using the trapezoidal rule?

Solution

From Example 6.2.2 you know that $K = 2$, $a = 1$, and $b = 2$, so that

$$|E_n| \leq \frac{2(2-1)^3}{12n^2} = \frac{1}{6n^2}$$

The goal is to find the smallest positive integer n for which

$$\frac{1}{6n^2} < 0.00005$$

Equivalently,

$$n^2 > \frac{1}{6(0.00005)}$$

or

$$n > \sqrt{\frac{1}{6(0.00005)}} \approx 57.74$$

The smallest such integer is $n = 58$, so 58 subintervals are required to ensure the desired accuracy.

Approximation Using Parabolas: Simpson's Rule

The relatively large number of subintervals required in Example 6.2.3 to ensure accuracy to within 0.00005 suggests that approximation by trapezoids may not be efficient enough for some applications. There is another approximation formula, called **Simpson's rule**, which is no harder to use than the trapezoidal rule but often requires substantially fewer calculations to achieve a specified degree of accuracy. Like the trapezoidal rule, it is based on the approximation of the area under a curve by columns, but unlike the trapezoidal rule, it uses parabolic arcs rather than line segments at the top of the columns.

To be more specific, the approximation of a definite integral using parabolas is based on the following construction (illustrated in Figure 6.5 for $n = 6$). Divide the interval $a \leq x \leq b$ into an **even** number of subintervals so that adjacent subintervals can be paired with none left over. Approximate the portion of the graph that lies above the first pair of subintervals by the (unique) parabola that passes through the three points $(x_1, f(x_1))$, $(x_2, f(x_2))$, and $(x_3, f(x_3))$, and use the area under this parabola between x_1 and x_3 to approximate the corresponding area under the curve. Do the same for the remaining pairs of subintervals, and use the sum of the resulting areas to approximate the total area under the graph. Here is the approximation formula that results from this construction.

Simpson's Rule ■ For an even integer n,

$$\int_a^b f(x)\,dx \approx \frac{\Delta x}{3}[f(x_1) + 4f(x_2) + 2f(x_3) + 4f(x_4) + \cdots + 2f(x_{n-1}) + 4f(x_n) + f(x_{n+1})]$$

Notice that the first and last function values in the approximating sum in Simpson's rule are multiplied by 1, while the others are multiplied alternately by 4 and 2.

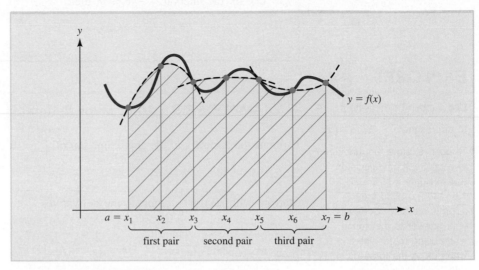

FIGURE 6.5 Approximation using parabolas.

The proof of Simpson's rule is based on the fact that the equation of a parabola is a polynomial of the form $y = Ax^2 + Bx + C$. For each pair of subintervals, the three given points are used to find the coefficients A, B, and C, and the resulting polynomial is then integrated to get the corresponding area. The details of the proof are straightforward but tedious and will be omitted.

EXPLORE!

Refer to Example 6.2.4, where $a = 1$, $b = 2$, and $n = 10$. The list feature of the graphing calculator can be used to facilitate the computations in Simpson's rule. Set $Y1 = \dfrac{1}{x}$. Place the x values, $x = 1.0$, $1.1, \ldots, 1.9, 2.0$, into list L1, and the Simpson's rule coefficients $1, 4, 2, \ldots, 4, 1$ into L2. For $H = \dfrac{b - a}{n}$, write L3 = Y1(L1)∗L2∗H/3. Confirm the result obtained in Example 6.2.4. Refer to the EXPLORE! box on page 496 for trapezoidal rule details.

EXAMPLE 6.2.4   Using Simpson's Rule

Use Simpson's rule with $n = 10$ to approximate $\displaystyle\int_1^2 \frac{1}{x}\,dx$.

Solution

As in Example 6.2.1, $\Delta x = 0.1$, and hence the interval $1 \leq x \leq 2$ is divided into the 10 subintervals by

$$x_1 = 1, x_2 = 1.1, x_3 = 1.2, \ldots, x_{10} = 1.9, x_{11} = 2$$

Then, by Simpson's rule,

$$\int_1^2 \frac{1}{x}\,dx \approx \frac{0.1}{3}\left(\frac{1}{1} + \frac{4}{1.1} + \frac{2}{1.2} + \frac{4}{1.3} + \frac{2}{1.4} + \frac{4}{1.5} + \frac{2}{1.6} + \frac{4}{1.7} + \frac{2}{1.8} + \frac{4}{1.9} + \frac{1}{2}\right)$$
$$\approx 0.693150$$

Notice that this is an excellent approximation to 5 decimal places of the true value, $\ln 2 = 0.693147\ldots$.

Accuracy of Simpson's Rule

The error estimate for Simpson's rule uses the 4th derivative $f^{(4)}(x)$ in much the same way the second derivative $f''(x)$ was used in the error estimate for the trapezoidal rule. Here is the estimation formula.

> **Error Estimate for Simpson's Rule** ∎ If M is the maximum value of $|f^{(4)}(x)|$ on the interval $a \le x \le b$, then
>
> $$|E_n| \le \frac{M(b-a)^5}{180n^4}$$

EXPLORE!

Use Simpson's rule with $n = 4$ to approximate $\int_0^2 (3x^2 + 1)\, dx$. To do this, store the function in your graphing calculator and evaluate it at 0, 0.5, 1.0, 1.5, and 2.0. Find the appropriate sum. Compare your answer to that found when using rectangles and when using the trapezoidal rule.

EXAMPLE 6.2.5 Using Error Estimates for Simpson's Rule

Estimate the accuracy of the approximation of $\int_1^2 \frac{1}{x}\, dx$ by Simpson's rule with $n = 10$.

Solution

Starting with $f(x) = \dfrac{1}{x}$, compute the derivatives

$$f'(x) = -\frac{1}{x^2} \qquad f''(x) = \frac{2}{x^3} \qquad f^{(3)}(x) = -\frac{6}{x^4} \qquad f^{(4)}(x) = \frac{24}{x^5}$$

and observe that the largest value of $|f^{(4)}(x)|$ on the interval $1 \le x \le 2$ is $|f^{(4)}(1)| = 24$.

Now apply the error formula with $M = 24$, $a = 1$, $b = 2$, and $n = 10$ to get

$$|E_{10}| \le \frac{24(2-1)^5}{180(10)^4} \approx 0.000013$$

Thus, the error in the approximation in Example 6.2.4 is guaranteed to be no greater than 0.000013.

In Example 6.2.6, the error estimate is used to determine the number of subintervals that are required to ensure a specified degree of accuracy.

EXAMPLE 6.2.6 Ensuring a Certain Level of Accuracy

How many subintervals are required to ensure accuracy to within 0.00005 in the approximation of $\int_1^2 \frac{1}{x}\, dx$ by Simpson's rule?

Solution

From Example 6.2.5 you know that $M = 24$, $a = 1$, and $b = 2$. Hence,

$$|E_n| \le \frac{24(2-1)^5}{180n^4} = \frac{2}{15n^4}$$

The goal is to find the smallest positive (even) integer for which

$$\frac{2}{15n^4} < 0.00005$$

Equivalently, $\qquad\qquad\qquad n^4 > \dfrac{2}{15(0.00005)}$

or
$$n > \left[\frac{2}{15(0.00005)}\right]^{1/4} \approx 7.19$$

The smallest such (even) integer is $n = 8$, so 8 subintervals are required to ensure the desired accuracy. Compare with the result of Example 6.2.3, where we found that 58 subintervals are required to ensure the same degree of accuracy using the trapezoidal rule.

Interpreting Data with Numerical Integration

Numerical integration can often be used to estimate a quantity $\int_a^b f(x)\,dx$ when all that is known about $f(x)$ is a set of experimentally determined data. Example 6.2.7 shows how to apply numerical integration to find an area.

EXAMPLE 6.2.7   Finding an Area Using Numerical Integration

Jack needs to know the area of his swimming pool to buy a pool cover, but this is difficult because of the pool's irregular shape. Suppose Jack makes the measurements shown in Figure 6.6 at 4-ft intervals along the base of the pool (all measurements are in feet). How can he use the trapezoidal rule to estimate the area?

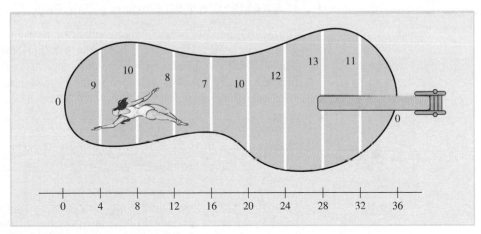

FIGURE 6.6   Measurements across a pool.

Solution

If Jack could find functions $f(x)$ for the far rim of the pool and $g(x)$ for the near rim, then the area would be given by the definite integral $A = \int_0^{36} [f(x) - g(x)]\,dx$. The irregular shape makes it impossible or at least impractical to find formulas for f and g, but Jack's measurements tell him that

$$f(0) - g(0) = 0 \qquad f(4) - g(4) = 9 \qquad f(8) - g(8) = 10 \ldots f(36) - g(36) = 0$$

Substituting this information into the trapezoidal rule approximation and using $\Delta x = \dfrac{36 - 0}{9} = 4$ (the width of each strip the pool was divided into), Jack obtains

$$A = \int_0^{36} [f(x) - g(x)]\,dx$$

$$= \frac{4}{2}[0 + 2(9) + 2(10) + 2(8) + 2(7) + 2(10) + 2(12) + 2(13) + 2(11) + 0]$$

$$= \frac{4}{2}(160) = 320$$

Thus, Jack estimates the area of the pool to be approximately 320 ft^2.

NOTE The curved shape of the pool in Example 6.2.7 suggests that using an approximation with parabolas might give a more accurate estimate of the area. Unfortunately, Simpson's rule cannot be used since the given measurements result in an odd number of subintervals. In Exercise 49, you are asked to use Simpson's rule to estimate the area of the same pool using a different set of measurements and an even number of subintervals. ■

EXAMPLE 6.2.8 Finding a Fair Price Using Numerical Integration

Carmen Chavez, the manager of a chain of pet supply stores, is planning to sell a 10-year franchise. Past records in similar localities suggest that t years from now the franchise will be generating income continuously at the rate of $f(t)$ thousand dollars per year, where $f(t)$ is as indicated in the following table for a typical decade:

Year t	0	1	2	3	4	5	6	7	8	9	10
Rate of income flow $f(t)$	510	580	610	625	654	670	642	610	590	573	550

If the prevailing rate of interest remains at 5% per year compounded continuously over the 10-year term, what would be a fair price for Carmen to charge for the franchise, based on the given information?

Solution

If the rate of income flow $f(t)$ were a continuous function, a fair price for the franchise might be determined by computing the present value of the income flow over the 10-year term. According to the formula developed in Section 5.5, this present value would be given by the definite integral

$$PV = \int_0^{10} f(t)e^{-0.05t}\,dt$$

since the prevailing rate of interest is 5% ($r = 0.05$). But we don't have such a continuous function $f(t)$, so we will use Simpson's rule with $n = 10$ and $\Delta t = 1$ to *estimate* the present value integral. We find that

$$PV = \int_0^{10} f(t)e^{-0.05t}\, dt$$

$$\approx \frac{\Delta t}{3}[f(0)e^{-0.05(0)} + 4f(1)e^{-0.05(1)} + 2f(2)e^{-0.05(2)} + \cdots + 4f(9)e^{-0.05(9)}$$

$$+ f(10)e^{-0.05(10)}]$$

$$\approx \frac{1}{3}[(510)e^{-0.05(0)} + 4(580)e^{-0.05(1)} + 2(610)e^{-0.05(2)} + 4(625)e^{-0.05(3)}$$

$$+ 2(654)e^{-0.05(4)} + 4(670)e^{-0.05(5)} + 2(642)e^{-0.05(6)} + 4(610)e^{-0.05(7)}$$

$$+ 2(590)e^{-0.05(8)} + 4(573)e^{-0.05(9)} + (550)e^{-0.05(10)}]$$

$$\approx \frac{1}{3}(14,387) \approx 4,796$$

Thus, the present value of the income stream over the 10-year term is approximately $4,796 thousand dollars ($4,796,000). The company may use this estimate as a fair asking price for the franchise.

EXERCISES ■ 6.2

In Exercises 1 through 14, approximate the given integral using (a) the trapezoidal rule and (b) Simpson's rule with the specified number of subintervals.

1. $\int_1^2 x^2\, dx;\ n = 4$

2. $\int_4^6 \frac{1}{\sqrt{x}}\, dx;\ n = 10$

3. $\int_0^1 \frac{1}{1 + x^2}\, dx;\ n = 4$

4. $\int_2^3 \frac{1}{x^2 - 1}\, dx;\ n = 4$

5. $\int_{-1}^0 \sqrt{1 + x^2}\, dx;\ n = 4$

6. $\int_0^3 \sqrt{9 - x^2}\, dx;\ n = 6$

7. $\int_0^1 e^{-x^2}\, dx;\ n = 4$

8. $\int_0^2 e^{x^2}\, dx;\ n = 10$

9. $\int_2^4 \frac{dx}{\ln x};\ n = 6$

10. $\int_1^2 \frac{\ln x}{x + 2}\, dx;\ n = 4$

11. $\int_0^1 \sqrt[3]{1 + x^2}\, dx;\ n = 4$

12. $\int_0^1 \frac{dx}{\sqrt{1 + x^3}};\ n = 6$

13. $\int_0^2 e^{-\sqrt{x}}\, dx;\ n = 8$

14. $\int_1^2 \frac{e^x}{x}\, dx;\ n = 4$

In Exercises 15 through 20, approximate the given integral and estimate the error $|E_n|$ using (a) the trapezoidal rule and (b) Simpson's rule with the specified number of subintervals.

15. $\int_1^2 \frac{1}{x^2}\, dx;\ n = 4$

16. $\int_0^2 x^3\, dx;\ n = 8$

17. $\int_1^3 \sqrt{x}\, dx;\ n = 10$

18. $\int_1^2 \ln x\, dx;\ n = 4$

19. $\int_0^1 e^{x^2}\, dx;\ n = 4$

20. $\int_0^{0.6} e^{x^3}\, dx;\ n = 6$

In Exercises 21 through 26, determine how many subintervals are required to guarantee accuracy to within 0.00005 in the approximation of the given integral by (a) the trapezoidal rule and (b) Simpson's rule.

21. $\int_1^3 \frac{1}{x}\, dx$

22. $\int_0^4 (x^4 + 2x^2 + 1)\, dx$

23. $\int_1^2 \frac{1}{\sqrt{x}}\, dx$

24. $\int_1^2 \ln(1 + x)\, dx$

25. $\int_{1.2}^{2.4} e^x\, dx$

26. $\int_0^2 e^{x^2}\, dx$

27. A quarter circle of radius 1 has the equation $y = \sqrt{1 - x^2}$ for $0 \le x \le 1$ and has area $\frac{\pi}{4}$.

Thus, $\int_0^1 \sqrt{1 - x^2}\, dx = \frac{\pi}{4}$. Use this formula to estimate π by applying:
a. The trapezoidal rule.
b. Simpson's rule.
In each case, use $n = 8$ subintervals.

28. Use the trapezoidal rule with $n = 8$ to estimate the area bounded by the curve $y = \sqrt{x^3 + 1}$, the x axis, and the lines $x = 0$ and $x = 1$.

29. Use the trapezoidal rule with $n = 10$ to estimate the average value of the function $f(x) = \dfrac{e^{-0.4x}}{x}$ over the interval $1 \le x \le 6$.

30. Use the trapezoidal rule with $n = 6$ to estimate the average value of the function $y = \sqrt{\ln x}$ over the interval $1 \le x \le 4$.

31. Use the trapezoidal rule with $n = 7$ to estimate the volume of the solid generated by rotating the region under the curve $y = \dfrac{x}{1 + x}$ between $x = 0$ and $x = 1$ about the x axis.

32. Use Simpson's rule with $n = 6$ to estimate the volume of the solid generated by rotating the region under the curve $y = \ln x$ between $x = 1$ and $x = 2$ about the x axis.

BUSINESS AND ECONOMICS APPLIED PROBLEMS

33. **FUTURE VALUE OF AN INVESTMENT**   An investment generates income continuously at the rate of $f(t) = \sqrt{t}$ thousand dollars per year at time t (years). If the prevailing rate of interest is 6% per year compounded continuously, use the trapezoidal rule with $n = 5$ to estimate the future value of the investment over a 10-year term. (See Example 5.5.2.)

34. **PRESENT VALUE OF A FRANCHISE**   The management of a national chain of fast-food restaurants is selling a 5-year franchise to operate its newest outlet in Tulare, California. Past experience in similar localities suggests that t years from now, the franchise will be generating profit at the rate of $f(t) = 12,000\sqrt{t}$ dollars per year. Suppose the prevailing annual interest rate remains fixed during the next 5 years at 5% compounded continuously. Use Simpson's rule with $n = 10$ to estimate the present value of the franchise.

35. **REVENUE FROM DEMAND DATA**   An economist studying the demand for a particular commodity gathers the data in the accompanying table, which lists the number of units q (in thousands) of the commodity that will be demanded (sold) at a price of p dollars per unit.

q (1,000 units)	0	4	8	12	16	20	24
p ($/unit)	49.12	42.90	31.32	19.83	13.87	10.58	7.25

Use this information together with Simpson's rule to estimate the total revenue

$$R = \int_0^{24} p(q)\, dq \quad \text{thousand dollars}$$

obtained as the level of production is increased from 0 to 24,000 units ($q = 0$ to $q = 24$).

36. **CONSUMERS' SURPLUS**   Jacob Lawrence is an economist who models the demand for a particular commodity by the function

$$p = D(q) = \frac{100}{q^2 + q + 1}$$

where q hundred units are sold when the price is p dollars per unit. He decides to use Simpson's rule with $n = 6$ to estimate the consumers' surplus for the commodity when 500 units ($q_0 = 5$) are manufactured and sold (see Example 5.5.5). What result does Jacob obtain?

37. **FUTURE VALUE OF AN INVESTMENT**   Marco has a small investment providing a variable income stream that is deposited continuously into an account earning interest at an annual rate of 4% compounded continuously. He spot-checks the monthly flow rate of the investment on the first day of every other month for a 1-year period, obtaining the results in this table:

Month	Jan.	Mar.	May	Jul.	Sep.	Nov.	Jan.
Rate of income flow	$437	$357	$615	$510	$415	$550	$593

For instance, income is entering the account at the rate of $615 per month on the first of May, but 2 months later, the rate of income flow is only $510 per month. Use this information together with Simpson's rule to estimate the future value of Marco's income flow during this 1-year period. [*Hint:* Recall Example 5.5.2.]

38. NET PROFIT FROM DATA   On the first day of each month, the manager of a small company estimates the rate at which profit is expected to increase during that month. The results are listed in the accompanying table for the first 6 months of the year, where $P'(t)$ is the rate of profit growth in thousands of dollars per month expected during the tth month ($t = 1$ for January, $t = 6$ for June). Use this information together with the trapezoidal rule to estimate the net profit earned by the company during this 6-month period (January through June).

t (month)	1	2	3	4	5	6
Rate of profit $P'(t)$	0.65	0.43	0.72	0.81	1.02	0.97

39. PRODUCERS' SURPLUS FROM DATA   An economist studying the supply for a particular commodity gathers the data in the accompanying table, which lists the number of units q (in thousands) of the commodity that will be supplied to the market by producers at a price of p dollars per unit. Use this information together with the trapezoidal rule to estimate the producers' surplus when 7,000 units are supplied ($q_0 = 7$).

q (1,000 units)	0	1	2	3	4	5	6	7
p (dollars per unit)	1.21	3.19	3.97	5.31	6.72	8.16	9.54	11.03

40. CONSUMERS' SURPLUS FROM DATA   An economist studying the demand for a particular commodity gathers the data in the accompanying table, which lists the number of units q (in thousands) of the commodity that will be demanded (sold) at a price of p dollars per unit. Use this information together with Simpson's rule to estimate the consumers' surplus when 24,000 units are produced, that is, when $q_0 = 24$.

q (1,000 units)	0	4	8	12	16	20	24
p (dollars per unit)	49.12	42.90	31.32	19.83	13.87	10.58	7.25

41. DISTRIBUTION OF INCOME   A sociologist studying the distribution of income in an industrial society compiles the data displayed in the accompanying table, where $L(x)$ denotes the fraction of the society's total wealth earned by the lowest-paid $100x\%$ of the wage-earners in the society. Use this information together with the trapezoidal rule to estimate the Gini index (GI) for this society; that is,

$$\text{GI} = 2 \int_0^1 [x - L(x)] \, dx$$

x	0	0.125	0.25	0.375	0.5	0.625	0.75	0.875	1
$L(x)$	0	0.0063	0.0631	0.1418	0.2305	0.3342	0.4713	0.6758	1

42. CONSTRUCTION   When excavation is being done for roads or buildings, the contractor will often pay nearby landowners to dump the excavated earth on their land, reducing the cost of hauling it away. One landowner, Dana Mays, is offered $0.20 per cubic foot to have dirt dumped to a depth of 2 feet over an oddly shaped portion of her land. The portion is 120 feet long, and Dana is provided with a survey map that has width measurements every 20 feet starting at the westernmost edge. The widths in feet are (going from west to east) 42, 61, 59, 54, 66, 88, and 54.

a. Use Simpson's rule to approximate the volume of dirt that will be dumped. [*Hint:* You can find volume by multiplying the area of the dump region and the depth.]

b. How much should Dana expect to be paid?

LIFE AND SOCIAL SCIENCE APPLIED PROBLEMS

43. SPREAD OF A DISEASE A new strain of influenza has just been declared an epidemic by health officials. Currently, 3,000 people have the disease and new victims are being added at the rate of $R(t) = 50\sqrt{t}$ people per week. Moreover, the fraction of infected people who still have the disease t weeks after contracting it is given by $S(t) = e^{-0.01t}$. Use Simpson's rule with $n = 8$ to estimate the number of people who have the flu 8 weeks from now. [*Hint:* Think of this as a survival/renewal problem, like Example 5.6.2.]

44. MENTAL HEALTH CARE A county mental health clinic has just opened. The clinic initially accepts 300 people for treatment and plans to accept new patients at the rate of 10 per month. Let $f(t)$ denote the fraction of people receiving treatment continuously for at least t days. For the first 60 days, records are kept and these values of $f(t)$ are obtained:

t (days)	0	5	10	15	20	25	30	35	40	45	50	55	60
$f(t)$	1	$\frac{3}{4}$	$\frac{3}{5}$	$\frac{1}{2}$	$\frac{1}{3}$	$\frac{3}{10}$	$\frac{1}{5}$	$\frac{1}{6}$	$\frac{1}{7}$	$\frac{1}{9}$	$\frac{1}{12}$	$\frac{1}{15}$	$\frac{1}{20}$

Use this information together with the trapezoidal rule to estimate the number of patients in the clinic at the end of the 60-day period. [*Hint:* This is a survival/renewal problem. Recall Example 5.6.1.]

45. POLLUTION CONTROL An industrial plant spills pollutant into a river. The pollutant spreads out as it is carried downstream by the current of the river, and 3 hours later, the spill forms the pattern shown in the accompanying figure. Measurements (in feet) across the spill are made at 5-foot intervals, as indicated in the figure. Use this information together with the trapezoidal rule to estimate the area of the spill.

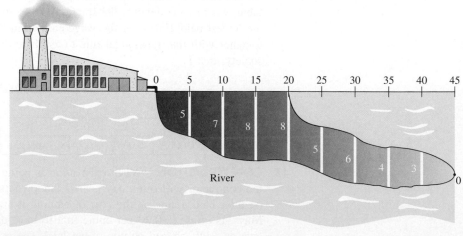

EXERCISE 45

46. **AVERAGE TEMPERATURE** One way to find the average temperature for a day is to take temperature readings at evenly spaced intervals for 24 hours and then find the average of those readings. A more sophisticated approach is to use Simpson's rule and the formula for finding the average value of a function from Section 5.4. The temperature readings for 1 day are provided in this table:

Time	12 (A.M.)	2	4	6	8	10	12 (P.M.)	2	4	6	8	10	12 (A.M.)
Temp (°F)	68	64	61	60	71	74	78	81	81	79	75	70	65

a. Approximate the average temperature for this day by finding the numeric average of the temperature readings. Round all answers to 2 decimal places.

b. Approximate the average temperature using Simpson's rule and the formula for finding the average value of a function. How do your answers compare? Which do you think is more representative of the average temperature? Explain.

c. Repeat parts (a) and (b), this time using only temperature readings taken every 6 hours starting at 12 A.M. Does this change your mind on which method you think is more effective?

47. **POPULATION DENSITY** A demographic study determines that the population density of a certain city at a distance of r miles from the city center is $D(r)$ people per square mile (mi^2), where D is as indicated in the following table for $0 \leq r \leq 10$ at 2-mile intervals.

Distance r (miles) from city center	0	2	4	6	8	10
Population density $D(r)$ (people/mi^2)	3,120	2,844	2,087	1,752	1,109	879

Use the trapezoidal rule to estimate the total population of the city that is located within a 10-mile radius of the city center. (See Example 5.6.3.)

48. **MORTALITY RATE FROM AIDS** The accompanying table gives the number of reported deaths due to AIDS, during the tth year after 1995, for the period 1995 to 2006. (Source: *Centers for Disease Control and Prevention, National Center for HIV, STD, and TB Prevention.*)

Year	t	Reported AIDS Deaths	Year	t	Reported AIDS Deaths
1995	0	51,670	2001	6	15,603
1996	1	38,396	2002	7	16,948
1997	2	22,245	2003	8	16,690
1998	3	18,823	2004	9	16,395
1999	4	18,249	2005	10	16,268
2000	5	16,672	2006	11	14,016

However, the table does not tell the complete story because many deaths actually due to AIDS go unreported or are attributed to other causes. Let $D(t)$ be the function that gives the cumulative number of AIDS deaths at time t. Then the data in the table can be thought of as rates of change of $D(t)$ at various times, that is, as mortality rates. For instance, the table tells us that in 1997 ($t = 2$), AIDS deaths were increasing at the rate of 22,245 per year.

a. Assuming that $D(t)$ is differentiable, explain why the total number of AIDS deaths N during the period 1995–2006 is given by the integral

$$N = \int_0^{11} D'(t)\, dt$$

b. Estimate N by using the data in the table together with the trapezoidal rule to approximate the integral in part (a).

c. Why is the mortality function $D(t)$ probably *not* differentiable? Does this invalidate the estimate of N made in part (b)? Write a paragraph either defending or challenging the procedure used to estimate N in this exercise.

49. **POOL AREA** Jack, the pool owner in Example 6.2.7, decides to use Simpson's rule to recalculate the surface area of his pool by making the following measurements across the pool in 6-ft intervals instead of the 4-ft intervals shown in Figure 6.6. What is the recalculated area?

Measurement point x_j	0	6	12	18	24	30	36
Width (in feet) of the pool at x_j	0	10	8	8	12	12	0

SECTION 6.3 Improper Integrals

Learning Objectives

1. Evaluate improper integrals with infinite limits of integration.
2. Use improper integrals in applied problems.

The definition of the definite integral $\int_a^b f(x)\, dx$, given in Section 5.3, requires the interval of integration $a \leq x \leq b$ to be bounded, but in certain applications, it is useful to consider integrals over *unbounded* intervals such as $x \geq a$. We will define such **improper integrals** in this section and examine a few properties and applications.

The Improper Integral $\int_a^{+\infty} f(x)\, dx$ We denote the improper integral of $f(x)$ over the unbounded interval $x \geq a$ by $\int_a^{+\infty} f(x)\, dx$. If $f(x) \geq 0$ for $x \geq a$, this integral can be interpreted as the area of the region under the curve $y = f(x)$ to the right of $x = a$, as shown in Figure 6.7a. Although this region has infinite extent, its area may be finite or infinite, depending on how quickly $f(x)$ approaches zero as x increases indefinitely. A reasonable strategy for finding the area of such a region is to first use a definite integral to compute the area from $x = a$ to some finite number $x = N$ and then to let N approach infinity in the resulting expression; that is,

$$\text{Total area} = \lim_{N \to +\infty} (\text{area from } a \text{ to } N) = \lim_{N \to +\infty} \int_a^N f(x)\, dx$$

This strategy is illustrated in Figure 6.7b and motivates the following definition of the improper integral.

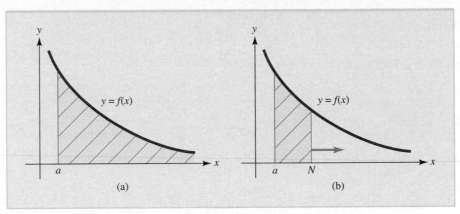

FIGURE 6.7 Area $= \displaystyle\int_{a}^{+\infty} f(x)\, dx = \lim_{N\to+\infty}\int_{a}^{N} f(x)\, dx.$

The Improper Integral $\displaystyle\int_{a}^{+\infty} f(x)\, dx$ ■ If $f(x)$ is continuous for $x \geq a$, then

$$\int_{a}^{+\infty} f(x)\, dx = \lim_{N\to+\infty}\int_{a}^{N} f(x)\, dx$$

If the limit exists, the improper integral is said to **converge** to the value of the limit. If the limit does not exist, the improper integral **diverges.**

EXAMPLE 6.3.1 Evaluating an Improper Integral

Either evaluate the improper integral

$$\int_{1}^{+\infty} \frac{1}{x^2}\, dx$$

or show that it diverges.

Solution

First compute the integral from 1 to N and then let N approach infinity. Arrange your work as follows:

$$\int_{1}^{+\infty} \frac{1}{x^2}\, dx = \lim_{N\to+\infty}\int_{1}^{N} \frac{1}{x^2}\, dx = \lim_{N\to+\infty}\left(-\frac{1}{x}\Big|_{1}^{N}\right) = \lim_{N\to+\infty}\left(-\frac{1}{N} + 1\right) = 1$$

EXAMPLE 6.3.2 Evaluating an Improper Integral

Either evaluate the improper integral

$$\int_{1}^{+\infty} \frac{1}{x}\, dx$$

or show that it diverges.

EXPLORE!

Refer to Example 6.3.1. Place $f(x) = \dfrac{1}{x^2}$ into Y1 of the equation editor and enter Y2 = **fnInt**(Y1, X, 1, X, 0.001), the numerical integration function. Set the table feature to start at X = 500 in increments of 500. Explain what you observe.

Solution

$$\int_1^{+\infty} \frac{1}{x}\, dx = \lim_{N\to +\infty} \int_1^N \frac{1}{x}\, dx = \lim_{N\to +\infty} \left(\ln |x| \bigg|_1^N \right) = \lim_{N\to +\infty} \ln N = +\infty$$

Since the limit does not exist (as a finite number), the improper integral diverges.

NOTE You have just seen that the improper integral $\int_1^{+\infty} \frac{1}{x^2}\, dx$ converges (Example 6.3.1) while $\int_1^{+\infty} \frac{1}{x}\, dx$ diverges (Example 6.3.2). In geometric terms, this says that the area under the curve $y = \frac{1}{x^2}$ to the right of $x = 1$ is *finite,* while the corresponding area under the curve $y = \frac{1}{x}$ to the right of $x = 1$ is *infinite.* The difference is due to the fact that as x increases without bound, $\frac{1}{x^2}$ approaches zero more quickly than does $\frac{1}{x}$. These observations are demonstrated in Figure 6.8. ■

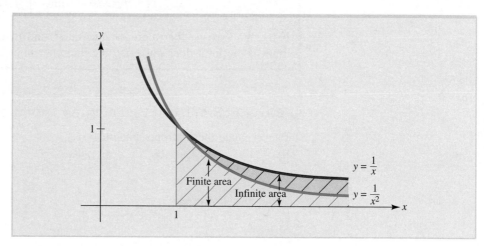

FIGURE 6.8 Comparison of the area under $y = \frac{1}{x}$ with that under $y = \frac{1}{x^2}$.

The evaluation of improper integrals arising from practical problems often involves limits of the form

$$\lim_{N\to\infty} \frac{N^p}{e^{kN}} = \lim_{N\to\infty} N^p e^{-kN} \quad (\text{for } k > 0)$$

In general, an exponential term e^{kN} will grow faster than *any* power term N^p, so

$$N^p e^{-kN} = \frac{N^p}{e^{kN}}$$

will become very small in the long run. To summarize:

> **A Useful Limit for Improper Integrals** ■ For any power p and positive number k,
>
> $$\lim_{N \to \infty} N^p e^{-kN} = 0$$

EXPLORE!

Refer to Example 6.3.3. Graph $f(x) = xe^{-2x}$ using the window $[0, 9.4]1$ by $[-0.05, 0.02]1$. Use the graphical integration feature of your calculator, **CALC (2nd TRACE)**, 7: $\int f(x)\,dx$, to show that the area under $f(x)$ from 0 to a relatively large x value approaches the value $\dfrac{1}{4}$.

■ EXAMPLE 6.3.3 **Evaluating an Improper Integral Using Parts**

Either evaluate the improper integral

$$\int_0^{+\infty} xe^{-2x}\,dx$$

or show that it diverges.

Solution

$$\int_0^{+\infty} xe^{-2x}\,dx = \lim_{N \to +\infty} \int_0^N xe^{-2x}\,dx \qquad \text{integration by parts}$$

$$= \lim_{N \to +\infty} \left(-\frac{1}{2}xe^{-2x}\Big|_0^N + \frac{1}{2}\int_0^N e^{-2x}\,dx \right) \quad \int e^{-2x}\,dx = \frac{-1}{2}e^{-2x} + C$$

$$= \lim_{N \to +\infty} \left(-\frac{1}{2}xe^{-2x} - \frac{1}{4}e^{-2x} \right)\Big|_0^N$$

$$= \lim_{N \to +\infty} \left(-\frac{1}{2}Ne^{-2N} - \frac{1}{4}e^{-2N} + 0 + \frac{1}{4} \right)$$

$$= \frac{1}{4}$$

since $e^{-2N} \to 0$ and $Ne^{-2N} \to 0$ as $N \to +\infty$.

There are other kinds of improper integrals of interest. For instance, if $f(x)$ is continuous on the unbounded interval $x \le b$, then the improper integral $\int_{-\infty}^b f(x)\,dx$ is defined by

$$\int_{-\infty}^b f(x)\,dx = \lim_{N \to -\infty} \int_N^b f(x)\,dx$$

More generally, if $f(x)$ is continuous for all x, then the improper integral over the entire x axis may be defined as follows.

> **The Improper Integral** $\displaystyle\int_{-\infty}^\infty f(x)\,dx$ ■ If $\int_c^\infty f(x)\,dx$ and $\int_{-\infty}^c f(x)\,dx$ both converge for some number c, then the improper integral of $f(x)$ over the unbounded interval $-\infty < x < \infty$ is defined by
>
> $$\int_{-\infty}^\infty f(x)\,dx = \int_c^\infty f(x)\,dx + \int_{-\infty}^c f(x)\,dx$$

> **EXAMPLE 6.3.4**   Evaluating an Improper Integral
> Over $-\infty < x < \infty$

Either evaluate the improper integral $\int_{-\infty}^{\infty} xe^{-0.1x^2} \, dx$ or show it diverges.

Solution

First, we find the indefinite integral $\int xe^{-0.1x^2} \, dx$.

$$\int xe^{-0.1x^2} \, dx = \int e^u \left[\frac{du}{-0.2} \right] \qquad \text{let } u = -0.1x^2, \, du = -0.2x \, dx$$

$$= -5e^u + C = -5e^{-0.1x^2} + C$$

Now, let $c = 0$ in the definition of $\int_{-\infty}^{\infty} f(x) \, dx$. We find that

$$\int_0^{\infty} xe^{-0.1x^2} \, dx = \lim_{N \to \infty} \int_0^N xe^{-0.1x^2} \, dx$$

$$= \lim_{N \to \infty} [-5e^{-0.1x^2}]_0^N$$

$$= \lim_{N \to \infty} [-5e^{-0.1N^2} - (-5e^0)] = 5$$

Similarly,

$$\int_{-\infty}^0 xe^{-0.1x^2} \, dx = \lim_{N \to -\infty} [-5e^{-0.1x^2}]_N^0$$

$$= \lim_{N \to -\infty} [(-5e^0) - (-5e^{-0.1N^2})] = -5$$

So

$$\int_{-\infty}^{\infty} xe^{-0.1x} \, dx = \int_0^{\infty} xe^{-0.1x} \, dx + \int_{-\infty}^0 xe^{-0.1x} \, dx$$

$$= 5 + (-5) = 0$$

Applications of Improper Integrals

Next, we examine two applications of improper integrals that generalize applications developed in Chapter 5. In each, the strategy is to express a quantity as a definite integral with a variable upper limit of integration that is then allowed to increase without bound. As you read through these applications, you may find it helpful to refer back to the analogous examples in Chapter 5.

Present Value of a Perpetual Income Flow

In Section 5.5, we showed how the present value of an investment that generates income continuously over a finite time period can be computed by a definite integral. If the generation of income continues in perpetuity, then an improper integral is required for computing its present value, as illustrated in Example 6.3.5.

> **EXAMPLE 6.3.5**   Finding Present Value of an Income Flow

Uday wishes to endow a scholarship at a local college with a gift that provides a continuous income stream at the rate of $25,000 + 1,200t$ dollars per year in perpetuity. Assuming the prevailing annual interest rate stays fixed at 5% compounded continuously, what donation is required to finance the endowment?

Solution

Uday's gift should equal the present value of the income stream in perpetuity. Recall from Section 5.5 that an income stream $f(t)$ deposited continuously for a term of T years into an account that earns interest at an annual rate r compounded continuously has a present value given by the integral

$$PV = \int_0^T f(t)e^{-rt}\, dt$$

For Uday's gift, we have $f(t) = 25{,}000 + 1{,}200t$ and $r = 0.05$, so for a term of T years, the present value is

*Present value of
the gift for T years*
$$PV = \int_0^T (25{,}000 + 1{,}200t)e^{-0.05t}\, dt$$

Integrating by parts, with

$$u = 25{,}000 + 1{,}200t \qquad dv = e^{-0.05t}\, dt$$

$$du = 1{,}200\, dt \qquad v = \frac{e^{-0.05t}}{-0.05} = -20e^{-0.05t}$$

we evaluate the present value as follows:

$$PV = \int_0^T (25{,}000 + 1{,}200t)e^{-0.05t}\, dt$$

$$= \left[(25{,}000 + 1{,}200t)(-20e^{-0.05t})\right]\Big|_0^T - \int_0^T 1{,}200(-20e^{-0.05t})\, dt$$

$$= \left[(-500{,}000 - 24{,}000t)e^{-0.05t}\right]\Big|_0^T + 24{,}000\left(\frac{e^{-0.05t}}{-0.05}\right)\Big|_0^T$$

$$= \left[(-980{,}000 - 24{,}000t)e^{-0.05t}\right]\Big|_0^T$$

$$= \left[(-980{,}000 - 24{,}000T)e^{-0.05T}\right] - \left[(-980{,}000 - 24{,}000(0))e^0\right]$$

$$= (-980{,}000 - 24{,}000T)e^{-0.05T} + 980{,}000$$

To find the present value in perpetuity, we take the limit as $T \to +\infty$; that is, we compute the improper integral

*Present value of
the gift in perpetuity*
$$\int_0^{+\infty} (25{,}000 + 1{,}200t)e^{-0.05t}\, dt$$

$$= \lim_{T \to +\infty} \left[(-980{,}000 - 24{,}000T)e^{-0.05T} + 980{,}000\right]$$

$$= 0 + 980{,}000 \qquad \text{since } e^{-0.05T} \to 0 \text{ and}$$
$$\qquad\qquad\qquad\qquad Te^{-0.05T} \to 0 \text{ as } T \to +\infty$$

$$= 980{,}000$$

Therefore, the endowment is established with a gift of \$980,000.

Survival/Renewal in Perpetuity

In Section 5.6, we examined a survival/renewal problem over a finite term in which renewal occurred at a constant rate. In Example 6.3.6, we examine a similar problem in which the survival/renewal process continues indefinitely.

EXAMPLE 6.3.6 Eventual Accumulation of Nuclear Waste

It is estimated that t years from now, a certain nuclear power plant will be producing radioactive waste at the rate of 400 pounds per year. The waste decays exponentially at the rate of 2% per year. How much radioactive waste will eventually accumulate?

Solution

The renewal function is $R(t) = 400$, and the survival function is $S(t) = e^{-0.02t}$. Suppose the current accumulation of waste is A_0 pounds. Then, according to the formula obtained in Section 5.6, the amount of waste accumulated over a period of T years is

$$A(T) = A_0 S(T) + \int_0^T R(t)S(T - t)\, dt$$

$$= A_0 e^{-0.02T} + \int_0^T 400 e^{-0.02(T - t)}\, dt$$

$$= A_0 e^{-0.02T} + 400 e^{-0.02T} \int_0^T e^{0.02t}\, dt$$

$$= A_0 e^{-0.02T} + 400 e^{-0.02T} \left[\frac{e^{0.02t}}{0.02} \right]_0^T$$

$$= A_0 e^{-0.02T} + 20{,}000 e^{-0.02T}[e^{0.02T} - 1]$$

$$= A_0 e^{-0.02T} + 20{,}000[1 - e^{-0.02T}]$$

To determine the amount of waste W that eventually accumulates, we take the limit of $A(T)$ as $T \to \infty$:

$$W = \lim_{T \to \infty} A(T) = \lim_{T \to \infty} [A_0 e^{-0.02T} + 20{,}000(1 - e^{-0.02T})] \quad e^{-0.02T} \to 0 \text{ as } T \to \infty$$

$$= 0 + 20{,}000 - 20{,}000(0) = 20{,}000$$

So we would expect a total of 20,000 pounds of waste to accumulate if the process continues indefinitely.

Population of an Urban Area

In Section 5.6, we showed that if an urban area has population density $p(r)$ people per square mile at a distance r miles from the city center, then the population within R miles of the city center is given by

$$P(R) = \int_0^R 2\pi r p(r)\, dr$$

In Example 6.3.7, we extend the radius R indefinitely to obtain an improper integral that estimates the total population of the urban area.

EXAMPLE 6.3.7 Estimating the Population of an Urban Area

A certain urban area has population density $p(r) = 1{,}100 e^{-0.002r^2}$ people per square mile at a distance of r miles from the city center. Estimate the total population of the urban area.

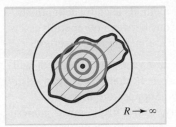

FIGURE 6.9 An urban area (red boundary) contained inside a circle of large radius R.

Solution

We can estimate the total population of the urban area by finding the area within a radius R of the city center and extending R indefinitely, as indicated in Figure 6.9. In other words, the total population TP is estimated by the improper integral

$$TP = \int_0^\infty 2\pi r p(r)\, dr = \lim_{R \to \infty} \int_0^R 2\pi r (1{,}100 e^{-0.002 r^2})\, dr$$

For simplicity, compute the indefinite integral separately

$$\int 2\pi r (1{,}100 e^{-0.002 r^2})\, dr \qquad \substack{\text{substitute } u = -0.002 r^2,\ du = -0.004 r\, dr}$$

$$\text{so } r\, dr = \frac{1}{-0.004}\, du = -250\, du$$

$$= 2{,}200 \pi \int e^u [-250\, du] \qquad \text{integrate and substitute } u = -0.002 r^2$$

$$= -550{,}000 \pi e^u + C = -550{,}000 \pi e^{-0.002 r^2} + C$$

and then evaluate the improper integral:

$$TP = \int_0^\infty 2\pi r (1{,}100 e^{-0.002 r^2})\, dr = \lim_{R \to \infty} \int_0^R 2{,}200 \pi r e^{-0.002 r^2}\, dr$$

$$= \lim_{R \to \infty} [-550{,}000 \pi e^{-0.002 r^2}]_0^R = -550{,}000 \pi \lim_{R \to \infty} [e^{-0.002 R^2} - e^0]$$

$$= -550{,}000 \pi [0 - 1] \qquad \lim_{R \to \infty} e^{-0.002 R^2} = 0 \text{ and } e^0 = 1$$

$$\approx 1{,}727{,}876$$

So the total population of the urban area is approximately 1,727,876 people.

EXERCISES ▪ 6.3

In Exercises 1 through 30, either evaluate the given improper integral or show that it diverges.

1. $\displaystyle \int_1^{+\infty} \frac{1}{x^3}\, dx$

2. $\displaystyle \int_1^{+\infty} x^{-3/2}\, dx$

3. $\displaystyle \int_1^{+\infty} \frac{1}{\sqrt{x}}\, dx$

4. $\displaystyle \int_1^{+\infty} x^{-2/3}\, dx$

5. $\displaystyle \int_3^{+\infty} \frac{1}{2x-1}\, dx$

6. $\displaystyle \int_3^{+\infty} \frac{1}{\sqrt[3]{2x-1}}\, dx$

7. $\displaystyle \int_3^{+\infty} \frac{1}{(2x-1)^2}\, dx$

8. $\displaystyle \int_0^{+\infty} e^{-x}\, dx$

9. $\displaystyle \int_0^{+\infty} 5e^{-2x}\, dx$

10. $\displaystyle \int_1^{+\infty} e^{1-x}\, dx$

11. $\displaystyle \int_1^{+\infty} \frac{x^2}{(x^3+2)^2}\, dx$

12. $\displaystyle \int_1^{+\infty} \frac{x^2}{x^3+2}\, dx$

13. $\displaystyle \int_1^{+\infty} \frac{x^2}{\sqrt{x^3+2}}\, dx$

14. $\displaystyle \int_0^{+\infty} x e^{-x^2}\, dx$

15. $\displaystyle \int_1^{+\infty} \frac{e^{-\sqrt{x}}}{\sqrt{x}}\, dx$

16. $\displaystyle \int_0^{+\infty} x e^{-x}\, dx$

17. $\displaystyle \int_0^{+\infty} 2x e^{-3x}\, dx$

18. $\displaystyle \int_0^{+\infty} x e^{1-x}\, dx$

19. $\displaystyle \int_1^{+\infty} \frac{\ln x}{x}\, dx$

20. $\displaystyle \int_1^{+\infty} \frac{\ln x}{x^2}\, dx$

21. $\displaystyle \int_2^{+\infty} \frac{1}{x \ln x}\, dx$

22. $\displaystyle \int_2^{+\infty} \frac{1}{x \sqrt{\ln x}}\, dx$

23. $\displaystyle \int_0^{+\infty} x^2 e^{-x}\, dx$

24. $\displaystyle \int_1^{+\infty} \frac{e^{1/x}}{x^2}\, dx$

25. $\displaystyle \int_{-\infty}^0 3e^{4x}\, dx$

26. $\displaystyle \int_{-\infty}^1 e^{1-x}\, dx$

27. $\displaystyle \int_{-\infty}^{-1} \frac{1}{x^2}\, dx$

28. $\displaystyle \int_{-\infty}^0 \frac{1}{(2x-1)^2}\, dx$

29. $\displaystyle \int_{-\infty}^\infty x e^{-x}\, dx$

30. $\displaystyle \int_{-\infty}^\infty \frac{x}{(x^2+1)^{3/2}}\, dx$

BUSINESS AND ECONOMICS APPLIED PROBLEMS

31. PRESENT VALUE OF AN INVESTMENT
An investment will generate $2,400 per year in perpetuity. If the money is dispensed continuously throughout the year and if the prevailing annual interest rate remains fixed at 4% compounded continuously, what is the present value of the investment?

32. PRESENT VALUE OF RENTAL PROPERTY
It is estimated that t years from now an apartment complex will be generating profit for its owner at the rate of $f(t) = 80,000 + 500t$ dollars per year, dispensed continuously. If the profit is generated in perpetuity and the prevailing annual interest rate remains fixed at 5% compounded continuously, what is the present value of the apartment complex?

33. PRESENT VALUE OF A FRANCHISE   The management of a national chain of fast-food outlets is selling a permanent franchise in Seattle, Washington. Past experience suggests that t years from now, the franchise will be generating profit in perpetuity at the rate of $f(t) = 100,000 + 900t$ dollars per year. If the prevailing interest rate remains fixed at 5% compounded continuously, what is the present value of the franchise?

34. ENDOWMENT   Harriet Truscott, a wealthy patron of a small private college, wishes to endow a chair in mathematics. Suppose the mathematician who occupies the chair is to receive $70,000 per year in salary and benefits. If the gift can be invested at an annual interest rate of 8% compounded continuously, how much money must Harriet donate to sustain the endowed chair?

35. CAPITALIZED COST OF AN ASSET   The **capitalized cost** of an asset is the sum of the original cost of the asset and the present value of maintaining the asset. Suppose a company is considering the purchase of two different machines. Machine 1 costs $10,000 and t years from now will cost $M_1(t) = 1,000(1 + 0.06t)$ dollars per year to maintain. Machine 2 costs only $8,000, but its maintenance cost at time t is $M_2(t) = 1,100$ dollars per year.
a. If the cost of money is 9% per year compounded continuously, what is the capitalized cost of each machine? Which machine should the company purchase?

b. Research various methods used by economists to make comparisons between competing assets. Write a paragraph comparing these methods.

36. PRESENT VALUE   In t years, an investment will be generating $f(t) = A + Bt$ dollars per year, where A and B are constants. Suppose the income is generated in perpetuity, with a fixed annual interest rate r compounded continuously. Show that the present value of this investment is $\dfrac{A}{r} + \dfrac{B}{r^2}$ dollars.

37. PRESENT VALUE   An investment will generate income continuously at the constant rate of Q dollars per year in perpetuity. Assuming a fixed annual interest rate r compounded continuously, use an improper integral to show that the present value of the investment is $\dfrac{Q}{r}$ dollars.

38. PRODUCT RELIABILITY   The manager of an electronics firm estimates that the proportion of components that last longer than t months is given by the improper integral

$$\int_t^\infty 0.008 e^{-0.008x} \, dx$$

Which is larger, the proportion of components that last longer than 5 years (60 months) or the proportion that fail in less than 10 years?

LIFE AND SOCIAL SCIENCE APPLIED PROBLEMS

39. HEALTH CARE   The fraction of patients who will still be receiving treatment at a certain health clinic t months after their initial visit is $f(t) = e^{-t/20}$. If the clinic accepts new patients at the rate of 10 per month, approximately how many patients will be receiving treatment at the clinic in the long run?

40. POPULATION GROWTH   Demographic studies conducted in a certain city indicate that the fraction of the residents that will remain in the city for at least t years is $f(t) = e^{-t/20}$. The current population of the city is 200,000, and it is estimated that new residents will be arriving at the rate of 100 people per year. If this estimate is correct, what will happen to the population of the city in the long run?

41. MEDICINE Marta receives 5 units of a certain drug per hour intravenously. The drug is eliminated exponentially, so that the fraction that remains in Marta's body for t hours is $f(t) = e^{-t/10}$. If the treatment is continued indefinitely, approximately how many units of the drug will be in Marta's body in the long run?

42. NUCLEAR WASTE A certain nuclear power plant produces radioactive waste at the rate of 600 pounds per year. The waste decays exponentially at the rate of 2% per year. How much radioactive waste from the plant will be present in the long run?

43. EPIDEMIOLOGY The proportion P of susceptible people who are infected t weeks after the outbreak of an epidemic is given by the integral

$$\int_0^t C(e^{-ax} - e^{-bx})\, dx$$

where a and b are parameters that depend on the disease and C is a constant. Assuming that all susceptible people are eventually infected, find C (in terms of a and b).

POPULATION OF AN URBAN AREA *In Exercises 44 and 45, the population density $p(r)$ of an urban area is given. In each case, use an improper integral to estimate the total population of the urban area.*

44. $p(r) = 100e^{-0.02r}$

45. $p(r) = 100e^{-0.02r} + 2{,}000e^{-0.001r^2}$

Important Terms, Symbols, and Formulas

Integration by parts: (480)

$$\int u\, dv = uv - \int v\, du$$

Table of integrals: (488–489)
  Forms involving $a + bu$ (488)
  Forms involving $a^2 + u^2$ (488)
  Forms involving $a^2 - u^2$ (489)
  Forms involving $u^2 - a^2$ (489)
  Forms involving e^{au} and $\ln u$ (489)

Reduction formulas (489)

Trapezoidal rule: (496)

$$\int_a^b f(x)\, dx$$

$$\approx \frac{\Delta x}{2}[f(x_1) + 2f(x_2) + \cdots + 2f(x_n) + f(x_{n+1})]$$

Error estimate: (497)

$$|E_n| \le \frac{K(b-a)^3}{12n^2}$$

where K is the maximum value of $|f''(x)|$ on $[a, b]$

Simpson's rule: for n even, (498)

$$\int_a^b f(x)\, dx \approx \frac{\Delta x}{3}[f(x_1) + 4f(x_2) + 2f(x_3) + \cdots$$
$$+ 2f(x_{n-1}) + 4f(x_n) + f(x_{n+1})]$$

Error estimate: (500)

$$|E_n| \le \frac{M(b-a)^5}{180n^4}$$

where M is the maximum value of $|f^{(4)}(x)|$ on $[a, b]$

Improper integral: (509)

$$\int_a^{+\infty} f(x)\, dx = \lim_{N \to +\infty} \int_a^N f(x)\, dx$$

$$\int_{-\infty}^{+\infty} f(x)\, dx = \lim_{N \to +\infty} \int_{-N}^0 f(x)\, dx + \lim_{N \to +\infty} \int_0^N f(x)\, dx$$

A useful limit: (511)
  For any power p and $k > 0$,

$$\lim_{N \to +\infty} N^p e^{-kN} = 0$$

Present value of a perpetual income flow (513)

$$PV = \int_0^{\infty} f(t)e^{-rt}\, dt$$

CHAPTER SUMMARY

Checkup for Chapter 6

1. Use integration by parts to find each of these indefinite and definite integrals.

   a. $\displaystyle\int \sqrt{2x}\, \ln x^2 \, dx$ **b.** $\displaystyle\int_0^1 xe^{0.2x} \, dx$

   c. $\displaystyle\int_{-4}^0 x\sqrt{1 - 2x}\, dx$ **d.** $\displaystyle\int \frac{x - 1}{e^x} \, dx$

2. In each case, either evaluate the given improper integral or show that it diverges.

   a. $\displaystyle\int_1^{+\infty} \frac{1}{x^{1.1}} \, dx$ **b.** $\displaystyle\int_1^{+\infty} xe^{-2x} \, dx$

   c. $\displaystyle\int_1^{+\infty} \frac{x}{(x + 1)^2} \, dx$ **d.** $\displaystyle\int_{-\infty}^{+\infty} xe^{-x^2} \, dx$

3. Use the integral table (Table 6.1) to find these integrals.

   a. $\displaystyle\int (\ln \sqrt{3x})^2 \, dx$ **b.** $\displaystyle\int \frac{dx}{x\sqrt{4 + x^2}}$

   c. $\displaystyle\int \frac{dx}{x^2\sqrt{x^2 - 9}}$ **d.** $\displaystyle\int \frac{dx}{3x^2 - 4x}$

4. **NUCLEAR WASTE** After t years of operation, a certain nuclear power plant produces radioactive waste at the rate of $R(t)$ pounds per year, where
$$R(t) = 300e^{0.001t}$$
   The waste decays exponentially at the rate of 3% per year. How much radioactive waste from the plant will be present in the long run?

5. **PRESENT VALUE OF AN ASSET** It is estimated that t years from now an office building will be generating profit for its owner at the rate of $R(t) = 50 + 3t$ thousand dollars per year. If the profit is generated in perpetuity and the prevailing annual interest rate remains fixed at 6% compounded continuously, what is the present value of the office building?

6. **DRUG CONCENTRATION** Zain, a patient in a hospital, receives 0.7 mg of a certain drug intravenously every hour. The drug is eliminated exponentially in such a way that the fraction of drug that remains in Zain's body after t hours is $f(t) = e^{-0.2t}$. If the treatment is continued indefinitely, approximately how much of the drug will remain in Zain's bloodstream in the long run (as $t \to \infty$)?

7. **WATER POLLUTION** Suppose that pollutant is being discharged into a river by an industrial plant in such a way that at time t (hours), the rate of discharge is $R(t) = 800e^{-0.05t}$ units per hour. If the discharge continues indefinitely, what is the total amount of pollutant that will be discharged into the river?

8. Use the trapezoidal rule with $n = 8$ to estimate the value of the integral
$$\int_3^4 \frac{\sqrt{25 - x^2}}{x} \, dx$$
   Then use Table 6.1 to compute the exact value of the integral and compare with your approximation.

9. **THE SPREAD OF A RUMOR** The rate at which a rumor spreads through a community of 2,000 people is jointly proportional to the number of people $N(t)$ who have heard it and the number of people $2,000 - N(t)$ who have not, so
$$\frac{dN}{dt} = kN(2,000 - N)$$
   Originally (at time $t = 0$), 50 people know the rumor, and 1 day later, 75 people have heard it.

   a. Use a formula from the integral table to find $N(t)$.

   b. How many people know the rumor after 1 week?

   c. How long does it take before 1,000 people know the rumor?

Review Exercises

In Exercises 1 through 10, use integration by parts to find the given integral.

1. $\displaystyle\int te^{1-t} \, dt$ **2.** $\displaystyle\int (5 + 3x)e^{-x/2} \, dx$

3. $\displaystyle\int x\sqrt{2x + 3} \, dx$ **4.** $\displaystyle\int_{-9}^{-1} \frac{y \, dy}{\sqrt{4 - 5y}}$

5. $\displaystyle\int_1^4 \frac{\ln \sqrt{s}}{\sqrt{s}} \, ds$ **6.** $\displaystyle\int (\ln x)^2 \, dx$

7. $\displaystyle\int_{-2}^1 (2x + 1)(x + 3)^{3/2} \, dx$

8. $\displaystyle\int \frac{w^3}{\sqrt{1 + w^2}} \, dw$

9. $\int x^3 \sqrt{3x^2 + 2}\, dx$ **10.** $\int_0^1 \dfrac{x + 2}{e^{3x}}\, dx$

In Exercises 11 through 16, use Table 6.1 to find the given integral.

11. $\int \dfrac{5\, dx}{8 - 2x^2}$ **12.** $\int \dfrac{2\, dt}{\sqrt{9t^2 + 16}}$

13. $\int w^2 e^{-w/3}\, dw$ **14.** $\int \dfrac{4\, dx}{x(9 + 5x)}$

15. $\int (\ln 2x)^3\, dx$ **16.** $\int \dfrac{dx}{x\sqrt{4 - x^2}}$

In Exercises 17 through 20, solve the given initial value problem using either integration by parts or a formula from Table 6.1. Note that Exercises 19 and 20 involve separable differential equations.

17. $\dfrac{dy}{dx} = x \ln \sqrt{x}$, where $y = 3$ when $x = 1$

18. $\dfrac{dy}{dx} = \dfrac{4}{x^2 + 2x - 3}$, where $y = 1$ when $x = 0$

19. $\dfrac{dy}{dx} = \dfrac{e^y}{xy}$, where $y = 0$ when $x = 1$

20. $\dfrac{dy}{dx} = \dfrac{xy}{3 + x}$, where $y = 1$ when $x = 1$

In Exercises 21 through 34, either evaluate the given improper integral or show that it diverges.

21. $\int_0^{+\infty} \dfrac{1}{\sqrt[3]{1 + 2x}}\, dx$ **22.** $\int_0^{+\infty} (1 + 2x)^{-3/2}\, dx$

23. $\int_0^{+\infty} \dfrac{3t}{t^2 + 1}\, dt$ **24.** $\int_0^{+\infty} 3e^{-5x}\, dx$

25. $\int_0^{+\infty} xe^{-2x}\, dx$ **26.** $\int_0^{+\infty} 2x^2 e^{-x^3}\, dx$

27. $\int_0^{+\infty} x^2 e^{-2x}\, dx$ **28.** $\int_2^{+\infty} \dfrac{1}{t(\ln t)^2}\, dt$

29. $\int_1^{+\infty} \dfrac{\ln x}{\sqrt{x}}\, dx$ **30.** $\int_0^{+\infty} \dfrac{x - 1}{x + 2}\, dx$

31. $\int_{-\infty}^{-1} xe^{x+1}\, dx$ **32.** $\int_{-\infty}^{0} e^{2x} + \dfrac{1}{(x - 1)^2}\, dx$

33. $\int_{-\infty}^{\infty} x^3 e^{-x^2}\, dx$ **34.** $\int_{-\infty}^{\infty} (e^x + e^{-x})\, dx$

35. PRESENT VALUE OF AN INVESTMENT
It is estimated that t years from now, a certain investment will be generating income at the rate of $f(t) = 8{,}000 + 400t$ dollars per year. If the income is generated in perpetuity and the prevailing annual interest rate remains fixed at 5% compounded continuously, find the present value of the investment.

36. PRODUCTION After t hours on the job, a factory worker can produce $100te^{-0.5t}$ units per hour. How many units does a worker who arrives on the job at 8:00 A.M. produce between 10:00 A.M. and noon?

37. DEMOGRAPHICS Demographic studies indicate that the fraction of the residents that will remain in a certain city for at least t years is $f(t) = e^{-t/20}$. The current population of the city is 100,000, and it is estimated that t years from now, new people will be arriving at the rate of $100t$ people per year. If this estimate is correct, what will happen to the population of the city in the long run?

38. SUBSCRIPTION GROWTH The publishers of a national magazine have found that the fraction of subscribers who remain subscribers for at least t years is $f(t) = e^{-t/10}$. Currently, the magazine has 20,000 subscribers and it is estimated that new subscriptions will be sold at the rate of 1,000 per year. Approximately how many subscribers will the magazine have in the long run?

39. NUCLEAR WASTE After t years of operation, a certain nuclear power plant produces radioactive waste at the rate of $R(t)$ pounds per year, where

$$R(t) = 300 - 200e^{-0.03t}$$

The waste decays exponentially at the rate of 2% per year. How much radioactive waste from the plant will be present in the long run?

40. PSYCHOLOGICAL TESTING In a psychological experiment, it is found that the proportion of participants who require more than t minutes to finish a particular task is given by

$$\int_t^{+\infty} 0.07e^{-0.07u}\, du$$

a. Find the proportion of participants who require more than 5 minutes to finish the task.

b. What proportion requires between 10 and 15 minutes to finish?

In Exercises 41 through 44, approximate the given integral and estimate the error with the specified number of subintervals using:

(a) The trapezoidal rule.

(b) Simpson's rule.

41. $\int_1^3 \frac{1}{x}\, dx;\ n = 10$

42. $\int_0^2 e^{x^2}\, dx;\ n = 8$

43. $\int_1^2 \frac{e^x}{x}\, dx;\ n = 10$

44. $\int_1^2 xe^{1/x}\, dx;\ n = 8$

In Exercises 45 and 46, determine how many subintervals are required to guarantee accuracy to within 0.00005 of the actual value of the given integral using:

(a) *The trapezoidal rule.*

(b) *Simpson's rule.*

45. $\int_1^3 \sqrt{x}\, dx$

46. $\int_{0.5}^1 e^{-1.1x}\, dx$

47. TOTAL COST FROM MARGINAL COST A manufacturer determines that the marginal cost of producing q units of a particular commodity is $C'(q) = \sqrt{q}e^{0.01q}$ dollars per unit.

a. Express the total cost of producing the first 8 units as a definite integral.

b. Estimate the value of the total cost integral in part (a) using the trapezoidal rule with $n = 8$ subintervals.

48. A child standing at the origin of a coordinate plane, holds a 5-foot length of rope that is attached to a sled. She walks along the x axis keeping the rope taut, as shown in the figure (units are in feet).

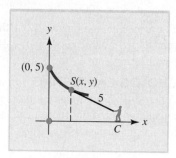

If the sled begins at $(0, 5)$ and is at $S(x, y)$ when the child is at C, it can be shown that

$$\frac{dy}{dx} = \frac{-y}{\sqrt{25 - y^2}}$$

a. Find the equation of the path following by the sled. (The path is called a **tractrix.**) [*Hint:* Use an integral formula from Table 6.1.]

b. Use the graphing utility of your calculator to draw the path. Then use **ZOOM** and **TRACE** or other features of your calculator to find the point on the path that is exactly 3 feet from the y axis.

49. Use the graphing utility of your calculator to draw the graphs of the curves $y = -x^3 - 2x^2 + 5x - 2$ and $y = x \ln x$ for $x > 0$ on the same screen. Use **ZOOM** and **TRACE** or some other feature of your calculator to find where the curves intersect, and then compute the area of the region bounded by the curves.

50. Repeat Exercise 49 for the curves

$$y = \frac{x - 2}{x + 1} \quad \text{and} \quad y = \sqrt{25 - x^2}$$

If your calculator has a numeric integration feature, use it to evaluate the integrals in Exercises 51 and 52. In each case, verify your result by applying an appropriate integration formula from Table 6.1.

51. $\int_{-1}^1 \frac{2}{9 - x^2}\, dx$

52. $\int_0^1 x^2\sqrt{9 + 4x^2}\, dx$

53. Use the numeric integration feature of your calculator to compute

$$I(N) = \int_0^N \frac{1}{\sqrt{\pi}} e^{-x^2}\, dx$$

for $N = 1, 10, 50$. Based on your results, do you think the improper integral

$$\int_0^{+\infty} \frac{1}{\sqrt{\pi}} e^{-x^2}\, dx$$

converges? If so, to what value?

54. Use the numeric integration feature of your calculator to compute

$$I(N) = \int_1^N \frac{\ln(x + 1)}{x}\, dx$$

for $N = 10, 100, 1,000, 10,000$. Based on your results, do you think the improper integral

$$\int_1^{+\infty} \frac{\ln(x + 1)}{x}\, dx$$

converges? If so, to what value?

EXPLORE! UPDATE

Complete solutions for all EXPLORE! boxes throughout the text can be accessed at the book-specific website, www.mhhe.com/hoffmann.

Solution for Explore!
on Page 496

The list features of the graphing calculator can be used to facilitate the numerical integration computations required by the trapezoidal rule. Following Example 6.2.1, we set $f(x) = \dfrac{1}{x}$, $a = 1$, $b = 2$, and $n = 10$. Write $Y1 = \dfrac{1}{x}$. On a cleared home screen, store values 1 into A, 2 into B, and 10 into N and define $(B - A)/N$ as H. A quick way to generate the numbers 1.0, 1.1, . . . , 1.9, 2.0 in L1 (accessed through **STAT, EDIT, 1:Edit**) is to write **L1 = seq(A + HX, X, 0, N)**, where the sequence command can be found in **LIST (2nd STAT), OPS, 5: seq(**. The following three screens display these steps.

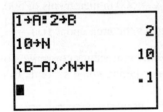

Now place the trapezoidal rule coefficients 1, 2, . . . , 2, 1 into L2. Write **L3 = Y1(L1)*L2*H/2,** recalling that you must place the cursor at the top heading line of L3 to write the desired command. The following screens in the middle and on the right show all the data.

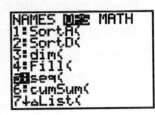

Finally, the sum of list L3 yields the desired trapezoidal rule approximation for $f(x) = \dfrac{1}{x}$ over [1, 2] with $n = 10$. The sum command can be found in **LIST (2nd STAT), MATH, 5:sum(**, as shown in the following screens. The answer coincides with that in Example 6.2.1 and is a good approximation of the answer obtained by direct evaluation of the integral.

EXPLORE! UPDATE

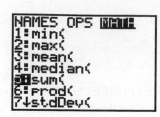

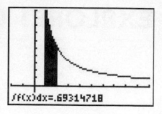

The graphing calculator steps shown here provide the key commands for writing a program to compute the trapezoidal rule of a function $f(x)$, over the interval $[a, b]$ with n subdivisions. As an exercise, carry out the same computations with $n = 20$ instead of $n = 10$.

Solution for Explore! on Page 509

Refer to Example 6.3.1. Place $f(x) = \dfrac{1}{x^2}$ into Y1 of the equation editor, and write Y2 = **fnInt(Y1, X, 1, X, 0.001)**, using the numerical integration function, found in the **MATH** key menu, **9:fnInt(**. We deselect Y1 to only display the values of Y2. Set the table feature to start at $X = 500$ in increments of 500, as shown in the middle screen.

The table on the right displays the area under $f(x) = \dfrac{1}{x^2}$ from $x = 1$ to the designated X value. It appears that the integral values converge slowly to the value of 1. You may find your graphing calculator taking some seconds to compute the Y2 values.

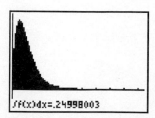

Solution for Explore! on Page 511

Refer to Example 6.3.3. Store $f(x) = xe^{-2x}$ into Y1, and graph using the window $[0, 9.4]$ 1 by $[-0.05, 0.2]$1, as shown on the following screen on the left. Use the graphical integration feature of your calculator, **CALC (2nd TRACE), 7: $\int f(x)\, dx$,** to display the area under $f(x)$. It appears that most of the area under the curve $f(x)$ for the nonnegative real numbers will occur over the interval $[0, x]$, for $x < 10$. So set the lower limit as 0 and the upper limit as an x value, say, $x = 6$, as shown in the following middle screen. We find that over this interval the area under the curve is approximately 0.24998, while the area over the interval $[0, \infty)$ is $\dfrac{1}{4}$, implying that a very small amount of area lies beyond $x > 6$. Use this method for an upper limit of $x = 7$ or 9.

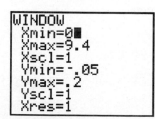
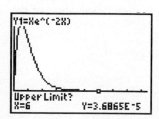
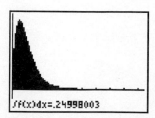

THINK ABOUT IT

MEASURING CARDIAC OUTPUT

The **cardiac output** F of a person's heart is the rate at which the heart pumps blood over a specified period of time. Measuring cardiac output helps physicians treat heart attack patients by diagnosing the extent of heart damage. Specifically, for a healthy person at rest, cardiac output is between 4 and 5 liters per minute and is relatively steady, but the cardiac output of someone with serious cardiovascular disease may be as low as 2 to 3 liters per minute. Because cardiac output is so important in cardiology, medical researchers have devised a variety of ways for measuring it. In this essay we examine two methods for using integration to measure cardiac output, one of which involves improper integration. Our coverage is adapted from *The UMAP Journal* article and the UMAP module listed in the references.

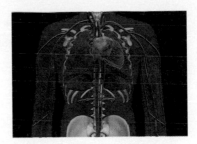

A traditional method for measuring cardiac output is a technique known as **dye dilution.** In this method, dye is injected into a main vein near the heart. Blood carrying this dye circulates through the right ventricle, to the lungs, through the left auricle and left ventricle, and then into the arterial system, returning to the left auricle and left ventricle. The concentration of dye passing through an artery or vein is measured at specific time intervals, and from these concentrations, cardiac output is computed. The dye that is typically used in dye dilution is indocyanine green. Very little of this dye is recirculated back to the heart after it passes through the body because it is metabolized by the liver.

Now suppose that X_0 units of dye is inserted into a vein near the heart where it mixes with a patient's blood. Let $C(t)$ denote the concentration of dye at time t at the beginning of the pulmonary artery. Note that $C(0) = 0$ and that $C(t)$ rises relatively quickly to a maximum, and then decreases to zero after a relatively short time period, say T seconds. The rate at which dye leaves the heart at time t equals $F \cdot C(t)$.

To derive a formula for F, we begin by splitting the interval between $t = 0$ and $t = T$ into n subintervals of equal length, using the points $0 = t_0$, $t_1 = \dfrac{T}{n}$, $t^2 = \dfrac{2T}{n}, \ldots, t_n = T$. Note that the $C_j(t)$, the concentration of dye during the jth subinterval, is nearly constant, so the amount of dye that leaves the heart during the jth subinterval is approximately

$$F \cdot C_j(t)\Delta t$$

To find the total amount of dye leaving the heart, we sum the terms for all n subintervals and obtain

$$X_0 = \sum_{j=0}^{n-1} FC_j(t)\Delta t$$

Taking the limit as n approaches infinity, we derive the estimate

$$X_0 = F\int_0^T C(t)\, dt$$

which means that

$$F = \frac{X_0}{\displaystyle\int_0^T C(t)\, dt}$$

To see how this formula for F is used, suppose that $X_0 = 5$ milligrams (mg) of indocyanine green dye is injected into a major vein near the heart of a patient who has been exercising and that the concentration of dye $C(t)$ (in milligrams per liter) measured each 4 seconds is as shown in Table 1.

TABLE 1   Dye Dilution Data for Measuring Cardiac Output

Time t (seconds)	Concentration $C(t)$ (mg/L)
0	0
4	1.2
8	4.5
12	3.1
16	1.4
20	0.5
24	0

Applying Simpson's rule to estimate the integral of the concentration $C(t)$ over the time interval $0 \le t \le 24$, we find that

$$\int_0^{24} C(t)\, dt \approx \frac{4}{3}[0 + 4(1.2) + 2(4.5) + 4(3.1) + 2(1.4) + 4(0.5) + 0] \approx 41.33$$

so that

$$F = \frac{X_0}{\displaystyle\int_0^{24} C(t)\, dt} \approx \frac{5}{41.33} \approx 0.121$$

Thus, the cardiac output F is approximately 0.121 liters per second or, equivalently,

$$(60 \text{ seconds/minute})(0.121 \text{ liters/second}) \approx 7.26 \text{ liters/minute}$$

which is normal for someone who has been exercising.

Because of the need to sample blood repeatedly, measuring cardiac output using blood flow can be difficult. Dye dilution may also produce inaccurate results because dyes can be unstable and a considerable amount of a dye may recirculate through the body or dye may dissipate slowly. Consequently, other methods of determining cardiac output may be preferable. In the 1970s, a method for measuring cardiac output known as **thermodilution** was introduced. To use this method, a catheter with a balloon tip is inserted into the arm vein of a patient. This catheter is guided to the superior vena cava. Then a thermistor, a device that measures temperatures, is guided through the right side of the heart and positioned several centimeters inside the pulmonary artery. Once the thermistor is in place, a cold solution is injected, typically 10 milliliters of a 5% solution of dextrose in water at 0°C. As this cold solution mixes with blood in the right side of the heart, the blood temperature decreases. The temperature of the blood mixed with this cold solution is detected by the thermistor. The thermistor is connected to a computer that records the temperatures and displays the graph of these temperatures on a monitor. Figure 1 shows the typical temperature variation of the blood at the beginning of the pulmonary artery t seconds after the injection of 10 milliliters of dextrose solution at 0°C.

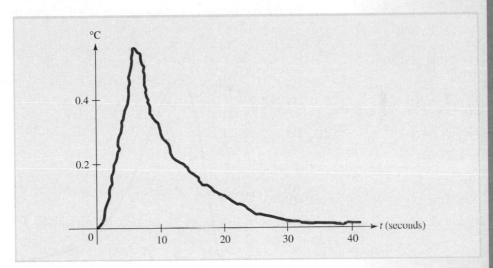

FIGURE 1   Variation in blood temperature T_{var} measured in the pulmonary artery t seconds after injection of cold solution into the right atrium of the heart.

Next, we show how to determine cardiac output using the temperature data collected during thermodilution. Let T_i denote the temperature of the injectant, and let T_b denote the body temperature, that is, the temperature of the blood before the cold solution is injected. Then Q_{in}, the cooling effect of the injectant, may be measured by the product of the volume V of the injectant and the temperature difference $T_b - T_i$ between the body and the injectant; that is,

$$Q_{in} = V(T_b - T_i)$$

Over a short time interval Δt, the cooling effect from the chilled blood passing the thermistor is given by the product of the temperature variation T_{var} that we recorded and the total amount of blood $F\Delta t$ that flows through the pulmonary artery during this interval. Thus, the *total* cooling effect Q_{out} (for all time $t \geq 0$) may be measured by the improper integral

$$Q_{out} = \int_0^\infty F\, T_{var}\, dt = F \int_0^\infty T_{var}\, dt$$

The key fact for computing cardiac output is that $Q_{out} = rQ_{in}$, where r is a correction factor related to how much cooling effect is lost as the chilled solution passes into the body. (The value of r for a particular way the injectant enters the bloodstream can be calculated through data collection.) Inserting the formulas we have found for Q_{in} and Q_{out}, we obtain

$$rV(T_b - T_i) = F \int_0^\infty T_{var}\, dt$$

Solving for the cardiac output rate F, we find that

$$F = \frac{rV(T_b - T_i)}{\int_0^\infty T_{var}\, dt}$$

To illustrate how this method can be used, we will model the temperature variation by the continuous function $T_{var} = 0.1t^2 e^{-0.31t}$, whose graph is shown in Figure 2.

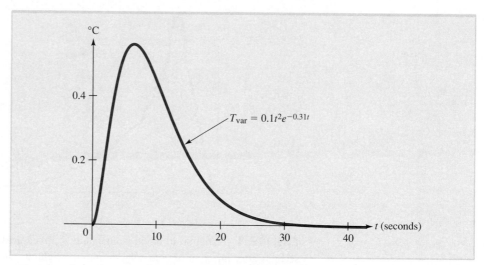

FIGURE 2   Approximation of temperature variation using $T_{var} = 0.1t^2 e^{-0.31t}$.

Using this approximation, suppose that $T_b = 37°C$, $T_i = 0°C$, $V = 10$, and $r = 0.9$ (so that 10% of the cooling effect of the injectant is lost as it passes into the bloodstream). Substituting into the equation for F, we get

$$F = \frac{(0.9)(10)(37 - 0)}{\int_0^\infty 0.1t^2 e^{-0.31t}\, dt}$$

Computing this improper integral using integration by parts (the details of this computation are left to the reader) shows that

$$F = \frac{333}{0.1[2/(0.31)^3]} \approx 49.6 \text{ milliliters/second} = 2.98 \text{ liters/minute}$$

Note that in our computations the numerator in the formula for F has units of milliliters times degrees Celsius and the denominator has units of degrees Celsius times seconds, so that F has units of milliliters per second. To convert this to units of liters per minute we multiply by $\dfrac{60}{1{,}000} = 0.06$. A cardiac output rate of 2.98 liters/minute indicates that our patient has a serious cardiovascular problem.

Questions

1. A physician injects 5 mg of dye into a vein near the heart of a patient and determines that the concentration of dye leaving the heart after t seconds is given by the function

    $$C(t) = 1.54te^{-0.12t} - 0.007t^2 \quad \text{mg/L}$$

    Assuming it takes 20 seconds for all the dye to clear the heart, what is the patient's cardiac output?

2. A physician injects 5 mg of dye into a vein near the heart of a patient. By monitoring the concentration of dye in the blood over a 24-second period, the physician determines that the concentration of dye leaving the heart after t seconds ($0 \leq t \leq 24$) is given by the function

    $$C(t) = -0.028t^2 + 0.672t \quad \text{mg/L}$$

    a. Use this information to find the patient's cardiac output.
    b. Sketch the graph of $C(t)$, and compare it to the graph in Figure 2. How are the graphs alike? How are they different?

3. Answer the questions in Exercise 2 for the dye concentration function

    $$C(t) = \begin{cases} 0 & \text{for } 0 \leq t \leq 2 \\ -0.034(t^2 - 26t + 48) & \text{for } 2 \leq t \leq 24 \end{cases}$$

4. Cardiac output is measured by injecting 10 milligrams of dye into a vein near the heart of a patient, and the dye concentration after t seconds is C mg/L, where C is modeled by the function $C(t) = 4te^{-0.15t}$ for $0 \leq t \leq 30$. Use integration by parts to find the cardiac output F (in liters per minute). Based on your result, do you think this patient is a healthy person at rest, a healthy person exercising, or a person with a serious cardiovascular problem?

5. Use Simpson's rule to estimate a patient's cardiac output if 7 milligrams of indocyanine green dye is injected into a major vein near the heart and the concentration of dye measured each 3 seconds is as shown in the following table.

Time t (seconds)	Concentration C(t) (mg/L)
0	0
3	1
6	3.5
9	5.8
12	4.4
15	3.2
18	2.1
21	1.2
24	0.5

Based on your result, do you think this patient is a healthy person at rest, a healthy person exercising, or a person with a serious cardiovascular problem?

6. Suppose that when the thermodilution method is applied to a patient, we find that $T_{var} = 0.2t^2 e^{-0.43t}$ can be used to model the temperature variation. Suppose that the patient's body temperature is 36°C, that 12 mL of injectant at temperature 5°C is inserted into a major artery near the patient's heart, and that the correction factor is $r = 0.9$. Estimate the cardiac output F for this patient.

7. The dye dilution approach to measuring cardiac output can be modified to apply to other flow problems. Use this approach to find a formula for P, the total daily discharge of a pollutant into a river flowing at the constant rate of F m³/hr, if it is known that at a fixed point downstream from the industrial plant where the discharge takes place, the concentration of pollutant at time t (hours) is given by the function $C(t)$.

8. A battery plant begins discharging cadmium into a river at midnight on a certain day and the concentration C of cadmium at a point 1,000 meters downriver is subsequently measured at 3-hour intervals over the next 24 hours, as indicated in the accompanying table. If the flow rate of the river is $F = 2.3 \times 10^5$ m³/hr, use the formula you found in Question 7 to estimate the total amount of cadmium discharged from the plant during this 24-hour period. You may assume that the concentration is 0 when the cadmium is first discharged; that is, $C(0) = 0$.

Time	Concentration (mg/m³)
3 A.M.	8.5
6 A.M.	12.0
9 A.M.	14.5
Noon	13.0
3 P.M.	12.5
6 P.M.	10.0
9 P.M.	7.5
Midnight	6.0

References

M. R. Cullen, *Linear Models in Biology*, Chichester: Ellis Horwood Ltd., 1985.

Arthur Segal, "Flow System Integrals," *The UMAP Journal*, Vol. III, No. 1, 1982.

Brindell Horelick and Sinan Koont, "Measuring Cardiac Output," UMAP Module 71, Birkhauser-Boston Inc., 1979.

William Simon, *Mathematical Techniques for Biology and Medicine*, Mineola, N.Y.: Dover Publications, 1986; pp. 194–210 (Chapter IX).

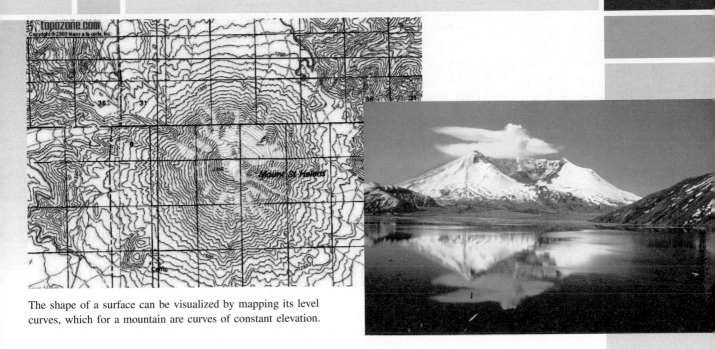

The shape of a surface can be visualized by mapping its level curves, which for a mountain are curves of constant elevation.

Calculus of Several Variables

SECTION 7.1   Functions of Several Variables

Learning Objectives

1. Define and examine functions of two or more variables.
2. Explore graphs and level curves of functions of two variables.
3. Study the Cobb-Douglas production function, isoquants, and indifference curves in economics.

In business, if a manufacturer determines that x units of a particular commodity can be sold domestically for \$90 per unit, and y units can be sold to foreign markets for \$110 per unit, then the total revenue obtained from all sales is given by

$$R = 90x + 110y$$

In psychology, a person's intelligence quotient (IQ) is measured by the ratio

$$\text{IQ} = \frac{100m}{a}$$

where a and m are the person's actual age and mental age, respectively. A carpenter constructing a storage box x feet long, y feet wide, and z feet deep knows that the box will have volume V and surface area S, where

$$V = xyz \quad \text{and} \quad S = 2xy + 2xz + 2yz$$

These are typical of practical situations in which a quantity of interest depends on the values of two or more variables. Other examples include the volume of water in a community's reservoir, which may depend on the amount of rainfall as well as the population of the community, and the output of a factory, which may depend on the amount of capital invested in the plant, the size of the labor force, and the cost of raw materials.

In this chapter, we will extend the methods of calculus to include functions of two or more independent variables. Most of our work will be with functions of two variables, which you will find can be represented geometrically by surfaces in three-dimensional space instead of curves in the plane. We begin with a definition and some terminology.

Function of Two Variables ▪ A function f of the two independent variables x and y is a rule that assigns to each ordered pair (x, y) in a given set D (the **domain** of f) exactly one real number, denoted by $f(x, y)$.

NOTE **Domain Convention:** Unless otherwise stated, we assume that the domain of f is the set of all (x, y) for which the expression $f(x, y)$ is defined. ▪

As in the case of a function of one variable, a function of two variables $f(x, y)$ can be thought of as a "machine" in which there is a unique "output" $f(x, y)$ for each "input" (x, y), as illustrated in Figure 7.1. The domain of f is the set of all possible inputs, and the set of all possible corresponding outputs is the **range** of f. Functions of three independent variables $f(x, y, z)$ or four independent variables $f(x, y, z, t)$, and so on can be defined in a similar fashion.

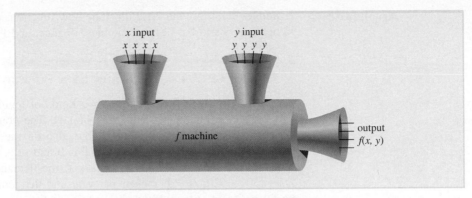

FIGURE 7.1   A function of two variables as a "machine."

EXPLORE!

Graphing calculators can naturally represent functions of a single independent variable but require additional capabilities to portray multivariate functions, such as three-dimensional graphing features. A simple method to deal with multivariate functions is to graph one variable at specific values of the other variables. For example, store $f(x, y) = x^3 - x^2y^2 - xy^3 - y^4$ into Y1 as X^3 − X^2*L1^2 − X*L1^3 − L1^4, where L1 is the list of values {0, 1.5, 2.0, 2.25, 2.5}. Graph using the modified decimal window [−9.4, 9.4]1 by [−150, 100]20. How does varying the y values in L1 affect the shape of the graph?

EXAMPLE 7.1.1   Evaluating a Function of Two Variables

Suppose $f(x, y) = \dfrac{3x^2 + 5y}{x - y}$.

a. Find the domain of f.

b. Compute $f(1, -2)$.

Solution

a. Since division by any real number except zero is possible, the expression $f(x, y)$ can be evaluated for all ordered pairs (x, y) with $x - y \neq 0$ or $x \neq y$. Geometrically, this is the set of all points in the xy plane except for those on the line $y = x$.

b. $f(1, -2) = \dfrac{3(1)^2 + 5(-2)}{1 - (-2)} = \dfrac{3 - 10}{1 + 2} = -\dfrac{7}{3}$.

EXAMPLE 7.1.2   Evaluating a Function of Two Variables

Suppose $f(x, y) = xe^y + \ln x$.

a. Find the domain of f.

b. Compute $f(e^2, \ln 2)$

Solution

a. Since xe^y is defined for all real numbers x and y and since $\ln x$ is defined only for $x > 0$, the domain of f consists of all ordered pairs (x, y) of real numbers for which $x > 0$.

b. $f(e^2, \ln 2) = e^2 e^{\ln 2} + \ln e^2 = 2e^2 + 2 = 2(e^2 + 1) \approx 16.78$

EXAMPLE 7.1.3   Evaluating a Function of Three Variables

Given the function of three variables $f(x, y, z) = xy + xz + yz$, evaluate $f(-1, 2, 5)$.

Solution

Substituting $x = -1$, $y = 2$, $z = 5$ into the formula for $f(x, y, z)$, we get

$$f(-1, 2, 5) = (-1)(2) + (-1)(5) + (2)(5) = 3$$

Applications

Examples 7.1.4 through 7.1.7 illustrate how functions of several variables can be used in applications to business, economics, finance, and life science.

EXAMPLE 7.1.4   Revenue as a Function of Two Variables

A sports store in St. Louis carries two kinds of tennis rackets, the Serena Williams and the Maria Sharapova autograph brands. The consumer demand for each brand depends not only on its own price, but also on the price of the competing brand. Sales figures indicate that if the Williams brand sells for x dollars per racket and the Sharapova brand for y dollars per racket, the demand for Williams rackets will be $D_1 = 300 - 20x + 30y$ rackets per year and the demand for Sharapova rackets will be $D_2 = 200 + 40x - 10y$ rackets per year. Express the store's total annual revenue from the sale of these rackets as a function of the prices x and y.

Solution

Let R denote the total monthly revenue. Then

$$R = \text{(number of Williams rackets sold)(price per Williams racket)}$$
$$+ \text{(number of Sharapova rackets sold)(price per Sharapova racket)}$$

Hence,

$$R(x, y) = (300 - 20x + 30y)(x) + (200 + 40x - 10y)(y)$$
$$= 300x + 200y + 70xy - 20x^2 - 10y^2$$

Output Q at a factory is often regarded as a function of the amount K of capital investment and the size L of the labor force. Output functions of the form

$$Q(K, L) = AK^\alpha L^\beta$$

where A, α, and β are positive constants with $\alpha + \beta = 1$, have proved to be especially useful in economic analysis and are known as **Cobb-Douglas production functions.** Example 7.1.5 involves such a function.

EXPLORE!

Refer to Example 7.1.5. Store the equation

  $0 = Q - 60K^(1/3)*L^(2/3)$

in the equation solver of your graphing calculator. Calculate the value of Q for $K = 512$ and $L = 1,000$. What happens to Q if the labor force L is doubled?

EXAMPLE 7.1.5   Production as a Function of Two Variables

Suppose that at a certain factory, output is given by the Cobb-Douglas production function $Q(K, L) = 60K^{1/3}L^{2/3}$ units, where K is the capital investment measured in units of $\$1,000$ and L the size of the labor force measured in worker-hours.

a. Compute the output if the capital investment is $\$512,000$ and 1,000 worker-hours of labor are used.

b. Show that the output in part (a) will double if both the capital investment and the size of the labor force are doubled.

Solution

a. Evaluate $Q(K, L)$ with $K = 512$ (thousand) and $L = 1,000$ to get

$$Q(512, 1,000) = 60(512)^{1/3}(1,000)^{2/3}$$
$$= 60(8)(100) = 48,000 \text{ units}$$

b. Evaluate $Q(K, L)$ with $K = 2(512)$ and $L = 2(1,000)$ as follows to get

$$Q[2(512), 2(1,000)] = 60[2(512)]^{1/3}[2(1,000)]^{2/3}$$
$$= 60(2)^{1/3}(512)^{1/3}(2)^{2/3}(1,000)^{2/3} = 96,000 \text{ units}$$

which is twice the output when $K = 512$ and $L = 1,000$.

EXAMPLE 7.1.6   Present Value as a Function of Four Variables

Recall (from Section 4.1) that the present value of B dollars in t years invested at the annual rate r compounded k times per year is given by

$$P(B, r, k, t) = B\left(1 + \frac{r}{k}\right)^{-kt}$$

Find the present value of \$10,000 in 5 years invested at 6% per year compounded quarterly.

Solution

We have $B = 10,000$, $r = 0.06$ (6% per year), $k = 4$ (compounded 4 times per year), and $t = 5$, so the present value is

$$P(10,000, 0.06, 4, 5) = 10,000\left(1 + \frac{0.06}{4}\right)^{-4(5)}$$
$$\approx 7,424.7$$

or approximately \$7,425.

EXAMPLE 7.1.7   Population as a Function of Three Variables

A population that grows exponentially satisfies

$$P(A, k, t) = Ae^{kt}$$

where P is the population at time t, A is the initial population (when $t = 0$), and k is the relative (per capita) growth rate. The population of a certain country is currently 5 million people and is growing at the rate of 3% per year. What will the population be in 7 years?

Solution

Let P be measured in millions of people. Substituting $A = 5$, $k = 0.03$ (3% annual growth), and $t = 7$ into the population function, we find that

$$P(5, 0.03, 7) = 5e^{0.03(7)} \approx 6.16839$$

Therefore, in 7 years, the population will be approximately 6,168,400 people.

Graphs of Functions of Two Variables

The **graph** of a function of two variables $f(x, y)$ is the set of all triples (x, y, z) such that (x, y) is in the domain of f and $z = f(x, y)$. To "picture" such graphs, we need to construct a **three-dimensional coordinate system.** The first step in this construction is to add a third coordinate axis (the z axis) perpendicular to the familiar xy coordinate plane, as shown in Figure 7.2. Note that the xy plane is taken to be horizontal, and the positive z axis is "up."

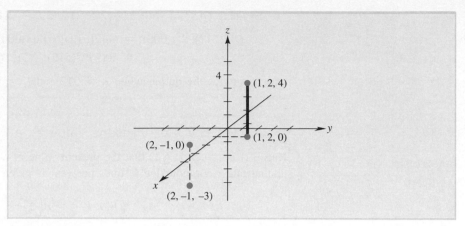

FIGURE 7.2   A three-dimensional coordinate system.

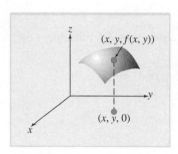

FIGURE 7.3   The graph of $z = f(x, y)$.

You can describe the location of a point in three-dimensional space by specifying three coordinates. For example, the point that is 4 units above the xy plane and lies directly above the point with xy coordinates $(x, y) = (1, 2)$ is represented by the ordered triple $(x, y, z) = (1, 2, 4)$. Similarly, the ordered triple $(2, -1, -3)$ represents the point that is 3 units directly below the point $(2, -1)$ in the xy plane. These points are shown in Figure 7.2.

To graph a function $f(x, y)$ of the two independent variables x and y, it is customary to introduce the letter z to stand for the dependent variable and to write $z = f(x, y)$ (Figure 7.3). The ordered pairs (x, y) in the domain of f are thought of as points in the xy plane, and the function f assigns a "height" z to each such point ("depth" if z is negative). Thus, if $f(1, 2) = 4$, you would express this fact geometrically by plotting the point $(1, 2, 4)$ in a three-dimensional coordinate space. The function may assign different heights to different points in its domain, and in general, its graph will be a surface in three-dimensional space.

Four such surfaces are shown in Figure 7.4 on page 535. The surface in Figure 7.4a is a **cone,** Figure 7.4b shows a **paraboloid,** Figure 7.4c shows an **ellipsoid,** and Figure 7.4d shows what is commonly called a **saddle surface.** Surfaces such as these play an important role in examples and exercises in this chapter.

Level Curves   It is usually not easy to sketch the graph of a function of two variables. One way to visualize a surface is shown in Figure 7.5 on page 535. Notice that when the plane $z = C$ intersects the surface $z = f(x, y)$, the result is a curve in space. The corresponding set of points (x, y) in the xy plane that satisfy $f(x, y) = C$ is called the **level curve** of f at C, and an entire family of level curves is generated as C varies over a set of numbers. By sketching members of this family in the xy plane, you can obtain a useful representation of the surface $z = f(x, y)$.

For instance, imagine the surface $z = f(x, y)$ as a "mountain" whose "elevation" at the point (x, y) is given by $f(x, y)$, as shown in Figure 7.6a on page 536. The level curve $f(x, y) = C$ lies directly below a path on the mountain where the elevation is always C. To graph the mountain, you can indicate the paths of constant elevation by sketching the family of level curves in the plane and pinning a "flag" to

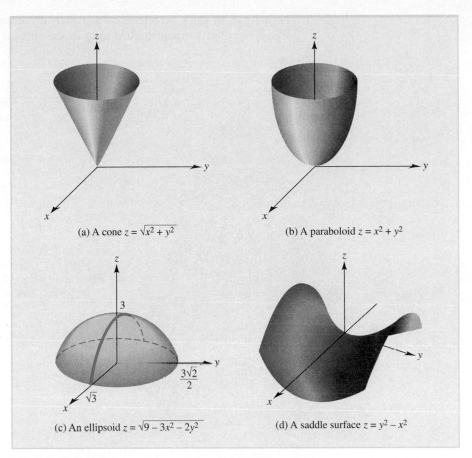

(a) A cone $z = \sqrt{x^2 + y^2}$

(b) A paraboloid $z = x^2 + y^2$

(c) An ellipsoid $z = \sqrt{9 - 3x^2 - 2y^2}$

(d) A saddle surface $z = y^2 - x^2$

FIGURE 7.4   Several surfaces in three-dimensional space.

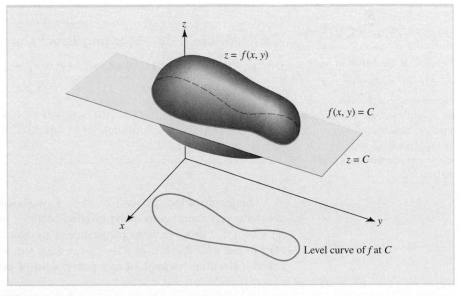

FIGURE 7.5   A level curve of the surface $z = f(x, y)$.

each curve to show the elevation to which it corresponds (Figure 7.6b). This "flat" figure is called a **topographical map** of the surface $z = f(x, y)$.

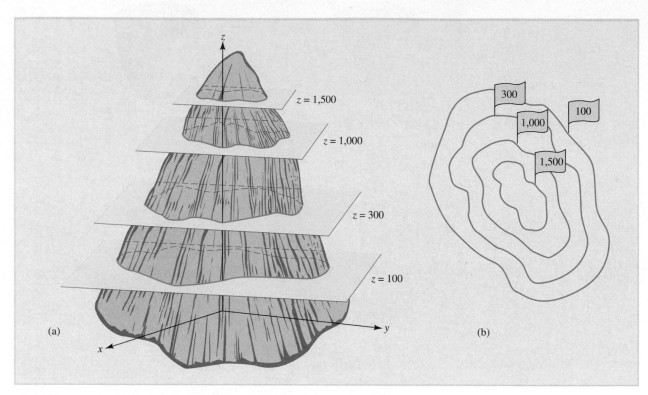

FIGURE 7.6   (a) The surface $z = f(x, y)$ as a mountain, and (b) level curves provide a topographical map of $z = f(x, y)$.

EXAMPLE 7.1.8   Studying Level Curves

Discuss the level curves of the function $f(x, y) = x^2 + y^2$.

Solution

The level curve $f(x, y) = C$ has the equation $x^2 + y^2 = C$. If $C = 0$, this is the point $(0, 0)$, and if $C > 0$, it is a circle of radius $\sqrt{C}$. If $C < 0$, there are no points that satisfy $x^2 + y^2 = C$.

The graph of the surface $z = x^2 + y^2$ is shown in Figure 7.7. The level curves you have just found correspond to cross sections perpendicular to the z axis. It can be shown that cross sections perpendicular to the x axis and the y axis are parabolas. (Try to see why this is true.) For this reason, the surface is shaped like a bowl. It is called a **circular paraboloid** or a **paraboloid of revolution**.

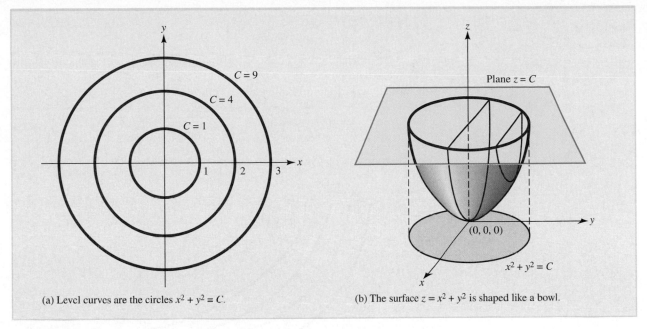

(a) Level curves are the circles $x^2 + y^2 = C$.

(b) The surface $z = x^2 + y^2$ is shaped like a bowl.

FIGURE 7.7   Level curves help to visualize the shape of a surface.

Level Curves in Economics: Isoquants and Indifference Curves

Level curves appear in many different applications. For instance, in economics, if the output $Q(x, y)$ of a production process is determined by two inputs x and y (say, hours of labor and capital investment), then the level curve $Q(x, y) = C$ is called the **curve of constant product** C or, more briefly, an **isoquant** (*iso* means "equal").

Another application of level curves in economics involves the concept of indifference curves. A consumer who is considering the purchase of a number of units of each of two commodities is associated with a **utility function** $U(x, y)$, which measures the total satisfaction (or **utility**) the consumer derives from having x units of the first commodity and y units of the second. A level curve $U(x, y) = C$ of the utility function is called an **indifference curve** and gives all the combinations of x and y that lead to the same level of consumer satisfaction. These terms are illustrated in Example 7.1.9.

EXPLORE!

Refer to Example 7.1.9. Represent the indifference curves $U(x, y) = x^{3/2}y = C$ by solving for y to get $y = Cx^{-3/2}$. Put X^(−3/2)*L1 into Y1 of the equation editor, where L1 = {800, 1,280, 2,000, 3,000} lists a few levels of constant utility C. Graph with the window [0, 37.6]5 by [0, 150]10. What effect does changing C have on the graph? Locate the point (16, 20) on the indifference curve $x^{3/2}y = 1,280$.

EXAMPLE 7.1.9 Applying Level Curves to Economics

Suppose the utility derived by a consumer from x units of one commodity and y units of a second commodity is given by the utility function $U(x, y) = x^{3/2}y$. If the consumer currently owns $x = 16$ units of the first commodity and $y = 20$ units of the second, find the consumer's current level of utility and sketch the corresponding indifference curve.

Solution

The current level of utility is

$$U(16, 20) = (16)^{3/2}(20) = 1,280$$

and the corresponding indifference curve is

$$x^{3/2}y = 1,280$$

or $y = 1,280x^{-3/2}$. This curve consists of all points (x, y) where the level of utility $U(x, y)$ is 1,280. The curve $x^{3/2}y = 1,280$ and several other curves of the family $x^{3/2}y = C$ are shown in Figure 7.8.

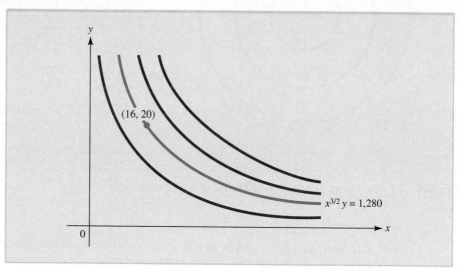

FIGURE 7.8 Indifference curves for the utility function $U(x, y) = x^{3/2}y$.

Computer Graphics   In practical work in the social, managerial, or life sciences, you will rarely, if ever, have to graph a function of two variables. Hence, we will spend no more time developing graphing procedures for such functions.

Computer software is now available for graphing functions of two variables. Such software often allows you to choose different scales along each coordinate axis and may also enable you to visualize a given surface from different viewpoints. These features permit you to obtain a detailed picture of the graph. A variety of computer-generated graphs are displayed in Figure 7.9.

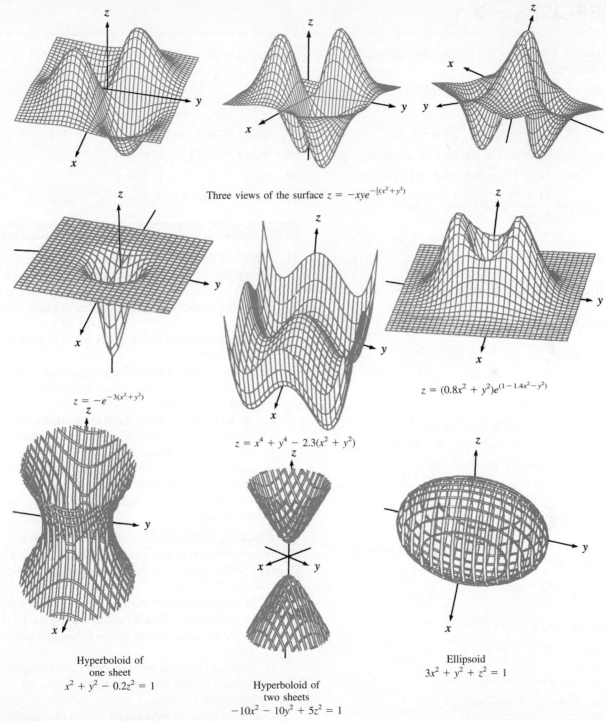

Three views of the surface $z = -xye^{-\frac{1}{2}(x^2+y^2)}$

$z = -e^{-3(x^2+y^2)}$

$z = (0.8x^2 + y^2)e^{(1-1.4x^2-y^2)}$

$z = x^4 + y^4 - 2.3(x^2 + y^2)$

Hyperboloid of
one sheet
$x^2 + y^2 - 0.2z^2 = 1$

Hyperboloid of
two sheets
$-10x^2 - 10y^2 + 5z^2 = 1$

Ellipsoid
$3x^2 + y^2 + z^2 = 1$

FIGURE 7.9   Some computer-generated surfaces.

EXERCISES ▪ 7.1

In Exercises 1 through 16, compute the indicated functional value.

1. $f(x, y) = 5x + 3y$; $f(-1, 2)$, $f(3, 0)$

2. $f(x, y) = x^2 + x - 4y$; $f(1, 3)$, $f(2, -1)$

3. $g(x, y) = x(y - x^3)$; $g(1, 1)$, $g(-1, 4)$

4. $g(x, y) = xy - x(y + 1)$; $g(1, 0)$, $g(-2, 3)$

5. $f(x, y) = (x - 1)^2 + 2xy^3$; $f(2, -1)$, $f(1, 2)$

6. $f(x, y) = \dfrac{3x + 2y}{2x + 3y}$; $f(1, 2)$, $f(-4, 6)$

7. $g(x, y) = \sqrt{y^2 - x^2}$; $g(4, 5)$, $g(-1, 2)$

8. $g(u, v) = 10u^{1/2}v^{2/3}$; $g(16, 27)$, $g(4, -1,331)$

9. $f(r, s) = \dfrac{s}{\ln r}$; $f(e^2, 3)$, $f(\ln 9, e^3)$

10. $f(x, y) = xye^{xy}$; $f(1, \ln 2)$, $f(\ln 3, \ln 4)$

11. $g(x, y) = \dfrac{y}{x} + \dfrac{x}{y}$; $g(1, 2)$, $g(2, -3)$

12. $f(s, t) = \dfrac{e^{st}}{2 - e^{st}}$; $f(1, 0)$, $f(\ln 2, 2)$

13. $f(x, y, z) = xyz$; $f(1, 2, 3)$, $f(3, 2, 1)$

14. $g(x, y, z) = (x + y)e^{yz}$; $g(1, 0, -1)$, $g(1, 1, 2)$

15. $F(r, s, t) = \dfrac{\ln(r + t)}{r + s + t}$; $F(1, 1, 1)$, $F(0, e^2, 3e^2)$

16. $f(x, y, z) = xye^z + xze^y + yze^x$; $f(1, 1, 1)$, $f(\ln 2, \ln 3, \ln 4)$

In Exercises 17 through 22, describe the domain of the given function.

17. $f(x, y) = \dfrac{5x + 2y}{4x + 3y}$

18. $f(x, y) = \sqrt{9 - x^2 - y^2}$

19. $f(x, y) = \sqrt{x^2 - y}$

20. $f(x, y) = \dfrac{x}{\ln(x + y)}$

21. $f(x, y) = \ln(x + y - 4)$

22. $f(x, y) = \dfrac{e^{xy}}{\sqrt{x - 2y}}$

In Exercises 23 through 30, sketch the indicated level curve $f(x, y) = C$ for each choice of constant C.

23. $f(x, y) = x + 2y$; $C = 1$, $C = 2$, $C = -3$

24. $f(x, y) = x^2 + y$; $C = 0$, $C = 4$, $C = 9$

25. $f(x, y) = x^2 - 4x - y$; $C = -4$, $C = 5$

26. $f(x, y) = \dfrac{x}{y}$; $C = -2$, $C = 2$

27. $f(x, y) = xy$; $C = 1$, $C = -1$, $C = 2$, $C = -2$

28. $f(x, y) = ye^x$; $C = 0$, $C = 1$

29. $f(x, y) = xe^y$; $C = 1$, $C = e$

30. $f(x, y) = \ln(x^2 + y^2)$; $C = 4$, $C = \ln 4$

BUSINESS AND ECONOMICS APPLIED PROBLEMS

31. **PRODUCTION** Using x skilled workers and y unskilled workers, a manufacturer can produce $Q(x, y) = 10x^2y$ units per day. Currently there are 20 skilled workers and 40 unskilled workers on the job.
    a. How many units are currently being produced each day?
    b. By how much will the daily production level change if 1 more skilled worker is added to the current workforce?
    c. By how much will the daily production level change if 1 more unskilled worker is added to the current workforce?
    d. By how much will the daily production level change if 1 more skilled worker *and* 1 more unskilled worker are added to the current workforce?

32. **PRODUCTION COST** Victor Murray's firm manufactures scientific graphing calculators that cost $40 each to produce and business calculators that cost $20 each.
    a. If x graphing calculators and y business calculators are produced each month, express the total monthly cost of production as a function of x and y.
    b. Compute the total monthly cost if 500 graphing calculators and 800 business calculators are produced.
    c. Victor wants to increase the output of graphing calculators by 50 each month from the level in part (b) but wants the total cost of production to remain the same. What change should he make in the monthly production of business calculators to achieve this goal?

33. RETAIL SALES A paint store carries two brands of latex paint. Sales figures indicate that if the first brand is sold for x_1 dollars per gallon and the second for x_2 dollars per gallon, the demand for the first brand will be $D_1(x_1, x_2) = 200 - 10x_1 + 20x_2$ gallons per month and the demand for the second brand will be $D_2(x_1, x_2) = 100 + 5x_1 - 10x_2$ gallons per month.

  a. Express the paint store's total monthly revenue from the sale of the paint as a function of the prices x_1 and x_2.

  b. Compute the revenue in part (a) if the first brand is sold for $21 per gallon and the second for $16 per gallon.

34. PRODUCTION The output at a certain factory is $Q(K, L) = 120K^{2/3}L^{1/3}$ units, where K is the capital investment measured in units of $1,000 and L the size of the labor force measured in worker-hours.

  a. Compute the output if the capital investment is $125,000 and the size of the labor force is 1,331 worker-hours.

  b. What will happen to the output in part (a) if both the level of capital investment and the size of the labor force are cut in half?

35. PRODUCTION The Easy-Gro agricultural company estimates that when $100x$ worker-hours of labor are employed on y acres of land, the number of bushels of wheat produced is $f(x, y) = Ax^a y^b$, where A, a, and b are positive constants. Suppose the company decides to double the production factors x and y. Determine how this decision affects the production of wheat in each of these cases:

  a. $a + b > 1$
  b. $a + b < 1$
  c. $a + b = 1$

36. PRODUCTION Suppose that when x machines and y worker-hours are used each day, a certain factory will produce $Q(x, y) = 10xy$ cell phones. Describe the relationship between the inputs x and y that results in an output of 1,000 phones each day. (Note that you are finding a level curve of Q.)

37. RETAIL SALES A manufacturer with exclusive rights to a sophisticated new industrial machine is planning to sell a limited number of the machines to both foreign and domestic firms. The price the manufacturer can expect to receive for the machines will depend on the number of machines made available. It is estimated that if the

manufacturer supplies x machines to the domestic market and y machines to the foreign market, the machines will sell for

$$60 - \frac{x}{5} + \frac{y}{20} \text{ thousand dollars apiece domestically}$$

and $50 - \dfrac{y}{10} + \dfrac{x}{20}$ thousand dollars apiece abroad. Express the revenue R as a function of x and y.

38. RETAIL SALES A manufacturer is planning to sell a new product at the price of A dollars per unit and estimates that if x thousand dollars is spent on development and y thousand dollars on promotion, consumers will buy approximately $\dfrac{320y}{y + 2} + \dfrac{160x}{x + 4}$ units of the product. If manufacturing costs are $50 per unit, express profit in terms of x, y, and A. [*Hint:* profit = revenue − total cost of manufacture, development, and promotion.]

39. CONSTANT PRODUCTION CURVES Using x skilled and y unskilled workers, a manufacturer can produce $Q(x, y) = 3x + 2y$ units per day. Currently the workforce consists of 10 skilled workers and 20 unskilled workers.

  a. Compute the current daily output.

  b. Find an equation relating the levels of skilled and unskilled labor if the daily output is to remain at its current level.

  c. On a two-dimensional coordinate system, draw the isoquant (constant production curve) that corresponds to the current level of output.

  d. What change should be made in the level of unskilled labor y to offset an increase in skilled labor x of two workers so that the output will remain at its current level?

40. INDIFFERENCE CURVES Suppose the utility derived by a consumer from x units of one commodity and y units of a second commodity is given by the utility function $U(x, y) = 2x^3 y^2$. The consumer currently owns $x = 5$ units of the first commodity and $y = 4$ units of the second. Find the consumer's current level of utility, and sketch the corresponding indifference curve.

41. INDIFFERENCE CURVES Beverly estimates that when she is corresponding on a regular basis with x close friends and is working on y interesting projects, her total satisfaction is measured by the utility function $U(x, y) = (x + 1)(y + 2)$. What is Beverly's level of satisfaction if she is currently communicating with 25 close friends and working on 8 interesting projects? Sketch the corresponding indifference curve. Explain what *indifference* means in this situation.

42. **CONSTANT RETURNS TO SCALE**   Suppose output Q is given by the Cobb-Douglas production function $Q(K, L) = AK^{\alpha}L^{1-\alpha}$, where A and α are positive constants and $0 < \alpha < 1$. Show that if K and L are both multiplied by the same positive number m, then the output Q will also be multiplied by m; that is, show that $Q(mK, mL) = mQ(K, L)$. A production function with this property is said to have *constant returns to scale*.

43. **AMORTIZATION OF DEBT**   Suppose a loan of A dollars is amortized over n years at an annual interest rate r compounded monthly. Let $i = \dfrac{r}{12}$ be the equivalent monthly rate of interest. Then the monthly payments will be M dollars, where

$$M(A, n, i) = \frac{Ai}{1 - (1 + i)^{-12n}}$$

(See Exercises 56 through 59 in Section 4.1.)

a. Allison has a home mortgage of $250,000 at the fixed rate of 5.2% per year for 15 years. What are her monthly payments? How much total interest does she pay for this loan?

b. Nathan also has a mortgage of $250,000 but at the fixed rate of 5.6% per year for 30 years. What are his monthly payments? How much total interest does he pay?

44. **INVENTORY**   Suppose a company requires N units per year of a certain commodity. Suppose further that it costs D dollars to order a shipment of the commodity and that the storage cost per unit is S dollars per year. The units are used (or sold) at a constant rate throughout the year, and each shipment is used up just as a new shipment arrives.

a. Show that the total cost $C(x)$ of maintaining inventory when x units are ordered in each shipment is minimized when $x = Q$ where

$$Q(N, D, S) = \sqrt{\frac{2DN}{S}}$$

(See Example 3.5.7 in Section 3.5.)

b. The optimum order size Q found in part (a) is called the *economic order quantity* (EOQ). What is the EOQ when 9,720 units per year are ordered with an ordering cost of $35 per shipment and a storage cost of 84 cents per unit per year?

45. **CES PRODUCTION**   A **constant elasticity of substitution** (CES) production function is one with the general form

$$Q(K, L) = A[\alpha K^{-\beta} + (1 - \alpha)L^{-\beta}]^{-1/\beta}$$

where K is capital expenditure; L is the level of labor; and A, α, β are constants that satisfy $A > 0$, $0 < \alpha < 1$, and $\beta > -1$. Show that such a function has *constant returns to scale*: that is,

$$Q(sK, sL) = sQ(K, L)$$

for any constant multiplier s. (Compare with Exercise 42.)

46. **PRODUCTION**   Output in a certain factory is given by the Cobb-Douglas production function

$$Q(K, L) = 57K^{1/4}L^{3/4}$$

where K is the capital in $1,000 and L is the size of the labor force, measured in worker-hours.

a. Use your calculator to obtain $Q(K, L)$ for the values of K and L in this table:

K($1,000)	277	311	493	554	718
L	743	823	1,221	1,486	3,197
$Q(K, L)$					

b. Note that the output $Q(277, 743)$ is doubled when K is doubled from 277 to 554 and L is doubled from 743 to 1,486. In a similar manner, verify that output is tripled when K and L are both tripled, and that output is halved when K and L are both halved. What do you think happens if K is doubled and L is halved? Verify your response with your calculator.

LIFE AND SOCIAL SCIENCE APPLIED PROBLEMS

47. **SURFACE AREA OF THE HUMAN BODY** Pediatricians and medical researchers sometimes use the following empirical formula* relating the surface area S (m²) of a person to the person's weight W (kg) and height H (cm):

$$S(W, H) = 0.0072W^{0.425}H^{0.725}$$

a. Find $S(15.83, 87.11)$. Sketch the level curve of $S(W, H)$ that passes through (15.83, 87.11). Sketch several additional level curves of $S(W, H)$. What do these level curves represent?

b. If Marc weighs 18.37 kg and has surface area 0.648 m², approximately how tall would you expect him to be?

---

*J. Routh, *Mathematical Preparation for Laboratory Technicians*, Philadelphia: Saunders Co., 1971, p. 92.

c. Suppose at some time in her life, Jenny weighs six times as much and is twice as tall as she was at birth. What is the corresponding percentage change in the surface area of her body?

48. **WIND POWER** A machine based on wind energy generally converts the kinetic energy of moving air to mechanical energy by means of a device such as a rotating shaft, as in a windmill. Suppose we have wind of velocity v traveling through a wind-collecting machine with cross-sectional area A. Then, in physics* it is shown that the total power generated by the wind is given by a formula of the form

$$P(v, A) = abAv^3$$

where $b = 1.2$ kg/m³ is the density of the air and a is a positive constant.

a. If a wind machine were perfectly efficient, then $a = \dfrac{1}{2}$ in the formula for $P(v, A)$. How much power would be produced by such an ideal windmill with blade radius 15 meters if the wind speed is 22 m/sec?

b. No wind machine is perfectly efficient. In fact, it has been shown that the best we can hope for is about 59% of ideal efficiency, and a good empirical formula for power is obtained by taking $a = \dfrac{8}{27}$. Compute $P(v, A)$ using this value of a if the blade radius of the windmill in part (a) is doubled and the wind velocity is halved.

c. Humankind has been trying to harness the wind for at least 4,000 years, sometimes with interesting or bizarre consequences. For example, in the source quoted in the footnote, the author notes that during World War II, a windmill was built in Vermont with a blade radius of 175 feet! Read an article on windmills and other devices using wind power. Do you think these devices have any place in the modern technological world? Explain.

49. **DAILY ENERGY EXPENDITURE** Suppose a person of age A years has weight w in kilograms (kg) and height h in centimeters (cm). Then, the

Harris-Benedict equations say that the daily basal energy expenditure in kilocalories is

$$B_m(w, h, A) = 66.47 + 13.75w + 5.00h - 6.77A$$

for a male and

$$B_f(w, h, A) = 655.10 + 9.60w + 1.85h - 4.68A$$

for a female.

a. Find the basal energy expenditure of a man who weighs 90 kg, is 190 cm tall, and is 22 years old.

b. Find the basal energy expenditure of a woman who weighs 61 kg, is 170 cm tall, and is 27 years old.

c. A man maintains a weight of 85 kg and a height of 193 cm throughout his adult life. At what age will his daily basal energy expenditure be 2,018 kilocalories?

d. A woman maintains a weight of 67 kg and a height of 173 cm throughout her adult life. At what age will her daily basal energy expenditure be 1,504 kilocalories?

50. **AIR POLLUTION** Dumping and other material-handling operations near a landfill may result in contaminated particles being emitted into the surrounding air. To estimate such particulate emission, the following empirical formula* can be used:

$$E(V, M) = k(0.0032)\left(\frac{V}{5}\right)^{1.3}\left(\frac{M}{2}\right)^{-1.4}$$

where E is the emission factor (pounds of particles released into the air per ton of soil moved), V is the mean speed of the wind (mph), M is the moisture content of the material (given as a percentage), and k is a constant that depends on the size of the particles.

a. For a small particle (diameter 5 mm), it turns out that $k = 0.2$. Find $E(10, 13)$.

b. The emission factor E can be multiplied by the number of tons of material handled to obtain a measure of total emissions. Suppose 19 tons of the material in part (a) is handled. How many tons of a second kind of material with $k = 0.48$ (diameter 15 mm) and moisture content 27% must be handled to achieve the same level of total emissions if the wind velocity stays the same?

---

*Raymond A. Serway, *Physics,* 3rd ed., Philadelphia: PA Saunders, 1992, pp. 408–410.

*M. D. LaGrega, P. L. Buckingham, and J. C. Evans, *Hazardous Waste Management,* New York: McGraw-Hill, 1994, p. 140.

c. Sketch several level curves of $E(V, M)$, assuming the size of the particle stays fixed. What is represented by these curves?

51. FLOW OF BLOOD   One of Poiseuille's laws[†] says that the speed of blood V (cm/sec) flowing at a distance r (cm) from the axis of a blood vessel of radius R (cm) and length L (cm) is given by

$$V(P, L, R, r) = \frac{9.3P}{L}(R^2 - r^2)$$

where P (dynes/cm^2) is the pressure in the vessel. Suppose a particular vessel has radius 0.0075 cm and is 1.675 cm long.

a. How fast is the blood flowing at a distance 0.004 cm from the axis of this vessel if the pressure in the vessel is 3,875 dynes/cm^2?

b. Since R and L are fixed for this vessel, V is a function of P and r alone. Sketch several level curves of $V(P, r)$. Explain what they represent.

52. PSYCHOLOGY   A person's intelligence quotient (IQ) is measured by the function

$$I(m, a) = \frac{100m}{a}$$

where a is the person's actual age and m is his or her mental age.

a. Find $I(12, 11)$ and $I(16, 17)$.

b. Sketch the graphs of several level curves of $I(m, a)$. How would you describe these curves?

MISCELLANEOUS PROBLEMS

53. REVERSE OSMOSIS   In manufacturing semiconductors, it is necessary to use water with an extremely low mineral content, and to separate water from contaminants, it is common practice to use a membrane process called **reverse osmosis.** A key to the effectiveness of such a process is the **osmotic pressure,** which may be determined by the **van't Hoff equation:**[*]

$$P(N, C, T) = 0.075NC(273.15 + T)$$

where P is the osmotic pressure (in atmospheres), N is the number of ions in each molecule of solute, C is the concentration of solute (gram-mole/ liter), and T is the temperature of the solute (°C). Find the osmotic pressure for a sodium chloride brine solution with concentration 0.53 gram-mole/ liter at a temperature of 23°C. (You will need to know that each molecule of sodium chloride contains two ions: $NaCl = Na^+ + Cl^-$.)

54. OPTICS   The lens equation in optics says that

$$\frac{1}{d_o} + \frac{1}{d_i} = \frac{1}{F}$$

where d_o is the distance of an object from a thin, spherical lens; d_i is the distance of its image on the other side of the lens; and F is the focal length of the lens (see the accompanying figure). Express F in terms of d_o and d_i, and sketch several level curves of this function (curves of constant focal length).

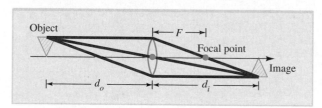

EXERCISE 54

55. CHEMISTRY   **Van der Waal's equation** of state says that 1 mole of a confined gas satisfies the equation

$$T(P, V) = 0.0122\left(P + \frac{a}{V^2}\right)(V - b) - 273.15$$

where T (°C) is the temperature of the gas, V (cm^3) is its volume, P (atmospheres) is the pressure of the gas on the walls of its container, and a and b are constants that depend on the nature of the gas.

a. Sketch the graphs of several level curves of T. These curves are called **curves of constant temperature** or **isotherms.**

b. If the confined gas is chlorine, experiments show that $a = 6.49 \times 10^6$ and $b = 56.2$. Find $T(1.13, 31.275 \times 10^3)$, that is, the temperature that corresponds to 31,275 cm^3 of chlorine under 1.13 atmospheres of pressure.

[†]E. Batschelet, *Introduction to Mathematics for Life Scientists,* 2nd ed., New York: Springer-Verlag, 1979, pp. 102–103.

[*]M. D. LaGrega, P. L. Buckingham, and J. C. Evans, *Hazardous Waste Management,* New York: McGraw-Hill, 1994, pp. 530–543.

56. ICE AGE PATTERNS OF ICE AND TEMPERATURE   The level curves in the land areas of the accompanying figure indicate ice elevations above sea level (in meters) during the last major ice age (approximately 18,000 years ago). The level curves in the sea areas indicate sea surface temperature. For instance, the ice was 1,000 meters thick above New York City, and the sea temperature near the Hawaiian Islands was about 24°C. Where on earth was the ice pack the thickest? What ice-bound land area was adjacent to the warmest sea?

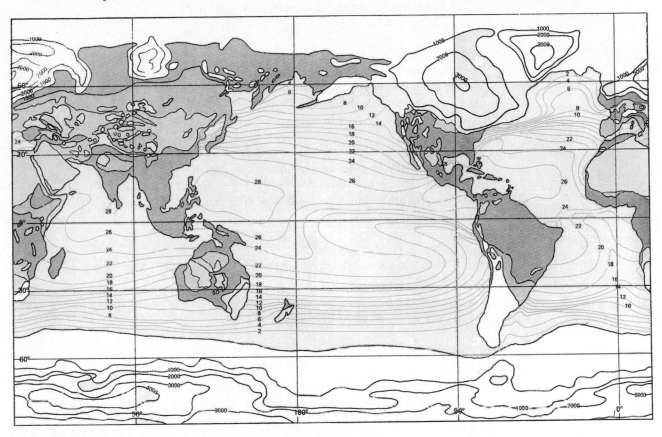

EXERCISE 56

Source: The Cambridge Encyclopedia of Earth Sciences, New York: Crown/Cambridge Press, 1981, p. 302.

SECTION 7.2 Partial Derivatives

Learning Objectives

1. Compute and interpret partial derivatives.
2. Apply partial derivatives to study marginal analysis problems in economics.
3. Compute second-order partial derivatives.
4. Use the chain rule for partial derivatives to find rates of change and make incremental approximations.

In many problems involving functions of two variables, the goal is to find the rate of change of the function with respect to one of its variables when the other is held constant. That is, the goal is to differentiate the function with respect to the particular variable in question while keeping the other variable fixed. This process is known as **partial differentiation,** and the resulting derivative is said to be a **partial derivative** of the function.

For example, suppose a manufacturer finds that

$$Q(x, y) = 5x^2 + 7xy$$

units of a certain commodity will be produced when x skilled workers and y unskilled workers are employed. Then if the number of unskilled workers remains fixed, the production rate with respect to the number of skilled workers is found by differentiating $Q(x, y)$ with respect to x while holding y constant. We call this the **partial derivative of Q with respect to x** and denote it by $Q_x(x, y)$; thus,

$$Q_x(x, y) = 5(2x) + 7(1)y = 10x + 7y$$

Similarly, if the number of skilled workers remains fixed, the production rate with respect to the number of unskilled workers is given by the **partial derivative of Q with respect to y,** which is obtained by differentiating $Q(x, y)$ with respect to y, holding x constant; that is, by

$$Q_y(x, y) = (0) + 7x(1) = 7x$$

Here is a general definition of partial derivatives and some alternative notation.

Partial Derivatives ■ Suppose $z = f(x, y)$. The partial derivative of f with respect to x is denoted by

$$\frac{\partial z}{\partial x} \quad \text{or} \quad f_x(x, y)$$

and is the function obtained by differentiating f with respect to x, treating y as a constant. The partial derivative of f with respect to y is denoted by

$$\frac{\partial z}{\partial y} \quad \text{or} \quad f_y(x, y)$$

and is the function obtained by differentiating f with respect to y, treating x as a constant.

NOTE   Recall from Chapter 2 that the derivative of a function of one variable $f(x)$ is defined by the limit of a difference quotient; namely,

$$f'(x) = \lim_{h \to 0} \frac{f(x + h) - f(x)}{h}$$

With this definition in mind, the partial derivative $f_x(x, y)$ is given by

$$f_x(x, y) = \lim_{h \to 0} \frac{f(x + h, y) - f(x, y)}{h}$$

and the partial derivative $f_y(x, y)$ by

$$f_y(x, y) = \lim_{h \to 0} \frac{f(x, y + h) - f(x, y)}{h}$$

Computation of Partial Derivatives

No new rules are needed for the computation of partial derivatives. To compute f_x, simply differentiate f with respect to the single variable x, pretending that y is a constant. To compute f_y, differentiate f with respect to y, pretending that x is a constant. Here are some examples.

EXAMPLE 7.2.1 Finding Partial Derivatives

Find the partial derivatives f_x and f_y if $f(x, y) = x^2 + 2xy^2 + \dfrac{2y}{3x}$.

Solution

To simplify the computation, begin by rewriting the function as

$$f(x, y) = x^2 + 2xy^2 + \frac{2}{3}yx^{-1}$$

To compute f_x, think of f as a function of x and differentiate the sum term by term, treating y as a constant to get

$$f_x(x, y) = 2x + 2(1)y^2 + \frac{2}{3}y(-x^{-2}) = 2x + 2y^2 - \frac{2y}{3x^2}$$

To compute f_y, think of f as a function of y and differentiate term by term, treating x as a constant to get

$$f_y(x, y) = 0 + 2x(2y) + \frac{2}{3}(1)x^{-1} = 4xy + \frac{2}{3x}$$

EXPLORE!

The graphing calculator can be used to calculate and visualize partial derivatives at specific points. Following Example 7.2.1, let

$f(x, y) = x^2 + 2xy^2 + \dfrac{2y}{3x}$

and store into Y1 the corresponding expression X^2 + 2X*L1^2 + 2L1/(3X). Suppose we wish to compute the partial derivative $f_x(-2, -1)$. Store the y value -1 into L1. Graph Y1 employing a decimal window, and determine the value of $f_x(-2, -1)$, using the derivative-finding feature of your graphing calculator.

EXAMPLE 7.2.2 Finding Partial Derivatives

Find the partial derivatives $\dfrac{\partial z}{\partial x}$ and $\dfrac{\partial z}{\partial y}$ if $z = (x^2 + xy + y)^5$.

Solution

Holding y fixed and using the chain rule to differentiate z with respect to x, you get

$$\frac{\partial z}{\partial x} = 5(x^2 + xy + y)^4 \frac{\partial}{\partial x}(x^2 + xy + y)$$

$$= 5(x^2 + xy + y)^4(2x + y)$$

Holding x fixed and using the chain rule to differentiate z with respect to y, you get

$$\frac{\partial z}{\partial y} = 5(x^2 + xy + y)^4 \frac{\partial}{\partial y}(x^2 + xy + y)$$

$$= 5(x^2 + xy + y)^4(x + 1)$$

EXAMPLE 7.2.3 Finding Partial Derivatives

Find the partial derivatives f_x and f_y if $f(x, y) = xe^{-2xy}$.

Solution

Using the product rule and the rule for exponential functions, we find

$$f_x(x, y) = x(-2ye^{-2xy}) + e^{-2xy} = (-2xy + 1)e^{-2xy}$$

Then, using the constant multiple rule and the rule for exponential functions, we find

$$f_y(x, y) = x(-2xe^{-2xy}) = -2x^2e^{-2xy}$$

Geometric Interpretation of Partial Derivatives

Recall from Section 7.1 that functions of two variables can be represented graphically as surfaces drawn on three-dimensional coordinate systems. In particular, if $z = f(x, y)$, an ordered pair (x, y) in the domain of f can be identified with a point in the xy plane and the corresponding function value $z = f(x, y)$ can be thought of as assigning a "height" to this point. The graph of f is the surface consisting of all points (x, y, z) in three-dimensional space whose height z is equal to $f(x, y)$.

The partial derivatives of a function of two variables can be interpreted geometrically as follows. For each fixed number y_0, the points (x, y_0, z) form a vertical plane whose equation is $y = y_0$. If $z = f(x, y)$ and if y is kept fixed at $y = y_0$, then the corresponding points $(x, y_0, f(x, y_0))$ form a curve in a three-dimensional space that is the intersection of the surface $z = f(x, y)$ with the plane $y = y_0$. At each point on this curve, the partial derivative $\dfrac{\partial z}{\partial x}$ is simply the slope of the line in the plane $y = y_0$ that is tangent to the curve at the point in question. That is, $\dfrac{\partial z}{\partial x}$ is the slope of the tangent line in the x direction. The situation is illustrated in Figure 7.10a.

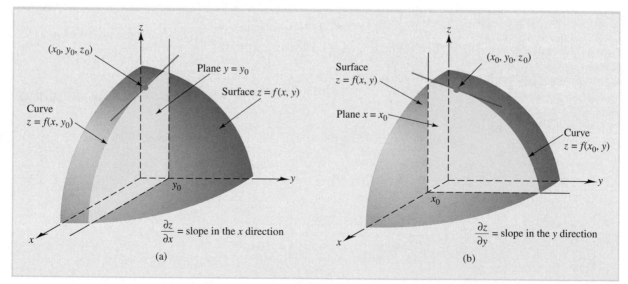

FIGURE 7.10 Geometric interpretation of partial derivatives.

Similarly, if x is kept fixed at $x = x_0$, the corresponding points $(x_0, y, f(x_0, y))$ form a curve that is the intersection of the surface $z = f(x, y)$ with the vertical plane $x = x_0$. At each point on this curve, the partial derivative $\dfrac{\partial z}{\partial y}$ is the slope of the tangent line in the plane $x = x_0$. That is, $\dfrac{\partial z}{\partial y}$ is the slope of the tangent line in the y direction.

The situation is illustrated in Figure 7.10b.

Marginal Analysis

In economics, the term **marginal analysis** refers to the practice of using a derivative to estimate the change in the value of a function resulting from a 1-unit increase in one of its variables. In Section 2.5, you saw some examples of marginal analysis involving ordinary derivatives of functions of one variable. Example 7.2.4 shows how partial derivatives can be used in a similar fashion.

EXAMPLE 7.2.4 Using Marginal Analysis to Study Output

It is estimated that the weekly output of a certain plant is given by the function $Q(x, y) = 1{,}200x + 500y + x^2y - x^3 - y^2$ units, where x is the number of skilled workers and y is the number of unskilled workers employed at the plant. Currently the workforce consists of 30 skilled workers and 60 unskilled workers. Use marginal analysis to estimate the change in the weekly output that will result from the addition of 1 more skilled worker if the number of unskilled workers is not changed.

Solution

The partial derivative

$$Q_x(x, y) = 1{,}200 + 2xy - 3x^2$$

is the rate of change of output with respect to the number of skilled workers. For any values of x and y, this is an approximation of the number of additional units that will be produced each week if the number of skilled workers is increased from x to $x + 1$ while the number of unskilled workers is kept fixed at y. In particular, if the workforce is increased from 30 skilled and 60 unskilled workers to 31 skilled and 60 unskilled workers, the resulting change in output is approximately

$$Q_x(30, 60) = 1{,}200 + 2(30)(60) - 3(30)^2 = 2{,}100 \text{ units}$$

For practice, compute the exact change $Q(31, 60) - Q(30, 60)$. Is the approximation a good one?

If $Q(K, L)$ is the output of a production process involving the expenditure of K units of capital and L units of labor, then the partial derivative $Q_K(K, L)$ is called the **marginal productivity of capital** and measures the rate at which output Q changes with respect to capital expenditure when the labor force is held constant. Similarly, the partial derivative $Q_L(K, L)$ is called the **marginal productivity of labor** and measures the rate of change of output with respect to the labor level when capital expenditure is held constant. Example 7.2.5 illustrates one way these partial derivatives can be used in economic analysis.

EXAMPLE 7.2.5 Marginal Productivity of Capital and Labor

A manufacturer estimates that the monthly output at a certain factory is given by the Cobb-Douglas function

$$Q(K, L) = 50K^{0.4}L^{0.6}$$

where K is the capital expenditure in units of $1,000 and L is the size of the labor force, measured in worker-hours.

a. Find the marginal productivity of capital Q_K and the marginal productivity of labor Q_L when the capital expenditure is $750,000, and the level of labor is 991 worker-hours.

b. Should the manufacturer consider adding capital or increasing the labor level to increase output?

Solution

a.
$$Q_K(K, L) = 50(0.4K^{-0.6})L^{0.6} = 20K^{-0.6}L^{0.6}$$

and

$$Q_L(K, L) = 50K^{0.4}(0.6L^{-0.4}) = 30K^{0.4}L^{-0.4}$$

so with $K = 750$ ($750,000) and $L = 991$

$$Q_K(750, 991) = 20(750)^{-0.6}(991)^{0.6} \approx 23.64$$

and

$$Q_L(750, 991) = 30(750)^{0.4}(991)^{-0.4} \approx 26.84$$

b. From part (a), you see that an increase in 1 unit of capital (that is, $1,000) results in an increase in output of 23.64 units, which is less than the 26.84 unit increase in output that results from a unit increase in the labor level. Therefore, the manufacturer should increase the labor level by 1 worker-hour (from 991 worker-hours to 992) to increase output as quickly as possible from the current level.

Substitute and Complementary Commodities

Two commodities are said to be **substitute commodities** if an increase in the demand for either results in a decrease in demand for the other. Substitute commodities are competitive, like butter and margarine.

On the other hand, two commodities are said to be **complementary commodities** if a decrease in the demand of either results in a decrease in the demand of the other. An example is provided by digital cameras and memory cards. If consumers buy fewer digital cameras, they will likely buy fewer memory cards, too.

We can use partial derivatives to obtain criteria for determining whether two commodities are substitute or complementary. Suppose $D_1(p_1, p_2)$ units of the first commodity and $D_2(p_1, p_2)$ of the second are demanded when the unit prices of the commodities are p_1 and p_2, respectively. It is reasonable to expect demand to decrease with increasing price, so

$$\frac{\partial D_1}{\partial p_1} < 0 \quad \text{and} \quad \frac{\partial D_2}{\partial p_2} < 0$$

For substitute commodities, the demand for each commodity increases with respect to the price of the other, so

$$\frac{\partial D_1}{\partial p_2} > 0 \quad \text{and} \quad \frac{\partial D_2}{\partial p_1} > 0$$

However, for complementary commodities, the demand for each decreases with respect to the price of the other, and

$$\frac{\partial D_1}{\partial p_2} < 0 \quad \text{and} \quad \frac{\partial D_2}{\partial p_1} < 0$$

Example 7.2.6 illustrates how these criteria can be used to determine whether a given pair of commodities are complementary, substitute, or neither.

EXAMPLE 7.2.6 Substitute and Complementary Commodities

Suppose the demand function for flour in a certain community is given by

$$D_1(p_1, p_2) = 500 + \frac{10}{p_1 + 2} - 5p_2$$

while the corresponding demand for bread is given by

$$D_2(p_1, p_2) = 400 - 2p_1 + \frac{7}{p_2 + 3}$$

where p_1 is the dollar price of a pound of flour and p_2 is the price of a loaf of bread. Determine whether flour and bread are substitute or complementary commodities or neither.

Solution

The partial derivatives of the demand for flour with respect to bread and the demand for bread with respect to flour are

$$\frac{\partial D_1}{\partial p_2} = -5 < 0 \quad \text{and} \quad \frac{\partial D_2}{\partial p_1} = -2 < 0$$

Since both partial derivatives are negative for all p_1 and p_2, it follows that flour and bread are complementary commodities.

Second-Order Partial Derivatives

Partial derivatives can themselves be differentiated. The resulting functions are called **second-order partial derivatives.** Here is a summary of the definition and notation for the four possible second-order partial derivatives of a function of two variables.

Second-Order Partial Derivatives ■ If $z = f(x, y)$, the partial derivative of f_x with respect to x is

$$f_{xx} = (f_x)_x \quad \text{or} \quad \frac{\partial^2 z}{\partial x^2} = \frac{\partial}{\partial x}\left(\frac{\partial z}{\partial x}\right)$$

The partial derivative of f_x with respect to y is

$$f_{xy} = (f_x)_y \quad \text{or} \quad \frac{\partial^2 z}{\partial y \, \partial x} = \frac{\partial}{\partial y}\left(\frac{\partial z}{\partial x}\right)$$

The partial derivative of f_y with respect to x is

$$f_{yx} = (f_y)_x \quad \text{or} \quad \frac{\partial^2 z}{\partial x \, \partial y} = \frac{\partial}{\partial x}\left(\frac{\partial z}{\partial y}\right)$$

The partial derivative of f_y with respect to y is

$$f_{yy} = (f_y)_y \quad \text{or} \quad \frac{\partial^2 z}{\partial y^2} = \frac{\partial}{\partial y}\left(\frac{\partial z}{\partial y}\right)$$

The computation of second-order partial derivatives is illustrated in Example 7.2.7.

EXAMPLE 7.2.7 Computing Second-Order Partial Derivatives

Compute the four second-order partial derivatives of the function

$$f(x, y) = xy^3 + 5xy^2 + 2x + 1$$

Solution

Since
$$f_x = y^3 + 5y^2 + 2$$

it follows that

$$f_{xx} = 0 \quad \text{and} \quad f_{xy} = 3y^2 + 10y$$

Since
$$f_y = 3xy^2 + 10xy$$

we have

$$f_{yy} = 6xy + 10x \quad \text{and} \quad f_{yx} = 3y^2 + 10y$$

NOTE   The two partial derivatives f_{xy} and f_{yx} are sometimes called the **mixed second-order partial derivatives** of f. Notice that the mixed partial derivatives in Example 7.2.7 are equal. This is not an accident. It turns out that for virtually all functions $f(x, y)$ you will encounter in practical work, the mixed partials will be equal; that is,

$$f_{xy} = f_{yx}$$

This means you will get the same answer if you first differentiate $f(x, y)$ with respect to x and then differentiate the resulting function with respect to y as you would if you performed the differentiation in the reverse order. ■

Example 7.2.8 illustrates how a second-order partial derivative can convey useful information in a practical situation.

EXAMPLE 7.2.8 Interpreting a Second-Order Partial Derivative

Suppose the output Q at a factory depends on the amount K of capital invested in the plant and equipment and also on the size L of the labor force, measured in worker-hours.

Give an economic interpretation of the sign of the second-order partial derivative $\dfrac{\partial^2 Q}{\partial L^2}$.

Solution

If $\dfrac{\partial^2 Q}{\partial L^2} < 0$, the marginal productivity of labor $\dfrac{\partial Q}{\partial L}$ decreases as L increases. This means that for a fixed level of capital investment, the effect on output of one additional worker-hour of labor is greater when the workforce is small than when the workforce is large.

Similarly, if $\dfrac{\partial^2 Q}{\partial L^2} > 0$, it follows that for a fixed level of capital investment, the effect on output of one additional worker-hour of labor is greater when the workforce is large than when it is small.

Typically, for a factory operating with an adequate workforce, the derivative $\dfrac{\partial^2 Q}{\partial L^2}$ will be negative. Can you give an economic explanation for this fact?

The Chain Rule for Partial Derivatives

In many practical situations, a particular quantity is given as a function of two or more variables, each of which can be thought of as a function of yet another variable, and the goal is to find the rate of change of the quantity with respect to this other variable. For example, the demand for a certain commodity may depend on the price of the commodity itself and on the price of a competing commodity, both of which are increasing with time, and the goal may be to find the rate of change of the demand with respect to time. You can solve problems of this type by using the following generalization of the chain rule

$$\frac{dz}{dt} = \frac{dz}{dx}\frac{dx}{dt}$$

obtained in Section 2.4.

> **Chain Rule for Partial Derivatives** ■ Suppose z is a function of x and y, each of which is a function of t. Then z can be regarded as a function of t and
>
> $$\frac{dz}{dt} = \frac{\partial z}{\partial x}\frac{dx}{dt} + \frac{\partial z}{\partial y}\frac{dy}{dt}$$

Observe that the expression for $\dfrac{dz}{dt}$ is the sum of two terms, each of which can be interpreted using the chain rule for a function of one variable. In particular,

$$\frac{\partial z}{\partial x}\frac{dx}{dt} = \text{rate of change of } z \text{ with respect to } t \text{ fixed } y$$

and

$$\frac{\partial z}{\partial y}\frac{dy}{dt} = \text{rate of change of } z \text{ with respect to } t \text{ for fixed } x$$

The chain rule for partial derivatives says that the total rate of change of z with respect to t is the sum of these two "partial" rates of change. Example 7.2.9 is a practical illustration of the use of the chain rule for partial derivatives.

EXAMPLE 7.2.9 Using the Chain Rule to Compute a Demand Rate

Divya manages a health store that carries two kinds of vitamin water, brand A and brand B. Sales figures indicate that if brand A is sold for x dollars per bottle and brand B for y dollars per bottle, the demand for brand A will be

$$Q(x, y) = 300 - 20x^2 + 30y \quad \text{bottles per month}$$

She estimates that t months from now the price of brand A will be

$$x = 2 + 0.05t \quad \text{dollars per bottle}$$

and the price of brand B will be

$$y = 2 + 0.1\sqrt{t} \quad \text{dollars per bottle}$$

At what rate should Divya expect the demand for brand A to be changing with respect to time 4 months from now? Should she expect the demand to be increasing or decreasing at that time?

Solution

Divya's goal is to find $\dfrac{dQ}{dt}$ when $t = 4$. Using the chain rule, she gets

$$\frac{dQ}{dt} = \frac{\partial Q}{\partial x}\frac{dx}{dt} + \frac{\partial Q}{\partial y}\frac{dy}{dt} = -40x(0.05) + 30(0.05t^{-1/2})$$

When $t = 4$, $\qquad\qquad\qquad x = 2 + 0.05(4) = 2.2$

and hence,

$$\frac{dQ}{dt} = -40(2.2)(0.05) + 30(0.05)(0.5) = -3.65$$

That is, 4 months from now the monthly demand for brand A will be decreasing at the rate of 3.65 bottles per month.

In Section 2.5, you learned how to use increments to approximate the change in a function resulting from a small change in its independent variable. In particular, if y is a function of x, then

$$\Delta y \approx \frac{dy}{dx}\Delta x$$

where Δx is a small change in the variable x and Δy is the corresponding change in y. Here is the analogous approximation formula for functions of two variables, based on the chain rule for partial derivatives.

Incremental Approximation Formula for Functions of Two Variables
■ Suppose z is a function of x and y. If Δx denotes a small change in x and Δy a small change in y, the corresponding change in z is approximated by

$$\Delta z \approx \frac{\partial z}{\partial x}\Delta x + \frac{\partial z}{\partial y}\Delta y$$

The marginal analysis approximations made earlier in Examples 7.2.4 and 7.2.5 involved unit increments. However, the incremental approximation formula allows a more flexible range of marginal analysis computations, as illustrated in Example 7.2.10.

EXAMPLE 7.2.10 Incremental Approximation of Output

At a certain factory, the daily output is $Q = 60K^{1/2}L^{1/3}$ units, where K denotes the capital investment measured in units of \$1,000 and L the size of the labor force measured in worker-hours. The current capital investment is \$900,000, and 1,000 worker-hours of labor are used each day. Estimate the change in output that will result if capital investment is increased by \$1,000 and labor is increased by 2 worker-hours.

Solution

Apply the approximation formula with $K = 900$, $L = 1,000$, $\Delta K = 1$ ($1,000), and $\Delta L = 2$ to get

$$\Delta Q \approx \frac{\partial Q}{\partial K} \Delta K + \frac{\partial Q}{\partial L} \Delta L$$

$$= 30K^{-1/2}L^{1/3}\Delta K + 20K^{1/2}L^{-2/3}\Delta L$$

$$= 30\left(\frac{1}{30}\right)(10)(1) + 20(30)\left(\frac{1}{100}\right)(2)$$

$$= 22 \text{ units}$$

That is, output will increase by approximately 22 units.

EXERCISES ■ 7.2

In Exercises 1 through 20, compute all first-order partial derivatives of the given function.

1. $f(x, y) = 7x - 3y + 4$

2. $f(x, y) = x - xy + 3$

3. $f(x, y) = 4x^3 - 3x^2y + 5x$

4. $f(x, y) = 2x(y - 3x) - 4y$

5. $f(x, y) = 2xy^5 + 3x^2y + x^2$

6. $z = 5x^2y + 2xy^3 + 3y^2$

7. $z = (3x + 2y)^5$

8. $f(x, y) = (x + xy + y)^3$

9. $f(s, t) = \dfrac{3t}{2s}$

10. $z = \dfrac{t^2}{s^3}$

11. $z = xe^{xy}$

12. $f(x, y) = xye^x$

13. $f(x, y) = \dfrac{e^{2-x}}{y^2}$

14. $f(x, y) = xe^{x+2y}$

15. $f(x, y) = \dfrac{2x + 3y}{y - x}$

16. $z = \dfrac{xy^2}{x^2y^3 + 1}$

17. $z = u \ln v$

18. $f(u, v) = u \ln uv$

19. $f(x, y) = \dfrac{\ln(x + 2y)}{y^2}$

20. $z = \ln\left(\dfrac{x}{y} + \dfrac{y}{x}\right)$

In Exercises 21 through 28, evaluate the partial derivatives $f_x(x, y)$ and $f_y(x, y)$ at the given point (x_0, y_0).

21. $f(x, y) - x^2 + 3y$ at $(1, -1)$

22. $f(x, y) = x^3y - 2(x + y)$ at $(1, 0)$

23. $f(x, y) = \dfrac{y}{2x + y}$ at $(0, -1)$

24. $f(x, y) = x + \dfrac{x}{y - 3x}$ at $(1, 1)$

25. $f(x, y) = 3x^2 - 7xy + 5y^3 - 3(x + y) - 1$; at $(-2, 1)$

26. $f(x, y) = (x - 2y)^2 + (y - 3x)^2 + 5$; at $(0, -1)$

27. $f(x, y) = xe^{-2y} + ye^{-x} + xy^2$; at $(0, 0)$

28. $f(x, y) = xy \ln\left(\dfrac{y}{x}\right) + \ln(2x - 3y)^2$; at $(1, 1)$

In Exercises 29 through 34, find the second partials (including the mixed partials).

29. $f(x, y) = 5x^4y^3 + 2xy$

30. $f(x, y) = \dfrac{x + 1}{y - 1}$

31. $f(x, y) = e^{x^2y}$

32. $f(u, v) = \ln(u^2 + v^2)$

33. $f(s, t) = \sqrt{s^2 + t^2}$

34. $f(x, y) = x^2ye^x$

In Exercises 35 through 40, use the chain rule to find
$\dfrac{dz}{dt}$. *Express your answer in terms of x, y, and t.*

35. $z = 2x + 3y;\ x = t^2,\ y = 5t$

36. $z = x^2y;\ x = 3t + 1,\ y = t^2 - 1$

37. $z = \dfrac{3x}{y};\ x = t,\ y = t^2$

38. $z = x^{1/2}y^{1/3};\ x = 2t,\ y = 2t^2$

39. $z = xy;\ x = e^{2t},\ y = e^{-3t}$

40. $z = \dfrac{x + y}{x - y};\ x = t^3 + 1,\ y = 1 - t^2$

BUSINESS AND ECONOMICS APPLIED PROBLEMS

SUBSTITUTE AND COMPLEMENTARY COMMODITIES   *In Exercises 41 through 46, the demand functions for a pair of commodities are given. Use partial derivatives to determine whether the commodities are substitute, complementary, or neither.*

41. $D_1 = 500 - 6p_1 + 5p_2;\ D_2 = 200 + 2p_1 - 5p_2$

42. $D_1 = 1{,}000 - 0.02p_1^2 - 0.05p_2^2;$
$D_2 = 800 - 0.001p_1^2 - p_1p_2$

43. $D_1 = 3{,}000 + \dfrac{400}{p_1 + 3} + 50p_2;$
$D_2 = 2{,}000 - 100p_1 + \dfrac{500}{p_2 + 4}$

44. $D_1 = 2{,}000 + \dfrac{100}{p_1 + 2} - 25p_2;$
$D_2 = 1{,}500 - \dfrac{p_2}{p_1 + 7}$

45. $D_1 = \dfrac{7p_2}{1 + p_1^2};\ D_2 = \dfrac{p_1}{1 + p_2^2}$

46. $D_1 = 200p_1^{-1/2}p_2^{-1/2};\ D_2 = 300p_1^{-1/2}p_2^{-3/2}$

47. MARGINAL ANALYSIS   At a certain factory, the daily output is $Q(K, L) = 60K^{1/2}L^{1/3}$ units, where K denotes the capital investment measured in units of $1,000 and L the size of the labor force measured in worker-hours. Suppose that the current capital investment is $900,000 and that 1,000 worker-hours of labor are used each day. Use marginal analysis to estimate the effect of an additional capital investment of $1,000 on the daily output if the size of the labor force is not changed.

48. MARGINAL PRODUCTIVITY   A manufacturer estimates that the annual output at a certain factory is given by
$$Q(K, L) = 30K^{0.3}L^{0.7}$$
units, where K is the capital expenditure in units of $1,000 and L is the size of the labor force in worker-hours.

a. Find the marginal productivity of capital and the marginal productivity of labor when the capital expenditure is $630,000 and the labor level is 830 worker-hours.

b. Should the manufacturer consider adding a unit of capital or a unit of labor to increase output more rapidly?

49. NATIONAL PRODUCTIVITY   The annual productivity of a certain country is
$$Q(K, L) = 150[0.4K^{-1/2} + 0.6L^{-1/2}]^{-2}$$
units, where K is capital expenditure in millions of dollars and L measures the labor force in thousands of worker-hours.

a. Find the marginal productivity of capital and the marginal productivity of labor.

b. Currently, capital expenditure is 5.041 billion dollars ($K = 5{,}041$) and 4,900,000 worker-hours ($L = 4{,}900$) are being employed. Find the marginal productivities at these levels.

c. Should the government of the country encourage capital investment or additional labor employment to increase productivity as rapidly as possible?

50. MARGINAL ANALYSIS   A grocer's daily profit from the sale of two brands of cat food is
$$P(x, y) = (x - 30)(70 - 5x + 4y)$$
$$+ (y - 40)(80 + 6x - 7y)$$
cents, where x is the price per can of the first brand and y is the price per can of the second. Currently the first brand sells for 50 cents per can and the second for 52 cents per can. Use marginal analysis to estimate the change in daily profit that will result if the grocer raises the price of the second brand by 1 cent per can but keeps the price of the first brand unchanged.

51. CONSUMER DEMAND   A bicycle dealer has found that if 10-speed bicycles are sold for x dollars each and the price of gasoline is y cents per gallon, approximately $F(x, y)$ bicycles will be sold each month, where
$$F(x, y) = 200 - 24\sqrt{x} + 4(0.1y + 3)^{3/2}$$

Currently, the bicycles sell for $324 apiece and gasoline sells for $3.80 per gallon. Use marginal analysis to estimate the change in the demand for bicycles that results when the price of bicycles is kept fixed but the price of gasoline decreases by 1 cent per gallon.

52. **CONSUMER DEMAND** The monthly demand for a certain brand of toasters is given by a function $f(x, y)$, where x is the amount of money (measured in units of $1,000) spent on advertising and y is the selling price (in dollars) of the toasters. Give economic interpretations of the partial derivatives f_x and f_y. Under normal economic conditions, what will be the sign of each of these derivatives?

53. **CONSUMER DEMAND** Two competing brands of power lawnmowers are sold in the same town. The price of the first brand is x dollars per mower, and the price of the second brand is y dollars per mower. The local demand for the first brand of mower is given by a function $D(x, y)$.
    a. How would you expect the demand for the first brand of mower to be affected by an increase in x? By an increase in y?
    b. Translate your answers in part (a) into conditions on the signs of the partial derivatives of D.
    c. If $D(x, y) = a + bx + cy$, what can you say about the signs of the coefficients b and c if your conclusions in parts (a) and (b) are to hold?

54. **MARGINAL PRODUCTIVITY** Suppose the output Q of a factory depends on the amount K of capital investment measured in units of $1,000 and on the size L of the labor force measured in worker-hours. Give an economic interpretation of the second-order partial derivative $\dfrac{\partial^2 Q}{\partial K^2}$.

55. **MARGINAL PRODUCTIVITY** At a certain factory, the output is $Q = 120K^{1/2}L^{1/3}$ units, where K denotes the capital investment measured in units of $1,000 and L the size of the labor force measured in worker-hours.
    a. Determine the sign of the second-order partial derivative $\dfrac{\partial^2 Q}{\partial L^2}$, and give an economic interpretation.
    b. Determine the sign of the second-order partial derivative $\dfrac{\partial^2 Q}{\partial K^2}$, and give an economic interpretation.

56. **SUBSTITUTE AND COMPLEMENTARY COMMODITIES** The demand function for peanut butter is
$$D_1(p_1, p_2) = 800 - 0.03p_1^2 - 0.04p_2^2$$
while that for a second commodity is
$$D_2(p_1, p_2) = 500 - 0.002p_1^3 - p_1p_2$$
Is the second commodity more likely to be jelly or bread? Explain.

57. **SUBSTITUTE AND COMPLEMENTARY COMMODITIES** The demand function for a certain brand of gel pens is
$$D_1(p_1, p_2) = 700 - 4p_1^2 + 7p_1p_2$$
while that of a second commodity is
$$D_2(p_1, p_2) = 300 - 2\sqrt{p_2} + 5p_1p_2$$
Is the second commodity more likely to be pencils or paper? Explain.

58. **LAW OF DIMINISHING RETURNS** Suppose the daily output Q of a factory depends on the amount K of capital investment and on the size L of the labor force. A **law of diminishing returns** states that in certain circumstances, there is a value L_0 such that the marginal product of labor will be increasing for $L < L_0$ and decreasing for $L > L_0$.
    a. Translate this law of diminishing returns into statements about the sign of a certain second-order partial derivative.
    b. Read about the principle of diminishing returns in an economics text. Then write a paragraph discussing the economic factors that might account for this phenomenon.

59. **MARGINAL ANALYSIS** It is estimated that the weekly output at a certain plant is given by
$$Q(x, y) = 1,175x + 483y + 3.1x^2y - 1.2x^3 - 2.7y^2$$
units, where x is the number of skilled workers and y is the number of unskilled workers employed at the plant. Currently the workforce consists of 37 skilled and 71 unskilled workers.
    a. Store the output function as
$$1,175X + 483Y + 3.1(X\text{^}2)*Y$$
$$- 1.2(X\text{^}3) - 2.7(Y\text{^}2)$$
Store 37 as X and 71 as Y and evaluate to obtain $Q(37, 71)$. Repeat for $Q(38, 71)$ and $Q(37, 72)$.
    b. Store the partial derivative $Q_x(x, y)$ in your calculator, and evaluate $Q_x(37, 71)$. Use the result to estimate the change in output resulting when the workforce is increased from 37

skilled workers to 38 and the unskilled work-force stays fixed at 71. Then compare with the actual change in output, given by the difference $Q(38, 71) - Q(37, 71)$.

c. Use the partial derivative $Q_y(x, y)$ to estimate the change in output that results when the number of unskilled workers is increased from 71 to 72 while the number of skilled workers stays at 37. Compare with the actual change $Q(37, 72) - Q(37, 71)$.

60. MARGINAL ANALYSIS Repeat Exercise 59 with the output function
$$Q(x, y) = 1{,}731x + 925y + x^2y - 2.7x^2 - 1.3y^{3/2}$$
and initial employment levels of $x = 43$ and $y = 85$.

Each of Exercises 61 through 68 involves either the chain rule for partial derivatives or the incremental approximation formula for functions of two variables.

61. ALLOCATION OF LABOR Using x hours of skilled labor and y hours of unskilled labor, a manufacturer can produce $Q(x, y) = 10xy^{1/2}$ units. Currently 30 hours of skilled labor and 36 hours of unskilled labor are being used. Suppose the manufacturer reduces the skilled labor level by 3 hours and increases the unskilled labor level by 5 hours. Use calculus to determine the approximate effect of these changes on production.

62. DEMAND FOR HYBRID CARS A car dealer determines that if gasoline-electric hybrid automobiles are sold for x dollars apiece and the price of gasoline is y cents per gallon, then approximately H hybrid cars will be sold each year, where
$$H(x, y) = 3{,}500 - 19x^{1/2} + 6(0.1y + 16)^{3/2}$$
She estimates that t years from now, the hybrid cars will be selling for
$$x(t) = 35{,}050 + 350t$$
dollars apiece and that gasoline will cost
$$y(t) = 300 + 10(3t)^{1/2}$$
cents per gallon. At what rate will the annual demand for hybrid cars be changing with respect to time 3 years from now? Will it be increasing or decreasing?

63. CONSUMER DEMAND The demand for a certain product is
$$Q(x, y) = 200 - 10x^2 + 20xy$$

units per month, where x is the price of the product and y is the price of a competing product. It is estimated that t months from now, the price of the product will be
$$x(t) = 10 + 0.5t$$
dollars per unit while the price of the competing product will be
$$y(t) = 12.8 + 0.2t^2$$
dollars per unit.

a. At what rate will the demand for the product be changing with respect to time 4 months from now?

b. At what percentage rate $\dfrac{100Q'(t)}{Q(t)}$ will the demand for the product be changing with respect to time 4 months from now?

64. ALLOCATION OF RESOURCES At a certain factory, when the capital expenditure is K thousand dollars and L worker-hours of labor are employed, the daily output will be $Q = 120K^{1/2}L^{1/3}$ units. Currently capital expenditure is \$400,000 ($K = 400$) and is increasing at the rate of \$9,000 per day, while 1,000 worker-hours are being employed and labor is being decreased at the rate of 4 worker-hours per day. At what rate is production currently changing? Is it increasing or decreasing?

65. ALLOCATION OF LABOR The output at a certain plant is
$$Q(x, y) = 0.08x^2 + 0.12xy + 0.03y^2$$
units per day, where x is the number of hours of skilled labor used and y is the number of hours of unskilled labor used. Currently, 80 hours of skilled labor and 200 hours of unskilled labor are used each day. Use calculus to estimate the change in output that will result if an additional $\dfrac{1}{2}$ hour of skilled labor is used each day, along with an additional 2 hours of unskilled labor.

66. PUBLISHING SALES An editor estimates that if x thousand dollars are spent on development and y thousand dollars are spent on promotion, approximately $Q(x, y) = 20x^{3/2}y$ copies of a new book will be sold. Current plans call for the expenditure of \$36,000 on development and \$25,000 on promotion. Use calculus to estimate how sales will be affected if the amount spent on development is increased by \$500 and the amount on promotion is decreased by \$1,000.

67. RETAIL SALES A grocer's daily profit from the sale of two brands of flavored iced tea is

$$P(x, y) = (x - 40)(55 - 4x + 5y)$$
$$+ (y - 45)(70 + 5x - 7y)$$

cents, where x is the price per bottle of the first brand and y is the price per bottle of the second, both in cents. Currently the first brand sells for 70 cents per bottle and the second, for 73 cents per bottle.

a. Find the marginal profit functions, P_x and P_y.

b. Evaluate P_x and P_y for the current values of x and y.

c. Use calculus to estimate the change in daily profit that will result if the grocer decides to raise the price of the first brand by 1 cent and the price of the second by 2 cents.

d. Estimate the change in profit if the price of the first brand is increased by 2 cents and the price of the second is decreased by 1 cent.

68. INVESTMENT SATISFACTION Ethan is an investor who derives $U(x, y)$ units of satisfaction from owning x stock units and y bond units, where

$$U(x, y) = (2x + 3)(y + 5)$$

He currently owns $x = 27$ stock units and $y = 12$ bond units.

a. Find the marginal utilities U_x and U_y.

b. Evaluate U_x and U_y for the current values of x and y.

c. Use calculus to estimate how Ethan's satisfaction changes if he adds 3 stock units and removes 2 bond units from his portfolio.

d. Estimate how many bond units Ethan could substitute for 1 stock unit without affecting his total satisfaction with his portfolio.

LIFE AND SOCIAL SCIENCE APPLIED PROBLEMS

69. FLOW OF BLOOD The smaller the resistance to flow in a blood vessel, the less energy is expended by the pumping heart. One of Poiseuille's laws* says that the resistance to the flow of blood in a blood vessel satisfies

$$F(L, r) = \frac{kL}{r^4}$$

---

*E. Batschelet, *Introduction to Mathematics for Life Scientists*, 2nd ed., New York: Springer-Verlag, 1979, p. 279.

where L is the length of the vessel, r is its radius, and k is a constant that depends on the viscosity of blood.

a. Find $F, \dfrac{\partial F}{\partial L}$, and $\dfrac{\partial F}{\partial r}$ in the case where $L = 3.17$ cm and $r = 0.085$ cm. Leave your answer in terms of k.

b. Suppose the vessel in part (a) is constricted and lengthened so that its new radius is 20% smaller than before and its new length is 20% greater. How do these changes affect the flow $F(L, r)$? How do they affect the values of $\dfrac{\partial F}{\partial L}$ and $\dfrac{\partial F}{\partial r}$?

70. SURFACE AREA OF THE HUMAN BODY Recall from Exercise 47 of Section 7.1 that the surface area of a person's body may be measured by the empirical formula

$$S(W, H) = 0.0072W^{0.425}H^{0.725}$$

where W (kg) and H (cm) are the person's weight and height, respectively. Currently, a child weighs 34 kg and is 120 cm tall.

a. Compute the partial derivatives $S_W(34, 120)$ and $S_H(34, 120)$, and interpret each as a rate of change.

b. Estimate the change in surface area that results if the child's height stays constant but her weight increases by 1 kilogram.

71. BLOOD CIRCULATION The flow of blood from an artery into a small capillary is given by the formula

$$F(x, y, z) = \frac{c\pi x^2}{4}\sqrt{y - z} \quad \text{cm}^3/\text{sec}$$

where c is a positive constant, x is the diameter of the capillary, y is the pressure in the artery, and z is the pressure in the capillary. What function gives the rate of change of blood flow with respect to capillary pressure, assuming fixed arterial pressure and capillary diameter? Is this rate increasing or decreasing?

72. CARDIOLOGY To estimate the amount of blood that flows through a patient's lung, cardiologists use the empirical formula

$$P(x, y, u, v) = \frac{100xy}{xy + uv}$$

where P is a percentage of the total blood flow, x is the carbon dioxide output of the lung, y is the arteriovenous carbon dioxide difference in the lung, u is the carbon dioxide output of the lung, and v is the arteriovenous carbon dioxide difference in the other lung.

It is known that blood flows into the lungs to pick up oxygen and dump carbon dioxide, so the arteriovenous carbon dioxide difference measures the extent to which this exchange is accomplished. (The actual measurement is accomplished by a device called a **cardiac shunt.**) The carbon dioxide is then exhaled from the lungs so that oxygen-bearing air can be inhaled.

Compute the partial derivatives P_x, P_y, P_u, and P_v, and give a physiological interpretation of each derivative.

MISCELLANEOUS PROBLEMS

*The function $z = f(x, y)$ is said to satisfy **Laplace's equation** if $z_{xx} + z_{yy} = 0$. Functions that satisfy such an equation play an important role in a variety of applications in the physical sciences, especially in the theory of electricity and magnetism. In Exercises 73 through 76, determine whether the given function satisfies Laplace's equation.*

73. $z = x^2 - y^2$

74. $z = xy$

75. $z = xe^y - ye^x$

76. $z = [(x - 1)^2 + (y + 3)^2]^{-1/2}$

77. PACKAGING A soft drink can is a cylinder H cm tall with radius R cm. Its volume is given by the formula $V = \pi R^2 H$. A particular can is 12 cm tall with radius 3 cm. Use calculus to estimate the change in volume that results if the radius is increased by 1 cm while the height remains at 12 cm.

78. PACKAGING For the soft drink can in Exercise 77, the surface area is given by $S = 2\pi R^2 + 2\pi RH$. Use calculus to estimate the change in surface that results if:
a. The radius is increased from 3 to 4 cm while the height stays at 12 cm.
b. The height is decreased from 12 to 11 cm while the radius stays at 3 cm.

79. CHEMISTRY The **ideal gas law** says that for n moles of an ideal gas, $PV = nRT$, where P is the pressure exerted by the gas, V is the volume of the gas, T is the temperature of the gas, and R is a constant (the **gas constant**). Compute the product

$$\frac{\partial V}{\partial T} \frac{\partial T}{\partial P} \frac{\partial P}{\partial V}$$

80. ELECTRIC CIRCUIT In an electric circuit with two resistors of resistance R_1 and R_2 connected in parallel, the total resistance R is given by the formula

$$\frac{1}{R} = \frac{1}{R_1} + \frac{1}{R_2}$$

Show that

$$R_1 \frac{\partial R}{\partial R_1} + R_2 \frac{\partial R}{\partial R_2} = R$$

Each of Exercises 81 through 85 involves either the chain rule for partial derivatives or the incremental approximation formula for functions of two variables.

81. PACKAGING A soft drink can is H centimeters (cm) tall and has a radius of R cm. The cost of material in the can is 0.0005 cents per cm^2, and the soda itself costs 0.001 cents per cm^3.
a. Find a function $C(R, H)$ for the cost of the materials and contents of a can of soda. (You will need the formulas for volume and surface area given in Exercises 77 and 78.)
b. The cans are currently 12 cm tall and have a radius of 3 cm. Use calculus to estimate the effect on cost of increasing the radius by 0.3 cm and decreasing the height by 0.2 cm.

82. LANDSCAPING A rectangular garden that is 30 yards long and 40 yards wide is bordered by a concrete path that is 0.8 yard wide. Use calculus to estimate the area of the concrete path.

83. CONSTRUCTION A cylindrical silo set on a concrete block has inside diameter 12 ft and height 80 ft without the lid. If the silo's lid and curved walls are 6 in and 4 in thick, respectively, use calculus to estimate the volume of the material used for the silo.

84. Suppose $y = h(x)$ is a differentiable function of x and that $F(x, y) = C$ for some constant C. Use the chain rule (with x taking the role of t) to show that

$$\frac{\partial F}{\partial x} + \frac{\partial F}{\partial y} \frac{dy}{dx} = 0$$

Conclude that the slope at each point (x, y) on the level curve $F(x, y) = C$ is given by

$$\frac{dy}{dx} = -\frac{F_x}{F_y}$$

85. Use the formula obtained in Exercise 84 to find the slope of the level curve
$$x^2 + xy + y^3 = 1$$
at the point $(-1, 1)$. What is the equation of the tangent line to the level curve at this point?

86. Use the formula obtained in Exercise 84 to find the slope of the level curve
$$x^2y + 2y^3 - 2e^{-x} = 14$$
at the point $(0, 2)$. What is the equation of the tangent line to the level curve at this point?

SECTION 7.3 Optimizing Functions of Two Variables

Learning Objectives

1. Locate and classify relative extrema for a function of two variables using the second partials test.
2. Examine applied problems involving optimization of functions of two variables.
3. Discuss and apply the extreme value property for functions of two variables to find absolute extrema on a closed, bounded region.

Suppose a manufacturer produces two models of Blu-ray player, the deluxe and the standard, and that the total cost of producing x units of the deluxe and y units of the standard is given by the function $C(x, y)$. How would you find the level of production $x = a$ and $y = b$ that results in minimal cost? Or perhaps the output of a certain production process is given by $Q(K, L)$, where K and L measure capital and labor expenditure, respectively. What levels of expenditure K_0 and L_0 result in maximum output?

In Section 3.4, you learned how to use the derivative $f'(x)$ to find the largest and smallest values of a function of a single variable $f(x)$, and the goal of this section is to extend those methods to functions of two variables $f(x, y)$. We begin with a definition.

Relative Extrema ■ The function $f(x, y)$ is said to have a **relative maximum** at the point (a, b) in the domain of f if $f(a, b) \geq f(x, y)$ for all points (x, y) in a circular disk centered at (a, b). Similarly, if $f(c, d) \leq f(x, y)$ for all points (x, y) in a circular disk centered at (c, d), then $f(x, y)$ has a **relative minimum** at (c, d).

In geometric terms, there is a relative maximum of $f(x, y)$ at (a, b) if the surface $z = f(x, y)$ has a peak at the point $(a, b, f(a, b))$, that is, if $(a, b, f(a, b))$ is at least as high as any nearby point on the surface. Similarly, a relative minimum of $f(x, y)$ occurs at (c, d) if the point $(c, d, f(c, d))$ is at the bottom of a valley, so $(c, d, f(c, d))$ is at least as low as any nearby point on the surface. These features are illustrated in Figure 7.11.

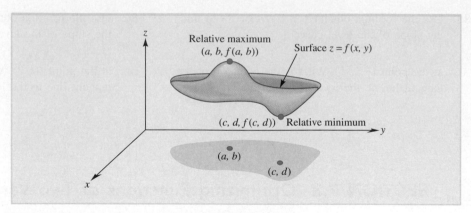

FIGURE 7.11   Relative extrema of the function $f(x, y)$.

Critical Points   The points (a, b) in the domain of $f(x, y)$ for which both $f_x(a, b) = 0$ and $f_y(a, b) = 0$ are said to be **critical points** of f. Like the critical numbers for functions of one variable, these critical points play an important role in the study of relative maxima and minima.

To see the connection between critical points and relative extrema, suppose $f(x, y)$ has a relative maximum at (a, b). Then the curve formed by intersecting the surface $z = f(x, y)$ with the vertical plane $y = b$ has a relative maximum and hence a horizontal tangent line when $x = a$ (Figure 7.12a). Since the partial derivative $f_x(a, b)$ is the slope of this tangent line, it follows that $f_x(a, b) = 0$. Similarly, the curve formed by intersecting the surface $z = f(x, y)$ with the plane $x = a$ has a relative maximum when $y = b$ (Figure 7.12b), and so $f_y(a, b) = 0$. This shows that a point at which a function of two variables has a relative maximum must be a critical point. A similar argument shows that a point at which a function of two variables has a relative minimum must also be a critical point.

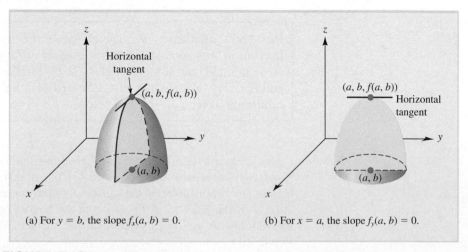

(a) For $y = b$, the slope $f_x(a, b) = 0$.   (b) For $x = a$, the slope $f_y(a, b) = 0$.

FIGURE 7.12   The partial derivatives are zero at a relative extremum.

Here is a more precise statement of the situation.

Critical Points and Relative Extrema ▪ A point (a, b) in the domain of $f(x, y)$ for which the partial derivatives f_x and f_y both exist is called a *critical point* of f if both

$$f_x(a, b) = 0 \qquad \text{and} \qquad f_y(a, b) = 0$$

If the first-order partial derivatives of f exist at all points in some region R in the xy plane, then the relative extrema of f in R can occur only at critical points.

Saddle Points

Although all the relative extrema of a function must occur at critical points, not every critical point of a function corresponds to a relative extremum. For example, if $f(x, y) = y^2 - x^2$, then

$$f_x(x, y) = -2x \qquad \text{and} \qquad f_y(x, y) = 2y$$

so $f_x(0, 0) = f_y(0, 0) = 0$. Thus, the origin $(0, 0)$ is a critical point for $f(x, y)$, and the surface $z = y^2 - x^2$ has horizontal tangents at the origin along both the x axis and the y axis. However, in the xz plane (where $y = 0$) the surface has the equation $z = -x^2$, which is a downward opening parabola, while in the yz plane (where $x = 0$), we have the upward opening parabola $z = y^2$. This means that at the origin, the surface $z = y^2 - x^2$ has a *relative maximum* in the x direction and a *relative minimum* in the y direction.

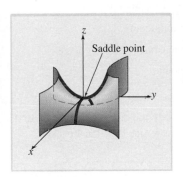

FIGURE 7.13 The saddle surface $z = y^2 - x^2$.

Instead of having a peak or a valley above the critical point $(0, 0)$, the surface $z = y^2 - x^2$ is shaped like a saddle, as shown in Figure 7.13, and for this reason is called a **saddle surface.** For a critical point to correspond to a relative extremum, the same extreme behavior (maximum or minimum) must occur in *all directions*. Any critical point (like the origin in this example) where there is a relative maximum in one direction and a relative minimum in another direction is called a **saddle point.**

The Second Partials Test

Here is a procedure involving second-order partial derivatives that you can use to decide whether a given critical point is a relative maximum, a relative minimum, or a saddle point. This procedure is the two-variable version of the second derivative test for functions of a single variable that you saw in Section 3.2.

The Second Partials Test

Let $f(x, y)$ be a function of x and y whose partial derivatives $f_x, f_y, f_{xx}, f_{yy},$ and f_{xy} all exist, and let $D(x, y)$ be the function

$$D(x, y) = f_{xx}(x, y) f_{yy}(x, y) - [f_{xy}(x, y)]^2$$

Step 1. Find all critical points of $f(x, y)$, that is, all points (a, b) so that

$$f_x(a, b) = 0 \qquad \text{and} \qquad f_y(a, b) = 0$$

Step 2. For each critical point (a, b) found in step 1, evaluate $D(a, b)$.

Step 3. If $D(a, b) < 0$, there is a **saddle point** at (a, b).

Step 4. If $D(a, b) > 0$, compute $f_{xx}(a, b)$, then

     if $f_{xx}(a, b) > 0$, there is a **relative minimum** at (a, b).
     if $f_{xx}(a, b) < 0$, there is a **relative maximum** at (a, b).

If $D(a, b) = 0$, the test is inconclusive and f may have either a relative extremum or a saddle point at (a, b).

Notice that there is a saddle point at the critical point (a, b) only when the quantity D in the second partials test is negative. If D is positive, there is either a relative maximum or a relative minimum *in all directions*. To decide which, you can restrict your attention to any one direction (say, the x direction) and use the sign of the second partial derivative f_{xx} in exactly the same way as the single variable second derivative was used in the second derivative test given in Chapter 3, namely,

$$\text{a relative minimum if } f_{xx}(a, b) > 0$$
$$\text{a relative maximum if } f_{xx}(a, b) < 0$$

You may find the following tabular summary a convenient way of remembering the conclusions of the second partials test:

Sign of D	Sign of f_{xx}	Behavior at (a, b)
+	+	Relative minimum
+	−	Relative maximum
−		Saddle point

The proof of the second partials test involves ideas beyond the scope of this text and is omitted. Examples 7.3.1 through 7.3.3 illustrate how the test can be used.

EXAMPLE 7.3.1   Classifying Critical Points

Find all critical points for the function $f(x, y) = x^2 + y^2$, and classify each as a relative maximum, relative minimum, or saddle point.

Solution

Since

$$f_x = 2x \qquad \text{and} \qquad f_y = 2y$$

the only critical point of f is $(0, 0)$. To test this point, use the second-order partial derivatives

$$f_{xx} = 2 \qquad f_{yy} = 2 \qquad \text{and} \qquad f_{xy} = 0$$

to get

$$D(x, y) = f_{xx}f_{yy} - (f_{xy})^2 = (2)(2) - 0^2 = 4$$

That is, $D(x, y) = 4$ for *all* points (x, y) and, in particular,

$$D(0, 0) = 4 > 0$$

Hence, f has a relative extremum at $(0, 0)$. Moreover, since

$$f_{xx}(0, 0) = 2 > 0$$

it follows that the relative extremum at $(0, 0)$ is a relative minimum. For reference, the graph of f is sketched in Figure 7.14.

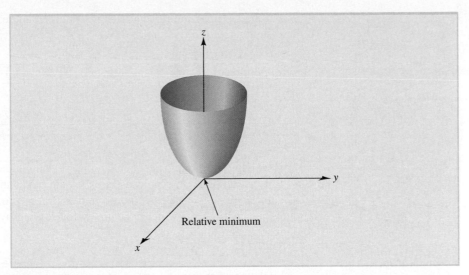

FIGURE 7.14   The surface $z = x^2 + y^2$ with a relative minimum at $(0, 0)$.

EXAMPLE 7.3.2 Classifying Critical Points

Find all critical points for the function $f(x, y) = 12x - x^3 - 4y^2$, and classify each as a relative maximum, relative minimum, or saddle point.

Solution

Since

$$f_x = 12 - 3x^2 \quad \text{and} \quad f_y = -8y$$

you find the critical points by solving simultaneously the two equations

$$12 - 3x^2 = 0$$
$$-8y = 0$$

From the second equation, you get $y = 0$ and from the first,

$$3x^2 = 12$$
$$x = 2 \quad \text{or} \quad -2$$

Thus, there are two critical points, $(2, 0)$ and $(-2, 0)$.

To determine the nature of these points, you first compute

$$f_{xx} = -6x \quad f_{yy} = -8 \quad \text{and} \quad f_{xy} = 0$$

and then form the function

$$D = f_{xx}f_{yy} - (f_{xy})^2 = (-6x)(-8) - 0 = 48x$$

Applying the second partials test to the two critical points, you find

$$D(2, 0) = 48(2) = 96 > 0 \quad \text{and} \quad f_{xx}(2, 0) = -6(2) = -12 < 0$$

and

$$D(-2, 0) = 48(-2) = -96 < 0$$

so a relative maximum occurs at $(2, 0)$ and a saddle point at $(-2, 0)$. These results are summarized in this table.

Critical Point (a, b)	Sign of $D(a, b)$	Sign of $f_{xx}(a, b)$	Behavior at (a, b)
$(2, 0)$	$+$	$-$	Relative maximum
$(-2, 0)$	$-$		Saddle point

Solving the equations $f_x = 0$ and $f_y = 0$ simultaneously to find the critical points of a function of two variables is rarely as simple as in Examples 7.3.1 and 7.3.2. The algebra in Example 7.3.3 is more typical. Before proceeding, you may wish to refer to Appendix A.2, in which techniques for solving systems of two equations in two unknowns are discussed.

EXAMPLE 7.3.3 Classifying Critical Points

Find all critical points for the function $f(x, y) = x^3 - y^3 + 6xy$, and classify each as a relative maximum, relative minimum, or saddle point.

Solution
Since

$$f_x = 3x^2 + 6y \qquad \text{and} \qquad f_y = -3y^2 + 6x$$

you find the critical points of f by solving simultaneously the two equations

$$3x^2 + 6y = 0 \qquad \text{and} \qquad -3y^2 + 6x = 0$$

Just-In-Time REVIEW

Recall that
$a^3 - b^3 = (a - b)(a^2 + ab + b^2)$
so that
$x^3 - 8 = (x - 2)(x^2 + 2x + 4)$.
Because $x^2 + 2x + 4 = 0$ has no real solutions (as can be seen using the quadratic formula), the only real solution of $x^3 - 8 = 0$ is $x = 2$.

Solving the first equation for y, you get $y = -\dfrac{x^2}{2}$, which you can substitute into the second equation to get

$$-3\left(\frac{-x^2}{2}\right)^2 + 6x = 0$$

$$-\frac{3x^4}{4} + 6x = 0 \qquad \begin{array}{l}\text{multiply both sides by} \\ \frac{4}{3} \text{ and factor out } -x\end{array}$$

$$-x(x^3 - 8) = 0$$

The solutions of this equation are $x = 0$ and $x = 2$. These are the x coordinates of the critical points of f. To get the corresponding y coordinates, substitute these values of x into the equation $y = -\dfrac{x^2}{2}$ (or into either one of the two original equations). You will find that $y = 0$ when $x = 0$ and $y = -2$ when $x = 2$. It follows that the critical points of f are $(0, 0)$ and $(2, -2)$.

The second-order partial derivatives of f are

$$f_{xx} = 6x \qquad f_{yy} = -6y \qquad \text{and} \qquad f_{xy} = 6$$

Hence,

$$D(x, y) = f_{xx}f_{yy} - (f_{xy})^2 = -36xy - 36 = -36(xy + 1)$$

Since

$$D(0, 0) = -36[(0)(0) + 1] = -36 < 0$$

it follows that f has a saddle point at $(0, 0)$. Since

$$D(2, -2) = -36[2(-2) + 1] = 108 > 0$$

and

$$f_{xx}(2, -2) = 6(2) = 12 > 0$$

you see that f has a relative minimum at $(2, -2)$. To summarize:

Critical Point (a, b)	$D(a, b)$	$f_{xx}(a, b)$	Behavior at (a, b)
$(0, 0)$	$-$		Saddle point
$(2, -2)$	$+$	$+$	Relative minimum

Examples 7.3.4 and 7.3.5 illustrate how the methods we have developed can be applied to practical optimization problems.

EXAMPLE 7.3.4 Maximizing Profit

Jamaal manages a grocery store that carries two brands of cat food, a local brand obtained at the cost of 30 cents per can and a well-known national brand obtained for 40 cents per can. He estimates that if the local brand is sold for x cents per can and the national brand for y cents per can, then approximately $70 - 5x + 4y$ cans of the local brand and $80 + 6x - 7y$ of the national brand will be sold each day. How should Jamaal price each brand to maximize total daily profit from the sale of cat food? (Assume that the largest daily profit occurs at a relative maximum.)

Solution
Since

$$\begin{pmatrix} \text{Total} \\ \text{profit} \end{pmatrix} = \begin{pmatrix} \text{profit from the sale} \\ \text{of the local brand} \end{pmatrix} + \begin{pmatrix} \text{profit from the sale} \\ \text{of the national brand} \end{pmatrix}$$

it follows that the total daily profit from the sale of the cat food is given by the function

$$f(x, y) = \underbrace{(70 - 5x + 4y)}_{\text{items sold}} \cdot \underbrace{(x - 30)}_{\text{profit per item}} + \underbrace{(80 + 6x - 7y)}_{\text{items sold}} \cdot \underbrace{(y - 40)}_{\text{profit per item}}$$

$$\underbrace{\qquad\qquad\qquad\qquad}_{\text{local brand profit}} \qquad \underbrace{\qquad\qquad\qquad\qquad}_{\text{national brand profit}}$$

$$= -5x^2 + 10xy - 20x - 7y^2 + 240y - 5{,}300$$

Compute the partial derivatives

$$f_x = -10x + 10y - 20 \quad \text{and} \quad f_y = 10x - 14y + 240$$

and set them equal to zero to get

$$-10x + 10y - 20 = 0 \quad \text{and} \quad 10x - 14y + 240 = 0$$

or

$$-x + y = 2 \quad \text{and} \quad 5x - 7y = -120$$

Then solve these equations simultaneously to get

$$x = 53 \quad \text{and} \quad y = 55$$

It follows that (53, 55) is the only critical point of f.

Next apply the second partials test. Since

$$f_{xx} = -10 \quad f_{yy} = -14 \quad \text{and} \quad f_{xy} = 10$$

you get

$$D(x, y) = f_{xx}f_{yy} - (f_{xy})^2 = (-10)(-14) - (10)^2 = 40$$

Because you have

$$D(53, 55) = 40 > 0 \quad \text{and} \quad f_{xx}(53, 55) = -10 < 0$$

it follows that f has a relative maximum when $x = 53$ and $y = 55$, which can be shown to also be an absolute maximum. So the grocer can maximize profit by selling the local brand of cat food for 53 cents per can and the national brand for 55 cents per can.

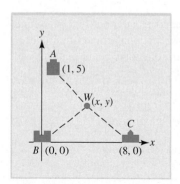

FIGURE 7.15 Locations of customers A, B, and C and warehouse W.

EXAMPLE 7.3.5   Finding Optimal Location for a Warehouse

The business manager for Acme Corporation plots a grid on a map of the region Acme serves and determines that the company's three most important customers are located at points A, B, and C, with coordinates (1, 5), (0, 0), and (8, 0), respectively, where units are in miles. What are the coordinates (x, y) of the point W where a warehouse should be located to minimize the sum of the squares of the distances from W to A, B, and C (see Figure 7.15)?

Solution

The sum of the squares of the distances from W to A, B, and C is given by the function

$$\underbrace{S(x, y)}_{\substack{\text{sum of squares} \\ \text{of distances}}} = \underbrace{[(x - 1)^2 + (y - 5)^2]}_{\substack{\text{square of distance} \\ \text{from } W \text{ to } A}} + \underbrace{(x^2 + y^2)}_{\substack{\text{square of distance} \\ \text{from } W \text{ to } B}} + \underbrace{[(x - 8)^2 + y^2]}_{\substack{\text{square of distance} \\ \text{from } W \text{ to } C}}$$

To minimize $S(x, y)$, you begin by computing the partial derivatives

$$S_x = 2(x - 1) + 2x + 2(x - 8) = 6x - 18$$
$$S_y = 2(y - 5) + 2y + 2y = 6y - 10$$

Then $S_x = 0$ and $S_y = 0$ when

$$6x - 18 = 0$$
$$6y - 10 = 0$$

or $x = 3$ and $y = \dfrac{5}{3}$. Since $S_{xx} = 6$, $S_{xy} = 0$, and $S_{yy} = 6$, you get

$$D = S_{xx}S_{yy} - S_{xy}^2 = (6)(6) - 0^2 = 36 > 0$$

and

$$S_{xx}\left(3, \frac{5}{3}\right) = 6 > 0$$

So the sum of squares is minimized at the map point W with coordinates $\left(3, \frac{5}{3}\right)$.

Finding Extreme Values on a Closed, Bounded Region

So far, we have discussed only the relative extrema of a function of two variables. The function $f(x, y)$ is said to have an **absolute maximum** at (x_0, y_0) on a region R in the xy plane if $f(x_0, y_0) \geq f(x, y)$ for every point (x, y) in R. Likewise, an **absolute minimum** occurs at (x_0, y_0) if $f(x_0, y_0) \leq f(x, y)$ for every point (x, y) in R. In Section 3.4, we found such absolute extrema by using the extreme value property, which says:

> A function $f(x)$ that is continuous on the closed interval $a \leq x \leq b$ attains both its absolute maximum and its absolute minimum values on $a \leq x \leq b$. These extreme values can occur either at an endpoint of the interval (a or b) or at a critical number c inside the interval $a < c < b$.

The extreme value property can be extended to functions of two variables and takes the following form:

> **Extreme Value Property for a Function of Two Variables** ■ A function $f(x, y)$ that is continuous on the closed, bounded region R in the xy plane attains both its absolute maximum and absolute minimum values on R. These extreme values occur either on the boundary of R or at a critical point in the interior of R.

But what exactly do we mean by a "closed, bounded region R" and the "boundary of R"? First, a **boundary point** of R is a point (c, d) with the property that every circle centered at (c, d) contains points both inside R and outside R, regardless of how small the radius may be, and the set of all boundary points of R is called its **boundary.** Then R is a closed, bounded region if it contains its boundary and can itself be contained in a circle of finite radius (Figure 7.16).

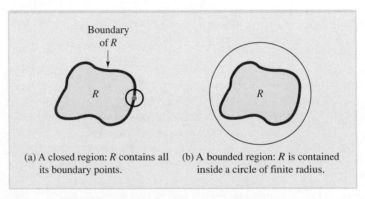

(a) A closed region: R contains all its boundary points.   (b) A bounded region: R is contained inside a circle of finite radius.

FIGURE 7.16   A closed, bounded region R in the xy plane.

With the extreme value property in mind, we can locate the absolute extrema of a continuous function on a closed, bounded region R using the following procedure:

> **A Procedure for Finding the Absolute Extrema of a Function $f(x, y)$ on a Closed, Bounded Region R**
>
> **Step 1.** Find the critical points of $f(x, y)$ in the region R.
> **Step 2.** Find all points on the boundary of R where extreme values can occur.
> **Step 3.** Evaluate $f(x_0, y_0)$ for each of the points (x_0, y_0) found in steps 1 and 2. The largest of these values is the absolute maximum of $f(x, y)$ on R, and the smallest of these values is the absolute minimum.

In general, closed bounded regions can be quite complicated, and it may be difficult to determine the boundary points where extreme values can occur (step 2). However, boundaries that occur in practical applications are often straight line segments or curves with simple equations (like circles). Example 7.3.6 illustrates the step-by-step procedure for finding extreme values when R is a triangular region. The procedure is then used in Example 7.3.7 to solve an applied optimization problem.

EXAMPLE 7.3.6   Finding the Absolute Extrema of a Function

Find the absolute maximum and minimum values of the function

$$f(x, y) = 4xy - x^2 - 4y + 9$$

on the triangular region R with vertices $(0, 0)$, $(8, 0)$, and $(0, 16)$. The graph of R is shown in Figure 7.17.

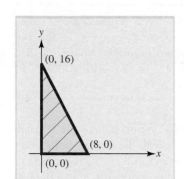

Solution
Step 1. Find the critical points of $f(x, y)$ in the region R.
The partial derivatives of f are

$$f_x(x, y) = 4y - 2x \qquad \text{and} \qquad f_y(x, y) = 4x - 4$$

These partial derivatives are defined for all x and y, and

$$f_x(x, y) = 4y - 2x = 0 \qquad \text{and} \qquad f_y(x, y) = 4x - 4 = 0$$

only when $x = 1$ and $y = \dfrac{1}{2}$, so $\left(1, \dfrac{1}{2}\right)$ is the only internal critical point of R. [Do you see why $\left(1, \dfrac{1}{2}\right)$ is inside R?]

Step 2. Find all points on the boundary of R where extreme values can occur.
The boundary lines of R have the equations $y = -2x + 16$, $x = 0$, and $y = 0$. The boundary consists of three line segments, and we examine each segment separately:
The horizontal line segment between $(0, 0)$ and $(8, 0)$. On this segment, $y = 0$ and $f(x, y)$ becomes a function of x alone:

$$u(x) = f(x, 0) = -x^2 + 9 \quad \text{for } 0 \le x \le 8$$

Since $u'(x) = -2x = 0$ on this interval only when $x = 0$, which corresponds to the vertex $(0, 0)$, the extreme values of $u(x)$ can occur only at the endpoints $(0, 0)$ and $(8, 0)$ in this case.

FIGURE 7.17   The region R in Example 7.3.6.

The vertical line segment between (0, 0) and (0, 16). Since $x = 0$ on this segment, the function $f(x, y)$ becomes a function of y alone:

$$v(y) = -4y + 9 \quad \text{for } 0 \leq y \leq 16$$

Since $v'(y) = -4 \neq 0$ for all y, the extreme values of $f(x, y)$ on this line segment can occur only at the endpoints $(0, 0)$ and $(0, 16)$.

The line segment between (0, 16) and (8, 0). We have $y = -2x + 16$ on this segment, and by substituting for y in the formula for $f(x, y)$, we get

$$w(x) = 4x(-2x + 16) - x^2 - 4(-2x + 16) + 9$$
$$= -9x^2 + 72x - 55 \quad \text{for } 0 \leq x \leq 8$$

We find that $w'(x) = -18x + 72$ and $w'(x) = 0$ when

$$x = 4 \quad \text{and} \quad y = -2(4) + 16 = 8$$

so the extreme values of $w(x)$ occur at either the boundary critical point $(4, 8)$ or at one of the endpoints $(0, 16)$ or $(8, 0)$.

Step 3. Evaluate $f(x_0, y_0)$ for each of the points (x_0, y_0) found in steps 1 and 2 and compare.

In the following table, the top line lists all the interior and boundary critical points found in steps 1 and 2 along with the three vertices of the triangular region R, while the bottom line lists the value of the function $f(x, y) = 4xy - x^2 - 4y + 9$ at each such point.

Point (x_0, y_0)	$\left(1, \dfrac{1}{2}\right)$	$(4, 8)$	$(0, 0)$	$(8, 0)$	$(0, 16)$
Value $f(x_0, y_0)$	8	89	9	-55	-55

Comparing the values in the second line in the table, we see that on the triangular region R, the function $f(x, y)$ attains the absolute maximum value 89 at the point $(4, 8)$ and the absolute minimum value -55 at both $(8, 0)$ and $(0, 16)$.

EXAMPLE 7.3.7 Applying the Extreme Value Property to a Business Problem

Paul Johnson is a salesman whose territory borders on a lake and can be described in terms of a rectangular grid as the region bounded by the curve $y = x^2$ (the lakefront) and the lines $y = 0$ and $x = 3$ as shown in Figure 7.18, where x and y are in miles. He determines that the number of units $S(x, y)$ he can sell at each grid point (x, y) in his region is given by the function

$$S(x, y) = 4x^2 - 16x + 4y^2 - 4y + 20$$

At what point(s) in his sales territory should Paul expect maximum sales to occur, and what are his maximum expected sales? Answer the same question for minimum sales.

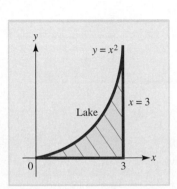

FIGURE 7.18 The sales territory in Example 7.3.7.

Solution

The partial derivatives of S are

$$S_x(x, y) = 8x - 16 \quad \text{and} \quad S_y(x, y) = 8y - 4$$

These partial derivatives are defined for all x and y, and

$$S_x(x, y) = 8x - 16 = 0 \quad \text{and} \quad S_y(x, y) = 8y - 4 = 0$$

when $x = 2$ and $y = \dfrac{1}{2}$, so $\left(2, \dfrac{1}{2}\right)$ is the only internal critical point of R. The boundary of the sales region consists of the horizontal line segment between $(0, 0)$ and $(3, 0)$, the vertical segment between $(3, 0)$ and $(3, 9)$, and the portion of the curve $y = x^2$ between $(0, 0)$ and $(3, 9)$.

On the horizontal boundary line $y = 0$, the sales function becomes

$$u(x) = 4x^2 - 16x + 20$$

Since $u'(x) = 8x - 16 = 0$ when $x = 2$, the extreme values of $S(x, y)$ on $y = 0$ can occur only at $(2, 0)$ or at the endpoints of the segment $(0, 0)$ and $(3, 0)$.

Similarly, on the vertical line segment between $(3, 0)$ and $(3, 9)$, we have $x = 3$ and the sales function becomes

$$v(y) = 4(3)^2 - 16(3) + 4y^2 - 4y + 20 = 4y^2 - 4y + 8$$

and $v'(y) = 8y - 4 = 0$ only when $y = \dfrac{1}{2}$. On this part of the boundary, extreme values of $S(x, y)$ can occur only at the boundary critical point $\left(3, \dfrac{1}{2}\right)$ or at the endpoints $(3, 0)$ and $(3, 9)$.

On the curved portion of the boundary where $y = x^2$, we substitute $y = x^2$ in the formula for the sales function to obtain

$$w(x) = 4x^2 - 16x + 4(x^2)^2 - 4(x^2) + 20 = 4x^4 - 16x + 20$$

The only real number that satisfies $w'(x) = 16x^3 - 16 = 0$ is $x = 1$, so on this part of the boundary, extreme values of $S(x, y)$ can occur only at the boundary critical point $(1, 1)$ or at the endpoints of the curved portion, $(0, 0)$ and $(3, 9)$.

Finally, we compute $S(x, y)$ at each of the points where extreme values can occur:

Point (x_0, y_0)	$\left(2, \dfrac{1}{2}\right)$	$(0, 0)$	$(2, 0)$	$(3, 0)$	$\left(3, \dfrac{1}{2}\right)$	$(3, 9)$	$(1, 1)$
Value $S(x_0, y_0)$	3	20	4	8	7	296	8

Comparing the values in the second line in the table, we see that Paul will be able to sell a maximum of 296 units at grid location $(3, 9)$. The minimum number of sales is 3 units at location $\left(2, \dfrac{1}{2}\right)$.

EXERCISES ▪ 7.3

In Exercises 1 through 22, find the critical points of the given functions and classify each as a relative maximum, a relative minimum, or a saddle point. (Note: The algebra in Exercises 19 through 22 is challenging.)

1. $f(x, y) = 5 - x^2 - y^2$

2. $f(x, y) = 2x^2 - 3y^2$

3. $f(x, y) = xy$

4. $f(x, y) = x^2 + 2y^2 - xy + 14y$

5. $f(x, y) = \dfrac{16}{x} + \dfrac{6}{y} + x^2 - 3y^2$

6. $f(x, y) = xy + \dfrac{8}{x} + \dfrac{8}{y}$

7. $f(x, y) = 2x^3 + y^3 + 3x^2 - 3y - 12x - 4$

8. $f(x, y) = (x - 1)^2 + y^3 - 3y^2 - 9y + 5$

9. $f(x, y) = x^3 + y^2 - 6xy + 9x + 5y + 2$

10. $f(x, y) = -x^4 - 32x + y^3 - 12y + 7$

11. $f(x, y) = xy^2 - 6x^2 - 3y^2$

12. $f(x, y) = x^2 - 6xy - 2y^3$

13. $f(x, y) = (x^2 + 2y^2)e^{1-x^2-y^2}$

14. $f(x, y) = e^{-(x^2+y^2-6y)}$

15. $f(x, y) = x^3 - 4xy + y^3$

16. $f(x, y) = (x - 4)\ln(xy)$

17. $f(x, y) = 4xy - 2x^4 - y^2 + 4x - 2y$

18. $f(x, y) = 2x^4 + x^2 + 2xy + 3x + y^2 + 2y + 5$

19. $f(x, y) = \dfrac{1}{x^2 + y^2 + 3x - 2y + 1}$

20. $f(x, y) = xye^{-(16x^2+9y^2)/288}$

21. $f(x, y) = x\ln\left(\dfrac{y^2}{x}\right) + 3x - xy^2$

22. $f(x, y) = \dfrac{x}{x^2 + y^2 + 4}$

In Exercises 23 through 28, find all interior and boundary critical points and determine the largest and smallest values of the function f(x, y) over the given closed, bounded region R.

23. $f(x, y) = xy - x - 3y$ on the triangular region R with vertices $(0, 0)$, $(5, 0)$, $(5, 5)$.

24. $f(x, y) = 4xy - 8x - 4y + 5$ on the triangular region R with vertices $(0, 0)$, $(2, 0)$, $(0, 3)$.

25. $f(x, y) = 2x^2 + y^2 + xy^2 - 2$ on the square region R with vertices $(5, 5)$, $(-5, 5)$, $(5, -5)$, $(-5, -5)$.

26. $f(x, y) = x^2 + 3y^2 - 4x + 6y - 3$ on the square region R with vertices $(0, 0)$, $(3, 0)$, $(3, -3)$, $(0, -3)$.

27. $f(x, y) = x^4 + 2y^3$ on the circular region R bounded by $x^2 + y^2 = 1$.

28. $f(x, y) = xy^2$ on the quarter circular region R bounded by $x^2 + y^2 = 12$ with $x \geq 0$, $y \geq 0$.

BUSINESS AND ECONOMICS APPLIED PROBLEMS

In Exercises 29 through 35, assume that the required extreme value is a relative extremum.

29. **RETAIL SALES** A T-shirt shop carries two competing shirts, one endorsed by Tim Duncan and the other by LeBron James. The owner of the store can obtain both types at a cost of $2 per shirt and estimates that if Duncan shirts are sold

for x dollars apiece and James shirts for y dollars apiece, consumers will buy $40 - 50x + 40y$ Duncan shirts and $20 + 60x - 70y$ James shirts each day. How should the owner price the shirts to generate the largest possible profit?

30. **PRICING** The telephone company is planning to introduce two new types of executive communications systems that it hopes to sell to its largest commercial customers. It is estimated that if the first type of system is priced at x hundred dollars per system and the second type at y hundred dollars per system, approximately $40 - 8x + 5y$ consumers will buy the first type and $50 + 9x - 7y$ will buy the second type. If the cost of manufacturing the first type is $1,000 per system and the cost of manufacturing the second type is $3,000 per system, how should the telephone company price the systems to generate the largest possible profit?

31. **RETAIL SALES** A company produces x units of commodity A and y units of commodity B. All the units can be sold for $p = 100 - x$ dollars per unit of A and $q = 100 - y$ dollars per unit of B. The cost (in dollars) of producing these units is given by the joint-cost function $C(x, y) = x^2 + xy + y^2$. What should x and y be to maximize profit?

32. **RETAIL SALES** Repeat Exercise 31 for the case where $p = 20 - 5x$, $q = 4 - 2y$, and $C = 2xy + 4$.

33. **PROFIT UNDER MONOPOLY** A manufacturer with exclusive rights to a sophisticated new industrial machine is planning to sell a limited number of the machines to both foreign and domestic firms. The price the manufacturer can expect to receive for the machines will depend on the number of machines made available. (For example, if only a few of the machines are placed on the market, competitive bidding among prospective purchasers will tend to drive the price up.) It is estimated that if the manufacturer supplies x machines to the domestic market and y machines to the foreign market, the machines will sell for

$$60 - \frac{x}{5} + \frac{y}{20} \text{ thousand dollars each domestically}$$

and for $50 - \dfrac{y}{10} + \dfrac{x}{20}$ thousand dollars each abroad. If the manufacturer can produce the machines at the cost of $10,000 each, how many should be supplied to each market to generate the largest possible profit?

34. **PROFIT UNDER MONOPOLY** A manufacturer with exclusive rights to a new industrial machine is planning to sell a limited number of them and estimates that if x machines are supplied to the domestic market and y to the foreign market, the machines will sell for $150 - \dfrac{x}{6}$ thousand dollars each domestically and for $100 - \dfrac{y}{20}$ thousand dollars each abroad. The operating costs are assumed to be constant.

   a. How many machines should the manufacturer supply to the domestic market to generate the largest possible profit at home?

   b. How many machines should the manufacturer supply to the foreign market to generate the largest possible profit abroad?

   c. How many machines should the manufacturer supply to each market to generate the largest possible *total* profit?

   d. Is the relationship between the answers in parts (a), (b), and (c) accidental? Explain. Does a similar relationship hold in Exercise 33? What accounts for the difference between these two problems in this respect?

35. **ALLOCATION OF FUNDS** A manufacturer is planning to sell a new product at the price of $210 per unit and estimates that if x thousand dollars is spent on development and y thousand dollars is spent on promotion, consumers will buy approximately $\dfrac{640y}{y + 3} + \dfrac{216x}{x + 5}$ units of the product. If manufacturing costs for this product are $135 per unit, how much should the manufacturer spend on development and how much on promotion to generate the largest possible profit from the sale of this product? [*Hint:* Profit = (number of units)(price per unit − cost per unit) − total amount spent on development and promotion.]

Exercises 36 through 39 require the extreme value property.

36. **SALES** Carmen Ramos, a colleague of the salesman in Example 7.3.7, has a territory that can be described in terms of a rectangular grid as the region bounded by the curve $y = x^2$ and the line $y = 16$, where x and y are in miles. She determines

that the number of units $S(x, y)$ she can sell at each grid point (x, y) in her region is given by the function

$$S(x, y) = 6x^2 - 36x + 9y^2 - 6y + 60$$

At what point(s) in Carmen's sales territory should she expect maximum sales to occur, and what are her maximum expected sales? Answer the same question for minimum sales.

37. **BUSINESS MANAGEMENT** McKinley Martin is the business manager of an electronics store. He determines that when he charges x dollars per unit for the standard Kindle eBook reader and y dollars per unit for the enhanced model, he will sell $400 - x - y$ standard units and $600 - 2x - y$ enhanced units. He further determines that the total cost of marketing and maintaining inventory for the Kindle readers will be C dollars, where

$$C(x, y) = x^2 - 280x + y^2 - 380y + 60,000$$

   a. Note that x and y must be nonnegative and must also satisfy $400 - x - y \geq 0$ and $600 - 2x - y \geq 0$. Describe the closed, bounded region R in the xy plane that meets these requirements.

   b. What prices x and y should McKinley charge to minimize cost?

38. **PROFIT** In this exercise, you are asked to re-examine Example 7.3.4 as an optimization problem over a closed, bounded region. [*Warning:* This problem involves some nasty algebra.]

   a. In Example 7.3.4, explain why the variables x and y must satisfy

$$x \geq 30, y \geq 40$$
$$70 - 5x + 4y \geq 0$$
$$80 + 6x - 7y \geq 0$$

   Show that the set of all points (x, y) that satisfy these inequalities is a triangular region, and find its vertices.

   b. Solve Example 7.3.4 by finding the largest value of the profit function

$$f(x, y) = -5x^2 + 10xy - 20x - 7y^2 + 240y - 5,300$$

   over the triangular region found in part (a). Does this change the solution provided? For what values of x and y is the *smallest* profit obtained?

39. **PROFIT** In this exercise, you are asked to re-examine Exercise 29 as an optimization problem over a closed, bounded region. [*Warning:* This problem involves some nasty algebra.]

a. In Exercise 29, explain why the variables x and y must satisfy

$$x \geq 2, \; y \geq 2$$
$$40 - 50x + 40y \geq 0$$
$$20 + 60x - 70y \geq 0$$

Show that the set of all points (x, y) that satisfy these inequalities is a triangular region, and find its vertices.

b. Solve Exercise 29 by finding the largest value of the profit function

$$P(x, y) = -50x^2 + 100xy + 20x - 70y^2 + 80y - 120$$

over the triangular region found in part (a). Does this change the solution found in Exercise 29? For what values of x and y is the *smallest* profit obtained?

LIFE AND SOCIAL SCIENCE APPLIED PROBLEMS

In Exercises 40 through 49, assume that the required extreme value is a relative extremum.

40. **SOCIAL CHOICES** The social desirability of an enterprise often involves making a choice between the commercial advantage of the enterprise and the social or ecological loss that may result. For instance, the lumber industry provides paper products to society and income to many workers and entrepreneurs, but the gain may be offset by the destruction of habitable territory for spotted owls and other endangered species. Suppose the social desirability of a particular enterprise is measured by the function

$$D(x, y) = (16 - 6x)x - (y^2 - 4xy + 40)$$

where x measures commercial advantage (profit and jobs) and y measures ecological disadvantage (species displacement, as a percentage) with $x \geq 0$ and $y \geq 0$. The enterprise is deemed desirable if $D \geq 0$ and undesirable if $D < 0$.

a. What values of x and y will maximize social desirability? Interpret your result. Is it possible for this enterprise to be desirable?

b. The function given in part (a) is artificial, but the ideas are not. Research the topic of ethics in industry, and write a paragraph on how you feel these choices should be made.*

41. **RESPONSE TO STIMULI** Consider an experiment in which a subject performs a task while being exposed to two different stimuli (for

*Start with the article by K. R. Stollery, "Environmental Controls in Extractive Industries," *Land Economics,* Vol. 61, 1985, p. 169.

example, sound and light). For low levels of the stimuli, the subject's performance might actually improve, but as the stimuli increase, they eventually become a distraction and the performance begins to deteriorate. Suppose in a certain experiment in which x units of stimulus A and y units of stimulus B are applied, the performance of a subject is measured by the function

$$f(x, y) = C + xye^{1-x^2-y^2}$$

where C is a positive constant. How many units of each stimulus result in maximum performance?

42. **MAINTENANCE** In relation to a rectangular map grid, four oil rigs are located at the points $(-300, 0)$, $(-100, 500)$, $(0, 0)$, and $(400, 300)$ where units are in feet. Where should a maintenance shed $M(a, b)$ be located to minimize the sum of squares of the distances from the rigs?

43. **CITY PLANNING** Four small towns in a rural area wish to pool their resources to build a television station. If the towns are located at the points $(-5, 0)$, $(1, 7)$, $(9, 0)$, and $(0, -8)$ on a rectangular map grid, where units are in miles, at what point $S(a, b)$ should the station be located to minimize the sum of squares of the distances from the towns?

44. **LEARNING** In a learning experiment, a subject is first given x minutes to examine a list of facts. The fact sheet is then taken away and the subject is allowed y minutes to prepare mentally for an exam based on the fact sheet. Suppose it is found that the score achieved by a particular subject is related to x and y by the formula

$$S(x, y) = -x^2 + xy + 10x - y^2 + y + 15$$

a. What score does the subject achieve if he takes the test "cold" (with no study or contemplation)?

b. How much time should the subject spend in study and contemplation to maximize his score? What is the maximum score?

45. **GENETICS** Alternative forms of a gene are called *alleles.* Three alleles, designated A, B, and O, determine the four human blood types A, B, O, and AB. Suppose that p, q, and r are the proportions of A, B, and O in a particular population, so that $p + q + r = 1$. Then according to the Hardy-Weinberg law in genetics, the proportion of individuals in the population who carry two different alleles is given by $P = 2pq + 2pr + 2rq$. What is the largest value of P?

46. BUTTERFLY WING PATTERNS   The beautiful patterns on the wings of butterflies have long been a subject of curiosity and scientific study. Mathematical models used to study these patterns often focus on determining the level of morphogen (a chemical that effects change). In a model dealing with eyespot patterns,* a quantity of morphogen is released from an eyespot and the morphogen concentration t days later is given by

$$S(r, t) = \frac{1}{\sqrt{4\pi t}} e^{-\left(\gamma kt + \frac{r^2}{4t}\right)} \qquad t > 0$$

where r measures the radius of the region on the wing affected by the morphogen, and k and γ are positive constants.

a. Find t_m so that $\dfrac{\partial S}{\partial t} = 0$. Show that the function $S_m(t)$ formed from $S(r, t)$ by fixing r has a relative maximum at t_m. Is this the same as saying that the function of two variables $S(r, t)$ has a relative maximum?

b. Let $M(r)$ denote the maximum found in part (a); that is, $M(r) = S(r, t_m)$. Find an expression for M in terms of $z = (1 + 4\gamma kr^2)^{1/2}$.

c. It turns out that $M(z)$ is what is really needed to analyze the eyespot wing pattern. Read pages 461–468 in the text cited with this problem, and write a paragraph on how biology and mathematics are blended in the study of butterfly wing patterns.

47. LIVABLE SPACE   Define the livable space of a building to be the volume of space in the building where a person 6 feet tall can walk upright. An A-frame cabin is y feet long and has equilateral triangular ends x feet on a side, as shown in the accompanying figure. If the surface area of the cabin (roof and two ends) is to be 500 ft², what dimensions x and y will maximize the livable space?

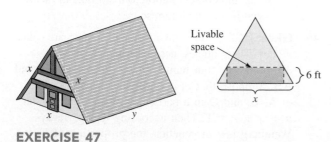

Livable space

6 ft

EXERCISE 47

*J. D. Murray, *Mathematical Biology*, 2nd ed., New York: Springer-Verlag, 1993, pp. 461–468.

48. CANCER THERAPY   Certain malignant tumors that do not respond to conventional methods of treatment such as surgery or chemotherapy may be treated by **hyperthermia**, which involves applying extreme heat to tumors using microwave transmissions.[†] One particular kind of microwave applicator used in this therapy produces an absorbed energy density that falls off exponentially. Specifically, the temperature at each point located r units from the central axis of a tumor and z units inside the tumor is given by a formula of the form

$$T(r, z) = Ae^{-pr^2}(e^{-qz} - e^{-sz})$$

where A, p, q, and s are positive constants that depend on properties of both blood and the heating appliance (see the accompanying figure). At what depth does maximum temperature occur? Express your answer in terms of A, p, q, and s. You are not required to apply the second partials test.

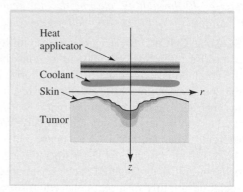

Heat applicator

Coolant

Skin

Tumor

r

z

EXERCISE 48

49. TIME OF TRAVEL   Tom, Dick, and Mary are participating in a cross-country relay race. Tom will trudge as fast as he can through thick woods to the edge of a river. Then Dick will take over and row to the opposite shore, where Mary will take the baton and run along the river road to the finish line. The course is shown in the accompanying figure. Teams must start at point S and finish at F, but they may position one member anywhere along the shore of the river and another anywhere along the river road.

a. Suppose Tom can trudge at 2 mph; Dick can row at 4 mph; and Mary can run at 6 mph. Where should Dick and Mary receive the baton for the team to finish the course as quickly as possible?

[†]The ideas in this exercise are based on the article by Leah Edelstein-Keshet, "Heat Therapy for Tumors," *UMAP Modules 1991: Tools for Teaching*, Lexington, MA: Consortium for Mathematics and Its Applications, Inc., 1992, pp. 73–101.

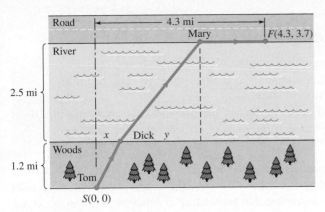

EXERCISE 49

b. The main competition for Tom, Dick, and Mary is the team of Nan, Fran, and Phineas. If Nan can trudge at 1.7 mph, Fran can row at 3.5 mph, and Phineas can run at 6.3 mph, which team should win? By how much time?

c. This exercise may remind you of the spy story in Exercise 49 of Section 3.5. Create your own spy story problem based on the mathematical ideas in this exercise.

MISCELLANEOUS PROBLEMS

In Exercises 50 through 52, assume that the required extreme value is a relative extremum.

50. **CONSTRUCTION** A farmer wishes to fence off a rectangular pasture along the bank of a river. The area of the pasture is to be 6,400 yd², and no fencing is needed along the river bank. Find the dimensions of the pasture that will require the least amount of fencing.

51. **CONSTRUCTION** Suppose you wish to construct a rectangular box with a volume of 32 ft³. Three different materials will be used in the construction. The material for the sides costs $1 per square foot, the material for the bottom costs $3 per square foot, and the material for the top costs $5 per square foot. What are the dimensions of the least expensive such box?

52. **PARTICLE PHYSICS** A particle of mass m in a rectangular box with dimensions x, y, and z has ground state energy

$$E(x, y, z) = \frac{k^2}{8m}\left(\frac{1}{x^2} + \frac{1}{y^2} + \frac{1}{z^2}\right)$$

where k is a physical constant. If the volume of the box satisfies $xyz = V_0$ for constant V_0, find the values of x, y, and z that minimize the ground state energy.

53. Let $f(x, y) = x^2 + y^2 - 4xy$. Show that f does *not* have a relative minimum at its critical point $(0, 0)$, even though it does have a relative minimum at $(0, 0)$ in both the x and y directions. [*Hint:* Consider the direction defined by the line $y = x$. That is, substitute x for y in the formula for f and analyze the resulting function of x.]

In Exercises 54 through 57, find the partial derivatives f_x and f_y and then use your graphing utility to determine the critical points of each function.

54. $f(x, y) = (x^2 + 3y - 5)e^{-x^2 - 2y^2}$

55. $f(x, y) = \dfrac{x^2 + xy + 7y^2}{x \ln y}$

56. $f(x, y) = 6x^2 + 12xy + y^4 + x - 16y - 3$

57. $f(x, y) = 2x^4 + y^4 - x^2(11y - 18)$

58. Sometimes you can classify the critical points of a function by inspecting its level curves. In each of the following cases, determine the nature of the critical point of f at $(0, 0)$.

a.

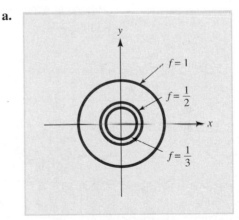

b.

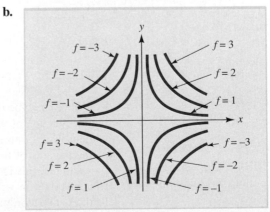

SECTION 7.4   The Method of Least-Squares

Learning Objectives

1. Explore least-squares approximation of data as an optimization problem involving a function of two variables.
2. Examine several applied problems using least-squares approximation of data.
3. Discuss nonlinear curve-fitting techniques using least-squares approximation.

Throughout this text, you have seen applied functions, many of which are derived from published research, and you may have wondered how researchers come up with such functions. A common procedure for associating a function with an observed physical phenomenon is to gather data, plot it on a graph, and then find a function whose graph "best fits" the data in some mathematically meaningful way. We will now develop such a procedure, called the **method of least-squares** or **regression analysis,** which was first mentioned in Example 1.3.8 of Section 1.3, in connection with fitting a line to unemployment data.

The Least-Squares Procedure

Suppose you wish to find a function $y = f(x)$ that fits a particular data set reasonably well. The first step is to decide what type of function to try. Sometimes this can be done by a theoretical analysis of the observed phenomenon and sometimes by inspecting the plotted data. Two sets of data are plotted in Figure 7.19. These are called **scatter diagrams.** In Figure 7.19a, the points lie roughly along a straight line, suggesting that a linear function $y = mx + b$ be used. However, in Figure 7.19b, the points appear to follow an exponential curve, and a function of the form $y = Ae^{-kx}$ would be more appropriate.

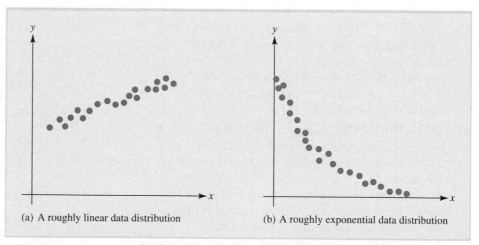

(a) A roughly linear data distribution   (b) A roughly exponential data distribution

FIGURE 7.19   Two scatter diagrams.

Once the type of function has been chosen, the next step is to determine the particular function of this type whose graph is "closest" to the given set of points. A convenient way to measure how close a curve is to a set of points is to compute the sum of the squares of the vertical distances from the points to the curve. In Figure 7.20, for example, this is the sum $d_1^2 + d_2^2 + d_3^2$. The closer the curve is to the points, the smaller this sum will be, and the curve for which this sum is smallest is said to best fit the data according to the **least-squares criterion.**

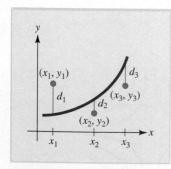

FIGURE 7.20   Sum of squares of the vertical distances $d_1^2 + d_2^2 + d_3^2$.

The use of the least-squares criterion to fit a linear function to a set of points is illustrated in Example 7.4.1. The computation involves the technique from Section 7.3 for minimizing a function of two variables.

EXAMPLE 7.4.1   Using the Least-Squares Criterion

Use the least-squares criterion to find the equation of the line that is closest to the three points $(1, 1)$, $(2, 3)$, and $(4, 3)$.

Solution

As indicated in Figure 7.21, the sum of the squares of the vertical distances from the three given points to the line $y = mx + b$ is

$$d_1^2 + d_2^2 + d_3^2 = (m + b - 1)^2 + (2m + b - 3)^2 + (4m + b - 3)^2$$

This sum depends on the coefficients m and b that define the line, and so the sum can be thought of as a function $S(m, b)$ of the two variables m and b. The goal, therefore, is to find the values of m and b that minimize the function

$$S(m, b) = (m + b - 1)^2 + (2m + b - 3)^2 + (4m + b - 3)^2$$

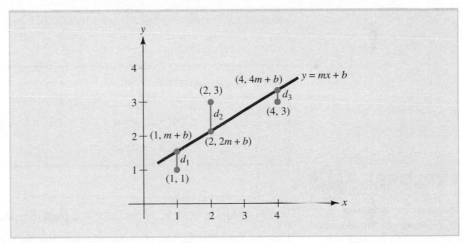

FIGURE 7.21   Minimize the sum $d_1^2 + d_2^2 + d_3^2$.

You do this by setting the partial derivatives $\dfrac{\partial S}{\partial m}$ and $\dfrac{\partial S}{\partial b}$ equal to zero to get

$$\frac{\partial S}{\partial m} = 2(m + b - 1) + 4(2m + b - 3) + 8(4m + b - 3)$$

$$= 42m + 14b - 38 = 0$$

and $\dfrac{\partial S}{\partial b} = 2(m + b - 1) + 2(2m + b - 3) + 2(4m + b - 3)$

$$= 14m + 6b - 14 = 0$$

Solving the resulting equations

$$42m + 14b = 38$$
$$14m + 6b = 14$$

simultaneously for m and b you conclude that

$$m = \frac{4}{7} \quad \text{and} \quad b = 1$$

It can be shown that the critical point $(m, b) = \left(\frac{4}{7}, 1\right)$ does indeed minimize the function $S(m, b)$, and so it follows that

$$y = \frac{4}{7}x + 1$$

is the equation of the line that is closest to the three given points.

The Least-Squares Line

The line that is closest to a set of points according to the least-squares criterion is called the **least-squares line** for the points. (The term **regression line** is also used, especially in statistical work.) The procedure used in Example 7.4.1 can be generalized to give formulas for the slope m and the y intercept b of the least-squares line for an arbitrary set of n points $(x_1, y_1), (x_2, y_2), \ldots, (x_n, y_n)$. The formulas involve sums of the x and y values. All the sums run from $j = 1$ to $j = n$, and to simplify the notation, the indices are omitted. For example, Σx is used instead of $\displaystyle\sum_{j=1}^{n} x_j$.

The Least-Squares Line ■ The equation of the least-squares line for the n points $(x_1, y_1), (x_2, y_2), \ldots, (x_n, y_n)$ is $y = mx + b$, where

$$m = \frac{n\Sigma xy - \Sigma x \Sigma y}{n\Sigma x^2 - (\Sigma x)^2} \quad \text{and} \quad b = \frac{\Sigma x^2 \Sigma y - \Sigma x \Sigma xy}{n\Sigma x^2 - (\Sigma x)^2}$$

EXPLORE!

Your graphing calculator can easily assist you in producing and displaying the lists and summations needed to calculate the coefficients of the least-squares equation. Using the data in Example 7.4.2, place the x values into L1 and y values into L2, and write L3 = L1∗L2 and L4 = L1². Use the summation features of your calculator to obtain the column totals needed to calculate the slope and y intercept formulas displayed in the box on this page.

EXAMPLE 7.4.2   Finding a Least-Squares Line

Use the formulas to find the least-squares line for the points $(1, 1)$, $(2, 3)$, and $(4, 3)$ from Example 7.4.1.

Solution

Arrange your calculations as follows:

x	y	xy	x^2
1	1	1	1
2	3	6	4
4	3	12	16
$\Sigma x = 7$	$\Sigma y = 7$	$\Sigma xy = 19$	$\Sigma x^2 = 21$

Then use the formulas with $n = 3$ to get

$$m = \frac{3(19) - 7(7)}{3(21) - (7)^2} = \frac{4}{7} \quad \text{and} \quad b = \frac{21(7) - 7(19)}{3(21) - (7)^2} = 1$$

from which it follows that the equation of the least-squares line is

$$y = \frac{4}{7}x + 1$$

This matches the result of Example 7.4.1.

**Least-Squares
Prediction**
The least-squares line (or curve) that best fits the data collected in the past can be used to make rough predictions about the future. This is illustrated in Example 7.4.3.

EXAMPLE 7.4.3 Making a Least-Squares Prediction of GPA

In his role as admissions officer for a small liberal arts college, Leonard Chapman has compiled these data relating a group of students' high school grade-point average (GPA) to the GPA they earned in college:

High school GPA	2.0	2.5	3.0	3.0	3.5	3.5	4.0	4.0
College GPA	1.5	2.0	2.5	3.5	2.5	3.0	3.0	3.5

Find the equation of the least-squares line for Leonard's data, and use it to help him predict the college GPA of a student whose high school GPA is 3.7.

Solution

Let x denote the high school GPA and y the college GPA, and arrange the calculations as follows:

x	y	xy	x^2
2.0	1.5	3.0	4.0
2.5	2.0	5.0	6.25
3.0	2.5	7.5	9.0
3.0	3.5	10.5	9.0
3.5	2.5	8.75	12.25
3.5	3.0	10.5	12.25
4.0	3.0	12.0	16.0
4.0	3.5	14.0	16.0
$\Sigma x = 25.5$	$\Sigma y = 21.5$	$\Sigma xy = 71.25$	$\Sigma x^2 = 84.75$

Use the least-squares formulas with $n = 8$ to get

$$m = \frac{8(71.25) - 25.5(21.5)}{8(84.75) - (25.5)^2} \approx 0.78$$

and

$$b = \frac{84.75(21.5) - 25.5(71.25)}{8(84.75) - (25.5)^2} \approx 0.19$$

The equation of the least-squares line is therefore

$$y = 0.78x + 0.19$$

To predict the college GPA y of a student whose high school GPA x is 3.7, substitute $x = 3.7$ into the equation of the least-squares line. This gives

$$y = 0.78(3.7) + 0.19 \approx 3.08$$

which suggests that the student's college GPA might be about 3.1.

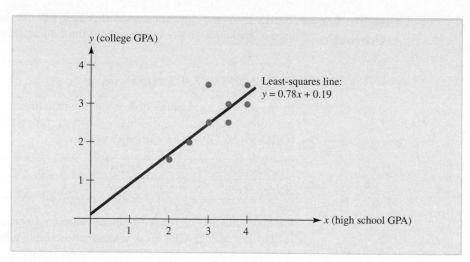

FIGURE 7.22   The least-squares line for high school and college GPAs.

The original data are plotted in Figure 7.22, together with the least-squares line $y = 0.78x + 0.19$. Actually, in practice, it is a good idea to plot the data before proceeding with the calculations. By looking at the graph you will usually be able to tell whether approximation by a straight line is appropriate or whether a better fit might be possible with a curve of some sort.

Nonlinear Curve-Fitting

In each of Examples 7.4.1, 7.4.2, and 7.4.3, the least-squares criterion was used to fit a linear function to a set of data. With appropriate modifications, the procedure can also be used to fit nonlinear functions to data. One kind of modified curve-fitting procedure is illustrated in Example 7.4.4.

EXPLORE!

Some graphing calculators can find best-fit equations for nonlinear data. Following Example 7.4.4, store the production and demand price data into lists L1 and L2, respectively. Then find and graph the nonlinear equation best fitting these data, using the Regression and Stat Plot techniques explained in the Calculator Introduction at the book-specific website, www.mhhe.com/hoffmann.

EXAMPLE 7.4.4   Finding an Exponential Demand Function

A manufacturer gathers these data relating the level of production x (hundred units) of a particular commodity to the demand price p (dollars per unit) at which all x units will be sold:

Production x (hundred units)	Demand Price p (dollars per unit)
6	743
10	539
17	308
22	207
28	128
35	73

a. Plot a scatter diagram for the data on a graph with production level on the x axis and demand price on the y axis.

b. Notice that the scatter diagram in part (a) suggests that the demand function is exponential. Modify the least-squares procedure to find a curve of the form $p = Ae^{mx}$ that best fits the data in the table.

c. Use the exponential demand function you found in part (b) to predict the revenue the manufacturer should expect if 4,000 ($x = 40$) units are produced.

Solution

a. The scatter diagram is plotted in Figure 7.23.

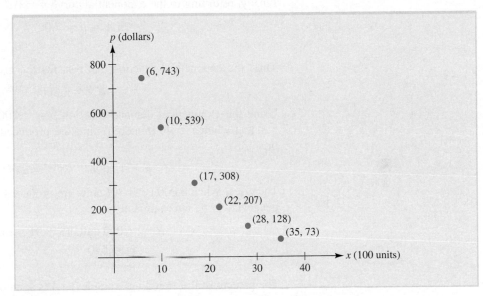

FIGURE 7.23   A scatter diagram for the demand data in Example 7.4.4.

Just-In-Time **REVIEW**

Recall the product rule for logarithms:

$\ln(ab) = \ln a + \ln b$

b. Taking logarithms on both sides of the equation $p = Ae^{mx}$, we find that

$$\begin{aligned}\ln p &= \ln(Ae^{mx}) &\text{product rule for logarithms}\\ &= \ln A + \ln(e^{mx}) &\ln e^u = u\\ &= \ln A + mx\end{aligned}$$

or equivalently, $y = mx + b$, where $y = \ln p$ and $b = \ln A$. Thus, to find the curve of the form $p = Ae^{mx}$ that best fits the given data points (x_k, p_k) for $k = 1, \ldots, 6$, we first find the least-squares line $y = mx + b$ for the data points $(x_k, \ln p_k)$. Arrange the calculations as follows:

k	x_k	p_k	$y_k = \ln p_k$	$x_k y_k$	x_k^2
1	6	743	6.61	39.66	36
2	10	539	6.29	62.90	100
3	17	308	5.73	97.41	289
4	22	207	5.33	117.26	484
5	28	128	4.85	135.80	784
6	35	73	4.29	150.15	1,225
	$\Sigma x = 118$		$\Sigma y = 33.10$	$\Sigma xy = 603.18$	$\Sigma x^2 = 2{,}918$

Use the least-squares formulas with $n = 6$ to get

$$m = \frac{6(603.18) - (118)(33.10)}{6(2{,}918) - (118)^2} = -0.08$$

and

$$b = \frac{2{,}918(33.10) - (118)(603.18)}{6(2{,}918) - (118)^2} = 7.09$$

which means the least-squares line has the equation

$$y = -0.08x + 7.09$$

Finally, returning to the exponential curve $p = Ae^{mx}$, recall that $\ln A = b$, so that

$$\ln A = b = 7.09$$
$$A = e^{7.09} = 1{,}200$$

Thus, the exponential function that best fits the given demand data is

$$p = Ae^{mx} = 1{,}200e^{-0.08x}$$

c. Using the exponential demand function $p = 1{,}200e^{-0.08x}$ found in part (b), we find that when $x = 40$ (hundred) units are produced, they can all be sold at a unit price of

$$p = 1{,}200e^{-0.08(40)} = \$48.91$$

Therefore, when 4,000 ($x = 40$) units are produced, we would expect the revenue generated to be approximately

$$R = (4{,}000 \text{ units})(\$48.91 \text{ per unit})$$
$$= \$195{,}640$$

The procedure illustrated in Example 7.4.4 is sometimes called **log-linear regression.** A curve of the form $y = Ax^k$ that best fits given data can also be found by log-linear regression. The specific procedure is outlined in Exercise 32, where it is used to verify an allometric formula first mentioned in the Think About It essay at the end of Chapter 1.

The least-squares procedure can also be used to fit other nonlinear functions to data. For instance, to find the quadratic function $y = Ax^2 + Bx + C$ whose graph (a parabola) best fits a particular set of data, you would proceed as in Example 7.4.1, minimizing the sum of squares of vertical distances from the given points to the graph. Such computations are algebraically complicated and usually require the use of a computer or graphing calculator. Most graphing calculators have built-in commands to find a variety of regression functions, allowing you to pick the type of function that best fits a given data set.

EXERCISES ■ 7.4

In Exercises 1 through 4, use the method of Example 7.4.1 to find the corresponding least-squares line.

1. (0, 1), (2, 3), (4, 2)

2. (1, 1), (2, 2), (6, 0)

3. (1, 2), (2, 4), (4, 4), (5, 2)

4. (1, 5), (2, 4), (3, 2), (6, 0)

In Exercises 5 through 12, use the formulas to find the corresponding least-squares line.

5. (1, 2), (2, 2), (2, 3), (5, 5)

6. (−4, −1), (−3, 0), (−1, 0), (0, 1), (1, 2)

7. (−2, 5), (0, 4), (2, 3), (4, 2), (6, 1)

8. (−6, 2), (−3, 1), (0, 0), (0, −3), (1, −1), (3, −2)

9. (0, 1), (1, 1.6), (2.2, 3), (3.1, 3.9), (4, 5)

10. (3, 5.72), (4, 5.31), (6.2, 5.12), (7.52, 5.32), (8.03, 5.67)

11. (−2.1, 3.5), (−1.3, 2.7), (1.5, 1.3), (2.7, −1.5)

12. (−1.73, −4.33), (0.03, −2.19), (0.93, 0.15), (3.82, 1.61)

In Exercises 13 through 16, modify the least-squares procedure as illustrated in Example 7.4.4 to find a curve of the form $y = Ae^{mx}$ that best fits the given data.

13. (1, 15.6), (3, 17), (5, 18.3), (7, 20), (10, 22.4)

14. (5, 9.3), (10, 10.8), (15, 12.5), (20, 14.6), (25, 17)

15. (2, 13.4), (4, 9), (6, 6), (8, 4), (10, 2.7)

16. (5, 33.5), (10, 22.5), (15, 15), (20, 10), (25, 6.8), (30, 4.5)

BUSINESS AND ECONOMICS APPLIED PROBLEMS

17. **DEMAND AND REVENUE** A manufacturer gathers the data listed in the accompanying table relating the level of production x (hundred units) of a particular commodity to the demand price p (dollars per unit) at which all the units will be sold:

Production x (hundreds of units)	5	10	15	20	25	30	35
Demand price p (dollars per unit)	44	38	32	25	18	12	6

   a. Plot these data on a graph.
   b. Find the equation of the least-squares line for the data.
   c. Use the linear demand equation you found in part (b) to predict the revenue the manufacturer should expect if 4,000 units ($x = 40$) are produced.

18. **SALES** A company's annual sales (in units of 1 billion dollars) for its first 5 years of operation are shown in this table:

Year	1	2	3	4	5
Sales	0.9	1.5	1.9	2.4	3.0

   a. Plot these data on a graph.
   b. Find the equation of the least-squares line.
   c. Use the least-squares line to predict the company's sixth-year sales.

19. **INVESTMENT ANALYSIS** Jennifer has several different kinds of investments, whose total value $V(t)$ (in thousands of dollars) at the beginning of the tth year after she began investing is given in this table, for $1 \le t \le 10$:

Year t	1	2	3	4	5	6	7	8	9	10
Value $V(t)$ of all investments	57	60	62	65	62	65	70	75	79	85

   a. Modify the least-squares procedure, as illustrated in Example 7.4.4, to find a function of the form $V(t) = Ae^{rt}$ whose graph best fits these data. Roughly at what annual rate, compounded continuously, is her account growing?
   b. Use the function you found in part (a) to predict the total value of her investments at the beginning of the 20th year after she began investing.
   c. Jennifer estimates she will need $300,000 to buy a vacation home. Use the function from part (a) to determine how long it will take her to attain this goal.
   d. Jennifer's friend, Frank Kornerkutter, looks over her investment analysis and snorts, "What a waste of time! You can find the A and r in your function $V(t) = Ae^{rt}$ by just using $V(1) = 57$ and $V(10) = 85$ and a little algebra." Find A and r using Frank's method, and comment on the relative merits of the two approaches.

20. DISPOSABLE INCOME AND CONSUMPTION The accompanying table gives the personal consumption expenditure and the corresponding disposable income (in billions of dollars) for the United States in the period 2003–2008:

Year	2003	2004	2005	2006	2007	2008
Disposable income	9,759.5	10,436.9	11,158.9	11,929.7	12,321.6	12,493.8
Personal consumption	7,804.0	8,285.1	8,819.0	9,322.7	9,806.3	10,104.5

SOURCE: U.S. Department of Commerce, Bureau of Economic Analysis, "Personal Consumption Expenditures by Major Type of Product" (http://www.bea.gov).

a. Plot these data on a graph, with disposable income on the x axis and consumption expenditure on the y axis.

b. Find the equation of the least-squares line for the data.

c. Use the least-squares line to predict the consumption that would correspond to $13 trillion ($13,000 billion) of disposable income.

d. Write a paragraph on the relationship between disposable income and consumption.

21. GASOLINE PRICES The average retail price per gallon (in cents) of regular unleaded gasoline at 3-year intervals from 1992 to 2010 is given in this table:

Year	1992	1995	1998	2001	2004	2007	2010
Price per gallon (cents)	109	111	103	142	185	280	276

SOURCE: U.S. Department of Energy (http://www.eia.doe.gov).

a. Plot these data on a graph, with the number of years after 1992 on the t axis and the average price of gasoline on the y axis.

b. Find the equation of the least-squares line for the data. Is the line a good fit?

c. What price per gallon does the least-squares line predict you will pay for a gallon of regular unleaded gasoline in 2015?

22. STOCK MARKET AVERAGE The accompanying table gives the Dow Jones Industrial Average (DJIA) at the close of the first trading day of the year shown:

Year	2001	2002	2003	2004	2005	2006
DJIA	10,646	10,073	10,454	10,783	10,178	12,463

SOURCE: Dow Jones (http://www.djindexes.com).

a. Plot these data on a graph, with the number of years after 2001 on the x axis and the DJIA on the y axis.

b. Find the equation of the least-squares line for the data.

c. What does the least-squares line predict for the DJIA on the first day of trading in 2008? Use the Internet to find where the DJIA actually closed on that day (January 2, 2008), and compare with the predicted value.

d. Write a paragraph on whether you think it is possible to find a curve that fits the DJIA well enough to usefully predict future market behavior.

23. GROSS DOMESTIC PRODUCT This table lists the gross domestic product (GDP) figures for China (billions of yuan) for the period 2004–2009:

Year	2004	2005	2006	2007	2008	2009
GDP	15,988	18,494	21,631	26,581	31,405	34,051

SOURCE: Chinese government website (http://www.china.org.cn).

a. Find the least-squares line $y = mt + b$ for these data, where y is the GDP of China t years after 2004.

b. Use the least-squares line found in part (a) to predict the GDP of China for the year 2020.

LIFE AND SOCIAL SCIENCE APPLIED PROBLEMS

24. **DRUG ABUSE** For each of five different years, the accompanying table gives the percentage of high school students who had used cocaine at least once in their lives up to that year:

Year	1991	1993	1995	1997	1999	2001	2003	2005	2007
Percentage who had used cocaine at least once	6.0	4.9	7.0	8.2	9.5	9.4	8.7	7.6	7.2

SOURCE: The White House Office of National Drug Control Policy, "2002 National Drug Control Strategy," (http://www.whitehousedrugpolicy.gov).

a. Plot these data on a graph, with the number of years after 1991 on the x axis and the percentage of cocaine users on the y axis.

b. Find the equation of the least-squares line for the data.

c. Use the least-squares line to predict the percentage of high school students who used cocaine at least once by the year 2013.

25. **COLLEGE ADMISSIONS** Over the past 4 years, a college admissions officer has compiled the following data (measured in units of 1,000) relating the number of college catalogs requested by high school students by December 1 to the number of completed applications received by March 1:

Catalogs requested	4.5	3.5	4.0	5.0
Applications received	1.0	0.8	1.0	1.5

a. Plot these data on a graph.

b. Find the equation of the least-squares line.

c. Use the least-squares line to predict how many completed applications will be received by March 1 if 4,800 catalogs are requested by December 1.

26. **POPULATION PREDICTION** The accompanying table gives the U.S. decennial census figures (in millions) for the period 1950–2000:

Year	1950	1960	1970	1980	1990	2000
Population	150.7	179.3	203.2	226.5	248.7	291.4

SOURCE: U.S. Census Bureau (http://www.census.gov).

a. Find the least-squares line $y = mt + b$ for these data, where y is the U.S. population t decades after 1950.

b. Use the least-squares line found in part (a) to predict the U.S. population for the year 2010. Use the Internet to find the actual U.S. population in 2010 and compare with the predicted value.

27. **POPULATION PREDICTION** Modify the least-squares procedure, as illustrated in Example 7.4.4, to find a function of the form $P(t) = Ae^{rt}$ whose graph best fits the population data in Exercise 26, where $P(t)$ is the U.S. population t decades after 1950.

a. Roughly at what percentage rate is the U.S. population growing?

b. Based on your population function, what would you expect the U.S. population to be in the year 2005? In 2010?

28. **PUBLIC HEALTH**   In a study of five industrial areas, a researcher obtained these data relating the average number of units of a certain pollutant in the air and the incidence (per 100,000 people) of a certain disease:

Units of pollutant	3.4	4.6	5.2	8.0	10.7
Incidence of disease	48	52	58	76	96

   a. Plot these data on a graph.
   b. Find the equation of the least-squares line.
   c. Use the least-squares line to estimate the incidence of the disease in an area with an average pollution level of 7.3 units.

29. **VOTER TURNOUT**   On election day, the polls in a certain state open at 8:00 A.M. Every 2 hours after that, an election official determines what percentage of the registered voters have already cast their ballots. The data through 6:00 P.M. are shown here:

Time	10:00	12:00	2:00	4:00	6:00
Percentage turnout	12	19	24	30	37

   a. Plot these data on a graph.
   b. Find the equation of the least-squares line. (Let x denote the number of hours after 8:00 A.M.)
   c. Use the least-squares line to predict what percentage of the registered voters will have cast their ballots by the time the polls close at 8:00 P.M.

30. **BACTERIAL GROWTH**   A biologist studying a bacterial colony measures its population each hour and records these data:

Time t (hours)	1	2	3	4	5	6	7	8
Population $P(t)$ (thousands)	280	286	292	297	304	310	316	323

   a. Plot these data on a graph. Does the scatter diagram suggest that the population growth is linear or exponential?
   b. If you think the scatter diagram in part (a) suggests linear growth, find a population function of the form $P(t) = mt + b$ that best fits the data. However, if you think the scatter diagram suggests exponential growth, modify the least-squares procedure, as illustrated in Example 7.4.4, to obtain a best-fitting population function of the form $P(t) = Ae^{kt}$.
   c. Use the population function you obtained in part (b) to predict how long it will take for the population to reach 400,000. How long will it take for the population to double from 280,000?

31. **SPREAD OF AIDS**   The number of reported cases of AIDS in the United States by year of reporting at 4-year intervals since 1980 is given in this table:

Year	1980	1984	1988	1992	1996	2000	2004	2008
Reported cases of AIDS	99	6,360	36,064	79,477	61,109	42,156	37,726	37,991

SOURCE: World Health Organization and the United Nations (http://www.unaids.org).

   a. Plot these data on a graph with time t (years after 1980) on the x axis.
   b. Find the equation of the least-squares line for the given data.
   c. How many cases of AIDS does the least-squares line in part (b) predict will be reported in 2012?

d. Do you think the least-squares line fits the given data well? If not, write a paragraph explaining which (if any) of the following four curve types would fit the data better:
1. (quadratic) $y = At^2 + Bt + C$
2. (cubic) $y = At^3 + Bt^2 + Ct + D$
3. (exponential) $y = Ae^{kt}$
4. (power-exponential) $y = Ate^{kt}$

(You may find it helpful to review the Think About It essay at the end of Chapter 3, which involves a similar analysis of the number of deaths due to AIDS.)

32. **ALLOMETRY** The determination of relationships between measurements of various parts of a particular organism is a topic of interest in the branch of biology called *allometry*.* (Recall the Think About It essay at the end of Chapter 1.) Suppose a biologist observes that the shoulder height h and antler size w of an elk, both in centimeters (cm), are related as indicated in this table:

Shoulder Height h (cm)	Antler Size w (cm)
87.9	52.4
95.3	60.3
106.7	73.1
115.4	83.7
127.2	98.0
135.8	110.2

a. For each data point (h, w) in the table, plot the point $(\ln h, \ln w)$ on a graph. Note that the scatter diagram suggests that $y = \ln w$ and $x = \ln h$ are linearly related.
b. Find the least-squares line $y = mx + b$ for the data $(\ln h, \ln w)$ you obtained in part (a).
c. Find numbers a and c so that $w = ah^c$. [*Hint:* Substitute $y = \ln w$ and $x = \ln h$ from part (a) into the least-squares equation found in part (b).]

33. **ALLOMETRY** The accompanying table relates the weight C of the large claw of a fiddler crab to the weight W of the rest of the crab's body, both measured in milligrams (mg).

Weight W (mg) of the body	57.6	109.2	199.7	300.2	355.2	420.1	535.7	743.3
Weight C (mg) of the claw	5.3	13.7	38.3	78.1	104.5	135.0	195.6	319.2

a. For each data point (W, C) in the table, plot the point $(\ln W, \ln C)$ on a graph. Note that the scatter diagram suggests that $y = \ln C$ and $x = \ln W$ are linearly related.
b. Find the least-squares line $y = mx + b$ for the data $(\ln W, \ln C)$ you obtained in part (a).
c. Find positive numbers a and k so that $C = aW^k$. [*Hint:* Substitute $y = \ln C$ and $x = \ln W$ from part (a) into the least-squares equation found in part (b).]

---

*Roger V. Jean, "Differential Growth, Huxley's Allometric Formula, and Sigmoid Growth," *UMAP Modules 1983: Tools for Teaching,* Lexington, MA: Consortium for Mathematics and Its Applications, Inc., 1984.

SECTION 7.5 Constrained Optimization: The Method of Lagrange Multipliers

Learning Objectives

1. Study the method of Lagrange multipliers as a procedure for locating points on a graph where constrained optimization can occur.
2. Use the method of Lagrange multipliers in a number of applied problems, including utility and allocation of resources.
3. Discuss the significance of the Lagrange multiplier λ.

In many applied problems, a function of two variables is to be optimized subject to a restriction or **constraint** on the variables. For example, an editor, constrained to stay within a fixed budget of \$60,000, may wish to decide how to divide this money between development and promotion to maximize the future sales of a new book. If x denotes the amount of money allocated to development, y the amount allocated to promotion, and $f(x, y)$ the corresponding number of books that will be sold, the editor would like to maximize the sales function $f(x, y)$ subject to the budgetary constraint that $x + y = 60,000$.

For a geometric interpretation of the process of optimizing a function of two variables subject to a constraint, think of the function itself as a surface in three-dimensional space and of the constraint (which is an equation involving x and y) as a curve in the xy plane. When you find the maximum or minimum of the function subject to the given constraint, you are restricting your attention to the portion of the surface that lies directly above the constraint curve. The highest point on this portion of the surface is the constrained maximum, and the lowest point is the constrained minimum. The situation is illustrated in Figure 7.24.

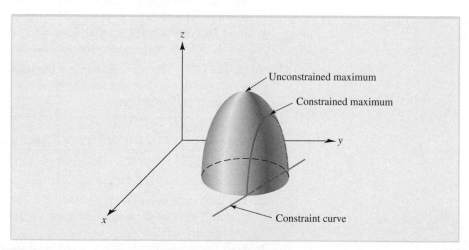

FIGURE 7.24 Constrained and unconstrained extrema.

You have already seen some constrained optimization problems in Chapter 3. (For instance, recall Example 3.5.1 of Section 3.5.) The technique you used in Chapter 3 to solve such a problem involved reducing it to a problem of a single variable by solving the constraint equation for one of the variables and then substituting the resulting expression into the function to be optimized. The success of this technique depended on solving the constraint equation for one of the variables, which is often difficult or even impossible to do in practice. In this section, you will see a more versatile technique called the **method of Lagrange multipliers,** in which the introduction of a *third* variable (the multiplier) enables you to solve constrained optimization problems without first solving the constraint equation for one of the variables.

More specifically, the method of Lagrange multipliers uses the fact that any relative extremum of the function $f(x, y)$ subject to the constraint $g(x, y) = k$ must occur at a critical point (a, b) of the function

$$F(x, y) = f(x, y) - \lambda[g(x, y) - k]$$

where λ is a new variable (the **Lagrange multiplier**). To find the critical points of F, compute its partial derivatives

$$F_x = f_x - \lambda g_x \qquad F_y = f_y - \lambda g_y \qquad F_\lambda = -(g - k)$$

and solve the equations $F_x = 0$, $F_y = 0$, and $F_\lambda = 0$ simultaneously, as follows:

$$
\begin{aligned}
F_x = f_x - \lambda g_x = 0 & \qquad \text{or} \qquad f_x = \lambda g_x \\
F_y = f_y - \lambda g_y = 0 & \qquad \text{or} \qquad f_y = \lambda g_y \\
F_\lambda = -(g - k) = 0 & \qquad \text{or} \qquad g = k
\end{aligned}
$$

Finally, evaluate $f(a, b)$ at each critical point (a, b) of F.

> **NOTE** The method of Lagrange multipliers tells you only that any constrained extrema must occur at critical points of the function $F(x, y)$. The method cannot be used to show that constrained extrema exist or to determine whether any particular critical point (a, b) corresponds to a constrained maximum, a minimum, or neither. However, *for the functions considered in this text, you can assume that if f has a constrained maximum (minimum) value, it will be given by the largest (smallest) of the critical values f(a, b).* ∎

Here is a summary of the Lagrange multiplier procedure for finding the largest and smallest values of a function of two variables subject to a constraint.

The Method of Lagrange Multipliers

Step 1. (*Formulation*) The goal is to find the largest (or smallest) value of $f(x, y)$ subject to the constraint $g(x, y) = k$, assuming that this extreme value exists.

Step 2. Compute the partial derivatives f_x, f_y, g_x, and g_y, and find all numbers $x = a$, $y = b$, and λ that satisfy the system of equations

$$
\begin{aligned}
f_x(a, b) &= \lambda g_x(a, b) \\
f_y(a, b) &= \lambda g_y(a, b) \\
g(a, b) &= k
\end{aligned}
$$

These are the *Lagrange equations.*

Step 3. Evaluate f at each point (a, b) that satisfies the system of equations in step 2.

Step 4. (*Interpretation*) If $f(x, y)$ has a largest (smallest) value subject to the constraint $g(x, y) = k$, it will be the largest (smallest) of the values found in step 3.

A geometric justification of the multiplier method is given at the end of this section. The method is illustrated in Examples 7.5.1 and 7.5.2. In particular, Example 7.5.1 shows how the multiplier method can be used to solve a constrained optimization problem we solved earlier (Example 3.5.1) by using the constraint equation to eliminate a variable.

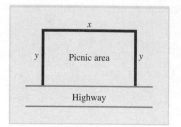

FIGURE 7.25   Rectangular picnic area.

EXAMPLE 7.5.1   Using Lagrange Multipliers in Construction

The highway department is planning to build a picnic area for motorists along a major highway. It is to be rectangular with an area of 5,000 square yards and is to be fenced off on the three sides not adjacent to the highway. What is the least amount of fencing that will be needed to complete the job?

Solution

Label the sides of the picnic area as indicated in Figure 7.25, and let f denote the amount of fencing required. Then,

$$f(x, y) = x + 2y$$

The goal is to minimize f given the requirement that the area must be 5,000 square yards, that is, subject to the constraint

$$g(x, y) = xy = 5,000$$

Find the partial derivatives

$$f_x = 1 \qquad f_y = 2 \qquad g_x = y \qquad \text{and} \qquad g_y = x$$

and obtain the three Lagrange equations

$$1 = \lambda y \qquad 2 = \lambda x \qquad \text{and} \qquad xy = 5,000$$

From the first and second equations you get

$$\lambda = \frac{1}{y} \qquad \text{and} \qquad \lambda = \frac{2}{x}$$

(since $y \neq 0$ and $x \neq 0$), which means that

$$\frac{1}{y} = \frac{2}{x} \qquad \text{or} \qquad x = 2y$$

Now substitute $x = 2y$ into the third Lagrange equation to get

$$2y^2 = 5,000 \qquad \text{or} \qquad y = \pm 50$$

and use $y = 50$ in the equation $x = 2y$ to get $x = 100$. Thus, $x = 100$ and $y = 50$ are the values that minimize the function $f(x, y) = x + 2y$ subject to the constraint $xy = 5,000$. The optimal picnic area is 100 yards wide (along the highway), extends 50 yards back from the road, and requires $100 + 50 + 50 = 200$ yards of fencing.

EXAMPLE 7.5.2   Using Lagrange Multipliers

Find the maximum and minimum values of the function $f(x, y) = xy$ subject to the constraint $x^2 + y^2 = 8$.

Solution

Let $g(x, y) = x^2 + y^2$ and use the partial derivatives

$$f_x = y \qquad f_y = x \qquad g_x = 2x \qquad \text{and} \qquad g_y = 2y$$

to get the three Lagrange equations

$$y = 2\lambda x \qquad x = 2\lambda y \qquad \text{and} \qquad x^2 + y^2 = 8$$

Neither x nor y can be zero if all three of these equations are to hold (do you see why?), and so you can rewrite the first two equations as

$$2\lambda = \frac{y}{x} \qquad \text{and} \qquad 2\lambda = \frac{x}{y}$$

which means that $\qquad \dfrac{y}{x} = \dfrac{x}{y} \qquad$ or $\qquad x^2 = y^2$

Now substitute $x^2 = y^2$ into the third equation to get

$$2x^2 = 8 \qquad \text{or} \qquad x = \pm 2$$

If $x = 2$, it follows from the equation $x^2 = y^2$ that $y = 2$ or $y = -2$. Similarly, if $x = -2$, we have either $y = 2$ or $y = -2$. Hence, the four points at which the constrained extrema can occur are $(2, 2)$, $(2, -2)$, $(-2, 2)$, and $(-2, -2)$. Since

$$f(2, 2) = f(-2, -2) = 4 \qquad \text{and} \qquad f(2, -2) = f(-2, 2) = -4$$

it follows that when $x^2 + y^2 = 8$, the maximum value of $f(x, y)$ is 4, which occurs at the points $(2, 2)$ and $(-2, -2)$, and the minimum value is -4, which occurs at $(2, -2)$ and $(-2, 2)$.

For practice, check these answers by solving the optimization problem using the methods of Chapter 3.

NOTE   In Examples 7.5.1 and 7.5.2, the first two Lagrange equations were used to eliminate the new variable λ, and then the resulting expression relating x and y was substituted into the constraint equation. For most constrained optimization problems you encounter, this particular sequence of steps will often lead quickly to the desired solution. ■

Maximization of Utility

A *utility function* $U(x, y)$ measures the total satisfaction or *utility* a consumer receives from having x units of one particular commodity and y units of another. Example 7.5.3 illustrates how the method of Lagrange multipliers can be used to determine how many units of each commodity the consumer should purchase to maximize utility while staying within a fixed budget.

EXAMPLE 7.5.3   Maximizing Utility

Esteban has $600 to spend on two commodities, the first of which costs $20 per unit and the second $30 per unit. Suppose that the utility he derives from x units of the first commodity and y units of the second commodity is given by the **Cobb-Douglas utility function** $U(x, y) = 10x^{0.6}y^{0.4}$. How many units of each commodity should Esteban buy to maximize utility?

Solution

The total cost of buying x units of the first commodity at \$20 per unit and y units of the second commodity at \$30 per unit is $20x + 30y$. Since Esteban has only \$600 to spend, his goal is to maximize utility $U(x, y)$ subject to the budgetary constraint $20x + 30y = 600$.

The three Lagrange equations are

$$6x^{-0.4}y^{0.4} = 20\lambda \qquad 4x^{0.6}y^{-0.6} = 30\lambda \qquad \text{and} \qquad 20x + 30y = 600$$

From the first two equations he gets

$$\frac{6x^{-0.4}y^{0.4}}{20} = \frac{4x^{0.6}y^{-0.6}}{30} = \lambda$$

$$9x^{-0.4}y^{0.4} = 4x^{0.6}y^{-0.6}$$

$$9y = 4x \qquad \text{or} \qquad y = \frac{4}{9}x$$

Substituting $y = \dfrac{4x}{9}$ into the third Lagrange equation, he gets

$$20x + 30\left(\frac{4}{9}x\right) = 600$$

$$\left(\frac{100}{3}\right)x = 600$$

so that

$$x = 18 \qquad \text{and} \qquad y = \frac{4}{9}(18) = 8$$

That is, to maximize utility, Esteban should buy 18 units of the first commodity and 8 units of the second.

Recall from Section 7.1 that the level curves of a utility function are known as *indifference curves*. A graph showing the relationship between the optimal indifference curve $U(x, y) = C$, where $C = U(18, 8)$ and the budgetary constraint $20x + 30y = 600$, is sketched in Figure 7.26.

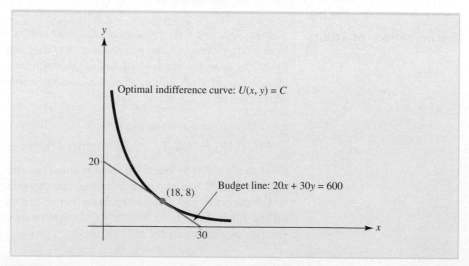

FIGURE 7.26 Budgetary constraint and optimal indifference curve.

Allocation of Resources

An important class of problems in business and economics involves determining an optimal allocation of resources subject to a constraint on those resources, as illustrated in Example 7.5.4.

EXAMPLE 7.5.4 Optimally Allocating Resources

A manufacturer has $600,000 to spend on the production of a certain product and determines that if x units of capital and y units of labor are allocated to production, then P units will be produced, where P is given by the Cobb-Douglas production function

$$P(x, y) = 120x^{4/5}y^{1/5}$$

Suppose each unit of labor costs $3,000 and each unit of capital costs $5,000. How many units of labor and capital should be allocated to maximize production?

Solution

The cost of capital is $3,000x$ and the cost of labor is $5,000y$, so the total cost of resources is $g(x, y) = 3,000x + 5,000y$. The goal is to maximize the production function $P(x, y) = 120x^{4/5}y^{1/5}$ subject to the cost constraint $g(x, y) = 600,000$. The corresponding Lagrange equations are

$$120\left(\frac{4}{5}\right)x^{-1/5}y^{1/5} = 3,000\lambda \qquad 120\left(\frac{1}{5}\right)x^{4/5}y^{-4/5} = 5,000\lambda$$

and

$$3,000x + 5,000y = 600,000$$

or, equivalently,

$$96x^{-1/5}y^{1/5} = 3,000\lambda \qquad 24x^{4/5}y^{-4/5} = 5,000\lambda \qquad \text{and} \qquad 3x + 5y = 600$$

Solving for λ in the first two equations, we get

$$\lambda = 0.032x^{-1/5}y^{1/5} = 0.0048x^{4/5}y^{-4/5}$$

Multiply both sides of this equation by $x^{1/5}y^{4/5}$ to obtain

$$[0.032x^{-1/5}y^{1/5}]x^{1/5}y^{4/5} = [0.0048x^{4/5}y^{-4/5}]x^{1/5}y^{4/5}$$
$$0.032y = 0.0048x$$

so

$$y = 0.15x$$

Substituting into the cost constraint equation $3x + 5y = 600$, we get

$$3x + 5(0.15x) = 600$$

Thus,

$$x = 160$$

and

$$y = 0.15x = 0.15(160) = 24$$

That is, to maximize production, the manufacturer should allocate 160 units to capital and 24 units to labor. If this is done,

$$P(160, 24) = 120(160)^{4/5}(24)^{1/5} \approx 13,138 \text{ units}$$

will be produced.

A graph showing the relationship between the cost constraint and the level curve for optimal production is shown in Figure 7.27.

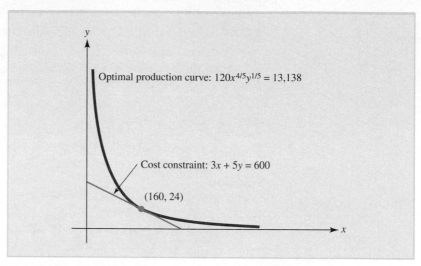

FIGURE 7.27   Optimal production curve and cost constraint.

The Significance of the Lagrange Multiplier λ

Usually, solving a constrained optimization problem by the method of Lagrange multipliers does not require actually finding a numerical value for the Lagrange multiplier λ. In some problems, however, you may want to compute λ, which has this useful interpretation.

The Lagrange Multiplier as a Rate ■ Suppose M is the maximum (or minimum) value of $f(x, y)$, subject to the constraint $g(x, y) = k$. The Lagrange multiplier λ is the rate of change of M with respect to k. That is,

$$\lambda = \frac{dM}{dk}$$

Hence,

$$\lambda \approx \text{change in } M \text{ resulting from a 1-unit increase in } k$$

EXAMPLE 7.5.5   Using the Lagrange Multiplier as a Rate

Suppose the manufacturer is given an extra $1,000 to spend on capital and labor for the production of the commodity in Example 7.5.4, that is, a total of $601,000. Estimate the effect on the maximum production level.

Solution

In Example 7.5.4, we found the maximum value M of the production function $P(x, y) = 120x^{4/5}y^{1/5}$ subject to the cost constraint $3,000x + 5,000y = 600,000$ by solving the three Lagrange equations

$$96x^{-1/5}y^{1/5} = 3,000\lambda \qquad 24x^{4/5}y^{-4/5} = 5,000\lambda \qquad \text{and} \qquad 3x + 5y = 600$$

to obtain $x = 160$ and $y = 24$ and the maximum production level

$$P(160, 24) \approx 13{,}138 \text{ units}$$

The multiplier λ can be found by substituting these values of x and y into either the first or second Lagrange equation. Using the first equation, we find that

$$\lambda = 0.032x^{-1/5}y^{1/5} = 0.032(160)^{-1/5}(24)^{1/5} \approx 0.0219$$

This means that the maximum production with the new cost constraint increases by approximately 0.0219 units for each \$1 increase in the constraint. Since the constraint increases by \$1,000, the maximum production increases by approximately

$$(0.0219)(1{,}000) = 21.9 \text{ units}$$

that is, to

$$13{,}138 + 21.9 = 13{,}159.9 \text{ units}$$

To check, if we repeat Example 7.5.4 with the revised constraint,

$$3{,}000x + 5{,}000y = 601{,}000$$

it can be shown that the maximum occurs when $x = 160.27$ and $y = 24.04$ (verify this result), so the maximum productivity is

$$P(160.27, 24.04) = 120(160.27)^{4/5}(24.04)^{1/5} \approx 13{,}159.82$$

This is essentially the estimate obtained using the Lagrange multiplier.

NOTE   A problem like Example 7.5.4, where production is maximized subject to a cost constraint, is called a **fixed budget problem** (see Exercise 29). In the context of such a problem, the Lagrange multiplier λ is called the **marginal productivity of money.** Similarly, the multiplier in a utility problem like that in Example 7.5.3 is called the **marginal utility of money** (see Exercise 25). ∎

Lagrange Multipliers for Functions of Three Variables

The method of Lagrange multipliers can be extended to constrained optimization problems involving functions of more than two variables and more than one constraint. For instance, to optimize $f(x, y, z)$ subject to the constraint $g(x, y, z) = k$, you solve

$$f_x = \lambda g_x \qquad f_y = \lambda g_y \qquad f_z = \lambda g_z \qquad \text{and} \qquad g = k$$

Example 7.5.6 features a problem involving this kind of constrained optimization.

EXAMPLE 7.5.6   Minimizing Cost of Construction

Gwyneth is constructing an ornate museum display box from material that costs \$1 per square inch for the bottom, \$2 per square inch for the sides, and \$5 per square inch for the top. If the total volume is to be 96 in.^{2}, what dimensions should Gwyneth choose to minimize the total cost of construction? What is the minimal cost of construction?

Solution

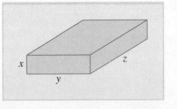

Let the box be x inches deep, y inches long, and z inches wide where x, y, and z are all positive, as indicated in the accompanying figure. Then the volume of the box is $V = xyz$ and the total cost of construction is given by

$$C = \underbrace{1yz}_{\text{bottom}} + \underbrace{2(2xy + 2xz)}_{\text{sides}} + \underbrace{5yz}_{\text{top}} = 6yz + 4xy + 4xz$$

Gwyneth wants to minimize $C = 6yz + 4xy + 4xz$ subject to $V = xyz = 96$. The Lagrange equations are

$$C_x = \lambda V_x \quad \text{or} \quad 4y + 4z = \lambda(yz)$$
$$C_y = \lambda V_y \quad \text{or} \quad 6z + 4x = \lambda(xz)$$
$$C_z = \lambda V_z \quad \text{or} \quad 6y + 4x = \lambda(xy)$$

and $xyz = 96$. Solving each of the first three equations for λ, she gets

$$\frac{4y + 4z}{yz} = \frac{6z + 4x}{xz} = \frac{6y + 4x}{xy} = \lambda$$

By multiplying each expression by xyz, she obtains

$$4xy + 4xz = 6yz + 4yx$$
$$4xy + 4xz = 6yz + 4xz$$
$$6yz + 4yx = 6yz + 4xz$$

which can be further simplified by canceling common terms on both sides of each equation to get

$$4xz = 6yz$$
$$4xy = 6yz$$
$$4yx = 4xz$$

By dividing z from both sides of the first equation, y from the second, and x from the third, she obtains

$$4x = 6y \quad \text{and} \quad 4x = 6z \quad \text{and} \quad 4y = 4z$$

so that $y = \frac{2}{3}x$ and $z = \frac{2}{3}x$. Substituting these values into the constraint equation $xyz = 96$, she first finds that

$$x\left(\frac{2}{3}x\right)\left(\frac{2}{3}x\right) = 96$$
$$\frac{4}{9}x^3 = 96$$
$$x^3 = 216 \quad \text{so} \quad x = 6$$

and then
$$y = z = \frac{2}{3}(6) = 4$$

Thus, Gwyneth can minimize the cost of constructing the display box by making it 6 inches deep with a square base, 4 inches on a side. With those dimensions, her minimal cost is

$$C_{\min} = 6(4)(4) + 4(6)(4) + 4(6)(4) = \$228$$

Why the Method of Lagrange Multipliers Works

Although a rigorous explanation of why the method of Lagrange multipliers works involves advanced ideas beyond the scope of this text, there is a rather simple geometric argument that you may find convincing. This argument depends on the fact that for any function $F(x, y)$, the level curve $F(x, y) = C$ has slope at each point (x, y) given by

$$\frac{dy}{dx} = -\frac{F_x}{F_y}$$

provided $F_y \neq 0$. We obtained this formula using the chain rule for partial derivatives in Exercise 84 in Section 7.2. Exercises 56 and 57 illustrate the use of the formula.

Now, consider the constrained optimization problem:

$$\text{Maximize } f(x, y) \text{ subject to } g(x, y) = k$$

Geometrically, this means you must find the highest level curve of f that intersects the constraint curve $g(x, y) = k$. As Figure 7.28 suggests, the critical intersection will occur at a point where the constraint curve is tangent to a level curve, that is, where the slope of the constraint curve $g(x, y) = k$ is equal to the slope of a level curve $f(x, y) = C$.

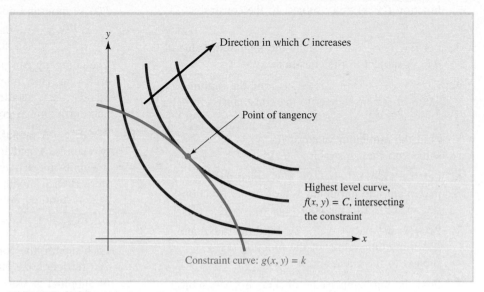

FIGURE 7.28 Increasing level curves and the constraint curve.

According to the formula stated at the beginning of this discussion, you have

$$\text{slope of constraint curve} = \text{slope of level curve}$$

$$-\frac{g_x}{g_y} = -\frac{f_x}{f_y}$$

or, equivalently,

$$\frac{f_x}{g_x} = \frac{f_y}{g_y}$$

If you let λ denote this common ratio, then

$$\frac{f_x}{g_x} = \lambda \qquad \text{and} \qquad \frac{f_y}{g_y} = \lambda$$

from which you get the first two Lagrange equations

$$f_x = \lambda g_x \qquad \text{and} \qquad f_y = \lambda g_y$$

The third Lagrange equation

$$g(x, y) = k$$

is simply a statement of the fact that the point of tangency actually lies on the constraint curve.

EXERCISES ▪ 7.5

In Exercises 1 through 16, use the method of Lagrange multipliers to find the indicated extremum. You may assume the extremum exists.

1. Find the maximum value of $f(x, y) = xy$ subject to the constraint $x + y = 1$.

2. Find the maximum and minimum values of the function $f(x, y) = xy$ subject to the constraint $x^2 + y^2 = 1$.

3. Let $f(x, y) = x^2 + y^2$. Find the minimum value of $f(x, y)$ subject to the constraint $xy = 1$.

4. Let $f(x, y) = x^2 + 2y^2 - xy$. Find the minimum value of $f(x, y)$ subject to the constraint $2x + y = 22$.

5. Find the minimum value of $f(x, y) = x^2 - y^2$ subject to the constraint $x^2 + y^2 = 4$.

6. Let $f(x, y) = 8x^2 - 24xy + y^2$. Find the maximum and minimum values of the function $f(x, y)$ subject to the constraint $8x^2 + y^2 = 1$.

7. Let $f(x, y) = x^2 - y^2 - 2y$. Find the maximum and minimum values of the function $f(x, y)$ subject to the constraint $x^2 + y^2 = 1$.

8. Find the maximum value of $f(x, y) = xy^2$ subject to the constraint $x + y^2 = 1$.

9. Let $f(x, y) = 2x^2 + 4y^2 - 3xy - 2x - 23y + 3$. Find the minimum value of the function $f(x, y)$ subject to the constraint $x + y = 15$.

10. Let $f(x, y) = 2x^2 + y^2 + 2xy + 4x + 2y + 7$. Find the minimum value of the function $f(x, y)$ subject to the constraint $4x^2 + 4xy = 1$.

11. Find the maximum and minimum values of $f(x, y) = e^{xy}$ subject to $x^2 + y^2 = 4$.

12. Find the maximum value of $f(x, y) = \ln(xy^2)$ subject to $2x^2 + 3y^2 = 8$ for $x > 0$ and $y > 0$.

13. Find the maximum value of $f(x, y, z) = xyz$ subject to $x + 2y + 3z = 24$.

14. Find the maximum and minimum values of $f(x, y, z) = x + 3y - z$ subject to $z = 2x^2 + y^2$.

15. Let $f(x, y, z) = x + 2y + 3z$. Find the maximum and minimum values of $f(x, y, z)$ subject to the constraint $x^2 + y^2 + z^2 = 16$.

16. Find the minimum value of $f(x, y, z) = x^2 + y^2 + z^2$ subject to $4x^2 + 2y^2 + z^2 = 4$.

17. **PROFIT**   A manufacturer of television sets makes two models, the Deluxe and the Standard. The manager estimates that when x hundred Deluxe sets and y hundred Standard sets are produced each year, the annual profit will be $P(x, y)$ thousand dollars, where
$$P(x, y) = -0.3x^2 - 0.5xy - 0.4y^2 + 85x + 125y - 2,500$$
The company can produce exactly 30,000 sets each year. How many Deluxe and how many Standard sets should be produced each year to maximize annual profit?

18. **PROFIT**   A manufacturer supplies refrigerators to two stores, A and B. The manager estimates that if x units are delivered to store A and y units to store B each month, the monthly profit will be $P(x, y)$ hundred dollars, where
$$P(x, y) = -0.02x^2 - 0.03xy - 0.05y^2 + 15x + 40y - 3,000$$
Each month, the company can produce exactly 700 refrigerators. How many refrigerators should be supplied to store A and how many to store B to maximize monthly profit?

19. **SALES**   Rajit is an editor who has been allotted $60,000 to spend on the development and promotion of a new book. He estimates that if x thousand dollars are spent on development and y thousand dollars on promotion, approximately
$$S(x, y) = 20x^{3/2}y$$
copies of the book will be sold.
    a. How much money should Rajit allocate to development and how much to promotion to maximize sales? What is the maximum sales level?
    b. If Rajit is allotted an extra $1,000 for development and promotion, how much should he expect the maximum sales level to change? Base your answer on the Lagrange multiplier λ.

20. **SALES**   A manager has been allotted $8,000 to spend on the development and promotion of a new product. It is estimated that if x thousand dollars are spent on development and y thousand dollars on promotion, approximately
$$f(x, y) = 50x^{1/2}y^{3/2}$$
units of the product will be sold.

a. How much money should the manager allocate to development and how much to promotion to maximize sales?

b. Suppose the manager is allotted an extra $1,000 for development and promotion. Use the Lagrange multiplier λ to estimate the change in the maximum sales level.

21. ALLOCATION OF FUNDS   When x thousand dollars are spent on labor and y thousand on equipment, the output of a certain factory is Q units, where

$$Q(x, y) = 60x^{1/3}y^{2/3}$$

Suppose $120,000 is available for labor and equipment.

a. How should the money be allocated between labor and equipment to generate the largest possible output?

b. Use the Lagrange multiplier λ to estimate the change in the maximum output of the factory that will result if the money available for labor and equipment is increased to $121,000.

22. ALLOCATION OF FUNDS   A manufacturer is planning to sell a new product at the price of $150 per unit and estimates that if x thousand dollars is spent on development and y thousand dollars is spent on promotion, approximately

$$\frac{320y}{y + 2} + \frac{160x}{x + 4}$$ units of the product will be sold.

The cost of manufacturing the product is $50 per unit. If the manufacturer has a total of $8,000 to spend on development and promotion, how should this money be allocated to generate the largest possible profit? [*Hint:* Profit = (number of units)(price per unit − cost per unit) − amount spent on development and production.]

23. MARGINAL ANALYSIS   Suppose the manufacturer in Exercise 22 decides to spend $8,100 instead of $8,000 on the development and promotion of the new product. Use the Lagrange multiplier λ to estimate how this change will affect the maximum possible profit.

24. ALLOCATION OF UNRESTRICTED FUNDS

a. If unlimited funds are available, how much should the manufacturer in Exercise 22 spend on development and how much on promotion to generate the largest possible profit? [*Hint:* Use the methods of Section 7.3.]

b. Suppose the allocation problem in part (a) is solved by the method of Lagrange multipliers. What is the value of λ that corresponds to the optimal budget? Interpret your answer in terms of $\dfrac{dM}{dk}$.

c. Your answer to part (b) should suggest another method for solving the problem in part (a). Solve the problem using this new method.

25. UTILITY   A consumer has $280 to spend on two commodities, the first of which costs $2 per unit and the second $5 per unit. Suppose that the utility derived by the consumer from x units of the first commodity and y units of the second is given by $U(x, y) = 100x^{0.25}y^{0.75}$.

a. How many units of each commodity should the consumer buy to maximize utility?

b. Compute the Lagrange multiplier λ and interpret in economic terms. (In the context of maximizing utility, λ is called the **marginal utility of money.**)

26. UTILITY   A consumer has k dollars to spend on two commodities, the first of which costs a dollars per unit and the second b dollars per unit. Suppose that the utility derived by the consumer from x units of the first commodity and y units of the second commodity is given by the Cobb-Douglas utility function $U(x, y) = x^\alpha y^\beta$, where $0 < \alpha < 1$ and $\alpha + \beta = 1$. Show that utility is maximized when $x = \dfrac{k\alpha}{a}$ and $y = \dfrac{k\beta}{b}$.

27. UTILITY   In Exercise 26, how does the maximum utility change if k is increased by 1 dollar?

In Exercises 28 through 30, let $Q(x, y)$ be a production function, where x and y represent units of labor and capital, respectively. If unit costs of labor and capital are given by p and q, respectively, then $px + qy$ represents the total cost of production.

28. MINIMUM COST   Use Lagrange multipliers to show that subject to a fixed production level c, the total cost is minimized when

$$\frac{Q_x}{p} = \frac{Q_y}{q} \qquad \text{and} \qquad Q(x, y) = c$$

provided Q_x and Q_y are not both 0 and $p \neq 0$ and $q \neq 0$. (This is often referred to as the **minimum-cost problem,** and its solution is called the **least-cost combination of inputs.)**

29. **FIXED BUDGET**   Show that the inputs x and y that maximize the production level $Q(x, y)$ subject to a fixed cost k satisfy

$$\frac{Q_x}{p} = \frac{Q_y}{q} \qquad \text{with } px + qy = k$$

(Assume that neither p nor q is 0.) This is called a **fixed-budget problem.**

30. **MINIMUM COST**   Show that with the fixed production level $Ax^\alpha y^\beta = k$, where k is a constant and α and β are positive with $\alpha + \beta = 1$, the joint cost function $C(x, y) = px + qy$ is minimized when

$$x = \frac{k}{A}\left(\frac{\alpha q}{\beta p}\right)^\beta \qquad y = \frac{k}{A}\left(\frac{\beta p}{\alpha q}\right)^\alpha$$

CES PRODUCTION   *A constant elasticity of substitution (CES) production function is one with the general form*

$$Q(K, L) = A[\alpha K^{-\beta} + (1 - \alpha)L^{-\beta}]^{-1/\beta}$$

where K is capital expenditure; L is the level of labor; and A, α, and β are constants that satisfy $A > 0$, $0 < \alpha < 1$, and $\beta > -1$. Exercises 31 through 33 involve such production functions.

---

31. Use the method of Lagrange multipliers to maximize the CES production function

$$Q = 55[0.6K^{-1/4} + 0.4L^{-1/4}]^{-4}$$

subject to the constraint

$$2K + 5L = 150$$

32. Use the method of Lagrange multipliers to maximize the CES production function

$$Q = 50[0.3K^{-1/5} + 0.7L^{-1/5}]^{-5}$$

subject to the constraint

$$5K + 2L = 140$$

33. Suppose you wish to maximize the CES production function

$$Q(K, L) = A[\alpha K^{-\beta} + (1 - \alpha)L^{-\beta}]^{-1/\beta}$$

subject to the linear constraint $c_1 K + c_2 L = B$. Show that the values of K and L at the maximum must satisfy

$$\left(\frac{K}{L}\right)^{\beta + 1} = \frac{c_2}{c_1}\left(\frac{\alpha}{1 - \alpha}\right)$$

34. **MARGINAL ANALYSIS**   Let $P(K, L)$ be a production function, where K and L represent the capital and labor required for a certain manufacturing procedure. Suppose we wish to

maximize $P(K, L)$ subject to a cost constraint, $C(K, L) = A$, for constant A. Use the method of Lagrange multipliers to show that optimal production is attained when

$$\frac{\dfrac{\partial P}{\partial K}}{\dfrac{\partial C}{\partial K}} = \frac{\dfrac{\partial P}{\partial L}}{\dfrac{\partial C}{\partial L}}$$

that is, when the ratio of marginal production from capital to the marginal cost of capital equals the ratio of marginal production of labor to the marginal cost of labor.

LIFE AND SOCIAL SCIENCE APPLIED PROBLEMS

35. **HAZARDOUS WASTE MANAGEMENT**   A study conducted at a waste disposal site reveals soil contamination over a region that may be described roughly as the interior of the ellipse

$$\frac{x^2}{4} + \frac{y^2}{9} = 1$$

where x and y are in miles. The manager of the site plans to build a circular enclosure to contain all polluted territory.

 a.   If the office at the site is at the point $S(1, 1)$, what is the radius of the smallest circle centered at S that contains the entire contaminated region? [*Hint:* The function

$$f(x, y) = (x - 1)^2 + (y - 1)^2$$

measures the square of the distance from $S(1, 1)$ to the point $P(x, y)$. The required radius can be found by maximizing $f(x, y)$ subject to a certain constraint.]

 b.   Read an article on waste management, and write a paragraph on how management decisions are made regarding landfills and other disposal sites.*

36. **SURFACE AREA OF THE HUMAN BODY**   Recall from Exercise 47 of Section 7.1 that an empirical formula for the surface area of a person's body is

$$S(W, H) = 0.0072W^{0.425}H^{0.725}$$

where W (kg) is the person's weight and H (cm) is his or her height. Suppose for a short period of

---

*An excellent case study may be found in M. D. LaGrega, P. L. Buckingham, and J. C. Evans, *Hazardous Waste Management*, New York: McGraw-Hill, 1994, pp. 946–955.

time, Maria's weight adjusts as she grows taller so that $W + H = 160$. With this constraint, what height and weight will maximize the surface area of Maria's body?

In Exercises 37 and 38, you will need to know that a closed cylinder of radius R and length L has volume $V = \pi R^2 L$ and surface area $S = 2\pi RL + 2\pi R^2$. The volume of a hemisphere of radius R is $V = \frac{2}{3}\pi R^3$ and its surface area is $S = 2\pi R^2$.

37. **MICROBIOLOGY** A bacterium is shaped like a cylindrical rod. If the volume of the bacterium is fixed, what relationship between the radius R and length H of the bacterium will result in minimum surface area?

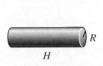
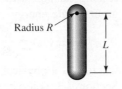

EXERCISE 37   **EXERCISE 38**

38. **MICROBIOLOGY** A bacterium is shaped like a cylindrical rod with two hemispherical "caps" on the ends. If the volume of the bacterium is fixed, what must be true about its radius R and length L to achieve minimum surface area?

39. **ECOLOGY** There are F hundred foxes and R hundred rabbits on a large island. An ecologist determines that the populations F and R are related by the formula

$$(R - 20)^2 + 25(F - 5)^2 = 234$$

What is the largest total number $F + R$ of foxes and rabbits that can be on the island at any one time?

40. **GENETICS** Alternative forms of a gene are called alleles. Three alleles, designated A, B, and O, determine the four human blood types, A, B, O, and AB. Suppose that p, q, and r are the proportions of A, B, and O in a particular population, so that $p + q + r = 1$. Then, according to the Hardy-Weinberg law in genetics, the proportion of individuals in the population who carry two different alleles is given by $P = 2pq + 2pr + 2qr$. In Exercise 45 of Section 7.3, you were asked to maximize P subject to the requirement $p + q + r = 1$. Solve the same constrained optimization problem using the method of Lagrange multipliers.

MISCELLANEOUS PROBLEMS

41. **CONSTRUCTION** A farmer wishes to fence off a rectangular pasture along the bank of a river. The area of the pasture is to be 3,200 square meters, and no fencing is needed along the river bank. Find the dimensions of the pasture that will require the least amount of fencing.

42. **CONSTRUCTION** There are 320 meters of fencing available to enclose a rectangular field. How should the fencing be used so that the enclosed area is as large as possible?

43. **POSTAL PACKAGING** According to postal regulations, the girth plus length of parcels sent by fourth-class mail may not exceed 108 inches. What is the largest possible volume of a rectangular parcel with two square sides that can be sent by fourth-class mail? (Refer to the accompanying figure.)

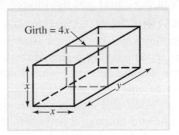

EXERCISE 43

44. **POSTAL PACKAGING** According to the postal regulation given in Exercise 43, what is the largest volume of a cylindrical can that can be sent by fourth-class mail? (A cylinder of radius R and length H has volume $\pi R^2 H$.)

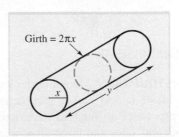

EXERCISE 44

45. **PACKAGING** Use the fact that 12 fluid ounces is approximately 6.89π cubic inches to find the dimensions of the 12-ounce soda can that can be constructed using the least amount of metal.

(Recall that the volume of a cylinder of radius r and height h is $\pi r^2 h$, and that a circle of radius r has area πr^2 and circumference $2\pi r$.)

46. PACKAGING   A cylindrical can is to hold 4π cubic inches of frozen orange juice. The cost per square inch of constructing the metal top and bottom is twice the cost per square inch of constructing the cardboard side. What are the dimensions of the least expensive can? (See the measurement information in Exercise 45.)

47. OPTICS   The thin lens formula in optics says that the focal length L of a thin lens is related to the object distance d_o and image distance d_i by the equation

$$\frac{1}{d_o} + \frac{1}{d_i} = \frac{1}{L}$$

If L remains constant while d_o and d_i are allowed to vary, what is the maximum distance $s = d_o + d_i$ between the object and the image?

48. CONSTRUCTION   A jewelry box is constructed by partitioning a box with a square base as shown in the accompanying figure. If the box is designed to have volume 800 cm³, what dimensions should it have to minimize its total surface area (top, bottom, sides, and interior partitions)? Notice that we have said nothing about where the partitions are located. Does it matter?

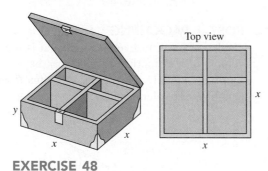

EXERCISE 48

49. CONSTRUCTION   Suppose the jewelry box in Exercise 48 is designed so that the material in the top costs twice as much as the material in the bottom and sides and three times as much as the material in the interior partitions. What dimensions minimize the total cost of constructing the box?

50. SPY STORY   Having escaped from Blabba's village before being detected (Exercise 63, Section 6.1), the spy sneaks into Scelerat's château. He enters a room and the door slams shut behind

him. He begins to feel warm and realizes he is trapped inside Scelerat's dreaded broiler room. Searching desperately for a way to survive, he notices that the room is shaped like the circle $x^2 + y^2 = 60$ and that he is standing at the center $(0, 0)$. He presses the stem on his special heat-detecting wristwatch and sees that the temperature at each point (x, y) in the room is given by

$$T(x, y) = x^2 + y^2 + 3xy + 5x + 15y + 130$$

From an informant's report, he knows that somewhere in the walls of this room there is a trap door leading outside the château, and he reasons that it must be located at the coolest point. Where is it? Just how cool will the spy be when he gets there?

51. PARTICLE PHYSICS   A particle of mass m in a rectangular box with dimensions x, y, and z has ground state energy

$$E(x, y, z) = \frac{k^2}{8m}\left(\frac{1}{x^2} + \frac{1}{y^2} + \frac{1}{z^2}\right)$$

where k is a physical constant. In Exercise 52 of Section 7.3, you were asked to minimize the ground state energy subject to the fixed volume constraint $V_0 = xyz$ using substitution. Solve the same constrained optimization problem using the method of Lagrange multipliers.

52. CONSTRUCTION   A rectangular building is to be constructed of material that costs \$31 per square foot for the roof, \$27 per square foot for the sides and the back, and \$55 per square foot for the facing and glass used in constructing the front. If the building is to have a volume of 16,000 ft³, what dimensions will minimize the total cost of construction?

53. CONSTRUCTION   A storage shed is to be constructed of material that costs \$15 per square foot for the roof, \$12 per square foot for the two sides and back, and \$20 per square foot for the front. What are the dimensions of the largest shed (in volume) that can be constructed for \$8,000?

54. SATELLITE CONSTRUCTION   A space probe has the shape of the surface

$$4x^2 + y^2 + 4z^2 = 16$$

where x, y, and z are in feet. When it reenters earth's atmosphere, the probe begins to heat up in such a way that the temperature at each point $P(x, y, z)$ on the probe's surface is given by

$$T(x, y, z) = 8x^2 + 4yz - 16z + 600$$

where T is in degrees Fahrenheit. Use the method of Lagrange multipliers to find the hottest and coolest points on the probe's surface. What are the extreme temperatures?

55. Use Lagrange multipliers to find the possible maximum or minimum points on that part of the surface $z = x - y$ for which $y = x^5 + x - 2$. Then use your calculator to sketch the curve $y = x^5 + x - 2$ and the level curves to the surface $f(x, y) = x - y$ and show that the points you have just found do not represent relative maxima or minima. What do you conclude from this observation?

56. Let $F(x, y) = x^2 + 2xy - y^2$.

a. If $F(x, y) = k$ for constant k, use the method of implicit differentiation developed in Chapter 2 to find $\dfrac{dy}{dx}$.

b. Find the partial derivatives F_x and F_y and verify that

$$\frac{dy}{dx} = -\frac{F_x}{F_y}$$

57. Repeat Exercise 56 for the function

$$F(x, y) = xe^{xy^2} + \frac{y}{x} + x\ln(x + y)$$

In Exercises 58 through 61, use the method of Lagrange multipliers to find the indicated maximum or minimum. You will need to use the graphing utility or the solve application on your calculator.

58. Maximize $f(x, y) = e^{x+y} - x\ln\left(\dfrac{y}{x}\right)$ subject to $x + y = 4$.

59. Minimize $f(x, y) = \ln(x + 2y)$ subject to $xy + y = 5$.

60. Minimize $f(x, y) = \dfrac{1}{x^2} + \dfrac{3}{xy} + \dfrac{1}{y^2}$ subject to $x + 2y = 7$.

61. Maximize $f(x, y) = xe^{x^2 - y}$ subject to $x^2 + 2y^2 = 1$.

SECTION 7.6 Double Integrals

Learning Objectives

1. Define and compute double integrals over rectangular and nonrectangular regions in the xy plane.
2. Use double integrals in problems involving area, volume, average value, and population density.

In Chapters 5 and 6, you integrated a function of one variable $f(x)$ by reversing the process of differentiation; a similar procedure can be used to integrate a function of two variables $f(x, y)$. However, since two variables are involved, we shall integrate $f(x, y)$ by holding one variable fixed and integrating with respect to the other.

For instance, to evaluate the partial integral $\displaystyle\int_1^2 xy^2\, dx$ you would integrate with respect to x, using the fundamental theorem of calculus with y held constant:

$$\int_1^2 xy^2\, dx = \frac{1}{2}x^2 y^2 \Big|_{x=1}^{x=2}$$

$$= \left[\frac{1}{2}(2)^2 y^2\right] - \left[\frac{1}{2}(1)^2 y^2\right] = \frac{3}{2}y^2$$

Similarly, to evaluate $\int_{-1}^{1} xy^2 \, dy$, you integrate with respect to y, holding x constant:

$$\int_{-1}^{1} xy^2 \, dy = x\left(\frac{1}{3}y^3\right)\Big|_{y=-1}^{y=1}$$

$$= \left[x\left(\frac{1}{3}(1)^3\right)\right] - \left[x\left(\frac{1}{3}(-1)^3\right)\right] = \frac{2}{3}x$$

In general, partially integrating a function $f(x, y)$ with respect to x results in a function of y alone, which can then be integrated as a function of a single variable, thus producing what we call an **iterated integral** $\int\left[\int f(x, y) \, dx\right] dy$. Similarly, the iterated integral $\int\left[\int f(x, y) \, dy\right] dx$ is obtained by first integrating with respect to y, holding x constant, and then with respect to x. Returning to our example, you find,

$$\int_{-1}^{1}\left(\int_{1}^{2} xy^2 \, dx\right) dy = \int_{-1}^{1} \frac{3}{2}y^2 \, dy = \frac{1}{2}y^3\Big|_{y=-1}^{y=1} = 1$$

and

$$\int_{1}^{2}\left(\int_{-1}^{1} xy^2 \, dy\right) dx = \int_{1}^{2} \frac{2}{3}x \, dx = \frac{1}{3}x^2\Big|_{x=1}^{x=2} = 1$$

In our example, the two iterated integrals turned out to have the same value, and you can assume this will be true for all iterated integrals considered in this text. The double integral of $f(x, y)$ over a rectangular region in the xy plane has the following definition in terms of iterated integrals.

The Double Integral over a Rectangular Region ■ The **double integral** $\iint_R f(x, y) \, dA$ over the rectangular region

$$R: a \le x \le b, c \le y \le d$$

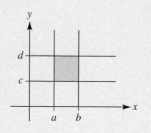

is given by the common value of the two iterated integrals

$$\int_{a}^{b}\left[\int_{c}^{d} f(x, y) \, dy\right] dx \quad \text{and} \quad \int_{c}^{d}\left[\int_{a}^{b} f(x, y) \, dx\right] dy$$

that is,

$$\iint_R f(x, y) \, dA = \int_{a}^{b}\left[\int_{c}^{d} f(x, y) \, dy\right] dx = \int_{c}^{d}\left[\int_{a}^{b} f(x, y) \, dx\right] dy$$

Example 7.6.1 illustrates the computation of this kind of double integral.

EXAMPLE 7.6.1   Evaluating a Double Integral

Evaluate the double integral

$$\iint_R xe^{-y}\, dA$$

where R is the rectangular region $-2 \le x \le 1,\, 0 \le y \le 5$ using:

a. x integration first

b. y integration first

Solution

a. Integrating with respect to x first:

$$\iint_R xe^{-y}\, dA = \int_0^5 \left(\int_{-2}^1 xe^{-y}\, dx \right) dy$$

$$= \int_0^5 \frac{1}{2} x^2 e^{-y} \Big|_{x=-2}^{x=1}\, dy$$

$$= \int_0^5 \frac{1}{2} e^{-y}[(1)^2 - (-2)^2]\, dy = \int_0^5 -\frac{3}{2} e^{-y}\, dy$$

$$= -\frac{3}{2}(-e^{-y}) \Big|_{y=0}^{y=5} = \frac{3}{2}(e^{-5} - e^0) = \frac{3}{2}(e^{-5} - 1)$$

b. Integrating with respect to y first:

$$\iint_R xe^{-y}\, dA = \int_{-2}^1 \left(\int_0^5 xe^{-y}\, dy \right) dx$$

$$= \int_{-2}^1 x(-e^{-y}) \Big|_{y=0}^{y=5}\, dx = \int_{-2}^1 [-x(e^{-5} - e^0)]\, dx$$

$$= \left[-(e^{-5} - 1)\left(\frac{1}{2} x^2\right) \right] \Big|_{x=-2}^{x=1}$$

$$= -\frac{1}{2}(e^{-5} - 1)[(1)^2 - (-2)^2] = \frac{3}{2}(e^{-5} - 1)$$

In Example 7.6.1, the order of integration made no difference. Not only do the computations yield the same result, but the integrations are essentially of the same level of difficulty. However, sometimes the order does matter, as illustrated in Example 7.6.2.

EXAMPLE 7.6.2   Evaluating a Double Integral

Evaluate the double integral

$$\iint_R xe^{xy}\, dA$$

where R is the rectangular region $0 \le x \le 2,\, 0 \le y \le 1$.

Solution

If you evaluate the integral in the order

$$\int_0^1 \left(\int_0^2 xe^{xy}\, dx \right) dy$$

it will be necessary to use integration by parts for the inner integration:

$$u = x \qquad dv = e^{xy}\, dx$$

$$du = dx \qquad v = \frac{1}{y}e^{xy}$$

$$\int_0^2 xe^{xy}\, dx = \frac{x}{y}e^{xy} \Big|_{x=0}^{x=2} - \int_0^2 \frac{1}{y}e^{xy}\, dx$$

$$= \left(\frac{x}{y} - \frac{1}{y^2} \right)e^{xy} \Big|_{x=0}^{x=2} = \left(\frac{2}{y} - \frac{1}{y^2} \right)e^{2y} - \left(\frac{-1}{y^2} \right)$$

Then the outer integration becomes

$$\int_0^1 \left[\left(\frac{2}{y} - \frac{1}{y^2} \right)e^{2y} + \frac{1}{y^2} \right] dy$$

Now what? Any ideas?

On the other hand, if you use y integration first, both computations are easy:

$$\int_0^2 \left(\int_0^1 xe^{xy}\, dy \right) dx = \int_0^2 \frac{xe^{xy}}{x} \Big|_{y=0}^{y=1} dx$$

$$= \int_0^2 (e^x - 1)\, dx = (e^x - x) \Big|_{x=0}^{x=2}$$

$$= (e^2 - 2) - e^0 = e^2 - 3$$

Double Integrals over Nonrectangular Regions

In each of the preceding examples, the region of integration is a rectangle, but double integrals can also be defined over nonrectangular regions. Before doing so, however, we will introduce an efficient procedure for describing certain such regions in terms of inequalities.

Vertical Cross Sections

The region R shown in Figure 7.29 on page 609 is bounded below by the curve $y = g_1(x)$, above by the curve $y = g_2(x)$, and on the sides by the vertical lines $x = a$ and $x = b$. This region can be described by the inequalities

$$R: a \le x \le b, g_1(x) \le y \le g_2(x)$$

The first inequality specifies the interval in which x must lie, while the second indicates the lower and upper bounds of the vertical cross section of R for each x in this interval. In words:

R is the region such that for each x between a and b,
y varies from $g_1(x)$ to $g_2(x)$.

This method for describing a region is illustrated in Example 7.6.3.

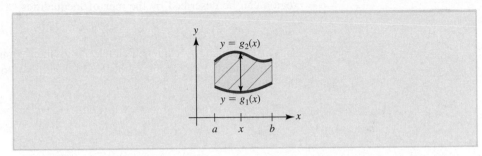

FIGURE 7.29 Vertical cross sections. The region R: $a \leq x \leq b$, $g_1(x) \leq y \leq g_2(x)$

EXAMPLE 7.6.3 Describing a Region by Vertical Cross Sections

Let R be the region bounded by the curve $y = x^2$ and the line $y = 2x$. Use inequalities to describe R in terms of its vertical cross sections.

Solution
Begin with a sketch of the curve and line as shown in Figure 7.30. Identify the region R, and, for reference, draw a vertical cross section. Solve the equations $y = x^2$ and $y - 2x$ simultaneously to find the points of intersection, $(0, 0)$ and $(2, 4)$. Observe that in the region R, the variable x takes on all values from $x = 0$ to $x = 2$ and that for each such value of x, the vertical cross section is bounded below by $y = x^2$ and above by $y = 2x$. Hence, R can be described by the inequalities

$$0 \leq x \leq 2 \qquad \text{and} \qquad x^2 \leq y \leq 2x$$

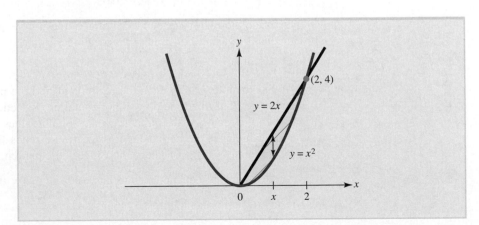

FIGURE 7.30 The region R between $y = x^2$ and $y = 2x$ described by vertical cross sections as R: $0 \leq x \leq 2$, $x^2 \leq y \leq 2x$.

Horizontal Cross Sections

The region R in Figure 7.31 is bounded on the left by the curve $x = h_1(y)$, on the right by $x = h_2(y)$, below by the horizontal line $y = c$, and above by $y = d$. This region can be described by the pair of inequalities

$$R: c \le y \le d, h_1(y) \le x \le h_2(y)$$

The first inequality specifies the interval in which y must lie, and the second indicates the left-hand ("trailing") and right-hand ("leading") bounds of a horizontal cross section. In words:

R is the region such that for each y between c and d,
x varies from $h_1(y)$ to $h_2(y)$.

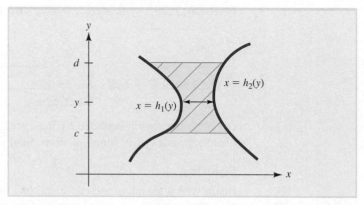

FIGURE 7.31   Horizontal cross sections. The region $R: c \le y \le d, h_1(y) \le x \le h_2(y)$

This method of description is illustrated in Example 7.6.4 for the same region described using vertical cross sections in Example 7.6.3.

EXAMPLE 7.6.4   Describing a Region by Horizontal Cross Sections

Describe the region R bounded by the curve $y = x^2$ and the line $y = 2x$ in terms of inequalities using horizontal cross sections.

Solution

As in Example 7.6.3, sketch the region and find the points of intersection of the line and curve, but this time draw a horizontal cross section (Figure 7.32).

In the region R, the variable y takes on all values from $y = 0$ to $y = 4$. For each such value of y, the horizontal cross section extends from the line $y = 2x$ on the left to the curve $y = x^2$ on the right. Since the equation of the line can be rewritten as $x = \frac{1}{2}y$ and the equation of the curve as $x = \sqrt{y}$, the inequalities describing R in terms of its horizontal cross sections are

$$0 \le y \le 4 \qquad \text{and} \qquad \frac{1}{2}y \le x \le \sqrt{y}$$

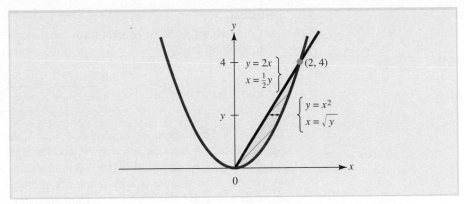

FIGURE 7.32 The region R between $y = x^2$ and $y = 2x$ described by horizontal cross sections as $R: 0 \leq y \leq 4, \frac{1}{2}y \leq x \leq \sqrt{y}$.

To evaluate a double integral over a region R described using either vertical or horizontal cross sections, you use an iterated integral whose limits of integration come from the inequalities describing the region. Here is a more precise description of how the limits of integration are determined.

Limits of Integration for Double Integrals ▪ If R can be described by the inequalities

$$a \leq x \leq b \qquad \text{and} \qquad g_1(x) \leq y \leq g_2(x)$$

then

$$\iint\limits_{R} f(x, y) \, dA = \int_a^b \left[\int_{g_1(x)}^{g_2(x)} f(x, y) \, dy \right] dx$$

If R can be described by the inequalities

$$c \leq y \leq d \qquad \text{and} \qquad h_1(y) \leq x \leq h_2(y)$$

then

$$\iint\limits_{R} f(x, y) \, dA = \int_c^d \left[\int_{h_1(y)}^{h_2(y)} f(x, y) \, dx \right] dy$$

NOTE When evaluating a double integral involving variable limits of integration, it is often vitally important to choose the order of integration carefully. For instance, in Example 7.6.5, evaluating the given integral using one order of integration is considerably easier than evaluating the same integral with the reverse order of integration. ▪

EXAMPLE 7.6.5 Evaluating a Double Integral over a Region

Let I be the double integral

$$I = \int_0^1 \int_0^y y^2 e^{xy} \, dx \, dy$$

a. Sketch the region of integration, and rewrite the integral with the order of integration reversed.

b. Evaluate I using either order of integration.

Solution

a. Comparing I with the general form for the order $dx\,dy$, we see that the region of integration is

$$R: \quad \underbrace{0 \le y \le 1}_{\substack{\text{outer limits} \\ \text{of integration}}}, \quad \underbrace{0 \le x \le y}_{\substack{\text{inner limits} \\ \text{of integration}}}$$

Thus, if y is a number in the interval $0 \le y \le 1$, the horizontal cross section of R at y extends from $x = 0$ on the left to $x = y$ on the right. The region is the triangle shown in Figure 7.33a. As shown in Figure 7.33b, the same region R can be described by taking vertical cross sections at each number x in the interval $0 \le x \le 1$ that are bounded below by $y = x$ and above by $y = 1$. Expressed in terms of inequalities, this means that

$$R: 0 \le x \le 1, x \le y \le 1$$

so the integral can also be written as

$$I = \int_0^1 \int_x^1 y^2 e^{xy}\, dy\, dx$$

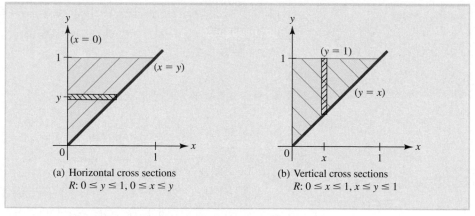

(a) Horizontal cross sections
$R: 0 \le y \le 1, 0 \le x \le y$

(b) Vertical cross sections
$R: 0 \le x \le 1, x \le y \le 1$

FIGURE 7.33   The region of integration for $I = \int_0^1 \int_0^y y^2 e^{xy}\, dx\, dy$.

b. Here is the evaluation of I using the given order of integration:

$$\int_0^1 \int_0^y y^2 e^{xy}\, dx\, dy = \int_0^1 \left(y e^{xy} \Big|_{x=0}^{x=y} \right) dy \qquad\qquad \text{since } \int e^{xy}\, dx = \frac{1}{y} e^{xy}$$

$$= \int_0^1 (y e^{y^2} - y)\, dy$$

$$= \left(\frac{1}{2} e^{y^2} - \frac{1}{2} y^2 \right) \Big|_0^1$$

$$= \left(\frac{1}{2} e - \frac{1}{2} \right) - \left(\frac{1}{2} - 0 \right) = \frac{1}{2} e - 1$$

Try to compute I by using the reverse order of integration found in part (a). What happens?

Applications   Next, we will examine a few applications of double integrals, all of which are generalizations of familiar applications of definite integrals of functions of one variable. Specifically, we will see how double integration can be used to compute area, volume, average value, and population from population density.

The Area of a Region in the Plane   The area of a region R in the xy plane can be computed as the double integral over R of the constant function $f(x, y) = 1$.

> **Area Formula** ■ The area of a region R in the xy plane is given by the formula
>
> $$\text{Area of } R = \iint_R 1 \, dA$$

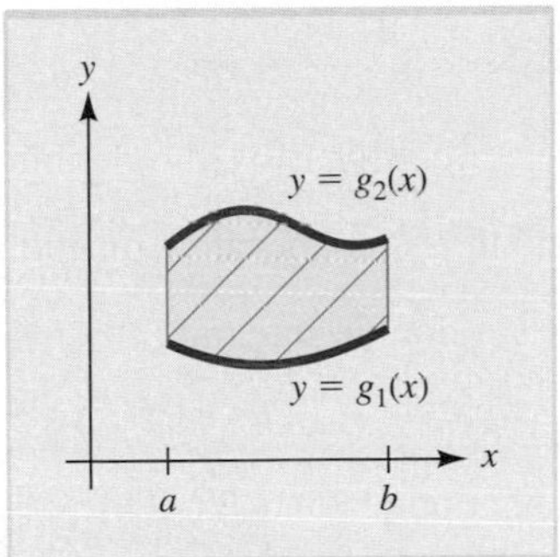

To get a feeling for why the area formula holds, consider the elementary region R shown in Figure 7.34, which is bounded above by the curve $y = g_2(x)$ and below by the curve $y = g_1(x)$, and which extends from $x = a$ to $x = b$. According to the double-integral formula for area,

$$\text{Area of } R = \iint_R 1 \, dA$$

$$= \int_a^b \int_{g_1(x)}^{g_2(x)} 1 \, dy \, dx$$

$$= \int_a^b \left[y \Big|_{y=g_1(x)}^{y=g_2(x)} \right] dx$$

$$= \int_a^b [g_2(x) - g_1(x)] \, dx$$

which is precisely the formula for the area between two curves that you saw in Section 5.4. Example 7.6.6 illustrates the use of the area formula.

FIGURE 7.34 Area of
$$R = \iint_R 1 \, dA$$

EXPLORE!

Find the area of the region R by using the numeric integration feature of your graphing utility to evaluate

$$\int_0^1 (x^2 - x^3) \, dx.$$

Explain why this gives the same answer as the method used in the solution to Example 7.6.6.

EXAMPLE 7.6.6 Finding Area Using a Double Integral

Find the area of the region R bounded by the curves $y = x^3$ and $y = x^2$.

Solution

The region is shown in Figure 7.35. Using the area formula, you get

$$\text{Area of } R = \iint_R 1 \, dA = \int_0^1 \int_{x^3}^{x^2} 1 \, dy \, dx$$

$$= \int_0^1 \left(y \Big|_{y=x^3}^{y=x^2} \right) dx$$

$$= \int_0^1 (x^2 - x^3) \, dx$$

$$= \left[\frac{1}{3} x^3 - \frac{1}{4} x^4 \right] \Big|_0^1$$

$$= \frac{1}{12}$$

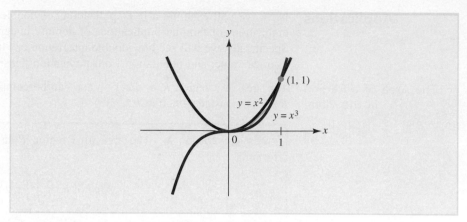

FIGURE 7.35   The region bounded by $y = x^2$ and $y = x^3$.

Volume as a
Double Integral

Recall from Section 5.3 that the region under the curve $y = f(x)$ over an interval $a \leq x \leq b$, where $f(x)$ is continuous and $f(x) \geq 0$, has area given by the definite integral $A = \displaystyle\int_a^b f(x)\, dx$. An analogous argument for a continuous, nonnegative function of two variables $f(x, y)$ yields this formula for volume as a double integral.

Volume as a Double Integral ▪ If $f(x, y)$ is continuous and $f(x, y) \geq 0$ on the region R, then the solid region under the surface $z = f(x, y)$ over R has volume given by

$$V = \iint\limits_R f(x, y)\, dA$$

EXAMPLE 7.6.7   Finding Volume of a Biomass Using a Double Integral

A biomass covers the triangular bottom of a container with vertices $(0, 0)$, $(6, 0)$, and $(3, 3)$, to a depth $h(x, y) = \dfrac{x}{y + 2}$ at each point (x, y) in the region, where all dimensions are in centimeters. What is the total volume of the biomass?

Solution

The volume is given by the double integral $V = \displaystyle\iint\limits_R h(x, y)\, dA$, where R is the triangular region R shown in Figure 7.36.

Note that this region is bounded by the x axis $(y = 0)$, and the lines $x = y$ and $x + y = 6$, so it can be described as

$$R\colon 0 \leq y \leq 3,\ y \leq x \leq 6 - y$$

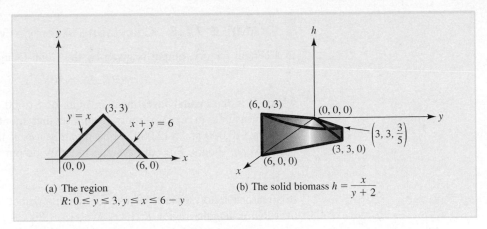

(a) The region
$R: 0 \le y \le 3, y \le x \le 6 - y$

(b) The solid biomass $h = \dfrac{x}{y + 2}$

FIGURE 7.36 The volume of a biomass.

Therefore, the volume of the biomass is given by

$$V = \int_0^3 \int_y^{6-y} \frac{x}{y+2}\, dx\, dy$$

$$= \int_0^3 \frac{1}{y+2}\left(\frac{x^2}{2}\right)\Big|_y^{6-y} dy = \int_0^3 \frac{1}{2(y+2)}[(6-y)^2 - y^2]\, dy$$

$$= \int_0^3 \frac{1}{2(y+2)}[36 - 12y]\, dy \qquad \text{divide } 2y + 4 \text{ into } -12y + 36$$

$$= \int_0^3 \left[-6 + \left(\frac{60}{2y+4}\right)\right] dy$$

$$= -6y + 30 \ln|2y + 4|\ \Big|_0^3$$

$$= [-6(3) + 30\ln(2(3) + 4)] - [-6(0) + 30\ln(0 + 4)]$$

$$\approx 9.489$$

We conclude that the volume of the biomass is approximately 9.5 cm^3.

Average Value of a Function f(x, y)

In Section 5.4, you saw that the average value of a function $f(x)$ over an interval $a \le x \le b$ is given by the integral formula

$$\text{AV} = \frac{1}{b-a}\int_a^b f(x)\, dx$$

That is, to find the average value of a function of one variable over an interval, you integrate the function over the interval and divide by the length of the interval. The two-variable procedure is similar. In particular, to find the average value of a function of two variables $f(x, y)$ over a region R, you integrate the function over R and divide by the area of R.

Average Value Formula ■ The average value of the function $f(x, y)$ over the region R is given by the formula

$$\text{AV} = \frac{1}{\text{area of } R}\iint_R f(x, y)\, dA$$

EXAMPLE 7.6.8   Calculating Average Monthly Output

In a certain factory, output is given by the Cobb-Douglas production function

$$Q(K, L) = 50K^{3/5}L^{2/5}$$

where K is the capital investment in units of $1,000 and L is the size of the labor force measured in worker-hours. Suppose that monthly capital investment varies between $10,000 and $12,000, while monthly use of labor varies between 2,800 and 3,200 worker-hours. Find the average monthly output for the factory.

Solution

It is reasonable to estimate the average monthly output by the average value of $Q(K, L)$ over the rectangular region R: $10 \leq K \leq 12$, $2,800 \leq L \leq 3,200$. The region has area

$$A = \text{area of } R = (12 - 10) \times (3,200 - 2,800)$$
$$= 800$$

so the average output is

$$AV = \frac{1}{800} \iint\limits_{R} 50K^{3/5}L^{2/5} \, dA$$

$$= \frac{1}{800} \int_{2,800}^{3,200} \left(\int_{10}^{12} 50K^{3/5}L^{2/5} \, dK \right) dL$$

$$= \frac{1}{800} \int_{2,800}^{3,200} 50L^{2/5} \left(\frac{5}{8} K^{8/5} \right) \Bigg|_{K=10}^{K=12} dL$$

$$= \frac{1}{800}(50)\left(\frac{5}{8} \right) \int_{2,800}^{3,200} L^{2/5}(12^{8/5} - 10^{8/5}) \, dL$$

$$= \frac{1}{800}(50)\left(\frac{5}{8} \right)(12^{8/5} - 10^{8/5})\left(\frac{5}{7} L^{7/5} \right) \Bigg|_{L=2,800}^{L=3,200}$$

$$= \frac{1}{800}(50)\left(\frac{5}{8} \right)\left(\frac{5}{7} \right)(12^{8/5} - 10^{8/5})[(3,200)^{7/5} - (2,800)^{7/5}]$$

$$\approx 5,181.23$$

Thus, the average monthly output is approximately 5,181 units.

Population Density

In Section 5.6, we showed how the population in a circular region can be obtained by integrating population density, that is, the function $p(r)$ that gives the number of people per square mile at a radial distance r from a fixed central point. More generally, if we know the population density $p(x, y)$ at each point $p(x, y)$ in a certain region R, then the total population ΔP in a small part of the region with area ΔA is given by the product

$$\Delta P \;=\; p(x, y) \;\cdot\; \Delta A$$

number   people per   area of
of people   square unit   the region
in region   area

Using double integration to "add up" the population of all such small regions, we can compute the total population P of the region R by the formula

$$P = \iint\limits_{R} p(x, y) \, dA$$

The same formula can also be used to compute more general populations, such as the number of people in a test region who have been influenced by an advertising campaign or the number of people in a community that are susceptible to a contagious disease. In Example 7.6.9, we apply the formula to predict the outcome of a vote.

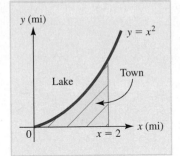

FIGURE 7.37 A lakeside community.

EXAMPLE 7.6.9 Finding Population from Population Density

A lakeside community whose boundaries are shown in Figure 7.37 is about to vote on a tax assessment to build a new city park. Based on a poll, a consultant estimates that the density of voters favorable to the assessment is $p(x, y) = 50xe^{-0.04y}$ hundred people per square mile at grid point (x, y) in the figure, where x and y are in miles. If 35,000 people vote on the assessment, does it pass or fail?

Solution

The community occupies the region R bounded above by $y = x^2$, below by the x axis, and on the right by $x = 2$. We will compute the total number N of favorable voters by the integral $\iint_R p(x, y)\, dA$ over R, integrating first with respect to y between $y = 0$ and $y = x^2$, and then with respect to x between $x - 0$ and $x = 2$. We find that

$$N = \iint_R p(x, y)\, dA$$

$$= \int_0^2 \int_0^{x^2} 50xe^{-0.04y}\, dy\, dx \qquad \text{exponential rule}$$

$$= \int_0^2 50x\left[\frac{e^{-0.04y}}{-0.04}\right]_{y=0}^{y=x^2} dx$$

$$= \int_0^2 -1{,}250x[e^{-0.04x^2} - e^0]\, dx \qquad \begin{array}{l}\text{substitute } u = -0.04x^2,\ du = -0.08x\, dx \\ \text{and use the rule for exponentials}\end{array}$$

$$= -1{,}250\left[\frac{e^{-0.04x^2}}{-0.08} - \frac{1}{2}x^2\right]_0^2$$

$$= -1{,}250[-12.5e^{-0.16} - 2] - (-1{,}250)[-12.5 - 0]$$

$$= 189.75$$

That is, 18,975 people (189.75 hundred) are expected to vote for the assessment and

$$35{,}000 - 18.975 = 16{,}025 \text{ people}$$

should vote against it, so the measure should pass.

EXERCISES ■ 7.6

Evaluate the double integrals in Exercises 1 through 18.

1. $\displaystyle\int_0^1 \int_1^2 x^2 y\, dx\, dy$

2. $\displaystyle\int_1^2 \int_0^1 x^2 y\, dy\, dx$

5. $\displaystyle\int_1^3 \int_0^1 \frac{2xy}{x^2 + 1}\, dx\, dy$

6. $\displaystyle\int_0^1 \int_0^1 x^2 e^{xy}\, dy\, dx$

3. $\displaystyle\int_0^{\ln 2} \int_{-1}^0 2xe^y\, dx\, dy$

4. $\displaystyle\int_2^3 \int_{-1}^1 (x + 2y)\, dy\, dx$

7. $\displaystyle\int_0^4 \int_{-1}^1 x^2 y\, dy\, dx$

8. $\displaystyle\int_0^1 \int_1^5 y\sqrt{1 - y^2}\, dx\, dy$

9. $\int_{2}^{3}\int_{1}^{2}\frac{x+y}{xy}\,dy\,dx$ 10. $\int_{1}^{2}\int_{2}^{3}\left(\frac{y}{x}+\frac{x}{y}\right)dy\,dx$

11. $\int_{0}^{4}\int_{0}^{\sqrt{x}}x^{2}y\,dy\,dx$ 12. $\int_{0}^{1}\int_{1}^{5}xy\sqrt{1-y^{2}}\,dx\,dy$

13. $\int_{0}^{1}\int_{y-1}^{1-y}(2x+y)\,dx\,dy$ 14. $\int_{0}^{1}\int_{x^{2}}^{x}2xy\,dy\,dx$

15. $\int_{0}^{1}\int_{0}^{4}\sqrt{xy}\,dy\,dx$ 16. $\int_{0}^{1}\int_{x}^{2x}e^{y-x}\,dy\,dx$

17. $\int_{1}^{e}\int_{0}^{\ln x}xy\,dy\,dx$ 18. $\int_{0}^{3}\int_{y^{2}/4}^{\sqrt{10-y^{2}}}xy\,dx\,dy$

In Exercises 19 through 24, use inequalities to describe R in terms of its vertical and horizontal cross sections.

19. R is the region bounded by $y=x^{2}$ and $y=3x$.

20. R is the region bounded by $y=\sqrt{x}$ and $y=x^{2}$.

21. R is the rectangle with vertices $(-1,1)$, $(2,1)$, $(2,2)$, and $(-1,2)$.

22. R is the triangle with vertices $(1,0)$, $(1,1)$, and $(2,0)$.

23. R is the region bounded by $y=\ln x$, $y=0$, and $x=e$.

24. R is the region bounded by $y=e^{x}$, $y=2$, and $x=0$.

In Exercises 25 through 36, evaluate the given double integral for the specified region R.

25. $\iint\limits_{R}3xy^{2}\,dA$, where R is the rectangle bounded by the lines $x=-1$, $x=2$, $y=-1$, and $y=0$.

26. $\iint\limits_{R}(x+2y)\,dA$, where R is the triangle with vertices $(0,0)$, $(1,0)$, and $(0,2)$.

27. $\iint\limits_{R}xe^{y}\,dA$, where R is the triangle with vertices $(0,0)$, $(1,0)$, and $(1,1)$.

28. $\iint\limits_{R}48xy\,dA$, where R is the region bounded by $y=x^{3}$ and $y=\sqrt{x}$.

29. $\iint\limits_{R}(2y-x)\,dA$, where R is the region bounded by $y=x^{2}$ and $y=2x$.

30. $\iint\limits_{R}12x\,dA$, where R is the region bounded by $y=x^{2}$ and $y=6-x$.

31. $\iint\limits_{R}(2x+1)\,dA$, where R is the triangle with vertices $(-1,0)$, $(1,0)$, and $(0,1)$.

32. $\iint\limits_{R}2x\,dA$, where R is the region bounded by $y=\frac{1}{x^{2}}$, $y=x$, and $x=2$.

33. $\iint\limits_{R}\frac{1}{y^{2}+1}\,dA$, where R is the triangle bounded by the lines $y=\frac{1}{2}x$, $y=-x$, and $y=2$.

34. $\iint\limits_{R}e^{y^{3}}\,dA$, where R is the region bounded by $y=\sqrt{x}$, $y=1$, and $x=0$.

35. $\iint\limits_{R}12x^{2}e^{y^{2}}\,dA$, where R is the region in the first quadrant bounded by $y=x^{3}$ and $y=x$.

36. $\iint\limits_{R}y\,dA$, where R is the region bounded by $y=\ln x$, $y=0$, and $x=e$.

In Exercises 37 through 44, sketch the region of integration for the given integral and set up an equivalent integral with the order of integration reversed.

37. $\int_{0}^{2}\int_{0}^{4-x^{2}}f(x,y)\,dy\,dx$ 38. $\int_{0}^{1}\int_{0}^{2y}f(x,y)\,dx\,dy$

39. $\int_{0}^{1}\int_{x^{3}}^{\sqrt{x}}f(x,y)\,dy\,dx$ 40. $\int_{0}^{4}\int_{y/2}^{\sqrt{y}}f(x,y)\,dx\,dy$

41. $\displaystyle\int_1^{e^2}\int_{\ln x}^2 f(x, y)\, dy\, dx$ **42.** $\displaystyle\int_0^{\ln 3}\int_{e^x}^3 f(x, y)\, dy\, dx$

43. $\displaystyle\int_{-1}^1\int_{x^2+1}^2 f(x, y)\, dy\, dx$ **44.** $\displaystyle\int_{-1}^1\int_{-\sqrt{y+1}}^{\sqrt{y+1}} f(x, y)\, dy\, dx$

In Exercises 45 through 54, use a double integral to find the area of R.

45. *R* is the triangle with vertices $(-4, 0)$, $(2, 0)$, and $(2, 6)$.

46. *R* is the triangle with vertices $(0, -1)$, $(-2, 1)$, and $(2, 1)$.

47. *R* is the region bounded by $y = \dfrac{1}{2}x^2$ and $y = 2x$.

48. *R* is the region bounded by $y = \sqrt{x}$ and $y = x^2$.

49. *R* is the region bounded by $y = x^2 - 4x + 3$ and the *x* axis.

50. *R* is the region bounded by $y = x^2 + 6x + 5$ and the *x* axis.

51. *R* is the region bounded by $y = \ln x$, $y = 0$, and $x = e$.

52. *R* is the region bounded by $y = x$, $y = \ln x$, $y = 0$, and $y = 1$.

53. *R* is the region in the first quadrant bounded by $y = 4 - x^2$, $y = 3x$, and $y = 0$.

54. *R* is the region bounded by $y = \dfrac{16}{x}$, $y = x$, and $x = 8$.

In Exercises 55 through 64, find the volume of the solid under the surface $z = f(x, y)$ and over the given region R.

55. $f(x, y) = 6 - 2x - 2y$;
$R: 0 \le x \le 1, 0 \le y \le 2$

56. $f(x, y) = 9 - x^2 - y^2$;
$R: -1 \le x \le 1, -2 \le y \le 2$

57. $f(x, y) = \dfrac{1}{xy}$;
$R: 1 \le x \le 2, 1 \le y \le 3$

58. $f(x, y) = e^{x+y}$;
$R: 0 \le x \le 1, 0 \le y \le \ln 2$

59. $f(x, y) = xe^{-y}$;
$R: 0 \le x \le 1, 0 \le y \le 2$

60. $f(x, y) = (1 - x)(4 - y)$;
$R: 0 \le x \le 1, 0 \le y \le 4$

61. $f(x, y) = 2x + y$; *R* is bounded by $y = x$, $y = 2 - x$, and $y = 0$.

62. $f(x, y) = e^{y^2}$; *R* is bounded by $x = 2y$, $x = 0$, and $y = 1$.

63. $f(x, y) = x + 1$; *R* is bounded by $y = 8 - x^2$ and $y = x^2$.

64. $f(x, y) = 4xe^y$; *R* is bounded by $y = 2x$, $y = 2$, and $x = 0$.

In Exercises 65 through 72, find the average value of the function $f(x, y)$ over the given region R.

65. $f(x, y) = xy(x - 2y)$;
$R: -2 \le x \le 3, -1 \le y \le 2$

66. $f(x, y) = \dfrac{y}{x} + \dfrac{x}{y}$;
$R: 1 \le x \le 4, 1 \le y \le 3$

67. $f(x, y) = xye^{x^2y}$;
$R: 0 \le x \le 1, 0 \le y \le 2$

68. $f(x, y) = \dfrac{\ln x}{xy}$;
$R: 1 \le x \le 2, 2 \le y \le 3$

69. $f(x, y) = 6xy$; *R* is the triangle with vertices $(0, 0)$, $(0, 1)$, $(3, 1)$.

70. $f(x, y) = e^{x^2}$; *R* is the triangle with vertices $(0, 0)$, $(1, 0)$, $(1, 1)$.

71. $f(x, y) = x$; *R* is the region bounded by $y = 4 - x^2$ and $y = 0$.

72. $f(x, y) = e^x y^{-1/2}$; *R* is the region bounded by $x = \sqrt{y}$, $y = 0$, and $x = 1$.

In Exercises 73 through 76, evaluate the double integral over the specified region R. Choose the order of integration carefully.

73. $\displaystyle\iint_R \dfrac{\ln(xy)}{y}\, dA$; $R: 1 \le x \le 3, 2 \le y \le 5$

74. $\displaystyle\iint_R ye^{xy}\, dA$; $R: -1 \le x \le 1, 1 \le y \le 2$

75. $\displaystyle\iint_R x^3 e^{x^2y}\, dA$; $R: 0 \le x \le 1, 0 \le y \le 1$

76. $\displaystyle\iint_R e^{x^3}\, dA$; $R: \sqrt{y} \le x \le 1, 0 \le y \le 1$

BUSINESS AND ECONOMICS APPLIED PROBLEMS

77. **PRODUCTION** At a certain factory, output Q is related to inputs x and y by the expression
$$Q(x, y) = 2x^3 + 3x^2y + y^3$$
If $0 \le x \le 5$ and $0 \le y \le 7$, what is the average output of the factory?

78. **PRODUCTION** A bicycle dealer has found that if 10-speed bicycles are sold for x dollars apiece and the price of gasoline is y cents per gallon, then approximately
$$Q(x, y) = 200 - 24\sqrt{x} + 4(0.1y + 3)^{3/2}$$
bicycles will be sold each month. If the price of bicycles varies between \$289 and \$324 during a typical month, and the price of gasoline varies between \$2.96 and \$3.05, approximately how many bicycles will be sold each month on average?

79. **AVERAGE PROFIT** A manufacturer estimates that when x units of a particular commodity are sold domestically and y units are sold to foreign markets, the profit is given by
$$P(x, y) = (x - 30)(70 + 5x - 4y)$$
$$+ (y - 40)(80 - 6x + 7y)$$
hundred dollars. If monthly domestic sales vary between 100 and 125 units and foreign sales between 70 and 89 units, what is the average monthly profit?

80. **PROPERTY VALUE** A community is laid out as a rectangular grid in relation to two main streets that intersect at the city center. Each point in the community has coordinates (x, y) in this grid, for $-10 \le x \le 10$, $-8 \le y \le 8$ with x and y measured in miles. Suppose the value of the land located at the point (x, y) is V thousand dollars, where
$$V(x, y) = (250 + 17x)e^{-0.01x - 0.05y}$$
Estimate the value of the block of land occupying the rectangular region $1 \le x \le 3$, $0 \le y \le 2$.

81. **PROPERTY VALUE** Repeat Exercise 80 for
$$V(x, y) = (300 + x + y)e^{-0.01x}$$
and the region $-1 \le x \le 1$, $-1 \le y \le 1$.

82. **PROPERTY VALUE** Repeat Exercise 80 for
$$V(x, y) = 400xe^{-y}$$
and the region R: $0 \le y \le x$, $0 \le x \le 1$.

83. **EFFECT OF ADVERTISING** A consultant determines that the density of people positively influenced by an advertising campaign in the region shown in the accompanying figure is $p(x, y) = x \ln y$ thousand people per square mile at the grid point (x, y) with x and y in miles. How many people in the test region are positively influenced by the campaign?

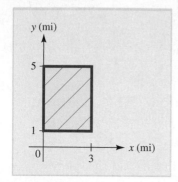

EXERCISE 83

84. **INVESTMENT** Sandy Gibbons is an investment advisor with a new product she has been actively promoting. She estimates that in the region shown in the accompanying region, the density of interest is $p(x, y) = y^2e^{-0.5x}$ people per square mile at the grid point (x, y) with x and y in miles. How many people in the sales region should Sandy expect to show interest in her product?

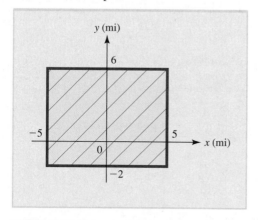

EXERCISE 84

LIFE AND SOCIAL SCIENCE APPLIED PROBLEMS

85. **AVERAGE ELEVATION** A map of a small regional park is a rectangular grid, bounded by the lines $x = 0$, $x = 4$, $y = 0$, and $y = 3$, where units are in miles. It is found that the elevation above sea level at each point (x, y) in the park is given by
$$E(x, y) = 90(2x + y^2) \text{ feet}$$
Find the average elevation in the park. (Remember, 1 mi = 5,280 feet.)

86. AVERAGE RESPONSE TO STIMULI In a psychological experiment, x units of stimulus A and y units of stimulus B are applied to a subject, whose performance on a certain task is then measured by the function

$$P(x, y) = 10 + xye^{1-x^2-y^2}$$

Suppose x varies between 0 and 1 while y varies between 0 and 3. What is the subject's average response to the stimuli?

87. CONSTRUCTION A storage bin is to be constructed in the shape of the solid bounded above by the surface

$$z = 20 - x^2 - y^2$$

below by the xy plane, and on the sides by the plane $y = 0$ and the parabolic cylinder $y = 4 - x^2$, where x, y, and z are in meters. Find the volume of the bin.

88. ARCHITECTURAL DESIGN A building is to have a curved roof above a rectangular base. In relation to a rectangular grid, the base is the rectangular region $-30 \leq x \leq 30$, $-20 \leq y \leq 20$, where x and y are measured in meters. The height of the roof above each point (x, y) in the base is given by

$$h(x, y) = 12 - 0.003x^2 - 0.005y^2$$

a. Find the volume of the building.
b. Find the average height of the roof.

89. EXPOSURE TO DISEASE The likelihood that a person with a contagious disease will infect others in a social situation may be assumed to be a function $f(s)$ of the distance s between individuals. Suppose contagious individuals are uniformly distributed throughout a rectangular region R in the xy plane. Then the likelihood of infection for someone at the origin $(0, 0)$ is proportional to the exposure index E, given by the double integral

$$E = \iint_R f(s)\, dA$$

where $s = \sqrt{x^2 + y^2}$ is the distance between $(0, 0)$ and (x, y). Find E for the case where

$$f(s) = 1 - \frac{s^2}{9}$$

and R is the square

$$R: -2 \leq x \leq 2, -2 \leq y \leq 2$$

90. CONSTRUCTION A box has the shape of the solid bounded above by the plane

$$3x + 4y + 2z = 12$$

below by the xy plane, and on the sides by the planes $x = 0$ and $y = 0$, where x, y, and z are in inches. Find the volume of the box.

91. POPULATION The population density is $f(x, y) = 2{,}500e^{-0.01x-0.02y}$ people per square mile at each point (x, y) within the triangular region R with vertices $(-5, -2)$, $(0, 3)$, and $(5, -2)$. Find the total population in the region R.

92. POPULATION The population density is $f(x, y) = 1{,}000y^2e^{-0.01x}$ people per square mile at each point (x, y) within the region R bounded by the parabola $x = y^2$ and the vertical line $x = 4$. Find the total population in the region R.

93. HEALTH TESTING A health official wants to estimate the number of people susceptible to a new strain of influenza. Examinations conducted in the test region R shown in the accompanying figure suggest that the density of susceptible people in the region is $p(x, y) = xy$ thousand people per square mile. How many people in the region are susceptible to the disease?

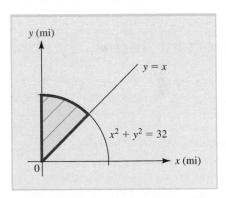

EXERCISE 93

94. REDISTRICTING A state legislature decides to create a new district from the region shown in the accompanying figure. It is found that the population density function for the region is $p(x, y) = ye^{-0.2x}$ thousand people per square mile.
a. If $c = 9$, what is the total population in the region?

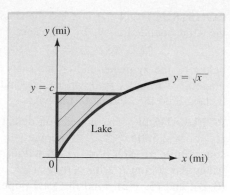

EXERCISE 94

 b. The committee responsible for redistricting decides there must be at least 800,000 people in the new district. What is the smallest value of c that achieves this goal?

95. AVERAGE SURFACE AREA OF THE HUMAN BODY Recall from Exercise 47, Section 7.1, that the surface area of a person's body may be estimated by the empirical formula

$$S(W, H) = 0.0072W^{0.425}H^{0.725}$$

where W is the person's weight in kilograms, H is the person's height in centimeters, and the surface area S is measured in square meters.

a. Find the average value of the function $S(W, H)$ over the region

$$R: 3.2 \leq W \leq 80, 38 \leq H \leq 180$$

b. A child weighs 3.2 kg and is 38 cm tall at birth and as an adult, has a stable weight of 80 kg and height of 180 cm. Can the average value in part (a) be interpreted as the average lifetime surface area of this person's body? Explain.

MISCELLANEOUS PROBLEMS

 In Exercises 96 through 98, use double integration to find the required quantity. In some cases, you may need to use the numeric integration feature of your calculator.

96. Find the area of the region bounded above by the curve (ellipse) $4x^2 + 3y^2 = 7$ and below by the parabola $y = x^2$.

97. Find the volume of the solid bounded above by the graph of $f(x, y) = x^2e^{-xy}$ and below by the rectangular region $R: 0 \leq x \leq 2, 0 \leq y \leq 3$.

98. Find the average value of $f(x, y) = xy \ln\left(\dfrac{y}{x}\right)$ over the rectangular region bounded by the lines $x = 1$, $x = 2$, $y = 1$, and $y = 3$.

Important Terms, Symbols, and Formulas

CHAPTER SUMMARY

Function of two variables: $z = f(x, y)$ (530)
Domain convention   (530)
Cobb-Douglas production function   (532)
Three-dimensional coordinate system   (533)
Level curve: $f(x, y) = C$ (534)
Topographical map   (536)
Utility   (537)
Indifference curve   (537)
Partial derivatives of $z = f(x, y)$:   (546)

$$f_x = \frac{\partial z}{\partial x} \quad f_y = \frac{\partial z}{\partial y}$$

Marginal productivity   (549)
Complementary and substitute commodities   (550)
Second-order partial derivatives: (551)

$$f_{xx} = \frac{\partial^2 z}{\partial x^2} \quad f_{xy} = \frac{\partial^2 z}{\partial y \, \partial x} \quad f_{yx} = \frac{\partial^2 z}{\partial x \, \partial y} \quad f_{yy} = \frac{\partial^2 z}{\partial y^2}$$

Equality of mixed second-order partial derivatives:

$$f_{xy} = f_{yx} \quad (552)$$

Chain rule for partial derivatives: (553)

$$\frac{dz}{dt} = \frac{\partial z}{\partial x}\frac{dx}{dt} + \frac{\partial z}{\partial y}\frac{dy}{dt}$$

Incremental approximation formula for a function of two variables $z = f(x, y)$:   (554)

$$\Delta z \approx \frac{\partial z}{\partial x}\Delta x + \frac{\partial z}{\partial y}\Delta y$$

Relative maximum; relative minimum   (561)
Critical point: $f_x = f_y = 0$ (562)
Saddle point   (563)
Second partials test at a critical point (a, b):   (563)
   Let $D(a, b) = f_{xx}f_{yy} - (f_{xy})^2$.
   If $D < 0$, f has a saddle point at (a, b).
   If $D > 0$, and $f_{xx} < 0$, f has a relative maximum at (a, b).
   If $D > 0$, and $f_{xx} > 0$, f has a relative minimum at (a, b).
   If $D = 0$, the test is inconclusive.

Absolute maximum and minimum   (569)

Extreme value property: (569)

A function $f(x, y)$ that is continuous on the closed, bounded region R in the xy plane attains both its absolute maximum and absolute minimum values on R. These extreme values occur either on the boundary of R or at a critical point in the interior of R.

Scatter diagram   (578)

Least-squares criterion   (578)

Least-squares line: $y = mx + b$, where

$$m = \frac{n\Sigma xy - \Sigma x\Sigma y}{n\Sigma x^2 - (\Sigma x)^2} \quad \text{and} \quad b = \frac{\Sigma x^2\Sigma y - \Sigma x\Sigma xy}{n\Sigma x^2 - (\Sigma x)^2} \quad (580)$$

Log-linear regression   (584)

Method of Lagrange multipliers: (591)

To find extreme values of $f(x, y)$ subject to $g(x, y) = k$, solve the equations

$$f_x = \lambda g_x \qquad f_y = \lambda g_y \qquad \text{and} \qquad g = k$$

The Lagrange multiplier: (596)

$$\lambda = \frac{dM}{dk}, \text{ where } M \text{ is the optimal value of } f(x, y)$$

subject to $g(x, y) = k$.

Double integral   (611)

over the region R: $a \le x \le b$, $g_1(x) \le y \le g_2(x)$

$$\iint_R f(x, y)\, dA = \int_a^b \left[\int_{g_1(x)}^{g_2(x)} f(x, y)\, dy \right] dx$$

over the region R: $c \le y \le d$, $h_1(y) \le x \le h_2(y)$

$$\iint_R f(x, y)\, dA = \int_c^d \left[\int_{h_1(y)}^{h_2(y)} f(x, y)\, dx \right] dy$$

Area of the region R in the xy plane is

$$\text{Area of } R = \iint_R 1\, dA \quad (613)$$

Volume under $z = f(x, y)$ over a region R where $f(x, y) \ge 0$ is

$$V = \iint_R f(x, y)\, dA \quad (614)$$

Average value of $f(x, y)$ over the region R: (615)

$$AV = \frac{1}{\text{area of } R} \iint_R f(x, y)\, dA$$

Population P of a region R from population density $p(x, y)$:   (616)

$$P = \iint_R p(x, y)\, dA$$

Checkup for Chapter 7

1. In each case, first describe the domain of the given function and then find the partial derivatives f_x, f_y, f_{xx}, and f_{yx}.

   a. $f(x, y) = x^3 + 2xy^2 - 3y^4$

   b. $f(x, y) = \dfrac{2x + y}{x - y}$

   c. $f(x, y) = e^{2x-y} + \ln (y^2 - 2x)$

2. Describe the level curves of each of these functions:

   a. $f(x, y) = x^2 + y^2$

   b. $f(x, y) = x + y^2$

3. In each case, find all critical points of the given function $f(x, y)$ and use the second partials test to classify each as a relative maximum, a relative minimum, or a saddle point.

   a. $f(x, y) = 4x^3 + y^3 - 6x^2 - 6y^2 + 5$

   b. $f(x, y) = x^2 - 4xy + 3y^2 + 2x - 4y$

   c. $f(x, y) = xy - \dfrac{1}{y} - \dfrac{1}{x}$

4. Use the method of Lagrange multipliers to find these constrained extrema:

   a. The smallest value of $f(x, y) = x^2 + y^2$ subject to $x + 2y = 4$.

   b. The largest and the smallest values of the function $f(x, y) = xy^2$ subject to $2x^2 + y^2 = 6$.

5. Evaluate each of these double integrals:

   a. $\displaystyle\int_{-1}^{3} \int_{0}^{2} x^3 y\, dx\, dy$ b. $\displaystyle\int_{0}^{2} \int_{-1}^{1} x^2 e^{xy}\, dx\, dy$

   c. $\displaystyle\int_{1}^{2} \int_{1}^{y} \frac{y}{x}\, dx\, dy$ d. $\displaystyle\int_{0}^{2} \int_{0}^{2-x} xe^{-y}\, dy\, dx$

6. **MARGINAL PRODUCTIVITY**   A company will produce $Q(K, L) = 120K^{3/4}L^{1/4}$ hundred units of a particular commodity when the capital expenditure is K thousand dollars and the size of the workforce is L worker-hours. Find the marginal productivity of capital Q_K and the marginal productivity of labor Q_L when the capital expenditure is $1,296,000 dollars and the labor level is 20,736 worker-hours.

7. **UTILITY** Everett has just received $500 as a birthday gift and has decided to spend it on DVDs and video games. He has determined that the utility (satisfaction) derived from the purchase of x DVDs and y video games is

$$U(x, y) = \ln (x^2 \sqrt{y})$$

If each DVD costs $20 and each video game costs $50, how many DVDs and video games should he purchase to maximize utility?

8. **MEDICINE** A certain disease can be treated by administering at least 70 units of drug C, but that level of medication sometimes results in serious side effects. Looking for a safer approach, a physician decides instead to use drugs A and B, which result in no side effects as long as their combined dosage is less than 60 units. Moreover, she determines that when x units of drug A and y units of drug B are administered to a patient, the effect is equivalent to administering E units of drug C, where

$$E = 0.05(xy - 2x^2 - y^2 + 95x + 20y)$$

What dosages of drugs A and B will maximize the equivalent level E of drug C? If the physician administers the optimum dosages of drugs A and B,

will the combined effect be enough to help the patient without running the risk of side effects?

9. **AVERAGE TEMPERATURE** A flat metal plate lying in the xy plane is heated in such a way that the temperature at the point (x, y) is T (°C), where

$$T(x, y) = 10ye^{-xy}$$

Find the average temperature over a rectangular portion of the plate for which $0 \le x \le 2$ and $0 \le y \le 1$.

10. **LEAST-SQUARES APPROXIMATION OF PROFIT DATA** A company's annual profit (in millions of dollars) for the first 5 years of operation is shown in this table:

Year	1	2	3	4	5
Profit (millions of dollars)	1.03	1.52	2.03	2.41	2.84

   a. Plot these data on a graph.

   b. Find the equation of the least-squares line through the data.

   c. Use the least-squares line to predict the company's 6th year profit.

Review Exercises

In Exercises 1 through 10, find the partial derivatives f_x and f_y.

1. $f(x, y) = 2x^3y + 3xy^2 + \dfrac{y}{x}$

2. $f(x, y) = (xy^2 + 1)^5$

3. $f(x, y) = \sqrt{x}(x - y^2)$

4. $f(x, y) = xe^{-y} + ye^{-x}$

5. $f(x, y) = \sqrt{\dfrac{x}{y}} + \sqrt{\dfrac{y}{x}}$

6. $f(x, y) = x \ln(x^2 - y) + y \ln(y - 2x)$

7. $f(x, y) = \dfrac{x^3 - xy}{x + y}$

8. $f(x, y) = xye^{xy}$

9. $f(x, y) = \dfrac{x^2 - y^2}{2x + y}$

10. $f(x, y) = \ln\left(\dfrac{xy}{x + 3y}\right)$

For each of the functions in Exercises 11 through 14, compute the second-order partial derivatives $f_{xx}, f_{yy}, f_{xy},$ and f_{yx}.

11. $f(x, y) = e^{x^2 + y^2}$

12. $f(x, y) = x^2 + y^3 - 2xy^2$

13. $f(x, y) = x \ln y$

14. $f(x, y) = (5x^2 - y)^3$

15. For each of these functions, sketch the indicated level curves:

   a. $f(x, y) = x^2 - y; f = 2, f = -2$

   b. $f(x, y) = 6x + 2y; f = 0, f = 1, f = 2$

16. For each of these functions, find the slope of the indicated level curve at the specified value of x:

   a. $f(x, y) = x^2 - y^3; f = 2; x = 1$

   b. $f(x, y) = xe^y; f = 2; x = 2$

In Exercises 17 through 24, find all critical points of the given function and use the second partials test to classify each as a relative maximum, a relative minimum, or a saddle point.

17. $f(x, y) = (x + y)(2x + y - 6)$

18. $f(x, y) = (x + y + 3)^2 - (x + 2y - 5)^2$

19. $f(x, y) = x^3 + y^3 + 3x^2 - 3y^2$

20. $f(x, y) = x^3 + y^3 + 3x^2 - 18y^2 + 81y + 5$

21. $f(x, y) = x^2 + y^3 + 6xy - 7x - 6y$

22. $f(x, y) = 3x^2y + 2xy^2 - 10xy - 8y^2$

23. $f(x, y) = xe^{2x^2 + 5xy + 2y^2}$

24. $f(x, y) = 8xy - x^4 - y^4$

In Exercises 25 through 30, find all interior and boundary critical points and determine the largest and smallest values of the function f(x, y) over the given closed, bounded region R.

25. $f(x, y) = x^2 + 2x + y^2 - 4y + 12$ on the triangular region R with vertices $(-4, 0)$, $(1, 0)$, and $(0, 4)$.

26. $f(x, y) = x^2 - 2x + 4y^2 - 6y + 15$ on the triangular region R with vertices $(0, 0)$, $(5, 5)$, and $(-5, 5)$.

27. $f(x, y) = x^3 - 4xy + 4x + y^2$ on the square region R with vertices $(1, 2)$, $(4, 2)$, $(1, 5)$, and $(4, 5)$.

28. $f(x, y) = ye^{x^2 - y}$ on the circular region R bounded by $x^2 + y^2 = 2$.

29. $f(x, y) = e^{x^2 + 4x + y^2}$ on the circular region R bounded by $x^2 + 4x + y^2 = 0$.

30. $f(x, y) = (y - 1)e^x - y^2$ on the square region R with vertices $(0, 0)$, $(1, 0)$, $(1, 1)$, and $(0, 1)$.

In Exercises 31 through 34, use the method of Lagrange multipliers to find the maximum and minimum values of the given function f(x, y) subject to the indicated constraint.

31. $f(x, y) = x^2 + 2y^2 + 2x + 3; x^2 + y^2 = 4$

32. $f(x, y) = 4x + y; \dfrac{1}{x} + \dfrac{1}{y} = 1$

33. $f(x, y) = x + 2y; 4x^2 + y^2 = 68$

34. $f(x, y) = x^2 + y^3; x^2 + 3y = 4$

35. **MARGINAL ANALYSIS** At a certain factory, the daily output is approximately $40K^{1/3}L^{1/2}$ units, where K denotes the capital investment measured in units of \$1,000 and L denotes the size of the labor force measured in worker-hours. Suppose that the current capital investment is \$125,000 and that 900 worker-hours of labor are used each day. Use marginal analysis to estimate the effect that an additional capital investment of \$1,000 will have on the daily output if the size of the labor force is not changed.

36. **MARGINAL ANALYSIS** In economics, the marginal product of labor is the rate at which output Q changes with respect to labor L for a fixed level of capital investment K. An economic law states that under certain circumstances, the marginal product of labor increases as the level of capital investment increases. Translate this law into a mathematical statement involving a second-order partial derivative.

37. **MARGINAL ANALYSIS** Using x skilled workers and y unskilled workers, a manufacturer can produce $Q(x, y) = 60x^{1/3}y^{2/3}$ units per day. Currently the manufacturer employs 10 skilled workers and 40 unskilled workers and is planning to hire 1 additional skilled worker. Use calculus to estimate the corresponding change that the manufacturer should make in the level of unskilled labor so that the total output will remain the same.

38. Use the method of Lagrange multipliers to prove that of all isosceles triangles with a given perimeter, the equilateral triangle has the largest area.

39. Use the method of Lagrange multipliers to prove that of all rectangles with a given perimeter, the square has the largest area.

40. **ALLOCATION OF FUNDS** A manufacturer is planning to sell a new product at the price of \$350 per unit and estimates that if x thousand dollars is spent on development and y thousand dollars is spent on promotion, consumers will buy approximately $\dfrac{250y}{y + 2} + \dfrac{100x}{x + 5}$ units of the product. If manufacturing costs for the product are \$150 per unit, how much should the manufacturer spend on development and how much on promotion to generate the largest possible profit if unlimited funds are available?

41. **ALLOCATION OF FUNDS** Suppose the manufacturer in Exercise 40 has only \$11,000 to spend on the development and promotion of the new product. How should this money be allocated to generate the largest possible profit?

42. ALLOCATION OF FUNDS Suppose the manufacturer in Exercise 41 decides to spend $12,000 instead of $11,000 on the development and promotion of the new product. Use the Lagrange multiplier λ to estimate how this change will affect the maximum possible profit.

43. Let $f(x, y) = \dfrac{12}{x} + \dfrac{18}{y} + xy$, where $x > 0$, $y > 0$.

How do you know that f must have a minimum in the region $x > 0$, $y > 0$? Find the minimum.

In Exercises 44 through 51, evaluate the double integral. You may need to exchange the order of integration.

44. $\displaystyle\int_0^1 \int_{-2}^0 (2x + 3y)\, dy\, dx$

45. $\displaystyle\int_0^1 \int_0^2 e^{-x-y}\, dy\, dx$

46. $\displaystyle\int_0^1 \int_0^2 x\sqrt{1-y}\, dx\, dy$

47. $\displaystyle\int_0^1 \int_{-1}^1 xe^{2y}\, dy\, dx$

48. $\displaystyle\int_0^2 \int_{-1}^1 \dfrac{6xy^2}{x^2+1}\, dy\, dx$

49. $\displaystyle\int_1^e \int_1^e \ln(xy)\, dy\, dx$

50. $\displaystyle\int_0^1 \int_0^{1-x} x(y-1)^2\, dy\, dx$

51. $\displaystyle\int_1^2 \int_0^x e^{y/x}\, dy\, dx$

In Exercises 52 and 53, evaluate the given double integral for the specified region R.

52. $\displaystyle\iint_R 6x^2y\, dA$, where R is the rectangle with vertices $(-1, 0)$, $(2, 0)$, $(2, 3)$, and $(-1, 3)$.

53. $\displaystyle\iint_R (x + 2y)\, dA$, where R is the rectangular region bounded by $x = 0$, $x = 1$, $y = -2$, and $y = 2$.

54. Find the volume under the surface $z = 2xy$ and above the rectangle with vertices $(0, 0)$, $(2, 0)$, $(0, 3)$, and $(2, 3)$.

55. Find the volume under the surface $z = xe^{-y}$ and above the rectangle bounded by the lines $x = 1$, $x = 2$, $y = 2$, and $y = 3$.

56. Find the average value of $f(x, y) = xy^2$ over the rectangular region with vertices $(-1, 3)$, $(-1, 5)$, $(2, 3)$, and $(2, 5)$.

57. Find three positive numbers x, y, and z so that $x + y + z = 20$ and the product $P = xyz$ is a maximum. [*Hint:* Use the fact that $z = 20 - x - y$ to express P as a function of only two variables.]

58. Find three positive numbers x, y, and z so that $2x + 3y + z = 60$ and the sum $S = x^2 + y^2 + z^2$ is minimized. (See the hint to Exercise 57.)

59. Find the shortest distance from the origin to the surface $y^2 - z^2 = 10$. [*Hint:* Express the distance $\sqrt{x^2 + y^2 + z^2}$ from the origin to a point (x, y, z) on the surface in terms of the two variables x and y, and minimize the *square* of the resulting distance function.]

60. Plot the points $(1, 1)$, $(1, 2)$, $(3, 2)$, and $(4, 3)$, and use partial derivatives to find the corresponding least-squares line.

61. SALES The marketing manager for a certain company has compiled these data relating monthly advertising expenditure and monthly sales (both measured in units of $1,000):

Advertising	3	4	7	9	10
Sales	78	86	138	145	156

a. Plot these data on a graph.

b. Find the least-squares line, and add it to the graph in part (a).

c. Use the least-squares line to predict monthly sales if the monthly advertising expenditure is $5,000.

62. UTILITY Suppose the utility derived by a consumer from x units of one commodity and y units of a second commodity is given by the utility function $U(x, y) = x^3y^2$. The consumer currently owns $x = 5$ units of the first commodity and $y = 4$ units of the second. Use calculus to estimate how many units of the second commodity the consumer could substitute for 1 unit of the first commodity without affecting total utility.

63. CONSUMER DEMAND A paint company makes two brands of latex paint. Sales figures indicate that if the first brand is sold for x dollars per quart and the second for y dollars per quart, the demand for the first brand will be Q quarts per month, where

$$Q(x, y) = 200 + 10x^2 - 20y$$

It is estimated that t months from now the price of the first brand will be $x(t) = 18 + 0.02t$ dollars per quart and the price of the second will be $y(t) = 21 + 0.4\sqrt{t}$ dollars per quart. At what rate will the demand for the first brand of paint be changing with respect to time 9 months from now?

64. **COOLING AN ANIMAL'S BODY** The difference between an animal's surface temperature and that of the surrounding air causes a transfer of energy by convection. The coefficient of convection h is given by

$$h = \frac{kV^{1/3}}{D^{2/3}}$$

where V (cm/sec) is wind velocity, D (cm) is the diameter of the animal's body, and k is a constant.

   a. Find the partial derivatives h_V and h_D. Interpret these derivatives as rates.

   b. Compute the ratio $\dfrac{h_V}{h_D}$.

65. **CONSUMER DEMAND** Suppose that when apples sell for x cents per pound and bakers earn y dollars per hour, the price of apple pies at a certain supermarket chain is

$$p(x, y) = \frac{1}{4}x^{1/3}y^{1/2}$$

dollars per pie. Suppose also that t months from now, the price of apples will be

$$x = 129 - \sqrt{8t}$$

cents per pound and bakers' wages will be

$$y = 15.60 + 0.2t$$

dollars per hour. If the supermarket chain can sell

$$Q = \frac{4,184}{p}$$ pies per week when the price is p

dollars per pie, at what rate will the weekly demand Q for pies be changing with respect to time 2 months from now?

66. **MARINE LIFE** Arnold, the heat-seeking mussel, is the world's smartest mollusk. Arnold likes to stay warm, and by using the crustacean coordinate system he learned from a passing crab, he has determined that at each nearby point (x, y) on the ocean floor the temperature (°C) is

$$T(x, y) = 2x^2 - xy + y^2 - 2y + 1$$

Arnold's world consists of a rectangular portion of ocean bed with vertices $(-1, -1)$, $(-1, 1)$, $(1, -1)$, and $(1, 1)$, and since it is very hard for him to move, he plans to stay where he is as long

as the average temperature of this region is at least 5°C. Does Arnold move or stay put?

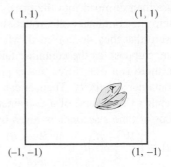

EXERCISE 66

67. **AIR POLLUTION** At a certain factory, the amount of air pollution generated each day is measured by the function $Q(E, T) = 125E^{2/3}\,T^{1/2}$, where E is the number of employees and T (°C) is the average temperature during the workday. Currently, there are $E = 151$ employees and the average temperature is $T = 10$°C. If the average daily temperature is falling at the rate of 0.21°C per day and the number of employees is increasing at the rate of 2 per month, use calculus to estimate the corresponding effect on the rate of pollution. Express your answer in units per day. You may assume that there are 22 workdays per month.

68. **POPULATION** A demographer sets up a grid to describe location within a suburb of a major metropolitan area. In relation to this grid, the population density at each point (x, y) is given by

$$f(x, y) = 1 + 3y^2$$

hundred people per square mile, where x and y are in miles. A housing project occupies the region R bounded by the curve $y^2 = 4 - x$ and the y axis $(x = 0)$. What is the total population within the project region R?

69. **POLLUTION** There are two sources of air pollution that affect the health of a certain community. Health officials have determined that at a point located r miles from source A and s miles from source B, there will be

$$N(r, s) = 40e^{-r/2}\,e^{-s/3}$$

units of pollution. A housing project lies in a region R for which

$$2 \le r \le 3 \qquad \text{and} \qquad 1 \le s \le 2$$

What is the total pollution within the region R?

CHAPTER SUMMARY

70. NUCLEAR WASTE DISPOSAL   Nuclear waste is often disposed of by sealing it into containers that are then dumped into the ocean. It is important to dump the containers into water shallow enough to ensure that they do not break when they hit bottom. Suppose as the container falls through the water, there is a drag force that is proportional to the container's velocity. Then, it can be shown that the depth s (in meters) of a container of weight W newtons at time t seconds is given by the formula

$$s(W, t) = \left(\frac{W - B}{k}\right)t + \frac{W(W - B)}{k^2 g}[e^{-(kgt/W)} - 1]$$

where B is a (constant) buoyancy force, k is the drag constant, and $g = 9.8$ m/sec^2 is the constant acceleration due to gravity.

a. Find $\dfrac{\partial s}{\partial W}$ and $\dfrac{\partial s}{\partial t}$. Interpret these derivatives as rates. Do you think it is possible for either partial derivative to ever be zero?

b. For a fixed weight, the speed of the container is $\dfrac{\partial s}{\partial t}$. Suppose the container will break when its speed on impact with the ocean floor exceeds 10 m/sec. If $B = 1{,}983$ newtons and $k = 0.597$ kg/sec, what is the maximum depth for safely dumping a container of weight $W = 2{,}417$ newtons?

c. Research the topic of nuclear waste disposal, and write a paragraph on whether you think it is best done on land or at sea.

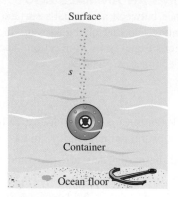

Surface

s

Container

Ocean floor

EXERCISE 70

71. PRODUCTION   For the production function given by $Q = x^a y^b$, where $a > 0$ and $b > 0$, show that

$$x\frac{\partial Q}{\partial x} + y\frac{\partial Q}{\partial y} = (a + b)Q$$

In particular, if $b = 1 - a$ with $0 < a < 1$, then

$$x\frac{\partial Q}{\partial x} + y\frac{\partial Q}{\partial y} = Q$$

EXPLORE! UPDATE

Complete solutions for all EXPLORE! boxes throughout the text can be accessed at the book-specific website, www.mhhe.com/hoffmann.

Solution for Explore! on Page 531

Store $f(x, y) = x^3 - x^2y^2 - xy^3 - y^4$ into Y1 as X^3 − X^2*L1^2 − X*L1^3 − L1^4, where L1 is the list of values {0, 1.5, 2.0, 2.25, 2.5}. Graph using the modified decimal window $[-9.4, 9.4]1$ by $[-150, 100]20$. Press the **TRACE** key and arrow right to $x = 2$ to observe the different $Y = f(x, L1)$ values that occur for varying L1 values. For larger L1 values the curves take on larger cubic dips in the positive x domain.

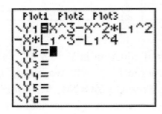 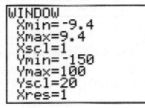 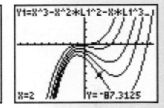

Solution for Explore! on Page 580

Using the data in Example 7.4.2, place the x values into L1 and the y values into L2. You can write L3 = L1*L2 and L4 = L1^2 if you wish to see the lists of values. To obtain all the sums needed to compute formulas for the slope m and y intercept b, press the **STAT** key, arrow right to **CALC,** select option **2:2-Var Stats,** and insert symbols for list L1 and L2, shown next in the middle screen. Pressing **ENTER** and arrowing up or down this screen gives all the desired sums, as shown in the far right screen.

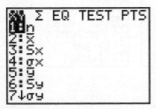 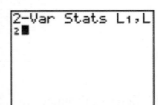 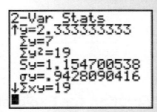

However, a more symbolic, presentable method on your calculator is to use the statistics symbolism available through the **VARS** key, **5:Statistics,** using both the **XY** and the **Σ** menus, shown next in the left and middle screens. The formulas for the slope m and y intercept b are computed in the screen next on the right, yielding $m = 0.5714 = \dfrac{4}{7}$ and $b = 1$.

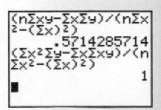

Solution for Explore! on Page 582

Following Example 7.4.4, store the production and demand price data into lists L1 and L2, respectively. Using the **STAT PLOT** procedure explained in the Calculator Introduction, obtain the scatterplot of decreasing prices shown in the following middle screen, which suggests an exponential curve with a negative exponent. Press **STAT,** arrow right to **CALC,** and arrow down to **0:ExpReg,** making sure to indicate the desired lists and function location. Specifically, write **ExpReg L1, L2, Y1** before completing the final keystroke. Recall that the symbol **Y1** is found through the keystroke sequence, **VARS, Y-VARS, 1:Function, 1:Y1.**

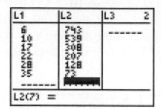 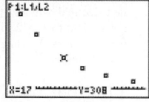 

The form of the exponential equation is $Y = a*b^x$, and we find that $a = 1,200$, with base $b = 0.92315$. If we write $b^x = e^{mx}$, we would find that $m = -0.079961$; that is, $0.92315 = e^{-0.079961}$. Pressing **ZOOM, 9:ZoomStat,** yields the following right screen, showing an almost perfect fit to the data for an exponential curve, with an equation $Y = 1,200(0.92315)^x = 1,200e^{(-0.079961)x}$. This equation coincides with the solution shown on page 584. We have computed a log-linear regression without having to take logarithms of the production and demand price variables, by directly selecting the Exponential Regression model.

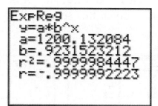 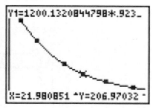

Solution for Explore! on Page 609

Refer to Example 7.6.3. Store $y = x^2$ into Y1 and $y = 2x$ into Y2 of the equation editor and graph using the window $[-0.15, 2.2]1$ by $[-0.5, 4.5]1$. By tracing, the intersection points of Y1 and Y2 are easily located. The vertical line feature can be found using the **DRAW** key **(2nd PRGM), 4:Vertical.** By arrowing left or right, cross sections of the area between Y1 and Y2 can be shown.

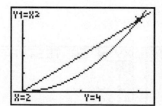 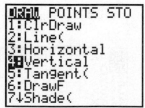 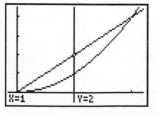

THINK ABOUT IT

MODELING POPULATION DIFFUSION

In 1905, five muskrats were accidentally released near Prague in the current Czech Republic. Subsequent to 1905, the range of the muskrat population expanded and the front (the outer limit of the population) moved as indicated in Figure 1. In the figure, the closed curves labeled with dates are equipopulation contours, that is, curves of constant, minimally detectable populations of muskrats. For instance, the curve labeled 1920 indicates that the muskrat population had expanded from Prague to the gates of Vienna in the 15 years after their release. A population dispersion such as this can be studied using mathematical models based on *partial differential equations*, that is, equations involving functions of two or more variables and their partial derivatives. We will examine such a model and then return to our illustration to see how well the model can be used to describe the dispersion of the muskrats.

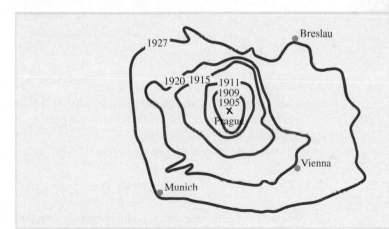

FIGURE 1   Equipopulation curves for a muskrat population in Europe.
Source: Leah Edelstein-Keshet, *Mathematical Models in Biology,* Boston: McGraw-Hill, 1988, p. 439.

The model we will discuss is based on the **diffusion equation,** an extremely versatile partial differential equation with important applications in the physical and life sciences as well as economics. *Diffusion* is the name used for the process by which particles spread out as they continually collide and randomly change direction after being inserted at a source. Suppose the particles can only move in one spatial direction (say, along a thin rod or tube). Then $C(x, t)$, the concentration of particles at time t located x units from the source (the point of insertion), satisfies the one-dimensional diffusion equation

$$\frac{\partial C}{\partial t} = \alpha \frac{\partial^2 C}{\partial x^2}$$

where α is a positive constant called the *diffusion coefficient*. Similarly, the two-dimensional diffusion equation

$$\frac{\partial C}{\partial t} = \alpha\left(\frac{\partial^2 C}{\partial x^2} + \frac{\partial^2 C}{\partial y^2}\right)$$

is used to model the dispersion of particles moving randomly in a plane, where $C(x, y, t)$ is the concentration of particles at the point (x, y) at time t.

Mathematical biologists have adapted the diffusion equation to model the spread of living organisms, including both plants and animals. We will examine such a model, due to J. G. Skellam. First, suppose that at a particular time ($t = 0$), an organism is introduced at a point (called a "source"), where it had previously not been present. Skellam assumed that the population of the organism disperses from the source in two ways:

    a. By growing exponentially at the continuous reproduction rate r.

    b. By moving randomly in an xy coordinate plane, with the source at the origin.

Based on these assumptions, he then modeled the dispersion of the population by the modified two-dimensional diffusion equation

$$(1) \quad \frac{\partial N}{\partial t} = D\underbrace{\left(\frac{\partial^2 N}{\partial x^2} + \frac{\partial^2 N}{\partial y^2}\right)}_{\substack{\text{expansion} \\ \text{by random} \\ \text{movement}}} + \underbrace{rN}_{\substack{\text{growth by} \\ \text{exponential} \\ \text{reproduction}}}$$

where $N(x, y, t)$ is the population density at the point (x, y) at time t, and D is a positive constant, called the *dispersion coefficient*, that is analogous to the diffusion coefficient.

It can be shown that one solution of Skellam's equation is

$$(2) \quad N(x, y, t) = \frac{M}{4\pi Dt} e^{rt - (x^2 + y^2)/(4Dt)}$$

where M is the number of individuals initially introduced at the source (see Question 5). The *asymptotic rate of population expansion*, V, is the distance between locations with equal population densities in successive years, and Skellam's model can be used to show that

$$(3) \quad V = \sqrt{4rD}$$

(see Question 4). Likewise, the *intrinsic rate of growth*, r, can be estimated using data of the growth of existing populations, and the dispersion coefficient, D, can be estimated using the formula

$$(4) \quad D \approx \frac{2A^2(t)}{\pi t}$$

where $A(t)$ is average distance organisms have traveled at time t.

Skellam's model has been used to study the spread of a variety of organisms, including oak trees, cereal leaf beetles, and cabbage butterflies. As an illustration of how the model can be applied, we return to the population of Central European muskrats introduced in the opening paragraph and Figure 1. Population studies indicate that r, the intrinsic rate of muskrat population increase, was no greater than 1.1 per year, and that D, the dispersion coefficient, was no greater than 230 km^2/year. Consequently, the solution to Skellam's model stated in Equation (2) predicts that the distribution of muskrats, under the best circumstances for the species, is given by

$$(5) \quad N(x, y, t) = \frac{5}{4\pi(230)t} e^{1.1t - (x^2 + y^2)/(920t)}$$

where (x, y) is the point x km east and y km north from the release point near Prague and t is the time in years (after 1905). Formula (3) predicts that the maximum rate of population expansion is

$$V = \sqrt{4rD} = \sqrt{4(1.1)(230)} \approx 31.8 \text{ km/yr}$$

which is a little greater than the observed rate of 25.4 km/yr.

The derivation of the diffusion equation may be found in many differential equations texts, or see *Introduction to Mathematics for Life Scientists*, 3rd ed., by Edward Batschelet, Springer-Verlag, New York, pages 392–395. Skellam's model and several variations are discussed in *Mathematical Models in Biology* by Leah Edelstein-Keshet, McGraw-Hill, Boston, 1988, pages 436–441. It is important to emphasize that Skellam's modified diffusion equation given in formula (1) has solutions other than formula (2). In general, solving partial differential equations is very difficult, and often the best that can be done is to focus on finding solutions with certain specified forms. Such solutions can then be used to analyze practical situations, as we did with the muskrat problem.

Questions

1. Verify that $C(x, t) = \dfrac{M}{2\sqrt{\pi Dt}} e^{-(x^2/4Dt)}$ satisfies the diffusion equation

$$\frac{\partial C}{\partial t} - \alpha \frac{\partial^2 C}{\partial x^2}$$

    Do this by calculating the partial derivatives and inserting them into the equation.

2. Find a relationship that must hold between the coefficients a and b for $C(x, t) = e^{ax + bt}$ to be a solution of the diffusion equation

$$\frac{\partial C}{\partial t} = \alpha \frac{\partial^2 C}{\partial x^2}$$

3. Suppose that a population of organisms spreads out along a one-dimensional line according to the partial differential equation

$$\frac{\partial N}{\partial t} = D \frac{\partial^2 N}{\partial x^2} + rN$$

    Show that the function $N(x, t) = \dfrac{M}{2\sqrt{\pi Dt}} e^{rt - (x^2/4Dt)}$ is a solution to this partial differential equation where M is the initial population of organisms located at the point $x = 0$ when $t = 0$.

4. Show that on the contours of equal population density (that is, the curves of the form $N(x, t) = A$ where A is a constant), the ratio $\dfrac{x}{t}$ equals

$$\frac{x}{t} = \pm \left[4rD - \frac{2D}{t} \ln t - \frac{4D}{t} \ln\left(\sqrt{2\pi D}\frac{A}{M} \right) \right]^{1/2}$$

THINK ABOUT IT

Using this formula, it can be shown that the ratio $\dfrac{x}{t}$ can be approximated by $\dfrac{x}{t} \approx \pm 2\sqrt{rD}$, which gives us a formula for the rate of population expansion.

5. Verify that $N(x, y, t) = \dfrac{M}{4\pi Dt}\, e^{rt - (x^2 + y^2)/(4Dt)}$ is a solution of the partial differential equation

$$\frac{\partial N}{\partial t} = D\left(\frac{\partial^2 N}{\partial x^2} + \frac{\partial^2 N}{\partial y^2}\right) + rN$$

6. Use Equation (5), which we obtained using Skellam's model, to find the population density for muskrats in 1925 at the location 50 km north and 50 km west of the release point near Prague.

7. Use Skellam's model to construct a function that estimates the population density of the small cabbage white butterfly if the largest diffusion coefficient observed is 129 km²/year and the largest intrinsic rate of increase observed is 31.5/year. What is the predicted maximum rate of population expansion? How does this compare with the largest observed rate of population expansion of 170 km/year?

8. In this question, you are asked to explore an alternative approach to analyzing the muskrat problem using Skellam's model. Recall that the intrinsic rate of growth of the muskrat population was $r = 1.1$ and that the maximum rate of dispersion was observed to be $V = 25.4$.
   a. Use these values for r and V in formula (3) to estimate the dispersion coefficient D.
   b. By substituting $r = 1.1$ into formula (2) along with the value for D you obtained in part (a), find the population density in 1925 for the muskrats at the location 50 km north and 50 km west of the release point (source) near Prague. Compare your answer with the answer to Question 6.
   c. Use formula (4) to estimate the average distance A of the muskrat population in 1925 from its source near Prague.

References

D. A. Andow, P. M. Kareiva, Simon A. Levin, and Akira Okubo, "Spread of Invading Organisms," *Landscape Ecology,* Vol. 4, nos. 2/3, 1990, pp. 177–188.

Leah Edelstein-Keshet, *Mathematical Models in Biology,* Boston: McGraw-Hill, 1988.

J. G. Skellam, "The Formulation and Interpretation of Mathematical Models of Diffusionary Processes in Population Biology," in *The Mathematical Theory of the Dynamics of Biological Populations,* edited by M. S. Bartlett and R. W. Hiorns, New York: Academic Press, 1973, pp. 63–85.

J. G. Skellam, "Random Dispersal in Theoretical Populations," *Biometrika,* Vol. 28, 1951, pp. 196–218.

CHAPTER **8**

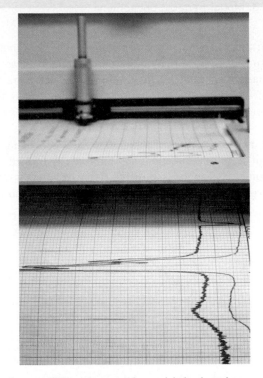

Periodic behavior, such as an EKG pattern, can be modeled using trigonometric functions.

Trigonometric Functions

SECTION 8.1   Angle Measurement; Trigonometric Functions

Learning Objectives

1. Examine angle measurement.
2. Explore the sine and cosine and other trigonometric functions.
3. Model periodic phenomena.

In this chapter, you will be introduced to some functions that are widely used to study periodic or rhythmic phenomena such as seasonal supply and demand, fluctuations in weather, the motion of planets, and the respiratory cycle and heartbeat of animals. These functions are also related to the measurement of angles and hence play an important role in such fields as architecture and surveying.

Angles   A **ray** is a portion of a line that starts at a point O (the endpoint) and extends indefinitely in one direction, and an **angle** is formed when a ray is rotated about its endpoint. The angle is **positive** if the rotation is in a counterclockwise direction and **negative** if the rotation is in a clockwise direction, as shown in Figure 8.1.

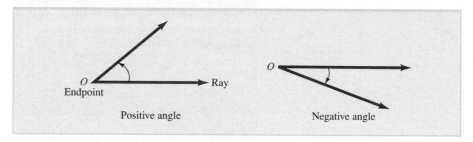

FIGURE 8.1   Positive and negative angles.

Measurement of Angles   You are probably already familiar with the use of degrees to measure angles. A **degree** is the amount by which a line segment must be rotated so that its free endpoint traces out $\frac{1}{360}$ of one complete rotation. Thus, for example, a complete counterclockwise rotation generating an entire circle contains 360°, one-half of a complete counterclockwise rotation contains 180°, and one-sixth of a complete counterclockwise rotation contains 60°. Some important angles are shown in Figure 8.2.

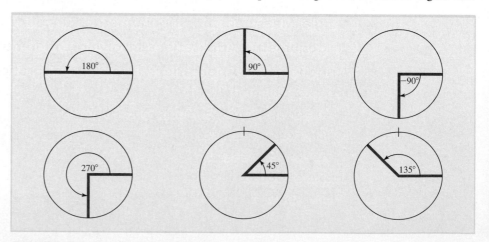

FIGURE 8.2   Angles measured in degrees.

Radian Measure

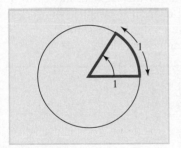

FIGURE 8.3 An angle of 1 radian.

Although measurement of angles in degrees is convenient for many geometric applications, there is another unit of angle measurement called a **radian** that leads to simpler rules for the differentiation and integration of trigonometric functions. One radian is defined to be the amount by which a line segment of length 1 must be rotated so that its free endpoint traces out a circular arc of length 1. The situation is illustrated in Figure 8.3.

The total circumference of a circle of radius 1 is 2π. Hence, 2π radians are equal to $360°$, π radians are equal to $180°$, and, in general, the relationship between radians and degrees is given by the following useful conversion formula.

Conversion Formula

$$\frac{\text{degrees}}{180} = \frac{\text{radians}}{\pi}$$

EXPLORE!

Use your graphing utility to convert 43°32′ to radians and to convert $\dfrac{\pi}{7}$ radian to degrees.

The use of this formula is illustrated in Example 8.1.1.

EXAMPLE 8.1.1 Angle Conversion

a. Convert 45° to radians.   **b.** Convert $\dfrac{\pi}{6}$ radian to degrees.

Solution

a. From the proportion $\dfrac{45}{180} = \dfrac{\text{radians}}{\pi}$, it follows that 45° is equivalent to $\dfrac{\pi}{4}$ radian.

b. From the proportion $\dfrac{\text{degrees}}{180} = \dfrac{\pi/6}{\pi}$, it follows that $\dfrac{\pi}{6}$ radian is equivalent to $\dfrac{180°}{6} = 30°$.

For reference, six of the most important angles are shown in Figure 8.4 along with their measurements in degrees and in radians.

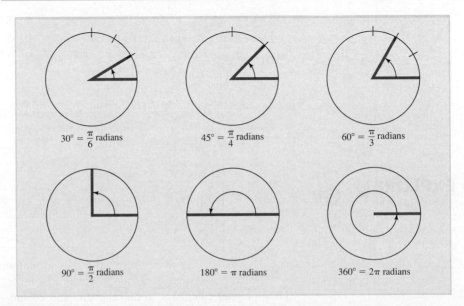

FIGURE 8.4 Six important angles in degrees and radians.

The Sine and Cosine

Suppose the line segment joining the points $(0, 0)$ and $(1, 0)$ on the x axis is rotated through an angle of θ radians so that the free endpoint of the segment moves from $(1, 0)$ to a point (x, y) as in Figure 8.5. The x and y coordinates of the point (x, y) are known, respectively, as the **cosine** and **sine** of the angle θ. The symbol $\cos \theta$ is used to denote the cosine of θ, and $\sin \theta$ is used to denote its sine.

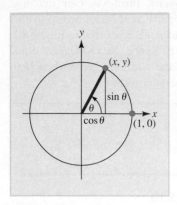

FIGURE 8.5   The sine and cosine of an angle.

The Sine and Cosine   ■   For any angle θ,

$$\cos \theta = x \quad \text{and} \quad \sin \theta = y$$

where (x, y) is the point to which $(1, 0)$ is carried by a rotation of θ radians about the origin.

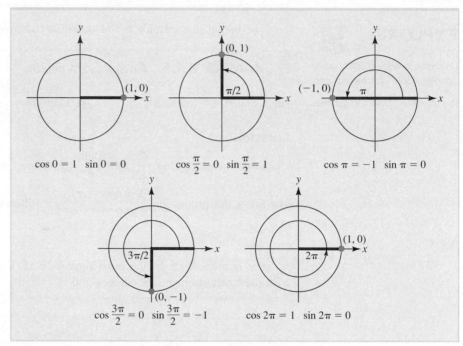

$$\cos 0 = 1 \quad \sin 0 = 0$$

$$\cos \frac{\pi}{2} = 0 \quad \sin \frac{\pi}{2} = 1$$

$$\cos \pi = -1 \quad \sin \pi = 0$$

$$\cos \frac{3\pi}{2} = 0 \quad \sin \frac{3\pi}{2} = -1$$

$$\cos 2\pi = 1 \quad \sin 2\pi = 0$$

FIGURE 8.6   The sine and cosine of multiples of $\dfrac{\pi}{2}$.

The values of the sine and cosine for multiples of $\dfrac{\pi}{2}$ can be read from Figure 8.6 and are summarized in the following table:

θ	0	$\dfrac{\pi}{2}$	π	$\dfrac{3\pi}{2}$	2π
$\cos \theta$	1	0	-1	0	1
$\sin \theta$	0	1	0	-1	0

The sine and cosine of a few other important angles are also easy to obtain geometrically, as you will see shortly. For other angles, use Table III at the back of the book or your calculator to find the sine and cosine.

Basic Properties of the Sine and Cosine

Since there are 2π radians in a complete rotation, it follows that

$$\sin(\theta + 2\pi) = \sin\theta \qquad \text{and} \qquad \cos(\theta + 2\pi) = \cos\theta$$

That is, the sine and cosine functions are **periodic** with period 2π. The situation is illustrated in Figure 8.7.

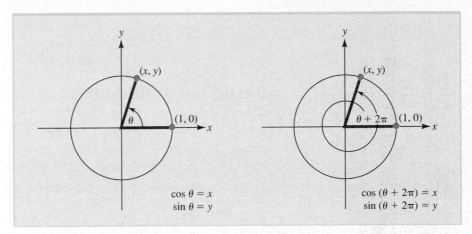

FIGURE 8.7 The periodicity of $\sin\theta$ and $\cos\theta$.

Since negative angles correspond to clockwise rotations, it follows that

$$\sin(-\theta) = -\sin\theta \qquad \text{and} \qquad \cos(-\theta) = \cos\theta$$

This is illustrated in Figure 8.8.

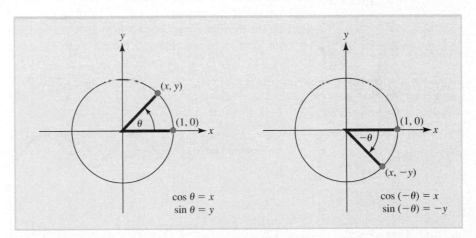

FIGURE 8.8 The sine and cosine of negative angles.

Properties of the Sine and Cosine

$$\sin(\theta + 2\pi) = \sin\theta \qquad \text{and} \qquad \cos(\theta + 2\pi) = \cos\theta$$
$$\sin(-\theta) = -\sin\theta \qquad \text{and} \qquad \cos(-\theta) = \cos\theta$$

The use of these properties is illustrated in Example 8.1.2.

EXAMPLE 8.1.2   Evaluation of Sine and Cosine

Evaluate the following expressions.

a. $\cos(-\pi)$ **b.** $\sin\left(-\dfrac{\pi}{2}\right)$ **c.** $\cos 3\pi$ **d.** $\sin\dfrac{5\pi}{2}$

Solution

a. Since $\cos \pi = -1$, it follows that

$$\cos(-\pi) = \cos \pi = -1$$

b. Since $\sin\dfrac{\pi}{2} = 1$, it follows that

$$\sin\left(-\frac{\pi}{2}\right) = -\sin\frac{\pi}{2} = -1$$

c. Since $3\pi = \pi + 2\pi$ and $\cos \pi = -1$, we have

$$\cos 3\pi = \cos(\pi + 2\pi) = \cos \pi = -1$$

d. Since $\dfrac{5\pi}{2} = \dfrac{\pi}{2} + 2\pi$ and $\sin\dfrac{\pi}{2} = 1$, it follows that

$$\sin\frac{5\pi}{2} = \sin\left(\frac{\pi}{2} + 2\pi\right) = \sin\frac{\pi}{2} = 1$$

The Graphs of sin θ and cos θ

We can see from the definitions of the sine and cosine that as θ goes from 0 to 2π, the function $\sin \theta$ oscillates between 1 and -1, starting with $\sin 0 = 0$, and the function $\cos \theta$ oscillates between 1 and -1, starting with $\cos 0 = 1$. This observation, together with the elementary properties previously derived, suggests that the graphs of the functions $\sin \theta$ and $\cos \theta$ resemble the curves in Figures 8.9 and 8.10, respectively.

EXPLORE!

Graph the function $f(x) = \sin x$ using the **ZTrig** window on the **ZOOM** key menu. Trace $f(x)$ for x close to but less than $\dfrac{\pi}{2}$.

Also determine $\lim\limits_{x \to \pi} \sin x$.

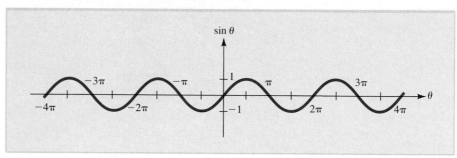

FIGURE 8.9   The graph of sin θ.

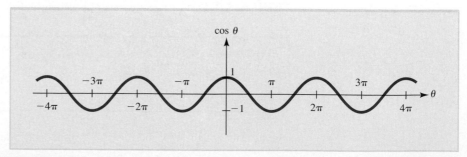

FIGURE 8.10   The graph of cos θ.

Other Trigonometric Functions

Four other useful trigonometric functions can be defined in terms of the sine and cosine as follows.

The Tangent, Cotangent, Secant, and Cosecant ■ For any angle θ,

$$\tan \theta = \frac{\sin \theta}{\cos \theta} \qquad \cot \theta = \frac{1}{\tan \theta} = \frac{\cos \theta}{\sin \theta}$$

$$\sec \theta = \frac{1}{\cos \theta} \qquad \csc \theta = \frac{1}{\sin \theta}$$

provided the denominators are not zero.

EXPLORE!

Graph $f(x) = \tan x$ using the **ZTrig** window on the **ZOOM** key menu. Trace $f(x)$ for x close to but not equal to $\dfrac{\pi}{2}$ and determine $\lim\limits_{x \to \pi/2^-} \tan x$.

EXAMPLE 8.1.3 Evaluation of Tangent, Cotangent, Secant, and Cosecant

Evaluate the following expressions.

    a. $\tan \pi$ **b.** $\cot \dfrac{\pi}{2}$ **c.** $\sec(-\pi)$ **d.** $\csc\left(-\dfrac{5\pi}{2}\right)$

Solution

   a. Since $\sin \pi = 0$ and $\cos \pi = -1$, it follows that

$$\tan \pi = \frac{\sin \pi}{\cos \pi} = \frac{0}{-1} = 0$$

   b. Since $\sin \dfrac{\pi}{2} = 1$ and $\cos \dfrac{\pi}{2} = 0$, we have

$$\cot \frac{\pi}{2} = \frac{\cos(\pi/2)}{\sin(\pi/2)} = \frac{0}{1} - 0$$

   c. Since $\cos \pi = -1$, then

$$\sec(-\pi) = \frac{1}{\cos(-\pi)} = \frac{1}{\cos \pi} = -1$$

   d. Since $\dfrac{5\pi}{2} = \dfrac{\pi}{2} + 2\pi$ and $\sin \dfrac{\pi}{2} = 1$, it follows that

$$\csc\left(-\frac{5\pi}{2}\right) = \frac{1}{\sin(-5\pi/2)} = \frac{1}{-\sin(5\pi/2)} = \frac{1}{-\sin(\pi/2)} = -1$$

Right Triangles

If you have had a course in trigonometry, you may remember the following definitions of the sine, cosine, and tangent involving the lengths of the sides of a right triangle such as the one in Figure 8.11.

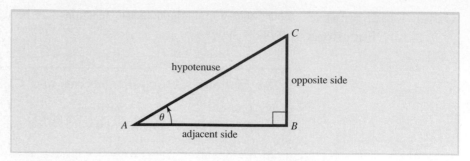

FIGURE 8.11 Triangle used to define trigonometric functions.

The Trigonometry of Right Triangles

$$\sin \theta = \frac{\text{opposite side}}{\text{hypotenuse}} \qquad \cos \theta = \frac{\text{adjacent side}}{\text{hypotenuse}} \qquad \tan \theta = \frac{\text{opposite side}}{\text{adjacent side}}$$

The definitions of trigonometric functions that you have seen in this section involving the coordinates of points on a circle of radius 1 are equivalent to the definitions from trigonometry. To see this, superimpose an *xy* coordinate system over the triangle *ABC* as shown in Figure 8.12 and draw the circle of radius 1 that is centered at the origin.

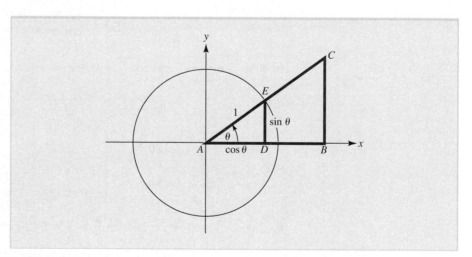

FIGURE 8.12 Similar triangles.

Since *ABC* and *ADE* are similar triangles, it follows that

$$\frac{AD}{AE} = \frac{AB}{AC}$$

But the length *AD* is the *x* coordinate of point *E* on the circle, and so, by the definition of the cosine, *AD* = cos θ. Moreover,

$$AE = 1 \qquad AB = \text{adjacent side} \qquad \text{and} \qquad AC = \text{hypotenuse}$$

so

$$\frac{\cos \theta}{1} = \frac{\text{adjacent side}}{\text{hypotenuse}} \quad \text{or} \quad \cos \theta = \frac{\text{adjacent side}}{\text{hypotenuse}}$$

For practice, convince yourself that similar arguments lead to the formulas for the sine and tangent.

Calculations with Right Triangles

Many calculations involving trigonometric functions can be performed easily and quickly with the aid of appropriate right triangles. For example, from the well-known 30-60-90 triangle in Figure 8.13, you see immediately that

$$\sin 30° = \frac{1}{2} \quad \text{and} \quad \cos 30° = \frac{\sqrt{3}}{2}$$

or, using radian measure,

$$\sin \frac{\pi}{6} = \frac{1}{2} \quad \text{and} \quad \cos \frac{\pi}{6} = \frac{\sqrt{3}}{2}$$

Example 8.1.4 further illustrates the use of right triangles.

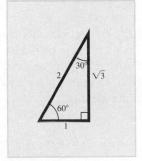

FIGURE 8.13 A 30-60-90 triangle.

EXAMPLE 8.1.4 Using a Right Triangle to Evaluate a Tangent

Find $\tan \theta$ if $\sec \theta = \dfrac{3}{2}$.

Solution

Since

$$\sec \theta = \frac{1}{\cos \theta} = \frac{\text{hypotenuse}}{\text{adjacent side}}$$

begin by drawing a right triangle (Figure 8.14) in which the hypotenuse has length 3 and the side adjacent to the angle θ has length 2.

By the Pythagorean theorem,

$$\text{Length of } BC = \sqrt{3^2 - 2^2} = \sqrt{5}$$

and so

$$\tan \theta = \frac{\text{opposite side}}{\text{adjacent side}} = \frac{\sqrt{5}}{2}$$

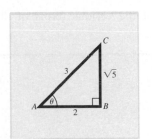

FIGURE 8.14 Right triangle for Example 8.1.4.

Using appropriate right triangles and the definitions and elementary properties of sine, cosine, and tangent, you should now be able to find the frequently used values listed in Table 8.1.

TABLE 8.1 Frequently Used Values of Sine, Cosine, and Tangent

θ	0	$\dfrac{\pi}{6}$	$\dfrac{\pi}{4}$	$\dfrac{\pi}{3}$	$\dfrac{\pi}{2}$	$\dfrac{2\pi}{3}$	$\dfrac{3\pi}{4}$	$\dfrac{5\pi}{6}$	π
$\sin \theta$	0	$\dfrac{1}{2}$	$\dfrac{\sqrt{2}}{2}$	$\dfrac{\sqrt{3}}{2}$	1	$\dfrac{\sqrt{3}}{2}$	$\dfrac{\sqrt{2}}{2}$	$\dfrac{1}{2}$	0
$\cos \theta$	1	$\dfrac{\sqrt{3}}{2}$	$\dfrac{\sqrt{2}}{2}$	$\dfrac{1}{2}$	0	$-\dfrac{1}{2}$	$-\dfrac{\sqrt{2}}{2}$	$-\dfrac{\sqrt{3}}{2}$	-1
$\tan \theta$	0	$\dfrac{\sqrt{3}}{3}$	1	$\sqrt{3}$	Undefined	$-\sqrt{3}$	-1	$-\dfrac{\sqrt{3}}{3}$	0

The Pythagorean Identity

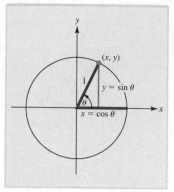

FIGURE 8.15   The identity $\sin^2 \theta + \cos^2 \theta = 1$.

An **identity** is an equation that is true for all values of its variable or variables when both sides are defined. Identities involving trigonometric functions are called **trigonometric identities** and can be used to simplify trigonometric expressions. You have already seen some elementary trigonometric identities in this section. For example, the formulas

$$\cos (-\theta) = \cos \theta \quad \text{and} \quad \sin (-\theta) = -\sin \theta$$

are identities because they are true for all values of θ. Most trigonometry texts list dozens of trigonometric identities. Fortunately, only a few of them will be needed in this book.

One of the most important and well-known trigonometric identities is a simple consequence of the Pythagorean theorem. It states that

$$\sin^2 \theta + \cos^2 \theta = 1$$

where $\sin^2 \theta$ stands for $(\sin \theta)^2$ and $\cos^2 \theta$ stands for $(\cos \theta)^2$. To see why this identity is true, look at Figure 8.15, which shows a point (x, y) on a circle of radius 1. By definition, $y = \sin \theta$ and $x = \cos \theta$, and since each point (x, y) on the unit circle satisfies $y^2 + x^2 = 1$, we have $\sin^2 \theta + \cos^2 \theta = 1$ as claimed. Note that if we divide each term in the Pythagorean identity by $\cos^2 \theta$, we obtain the identity

$$\frac{\sin^2 \theta}{\cos^2 \theta} + \frac{\cos^2 \theta}{\cos^2 \theta} = \frac{1}{\cos^2 \theta}$$

or, equivalently,

$$\tan^2 \theta + 1 = \sec^2 \theta$$

Similarly, $1 + \cot^2 \theta = \csc^2 \theta$. Here is a summary of the various forms of the Pythagorean identity.

The Pythagorean Identity ■ For any angle θ.

$$\sin^2 \theta + \cos^2 \theta = 1$$

In addition,

$$\tan^2 \theta + 1 = \sec^2 \theta \quad \text{whenever } \cos \theta \neq 0$$

and

$$1 + \cot^2 \theta = \csc^2 \theta \quad \text{whenever } \sin \theta \neq 0$$

EXPLORE!

Refer to Example 8.1.5. Set Y1=sin(X)²−cos(X)²+sin(X) in the equation editor of your graphing utility. Graph using the window [0, 2π]$\frac{\pi}{2}$ by [−2, 2]0.5. Use the **TRACE** key or the root-finding features of your calculator to find the x intercepts of the graph of the function. Compare your results with those determined in the example.

EXAMPLE 8.1.5 Using an Identity to Solve a Trigonometric Equation

Find all the values of θ in the interval $0 \leq \theta \leq 2\pi$ that satisfy the equation $\sin^2 \theta - \cos^2 \theta + \sin \theta = 0$.

Solution

Use the Pythagorean identity

$$\sin^2 \theta + \cos^2 \theta = 1 \quad \text{or} \quad \cos^2 \theta = 1 - \sin^2 \theta$$

to rewrite the original equation without any cosine terms as

$$\sin^2 \theta - (1 - \sin^2 \theta) + \sin \theta = 0$$

or

$$2 \sin^2 \theta + \sin \theta - 1 = 0$$

and factor to get

$$(2 \sin \theta - 1)(\sin \theta + 1) = 0$$

The only solutions of

$$2 \sin \theta - 1 = 0 \quad \text{or} \quad \sin \theta = \frac{1}{2}$$

in the interval $0 \le \theta \le 2\pi$ are

$$\theta = \frac{\pi}{6} \quad \text{and} \quad \theta = \frac{5\pi}{6}$$

and the only solution of

$$\sin \theta + 1 = 0 \quad \text{or} \quad \sin \theta = -1$$

in the interval is

$$\theta = \frac{3\pi}{2}$$

Hence, the solutions of the original equation in the specified interval are $\theta = \frac{\pi}{6}, \theta = \frac{5\pi}{6},$ and $\theta = \frac{3\pi}{2}$.

Modeling Periodic Phenomena

A *periodic phenomenon* is one that recurs repeatedly in cycles. The motion of a pendulum, the rise and fall of tides, and the circadian rhythms of human metabolism are examples of periodic phenomena.

Periodic phenomena can often be modeled by trigonometric functions of one of these general forms:

$$s(t) = a + b \sin \left[\frac{2\pi}{p}(t - d) \right] \quad \text{or} \quad c(t) = a + b \cos \left[\frac{2\pi}{p}(t - d) \right]$$

The graph of such a function $f(t)$ repeats itself every p units along the t axis in the sense that $f(t + p) = f(t)$ for all t (see Exercise 77). The number p in the form is called the **period** of the function. The reciprocal of the period $\frac{1}{p}$ is called the **frequency** and provides a measure of how often the cycles recur. The number b is called the **amplitude** and measures the vertical spread in the graph of $f(t)$ in the sense that $|2b|$ is the difference between the largest and smallest value of $f(t)$. The **horizontal shift** d determines the horizontal displacement of the graph of $f(t)$ from the position of the standard sine and cosine curves in Figures 8.9 and 8.10, while the **vertical shift** a determines the vertical displacement of the graph from the standard sine and cosine curves.

For instance, the function

$$f(t) = 3 + 0.5 \sin \pi(t - 1)$$

has period $p = 2$, frequency $\frac{1}{2}$, amplitude 0.5, horizontal shift 1, and vertical shift 3.

The graph of one period of this function is shown in Figure 8.16.

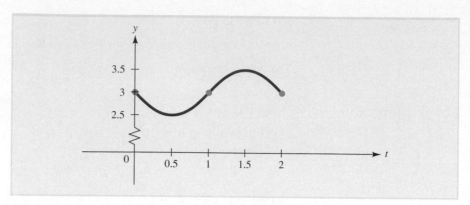

FIGURE 8.16   The graph of one period of the function $f(t) = 3 + 0.5 \sin \pi(t - 1)$.

We close this introductory section with Examples 8.1.6 and 8.1.7, which illustrate how periodic functions can be used for modeling important phenomena. Additional applications will be examined in the remaining sections of this chapter.

EXAMPLE 8.1.6   Studying a Trigonometric Model in Medicine

A device monitoring the vital signs of a hospital patient provides data that suggests a patient's blood pressure can be modeled by the trigonometric function

$$P(t) = 108 + 27 \cos 8t$$

where time t is measured in seconds and P is in millimeters of mercury.

a. What is the maximum (systolic) pressure P_s of the patient? What is the minimum (diastolic) pressure P_d?

b. A heartbeat is the interval between successive systolic peaks. What is the patient's pulse rate (heartbeats per minute)?

Solution

a. The largest $\cos 8t$ can be is 1, and when $\cos 8t = 1$, the blood pressure has the (maximum) value of

$$P_s = 108 + 27(1) = 135$$

Likewise, when $\cos 8t = -1$, the blood pressure function has the minimum value of

$$P_d = 108 + 27(-1) = 81$$

b. Systolic pressure occurs whenever $\cos 8t = 1$. Since the cosine function repeats itself every 2π radians, it follows that each heartbeat takes t_h seconds, where

$$8t_h = 2\pi$$

$$t_h = \frac{\pi}{4}$$

Therefore, the number of heartbeats that occur each minute is given by the product

$$\left(\frac{1 \text{ heartbeat}}{\pi/4 \text{ seconds}} \right) \left(\frac{60 \text{ seconds}}{\text{minute}} \right) = \frac{4}{\pi}(60) \approx 76.4$$

We conclude that the pulse rate is approximately 76 heartbeats per minute.

EXAMPLE 8.1.7 Optimizing Periodic Sales

Suppose the number of bathing suits sold in a store at the New Jersey shore during week t of a particular year is modeled by the function

$$B(t) = 40 + 36 \cos\left[\frac{\pi}{52}(t - 24)\right] \quad \text{for } 1 \leq t \leq 52$$

During what week are sales the largest? What is the maximum sales level?

Solution

The graph of the sales function is shown in Figure 8.17. Sales are largest when $\cos\left[\dfrac{\pi}{52}(t - 24)\right] = 1$, and for the time interval $1 \leq t \leq 52$, this occurs only when $t = 24$. Thus, the largest sales occur during the 24th week, when

$$B(24) = 40 + 36 = 76 \text{ suits}$$

are sold.

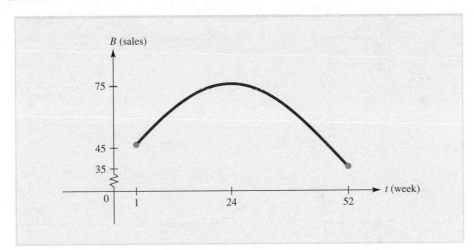

FIGURE 8.17   The graph of the sales function $B(t) = 40 + 36 \cos\left[\dfrac{\pi}{52}(t - 24)\right]$.

EXERCISES ▪ 8.1

In Exercises 1 through 4, specify the number of degrees in each indicated angle.

1.

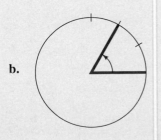

a. **b.**

2.

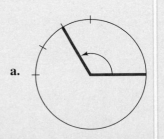

a.

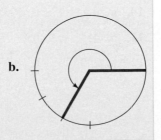

b.

3.
4.

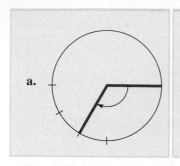

In Exercises 5 through 8, specify the number of radians in each indicated angle.

5.
6.

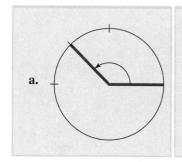
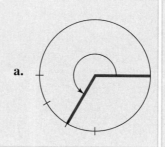

7.
8.

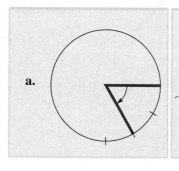
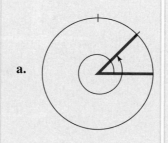
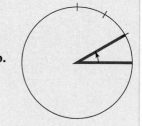

In Exercises 9 through 12, draw each angle with the indicated degree measure in a coordinate plane with the ray initially along the positive x axis and the endpoint at the origin.

9. a. $60°$ **b.** $120°$

10. a. $-45°$ **b.** $135°$

11. a. $45°$ **b.** $-150°$

12. a. $240°$ **b.** $405°$

In Exercises 13 through 16, draw each angle with the indicated radian measure in a coordinate plane with the ray initially along the positive x axis and the endpoint at the origin.

13. a. $\dfrac{3\pi}{4}$ **b.** $-\dfrac{2\pi}{3}$

14. a. $-\dfrac{\pi}{6}$ **b.** $\dfrac{5\pi}{6}$

15. a. $-\dfrac{5\pi}{3}$ **b.** $\dfrac{3\pi}{2}$

16. a. $-\dfrac{5\pi}{4}$ **b.** $\dfrac{7\pi}{6}$

In Exercises 17 through 20, convert the indicated angles from degrees to radians.

17. a. $15°$ **b.** $-240°$

18. a. $30°$ **b.** $270°$

19. a. $135°$ **b.** $540°$

20. a. $-150°$ **b.** $1°$

In Exercises 21 through 24, convert the indicated angle from radians to degrees.

21. a. $\dfrac{5\pi}{6}$ **b.** $-\dfrac{\pi}{12}$

22. a. $\dfrac{2\pi}{3}$ **b.** $-\dfrac{3\pi}{4}$

23. a. 3π **b.** 1

24. a. 3 **b.** $\dfrac{13\pi}{12}$

In Exercises 25 through 44, evaluate the given expression without using a calculator.

25. $\cos \dfrac{7\pi}{2}$ **26.** $\sin 5\pi$

27. $\sin\left(-\dfrac{7\pi}{2}\right)$ **28.** $\cos\left(-\dfrac{5\pi}{2}\right)$

29. $\cot \dfrac{5\pi}{2}$ **30.** $\sec 3\pi$

31. $\csc\left(-\dfrac{7\pi}{2}\right)$ **32.** $\tan(-\pi)$

33. $\cos\left(-\dfrac{2\pi}{3}\right)$ **34.** $\sin\left(-\dfrac{\pi}{6}\right)$

35. $\sin\left(-\dfrac{7\pi}{6}\right)$ **36.** $\cos \dfrac{7\pi}{3}$

37. $\cot \dfrac{\pi}{3}$ **38.** $\sec \dfrac{\pi}{3}$

39. $\tan\left(-\dfrac{\pi}{4}\right)$ **40.** $\cot\left(-\dfrac{5\pi}{3}\right)$

41. $\csc \dfrac{2\pi}{3}$ **42.** $\tan \dfrac{\pi}{6}$

43. $\sec \dfrac{5\pi}{4}$ **44.** $\sin\left(-\dfrac{3\pi}{4}\right)$

45. Use a right triangle to find $\cos \dfrac{\pi}{3}$ and $\sin \dfrac{\pi}{3}$.

46. Use a right triangle to find $\cos \dfrac{\pi}{4}$ and $\sin \dfrac{\pi}{4}$.

47. Complete the following table:

θ	$\dfrac{7\pi}{6}$	$\dfrac{5\pi}{4}$	$\dfrac{4\pi}{3}$	$\dfrac{3\pi}{2}$	$\dfrac{5\pi}{3}$	$\dfrac{7\pi}{4}$	$\dfrac{11\pi}{6}$	2π
$\sin \theta$								
$\cos \theta$								

48. Find $\tan \theta$ if $\sec \theta = \dfrac{5}{4}$.

49. Find $\tan \theta$ if $\csc \theta = \dfrac{5}{3}$.

50. Find $\cos \theta$ if $\tan \theta = \dfrac{3}{4}$.

51. Find $\tan \theta$ if $\cos \theta = \dfrac{3}{5}$.

52. Find $\sin \theta$ if $\tan \theta = \dfrac{2}{3}$.

53. Find $\sec \theta$ if $\cot \theta = \dfrac{4}{3}$.

In Exercises 54 through 57, determine the period p, the amplitude b, the horizontal shift d, and the vertical shift a of the given trigonometric function f(t). Then sketch the graph of f(t).

54. $f(t) = -1 + 2\cos\left[\dfrac{\pi}{3}(t - 3)\right]$

55. $f(t) = 2 + 1.5\sin\left[\dfrac{\pi}{2}(t - 1)\right]$

56. $f(t) = 5 - 3\sin\left[\dfrac{\pi}{4}(t + 1)\right]$

57. $f(t) = 3 - \cos\left[2\left(t - \dfrac{\pi}{2}\right)\right]$

In Exercises 58 through 61, find all the values of θ in the specified interval that satisfy the given equation.

58. $2\cos^2\theta - 2\sin\theta\cos\theta = 0; \; 0 \le \theta \le \pi$

59. $3\cos^2\theta - \sin^2\theta = 2; \; 0 \le \theta \le \pi$

60. $2\cos^2\theta + \sin\theta = 1; \; 0 \le \theta \le \pi$

61. $\cos^2\theta - \sin^2\theta + \cos\theta = 0; \; 0 \le \theta \le \pi$

BUSINESS AND ECONOMICS APPLIED PROBLEMS

62. DEMAND When q hundred units of a commodity are produced, they can all be sold at a price of p dollars per unit for $0 \le q \le 6$, where

$$p(q) = 3 + 2\cos\left(\frac{\pi q}{12}\right)$$

   a. Sketch the graph of the demand function $p(q)$.
   b. What is the net change in revenue as the level of production increases from 200 to 300 units?

63. SUPPLY When a commodity is priced at p dollars per unit, producers will supply q thousand units for $0 \le p \le 6$, where

$$q(p) = 5 + 2\sin\left(\frac{\pi p}{6}\right)$$

   a. Sketch the graph of the supply function $q(p)$.
   b. What is the smallest number of units supplied? The largest?
   c. For what price will 6,000 units be supplied?

64. PROFIT Harish's company determines that the profit P (in hundreds of dollars) obtained from sales during week t (for $1 \le t \le 52$) of a particular year may be modeled by the trigonometric function

$$P(t) = 75 + 55\cos\left[\frac{\pi(t - 15)}{26}\right]$$

   a. What is Harish's largest weekly profit? When does it occur?
   b. What is his smallest weekly profit? When does it occur?
   c. Sketch the graph of $P(t)$.

65. SALES Suppose that the number of pairs of skis sold by a sporting goods store during week t after the beginning of a particular year is modeled by the function

$$S(t) = 125 - 120\cos\left[\frac{\pi(t - 18)}{26}\right]$$

for $1 \le t \le 12$.
   a. During what week are the maximum number of skis sold? What is the maximum number?
   b. Sketch the graph of $S(t)$.

66. COST OF ENERGY Margaret's monthly gas and electrical bills vary periodically throughout a typical year. She finds that at month t for $1 \le t \le 12$, her gas bill $G(t)$ and electrical bill $E(t)$ satisfy

$$G(t) = 80 + 50\cos\left(\frac{\pi t}{6}\right)$$

$$E(t) = 150 - 70\sin\left[\frac{\pi(t - 12)}{8}\right]$$

Find her total utility bill in January ($t = 1$), April, August, and November.

67. STOCK PRICES The closing price of the stock of a small company is $S(t)$ dollars per share at the end of week t, where

$$S(t) = 12 + 3\cos\left(\frac{\pi t}{6}\right) + 2\sin\left[\frac{\pi(t - 7)}{10}\right]$$

for $1 \le t \le 40$.
   a. Why can the share prices never sink lower than \$7 per share? What is the upper bound on share prices?
   b. Does the share price ever actually attain the lowest or highest levels found in part (a)? If so, when?

LIFE AND SOCIAL SCIENCE APPLIED PROBLEMS

68. ENERGY CONSUMPTION Let $H(t)$ be the number of gallons of heating oil used to heat a particular house in Buffalo, New York, during month t after February 2008. Suppose that $H(t)$ is modeled by the function.

$$H(t) = 200 + 165\cos\left(\frac{\pi t}{6}\right)$$

   a. During what month is oil consumption the largest? What is this maximum amount?
   b. During what month is oil consumption the smallest? What is this minimum amount?

69. HUMAN RESPIRATION A typical adult male inhales and exhales approximately every 3.5 seconds. Let $B(t)$ be the volume (in liters) of the air in the lungs of an adult male subject t seconds after exhaling. Suppose $B(t)$ can be modeled by the function

$$B(t) = 0.5 - 0.4\cos\left(\frac{\pi t}{1.75}\right)$$

   a. Find the maximum volume of air in the lungs of the subject and the times when his lungs contain the maximum volume of air.

b. Answer the questions in part (a) for the minimum volume of air in the subject's lungs.

c. Sketch the graph of $B(t)$.

70. **MEDICINE** A device monitoring a patient's blood pressure (measured in millimeters of mercury) indicates that the *systolic* (maximum) pressure is 130, while the *diastolic* (minimum) pressure is 82. The same patient has a pulse rate of 79 heartbeats per minute. Suppose that t seconds after a systolic reading, the blood pressure is $P(t)$, where P is modeled by a function of the form

$$P(t) = A + B \cos kt$$

where A, B, and k are positive constants.

a. Use the given information to find A, B, and k.

b. What is the patient's blood pressure after 1 minute?

71. **BIORHYTHMS** Jake believes that his emotional well-being varies periodically with time t in such a way that the pattern is repeated every 28 days. That is, assuming his emotional state was neutral at birth, his emotional level t days later will be

$$E(t) = E_0 \sin\left(\frac{2\pi t}{28}\right)$$

where E_0 is his peak emotional level.

a. What will Jake's emotional level be on his 21st birthday? (Ignore leap years.)

b. Jake also believes that his state of physical well-being follows a similar periodic pattern, only this time the pattern repeats itself every 23 days. If P_0 is Jake's peak level of physical well-being, find a formula for his physical state at time t and determine the level on his 21st birthday.

c. How many days after Jake's 21st birthday should he expect his emotional and physical cycles to *both* be at peak levels?

72. **PREDATOR-PREY** A **predator-prey model** in ecology involves the study of interactions between two species occupying the same environment. One species (the predator) feeds on the other (the prey), which feeds on vegetation or some other food source obtained from the environment. The populations of the two species often behave periodically in relation to one another over time.

Suppose a fox population (predator) $F(t)$ and a rabbit population $R(t)$ at time t (years) are modeled by

$$F(t) = A + B \sin\left(\frac{\pi t}{24}\right)$$

$$R(t) = C + D \cos\left(\frac{\pi t}{24}\right)$$

where A, B, C, and D are constants. Further suppose that the largest and smallest fox populations are 54 and 16, respectively, and that the corresponding largest and smallest rabbit populations are 581 and 261.

a. Use the given information to find A, B, C, and D. How long is the population cycle (period) for each species?

b. What is the rabbit population when the fox population is largest? What is the rabbit population when the fox population is smallest?

c. What is the fox population when the rabbit population is largest? What is the fox population when the rabbit population is smallest?

d. Sketch the graphs of the two population functions on the same set of axes.

73. **PREDATOR-PREY** Answer the questions in Exercise 72 for the case where

$$F(t) = A + B \sin\left(\frac{\pi t}{12}\right)$$

$$R(t) = C + D \cos\left(\frac{\pi t}{12}\right)$$

and the largest and smallest fox populations are 70 and 22, while the largest and smallest rabbit populations are 635 and 227.

MISCELLANEOUS PROBLEMS

74. **REFRACTION OF LIGHT BY WATER** Lloyd is swimming at a depth d under water, but because light is refracted by the water, his *apparent* depth s is less than d. In physics,* it is shown that if Lloyd is viewed from an angle of incidence θ, then

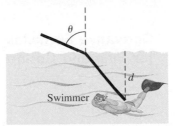

EXERCISE 74

$$s = \frac{3d \cos \theta}{\sqrt{7 + 9 \cos^2 \theta}}$$

*R. A. Serway, *Physics*, 3rd ed., Philadelphia: Saunders, 1992, p. 1007.

a. If $d = 3$ meters and $\theta = 31°$, what is the apparent depth of the man?

 b. If the actual depth is $d = 1.9$ meters, what angle of incidence yields an apparent depth of $s = 1.12$ meters?

ADDITION FORMULAS FOR SINE AND COSINE

For any angles A and B,

$$\sin(A + B) = \sin A \cos B + \cos A \sin B$$

and

$$\cos(A + B) = \cos A \cos B - \sin A \sin B$$

These formulas are used in Exercises 75 through 78.

---

75. Use the addition formula for sine to derive the identity

$$\sin\left(\frac{\pi}{2} - \theta\right) = \cos\theta$$

76. Use the addition formula for cosine to derive the identity

$$\cos\left(\frac{\pi}{2} - \theta\right) = \sin\theta$$

77. Show that $f(t + p) = f(t)$ for all t, where

a. $f(t) = a + b \sin\left[\dfrac{2\pi}{p}(t - d)\right]$

b. $f(t) = a + b \cos\left[\dfrac{2\pi}{p}(t - d)\right]$

78. Starting with the addition formulas for sine and cosine, derive the addition formula

$$\tan(A + B) = \frac{\tan A + \tan B}{1 - \tan A \tan B}$$

for the tangent.

79. Give geometric arguments involving the coordinates of points on a circle of radius 1 to show why the identities in Exercises 75 and 76 are valid.

80. Starting with the identity $\sin^2\theta + \cos^2\theta = 1$, derive the identity $1 + \tan^2\theta = \sec^2\theta$.

SECTION 8.2 Trigonometric Applications Involving Differentiation

Learning Objectives

1. Derive and use differentiation formulas for trigonometric functions.
2. Study periodic rate and optimization problems using derivatives of trigonometric functions.

In this section, we obtain derivatives for the trigonometric functions and then illustrate the use of these derivative formulas by finding rates of change and solving optimization problems.

Derivative Formulas for Sine and Cosine

Here are the formulas for the derivatives of sine and cosine. Since these formulas are often used in conjunction with the chain rule, we also give the generalized version of each formula. The proof of the sine formula will be given at the end of this section, after we have had a chance to see how the formulas are used in practice.

Derivatives for Sine and Cosine ■ If t is measured in radians, then

$$\frac{d}{dt}(\sin t) = \cos t \qquad \text{and} \qquad \frac{d}{dt}(\cos t) = -\sin t$$

and according to the chain rule, if $u = u(t)$ is a differentiable function

$$\frac{d}{dt}(\sin u) = \cos u \, \frac{du}{dt} \qquad \text{and} \qquad \frac{d}{dt}(\cos u) = -\sin u \, \frac{du}{dt}$$

EXPLORE!

Refer to Example 8.2.1. Set Y1 = sin(3X + 1) in the equation editor of your graphing utility, and graph using the window $[-\pi, \pi]\frac{\pi}{4}$ by $[-2, 2]0.25$. Use the numeric derivative feature of your utility to find $f'(x)$ at $x = \frac{\pi}{4}$. Why is this derivative negative? Also find and display the equation of the tangent line to $f(x)$ at $x = \frac{\pi}{4}$.

Just-In-Time REVIEW

The nth power of $\cos x$ is written as $\cos^n x$; that is,

$$(\cos x)^n = \cos^n x$$

Similarly,

$$(\sin x)^n = \sin^n x$$

and

$$(\tan x)^n = \tan^n x$$

Examples 8.2.1 and 8.2.2 illustrate the use of these formulas.

EXAMPLE 8.2.1 Differentiating a Sine Function

Differentiate the function $f(t) = \sin(3t + 1)$.

Solution

Using the chain rule for the sine function with $u(t) = 3t + 1$, you get

$$f'(t) = \cos(3t + 1)\frac{d}{dt}(3t + 1) = \cos(3t + 1)(3)$$

EXAMPLE 8.2.2 Differentiating a Cosine Function

Differentiate the function $f(t) = \cos^3 t$.

Solution

Since

$$f(t) = \cos^3 t = (\cos t)^3$$

you use the chain rule for powers and the formula for the derivative of the cosine to get

$$f'(t) = 3(\cos t)^2 \frac{d}{dt}(\cos t) = 3(\cos t)^2(-\sin t)$$
$$= -3\cos^2 t \sin t$$

The production cost of certain commodities varies with seasonal availability and may exhibit periodic behavior. In Example 8.2.3, we find the marginal cost of such a commodity.

EXAMPLE 8.2.3 Computing Marginal Cost for a Periodic Cost Function

The cost of producing x units of a particular commodity is $C(x)$ hundred dollars, where

$$C(x) = 400 + 15\sin\left[\frac{\pi}{24}(x - 11)\right] \quad \text{for } 0 \le x \le 30$$

Find the marginal cost of producing 3 units.

Solution

The marginal cost is given by the derivative of $C(x)$ evaluated at $x = 3$. Using the chain rule, we find that

$$C'(x) = 15\cos\left[\frac{\pi}{24}(x - 11)\right]\left[\frac{\pi}{24}\right]$$
$$= \frac{5\pi}{8}\cos\left[\frac{\pi}{24}(x - 11)\right]$$

So the marginal cost when $x = 3$ is

$$C'(3) = \frac{5\pi}{8} \cos\left[\frac{\pi}{24}(3-11)\right] = \frac{5\pi}{8}\cos\left(-\frac{\pi}{3}\right) = \frac{5\pi}{16} \approx 0.9817$$

Since $C(x)$ is given in hundreds of dollars, the marginal cost is $98.17 per unit.

In Example 8.2.4, we study the motion of an object moving periodically along a straight line.

EXAMPLE 8.2.4 Studying Periodic Movement Along a Line

An object moves along a straight line in such a way that after t seconds, it is located at position x on the line where x is in meters and

$$x(t) = \sin(\pi - 3t) + \cos(\pi t)$$

Where is the object and what are its velocity and acceleration at time $t = 0$?

Solution

At time $t = 0$, the object is located at

$$x(0) = \sin[\pi - 3(0)] + \cos[\pi(0)] = 0 + 1 = 1$$

that is, at the point 1 meter to the right of the 0 point on the line.
The velocity of the object is

$$v(t) = x'(t) = \cos(\pi - 3t)(-3) + [-\sin(\pi t)(\pi)]$$
$$= -3\cos(\pi - 3t) - \pi\sin(\pi t)$$

and its acceleration is

$$a(t) = v'(t) = -3[-\sin(\pi - 3t)(-3)] - \pi\cos(\pi t)(\pi)$$
$$= -9\sin(\pi - 3t) - \pi^2\cos(\pi t)$$

So at time $t = 0$, the velocity is

$$v(0) = -3\cos[\pi - 3(0)] - \pi\sin[\pi(0)] = -3(-1) - \pi(0)$$
$$= 3 \text{ meters per second}$$

and the acceleration is

$$a(0) = -9\sin[\pi - 3(0)] - \pi^2\cos[\pi(0)] = -9(0) - \pi^2(1)$$
$$= -\pi^2 \text{ meters per second per second}$$

Using calculus to analyze applied optimization problems involving periodic functions follows the basic procedure we used successfully in Chapter 3:

1. Construct a function representing the quantity to be optimized in terms of a convenient variable.
2. Identify an interval on which the function has a practical interpretation.
3. Differentiate the function and determine all critical numbers (where the derivative is equal to zero or does not exist).
4. Compare the values of the function at the critical numbers in the interval and at the endpoints (or use the second derivative test) to verify that the desired absolute extremum has been found.

This procedure is illustrated for various periodic functions in Examples 8.2.5 through 8.2.7.

EXAMPLE 8.2.5 Studying Worker Efficiency

Thelma, a factory manager, determines that t hours after the second shift begins at noon, the average worker will have produced $W(t)$ units, where

$$W(t) = 4(6t + \sin t - \cos t + 1) \quad \text{for } 0 \le t \le 4$$

a. At what rate is the average worker producing units at 3 P.M.?
b. When should Thelma expect the average worker to be producing units most efficiently?

Solution

a. The rate of production at time t is given by the derivative

$$W'(t) = \frac{d}{dt}[4(6t + \sin t - \cos t + 1)]$$

$$= 4[6 + \cos t - (-\sin t) + 0] = 4[6 + \cos t + \sin t]$$

At 3 P.M., when $t = 3$, the rate is

$$W'(3) = 4[6 + \cos 3 + \sin 3] \approx 20.6$$

so the worker is producing about 21 units per hour.

b. The average worker is producing units most efficiently when the rate of production $R(t) = W'(t)$ is maximized. Differentiating $R(t)$ and setting it equal to 0, we find that

$$R'(t) = 4[0 + (-\sin t) + \cos t] = 4[-\sin t + \cos t]$$
$$= 0$$

when $\sin t = \cos t$, that is, when $t = \dfrac{\pi}{4}$ or $\dfrac{5\pi}{4}$ on the time interval $0 \le t \le 4$. Since $R(t)$ is continuous on the closed, bounded interval $0 \le t \le 4$, the extreme value property given in Section 3.4 tells us that its largest value must occur either at $\dfrac{\pi}{4}$ or $\dfrac{5\pi}{4}$ or at one of the endpoints 0 or 4. We find that

$$R(0) = 28 \qquad R\left(\frac{\pi}{4}\right) = 29.66 \qquad R\left(\frac{5\pi}{4}\right) = 18.34 \qquad R(4) = 18.36$$

so the greatest rate occurs when $t = \dfrac{\pi}{4} \approx 0.79$, that is, at about 47 minutes after noon, when the average worker is producing approximately 30 units per hour.

EXAMPLE 8.2.6 Maximizing the Area of a Triangular-Shaped Garden

A landscape architect is designing a garden that is to be triangular in shape with two equal sides, each 4 yards in length. What should be the angle between the two equal sides to make the area of the garden as large as possible?

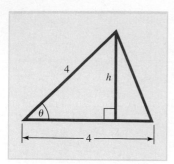

FIGURE 8.18 Triangle for Example 8.2.6.

Solution

The triangle is shown in Figure 8.18. In general, the area of a triangle is given by the formula

$$\text{Area} = \frac{1}{2} \, (\text{base})(\text{height})$$

In this case, the base is $b = 4$, and since $\sin \theta = \frac{h}{4}$, the height is $h = 4 \sin \theta$. Hence, the area of the triangle is given by the function

$$A(\theta) = \frac{1}{2} \, (4)(4 \sin \theta) = 8 \sin \theta$$

Since only values of θ between 0 and π radians are meaningful in the context of this problem, the goal is to find the absolute maximum of the function $A(\theta)$ on the interval $0 \le \theta \le \pi$.

The derivative of $A(\theta)$ is

$$A'(\theta) = 8 \cos \theta$$

which is zero on the interval $0 \le \theta \le \pi$ only when $\theta = \frac{\pi}{2}$. Comparing

$$A(0) = 8 \sin 0 = 0$$

$$A\left(\frac{\pi}{2}\right) = 8 \sin \frac{\pi}{2} = 8(1) = 8$$

$$A(\pi) = 8 \sin \pi = 0$$

we conclude that the area is maximized when the angle θ measures $\frac{\pi}{2}$ radians (or 90°), that is, when the garden has the shape of a right triangle.

EXAMPLE 8.2.7 Minimizing a Supply Route

Two off-shore oil wells A and B are, respectively, a miles and b miles out to sea. A motorboat traveling at a constant speed s is to shuttle workers from well A to a depot on the shore and then on to well B. To make this operation cost-efficient, the location of the depot is chosen to minimize the total travel time. Show that at this optimum location P, the angle α between the motorboat's path of arrival and the shoreline will be equal to the angle β between the shoreline and the path of departure.

Solution

Even though the goal is to prove that two angles are equal, it turns out that the easiest way to solve this problem is to let the variable x represent a convenient distance and to introduce trigonometry only at the end.

Begin with a sketch as in Figure 8.19 and define the variable x and the constant distance d as indicated. Then, by the Pythagorean theorem,

$$\text{Distance from first well to } P = \sqrt{a^2 + x^2}$$

and

$$\text{Distance from second well to } P = \sqrt{b^2 + (d - x)^2}$$

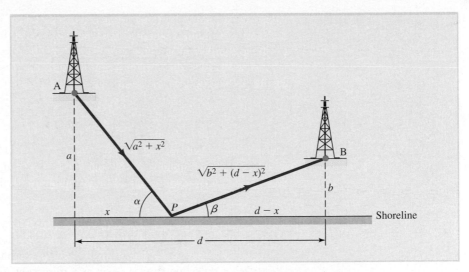

FIGURE 8.19 Path of motorboat from oil well to shore to oil well.

Since (for constant speeds),

$$\text{Time} = \frac{\text{distance}}{\text{speed}}$$

the total travel time is given by the function

$$T(x) = \frac{\sqrt{a^2 + x^2}}{s} + \frac{\sqrt{b^2 + (d-x)^2}}{s}$$

where $0 \le x \le d$. The derivative of $T(x)$ is

$$T'(x) = \frac{x}{s\sqrt{a^2 + x^2}} - \frac{d-x}{s\sqrt{b^2 + (d-x)^2}}$$

and by setting $T'(x) = 0$ and multiplying by s, we get

$$\frac{x}{\sqrt{a^2 + x^2}} = \frac{d-x}{\sqrt{b^2 + (d-x)^2}}$$

Now look again at the right triangles in Figure 8.19 and observe that

$$\frac{x}{\sqrt{a^2 + x^2}} = \cos \alpha \quad \text{and} \quad \frac{d-x}{\sqrt{b^2 + (d-x)^2}} = \cos \beta$$

Hence, $T'(x) = 0$ when

$$\cos \alpha = \cos \beta$$

In the context of this problem, both α and β are between 0 and $\dfrac{\pi}{2}$ radians, and for angles in this range, equality of the cosines indicates equality of the angles themselves. (See, for example, the graph of the cosine function in Figure 8.10.) Hence, $T'(x) = 0$ when

$$\alpha = \beta$$

Moreover, it should be clear from Figure 8.19 that no matter what a, b, and d may be, there is a (unique) point P for which $\alpha = \beta$. It follows that the function $T(x)$ has a critical number in the interval $0 \le x \le d$ where $\alpha = \beta$.

To verify that this critical point is really the absolute minimum, we can use the second derivative test. A routine calculation gives

$$T''(x) = \frac{a^2}{s(a^2 + x^2)^{3/2}} + \frac{b^2}{s[b^2 + (d-x)^2]^{3/2}}$$

(For practice, check this calculation.) Since $T''(x) > 0$ for all values of x and since there is only one critical number in the interval $0 \le x \le d$, it follows that this critical point does indeed correspond to the absolute minimum of the total time T on this interval.

NOTE According to **Fermat's principle** in optics, light traveling from one point to another takes the path that requires the least amount of time. Suppose, as illustrated in Figure 8.20, that a ray of light is transmitted from a source at a point A, strikes a reflecting surface (such as a mirror) at P, and is subsequently received by an observer at point B. Since by Fermat's principle, the path from A to P to B minimizes time, it follows from the calculation in Example 8.2.7 that the angles α and β are equal. This, in turn, gives the **law of reflection,** which states that *the angle of incidence must equal the angle of reflection* ($\theta_1 = \theta_2$ in Figure 8.20). ∎

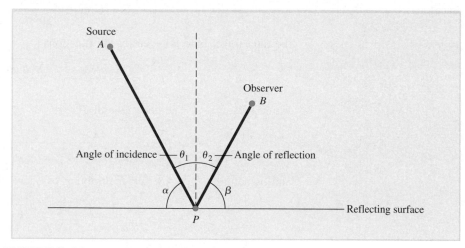

FIGURE 8.20   The reflection of light: $\theta_1 = \theta_2$.

Derivatives of Other Trigonometric Functions

The differentiation formulas for sine and cosine can be used to obtain formulas for the derivatives of other trigonometric functions. For instance, the differentiation formula for tangent may be obtained by using the quotient rule as follows:

$$\frac{d}{dt}(\tan t) = \frac{d}{dt}\left(\frac{\sin t}{\cos t}\right) \qquad \text{use the quotient rule}$$
$$= \frac{\cos t\,(\cos t) - \sin t\,(-\sin t)}{\cos^2 t} \qquad \sin^2 t + \cos^2 t = 1$$
$$= \frac{\cos^2 t + \sin^2 t}{\cos^2 t} = \frac{1}{\cos^2 t} \qquad \sec t = \frac{1}{\cos t}$$
$$= \sec^2 t$$

Similarly, it can be shown that

$$\frac{d}{dt} (\sec t) = \sec t \tan t$$

(see Exercise 71).

> **Derivatives for Tangent and Secant** ■ If t is measured in radians, then
>
> $$\frac{d}{dt} (\tan t) = \sec^2 t \quad \text{and} \quad \frac{d}{dt} (\sec t) = \sec t \tan t$$
>
> and if $u = u(t)$ is a differentiable function of t, then
>
> $$\frac{d}{dt} (\tan u) = \sec^2 u \, \frac{du}{dt} \quad \text{and} \quad \frac{d}{dt} (\sec u) = \sec u \tan u \, \frac{du}{dt}$$

You are asked to obtain derivatives for the remaining two trigonometric functions, cotangent and cosecant, in Exercise 71.

EXAMPLE 8.2.8 Differentiating a Function Involving Tangent and Secant

Differentiate the function $f(t) = \tan (t^2) + \sec (1 - 3t)$.

Solution

Using the chain rule, we get

$$
\begin{aligned}
f'(t) &= \frac{d}{dt} [\tan (t^2)] + \frac{d}{dt} [\sec (1 - 3t)] \\
&= [\sec^2 (t^2)] \frac{d}{dt} (t^2) + [\sec (1 - 3t) \tan (1 - 3t)] \frac{d}{dt} (1 - 3t) \\
&= [\sec^2 (t^2)](2t) + [\sec (1 - 3t) \tan (1 - 3t)](-3) \\
&= 2t \sec^2 (t^2) - 3 \sec (1 - 3t) \tan (1 - 3t)
\end{aligned}
$$

In Example 8.2.9, the derivative of the tangent is used in a related rates problem.

EXAMPLE 8.2.9 Solving a Periodic Related Rates Problem

Alonzo watches a plane approach at a speed of 400 miles per hour and at an altitude of 4 miles. At what rate is the angle of elevation of his line of sight changing with respect to time when the horizontal distance between the plane and Alonzo is 3 miles?

Solution

Let x denote the horizontal distance between the plane and Alonzo, let t denote time (in hours), and draw a diagram representing the situation as in Figure 8.21.

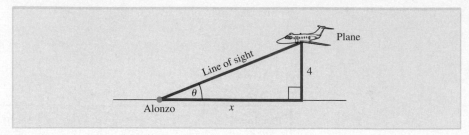

FIGURE 8.21 Observation of an approaching plane.

We know that $\dfrac{dx}{dt} = -400$ (the minus sign indicates that the distance x is decreasing), and the goal is to find $\dfrac{d\theta}{dt}$ when $x = 3$. From the right triangle in Figure 8.21 we see that

$$\tan \theta = \frac{4}{x}$$

Use the chain rule (remember x is a function of t) to differentiate both sides of this equation with respect to t to get

$$\sec^2 \theta \, \frac{d\theta}{dt} = -\frac{4}{x^2}\frac{dx}{dt} \qquad \text{divide both sides by } \sec^2 \theta: \\ \frac{1}{\sec^2 \theta} = \cos^2 \theta$$

or

$$\frac{d\theta}{dt} = -\frac{4}{x^2}\cos^2 \theta \, \frac{dx}{dt}$$

The values of x and $\dfrac{dx}{dt}$ are given. We get the value of $\cos^2 \theta$ when $x = 3$ from the right triangle in Figure 8.22:

$$\cos^2 \theta = \left(\frac{3}{5}\right)^2 = \frac{9}{25}$$

Substituting $x = 3, \dfrac{dx}{dt} = -400$, and $\cos^2 \theta = \dfrac{9}{25}$ into the formula for $\dfrac{d\theta}{dt}$, we get

$$\frac{d\theta}{dt} = -\frac{4}{9}\left(\frac{9}{25}\right)(-400) = 64 \text{ radians per hour}$$

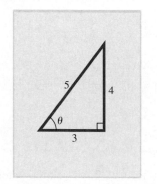

FIGURE 8.22 Triangle for the computation of $\cos \theta$ when $x = 3$.

The human vascular system operates in such a way that the circulation of blood from the heart, through the organs, and back to the heart is accomplished with as little expenditure of energy as possible. Thus, it is reasonable to expect that when an artery branches, the angle between the "parent" and "daughter" arteries should minimize the total resistance to the flow of blood. In Example 8.2.10, we use calculus to find this optimal branching angle.

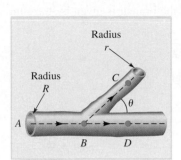

FIGURE 8.23 Vascular branching.

EXAMPLE 8.2.10   Vascular Branching: Minimizing Resistance to Blood Flow

Figure 8.23 shows a small artery of radius r branching at an angle $\theta \left(0 < \theta < \dfrac{\pi}{2} \right)$ from a larger artery of radius R ($R > r$).* Blood flows in the direction of the arrows from point A to the branch at B and then to points C and D. Show that the resistance of the blood flow is minimized when θ satisfies

$$\cos \theta = \frac{r^4}{R^4}$$

Solution

Figure 8.24 is a geometric version of Figure 8.23. For simplicity, we assume that C and D are located so that CD is perpendicular to the main line through A and B, with C located h units above D. We wish to find the value of the branching angle θ that minimizes the total resistance to the flow of blood as it moves from A to B and then to point C.

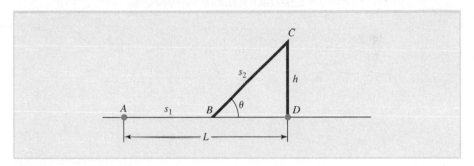

FIGURE 8.24 Minimizing resistance to the flow of blood.

Before we can solve this problem, we must have a way of measuring the resistance to blood flow, and this is provided by the following result, which was discovered empirically by the French physiologist and physician Jean Louis Poiseuille (1799–1869).

> **Poiseuille's Resistance Law** ■ The resistance to flow of blood in an artery is directly proportional to the artery's length and inversely proportional to the fourth power of its radius.

According to Poiseuille's law, the resistance to the flow of blood from point A to point B in Figure 8.24 is

$$F_1 = \frac{ks_1}{R^4}$$

and the resistance from point B to point C is

$$F_2 = \frac{ks_2}{r^4}$$

---

*This application is adapted from an example in the text by E. Batschelet, *Introduction to Mathematics for Life Scientists*, 2nd ed., New York: Springer-Verlag, 1975, pp. 278–280. As the original source of the material, the author cites the text by R. Rosen, *Optimality Principles in Biology*, London, England: Butterworths, 1967.

Just-In-Time **REVIEW**

Recall that

$$\frac{1}{\sin q} = \csc q$$

and

$$\frac{1}{\tan q} = \cot q$$

and note that $\csc q$ is never 0.

where k is a constant that depends on the viscosity of blood, and s_1 and s_2 are the lengths of the portions of artery from A to B and from B to C, respectively. Thus, the total resistance to the flow of blood up to and through the branch is given by the sum

$$F = F_1 + F_2 = \frac{ks_1}{R^4} + \frac{ks_2}{r^4} = k\left(\frac{s_1}{R^4} + \frac{s_2}{r^4}\right)$$

Next, we write F as a function of the branching angle θ. To do this, in Figure 8.24, note that

$$\sin \theta = \frac{h}{s_2} \qquad \text{and} \qquad \tan \theta = \frac{h}{L - s_1}$$

where L is the length of the main artery from A to D. Solving these equations for s_1 and s_2, we get

$$s_2 = \frac{h}{\sin \theta} = h \csc \theta \qquad \text{and} \qquad s_1 = L - \frac{h}{\tan \theta} = L - h \cot \theta$$

so that F can be expressed as

$$F = k\left(\frac{s_1}{R^4} + \frac{s_2}{r^4}\right) = k\left(\frac{L - h \cot \theta}{R^4} + \frac{h \csc \theta}{r^4}\right)$$

In Exercise 71, you are asked to show that

$$\frac{d}{dx}(\cot u) = -\csc^2 u \, \frac{du}{dx} \qquad \text{and} \qquad \frac{d}{dx}(\csc u) = -\cot u \csc u \, \frac{du}{dx}$$

and by using these formulas to differentiate $F(\theta)$, you get

$$F'(\theta) = k\left[\frac{-h(-\csc^2 \theta)}{R^4} + \frac{h(-\csc \theta \cot \theta)}{r^4}\right]$$

$$= kh \csc \theta \left(\frac{\csc \theta}{R^4} - \frac{\cot \theta}{r^4}\right)$$

Solving $F'(\theta) = 0$ to obtain all critical numbers for F, you get

$$\csc \theta = 0 \qquad \text{or} \qquad \frac{\csc \theta}{R^4} - \frac{\cot \theta}{r^4} = 0$$

Because $\csc \theta$ is never 0, it follows that $F'(\theta) = 0$ only when

$$\frac{\csc \theta}{R^4} - \frac{\cot \theta}{r^4} = 0$$

$$\frac{\csc \theta}{R^4} = \frac{\cot \theta}{r^4}$$

$$\frac{1}{R^4 \sin \theta} = \frac{\cos \theta}{r^4 \sin \theta}$$

$$\frac{r^4}{R^4} = \cos \theta$$

Finally, if θ_0 is the acute angle that satisfies this equation, it can be shown that the resistance F has an absolute minimum when $\theta = \theta_0$ (see Exercise 45). Thus, θ_0 is the optimal angle of vascular branching.

The Proof That

$$\frac{d}{dt}\sin t = \cos t$$

Since the function $f(t) = \sin t$ is unrelated to any of the functions whose derivatives were obtained in previous chapters, it will be necessary to return to the definition of the derivative (from Section 2.1) to find the derivative in this case. In terms of t, the definition states that

$$f'(t) = \lim_{h \to 0} \frac{f(t + h) - f(t)}{h}$$

In performing the required calculations when $f(t) = \sin t$, we need to use the addition formula

$$\sin (A + B) = \sin A \cos B + \cos A \sin B$$

(see Exercises 75 to 78, Section 8.1) as well as the following two special limit formulas:

> **Two Important Limits Involving Sine and Cosine** ■ If t is measured in radians, then
>
> $$\lim_{t \to 0} \frac{\sin t}{t} = 1$$
>
> and
>
> $$\lim_{t \to 0} \frac{\cos t - 1}{t} = 0$$

EXPLORE!

Graph $f(x) = \dfrac{\sin x}{x}$ using the

viewing window $[-\pi, \pi]\dfrac{\pi}{4}$ by

$[-1.5, 1.5]0.5$. Use **ZOOM** and

TRACE to investigate $\lim\limits_{x \to 0} \dfrac{\sin x}{x}$.

Now we are ready to establish the derivative formula for $\sin t$:

$$\frac{d}{dt}(\sin t) = \lim_{h \to 0} \frac{\sin (t + h) - \sin t}{h} \qquad \text{addition formula for sine}$$

$$= \lim_{h \to 0} \frac{[\sin t \cos h + \cos t \sin h] - \sin t}{h} \qquad \text{factor sin } t \text{ from the first and third terms}$$

$$= \lim_{h \to 0} \frac{(\sin t)[\cos h - 1] + (\cos t) \sin h}{h} \qquad \text{properties of limits}$$

$$= (\sin t)\lim_{h \to 0} \frac{\cos h - 1}{h} + (\cos t)\lim_{h \to 0} \frac{\sin h}{h} \qquad \begin{array}{l}\text{these limits affect}\\ \text{only } h, \text{ not } t\end{array}$$

$$= (\sin t)(0) + (\cos t)(1) = \cos t$$

That is,

$$\frac{d}{dt}(\sin t) = \cos t$$

The derivative formula for the cosine can be derived in a similar manner (see Exercise 69).

EXERCISES ■ 8.2

In Exercises 1 through 20, differentiate the given function.

1. $f(t) = \sin 3t$

2. $f(t) = \cos 2t$

3. $f(t) = \sin (1 - 2t)$

4. $f(t) = \sin t^2$

5. $f(t) = \cos (t^3 + 1)$

6. $f(t) = \sin^2 t$

7. $f(t) = \cos^2\left(\dfrac{\pi}{2} - t\right)$

8. $f(t) = \sin (2t + 1)^2$

9. $f(x) = \cos (1 + 3x)^2$

10. $f(x) = e^{-x} \sin x$

11. $f(u) = e^{-u/2} \cos 2\pi u$

12. $f(u) = \dfrac{\cos u}{1 - \cos u}$

13. $f(t) = \dfrac{\sin t}{1 + \sin t}$

14. $f(t) = \tan (5t + 2)$

15. $f(t) = \tan (1 - t^3)$

16. $f(t) = \tan^2 t$

17. $f(t) = \sec \left(\dfrac{\pi}{2} - 2\pi t \right)$

18. $f(t) = \sec (\pi - 4t)^2$

19. $f(t) = \ln \sin^2 t$

20. $f(t) = \ln \tan^2 t$

An object moving back and forth along a straight line is said to undergo **vibrational motion** *if its displacement from equilibrium at time t is given by an equation of the form*

$$x(t) = e^{-kt}(A \cos \omega t + B \sin \omega t)$$

where A, B, k, and ω are constants. In Exercises 21 through 26, answer the following questions about the given function x(t) with this form for the indicated time interval a ≤ x ≤ b:

  (a) Sketch the graph of x(t). You may wish to use the graphing utility of your calculator.

  (b) Find the values of t for which the largest and smallest values of x(t) occur for a ≤ t ≤ b.

21. $x(t) = \sin 2t + \cos 2t$ for $0 \le t \le \dfrac{\pi}{2}$

22. $x(t) = 2 \sin 2t + 3 \cos 2t$ for $0 \le t \le \dfrac{\pi}{2}$

23. $x(t) = 5 \sin 3t + 2 \cos t$ for $0 \le t \le \pi$

24. $x(t) = 2 \sin t - 3 \cos 2t$ for $0 \le t \le \pi$

25. $x(t) = e^{-t}(\sin t + \cos t)$ for $\dfrac{\pi}{2} \le t \le \dfrac{3\pi}{2}$

26. $x(t) = e^{-2t}(3 \sin 2t + 2 \cos 3t)$ for $0 \le t \le \pi$

27. Find the largest value of

$$f(t) = \cot t - \sqrt{2} \csc t \quad \text{for } 0 < t < \pi$$

Does $f(t)$ have a smallest value on this interval?

28. Find the smallest value of

$$f(t) = 27 \sec t + 8 \csc t \quad \text{for } 0 < t < \dfrac{\pi}{2}$$

Does $f(t)$ have a largest value on this interval?

29. SALES   The sales of a certain product satisfy

$$S(t) = 27.5 + 11.2 \sin \left(\dfrac{\pi t}{5} \right) + 2.3 \cos \left(\dfrac{\pi t}{3} \right)$$

where t is the time in months measured from January, and S is measured in thousands of units.

Find the rate of change of sales at midyear ($t = 6$). Are sales increasing or decreasing at that time?

30. PROFIT   When x hundred units of a certain commodity are produced, they can all be sold at a price of

$$p(x) = 3 - \sin (2x)$$

dollars per unit. It costs $2.50 to produce each unit.
  a. Express the profit $P(x)$ as a function of x.
  b. What is the profit or loss when 100 units are produced? 200 units?
  c. At what rate is the profit changing when 50 units are being produced? Is the profit increasing or decreasing at that level of production?

31. WORKER EFFICIENCY   The manager of a factory estimates that t hours after the morning shift begins at 8 A.M., the average worker will have produced

$$W(t) = 20(t - \sin t - \cos t + 1)$$

units, for $0 \le t \le 4$.
  a. How many units will an average worker produce by 10 A.M.?
  b. What is the average worker's production rate at noon? Is production increasing or decreasing at that time?
  c. At what time is an average worker producing units most efficiently? What is the most efficient rate?

32. FARM INCOME   A farmer determines that t weeks into a typical year, his weekly income will be $I(t)$ thousand dollars where

$$I(t) = 20 + 15 \cos \left[\dfrac{2\pi(t - 20)}{52} \right] + 8 \sin \left[\dfrac{\pi(t - 35)}{52} \right]$$

for $0 \le t \le 52$.
  a. What is the weekly income when $t = 26$ (halfway through the year)?
  b. When is the weekly income the greatest? What is the largest weekly income?
  c. When is the weekly income the least? What is the least weekly income?

33. INVENTORY MANAGEMENT   Stella, the owner of a tire store, determines that t months from now, $P(t)$ percent of her available storage space will be utilized, where

$$P(t) = 43 + 52 \cos^2 \left[\dfrac{\pi(t - 3)}{8} \right]$$

for $0 \le t \le 12$.

a. When does the largest percentage of utilization occur? The smallest?

b. Stella reorders when the percentage of storage utilization is increasing most rapidly. When does this occur? What percentage of storage space is in use at this time?

34. APARTMENT VACANCY   The manager of an apartment complex determines that t months from now, $V(t)$ percent of her apartments will be vacant, where

$$V(t) = 8 + 15 \sin^2\left(\frac{\pi t}{12}\right)$$

for $0 \le t \le 12$.

a. When does the largest percentage of vacancy occur? The smallest?

b. When is the percentage of vacancy increasing most rapidly? What is the percentage of vacancy at this time?

LIFE AND SOCIAL SCIENCE APPLIED PROBLEMS

35. MALE METABOLISM   Suppose the body temperature (in °F) of a particular man t hours after midnight (for $0 \le t \le 24$) is given by the function

$$T(t) = 98.1 + 0.6 \cos\left[\frac{\pi(t - 15)}{24}\right]$$

a. What are the maximum and minimum body temperatures and the times when they occur?

b. At what rate is the man's body temperature changing at noon?

c. When is the man's body temperature increasing most rapidly?

36. FEMALE METABOLISM   The body temperature of a woman is affected by two different cycles, a daily cycle and a monthly cycle. Suppose the body temperature $T(x)$ in °C of a woman x days after the start of a monthly cycle is modeled by the formula

$$T(x) = 37 + 0.3 \cos\left[\frac{2\pi(x - 15)}{28}\right] + 0.4 \cos\left[2\pi(x - 0.6)\right]$$

for $0 \le x \le 28$. At what rate is the body temperature changing when $x = 15$? When $x = 28$?

37. WEATHER   The average temperature t hours after midnight in a certain city is given by the formula

$$T(t) = 60 + 12 \sin\left(\frac{\pi t}{10} - \frac{5}{6}\right)$$

At what rate is the temperature changing at noon? Is it getting hotter or cooler at that time?

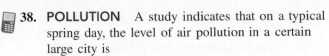

38. POLLUTION   A study indicates that on a typical spring day, the level of air pollution in a certain large city is

$$P(t) = 1.5 + 18t - 0.75t^2 + 0.92 \sin(2t + 1)$$

units, where t ($0 \le t \le 24$) is the number of hours after midnight.

a. Find the rate of change of the pollution level at each of the following times: $t = 0$, $t = 9$, $t = 12$, and $t = 24$.

b. Use your calculator to show that the pollution level is highest at about 12:36 P.M. What is the highest pollution level?

c. Show that the pollution rate is maximized when

$$\sin(2t + 1) = -0.4076$$

Then use your calculator to show that the pollution is increasing most rapidly at approximately 2:26 A.M.

39. HEALTH   Health officials in a community determine that the number of people in the community who are infected with influenza fluctuates periodically in such a way that t months after the beginning of the year, $P(t)$ percent of the total population have the disease, where

$$P(t) = 25 + 10 \cos^2\left[\frac{\pi(t - 3)}{14}\right]$$

for $0 \le t \le 12$.

a. What is the highest percentage of infected people? The lowest percentage?

b. When is the percentage of infected people increasing most rapidly?

40. MEDICINE   Human metabolism excretes phosphate periodically. Suppose the excretion rate is given by

$$\frac{dP}{dt} = -\frac{1}{3} + \frac{1}{6} \cos\left[\frac{\pi}{12}(t - 6)\right]$$

grams per hour at time t (hours) over a 24-hour period ($0 \le t \le 24$), where $P(t)$ is the amount of phosphate in the body at time t.

a. When does the maximum level of phosphate occur? The minimum level?

b. When is the phosphate level decreasing least rapidly?

41. POPULATION GROWTH   The number $P(t)$ of deer ticks per acre in a Maryland park can be modeled by the function

$$P(t) = 1{,}250e^{0.3t}\left\{1.47 + 1.38 \cos\left[\frac{2\pi}{365}(t - 145)\right]\right\}$$

where t is the number of years after the base year 2000.

a. Find the number of deer ticks per acre in the park in the years 2000, 2005, and 2010.

b. Find the rate of change of the number of deer ticks per acre in the park in the years 2000, 2005, and 2010.

c. Draw the graph of the number of deer ticks per acre in the park.

 d. Write a paragraph explaining what there is about the function $P(t)$ that makes it a good model for the number of deer ticks per acre.

42. **WEATHER**   Suppose the weather bureau determines that the average temperature (°F) of a certain city is given by

$$F(t) = 45 \sin\left[\frac{2\pi}{365}(t - 97)\right] + 38$$

where t is the number of days after January 1.

a. Sketch the graph of the temperature function $F(t)$. You may wish to use the graphing utility of your calculator.

b. What are the highest and lowest temperatures of the year in this city? When do these extreme temperatures occur?

43. **AVIAN BEHAVIOR**   Experiments indicate that homing pigeons will avoid flying over large areas of water whenever possible, perhaps because it takes more effort to fly through the heavy, cool air over the water than the relatively light, warm air over land.*

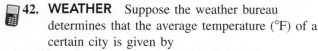

EXERCISE 43

Suppose a pigeon is released from a boat at point P on a lake and flies to its loft at point L on the lakeshore, as shown in the accompanying figure. Assume that E_W units of energy per mile are required to fly over water and E_L units of energy per mile are required over land. If the boat is s_1 miles from the nearest point A on the shore and A is s_2 miles from the loft L along the shore, what heading θ should the pigeon take from the boat to minimize the total energy expended in getting to L? (Express your answer in terms of $\sin \theta$.)

44. **SURFACE AREA OF A BEE'S CELL**   A beehive cell is a regular hexagonal prism that is open at one end and has a trihedral angle θ at the other, as shown in the accompanying figure.

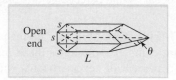

EXERCISE 44

Using trigonometry[†] (quite a lot!), it can be shown that the cell has surface area

$$S(\theta) = 6sL + 1.5s^2(-\cot \theta + \sqrt{3} \csc \theta)$$

for $0 < \theta < \frac{\pi}{2}$, where s is the length of each side in the open hexagonal base, and L is the height of the prism.

a. Assuming that s and L are fixed (constant), what angle minimizes the surface area $S(\theta)$? Express the minimum area in terms of s and L.

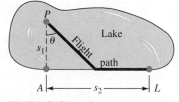 b. Read the article cited with this exercise, and write a paragraph on whether or not you think bees actually do conserve wax by constructing their cells with the "bee angle" found in part (a).

45. **BLOOD FLOW**   Let $F(\theta)$ be the resistance to blood flow function obtained in Example 8.2.10, namely,

$$F(\theta) = k\left[\frac{L - h \cot \theta}{R^4} + \frac{h \csc \theta}{r^4}\right]$$

We found the derivative $F'(\theta)$ in Example 8.2.10 and showed that $F(\theta)$ has its only critical number at θ_0 where $\cos \theta_0 = \frac{r^4}{R^4}$. Show that $F'(\theta) < 0$ if $0 < \theta < \theta_0$ and that $F'(\theta) > 0$ if $\theta > \theta_0$. Why does this mean that $F(\theta)$ has an absolute minimum at θ_0?

MISCELLANEOUS PROBLEMS

46. **LANDSCAPE DESIGN**   A landscape architect wishes to construct a garden in the shape of an isosceles triangle. The two equal sides are to each have length 20 feet. How long should the third side be for the area of the garden to be as large as possible? What is the maximum area of a garden that is designed with these specifications?

47. VELOCITY AND ACCELERATION An object moves along a straight line in such a way that its position at time t is

$$x(t) = \sin \pi(t - 1) + \sqrt{t}$$

  a. What is the object's velocity $v(t)$ at time t? What is its acceleration $a(t)$?

  b. Where is the object when $t = 1$, and what are its velocity and acceleration at that time?

48. VELOCITY AND ACCELERATION An object moves along a straight line in such a way that its position at time t is

$$x(t) = t \cos \pi t$$

  a. What is the object's velocity $v(t)$ at time t? What is its acceleration $a(t)$?

  b. Where is the object when $t = 2$, and what are its velocity and acceleration at that time?

49. RELATED RATES A man 6 feet tall is watching a streetlight 18 feet high while walking toward it at a speed of 5 feet per second. At what rate is the angle of elevation of the man's line of sight changing with respect to time when he is 9 feet from the base of the light?

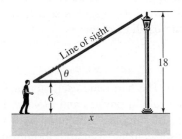

EXERCISE 49

50. RELATED RATES An attendant is standing at the end of a pier 12 feet above a rowboat and is pulling a rope attached to the boat at the rate of 4 feet of rope per minute. At what rate is the angle that the rope makes with the surface of the water changing with respect to time when the boat is 16 feet from the pier?

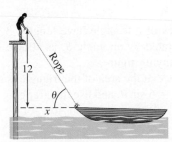

EXERCISE 50

51. RELATED RATES An observer watches a plane approach at a speed of 500 miles per hour and at an altitude of 3 miles. At what rate is the angle of elevation of the observer's line of sight changing with respect to time t when the horizontal distance between the plane and the observer is 4 miles?

52. OPTIMAL ANGLE OF OBSERVATION An 8-foot-high painting is hung on the wall of an art museum, 2 feet above the eye level of an observer, as shown in the accompanying figure.

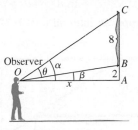

EXERCISE 52

Let θ be the angle subtended by the painting at the observer's eye when he stands x feet from the wall. It is reasonable to assume that the observer gets the best view of the painting when θ is as large as possible.

  a. Let β be the angle $\angle AOB$ and let α be $\angle AOC$. Use the trigonometric identity

$$\tan(\alpha - \beta) = \frac{\tan \alpha - \tan \beta}{1 + \tan \alpha \tan \beta}$$

  to express $\tan \theta$ in terms of x.

  b. How far back from the wall should the observer stand to maximize the angle θ? [*Hint:* Use implicit differentiation with the equation for $\tan \theta$ you found in part (a).]

53. CONSTRUCTION You have a piece of metal that is 20 meters long and 6 meters wide, which you are going to bend to form a trough as indicated in the accompanying figure. At what angle should the sides meet so that the volume of the trough is as large as possible? [*Hint:* The volume is the length of the trough times the area of its triangular cross section.]

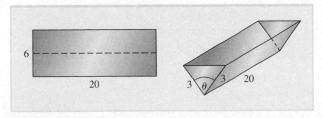

EXERCISE 53

54. CONSTRUCTION In a building under construction, there is a narrow hallway only $3\sqrt{2}$ feet wide that turns at a corner into another hallway that is 6 feet wide, as shown in the accompanying figure. What is the length of the longest pipe that workers can carry horizontally around this corner? Ignore the thickness of the pipe. [*Hint:* Why is this the same as minimizing the horizontal clearance

$$C(\theta) = \frac{6}{\sin \theta} + \frac{3\sqrt{2}}{\cos \theta}$$

on an appropriate interval $a \le \theta \le b$?]

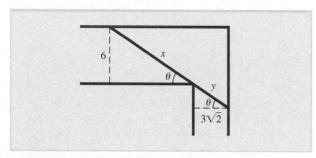

EXERCISE 54

55. REFRACTION OF LIGHT Fermat's principle in physics states that light takes the path requiring the least amount of time. The accompanying figure shows a ray of light that is emitted from a source at point A under water and, subsequently, received by an observer at point B above the surface of the water. If v_1 is the speed of light in water and v_2 is the speed of light in air, use Fermat's principle to show that

$$\frac{\sin \theta_1}{\sin \theta_2} = \frac{v_1}{v_2}$$

where θ_1 is the **angle of incidence** and θ_2 is the **angle of refraction.** [*Hint:* Use the solution to Example 8.2.7 as a guide. You may assume without proof that the total time is minimized when its derivative is equal to zero.]

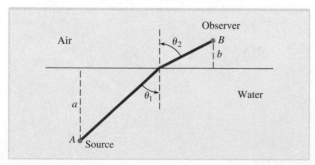

EXERCISE 55

56. SATELLITE ORBIT A telecommunications satellite is located

$$r(t) = \frac{4{,}831}{1 + 0.15 \cos (0.06t)}$$

miles above the center of Earth t minutes after achieving orbit, as indicated in the accompanying figure.
   a. Sketch the graph of $r(t)$.
   b. What are the lowest and highest points on the satellite's orbit? (These are called the *perigee* and *apogee* positions, respectively.)

EXERCISE 56

57. HARMONIC OSCILLATOR An object that moves along the x axis in such a way that its acceleration is proportional to its position and is directed toward the origin is said to be an **harmonic oscillator;** that is,

$$\frac{d^2x}{dt^2} = -kx(t)$$

for $k > 0$. The harmonic oscillator plays an important role in physics and biology. For instance, a frictionless pendulum with small initial displacement from equilibrium may be modeled as an harmonic oscillator.

   It can be shown that the solution of the harmonic oscillator equation has the general form

$$x(t) = A \cos ct + B \sin ct$$

for constants A, B, and c. Find a solution $x(t)$ with $x(0) = 0$ and $x'(0) = v_0$. Your answer will be expressed in terms of k and the initial velocity v_0.

58. Two sides of a triangle have lengths 3 and 4 and are separated by an angle θ, as shown in the accompanying figure.
   a. Show that the area of the triangle is given by $A(\theta) = 6 \sin \theta$, and use the graphing utility of your calculator to graph $A(\theta)$ for $0 \le \theta \le \pi$.

b. Use **TRACE** and **ZOOM** to find the angle for which $A(\theta)$ is maximized for $0 \leq \theta \leq \pi$.

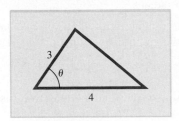

EXERCISE 58

MOTION OF A PROJECTILE *Suppose a projectile is fired from the top of a cliff s_0 feet high at an angle of θ radians with an initial (muzzle) speed of v_0 ft/sec. In physics, it is shown that t seconds later, the projectile will be at the point (x, y), where x and y are in feet and*

$$x = (v_0 \cos \theta)t$$
$$y = -16t^2 + (v_0 \sin \theta)t + s_0$$

These formulas are used in Exercises 59 through 62.

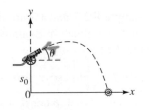

59. MOTION OF A PROJECTILE A projectile is fired from the top of a cliff. Sketch the path of the projectile for each of the cases described in the following table. You may wish to use the graphing utility of your calculator.

s_0 (feet)	100	100	0	0
v_0 (ft/sec)	50	150	50	200
θ (radians)	0	$\dfrac{\pi}{6}$	$\dfrac{\pi}{3}$	$\dfrac{\pi}{8}$

a. Based on these results, what kind of curve do you think the path of a projectile will be?

b. In each case listed in the table, find the highest point on the path and the point of impact (where the path intersects the *x* axis).

60. HEIGHT AND RANGE OF A PROJECTILE A projectile is fired at ground level ($s_0 = 0$) at an angle θ with initial speed v_0 ft/sec.

a. Find a formula for the maximum height reached by the projectile during its flight. (Your answer will involve θ and v_0.)

b. The **range** of the projectile is the horizontal distance measured from its firing position to the point of impact. Find a formula for the range (in terms of θ and v_0). For what value of θ is the range the largest?

61. WATER-SKIING A water-skier is being pulled by a motorboat toward a jump ramp. The angle of the ramp is 12° and its end is 3 feet above the water, as shown in the accompanying figure.

a. If the skier is traveling at 40 ft/sec as she leaves the ramp, how far from the ramp does she land?

b. The water-skier makes a second run, with the boat traveling at a higher speed. If she lands 10 feet farther out than in part (a), how fast is she now traveling as she leaves the ramp?

EXERCISE 61

62. SPY STORY The spy escapes from Scelerat's broiler room (Exercise 50, Section 7.5) and runs for his life, with the bad guys in hot pursuit. To his horror, he sees a grenade coming toward him and estimates that it was released at an angle of 45° from a height of 6 ft with an initial speed of 50 ft/sec. Once the grenade lands, the explosion will damage everything within a 10-ft radius of its point of impact. How far away from the thug who threw the grenade must the spy be when the grenade lands if he wants to escape unharmed?

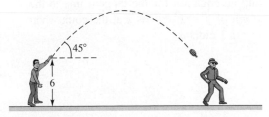

EXERCISE 62

63. ILLUMINATION   A lamp with adjustable height hangs directly above the center of a circular kitchen table that is 8 feet in diameter. The illumination at the edge of the table is directly proportional to the cosine of the angle θ shown in the accompanying figure and inversely proportional to the square of the distance d, also shown, so that

$$L = \frac{k \cos \theta}{d^2}$$

How close to the table should the lamp be lowered to maximize the illumination at the edge of the table?

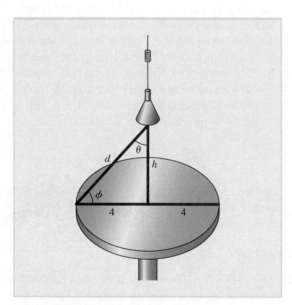

EXERCISES 63 and 64

64. ILLUMINATION   Rework Exercise 63 assuming that the illumination is directly proportional to $\sin \varphi$ instead of $\cos \theta$, where φ is the angle at which the ray of light meets the table. Explain why your answer is the same as in Exercise 63.

 65. Use the graphing utility of your calculator to find an equation for the tangent line to the graph of $y = x^2 - \cos x$ at the point where $x = 1.2$ radians.

 66. Use the graphing utility of your calculator to find an equation for the tangent line to the graph of $y = x + \sin x$ at the point where $x = 1.5$ radians.

 67. Use the graphing utility of your calculator to sketch the graph of the function

$$f(x) = \frac{\cos x - 1}{x}$$

Then use **TRACE** and **ZOOM** to investigate the limit

$$\lim_{x \to 0} \frac{\cos x - 1}{x}$$

(*Hint:* Use the viewing rectangle $[-\pi, \pi] \frac{\pi}{4}$ by $[-1, 1]\frac{1}{2}$.)

68. Repeat Exercise 67 for the function

$$f(x) = \frac{\tan x}{x}.$$

69. Use the definition of the derivative to show that

$$\frac{d}{dt} (\cos t) = -\sin t$$

70. Refer to Example 8.2.7 and Figure 8.19.
   a. Show that the problem in Example 8.2.7 can be restated in the following form:

   Minimize $T = \frac{1}{s}(a \csc \alpha + b \csc \beta)$

   subject to $a \cot \alpha + b \cot \beta = d$

   b. Solve the constrained optimization problem in part (a) using the method of Lagrange multipliers (Section 7.5) to confirm that the minimum time of travel occurs when $\cos \alpha = \cos \beta$.

71. Use the derivative formulas for $\sin t$ and $\cos t$ and the quotient rule to show that:
   a. $\dfrac{d}{dt} \sec t = \sec t \tan t$
   b. $\dfrac{d}{dt} \cot t = -\csc^2 t$
   c. $\dfrac{d}{dt} \csc t = -\csc t \cot t$

SECTION 8.3   Trigonometric Applications Involving Integration

Learning Objectives

1. Derive and use integration formulas for trigonometric functions.
2. Apply integrals of periodic functions.

The applications of integration developed in Chapters 5 and 6, such as area, volume, average value, consumers' surplus, future and present value of a continuous income flow, survival/renewal, and modeling with differential equations can be applied to periodic phenomena. We will examine several such applications in this section, after first showing how various rules and techniques of integration apply to trigonometric functions.

Integrals Involving Trigonometric Functions

Since the indefinite integral of a function $f(t)$ is determined by an antiderivative of $f(t)$, each of the differentiation formulas for trigonometric functions obtained in Section 8.2 corresponds to an integration formula. For instance, since the derivative of $\sin u$ is $\cos u$, we have

$$\int \cos u \, du = \sin u + C$$

Integration Formulas Involving Trigonometric Functions

$$\int \cos u \, du = \sin u + C$$

$$\int \sin u \, du = -\cos u + C$$

$$\int \sec^2 u \, du = \tan u + C$$

$$\int \sec u \tan u \, du = \sec u + C$$

EXAMPLE 8.3.1   Computing Trigonometric Integrals

Find the following integrals:

a. $\displaystyle\int \sec t \, (\sec t + \tan t) \, dt$ **b.** $\displaystyle\int x \sin (x^2) \, dx$

Solution

a. We have

$$\int \sec t \, (\sec t + \tan t) \, dt = \int \sec^2 t \, dt + \int \sec t \tan t \, dt$$

$$= \tan t + \sec t + C$$

b. Use the substitution $u = x^2$, $du = 2x\,dx$. Then

$$\int x \sin (x^2)\, dx = \int \sin (x^2)(x\,dx)$$

$$= \int \sin u \left(\frac{1}{2}\,du\right) = \frac{1}{2}[-\cos u] + C$$

$$= -\frac{1}{2} \cos (x^2) + C$$

Integration is used in Examples 8.3.2 and 8.3.3 to compute area and volume involving trigonometric functions and in Example 8.3.4 to find average sales for a seasonal product.

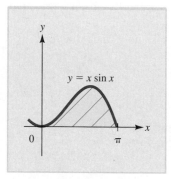

FIGURE 8.25   Area under $y = x \sin x$ between $x = 0$ and $x = \pi$.

EXPLORE!

Refer to Example 8.3.2. Set Y1 = X sin(X) in the equation editor of your graphing utility, and graph using the window $\left[-\dfrac{\pi}{2}, \dfrac{3\pi}{2}\right]\dfrac{\pi}{2}$ by $[-1, 2]$ 0.25. Use the numerical integration features of your utility to compute the area under the curve between $x = 0$ and $x = \pi$.

Just-In-Time **REVIEW**

Recall that $\sin 0 = 0$ and $\cos 0 = 1$, and that $\sin\left(\dfrac{\pi}{2}\right) = 1$ and $\cos\left(\dfrac{\pi}{2}\right) = 0$.

EXAMPLE 8.3.2   Computing Area by Integration

Find the area of the region under the curve $y = x \sin x$ between the lines $x = 0$ and $x = \pi$ (Figure 8.25).

Solution

The area is given by the definite integral $A = \displaystyle\int_0^{\pi} x \sin x\,dx$. To evaluate this integral, use integration by parts, as follows (see Section 6.1):

$$u = x \qquad dv = \sin x\,dx$$
$$du = dx \qquad v = -\cos x$$

so that

$$\int x \sin x\,dx = x(-\cos x) - \int (-\cos x)\, dx$$

$$= -x \cos x + \sin x + C$$

Therefore, the area is

$$A = \int_0^{\pi} x \sin x\,dx = (-x \cos x + \sin x)\Big|_0^{\pi}$$
$$= [-\pi \cos (\pi) + \sin (\pi)] - [-(0) \cos (0) + \sin (0)]$$
$$= [-\pi (-1) + 0] - [0] = \pi \quad \text{square units}$$

EXAMPLE 8.3.3   Computing Volume of Revolution by Integration

Find the volume of the solid S formed by revolving the region under the curve $y = \sqrt{\cos x}$ from $x = 0$ to $x = \dfrac{\pi}{2}$ about the x axis.

Solution

According to the formula obtained in Section 5.6, the volume is given by

$$V = \int \pi y^2 \, dx = \int_0^{\pi/2} \pi (\sqrt{\cos x})^2 \, dx$$

$$= \int_0^{\pi/2} \pi \cos x \, dx$$

$$= \pi \sin x \Big|_0^{\pi/2}$$

$$= \pi \left[\sin\left(\frac{\pi}{2}\right) - \sin(0) \right] = \pi[1 - 0]$$

$$= \pi$$

EXAMPLE 8.3.4 Studying Average Periodic Sales

The sales of baseball equipment in a store in California during week t of a particular year is $S(t)$ hundred dollars, where

$$S(t) = 35 + 18 \cos\left[\frac{\pi}{52}(t - 16)\right] \quad \text{for } 1 \le t \le 52$$

a. Find the average weekly sales of the equipment over the year.

b. During what week (or weeks) are average weekly sales actually attained?

Solution

a. The average sales is given by the integral

$$AV = \frac{1}{52 - 1} \int_t^{52} 35 + 18 \cos\left[\frac{\pi}{52}(t - 16)\right] dt \qquad \begin{array}{c} \text{substitute} \\ u = \frac{\pi}{52}(t - 16), \, du = \frac{\pi}{52} dt \end{array}$$

$$= \frac{1}{51}\left\{ 35t + 18\left(\frac{52}{\pi}\right) \sin\left[\frac{\pi}{52}(t - 16)\right] \right\}\Big|_1^{52}$$

$$= \frac{1}{51}\left\{ 35(52) + 18\left(\frac{52}{\pi}\right) \sin\left[\frac{\pi}{52}(52 - 16)\right] \right\}$$

$$\quad - \frac{1}{51}\left\{ 35(1) + 18\left(\frac{52}{\pi}\right) \sin\left[\frac{\pi}{52}(1 - 16)\right] \right\}$$

$$= \frac{1}{51}[2,065.20 - (-199.53)] \approx 44.41$$

Since sales are given in hundreds of dollars, sales will average approximately $4,441 per week.

b. The average sales are attained when

$$35 + 18 \cos\left[\frac{\pi}{52}(t - 16)\right] = 44.41$$

That is, when

$$\cos\left[\frac{\pi}{52}(t - 16)\right] = \frac{44.41 - 35}{18} \approx 0.5228$$

Solving this equation for t (with a calculator), we find that $t = 32.9$ weeks, so the average sales occur sometime during August.

Differential equations and initial value problems were introduced in Section 5.1. Example 8.3.5 is an applied initial value problem involving a periodic function.

EXAMPLE 8.3.5   Using Integration to Study an Animal Population

The size of an animal population often varies with the seasons. Suppose $P(t)$ is the population of a herd of large mammals at time t (months) and that

$$\frac{dP}{dt} = 12 \sin (0.1\pi t)$$

If the population is initially 3,000, what is it 1 year later?

Solution

Separating the variables and integrating, we get

$$\int dP = \int 12 \sin (0.1\pi t)\, dt \qquad\qquad \text{substitute } u = 0.1\pi t,\ du = 0.1\pi\, dt$$

$$P(t) = 12\left(\frac{10}{\pi}\right)[-\cos (0.1\pi t)] + C \approx -38.2 \cos (0.1\pi t) + C$$

Since $P(0) = 3,000$, we find that

$$P(0) = 3,000 \approx -38.2 \cos (0) + C = -38.2 + C$$
$$C \approx 3,000 + 38.2 = 3,038.2$$

and the population at time t is

$$P(t) \approx -38.2 \cos (0.1\pi t) + 3,038.2$$

Thus, after 1 year ($t = 12$ months), we have

$$P(12) \approx -38.2 \cos [0.1\pi(12)] + 3,038.2$$
$$\approx 3,069.1$$

so the population increases to about 3,069 during the year.

The price, supply, and demand of crops and certain other commodities fluctuate naturally and can often be represented by trigonometric functions. In Example 8.3.6, we find the consumers' surplus for a periodic demand.

EXAMPLE 8.3.6   Computing Consumers' Surplus for Periodic Demand

Consumers will demand (buy) q hundred units of a certain commodity when the price is $p = D(q)$ dollars per unit, where

$$D(q) = 15 - 4q + \sin\left(\frac{\pi q}{2}\right)$$

Find the consumers' surplus for this commodity when 300 units are demanded and produced.

Solution

When 300 units are demanded, we have $q_0 = 3$ and the corresponding price is p_0 dollars per unit, where

$$p_0 = D(3) = 15 - 4(3) + \sin\left[\frac{\pi(3)}{2}\right] \quad \text{since } \sin\left(\frac{3\pi}{2}\right) = -1$$

$$= 15 - 12 + (-1)$$

$$= 2$$

Recall from Section 5.5 that the consumers' surplus for this commodity at the specified level of demand is given by

$$CS = \int_0^3 D(q)\, dq - p_0 q_0$$

$$= \int_0^3 \left[15 - 4q + \sin\left(\frac{\pi q}{2}\right)\right] dq - (2)(3) \qquad \begin{array}{c}\text{substitute}\\ u = \frac{\pi}{2}q,\ du = \frac{\pi}{2}\,dq\end{array}$$

$$= \left[15q - 4\left(\frac{q^2}{2}\right) + \left(-\cos\left(\frac{\pi q}{2}\right)\right)\left(\frac{1}{\pi/2}\right)\right]\Big|_0^3 - 6$$

$$= \left[15q - 2q^2 - \left(\frac{2}{\pi}\right)\cos\left(\frac{\pi q}{2}\right)\right]\Big|_0^3 - 6$$

$$= \left[15(3) - 2(3)^2 - \left(\frac{2}{\pi}\right)\cos\left(\frac{3\pi}{2}\right)\right] - \left[15(0) - 2(0) - \left(\frac{2}{\pi}\right)\cos(0)\right] - 6$$

$$= 21 + \frac{2}{\pi} \approx 21.64$$

Thus, the consumers' surplus is approximately $2,164 (since $q_0 = 3$ is really 300 units).

The rate of flow of income into an interest-earning account can fluctuate periodically and thus can often be modeled by trigonometric functions. In this context, positive values of the flow rate function correspond to deposits into the account, while negative values correspond to withdrawals from the account.

In Section 5.5, we showed that if income is transferred continuously at the rate $f(t)$ into an account earning interest at the annual rate r compounded continuously, then the accumulated (future) value of the account after a term of T years is given by the integral

$$FV = \int_0^T f(t)\, e^{r(T-t)}\, dt$$

and the present value of the same income flow is given by

$$PV = \int_0^T f(t)\, e^{-rt}\, dt$$

Recall that present value PV represents the amount that must be deposited now at the prevailing rate for the same term of T years to generate the accumulated value FV of the income flow. Present value is used to assess the value of a periodic income flow in Example 8.3.7. For this example and also for the survival/renewal problem in Example 8.3.8, we will need the following two integration formulas, which may be obtained by two applications of integration by parts (see Exercise 66).

> **Integration Formulas Involving $e^{au} \sin{(bu)}$ and $e^{au} \cos{(bu)}$**
>
> $$\int e^{au} \sin{(bu)} \, du = \frac{e^{au}}{a^2 + b^2} [a \sin{(bu)} - b \cos{(bu)}] + C$$
>
> $$\int e^{au} \cos{(bu)} \, du = \frac{e^{au}}{a^2 + b^2} [a \cos{(bu)} + b \sin{(bu)}] + C$$

EXAMPLE 8.3.7   Present Value of a Continuous Periodic Income Flow

Martha is considering an investment that is projected to be generating income continuously at the rate of $f(t)$ dollars per year t years after it is initiated, where

$$f(t) = 2{,}000 \cos\left(\frac{t}{4}\right)$$

She plans to hold the investment for a term of 4 years and expects the prevailing annual interest rate to remain fixed at 5% compounded continuously during that period. What is a fair price for her to pay for the investment?

Solution

A fair price for the investment would be the present value

$$PV = \int_0^4 2{,}000 \cos\left(\frac{t}{4}\right) e^{-0.05t} \, dt$$

$$= 2{,}000 \int_0^4 e^{-0.05t} \cos\left(\frac{t}{4}\right) dt$$

Applying the integral formula for $\displaystyle\int e^{au} \cos{(bu)} \, du$, with $u = t$, $a = -0.05$, and $b = \dfrac{1}{4}$, we get

$$\int_0^4 e^{-0.05t} \cos\left(\frac{t}{4}\right) dt$$

$$= \frac{e^{-0.05t}}{(-0.05)^2 + (1/4)^2} \left[-0.05 \cos\left(\frac{t}{4}\right) + \frac{1}{4} \sin\left(\frac{t}{4}\right) \right]\Bigg|_0^4$$

$$= \frac{e^{-0.2}}{0.065} \left[-0.05 \cos{(1)} + \frac{1}{4} \sin{(1)} \right] - \frac{e^0}{0.065} \left[-0.05 \cos{(0)} - \frac{1}{4} \sin{(0)} \right]$$

$$\approx 3.0787$$

Therefore, the present value of the investment is

$$PV = 2{,}000 \int_0^4 e^{-0.05t} \cos\left(\frac{t}{4}\right) dt$$

$$= 2{,}000(3.0787)$$

$$\approx 6{,}157$$

so a fair price is roughly $6,157.

EXAMPLE 8.3.8 Studying Water Pollution

A large lake contains 500 tons of organic pollutant that degrades exponentially at the rate of 3.2% per year. If waste is added continuously at the rate of $10 + 2 \cos t$ tons per year, how much waste remains after 10 years?

Solution

Think of this as a survival/renewal problem, with an initial population of $P_0 = 500$ tons of waste, a survival rate of $S(t) = e^{-0.032t}$, and a renewal rate of $R(t) = 10 + 2 \cos t$. According to the formula developed in Section 5.6, the amount of waste that remains after $T = 10$ years is

$$W(10) = 500e^{-0.032(10)} + \int_0^{10} (10 + 2 \cos t)e^{-0.032(10-t)} \, dt$$

$$= 500e^{-0.032(10)} + e^{-0.032(10)} \int_0^{10} (10 + 2 \cos t)e^{0.032t} \, dt \qquad \begin{array}{l} \text{integration formula} \\ \text{for } \int e^{au} \cos (bu) \, du \end{array}$$

$$= 363.07 + e^{-0.032(10)} \left[\frac{10e^{-0.032t}}{0.032} + \frac{2e^{0.032t}}{(0.032)^2 + 1^2} (0.032 \cos t + \sin t) \right]_0^{10}$$

$$= 363.07 + 0.726 \left\{ 312.5e^{0.032(10)} + \frac{2e^{0.032(10)}}{(0.032)^2 + 1^2} (0.032 \cos 10 + \sin 10) \right\}$$

$$\qquad - 0.726 \left\{ 312.5e^{0.032(0)} + \frac{2e^{0.032(0)}}{(0.032)^2 + 1^2} (0.032 \cos 0 + \sin 0) \right\}$$

$$= 447.47$$

So approximately 447 tons of waste material will remain.

EXERCISES ▪ 8.3

In Exercises 1 through 14, find the indicated integral.

1. $\int \sin \left(\frac{t}{2} \right) dt$

2. $\int [\sin t + \cos 2t] \, dt$

3. $\int x \cos (1 - 3x^2) \, dx$

4. $\int [\tan^2 x + \sec^2 x] \, dx$

  [*Hint:* Use the identity $\tan^2 x + 1 = \sec^2 x$.]

5. $\int x \cos (2x) \, dx$ **6.** $\int \frac{\sin x}{1 + \cos x} \, dx$

7. $\int \sin x \cos x \, dx$ **8.** $\int \sqrt{\cos x} \sin x \, dx$

9. $\int x \sin x \, dx$ **10.** $\int \cos^4 x \sin x \, dx$

11. $\int_0^{\pi/3} \cos (3t) \, dt$ **12.** $\int_0^{\pi/2} \sin^2 t \cos t \, dt$

13. $\int_0^{\pi} (\sin x + \cos x)^2 \, dx$ **14.** $\int_0^{\pi} x \sin x \, dx$

In Exercises 15 through 18, find the area between the given curves over the specified interval.

15. $y = \sin \left(\frac{\pi x}{2} \right)$ and $y = \cos (\pi x)$; $0 \le x \le \frac{1}{3}$

16. $y = \sin x$ and $y = \cos (\pi - x)$; $0 \le x \le \frac{\pi}{2}$

17. $y = x^2 - x$ and $y = \sin (\pi x)$; $0 \le x \le 1$

18. $y = x^2 - 1$ and $y = \cos \left(\frac{\pi x}{2} \right)$; $0 \le x \le 1$

In Exercises 19 through 22, find the volume of the solid generated when the region under the given curve $y = f(x)$ over the specified region $a \leq x \leq b$ is rotated about the x axis.

19. $y = \sqrt{\sin x}; 0 \leq x \leq \pi$

20. $y = \sin x \sqrt{\cos x}; 0 \leq x \leq \dfrac{\pi}{2}$

21. $y = \sec x; -\dfrac{\pi}{3} \leq x \leq \dfrac{\pi}{3}$

22. $y = \tan x; 0 \leq x \leq \dfrac{\pi}{4}$

In Exercises 23 through 28, solve the given initial value problem (differential equation subject to a specified condition). Note that Exercises 26 through 28 involve separable differential equations.

23. $\dfrac{dy}{dx} = \cos x - \sin x; y(0) = 3$

24. $\dfrac{dy}{dx} = e^{-x} \sin (e^{\pi - x}); y(\pi) = 0$

25. $\dfrac{dx}{dt} = \dfrac{\sin (\sqrt{t})}{\sqrt{t}}; x(0) = -1$

26. $\dfrac{dy}{dt} = \dfrac{\sin (2t)}{y + 1}; y(0) = 1$

27. $\dfrac{dx}{dt} = x^2 \cos t; x(\pi) = 1$

28. $\dfrac{dy}{dx} = \dfrac{\sin x}{\cos y}; y(\pi) = 0$

BUSINESS AND ECONOMICS APPLIED PROBLEMS

CONSUMERS' SURPLUS In Exercises 29 through 32, a periodic demand function $D(q)$ is given along with a level of production (demand) q_0. In each case, compute the corresponding consumers' surplus (see Section 5.5).

29. $D(q) = 10 - 2q + \sin \left(\dfrac{\pi q}{3}\right); q_0 = 3$

30. $D(q) = 8 - q + \cos \left(\dfrac{\pi q}{10}\right); q_0 = 5$

31. $D(q) = 15 - 5q + \sin (\pi q) \cos (\pi q); q_0 = 2$

32. $D(q) = 20 - q^2 + \sin \left(\dfrac{\pi q}{2}\right); q_0 = 3$

PRODUCERS' SURPLUS In Exercises 33 through 36, a periodic supply function $S(q)$ is given along with a level of production q_0. In each case, compute the corresponding producers' surplus (see Section 5.5).

33. $S(q) = 3 + 2q + 0.5 \cos (\pi q); q_0 = 4$

34. $S(q) = 2 + 5q + \sin (\pi q); q_0 = 2$

35. $S(q) = 1 + 5q + 2 \sin \left(\dfrac{\pi q}{2}\right) \cos \left(\dfrac{\pi q}{2}\right); q_0 = 3$

36. $S(q) = 1 + q^2 - \cos (\pi q); q_0 = 1$

FUTURE AND PRESENT VALUE In each of Exercises 37 through 40, find the future value FV and the present value PV of a continuous income flow $f(t)$ into an account earning interest at the specified annual rate r over the specified term T.

37. $f(t) = 2,000 \cos \left(\dfrac{t}{5}\right); 4\%; T = 5$

38. $f(t) = 1,000 \sin \left(\dfrac{\pi t}{3}\right); 5\%; T = 6$

39. $f(t) = 3,000 \sin (\pi t); 3\%; T = 3$

40. $f(t) = 10,000 (\sin t + \cos t); 6\%; T = 5$

41. CONSUMERS' AND PRODUCERS' SURPLUS AT EQUILIBRIUM A manufacturer plans to supply q hundred units of a certain commodity to the market when the price is $p = S(q)$ dollars per unit and expects all q hundred units to sell (be demanded) at a price of $p = D(q)$ dollars per unit, where

$$D(q) = 9 - 6 \sin \left(\dfrac{\pi q}{20}\right)$$

and

$$S(q) = 5 + 2 \sin \left(\dfrac{\pi q}{20}\right)$$

are the demand and supply functions, for $0 \leq q \leq 10$.
   a. Find the equilibrium price p_e and the corresponding level of production q_e.
   b. Find the consumers' surplus and producers' surplus for this commodity at equilibrium.

42. CONSUMERS' AND PRODUCERS' SURPLUS AT EQUILIBRIUM A manufacturer plans to supply q hundred units of a certain commodity to the market when the price is $p = S(q)$ dollars per unit and expects all q hundred units to sell (be demanded) at a price of $p = D(q)$ dollars per unit, where

$$D(q) = 7 + 6 \cos\left(\frac{\pi q}{15}\right)$$

and

$$S(q) = 3 - 2 \cos\left(\frac{\pi q}{15}\right)$$

are the demand and supply functions, for $0 \le q \le 15$.

a. Find the equilibrium price p_e and the corresponding level of production q_e.

b. Find the consumers' surplus and producers' surplus for this commodity at equilibrium.

43. **PERIODIC INVESTMENT** Hank invests $10,000 in an account that pays 4% annual interest, compounded continuously. After that, he makes periodic adjustments to the account in the amount of $F(t)$ dollars at time t (years), where $F(t) = 50 \cos t$ (positive values of F are deposits and negative values are withdrawals). What is Hank's account worth in 10 years? [*Hint:* Think of this as a survival/renewal problem.]

44. **SALES** The monthly sales of buffalo wings in Los Angeles can be estimated using the function

$$B(t) = 4{,}128{,}500 + F_1(t) + F_2(t)$$

where

$$F_1(t) = -841{,}000 \cos\left(\frac{\pi t}{2}\right) - 111{,}500 \sin\left(\frac{\pi t}{2}\right)$$

$$F_2(t) = 234{,}500 \cos\left(\frac{\pi t}{3}\right) - 88{,}000 \sin\left(\frac{\pi t}{3}\right)$$

and t is measured in months, with $t = 1$ corresponding to January.

a. Use the formula for $B(t)$ to estimate the number of buffalo wings sold in Los Angeles during June.

b. Find the average monthly sales of buffalo wings during the first year ($0 \le t \le 12$).

45. **PROFIT** The manager of a small company in Montana finds that her profit varies periodically throughout the year and at week t is $P(t)$ hundred dollars, where

$$P(t) = 19 - 10 \sin\left[\frac{\pi}{52}(2 - t)\right] \quad \text{for } 1 \le t \le 52$$

a. Find the average weekly profit over the year.

b. During what week (or weeks) is average weekly profit actually attained?

46. **STOCK PRICES** The closing price of the stock of DeBest Ecology Corporation is $S(t)$ dollars per share at the end of week t, where

$$S(t) = 35 - 9 \cos\left(\frac{\pi t}{5}\right) + 4 \sin\left[\frac{\pi(t - 3)}{12}\right]$$

for $1 \le t \le 52$. What is the average weekly closing price of the stock during the year?

LIFE AND SOCIAL SCIENCE APPLIED PROBLEMS

47. **BIOLOGY** Suppose that a biomass $B(t)$ changes with respect to time t (hours) at the rate

$$\frac{dB}{dt} = \sqrt{t} + \sin\left(\frac{\pi t}{4}\right)$$

over a 24-hour period, $0 \le t \le 24$. If the biomass is B_0 at time $t = 0$, how large is it at time $t = 24$?

48. **SEASONAL POPULATIONS** The population $P(t)$ of a species fluctuates periodically with the seasons in such a way that

$$\frac{dP}{dt} = 0.03P + 0.004 \cos 2\pi t$$

where t is time, in years. If the population is initially $P(0) = 10{,}000$, what is it 10 years later?

49. **SEASONAL POPULATIONS** The death rate and birthrate of many animal and plant species fluctuate periodically with the seasons. The population $P(t)$ of such a species at time t changes at a rate that may be modeled by a differential equation of the form

$$\frac{dP}{dt} = (b + a \cos 2\pi t)P$$

where a and b are constants. Solve this equation to show that

$$P(t) = P_0 e^{bt + \frac{a}{2\pi} \sin 2\pi t}$$

where $P_0 = P(0)$ is the initial population.

50. **SEASONAL POPULATIONS** A herd of large mammals has a population $P(t)$ that fluctuates periodically with the seasons and satisfies a differential equation of the form given in Exercise 49, where t is measured in years. Initially, there are $P(0) = 3{,}000$ animals in the herd, and it is observed that the populations after 3 months and 6 months, respectively, are

$$P\left(\frac{1}{4}\right) = 2{,}800 \quad \text{and} \quad P\left(\frac{1}{2}\right) = 3{,}200$$

a. Use this information to determine $P(t)$. That is, find the parameters a and b in the formula obtained in Exercise 49.

b. What will the population be after 9 months $\left(t = \dfrac{3}{4}\right)$? After 1 year?

c. When during the first year ($0 \le t \le 1$) is the population of the herd the largest? What is the maximum population?

d. Answer the questions in part (c) for the minimum population of the herd.

51. ENERGY CONSUMPTION The amount of household energy consumption for space heating is sometimes measured by so-called *heating degree days* (hdd), which are defined as $65 - T_{ave}$, where T_{ave} is the average of the high and low daily temperatures, in °F.

Suppose the number of hdd in Bangor, Maine, on day t of a particular year is modeled by the function

$$H(t) = 25 + 22 \cos\left[\frac{2\pi}{365}(t - 35)\right]$$

for $0 \le t \le 365$, where $t = 0$ corresponds to the beginning of the day on January 1.

a. On what day of the year does this model predict the maximum number of hdd in Bangor will be experienced? What is this maximum number?

b. Answer the questions in part (a) for the minimum number of hdd.

c. At what rate is the energy consumption function $H(t)$ changing on January 1? Is $H(t)$ increasing or decreasing at this time? Repeat for April 1 ($t = 91$) and June 1 ($t = 152$).

d. Sketch the graph of $H(t)$ over the entire year.

e. What is the average hdd over the first 90 days of the year (January through March)? This provides a measure of the average daily energy consumption of a household in Bangor during the first quarter of the year.

52. ENERGY CONSUMPTION Let $H(t)$ be the number of gallons of heating oil used to heat a particular house in Erie, Pennsylvania, during month t after January 1, 2013. Suppose

$$H(t) = 175 + 153 \cos\left(\frac{\pi t}{12}\right) \quad \text{for } 0 \le t \le 12$$

a. What is the total amount of heating oil that will be consumed during the year 2013?

b. What is the average monthly consumption of heating oil during the year 2013?

53. NUCLEAR WASTE A nuclear power plant produces radioactive waste with a half-life of 47 years at the rate of

$$200e^{-0.1t}(\cos t + 1)$$

pounds per year. Currently, there are 300 pounds of waste.

a. How much waste will there be 100 years from now? [*Hint:* Think of this as a survival/renewal problem.]

b. How much residual waste will there be (amount of waste as $t \to \infty$)?

54. POPULATION GROWTH Demographers estimate that t years from now, the number of new people who will take up residence in a particular town is $500 + 12 \sin t$. If the current population of the town is 154,000 and the population decreases exponentially at the rate of 2.5% per year, how many people will live in the town 10 years from now? [*Hint:* Think of this as a survival/renewal problem.]

55. MEDICINE A *pneumotachograph* is a device used to measure the rate of air flow in and out of the lungs. Let $V(t)$ be the volume of air in the lungs (in liters) at time t (seconds). Suppose the rate of air flow is modeled by

$$V'(t) = 0.87 \sin (0.65t)$$

a. Inhalation occurs when $V'(t) > 0$ and exhalation when $V'(t) < 0$. The *respiratory period* is the amount of time L required to inhale and then exhale exactly once. What is L?

b. What volume of air is inhaled during the time interval $0 \le t \le L$? Is this the same as the volume exhaled during the same time period?

c. Answer the questions in parts (a) and (b) if the rate of air flow is modeled by

$$V'(t) = 0.87t \sin (0.65t)$$

[*Hint:* Use integration by parts.]

56. MEDICINE Repeat Exercise 51 for the rate function

$$V'(t) = 0.31e^{-0.1t} \sin (0.81t)$$

57. ALZHEIMER'S DISEASE Research* indicates that the body temperature $T(t)$ (in °C) of patients

---

*L. Volicer et al., "Sundowning and Circadian Rhythms in Alzheimer's Disease," *American Journal of Psychiatry*, Vol. 158, No. 5, May 2001, pp. 704–711.

with Alzheimer's disease fluctuates periodically over a 24-hour period according to the formula

$$T(t) = 37.29 + 0.46 \cos\left[\frac{\pi(t - 16.37)}{12}\right]$$

where t ($0 \le t \le 24$) is the number of hours past midnight.

a. Find the derivative $T'(t)$.

b. At what time (or times) during the 24-hour period does the maximum body temperature occur? What is the maximum temperature?

c. Answer the questions in part (b) for the minimum body temperature during the 24-hour period.

d. Suppose that patients in a certain control group are awake from 7 A.M. to 10 P.M. What is the average body temperature of such a patient over this wakeful period?

58. **POLLUTION** Air pollution fluctuates throughout the day. Suppose on a typical summer day in a certain large city, the level of pollution t hours after midnight is $P(t)$ units, where

$$P(t) = 2.1 + 16t - 0.63t^2 + 0.84 \sin(2t - 1)$$

for $0 \le t \le 24$

a. Use your calculator to sketch the graph of $P(t)$. Notice how the graph "wiggles" along the graph of a quadratic function.

b. If dawn is at 5 A.M. and dusk is at 8 P.M., what is the average hourly level of pollution from dawn to dusk?

c. Use your calculator to determine when the average level of pollution found in part (b) actually occurs.

MISCELLANEOUS PROBLEMS

59. **LANDSCAPING** Celine's garden is bounded by the positive x and y axes and a cobbled path that follows the curve

$$y = 50 \sin(0.3x + 1) + 10 \cos(0.6x + 2)$$

where x and y are measured in feet.

a. Use the graphing utility of your calculator to sketch the graph of the path. What are the endpoints of the path (the points where it intersects the x and y axes)?

b. Find the area of Celine's garden.

60. **AREA AND VOLUME** Let R be the region bounded by the curves $y = \sin t$ and $y = \cos t$ for $0 \le t \le \dfrac{\pi}{4}$.

a. Find the area of R.

b. Find the volume of the solid formed by revolving R about the x axis. [*Hint:* This is easier than it may seem.]

61. **MOTION OF AN OBJECT** An object begins at rest (initial velocity 0) at the position $x(0) = -2$ and moves along a straight line in such a way that its acceleration at time t is given by $a(t) = 3 \sin(2t)$ units per second, per second.

a. Find the velocity $v(t)$ of the object at time t.

b. Find the position $x(t)$ of the object at time t.

c. What is the average velocity of the object over the time period $0 \le t \le 3$?

62. **MOTION OF AN OBJECT** An object begins at the position $x(0) = 4$ and moves along a straight line in such a way that its velocity at time t is given by $v(t) = 4 + 2 \cos t$ units per second.

a. Find the position $x(t)$ of the object at time t.

b. How far does the object travel between time $t = 1$ and $t = 5$?

c. What is the average velocity of the object over the time period $0 \le t \le 5$?

63. The definite integral

$$\int_0^{\pi/2} \sin(x^2)\, dx$$

is associated with the diffraction of light, but the integrand $f(x) = \sin(x^2)$ has no simple antiderivative.

a. Use Simpson's rule (Section 6.2) with $n = 10$ to estimate the value of the given definite integral.

b. Check the result in part (a) by using the numeric integration feature of your calculator.

64. Repeat the steps in Exercise 63 for the definite integral

$$\int_0^{\pi/2} \cos(x^2)\, dx$$

65. a. Evaluate $\displaystyle\int \tan x\, dx$. [*Hint:* Use the substitution $u = \cos x$.]

b. Evaluate $\displaystyle\int \cot x\, dx$.

66. Let I denote the integral $I = \displaystyle\int e^{au} \sin(bu)\, du$.

a. Use integration by parts to show that

$$I = \frac{1}{b}\left[-e^{au} \cos(bu) + a \int e^{au} \cos(bu)\, du\right]$$

b. Use integration by parts a second time to show that

$$I = \frac{1}{b}\left(-e^{au}\cos(bu)\right) + \frac{a}{b}\left(\frac{1}{b}e^{au}\sin(bu) - \frac{a}{b}I\right)$$

c. Solve the equation in part (b) for I to obtain

$$I = \int e^{au}\sin(bu)\,du = \frac{e^{au}}{a^2+b^2}[a\sin bu - b\cos bu]$$

d. Substitute the integration formula obtained in part (c) into the equation in part (a) and solve to obtain a formula for $\int e^{au}\cos(bu)\,du$.

67. a. Find $\dfrac{dy}{dx}$, where $y = \ln(\sec x + \tan x)$.

b. Find $\int \sec x\,dx$.

✎ **c.** The **Mercator projection*** in cartography is a procedure for drawing the map of Earth on a flat

surface in which the paths of constant compass direction are all straight lines. It turns out that drawing such a map requires the values of the integral

$$\int_0^x \sec\theta\,d\theta$$

for all $\dfrac{0 \le x \le \pi}{2}$. Read the article cited in the footnote to this exercise, and write a paragraph on the role played by this integral in the development of the Mercator projection.

---

*Philip M. Tuchinsky, "Mercator's World Map and Calculus," *UMAP Modules 1977: Tools for Teaching*, Lexington, MA: Consortium for Mathematics and Its Applications, Inc., 1978.

Important Terms, Symbols, and Formulas

CHAPTER SUMMARY

Angle measurement:
  Degree measure (636)
  Radian measure (637)
  Conversion formula (637)

$$\frac{\text{degrees}}{180} = \frac{\text{radians}}{\pi}$$

The sine and cosine (638)
Table of frequently used values of sine and cosine (643)

θ	0	$\dfrac{\pi}{6}$	$\dfrac{\pi}{4}$	$\dfrac{\pi}{3}$	$\dfrac{\pi}{2}$	$\dfrac{2\pi}{3}$	$\dfrac{3\pi}{4}$	$\dfrac{5\pi}{6}$	π
$\sin\theta$	0	$\dfrac{1}{2}$	$\dfrac{\sqrt{2}}{2}$	$\dfrac{\sqrt{3}}{2}$	1	$\dfrac{\sqrt{3}}{2}$	$\dfrac{\sqrt{2}}{2}$	$\dfrac{1}{2}$	0
$\cos\theta$	1	$\dfrac{\sqrt{3}}{2}$	$\dfrac{\sqrt{2}}{2}$	$\dfrac{1}{2}$	0	$\dfrac{-1}{2}$	$\dfrac{-\sqrt{2}}{2}$	$\dfrac{-\sqrt{3}}{2}$	-1

Properties of the sine and cosine: (639)

$$\sin(\theta + 2\pi) = \sin\theta \qquad \cos(\theta + 2\pi) = \cos\theta$$
$$\sin(-\theta) = -\sin\theta \qquad \cos(-\theta) = \cos\theta$$

Graphs of $\sin\theta$ and $\cos\theta$ (640)
Other trigonometric functions: (641)

$$\tan\theta = \frac{\sin\theta}{\cos\theta} \qquad \cot\theta = \frac{\cos\theta}{\sin\theta}$$

$$\sec\theta = \frac{1}{\cos\theta} \qquad \csc\theta = \frac{1}{\sin\theta}$$

The trigonometry of right angles: (642)

$$\sin\theta = \frac{\text{opposite side}}{\text{hypotenuse}}$$

$$\cos\theta = \frac{\text{adjacent side}}{\text{hypotenuse}}$$

$$\tan\theta = \frac{\text{opposite side}}{\text{adjacent side}}$$

Pythagorean identity: (644)

$$\sin^2\theta + \cos^2\theta = 1$$
$$\text{or}\qquad \tan^2\theta + 1 = \sec^2\theta$$

Models for periodic phenomena: (645)

$$s(t) = a + b\sin\left[\frac{2\pi}{p}(t - d)\right]$$

$$c(t) = a + b\cos\left[\frac{2\pi}{p}(t - d)\right]$$

where
  p is the period
  $\dfrac{1}{p}$ is the frequency
  b is the amplitude
  d is the horizontal shift
  a is the vertical shift

Differentiation formulas: (652, 659)

$$\frac{d}{dt}(\sin u(t)) = \cos u \frac{du}{dt}$$

$$\frac{d}{dt}(\cos u(t)) = -\sin u \frac{du}{dt}$$

$$\frac{d}{dt}(\tan u(t)) = \sec^2 u \frac{du}{dt}$$

$$\frac{d}{dt}(\sec u(t)) = \sec u \tan u \frac{du}{dt}$$

Fermat's principle (658)

Poiseuille's resistance law (661)

Integration formulas: (671)

$$\int \cos u \, du = \sin u + C$$

$$\int \sin u \, du = -\cos u + C$$

$$\int \sec^2 u \, du = \tan u + C$$

$$\int \sec u \tan u \, du = \sec u + C$$

Checkup for Chapter 8

1. Evaluate each of the following expressions without using tables or a calculator:

   a. $\sin\left(\frac{5\pi}{6}\right)$ b. $\cos\left(-\frac{4\pi}{3}\right)$

   c. $\tan\left(\frac{3\pi}{4}\right)$ d. $\sec\left(-\frac{5\pi}{6}\right)$

2. Find $\cos\theta$ for $0 \le \theta \le \frac{\pi}{2}$ given that $\tan\theta = \frac{2}{3}$.

3. Differentiate each of the following functions:

   a. $f(x) = \sin(2x)$ b. $f(x) = x\cos x$

   c. $f(x) = \frac{\tan x}{x}$ d. $f(x) = x\sin(x^2)$

4. Find each of the following integrals:

   a. $\int \sin(2x)\,dx$ b. $\int x\cos(x^2)\,dx$

   c. $\int_0^{\pi/2} \sec^2(2t)\,dt$ d. $\int_0^{\pi/6} \sin^2 t \cos t\,dt$

5. Find all values of θ such that $\sin(2\theta) = \cos\theta$ for $0 \le \theta \le \pi$.

6. Find the area of the region bounded above by $y = \cos x$, below by $y = \sin x$, and on the left by the y axis.

7. Starting with the identity $\sin^2 x + \cos^2 x = 1$, derive the identity $1 + \cot^2 x = \csc^2 x$.

8. **SALES** The monthly sales of a certain product satisfy

$$S(t) = 19.3 + 9.4\sin\left(\frac{\pi t}{6}\right) + 3.4\cos\left(\frac{\pi t}{3}\right)$$

where t ($0 \le t \le 12$) is the time in months measured from the beginning of January and S is measured in thousands of units.

   a. At what rate are sales changing at midyear ($t = 6$)? Are sales increasing or decreasing at this time?

   b. What are the largest and smallest sales levels for the year? In what months do these extreme sales levels occur? [*Hint:* You may need the double-angle identity $\sin(2A) = 2\sin A \cos A$.]

9. **MARGINAL COST** Suppose $C(x)$ is the cost (in hundreds of dollars) of producing x units of a particular commodity, and that the marginal cost is given by

$$C'(x) = 3x + 2.5\sin(2\pi x)$$

for $0 \le x \le 10$.

   a. What is the net cost of producing the first 5 units?

   b. At what rate is the cost changing when 5 units are produced? By how much is the marginal cost changing at this level of production?

10. **FEMALE METABOLISM** Suppose Lois' body temperature (in °F) after t days in the month of June is given by

$$T(t) = 98.3 + 0.3\cos\left[\frac{2\pi(t-15)}{28}\right] + 0.2\cos[2\pi(t-0.6)]$$

for $0 \le t \le 30$.

   a. At what rate is Lois' body temperature changing when $t = 15$?

   b. What is Lois' average body temperature during the month of June?

CHAPTER SUMMARY

11. BLOOD PRESSURE The blood pressure of a particular patient at time t (seconds) is modeled by the function

$$B(t) = 105 + 31 \cos kt$$

where k is a positive constant and B is measured in millimeters of mercury. Suppose the patient has a pulse rate of 80 heartbeats per minute.

a. Find k.

b. Using the value of k you found in part (a), sketch the graph of $B(t)$.

c. Determine the systolic (largest) and diastolic (smallest) pressures for the patient.

d. What is the patient's blood pressure when it is increasing most rapidly?

Review Exercises

1. Specify the radian measurement and degree measurement for each of the following angles:

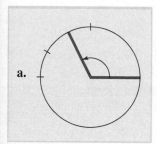
a.

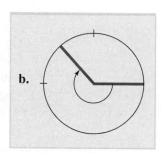
b.

2. Convert each of the following degree measurements to radians:

a. $50°$ **b.** $120°$ **c.** $-15°$

3. Convert each of the following radian measurements to degrees:

a. 0.25 radian   **b.** 1 radian   **c.** -1.5 radians

4. Evaluate each of the following expressions:

a. $\sin\left(-\dfrac{5\pi}{3}\right)$ **b.** $\cos\left(\dfrac{15\pi}{4}\right)$

c. $\sec\left(\dfrac{7\pi}{3}\right)$ **d.** $\cot\left(\dfrac{2\pi}{3}\right)$

5. Find $\tan \theta$ if $\sin \theta = \dfrac{4}{5}$ and $0 \le \theta \le \dfrac{\pi}{2}$.

6. Find $\csc \theta$ if $\cot \theta = \dfrac{\sqrt{5}}{2}$ and $0 \le \theta \le \dfrac{\pi}{2}$.

Exercises 7 and 8 use the following addition formulas for the sine and cosine:

$$\sin (A + B) = \sin A \cos B + \sin B \cos A$$

and

$$\cos (A + B) = \cos A \cos B - \sin A \sin B$$

7. Starting with the addition formulas for the sine and cosine, derive these identities:

$$\cos\left(\frac{\pi}{2} + \theta\right) = -\sin \theta \quad \text{and} \quad \sin\left(\frac{\pi}{2} + \theta\right) = \cos \theta$$

Give geometric arguments to justify the identities.

8. Use the addition formulas for sine and cosine to derive the double-angle formulas

$$\sin (2A) = 2 \sin A \cos A$$
$$\cos (2A) = \cos^2 A - \sin^2 A$$
$$= 2 \cos^2 A - 1$$
$$= 1 - 2 \sin^2 A$$

9. a. Use the double-angle formulas along with the Pythagorean identity $\sin^2 A + \cos^2 A = 1$ to show that

$$\cos^2 \theta = \frac{1}{2}(1 + \cos 2\theta) \quad \text{and} \quad \sin^2 \theta = \frac{1}{2}(1 - \cos 2\theta)$$

b. Use the identities in part (a) to show that

$$\int \cos^2 x \, dx = \frac{1}{2}x + \frac{1}{4}\sin (2x) + C$$

and

$$\int \sin^2 x \, dx = \frac{1}{2}x - \frac{1}{4}\sin (2x) + C$$

c. An object moves along a straight line in such a way that after t seconds, its velocity is given by

$$v(t) = 2t + \sin^2\left(\frac{\pi t}{6}\right)$$

meters per second. Find the average velocity of the object over the time period $0 \le t \le 3$.

10. Use the double-angle formulas for sine and cosine in Exercise 8 to derive a double-angle formula for $\tan \theta$.

In Exercises 11 through 18, differentiate the given function.

11. $f(x) = \cos (1 - 5x)$

12. $f(x) = \sin (3x + 1) \cos x$

13. $f(x) = \cos^2 x$

14. $f(x) = \tan(3x^2 + 1)$

15. $f(x) = \tan^7(3x + 1)$ **16.** $f(x) = \dfrac{\sin x}{1 - \cos x}$

17. $f(x) = \ln(\cos^2 x)$ **18.** $f(x) = e^{-2x} \cos 3x$

In Exercises 19 through 24, find the indicated integral.

19. $\displaystyle\int (\sin 2t + \cos 2t)\, dt$

20. $\displaystyle\int \cos(1 - 2t)\, dt$ **21.** $\displaystyle\int \sin x \cos x\, dx$

22. $\displaystyle\int x \sin x\, dx$ **23.** $\displaystyle\int \dfrac{\sec^2 t}{\tan t}\, dt$

24. $\displaystyle\int \tan^2 t\, dt$ [*Hint:* Use $1 + \tan^2 t = \sec^2 t$.]

25. $\displaystyle\int_0^\pi \cos\left(\dfrac{x}{3}\right) dx$ **26.** $\displaystyle\int_0^1 x \sin(x^2)\, dx$

27. Angle θ has a terminal side in quadrant IV, as shown in the accompanying figure. Find the reference angle A in degree measure to the nearest second. Then find the smallest positive angle θ so that $\csc \theta = \csc A$.

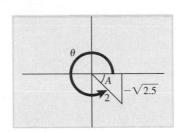

EXERCISE 27

28. In each of the following cases, use the graphing utility of your calculator to draw the graphs of the given pair of functions $f(x)$ and $g(x)$ on the same screen. Describe the relationship between the graphs of $f(x)$ and $g(x)$.

   a. $f(x) = \sin x$ and $g(x) = 2 \sin x$

   b. $f(x) = \cos x$ and $g(x) = 2 \cos 2x$

   c. $f(x) = \sin x$ and $g(x) = \sin\left(x + \dfrac{\pi}{2}\right)$

   d. $f(x) = \cos x$ and $g(x) = 2 + \cos x$

 29. Use your calculator to solve the equation

$$2 \tan 3x - 5.87 = 2 \sin 2x \quad \text{for } 0 \le x \le \dfrac{\pi}{2}$$

to three decimal places.

30. Find the area of the region bounded by the curves $y = \sin 2x$ and $y = \cos x$ over the interval $\dfrac{\pi}{6} \le x \le \dfrac{\pi}{2}$.

31. Let R be the region bounded by the x axis, the curve $y = \cos x + \sin x$, and the lines $x = -\dfrac{\pi}{2}$ and $x = \dfrac{\pi}{6}$. Find the volume of the solid generated by rotating R about the x axis.

32. a. Find the period p, the amplitude b, the horizontal shift d, and the vertical shift a of the function

$$f(x) = 5.0 + 3.0 \cos\left[\dfrac{\pi}{4}(x - 1.5)\right]$$

   b. Sketch the graph of the function $f(x)$ in part (a).

33. a. Find the period p, the amplitude b, the horizontal shift d, and the vertical shift a of the function

$$f(x) = 33 + 27 \cos\left[\dfrac{2\pi}{25}(x - 11)\right]$$

   b. Sketch the graph of the function $f(x)$ in part (a).

34. WEATHER The maximum daily temperature $T(x)$ in degrees Celsius in Minneapolis on day x of the year can be modeled as

$$T(x) = 13 + 33 \cos\left[\dfrac{2\pi}{365}(x - 271)\right]$$

where $x = 0$ corresponds to January 1.

   a. Using a calculator, find the maximum daily temperature in Minneapolis on the first day of January. Repeat for the first days of March, May, July, September, and November.

   b. Find the largest and smallest maximum daily temperature in Minneapolis during the year.

   c. Draw the graph of the maximum daily temperature function $T(x)$.

35. CONSTRUCTION COST A cable is run in a straight line from a power plant on one side of a river 900 meters wide to a point P on the other side and then along the river bank to a factory,

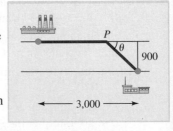

EXERCISE 35

3,000 meters downstream from the power plant, as shown in the accompanying figure. The cost of running the cable under the water is \$5 per meter, while the cost over land is \$4 per meter. If θ is the (smaller) angle between the segment of cable under the river and the opposite bank, show that

$$\cos \theta = \dfrac{4}{5} \text{ (the ratio of the per-meter costs) for the}$$

CHAPTER SUMMARY

route that minimizes the total installation cost. (You may assume without proof that the absolute minimum occurs when the derivative of the cost function is zero.)

36. **SEASONAL POPULATIONS** We can model the population $P(t)$ of an animal or plant species that fluctuates seasonally using a function of the form $P(t) = P_0 e^{a \sin (2\pi t)}$, where t is the time in years and $P_0 = P(0)$ is the initial population.

    The population function $P(t)$ satisfies a differential equation of the form

    $$\frac{dP}{dt} = R(t)P(t)$$

    where $R(t)$ is the relative growth rate function for the species (birthrate − death rate). What is $R(t)$?

37. **SEASONAL POPULATIONS** Suppose the population of mosquitoes in a swamp can be modeled by $P(t) = 500e^{7 \sin 2\pi t}$, where time t is in years and $t = 0$ corresponds to May 1.

    a. Find the number of mosquitoes in the swamp at time $t = 0.125$. Repeat for times $t = 0.25$, $t = 0.325$, and $t = 0.5$.

    b. Find the rate of change of the population of mosquitoes in the swamp at time $t = 0.125$. Repeat for times $t = 0.25$, $t = 0.325$, and $t = 0.5$.

    c. Draw the graph of $P(t)$.

38. **POLLUTION** The ozone levels in parts per million (ppm) in a city can be modeled by the function $F(t) = 0.01t^3 + 0.05t^2 + 1.1t + 56 + 22 \sin (2\pi t)$ where t is the time in years after 1990.

    a. Find the levels of ozone on July 1, 1990. Repeat for January 1, 2000, and March 1, 2005.

    b. Find the rate of change of the level of ozone on the three dates in part (a).

    c. Graph $F(t)$ for the time period from 1990 to 2010 ($0 \le t \le 20$).

    d. Describe the behavior of $F(t)$ as t increases from 0 to 20. Interpret the roles of the polynomial part of $F(t)$ and the periodic part.

39. **POPULATION GROWTH** Suppose that caribou enter the Arctic National Wildlife Reserve at the rate modeled by the derivative

    $$C'(t) = 275 + 275 \cos \left[\frac{\pi}{6} (t - 3) \right]$$

    where $C(t)$ is the number of caribou in the reserve at time t (months).

    a. By how much does the caribou population in the reserve grow during the first year ($0 \le t \le 12$)? What about during the second year ($12 \le t \le 24$)?

    b. During what month of the year is the caribou population the largest? When is it smallest?

40. **OPTICS** In the study of Fraunhofer diffraction in optics, a light beam of intensity I_0 from a source L passes through a narrow slit and is diffracted onto a screen. Experiments show that the intensity $I(\theta)$ of light on the screen depends on the angle θ shown in the accompanying figure in such a way that

    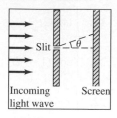

    EXERCISE 40

    $$I = I_0 \left(\frac{\sin (\beta/2)}{\beta/2} \right)^2$$

    where $\beta = 2\pi a \sin \left(\dfrac{\pi}{\lambda} \right)$, λ is the wavelength of the light, and a is a positive constant related to the width of the slit.

    a. Find $\dfrac{dI}{d\beta}$; then use the chain rule to find $\dfrac{dI}{d\theta}$.

    b. Sketch the graph of $I(\beta)$. For what values of β is the intensity $I(\beta)$ zero? (The corresponding angles result in blank bands in the diffraction pattern.)

    c. In Fraunhofer diffraction, it is assumed that the incident rays of light are parallel as they come toward the slit. Read an article on Fraunhofer diffraction and a related phenomenon called **Fresnel diffraction,** where the incident rays are not parallel. Then write a paragraph on how diffraction affects optical systems.

41. **HOURS OF DAYLIGHT** The number of hours of daylight in New York for day t of the year can be modeled by the function

    $$D(t) = 12.2 + 3.09 \cos \left[\frac{2\pi}{365} (t - 185) \right]$$

    where $t = 0$ corresponds to January 1.

    a. How many hours of daylight are there on January 1? On March 15 ($t = 74$)? On June 21 ($t = 172$)?

    b. On which day of the year is the number of daylight hours the greatest? When does the least number of daylight hours occur?

c. What is the average number of daylight hours per day over the entire year ($0 \le t \le 365$)?

42. INVENTORY MANAGEMENT The manager of a company determines that t months from now, $P(t)$ percent of his available storage space will be utilized, where

$$P(t) = 15 + 83 \cos^2\left[\frac{\pi(t-5)}{7}\right] \quad \text{for } 0 \le t \le 12$$

What is the average monthly utilization over the period $0 \le t \le 12$? [*Hint:* You may need one of the formulas in Review Exercise 9.]

43. POLITICAL POLLING A poll is conducted to determine public reaction to a new bill pending in the state legislature. It is found that the number of people favoring the bill fluctuates in such a way that t weeks after polling begins, the percentage of approval $P(t)$ satisfies

$$P(t) = 43 + 14 \sin^2\left[\frac{\pi(t-3)}{28}\right]$$

What is the average weekly percentage of public approval over the period $0 \le t \le 10$? [*Hint:* You may need one of the formulas in Review Exercise 9.]

44. RELATIVE RATE OF CHANGE A revolving searchlight in a lighthouse 2 miles off shore is following a jogger running along the shore, as shown in the accompanying figure. When the jogger is 1 mile from the point A on the shore that is

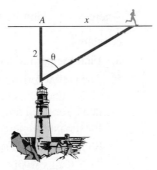

EXERCISE 44

closest to the lighthouse, the searchlight is turning at the rate of 0.25 revolutions per hour. How fast is the jogger running at this moment?

$$\left[\text{\emph{Hint:} Since 0.25 revolutions per hour is } \frac{\pi}{2} \text{ radians}\right.$$

per hour, the problem is to find $\dfrac{dx}{dt}$ when $x = 1$

$$\left. \text{and } \frac{d\theta}{dt} = \frac{\pi}{2}.\right]$$

45. RELATIVE RATE OF CHANGE On New Year's Eve, Zain is watching the descent of a lighted ball from atop a tall building that is 600 feet away. The ball is falling at the rate of 20 feet per minute. At what rate is the angle of elevation of Zain's line

of sight changing with respect to time when the ball is 800 feet from the ground?

46. CONSTRUCTION A trough 9 meters long is to have a cross section consisting of an isosceles trapezoid in which the base and two sides are all 4 meters long, as shown in the accompanying figure. At what angle should the sides of the trapezoid meet the horizontal top to maximize the capacity of the trough?

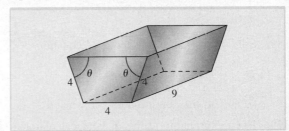

EXERCISE 46

47. CONSTRUCTION A piece of paper measuring 8.5 in. by 11 in. is folded so that the lower right-hand corner reaches the left edge, forming an angle θ at the crease, as shown in the accompanying figure.

a. Express the length of the crease L as a function of θ.

b. Find the crease angle θ for which L is minimized. What is the minimum length?

c. What is the smallest area of the triangle $\triangle ABC$ shown in the figure?

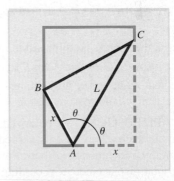

EXERCISE 47

48. ARCHITECTURE An architect is asked to construct a game room in the shape of an isosceles trapezoid, as shown in the accompanying figure. The two equal (nonparallel) walls of the room and the shorter of the two parallel walls are each to be 20 feet long. How long should the fourth wall be to maximize the area of the room? What is the maximum area of a room designed with these specifications? [*Hint:* Express the area in terms of the angle θ shown in the figure.]

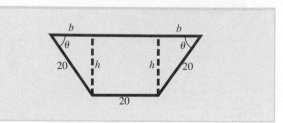

EXERCISE 48

49. Solve the separable differential equation
$$\frac{dy}{dx} = \sin x \sec y$$
subject to the condition $y = 1$ when $x = 0$.

50. Use the graphing utility of your calculator to draw the curves $y = \sin x$ and $y = e^{x-2}$ for $x \geq 0$ on the same screen. Find all points of intersection of the two curves. Let R be the region enclosed by the two curves.
   a. Find the area of the region R.
   b. Find the volume of the solid formed by revolving the region R around the x axis. $\left[\text{*Hint:* It may help to recall the identity } \sin^2 x = \dfrac{1 - \cos 2x}{2}.\right]$
   c. Check the integration in part (b) by using the numeric integration feature of your calculator.

Exercises 51 through 56 deal with topics developed in Chapter 7.

51. What is the largest possible value of the product
$$f(A, B, C) = \sin A \sin B \sin C$$
given that A, B, and C are the angles in a triangle? [*Hint:* It may help to note that $A + B + C = \pi$.]

52. **POLARIZED LIGHT** A polarized light wave travels in such a way that its vertical displacement y at time t is a function of both t and its horizontal displacement x according to the formula
$$y(x, t) = 0.27 \sin\left(10\pi t - 3\pi x + \frac{\pi}{4}\right)$$
   a. Find $\dfrac{\partial y}{\partial x}$ and $\dfrac{\partial y}{\partial t}$.
   b. For what points (x, t) is $y(x, t)$ maximized? For what points is $y(x, t)$ minimized?

53. **THE DIFFUSION EQUATION** The following equation involving partial derivatives of the function $u(x, t)$ is called the *diffusion equation*:
$$u_t = c^2 u_{xx}$$
The diffusion equation is used in modeling a large variety of physical phenomena. For instance, in biology it is used to model the mechanism for butterfly wing patterns, the effects of genetic drift, and macrophage response to bacteria in the lungs, while in physics, it is used to study the motion of molecules and heat conduction.
   a. Show that the function $u = e^{-c^2 k^2 t} \sin kx$ satisfies the diffusion equation.
   b. Read an article on the diffusion equation, and write a paragraph on one of its applications.

54. **ECOLOGY** Ground temperature models are important in ecology, where they are used to study phenomena such as frost penetration. Suppose ground temperature T at time t (months) and depth x (centimeters) is modeled by a function of the form
$$T(x, t) = A + Be^{-kx} \sin (at - kx)$$
where $a = \dfrac{\pi}{6}$ and A, B, and k are positive constants.
   a. Find the partial derivatives T_x and T_t.
   b. The partial derivative T_x measures the rate at which the ground temperature drops with increasing depth for fixed time. Give a similar interpretation for the partial derivative T_t.
   c. Show that $T(x, t)$ satisfies the diffusion equation $T_t = c^2 T_{xx}$, where c is a constant involving B and k.

55. Use the method of Lagrange multipliers to find the maximum value of
$$f(x, y) = \cos x + \cos y$$
subject to the constraint condition
$$y - x = \frac{\pi}{4}$$

56. Find the largest and the smallest values of the function
$$f(x, y) = 2 \sin x + 5 \cos y$$
over the rectangle R with vertices $(0, 0)$, $(2, 0)$, $(2, 5)$, and $(0, 5)$.

EXPLORE! UPDATE

Complete solutions for all EXPLORE! boxes throughout the text can be accessed at the book-specific website, www.mhhe.com/hoffmann.

Solution for Explore! on Page 640

Store the function $f(x) = \sin x$ into Y1 of the equation editor of your graphing calculator. Graph by using the **ZOOM** key, option **7:ZTrig,** which employs a window approximately $[-2\pi, 2\pi]\frac{\pi}{2}$ by $[-4, 4]1$, as displayed in the following left screen. Using the **TRACE** key, we see that $f(x)$ approaches 1 for x values close to but less than $\frac{\pi}{2}$. Also by tracing, we determine $\lim\limits_{x \to \pi} \sin x = 0$ (the negligible calculator value of 5.898×10^{-10}).

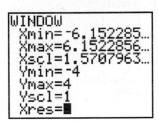 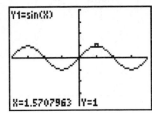 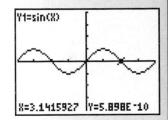

Solution for Explore! on Page 641

Store $f(x) = \tan x$ into Y1 of the equation editor of your graphing calculator. Graph by using the **ZOOM** key, option **7:ZTrig.** As x approaches $\frac{\pi}{2} \approx 1.571$, $\tan x$ gets asymptotically large (approaching $+\infty$). We can write $\lim\limits_{x \to \pi/2^-} \tan x = +\infty$.

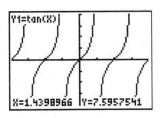

Solution for Explore! on Page 644

Refer to Example 8.1.5. Set Y1=sin(X)^2−cos(X)^2+sin(X) in the equation editor of your graphing calculator. Graph using the window $[0, 2\pi]\frac{\pi}{2}$ by $[-2, 2]0.5$. Using the root-finding feature of the **CALC** key **(2nd TRACE), 2:zero,** the x intercepts of $\frac{\pi}{6} \approx 0.5236$, $\frac{5\pi}{6} \approx 2.6180$, and $\frac{3\pi}{2} \approx 4.7124$ can be determined, confirming results found in the example.

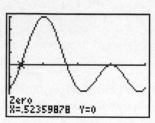
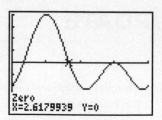
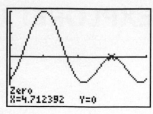

**Solution for Explore!
on Page 672**

Refer to Example 8.3.2. Set Y1=Xsin(X) in the equation editor of your graphing calculator, and graph using the window $\left[-\dfrac{\pi}{2}, \dfrac{3\pi}{2}\right]\dfrac{\pi}{2}$ by $[-1, 2]0.25$. Access the graphical integral display using the **CALC** key **(2nd TRACE), 7: ∫f(x)dx,** setting the lower limit at $x = 0$ and the upper limit at $x = \pi$, entered directly. As viewed in the next screen on the far right, the area under the curve over the interval $[0, \pi]$ is π.

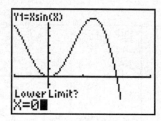
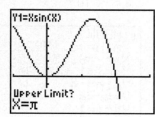
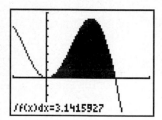

THINK ABOUT IT

WHAT ANGLE IS BEST FOR VIEWING A RAINBOW?

Rainbows have been observed and admired by philosophers, poets, and scientists for thousands of years. Most people would agree that experiencing the ephemeral beauty of a rainbow is one of the great joys of life, but at what viewing angle will the rainbow appear most vivid?

Rainbows are formed when sunlight traveling through the air is both reflected and refracted by raindrops. Figure 1a shows a raindrop, which for simplicity, is assumed to be spherical in shape. An incoming beam of sunlight strikes the raindrop at point A with angle of incidence α. Some of the light is reflected and some is refracted through angle β. The refracted beam then continues to point B on the other side of the raindrop, where part of it is reflected through angle β back through the raindrop to point C, while part of what remains is refracted through angle α back into the air.

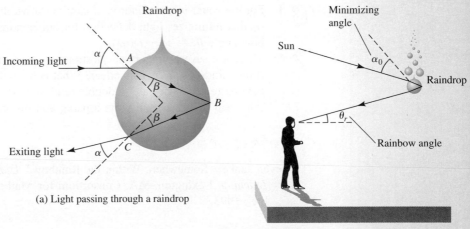

(a) Light passing through a raindrop

(b) A rainbow is brightest when viewed at the rainbow angle.

FIGURE 1   Analysis of light passing through a raindrop.

At each interface, the light beam is deflected (from a straight line path), which causes its intensity to be reduced. In particular, at A, the incoming beam is deflected through a clockwise rotation of $\alpha - \beta$ radians, while at B, the deflection is $\pi - 2\beta$ radians and at C, it is again $\alpha - \beta$ radians. Thus, the total deflection is D radians, where

$$D = (\alpha - \beta) + \pi - 2\beta + \alpha - \beta = \pi + 2\alpha - 4\beta$$

The intensity of the light exiting from the raindrop is greatest for the angle of incidence α_0 that minimizes the total deflection D. As indicated in Figure 1b, this minimizing angle α_0 is directly related to the angle of observation θ_r for which the rainbow is brightest. The angle θ_r is appropriately called the **rainbow angle.** You will determine the value of the rainbow angle by using calculus to answer Questions 1–3 in the following list.

This essay was adapted from the article, *Somewhere Within the Rainbow,* by Steven Janke, *UMAP Modules 1992: Tools for Teaching,* Lexington, MA: Consortium for Mathematics and Its Applications, Inc., 1993. The article explores a variety of issues regarding the observation of a rainbow. You are asked to address two such issues in Question 4.

Questions

1. The **law of refraction** in optics says that the angle of incidence α is related to the angle of refraction β by the equation $\sin \alpha = k \sin \beta$, where k is a constant that depends on the refracting medium (recall Exercise 55 of Section 8.2). Use this law to express the light deflection D in terms of α and then differentiate to show that

$$\frac{d\beta}{d\alpha} = 2 - \frac{4 \cos \alpha}{k \cos \beta}$$

2. Show that $D'(\alpha_0) = 0$, where α_0 satisfies $\cos (\alpha_0) = \sqrt{\dfrac{k^2 - 1}{3}}$. Use the second derivative test to show that $D(\alpha)$ is minimized at $\alpha = \alpha_0$.

3. For the water in our raindrop, it can be shown that $k \approx 1.33$. Compute the angle α_0 that minimizes light deflection for our raindrop. Then use α_0 to find the rainbow angle $\theta_r = \pi - D(\alpha_0)$.

4. There are several important questions we have not explored. For instance, what if the raindrops are not spherical? What is the optimum viewing angle if the rainbow has a secondary arc (a double rainbow)? Read the UMAP module on which this essay is based, and write a paragraph on one of these topics.

Reference

Steven Janke, "Somewhere Within the Rainbow," *UMAP Modules 1992: Tools for Teaching,* Lexington, MA: Consortium for Mathematics and Its Applications, Inc., 1993.

CHAPTER

9

The concentration of pollutant in a lake can be determined using differential equations.

Differential Equations

SECTION 9.1   Modeling with Differential Equations

Learning Objectives

1. Solve separable differential equations and initial value problems.
2. Construct and use mathematical models involving differential equations.
3. Explore learning and population models, including exponential and logistic growth.

In Section 1.4, we introduced mathematical modeling as a dynamic process involving three stages:

1. A real-world problem is given a mathematical formulation, called a *mathematical model.* This often involves making simplifying assumptions about the problem to make it more accessible.
2. The mathematical model is analyzed or solved by using tools from areas such as algebra, calculus, or statistics, among others, or by numerical methods involving graphing calculators or computers.
3. The solution of the mathematical model is interpreted in terms of the original real-world problem. This often leads to adjustments in the assumptions of the model.

The process may then be repeated, each time with a more refined model, until a satisfactory understanding of the real-world problem is attained.

We have used calculus to analyze mathematical models throughout this text, and many of these models have involved rates of change. Sometimes the mathematical formulation of a problem involves an equation in which a quantity and the rate of change of that quantity are related by an equation. Since rates of change are expressed in terms of derivatives or differentials, such an equation is appropriately called a **differential equation.** For example,

$$\frac{dy}{dx} = 3x^2 + 5 \qquad \text{and} \qquad \frac{dP}{dt} = kP \qquad \text{and} \qquad \left(\frac{dy}{dx}\right)^2 + 3\frac{dy}{dx} + 2y = e^x$$

are all differential equations.

We first encountered differential equations in Section 5.1 as part of the introduction to indefinite integration. In that section and in Section 5.2, we showed how to set up and solve elementary differential equations and examined a few applications. In this chapter, we will introduce additional modeling and solution techniques for differential equations and will explore selected applications to business, economics, and the life, social, and physical sciences.

We begin by reviewing some basic terminology. A function that satisfies a given differential equation is called a **solution** of the equation, and a **general solution** is a characterization of a family of solutions. A differential equation coupled with a side condition is referred to as an **initial value problem,** and a solution that satisfies both the differential equation and the side condition is called a **particular solution** of the initial value problem. This terminology is illustrated in Examples 9.1.1 and 9.1.2 for equations of the form

$$\frac{dy}{dx} = g(x)$$

which can be solved by simply finding the indefinite integral of the function $g(x)$.

EXAMPLE 9.1.1　Solving an Initial Value Problem

Find the general solution of the differential equation

$$\frac{dy}{dx} = x^2 + 3x$$

and the particular solution that satisfies $y = 2$ when $x = 1$.

Solution

Integrating, you get

$$y = \int \left(\frac{dy}{dx}\right) dx = \int (x^2 + 3x)\, dx$$

$$= \frac{1}{3}x^3 + \frac{3}{2}x^2 + C$$

This is the general solution since all solutions can be expressed in this form. For the particular solution, substitute $x = 1$ and $y = 2$ into the general solution:

$$2 = \frac{1}{3}(1)^3 + \frac{3}{2}(1)^2 + C$$

$$C = 2 - \frac{1}{3} - \frac{3}{2} = \frac{1}{6}$$

Thus, the required particular solution is $y = \frac{1}{3}x^3 + \frac{3}{2}x^2 + \frac{1}{6}$.

EXAMPLE 9.1.2　Determining the Resale Value of a Machine

The resale value of a certain industrial machine decreases over a 10-year period at a rate that depends on the age of the machine. When the machine is x years old, the rate at which its value is changing is $220(x - 10)$ dollars per year. Express the value of the machine as a function of its age and initial value. If the machine was originally worth \$12,000, how much will it be worth when it is 10 years old?

Solution

Let $V(x)$ denote the value of the machine when it is x years old. The derivative $\frac{dV}{dx}$ is equal to the rate $220(x - 10)$ at which the value of the machine is changing. Hence, to model this problem, we can use the differential equation

$$\frac{dV}{dx} = 220(x - 10) = 220x - 2{,}200$$

To find V, solve this differential equation by integration:

$$V(x) = \int (220x - 2{,}200)\, dx = 110x^2 - 2{,}200x + C$$

Notice that C is equal to $V(0)$, the initial value of the machine. A more descriptive symbol for this constant is V_0. Using this notation, you can write the general solution as

$$V(x) = 110x^2 - 2{,}200x + V_0$$

If $V_0 = 12{,}000$, the corresponding particular solution is

$$V(x) = 110x^2 - 2{,}200x + 12{,}000$$

Thus, when the machine is $x = 10$ years old, its value is

$$V(10) = 110(10)^2 - 2{,}200(10) + 12{,}000 = \$1{,}000$$

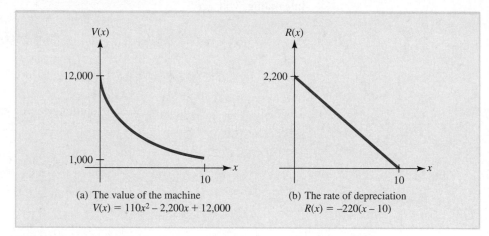

(a) The value of the machine
$V(x) = 110x^2 - 2{,}200x + 12{,}000$

(b) The rate of depreciation
$R(x) = -220(x - 10)$

FIGURE 9.1   The value of the machine and its rate of depreciation.

The negative of the rate of change of resale value of the machine

$$R = -\frac{dV}{dx} = -220(x - 10)$$

is called the **rate of depreciation.** Graphs of the resale value V of the machine and the rate of depreciation R are shown in Figure 9.1.

In Section 5.1 we noted that a differential equation is said to be **separable** if it can be formally rewritten so that all terms containing the independent variable appear on one side of the equation and all terms involving the dependent variable appear on the other. Recall that such an equation can be solved by performing two separate integrations. The procedure is summarized here:

Separable Differential Equations   ■   A differential equation that can be written in the form

$$\frac{dy}{dx} = \frac{h(x)}{g(y)}$$

is said to be **separable.** The general solution of such an equation can be obtained by separating the variables and integrating both sides; that is,

$$\int g(y)\, dy = \int h(x)\, dx$$

EXAMPLE 9.1.3   Using Velocity to Study the Motion of an Object

An object moves along the x axis in such a way that at each time t, its velocity is given by the differential equation

$$\frac{dx}{dt} = x^2 \ln t$$

If the object is at $x = -2$ when $t = 1$, where is it when $t = 3$?

Solution

Separating the variables in the given differential equation and integrating, we get

$$\int \frac{1}{x^2}\, dx = \int \ln t\, dt \qquad \text{integration by parts with}$$
$$u = \ln t \qquad dv = dt$$
$$-\frac{1}{x} + C_1 = t \ln t - \int \frac{1}{t}(t)\, dt \qquad du = \frac{1}{t}\, dt \qquad v = t$$
$$= t \ln t - t + C_2$$

so the general solution is

$$-\frac{1}{x} = t \ln t - t + C \quad \text{where } C = C_2 - C_1$$

Since $x = -2$ when $t = 1$, we have

$$-\frac{1}{(-2)} = (1)\ln(1) \quad (1) + C \quad \ln 1 - 0$$

$$\frac{1}{2} = 0 - 1 + C \qquad \text{isolate } C$$

$$C = \frac{1}{2} + 1 = \frac{3}{2}$$

so that

$$-\frac{1}{x} = t \ln t - t + \frac{3}{2}$$

In particular, when $t = 3$, we have

$$-\frac{1}{x} = (3)\ln(3) - (3) + \frac{3}{2} \approx 1.80$$

Solving this equation for x, we get

$$x \approx -\frac{1}{1.8} \approx -0.56$$

which means that the object is at $x = -0.56$ when $t = 3$.

Modeling with Differential Equations

Quantities of interest in business, economics, or life science often change at rates proportional to expressions involving their own current value. In Example 9.1.4, we show how several such rate relationships can be expressed in terms of separable differential equations. Then in Examples 9.1.5 through 9.1.7, we explore models from three separate applied areas.

Just-In-Time **REVIEW**

Recall from page 50 that the quantity Q is:

Directly proportional to x if $Q = kx$

Inversely proportional to x if $Q = \dfrac{k}{x}$

Jointly proportional to x and y if $Q = kxy$

EXAMPLE 9.1.4 Modeling with Separable Differential Equations

In each of the following cases, write a differential equation describing the given situation:

a. The marginal cost of producing a certain commodity is proportional to the square root of cost.

b. A fish population grows at a rate inversely proportional to the square of the population.

c. The rate at which a rumor spreads through a social group is jointly proportional to the number of people in the group who have heard the rumor and the number who have not.

Solution

a. Let $C(x)$ be the total cost of producing x units. Then the derivative $\dfrac{dC}{dx}$ is the marginal cost, which must be proportional to the square root of $C(x)$. Therefore, there is a constant k for which

$$\frac{dC}{dx} = k\sqrt{C}$$

b. Let $P(t)$ be the fish population at time t. Since the growth rate of the population, $\dfrac{dP}{dt}$, is inversely proportional to the square of the population, we have

$$\frac{dP}{dt} = \frac{k}{P^2}$$

c. Let $N(t)$ be the number of people who have heard the rumor at time t, and let N_0 be the total number of people in the group. Then the number of people who have not yet heard the rumor at time t is $N_0 - N$, and the rumor is spreading at the rate

$$\frac{dN}{dt} = kN(N_0 - N)$$

Learning Models

A **learning model** is one in which a quantity $Q(t)$ grows at a rate proportional to the difference between its current size and a fixed upper limit Q_m, so that

$$\frac{dQ}{dt} = k(Q_m - Q)$$

In Example 9.1.5, we examine a learning model from psychology.

EXAMPLE 9.1.5 Studying a Learning Model

The rate at which a person correctly performs a series of tasks is proportional to the number of possible tasks not yet performed. Ross is given a list of 20 tasks and is able to perform 3 of them immediately. If Ross is able to perform half the tasks after 15 minutes, how long does it take him to perform 17 tasks?

Solution

If $N(t)$ is the number of tasks correctly performed after t minutes, the given information tells us that

$$\frac{dN}{dt} = k(20 - N) \quad \text{where } \underbrace{N(0) = 3}_{\substack{\text{3 tasks} \\ \text{initially}}} \quad \text{and} \quad \underbrace{N(15) = 10}_{\substack{\text{half (10) the tasks} \\ \text{after 15 minutes}}}$$

Separating the variables in this differential equation and integrating, we get

$$\int \frac{dN}{20 - N} = \int k \, dt \qquad \text{let } u = 20 - N, \, du = -dN$$

$$-\ln|20 - N| = kt + C_1 \quad \text{multiply by } -1 \text{ and apply exponential base } e$$

$$20 - N = Ce^{-kt} \qquad \text{since } 20 - N > 0 \text{ (there are only 20 tasks)}$$

where $C = e^{-C_1}$. So

$$N(t) = 20 - Ce^{-kt}$$

Since $N(0) = 3$, we have

$$3 = N(0) = 20 - Ce^{(0)} = 20 - C$$
$$C = 20 - 3 = 17$$

and

$$N(t) = 20 - 17e^{-kt}$$

To evaluate k, we use the fact that $N(15) = 10$:

$$10 = N(15) = 20 - 17e^{-15k}$$

$$e^{-15k} = \frac{20 - 10}{17} = \frac{10}{17} \qquad \text{take logarithms}$$

$$-15k = \ln\left(\frac{10}{17}\right)$$

$$k = -\frac{1}{15}\ln\left(\frac{10}{17}\right) \approx 0.0354$$

Therefore, at any time t for $t \geq 0$

$$N(t) = 20 - 17e^{-0.0354t}$$

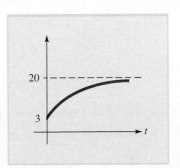

FIGURE 9.2 The learning curve $N(t) = 20 - 17e^{-0.0354t}$.

(see Figure 9.2). Finally, to find how long it takes for 17 tasks to be performed, we solve

$$N(t) = 17$$
$$20 - 17e^{-0.0354t} = 17$$
$$e^{-0.0354t} = \frac{17 - 20}{-17} = \frac{3}{17} \qquad \text{take logarithms}$$
$$-0.0354t = \ln\left(\frac{3}{17}\right)$$
$$t = \frac{1}{-0.0354}\ln\left(\frac{3}{17}\right) \approx 49$$

so it takes about 49 minutes for Tom to perform 17 of the 20 tasks.

Not all learning models involve learning in the traditional sense. In Example 9.1.6, we examine a learning model in which the learning process involves an adjustment to inventory.

EXAMPLE 9.1.6   Studying an Inventory Model

A retailer can maintain as many as 100 units of a particular product in inventory and determines that the product is being sold at a weekly rate equal to 10% of the available capacity. Suppose the retailer initially has 75 units in inventory.

a. Write an initial value problem for the number of units $N(t)$ in inventory after t weeks.

b. Solve the initial value problem in part (a) to find $N(t)$.

c. How long does it take for all the inventory to be used up?

Solution

a. Since the available capacity at any time t is $100 - N(t)$, we have

$$\underbrace{\frac{dN}{dt}}_{\substack{\text{rate of change} \\ \text{of inventory}}} = -\overbrace{0.10}^{\substack{\text{decreasing}}}\underbrace{}_{10\%}\underbrace{(100 - N)}_{\substack{\text{available} \\ \text{capacity}}} \quad \text{with } N(0) = 75$$

b. Separating the variables in the differential equation in part (a), and integrating, we find that

$$\frac{dN}{dt} = -0.10(100 - N)$$
$$\int \frac{dN}{100 - N} = \int -0.1 \, dt \qquad |100 - N| = 100 - N \text{ since } N \text{ is at most } 100$$
$$-\ln(100 - N) = -0.1t + C_1 \qquad \text{multiply by } -1 \text{ and raise to the exponent base } e$$
$$100 - N = C \, e^{0.1t} \qquad C = e^{-C_1}; \text{ solve for } N$$
$$N = 100 - C \, e^{0.1t}$$

Finally, since $N(0) = 75$, we have

$$75 = N(0) = 100 - C \, e^{0} = 100 - C \qquad \text{solve for } C$$
$$C = 100 - 75 = 25$$
$$N(t) = 100 - 25e^{0.1t} \quad \text{for all } t \geq 0$$

The graph of $N(t)$ is shown in Figure 9.3.

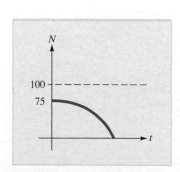

FIGURE 9.3   The inventory curve $N(t) = 100 - 25e^{0.1t}$.

c. The inventory is used up when $N(t) = 0$. Solving, we find that

$$0 = N(t) = 100 - 25e^{0.1t} \qquad \text{add } 25e^{0.1t} \text{ to both sides}$$
$$25e^{0.1t} = 100$$
$$e^{0.1t} = 4 \qquad\qquad \text{take logarithms}$$
$$0.1t = \ln 4 \qquad\qquad \text{solve for } t$$
$$t = \frac{\ln 4}{0.1} \approx 13.86$$

So in about 14 weeks, the inventory will be used up.

Population Models; Logistic Growth

While the population of a species can change only one unit at a time (a birth or death), useful information about the population can be gained by assuming the population is a continuous function of time $P(t)$. To model the population, we will assume that the relative rate of change of $P(t)$ (time rate of change in P per person) is equal to the difference between the birth rate $B(t)$ per unit time and the death rate $D(t)$; that is,

$$\underbrace{\frac{\dfrac{dP}{dt}}{P}}_{\substack{\text{rate of change} \\ \text{in } P \text{ per individual}}} = \underbrace{B(t)}_{\substack{\text{birth} \\ \text{rate}}} - \underbrace{D(t)}_{\substack{\text{death} \\ \text{rate}}}$$

or, equivalently

$$\frac{dP}{dt} = P(t)[B(t) - D(t)]$$

We refer to this as the **basic population equation.**

As might be expected, the character of the model depends on the nature of the birth and death rates. For instance, when $B(t) = B$ and $D(t) = D$ are both constant, the population equation becomes

$$\frac{dP}{dt} = kP \quad \text{where } k = B - D$$

which we solved in Section 5.1 (Example 5.1.9) to obtain

$$P(t) = P_0 e^{kt}$$

In this case, the population is said to **grow exponentially** (see Figure 9.4a).

Exponential growth often occurs if a species lives in an environment with no restrictions on food supply or space for expansion. When such limitations are present, studies show that the birth rate tends to decrease linearly with increasing population, so that $B(t) = B_0 - kP(t)$ for positive constants B_0 and k. If the death rate D remains constant, the population equation becomes

$$\frac{dP}{dt} = P(t)[(B_0 - kP(t)) - D]$$
$$= kP(M - P)$$

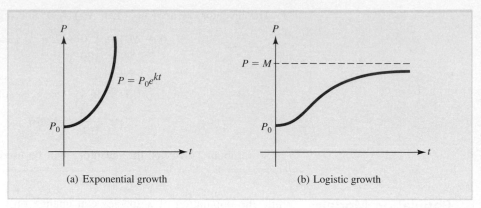

FIGURE 9.4   Exponential and logistic growth.

where $M = \dfrac{B_0 - D}{k}$ is called the **carrying capacity** of the environment. This modification of the exponential population model is called the **logistic model** for population growth. A logistic curve has a characteristic S shape and approaches the carrying capacity M as $t \to \infty$, as illustrated in Figure 9.4b and verified in Example 9.1.7.

EXAMPLE 9.1.7   Studying Logistic Growth

A population grows at a rate given by the logistic equation

$$\frac{dP}{dt} = kP(M - P) \quad \text{with } M > P(t) \text{ for all } t$$

Show that the equation has the general solution

$$P(t) = \frac{MP_0}{P_0 + (M - P_0)e^{-Mkt}}$$

where $P_0 = P(0)$. Then explain why the graph of $P(t)$ approaches the value M as $t \to \infty$.

Solution

Note that the logistic equation is separable. Separating the variables and integrating, we find that

$$\int \frac{dP}{P(M - P)} = \int k \, dt$$

$$\frac{1}{M} \ln\left|\frac{P}{M - P}\right| = kt + C_1$$

where the integration was performed using integration formula 6 in Table 6.1 (Section 6.1):

$$\int \frac{du}{u(a + bu)} = \frac{1}{a} \ln\left|\frac{u}{a + bu}\right| + C$$

with $u = P$, $a = M$, and $b = -1$.

Since $P > 0$ and $M > P$, we can remove the absolute value bars from the solution and write

$$\frac{1}{M} \ln\left(\frac{P}{M - P}\right) = kt + C_1 \qquad \text{multiply both sides by } M$$

$$\ln\left(\frac{P}{M - P}\right) = Mkt + MC_1 \qquad \text{apply exponential base } e$$

$$\frac{P}{M - P} = e^{Mkt + MC_1} = e^{Mkt} e^{MC_1}$$

$$\frac{P}{M - P} = Ce^{Mkt} \qquad \text{where } C = e^{MC_1}$$

Solving this equation for P, we find that

$$P(t) = \frac{MC}{C + e^{-Mkt}}$$

and since $P(0) = P_0$, we have

$$P_0 = P(0) = \frac{MC}{C + e^0} = \frac{MC}{C + 1}$$

so that

$$C = \frac{P_0}{M - P_0}$$

Substituting into our formula for $P(0)$, we get

$$P(t) = \frac{\dfrac{MP_0}{M - P_0}}{\dfrac{P_0}{M - P_0} + e^{-Mkt}} \qquad \text{simplify the complex fraction}$$

so

$$P(t) = \frac{MP_0}{P_0 + (M - P_0)e^{-Mkt}}$$

as claimed.

Finally, as $t \to \infty$, note that $e^{-Mkt} \to 0$, so that

$$\lim_{t \to \infty} P(t) = \lim_{t \to \infty} \frac{MP_0}{P_0 + (M - P_0)e^{-Mkt}} = \frac{MP_0}{P_0 + 0}$$

$$= M$$

which means that $P(t)$ tends toward the value M as $t \to \infty$.

EXPLORE!

Suppose the membership of a private club increases logistically over a 7-year period as indicated in the following table:

Weeks	Membership
1	45
2	75
3	200
4	450
5	595
6	700
7	760

Use the statistical modeling feature of your graphing calculator to fit a logistic curve to the data, and determine the projected upper limit of the membership based on the logistic model.

As a specific example of logistic growth, let $P(t)$ be the population of the United States, where t is the number of years after 1900. The population of the United States in 1900 was 76 million, so $P_0 = P(0) = 76$. There are still two parameters, k and M, in the formula for $P(t)$ to be evaluated, so we need two additional side conditions. The population of the United States was 150.7 million in 1950 and 281.4 million in

2000, so $P(50) = 150.7$ and $P(100) = 281.4$. With this information (and a lot of algebra!), we find that

$$M = 1{,}262 \quad \text{and} \quad k = 0.015$$

and obtain the formula

$$P(t) = \frac{95{,}904}{76 + 1{,}106e^{-0.015t}}$$

which can then be used to predict future populations. For instance, the model predicts that the population of the United States in 2015 ($t = 115$) will be

$$P(115) = \frac{95{,}904}{76 + 1{,}106e^{-0.015(115)}} \approx 333.8 \text{ million people}$$

Perhaps more important, the model predicts that the carrying capacity of the United States is $M = 1{,}262$ million people. This suggests that the environment of the United States can support a maximum of 1 billion 262 million people.

Population models actually apply in any situation where the number of individuals in a group satisfies the basic population equation. For instance, the future value of a bank account growing with continuous compounding of interest may be regarded as a population model in which the population is the number of dollars in the account. In Section 6.1, we used a logistic model to study the spread of a rumor (Example 6.1.10), where the population was the number of people who had heard the rumor, at time t. Logistic models can also be used to study the spread of an infection, the effect of radioactive decay, the growth of bacteria, and the concentration of a drug in a person's bloodstream. Several such applications are featured in the exercises.

EXERCISES ■ 9.1

In Exercises 1 through 20, find the general solution of the given differential equation. You will need integration by parts in Exercises 17 through 20.

1. $\dfrac{dy}{dx} = 3x^2 + 5x - 6$

2. $\dfrac{dP}{dt} = \sqrt{t} + e^{-t}$

3. $\dfrac{dy}{dx} = 3y$

4. $\dfrac{dy}{dx} = y^2$

5. $\dfrac{dy}{dx} = e^y$

6. $\dfrac{dy}{dx} = e^{x+y}$

7. $\dfrac{dy}{dx} = \dfrac{x}{y}$

8. $\dfrac{dy}{dx} = \dfrac{y}{x}$

9. $\dfrac{dy}{dx} = \sqrt{xy}$

10. $\dfrac{dy}{dx} = \dfrac{y^2 + 4}{xy}$

11. $\dfrac{dy}{dx} = \dfrac{y}{x - 1}$

12. $\dfrac{dy}{dx} = e^y\sqrt{x + 1}$

13. $\dfrac{dy}{dx} = \dfrac{y + 3}{(2x - 5)^6}$

14. $\dfrac{dy}{dx} = e^{-2y}(x - 2)^9$

15. $\dfrac{dx}{dt} = \dfrac{xt}{2t + 1}$

16. $\dfrac{dy}{dt} = \dfrac{te^y}{2t - 1}$

17. $\dfrac{dy}{dx} = xe^{x-y}$

18. $\dfrac{dw}{ds} = \dfrac{se^{2w}}{w}$

19. $\dfrac{dy}{dt} = y \ln\sqrt{t}$

20. $\dfrac{dx}{dt} = \dfrac{\ln t}{\ln x}$

In Exercises 21 through 28, find the particular solution of the differential equation satisfying the indicated condition.

21. $\dfrac{dy}{dx} = e^{5x}$; $y = 1$ when $x = 0$

22. $\dfrac{dy}{dx} = 5x^4 - 3x^2 - 2$; $y = 4$ when $x = 1$

23. $\dfrac{dy}{dx} = \dfrac{x}{y^2}$; $y = 3$ when $x = 2$

24. $\dfrac{dy}{dx} = 4x^3y^2$; $y = 2$ when $x = 1$

25. $\dfrac{dy}{dx} = y^2\sqrt{4-x}$; $y = 2$ when $x = 4$

26. $\dfrac{dy}{dx} = xe^{y-x^2}$; $y = 0$ when $x = 1$

27. $\dfrac{dy}{dt} = \dfrac{y+1}{t(y-1)}$; $y = 2$ when $t = 1$

$$\left[\text{Hint: } \frac{y-1}{y+1} = 1 - \frac{2}{y+1}\right]$$

28. $\dfrac{dx}{dt} = xt\sqrt{t+1}$; $x = 1$ when $t = 0$

In Exercises 29 through 40, write a differential equation describing the given situation. Define all variables you introduce. (Do not try to solve the differential equation at this time.)

29. INVESTMENT GROWTH An investment grows at a rate equal to 7% of its size.

30. MARGINAL COST A manufacturer's marginal cost is $60 per unit.

31. GROWTH OF BACTERIA The number of bacteria in a culture grows at a rate that is proportional to the number present.

32. CONCENTRATION OF DRUGS The rate at which the concentration of a drug in the bloodstream decreases is proportional to the concentration.

33. POPULATION GROWTH The population of a certain town increases at the constant rate of 500 people per year.

34. RADIOACTIVE DECAY A sample of radium decays at a rate that is proportional to its size.

35. TEMPERATURE CHANGE The rate at which the temperature of an object changes is proportional to the difference between its own temperature and the temperature of the surrounding medium.

36. DISSOLUTION OF SUGAR After being placed in a container of water, sugar dissolves at a rate proportional to the amount of undissolved sugar remaining in the container.

37. RECALL FROM MEMORY When a person is asked to recall a set of facts, the rate at which the facts are recalled is proportional to the number of relevant facts in the person's memory that have not yet been recalled.

38. MARKET SHARE The rate at which a new product is replacing an old, obsolete product is proportional to the market share of the old product.

39. CORRUPTION IN GOVERNMENT The rate at which people are implicated in a government scandal is jointly proportional to the number of people already implicated and the number of people involved who have not yet been implicated.

40. SPREAD OF A RUMOR The rate at which a rumor spreads through a community is jointly proportional to the number of people in the community who have heard the rumor and the number who have not.

41. Verify that the function $y = Ce^{kx}$ is a solution of the differential equation $\dfrac{dy}{dx} = ky$.

42. Verify that the function $Q = B - Ce^{-kt}$ is a solution of the differential equation $\dfrac{dQ}{dt} = k(B - Q)$.

43. Verify that $y = C_1e^x + C_2xe^x$ is a solution of the differential equation $\dfrac{d^2y}{dx^2} - 2\dfrac{dy}{dx} + y = 0$.

44. Verify that the function $y = \dfrac{1}{20}x^4 - \dfrac{C_1}{x} + C_2$ is a solution of the differential equation $x\dfrac{d^2y}{dx^2} + 2\dfrac{dy}{dx} = x^3$.

45. INVENTORY MANAGEMENT Jill owns an electronics store with storage capacity for 70 iPads. She currently has 50 iPads in inventory and determines that they are selling at a daily rate equal to 15% of the available capacity. When will Jill sell out of iPads?

46. INVENTORY MANAGEMENT Suppose Jill, the store owner in Exercise 45, finds that her iPads are actually selling at a rate jointly proportional to the number of units in inventory and the available storage capacity. She plans to reorder when only 5 units remain unsold. If half her original inventory of 50 iPads has been sold after 15 days, when should she reorder?

47. MARKET SHARE A new product is replacing an old, obsolete product at a rate jointly proportional to the market share of the new product and that of the old product. If the new product has only 10% of the market at time $t = 0$ and the two products have equal market share after 6 months, how long does it take before the new product controls 95% of the total market?

48. STOCK INVESTMENT A new offering of 270,000 shares of a growth stock is presented to the market, and 100,000 shares are purchased when the market opens at 6.30 A.M. It is found that t hours later, sales of the shares are changing at a rate jointly proportional to the number already sold and the number unsold. If half the shares are sold by noon, how many will be sold by the end of the trading day at 2.00 P.M.? (You may assume that shares purchased during the day are retained by the buyer and not resold.)

49. VALUE OF A DEPRECIATING ASSET The resale value of an industrial machine decreases at a rate proportional to the difference between its current value and its scrap value of $5,000. The machine was bought new for $40,000 and was worth $30,000 after 4 years. How much will it be worth when it is 8 years old?

50. VALUE OF A DEPRECIATING ASSET Repeat Exercise 49 for a machine initially bought for $100,000, if the machine is worth $60,000 after 4 years, and its scrap value is $12,000.

51. WORKER EFFICIENCY For $0 \le p \le 1$, let $p(t)$ be the likelihood that an assembly line worker will make a mistake t hours into the worker's 8-hour shift. A particular worker, Stefano, never makes a mistake at the beginning of his shift and is only 5% likely to make a mistake at the end. Set up and solve a differential equation assuming that at each time t, the likelihood of Stefano making an error increases at a rate proportional to the likelihood $1 - p(t)$ that a mistake has not already been made.

52. WORKER EFFICIENCY Lou, a coworker of Stefano in Exercise 51, has a 1% likelihood of making an error at the beginning of her shift but only a 3% likelihood of making an error at the end. Set up and solve a differential equation assuming that at each time t, the likelihood $p(t)$ of Lou making an error increases at a rate jointly proportional to $p(t)$ and the likelihood $1 - p(t)$ that a mistake has not already been made.

53. OIL PRODUCTION An oil well that yields 400 barrels of crude oil per month will run dry in 2 years. The price of crude oil is currently $130 per barrel and is expected to rise at the constant rate of 4 cents per barrel per month. If the oil is sold as soon as it is extracted from the ground, how much total revenue will be obtained from the well?

54. OIL PRODUCTION Suppose the owner of the oil well in Exercise 53 decides to step up

production so that 600 barrels per month are extracted but everything else remains the same.
a. How many months pass before the well runs dry?
b. How much total revenue will be obtained from the well?

55. SUPPLY The supply $S(p)$ of a commodity increases with respect to price p at a rate proportional to the difference between the current supply and a limiting value of 10,000 units imposed by market conditions. Suppose 1,000 units are supplied when the price is $4 per unit and 2,000 units are supplied when the price is $6 per unit.
a. Set up and solve an initial value problem for $S(p)$.
b. How many units are supplied when the price is $8 per unit?
c. At what price are 5,000 units supplied?

56. DEMAND The demand $D(p)$ of a commodity changes with respect to price p at a rate inversely proportional to the price. Suppose 8,000 units are demanded at a price of $1 per unit and 4,000 units are demanded when $p = $3 per unit.
a. Set up and solve an initial value problem for $D(p)$.
b. How many units are demanded when the price is $5 per unit?
c. At what price are 2,000 units demanded?

LIFE AND SOCIAL SCIENCE APPLIED PROBLEMS

57. SPREAD OF AN EPIDEMIC The rate at which an epidemic spreads through a community is jointly proportional to the number of residents who have been infected and the number of susceptible residents who have not. Show that the epidemic is spreading most rapidly when one-half of the susceptible residents have been infected. [*Hint:* You do not have to solve a differential equation to do this. Just start with a formula for the *rate* at which the epidemic is spreading and use calculus to maximize this rate.]

58. SPREAD OF AN EPIDEMIC The rate at which an epidemic spreads through a community with 2,000 susceptible residents is jointly proportional to the number of residents who have been infected and the number of susceptible residents who have not. Express the number of residents who have been infected as a function of time (in weeks) if 500 residents had the disease initially and 855 residents had been infected by the end of the first week.

59. CORRUPTION IN GOVERNMENT The number of people implicated in a certain major government scandal increases at a rate jointly proportional to the number of people already implicated and the number involved who have not yet been implicated. Suppose that 7 people were implicated when a Washington newspaper first made the scandal public, that 9 more were implicated over the next 3 months, and that another 12 were implicated during the following 3 months. Approximately how many people are involved in the scandal? [*Warning:* This problem will test your algebraic ingenuity!]

60. POPULATION GROWTH Demographers estimate that the environment of a certain country can support no more than 10 million people and that the population is growing at a rate proportional to the difference between this upper limit and the current population. The current population is 2 million and 5 years ago, it was 1.7 million.
 a. Set up and solve an initial value problem for the population $P(t)$ (in millions).
 b. What will the population be 10 years from now?

61. POPULATION GROWTH Demographers estimate that environmental factors place an upper bound of 10 million people on the population of a certain country. At each time t, the population $P(t)$ is growing at a rate that is jointly proportional to the current population P and the difference between the upper bound and P. Suppose the population of this country was 4 million in 2005 and 7.4 million in 2009. Set up and solve a differential equation to find $P(t)$. What will the population be in 2015?

62. AVIAN POPULATION A bird species is introduced to a favorable environment in which its birth rate $B(t)$ and death rate $D(t)$ are both inversely proportional to the **square root** of the population $P(t)$. Initially, there are 100 birds and the population is increasing at the rate of 2 birds per month.
 a. Set up and solve an initial value problem for $P(t)$. [*Hint:* What are P and P' when $t = 0$?]
 b. How many birds are there after 1 year?
 c. How long does it take for the bird population to exceed 400?

63. RUNAWAY POPULATION GROWTH Consider a prolific rodent species whose birth rate $B(t)$ and death rate $D(t)$ are both proportional to the population $P(t)$. Initially, there are 100 rodents and the population is increasing at the rate of 2 rodents per month.
 a. According to the basic population equation, which of the following differential equations models the rodent population for some constant k?
 (1) $\dfrac{dP}{dt} = kP$ or (2) $\dfrac{dP}{dt} = kP^2$

 b. Find k. [*Hint:* What are P and P' when $t = 0$?]
 c. Solve an appropriate initial value problem to find $P(t)$.
 d. What is the rodent population after 1 year? After 4 years?
 e. Sketch the graph of $P(t)$. How long does it take for the rodent population to become effectively infinite? That is, for what value c does $\lim\limits_{t \to c} P(t) = \infty$?

64. POPULATION EXPLOSION OR EXTINCTION The birth rate of a certain animal species is 0.04% of the population $P(t)$ at each time t (months) and the death rate is 50 individuals per month.
 a. Which of the following differential equations models the population $P(t)$ in thousands?
 (1) $\dfrac{dP}{dt} = 0.0004P(P - 125)$

 (2) $\dfrac{dP}{dt} = 0.0004P(P - 50)$

 b. Solve the differential equation in part (a) for $P(t)$ in terms of the initial population $P_0 = P(0)$.
 c. Suppose the initial population is 150,000. Sketch the graph of $P(t)$. What is the population after 2 years (24 months)? After 35 months? This is called **exploding** or **runaway** population growth.
 d. Suppose, instead, the initial population is 100,000. Sketch the graph of $P(t)$. What is the population after 2 years? Show that the species is virtually extinct after 20 years.

Exercises 65 and 66 use trigonometric functions introduced in Chapter 8.

65. HUMAN HORMONAL EXCRETION The human body excretes certain hormones periodically. Suppose the rate of such an excretion in a particular patient is

$$H'(t) = \cos\left(\frac{\pi t}{12}\right)H \quad \text{for } 0 \le t \le 24$$

grams per hour t hours after midnight. The patient's body originally contains 300 grams of the hormone, and her diet is restricted so that no new hormone is produced during a 24-hour observational period.
 a. Find $H(t)$.
 b. How much hormone remains at the end of the period (that is, at $t = 24$)?

66. HUMAN HORMONAL EXCRETION Answer the questions in Exercise 65 for the excretion rate

$$H'(t) = \sin\left(\frac{\pi(t - 6)}{12}\right)H \quad \text{for } 0 \le t \le 24$$

67. IRRIGATION   A reservoir is filled with 216,000 ft^3 of water. Water flows into the reservoir at the rate of 5,000 ft^3/week and is used for irrigation at the same rate. Water also evaporates from the reservoir at a weekly rate proportional to $V^{2/3}$, where $V(t)$ is the volume of water in the reservoir at time t (weeks). After 5 weeks, the reservoir contains 180,000 ft^3 of water.

  a. Set up and solve an initial value problem to find $V(t)$. For simplicity, assume that V is expressed in thousands of cubic feet.

  b. How long does it take for the reservoir to drain?

68. IRRIGATION   A reservoir in the shape of a cylinder with radius 25 feet is filled with water to a depth of 40 feet. Water flows into the reservoir at the rate of 3,000 ft^3/week and is used for irrigation at the same rate. Water also evaporates at a rate proportional to the surface area of the water. After 1 week, the reservoir contains 75,000 ft^3 of water.

  a. Set up and solve an initial value problem for the volume of water in the reservoir. It will help to know that a cylinder of radius r and height h has volume $\pi r^2 h$ and top surface area πr^2.

  b. How long does it take for the reservoir to drain?

69. LOGISTIC CURVES   Show that if a quantity Q satisfies the differential equation $\dfrac{dQ}{dt} = kQ(B - Q)$, where k and B are positive constants, then the rate of change $\dfrac{dQ}{dt}$ is greatest when $Q(t) = \dfrac{B}{2}$. What does this result tell you about the inflection point of a logistic curve? Explain. [*Hint:* See the hint for Exercise 57.]

70. FICK'S LAW   When a cell is placed in a liquid containing a solute, the solute passes through the cell wall by diffusion. As a result, the concentration of the solute inside the cell changes, increasing if the concentration of the solute outside the cell is greater than the concentration inside and decreasing if the opposite is true. In biology, Fick's law asserts that the concentration of the solute inside the cell changes at a rate that is jointly proportional to the area of the cell wall and the difference between the concentrations of the solute inside and outside the cell. Assuming that the concentration of the solute outside the cell is constant and greater than the concentration inside, derive a formula for the concentration of the solute inside the cell.

SECTION 9.2   First-Order Linear Differential Equations

Learning Objectives

  1. Solve first-order linear differential equations and initial value problems.

  2. Explore compartmental analysis with applications to finance, drug administration, and dilution models.

A first-order linear differential equation is one with the general form

$$\frac{dy}{dx} + p(x)y = q(x)$$

For instance,

$$\frac{dy}{dx} + \frac{y}{x} = e^{-x} \qquad \text{for } x > 0$$

is a first-order linear differential equation with $p(x) = \dfrac{1}{x}$ and $q(x) = e^{-x}$. In its present form, this equation cannot be solved by simple integration or by separation of variables. However, if you multiply both sides of the equation by x, it becomes

$$x\frac{dy}{dx} + y = xe^{-x}$$

Notice that the left side of this equation is the derivative (with respect to x) of the product xy since

$$\frac{d}{dx}(xy) = x\frac{dy}{dx} + y\frac{dx}{dx} = x\frac{dy}{dx} + y$$

Thus, the given differential equation now reads

$$\frac{d}{dx}(xy) = xe^{-x}$$

which you can solve by integrating both sides with respect to x:

$$\int \frac{d}{dx}(xy)\,dx = \int xe^{-x}\,dx$$

$$xy = \int xe^{-x}\,dx = -(x+1)e^{-x} + C \quad \text{using integration by parts}$$

Solving for y, you conclude that the given first-order linear differential equation has the general solution

$$y = \frac{1}{x}[e^{-x}(-x-1) + C]$$

In this example, multiplication by the function x converted the given first-order linear differential equation into one that could be solved by simple integration. Such a function is called an **integrating factor** of the differential equation. In more advanced texts, it is shown that every first-order linear differential equation $\frac{dy}{dx} + p(x)y = q(x)$ has an integrating factor equal to $I(x) = e^{\int p(x)\,dx}$. For instance, in the introductory example, where $p(x) = \frac{1}{x}$, the integrating factor is

$$I(x) = e^{\int(1/x)\,dx} = e^{\ln x} = x \quad \text{since } x > 0$$

To summarize:

Solution of a First-Order Linear Differential Equation ■ The first-order linear differential equation $\dfrac{dy}{dx} + p(x)y = q(x)$ has the general solution

$$y = \frac{1}{I(x)}\left[\int I(x)q(x)\,dx + C\right]$$

where C is an arbitrary constant and $I(x)$ is the integrating factor

$$I(x) = e^{\int p(x)\,dx}$$

EXAMPLE 9.2.1 Solving a First-Order Linear Differential Equation

Find the general solution of the differential equation

$$\frac{dy}{dx} + 2y = 2x$$

Solution

This is a first-order linear differential equation with $p(x) = 2$ and $q(x) = 2x$. The integrating factor is

$$I(x) = e^{\int 2\, dx} = e^{2x}$$

and the general solution of the first-order linear equation is

$$y = \frac{1}{e^{2x}}\left[\int 2x\, e^{2x}\, dx + C\right]$$

The integral on the right can be found using integration by parts:

$$\int 2x\, e^{2x}\, dx = \left(x - \frac{1}{2}\right)e^{2x} = xe^{2x} - \frac{1}{2}e^{2x}$$

(Verify the details.) Thus, the general solution is

$$y = \frac{1}{e^{2x}}\left[xe^{2x} - \frac{1}{2}e^{2x} + C\right] = x - \frac{1}{2} + Ce^{-2x}$$

EXAMPLE 9.2.2 Solving a First-Order Linear Initial Value Problem

Find the general solution of the differential equation

$$\frac{dy}{dx} = \frac{2x + y}{x}$$

and the particular solution that satisfies $y = -2$ when $x = 1$.

Solution

The given differential equation can be rewritten as

$$\frac{dy}{dx} = \frac{2x + y}{x} = 2 + \frac{y}{x} \qquad \text{divide } 2x + y \text{ by } x$$

$$\frac{dy}{dx} - \frac{y}{x} = 2 \qquad \text{subtract } \frac{y}{x} \text{ from both sides}$$

We recognize this as a first-order linear differential equation with $p(x) = -\frac{1}{x}$ and $q(x) = 2$. The integrating factor is

$$I(x) = e^{\int(-1/x)\, dx} = e^{-\ln x} \qquad -\ln x = \ln x^{-1}$$
$$= e^{\ln x^{-1}} = x^{-1} = \frac{1}{x}$$

and the general solution of the equation is

$$y = \frac{1}{1/x}\left[\int \frac{1}{x}(2)\, dx + C\right] = x\left[\int \frac{2}{x}\, dx + C\right]$$
$$= x[2\ln x + C]$$

To find the particular solution, substitute $x = 1$ and $y = -2$ into the general solution to get

$$-2 = (1)[2\ln(1) + C] = 2(0) + C = C$$

Thus, the particular solution is

$$y = x[2 \ln x - 2] = 2x[\ln x - 1]$$

Compartmental Analysis

Complex processes in business, economics or the life, social, and physical sciences can often be split up into separate stages and then analyzed by studying the individual components. The stages in this procedure are called blocks or compartments, and the procedure itself is known as **compartmental** (or "black box") **analysis.**

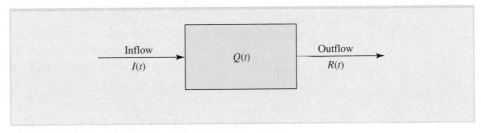

FIGURE 9.5 A diagram illustrating compartmental analysis.

In this section, we will consider only one-compartment systems like the one illustrated in Figure 9.5. In such a system, a quantity $Q(t)$ flows into the system at the input rate $I(t)$, is modified in some way, and then flows out at an output (removal) rate $R(t)$. The net rate of change of $Q(t)$ is then the difference between the input rate and the output rate, so

$$\underset{\substack{\text{input}\\\text{rate}}}{\frac{dQ}{dt}} = \underset{\substack{\text{input}\\\text{rate}}}{I(t)} - \underset{\substack{\text{output}\\\text{rate}}}{R(t)}$$

Compartmental analysis is illustrated in Examples 9.2.3 through 9.2.5. In Example 9.2.3, we use the procedure to examine a financial model.

EXAMPLE 9.2.3 Studying a Personal Finance Model

David deposits $20,000 into an account in which interest accumulates at the rate of 5% per year, compounded continuously. He plans to withdraw $3,000 per year.

a. Set up and solve a differential equation to determine the value $Q(t)$ of his account t years after the initial deposit.

b. How long does it take for his account to be exhausted?

Solution

a. If no withdrawals were made, the value of the account would change at a percentage rate equal to the annual interest rate; that is,

$$\frac{100Q'(t)}{Q(t)} = 5$$

or, equivalently, $Q'(t) = 0.05Q(t)$. This is the rate at which interest is added to the account, and by subtracting the annual withdrawal rate of \$3,000, we obtain the net rate of change of the account; that is,

$$\underbrace{\frac{dQ}{dt}}_{\substack{\text{net rate of} \\ \text{change of } Q}} = \underbrace{0.05Q}_{\substack{\text{rate at which} \\ \text{interest is added}}} - \underbrace{3{,}000}_{\substack{\text{rate at which} \\ \text{money is withdrawn}}}$$

Rewriting this equation as

$$\frac{dQ}{dt} - 0.05Q = -3{,}000$$

we recognize it as a first-order linear differential equation with $p(t) = -0.05$ and $q(t) = -3{,}000$ that we wish to solve subject to the initial condition that $Q(0) = 20{,}000$. The integrating factor for this equation is

$$I(t) = e^{\int -0.05\, dt} = e^{-0.05t}$$

so the general solution is

$$Q(t) = \frac{1}{e^{-0.05t}}\left[\int e^{-0.05t}(-3{,}000)\, dt + C\right] \quad \text{substitute } u = -0.05t$$

$$= e^{0.05t}\left[-3{,}000\left(\frac{e^{-0.05t}}{-0.05}\right) + C\right]$$

$$= 60{,}000 + Ce^{0.05t}$$

Since $Q(0) = 20{,}000$, we have

$$Q(0) = 20{,}000 = 60{,}000 + Ce^0 = 60{,}000 + C$$

so that

$$C = -40{,}000$$

and therefore,

$$Q(t) = 60{,}000 - 40{,}000e^{0.05t}$$

b. The account becomes exhausted when $Q(t) = 0$. Solving the equation

$$Q(t) = 0 = 60{,}000 - 40{,}000e^{0.05t}$$

we get

$$40{,}000e^{0.05t} = 60{,}000 \qquad \text{divide both sides by 40,000}$$

$$e^{0.05t} = \frac{60{,}000}{40{,}000} = 1.5 \qquad \text{take logarithms on both sides}$$

$$0.05t = \ln 1.5 \qquad \text{divide both sides by 0.05}$$

$$t = \frac{\ln 1.5}{0.05}$$

$$\approx 8.11$$

Thus, the account is exhausted in approximately 8 years.

In Example 9.2.4, we show how compartmental analysis can be used to assist a physician in making a decision regarding the medication administered to a patient.

EXAMPLE 9.2.4 Studying Drug Administration

A drug is infused into a patient's bloodstream at a constant rate r (the *infusion rate*) and is eliminated from the bloodstream at a rate proportional to the amount of drug present at time t. Initially (when $t = 0$), the patient's blood contains no drug.

a. Set up and solve a differential equation for the amount of drug $A(t)$ present in the patient's bloodstream at time t.

b. Show that $A(t)$ approaches a constant value as $t \to \infty$. This is the *steady-state* level of medication.

c. Suppose an attending physician sets the infusion rate for a particular patient at 2 mg per hour and that a sample is taken indicating there are 1.69 mg of the drug in the patient's blood after 1 hour. How much medication will be present after 6 hours? What is the steady-state level of medication?

d. Suppose the physician in part (c) wants the steady-state level of medication in the patient's bloodstream to be 8 mg. How should the infusion rate be adjusted to achieve this goal?

Solution

a. The amount of drug $A(t)$ in the bloodstream at time t changes at the rate

$$\frac{dA}{dt} = \underbrace{r}_{\text{infusion}} - \underbrace{kA}_{\text{elimination}}$$

where k is a constant (the *elimination coefficient*). Rewriting this equation as

$$\frac{dA}{dt} + kA = r$$

we recognize it as a first-order linear differential equation with $p(t) = k$ and $q(t) = r$. The integrating factor is

$$I(t) = e^{\int k\, dt} = e^{kt}$$

so the general solution is

$$A(t) = \frac{1}{e^{kt}}\left[\int re^{kt}\, dt + C\right]$$
$$= e^{-kt}\left[\frac{re^{kt}}{k} + C\right]$$
$$= \frac{r}{k} + Ce^{-kt}$$

Since there is no drug initially present in the blood, we have $A(0) = 0$ so that

$$A(0) = 0 = \frac{r}{k} + Ce^0$$

Thus, $C = -\dfrac{r}{k}$, and

$$A(t) = \frac{r}{k} - \frac{r}{k}e^{-kt} = \frac{r}{k}[1 - e^{-kt}]$$

b. Since

$$\lim_{t\to\infty} A(t) = \lim_{t\to\infty} \frac{r}{k}[1 - e^{-kt}] = \frac{r}{k}(1 - 0) = \frac{r}{k}$$

the steady-state level of medication is $\dfrac{r}{k}$.

c. In this particular case, we have $r = 2$ and $A(1) = 1.69$. Substituting this information into the formula derived for $A(t)$ in part (a), we get

$$A(1) = 1.69 = \frac{2}{k}[1 - e^{-k(1)}]$$

Using a graphing calculator, we solve this equation to obtain $k = 0.35$. Thus, after 6 hours, the amount of drug in the patient's bloodstream is

$$A(6) = \frac{2}{0.35}[1 - e^{-0.35(6)}] \approx 5.01 \text{ mg}$$

The steady-state level of medication is

$$\frac{r}{k} = \frac{2}{0.35} \approx 5.71 \text{ mg}$$

d. We assume that the elimination coefficient k depends on the patient's metabolism and not on the infusion rate r. (Does this assumption seem reasonable to you?) Therefore, with k fixed at 0.35, a steady-state level of 8 mg is achieved by adjusting the infusion rate r to satisfy

$$\frac{r}{0.35} = 8$$

so that

$$r = 2.8 \text{ mg/hr}$$

The outflow rate in a compartmental analysis model can often be written in the form $R(t) = k\,Q(t)$ since this rate represents the proportion of $Q(t)$ that is being removed from the compartment. In this context, the multiplier k is called the **transfer coefficient.** This coefficient was constant in Examples 9.2.3 and 9.2.4, but in Example 9.2.5, we examine a dilution problem in which the transfer coefficient varies with time.

EXAMPLE 9.2.5 Studying a Dilution Model

A 70-gallon tank initially contains 20 pounds of salt dissolved in 50 gallons of water. Suppose that 3 gallons of brine containing 2 pounds of dissolved salt per gallon runs into the tank every minute and that the mixture (kept uniform by stirring) runs out of the tank at the rate of 2 gallons per minute.

a. Set up and solve a differential equation to find an expression for the amount of salt in the tank after t minutes.

b. How much salt will be in the tank at the instant it begins to overflow?

Solution

a. Let $S(t)$ be the amount of salt in the tank at time t. Since 3 gallons of brine flow into the tank each minute and each gallon contains 2 pounds of salt, then $(3)(2) = 6$ pounds of salt flow into the tank each minute. To find the number of pounds of salt flowing out of the tank each minute, note that at time t, there are $S(t)$ pounds of salt and $50 + (3 - 2)t = 50 + t$ gallons of solution in the tank (because there is a net increase of $3 - 2 = 1$ gallon of solution every minute). Thus, the concentration of salt in solution at time t is $\dfrac{S(t)}{50 + t}$ pounds per gallon, and salt is leaving the tank at the rate

$$\left[\frac{S(t)}{50 + t} \text{ pounds/gallon}\right][2 \text{ gallons/minute}] = \frac{2S(t)}{50 + t} \text{ pounds/minute}$$

It follows that the net rate of change $\dfrac{dS}{dt}$ of salt in the tank is given by

$$\underbrace{\frac{dS}{dt}}_{\text{net rate of change}} = \underbrace{6}_{\text{inflow}} - \underbrace{\frac{2S}{50 + t}}_{\text{outflow}}$$

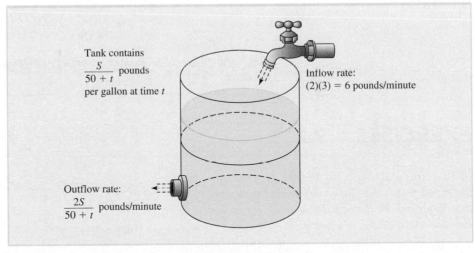

Tank contains
$\dfrac{S}{50 + t}$ pounds
per gallon at time t

Inflow rate:
$(2)(3) = 6$ pounds/minute

Outflow rate:
$\dfrac{2S}{50 + t}$ pounds/minute

FIGURE 9.6 The net rate of change of salt in solution equals inflow rate minus outflow rate.

(see Figure 9.6). This equation can be written as

$$\frac{dS}{dt} + \frac{2S}{50 + t} = 6$$

which is a first-order linear differential equation with $p(t) = \dfrac{2}{50 + t}$ and $q(t) = 6$.

The integrating factor is

$$I(t) = e^{\int p(t)\, dt} = e^{\int [2/(50+t)]\, dt}$$
$$= e^{2 \ln|50 + t|} = e^{\ln|50 + t|^2}$$
$$= (50 + t)^2$$

Hence, the general solution is

$$S(t) = \frac{1}{(50 + t)^2}\left[\int (6)(50 + t)^2\, dt + C\right]$$

$$= \frac{1}{(50 + t)^2}[2(50 + t)^3 + C]$$

$$= 2(50 + t) + \frac{C}{(50 + t)^2}$$

To find C, note that there are 20 pounds of salt initially in the tank. Thus,

$$20 = S(0) = 2(50 + 0) + \frac{C}{(50 + 0)^2}$$

so that $C = -80(50)^2$. It follows that the amount of salt at time t is

$$S(t) = 2(50 + t) - \frac{80(50)^2}{(50 + t)^2}$$

b. After t minutes, the tank contains $50 + t$ gallons of fluid, and it begins to overflow when the amount of fluid reaches its capacity, 70 gallons. Solving the equation $50 + t = 70$, we see that this occurs when $t = 20$ minutes. By substituting $t = 20$ into the formula obtained in part (a), we find that

$$S(20) = 2(50 + 20) - \frac{80(50)^2}{(50 + 20)^2}$$

$$\approx 99.18$$

Thus, the tank contains approximately 99 pounds of salt when it begins to overflow.

EXERCISES ■ 9.2

In Exercises 1 through 18, find the general solution of the given first-order linear differential equation. Notice that Exercises 15 through 18 involve trigonometric functions introduced in Chapter 8.

1. $\dfrac{dy}{dx} + \dfrac{3y}{x} = x$

2. $\dfrac{dy}{dx} + \dfrac{2y}{x} = \sqrt{x} + 1$

3. $\dfrac{dy}{dx} + \dfrac{y}{2x} = \sqrt{x}\, e^x$

4. $x^4\dfrac{dy}{dx} + 2x^3y = 5$

5. $x^2\dfrac{dy}{dx} + xy = 2$

6. $x\dfrac{dy}{dx} + 2y = xe^{x^3}$

7. $\dfrac{dy}{dx} + \left(\dfrac{2x + 1}{x}\right)y = e^{-2x}$

8. $\dfrac{dy}{dx} + \dfrac{xy}{x^2 + 1} = x$

9. $\dfrac{dy}{dx} = \dfrac{1 + xy}{1 + x}$

10. $\dfrac{dy}{dx} = \dfrac{x + y}{2x + 1}$

11. $\dfrac{dx}{dt} + \dfrac{x}{1 + t} = t$

12. $\dfrac{dy}{dt} - \dfrac{y}{t} = te^{-t}$

13. $\dfrac{dy}{dt} + \dfrac{ty}{t + 1} = t + 1$

14. $\dfrac{dx}{dt} + \dfrac{x}{1 + 2t} = 5$

15. $\dfrac{dy}{dx} = \dfrac{\sin x - y}{x}$

16. $\dfrac{dy}{dx} - \dfrac{2y}{x} = x^2 \cos x$

17. $\dfrac{dy}{dx} + y \tan x = \sin x$

18. $\dfrac{dy}{dx} + y = e^{-x} \cos x$

In Exercises 19 through 24, find the particular solution of the given differential equation that satisfies the given condition.

19. $x\dfrac{dy}{dx} - 2y = 2x^3$; $y = 2$ when $x = 1$

20. $\dfrac{dy}{dx} + y = x^2$; $y = 2$ when $x = 0$

21. $\dfrac{dy}{dx} + xy = x + e^{-x^2/2}$; $y = -1$ when $x = 0$

22. $\dfrac{dy}{dx} + y = x$; $y = 4$ when $x = 0$

23. $\dfrac{dy}{dx} + \dfrac{y}{x} = \dfrac{1}{x^2}$; $y = -2$ when $x = 1$

24. $\dfrac{dy}{dx} - \dfrac{y}{x} = \ln x$; $y = 3$ when $x = 1$

Differential equations can often be solved in several different ways. In Exercises 25 through 28, solve the given differential equation by regarding it as (a) separable and (b) first-order linear.

25. $\dfrac{dy}{dx} + 3y = 5$

26. $\dfrac{dy}{dx} + \dfrac{y}{x} = 0$

27. $\dfrac{dy}{dx} + \dfrac{y}{x+1} = \dfrac{2}{x+1}$

28. $\dfrac{dy}{dx} = 1 + x + y + xy$ (*Hint:* Factor.)

BUSINESS AND ECONOMICS APPLIED PROBLEMS

29. **PRICE ADJUSTMENT FOR SUPPLY AND DEMAND** At time t (months), the price $p(t)$ of a certain commodity changes at a rate equal to 20% of the shortage $D - S$, where $S(t) = 2 + p$ and $D(t) = 14 - t$ are the supply and demand for the commodity, respectively, both in thousands of units. If the price is \$2 when $t = 0$, what is the price after 4 months?

30. **PRICE ADJUSTMENT FOR SUPPLY AND DEMAND** A commodity is introduced with an initial price of \$5 per unit, and t months later, the price is $p(t)$ dollars per unit. A study indicates that at time t, the price changes at a rate equal to 2% of the shortage $D - S$, where

$S(t) = 2 + p$ and $D(t) = 3 + 10e^{-0.01t}$ are, respectively, the supply and demand for the commodity, both in thousands of units.
  a. Set up and solve an initial value problem for the unit price $p(t)$.
  b. What is the unit price after 6 months?
  c. At what time is the unit price the largest? What is the maximum unit price and the corresponding supply and demand?
  d. As $t \to \infty$, what happens to $p(t)$, $S(t)$, and $D(t)$?

31. **INVENTORY MANAGEMENT** A retailer has the storage capacity for 80 units of a certain popular product. Every day, 5 units enter inventory, either from purchase or return, and 10% of the current inventory is sold. Currently ($t = 0$), there are 60 units in inventory.
  a. Set up an initial value problem to model the number of units $N(t)$ in inventory at time t (days).
  b. Solve the initial value problem in part (a).
  c. How many units are in inventory after 30 days? Does the inventory ever reach capacity (80 units)? Explain.

32. **STOCK INVESTMENT** A highly anticipated offering of 300,000 shares of a growth stock is presented to the market, and half the offered shares are purchased when the market opens at 6:30 A.M. Each hour during the trading day, 65,000 previously uncirculated shares are purchased and 20% of the shares currently in circulation are resold.
  a. Set up an initial value problem for the number of shares $S(t)$ in circulation t hours into the trading day.
  b. How many shares are in circulation when the market closes at 2:00 P.M.?

33. **REAL ESTATE** The price of a certain house is currently \$200,000. Suppose it is estimated that after t months, the price $p(t)$ will be increasing at the rate of $0.01p(t) + 1,000t$ dollars per month. How much will the house cost 9 months from now?

34. **RETAIL PRICES** The price of a certain commodity is currently \$3 per unit. It is estimated that t weeks from now, the price $p(t)$ will be increasing at the rate of $0.02p(t) + e^{0.1t}$ cents per week. How much will the commodity cost in 10 weeks?

35. **INVESTMENT PLAN** An investor makes regular deposits totaling D dollars each year into an account that earns interest at the annual rate r compounded continuously.
   a. Explain why the account grows at the rate
   $$\frac{dV}{dt} = rV + D$$
   where $V(t)$ is the value of the account t years after the initial deposit. Solve this differential equation to express $V(t)$ in terms of r and D.
   b. Amanda wants to retire in 20 years. To build up a retirement fund, she makes regular annual deposits of $8,000. If the prevailing interest rate stays constant at 4% compounded continuously, how much will she have in her account at the end of the 20-year period?
   c. Ray estimates he will need $800,000 to retire. If the prevailing rate of interest is 5% compounded continuously, how large should his regular annual deposits be so that he can retire in 30 years?

36. **SAVINGS** Chris has a starting salary of $47,000 per year and figures that with salary increases and bonuses, her compensation will increase at an average annual rate of 9%. She regularly deposits 5% of her salary in a savings account that earns interest at the annual rate of 8% compounded continuously. How much will be in her account in 20 years? (See Exercise 35.)

37. **SAVINGS** Vijay is saving to take an $8,000 tour of Europe in 4 years. His parents want to help by making a lump sum deposit of A dollars into a savings account that earns 4% interest per year compounded continuously. Vijay plans to add to this amount by making frequent deposits into the account totaling $800 per year. What must A be for him to meet his goal? (See Exercise 35.)

38. **SAVINGS** Ella deposits $10,000 into an account in which interest accumulates at the rate of 4% per year, compounded continuously. She plans to withdraw $2,000 per year.
   a. Set up and solve a differential equation to determine the value $Q(t)$ of her account t years after the initial deposit.
   b. How long does it take for her account to be exhausted?

39. **SAVINGS** Alonzo plans to make an initial deposit into an account that earns 5% interest compounded continuously and then to make $5,000 withdrawals each year. How much must he deposit if it takes 10 years for his account to be exhausted?

40. **SAVINGS** Janine has a starting salary of $40,000 and figures that with salary increases and bonuses her compensation will increase at an average annual rate of 5%. She regularly deposits 10% of her salary into a savings account that earns interest at the rate of 4% per year compounded continuously and withdraws $3,000 per year for expenses.
   a. Set up and solve a differential equation for the amount of money $A(t)$ in her account after t years.
   b. How much will Janine have in her account when she retires in 40 years?

41. **CURRENCY RENEWAL** A nation has 5 billion dollars in currency. Each day about 18 million dollars comes into the banks and the same amount is paid out. Suppose the government decides that whenever old currency comes into a bank, it is destroyed and replaced by a new style of currency.
   a. Set up an initial value problem for the amount of new currency $A(t)$ in circulation after t days.
   b. Solve the initial value problem in part (a).
   c. How long does it take for 90% of the currency in circulation to be the new style?

42. **CURRENCY INFLATION** Suppose the officials of the nation in Exercise 41 decide to inflate the money supply from 5 billion dollars to 6 billion dollars in the new currency. They achieve this by having banks pay out 19 million dollars in new currency each day instead of just matching the 18 million dollars taken in daily.
   a. Set up an initial value problem for the amount of new currency $A(t)$ in circulation after t days.
   b. Solve the initial value problem in part (a).
   c. How long does it take for the goal of 6 billion dollars in new currency to be reached?

Exercises 43 and 44 use trigonometric functions introduced in Chapter 8.

43. **PRICE ADJUSTMENT FOR PERIODIC SUPPLY AND DEMAND** At time t (months), the price $p(t)$ of a commodity changes at a rate

equal to the shortage $D - S$, where $S(t) = \sin t$ and $D(t) = 9 - p$ are the supply and demand for the commodity, respectively, both in thousands of units. If the price is \$2 when $t = 0$, what is the price after 4 months?

44. **PRICE ADJUSTMENT FOR PERIODIC SUPPLY AND DEMAND** Answer the question in Exercise 43 for periodic supply $S(t) = \sin(\pi t/11)$ and demand $D(t) = p \tan(\pi t/11)$, both in thousands of units for $0 \le t \le 5$.

LIFE AND SOCIAL SCIENCE APPLIED PROBLEMS

45. **MEDICATION** A drug injected into a patient's bloodstream at a constant infusion rate of 3 mg per hour is eliminated from the blood at a rate proportional to the amount of the drug currently present. Initially (when $t = 0$), the patient's blood contains no drug.
    a. Set up and solve a differential equation for the amount of drug $A(t)$ present in the patient's bloodstream at time t.
    b. Suppose that a sample is taken indicating there are 2.3 mg of the drug in the patient's blood after 1 hour. How much medication will be present after 8 hours? How much medication is present as $t \to \infty$ (the steady-state level)?
    c. Suppose the patient's physician wants the steady-state level of medication in the patient's bloodstream to be 9 mg. How should the infusion rate be adjusted to achieve this goal?

46. **MEDICATION** A drug injected into a patient's bloodstream at a constant infusion rate of 2.2 mg per hour is eliminated from the blood at a rate proportional to the amount of the drug currently present. Initially (when $t = 0$), the patient's blood contains 0.5 mg of the drug.
    a. Set up and solve a differential equation for the amount of drug $A(t)$ present in the patient's bloodstream at time t.
    b. Suppose that a sample is taken indicating there are 2 mg of the drug in the patient's blood after 1 hour. How much medication will be present after 6 hours? How much medication is present as $t \to \infty$ (the steady-state level)?
    c. Suppose the patient's physician wants the steady-state level of medication in the

patient's bloodstream to be 8 mg. How should the infusion rate be adjusted to achieve this goal?

47. **VETERINARY MEDICINE** A drug used as an animal anesthesia is known to be eliminated exponentially from the bloodstream at the continuous rate of 14% per hour. The animal is anesthetized when there are at least 40 mg of drug in its bloodstream for every kilogram of the animal's body weight. The veterinarian is planning to operate on a 60-kg animal.
    a. Set up and solve an initial value problem for the amount of drug $D(t)$ in the animal's bloodstream after t hours. Express your answer in terms of the initial drug dose $D_0 = D(0)$. [*Hint:* Think of this as a dilution problem like Example 9.2.5.]
    b. If the veterinarian expects the operation to take 90 minutes, how much anesthesia should be administered to keep the animal anesthetized for the entire operation?

48. **AIR FILTRATION** A 2,400-cubic-foot room contains an activated charcoal air filter through which air passes at the rate of 400 cubic feet per minute. The ozone in the air is absorbed by the charcoal as the air flows through the filter, and the purified air is recirculated in the room. Assume that the ozone is evenly distributed throughout the room at all times.
    a. Set up an initial value problem for the amount of ozone $A(t)$ in the room after t minutes.
    b. Solve the differential equation in part (a), and express your answer in terms of the amount of ozone A_0 in the room at time $t = 0$.
    c. How long does it take for the filtration system to remove 50% of the ozone originally in the room?

49. **POLITICAL POLLING** Zamora Lopez, a candidate for the state senate, conducts weekly polls on her popularity from a pool of 50,000 voters selected at random. Suppose 5,000 people in the sample group favor Zamora's candidacy in the first poll (at time $t = 0$) and that in each successive poll, 1,200 new people per week favor her candidacy while 2% of the people who had previously favored her switch to her opponent.

a. Set up an initial value problem for the number of people $P(t)$ in the pool who favor Zamora after week t.

b. Solve the initial value problem in part (a).

c. The polling goes on for 21 weeks, with the final poll being conducted just before the election. If Zamora's polling model is accurate, should she expect to win or lose?

d. Suppose that instead of increasing at the constant rate of 1,200 people per week, Zamora receives $125t$ new supporters per week at time t. Answer the questions in parts (a) through (c) for this revised polling model.

50. POLITICAL POLLING *(This problem involves trigonometric functions introduced in Chapter 8.)* Suppose the candidate in Exercise 49 discovers that instead of increasing at the constant rate of 1,200 people per week, the number of new supporters varies periodically so that $1,500 \sin\left(\dfrac{\pi t}{52}\right)$ people per week are actually added. Everything else remains the same as in Exercise 49.

a. Set up and solve an initial value problem for the number of people $P(t)$ who favor the candidate after t weeks.

b. Based on the results of the poll conducted after 21 weeks ($t = 20$), should the candidate expect to win or lose?

51. GLACIAL MELTDOWN A glacier has a volume of 4 cubic miles. Suppose ice is added at the rate of 100 million cubic feet per year through freezing snow pack and is removed from the glacier (by melting or calving) at the rate of 0.5% of the total volume per year.

a. Set up and solve an initial value problem for the volume of ice in the glacier $V(t)$ (in millions of cubic feet) after t years. (Use 588,792 for the current volume of the glacier.)

b. How long does it take for half the glacier to be removed?

52. GLACIAL MELTDOWN *(This problem involves trigonometric functions introduced in Chapter 8.)* Suppose ice is added to the glacier in Exercise 51 periodically at the rate of $100 \cos(t)$ million cubic feet per year (instead of at the constant rate of 100 million cubic feet per year). If everything else stays the same, what proportion of the ice remains after 10 years?

53. DILUTION A tank contains 5 pounds of salt dissolved in 40 gallons of water. Pure water runs into the tank at the rate of 1 gal/min, and the mixture, kept uniform by stirring, runs out at the rate of 3 gal/min.

a. Set up and solve an initial value problem for the amount of salt $S(t)$ in the tank at time t.

b. How long does it take for the amount of salt in solution to reach 4 pounds?

c. How much salt is in the tank just as it drains empty?

54. DILUTION A tank contains 10 pounds of salt dissolved in 30 gallons of water. Suppose that 2 gallons of brine containing 1 pound of dissolved salt per gallon runs into the tank every minute and that the mixture, kept uniform by stirring, runs out at the rate of 1 gal/min.

a. Set up and solve an initial value problem for the amount of salt $S(t)$ in the tank at time t.

b. Suppose the tank has a capacity of 50 gallons. How much salt is in the tank when it is full?

55. GEOMETRY Find a function $f(x)$ whose graph passes through the point $(-1, 2)$ and has the property that at each point (x, y) on the graph, the slope of the tangent line equals the sum $x + y$ of the coordinates.

56. GEOMETRY Find a function $f(x)$ whose graph passes through the point $(1, 1)$ and has the property that at each point (x, y) on the graph, the slope of the tangent line equals $x^2 + 2y$.

57. SPY STORY The spy is shaken by his close call with the hand grenade (Exercise 62, Section 8.2). While gathering his nerve, he reflects on the dangerous life he has led since he and nine friends from college started their organization 15 years ago. For instance, he knows that each year, 10 new spies join the organization and 20% of all spies either resign or are eliminated.

a. Set up and solve an initial value problem for the number of people $P(t)$ in the spy's organization t years after it began.

b. How many spies are in the organization now?

c. Instead of using a differential equation, find the current size of the spy's organization by solving an appropriate survival/renewal problem with survival function $S(t) = (0.8)^t$. [*Hint:* Note that $(0.8)^t = e^{t \ln(0.8)}$.]

SECTION 9.3   Additional Applications of Differential Equations

Learning Objectives

1. Use differential equations to model applications involving public health, orthogonal trajectories, and finance.
2. Explore the predator–prey model.

In Sections 9.1 and 9.2, you saw applications of differential equations to finance, learning, population growth, the spread of an epidemic, and the amount of medication in a patient's bloodstream. In all these models, a specified rate of change is interpreted as a derivative that is then related to other features of the model by a differential equation. In this section, we examine several additional models involving separable and first-order linear differential equations. We begin with an applied model involving public health.

A Public Health Dilution Model

EXAMPLE 9.3.1   Modeling Defluoridation of a Water Supply

The residents of a certain community have voted to discontinue the fluoridation of their water supply. The local reservoir currently holds 200 million gallons of fluoridated water that contains 1,600 pounds of fluoride. The fluoridated water is flowing out of the reservoir at the rate of 4 million gallons per day and is being replaced at the same rate by unfluoridated water. At all times, the remaining fluoride is evenly distributed in the reservoir. Express the amount of fluoride in the reservoir as a function of time.

Solution

Let $Q(t)$ be the amount of fluoride (in pounds) in the reservoir t days after the fluoridation ends. To model the flow of fluoride, start with the rate relationship

$$\begin{bmatrix} \text{Net rate of change of fluoride} \\ \text{with respect to time} \end{bmatrix} = \begin{bmatrix} \text{daily rate of fluoride} \\ \text{flowing in} \end{bmatrix} - \begin{bmatrix} \text{daily rate of fluoride} \\ \text{flowing out} \end{bmatrix}$$

The net rate of change of the fluoride with respect to time is $\dfrac{dQ}{dt}$, and since the fluoridation has been terminated, the rate of fluoride flowing in is 0. Since the volume of fluoridated water in the reservoir stays fixed at 200 million gallons and the fluoride is always evenly distributed in the reservoir, the concentration of fluoride in the reservoir at time t is given by the ratio

$$\frac{Q(t) \text{ pounds of fluoride}}{200 \text{ million gallons of fluoridated water}}$$

Therefore, since 4 million gallons of fluoridated water are being removed each day, the daily outflow rate of fluoride is given by the product

$$\begin{array}{l} \text{Daily rate of fluoride} \\ \text{flowing out} \end{array} = \left(\frac{Q(t) \text{ lb}}{200 \text{ million gal}} \right) \left(\frac{4 \text{ million gal}}{\text{day}} \right) = \frac{4Q}{200} \text{ lb/day}$$

The flow rate relationships are summarized in Figure 9.7.

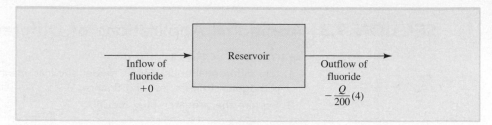

FIGURE 9.7   Flow rates for fluoride dilution in Example 9.3.1.

Since the rate of change of the amount of fluoride in the reservoir equals the inflow rate minus the outflow rate, it follows that

$$\frac{dQ}{dt} = \underbrace{0}_{\substack{\text{inflow} \\ \text{rate}}} - \underbrace{\frac{4Q}{200}}_{\substack{\text{outflow} \\ \text{rate}}} = -\frac{Q}{50}$$

Solving this differential equation by separation of variables, you get

$$\int \frac{1}{Q}\, dQ = -\int \frac{1}{50}\, dt \qquad \text{Q is never negative}$$

$$\ln Q = -\frac{t}{50} + C \qquad \text{apply exponential base e on both sides}$$

$$Q = e^{C - t/50} = e^C e^{-t/50}$$

$$= Q_0 e^{-t/50} \qquad \text{where } Q_0 = e^C$$

Initially, the reservoir contained 1,600 pounds of fluoride, so

$$1{,}600 = Q(0) = Q_0 e^0 = Q_0$$

and thus,

$$Q(t) = 1{,}600 e^{-t/50}$$

and the amount of fluoride decreases exponentially, as illustrated in Figure 9.8.

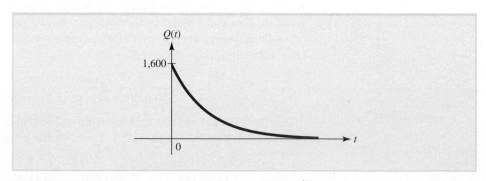

FIGURE 9.8   The amount of fluoride $Q(t) = 1{,}600 e^{-t/50}$.

Newton's Law of Cooling: A Forensics Model

Newton's law of cooling says that the rate of change of the temperature $T(t)$ of an object is proportional to the difference between $T(t)$ and the temperature T_m of the surrounding medium; that is,

$$\frac{dT}{dt} = k(T - T_m)$$

The law of cooling can be used to model a variety of situations involving temperature change. For instance, when a dead body is discovered relatively quickly (within 2 days of the time of death) and the temperature of the air surrounding the body is essentially constant, the time of death can often be determined using the law of cooling as illustrated in Example 9.3.2.

EXAMPLE 9.3.2   Studying a Forensic Investigation

A body is discovered at noon on Friday in a room where the air temperature is 70°F. The temperature of the body at the time of discovery is 76°F and one hour later is 75.3°F. Use this information to determine the time of death.

Solution

Since the temperature of the medium surrounding the body is $T_m = 70$, Newton's law of cooling tells us that

$$\frac{dT}{dt} = k(T - 70)$$

where t is the number of hours since the time of death. Separating the variables and integrating, we find that

$$\int \frac{dT}{T - 70} = \int k \, dt$$
$$\ln |T - 70| = kt + C_1$$

By taking exponentials on both sides, we get

$$T - 70 = e^{kt+C_1} = e^{kt} e^{C_1} \qquad T - 70 \text{ is never negative}$$

so that

$$T = 70 + Ce^{kt} \qquad \text{where } C = e^{C_1}$$

Assume that the body had a normal temperature of 98.6°F at the time of death. Then

$$98.6 = 70 + Ce^{k(0)} = 70 + C$$
$$C = 98.6 - 70 = 28.6$$

and

$$T = 70 + 28.6e^{kt}$$

Let t_0 be the number of hours from the time of death to the time of discovery. Then the given information tells us that $T(t_0) = 76$ and $T(t_0 + 1) = 75.3$. Since

$$T(t_0) = 76 = 70 + 28.6e^{kt_0}$$

we have

$$28.6e^{kt_0} = 76 - 70 = 6$$

We also have

$$T(t_0 + 1) = 75.3 = 70 + 28.6e^{k(t_0+1)}$$
$$= 70 + 28.6e^{kt_0}e^k$$

so that

$$(28.6e^{kt_0})e^k = 5.3 \qquad\qquad\qquad 28.6e^{kt_0} = 6$$
$$(6)e^k = 5.3$$
$$e^k = \frac{5.3}{6} = 0.8833 \qquad\qquad\qquad \text{take logarithms}$$
$$k = \ln(0.8833) \approx -0.1241$$

Since $28.6e^{kt_0} = 6$, it follows that

$$e^{kt_0} = \frac{6}{28.6} \approx 0.2098 \qquad\qquad\qquad k \approx -0.1241$$
$$e^{-0.1241t_0} \approx 0.2098 \qquad\qquad\qquad \text{take logarithms}$$
$$-0.1241t_0 \approx \ln 0.2098 \approx -1.5616$$
$$t_0 \approx \frac{-1.5616}{-0.1241} \approx 12.58$$

Thus, the body had been dead for approximately 12.58 hours (12 hours and 35 minutes) at the time of discovery at noon on Friday. This places the time of death at 11:25 P.M. on Thursday.

Orthogonal Trajectories: Chemotaxis

A curve that intersects every curve in a given family of curves at right angles is said to be an **orthogonal trajectory** for that family (Figure 9.9). Orthogonal trajectories appear in a variety of applications. For instance, in thermodynamics, heat flow lines are orthogonal to curves of constant temperature (isotherms) while in aerodynamics, the curves of airflow direction (streamline trajectories) are orthogonal to curves of equal velocity potential.

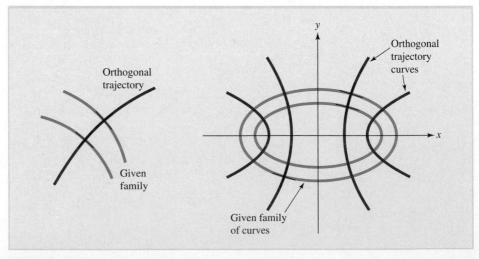

FIGURE 9.9 Orthogonal trajectories of a family of curves.

The general procedure for finding the orthogonal trajectories of a given family of curves involves solving a differential equation. This procedure is illustrated in Example 9.3.3 as we address an applied problem from biology.

EXAMPLE 9.3.3   Modeling Chemotaxis in Biology

In biology, an organism moving on a flat surface in a nutrient bed ideally follows a path along which the level of nutrient concentration increases. The organism receives cues to its progress by continually crossing curves of constant nutrient concentration; that is, curves along which the concentration stays the same. This type of movement, called *chemotaxis*, is performed most efficiently by following a path that is orthogonal to the curves of constant nutrient concentration. In other words, in its quest for greater and greater nutrient levels, the organism will proceed along an orthogonal trajectory of the family of curves of constant nutrient concentration.

Suppose a particular nutrient bed is distributed in the xy plane in such a way that the nutrient concentration is always the same along any curve of the form $x^2 + 2y^2 = C$. If a chemotactic organism is introduced to the nutrient bed at the point $(1, 2)$, what path does it follow?

Solution

We need to find an orthogonal trajectory of the family of curves $x^2 + 2y^2 = C$ that passes through the point $(1, 2)$. We will use uppercase letters X and Y when representing the orthogonal trajectory curves to distinguish those curves from curves in the given family $x^2 + 2y^2 = C$.

Differentiating implicitly with respect to x in the equation $x^2 + 2y^2 = C$, we find that

$$2x + 2(2y)\frac{dy}{dx} = 0$$

so that

$$\frac{dy}{dx} = -\frac{2x}{4y} = -\frac{x}{2y}$$

Just-In-Time REVIEW

When two curves intersect at right angles at a point where the tangent lines to the curves have slopes m_1 and m_2, respectively, then

$$m_2 = -\frac{1}{m_1}$$

This is called the *differential equation of the family* $x^2 + 2y^2 = C$. It tells us that at each point (x, y) on any curve of the form $x^2 + 2y^2 = C$, the slope is $\frac{dy}{dx} = -\frac{x}{2y}$. Since an orthogonal trajectory curve intersects such a curve at right angles, the slope of the orthogonal trajectory must satisfy

$$\frac{dY}{dX} = \frac{-1}{dy/dx} = \frac{-1}{-x/2y} = \frac{2y}{x}$$

$$= \frac{2Y}{X}$$

because $x = X$ and $y = Y$ at the point of intersection. Separating the variables in this differential equation and integrating, we get

$$\int \frac{dY}{Y} = \int \frac{2\,dX}{X} \qquad\qquad X \text{ and } Y \text{ both positive}$$

$$\ln Y = 2 \ln X + K_1 = \ln X^2 + K_1$$

for some constant K_1. Taking exponentials on each side of this equation, we find that

$$Y = KX^2 \quad \text{where } K = e^{K_1}$$

This is the formula for the family of orthogonal trajectories of the given family of curves $x^2 + 2y^2 = C$. To find the particular member of the orthogonal trajectory family that passes through the point $(1, 2)$ where the organism begins its search for food, we substitute $X = 1$ and $Y = 2$ into the formula $Y = KX^2$ and note that

$$2 = K(1)^2$$

so that

$$K = 2$$

Thus, the organism travels along the orthogonal trajectory curve $Y = 2X^2$. Members of the given family of curves (ellipses) are shown in Figure 9.10 along with the orthogonal trajectory curve $Y = 2X^2$ (a parabola).

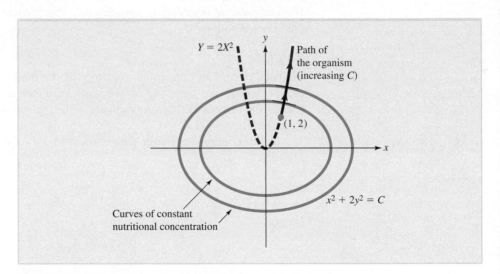

FIGURE 9.10 Motion of an organism in a nutrient bed.

A Financial Model: Linked Differential Equations

Certain applications involve variables whose rates of change are linked by a system of differential equations. In Example 9.3.4, we examine a model in which the flow of money between two components of a financial portfolio is determined by such a system.

EXAMPLE 9.3.4 Modeling Finance with Linked Differential Equations

Jessica currently has $15,000 invested in a money market fund earning 3% per year compounded continuously. Market conditions are improving, and she decides to continuously transfer 20% of her account into a stock fund earning 8% per year. How much will be in each account 4 years from now?

Solution

Let $M(t)$ be the value of the money market account after t years, and let $S(t)$ be the corresponding value of the stock account. According to the given information, the money market account changes at the rate

$$\frac{dM}{dt} = \underbrace{0.03M}_{\substack{\text{percentage} \\ \text{rate of growth}}} - \underbrace{0.20M}_{\substack{\text{transfer} \\ \text{rate to } S}} = -0.17M$$

while the rate of change of the stock account is

$$\frac{dS}{dt} = \underbrace{0.08S}_{\substack{\text{percentage} \\ \text{rate of growth}}} + \underbrace{0.20M}_{\substack{\text{transfer} \\ \text{rate from } M}}$$

The differential equation $\dfrac{dM}{dt} = -0.17M$ is an exponential decay equation. In Section 5.1 (Example 5.1.9), we showed that such an equation has the solution

$$M(t) = M_0 e^{-0.17t}$$

where $M_0 = M(0) = 15{,}000$ is the initial value of the money market account. Substituting $M(t) = 15{,}000e^{-0.17t}$ into the differential equation for $S(t)$, we obtain

$$\frac{dS}{dt} = 0.08S + 0.20[15{,}000e^{-0.17t}] = 0.08S + 3{,}000e^{-0.17t}$$

or, equivalently,

$$\frac{dS}{dt} - 0.08S = 3{,}000e^{-0.17t}$$

This is a first-order linear differential equation with $p(t) = -0.08$ and $q(t) = 3{,}000e^{-0.17t}$. The integrating factor is

$$I(t) = e^{\int -0.08\, dt} = e^{-0.08t}$$

and the general solution is

$$S(t) = \frac{1}{e^{-0.08t}}\left[\int e^{-0.08t}(3{,}000e^{-0.17t})\, dt + C\right]$$

$$= e^{0.08t}\left[\int 3{,}000e^{-0.25t}\, dt + C\right]$$

$$= e^{0.08t}\left[\frac{3{,}000e^{-0.25t}}{-0.25} + C\right]$$

$$= -12{,}000e^{-0.17t} + Ce^{0.08t}$$

Since the stock account initially contains no money, we have $S(0) = 0$ and

$$S(0) = 0 = -12{,}000e^0 + Ce^0 = -12{,}000 + C$$

so that $C = 12{,}000$ and

$$S(t) = 12{,}000(-e^{-0.17t} + e^{0.08t})$$

Therefore, when $t = 4$, we have

$$M(4) = 15,000e^{-0.17(4)} \approx 7,599.25$$

and

$$S(4) = 12,000(-e^{-0.17(4)} + e^{0.08(4)})$$
$$\approx 10,446.13$$

That is, after 4 years, the money market account contains approximately $7,599, while the stock account contains about $10,446.

Interaction of Species: The Predator-Prey Model

An important area of interest in ecology involves the study of interactions among various species occupying the same environment. We shall examine a classic case, called the **predator-prey model**, in which one species (the predator) feeds on a second species (the prey) that in turn has an adequate supply of food and space. To simplify the discussion of the model, we shall call the predators "foxes" and the prey "rabbits" and denote their respective populations at time t by $F(t)$ and $R(t)$.

If there were no foxes, then since rabbits have access to ample food and space, we would expect their population to grow exponentially; that is,

$$\frac{dR}{dt} = aR \qquad \text{where } a > 0$$

On the other hand, if there were no rabbits, the foxes would have no food supply and their population would be expected to decline exponentially, so that

$$\frac{dF}{dt} = -bF \qquad \text{where } b > 0$$

If both foxes and rabbits are present, then encounters between them result in a decline in the rabbit population (rabbit deaths) and growth of the fox population (more food). We assume such encounters are proportional to the product FR, since an increase in either population should increase the frequency of encounters. Combining these assumptions, we model the growth rate of rabbits by

$$\frac{dR}{dt} = \underbrace{aR}_{\substack{\text{natural} \\ \text{growth} \\ \text{rate}}} - \underbrace{cFR}_{\substack{\text{accelerated} \\ \text{death rate due} \\ \text{to encounters}}}$$

and the growth rate of foxes by

$$\frac{dF}{dt} = \underbrace{-bF}_{\substack{\text{starvation} \\ \text{rate}}} + \underbrace{dFR}_{\substack{\text{accelerated} \\ \text{growth rate due} \\ \text{to encounters}}}$$

where c and d are positive constants called *interaction coefficients*.

Solving this linked pair of differential equations would allow us to determine the two populations at each time t. Unfortunately, however, it is usually impossible to find explicit formulas for $F(t)$ and $R(t)$. Instead, we analyze the populations indirectly using the differential equations themselves, as illustrated in Example 9.3.5.

EXAMPLE 9.3.5   Studying a Predator-Prey Population Model

An ecologist studying a forested region models the dynamics of the fox and rabbit populations within the region by the predator-prey equations

$$\frac{dR}{dt} = 0.045R - 0.0015FR$$

$$\frac{dF}{dt} = -0.12F + 0.0003FR$$

a. An *equilibrium solution* is the constant populations $R(t) = R_e$ and $F(t) = F_e$ that satisfy the modeling equations. Find all equilibrium solutions of the given model, and interpret the meaning of each.

b. Find an expression for $\dfrac{dF}{dR}$ and separate variables to obtain an implicit solution for the predator-prey model.

c. Suppose the ecologist determines that there are 300 rabbits and 18 foxes at time $t = 0$. Use a calculator or computer along with the implicit solution found in part (b) to sketch the solution curve that passes through the point (300, 18). Then use your curve to describe how the two populations change for increasing time t.

Solution

a. An equilibrium solution (R_e, F_e) must satisfy

$$0.045R_e - 0.0015F_eR_e = 0$$
$$-0.12F_e + 0.0003F_eR_e = 0$$

One solution is $R_e = 0$ and $F_e = 0$, that is, no rabbits or foxes. Clearly, if this situation ever occurs, it will never change. If $R_e \neq 0$ and $F_e \neq 0$, then R_e can be divided out of the first equation and F_e out of the second to obtain

$$0.045 - 0.0015F_e = 0$$
$$-0.12 + 0.0003R_e = 0$$

so that

$$F_e = \frac{0.045}{0.0015} = 30$$

$$R_e = \frac{0.12}{0.0003} = 400$$

If the rabbit population is ever 400 at the same time the fox population is 30, then these populations will never change because the rates of change $R'(t)$ and $F'(t)$ are both zero.

b. According to the chain rule, we have

$$\frac{dF}{dt} = \frac{dF}{dR}\frac{dR}{dt}$$

so that

$$\frac{dF}{dR} = \frac{dF/dt}{dR/dt} = \frac{-0.12F + 0.0003FR}{0.045R - 0.0015FR} \qquad \text{factor and reduce}$$

$$= \frac{0.0003F(-400 + R)}{0.0015R(30 - F)} = \frac{F(-400 + R)}{5R(30 - F)}$$

Separating the variables and integrating, we get

$$\frac{dF}{dR} = \frac{F(-400 + R)}{5R(30 - F)}$$ cross multiply and integrate

$$\int \frac{5(30 - F)}{F}\, dF = \int \frac{-400 + R}{R}\, dR$$

$$5\int \left[\frac{30}{F} - 1\right] dF = \int \left[\frac{-400}{R} + 1\right] dR$$

$$5[30 \ln F - F] = -400 \ln R + R + C$$

This equation provides an implicit general solution of the given predator-prey system of equations.

c. Substituting the initial conditions $R = 300$ when $F = 18$ into the equation in part (b), we get

$$5[30 \ln(18) - 18] = -400 \ln(300) + 300 + C$$

$$C = 400 \ln(300) + 150 \ln(18) - 300 - 90$$

$$\approx 2{,}325$$

so the required solution curve has the equation

$$5[30 \ln F - F] = -400 \ln R + R + 2{,}325$$

or, equivalently,

$$150 \ln F + 400 \ln R - 5F - R = 2{,}325$$

A computer sketch of this solution curve is shown in Figure 9.11. Note that the solution is an oval-shaped closed curve that contains the equilibrium point (400, 30) in its interior.

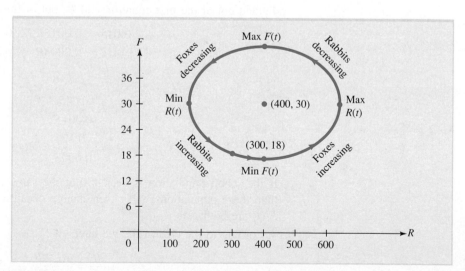

FIGURE 9.11   The predator-prey solution curve through the point (300, 18): $150 \ln F + 400 \ln R - 5F - R = 2{,}325$.

Next, we shall interpret the behavior of the fox and rabbit populations in terms of the solution curve shown in Figure 9.11. Toward this end, we need to know whether the curve is traversed clockwise or counterclockwise with increasing time. To make this determination, we substitute the initial values $R = 300$ and $F = 18$ into the given predator-prey equations and find that

$$\frac{dR}{dt} = 0.045(300) - 0.0015(18)(300) = 5.4$$

and

$$\frac{dF}{dt} = -0.12(18) + 0.0003(18)(300) = -0.54$$

Since $\frac{dR}{dt} > 0$ and $\frac{dF}{dt} < 0$ at the initial point $(300, 18)$ on the solution curve, it follows that as t increases, the rabbit population *increases* while the fox population *decreases*. That is, we initially move to the right (R increasing) and down (F decreasing) along the solution curve, so the curve is *traversed counterclockwise*, as indicated by the arrows in Figure 9.11.

Since the fox population is initially decreasing, 18 foxes are not enough to keep the rabbit population in check. As we move from the initial point $(300, 18)$ to the right along the solution curve (rabbits increasing), the curve keeps falling (fox population is decreasing) until the rabbit population reaches 400 $\left(\text{where } \dfrac{dR}{dt} = 0\right)$. At this time, there are so many rabbits that hunting becomes almost a sure thing, and the fox population begins to increase. For awhile, the curve rises as we move to the right (both populations are increasing), but eventually, there are so many foxes that the rabbits have trouble foraging for food without being killed. This corresponds to the extreme right end of the curve $\left(F = 30, \text{ where } \dfrac{dR}{dt} = 0\right)$. The curve then moves to the left and up (rabbits are decreasing while foxes are increasing) until it reaches its highest point. At this time, the rabbit population has fallen to the level where hunting ceases to be automatic $\left(R = 400, \text{ where } \dfrac{dF}{dt} = 0\right)$. After that, the curve falls to the left (both populations decreasing) until it reaches its extreme left end $\left(F = 30, \text{ where } \dfrac{dR}{dt} = 0\right)$, which corresponds to a fox population small enough to allow rabbits to again forage freely. Finally, from this extreme point, the curve moves again to the right and down, eventually returning to the initial point $(300, 18)$. The cycle then repeats indefinitely.

NOTE The predator-prey situation we have analyzed is only one of several possible kinds of interaction that can occur between species. For instance, some species actually cooperate, with interactions acting to increase both growth rates. Other species, such as squirrels and rabbits, compete for food within the same ecological space without one feeding on the other. Sometimes such species coexist within their environment, and sometimes one species dominates while the other becomes extinct. ∎

EXERCISES ▪ 9.3

ORTHOGONAL TRAJECTORIES   *In Exercises 1 through 4, find an equation for the orthogonal trajectories of the given family of curves. In each case, sketch a few members of both the given family and the family of orthogonal trajectories.*

1. $xy^2 = C$ **2.** $x^2 + y^2 = C$

3. $y = Cx^2$ **4.** $y = Ce^{-x}$

BUSINESS AND ECONOMICS APPLIED PROBLEMS

5. SAVINGS   Marcia currently has $30,000 invested in a money market fund earning 2% per year compounded continuously. Market conditions are improving, and she decides to continuously transfer 25% of her account into a stock fund earning 10% per year. How much will be in each account 5 years from now?

6. SAVINGS   Elvin currently has $40,000 invested in a savings account earning 3% per year compounded continuously. Market conditions are improving, and he decides to continuously transfer 15% of his account into a stock fund earning 12% per year. He also withdraws $5,000 from the savings account each year for expenses.

    a. Set up and solve a linked pair of differential equations to determine the amount of money in each account t years from now.

     b. How long does it take for the savings account to be exhausted? How much money is in the stock account at this time?

7. PARETO'S LAW   *Pareto's law* in economics says that the rate of change (decrease) of the number of people P in a stable economy who have an income of at least x dollars is directly proportional to the number of such people and inversely proportional to their income. Express this law as a differential equation, and solve for P in terms of x.

8. DOMAR DEBT MODEL   Let D and I denote the national debt and national income, and assume that both are functions of time t. One of several **Domar debt models** assumes that the time rates of change of D and I are both proportional to I, so that

$$\frac{dD}{dt} = aI \quad \text{and} \quad \frac{dI}{dt} = bI$$

where a and b are positive constants. Suppose $I(0) = I_0$ and $D(0) = D_0$.

a. Solve the equation for $I(t)$ first in terms of b and I_0; then substitute your expression for $I(t)$ into the equation for $D(t)$ and solve to express D in terms of a, b, I_0, and D_0.

b. The economist, Evsey Domar, who first studied this model, was interested in the ratio of national debt to national income. Determine what happens to this ratio in the long run by computing the limit

$$\lim_{t \to \infty} \frac{D(t)}{I(t)}$$

9. EQUILIBRIUM PRICE ADJUSTMENT   The supply S and demand D of a particular commodity are given by

$$S(p) = 34 + 3p + 2p'$$

and

$$D(p) = 25 - 2p + 5p'$$

where $p(t)$ is the price of the commodity as a function of time t (weeks). Suppose the price is $8 per unit at time $t = 0$ and adjusts continuously so that supply always equals demand; that is, $S(p) = D(p)$.

    The pricing of the commodity is *stable* if $p(t)$ approaches a finite value as $t \to \infty$ and *unstable* (or *inflated*) if $p(t) \to \infty$.

a. What is $p(t)$?

b. Is the pricing of this commodity stable or unstable?

10. EQUILIBRIUM PRICE ADJUSTMENT   Answer the questions in Exercise 9 for the supply and demand functions

$$S(p) = 6 + 5p + 3p'$$

and

$$D(p) = 30 - p + p'$$

11. TIME ADJUSTMENT OF SUPPLY AND DEMAND   The supply $S(t)$ and demand $D(t)$ of a certain commodity vary with time t (months) in such a way that

$$\frac{dD}{dt} = -kD \quad \text{and} \quad \frac{dS}{dt} = 2kS$$

for some constant $k > 0$. It is known that $D(0) = 50$ units and $S(0) = 5$ units and that equilibrium occurs when $t = 10$ months; that is, $D(10) = S(10)$.

a. Use this information to find k.
b. Find $D(t)$ and $S(t)$.
c. How many units are supplied and demanded at equilibrium?

LIFE AND SOCIAL SCIENCE APPLIED PROBLEMS

12. **AMOUNT OF DRUG IN AN ORGAN** A drug with concentration 0.08 g/cc is delivered into an organ at the rate of 12 cc/sec and is dispersed at the same rate. Suppose the organ has a volume of 600 cc and initially contains none of the drug.
    a. Set up and solve a differential equation for the amount of drug $A(t)$ in the organ at time t.
    b. What is the concentration of the drug in the organ after 30 seconds? After 1 minute?
    c. How long does it take for the concentration of drug in the organ to reach 0.06 g/cc?

13. **AMOUNT OF DRUG IN AN ORGAN** A drug with concentration 0.16 g/cc is delivered into an organ at the rate of 6 cc/sec and is dispersed at the same rate. Suppose the organ has volume 200 cc and initially contains none of the drug. Set up and solve a differential equation for the amount of drug $A(t)$ in the organ at time t.

14. **HAZARDOUS EMISSIONS** Noxious fumes from a garage enter an adjacent 2,000-ft^3 workroom through a vent at the rate of 0.2 ft^3/min, and the mixed air (fresh air and fumes) escapes through a crack under the door at the same rate.
    a. Set up and solve a differential equation for the amount of noxious fumes $g(t)$ in the room t minutes after it begins to enter the room [so $g(0) = 0$].
    b. The air in the room becomes hazardous when it contains 1% noxious fumes. How long does it take for this to occur?

15. **FORENSICS** A dead body is discovered at 3:00 P.M. on Monday in a storage room where the air temperature is 50°F. The temperature of the body at the time of discovery is 80°F and 20 minutes later, is 78°F. What was the time of death?

16. **ATMOSPHERIC PRESSURE** Assuming constant temperature, atmospheric pressure P decreases at a rate proportional to P with respect to altitude h above sea level. Suppose the pressure is 30 inches of mercury at sea level and at an altitude of 1 mile is 24.6 inches of mercury.
    a. Set up and solve a differential equation to express P as a function of h.

b. What is the pressure at the top of Mt. Everest at an altitude of roughly 29,000 feet?
c. A healthy human adult becomes distressed when the atmospheric pressure decreases to half its value at sea level. At what altitude does this occur?

17. **RESTAURANT MAINTENANCE** As part of its décor, a seafood restaurant features a 3,000-gallon aquarium filled with exotic fish. To keep the fish healthy and active, a solution containing $0.2te^{-t/40}$ ounces of soluble nutrient is pumped into the tank and circulated at the rate of 75 gallons per minute.
    a. Set up and solve a differential equation for the amount of nutrient $N(t)$ in solution at time t.
    b. What is the maximum concentration of nutrient in the tank and when does it occur?

18. **CRIMINAL INVESTIGATION** At midnight, an automatic camera located at an intersection photographs a man driving erratically enough to suggest that he may be under the influence of alcohol. Two hours later, the police locate the driver and determine that there is 0.06% alcohol in his blood. An hour later at police headquarters, a follow-up test detects a blood-alcohol level of 0.05%. Evidence indicates that the suspect consumed no alcohol after midnight. If the legal blood-alcohol level is 0.08%, can the suspect be charged with driving under the influence of alcohol (DUI)? (Assume that the percentage P of alcohol in the bloodstream satisfies the decay equation $P' = -kP$.)

EMIGRATION FROM A POPULATION *Suppose a population $P(t)$ grows exponentially at a natural rate r and that $E(t)$ individuals are emigrating from the population at time t so that*

$$\frac{dP}{dt} = rP - E$$

Solve this equation to determine the population at time t for each of the cases in Exercises 19 and 20.

19. $r = 0.03$, $E(t) = 10t$, and $P(0) = 100,000$

20. $r = 0.015$, $E(t) = 200e^{-t}$, and $P(0) = 250,000$

21. **IMMIGRATION TO A POPULATION** Suppose a population $P(t)$ grows exponentially at the rate r and that $I(t)$ individuals are immigrating into the population at time t. Set up and solve a differential equation for $P(t)$ in the case where t is in years, $r = 0.02$, $I(t) = 100e^{-t}$, and $P(0) = 300,000$.

22. CHEMOTAXIS Suppose a particular nutrient bed is distributed in the xy plane in such a way that the nutrient concentration is always the same along any curve of the form $x^2 - y^2 = C$. If a chemotactic organism is introduced to the nutrient bed at the point $(3, 1)$, what path does it follow?

23. CHEMOTAXIS Suppose a particular nutrient bed is distributed in the xy plane in such a way that the nutrient concentration is always the same along any curve of the form $3x^2 + y^2 = C$. If a chemotactic organism is introduced to the nutrient bed at the point $(1, 1)$, what path does it follow?

24. MANAGEMENT OF A FISHERY Suppose a population of fish in a fishery grows exponentially with natural growth rate k and that harvesting (fishing) is allowed at a constant, continuous rate h.

    a. Set up and solve a differential equation for the population $P(t)$ at time t. [Your answer will involve k, h, and the initial population $P_0 = P(0)$.]

    b. Find $P(t)$ for the case where t is in years, $k = 0.2$, the initial population is $P(0) = 2,000$, and the harvesting rate is $h = 300$ fish per year.

    c. Suppose the fishery in part (b) decides on a policy of harvesting just enough fish so the population remains constant at 2,000. What harvesting rate h will implement this policy?

25. SPREAD OF AN EPIDEMIC Let $N(t)$ be the number of people from a susceptible population N_s who are infected by a certain disease t days' after the onset of an epidemic. Researchers model the epidemic using the differential equation

$$\frac{dN}{dt} = k(N_s - N)(N - m)$$

where $k > 0$ and m are constants, with $0 < m < N_s$. Call m the *epidemic threshold* of the disease.

    Suppose that $k = 0.04$, that $N_s = 250,000$ people are susceptible to the disease, and that the epidemic threshold m is 1% of the susceptible population.

    a. Solve the differential equation for this epidemic in terms of $N_0 = N(0)$, the number of people who are initially infected. [*Hint:* Integration formula 6 in Table 6.1 may help.]

    b. Show that if $N_0 = 5,000$, then $N(t)$ approaches N_s in the long run. That is, the entire susceptible population is eventually infected.

    c. Show that if $N_0 = 1,000$, then there is a finite time T such that $N(T) = 0$. That is, the epidemic eventually dies out.

    d. In general, what relationship between N_0 and m determines whether an epidemic modeled by a differential equation of the given form dies out in finite time? What effect (if any) does changing the value of k have on this determination?

26. WATER POLLUTION A lake holds 4 billion ft^3 of water and, initially, its pollutant content is 0.19%. A river whose water contains 0.04% pollutant flows into the lake at the rate of 250 million ft^3/day and then flows out of the lake at the same rate. Assuming that the water in the lake and the two rivers is always well mixed, how long does it take for the pollutant content in the lake to be reduced to 0.1%?

27. WATER POLLUTION Two lakes are connected by a river as shown in the accompanying figure. The lakes both contain pure water until 2,000 lb of pollutant is dumped into the upper lake. Suppose the upper lake contains 700,000 gallons of water, the lower lake contains 400,000 gallons, and the river flows at the rate of 1,500 gallons per hour. Assume that the pollutant disperses rapidly enough so that the mixture of pollutant and water is well-mixed at all times.

    a. Set up and solve a differential equation for the amount of pollutant $P_1(t)$ in the upper lake at time t.

    b. If $P_2(t)$ is the amount of pollutant in the lower lake at time t, then there are constants a and b, such that

$$\frac{dP_2}{dt} = aP_1 - bP_2$$

    Use the given information to determine a and b. Then solve this equation for $P_2(t)$.

    c. What is the maximum amount of pollutant contained in the lower lake? When does this maximum amount occur?

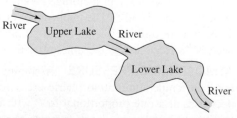

                                       EXERCISE 27

28. PREDATOR-PREY Suppose a large wildlife preserve contains W thousand wolves (predators) and E thousand elk (prey). An ecologist finds that there are currently 20,000 elk and 5,000 wolves and models the growth rates of these two populations with respect to time t by this pair of differential equations:

$$\frac{dE}{dt} = 96E - 8EW$$

$$\frac{dW}{dt} = -123W + 3EW$$

a. Find equilibrium populations for this model; that is, populations E_e and W_e that satisfy

$$\frac{dE}{dt} = \frac{dW}{dt} = 0$$

b. Note that

$$\frac{dE}{dW} = \frac{96E - 8EW}{-123W + 3EW}$$

   Separate the variables in this equation, and solve to obtain an implicit solution for the system.

29. **COMPETITION BETWEEN SPECIES** A certain ecological territory contains S thousand squirrels and R thousand rabbits. Currently, there are 4,000 of each species and the growth rates of the populations with respect to time t are given by this pair of differential equations:

$$\frac{dR}{dt} = 63R - 3RS$$

$$\frac{dS}{dt} = 26S - RS$$

a. Find equilibrium populations for this model, that is, populations R_e and S_e that satisfy

$$\frac{dR}{dt} = \frac{dS}{dt} = 0$$

b. Note that

$$\frac{dR}{dS} = \frac{63R - 3RS}{26S - RS}$$

   Separate the variables in this equation, and solve to obtain an implicit solution for the system.

30. **DEMOGRAPHY** The differential equation

$$\frac{dQ}{dt} = Q(a - b \ln Q)$$

where a and b are constants, is called the *Gompertz equation,* and a solution of the equation is called a *Gompertz function.* Such functions are used to describe restricted growth in populations as well as matters such as learning and growth within an organization.

a. Use the Gompertz equation to show that a Gompertz function is growing most rapidly when $\ln Q = \dfrac{a - b}{b}$. [*Hint:* What derivative measures how fast the growth rate $\dfrac{dQ}{dt}$ is changing?]

b. Solve the Gompertz equation. [*Hint:* Substitute $u = \ln Q$.]

c. Compute $\lim\limits_{t \to \infty} Q(t)$.

d. Sketch the graph of a typical Gompertz function.

31. **AMOUNT OF DRUG IN AN ORGAN** A certain drug is injected into an organ of a patient at the rate of 5 units/hr and is dispersed at the rate of 4 units/hr. This causes the organ to expand at the rate of 2 cm³/hr. Initially, the organ contains 50 units of the drug and has volume 800 cm³.

a. Set up and solve a differential equation to find the concentration $C(t)$ of drug in the organ after t hours.

b. The organ can safely accommodate a concentration of no more than 0.13 units/cm³. What is the maximum length of time the drug can be allowed to enter the organ?

32. **RUNAWAY GROWTH** For certain prolific species, the population $P(t)$ grows at a rate proportional to P^{1+c}, for $c > 0$; that is,

$$\frac{dP}{dt} = kP^{1+c}$$

a. Solve this differential equation in terms of k, c, and the initial population $P_0 = P(0)$.

b. Show that there is a finite time t_a such that $\lim\limits_{t \to t_a} P(t) = \infty$.

c. Suppose $c = 0.02$ for a certain population of rodents. Find $P(t)$ if the population begins with a mated pair and 2 months later has doubled [that is, $P(0) = 2$ and $P(2) = 4$].

d. How long does it take for the population in part (c) to become essentially infinite?

MISCELLANEOUS PROBLEMS

33. **DECOMPOSITION OF SUGAR** When placed in a container of water, sugar dissolves at a rate proportional to the amount $Q(t)$ of undissolved sugar remaining in the container at that time (see Exercise 36, Section 9.1). If half the sugar has dissolved after 3 minutes, how long does it take for $\dfrac{3}{4}$ of the sugar to dissolve?

34. NEWTON'S LAW OF COOLING When a cold drink is taken from the refrigerator, its temperature is 40°F. It is taken outside, where the air temperature is 70°F, and 20 minutes later, its temperature is 50°F.
  a. What will its temperature be 30 minutes after being removed from the refrigerator?
  b. How long will it take before the temperature of the drink is 60°F?

35. NEWTON'S LAW OF COOLING On a cold winter day, a cup of hot chocolate is taken outside, where the temperature is −5°C. After 10 minutes, its temperature is 70°C, and 10 minutes after that, its temperature is 50°C. What was the original temperature of the drink?

36. NEWTON'S LAW OF COOLING At noon, a roast is removed from the oven where its temperature was 170°F and is placed in a room where it takes 1 hour for it to cool to 125°F. If the roast is ready to be served at 2:00 P.M. when its temperature reaches 100°F, what is the air temperature in the room?

37. WATER CLOCK According to Torricelli's law in physics, water drains from a small hole in a tank at a rate proportional to the square root of the height h of water in the tank; that is,

$$\frac{dV}{dt} = -k\sqrt{h}$$

where $V(t)$ is the volume of water at time t.
  a. Suppose the tank is a cube, 9 cm on a side, and that at time $t = 0$, water is draining at the rate of 8.1 cm^3/min. Assuming the tank is full at time $t = 0$, how long does it take for the tank to drain? $\left[\textit{Hint:} \text{ Use the chain rule } \frac{dV}{dt} = \frac{dV}{dh}\frac{dh}{dt}. \right]$
  b. Do you see how the tank in part (a) could be used to tell time? Write a paragraph on how water clocks were constructed and used in ancient times.

38. ABSORPTION OF LIGHT The intensity of light $I(d)$ at a depth d below the surface of a body of water changes at a rate proportional to I. If the intensity at a depth of $d = 0.5$ meter is half of the surface intensity I_0, at what depth is the intensity 5% of I_0?

39. CHEMICAL CONVERSION Two substances, A and B, are being converted into a third substance C in such a way that the time rate of conversion is jointly proportional to the amount of unconverted amounts of A and B. Assume units are chosen so that one unit of C is formed from the combination of one unit of A and one unit of B.
  a. Let x be the amount of substance C at time t, and explain why the unconverted amounts of A and B at time t are $a - x$ and $b - x$, respectively, where a and b are the initial amounts of A and B. Thus, the conversion process may be modeled by the initial value problem

$$\frac{dx}{dt} = k(a - x)(b - x)$$

with $x(0) = 0$.
  b. Solve the problem in part (a) for the case where $a = b$.
  c. Solve the problem in part (a) for the general case. [*Hint:* You may need one of the integration formulas from Table 6.1.]

40. DIFFUSION ACROSS A MEMBRANE A chemical in a solution diffuses from a compartment with concentration $C_1(t)$ across a membrane to a second compartment where the concentration is $C_2(t)$. Experiments suggest that the rate of change of $C_2(t)$ with respect to time is proportional to the difference $C_1 - C_2$ in concentrations. Set up and solve a differential equation for $C_2(t)$ for the case where $C_2(0) = 0$, $C_1(t) = 100e^{-t}$, and $k = 1.5$ is the constant of proportionality.

SECTION 9.4 Approximate Solutions of Differential Equations

Learning Objectives
1. Analyze solutions of differential equations using slope fields.
2. Use Euler's method for approximating solutions of initial value problems.

Certain differential equations, including many that arise in practical applications, cannot be solved by any known method. However, it is often possible to approximate solutions to such equations. In this section, we will discuss two procedures for obtaining approximate solutions of differential equations.

Slope Fields   Suppose we wish to solve the initial value problem

$$\frac{dy}{dx} = f(x, y) \qquad y(x_0) = y_0$$

We may not be able to find the exact solution, but we can obtain a rough sketch by exploiting the fact that since $y' = f(x, y)$ at each point (x, y) on a solution curve $y = y(x)$, the slope of the tangent line to the curve at (x, y) is given by $f(x, y)$. Here is a description of the procedure we will follow.

Slope Field Procedure for Sketching a Solution Curve of $y' = f(x, y)$ That Contains the Point (x_0, y_0)

Step 1. At each point (x, y) on a grid in the xy plane, draw a short line segment with slope equal to $f(x, y)$. The collection of all such points and segments is called a **slope field** for the differential equation $y' = f(x, y)$.

Step 2. Starting with the point (x_0, y_0), use the slopes indicated by the slope field to sketch the required solution as demonstrated in this figure:

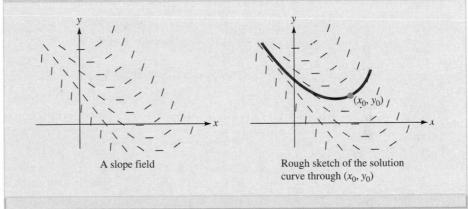

A slope field   Rough sketch of the solution
                             curve through (x_0, y_0)

The slope field procedure for visualizing a solution of a differential equation is illustrated in Example 9.4.1.

EXAMPLE 9.4.1 Using a Slope Field to Solve an Initial Value Problem

Use a slope field to sketch several solutions of the differential equation

$$\frac{dy}{dx} = x - 2y$$

In particular, sketch the solution that passes through the point $(0, 1)$.

Solution

Note that if (x, y) is a point on the line $x - 2y = m$, then any solution curve $y = y(x)$ of the differential equation $y' = x - 2y$ that passes through (x, y) will have slope $y' = m$. Thus, we can construct the slope field for this differential equation by drawing the family of parallel lines $x - 2y = m$ and placing a small line segment of slope m at selected points along each such line. This is done in Figure 9.12a.

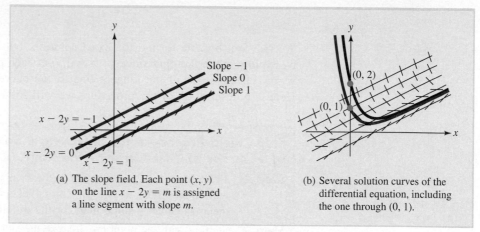

(a) The slope field. Each point (x, y) on the line $x - 2y = m$ is assigned a line segment with slope m.

(b) Several solution curves of the differential equation, including the one through $(0, 1)$.

FIGURE 9.12   Using a slope field to construct solutions for $y' = x - 2y$.

In Figure 9.12b, we use the slope field in Figure 9.12a to draw several solutions to the differential equation $y' = x - 2y$. As a check, notice that the given differential equation can be rewritten in the first-order linear form

$$y' + 2y = x$$

and then solved analytically using the method developed in Section 9.2. The integrating factor is

$$I(x) = e^{\int 2\,dx} = e^{2x}$$

so the general solution is

$$y = \frac{1}{e^{2x}}\left[\int x\, e^{2x}\, dx + C\right]$$

$$= e^{-2x}\left[\left(\frac{x}{2} - \frac{1}{4}\right)e^{2x} + C\right] \quad \begin{array}{l}\text{integration by parts:}\\ u = x \qquad dv = e^{2x}\, dx\end{array}$$

$$= \frac{x}{2} - \frac{1}{4} + Ce^{-2x}$$

For the particular solution with $y = 1$ when $x = 0$, we have

$$1 = 0 - \frac{1}{4} + Ce^{0} = -\frac{1}{4} + C$$

so that $C = \dfrac{5}{4}$ and

$$y = \frac{x}{2} - \frac{1}{4} + \frac{5}{4}e^{-2x}$$

This is the curve sketched in red in Figure 9.12b.

The slope field procedure was not really necessary in Example 9.4.1 because it was fairly easy to solve the differential equation using the methods of Section 9.2. However, in Example 9.4.2, we use the procedure to obtain approximate solutions to a differential equation that cannot be solved directly.

EXAMPLE 9.4.2   Using a Slope Field to Solve a Differential Equation

Use a slope field to sketch solutions of the differential equation

$$\frac{dy}{dx} = \sqrt{x + y}$$

Solution

If (x, y) is a point on the line $x + y = m$ for $m \geq 0$, then any solution curve $y = y(x)$ of the differential equation $y' = \sqrt{x + y}$ that passes through (x, y) will have slope $y' = \sqrt{m}$. We construct the slope field for this equation by drawing the family of parallel lines $x + y = m$ and assigning a small line segment of slope $\sqrt{m}$ at selected points along each such line. Figure 9.13a shows the slope field. Note that since $\sqrt{x + y}$ is not defined for $x + y < 0$, the entire slope field lies on or above the line $x + y = 0$. Several solution curves are sketched in Figure 9.13b. Notice that each solution curve is "flat" (has slope 0) on the line $x + y = 0$ and then rises rapidly to the right.

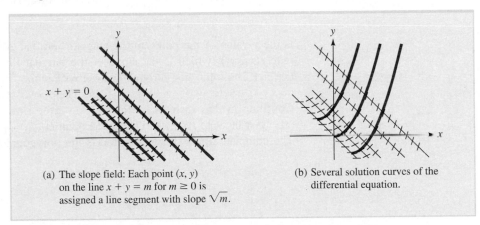

(a) The slope field: Each point (x, y) on the line $x + y = m$ for $m \geq 0$ is assigned a line segment with slope $\sqrt{m}$.

(b) Several solution curves of the differential equation.

FIGURE 9.13   A slope field construction of solutions for $y' = \sqrt{x + y}$.

Euler's Method   The slope field procedure we have just examined enables us to obtain a rough sketch of the solution to an initial value problem, and our next goal is to show how such a solution can be generated numerically. The great advantage of a numerical approximation procedure is that it can be programmed for solution by a high-speed computer. Indeed, using computers and numerical approximation schemes, it is now possible to analyze accurate models of complex situations such as weather patterns, ecosystems, and economic interactions among nations.

The particular numerical approximation procedure we will discuss is called Euler's method, after its originator, the great Swiss mathematician Leonhard Euler (1707–1783). Although Euler's method is much less accurate than the numerical procedures used in practical applications, it deals with the same issues as the more sophisticated procedures and its relatively simple structure makes it easier to demonstrate.

To illustrate the ideas behind Euler's method, suppose we wish to approximate the solution to the initial value problem

$$\frac{dy}{dx} = f(x, y) \qquad \text{for } y(x_0) = y_0$$

The key to Euler's approach is that once the solution value $y(c)$ is known for some number c, then we can compute $f(c, y(c))$ which equals $y'(c)$, the slope of the tangent line to the graph of the solution at $(c, y(c))$. Using this fact, we can approximate values of the solution $y = y(x)$ at successive, evenly spaced points.

To begin the approximation process, choose the *step size h*. This is the distance along the x axis between successive points on the approximate solution curve. That is, we will approximate values of the solution at the points

$$x_0, x_1 = x_0 + h, x_2 = x_0 + 2h, \ldots, x_n = x_0 + nh$$

Starting at (x_0, y_0) compute $f(x_0, y_0)$ and draw the line through (x_0, y_0) with slope $f(x_0, y_0)$. We know that this line is tangent to the solution curve $y = y(x)$ at (x_0, y_0), and we proceed along this line to the point (x_1, y_1), where

$$y_1 = y_0 + f(x_0, y_0)(x_1 - x_0) = y_0 + f(x_0, y_0)h$$

is the y value of the point on the tangent line that corresponds to x_1. Next, we compute $f(x_1, y_1)$, which is the slope of the tangent line to the solution curve through (x_1, y_1). Draw the line through (x_1, y_1) with slope $f(x_1, y_1)$, and proceed to the point (x_2, y_2), where $y_2 = y_1 + f(x_1, y_1)h$. These two steps are shown in Figure 9.14a. Continue in this fashion, at the $k + 1$st step drawing the line through the point (x_k, y_k) with slope $f(x_k, y_k)$ to locate the point (x_{k+1}, y_{k+1}) where $y_{k+1} = y_k + f(x_k, y_k)h$. The complete approximate solution is the polygonal path shown in Figure 9.14b.

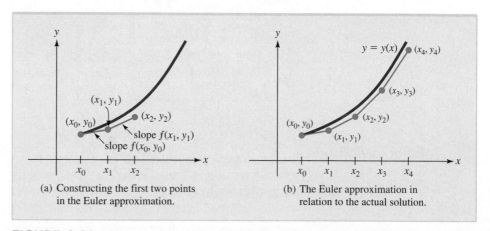

(a) Constructing the first two points in the Euler approximation.

(b) The Euler approximation in relation to the actual solution.

FIGURE 9.14 Euler's method for approximating the solution to the initial value problem $y'(x) = f(x, y)$ with $y(x_0) = y_0$.

Here is a step-by-step summary of how to use Euler's method to obtain an approximate solution of an initial value problem. The method is illustrated in Example 9.4.3.

Euler's Method for Approximating the Solution to the Initial Value Problem $y'(x) = f(x, y)$ with $y(x_0) = y_0$

Step 1. Choose the step size h.

Step 2. For $k = 1, 2, \ldots, n$, compute the successive points (x_k, y_k), where
$$x_k = x_{k-1} + h \quad \text{and} \quad y_k = y_{k-1} + f(x_{k-1}, y_{k-1})h.$$

Step 3. Construct the Euler approximation by plotting the points (x_k, y_k) for $k = 0, 1, 2, \ldots, n$ in the xy plane and connecting successive pairs of points with line segments.

EXPLORE!

Refer to Example 9.4.3. Create the slope field for the differential equation $\dfrac{dy}{dx} = x - y$, using the program appearing in the Explore! Update section of this chapter. Graphically display the solution $g(x) = x - 1 + 2e\^(-x)$, which passes through the point (0, 1).

EXAMPLE 9.4.3   Using Euler's Method to Solve an Initial Value Problem

Use Euler's method with step size $h = 0.1$ to approximate the solution of the initial value problem

$$\frac{dy}{dx} = x - y \qquad \text{with } y(0) = 1$$

Compare the approximation with the actual solution.

Solution

The given differential equation can be written in the first-order linear form

$$\frac{dy}{dx} + y = x$$

By using the methods of Section 9.2, it can be shown that the general solution is $y = x - 1 + Ce^{-x}$, and the particular solution with $y(0) = 1$ is

$$g(x) = x - 1 + 2e^{-x}$$

We will compare this solution with the approximate solution obtained by applying Euler's method with $h = 0.1$, $f(x, y) = x - y$, $x_0 = 0$, $y_0 = 1$, and 10 steps. The first step yields

$$y_1 = y_0 + f(x_0, y_0)h = 1 + (0 - 1)(0.1) = 0.9$$

and comparing this with the actual value of the solution $y = g(x)$ at $x_1 = 0.1$, we get $y_1 - g(x_1) = 0.9 - 0.90967 = -0.00967$. The rest of the computations are shown in the table on page 742:

k	x_k	y_k	Slope $f(x_k, y_k)$	y_{k+1}	Actual Solution $g(x_k)$	Difference $y_k - g(x_k)$
0	0	1	−1	0.9	1	0
1	0.1	0.9	−0.8	0.820	0.910	−0.010
2	0.2	0.82	−0.62	0.758	0.837	−0.017
3	0.3	0.758	−0.458	0.712	0.782	−0.024
4	0.4	0.712	−0.312	0.681	0.741	−0.029
5	0.5	0.681	−0.181	0.663	0.713	−0.032
6	0.6	0.663	−0.063	0.657	0.698	−0.035
7	0.7	0.657	0.043	0.661	0.693	−0.036
8	0.8	0.661	0.139	0.675	0.699	−0.038
9	0.9	0.675	0.225	0.698	0.713	−0.038
10	1.0	0.698			0.736	−0.038

In Figure 9.15, the Euler approximation (polygonal path) is shown along with the actual solution curve for comparison.

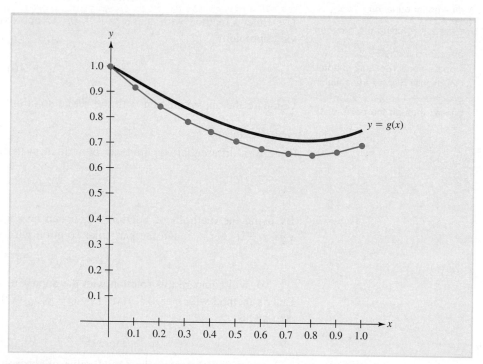

FIGURE 9.15   Comparing the solution curve for the initial value problem $y' = x - y$ with $y(0) = 1$ to the Euler approximation.

Applied models often involve initial value problems whose exact solution is either difficult or impossible to obtain. In Example 9.4.4, we approximate a solution to a population model using Euler's method.

EXAMPLE 9.4.4 Using Euler's Method to Study a Population Model

After t months of observation, the population $P(t)$ of an animal species isolated on an island is found to be changing at the rate

$$\frac{dP}{dt} = 0.03\sqrt{P} - 0.005t$$

Suppose the population was 1,000 when observations began ($t = 0$). Use Euler's method to estimate the population after 1 year.

Solution

The slope function is $f(t, P) = 0.03\sqrt{P} - 0.005t$, with $P = 1,000$ when $t = 0$. Since time is measured in months, we wish to estimate $P(12)$. Assume a step size $h = 2$. Then the population P_{k+1} at the $k+1$st step is related to the population at the kth step by the formula

$$\begin{aligned} P_{k+1} &= P_k + f(t_k, P_k)h \\ &= P_k + [0.03\sqrt{P_k} - 0.005t_k](2) \end{aligned}$$

where $t_k = 0 + 2k = 2k$. The table summarizes the computations for the approximation.

k	t_k	P_k	Slope $f(t_k, P_k)$	P_{k+1}
0	0	1,000.00	0.9487	1,001.90
1	2	1,001.90	0.9396	1,003.78
2	4	1,003.78	0.9305	1,005.64
3	6	1,005.64	0.9214	1,007.48
4	8	1,007.48	0.9122	1,009.30
5	10	1,009.30	0.9031	1,011.11
6	12	1,011.11		

Therefore, based on this Euler approximation, we would estimate the population at the end of 1 year to be approximately 1,011 animals.

EXERCISES ▪ 9.4

In Exercises 1 through 12, plot the slope field for the given differential equation and use it to sketch the graph of the solution that passes through the specified point.

1. $y' = x + y$; $(1, 1)$ **2.** $y' = x - y$; $(1, 1)$

3. $y' = y^2$; $(0, 1)$ **4.** $y' = \sqrt{x}$; $(4, 0)$

5. $y' = y(2 - y)$; $(0, -1)$ **6.** $y' = \dfrac{x}{y + 1}$; $(0, 0)$

7. $y' = x^2 + y^2$; $(1, -1)$

8. $y' = x - y^2$; $(1, 1)$

9. $y' = \dfrac{2y - 3x}{x + y}$; $(0, 1)$

10. $y' = 2 - x - y$; $(0, 0)$

11. $y' = e^{xy}$; $(0, 0)$

12. $y' = e^{x/y}$; $(1, 1)$

In Exercises 13 through 20, use Euler's method with the indicated step size h and number of increments n to obtain an approximate solution of the given initial value problem.

13. $y' = 2x + y$; $y(0) = 1$; $h = 0.2$; $n = 5$

14. $y' = x - 3y$; $y(0) = 0$; $h = 0.4$; $n = 5$

15. $y' = \dfrac{y}{x} + x$; $y(1) = 0$; $h = 0.2$; $n = 5$

16. $y' = xy + y$; $y(0) = 1$; $h = 0.5$; $n = 6$

17. $y' = \sqrt{x} - y$; $y(1) = 1$; $h = 0.5$; $n = 8$

18. $y' = \dfrac{e^x}{y}$; $y(0) = -1$; $h = 0.5$; $n = 8$

19. $y' = \dfrac{x^2 - y^2}{xy}$; $y(1) = 1$; $h = 0.4$; $n = 5$

20. $y' = \dfrac{x}{\sqrt{x^2 + y^2}}$; $y(0) = 1$; $h = 0.2$; $n = 5$

In Exercises 21 through 26, use Euler's method with the indicated step size h to estimate the required solution of the given initial value problem.

21. $y' = x + 2y$; $y(0) = 1$; $h = 0.2$; estimate $y(1)$

22. $y' = \dfrac{x}{y}$; $y(1) = 1$; $h = 0.2$; estimate $y(2)$

23. $y' = \dfrac{x - y}{2x + y}$; $y(0) = 1$; $h = 0.2$; estimate $y(1)$

24. $y' = x\sqrt{1 + y^2}$; $y(0) = 0$; $h = 0.5$; estimate $y(4)$

25. $y' = x^2 + 2y^2$; $y(0) = 0$; $h = 0.4$; estimate $y(2)$

26. $y' = \ln(x + y)$; $y(0) = 1$; $h = 0.5$; estimate $y(4)$

27. The length $L(t)$ of the curve $y = \ln x$ over the interval $1 \le x \le t$ satisfies

$$\frac{dL}{dt} = \frac{\sqrt{1 + t^2}}{t} \qquad L(1) = 0$$

Estimate the length $L(4)$ using Euler's method with step size $h = 0.5$.

28. SUPPLY Suppose the number of units $S(t)$ of a particular commodity that will be supplied by producers at time t increases at the rate

$$\frac{dS}{dt} = 0.3Se^{-0.1t}$$

If $S(0) = 3$, use Euler's method with $h = 0.5$ to estimate $S(4)$.

29. PRODUCTION The output Q at a certain factory changes at the rate

$$\frac{dQ}{dK} = 0.2Q + K^{1/2}$$

where K is the capital expenditure (in units of $1,000). Suppose when the capital expenditure is $2,000, there will be $Q(2) = 500$ units produced. Use Euler's method with $h = 1$ to estimate the number of units produced when the capital expenditure is $5,000.

30. GROWTH OF A SPECIES The fox population $F(t)$ on a large island satisfies the modified logistic initial value problem

$$\frac{dF}{dt} = 0.01(80 - F)(F - 76) \qquad F(0) = 70$$

Use Euler's method with $h = 1$ to estimate $F(10)$, the fox population after 10 years.

31. GROWTH OF A SPECIES The length $L(t)$ of a species of fish at age t is modeled by the von Bertalanffy initial value problem

$$\frac{dL}{dt} = 1.03(5.3 - L) \qquad L(0) = 0.2$$

where L is measured in inches and t in months.
a. Use Euler's method with $h = 0.4$ to estimate the length of the fish at time $t = 2$.
b. Solve the initial value problem and evaluate $L(2)$. Compare this exact value with the estimate you obtained in part (a).

32. POPULATION EXTINCTION When the population density of a sexually reproducing species falls below a certain threshold, individuals have difficulty finding a suitable mate, and this has the effect of depressing the population still further, eventually leading to extinction. In biology, this is called the *Allee effect,* after W. C. Allee, who first studied the effect in 1931. If the population at time t is $P(t)$, this effect can be modeled by the modified logistic equation

$$\frac{dP}{dt} = kP(P - A)\left(1 - \frac{P}{K}\right)$$

where k is the natural growth rate of the species, K is the carrying capacity, and A is the threshold population. Assume that $k > 0$ and $0 < A < K$.

For a particular species, suppose P is measured in thousands, t in years, and that $k = 0.2$, $K = 20$, and $A = 2$. Assume further that the initial population is $P(0) = 1$. Use Euler's method with $h = 0.25$ to estimate $P(1)$, the population after 1 year.

SECTION 9.5 Difference Equations; The Cobweb Model

Learning Objectives

1. Solve first-order linear difference equations.
2. Use difference equations to analyze models in population dynamics, finance (amortization of debt), and fishery management.
3. Explore the cobweb model in economics.
4. Compare difference equations with differential equations as modeling tools.

We have already seen how differential equations can be used to model a variety of dynamic situations involving continuous behavior. However, it is often more natural or convenient to describe certain situations in terms of behavior measured at a collection of discrete points instead of continuously. For example, the population of the United States is measured by a census taken every 10 years, a biologist interested in knowing the population of a bacterial colony might take measurements every hour, or an economist charting the supply and demand of a commodity might gather the pertinent data once a week.

A **difference equation** is an equation involving the differences between values of a variable at discrete times. For instance, if f is a function and $y_n = f(n)$ for $n = 0$, 1, 2, ..., then

$$y_n - y_{n-1} = 3$$

is a difference equation, as are

$$y_n - y_{n-1} = y_{n-2} \qquad \text{the Fibonacci equation}$$

and

$$y_n - y_{n-1} = ky_{n-1}\left(1 - \frac{1}{K}y_{n-1}\right) \qquad \text{the logistic equation}$$

Such equations play the same role in the study of discrete dynamical situations that differential equations play in modeling continuous phenomena. Our goal in this section is to examine a few properties and applications of difference equations. We begin with some terminology.

A **solution** of a given difference equation is an ordered list $\{y_n\}$ (called a *sequence*) whose terms y_n satisfy the equation, and a **general solution** is a formula that characterizes a collection of solutions. An **initial value problem** consists of a difference equation together with a collection of conditions that must be satisfied, and a solution of such a problem is called a **particular solution.**

For instance, suppose a bacterial colony initially containing 100,000 individuals is observed every hour. If it is found that the population P_{k+1} at time $k + 1$ is always 1,000 more than 3% of the population P_k at time k, we can express this information in terms of the initial value problem

$$P_{k+1} = 0.03P_k + 1,000 \qquad \text{with } P_0 = 100,000$$

Examples 9.5.1 and 9.5.2 further illustrate this notation and terminology.

EXPLORE!

Some graphing calculator models can handle difference equations through the recursive relations feature provided by the sequence mode. See the Calculator Introduction on the book-specific website www.mhhe.com/hoffmann for preliminary instructions on displaying such functions and their graphs. Following the example on this page involving a bacterial colony, construct the function

$u(n) = 0.03u(n - 1) + 1,000$

with $u(1) = 100,000$, as the initial value in the equation editor of your graphing calculator. Display the values for at least seven generations using the list editor as well as the graphical features of your calculator. Does it appear that the population of bacteria is approaching a limiting value?

> **EXAMPLE 9.5.1**   Finding Terms in a First-Order Difference Equation

Write out the first five terms of the sequence $y_1, y_2, \ldots$ that satisfies the initial value problem

$$y_n = -3y_{n-1} + 1 \qquad \text{with } y_0 = 2$$

Solution

The initial term is $y_0 = 2$, and we generate each successive term by substituting into the difference equation. We find that

$$
\begin{aligned}
y_0 &= 2 \\
y_1 &= -3y_0 + 1 = -3(2) + 1 = -5 \\
y_2 &= -3y_1 + 1 = -3(-5) + 1 = 16 \\
y_3 &= -3y_2 + 1 = -3(16) + 1 = -47 \\
y_4 &= -3y_3 + 1 = -3(-47) + 1 = 142
\end{aligned}
$$

Thus, the first five terms of the sequence are $2, -5, 16, -47,$ and 142.

Difference equations play an important role in mathematical modeling. For instance, such equations can be used to relate the values of a particular quantity at different discrete times or periods, such as generations. This is illustrated in Example 9.5.2 for an application involving genetics.

> **EXAMPLE 9.5.2**   Verifying a Solution to a Difference Equation

Population genetics is concerned with how the distribution of genotypes in a population changes from generation to generation. Suppose random mating occurs in a population, and it is assumed that individuals who inherit a particular recessive gene from both parents will themselves have no offspring. Under these conditions, the frequency f_n of the gene recurring in the nth generation may be modeled by the difference equation

$$f_{n+1} = \frac{f_n}{1 + f_n}$$

Verify that

$$f_n = \frac{f_0}{1 + nf_0}$$

satisfies this difference equation, where f_0 is the initial frequency.

Solution

To verify that $f_n = \dfrac{f_0}{1 + nf_0}$ satisfies the given difference equation, note that

$$f_{n+1} = \frac{f_0}{1 + (n+1)f_0}$$

while

$$\frac{f_n}{1 + f_n} = \frac{\dfrac{f_0}{1 + nf_0}}{1 + \left(\dfrac{f_0}{1 + nf_0}\right)} = \frac{\dfrac{f_0}{1 + nf_0}}{\dfrac{1 + nf_0 + f_0}{1 + nf_0}}$$

$$= \frac{f_0}{1 + f_0 + nf_0} = \frac{f_0}{1 + (n + 1)f_0}$$

Thus,

$$f_{n+1} = \frac{f_0}{1 + (n + 1)f_0} = \frac{f_n}{1 + f_n}$$

as required.

Solution of a Difference Equation

We can always find solutions of a given initial value problem by starting with the initial condition and using the difference equation to generate one value after another as we did in Example 9.5.1. However, what we really want is a formula for y_n in terms of n, which will enable us to find any particular term y_k in the solution sequence by simply substituting k for n in the formula without having to first find $y_1, y_2, \ldots, y_{k-1}$. For instance, the result in Example 9.5.2, tells us that the 20th term in the solution sequence for the difference equation

$$f_{n+1} = \frac{f_n}{1 + f_n}$$

can be found by substituting $n = 20$ into the formula

$$f_n = \frac{f_0}{1 + nf_0}$$

That is,

$$f_{20} = \frac{f_0}{1 + 20f_0}$$

A **first-order linear difference equation** is one of the general form

$$y_n = ay_{n-1} + b$$

where a and b are constants. Our next goal is to show how the general term y_n of the solution sequence of such an equation can be expressed in terms of the initial term y_0. We will use an iterative procedure in which y_n is expressed in terms of y_{n-1}, then y_{n-1} in terms of y_{n-2}, and so on until we have a formula for y_n in terms of y_0. Here are the specific steps:

EXPLORE!

Refer to the problem explored in the Explore! box on page 745. Return the **MODE** menu back to Func mode and write Y1 = 100,000(0.03)^X + 1000(1 − 0.03^X)/(1 − 0.03) and graph using the window [0, 7]1 by [0, 100,000]1,000. How is this different from the graph of individual points for the Explore! box on page 745? Which is a better representation of the problem situation?

$$y_0$$
$$y_1 = ay_0 + b$$
$$y_2 = ay_1 + b = a(ay_0 + b) + b = a^2y_0 + ab + b$$
$$y_3 = ay_2 + b = a(a^2y_0 + ab + b) + b = a^3y_0 + a^2b + ab + b$$
$$y_4 = ay_3 + b = a^4y_0 + a^3b + a^2b + ab + b$$

and, in general,

$$y_n = ay_{n-1} + b = a^n y_0 + a^{n-1}b + a^{n-2}b + \cdots + a^2b + ab + b$$
$$= a^n y_0 + (a^{n-1} + a^{n-2} + \cdots + a + 1)b$$

For $a = 1$, all powers of a also equal 1, and we have $y_n = y_0 + nb$. If $a \neq 1$, the expression $a^{n-1} + a^{n-2} + \cdots + a + 1$ can be summed by the **geometric series** formula

$$a^{n-1} + a^{n-2} + \cdots + a + 1 = \frac{1 - a^n}{1 - a}$$

(see Exercise 48 or the general discussion of geometric series in Section 11.1). In this case, our formula for y_n can be written as

$$y_n = a^n y_0 + \left(\frac{1 - a^n}{1 - a}\right)b$$

To summarize:

Solving a First-Order Linear Difference Equation with an Initial Condition

For constants a and b, consider the initial value problem

$$y_n = ay_{n-1} + b \qquad \text{with initial value } y_0$$

If $a = 1$, the general solution is

$$y_n = y_0 + nb$$

If $a \neq 1$, the general solution is

$$y_n = a^n y_0 + \left(\frac{1 - a^n}{1 - a}\right)b$$

EXAMPLE 9.5.3   Solving a First-Order Linear Difference Equation

Solve the first-order linear initial value problem

$$y_n = -\frac{1}{2}y_{n-1} + 5 \qquad \text{with initial value } y_0 = 3$$

Solution

Comparing the given difference equation with the first-order linear form $y_n = ay_{n-1} + b$, we see that $a = -\frac{1}{2}$ and $b = 5$. Since $a \neq 1$, the solution is

$$
\begin{aligned}
y_n &= a^n y_0 + \left(\frac{1 - a^n}{1 - a}\right)b \\[2mm]
&= \left(-\frac{1}{2}\right)^n (3) + \left[\frac{1 - (-1/2)^n}{1 - (-1/2)}\right](5) \\[2mm]
&= \left(-\frac{1}{2}\right)^n (3) + \frac{10}{3}\left[1 - \left(-\frac{1}{2}\right)^n\right] \\[2mm]
&= \frac{10}{3} + \left(3 - \frac{10}{3}\right)\left(-\frac{1}{2}\right)^n \\[2mm]
&= \frac{10}{3} - \frac{1}{3}\left(-\frac{1}{2}\right)^n
\end{aligned}
$$

For example, the fifth term in the solution sequence is

$$y_5 = \frac{10}{3} - \frac{1}{3}\left(-\frac{1}{2}\right)^5 = \frac{107}{32}$$

EXPLORE!

Refer to Example 9.5.4. Store the function P_n as Y1 in the function editor of your graphing calculator, using the variable x instead of n. Graph using the window [0, 300]10 by [10,000, 25,000]1,000. How does this illustrate the problem situation for this example? Also explain why the continuous function does not truly represent the problem situation. Use graphical features of your calculator to show that the population tends toward 24,000 over a long period of time.

EXAMPLE 9.5.4 Using a Difference Equation to Study a Population Model

There are 12,000 people in Styxville. Each year, 3% of the people in the city die and 720 babies are born.

a. What will be the population in 5 years?

b. What will be the population in the long run?

Solution

a. Let P_n be the population after n years. Since $0.03P_{n-1}$ people die during the nth year, it follows that $0.97P_{n-1}$ people survive, and to that number is added the 720 births so that

$$\underbrace{P_n}_{\substack{\text{current}\\\text{population}}} = \underbrace{0.97P_{n-1}}_{\substack{\text{surviving}\\\text{population}}} + \underbrace{720}_{\text{births}}$$

Substituting $a = 0.97$, $b = 720$, and $P_0 = 12{,}000$ into our formula for the general solution of the equation $y_n = ay_{n-1} + b$ we get

$$P_n = (0.97)^n(12{,}000) + \frac{[1 - (0.97)^n](720)}{1 - 0.97}$$

Thus, when $n = 5$, the population is

$$P_5 = (0.97)^5(12{,}000) + \frac{[1 - (0.97)^5](720)}{1 - 0.97}$$

$$= 13{,}695$$

b. To find the long-run population, we compute the limit

$$\lim_{n \to \infty} P_n = \lim_{n \to \infty}\left\{ (0.97)^n(12{,}000) + \frac{[1 - (0.97)^n](720)}{1 - 0.97} \right\}$$

$$= 0 + \frac{[1 - 0](720)}{0.03} = 24{,}000$$

Thus, over a long period of time, the population tends toward 24,000 people.

Applications of Difference Equations

Difference equations are used in models involving business and economics, psychology, sociology, biology, information theory, and a variety of other areas. Two such applications are examined in Examples 9.5.5 and 9.5.6.

EXAMPLE 9.5.5   Amortization of Debt

When money is borrowed at an interest rate r, the total debt increases in the same way as a bank account paying the same rate of interest. **Amortization** is a procedure for repaying the debt, including interest, by making a sequence of payments. Usually the payments are of equal size and are made at equal time intervals (annually, quarterly, or monthly).

Suppose a debt of D dollars at an annual interest rate r is amortized over a period of n years by making annual payments of A dollars. Set up and solve a difference equation to show that

$$A = \frac{rD}{1 - (1 + r)^{-n}}$$

Solution

Let y_n be the total debt after n years. Then $y_n - y_{n-1}$ is the change in debt during the nth year, which equals the interest ry_{n-1} charged during the nth year less the annual payment A. That is,

$$\underbrace{y_n - y_{n-1}}_{\substack{\text{change} \\ \text{in debt}}} = \underbrace{ry_{n-1}}_{\substack{\text{interest} \\ \text{charged}}} - \underbrace{A}_{\substack{\text{annual} \\ \text{payment}}}$$

By combining the y_{n-1} terms and using the fact that $y_0 = D$, the initial debt, we obtain the initial value problem

$$y_n = (1 + r)y_{n-1} - A \qquad \text{with } y_0 = D$$

This difference equation has the first-order linear form, with $a = 1 + r$ and $b = -A$, so the general solution is

$$
\begin{aligned}
y_n &= a^n y_0 + \left(\frac{1 - a^n}{1 - a}\right)b \\
&= (1 + r)^n D + \left[\frac{1 - (1 + r)^n}{1 - (1 + r)}\right](-A) \\
&= (1 + r)^n D + \left(\frac{-A}{-r}\right)[1 - (1 + r)^n] \\
&= (1 + r)^n D + \frac{A}{r}[1 - (1 + r)^n]
\end{aligned}
$$

Since the debt is to be paid off in n years, we want to find A so that $y_n = 0$; that is,

$$y_n = (1 + r)^n D + \frac{A}{r}[1 - (1 + r)^n] = 0$$

By subtracting $\dfrac{A}{r}[1 - (1 + r)^n]$ from both sides of this equation, we get

$$(1 + r)^n D = -\frac{A}{r}[1 - (1 + r)^n] = \frac{A}{r}[(1 + r)^n - 1]$$

so that

$$A = rD\left[\frac{(1 + r)^n}{(1 + r)^n - 1}\right]$$

Finally, multiplying both numerator and denominator by $(1 + r)^{-n}$, we obtain the amortization formula

$$A = \frac{rD}{1 - (1 + r)^{-n}}$$

EXAMPLE 9.5.6 Biology-Fishery Management

A biologist is employed by a fishery to help regulate the population of trout in a lake it owns and controls. The biologist estimates that the lake initially contains 150,000 trout, and that the trout population is growing at the rate of 15% per year. Currently, the fishery harvests trout at the rate of 30,000 fish per year.

  a. Set up and solve a difference equation for the fish population after n years.

  b. How long does it take before the lake contains no trout?

  c. Suppose the fishery decides to change the harvesting policy so that the lake will still contain trout 20 years from now. What is the maximum harvesting rate that will allow this to occur?

  d. What is the maximum harvesting rate that can be allowed if the fishery wants the lake to always contain some trout?

Solution

  a. Let p_n denote the trout population at the end of the nth year. Then the change in population $p_n - p_{n-1}$ during the nth year satisfies

$$\underbrace{p_n - p_{n-1}}_{\substack{\text{change in} \\ \text{population}}} = \underbrace{0.15 p_{n-1}}_{\substack{\text{natural} \\ \text{growth}}} - \underbrace{30{,}000}_{\substack{\text{harvested} \\ \text{fish}}} \qquad \text{where } p_0 = 150{,}000$$

Combining the p_{n-1} terms on the right, we see that this difference equation has the first-order linear form

$$p_n = (1 + 0.15) p_{n-1} - 30{,}000 = 1.15 p_{n-1} - 30{,}000$$

where $a = 1.15$ and $b = -30{,}000$. The solution of this equation is

$$p_n = (1.15)^n (150{,}000) + \left[\frac{1 - (1.15)^n}{1 - 1.15} \right] (-30{,}000)$$

$$= (1.15)^n (150{,}000) + \left(\frac{-30{,}000}{-0.15} \right)[1 - (1.15)^n]$$

$$= (1.15)^n (150{,}000) + 200{,}000[1 - (1.15)^n]$$

$$= (1.15)^n [150{,}000 - 200{,}000] + 200{,}000$$

$$= -50{,}000(1.15)^n + 200{,}000$$

  b. The lake contains no trout when $p_n = 0$, that is, when

$$-50{,}000(1.15)^n + 200{,}000 = 0$$

$$(1.15)^n = \frac{200{,}000}{50{,}000} = 4$$

Taking logarithms on both sides of this equation, we find that

$$\ln(1.15)^n = \ln 4$$

so that

$$n \ln(1.15) = \ln 4$$

$$n = \frac{\ln 4}{\ln(1.15)} = 9.92$$

Therefore, no fish will remain in the lake after roughly 10 years.

c. Suppose the harvesting rate is h. Then the difference equation becomes $p_n = (1.15)p_{n-1} - h$, and its solution is

$$p_n = (1.15)^n(150{,}000) + \left[\frac{1 - (1.15)^n}{1 - 1.15}\right](-h)$$

$$= (1.15)^n(150{,}000) + \left(\frac{-h}{-0.15}\right)[1 - (1.15)^n]$$

$$= (1.15)^n\left(150{,}000 - \frac{h}{0.15}\right) + \frac{h}{0.15}$$

We want to find h so that the lake will be "fished out" when $n = 20$. Setting $p_{20} = 0$ and solving for h, we find that

$$p_{20} = (1.15)^{20}\left(150{,}000 - \frac{h}{0.15}\right) + \frac{h}{0.15} = 0$$

$$\frac{h}{0.15}[1 - (1.15)^{20}] = -(1.15)^{20}(150{,}000)$$

$$h = \frac{-0.15(1.15)^{20}(150{,}000)}{1 - (1.15)^{20}} = 23{,}964$$

Therefore, approximately 24,000 trout may be harvested each year.

d. If the lake is always to contain some trout, the year n in which all the trout disappear must satisfy $n \to \infty$. According to the computation in part (c), the fish population will be 0 in year n if

$$p_n = (1.15)^n\left(150{,}000 - \frac{h}{0.15}\right) + \frac{h}{0.15} = 0$$

Solving this equation for h, we find that

$$h = \frac{-0.15(1.15)^n(150{,}000)}{1 - (1.15)^n} \qquad \text{multiply numerator and denominator by } -1$$

$$= \frac{0.15(1.15)^n(150{,}000)}{(1.15)^n - 1} \qquad \text{multiply numerator and denominator by } (1.15)^{-n}$$

$$= \frac{0.15(150{,}000)}{1 - (1.15)^{-n}}$$

To find the fishing rate for the case where the lake always contains some trout, we compute h as $n \to \infty$; that is,

$$h = \lim_{n \to \infty} \frac{0.15(150{,}000)}{1 - (1.15)^{-n}} = \frac{0.15(150{,}000)}{1 - 0} \qquad \text{since } \lim_{n \to \infty}(1.15)^{-n} = 0$$

$$= 22{,}500$$

Therefore, if the fishing rate is set at no more than 22,500 trout per year, the lake will always contain some trout.

The Cobweb Model   In Chapter 1, we introduced the economic concepts of supply and demand and defined market equilibrium as the situation that occurs when supply equals demand. In Section 9.1, we discussed a price adjustment model in which supply and demand are regarded as functions of price, which changes continuously with time as market conditions adjust. However, equilibrium does not occur instantaneously by mutual agreement between producers and consumers. Instead, it is reached by a sequence of discrete adjustments, and our next goal is to describe a model involving difference equations for dealing with discrete supply-demand adjustments in a dynamic market.

We will consider a situation involving a single commodity, which for purposes of illustration, we assume to be a crop. Suppose the price is currently high, inspiring farmers to plant a large crop. However, by the time the crop is harvested and brought to market, the demand may be less than the farmers were expecting. If so, the price drops, prompting farmers to plant a smaller crop the next year. This time, when the crop is harvested, the supply may be smaller than demand. The price will then rise, farmers will be encouraged to grow larger crops again, supply will exceed demand, the price will fall, and so on. The pattern continues, with supply in each period determined by the price of the *previous* period.

To describe this supply-demand adjustment mathematically, let p_n be the price of the crop during the nth period, and let S_n and D_n be the corresponding supply and demand. Assume for simplicity that the supply and demand functions are linear; that is,

$$D = a - bp \qquad \text{and} \qquad S = c + dp$$

for constants a, b, c, and d. During the nth period, the demand satisfies

$$D_n = a - bp_n$$

but since supply in the nth period is determined by the price p_{n-1} of the previous period, we have

$$S_n = c + dp_{n-1}$$

During each period, the price is determined by equating supply and demand, so we must have

$$\underset{\text{supply } S_n}{c + dp_{n-1}} \quad = \quad \underset{\text{demand } D_n}{a - bp_n}$$

We recognize this as a difference equation that can be rewritten in the first-order linear form as follows:

$$c + dp_{n-1} = a - bp_n$$
$$bp_n = a - (c + dp_{n-1})$$
$$p_n = \left(-\frac{d}{b}\right)p_{n-1} + \left(\frac{a-c}{b}\right)$$

Substituting into our formula for the general solution of a first-order linear difference equation, we find that

$$p_n = \left(-\frac{d}{b}\right)^n p_0 + \left[\frac{1 - (-d/b)^n}{1 - (-d/b)}\right]\left(\frac{a-c}{b}\right)$$
$$= \left(-\frac{d}{b}\right)^n p_0 + \left(\frac{a-c}{b+d}\right)\left[1 - \left(-\frac{d}{b}\right)^n\right]$$
$$= \left(-\frac{d}{b}\right)^n\left[p_0 - \left(\frac{a-c}{b+d}\right)\right] + \left(\frac{a-c}{b+d}\right)$$

where p_0 is the initial price of the commodity (the crop in our example).

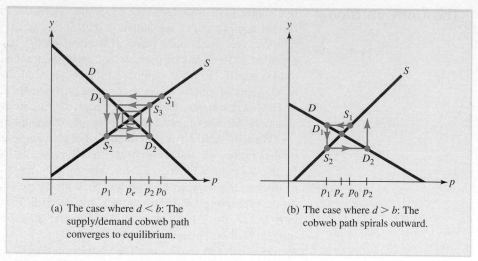

(a) The case where $d < b$: The supply/demand cobweb path converges to equilibrium.

(b) The case where $d > b$: The cobweb path spirals outward.

FIGURE 9.16 The progression of supply and demand in the cobweb model.

The model we have just described is called the **cobweb model.** To see why, look at Figure 9.16. The supply and demand lines, $y = c + dp$ and $y = a - bp$, are shown in two different cases. In both, the point of market equilibrium occurs where the lines intersect, that is, at the price p_e where

$$a - bp_e = c + dp_e$$

$$p_e = \frac{a - c}{b + d}$$

In Figure 9.16a, we plot the path of supply and demand over several time periods in the case where $d < b$. Notice that the initial price p_0 determines the supply S_1 in the first period, which in turn determines the demand D_1 in that period and the price p_1 such that $S_1 = D_1$. Then p_1 determines the supply S_2 in the second period, and so on. The path traced out in this fashion resembles a cobweb converging toward the equilibrium point. Figure 9.16b shows the case where $d > b$. This time, the "cobweb" spirals outward, diverging away from the equilibrium point.

Analytically, notice that if $d < b$, then $\left(-\dfrac{d}{b}\right)^n$ tends to 0 as $n \to \infty$. Therefore, no matter what the initial price p_0 may be, the price p_n will approach the equilibrium price $p_e = \dfrac{a - c}{b + d}$. However, if $d > b$, the term $\left(-\dfrac{d}{b}\right)^n$ will grow larger and larger in absolute value, while if $d = b$, then $\left(-\dfrac{d}{b}\right)^n = (-1)^n$ oscillates between 1 and -1. In both these latter cases, the price sequence $p_0, p_1, p_2, \ldots$ will not converge (approach a finite number as $n \to \infty$).

Since d is the slope of the supply line, it measures the rate of increase of supply, just as b measures the rate of decrease of demand. Thus, the result we have obtained can be interpreted as saying that the price sequence $p_0, p_1, p_2, \ldots$ converges to the equilibrium p_e if and only if the rate of decrease in demand exceeds the rate of increase in supply. In the language of economics, *market stability occurs in the long run only when suppliers are less sensitive to price changes than consumers.*

Difference Equations Versus Differential Equations

When modeling with difference equations, the difference $y_n - y_{n-1}$ between consecutive terms is used in much the same way as the derivative $\dfrac{dy}{dx}$ is used when modeling with a differential equation. Continuing the analogy, the ratio

$$\frac{y_n - y_{n-1}}{y_{n-1}}$$

is used to represent a relative or percentage rate of change in a discrete model.

For instance, suppose a species of fish is observed to reproduce at the rate of 15% per year. Then the fish population F_n after n years may be modeled by the difference equation

$$\frac{F_n - F_{n-1}}{F_{n-1}} = 0.15$$

or, equivalently, by

$$F_n - F_{n-1} = 0.15 F_{n-1}$$

On the other hand, if we wish to model the fish population as a continuous function $F(t)$ of time t, we could represent its growth by the analogous differential equation

$$\frac{dF}{dt} = 0.15F$$

Results obtained with difference equations may be quite different from those obtained using the analogous differential equations. For instance, in Example 9.5.5, we used the difference equation

$$y_n - y_{n-1} = ry_{n-1} - A$$

to model the total debt y_n accumulated after n years when interest is compounded continuously at an annual rate r and A dollars are spent annually to reduce debt. An analogous model in which the debt $Q(t)$ is regarded as a continuous function of time t would involve the differential equation

$$\frac{dQ}{dt} = rQ - A$$

Using the methods discussed in Section 9.2 to solve this equation subject to the initial condition that $Q(0) = D$ (the initial debt), we find that

$$Q(t) = \left(D - \frac{A}{r}\right)e^{rt} + \frac{A}{r}$$

(for practice, you should verify this result). If the debt is to be amortized over a term of n years, then we want $Q(n) = 0$, so that

$$0 = Q(n) = \left(D - \frac{A}{r}\right)e^{rt} + \frac{A}{r} \qquad \text{subtract } \frac{A}{r} \text{ from both sides}$$

$$\left(D - \frac{A}{r}\right)e^{rt} = -\frac{A}{r} \qquad \text{multiply both sides by } e^{-rt}$$

$$D - \frac{A}{r} = -\frac{A}{r}e^{-rt} \qquad \text{add } \frac{A}{r} \text{ to both sides and factor}$$

$$D = (1 - e^{-rt})\frac{A}{r} \qquad \text{multiply both sides by } \frac{r}{1 - e^{-rt}}$$

$$A = \frac{rD}{1 - e^{-rt}}$$

Notice the similarities and differences between this continuous amortization formula and the discrete formula found in Example 9.5.5. For instance, if the initial debt is $D = 10,000$ dollars and the annual interest rate is 8% ($r = 0.08$), then the discrete model tells us the debt will be amortized over a term of $n = 10$ years by making annual payments of

$$A = \frac{0.08(10,000)}{1 - (1 + 0.08)^{-10}} = 1,490.29$$

while the continuous model yields annual payments of

$$A = \frac{0.08(10,000)}{1 - e^{-0.08(10)}} = 1,452.77$$

The difference is that in the discrete case, the payment is made once a year, as a lump sum, while in the continuous case, the annual payment is assumed to be spread out continuously over the entire year.

EXERCISES ■ 9.5

In Exercises 1 through 8, write out the first five terms of the given initial value problem.

1. $y_n = 3y_{n-1}; y_0 = 1$

2. $y_n = y_{n-1} + 2; y_0 = 0$

3. $y_n = y_{n-1}^2; y_0 = 1$

4. $y_n = y_{n-1}^3 + (1 - y_{n-1})y_{n-1}; y_0 = -0.1$

5. $y_{n+1} = y_n + y_{n-1}; y_0 = 1, y_1 = 1$

6. $y_{n+1} = y_n^2 - 2y_{n-1}; y_0 = -1, y_1 = 0$

7. $y_{n+1} = \dfrac{y_n}{y_n + 2}; y_0 = 1$

8. $y_{n+1} = \dfrac{y_n}{y_{n-1} + 1}; y_0 = 0, y_1 = 1$

In Exercises 9 and 10, find constants A and B so that the given expression y_n satisfies the specified difference equation.

9. $ny_n + (n - 1)y_{n-1} = 2n - 3; y_n = A + \dfrac{B}{n}$

10. $y_{n+1} = y_n + n; y_n = An^2 + Bn$

In Exercises 11 through 17, solve the given first-order linear initial value problem.

11. $y_n = -y_{n-1}; y_0 = 1$

12. $y_n = 2y_{n-1} + 3; y_0 = 0$

13. $y_n = \dfrac{1}{2}y_{n-1} + 1; y_0 = 0$

14. $y_n = -\dfrac{1}{2}y_{n-1} - \dfrac{1}{4}; y_0 = 10$

15. $y_n - y_{n-1} = \dfrac{1}{8}y_{n-1} + \dfrac{1}{4}; y_0 = 0$

16. $y_n - y_{n-1} = -\dfrac{1}{2}y_{n-1} + \dfrac{1}{3}; y_0 = 1$

17. $y_n - y_{n-1} = \dfrac{1}{10}y_{n-1}; y_0 = 10$

BUSINESS AND ECONOMICS APPLIED PROBLEMS

18. **SAVINGS ACCOUNT** Jose deposits $1,000 in an account that pays 5% interest, compounded annually. Set up and solve an initial value problem for the amount of money A_n in Jose's account after n years.

19. **SAVINGS ACCOUNT** Like Jose in Exercise 18, Myra invests $1,000 in an account that pays 5% interest, compounded annually. However, she also makes additional deposits of $50 at the end of each year. Set up and solve an initial value problem for the amount of money A_n in Myra's account after n years.

20. **STOCK MARKET SPECULATION** Sanjay invests in a speculative stock whose price p_n at the end of each month satisfies the initial value problem

$$p_n = -0.1p_{n-1} + 300 \qquad \text{where } p_0 = 100$$

How much will Sanjay's investment be worth after n months?

21. **BUSINESS MANAGEMENT** A car rental company currently owns 300 cars. Each year, the company plans to dispose of 25% of its car fleet and add 50 new cars.
   a. Set up and solve a difference equation for the number of cars C_n in the fleet at the end of the nth year.
   b. How many cars will be in the fleet after 10 years? How large will the fleet be in the long run?

22. **AMORTIZATION OF DEBT** In Example 9.5.5, we obtained an amortization formula for the case where interest is compounded annually and annual payments are made. In practice, monthly compounding and monthly payments are more common.

      Set up and solve a difference equation to show that a debt of D dollars due in t years at an annual interest rate r compounded monthly will be amortized by making equal monthly payments of M dollars, where

$$M = \frac{iD}{1 - (1 + i)^{-12t}} \quad \text{with } i = \frac{r}{12}$$

23. **AMORTIZATION OF DEBT** Generalize the result in Exercise 22 to obtain a formula for amortizing a debt of D dollars due in t years at an annual interest rate r compounded k times per year by making equal payments of M dollars, also k times per year.

In Exercises 24 through 27, use the formula in Exercise 22 to answer the given question.

24. **MORTGAGE PAYMENTS** A homeowner puts $35,000 down on a $350,000 home and takes out a 30-year mortgage for the balance. If the interest rate is 6% per year, compounded monthly, what are the monthly mortgage payments?

25. **INSTALLMENT PAYMENTS** Marvin pays $3,000 down on a $25,000 new car and will pay off the balance by making monthly installment payments over the next 5 years. What are Marvin's monthly payments if the interest rate is 8% per year, compounded monthly?

26. **INSTALLMENT PAYMENTS** Joan borrows $5,000 for 24 months at 9% interest, compounded monthly. If the loan is to be repaid in monthly installments of $200 with a final balloon payment of A dollars at the end of the 24-month term, what is A?

27. **SINKING FUND** Shen has a payroll savings plan at work in which a certain fixed amount A is taken from his monthly paycheck and placed into an account earning 5% annual interest compounded monthly. What should A be for Shen to save $7,000 for a trip to China in 3 years?

28. **FUTURE VALUE OF AN ANNUITY** An annuity consists of equal deposits of D dollars made annually into an account that pays interest at an annual rate r compounded annually.
   a. Set up and solve a difference equation to obtain a formula for the accumulated value A_n of the annuity after n deposits.
   b. Tom deposits $1,000 annually into an account that pays 5% interest compounded annually. How much is his account worth after 20 years?

29. **CURRENCY RENEWAL** A nation has 5 billion dollars in currency. Each day, about $18 million comes into the nation's banks and the same amount is paid out. Suppose the country decides that whenever old currency comes into a bank, it is destroyed and replaced by a new style of currency.
   a. Set up and solve a difference equation for the amount of new currency C_n in circulation at the beginning of the nth day after the currency renewal begins. You may assume that $C_0 = 0$.
   b. How long will it take before 90% of the currency in circulation is new?
   c. Compare the results you have just obtained for your discrete currency model with the analogous continuous model analyzed in Exercise 41 in Section 9.2. In particular, do the two models yield the same time period for 90% renewal of the currency?
   d. Write a paragraph on which of the two models (discrete or continuous) you think is more appropriate for describing the renewal process.

30. **COLLEGE FUNDING** A department at a large state college was initially funded with a budget of $10 million. Each year, the department requests $300,000 more than it received the previous year, but the final budget approved by the board of trustees is always 5% less than the amount requested.
   a. Set up and solve a difference equation for the budget B_n of the department n years after its initial funding.
   b. How large will the department budget be 20 years after its initial funding? How large will it be in the long run?

31. VALUE OF AN INHERITANCE   Jaqui receives an inheritance in which she is paid $25,000 each year on her birthday until she retires. Suppose she deposits the money into a retirement annuity that pays annual interest at an effective rate of 7%. (Recall the definition of effective rate given in Section 4.1.)

a. Set up and solve a difference equation for the accumulated value A_n in her account n years after the payments begin. [*Note:* $A_0 = 25,000$.]

b. She decides she will retire on the first birthday when her retirement account is worth a million dollars. If the inheritance payments begin on her 21st birthday, when does she retire?

c. What is the equivalent cash value (that is, the present value) of Jaqui's inheritance?

COBWEB MODEL   *In Exercises 32 through 35, p is the price of a particular commodity and S(p) and D(p) are the corresponding supply and demand functions. Assume these quantities are recorded at a sequence of regular time periods, and let p_n, S_n, and D_n be the price, supply, and demand, respectively, during the nth period. In each case:*

(a) Find the equilibrium price p_e.

(b) Set up and solve a difference equation to find a formula for p_n.

(c) Find p_n, S_n, and D_n for n = 1, 2, and 3, and then sketch the corresponding cobweb graph for the first three time periods.

32. $S_n = 1.5p_{n-1} - 2; D_n = -2p_n + 3; p_0 = 1.5$

33. $S_n = 3p_{n-1} - 2; D_n = -p_n + 1.3; p_0 = 1$

34. $S_n = 1.5p_{n-1} + 3; D_n = -p_n + 4; p_0 = 3$

35. $S_n = p_{n-1} + 1; D_n = -2p_n + 7; p_0 = 3$

36. INSTALLMENT LOANS   A loan of P dollars is being paid off with monthly payments of M dollars to a bank that charges interest at the annual rate r compounded monthly. Set up and solve a difference equation for the balance owed B_n after n payments have been made.

37. NATIONAL INCOME   In the classic Harrod-Hicks model* of an expanding national economy, it is assumed that if I_n denotes the national

income at the end of the nth year, then the change in national income during that year is given by the difference equation

$$I_n - I_{n-1} = cI_{n-1} - h$$

where c and h are positive constants (c is the *marginal propensity to consume*). Solve this difference equation to find I_n in terms of c, h, and the initial national income $I_0 = I(0)$.

LIFE AND SOCIAL SCIENCE APPLIED PROBLEMS

38. SPREAD OF A DISEASE   A disease spreads in such a way that the number of people exposed during the month n is proportional to the number of susceptible people in the population who have not yet acquired the disease. That is, if S_n is the number of sick people in a susceptible population P, then

$$S_{n+1} - S_n = k(P - S_n)$$

Suppose 10,000 people in a community are susceptible to a new strain of influenza. If 500 people initially have the disease and 700 are newly infected during the first month, how many will have the disease n months after the outbreak? What happens to S_n in the long run (as $n \to \infty$)?

39. SPREAD OF A DISEASE WITH A CURE   In Exercise 38, suppose 20% of the people who have the disease at the end of the $n - 1$st month are cured during the nth month.

a. Rewrite the difference equation in Exercise 38 to take into account this new feature.

b. Now how many people will have the disease n months after the outbreak? What happens to S_n as $n \to \infty$ in this case?

40. RECYCLING   Gillian has a library of 1,200 paperback books. Each year, she gives away 15% of the books she has and buys 135 new books. Set up and solve a difference equation for the number of books B_n in her library at the end of the nth year. How many books will Gillian have at the end of the 10th year? How many will she have in the long run?

41. ANIMAL POPULATION WITH HUNTING   The fox population in a certain region is initially 50,000 and is found to grow at the rate of 8% per year. Hunters are allowed to "harvest" h foxes per year.

a. If $h = 2,000$, what is the size of the fox population after n years?

b. What should h be for the fox population to remain the same from year to year?

---

*J. R. Hicks, "Mr. Harrod's Dynamic Theory," *Economica*, New Series, Vol. 16, pp. 106–121. A readable treatment of applications of difference equations to economics may be found in the text, *Introduction to Difference Equations,* by Samuel Goldberg, New York: Wiley, 1958, pp. 93–96.

42. GENETICS Certain genetic diseases can be modeled by a difference equation of the form

$$f_{n+1} = \frac{f_n}{1 + f_n}$$

where f_n is the relative frequency of a recessive, disease-producing gene in the nth generation, assuming that individuals who carry the recessive gene do not reproduce (recall Example 9.5.2). This difference equation is not first-order linear, but you can solve it by proceeding as follows:

a. Show that if $g_n = \dfrac{1}{f_n}$, then g_n satisfies the first-order linear difference equation $g_n = Ag_{n-1} + B$ for certain constants A and B.

b. Solve the difference equation for g_n in part (a), and then use this result to find a formula for f_n.

c. Read an article on population genetics, and write a paragraph on mathematical methods of modeling genetic diseases.[†]

43. DRUG ELIMINATION A patient receives an intravenous injection of a certain drug which is then eliminated from his body at the rate of 6% per hour.

a. Set up and solve a difference equation for the amount D_n of the drug remaining in the patient's body after n hours.

b. If the patient receives 150 mg of the drug at 6 A.M., how much remains in his body at 6 P.M.?

44. POPULATION Suppose the number of females in the nth generation of the population of a certain species is given by P_n. Further assume that on average, each female member of the $n - 1$st generation gives birth to b female offspring and that in each generation, d females can be expected to die, where b and d are positive constants.

a. Set up and solve a difference equation to find P_n in terms of b, d, and the initial population of females $P_0 = P(0)$.

b. Suppose $P_0 = 100$, $b = 0.2$, and $d = 15$. What will be the female population of the 10th generation? What happens to the female population in the long run (as $n \to \infty$)?

45. LOGISTIC GROWTH An alternate form of the logistic differential equation discussed in Section 9.1 is

$$\frac{dP}{dt} = kP\left(1 - \frac{P}{K}\right)$$

Such equations are frequently used in continuous models for populations $P(t)$ that grow rapidly at first but then slow down and eventually approach the carrying capacity K as $t \to \infty$. The analogous difference equation

$$P_n - P_{n-1} = kP_{n-1}\left(1 - \frac{1}{K}P_{n-1}\right)$$

is called the *discrete logistic equation* and is a useful tool in discrete modeling, especially in biology.

a. Suppose a particular population is modeled by a discrete logistic equation. The population is said to reach an *equilibrium* level L if $P_m = P_{m+1} = L$ for some m. Show that a population at an equilibrium level stays there; that is, $P_n = L$ for all $n > m$.

b. The population of a certain bacterial colony is recorded at discrete time intervals. Let P_n be the population (in thousands) at the nth observation, and suppose that $P_0 = 50$ (that is, 50,000), $P_1 = 55$, and $P_2 = 60$. If the population is modeled by a discrete logistic equation, use this information to find K and k and the equilibrium level L.

c. Set up and solve an analogous logistic differential equation for the bacterial population in part (b), regarded as a continuous function $P(t)$ of time t, with $P(0) = 50$, $P(1) = 55$, and $P(2) = 60$. Compare with the results obtained for the discrete model in part (b).

46. LEARNING Consider an experiment in which a subject is repeatedly presented with a stimulus, which may be thought of as a "test." Each time the test is presented, the subject is required to respond, and if the response is "positive," a reward is given. For example, a rat may be presented with a maze (the stimulus). If it responds by choosing the "correct" path, it is rewarded with food, while an "incorrect" path leads to no food or a small electric shock.[*]

Suppose in our experiment, p_n (for $0 \le p_n \le 1$) is the likelihood that the subject "succeeds" on the test on the nth trial. Assume that the change in performance $p_n - p_{n-1}$ between the $n - 1$st and nth trials satisfies the difference equation

$$p_n - p_{n-1} = A(1 - p_{n-1}) - Bp_{n-1}$$

[†]A good place to start is the text, "*Introduction to Dynamic Systems*," by David G. Luenberger, New York: Wiley, 1979, pp. 382–389.

[*]The learning model in Exercise 46 is adapted from a classic model developed by R. R. Bush and F. Mosteller, as described in *Introduction to Difference Equations,* by Samuel Goldberg, New York: Wiley, 1958, pp. 103–107.

where A and B are parameters that satisfy $0 < A < 1$ and $0 < B < 1$. The parameter A can be thought of as a measure of the value of a reward to the subject, while $-B$ measures the negative impact of a punishment.

a. Solve the difference equation for p_n in terms of A, B, and the likelihood p_0 that the subject enjoys immediate success.

b. How likely is the subject to succeed on the test in the long run (as the number of trials increases indefinitely)?

c. Suppose the likelihood $p(t)$ varies continuously with time t. Set up and solve an analogous differential equation to find $p(t)$. Compare this solution with the solution to the discrete problem you found in part (a). In particular, do both models yield the same eventual likelihood of success?

d. Write a paragraph on which model (discrete or continuous) you think is more appropriate for this experiment.

MISCELLANEOUS PROBLEMS

47. **LOTTERY WINNINGS** Lacey wins a million dollar lottery that is to be paid in equal annual amounts over a period of years.

a. Set up and solve a difference equation to find the present value P_n of the payout if the prevailing annual interest rate is 5% and the payments are spread over n years.

b. Lacey does the math in part (a) and sues the lottery for false advertising, claiming that she was promised a million dollars but will receive less. The court agrees with her and orders the lottery to make an extra payment of A dollars each year in order to bring the present value up to the promised million dollar level. What must A be if the payout is spread over 20 years?

48. a. Show that for $a \neq 1$,

$$S_n = a^{n-1} + a^{n-2} + \cdots + a + 1 = \frac{1 - a^n}{1 - a}$$

[*Hint:* What is $S_n - aS_{n-1}$?]

b. Show that

$$1 + a + a^2 + a^3 + \cdots = \begin{cases} \dfrac{1}{1 - a} & \text{if } |a| < 1 \\ \infty & \text{if } a = 1 \end{cases}$$

What happens when $a = -1$?

Important Terms, Symbols, and Formulas

$$\frac{dy}{dx} = \frac{h(x)}{g(y)}$$

$$\int g(y)\, dy = \int h(x)\, dx$$

$$\frac{dQ}{dt} = k(B - Q)$$

$$\frac{dP}{dt} = kP(\text{birth rate} - \text{death rate})$$

$$\frac{dP}{dt} = kP \qquad \text{with } P(0) = P_0$$

Solution:

$$P(t) = P_0 e^{kt}$$

$$\frac{dP}{dt} = kP(M - P) \qquad \text{with } P(0) = P_0$$

where M is the carrying capacity
Solution:

$$P(t) = \frac{MP_0}{P_0 + (M - P_0)e^{-Mkt}}$$

$$\frac{dy}{dx} + p(x)y = q(x)$$

$$I(x) = e^{\int p(x)\, dx}$$

CHAPTER SUMMARY

Solution: (709)

$$y = \frac{1}{I(x)}\left[\int I(x)q(x)\,dx + C\right]$$

Compartmental analysis   (711)

$$\frac{dQ}{dt} = \text{input rate} - \text{output rate}$$

Dilution models   (714)
Newton's law of cooling   (723)
Orthogonal trajectories   (724)
Chemotaxis   (725)
Linked differential equations   (726)
Predator-prey model   (728)

Slope fields   (737)
Euler's method   (741)
Difference equations   (745)
First-order linear difference equation:
   Form: (747)

$$y_n = ay_{n-1} + b \qquad \text{with initial value } y_0$$

   Solution: (748)

$$y_n = a^n y_0 + \left(\frac{1 - a^n}{1 - a}\right)b$$

Amortization of debt   (750)
Cobweb model   (754)

CHAPTER SUMMARY

Checkup for Chapter 9

1. Find the general solution for each of the following differential equations:

   a. $\dfrac{dy}{dx} = \dfrac{xy}{x^2 + 1}$

   b. $\dfrac{dy}{dx} = \dfrac{2y}{x} + x^3$

   c. $\dfrac{dy}{dx} = xe^{y-x}$

2. In each case, find the particular solution of the given differential equation that satisfies the indicated condition.

   a. $\dfrac{dy}{dx} = -\dfrac{2}{x^2 y}$ where $y = 1$ when $x = -1$

   b. $\dfrac{dy}{dx} + \dfrac{y}{x} = 5$ where $y = 0$ when $x = 1$

   c. $\dfrac{dy}{dx} + \dfrac{xy}{x^2 + 1} = x$ where $y = 0$ when $x = 0$

3. Solve each of the following difference equations subject to the indicated initial condition.

   a. $y_n = -\dfrac{1}{2}y_{n-1};\ y_0 = -1$

   b. $y_n - y_{n-1} = \dfrac{1}{3}y_{n-1};\ y_0 = 2$

   c. $y_n - y_{n-1} = \dfrac{1}{4}y_{n-1} - \dfrac{1}{2};\ y_0 = 0$

4. Plot a slope field for each of the following differential equations, and use the field to sketch the graph of the solution that passes through the specified point.

   a. $y' = 2x - y;\ (0, 0)$

   b. $y' = x^2 - y^2;\ (1, 1)$

   c. $y' = \dfrac{y}{x + 1};\ (0, 0)$

In Exercises 5 and 6, write a differential equation describing the given situation. (Do not try to solve the equation.)

5. **WATER POLLUTION**   The amount $P(t)$ of pollution in a stream t days after an industrial spill decreases with respect to time at a rate equal to 11% of the amount present at that time.

6. **PRICE ADJUSTMENT**   The price $p(t)$ of a commodity changes with respect to time at a rate that is proportional to the surplus $S(p) - D(p)$, where S and D are the supply and demand, respectively.

7. **MEDICATION**   A child is injected with A_0 mg of an anesthetic drug, and t hours later, the amount $A(t)$ of drug still present in the child's bloodstream is decreasing with respect to time at a rate proportional to $A(t)$. It takes 4 hours to eliminate half the original dose of the drug, and the anesthetic becomes ineffective when the blood contains less than 1,800 mg. What dose A_0 should be administered to guarantee that the child will remain fully anesthetized during a 2-hour operation?

8. **GROWTH OF AN INVESTMENT**   An investment manager determines that the value $V(t)$ of a certain investment is growing with respect to time at a rate proportional to V^2. The initial value

of the investment was $10,000, and it took 10 years to double. How long will it take before the investment is worth $50,000?

9. **DILUTION** A 500-gallon tank initially contains 200 gallons of brine containing 75 pounds of dissolved salt. Brine containing 2 pounds of salt per gallon flows into the tank at the rate of 4 gallons per minute, and the well-stirred mixture flows out at the rate of 1 gallon per minute. Set up and solve a differential equation for the amount of salt $A(t)$ in the tank at time t. How much salt is in the tank when it is full?

10. **COBWEB MODEL** The supply S_n and demand D_n for a certain commodity during the nth time period for $n = 1, 2, \ldots$ are given by

$$S_n = 20 + p_{n-1} \quad \text{and} \quad D_n = 50 - 2p_n$$

where p_n is the price of the commodity during the nth period. The initial price of the commodity is $p_0 = 20$.

a. Find the equilibrium price p_e.
b. Set up and solve a difference equation to find a formula for p_n.
c. Sketch the cobweb graph for the first three time periods.

Review Exercises

In Exercises 1 through 12, find the general solution of the given differential equation.

1. $\dfrac{dy}{dx} = x^3 - 3x^2 + 5$

2. $\dfrac{dy}{dx} = 0.02y$

3. $\dfrac{dy}{dx} = k(80 - y)$

4. $\dfrac{dy}{dx} = y(1 - y)$

5. $\dfrac{dy}{dx} = e^{x+y}$ [*Hint:* $e^{x+y} = e^x e^y$]

6. $\dfrac{dy}{dx} = ye^{-2x}$

7. $\dfrac{dy}{dx} + \dfrac{4y}{x} = e^{-x}$

8. $\dfrac{dy}{dx} + \dfrac{y}{x} = \dfrac{2}{x+1}$

9. $\dfrac{dy}{dx} + \dfrac{2y}{x+1} = \dfrac{4}{x+2}$

10. $x^3 \dfrac{dy}{dx} + xy = 5$

11. $\dfrac{dy}{dx} = \dfrac{\ln x}{xy}$

12. $\dfrac{dy}{dx} = e^{2x-y}$ $\left[\text{Hint: } e^{2x-y} = \dfrac{e^{2x}}{e^y} \right]$

In Exercises 13 through 24, find the particular solution of the given differential equation that satisfies the specified condition.

13. $\dfrac{dy}{dx} = 5x^4 - 3x^2 - 2$; $y = 4$ when $x = 1$

14. $\dfrac{dy}{dx} = 3 - y$; $y = 2$ when $x = 0$

15. $\dfrac{dy}{dx} = 0.06y$; $y = 100$ when $x = 0$

16. $\dfrac{dy}{dx} - \dfrac{y}{x} = \dfrac{2}{x}$; $y = -1$ when $x = 1$

17. $\dfrac{dy}{dx} - \dfrac{5y}{x} = x^2$; $y = 4$ when $x = -1$

18. $\dfrac{dy}{dx} + \dfrac{y}{x+1} = x$; $y = 0$ when $x = 3$

19. $\dfrac{dy}{dx} - xy = e^{x^2/2}$; $y = 4$ when $x = 0$

20. $\dfrac{dy}{dx} + 3x^2 y = x^2$; $y = 2$ when $x = 0$

21. $\dfrac{dy}{dx} = y \ln \sqrt{x}$; $y = -4$ when $x = 1$

22. $\dfrac{dy}{dx} = \dfrac{\ln x}{y}$; $y = 100$ when $x = 1$

23. $\dfrac{dy}{dx} = \dfrac{xy}{\sqrt{1 - x^2}}$; $y = 2$ when $x = 0$

24. $\dfrac{d^2 y}{dx^2} = 2$; $y = 5$ and $\dfrac{dy}{dx} = 3$ when $x = 0$

In Exercises 25 through 30, find the solution of the given first-order linear difference equation that satisfies the specified initial condition.

25. $y_n = y_{n-1} + 2$; $y_0 = 1$

26. $y_n = -8y_{n-1}$; $y_0 = 1$

27. $y_n - y_{n-1} = 3$; $y_0 = 1$

28. $y_n - y_{n-1} = -1$; $y_0 = -2$

29. $y_n - 5y_{n-1} = 2$; $y_0 = 2$

30. $y_n - 2y_{n-1} = 8$; $y_0 = -1$

In Exercises 31 through 36, plot a slope field for the given differential equation and use the field to sketch the graph of the particular solution that passes through the specified point.

31. $y' = 3x - y$; $(1, 0)$

32. $y' = 3 - x + y;\ (0, 0)$

33. $y' = x + y^2;\ (1, 1)$

34. $y' = \dfrac{x - y}{x + y};\ (1, 1)$

35. $y' = \dfrac{y}{x} + \dfrac{x}{y};\ (1, 0)$

36. $y' = \dfrac{xy}{x^2 + y^2};\ (0, 1)$

In Exercises 37 through 40, use Euler's method with the indicated step size h and number of increments n to obtain an approximate solution of the given initial value problem.

37. $y' = 3x - y;\ y(0) = 0;\ h = 0.1;\ n = 5$

38. $y' = x + xy;\ y(0) = 1;\ h = 0.1;\ n = 4$

39. $y' = \dfrac{y}{x} + y;\ y(1) = 1;\ h = 0.2;\ n = 4$

40. $y' = \sqrt{x} + \sqrt{y};\ y(0) = 0;\ h = 0.2;\ n = 5$

In Exercises 41 through 44, use Euler's method with the indicated step size h to estimate the required solution of the given initial value problem.

41. $y' = x + 4y;\ y(0) = 1;$ estimate $y(1)$ using $h = 0.2$

42. $y' = \dfrac{y}{x};\ y(1) = 0;$ estimate $y(2)$ using $h = 0.2$

43. $y' = e^{xy};\ y(0) = 0;$ estimate $y(2)$ using $h = 0.4$

44. $y' = \ln(xy);\ y(1) = 1;$ estimate $y(3)$ using $h = 0.4$

45. RECALL FROM MEMORY Some psychologists believe that when a person is asked to recall a set of facts, the rate at which the facts are recalled is proportional to the number of relevant facts in the subject's memory that have not yet been recalled. Suppose a subject is presented with a list of A facts, and that at time $t = 0$, the subject can recall none of these facts. Set up and solve a differential equation to find the number of facts $N(t)$ that have been recalled as a function at time t.

46. DECAY OF A BIOMASS A researcher determines that a certain protein with mass $m(t)$ grams at time t (hours) disintegrates into amino acids at a rate jointly proportional to $m(t)$ and t. Experiments indicate that half of any given sample will disintegrate in 12 hours.

    a. Set up and solve a differential equation for $m(t)$.

    b. What fraction of a given sample remains after 9 hours?

47. DILUTION A tank currently holds 200 gallons of brine that contains 3 pounds of salt per gallon. Pure water flows into the tank at the rate of 4 gallons per minute, while the mixture, kept uniform by stirring, runs out of the tank at the same rate. How much salt is in the tank at the end of 100 minutes?

48. DILUTION A 200-gallon tank currently holds 30 pounds of salt dissolved in 50 gallons of water. Brine containing 2 pounds of salt per gallon runs into the tank at the rate of 2 gallons per minute, and the mixture, kept uniform by stirring, flows out at the rate of 1 gallon per minute.

    a. Set up and solve a differential equation for the amount of salt $S(t)$ in solution at time t.

    b. How many gallons of solution are in the tank when there are 40 pounds of salt in solution?

    c. How much salt is in the tank when it overflows?

49. AMOUNT OF DRUG IN AN ORGAN Suppose that a drug with concentration c_0 g/cc enters an organ of volume V cc at the rate of A cc/sec and is dispersed at the same rate. If the organ initially contains none of the drug and it can safely accommodate a concentration no greater than L g/cc, where $L < c_0$, what is the maximum length of time that the drug can be allowed to enter the organ? Your answer will be an expression involving V, c_0, L, and A.

50. RETIREMENT INCOME A retiree deposits S dollars into an account that earns at an annual rate r compounded continuously and annually withdraws W dollars.

    a. Explain why the account changes at the rate

$$\frac{dV}{dt} = rV - W$$

    where $V(t)$ is the value of the account t years after the account is initiated. Solve this differential equation to express $V(t)$ in terms of r, W, and S.

    b. Frank and Jessie Jones deposit \$500,000 into an account that pays 5% interest compounded continuously. If they withdraw \$50,000 annually, what is their account worth at the end of 10 years?

    c. What annual amount W can Frank and Jessie withdraw each year if their goal is to keep their account unchanged at \$500,000?

    d. If Frank and Jessie decide to withdraw \$80,000 annually, how long does it take to exhaust their account? [That is, when does $V(t)$ become 0?]

CHAPTER SUMMARY

51. CONCENTRATION OF BLOOD GLUCOSE
Glucose is infused into the bloodstream of a
patient at a constant rate R, and at each time t, is
converted and excreted at a rate proportional to
the present concentration of glucose $C(t)$.

  a. Set up and solve a differential equation for $C(t)$.
  Express your answer in terms of R, the constant of
  proportionality k (the elimination constant), and
  the initial concentration of glucose $C_0 = C(0)$.

  b. Show that in the long run (as $t \to \infty$) the glucose
  concentration approaches a level that depends only
  on r and k and not on the initial concentration C_0.

  c. Find an expression for the time T required for
  half the initial concentration C_0 to be eliminated. (Notice that this half-life depends on
  C_0 unless $C_0 = 0$.)

52. REDUCTION OF AIR POLLUTANT   The air
in a room with volume 3,000 ft^3 initially contains
0.21% carbon dioxide. Suppose the room receives
fresh air containing 0.04% carbon dioxide at the
rate of 800 ft^3 per minute. All the air in the room
is circulated by a fan, and the mixture leaves at
the same rate (800 ft^3/min).

  a. Set up and solve a differential equation for the
  percentage of carbon dioxide in the room after
  t minutes.

  b. How long does it take before the room contains
  only 0.1% carbon dioxide?

  c. What percentage of carbon dioxide is left in the
  room in the long run (as $t \to \infty$)?

53. PERSONNEL MANAGEMENT   A certain
company began operations with 100 employees,
and in subsequent years, the number of employees
E_n during the nth year always exceeded the number
E_{n-1} during the previous year by 5%. Express this
relationship as a first-order linear difference
equation, and solve the equation to find E_n.

54. MEDICATION   A drug administered
intravenously to a patient is eliminated from the
patient's bloodstream at the rate of 14% per hour.
Set up and solve a difference equation for the
amount D_n of a dose of D_0 milligrams of drug
that remains after n hours. How much of the dose
remains after 10 hours?

**55. PRICE-ADJUSTMENT FOR SUPPLY AND
DEMAND**   A commodity is introduced with an
initial price of $4 per unit, and t months later, the
price is $p(t)$ dollars per unit. A study indicates that

at time t, the price changes at a rate equal to 2%
of the shortage $D - S$, where $S(t) = 2 + p$ and
$D(t) = 3 + 7e^{-t}$ are, respectively, the supply and
demand for the commodity, both in thousands of
units.

  a. Set up and solve an initial value problem for
  the unit price $p(t)$.

  b. What is the unit price after 6 months?

  c. At what time is the unit price the largest?
  What is the maximum unit price and the
  corresponding supply and demand?

  d. As $t \to \infty$, what happens to $p(t)$, $S(t)$, and $D(t)$?

**56. PRICE-ADJUSTMENT FOR SUPPLY AND
DEMAND**   Answer the question in Exercise 55
for a commodity with initial price $2 per unit
whose price changes at a rate equal to 3% of the
shortage $D - S$, with supply $S(t) = 1 + p$ and
demand $D(t) = 2 + 8e^{-t/2}$.

In Exercises 57 and 58, you will need the formula
$$1 + a + a^2 + a^3 + \cdots = \frac{1}{1 - a} \qquad \text{for } |a| < 1$$
derived in Exercise 48 of Section 9.5.

---

57. QUALITY CONTROL   Suppose that in a
particular industrial assembly line, the likelihood
p_n that a defective unit receives its imperfection
during the nth stage of production is 30% of the
likelihood p_{n-1} of the imperfection being
introduced during the $n - 1$st stage.

  a. Set up and solve a difference equation for p_n in
  terms of the likelihood p_0 that the raw material
  of the unit is already defective.

  b. Why is it true that $p_0 + p_1 + p_2 + \cdots = 1$? Use
  this fact to find a formula for p_n. How likely is it
  that the imperfection occurred during the third
  stage?

58. POLITICAL INFLUENCE   Suppose the
likelihood p_n that a particular politician influences
n colleagues is 60% of the likelihood p_{n-1} that the
same politician influences only $n - 1$ colleagues.

  a. Set up and solve a difference equation for p_n in
  terms of the likelihood p_0 that the politician influences no colleagues.

  b. Why is it true that $p_0 + p_1 + p_2 + \cdots = 1$? Use
  this fact to find a formula for p_n.

  c. How likely is it that the politician influences as
  many as five colleagues?

59. **POPULATION GROWTH** A bacterial colony begins with population $P_0 > 0$ and grows in such a way that the population P_n after n hours is related to the population P_{n-1} after $n-1$ hours by the difference equation

$$P_n = \frac{bP_{n-1}}{a + P_{n-1}}$$

where a and b are positive constants.

a. Show that $P_n < \left(\dfrac{b}{a}\right)P_{n-1}$ and hence that

$$P_n < \left(\frac{b}{a}\right)^n P_0.$$

b. If $a > b$, show that the population eventually becomes extinct; that is, $\lim\limits_{n\to\infty} P_n = 0$.

c. If $a < b$, what happens to the population in the long run? You may assume that $\lim\limits_{n\to\infty} P_n$ exists and that $\lim\limits_{n\to\infty} P_n = \lim\limits_{n\to\infty} P_{n-1}$.

60. Solve the difference equation

$$P_n = \frac{bP_{n-1}}{a + P_{n-1}}$$

in Exercise 59 by following these steps:

a. Show that substituting $Q_n = \dfrac{1}{P_n}$ converts the given difference equation into the linear equation

$$Q_n = \frac{a}{b}Q_{n-1} + \frac{1}{b}$$

b. Solve the linear equation in Q_n obtained in step (a), and then use your solution to solve the given difference equation in terms of a, b, and P_0.

61. **THE FIBONACCI EQUATION** The 13th century mathematician Fibonacci (1170–1250), also known as Leonardo of Pisa, posed this problem in his book *Liber Abaci*:

Consider an island where rabbits live forever. Every month each pair of adult rabbits produces a pair of offspring which themselves become reproductive adults 2 months later. If the rabbit colony begins with one newborn pair, how many pairs will there be after n months?

a. Let R_n be the total number of rabbit pairs (adult, newborn, and adolescent) in the colony after n months. Explain why $R_1 = 1$, $R_2 = 1$, $R_3 = 2$, $R_4 = 3$, and in general,

$$R_n = R_{n-1} + R_{n-2}$$

This difference equation is called the *Fibonacci equation*.

b. Find all numbers λ such that $R_n = \lambda^n$ satisfies the Fibonacci equation.

c. If λ_1 and λ_2 are the numbers you found in part (b), it can be shown that the general solution of the Fibonacci equation has the form

$$R_n = A\lambda_1^n + B\lambda_2^n$$

for constants A and B. Find A and B so that the initial conditions $R_1 = 1$ and $R_2 = 1$ are satisfied.

d. The numbers 1, 1, 2, 3, 5, 8, 13, ... generated by the Fibonacci equation are called *Fibonacci numbers*. Such numbers occur in a remarkable variety of applications, ranging from population genetics to stock market analysis. For instance, in botany, the numbers appear in the study of phyllotaxis, the arrangement of leaves on the stems of plants. Write a paragraph on applications of the Fibonacci numbers, using the Internet as a research source.

62. A child, standing at the origin of a coordinate plane, holds a 5-foot length of rope that is attached to a sled. She walks along the x axis keeping the rope taut, as shown in the figure (units are in feet).

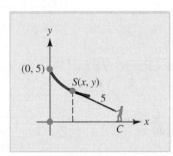

EXERCISE 62

a. If the sled begins at $(0, 5)$ and is at $S(x, y)$ when the child is at C, explain why

$$\frac{dy}{dx} = \frac{-y}{\sqrt{25 - y^2}}$$

b. Find the equation of the path followed by the sled. The path is called a **tractrix**. [*Hint:* Use an integral formula from Table 6.1.]

c. Use the graphing utility of your calculator to draw the path. Then use **ZOOM** and **TRACE** or other features of your calculator to find the point on the path that is exactly 3 feet from the y axis.

EXPLORE! UPDATE

Complete solutions for all EXPLORE! boxes throughout the text can be accessed at the book-specific website, www.mhhe.com/hoffmann.

Solution for Explore! on Page 703

Place the data into the **STAT** editor, as shown in the left screen. In the **CALC** menu, select option **B:Logistic** and write the command **Logistic L1, L2, Y1** to obtain the parameters of the logistic equation, displayed in the middle screen. The data points and logistic graph can be presented using the **STAT PLOT** command (as in the right screen). After a long period (as x approaches a relatively large number), the denominator of the logistic formula becomes arbitrarily close to 1, indicating that the club membership will level off at about $y = 766$. At approximately when did half of the susceptible community become infected?

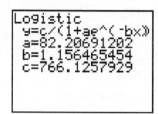
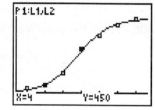

Solution for Explore! on Page 738

A relatively simple program (labeled SLPFIELD) appears in the table on page 767, which constructs a slope field for a first-order differential equation, $\frac{dy}{dx} = f(x, y)$.

It requires that the function $f(x, y)$ be entered as Y1 in the equation editor of your calculator. See the Calculator Introduction on the book-specific website, www.mhhe.com/hoffmann for preliminary instructions on how to construct programs in your graphing calculator. Line numbers are included for the benefit of the reader. Line 1 clears the homescreen of unwanted commands, graphs, and drawings. Lines 2 and 3 request the number of desired slope segments in the x and y directions, respectively, while lines 4 and 5 compute the length and width of the subrectangle housing each slope segment. Imagine that the graphing calculator screen will be divided into a gridwork of smaller rectangles, R by L of them. Lines 6 and 8 locate the coordinates of both the center point of the first subrectangle and midpoint of the initial line segment. Two loops (lines 7 to 20 and 9 to 18) will perform the calculation of slope segments within the confines of the subrectangles, first in the x direction and then in the y direction. The slope segment has a slope $f(X, Y) = S$ (line 10) and a length one-half that of the diagonal of the subrectangle (line 11). Lines 12 to 15 locate the coordinates of the endpoints, (E, G) and (F, H), of the slope segment and line 16 actually draws the slope segment.

Line	PROGRAM: SLPFIELD	Line	(continued from first column)
1	:ClrHome: ClrDraw: FnOff	11	$:\sqrt{(A^2 + B^2)}/2 \to C$
2	:Input "No. X Values", R	12	$:X - C/2*\cos(\tan^{-1}(S)) \to E$
3	:Input "No. Y Values", L	13	$:X + C/2*\cos(\tan^{-1}(S)) \to F$
4	$:(X_{max} - X_{min})/R \to A$	14	$:Y - C/2*\sin(\tan^{-1}(S)) \to G$
5	$:(Y_{max} - Y_{min})/L \to B$	15	$:Y + C/2*\sin(\tan^{-1}(S)) \to H$
6	$:X_{min} + A/2 \to X$	16	:Line (E, G, F, H)
7	:For (I, 1, R)	17	$:Y + B \to Y$
8	$:Y_{min} + B/2 \to Y$	18	:End
9	:For (J, 1, L)	19	$:X + A \to X$
10	$:Y1 \to S$	20	:End

Now refer to Example 9.4.1 and Figure 9.12. Enter Y1 = X − 2Y in the equation editor of your calculator. Since we eventually want to view the solution curve passing through the point (0, 1), set the window with modified decimal dimensions [−2, 7.4]1 by [−1, 5.2]1 and run the SLPFIELD program with 12 elements in the *x* direction and 8 in the *y* direction, as shown on the left. In the equation editor, enter Y2 = (X + 2)/2, Y3 = X/2, and Y4 = (X − 2)/2 and then press **GRAPH.** You will see straight lines, called **isoclines,** on which slope segments all have the same slope, namely, S = −2, 0, and 2, respectively. See the middle screen. Finally, write Y5 = 0.5X − 0.25 + 1.25e^(−2X) and graph in bold, as shown on the right screen, to view the solution curve passing through the point (0, 1).

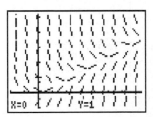 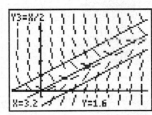 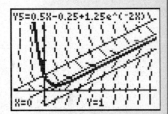

Solution for Explore! on Page 741

Refer to Example 9.4.3. Employing the **SLPFIELD** program listed earlier, construct the slope field for the differential equation $\dfrac{dy}{dx} = x - y$ using a window with modified decimal dimensions [−2, 7.4]1 by [−1, 5.2]1. The particular solution g(x) = x − 1 + 2e^(−x) passing through the point (0, 1) is also displayed.

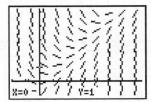 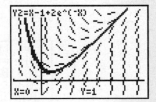

Solution for Explore! on Page 745

See the Calculator Introduction in the book-specific website, www.mhhe.com/hoffmann for preliminary instructions on displaying recursive functions, such as difference equations. Make sure that the **MODE** settings are as on the screen shown on the left. Following the example on page 745 involving a bacterial colony, enter the function u(n)=0.03u(n−1)+1,000, with initial value u(1)=100,000, in the equation editor of your graphing calculator (see the middle screen). The symbol u is found via the **2nd** function key above **7.** The sequence $u(n)$ begins with initial condition $u(1)=100,000$. In the list editor, write L2 = u(L1), where L1 is the sequence of natural numbers from 1 to 7. See the screen on the right, which displays seven generations of bacteria populations.

Does the population of bacteria appear to approach a limiting value? This question is answered graphically. Make sure the **FORMAT (2nd ZOOM)** is set for **Time** graphs, as shown in the left screen. Use a **WINDOW** with the time set as nMin = 1 and nMax = 10 and XY dimensions of [0, 10]1 by [−200, 5,000]100, as seen in the middle screen. Press **GRAPH** to obtain the dot sequence graph shown on the right.

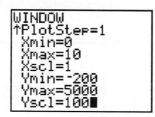

The population values quickly drop from 100,000 to a value approaching 1,000 bacteria.

Solution for Explore! on Page 754

Refer to the cobweb model (page 754). Enter u(n) = (−2/3)u(n−1) + 10/3 in sequence mode in the equation editor, with p_0 = u(0) = 3 ($300), as in the left screen below. In the list editor, store the numbers 0 to 6 in L1, and enter L2 = u(L1), L3 = 2L2 + 20, and L4 = −3L2 + 30 to obtain $u(n)$, S = 2p + 20, and D = −3p + 30, for n from 0 to 6. See the middle screen. Graphing $u(n)$ with window [0, 6]1 by [−3, 4]1, and notice that $u(n)$ appears to converge, indicating a convergent cobweb spiral. This is confirmed by d = 2 < 3 = b. Repeating with the initial price of $100 ($p_0$ = 1) also yields a convergent cobweb spiral but with a different starting place.

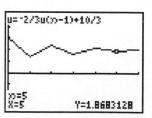

THINK ABOUT IT

MODELING EPIDEMICS

An outbreak of a disease affecting a large number of people is called an *epidemic* (from the Greek roots *epi* "upon" and *demos* "the people"). Mathematics plays an important role in the study of epidemics, a discipline known as *epidemiology*. Mathematical models have been developed for epidemics of influenza, bubonic plague, HIV, AIDS, smallpox, gonorrhea, and various other diseases. By making some simplifying assumptions, we can construct a model based on differential equations, called the **S-I-R model,** that is useful for making predictions about the course of an epidemic. Our model will be fairly straightforward, with enough simplifying assumptions to make the mathematics accessible. Later, we will examine the assumptions, with a view toward improving the model by making it more realistic.

To build our model, we consider an epidemic affecting a population of N people, each of whom falls into exactly one of the following groups at time t:

Susceptibles, $S(t)$, are people who have not yet become ill, but who could become sick later. When a susceptible person gets sick, it is assumed to happen with no time lag.

Infectives, $I(t)$, are people who have the disease and can infect others. Infectives are not quarantined and continue a normal pattern of interacting with other people.

Removeds, $R(t)$, are people who can no longer infect others. In particular, a recovered person cannot catch the disease again or infect others.

We assume that no people enter or leave the population (including births and deaths), and that at time $t = 0$, there are some susceptibles and infectives but no removeds; that is,

$$N = S(t) + I(t) + R(t) \qquad \text{for all } t \geq 0$$

where

$$S(0) > 0 \qquad I(0) > 0 \qquad R(0) = 0$$

so that $S(0) + I(0) = N$.

The key to modeling an epidemic or any other dynamic situation lies in the assumptions made about rates of change. In our model, we assume that the rate at which susceptible people are being infected at any time is proportional to the number of contacts between susceptibles and infectives, that is, to the product SI. Thus, the susceptible population $S(t)$ is *decreasing* (becoming infected) at a rate given by

$$(1) \qquad \frac{dS}{dt} = -aSI$$

We also assume that the rate people are being removed from the infected population is proportional to the number of people who are infective, so that

$$(2) \qquad \frac{dR}{dt} = bI$$

for some number $b > 0$. Finally, we assume that the rate at which the number of infectives changes is the rate at which susceptibles become infective minus the rate at which infective people are removed from the population of infectives. This means that

$$(3) \qquad \frac{dI}{dt} = aSI - bI = aI\left(S - \frac{b}{a}\right) = aI(S - c)$$

with $c = \dfrac{b}{a}$. [In Question 1, you are asked to obtain this last rate by differentiating both sides of the equation $I(t) = N - S(t) - R(t)$ with respect to t and substituting the rates in equations (1) and (2).]

The S-I-R model consists of the three simultaneous differential equations we have constructed; namely,

$$\frac{dS}{dt} = -aSI \qquad \frac{dR}{dt} = bI \qquad \frac{dI}{dt} = aI(S - c)$$

where $S(t) + I(t) + R(t) = N$ for all t, with $R(0) = 0$, $S(0) > 0$, and $I(0) > 0$. Although we cannot explicitly solve this system to obtain simple formulas for the functions $S(t)$, $I(t)$, and $R(t)$, we can use these differential equations to help us understand the progress of epidemics. We illustrate this by applying our S-I-R model to a historic case, an epidemic of bubonic plague in Eyam, a village near Sheffield, England, during 1665–1666. This analysis is possible because the village quarantined itself during the epidemic and kept sufficiently detailed records. We will use these records to estimate the constants a and b in equations (1) and (2).

First, note that the constant b in equation (2) can be interpreted as the rate at which people are removed from the population of infectives. The infective period of the bubonic plague in Eyam was 11 days or 0.367 months. Therefore, if we measure time t in months and assume, for simplicity, that infected people are removed from this population at a constant rate, we can estimate b by the ratio

$$b = \frac{1}{0.367} \approx 2.72$$

Next, records kept during the epidemic show that on July 19, 1666, there were 201 susceptibles and 22 infectives, while on August 19, 1666, there were 121 susceptibles and 21 infectives. Thus, the number of susceptibles changed by $121 - 201 = -80$

during this month-long period, and this change can be used to estimate $\dfrac{dS}{dt}$ on the day the period began. In other words, on July 19, 1666,

$$S = 201 \qquad I = 22 \qquad \text{and} \qquad \frac{dS}{dt} \approx -80$$

Substituting into equation (1) and solving for the constant a, we find that

$$a = \frac{-dS/dt}{SI} = \frac{-(-80)}{(201)(22)} = 0.018$$

It turns out that the resulting model, with the estimates we have just obtained for the values of a and b, does a good job of estimating the actual data. Figure 1 displays the values predicted by the model and the actual data. [This information is adapted from Raggett (1982) and from the discussion in Brauer and Castillo-Chavez (2001).]

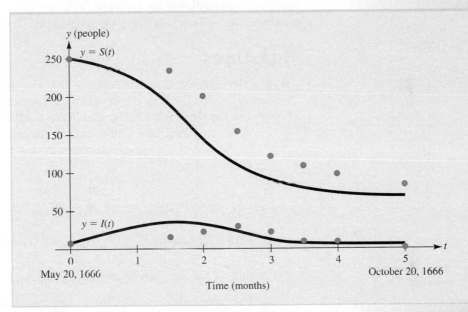

FIGURE 1   Comparison of Eyam plague data with value predicted by the S-I-R model.

The S-I-R model can be used to make some general observations about the course of epidemics. For example, unless the number of infectives increases initially, the epidemic will never get started (since the number of infectives does not grow beyond the initial number). In the language of calculus, an epidemic begins only if $\dfrac{dI}{dt} > 0$ at $t = 0$. Examining equation (3) of the S-I-R model, we see that this happens only if $S(0) > c$. For this reason, c is called the *threshold number of susceptibles*. When the initial number of susceptibles exceeds c, the epidemic spreads; otherwise, it dies out. For example, the records of the 1666 bubonic plague in Eyam indicate there were initially $S(0) = 254$ susceptibles. Using our estimates of $a \approx 0.018$ and $b \approx 2.72$, we estimate the threshold number for this model to be

$$c = \frac{b}{a} = \frac{2.72}{0.018} \approx 151$$

Since $S(0) = 254 > 151$, the model predicts that the epidemic should spread, which is exactly what happened historically.

You probably have noticed that in constructing our S-I-R model and applying it to the Eyam epidemic, we have made several fairly strong assumptions, not the least of which is that $S(t) + I(t) + R(t)$ is a constant. Is it realistic to assume that the size of a population subjected to an epidemic would never change? Is it any more realistic to assume that people become infective immediately with no time lag? Certainly not, since diseases usually have incubation periods, but inserting a time lag to account for the incubation period would make the resulting model too complicated to analyze without more sophisticated mathematical tools.

This illustrates the central issue facing anyone who uses mathematical modeling: *A completely realistic model is often difficult if not impossible to analyze, while making assumptions to simplify the analysis can lead to results that are not entirely realistic.* The S-I-R model is indeed based on several simplifying assumptions that are somewhat unrealistic, but we have also seen that the model can be used effectively to explore the dynamics of an epidemic. How do you think the model could be made more realistic without sacrificing too much of its simplicity?

Questions

1. Use the equation $N = S(t) + I(t) + R(t)$ and the rate relations in equations (1) and (2) of the S-I-R model to verify the rate relation in equation (3).

2. **a.** Use the chain rule to show that in the S-I-R model, the rate of change of the number of infectives with respect to the number of susceptibles is the ratio of the rate of change of infectives divided by the rate of change of susceptibles, that is

    $$\frac{dI}{dS} = \frac{dI/dt}{dS/dt}$$

    b. Find a formula for $\dfrac{dI}{dS}$ using equations (1) and (3) of the model. Use this to find $\dfrac{dI}{dS}$ for the 1666 Eyam bubonic plague epidemic.

    c. Use your answer in part (b) to find the number of susceptibles when the number of infected is a maximum.

3. An influenza epidemic broke out at a British boarding school in 1978. At the start of the epidemic there were 762 susceptible boys and 1 infective boy, and 1 day later, two more boys became ill. Assume that anyone sick with influenza is also infective.

    a. Use the data supplied to estimate the constant a in equation (1) of the S-I-R model for this epidemic with t representing time in days.

    b. Suppose that all infected boys are removed from the population the day after they become ill. Use this information to estimate the constant b in equation (2) of the S-I-R model for this epidemic.

    c. Use the values of a and b you found in parts (a) and (b) to find the threshold number of susceptibles needed for this epidemic to get started.

    d. Use the results of Question 2 to find the number of susceptibles when the number of students infected with influenza was the maximum.

4. Modify the S-I-R model for an epidemic to take into account vaccinations of people at a constant rate d, assuming that someone vaccinated immediately is no longer susceptible. Find a formula for $\dfrac{dI}{dS}$ for the resulting model.

5. An S-I model can be used to model an epidemic where everyone who gets the disease remains infective. (An S-I model is really the special case of the S-I-R model in which $b = 0$.) Answer these questions about such an epidemic.

   a. Show that $\dfrac{dS}{dt} = -aS(N - S)$ where N is the population size (which we assume is constant).

   b. Verify that $S(t) = \dfrac{N(N - 1)}{(N - 1) + e^{aNt}}$ is a solution of this differential equation, and conclude that $I(t) = \dfrac{Ne^{aNt}}{(N - 1) + e^{aNt}}$.

   c. Show that $\lim\limits_{t \to +\infty} S(t) = 0$ and $\lim\limits_{t \to +\infty} I(t) = N$. Draw a graph showing the susceptible and infective curves, $y = S(t)$ and $y = I(t)$, assuming that at time $t = 0$ there are I_0 infected people, where $0 < I_0 < N$.

References

W. O. Kermack and A. G. McKendrick, "A Contribution to the Mathematical Theory of Epidemics," *Proc. Royal Soc. London*, Vol. 115, 1927, pp. 700–721.

G. F. Raggett, "Modeling the Eyam Plague," *IMA Journal*, Vol. 18, 1982, pp. 221–226.

Fred Brauer and Carlos Castillo-Chavez, *Mathematical Models in Population Biology and Epidemiology*, New York: Springer-Verlag, 2001.

Leah Edelstein-Keshet, *Mathematical Models in Biology*, Birkhauser Mathematics Series, Boston: McGraw-Hill, 1988, pp. 243–256.

CHAPTER 10

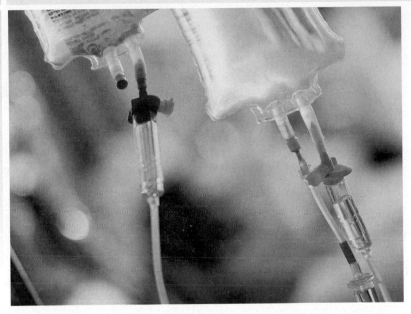

The accumulation of medication in a patient's bloodstream can be determined using infinite series.

Infinite Series and Taylor Series Approximations

SECTION 10.1   Infinite Series; Geometric Series

Learning Objectives

1. Determine convergence or divergence of an infinite series.
2. Examine and use geometric series.

The sum of infinitely many numbers may be finite. This statement, which may seem paradoxical at first, plays a central role in mathematics and has a variety of important applications. The purpose of this chapter is to explore its meaning and some of its consequences.

You are already familiar with the phenomenon of a finite-valued infinite sum. You know, for example, that the repeating decimal 0.333 . . . stands for the infinite sum $\frac{3}{10} + \frac{3}{100} + \frac{3}{1,000} + \cdots$ and that its value is the finite number $\frac{1}{3}$. What may not be familiar, however, is exactly what it means to say that this infinite sum "adds up to" $\frac{1}{3}$.

The situation will be clarified in this introductory section.

Convergence and Divergence of Infinite Series

An expression of the form

$$a_1 + a_2 + \cdots + a_n + \cdots$$

is called an **infinite series.** It is customary to use summation notation to write series compactly as follows:

$$a_1 + a_2 + \cdots + a_n + \cdots = \sum_{n=1}^{\infty} a_n$$

The use of this notation is illustrated in Example 10.1.1.

EXAMPLE 10.1.1   Using Summation Notation

a. Write out some representative terms of the series $\sum_{n=1}^{\infty} \frac{1}{2^n}$.

b. Use summation notation to write the series $1 - \frac{1}{4} + \frac{1}{9} - \frac{1}{16} + \cdots$ in compact form.

Solution

a. $\sum_{n=1}^{\infty} \frac{1}{2^n} = \frac{1}{2} + \frac{1}{4} + \frac{1}{8} + \cdots + \frac{1}{2^n} + \cdots$

b. The nth term of this series is

$$\frac{(-1)^{n+1}}{n^2}$$

where the factor $(-1)^{n+1}$ generates the alternating signs starting with a plus sign when $n = 1$. Thus,

$$1 - \frac{1}{4} + \frac{1}{9} - \frac{1}{16} + \cdots = \sum_{n=1}^{\infty} \frac{(-1)^{n+1}}{n^2}$$

Roughly speaking, an infinite series

$$a_1 + a_2 + a_3 + \cdots = \sum_{k=1}^{\infty} a_k$$

is said to *converge* if it "adds up" to a finite number and to *diverge* if it does not. A more precise statement of the criterion for convergence involves the sum

$$S_n = a_1 + a_2 + \cdots + a_n = \sum_{k=1}^{n} a_k$$

of the first n terms of the infinite series. This finite sum is called the **nth partial sum** of the infinite series, and its behavior as n approaches infinity determines the convergence or divergence of the series. Here is a summary of the criterion.

EXPLORE!

Evaluate

$$\sum_{n=1}^{7} \frac{n^3 - 2n}{n+1}$$

to at least three-decimal-place accuracy. To achieve this on your graphing calculator, enter

Y1 = (X^3 − 2X)/(X + 1)

into the equation editor and employ the sequence and cumulative summation features of the **LIST (2nd STAT)** key, as demonstrated in the Calculator Introduction on the book-specific website, www.mhhe.com/hoffmann. Explain the behavior of the sequence of partial sums.

Convergence or Divergence of an Infinite Series ■ An infinite series $\sum_{k=1}^{\infty} a_k$ with nth partial sum

$$S_n = a_1 + a_2 + \cdots + a_n = \sum_{k=1}^{n} a_k$$

is said to **converge with sum S** if S is a finite number such that

$$\lim_{n \to \infty} S_n = S$$

and in this case, we write

$$\sum_{k=1}^{\infty} a_k = S$$

The series is said to **diverge** if $\lim_{n \to \infty} S_n$ does not exist as a finite number.

According to this criterion, to determine the convergence or divergence of an infinite series, you start by finding an expression for the sum of the first n terms of the series and then take the limit of this finite sum as n approaches infinity. This procedure is similar to the one used in Section 6.3 to determine the convergence or divergence of an improper integral, as illustrated in Example 10.1.2.

EXAMPLE 10.1.2 Determining Convergence or Divergence

Investigate the possible convergence of the following series.

a. $\displaystyle\sum_{k=1}^{\infty} (-1)^{k+1} = 1 - 1 + 1 - 1 + \cdots$

b. $\displaystyle\sum_{k=1}^{\infty} \frac{3}{10^k} = \frac{3}{10} + \frac{3}{10^2} + \frac{3}{10^3} + \cdots$

Solution

a. The first three partial sums of the series $\displaystyle\sum_{k=1}^{\infty} (-1)^{k+1}$ are

$$S_1 = 1$$
$$S_2 = 1 - 1 = 0$$
$$S_3 = 1 - 1 + 1 = 1$$

This pattern continues, and in general, the even partial sums are 0 and the odd partial sums are 1. Therefore, $\lim\limits_{n \to \infty} S_n$ does not exist, so the series diverges.

b. Notice that each term of the series

$$\sum_{k=1}^{\infty} \frac{3}{10^k} = \frac{3}{10} + \frac{3}{10^2} + \frac{3}{10^3} + \cdots$$

is $\frac{1}{10}$ times the preceding term. This leads to the following trick for finding a compact formula for S_n.

Begin with the nth partial sum in the form

$$S_n = \frac{3}{10} + \frac{3}{10^2} + \frac{3}{10^3} + \cdots + \frac{3}{10^n}$$

and multiply both sides of this equation by $\frac{1}{10}$ to get

$$\frac{1}{10} S_n = \frac{3}{10^2} + \frac{3}{10^3} + \cdots + \frac{3}{10^n} + \frac{3}{10^{n+1}}$$

Now subtract the expression for $\frac{1}{10} S_n$ from the expression for S_n. Most of the terms cancel, and you are left with only the first term of S_n and the last term of $\frac{1}{10} S_n$; that is,

$$S_n - \frac{1}{10} S_n = \frac{9}{10} S_n = \frac{3}{10} - \frac{3}{10^{n+1}} = \frac{3}{10}\left(1 - \frac{1}{10^n}\right) \qquad \begin{array}{l} \text{multiply both} \\ \text{sides by } \dfrac{10}{9} \end{array}$$

or

$$S_n = \frac{1}{3}\left(1 - \frac{1}{10^n}\right)$$

Then

$$\lim_{n \to \infty} S_n = \lim_{n \to \infty} \frac{1}{3}\left(1 - \frac{1}{10^n}\right) = \frac{1}{3} \qquad \lim_{n \to \infty} \frac{1}{10^n} = 0$$

and so

$$\sum_{n=1}^{\infty} \frac{3}{10^n} = \frac{1}{3}$$

EXPLORE!

Calculate the first seven partial sums of the series $\sum\limits_{n=1}^{\infty} \dfrac{3}{10^n}$, using the sequence and cumulative summation features of your graphing calculator. Conjecture the value of the infinite sum S, and compare this with the calculations shown in Example 10.1.2b.

EXPLORE!

Use the sequence and cumulative summation features of your calculator to show that the nth partial sum of the series in Example 10.1.3 satisfies

$$S_n = 1 - \frac{1}{n+1}$$

EXAMPLE 10.1.3   Summing a Telescoping Series

Investigate the possible convergence of the series

$$\sum_{k=1}^{\infty} \frac{1}{k(k+1)} = \frac{1}{(1)(2)} + \frac{1}{(2)(3)} + \frac{1}{(3)(4)} + \cdots$$

Solution

First, note that

$$\frac{1}{k(k+1)} = \frac{1}{k} - \frac{1}{k+1}$$

Therefore, the nth partial sum of the given series can be expressed as

$$S_n = \sum_{n=1}^{n} \frac{1}{k(k+1)} = \sum_{k=1}^{n} \left[\frac{1}{k} - \frac{1}{k+1} \right]$$

$$= \left(\frac{1}{1} - \frac{1}{2} \right) + \left(\frac{1}{2} - \frac{1}{3} \right) + \left(\frac{1}{3} - \frac{1}{4} \right) + \cdots + \left(\frac{1}{n} - \frac{1}{n+1} \right)$$

$$= 1 + \left(-\frac{1}{2} + \frac{1}{2} \right) + \left(-\frac{1}{3} + \frac{1}{3} \right) + \cdots + \left(-\frac{1}{n} + \frac{1}{n} \right) - \frac{1}{n+1}$$

$$= 1 - \frac{1}{n+1}$$

Taking the limit of the partial sums as $n \to \infty$, we find that

$$\lim_{n \to \infty} S_n = \lim_{n \to \infty} \left(1 - \frac{1}{n+1} \right) = 1$$

Therefore, we conclude that the given series converges with sum $S = 1$.

NOTE A series like the one in Example 10.1.3 in which the terms of the partial sums cancel in pairs is called a *telescoping series*. ▪

Convergent infinite series can be added together and multiplied by constants in much the same way as ordinary sums of numbers. Here is a summary of these properties.

Sum and Multiple Rules for Convergent Infinite Series ▪ If the series $\sum_{n=1}^{\infty} a_n$ and $\sum_{n=1}^{\infty} b_n$ converge, then the series $\sum_{n=1}^{\infty} (a_n + b_n)$ and $\sum_{n=1}^{\infty} c a_n$ for constant c also converge and

$$\sum_{n=1}^{\infty} (a_n + b_n) = \sum_{n=1}^{\infty} a_n + \sum_{n=1}^{\infty} b_n$$

$$\sum_{n=1}^{\infty} c a_n = c \sum_{n=1}^{\infty} a_n$$

EXAMPLE 10.1.4 Using Sum and Multiple Rules

Investigate the possible convergence of the series

$$\sum_{k=1}^{\infty} \left[\frac{6}{10^k} - \frac{5}{k(k+1)} \right]$$

Solution

In Examples 10.1.2 and 10.1.3, we found that the series $\sum_{k=1}^{\infty} \frac{3}{10^k}$ and $\sum_{k=1}^{\infty} \frac{1}{k(k+1)}$ both converge and that

$$\sum_{k=1}^{\infty} \frac{3}{10^k} = \frac{1}{3} \qquad \text{and} \qquad \sum_{k=1}^{\infty} \frac{1}{k(k+1)} = 1$$

Therefore, by applying the sum and multiple rules for series, we find that the given series converges and that

$$\sum_{k=1}^{\infty}\left[\frac{6}{10^k} - \frac{5}{k(k+1)}\right] = \sum_{k=1}^{\infty} 2\left(\frac{3}{10^k}\right) + \sum_{k=1}^{\infty}(-5)\frac{1}{k(k+1)} \quad \text{sum rule}$$

$$= 2\sum_{k=1}^{\infty}\frac{3}{10^k} + (-5)\sum_{k=1}^{\infty}\frac{1}{k(k+1)} \quad \text{multiple rule}$$

$$= 2\left(\frac{1}{3}\right) + (-5)(1)$$

$$= -\frac{13}{3}$$

Geometric Series

Notice that the series featured in Example 10.1.2b has the property that the ratio of successive terms is a constant $\left(\text{namely, }\frac{1}{10}\right)$. Any series with this property is called a **geometric series** and must have the general form

$$\sum_{k=0}^{\infty} ar^k = a + ar + ar^2 + \cdots$$

where a is a constant and r is the ratio of successive terms. For example,

$$\frac{2}{9} - \frac{2}{27} + \frac{2}{81} - \frac{2}{243} + \cdots = \frac{2}{9}\left(1 - \frac{1}{3} + \frac{1}{3^2} - \frac{1}{3^3} + \cdots\right)$$

$$= \sum_{k=0}^{\infty} \frac{2}{9}\left(-\frac{1}{3}\right)^k$$

is a geometric series with $a = \frac{2}{9}$ and ratio $r = -\frac{1}{3}$.

A geometric series $\sum_{k=0}^{\infty} ar^k$ converges if $|r| < 1$ and diverges if $|r| \geq 1$. To show this, we proceed as in Example 10.1.2b. Begin with the nth partial sum

$$S_n = a + ar + ar^2 + \cdots + ar^{n-1} \quad (r \neq 0)$$

(Note that the power of r in the nth term is only $n - 1$ since the first term in the sum is $a = ar^0$.) Now multiply both sides of this equation by r to get

$$rS_n = ar + ar^2 + \cdots + ar^{n-1} + ar^n$$

and subtract the expression for rS_n from the expression for S_n to get

$$S_n - rS_n = a - ar^n \quad \text{all other terms cancel}$$

$$(1 - r)S_n = a(1 - r^n)$$

$$S_n = \frac{a(1 - r^n)}{1 - r}$$

If $|r| < 1$, we have $\lim_{n\to\infty} r^n = 0$, and it follows that

$$\lim_{n\to\infty} S_n = \lim_{n\to\infty} \frac{a(1 - r^n)}{1 - r} = \frac{a}{1 - r}$$

In other words, the given geometric series converges with sum $\dfrac{a}{1-r}$. If $|r| > 1$, then $\lim\limits_{n\to\infty} |r| = \infty$, and $\lim\limits_{n\to\infty} S_n$ cannot exist. The series also diverges when $|r| = 1$ (see Exercise 53). To summarize:

Convergence of a Geometric Series ■ The geometric series $\sum\limits_{n=0}^{\infty} ar^n$, with *common ratio* r, converges if $|r| < 1$ with sum

$$S = \sum_{n=0}^{\infty} ar^n = \frac{a}{1-r}$$

If $|r| \geq 1$, the series diverges.

EXPLORE!

Given the geometric series with $a = 2$ and $r = 0.85$, evaluate S_{15} and S_{50} to five-decimal-place accuracy. Compare S_{15} and S_{50} to S_∞. What is the error in each case?

For example, the following two series both diverge since $|r| > 1$:

$$2 + 2(1.5) + 2(1.5)^2 + \cdots \qquad (r = 1.5)$$

and

$$\frac{1}{2} - \frac{1}{2}\left(\frac{9}{7}\right) + \frac{1}{2}\left(\frac{9}{7}\right)^2 - \frac{1}{2}\left(\frac{9}{7}\right)^3 + \cdots \qquad \left(r = -\frac{9}{7}\right)$$

Examples 10.1.5, 10.1.6, and 10.1.7 illustrate the case where $|r| < 1$.

EXAMPLE 10.1.5 Finding the Sum of a Geometric Series

Find the sum of the series $\sum\limits_{n=0}^{\infty} \left(-\dfrac{2}{3}\right)^n$.

Solution

This is a geometric series with $r = -\dfrac{2}{3}$. Since $|r| < 1$, it converges with sum

$$\sum_{n=0}^{\infty} \left(-\frac{2}{3}\right)^n = \frac{1}{1 - \left(-\frac{2}{3}\right)} = \frac{1}{\frac{5}{3}} = \frac{3}{5} \qquad a = 1$$

EXAMPLE 10.1.6 Finding the Sum of a Geometric Series

Find the sum of the series $\sum\limits_{n=0}^{\infty} \dfrac{3}{2^n}$.

Solution

$$\sum_{n=0}^{\infty} \frac{3}{2^n} = \sum_{n=0}^{\infty} 3\left(\frac{1}{2}\right)^n = 3\left(\frac{1}{1 - \frac{1}{2}}\right) = 3\left(\frac{1}{\frac{1}{2}}\right) = 6$$

> **EXAMPLE 10.1.7   Finding the Sum of a Geometric Series**
>
> Find the sum of the series $\sum_{n=1}^{\infty} \frac{(-2)^n}{3^{n+1}}$.

Solution

$$\sum_{n=1}^{\infty} \frac{(-2)^n}{3^{n+1}} = \frac{-2}{3^2} + \frac{(-2)^2}{3^3} + \frac{(-2)^3}{3^4} + \cdots$$

$$= -\frac{2}{3^2}\left[1 + \left(-\frac{2}{3}\right) + \left(-\frac{2}{3}\right)^2 + \cdots\right]$$

$$= \left(-\frac{2}{9}\right)\sum_{n=0}^{\infty}\left(-\frac{2}{3}\right)^n$$

$$= -\frac{2}{9}\left(\frac{1}{1+\frac{2}{3}}\right) = -\frac{2}{9}\left(\frac{3}{5}\right) = -\frac{2}{15}$$

Applications of Geometric Series

Geometric series arise in many branches of mathematics and in a variety of applications. Four such applications are considered in Examples 10.1.8 through 10.1.11.

Repeating Decimals

It turns out that any number N whose decimal form contains a pattern that repeats infinitely often is rational; that is, $N = \frac{p}{q}$ for integers p and q ($q \neq 0$). Example 10.1.8 illustrates a general procedure for using geometric series to find p and q.

> **EXAMPLE 10.1.8   Writing a Repeating Decimal as a Fraction**
>
> Express the repeating decimal 0.232323 . . . as a fraction.

Solution

Write the decimal as a geometric series as follows:

$$0.232323\ldots = \frac{23}{100} + \frac{23}{10,000} + \frac{23}{1,000,000} + \cdots$$

$$= \frac{23}{100}\left[1 + \frac{1}{100} + \left(\frac{1}{100}\right)^2 + \cdots\right]$$

$$= \frac{23}{100}\sum_{n=0}^{\infty}\left(\frac{1}{100}\right)^n$$

$$= \frac{23}{100}\left(\frac{1}{1-\frac{1}{100}}\right) = \frac{23}{100}\left(\frac{100}{99}\right) = \frac{23}{99}$$

The Multiplier Effect in Economics

If a sum of money A is injected into the economy, say by a tax rebate, the total effect on the economy is to induce a level of spending that is much larger than A. The reason for this is that the portion of S that is spent by one individual becomes income for one or more others, who, in turn, spend some of it again, thus creating income for yet other individuals to spend. Economists refer to this kind of multistage spending as the multiplier effect. If the fraction of income that is spent at each stage of this process remains constant, then the total amount of spending is the sum of a geometric series, as illustrated in Example 10.1.9.

EXPLORE!

Following Example 10.1.9, enter the function 40(0.9)^X into Y1 of the equation editor. Use the sequence and cumulative summation features of your graphing calculator to determine how many terms of $\sum 40(0.9)^n$ need to be calculated to be accurate to the nearest billion.

EXAMPLE 10.1.9 Studying the Multiplier Effect

Suppose that nationwide, approximately 90% of all income is spent and 10% is saved. What is the total amount of spending generated by a $40 billion tax rebate if savings habits do not change?

Solution

The government begins by spending $40 billion, and then the original recipients of the rebate spend 0.9(40) billion dollars. This becomes new income, of which

$$0.9[0.9(40)] = 40(0.9)^2$$

billion dollars is spent. This, in turn, generates additional income of

$$0.9[40(0.9)^2] = 40(0.9)^3$$

and so on. If this process continues indefinitely, the total amount spent is

$$T = 40 + 40(0.9) + 40(0.9)^2 + 40(0.9)^3 + \cdots = \sum_{n=0}^{\infty} 40(0.9)^n$$

which we recognize as a geometric series with $u = 40$ and $r = 0.9$. Since $|r| < 1$, the geometric series converges with sum

$$T = \frac{40}{1 - 0.9} = 400$$

Thus, the total amount of spending generated is $400 billion.

In general, suppose A dollars are originally injected into the economy and that each time a person receives a dollar, that person will spend a fixed fraction c of the dollar. For example, $c = \dfrac{9}{10}$ in Example 10.1.9. Then c is called the **marginal propensity to consume,** and by proceeding as in Example 10.1.9, it can be shown that the total amount spent will be

$$T = \frac{A}{1 - c} = \frac{A}{s} = mA$$

where $s = 1 - c$ is the **marginal propensity to save** and $m = \dfrac{1}{s}$ is the **multiplier** for the spending process.

Present Value of a
Perpetual Annuity

EXAMPLE 10.1.10 Using Series to Study a Perpetual Annuity

Find the amount of money Ellis should invest today at an annual interest rate of 10% compounded continuously so that, starting next year, he can make annual withdrawals of $400 in perpetuity.

Solution

The amount Ellis should invest today to generate the desired sequence of withdrawals is the sum of the present values of the individual withdrawals. He computes the present value of each withdrawal using the formula $P = Be^{-rt}$ from Section 4.1, with $B = 400$, $r = 0.1$, and t being the time (in years) at which the withdrawal is made. Thus,

$$P_1 = \text{present value of first withdrawal} = 400e^{-0.1(1)} = 400e^{-0.1}$$
$$P_2 = \text{present value of second withdrawal} = 400e^{-0.1(2)} = 400e^{-0.2}$$

$$\vdots$$

$$P_n = \text{present value of } n\text{th withdrawal} = 400e^{-0.1(n)}$$

and so on. The situation is illustrated in Figure 10.1.

EXPLORE!

Following Example 10.1.10, enter the function 400e^(−0.1X) into Y1 of the equation editor. Using the sequence and cumulative summation features of your graphing calculator, as demonstrated in the Explore! Update section of this chapter, determine how many terms of the infinite series are needed to be within a penny of the actual total.

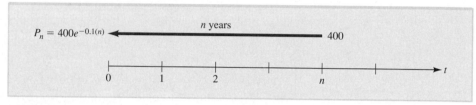

FIGURE 10.1   The present value of the nth withdrawal.

The amount that he should invest today is the sum of these infinitely many present values. That is,

$$\begin{aligned}\text{Amount to be invested} &= \sum_{n=1}^{\infty} P_n = \sum_{n=1}^{\infty} 400e^{-0.1(n)}\\ &= 400[e^{-0.1} + e^{-0.2} + e^{-0.3} + \cdots]\\ &= 400e^{-0.1}[1 + e^{-0.1} + (e^{-0.1})^2 + \cdots]\end{aligned}$$

This is a geometric series with $a = 400e^{-0.1}$ and $r = e^{-0.1} = 0.9048$. Since $|r| < 1$, the series converges with sum

$$\sum_{n=1}^{\infty} 400e^{-0.1(n)} = \frac{400e^{-0.1}}{1 - e^{-0.1}} \approx 3,803.33$$

Thus, Ellis should invest \$3,803.33 to achieve the stated objective.

Accumulation of Medication in the Body

In Example 6.2.5 we considered a nuclear power plant generating radioactive waste that was decaying exponentially. The problem was to determine the long-run accumulation of waste from the plant. Because the waste was being generated *continuously*, the model used to analyze this situation involved an (improper) integral. The situation in Example 10.1.11 is similar. A patient receives medication that is eliminated from the body exponentially, and the goal is to determine the long-run accumulation of medication in the patient's body. In this case, however, the medication is being administered in *discrete* doses, and the model involves an infinite series rather than an integral.*

*A source containing a variety of mathematical models for drug dosage is the module by Brindell Horelick and Sinan Koont, "Prescribing Safe and Effective Dosage," *UMAP Modules 1977: Tools for Teaching,* Lexington, MA: Consortium for Mathematics and Its Applications, Inc., 1978.

EXAMPLE 10.1.11 Studying Accumulation of Medication

A patient is given an injection of 10 units of a certain drug every 24 hours. The drug is eliminated exponentially so that the fraction that remains in the patient's body after t days is $f(t) = e^{-t/5}$. If the treatment is continued indefinitely, approximately how many units of the drug will eventually be in the patient's body just prior to an injection?

Solution

Of the original dose of 10 units, only $10e^{-1/5}$ units are left in the patient's body after 1 day (just prior to the second injection). That is,

$$\text{Amount in body after 1 day} = S_1 = 10e^{-1/5}$$

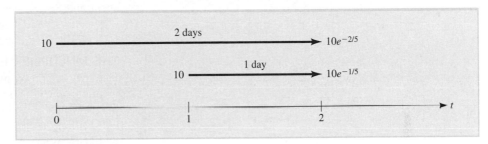

FIGURE 10.2 Amount of medication in the body after 2 days.

The medication in the patient's body after 2 days consists of what remains from the first *two* doses. Of the original dose, only $10e^{-2/5}$ units are left (since 2 days have elapsed), and of the second dose, $10e^{-1/5}$ units remain (Figure 10.2). Hence,

$$\text{Amount in body after 2 days} = S_2 = 10e^{-1/5} + 10e^{-2/5}$$

Similarly,

$$\begin{array}{c}\text{Amount in body}\\ \text{after } n \text{ days}\end{array} = S_n = 10e^{-1/5} + 10e^{-2/5} + \cdots + 10e^{-n/5}$$

The amount S of medication in the patient's body in the long run is the limit of S_n as n approaches infinity. That is,

$$S = \lim_{n \to \infty} S_n = \sum_{n=1}^{\infty} 10e^{-n/5}$$

$$= \sum_{n=1}^{\infty} 10(e^{-1/5})^n = 10e^{-1/5} \sum_{n=0}^{\infty} (e^{-1/5})^n$$

$$= 10e^{-1/5}\left(\frac{1}{1 - e^{1/5}}\right) \approx 45.17 \text{ units}$$

EXERCISES ▪ 10.1

In Exercises 1 through 6, use summation notation to write the given series in compact form.

1. $\dfrac{1}{3} + \dfrac{1}{9} + \dfrac{1}{27} + \dfrac{1}{81} + \cdots$

2. $1 + \dfrac{1}{8} + \dfrac{1}{27} + \dfrac{1}{64} + \cdots$

3. $\dfrac{1}{2} + \dfrac{2}{3} + \dfrac{3}{4} + \dfrac{4}{5} + \cdots$

4. $\dfrac{1}{3} + \dfrac{2}{5} + \dfrac{3}{7} + \dfrac{4}{9} + \cdots$

5. $\dfrac{1}{2} - \dfrac{4}{3} + \dfrac{9}{4} - \dfrac{16}{5} + \cdots$

6. $-3 + \dfrac{9}{4} - \dfrac{27}{9} + \dfrac{81}{16} - \cdots$

In Exercises 7 through 10, find the fourth partial sum S_4 of the given series.

7. $\displaystyle\sum_{n=1}^{\infty} \dfrac{1}{2^n}$ **8.** $\displaystyle\sum_{n=1}^{\infty} \dfrac{n}{n+1}$

9. $\displaystyle\sum_{n=1}^{\infty} \dfrac{(-1)^n}{n}$ **10.** $\displaystyle\sum_{n=1}^{\infty} \dfrac{(-1)^{n+1}}{n(n+1)}$

In Exercises 11 through 24, determine whether the given geometric series converges, and if so, find its sum.

11. $\displaystyle\sum_{n=0}^{\infty} \left(\dfrac{4}{5}\right)^n$ **12.** $\displaystyle\sum_{n=0}^{\infty} \left(-\dfrac{4}{5}\right)^n$

13. $\displaystyle\sum_{n=0}^{\infty} \dfrac{2}{3^n}$ **14.** $\displaystyle\sum_{n=0}^{\infty} \dfrac{2}{(-3)^n}$

15. $\displaystyle\sum_{n=1}^{\infty} \left(\dfrac{3}{2}\right)^n$ **16.** $\displaystyle\sum_{n=1}^{\infty} \dfrac{3}{2^n}$

17. $\displaystyle\sum_{n=2}^{\infty} \dfrac{3}{(-4)^n}$ **18.** $\displaystyle\sum_{n=2}^{\infty} \left(-\dfrac{4}{3}\right)^n$

19. $\displaystyle\sum_{n=1}^{\infty} 5(0.9)^n$ **20.** $\displaystyle\sum_{n=1}^{\infty} e^{-0.2n}$

21. $\displaystyle\sum_{n=1}^{\infty} \dfrac{3^n}{4^{n+2}}$ **22.** $\displaystyle\sum_{n=2}^{\infty} \dfrac{(-2)^{n-1}}{3^{n+1}}$

23. $\displaystyle\sum_{n=0}^{\infty} \dfrac{4^{n+1}}{5^{n-1}}$ **24.** $\displaystyle\sum_{n=2}^{\infty} (-1)^n \dfrac{2^{n+1}}{3^{n-3}}$

In each of Exercises 25 through 28, evaluate the specified series given that $\displaystyle\sum_{k=1}^{\infty} a_k = 5$ and $\displaystyle\sum_{k=1}^{\infty} b_k = -3$.

25. $\displaystyle\sum_{k=1}^{\infty} 2a_k$ **26.** $\displaystyle\sum_{k=1}^{\infty} (3a_k - 5b_k)$

27. $\displaystyle\sum_{k=1}^{\infty} \left[\dfrac{1}{2^k} - b_k\right]$ **28.** $\displaystyle\sum_{k=1}^{\infty} \left[a_k - \dfrac{(-1)^k}{3^{k+1}}\right]$

In each of Exercises 29 through 32, find the sum of the given telescoping series.

29. $\displaystyle\sum_{k=0}^{\infty} \dfrac{1}{(k+1)(k+2)}$ **30.** $\displaystyle\sum_{k=0}^{\infty} \dfrac{1}{(2k+1)(2k+3)}$

31. $\displaystyle\sum_{k=1}^{\infty} \dfrac{1}{(2k-1)(2k+1)}$ **32.** $\displaystyle\sum_{k=1}^{\infty} \dfrac{\sqrt{k+1} - \sqrt{k}}{\sqrt{k(k+1)}}$

In Exercises 33 through 36, express the given decimal as a fraction.

33. $0.4444\ldots$ **34.** $0.5555\ldots$

35. $0.252525\ldots$ **36.** $1.405405405\ldots$

BUSINESS AND ECONOMICS APPLIED PROBLEMS

37. THE MULTIPLIER EFFECT   Suppose that nationwide, approximately 92% of all income is spent and 8% is saved. What is the total amount of spending generated by a 50 billion dollar tax rebate if savings habits do not change?

38. THE MULTIPLIER EFFECT   Suppose that nationwide, approximately nine times as much income is spent as is saved and that economic planners wish to generate total spending of 270 billion dollars by instituting a tax cut of N dollars. What is N?

39. PRESENT VALUE   An investment guarantees annual payments of $1,000 in perpetuity, with the payments beginning immediately. Find the present value of this investment if the prevailing annual interest rate remains fixed at 4% compounded continuously. [*Hint:* The present value of the investment is the sum of the present values of the individual payments.]

40. PRESENT VALUE   How much should Leah invest today at an annual interest rate of 5% compounded continuously so that, starting next year, she can make annual withdrawals of $2,000 in perpetuity?

41. PRESENT VALUE   An investment guarantees annual payments of A dollars each, to be paid at the end of each year in perpetuity into an account earning a fixed annual interest rate r, compounded annually. Show that the present value of the investment is $\dfrac{A}{r}$.

42. PRESENT VALUE   Suppose in Exercise 41, payments of A dollars each are made k times per year (at the end of each period) into an account that earns interest at an annual rate r, compounded k times per year. Show that the present value of the investment is $\dfrac{kA}{r}$.

43. DEPLETION OF RESERVES A certain rare gas used in industrial processes had known reserves of 3×10^{11} m³ in 2000. In 2001, 1.7×10^9 m³ of the gas was consumed with an annual increase of 7.4% each year thereafter. When will all known reserves of the gas be consumed?

44. GROUP MEMBERSHIP Each January first, the administration of a certain private college adds six new members to its board of trustees. If the fraction of trustees who remain active for at least t years is $f(t) = e^{-0.02t}$, approximately how many active trustees will the college have on December 31 in the long run?

45. PUBLIC HEALTH A health problem, especially in emerging nations, is that the food available to people often contains toxins or poisons. Suppose that a person consumes a dose d of a certain toxin every day and that the proportion of accumulated toxin excreted by the person's body each day is q. (For example, if $q = 0.7$, then 70% of the accumulated toxin is excreted each day.) Find an expression for the total amount of toxin that will eventually accumulate in the person's body.

46. ACCUMULATION OF MEDICATION Jodie Michaels, a patient in a hospital, is being given an injection of A units of a certain drug every h hours. The drug is eliminated exponentially from her body so that t hours after each injection is administered, the fraction that remains is $f(t) = e^{-rt}$, where r is a positive constant (called the *elimination constant*).
  a. If this treatment is continued indefinitely, how many total units T of the drug will eventually accumulate in Jodie's body?
  b. Suppose a physician determines that the accumulated amount of drug T found in part (a) is 20% too much for Jodie's safety. Assuming that the elimination constant r is a fixed characteristic of Jodie's body, how should h be adjusted to guarantee that the amount of accumulated drug does not become unsafe?

47. ACCUMULATION OF MEDICATION A patient is given an injection of 20 units of a certain drug every 24 hours. The drug is eliminated exponentially so that the fraction that remains in the patient's body after t days is $f(t) = e^{-t/2}$. If the treatment is continued indefinitely, approximately how many units of the drug will eventually be in the patient's body just prior to an injection?

48. DISTANCE TRAVELED BY A BOUNCING BALL A ball has the property that each time it falls from a height h onto a level, rigid surface, it will rebound to a height rh, where r ($0 < r < 1$) is a number called the *coefficient of restitution*.
  a. A super ball with $r = 0.8$ is dropped from a height of 5 meters. Show that if the ball continues to bounce indefinitely, it will travel a total distance of 45 meters.
  b. Find a formula for the total distance traveled by a ball with coefficient of restitution r that is dropped from a height H.
  c. A certain ball that bounces indefinitely after being dropped from a height H travels a total distance of $2H$. What is r?

49. TIME OF TRAVEL BY A BOUNCING BALL Refer to Exercise 48. Assuming no air resistance, an object dropped from a height will fall $s = 16t^2$ feet in t seconds.
  a. A golf ball with coefficient of restitution $r = 0.36$ is dropped from a height of 16 feet. Show that if the ball continues to bounce indefinitely, its total time of travel will be 4 seconds.
  b. Find a formula for the total time of travel of a ball with coefficient of restitution r that is dropped from a height of H feet.
  c. A certain bouncing ball is dropped from a height of 9 feet and is allowed to bounce indefinitely. If its total time of travel is 2.25 seconds, how far does it travel?

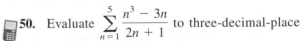**50.** Evaluate $\displaystyle\sum_{n=1}^{5} \frac{n^3 - 3n}{2n + 1}$ to three-decimal-place accuracy by following these steps on your calculator:
  a. Store the expression as $(X^3 - 3X)/(2X + 1)$ in the function list and store the value 1 as X. Evaluate the expression and store the result in a temporary location, say, A.
  b. Store 2 as X, evaluate the function, and add the result to the value in memory location A.
  c. Repeat step (b) with 3 as X, then 4, and finally, 5.

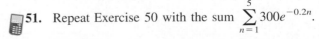**51.** Repeat Exercise 50 with the sum $\displaystyle\sum_{n=1}^{5} 300e^{-0.2n}$.

52. Find the sum S of the geometric series $\sum_{n=0}^{\infty} 2(0.85)^n$. Then calculate the partial sums S_5 and S_{10} with five-decimal-place accuracy and compare the values of these partial sums to the sum S of the series.

53. Show that the geometric series $\sum_{k=0}^{\infty} ar^k$ diverges when

a. $r = 1$ [*Hint:* Note that the nth partial sum is $S_n = na$.]

b. $r = -1$ [*Hint:* Note that the nth partial sum is $S_n = 0$ if n is odd or $S_n = a$ if n is even.]

54. ZENO'S PARADOX The Greek philosopher Zeno of Elea (495–435 B.C.E.) presented a number of paradoxes that perplexed many mathematicians of his day. Perhaps the most famous is the racecourse paradox, which may be stated as follows:

As a runner runs a racecourse, he must first run $\frac{1}{2}$ its length, then $\frac{1}{2}$ the remaining distance, then $\frac{1}{2}$ the distance that remains after that, and so on, indefinitely. Since there are an infinite number of such half segments and the runner requires finite time to complete each one, he can never finish the race.

Suppose the runner runs at a constant pace and takes T minutes to run the first half of the course. Set up an infinite series that gives the total time required to finish the course, and verify that the total time is $2T$, as intuition would suggest.

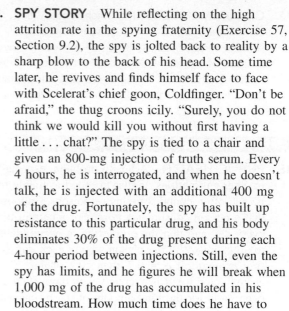 **55. SPY STORY** While reflecting on the high attrition rate in the spying fraternity (Exercise 57, Section 9.2), the spy is jolted back to reality by a sharp blow to the back of his head. Some time later, he revives and finds himself face to face with Scelerat's chief goon, Coldfinger. "Don't be afraid," the thug croons icily. "Surely, you do not think we would kill you without first having a little . . . chat?" The spy is tied to a chair and given an 800-mg injection of truth serum. Every 4 hours, he is interrogated, and when he doesn't talk, he is injected with an additional 400 mg of the drug. Fortunately, the spy has built up resistance to this particular drug, and his body eliminates 30% of the drug present during each 4-hour period between injections. Still, even the spy has limits, and he figures he will break when 1,000 mg of the drug has accumulated in his bloodstream. How much time does he have to plan and carry out an escape?

SECTION 10.2 Tests for Convergence

Learning Objective

1. Use the divergence test, integral test, and ratio test to determine convergence and divergence of series.

In Section 10.1, we introduced the notions of convergence and divergence of infinite series and found that a geometric series $\sum_{n=b}^{\infty} ar^n$ converges if $|r| < 1$ and diverges otherwise. This is a fairly easy criterion to apply, but what about series that are not geometric? In this section, we shall develop several procedures called "convergence tests" that will enable us to determine the convergence or divergence of a variety of series. We begin by examining a special series that will be used in subsequent examples and applications.

The Harmonic Series

The series

$$\sum_{k=1}^{\infty} \frac{1}{k} = 1 + \frac{1}{2} + \frac{1}{3} + \frac{1}{4} + \frac{1}{5} + \cdots$$

is called the **harmonic series** because it occurs in physics in the study of sound. In Example 10.2.1, we determine whether this series converges by examining its partial sums.

EXPLORE!

As proved in Example 10.2.1, the harmonic series diverges, but it does so very slowly. A graphing calculator exercise will illustrate this. Store the function 1/X into Y1 of the equation editor. Use the sequence and cumulative summation features of your calculator, as shown in the Explore! Update section of this chapter, to determine how many terms are needed for the partial sum of the harmonic series to reach the value 6.

EXAMPLE 10.2.1   Studying the Harmonic Series

Show that the harmonic series $\displaystyle\sum_{k=1}^{\infty} \frac{1}{k}$ diverges.

Solution

Let $S_n = \displaystyle\sum_{k=1}^{n} \frac{1}{k}$ be the nth partial sum of the harmonic series, and note that

$$S_1 = 1$$

$$S_2 = 1 + \frac{1}{2}$$

$$S_4 = 1 + \frac{1}{2} + \left(\frac{1}{3} + \frac{1}{4}\right)$$

$$> 1 + \frac{1}{2} + \left(\frac{1}{4} + \frac{1}{4}\right) = 1 + \frac{1}{2} + \frac{1}{2} = 1 + \frac{2}{2}$$

$$S_8 = 1 + \frac{1}{2} + \left(\frac{1}{3} + \frac{1}{4}\right) + \left(\frac{1}{5} + \frac{1}{6} + \frac{1}{7} + \frac{1}{8}\right)$$

$$> 1 + \frac{1}{2} + \left(\frac{1}{4} + \frac{1}{4}\right) + \left(\frac{1}{8} + \frac{1}{8} + \frac{1}{8} + \frac{1}{8}\right) = 1 + \frac{1}{2} + \frac{1}{2} + \frac{1}{2} = 1 + \frac{3}{2}$$

and in general,

$$S_{2^n} > 1 + \frac{n}{2}$$

This means that the partial sums S_n can be made as large as desired by taking n sufficiently large. Therefore, $\displaystyle\lim_{n\to\infty} S_n$ does not exist as a finite number, and the harmonic series must diverge.

We have just seen that the partial sums of the harmonic series grow without bound as n becomes large. However, the size of the partial sums grows very slowly. For instance, computations show that $S_{100} \approx 5.19$ and $S_{1,000} \approx 7.49$, and that to get a partial sum exceeding 20 requires more than 250 million terms. Example 10.2.2 features an application in which estimating the size of a partial sum of the harmonic series plays a crucial role.

EXAMPLE 10.2.2   Using Harmonic Series to Study Snowfall

Suppose records are kept over a 100-year period of the annual snowfall in a certain location. For how many years should we expect record-breaking snowfall in the sense that the snowfall in that winter exceeds the snowfall in all previous years during the period?

Solution

In modeling this situation, we assume that the snowfall in any given year of the 100-year period is not affected by snowfall in previous years and that the climate does not change in a systematic way. We also assume that the snowfall is measured quite accurately so that no ties occur. Our analysis will involve some basic (intuitive) notions from probability.

Consider the 100 years as a sequence of events in which the snowfall levels are recorded. The snowfall in the first year is a record, since records begin at that time. In the second year, the snowfall is equally likely to be more than as less than the snowfall in the first year. It follows that there is a probability of $\frac{1}{2}$ that the snowfall in the second year sets a record by exceeding the snowfall in the first year. Consequently, in the first 2 years of the 100-year period, we expect to have $1 + \frac{1}{2}$ records set. For the third year, the probability is $\frac{1}{3}$ that the snowfall that year exceeds the snowfall in both of the first 2 years, so the expected number of record snowfalls in 3 years is $1 + \frac{1}{2} + \frac{1}{3}$. Continuing in this fashion for n years, we see that the expected number of years with record snowfalls is

$$1 + \frac{1}{2} + \frac{1}{3} + \cdots + \frac{1}{n}$$

which we recognize as the nth partial sum of the harmonic series. In particular, the expected number of record snowfalls over the entire 100-year period is

$$1 + \frac{1}{2} + \frac{1}{3} + \cdots + \frac{1}{100} \approx 5.19$$

We conclude that on average, we would expect about 5 record snowfall years during the 100-year period.

If you were to test this result by checking a specific 100-year period, you may find less than 5 record years or more than 5, perhaps many more. The correct interpretation of our result is that if you look at a large collection (the larger the better) of 100-year periods, then on average, there would be about 5 years of record snowfall per 100-year period.

The Divergence Test

Next, we shall develop several general tests for determining whether a series $\sum_{k=1}^{\infty} a_k$ converges or diverges. The easiest such test to apply can only be used to establish divergence and is appropriately called the *divergence test*.

The Divergence Test ■ If $\lim_{n\to\infty} a_n$ does not exist or if $\lim_{n\to\infty} a_n \neq 0$, then $\sum_{k=1}^{\infty} a_k$ diverges.

We shall establish the divergence test by showing that if the series $\sum_{k=1}^{\infty} a_k$ converges, the terms a_n must tend to 0 as n tends to infinity; that is, $\lim_{n\to\infty} a_n = 0$. (Do you see why this is equivalent to the statement of the divergence test?) To see why this is true, let $S_n = \sum_{k=1}^{n} a_k$ be the nth partial sum of the series, and note that a_n is the difference between S_n and S_{n-1}; that is,

$$a_n = (a_1 + a_2 + \cdots + a_n) - (a_1 + a_2 + \cdots + a_{n-1}) = S_n - S_{n-1}$$

If $\sum_{k=1}^{\infty} a_k$ converges with sum S, then

$$\lim_{n \to \infty} S_n = \lim_{n \to \infty} S_{n-1} = S$$

and it follows that

$$\lim_{n \to \infty} a_n = \lim_{n \to \infty} (S_n - S_{n-1}) = \lim_{n \to \infty} S_n - \lim_{n \to \infty} S_{n-1}$$
$$= S - S = 0$$

as claimed.

Example 10.2.3 shows how the divergence test can be used to show that a given series diverges.

EXAMPLE 10.2.3   Using the Divergence Test

Show that each of the following series diverges:

    a. $\displaystyle\sum_{k=1}^{\infty} \sqrt{k}$ **b.** $\displaystyle\sum_{k=1}^{\infty} \dfrac{k}{k+1}$

Solution

    a. Since $\lim_{k \to \infty} \sqrt{k} = \infty$, it follows that $\lim_{k \to \infty} \sqrt{k}$ is not 0, so $\sum_{k=1}^{\infty} \sqrt{k}$ diverges by the divergence test.

    b. The series $\displaystyle\sum_{k=1}^{\infty} \dfrac{k}{k+1}$ diverges since

$$\lim_{k \to \infty} \frac{k}{k+1} = \lim_{k \to \infty} \frac{1}{1 + 1/k} = \frac{1}{1 + 0} = 1 \neq 0$$

WARNING   *The divergence test cannot be used to show that a series* $\sum_{n=1}^{\infty} a_n$ *converges.* Specifically, you cannot conclude that $\sum_{n=1}^{\infty} a_n$ converges just because $\lim_{n \to \infty} a_n = 0$.

For instance, for the harmonic series $\sum_{k=1}^{\infty} \dfrac{1}{k}$ we have

$$\lim_{k \to \infty} \frac{1}{k} = 0$$

yet the harmonic series diverges. ∎

The Integral Test   Our second test for convergence, the *integral test,* relates the convergence of the infinite series $\sum_{n=1}^{\infty} a_n$ to the convergence of the improper integral $\int_{1}^{\infty} f(x)\, dx$, where $f(n) = a_n$. Here is a precise statement of the test.

The Integral Test ■ Let $\sum_{n=1}^{\infty} a_n$ be a series with $a_n > 0$ for all n. If $f(x)$ is a function such that $f(n) = a_n$ for all n and if f is continuous, positive ($f(x) > 0$), and decreasing for $x \geq 1$, then $\sum_{n=1}^{\infty} a_n$ and $\int_{1}^{\infty} f(x)\, dx$ either both converge or both diverge. In particular,

$$\sum_{n=1}^{\infty} a_n \text{ converges if } \int_{1}^{\infty} f(x)\, dx \text{ converges}$$

and

$$\sum_{n=1}^{\infty} a_n \text{ diverges if } \int_{1}^{\infty} f(x)\, dx \text{ diverges}$$

NOTE   The integral test actually applies whenever the function f is *eventually* positive and decreasing, that is, if f is positive and decreasing for $x \geq N$ for some integer N. ■

We will provide a geometric justification of the integral test after illustrating its use in Example 10.2.4.

Just-In-Time **REVIEW**

Recall that ln $x > 1$ when $x > e$. Also, $e^x > 0$ and $e^{-x} > 0$ for all real numbers x.

EXAMPLE 10.2.4 Using the Integral Test

Determine whether each of the following series converges or diverges.

a. $\sum_{n=3}^{\infty} \dfrac{\ln n}{n}$ **b.** $\sum_{n=1}^{\infty} ne^{-n}$

Solution

a. Let $f(x) = \dfrac{\ln x}{x}$, so that $f(n) = \dfrac{\ln n}{n}$. Note that $f(x)$ is continuous and positive for $x \geq 2$, and it is decreasing since its derivative satisfies

$$f'(x) = \frac{x(1/x) - (1)\ln x}{x^2} = \frac{1 - \ln x}{x^2} < 0 \qquad \text{for } x > e > 2$$

Therefore, the integral test applies, and since

$$\int \frac{\ln x}{x}\, dx = \frac{1}{2}(\ln x)^2 + C$$

substitution
let $u = \ln x$
$du = \dfrac{1}{x}\, dx$

it follows that

$$\int_{2}^{\infty} \frac{\ln x}{x}\, dx = \lim_{t \to \infty} \int_{2}^{t} \frac{\ln x}{x}\, dx = \lim_{t \to \infty} \left[\frac{1}{2}(\ln t)^2 - \frac{1}{2}(\ln 2)^2 \right] = \infty$$

Since the improper integral diverges, it follows that the series $\displaystyle\sum_{n=2}^{\infty} \frac{\ln n}{n}$ diverges as well.

b. Let $f(x) = xe^{-x}$, so that $f(n) = ne^{-n}$. Then $f(x)$ is continuous and positive for $x \geq 1$, and it is decreasing since its derivative satisfies

$$f'(x) = x(-e^{-x}) + (1)e^{-x} = (1 - x)e^{-x} < 0 \qquad \text{for } x > 1$$

Therefore, the integral test applies, and since

$$\int xe^{-x}\, dx = (-x - 1)e^{-x} + C \qquad \begin{array}{l}\text{integration by parts with}\\ u = x \quad dv = e^{-x}\, dx\end{array}$$

it follows that

$$\int_1^{\infty} xe^{-x}\, dx = \lim_{t\to\infty} \int_1^{t} xe^{-x}\, dx = \lim_{t\to\infty} [(-t - 1)e^{-t} - (-1 - 1)e^{-1}] = 2e^{-1}$$

Since the improper integral converges, the series $\displaystyle\sum_{n=1}^{\infty} ne^{-n}$ also converges.

NOTE In Example 10.2.4, the integral test tells us that $\displaystyle\sum_{n=1}^{\infty} ne^{-n}$ converges because the improper integral $\displaystyle\int_1^{\infty} xe^{-x}\, dx$ converges, but the actual sum of the series is not the same as the value $2e^{-1}$ of the improper integral. ∎

Geometric Justification of the Integral Test

To see why the integral test must be true, first note that since the test requires $a_k > 0$ for all k, the series $\displaystyle\sum_{k=1}^{\infty} a_k$ will converge unless it becomes infinitely large; in particular, the series cannot "diverge by oscillation" as with the geometric series

$$\sum_{k=1}^{\infty} (-1)^k - 1 - 1 + 1 - 1 + \cdots$$

Similarly, since $f(x) > 0$ and $f(x)$ is decreasing for $x > 1$, the improper integral $\displaystyle\int_1^{\infty} f(x)\, dx$ will converge unless it becomes infinite. We shall show that if the improper integral is finite, then so is the series $\displaystyle\sum_{k=1}^{\infty} a_k$, and that if $\displaystyle\int_1^{\infty} f(x)\, dx$ is infinite, then $\displaystyle\sum_{k=1}^{\infty} a_k$ must also be infinite.

Refer to the two parts of Figure 10.3. Both parts show the graph of $f(x)$ together with a number of approximating rectangles built on the intervals $[1, 2], [2, 3], \ldots$. In Figure 10.3a, the rectangles are *inscribed* under the curve $y = f(x)$, and the height of the approximating rectangle built on the interval $[k - 1, k]$ is $f(k) = a_k$, the height of the curve $y = f(x)$ above the *right* endpoint of the interval. Since this approximating rectangle has height a_k and width 1, its area is $a_k \cdot 1 = a_k$ and the total area of the first n rectangles is $a_2 + a_3 + \cdots + a_n$. Since the rectangles are inscribed

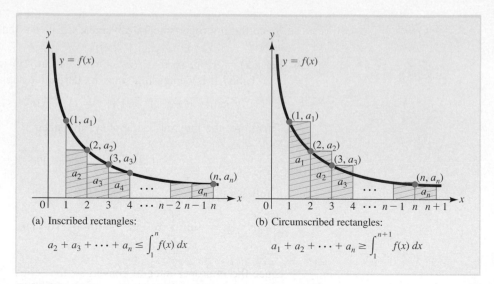

FIGURE 10.3 A geometric interpretation of the integral test.

under the curve $y = f(x)$, this area is no greater than the area under the curve above the interval $[1, n]$, so we have

$$a_2 + a_3 + \cdots + a_n \leq \int_1^n f(x)\, dx$$

which means that the nth partial sum of the series $\sum_{k=1}^{\infty} a_k$ satisfies

$$S_n = a_1 + a_2 + a_3 + \cdots + a_n \leq a_1 + \int_1^n f(x)\, dx$$

The rectangles in Figure 10.3b are *circumscribed* over the curve $y = f(x)$ on the interval $[1, n + 1]$. This time, the height of the rectangle based on the interval $[k - 1, k]$ is $f(k - 1) = a_{k-1}$, the height of the curve above the *left* endpoint of the subinterval. The area of this approximating rectangle is $a_{k-1} \cdot 1 = a_{k-1}$, and since the circumscribing rectangles cover an area no less than the area under the curve on $[1, n + 1]$, we have

$$S_n = a_1 + a_2 + \cdots + a_n \geq \int_1^{n+1} f(x)\, dx$$

Combining these results, we see that the nth partial sum S_n is "trapped" between two integrals:

$$\int_1^{n+1} f(x)\, dx \leq S_n \leq a_1 + \int_1^n f(x)\, dx$$

If the improper integral converges, then the partial sums S_n are bounded from above by the (finite) quantity $a_1 + \int_1^{\infty} f(x)\, dx$, which means that the series $\sum_{k=1}^{\infty} a_k$ converges to a finite value. On the other hand, if the improper integral diverges, the partial sums cannot converge since they are larger than a quantity $\int_1^{n+1} f(x)\, dx$ that is tending to infinity.

The p-Series Test   Infinite series of the general form $\displaystyle\sum_{n=1}^{\infty} \frac{1}{n^p}$, where p is a positive constant, are called **p-series.** For instance, the harmonic series is a p-series with $p = 1$. An important application of the integral test is to establish this simple rule for determining the convergence or divergence of p-series.

The p-Series Test ■ The p-series $\displaystyle\sum_{n=1}^{\infty} \frac{1}{n^p}$ converges if $p > 1$ and diverges if $p \leq 1$.

To establish the p-series test, let $f(x) = \dfrac{1}{x^p} = x^{-p}$, so that $f(n) = \dfrac{1}{n^p}$. Then $f(x)$ is continuous and positive for $x \geq 1$, and it is also decreasing since its derivative satisfies

$$f'(x) = -px^{-p-1} = \frac{-p}{x^{p+1}} < 0$$

Applying the integral test, we first note that if $p = 1$, then

$$\int_1^{\infty} \frac{1}{x}\, dx = \lim_{t \to \infty} \int_1^t \frac{1}{x}\, dx = \lim_{t \to \infty}\, (\ln t - \ln 1) = \infty$$

which confirms that the harmonic series $\displaystyle\sum_{n=1}^{\infty} \frac{1}{n}$ diverges (recall Example 10.2.1). For the case where $p \neq 1$, we have

$$\int_1^{\infty} \frac{1}{x^p}\, dx = \lim_{t \to \infty} \int_1^t x^{-p}\, dx = \lim_{t \to \infty} \left[\frac{t^{-p+1}}{(-p+1)} - \frac{1}{(-p+1)} \right]$$

$$= \begin{cases} \infty & \text{if } p < 1 \\ \dfrac{1}{p-1} & \text{if } p > 1 \end{cases}$$

Therefore, the improper integral and hence the p-series both converge for $p > 1$ and both diverge for $p \leq 1$.

EXAMPLE 10.2.5 Using the p-Series Test

Determine whether each of the following series converges or diverges:

a. $\displaystyle\sum_{n=1}^{\infty} \frac{1}{\sqrt{n}}$ **b.** $\displaystyle\sum_{n=1}^{\infty} \frac{1}{n\sqrt{n^3}}$ **c.** $\displaystyle\sum_{n=1}^{\infty} \frac{1}{n^{0.999}}$

Solution

a. $\displaystyle\sum_{n=1}^{\infty} \frac{1}{\sqrt{n}} = \sum_{n=1}^{\infty} \frac{1}{n^{1/2}}$ is a p-series with $p = \dfrac{1}{2} < 1$, so the series diverges.

b. $\displaystyle\sum_{n=1}^{\infty} \frac{1}{n\sqrt{n^3}} = \sum_{n=1}^{\infty} \frac{1}{n^{5/2}}$ is a p-series with $p = \dfrac{5}{2} > 1$. Thus, the series converges.

c. $\displaystyle\sum_{n=1}^{\infty} \frac{1}{n^{0.999}}$ is a p-series with $p = 0.999 < 1$, so the series diverges.

The Direct Comparison Test

Our next convergence test compares the behavior of a given series to that of another series whose convergence properties are known.

The Direct Comparison Test ■ Consider two series $\sum_{n=1}^{\infty} a_n$ and $\sum_{n=1}^{\infty} b_n$, where $0 < b_n \le a_n$ for all n. Then:

If $\sum_{n=1}^{\infty} a_n$ converges, so does $\sum_{n=1}^{\infty} b_n$.

If $\sum_{n=1}^{\infty} b_n$ diverges, so does $\sum_{n=1}^{\infty} b_n$.

To see why the direct comparison test works, note that a series of positive terms diverges only if it becomes infinitely large. If $b_n \le a_n$ for all n and $\sum_{n=1}^{\infty} a_n$ converges (exists as a finite number), then $\sum_{n=1}^{\infty} b_n \le \sum_{n=1}^{\infty} a_n$. It follows that the smaller series $\sum_{n=1}^{\infty} b_n$ must also be finite, so it converges.

Similarly, if $\sum_{n=1}^{\infty} b_n$ diverges, then $\sum_{n=1}^{\infty} b_n$ is infinitely large. Since $\sum_{n=1}^{\infty} b_n \le \sum_{n=1}^{\infty} a_n$, the series $\sum_{n=1}^{\infty} a_n$ must also be infinitely large and hence diverges.

EXAMPLE 10.2.6 Using the Direct Comparison Test

Determine whether each of the following series converges or diverges:

a. $\sum_{n=1}^{\infty} \dfrac{1}{n2^n}$ **b.** $\sum_{n=5}^{\infty} \dfrac{4}{\sqrt{n} - 2}$

Solution

a. $\sum_{n=1}^{\infty} \dfrac{1}{2^n}$ is a convergent geometric series $\left(r = \dfrac{1}{2} < 1 \right)$, and since $\dfrac{1}{n2^n} \le \dfrac{1}{2^n}$ for $n \ge 1$, it follows from the direct comparison test that the given series also converges.

b. Note that $\dfrac{4}{\sqrt{n} - 2} > \dfrac{4}{\sqrt{n}}$ for all $n \ge 5$ because the first fraction has a smaller denominator than the second. Since $\sum_{n=5}^{\infty} \dfrac{4}{\sqrt{n}}$ is a multiple of the divergent p-series $\sum_{n=5}^{\infty} \dfrac{1}{n^{1/2}} \left(p = \dfrac{1}{2} < 1 \right)$, it follows from the direct comparison test that $\sum_{n=5}^{\infty} \dfrac{4}{\sqrt{n} - 2}$ also diverges.

The Ratio Test

Applying the comparison test can be frustrating since it may require several steps of trial and error to find a suitable comparison. An alternative, the *ratio test,* enables a kind of internal comparison of terms of the given series and has the added advantage of being applicable to series that contain negative terms. Here is a statement of this test.

> **The Ratio Test** ■ For the series $\displaystyle\sum_{n=1}^{\infty} a_n$, let $L = \displaystyle\lim_{n\to\infty} \left| \dfrac{a_{n+1}}{a_n} \right|$. Then:
>
> **a.** If $L < 1$, the series converges.
> **b.** If $L > 1$, the series diverges.
> **c.** If $L = 1$, the test is inconclusive; that is, when this occurs, the series may converge or it may diverge.

The ratio test is especially useful when applied to a series $\displaystyle\sum_{n=1}^{\infty} a_n$ in which the terms a_n involve powers or factorials. This is illustrated in Example 10.2.7.

EXAMPLE 10.2.7 Using the Ratio Test

Determine whether each of the following series converges or diverges:

a. $\displaystyle\sum_{n=1}^{\infty} \frac{5^n}{n^2}$ **b.** $\displaystyle\sum_{n=1}^{\infty} \frac{(-2)^n}{n!}$ **c.** $\displaystyle\sum_{n=1}^{\infty} \frac{n}{n+1}$

Solution

a. To apply the ratio test, we compute the limit

$$L = \lim_{n\to\infty} \left| \frac{\dfrac{5^{n+1}}{(n+1)^2}}{\dfrac{5^n}{n^2}} \right|$$

$$= \lim_{n\to\infty} \frac{5^{n+1} n^2}{5^n (n+1)^2}$$

$$= \lim_{n\to\infty} 5\left(\frac{n}{n+1} \right)^2$$

$$= \lim_{n\to\infty} 5\left(\frac{1}{1 + 1/n} \right)^2 = 5\left(\frac{1}{1+0} \right)^2 = 5$$

Since $L = 5 > 1$, the series diverges.

b. Applying the ratio test, we find that

$$L = \lim_{n\to\infty} \left| \frac{\dfrac{(-2)^{n+1}}{(n+1)!}}{\dfrac{(-2)^n}{n!}} \right|$$

$$= \lim_{n\to\infty} \left| \frac{(-2)^{n+1} n!}{(-2)^n (n+1)!} \right| \qquad\qquad \frac{n!}{(n+1)!} = \frac{1 \cdot 2 \cdots n}{1 \cdot 2 \cdots n \cdot (n+1)}$$

$$= \lim_{n\to\infty} \left| \frac{-2}{n+1} \right| \qquad\qquad\qquad\qquad = \frac{1}{n+1}$$

$$= 0$$

Since $L = 0 < 1$, the series converges.

c. We have already shown that this series diverges in Example 10.2.3b. However, if we try to confirm this result using the ratio test, we get

$$L = \lim_{n \to \infty} \left| \frac{\dfrac{n+1}{(n+1)+1}}{\dfrac{n}{n+1}} \right|$$

$$= \lim_{n \to \infty} \left| \frac{(n+1)(n+1)}{n(n+2)} \right| = \lim_{n \to \infty} \left| \frac{n^2 + 2n + 1}{n^2 + 2n} \right|$$

$$= \lim_{n \to \infty} \left| \frac{1 + 2/n + 1/n^2}{1 + 2/n} \right| = \frac{1 + 0 + 0}{1 + 0}$$

$$= 1$$

so the ratio test is inconclusive.

Knowing what convergence test to apply to a given series takes experience, and to gain that experience, it helps to work as many practice exercises as possible. In the following exercise set, there is a group of exercises focused on each test we have discussed. In addition, there is a larger collection of exercises in which the test is not specified.

EXERCISES ■ 10.2

In Exercises 1 through 4, use the divergence test to show that the given series diverges.

1. $\displaystyle\sum_{k=1}^{\infty} \frac{2k}{k+5}$ **2.** $\displaystyle\sum_{k=1}^{\infty} \frac{k+3}{\sqrt{k}+1}$

3. $\displaystyle\sum_{k=1}^{\infty} [1 + (-1)^k]$ **4.** $\displaystyle\sum_{k=1}^{\infty} \frac{e^{-k}+1}{e^{-k}+2}$

In Exercises 5 through 8, use the integral test to determine whether the given series converges or diverges.

5. $\displaystyle\sum_{k=1}^{\infty} \frac{2}{3k+1}$ **6.** $\displaystyle\sum_{k=1}^{\infty} ke^{-k^2}$

7. $\displaystyle\sum_{k=1}^{\infty} \frac{k}{3+k^2}$ **8.** $\displaystyle\sum_{k=1}^{\infty} \frac{1}{\sqrt[3]{5k-1}}$

In Exercises 9 through 12, use the p-series test to determine whether the given series converges or diverges.

9. $\displaystyle\sum_{k=1}^{\infty} \frac{1}{k^{5/6}}$ **10.** $\displaystyle\sum_{k=1}^{\infty} \frac{1}{k\sqrt[3]{k^2}}$

11. $\displaystyle\sum_{k=1}^{\infty} \frac{1}{k^2\sqrt{k}}$ **12.** $\displaystyle\sum_{k=1}^{\infty} \frac{1}{k^{1.001}}$

In Exercises 13 through 16, use the comparison test to determine whether the given series converges or diverges.

13. $\displaystyle\sum_{k=1}^{\infty} \frac{3}{2^k+1}$ **14.** $\displaystyle\sum_{k=1}^{\infty} \frac{3^k+1}{2^{k-1}}$

15. $\displaystyle\sum_{k=2}^{\infty} \frac{1}{\sqrt{k}-1}$ **16.** $\displaystyle\sum_{k=2}^{\infty} \frac{1}{k\sqrt{k}+1}$

In Exercises 17 through 20, use the ratio test to determine whether the given series converges or diverges.

17. $\displaystyle\sum_{k=1}^{\infty} \frac{2^k}{k^3}$ **18.** $\displaystyle\sum_{k=1}^{\infty} k\left(\frac{1}{3}\right)^k$

19. $\displaystyle\sum_{k=1}^{\infty} \frac{(-3)^k}{k2^{2k}}$ **20.** $\displaystyle\sum_{k=1}^{\infty} \frac{2^k}{k!}$

In Exercises 21 through 36, determine whether the given series converges or diverges. You may use any test.

21. $\displaystyle\sum_{k=3}^{\infty} \frac{k}{\ln k}$ **22.** $\displaystyle\sum_{k=1}^{\infty} \frac{2^k+3^k}{4^k+1}$

23. $\displaystyle\sum_{k=1}^{\infty} \left(\frac{\pi}{2}\right)^k$ **24.** $\displaystyle\sum_{k=1}^{\infty} \frac{k^3+1}{k^3+2}$

25. $\displaystyle\sum_{k=1}^{\infty} \frac{1}{(3k)^{3/2}}$

26. $\displaystyle\sum_{k=1}^{\infty} \frac{1}{(3k+1)^2}$

27. $\displaystyle\sum_{k=3}^{\infty} \frac{1}{k\sqrt{\ln k}}$

28. $\displaystyle\sum_{k=3}^{\infty} \frac{1}{k(\ln k)^3}$

29. $\displaystyle\sum_{k=1}^{\infty} \frac{e^{1/k}}{k^2}$

30. $\displaystyle\sum_{k=1}^{\infty} \frac{k!}{k^3}$

31. $\displaystyle\sum_{k=2}^{\infty} \frac{\ln k}{e^k}$

32. $\displaystyle\sum_{k=1}^{\infty} \frac{3+(-1)^k}{2^k}$

33. $\displaystyle\sum_{k=1}^{\infty} \frac{1}{k^{\sqrt{2}}}$

34. $\displaystyle\sum_{k=3}^{\infty} \frac{1}{k^{3/4}-2}$

35. $\displaystyle\sum_{k=1}^{\infty} e^{1/k}$

36. $\displaystyle\sum_{k=1}^{\infty} \frac{e^k}{k!}$

In Exercises 37 and 38, show that the given series converges by applying the comparison test in two different ways. That is, find two different convergent series with terms greater than the terms of the given series.

37. $\displaystyle\sum_{k=1}^{\infty} \frac{1}{k^2\, 2^k}$

38. $\displaystyle\sum_{k=1}^{\infty} \frac{1}{k^2+3^k}$

In Exercises 39 and 40, use the integral test to determine the values of p for which the given series converges.

39. $\displaystyle\sum_{k=1}^{\infty} \frac{1}{(1+2k)^p}$

40. $\displaystyle\sum_{k=3}^{\infty} \frac{1}{k(\ln k)^p}$

41. For what values of the positive number c does this series converge?

$$\sum_{k=1}^{\infty} \frac{c^k}{k^c}$$

42. Show that the series

$$1 + \frac{1}{3} + \frac{1}{5} + \cdots = \sum_{n=0}^{\infty} \frac{1}{2n+1}$$

diverges.

43. Show that the series

$$1 + \frac{1}{4} + \frac{1}{7} + \cdots = \sum_{n=0}^{\infty} \frac{1}{3n+1}$$

diverges.

44. **TRAFFIC ENGINEERING** Suppose there is no passing on a single-lane road. You observe that a slow car is followed by a group of cars that want to drive faster but are blocked from doing so.

Explain why if n cars are driving in the same direction on this road, you would expect to see

$$1 + \frac{1}{2} + \frac{1}{3} + \cdots + \frac{1}{n}$$

bunches of cars that are more widely spaced. About how many bunches of cars will you observe if 20 cars are driving in the same direction? What if there are 100 cars driving in the same direction? [*Hint:* Look for record low speeds.]

45. **MATERIAL TESTING** Todd wants to estimate the minimum breaking strain of 100 wooden beams of approximately the same dimensions using a machine that applies a gradually increasing force to the center of a beam supported at its ends. By increasing the force until a beam breaks, he can find the breaking strain of the beam. Show that Todd can determine the minimum breaking strain of all 100 beams, so that on average, he will expect to break 5.19 beams in testing. [*Hint:* Find the breaking strain of the first beam. Next, test the second beam, increasing the force up to the breaking strain of the first beam. Then test the third beam, increasing the force up to the breaking strain of the first or second beam, whichever is smaller, and so on.]

46. **DESERT WARFARE** At a base in the desert, there are an unlimited number of jeeps available for action. The gas tanks of the jeeps are full, but no other gasoline is available. Show that it is possible to have one of the jeeps travel from the base to a destination at an arbitrary distance by transferring gas from other jeeps in such a way that all other jeeps return to base without running out of gas on the desert. [*Hint:* The key idea is to send out a sequence of jeeps, transferring gas in such a way that the first jeep gets to its destination, leaving all other jeeps enough gas to return to base. To develop this strategy, note that if you send out one jeep, it can travel only as far as it can go on one tank of gas, but you can get the first jeep farther out by sending a second jeep. Have both jeeps stop when they have used one-third of their gas. Then transfer one-third of the second jeep's tank of gas to the first jeep, thus leaving the second jeep enough gas to get back to base while providing the first jeep with enough gas to travel $1 + \dfrac{1}{3}$ times its usual range, that is,

the distance it can go on one tank of gas. Continue by considering how to use three jeeps, four jeeps, and so on. You will need the result of Exercise 42.]

47. **ZENO'S PARADOX REVISITED**   This is a continuation of Exercise 54, Section 10.1. A supporter of Zeno, called Summo (dates unknown), is not convinced that a racer who keeps running will always finish a racecourse of finite length. "Not so fast!" he challenges. "Suppose the runner takes T minutes for the first half of the course and $\dfrac{T}{2}$ minutes for the next quarter, but then he tires and it takes $\dfrac{T}{3}$ minutes for the next eighth, $\dfrac{T}{4}$ for the next sixteenth, and so on. Now how long does it take him to finish the race?"

*The **limit comparison test** says that if $\displaystyle\sum_{k=1}^{\infty} a_k$ and $\displaystyle\sum_{k=1}^{\infty} b_k$ are positive-term series ($a_k > 0$ and $b_k > 0$*

for all k) and the limit $L = \displaystyle\lim_{k\to\infty} \dfrac{a_k}{b_k}$ is a positive, finite real number ($0 < L < \infty$), then $\displaystyle\sum_{k=1}^{\infty} a_k$ and $\displaystyle\sum_{k=1}^{\infty} b_k$ either both converge or both diverge. Use this test in Exercises 48 through 51 to determine whether the given series converges or diverges.

48. $\displaystyle\sum_{k=1}^{\infty} \dfrac{2}{k^3 + 5k + 71}$ 49. $\displaystyle\sum_{k=1}^{\infty} \dfrac{k+2}{\sqrt[3]{k^5 + k^2}}$

50. $\displaystyle\sum_{k=2}^{\infty} \dfrac{3}{(2k+5)(\ln k)^2}$ 51. $\displaystyle\sum_{k=1}^{\infty} \dfrac{1 + 2^k}{3 + 5^k}$

It can be shown that $\ln n < n$ and $e^n > n$ for all $n \geq 1$. Use these facts together with the direct comparison test to determine whether the series in Exercises 52 and 53 converge or diverge.

52. $\displaystyle\sum_{n=1}^{\infty} \dfrac{e^n}{n^2}$ 53. $\displaystyle\sum_{n=3}^{\infty} \dfrac{1}{(\ln n)^2}$

SECTION 10.3   Functions as Power Series; Taylor Series

Learning Objectives

1. Find the radius and interval of convergence for a power series.
2. Study term-by-term differentiation and integration of power series.
3. Explore Taylor series representation of functions.

A series of the form

$$a_0 + a_1 x + a_2 x^2 + a_3 x^3 + \cdots = \sum_{n=0}^{\infty} a_n x^n$$

is called a **power series** in the variable x. In this section, we will show how functions can be represented by power series. Such representations are useful in a variety of ways, mainly because power series can be thought of as generalized polynomials and polynomials are easily evaluated and manipulated. For instance, the functional values for functions like $\sin x$ and $\ln x$ on calculators and computers are obtained by approximating functions by polynomials. A similar procedure is used to evaluate definite integrals and to solve certain differential equations.

Convergence of a Power Series   A power series is a function of x, and as with any function, it is important to know the domain of this function. In other words, what is the set of all x for which a given power series converges? Here is the answer to this question.

Convergence of a Power Series ■ For a power series

$$a_0 + a_1x + a_2x^2 + a_3x^3 + \cdots = \sum_{n=0}^{\infty} a_n x^n$$

exactly one of the following must be true:
   a. The power series converges only for $x = 0$.
   b. The series converges for all x.
   c. There is a number R such that the series converges for all $|x| < R$ and diverges for $|x| > R$. It may converge or diverge at the endpoints $x = -R$ and $x = R$.

The interval $|x| < R$ or, equivalently, $-R < x < R$ is called the **interval of absolute convergence** of the power series, and R is called the **radius of convergence**. If the power series converges only for $x = 0$, then $R = 0$, and $R = \infty$ in the case where the power series converges for all x. In practice, the radius of convergence of a power series is usually determined by applying the ratio test, as illustrated in Example 10.3.1. Convergence or divergence at the endpoints $x = -R$ and $x = R$ of the interval of absolute convergence must be checked by other tests, some of which are outside the scope of this text.

EXAMPLE 10.3.1 Studying Convergence of Power Series

For each of the following power series, find the radius of convergence and the interval of absolute convergence.

a. $\displaystyle\sum_{k=0}^{\infty} k! x^k$ **b.** $\displaystyle\sum_{k=0}^{\infty} \frac{x^k}{k!}$ **c.** $\displaystyle\sum_{k=0}^{\infty} \frac{x^{2k}}{4^k}$

Just-In-Time **REVIEW**

Recall that
$$\frac{(k+1)!}{k!} = \frac{1 \cdot 2 \cdot 3 \cdots k \cdot (k+1)}{1 \cdot 2 \cdot 3 \cdots k}$$
$$= k + 1$$

and
$$\frac{k!}{(k+1)!} = \frac{1}{k+1}$$

Solution

a. Applying the ratio test, we find that the limit of the absolute value of the ratio of consecutive terms $k! \, x^k$ and $(k+1)! \, x^{k+1}$ is

$$L = \lim_{k\to\infty} \left| \frac{(k+1)! \, x^{k+1}}{k! \, x^k} \right| = \lim_{k\to\infty} |(k+1)x|$$

$$= \infty \quad \text{unless } x = 0$$

Therefore, the power series converges only for $x = 0$. The radius of convergence is $R = 0$, and the interval of absolute convergence is the single point $x = 0$.

b. For this series, we have

$$L = \lim_{k\to\infty} \left| \frac{\dfrac{x^{k+1}}{(k+1)!}}{\dfrac{x^k}{k!}} \right| = \lim_{k\to\infty} \left| \frac{k! \, x^{k+1}}{(k+1)! \, x^k} \right|$$

$$= \lim_{k\to\infty} \left| \frac{x}{k+1} \right| = 0 \quad \text{for all } x$$

Since $L = 0 < 1$ for all x, it follows from the ratio test that the power series converges for all x. In this case, the radius of convergence is $R = \infty$, and the interval of absolute convergence is the entire x axis.

c. For this power series, we get

$$L = \lim_{k \to \infty} \left| \frac{\dfrac{x^{2(k+1)}}{4^{k+1}}}{\dfrac{x^{2k}}{4^k}} \right| = \lim_{k \to \infty} \left| \frac{4^k x^{2k+2}}{4^{k+1} x^{2k}} \right| = \frac{x^{2k+2}}{x^{2k}} = x^{2k+2-2k} = x^2$$

$$= \frac{x^2}{4}$$

According to the ratio test, the power series converges if $L < 1$ and diverges if $L > 1$. In other words, the series converges if $\dfrac{x^2}{4} < 1$ or, equivalently, if $x^2 < 4$.

Thus, the interval of absolute convergence is $-2 < x < 2$, and the radius of convergence is $R = 2$.

Power Series Representations of Functions

Next, we will see how to use geometric series and functions related to geometric series as power series. Recall that if x is any number such that $|x| < 1$, we have

$$1 + x + x^2 + x^3 + \cdots = \frac{1}{1 - x}$$

which may be interpreted as saying that the function $f(x) = \dfrac{1}{1 - x}$ can be represented by the power series $1 + x + x^2 + \cdots = \displaystyle\sum_{k=0}^{\infty} x^k$, with interval of convergence $-1 < x < 1$. As illustrated in Example 10.3.2, the geometric series formula can be modified by substitution to find power series representations for other functions as well.

EXAMPLE 10.3.2 Modifying a Geometric Series

In each case, find a power series for the given function and determine its interval of absolute convergence.

a. $f(x) = \dfrac{x}{1 + 3x^2}$ **b.** $g(x) = \dfrac{1}{2 + 3x}$

Solution

a. Start with the geometric power series

$$\frac{1}{1 - x} = \sum_{n=0}^{\infty} x^n \qquad \text{for } |x| < 1$$

and replace x by $-3x^2$ to get

$$\frac{1}{1 + 3x^2} = \sum_{n=0}^{\infty} (-3x^2)^n = \sum_{n=0}^{\infty} (-3)^n x^{2n}$$

The new interval of absolute convergence is $|-3x^2| < 1$, which can be rewritten as $|x^2| < \dfrac{1}{3}$ or, equivalently, as $-\dfrac{1}{\sqrt{3}} < x < \dfrac{1}{\sqrt{3}}$. Now multiply by x (which does not affect the interval of convergence) to get

$$\frac{x}{1 + 3x^2} = \sum_{n=0}^{\infty} (-3)^n x^{2n+1} \qquad \text{for } |x| < \frac{1}{\sqrt{3}}$$

b. The form of the geometric series $\dfrac{1}{1-x}$ requires the constant term in the denominator to be 1 not 2. Therefore, we begin by factoring 2 from the denominator of the given functional expression to obtain

$$\frac{1}{2 + 3x} = \frac{1}{2\left(1 + \dfrac{3x}{2}\right)} = \frac{1}{2}\left[\frac{1}{1 - \left(-\dfrac{3x}{2}\right)}\right]$$

$$= \frac{1}{2}\left[1 + \left(-\frac{3x}{2}\right) + \left(-\frac{3x}{2}\right)^2 + \cdots\right] = \frac{1}{2}\sum_{n=0}^{\infty}\left(-\frac{3x}{2}\right)^n$$

$$= \frac{1}{2}\sum_{n=0}^{\infty}(-1)^n \frac{3^n}{2^n} x^n$$

$$= \sum_{n=0}^{\infty} \frac{(-1)^n 3^n}{2^{n+1}} x^n$$

The series converges when

$$\left|-\frac{3x}{2}\right| < 1$$

$$|x| < \frac{2}{3}$$

so the interval of absolute convergence is $-\dfrac{2}{3} < x < \dfrac{2}{3}$.

Polynomials are particularly easy to differentiate and integrate because the operations involve term-by-term application of the power rule; that is,

$$\frac{d}{dx}[a_0 + a_1 x + a_2 x^2 + \cdots + a_n x^n] = a_1 + a_2(2x) + a_3(3x^2) + \cdots + a_n(nx^{n-1})$$

and

$$\int [a_0 + a_1 x + a_2 x^2 + \cdots + a_n x^n]\, dx$$

$$= a_0 x + a_1\left(\frac{x^2}{2}\right) + a_2\left(\frac{x^3}{3}\right) + \cdots + a_n\left(\frac{x^{n+1}}{n+1}\right) + C$$

Here is a statement of this principle as it applies to power series.

Term-by-Term Differentiation and Integration of Power Series ■

Suppose the power series $a_0 + a_1x + a_2x^2 + \cdots = \sum_{n=0}^{\infty} a_n x^n$ converges for $|x| < R$, and let f be the function defined by

$$f(x) = \sum_{n=0}^{\infty} a_n x^n \qquad \text{for } -R < x < R$$

Then $f(x)$ is differentiable for $-R < x < R$, and its derivative is given by

$$f'(x) = a_1 + a_2(2x) + a_3(3x^2) + \cdots = \sum_{n=1}^{\infty} n a_n x^{n-1}$$

The function $f(x)$ is also integrable for $-R < x < R$, and its integral is given by

$$\int f(x)\, dx = C + a_0 x + a_1\left(\frac{x^2}{2}\right) + a_2\left(\frac{x^3}{3}\right) + \cdots = \sum_{n=0}^{\infty} a_n\left(\frac{x^{n+1}}{n+1}\right) + C$$

The proof of this result is omitted. Note that the constant of integration C is placed in front of the expanded form of the power series to distinguish it from the actual terms of the series.

We can use term-by-term differentiation and integration when we need to generate power series representations of functions in much the same way that we used substitution in Example 10.3.2. This procedure is illustrated in Examples 10.3.3 and 10.3.4.

EXAMPLE 10.3.3   Term-by-Term Differentiation of a Series

Find a power series for the function

$$g(x) = \frac{1}{(1 - x)^2}$$

Solution

Note that if $f(x) = \dfrac{1}{1 - x} = (1 - x)^{-1}$, then

$$f'(x) = -1(1 - x)^{-2}(-1) = \frac{1}{(1 - x)^2} = g(x)$$

Therefore, since $f(x)$ is represented by the geometric series

$$f(x) = \frac{1}{1 - x} = 1 + x + x^2 + \cdots = \sum_{n=0}^{\infty} x^n$$

we can find its derivative $f'(x) = g(x)$ by differentiating the geometric series term by term:

$$g(x) = f'(x) = 1 + 2x + 3x^2 + \cdots = \sum_{n=1}^{\infty} n x^{n-1}$$

Since the geometric series converges for $|x| < 1$, it follows that the power series we have obtained for $g(x)$ by term-by-term differentiation also converges for $|x| < 1$.

EXAMPLE 10.3.4 Term-by-Term Integration of a Series

Find a power series for the logarithmic function $L(x) = \ln(1 + x)$.

Solution

First, note that

$$\ln(1 + x) = \int \frac{1}{1 + x}\, dx$$

which suggests that we can find a power series for $\ln(1 + x)$ by finding one for $\dfrac{1}{1 + x}$ and integrating term by term. Substituting $-x$ for x in the (geometric) power series for $\dfrac{1}{1 - x}$, we get

$$\frac{1}{1 + x} = \frac{1}{1 - (-x)} = 1 + (-x) + (-x)^2 + (-x)^3 + \cdots$$

$$= 1 - x + x^2 - x^3 + \cdots = \sum_{n=0}^{\infty} (-1)^n x^n$$

and by integrating term by term, we find that

$$\ln(1 + x) - \int \frac{1}{1 + x}\, dx = \int [1 - x + x^2 - x^3 + \cdots]\, dx$$

$$= C + x - \frac{x^2}{2} + \frac{x^3}{3} - \frac{x^4}{4} + \cdots$$

$$= \sum_{n=1}^{\infty} \frac{(-1)^{n+1} x^n}{n} + C$$

To determine the value of C, we put $x = 0$ into this equation and get

$$\ln(1 + 0) = \sum_{n=1}^{\infty} \frac{(-1)^{n+1}(0)^n}{n} + C = 0 + C$$

$$0 = C \qquad\qquad \text{since } \ln 1 = 0$$

Thus, we have

$$\ln(1 + x) = \sum_{n=1}^{\infty} \frac{(-1)^{n+1} x^n}{n}$$

Since the modified geometric series for $\dfrac{1}{1 + x}$ converges for $|-x| < 1$, it follows that the power series we have just obtained for $\ln(1 + x)$ converges on the interval $-1 < x < 1$.

Taylor Series   Imagine that a function $f(x)$ is given and that you would like to find the corresponding coefficients a_n such that the power series $\sum_{n=0}^{\infty} a_n x^n$ converges to $f(x)$ on some interval. To discover what these coefficients might be, suppose that

$$f(x) = \sum_{n=0}^{\infty} a_n x^n = a_0 + a_1 x + a_2 x^2 + a_3 x^3 + \cdots + a_n x^n + \cdots$$

If $x = 0$, only the first term in the sum is nonzero and so

$$a_0 = f(0)$$

Now differentiate the series term by term. It can be shown that if the original power series converges to $f(x)$ on the interval $-R < x < R$, then the differentiated series converges to $f'(x)$ on this interval. Hence,

$$f'(x) = a_1 + 2a_2 x + 3a_3 x^2 + \cdots + n a_n x^{n-1} + \cdots$$

and, if $x = 0$, it follows that

$$a_1 = f'(0)$$

Differentiate again to get

$$f''(x) = 2a_2 + 3 \cdot 2a_3 x + \cdots + n(n-1)a_n x^{n-2} + \cdots$$

and let $x = 0$ to conclude that

$$f''(0) = 2a_2 \qquad \text{or} \qquad a_2 = \frac{f''(0)}{2}$$

Similarly, the third derivative of f is

$$f^{(3)}(x) = 3 \cdot 2a_3 + \cdots + n(n-1)(n-2)a_n x^{n-3} + \cdots$$

and

$$f^{(3)}(0) = 3 \cdot 2a_3 \qquad \text{or} \qquad a_3 = \frac{f^{(3)}(0)}{3!}$$

and so on. In general,

$$f^{(n)}(0) = n! a_n \qquad \text{or} \qquad a_n = \frac{f^{(n)}(0)}{n!}$$

where $f^{(n)}(0)$ denotes the nth derivative of f evaluated at $x = 0$, and $f^{(0)}(0) = f(0)$.

The preceding argument shows that *if* there is any power series that converges to $f(x)$, it must be the one whose coefficients are obtained from the derivatives of f by the formula

$$a_n = \frac{f^{(n)}(0)}{n!}$$

This series is known as the **Taylor series of f** (about $x = 0$), and the corresponding coefficients a_n are called the **Taylor coefficients of f.**

Taylor Series ■ The Taylor series of $f(x)$ about $x = 0$ is the power series $\sum_{n=0}^{\infty} a_n x^n$, where

$$a_n = \frac{f^{(n)}(0)}{n!}$$

When we developed the Taylor series representation of a function, we noted that if there is a power series that converges to a given function $f(x)$, then it must be the Taylor series. For instance, since the geometric series $\sum_{n=0}^{\infty} x^n$ represents $g(x) = \dfrac{1}{1-x}$ for $|x| < 1$, we would expect the Taylor series for $g(x)$ to be $\sum_{n=0}^{\infty} x^n$. You are asked to verify this fact in Exercise 24.

In Examples 10.3.2 through 10.3.4, we showed how to modify geometric series to obtain power series representations for several rational functions and a logarithmic function. Sometimes, however, finding the Taylor series for a given function is the most direct way of obtaining a power series representation of the function. For instance, the power series representation for the exponential function $E(x) = e^x$ cannot be found by modifying or manipulating a geometric series, but as is shown in Example 10.3.5, the Taylor series for $E(x)$ can be found quite easily.

EXAMPLE 10.3.5 Finding a Taylor Series about x = 0

Show that the Taylor series about $x = 0$ for the function $f(x) = e^x$ is

$$e^x = \sum_{n=0}^{\infty} \frac{x^n}{n!}$$

Show also that this power series converges for all x.

Solution
Compute the Taylor coefficients as follows:

$$f(x) = e^x \qquad f(0) = 1 \qquad a_0 = \frac{f(0)}{0!} = \frac{1}{0!}$$

$$f'(x) = e^x \qquad f'(0) = 1 \qquad a_1 = \frac{f'(0)}{1!} = \frac{1}{1!}$$

$$f''(x) = e^x \qquad f''(0) = 1 \qquad a_2 = \frac{f''(0)}{2!} = \frac{1}{2!}$$

$$f^{(3)}(x) = e^x \qquad f^{(3)}(0) = 1 \qquad a_3 = \frac{f^{(3)}(0)}{3!} = \frac{1}{3!}$$

$$\vdots \qquad\qquad\qquad \vdots \qquad\qquad\qquad \vdots$$

$$f^{(n)}(x) = e^x \qquad f^{(n)}(0) = 1 \qquad a_n = \frac{f^{(n)}(0)}{n!} = \frac{1}{n!}$$

The corresponding Taylor series is

$$\sum_{n=0}^{\infty} a_n x^n = \sum_{n=0}^{\infty} \frac{x^n}{n!}$$

To determine the convergence set for this power series, we use the ratio test. We find that

$$L = \lim_{n \to \infty} \left| \frac{\dfrac{x^{n+1}}{(n+1)!}}{\dfrac{x^n}{n!}} \right| = \lim_{n \to \infty} \left| \frac{n!\, x^{n+1}}{(n+1)!\, x^n} \right|$$

$$= \lim_{n \to \infty} \frac{|x|}{n+1} = 0 \qquad \text{for all } x$$

Since $L = 0 < 1$, it follows that the Taylor series for e^x converges for all x. In addition, it can be shown that the Taylor series converges to e^x for all x; that is,

$$e^x = \sum_{n=0}^{\infty} \frac{x^n}{n!} = 1 + x + \frac{x^2}{2!} + \frac{x^3}{3!} + \cdots$$

for all x.

Sometimes it is inappropriate or even impossible to represent a given function by a power series of the form $\sum a_n x^n$, and you will have to use a series of the form $\sum a_n (x - a)^n$ instead, where a is a constant. This is the case, for instance, for the function $f(x) = \ln x$, which is undefined for $x \le 0$ and so cannot possibly be the sum of a series of the form $\sum a_n x^n$ on a symmetric interval about $x = 0$.

Using an argument similar to the one on page 806, you can show that if $f(x) = \sum_{n=0}^{\infty} a_n (x - a)^n$, the coefficients of the series are related to the derivatives of f by the formula

$$a_n = \frac{f^{(n)}(a)}{n!}$$

The power series in $(x - a)$ with these coefficients is called the **Taylor series of $f(x)$ about $x = a$.** To summarize:

Taylor Series about $x = a$ ■ The Taylor series of $f(x)$ about $x = a$ is the power series

$$\sum_{n=0}^{\infty} a_n (x - a)^n$$

where

$$a_n = \frac{f^{(n)}(a)}{n!}$$

Approximation by Taylor Polynomials

The partial sums of a Taylor series of a function are polynomials that can be used to approximate the function. In general, the more terms there are in the partial sum, the better the approximation will be. Let $P_n(x)$ denote the $(n + 1)$st partial sum, which is a polynomial of degree (at most) n. In particular,

$$P_n(x) = f(a) + f'(a)(x - a) + \frac{f''(a)}{2!}(x - a)^2 + \cdots + \frac{f^{(n)}(a)}{n!}(x - a)^n$$

The polynomial $P_n(x)$ is known as the ***n*th Taylor polynomial** of $f(x)$ about $x = a$. The approximation of $f(x)$ by its Taylor polynomials $P_n(x)$ is most accurate near $x = a$ and for large values of n. Here is a geometric argument that should give you some additional insight into the situation.

Observe first that at $x = a$, f and P_n are equal, as are, respectively, their first n derivatives. For example, with $n = 2$

$$P_2(x) = f(a) + f'(a)(x - a) + \frac{f''(a)}{2!}(x - a)^2 \quad \text{and} \quad P_2(a) = f(a)$$

$$P_2'(x) = f'(a) + f''(a)(x - a) \quad \text{and} \quad P_2'(a) = f'(a)$$

$$P_2''(x) = f''(a) \quad \text{and} \quad P_2''(a) = f''(a)$$

The fact that $P_n(a) = f(a)$ indicates that the graphs of P_n and f intersect at $x = a$. The fact that $P_n'(a) = f'(a)$ also shows that the graphs have the same slope at $x = a$. The fact that $P_n''(a) = f''(a)$ imposes the further restriction that the graphs have the same concavity at $x = a$. In general, as n increases, the number of matching derivatives increases, and the graph of P_n approximates more closely that of f near $x = a$. The situation is illustrated in Figure 10.4, which shows the graphs of e^x and its first three Taylor polynomials.

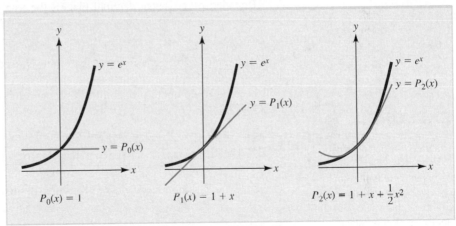

FIGURE 10.4 The graphs of $f(x) = e^x$ and its first three Taylor polynomials.

The use of Taylor polynomials to approximate functions is illustrated in Example 10.3.6.

EXAMPLE 10.3.6 Using a Taylor Polynomial to Estimate a Root

Use an appropriate Taylor polynomial of degree 3 to approximate $\sqrt{4.1}$.

Solution

The goal is to estimate $f(x) = \sqrt{x}$ when $x = 4.1$. Since 4.1 is close to 4 and since the values of f and its derivatives at $x = 4$ are easy to compute, we will use a Taylor polynomial about $x = 4$.

Compute the necessary Taylor coefficients about $x = 4$ as follows:

$$f(x) = \sqrt{x} \qquad\qquad f(4) = 2 \qquad\qquad a_0 = \frac{2}{0!} = 2$$

$$f'(x) = \frac{1}{2}x^{-1/2} \qquad\qquad f'(4) = \frac{1}{4} \qquad\qquad a_1 = \frac{1/4}{1!} = \frac{1}{4}$$

$$f''(x) = -\frac{1}{4}x^{-3/2} \qquad\qquad f''(4) = -\frac{1}{32} \qquad\qquad a_2 = \frac{-1/32}{2!} = -\frac{1}{64}$$

$$f^{(3)}(x) = \frac{3}{8}x^{-5/2} \qquad\qquad f^{(3)}(4) = \frac{3}{256} \qquad\qquad a_3 = \frac{3/256}{3!} = \frac{1}{512}$$

The corresponding Taylor polynomial of degree 3 is

$$P_3(x) = 2 + \frac{1}{4}(x - 4) - \frac{1}{64}(x - 4)^2 + \frac{1}{512}(x - 4)^3$$

and so

$$\sqrt{4.1} \approx P_3(4.1) = 2 + \frac{1}{4}(0.1) - \frac{1}{64}(0.1)^2 + \frac{1}{512}(0.1)^3 \approx 2.02485$$

Rounded off to seven decimal places, the true value of $\sqrt{4.1}$ is 2.0248457.

In Example 10.3.7, a Taylor polynomial is used to estimate a definite integral of a function that has no elementary antiderivative. In Chapter 11, a form of this integral will be used to compute probabilities involving the normal distribution.

EXPLORE!

Evaluate the definite integral in Example 10.3.7 using the numerical integration feature of your graphing calculator. Compare this value with the approximation obtained using a Taylor polynomial of degree 8.

EXAMPLE 10.3.7   Integration Using a Taylor Polynomial

Use a Taylor polynomial of degree 8 to approximate $\displaystyle\int_0^1 e^{-x^2}\, dx$.

Solution

The easiest way to get the Taylor series of e^{-x^2} is to start with the series

$$e^x = \sum_{n=0}^{\infty} \frac{x^n}{n!}$$

for e^x and replace x by $-x^2$ to get

$$e^{-x^2} = \sum_{n=0}^{\infty} \frac{(-1)^n x^{2n}}{n!}$$

Thus,

$$e^{-x^2} \approx 1 - x^2 + \frac{1}{2}x^4 - \frac{1}{6}x^6 + \frac{1}{24}x^8$$

and so

$$\int_0^1 e^{-x^2}\, dx \approx \int_0^1 \left(1 - x^2 + \frac{1}{2}x^4 - \frac{1}{6}x^6 + \frac{1}{24}x^8\right) dx$$

$$= \left(x - \frac{1}{3}x^3 + \frac{1}{10}x^5 - \frac{1}{42}x^7 + \frac{1}{216}x^9\right)\Big|_0^1$$

$$= 1 - \frac{1}{3} + \frac{1}{10} - \frac{1}{42} + \frac{1}{216} \approx 0.7475$$

This value is correct only to two decimal places. The true value of the integral to seven places is 0.7468241.

EXERCISES ▪ 10.3

In Exercises 1 through 8, determine the radius of convergence and the interval of absolute convergence for the given power series.

1. $\sum_{k=0}^{\infty} 5^k x^k$

2. $\sum_{k=1}^{\infty} \sqrt{k} x^k$

3. $\sum_{k=1}^{\infty} \frac{x^k}{\sqrt{k}}$

4. $\sum_{k=1}^{\infty} \frac{3^k x^k}{k}$

5. $\sum_{k=1}^{\infty} \frac{2^{2k} x^k}{k^2}$

6. $\sum_{k=1}^{\infty} \frac{x^k}{(2k)!}$

7. $\sum_{k=1}^{\infty} \frac{5^k x^k}{k!}$

8. $\sum_{k=0}^{\infty} \frac{k! x^k}{3^k}$

In Exercises 9 through 14, find a power series for the given function and determine its interval of absolute convergence.

9. $f(x) = \dfrac{x}{1 + x}$

10. $f(x) = \dfrac{1}{2 - x}$

11. $f(x) = \dfrac{x^2}{1 - x^2}$

12. $f(x) = \dfrac{x}{3 + 2x}$

13. $f(x) = \ln(2 + x)$

14. $f(x) = \ln(1 + 2x)$

In Exercises 15 through 20, find the Taylor series for the given function at the indicated point $x = a$.

15. $f(x) = e^{3x};\ a = 0$

16. $f(x) = e^{-2x};\ a = 0$

17. $f(x) = \dfrac{1}{2}(e^x + e^{-x});\ a = 0$

18. $f(x) = (1 + x)e^x;\ a = 0$

19. $f(x) = e^{1-x};\ a = -1$

20. $f(x) = \dfrac{1}{x};\ a = 1$

21. Find the Taylor series for $f(x) = 3x^2 e^{x^3}$ about $x = 0$ from the Taylor series for e^x obtained in Example 10.3.5 in two different ways:
  a. Substitute x^3 for x and then multiply the resulting series by $3x^2$.
  b. Substitute x^3 for x and then differentiate term by term.

22. Find a power series for $f(x) = \dfrac{x}{1 + x}$ in two different ways:
  a. Using an appropriate geometric series.
  b. Finding the Taylor coefficients.

23. Find a power series for $f(x) = \dfrac{1}{(1 - x)^3}$ by using term-by-term differentiation and the result of Example 10.3.3.

24. a. Find the Taylor coefficients for $f(x) = \dfrac{1}{1 - x}$, and verify that the Taylor series for $f(x)$ at $x = 0$ is the geometric series with $a = 1$ and $r = x$.
  b. Substitute $-x^2$ for x in the geometric series and integrate term by term to obtain the Taylor series for the indefinite integral
  $$\int \frac{1}{1 + x^2}\, dx$$

25. Find the Taylor series about $x = 0$ for the indefinite integral
$$\int x^2 e^{-x^2}\, dx$$
[*Hint:* Start by substituting $-x^2$ for x in the Taylor series for e^x obtained in Example 10.3.5. Then multiply the resulting series by x^2 and integrate term by term.]

In Exercises 26 through 31, use a Taylor polynomial of specified degree n to approximate the indicated quantity.

26. $\sqrt{3.8}$; $n = 3$ **27.** $\sqrt{1.2}$; $n = 3$

28. $\ln 1.1$; $n = 5$ **29.** $\ln 0.7$; $n = 5$

30. $e^{0.3}$; $n = 4$ **31.** $\dfrac{1}{\sqrt{e}}$; $n = 4$

In Exercises 32 through 35, use a Taylor polynomial of specified degree n together with term-by-term integration to estimate the indicated definite integral.

32. $\displaystyle\int_0^{1/2} e^{-x^2}\, dx$; $n = 6$ **33.** $\displaystyle\int_{-0.2}^{0.1} e^{-x^2}\, dx$; $n = 4$

34. $\displaystyle\int_0^{0.1} \dfrac{1}{1 + x^2}\, dx$; $n = 4$ **35.** $\displaystyle\int_{-1/2}^{0} \dfrac{1}{1 - x^3}\, dx$; $n = 9$

In Exercises 36 through 39, calculate the first four Taylor polynomials $P_0(x)$, $P_1(x)$, $P_2(x)$, and $P_3(x)$ for the given function about the specified point $x = a$. Then use the graphing capability of your calculator to draw the graphs of $f(x)$ and the four approximating polynomials on the same set of axes.

36. $f(x) = e^{2x}$; $a = 0$

37. $f(x) = 1 - e^{-x}$; $a = 0$

38. $f(x) = \ln x$; $a = 1$

39. $f(x) = \dfrac{\ln x}{x}$; $a = 1$

40. MARGINAL ANALYSIS The marginal cost of producing x units of a particular commodity is $C'(x)$ thousand dollars per unit, where

$$C'(x) = \ln(1 + 0.01x^2)$$

The net cost of producing the first eight units of the commodity is given by the integral

$$C(8) - C(0) = \int_0^8 C'(x)\, dx = \int_0^8 \ln(1 + 0.01x^2)\, dx$$

a. Find the Taylor polynomial of degree 6 for $\ln(1 + 0.01x^2)$ at $x = 0$. [*Hint:* Example 10.3.4 may help.]

b. Integrate the Taylor polynomial you found in part (a) to estimate the net change $C(8) - C(0)$. [*Note:* The actual net cost is about 1.452, that is, $1,452.]

41. MARGINAL ANALYSIS The marginal profit derived from producing x units of a particular commodity is $P'(x)$ thousand dollars per unit, where

$$P'(x) = 10x^2 e^{-x^2}$$

Find the Taylor polynomial of degree 8 for $P'(x)$ at $x = 0$ and integrate to obtain an estimate of the net profit $P(1) - P(0)$ obtained from producing the first unit.

POPULATION DENSITY   *Recall from Section 5.6 that if the population density x miles from the center of an urban area is D(x), then the total population P living within a radius of M miles of the city center is given by the integral*

$$P(M) = \int_0^M [2\pi\, x\, D(x)]\, dx$$

Use this formula in Exercises 42 and 43.

42. Suppose the population density is

$$D(x) = \dfrac{5{,}000}{1 + 0.5x^2}$$

a. Find the Taylor polynomial of degree 6 for $2\pi\, x\, D(x)$ at $x = 0$ and integrate to obtain an estimate of the total number of people $P(1)$ living within 1 mile of the city center.

b. Evaluate the integral for $P(1)$ directly using the substitution $u = 1 + 0.5x^2$. Compare this exact value of $P(1)$ with your estimate in part (a).

c. Repeat parts (a) and (b) for $P(10)$.

d. Are your two answers in part (c) even close to each other? Write a paragraph on what you think went wrong.

43. Suppose the population density is

$$D(x) = \dfrac{5{,}000}{1 + 0.5x^3}$$

Use a Taylor polynomial of degree 8 to estimate $P(1)$, the population within 1 mile of the city center.

44. GENETICS The Jukes-Cantor model* in genetics is used for studying mutations from an original DNA sequence. The Jukes-Cantor distance between DNA sequences S_0 and S_1 is the quantity

$$d = -\frac{3}{4}\ln\left(1 - \frac{4}{3}p\right)$$

where p is the fraction of sites that disagree between the two sequences. It provides a measure for the total number of substitutions per site that occur as S_0 evolves into S_1.

---

*For example, see E. S. Allman and J. A. Rhodes, *Mathematical Models in Biology*, Cambridge, UK: Cambridge Univ. Press, 2004, pp. 155–162.

a. Approximate d by a Taylor polynomial of degree 3 at $x = 0$.

b. A researcher finds that out of 43 sites in a particular DNA sequence S_0, 17 have undergone a substitution when the sequence evolves to S_1. Use the Taylor polynomial you obtained in part (a) to estimate the Jukes-Cantor distance for this mutation.

c. Read an article on how the Jukes-Cantor distance and other so-called *phylogenetic distances* are used to measure sequence similarity. Then write a paragraph describing the role played by such ideas in mathematical models of evolution.

45. **LOGISTIC GROWTH** The population of a city grows logistically in such a way that after t years, the population is P million, where

$$P(t) = \frac{1}{1 + e^{-1.2t}}$$

Thus, the average population over the next 10 years is given by the integral

$$P_{ave} = \frac{1}{10} \int_0^{10} P(t)\, dt$$

a. Estimate the value of P_{ave} by integrating the Taylor polynomial of degree 2 for $P(t)$ expanded at $t = c$ for $c = 0$, $c = 5$, and $c = 10$.

b. The actual average population is

$$P_{ave} = 0.942 \ (942{,}000 \text{ people})$$

Which choice of c in part (a) resulted in the best approximation?

Exercises 46 through 49 involve trigonometric functions introduced in Chapter 8.

46. a. Show that the Taylor series for $f(x) = \sin x$ at $x = 0$ is

$$\sin x = \sum_{n=0}^{\infty} \frac{(-1)^n x^{2n+1}}{(2n+1)!}$$

b. Show that the series in part (a) converges for all x.

47. Show that the Taylor series for $f(x) = \cos x$ at $x = 0$ is

$$\cos x = \sum_{n=0}^{\infty} \frac{(-1)^n x^{2n}}{(2n)!}$$

in two different ways:

a. By finding the Taylor coefficients.

b. By differentiating the series for $\sin x$ in Exercise 46 term by term.

48. The integral of the function $f(x) = \dfrac{\sin x}{x}$ appears in certain engineering applications, but $f(x)$ has no simple antiderivative.

a. Use the result of Exercise 46 to show that

$$\frac{\sin x}{x} = 1 - \frac{x^2}{3!} + \frac{x^4}{5!} - \frac{x^6}{7!} + \cdots = \sum_{n=0}^{\infty} \frac{(-1)^n x^{2n}}{(2n+1)!}$$

Then integrate this power series term by term to express the definite integral

$$\int_0^1 \frac{\sin x}{x}\, dx$$

as a numerical infinite series.

b. Estimate the value of the definite integral

$$\int_0^1 \frac{\sin x}{x}\, dx$$

by computing the sum of the first four terms of the numerical series in part (a).

49. a. Modify the series in Exercise 47 to obtain the Taylor series for $f(x) = \cos(x^2)$.

b. Use a Taylor polynomial of degree 16 to estimate the value of the definite integral

$$\int_0^1 \cos(x^2)\, dx$$

Important Terms, Symbols, and Formulas

Infinite series:

$$\sum_{k=1}^{\infty} a_k = a_1 + a_2 + a_3 + \cdots \quad (776)$$

The nth partial sum of a series:

$$S_n = \sum_{k=1}^{n} a_k = a_1 + a_2 + \cdots + a_n \quad (777)$$

Convergence and divergence of a series: (777)

$$\sum_{k=1}^{\infty} a_k \text{ converges with sum } S \text{ if } \lim_{n \to \infty} S_n = S$$

and diverges otherwise.

Geometric series with common ratio r: (781)

$$\sum_{n=0}^{\infty} ar^n \text{ converges to } \frac{a}{1-r} \text{ if } |r| < 1$$

and diverges if $|r| \geq 1$.

The harmonic series: $\displaystyle\sum_{k=1}^{\infty} \frac{1}{k}$ (789)

The divergence test: (790)

$$\sum_{k=1}^{\infty} a_k \text{ diverges if } \lim_{k \to \infty} a_k \neq 0.$$

The integral test: (792)

If f is decreasing, $f(x) > 0$, and $f(n) = a_n$,

then $\displaystyle\sum_{n=1}^{\infty} a_n$ and $\displaystyle\int_{1}^{\infty} f(x)\, dx$ either both converge or both diverge.

The p-series test: (795)

$\displaystyle\sum_{n=1}^{\infty} \frac{1}{n^p}$ converges if $p > 1$ and diverges if $0 \leq p \leq 1$.

The direct comparison test: (796)

Suppose $0 \leq b_n \leq a_n$ for all n. Then

if $\displaystyle\sum_{n=1}^{\infty} a_n$ converges, so does $\displaystyle\sum_{n=1}^{\infty} b_n$

if $\displaystyle\sum_{n=1}^{\infty} b_n$ diverges, so does $\displaystyle\sum_{n=1}^{\infty} a_n$.

The ratio test: (797)

For the series $\displaystyle\sum_{n=1}^{\infty} a_n$, let $L = \lim_{n \to \infty} \left| \dfrac{a_{n+1}}{a_n} \right|$.

The series converges if $L < 1$ and diverges if $L > 1$. The test is inconclusive if $L = 1$.

A power series $\displaystyle\sum_{n=0}^{\infty} a_n x^n$ either: (801)

Converges only for $x = 0$
Converges for all x
Converges for $|x| < R$ and diverges for $|x| > R$

Interval of absolute convergence   (801)
Radius of convergence   (801)
Term-by-term differentiation and integration of power series   (804)

Taylor series of $f(x)$ about $x = 0$:   (806)

$$\sum_{n=0}^{\infty} a_n x^n \text{ where } a_n = \frac{f^{(n)}(0)}{n!}$$

Taylor series of $f(x)$ about $x = a$:   (808)

$$\sum_{n=0}^{\infty} a_n (x - a)^n \text{ where } a_n = \frac{f^{(n)}(a)}{n!}$$

Important Taylor series:

$$\frac{1}{1 - x} = \sum_{n=0}^{\infty} x^n \qquad \text{for } |x| < 1 \quad (781)$$

$$e^x = \sum_{n=0}^{\infty} \frac{x^n}{n!} \qquad \text{for all } x \quad (807)$$

$$\ln x = \sum_{n=1}^{\infty} \frac{(-1)^{n+1}}{n} (x - 1)^n \qquad \text{for } 0 < x \leq 2 \quad (805)$$

Taylor polynomials   (808)

Checkup for Chapter 10

1. Determine whether each of the following geometric series converges or diverges. If the series converges, find its sum.

    a. $\displaystyle\sum_{n=0}^{\infty} \frac{(-3)^n}{5^{n+1}}$ **b.** $\displaystyle\sum_{n=1}^{\infty} \frac{2^{2n}}{3^{n-1}}$

2. In each case, show that the given series converges.

    a. $\displaystyle\sum_{n=1}^{\infty} \frac{1}{n^{3/2}}$ **b.** $\displaystyle\sum_{n=1}^{\infty} \frac{3^n}{n!}$

3. In each case, show that the given series diverges.

    a. $\displaystyle\sum_{n=1}^{\infty} \frac{n(n^2 + 1)}{100n^3 + 9}$ **b.** $\displaystyle\sum_{n=2}^{\infty} \frac{\ln n}{\sqrt{n}}$

4. Determine whether each of the following series converges or diverges.

    a. $\displaystyle\sum_{n=1}^{\infty} \frac{1}{\sqrt{n}}$ **b.** $\displaystyle\sum_{n=1}^{\infty} \sqrt{n}\, e^{-n}$

    c. $\displaystyle\sum_{n=1}^{\infty} \frac{n^2}{2^n}$ **d.** $\displaystyle\sum_{n=0}^{\infty} \frac{1}{n^2 + 1}$

5. For each of the following power series, find the interval of absolute convergence.

    a. $\displaystyle\sum_{n=1}^{\infty} \frac{x^n}{n + 1}$ **b.** $\displaystyle\sum_{n=1}^{\infty} \frac{(2x)^n}{n!}$

6. Find a power series for each of the following functions.

    a. $f(x) = \dfrac{x}{1 + x^2}$ **b.** $g(x) = \dfrac{5}{2 + 3x}$

7. Find the Taylor series about $x = 0$ for each of the following functions.

    a. $f(x) = e^x + e^{-x}$ **b.** $g(x) = \ln(x^2 + 1)$

8. **ACCUMULATION OF MEDICATION** A patient is given an injection of 25 units of a certain drug every 24 hours. The drug is eliminated exponentially so that the fraction that remains after t days is $f(t) = e^{-t/3}$. If the treatment is continued indefinitely, approximately how many units will eventually be in the patient's bloodstream just prior to an injection?

9. PRESENT VALUE How much should you invest today at an annual rate of 4% compounded continuously so that starting next year, you can make annual withdrawals of $5,000 in perpetuity?

10. Estimate the value of the integral

$$\int_0^1 \frac{1}{1 + x^2}\, dx$$

by finding a power series for $f(x) = \dfrac{1}{1 + x^2}$ and integrating term by term.

Review Exercises

In Exercises 1 through 8, determine whether the given geometric series converges or diverges. If the series converges, find its sum.

1. $\displaystyle\sum_{n=0}^{\infty} \frac{(-2)^n}{5^{n+1}}$

2. $\displaystyle\sum_{n=1}^{\infty} \frac{4^{n-1}}{3^{2n}}$

3. $\displaystyle\sum_{n=0}^{\infty} \frac{13}{(-5)^n}$

4. $\displaystyle\sum_{n=0}^{\infty} \left(-\frac{3}{2}\right)^n$

5. $\displaystyle\sum_{n=0}^{\infty} e^{-0.5n}$

6. $\displaystyle\sum_{n=1}^{\infty} \frac{2^{n+1}}{3^{n-1}}$

7. $\displaystyle\sum_{n=0}^{\infty} \left[\left(\frac{2}{3}\right)^n + \left(\frac{3}{2}\right)^n\right]$

8. $\displaystyle\sum_{n=0}^{\infty} \left(\frac{3}{2}\right)\left(\frac{2}{3}\right)^n$

In Exercises 9 through 18, determine whether the given series converges or diverges.

9. $\displaystyle\sum_{n=0}^{\infty} \frac{1}{2n + 1}$

10. $\displaystyle\sum_{n=1}^{\infty} \left(1 + \frac{1}{n}\right)^2$

11. $\displaystyle\sum_{n=1}^{\infty} \ln\left(2 + \frac{1}{n}\right)$

12. $\displaystyle\sum_{n=1}^{\infty} \frac{1}{n^2 \sqrt{n}}$

13. $\displaystyle\sum_{n=2}^{\infty} \frac{\ln \sqrt{n}}{\sqrt{n}}$

14. $\displaystyle\sum_{n=1}^{\infty} \frac{10^n}{n!}$

15. $\displaystyle\sum_{n=1}^{\infty} \frac{n^3}{3^n}$

16. $\displaystyle\sum_{n=1}^{\infty} \frac{(-2)^n}{n^2}$

17. $\displaystyle\sum_{n=1}^{\infty} \frac{(-3)^{2n}}{n!}$

18. $\displaystyle\sum_{n=2}^{\infty} \frac{1}{n^2 - 1}$

In Exercises 19 through 22, find the interval of absolute convergence for the given power series.

19. $\displaystyle\sum_{n=1}^{\infty} \frac{(2x)^{2n}}{3^{n-1}}$

20. $\displaystyle\sum_{n=0}^{\infty} (3x)^n$

21. $\displaystyle\sum_{n=1}^{\infty} e^n x^{n-1}$

22. $\displaystyle\sum_{n=1}^{\infty} \frac{(-2x)^n}{n!}$

In Exercises 23 through 26, find the Taylor series at $x = 0$ for the given function, either by using the definition or by manipulating a known series.

23. $f(x) = e^x - e^{-x}$

24. $f(x) = \dfrac{1}{(1 + 2x)^2}$

25. $f(x) = \ln\left(\dfrac{x + 1}{2x + 1}\right)$

26. $f(x) = x^2 e^{-2x}$

In Exercises 27 through 30, find the Taylor series for the given function at the specified value of $x = a$.

27. $f(x) = \dfrac{1}{(1 - x)^2}$; $a = 2$

28. $f(x) = \ln(2 + x)$; $a = -1$

29. $f(x) = x \ln x$; $a = 1$

30. $f(x) = \dfrac{1 - x}{1 + x}$; $a = 0$

31. Use a Taylor polynomial of degree 3 to approximate $\sqrt{0.9}$.

32. Use a Taylor polynomial of degree 10 to approximate

$$\int_0^{1/2} \frac{x}{1 + x^3}\, dx$$

33. Calculate the first four Taylor polynomials $P_0(x)$, $P_1(x)$, $P_2(x)$, and $P_3(x)$ at $x = 0$ for the function

$$f(x) = \frac{1}{\sqrt{1 - x^2}}$$

a. Use the graphing utility on your calculator to sketch the graphs of these four polynomials on the same set of axes.

b. Use $P_3(x)$ to estimate the value of the definite integral

$$\int_0^{1/2} \frac{dx}{\sqrt{1 - x^2}}$$

34. Use your calculator to compute the sum

$$S(N) = 1 - \frac{1}{2!} + \frac{1}{4!} - \frac{1}{6!} + \cdots + \frac{(-1)^N}{(2N)!}$$

for $N = 2, 7, 10, 15$, and 30. Based on your results, do you think the series

$$\sum_{k=0}^{\infty} \frac{(-1)^k}{(2k)!}$$

converges or diverges? If it converges, what do you think its sum will be?

In Exercises 35 and 36, express the repeating decimal as a fraction $\frac{p}{q}$.

35. 51.34747...

36. 0.1223535...

37. BOUNCING BALL A ball is dropped from a height of H feet and bounces indefinitely, repeatedly rebounding to 75% of its previous height. If it travels a total distance of 70 feet, what is H?

38. BOUNCING BALL A ball is dropped from a height of 6 feet and bounces indefinitely, repeatedly rebounding to 80% of its previous height. How far does the ball travel?

39. PRESENT VALUE An investment guarantees annual payments of $5,000 in perpetuity, with the payments beginning immediately. Find the present value of this investment if the prevailing annual interest rate remains fixed at 5% compounded continuously.

40. THE MULTIPLIER EFFECT Suppose that nationwide, approximately 91% of all income is spent and 9% is saved. What is the total amount of spending generated by a 60 billion dollar tax rebate if saving habits do not change?

41. ACCUMULATION OF MEDICATION A patient is given an injection of 25 units of a certain drug every 24 hours. The drug is eliminated exponentially so that the fraction that remains in the patient's body after t days is $f(t) = e^{-kt}$ for some constant k. If 70 units of drug eventually accumulate in the patient's bloodstream just prior to an injection, what is k?

42. SAVINGS How much should you invest today at an annual interest rate of 5% compounded continuously so that starting next year, you can make annual withdrawals of $4,000 in perpetuity?

43. DEMOGRAPHICS A developing country currently has 2,500 trained scientists. The government estimates that each year, 6% of the current number of scientists either retire, die, or emigrate, while 278 new scientists graduate from college. If these trends continue, how many scientists will there be in 20 years? How many in the long run?

44. GENETICS In a classic genetic model,[*] the average life span of a harmful gene is related to the infinite series

$$1 + 2r + 3r^2 + 4r^3 + \cdots = \sum_{k=1}^{\infty} kr^{k-1}$$

for $0 < r < 1$. Show that this series converges and find its sum in terms of r. [*Hint:* If S_n is the nth partial sum of the series, what is $S_n - rS_n$?]

45. LINGUISTICS Linguists and psychologists who are interested in the evolution of language have noticed an interesting pattern in the frequency of so-called rare words in certain literary works. According to one classic model,[†] if a book contains a total of T different such words, then approximately $\frac{T}{(1)(2)}$ words appear exactly once, $\frac{T}{(2)(3)}$ words appear exactly twice, and in general $\frac{T}{k(k+1)}$ words appear exactly k times.

a. Assuming the word frequency pattern in the model is accurate, why should you expect the series

$$\sum_{k=1}^{\infty} \frac{T}{k(k+1)}$$

to converge? What would you expect its sum to be?

b. Verify your conjecture in part (a) by actually summing the series. [*Hint:* See Example 10.1.3.]

c. Read an article on information and learning theory, and write a paragraph on mathematical methods in this subject.

46. LABOR MIGRATION Economists refer to the process of moving from one job to another as *labor migration*. Such movement is usually undertaken as a means for social or economic improvement, but it also involves costs, such as the loss of seniority in the old job and the psychological cost of disrupting relationships. Consider the function[‡]

$$V = \sum_{n=1}^{N} \frac{E_2(n) - E_1(n)}{(1+i)^n} - \sum_{n=1}^{N} \frac{C_m(n)}{(1+i)^n} - C_p$$

[*]C. C. Li, *Human Genetics: Principles and Methods,* New York: McGraw-Hill.
[†]G. K. Zipf, *Human Behavior and the Principle of Least Effort,* Cambridge, MA: Addison-Wesley. Another good source is C. E. Shannon and W. Weaver, *The Mathematical Theory of Communication,* Urbana, IL: Univ. of Illinois Press.
[‡]C. R. McConnell and S. L. Brue, *Contemporary Labor Economics,* New York: McGraw-Hill, 1992, pp. 440–444.

where $E_2(n)$ and $E_1(n)$ denote the earnings from the new and old jobs, respectively, in year n after the move is to be made; i is the prevailing annual interest rate (in decimal form); N is the number of years the person is expected to be on the new job; and C_m and C_p are the expected monetary and net psychological costs of the move (psychological gain minus psychological loss).

a. What does V represent? Why is the job move desirable if $V > 0$ and undesirable if $V < 0$?

b. For simplicity, assume that $E_2 - E_1$ and C_m are constant for all n and that the person expects to stay on the new job "forever" once he or she moves (that is, $N \to \infty$). Find a formula for V, and use it to obtain a criterion for whether or not the job move should be made. [*Hint:* Your criterion should be an inequality involving E_1, E_2, C_m, C_p, and i.]

c. Read an article on job mobility and labor migration, and write a paragraph on mathematical methods for modeling such issues.

47. GEOLOGICAL DATING   At the time we studied carbon dating in Chapter 4, we noted that radiocarbon methods are used mainly for dating specimens that are not too old.* To date rocks or artifacts that are older than 40,000 years, it is necessary to use other methods. Whether ^{14}C or a radioactive isotope of some other element is used for dating, it can be shown that

$$\frac{S(t) - S(0)}{R(t)} + 1 = e^{(\ln 2)t/\lambda}$$

where $R(t)$ is the number of atoms of radioactive isotope at time t, $S(t)$ is the number of atoms of the stable product of radioactive decay, $S(0)$ is the number of atoms of stable product initially present (at $t = 0$), and λ is the half-life of the radioactive isotope (the time it takes for half a sample to decay).

a. Approximate the time t in this formula using the Taylor polynomial of degree 2 for e^x at $x = 0$.

b. Suppose a piece of mica is analyzed, and it is found that 5% of the atoms in the rock are radioactive rubidium-87 and 0.04% are strontium-87. If all the strontium-87 was produced by decay of the rubidium-87 in the rock, how old is the rock? Use the approximation obtained in part (a). You will

need to know that the half-life of rubidium-87 is 48.6×10^9 years.

c. Read an article on geological dating methods, and write a paragraph on how these methods differ from radioactive dating and from each other.

48. GEOLOGICAL DATING   Reconsider the geological dating equation

$$\frac{S(t) - S(0)}{R(t)} + 1 = e^{(\ln 2)t/\lambda}$$

from Exercise 47.

a. Solve the equation for t. Then use the Taylor polynomial of degree 2 for $\ln(1 + x)$ at $x = 0$ to estimate the value of t that satisfies the equation.

b. What adjustment would be necessary if your approximation were used to measure a value of t as large or larger than the half-life of the radioactive substance used in the dating?

49. BEE STORY   The story goes that the famous mathematician, John von Neumann (1903–1957), was once challenged to solve a version of the following problem:

Two trains, each traveling at 30 ft/sec, approach each other on a straight track. When they are 1,000 feet apart, a bee begins flying from one train to the other and back again at the rate of 60 ft/sec, and continues to do so until the trains crash. How far does the bee fly before it is crushed by the crashing trains?

Von Neumann pondered the question only briefly before giving the correct answer. The poser of the problem chuckled appreciatively and said, "You saw the trick. I should have known better than to try to fool you, Professor." Von Neumann looked puzzled. "What trick?" he replied, "I summed the series."

a. Sum a series as von Neumann did to find the distance traveled by the ill-fated bee.

b. Unlike von Neumann, do you see an easy way to solve this problem? (Incidentally, a form of this question is sometimes used by Microsoft to test new employees.)

c. Try this kinder, gentler version of the same problem. Suppose the conductors of the two trains see each other and hit the brakes when they are 180 feet apart. If both trains decelerate at the rate of 5 ft/sec^2, how far does the bee fly before the trains come together?

---

*Paul J. Campbell, "*How Old Is the Earth?*" UMAP Modules 1992: Tools for Teaching, Lexington, MA: Consortium for Mathematics and Its Applications, Inc., 1993, pp. 105–137.

EXPLORE! UPDATE

Complete solutions for all EXPLORE! boxes throughout the text can be accessed at the book-specific website, www.mhhe.com/hoffmann.

Solution for Explore!
on Page 777

To evaluate the sum $\sum_{n=1}^{7} \dfrac{n^3 - 2n}{n + 1}$ to at least three-decimal-place accuracy, first enter Y1 = (X^3 − 2X)/(X + 1) into the equation editor. As demonstrated in the Calculator Introduction found at the website www.mhhe.com/hoffmann, employ the features of sequence, **5: seq(,** and cumulative summation, **6:sumSum(,** from the **OPS** menu of the **LIST** key **(2nd STAT),** shown in the following left screen. In the **STAT** editor, enter L1 = seq(X, X, 1, 7), L2 = Y1(L1), and L3 = cumSum(L2), making sure the cursor is at the very top line of each list column before entering the respective commands. The final result on the following right screen shows that the individual terms of the series in L2 appear to be getting larger and larger, creating the sequence of increasing cumulative partial sums in L3.

Solution for Explore!
on Page 781

To evaluate the geometric series with $a = 2$ and $r = 0.85$, first enter Y1 = 2(0.85)^X into the equation editor of your graphing calculator. Put the numbers 0 to 49 in List 1 of the **STAT** editor by using the sequence feature: L1 = seq(X, X, 0, 49). Next enter L2 = Y1(L1) at the top of the List 2 column. Finally enter L3 = cumSum(L2) and arrow down to rows 15 and 50 to obtain the partial sums $S_{15} = 12.1686104$ and $S_{50} = 13.3293898$, respectively. (Note that the L1 values are, respectively, 14 and 49.) The S values are similar to those obtained via the formula $Sn = \dfrac{a(1 - r^n)}{1 - r}$. In comparison, $S_{15} = 12.1686104$ is within two units of $S_\infty = 13\frac{1}{3}$ and $S_{50} = 13.3293898$ is within a hundredth of S_∞. Try comparing S_{100} with S_∞.

Solution for Explore!
on Page 783

Following Example 10.1.9, enter the function 40(0.9)^X into Y1 of the equation editor. For exploration purposes, put the sequence {0, 1, . . . , 120} in List 1 of the **STAT** editor, by entering L1 = seq(X, X, 0, 120). At the very top of List 2, enter L2 = Y1(L1) (see the following screen on the left) to generate the corresponding terms of the series, and in List 3 create the corresponding partial sums, using L3 = cumSum(L2). Scrolling down L3 (see the following right screen) we determine that 57 terms are needed to have the partial sum be within a billion of the infinite series total of 400 billion.

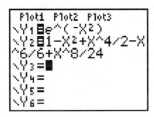 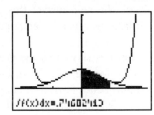 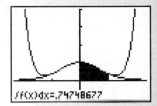

Solution for Explore! on Page 810

Referring to Example 10.3.7, store the function $f(x) = e^{\wedge}(-X^2)$ into Y1 and $P_8(x) = 1 - X^2 + X^4/2 - X^6/6 + X^8/24$ into Y2 of the equation editor of your graphing calculator. Use the modified decimal viewing window $[-2.35, 2.35]1$ by $[-1, 4]1$ and select **7: $\int$f(x)dx** from the **CALC** menu (**2nd TRACE**) of your calculator. The following middle screen displays the integral of $f(x) = e^{\wedge}(-X^2)$, with the lower limit of integration as $X = 0$ and the upper limit as $X = 1$. Before generating the following right screen for the integral of $P_8(x)$, make sure to clear the previous shading by entering **1:ClrDraw** on the **DRAW** (**2nd PRGM**) key. The integral of the Taylor polynomial of degree 8 approximates that of the original function to within 0.001.

THINK ABOUT IT

HOW MANY CHILDREN WILL A COUPLE HAVE?

In family planning, one consideration is that many couples want at least one boy and at least one girl. Call this the *gender requirement*. On average, how many children must be born to a couple before the gender requirement is met?

We will answer this question in terms of the proportion of male births among all births in a particular population. If this proportion is m, where $0 < m < 1$, then the proportion of female births will be $f = 1 - m$. For example, in the United States, birth statistics show that the proportions are $m = 0.51$ and $f = 0.49$ (51% male births versus 49% female). In the discussion that follows, we assume that the likelihood of a male child being born to a couple is m regardless of the gender of their previous children.

We first determine the possibilities when a couple continues to have children until the gender requirement is met. Suppose their first child is a boy. The couple then continues to have children until they have a girl. Consequently, the couple can have a boy followed by a girl, or two boys followed by a girl, or three boys followed by a girl, and so on. We denote these possibilities by BG, BBG, $BBBG$, . . . , where B denotes a boy child and G denotes a girl.

Our next step is to determine the proportion of each of these possibilities among all couples who continue to have children until the gender requirement is met. The proportion of such families who have a boy followed by a girl (BG) is

$$mf = m(1 - m)$$

while the proportion who have two boys followed by a girl (BBG) is

$$mmf = m^2(1 - m)$$

and, in general, the proportion of families that have $j - 1$ boys before having their first girl is

$$\underbrace{mm \cdots m}_{j\,-\,1\ \text{terms}} f = m^{j-1}(1 - m)$$

Similarly, the proportion of such families who have $j - 1$ girls before having their first boy is

$$\underbrace{ff \cdots f}_{j\,-\,1\ \text{terms}} m = f^{j-1}m = (1 - m)^{j-1}m$$

Putting things together, we see that the proportion of couples that have $j - 1$ children of the same gender before having a child of the opposite gender is given by the sum

$$p(m, j) = m^{j-1}(1 - m) + (1 - m)^{j-1}m$$

If we multiply the proportion $p(m, j)$ we have just found by the number of children j in the family at the time the gender requirement is met, we get the weighted proportion

$$p(m, j) = j[m^{j-1}(1 - m) + (1 - m)^{j-1}m]$$

Then the average size $A(m)$ of a family that satisfies the gender requirement is found by adding such weighted proportions for all possible family sizes; that is,

$$A(m) = \sum_{j=2}^{\infty} j[m^{j-1}(1 - m) + (1 - m)^{j-1}m]$$

where the sum begins at $j = 2$ since there must be at least two children in any family that satisfies the gender requirement. To see why such a sum represents an average, consider the simpler problem in which a class of students is asked to rate their instructor on a point scale from 1 (terrible) to 4 (excellent). Suppose the instructor receives 1 point from 10% of the students, 2 points from 40%, 3 points from 20%, and 4 points from 30%. Then the instructor's average score A is given by the sum of weighted proportions

$$A = 1(0.1) + 2(0.4) + 3(0.2) + 4(0.3) = 2.7$$

so the instructor is rated as slightly above average by the students.

Now we shall determine the average size $A(m)$ of a family that satisfies the gender requirement by finding the sum of the infinite series

$$A(m) = \sum_{j=2}^{\infty} j[m^{j-1}(1 - m) + (1 - m)^{j-1}m]$$

$$= \sum_{j=2}^{\infty} j[m^{j-1}(1 - m)] + \sum_{j=2}^{\infty} j[(1 - m)^{j-1}m] \qquad \begin{array}{l} \text{factor } 1 - m \text{ from the left series} \\ \text{and } m \text{ from the right series} \end{array}$$

$$= (1 - m) \sum_{j=2}^{\infty} jm^{j-1} + m \sum_{j=2}^{\infty} j(1 - m)^{j-1}$$

Notice that we have expressed the given series for $A(m)$ as the sum of two subseries, each of which has the form $\sum_{j=2}^{\infty} jx^{j-1}$, where $x = m$ for the first subseries and $x = 1 - m$ for the second. We can sum these series using the formula

$$\frac{1}{(1 - x)^2} = \sum_{j=1}^{\infty} jx^{j-1}$$

derived in Example 10.3.3. Applying this formula, we find that

$$\frac{1}{(1 - m)^2} = \sum_{j=1}^{\infty} jm^{j-1}$$

and

$$\frac{1}{m^2} = \frac{1}{[1 - (1 - m)]^2} = \sum_{j=1}^{\infty} j(1 - m)^{j-1}$$

Because both these sums are missing their first term (that is, each sum begins with $j = 2$), we subtract off the $j = 1$ term in each sum to compute its value. We find that

$$\sum_{j=2}^{\infty} jm^{j-1} = \left(\sum_{j=1}^{\infty} jm^{j-1} \right) - 1 \cdot m^{1-1}$$

$$= \frac{1}{(1-m)^2} - 1 = \frac{2m - m^2}{(1-m)^2} = \frac{m(2-m)}{(1-m)^2}$$

and

$$\sum_{j=2}^{\infty} j(1-m)^{j-1} = \sum_{j=1}^{\infty} j(1-m)^{j-1} - 1(1-m)^{1-1}$$

$$= \frac{1}{m^2} - 1 = \frac{1-m^2}{m^2} = \frac{(1+m)(1-m)}{m^2}$$

It follows that the average number of children a couple will have if they decide to have children until they have at least one boy and one girl is

$$A(m) = (1-m)m \frac{2-m}{(1-m)^2} + m(1-m) \frac{1+m}{m^2}$$

$$= \frac{m(2-m)}{1-m} + \frac{1-m^2}{m} = \frac{1-m+m^2}{m(1-m)}$$

So what is the average number of children of such families when boys and girls are born in the same proportion, that is, when $m = \dfrac{1}{2}$? We find that

$$A\left(\frac{1}{2}\right) = \frac{1 - \frac{1}{2} + \left(\frac{1}{2}\right)^2}{\left(\frac{1}{2}\right)^2} = \frac{\frac{3}{4}}{\frac{1}{4}} = 3$$

This essay was adapted from a discussion in Paul J. Nahin's fascinating little book, *Duelling Idiots and Other Probability Puzzlers*, Princeton, NJ: Princeton University Press, 2000. In our discussion, we have explicitly avoided using ideas and terminology from probability theory, but you may find it instructive to revisit this essay after encountering probability in Chapter 11.

Questions

1. Find the average number of children that will be born to a couple before the gender requirement is met if $m = 0.51$, the observed proportion of male births in the United States; that is, find $A(0.51)$.

2. Use calculus to show that the average size $A(m)$ of a family that satisfies the gender requirement is minimized when $m = f = \dfrac{1}{2}$.

3. What happens to the average size $A(m)$ of a family that satisfies the gender requirement as the proportion of male births m approaches 0? What happens as m approaches 1?

4. A couple decides to have children until they have a girl. On average, how many children will they have before this requirement is satisfied? Express your answer in terms of a formula involving m, the proportion of male births.

CHAPTER 11

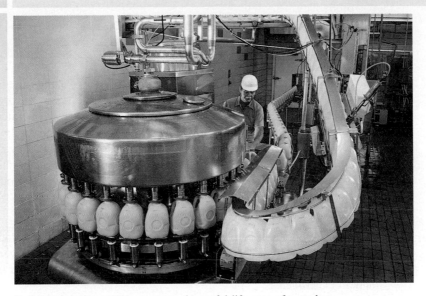

Probability can be used to measure the useful life span of a product.

Probability and Calculus

SECTION 11.1   Introduction to Probability; Discrete Random Variables

Learning Objectives

1. Define outcome, sample space, random variable, and other basic concepts of probability.
2. Study histograms, expected value, and variance of discrete random variables.
3. Examine and use geometric distributions.

Uncertainty is a daily feature of every aspect of life. We cannot predict exactly what will happen when we flip a coin, when we make an investment, or when an insurance company underwrites a policy. The price we pay for a computer may be specified exactly, but the length of time the computer operates effectively before the hard drive crashes is uncertain, a random occurrence. **Probability theory** is the mathematical area that deals with random behavior. The primary goal of this chapter is to explore the use of calculus in probability, but first we need to introduce some basic concepts and terminology.

Outcomes, Events, and Sample Spaces

Mathematical modeling with probability involves the study of **random experiments,** that is, actions with various possible outcomes that can be observed and recorded. Each repetition of a random experiment is called a **trial.** Here are a few examples of random experiments.

1. Rolling a die and recording the number on the top face.
2. Sampling from a randomly chosen production run and recording whether or not the item selected is defective.
3. Choosing a child at random from a particular school and recording the number of siblings of that child.

In describing an experiment, we use the following terminology.

> **Outcomes, Events, and Sample Space** ■ For a particular random experiment, an **outcome** is a result of the experiment. Any collection of outcomes is called an **event,** and a **simple event** is the event of a single outcome. The **empty set** $\varnothing$ is the event that contains no outcomes, and the collection of all possible outcomes is called the **sample space.**

EXAMPLE 11.1.1   Finding the Events for a Random Experiment

Suppose a box contains three balls, colored red, black, and yellow. You conduct an experiment by selecting one of the balls from the box without looking. List the events and the sample space for this experiment.

Solution

Denote the balls by R, B, and Y, for "red," "black," and "yellow," respectively. The sample space is the set of all outcomes $S = \{R, B, Y\}$. Events can be thought of as subsets of S. For instance, the simple events are the subsets $\{R\}$, $\{B\}$, and $\{Y\}$, and

the event "the ball picked is red or black" is the subset $\{R, B\}$. Listing the set of all events in this fashion, we obtain

$$\{\varnothing\}, \{R\}, \{B\}, \{Y\}, \{R, B\}, \{R, Y\}, \{B, Y\}, \{R, B, Y\}$$

Random Variables

A **random variable** is a function X that assigns a numerical value to each outcome of a random experiment. More specifically, a **discrete random variable** takes on values from a finite set of numbers or an infinite succession of numbers such as the positive integers, while a **continuous random variable** takes on values from an entire interval of numbers. We study discrete random variables in this section and will examine continuous random variables in subsequent sections. Example 11.1.2 illustrates how to distinguish between discrete and continuous random variables.

> ## EXAMPLE 11.1.2   Identifying Discrete and Continuous Random Variables

In each of the following cases, determine whether the given random variable X is discrete or continuous. If the random variable is discrete, state its value at each point of the sample space.

a. A manufacturing process involves daily production runs of N units. Let X be the number of defective units in a randomly selected production run.

b. Let X be the age of a cell chosen at random from a particular culture.

c. Let X denote the time the hard drive of a randomly selected personal computer works until it crashes.

d. Let X count the number of times a randomly chosen coin must be tossed until heads comes up.

Solution

a. The random variable X is discrete and finite because its possible values are 0, 1, 2, . . . , N.

b. Since the age of a cell can be any positive real number, this random variable is continuous.

c. The hard drive of a personal computer may be defective or may crash at any time, so the random variable X can take on any nonnegative real number. Therefore, X must be continuous.

d. When a coin is tossed repeatedly until it comes up heads, the possible outcomes consist of an immediate head or a string of tails terminating in a head. In other words, the sample space is $\{$H, TH, TTH, TTTH, TTTTH, . . .$\}$, and since X measures the number of times the coin is tossed before heads appear, we have $X(\text{H}) = 1$, $X(\text{TH}) = 2$, $X(\text{TTH}) = 3, ...$ Since the set of values of X is the set of all positive integers, it follows that X is a discrete random variable.

Probability Distributions

The probability of an outcome is a numerical measure of the likelihood of that outcome occurring. Suppose a particular random experiment has outcomes $s_1, s_2, . . . , s_n$, and that to each outcome s_k we assign a real number $P(s_k)$ in such a way that these two conditions are satisfied:

$$0 \leq P(s_k) \leq 1$$
$$P(s_1) + P(s_2) + \cdots + P(s_n) = 1$$

Then $P(s_k)$ is called the **probability** of outcome s_k. The first required condition says that the likelihood of any outcome must be somewhere between impossible ($P = 0$) and certain ($P = 1$). The second requirement says that whenever the random experiment is performed, it is certain that exactly one outcome will occur.

More generally, for any event E consisting of outcomes $t_1, t_2, \ldots, t_k$, we have

$$P(E) = P(t_1) + P(t_2) + \cdots + P(t_k)$$

It follows that the **complement** of E, the event $\overline{E}$ consisting of all points in the sample space that are not in E, has probability

$$P(\overline{E}) = 1 - P(E)$$

since every outcome is either in E or in $\overline{E}$, and the sum of the probabilities of all outcomes is 1.

EXAMPLE 11.1.3 Finding Probability

A fair die is thrown, and the number on the top face is observed. The possible outcomes are 1, 2, 3, 4, 5, 6, and we assign a probability of $\dfrac{1}{6}$ to each. Find the probability of the following events:

a. The event E_1 that the die shows either a 1 or a 2.
b. The event E_2 that the die shows a 3, 4, 5, or 6.

Solution

a. The probability that the die shows a 1 or 2 is

$$P(E_1) = \frac{1}{6} + \frac{1}{6} = \frac{1}{3}$$

b. The event that the die shows a 3, 4, 5, or 6 is the complement of the event that a 1 or 2 is shown. Thus, $E_2 = \overline{E_1}$ and we have

$$P(E_2) = P(\overline{E_1}) = 1 - P(E_1) = 1 - \frac{1}{3} = \frac{2}{3}$$

We shall write $(X = c)$ to denote the event that contains all points of the sample space of an experiment for which the random variable X has the value c. For instance, the sample space for the random experiment of tossing a coin twice is {HH, HT, TH, TT}. If X is the random variable that counts the number of heads that occur, then the event $(X = 2)$ contains the single sample point HH because this is the only way to get two heads. Likewise, the event $(X = 1)$ contains both HT and TH, while $(X = 0)$ contains only TT.

Once probabilities have been assigned to the points in the sample space S of a random experiment, we can examine how the probability varies in relation to a particular discrete random variable X defined on S. This is done by studying the *probability distribution* for X, which may be defined as follows.

> **Probability Distribution** ▪ Suppose a probability assignment has been made for the sample space S of a particular random experiment, and let X be a discrete random variable defined on S. Then the function p such that
>
> $$p(x) = P(X = x)$$
>
> for each value x assumed by X is called a probability function or **probability distribution** for X.

EXAMPLE 11.1.4 Finding a Probability Distribution

An industrial efficiency expert conducts a test in which a randomly selected worker is required to perform a particular task three times in succession. Each time the task is performed successfully, the worker is awarded 1 point. An additional bonus point is awarded if the worker succeeds all three times, and there is a 5-point penalty for three successive failures. Find the probability distribution for the random variable X that measure's the worker's test score.

Solution

Let "S" denote success and "F" denote failure in the performance of the task. The eight outcomes are

$$\{SSS, SSF, SFS, SFF, FSS, FSF, FFS, FFF\}$$

Assume success and failure are equally likely each time the task is performed. Then each of the eight possible outcomes has probability $\dfrac{1}{8}$. We find the probability for X by adding the probabilities of all the outcomes for which X has the value x:

$$p(-5) = P(X = -5) = P(FFF) - \frac{1}{8}$$

$$p(1) = P(X = 1) = P(SFF) + P(FSF) + P(FFS) = \frac{1}{8} + \frac{1}{8} + \frac{1}{8} = \frac{3}{8}$$

$$p(2) = P(X = 2) = P(SSF) + P(SFS) + P(FSS) = \frac{1}{8} + \frac{1}{8} + \frac{1}{8} = \frac{3}{8}$$

$$p(4) = P(X = 4) = P(SSS) = \frac{1}{8}$$

Note that

$$p(0) + p(1) + p(2) + p(3) = \frac{1}{8} + \frac{3}{8} + \frac{3}{8} + \frac{1}{8} = 1$$

as required.

NOTE   There is no such thing as a "natural" probability assignment. Consider the random experiment of flipping a coin, where the outcomes are heads (H) and tails (T). It may be reasonable to assign $P(\text{H}) = P(\text{T}) = \dfrac{1}{2}$ if the coin is fair. However, when we carry out an experiment of flipping a particular coin a large number of times, we may observe different relative frequencies for heads and tails. For example, if heads comes up 55% of the time in such an experiment, we may assign $P(\text{H}) = 0.55$ and $P(\text{T}) = 0.45$. ■

Histograms   One way to gain insight into the probability function for a discrete random variable X is to construct a graph showing the values of X and the probability assignments that correspond to those values. A **histogram** is a graphical representation obtained by first placing the values of the random variable X on a number line and then constructing above each such value n a vertical rectangle with width 1 and height equal to the probability $P(X = n)$. Note that

$$\begin{array}{c}\text{Area of the}\\ n\text{th rectangle}\end{array} = 1 \cdot P(X = n) = P(X = n)$$

so the total area bounded by all the rectangles in the histogram is

$$\begin{array}{c}\text{Total}\\ \text{area}\end{array} = \sum P(X = n) = 1$$

where the sum is over all positive integers n for which $P(X = n) > 0$. The histogram for the random variable in Example 11.1.4 is shown in Figure 11.1.

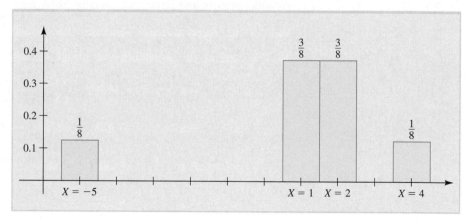

FIGURE 11.1   The histogram for the random variable in Example 11.1.4.

Example 11.1.5 provides a second illustration of how a histogram can be used to visually display a probability distribution, this time in an application involving health science.

EXAMPLE 11.1.5   Representing a Probability Distribution with a Histogram

A study of the effect of an influenza epidemic on the students living in a particular college dormitory determines that during the 30 days of April, there were:

4 days when no students had the flu
3 days when exactly one student had the flu
1 day when exactly two students had the flu
12 days when exactly six students had the flu
8 days when exactly seven students had the flu
2 days when exactly eight students had the flu

Let X be the random variable that represents the number of students in the dormitory who had the flu on a randomly chosen day in April. Describe the event $(X = x)$ and compute the probabilities $P(X = x)$ for $x = 0, 1, 2, \ldots , 8$. Then draw the histogram for X.

Solution

For each value of x, $(X = x)$ is the event that on a randomly selected day in April, exactly x students in the dormitory had the flu, and $P(X = x)$ is the probability assignment for that event. We find that

$$P(X = 0) = \frac{4}{30} \qquad \text{for 4 days of the 30, } x = 0 \text{ students had the flu}$$

$$P(X = 1) = \frac{3}{30} \qquad \text{for 3 days of the 30, } x = 1 \text{ student had the flu}$$

$$P(X = 2) = \frac{1}{30} \qquad \text{for 1 day of the 30, } x = 2 \text{ students had the flu}$$

$$P(X = 3) = P(X = 4) \qquad \text{there are no days when } x = 3, x = 4, \text{ or } x = 5$$
$$= P(X = 5) = 0 \qquad \text{students had the flu}$$

$$P(X = 6) = \frac{12}{30} \qquad \text{for 12 days of the 30, } x = 6 \text{ students had the flu}$$

$$P(X = 7) = \frac{8}{30} \qquad \text{for 8 days of the 30, } x = 7 \text{ students had the flu}$$

$$P(X = 8) = \frac{2}{30} \qquad \text{for 2 days of the 30, } x = 8 \text{ students had the flu}$$

The histogram for this random variable is shown in Figure 11.2.

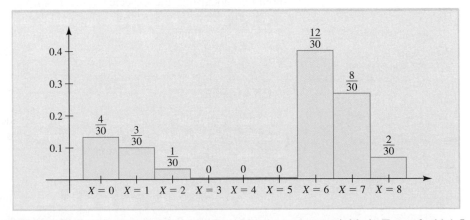

FIGURE 11.2   The histogram for the flu epidemic random variable in Example 11.1.5.

Expected Value

For the flu epidemic in Example 11.1.5, how many students would you expect to be ill on a randomly chosen day in April? From the calculation in the example, we know that 0 students were ill $\dfrac{4}{30}$ of the time during the month, 1 student was ill $\dfrac{3}{30}$ of the time, 2 students were ill $\dfrac{1}{30}$ of the time, and so on. Therefore, we compute the expected number of ill students on our randomly selected day by evaluating the weighted sum

$$0 \cdot P(X = 0) + 1 \cdot P(X = 1) + 2 \cdot P(X = 2) + 3 \cdot P(X = 3) + 4 \cdot P(X = 4)$$
$$+ 5 \cdot P(X = 5) + 6 \cdot P(X = 6) + 7 \cdot P(X = 7) + 8 \cdot P(X = 8)$$
$$= 0\left(\frac{4}{30}\right) + 1\left(\frac{3}{30}\right) + 2\left(\frac{1}{30}\right) + 3(0) + 4(0) + 5(0) + 6\left(\frac{12}{30}\right) + 7\left(\frac{8}{30}\right) + 8\left(\frac{2}{30}\right)$$
$$= \frac{149}{30} \approx 5$$

Thus, if we repeatedly pick days in April at random, on average we would expect to find about 5 students ill.

What we have computed in this illustration is the *expected value* of the random variable X that represents the number of students who were ill on a randomly selected day in April. Here is a definition of expected value.

Expected Value of a Discrete Random Variable ■ Let X be a discrete random variable, with values $x_1, x_2, \ldots, x_n$, and let p be the probability function of X, so that $p(x_k) = P(X = x_k)$ for $k = 1, 2, \ldots, n$. Then the expected value of X is given by the weighted sum

$$E(X) = x_1 p(x_1) + x_2 p(x_2) + \cdots + x_n p(x_n) = \sum_{k=1}^{n} x_k p(x_k)$$

The expected value is also referred to as the **mean** of X.

Intuitively, the expected value $E(X)$ can be thought of as a long-term probability-weighted average of all possible values that the random variable X can assume. It should be noted that the expected value $E(X)$ may not be a value actually assumed by X. For instance, if X measures the number of pets owned by a randomly chosen family in a particular community, it is quite possible for the expected value to be something like $E(X) = 3.1$ even though no family can actually own 3.1 pets.

EXAMPLE 11.1.6 Finding Expected Value

Continuing Example 11.1.4, find the expected value of the random variable X that measures the performance test score achieved by a randomly selected worker performing a task three times in succession.

Solution

Using the formula for $E(X)$ and the probabilities found in Example 11.1.4, we obtain

$$E(X) = -5 \cdot P(X = -5) + 1 \cdot P(X = 1) + 2 \cdot P(X = 2) + 4 \cdot P(X = 4)$$
$$= -5\left(\frac{1}{8}\right) + 1\left(\frac{3}{8}\right) + 2\left(\frac{3}{8}\right) + 4\left(\frac{1}{8}\right) = 1$$

We conclude that if the worker in Example 11.1.4 repeatedly performs the task three times in succession, then on average, the performance test score will be 1.

EXAMPLE 11.1.7 Using Expected Value for Pricing

The grand prize of a lottery is $1,000, and there are also three second prizes each worth $500 and five third prizes each worth $100. Suppose that 10,000 lottery tickets are sold, one of which is to you. How much should you be willing to pay for your ticket?

Solution

A fair price to pay would be the expected value $E(X)$ of the random variable X that represents the payoff of a randomly selected ticket. This is because if the lottery were held a large number of times and you paid $E(X)$ dollars for each ticket, then your average winnings or losses would be approximately 0.

Computing the payoff probabilities, we find that

$$p(1,000) = P(X = 1,000) = \frac{1}{10,000} \qquad \text{since only 1 ticket of the 10,000 wins \$1,000}$$

$$p(500) = P(X = 500) = \frac{3}{10,000} \qquad \text{since 3 of the 10,000 tickets win \$500}$$

$$p(100) = P(X = 100) = \frac{5}{10,000} \qquad \text{since 5 of the 10,000 tickets win \$100}$$

$$p(0) = P(X = 0) = \frac{9,991}{10,000} \qquad \text{since the other 9,991 tickets win nothing}$$

Using this information, we find the expected value of X is

$$E(X) = 0 \cdot p(0) + 100 \cdot p(100) + 500 \cdot p(500) + 1,000 \cdot p(1,000)$$

$$= 0\left(\frac{9,991}{10,000}\right) + 100\left(\frac{5}{10,000}\right) + 500\left(\frac{3}{10,000}\right) + 1,000\left(\frac{1}{10,000}\right)$$

$$= \frac{3}{10}$$

It follows that as more and more of the 10,000 available tickets are purchased, the average amount to be won gets closer and closer to $0.30. Therefore, a fair price for your ticket is 30 cents.

Variance and Standard Deviation

After we have found the mean (expected value) of a random variable X, we can learn more about X by examining how close its values are to the mean. Are the values clustered close to $E(X)$ or are they dispersed? The **variance** of X provides a measure of the tendency of the values of X to spread out from the mean $E(X)$.

For a discrete random variable X with values $x_1, x_2, \ldots, x_n$ and respective probabilities $p(x_1), p(x_2), \ldots, p(x_n)$, the variance is defined as

$$\text{Var}(X) = \sum_{k=1}^{n} (x_k - E(X))^2 p(x_k)$$

Since the variance is a sum of squares, the units in which the values x_j are expressed will be squared in the variance. For this reason, it is sometimes more convenient to measure the spread of X by the *square root* of the variance, which is called the **standard deviation** of X and is denoted by $\sigma(X)$; that is,

$$\sigma(X) = \sqrt{\text{Var}(X)}$$

Variance and Standard Deviation of a Discrete Random Variable ■
Let X be a discrete random variable with values $x_1, x_2, \ldots, x_n$ and expected value $E(X) = \mu$, and let p be the probability function for X, so that $p(x_k) = P(X = x_k)$ for $k = 1, 2, \ldots, n$. Then the **variance** of X, denoted by $\text{Var}(X)$, is the weighted sum

$$\text{Var}(X) = (x_1 - \mu)^2 \cdot p(x_1) + (x_2 - \mu)^2 \cdot p(x_2) + \cdots + (x_n - \mu)^2 \cdot p(x_n)$$

$$= \sum_{k=1}^{n} (x_k - \mu)^2 \cdot p(x_k)$$

The square root of the variance is called the **standard deviation** of X and is denoted by $\sigma(X)$; that is,

$$\sigma(X) = \sqrt{\text{Var}(X)}$$

Note that the variance $\text{Var}(X)$ of a discrete random variable X with values $x_1, x_2, \ldots, x_n$ and probability assignments $p(x_1), \ldots, p(x_n)$ is the weighted sum of squares $(x_k - \mu)^2 p(x_k)$ of the differences between each x_k and the mean $\mu = E(X)$. In this sense, the variance measures the deviation of X from its mean μ. You may wonder why this deviation is not measured by simply adding the weighted differences $(x_k - \mu)p(x_k)$ without squaring, but such a sum $\sum_{k=1}^{n} (x_k - \mu)p(x_k)$ will always be 0 (see Exercise 50).

EXAMPLE 11.1.8   Finding Variance and Standard Deviation

Continuing Examples 11.1.4 and 11.1.6, find the variance and standard deviation of the random variable X that measures the performance test score achieved by a randomly selected worker performing a task three times in succession.

Solution

In Example 11.1.6, we showed that the expected value of X satisfies $E(X) = 1$. Therefore, the variance of X is given by

$$\text{Var}(X) = (-5 - 1)^2 \cdot \frac{1}{8} + (1 - 1)^2 \cdot \frac{3}{8} + (2 - 1)^2 \cdot \frac{3}{8} + (4 - 1)^2 \cdot \frac{1}{8} = 6$$

and the standard deviation is $\sigma = \sqrt{\text{Var}(X)} = \sqrt{6} \approx 2.45$. The relatively large variance (compared to the mean) indicates that the values of the random variable X are not clustered especially near its expected value.

Just-In-Time **REVIEW**

If you need a quick review of infinite series, revisit Section 10.1.

An *infinite discrete random variable* X is one whose values $x_1, x_2, \ldots$ form an infinite sequence. If the probability assignment for such a random variable is $p(x_k) = P(X = x_k)$, then the expected value of X is given by the infinite series

$$E(X) = x_1 p(x_1) + x_2 p(x_2) + \cdots = \sum_{k=1}^{\infty} x_k p(x_k)$$

If we set $\mu = E(X)$, then the variance of such a random variable X is given by the infinite series

$$\text{Var}(X) = (x_1 - \mu)^2 \cdot p(x_1) + (x_2 - \mu)^2 \cdot p(x_2) + \cdots = \sum_{k=1}^{\infty} (x_k - \mu)^2 \cdot p(x_k)$$

and as before, the standard deviation of X is given by $\sigma(X) = \sqrt{\text{Var}(X)}$. These formulas will be used in the following discussion of geometric random variables.

Geometric Random Variables

Certain random experiments that occur in modeling can be characterized as having two possible outcomes, traditionally called **success** and **failure,** and are carried out repeatedly until the first success is achieved. For example, a coin may be tossed repeatedly until tails first appears, or newly manufactured computer chips may be examined until a defective chip is found. In such situations, the random variable X that counts the number of repetitions required to produce the first success is known as a **geometric random variable.**

A geometric random variable can take on any positive integer value, so it is discrete but infinite. For any positive integer n, the probability that $X = n$ is the probability that the first $n - 1$ trials of the experiment result in failure and the nth trial is a success:

$$\underbrace{\text{FF} \cdots \text{F}}_{\substack{n - 1 \\ \text{failures}}} \cdot \underbrace{\text{S}}_{\substack{\text{first} \\ \text{success}}}$$

If, on each trial, the probability of success is p, then the probability of failure must be $1 - p$, so the probability of $n - 1$ failures followed by 1 success is the product

$$[\underbrace{(1 - p)(1 - p) \cdots (1 - p)}_{n - 1 \text{ failures}}] \cdot \underbrace{p}_{\text{success}} = (1 - p)^{n-1} p$$

That is,

$$P(X = n) = (1 - p)^{n-1} p$$

If $p = 1$, success occurs immediately, while if $p = 0$, success is impossible. Otherwise, where $0 < p < 1$, we expect success to eventually occur if the experiment is repeated indefinitely, so $P(X \geq 1)$ ought to be 1. To verify that $P(X \geq 1) = 1$, we compute this probability by adding all the probabilities $P(X = n)$ for $n = 1, 2, \ldots$. This leads to the following infinite geometric series, which we sum using the geometric series formula of Section 10.1:

$$
\begin{aligned}
P(X \geq 1) &= \sum_{n=1}^{\infty} P(X = n) \\
&= \sum_{n=1}^{\infty} (1 - p)^{n-1} p \\
&= p + (1 - p)p + (1 - p)^2 p + \cdots \\
&= p[1 + (1 - p) + (1 - p)^2 + \cdots] \\
&= p \sum_{n=0}^{\infty} (1 - p)^n \\
&= \frac{p}{1 - (1 - p)} = \frac{p}{p} = 1
\end{aligned}
$$

Note that in the last step, we have used the geometric series formula $\displaystyle\sum_{n=0}^{\infty} ar^n = \frac{a}{1-r}$ with $a = p$ and $r = 1 - p$, which is valid since $0 < p < 1$ means that $0 < 1 - p < 1$.

The random variable X such that $P(X = n) = (1 - p)^{n-1}p$ is called a **geometric random variable with parameter p.** The histogram for the first seven values of a geometric random variable with parameter $p = 0.5$ is shown in Figure 11.3. Using techniques for working with power series developed in Chapter 10 (see Exercise 49), it can be shown that a geometric random variable X with parameter p has expected value

$$E(X) = \frac{1}{p}$$

Similar, but slightly more complicated computations can be used to show that X has variance

$$\mathrm{Var}(X) = \frac{1-p}{p^2}$$

Applications of geometric random variables are examined in Examples 11.1.9 and 11.1.10.

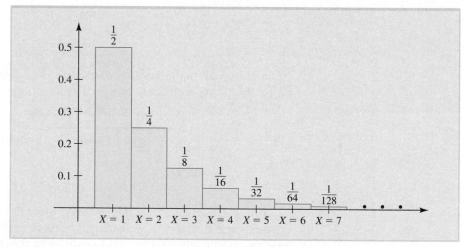

FIGURE 11.3   The histogram for the geometric random variable with parameter $p = 0.5$.

EXAMPLE 11.1.9   Studying Quality Control

In an industrial process, items are sampled repeatedly until a defective item is found. The company claims that any item of its product is 100 times more likely to be effective than defective. Assuming the company's claim is accurate, find the following:

a. The probability that a randomly chosen item is defective.

b. The probability that it takes exactly 100 samplings before a defective item is found.

c. The expected number of samplings before a defective item is found.

Solution

a. If p is the probability of a defective item, then $1 - p$ is the probability of an effective item, and the company claims that

$$1 - p = 100p$$

so that

$$p = \frac{1}{101} \approx 0.0099$$

b. Let X be the geometric random variable that counts the number of items that must be sampled before a defective item is found. The probability that it takes exactly 100 samplings before a defective item is found is

$$P(X = 100) = (1 - p)^{100-1} \cdot p \approx (1 - 0.0099)^{99}(0.0099) \approx 0.0037$$

c. Using the formula $E(X) = \frac{1}{p}$ for the expected value of a geometric random variable of parameter p, we find that since $p = \frac{1}{101}$, the number of expected samplings is

$$E(p) = \frac{1}{1/101} = 101$$

That is, if the random sampling is carried out repeatedly, on average we would expect to find the first defective item on the 101st sampling.

EXAMPLE 11.1.10 Finding Probability and Expected Value

Suppose that a fair coin is tossed repeatedly until it comes up heads. Find each of the following quantities:

a. The probability that we have to toss the coin exactly three times until heads comes up.

b. The probability that the coin comes up heads within the first three tosses.

c. The probability that the coin must be tossed more than three times to come up heads.

d. The expected number of times we have to toss the coin before a head appears.

EXPLORE!

Some graphing calculators have built-in probability density functions. On a TI-84 Plus, press **2nd VARS** to obtain the **DISTR** screen, which lists both continuous and discrete probability density functions. Find the probabilities in parts (a) and (b) of Example 11.1.10 using **E:geometpdf(** and **F:geometcdf(.** Do you get the same values as in the example?

Solution

Let X be the random variable that represents the number of times we have to toss the coin until it comes up heads. This is a geometric random variable with parameter $p = \frac{1}{2}$, so

$$P(X = n) = \left(1 - \frac{1}{2}\right)^{n-1}\left(\frac{1}{2}\right) = \left(\frac{1}{2}\right)^{n}$$

a. The required probability is $P(X = 3) = \left(\frac{1}{2}\right)^{3} = \frac{1}{8}$.

b. The coin comes up heads within the first three tosses if a head appears on toss 1, toss 2, or toss 3. This is the event $X \leq 3$, and we have

$$P(X \leq 3) = P(X = 1) + P(X = 2) + P(X = 3)$$
$$= \frac{1}{2} + \left(\frac{1}{2}\right)^{2} + \left(\frac{1}{2}\right)^{3} = \frac{1}{2} + \frac{1}{4} + \frac{1}{8}$$
$$= \frac{7}{8}$$

c. The event $X > 3$ that the coin must be tossed more than three times to come up heads is the complement of the event $X \leq 3$ that it comes up heads within the first three tosses, so we must have $P(X > 3) = 1 - P(X \leq 3)$. Since we found that $P(X \leq 3) = \dfrac{7}{8}$ in part (b), it follows that

$$P(X > 3) = 1 - P(X \leq 3) = 1 - \frac{7}{8} = \frac{1}{8}$$

d. Using the formula for the expected value of a geometric random variable with parameter $p = \dfrac{1}{2}$, we find that $E(X) = \dfrac{1}{p} = \dfrac{1}{1/2} = 2$. Consequently, if we carry out this random experiment repeatedly, on average we should expect to have to toss the coin only twice before a head appears.

EXAMPLE 11.1.11   Finding Probability in a Board Game

In a certain two-person board game played with 1 die, each player needs to roll a 6 to start play. You and your opponent take turns rolling the die. Find the probability that you will get the first 6 if you roll first.

Solution

In this context, *success* is getting a 6 and *failure* is getting any of the other five possible numbers. Hence, the probability of success is $p = \dfrac{1}{6}$ and the probability of failure is $1 - p = \dfrac{5}{6}$. Let X be the random variable that counts the number of the roll on which the first 6 comes up. The first 6 will occur on one of your rolls if X is odd and on one of your opponent's rolls if X is even. Hence, the probability that you get the first 6 is

$$P(X \text{ is odd}) = P(X = 1) + P(X = 3) + P(X = 5) + \cdots$$

$$= \frac{1}{6} + \left(\frac{5}{6}\right)^2\left(\frac{1}{6}\right) + \left(\frac{5}{6}\right)^4\left(\frac{1}{6}\right) + \cdots$$

$$= \frac{1}{6} + \left(\frac{25}{36}\right)\left(\frac{1}{6}\right) + \left(\frac{25}{36}\right)^2\left(\frac{1}{6}\right) + \cdots$$

$$= \frac{1}{6}\sum_{n=0}^{\infty}\left(\frac{25}{36}\right)^n$$

$$= \frac{1}{6}\left(\frac{1}{1 - \dfrac{25}{36}}\right) = \frac{1}{6}\left(\frac{36}{11}\right) = \frac{6}{11} \approx 0.5455$$

EXERCISES ■ 11.1

In each of Exercises 1 through 8, determine whether the given random variable X is discrete or continuous.

1. X counts the number of tails that come up when a coin is tossed 10 times.

2. X measures the time a randomly selected flashlight battery works before failing.

3. X measures the weight of a randomly selected member of your class.

4. X measures the income in dollars of a randomly selected person in your community.

5. X measures the living area of a randomly selected house in your neighborhood.

6. X measures the length of time a randomly selected traveler has to wait until the next airport shuttle arrives.

7. X counts the number of players on a hockey team who are injured in a randomly selected game.

8. X counts the number of birds in a randomly selected flock.

9. Consider the random experiment of tossing a fair coin three times, and let X be the random variable that counts the number of heads that appear. Find the probability distribution for X, and sketch its histogram.

10. A deck of 10 cards has four cards labeled with a 1, three cards with a 2, two cards with a 3, and a single card labeled 4. A card is picked at random from the deck, and X is the discrete random variable that measures the number on the card. Find the probability distribution for X, and sketch its histogram.

In Exercises 11 through 14, a die is rolled twice. In each case, describe the sample space for the indicated random experiment.

11. Observe how many 6s are thrown.

12. Observe how many 1s and 4s are thrown.

13. Observe how many even numbers are thrown.

14. Observe the sum of the numbers on the two throws.

15. A box contains 4 tickets numbered 1 through 4. A ticket is selected at random from the box, a die is rolled, and the numbers on the ticket and die are recorded.

a. List the sample space for this experiment.

b. List the outcomes that make up the event that the sum of the numbers on the ticket and die is even.

c. List the outcomes that make up the event that the number on the ticket is the same as the number on the die.

16. A number between 15 and 49, inclusive, is selected at random. List the outcomes that make up each of the following events.

a. The two digits of the selected number are the same.

b. The sum of the digits of the selected number is 8.

17. Consider the random experiment of tossing a fair coin four times, and let X denote the random variable that counts the number of times tails appears.

a. List all possible outcomes of this experiment, and the value that X assigns to each outcome.

b. List the outcomes that make up the event $(X = 3)$.

c. List the outcomes that make up the event $(X < 4)$.

d. Find each of these probabilities:

$$P(X = 0), P(X \le 3), \text{ and } P(X > 3)$$

18. Consider the random experiment of rolling a fair die three times, and let X denote the random variable that counts the number of times a 6 appears.

a. List all possible outcomes of this experiment and the value that X assigns to each outcome.

b. List the outcomes that make up the event $(X = 2)$.

c. List the outcomes that make up the event $(X < 3)$.

d. Find each of these probabilities:

$$P(X = 0), P(X \le 2), \text{ and } P(X > 2)$$

In Exercises 19 through 22, the outcomes and corresponding probability assignments for a discrete random variable X are given. In each case, find the expected value E(X), variance Var(X), and standard deviation σ(X).

19.

Outcomes for X	0	1	2	4
Probability	1/8	1/8	1/4	1/2

20.

Outcomes for X	1	2	3	5
Probability	1/4	1/3	1/6	1/4

21.

Outcomes for X	0	2	4	6
Probability	1/7	3/7	2/7	1/7

22.

Outcomes for X	0	1	2	$\cdots$	k	$\cdots$
Probability	1/2	1/4	1/8	$\cdots$	$1/2^{k+1}$	$\cdots$

23. Draw the probability histogram for the random variable X in Exercise 17. Then find the expected value, variance, and standard deviation for X.

24. Draw the probability histogram for the random variable X in Exercise 18. Then find the expected value, variance, and standard deviation for X.

BUSINESS AND ECONOMICS APPLIED PROBLEMS

25. PORTFOLIO MANAGEMENT   Louise estimates that the probability is 0.6 that a certain stock in her portfolio will go up 25% in price. She also figures that the probability is 0.25 that it will go up only 10%, and that the probability is 0.15 that it will go down 10%. What percentage return should she expect from this stock?

26. CAR SALES   Ken King, the owner of King's Kars, determines that the number of daily new car sales may be measured by the discrete random variable X with this probability distribution:

$$P(X = 0) = 0.05, P(X = 1) = 0.15,$$
$$P(X = 2) = 0.20, P(X = 3) = 0.25,$$
$$P(X = 4) = 0.12, P(X = 5) = 0.13,$$
$$P(X = 6) = 0.05, P(X = 7) = 0.02,$$
$$P(X = 8) = 0.03$$

a. Describe the event $(X = x)$.
b. Draw the histogram for this random variable.
c. How many new cars should Ken expect to sell on a randomly selected day?
d. Find the variance and standard deviation of Ken's daily new car sales.

27. RELIABILITY   Suppose that a particular printer at your school has a 99.95% chance of successfully printing out a randomly selected page in a document and a 0.05% chance of jamming.
a. Find the probability that the printer will jam on the tenth page of a 10-page document after printing out the first 9 pages without jamming.
b. On what page would you expect the printer to jam when a large document is printed out?
c. Determine the probability that the printer completes a document of 25 pages without jamming.

28. QUALITY CONTROL   Three inspectors take turns checking electronic components as they come off an assembly line. If 10% of all components produced on the assembly line are defective, find the probability that the inspector who checks the first (and the fourth, seventh, and

so on) component will be the one who finds the first defective component.

29. RELIABILITY   Suppose that 7% of the computer chips manufactured by a certain company are defective. A quality control inspector tests each chip before it goes out of the factory. Using testing equipment, this inspector can always detect a defective chip.
a. Find the probability that the inspector passes exactly four chips as good before detecting a defective chip.
b. Find the probability that the inspector passes at most four chips as good before detecting a defective chip.
c. Find the probability that the inspector passes at least four chips as good before detecting a defective chip.
d. Find the expected number of chips the inspector examines before detecting a defective chip.

30. WORKER EFFICIENCY   Mingxia, an industrial efficiency expert, conducts a test in which a worker is required to perform a particular task three times in succession. Each time the task is performed successfully, the worker is awarded 1 point, but if the task is performed incorrectly more than once, the worker loses 2 points for each unsuccessful performance. Assume success and failure are equally likely.
a. Find the probability distribution for the discrete random variable that measures the score of a randomly selected worker.
b. What score should Mingxia expect a randomly selected worker to receive on this test?

LIFE AND SOCIAL SCIENCE APPLIED PROBLEMS

31. HIGHWAY SAFETY   The highway department kept track of the number of accidents on a 10-mile stretch of a major highway every day during the month of September. They found that there were:

6 days with no accidents
8 days with one accident
7 days with two accidents
2 days with three accidents
3 days with four accidents
0 days with five accidents
2 days with six accidents
0 days with seven accidents
1 day with eight accidents
1 day with nine accidents

Let X be the random variable that represents the number of accidents on a randomly chosen day in September.
a. Construct a histogram for X.
b. Find the expected value $E(X)$, and give an interpretation of this number.
c. Find the variance and standard deviation of X.

32. **JURY SELECTION** Potential jurors from a large jury pool are examined one at a time by the defense and prosecution. Suppose that a person selected at random from the jury pool has probability $p = 0.25$ of being disqualified.
a. Find the probability that the third potential juror examined is the first to be disqualified.
b. Find the probability that at least one of the first three jurors is disqualified.
c. Find the probability that a jury of 12 people can be seated without any potential jurors being disqualified.

33. **AIRLINE TRAVEL** Let X represent the number of passengers bumped from a randomly chosen flight on a particular commercial airline. Suppose the probability $P(X = N)$ that N people (for $N = 0$ to 6) are bumped is given in the table.

N	0	1	2	3
$P(X = N)$	0.961	0.015	0.009	0.005

4	5	6
0.004	0.004	0.002

Based on this information, how many people on average would you expect to be bumped from a randomly chosen flight?

34. **EFFECT OF AN EPIDEMIC** In Example 11.1.5, the geometric random variable X measures the number of students in a particular dormitory who are ill with the flu in April. Find the variance and standard deviation of X.

35. **POPULATION STUDIES** Suppose 4% of the students at a large university have red hair. You interview a number of students, one at a time.
a. Find the probability that the fifth student interviewed is the first to have red hair.
b. Find the probability that at least one student with red hair is among the first five students interviewed.
c. Find the probability that there are no red-haired students among the first 12 interviewees.
d. On average, how many students would you expect to interview before finding one with red hair?

MISCELLANEOUS PROBLEMS

36. **GAMES OF CHANCE** You and a friend take turns rolling a die until one of you wins by rolling a 3 or a 4. If your friend rolls first, find the probability that you will win.

37. **NUMISMATICS** In 1959, the design on the reverse side of U.S. pennies was changed from one depicting a pair of wheat stalks to one depicting the Lincoln Memorial. Today, very few of the old pennies are still in circulation. If only 0.02 of the pennies now in circulation were minted before 1959, find the probability that you will have to examine at least 100 pennies to find one with the old design. (Incidentally, there is an easy way to do this exercise that does not involve infinite series. For practice, do it both ways.)

38. **BIASED COINS** A biased coin that comes up heads with probability 0.55 is tossed repeatedly until it comes up heads. Let X be the random variable that measures the number of times we have to toss the coin until it comes up heads. Find:
a. The probability that we have to toss the coin exactly three times until heads comes up.
b. The probability that the coin comes up heads within the first three tosses.
c. The probability that the coin must be tossed more than three times to come up heads.
d. The expected number of times we have to toss the coin until it comes up heads.
e. The variance of the number of times we have to toss the coin until it comes up heads.

39. **LOTTERY** Suppose that the grand prize in a lottery is $1,000,000. There are also 5 second prizes of $100,000 and 50 third prizes of $10,000. If 100,000 tickets are sold, what is a fair price to pay for a ticket to this lottery?

40. **CASINO GAMES** A roulette wheel has 18 red numbers, 18 black numbers, and 2 green numbers. Each time the wheel is spun, George bets that the ball will land on a red number. When the ball lands on red, he wins $1 from the casino for every $1 bet. When the ball lands on a black or green number, he loses his bet. Find his expected winnings from a $1 bet on red.

41. **CASINO GAMES** George, the roulette player in Exercise 40, finds another wheel with the same number of red and black numbers but with only 1 green number. Now what are his expected winnings from a $1 bet on red?

42. BOARD GAMES There are 100 tiles in the board game Scrabble; 98 of these tiles have letters of the English alphabet printed on them and two are blank. Each tile also has a value, depending on the letter printed on it. Blank tiles have zero value. This table shows the number of tiles having each value:

Value	Number of Tiles
0	2
1	68
2	7
3	8
4	10
5	1
6	0
7	0
8	2
9	0
10	2

a. Let X be the random variable that represents the value of a randomly selected Scrabble tile from all 100 tiles. Draw the histogram of X.

b. Find $E(X)$, the expected value of a randomly selected Scrabble tile.

FAIR BETS AND ODDS *A gaming bet is said to be* **fair** *if, on average, a player's expected winnings are 0 (no loss or gain). The bet has* m *to 1* **odds** *if for every $1 wagered, a winning player gains* m *dollars (in addition to the return of the $1 bet).* **Fair odds** *on a bet are the odds that make the bet fair. This terminology is used in Exercises 43 through 45.*

---

43. For a certain game, the odds are w to 1, and the probability of winning a particular bet is p.

a. Let X be the random variable representing a player's winnings when betting $1. Express the expected value $E(X)$ in terms of w.

b. Show that the fair odds for this bet are $\dfrac{1-p}{p}$ to 1.

44. Use the formula found in Exercise 43 to find fair odds for the game of roulette with two green numbers as described in Exercise 40.

45. Use the formula found in Exercise 43 to find fair odds for the game of roulette with one green number as described in Exercise 41.

46. Let X be a geometric random variable with parameter $p = 0.4$. Compute $P(X = n)$ for $n = 1$ to 5, and then draw the histogram for the first five values of X.

47. Let X be a geometric random variable with parameter $p = 0.8$. Compute $P(X = n)$ for $n = 1$ to 5, and then draw the histogram for the first five values of X.

48. Show that for a geometric random variable X, the most likely number of trial runs is one. That is, $P(X = n)$ takes on its largest value when $n = 1$, as n ranges over all positive integers.

49. In Section 10.3, it is shown that

$$1 + 2x + 3x^2 + \cdots = \frac{1}{(1-x)^2} \text{ for } -1 < x < 1$$

Use this fact to show that a geometric random variable X with parameter p has expected value $E(X) = \dfrac{1}{p}$.

50. Let X be a discrete random variable with values $x_1, x_2, \ldots, x_n$ and probability assignments $p(x_1) = p_1, \ldots, p(x_n) = p_n$. If $\mu = E(X)$ is the expected value of X, show that

$$(x_1 - \mu)p_1 + (x_2 - \mu)p_2 + \cdots + (x_n - \mu)p_n = 0$$

A random variable X follows a **zeta distribution** *if there are positive integers a and b with b > 1 such that the event (X = n) has probability*

$$P(X = n) = \frac{a}{n^b}$$

for all positive integers n. For instance, the frequency of occurrence of words in English ranked according to their use approximately follows a zeta distribution. This definition is used in Exercises 51 and 52.

---

51. Let X be a random variable with a zeta distribution. Use the requirement that

$$\sum_{n=1}^{\infty} P(X = n) = 1$$

to show that $a = \dfrac{1}{\zeta(b)}$, where

$$\zeta(b) = \sum_{n=1}^{\infty} \frac{1}{n^b}$$

52. THE INTERNET A website on the Internet is said to have *rank n* if it is the nth most popular site. Consider the random experiment in which someone currently surfing the Web selects a website, and let X be the random variable that measures the rank of the selected website. Recent studies suggest that X follows a zeta distribution with parameter $b = 2$.

a. It can be shown that

$$\zeta(2) = \sum_{n=1}^{\infty} \frac{1}{n^2} = \frac{\pi^2}{6}$$

Use this fact and the result of Exercise 51 to find a formula for the probability $P(X = n)$ that a

randomly chosen visit to the Internet is to the nth most popular site.

b. Find the probability that a random visit to a website is to the fifth most popular site.

c. Find the probability that a random visit to a website is to one of the three most popular sites.

d. Find the probability that a random visit to a website is to a site not ranked among the four most popular.

e. What expression represents the expected value of X? Use your calculator to estimate $E(X)$, and interpret your result.

SECTION 11.2 Continuous Probability Distributions

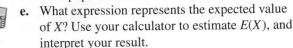

Learning Objectives

1. Define and examine continuous probability density functions.
2. Use uniform and exponential probability distributions.
3. Study joint probability distributions.

In Section 11.1, we studied discrete random variables as a means for introducing basic probabilistic concepts and procedures. Beginning with this section, our focus will be on continuous random variables and the use of calculus in the study of probability.

Recall that a continuous random variable is one that can take on any value in an interval. An important difference between discrete and continuous probability distributions is the manner in which probability is assigned. For instance, suppose X is a continuous random variable that measures the average weekly rainfall in a particular city. Assume for simplicity that X can assume any value between 0 (no rain) and 10 inches (storm). Listing probabilities for X as we did for the discrete random variables in Section 11.1 is unrealistic because there are too many values (possible rainfall levels) between 0 and 10 to list. It makes more sense to ask questions like: What is the probability that X lies between 4 and 5 inches? or "How likely is X to be greater than 6 inches?"

Suppose the interval $[0, 10]$ is divided into five subintervals of equal length and the probability that X lies in each subinterval is as follows:

$$P(0 \leq X \leq 2) = 0.15$$
$$P(2 \leq X \leq 4) = 0.25$$
$$P(4 \leq X \leq 6) = 0.30$$
$$P(6 \leq X \leq 8) = 0.20$$
$$P(8 \leq X \leq 10) = 0.10$$

We can construct a histogram with five vertical rectangles based on these five subintervals, each having area equal to the probability that X lies between the endpoints of the base subinterval. For instance, the third rectangle is based on the subinterval $[4, 6]$ and has area equal to $P(4 \leq X \leq 6) = 0.30$. The histogram is shown in Figure 11.4 along with a curve passing through the midpoint of the top bar of each rectangle.

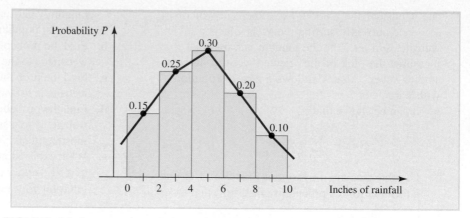

FIGURE 11.4   The histogram for the probability function of the continuous random variable measuring weekly rainfall.

If we increase the number of subintervals while retaining the general plan of construction, we obtain a histogram like the one shown in Figure 11.5a. Now the curve passing through the top of each approximating rectangle has a clear form, and if we allow the number of subdivisions to tend toward infinity, the curve approaches the graph of a function $f(x)$, which is called the **probability density function** of the random variable X.

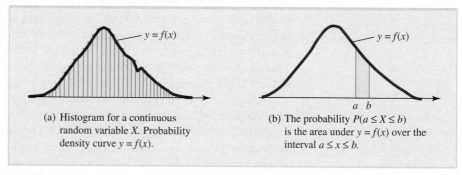

FIGURE 11.5   The histogram and probability density function for a continuous random variable X.

Since the area of each approximating rectangle is equal to the probability that X lies between the endpoints of the subinterval that forms its base, it is reasonable to expect the probability that X lies between any two values a and b to be given by the area under the density function over the interval $a \leq x \leq b$ (Figure 11.5b). In Chapter 5, we found that such an area is given by an integral. For instance, the probability that the weekly rainfall is between 4.12 and 5.3 inches is given by

$$P(4.12 \leq X \leq 5.3) = \int_{4.12}^{5.3} f(x)\, dx$$

A probability density function is something like the population density functions studied in Chapters 5 through 7, in which likelihood of occurrence plays the role of "population." If we know the population density function over a region, we can find the population in a portion of the region by integrating. Likewise, if the probability density function $f(x)$ for a particular continuous random variable X is known, we can find the probability of an event E by integrating $f(x)$ over the interval that defines E. Here is a summary of the general procedure for computing probability associated with a continuous random variable.

Continuous Probability Density Function and Distribution ■ A probability density function for the continuous random variable X is a function f that satisfies these conditions:

1. $f(x) \geq 0$ for all x

2. $\displaystyle\int_{-\infty}^{+\infty} f(x)\, dx = 1$

3. The probability that X lies in the interval $a \leq X \leq b$ is given by the integral

$$P(a \leq X \leq b) = \int_{a}^{b} f(x)\, dx$$

This probability assignment for X is called a **continuous probability distribution.**

The values of a and b in this formula need not be finite, and if either is infinite, the corresponding probability is given by an improper integral. For example, the probability that $X \geq a$ is

$$P(X > a) = P(a \leq X < +\infty) = \int_{a}^{+\infty} f(x)\, dx$$

The second condition in the definition of a probability density function expresses the fact that since X must have some value, the event $-\infty < X < +\infty$ is certain to occur. Thus, we have $P(-\infty < X < +\infty) = 1$, which means that

$$\int_{-\infty}^{+\infty} f(x)\, dx = 1$$

where the improper integral on the left is defined as

$$\int_{-\infty}^{\infty} f(x)\, dx = \lim_{N \to +\infty} \int_{-N}^{0} f(x)\, dx + \lim_{N \to +\infty} \int_{0}^{N} f(x)\, dx$$

and converges only if both limits on the right exist. Examples 11.2.1 and 11.2.2 illustrate how these criteria can be used to determine whether a given function is a probability density function.

EXAMPLE 11.2.1 Recognizing a Probability Density Function

Determine whether the following function is a probability density function:

$$f(x) = \begin{cases} \dfrac{3}{2}x(x - 1) & \text{for } 0 \leq x \leq 2 \\ 0 & \text{otherwise} \end{cases}$$

Solution

We find that

$$\int_{-\infty}^{\infty} f(x)\, dx = \int_{0}^{2} \frac{3}{2} x(x-1)\, dx = \frac{3}{2}\left[\frac{1}{3}x^3 - \frac{1}{2}x^2\right]_{0}^{2}$$

$$= \frac{3}{2}\left[\frac{1}{3}(2)^3 - \frac{1}{2}(2)^2\right] - \frac{3}{2}[0-0] = 1$$

However, the condition $f(x) \geq 0$ for all x is not satisfied [$f(x) < 0$ for $0 < x < 1$], so $f(x)$ is *not* a probability density function.

EXAMPLE 11.2.2 Constructing a Probability Density Function

Either find a number k such that the following function $f(x)$ is a probability density function or explain why no such number exists:

$$f(x) = \begin{cases} k(x^3 + x + 1) & \text{for } 1 \leq x \leq 3 \\ 0 & \text{otherwise} \end{cases}$$

Solution

First, note that if $k \geq 0$, then $f(x)$ is positive on the interval $1 \leq x \leq 3$, which means $f(x) \geq 0$ for all x. Next, we find that

$$\int_{-\infty}^{\infty} f(x)\, dx = \int_{1}^{3} k(x^3 + x + 1)\, dx = k\left[\frac{1}{4}x^4 + \frac{1}{2}x^2 + x\right]_{1}^{3}$$

$$= k\left\{\left[\frac{1}{4}(3)^4 + \frac{1}{2}(3)^2 + 3\right] - \left[\frac{1}{4}(1)^4 + \frac{1}{2}(1)^2 + 1\right]\right\} = 26k$$

Thus, $f(x)$ is a probability density function if $26k = 1$, that is, if $k = \dfrac{1}{26}$.

There are many different kinds of continuous probability density functions, and choosing an appropriate such function for a particular random experiment often involves special insights and techniques developed in courses in probability and statistics. We next examine uniform and exponential density functions, two special forms that prove to be particularly useful in modeling applications.

Uniform Density Functions

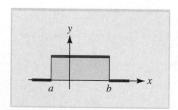

FIGURE 11.6 A uniform density function.

A **uniform probability density function** (Figure 11.6) is constant over a bounded interval $a \leq x \leq b$ and zero outside that interval. A random variable that has a uniform density function is said to be **uniformly distributed.** Roughly speaking, for a uniformly distributed random variable, all values in some bounded interval are "equally likely." More precisely, a continuous random variable is uniformly distributed if the probability that its value will be in a particular subinterval of the bounded interval is equal to the probability that it will be in any other subinterval that has the same length. An example of a uniformly distributed random variable is the waiting time of a motorist at a traffic light that remains red for, say, 40 seconds at a time. This random variable has a uniform distribution because all waiting times between 0 and 40 seconds are equally likely.

If k is the constant value of a uniform density function $f(x)$ on the interval $a \leq x \leq b$, its value may be determined using the requirement that the total area under the graph of f must be equal to 1. In particular,

$$1 = \int_{-\infty}^{+\infty} f(x)\, dx = \int_a^b f(x)\, dx \qquad \text{since } f(x) = 0 \text{ outside the interval } a \leq x \leq b$$

$$= \int_a^b k\, dx = kx \Big|_a^b = k(b - a)$$

and so

$$k = \frac{1}{b - a}$$

This observation leads to the following formula for a uniform density function.

Uniform Density Function

$$f(x) = \begin{cases} \dfrac{1}{b - a} & \text{if } a \leq x \leq b \\ 0 & \text{otherwise} \end{cases}$$

EXAMPLE 11.2.3 Using a Uniform Distribution

A certain traffic light remains red for 40 seconds at a time. Brie arrives (at random) at the light and finds it red. Use an appropriate uniform density function to find the probability that she will have to wait at least 15 seconds for the light to turn green.

Solution

Let X denote the random variable that measures the time (in seconds) that Brie must wait. Since all waiting times between 0 and 40 are equally likely, X is uniformly distributed over the interval $0 \leq x \leq 40$. The corresponding uniform density function is

$$f(x) = \begin{cases} \dfrac{1}{40} & \text{if } 0 \leq x \leq 40 \\ 0 & \text{otherwise} \end{cases}$$

and the desired probability is

$$P(15 \leq X \leq 40) = \int_{15}^{40} \frac{1}{40}\, dx = \frac{1}{40}x \Big|_{15}^{40} = \frac{40}{40} - \frac{15}{40} = \frac{5}{8}$$

Exponential Density Functions

An **exponential probability density function** is a function $f(x)$ that is zero for $x < 0$ and decreases exponentially for $x \geq 0$. That is,

$$f(x) = \begin{cases} Ae^{-\lambda x} & \text{for } x \geq 0 \\ 0 & \text{for } x < 0 \end{cases}$$

where A and λ are positive constants.

The value of A is determined by the requirement that the total area under the graph of f must be equal to 1. Thus,

$$1 = \int_{-\infty}^{+\infty} f(x)\, dx = \int_0^{+\infty} Ae^{-\lambda x}\, dx = \lim_{N \to +\infty} \int_0^N Ae^{-\lambda x}\, dx$$

$$= \lim_{N \to +\infty}\left(-\frac{A}{\lambda}e^{-\lambda x}\Big|_0^N\right) = \lim_{N \to +\infty}\left(-\frac{A}{\lambda}e^{-\lambda N} + \frac{A}{\lambda}\right) = \frac{A}{\lambda}$$

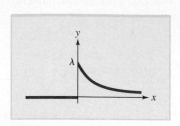

FIGURE 11.7   An exponential density function.

and so $A = \lambda$.

This calculation leads to the following general formula for an exponential density function. The corresponding graph is shown in Figure 11.7.

Exponential Density Function

$$f(x) = \begin{cases} \lambda e^{-\lambda x} & \text{if } x \geq 0 \\ 0 & \text{if } x < 0 \end{cases}$$

A random variable that has an exponential density function is said to be **exponentially distributed.** As you can see from the graph in Figure 11.7 the value of an exponentially distributed random variable is much more likely to be small than large. Such random variables include the life span of electronic components, the duration of telephone calls, and the interval between the arrivals of successive planes at an airport. Example 11.2.4 shows how to use an exponential distribution.

EXAMPLE 11.2.4   Using an Exponential Distribution

Let X be a random variable that measures the duration of cell phone calls in a certain city, and assume that X has an exponential distribution with density function

$$f(t) = \begin{cases} 0.5e^{-0.5t} & \text{if } t \geq 0 \\ 0 & \text{if } t < 0 \end{cases}$$

where t denotes the duration (in minutes) of a randomly selected call.

a. Find the probability that a randomly selected call lasts no more than 1 minute.

b. Find the probability that a randomly selected call will last at least 2 minutes.

Solution

a. The graph of the exponential density function $f(t)$ is shown in Figure 11.8. Since the probability density function X measures the duration of a cell phone call, the shape of the curve reflects the fact that most calls are relatively short. The event that a call lasts no more than 1 minute can be described as $0 \leq X \leq 1$. The probability $P(0 \leq X \leq 1)$ of this event is given by the area of the indicated shaded region in the figure and is computed by the integral

$$P(0 \leq X \leq 1) = \int_0^1 0.5e^{-0.5t}\, dt = -e^{-0.5t}\Big|_0^1$$

$$= -e^{-0.5(1)} - (-e^0) \approx 0.3935$$

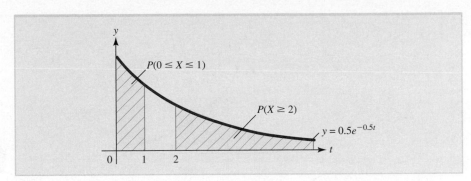

FIGURE 11.8   An exponential probability density function for the duration of cell phone calls.

b. There are two ways to compute this probability. The first method is to evaluate an improper integral.

$$P(X \geq 2) = P(2 \leq X < +\infty) = \int_{2}^{+\infty} 0.5e^{-0.5t}\, dt$$

$$= \lim_{N \to +\infty} \int_{2}^{N} 0.5e^{-0.5t}\, dt = \lim_{N \to +\infty} \left(-e^{-0.5t}\bigg|_{2}^{N} \right)$$

$$= \lim_{N \to +\infty} (-e^{-0.5N} + e^{-1}) = e^{-1} \approx 0.3679$$

Alternatively, since the duration of a call must be either less than 2 minutes or greater than or equal to 2 minutes, we have $P(X < 2) + P(X \geq 2) = 1$, so

$$P(X \geq 2) = 1 - P(X < 2)$$

$$= 1 - \int_{0}^{2} 0.5e^{-0.5t}\, dt = 1 - \left(-e^{-0.5t} \right)\bigg|_{0}^{2}$$

$$= 1 - (-e^{-1} + 1) = e^{-1} \approx 0.3679$$

Joint Probability Distributions

As we have seen, if $f(x)$ is the probability density function for the continuous random variable X, then the probability that X is between a and b is given by the formula

$$P(a \leq X \leq b) = \int_{a}^{b} f(x)\, dx$$

which can be interpreted geometrically as the area under the curve $y = f(x)$ over the interval $a \leq x \leq b$ (Figure 11.9a).

In situations involving two random variables X and Y, we compute probabilities by evaluating double integrals of a two-variable density function. In particular, we integrate a **joint probability density function** $f(x, y)$, which has these properties:

Just-In-Time **REVIEW**

If you need a quick review of working with double integrals, refer to Section 7.6.

1. $f(x, y) \geq 0$ for all points (x, y) in the xy plane

2. $\displaystyle\int_{-\infty}^{+\infty} \int_{-\infty}^{+\infty} f(x, y)\, dy\, dx = 1$

3. The probability that the ordered pair (X, Y) lies in a region R is given by

$$P[(X, Y) \text{ in } R] = \iint_{R} f(x, y)\, dA$$

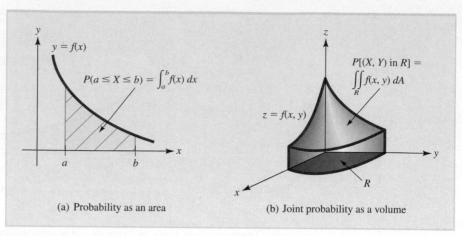

FIGURE 11.9   Geometric interpretation of probability.

This probability assignment is called the **joint probability distribution** for the density function $f(x, y)$.

Note that by property (3), the probability that X is between a and b and Y is between c and d is given by

$$P(a \leq X \leq b \text{ and } c \leq Y \leq d) = \int_c^d \int_a^b f(x, y)\, dx\, dy$$

$$= \int_a^b \int_c^d f(x, y)\, dy\, dx$$

In geometric terms, the probability $P[(X, Y) \text{ in } R]$ is the volume under the probability density surface $z = f(x, y)$ above the region R (Figure 11.9b), and condition (2) says that the total volume under the surface is 1.

The techniques for constructing joint probability density functions from experimental data are discussed in most texts on probability and statistics. Joint probability density functions are used to study product reliability, life expectancy depending on different factors, and a variety of other applications. Examples 11.2.5 and 11.2.6 illustrate how integration may be used to compute probability when the joint density function is known.

EXAMPLE 11.2.5   Using a Joint Distribution to Study Reliability

Smoke detectors manufactured by a certain firm contain two independent circuits, one manufactured at the firm's California plant and the other at the firm's plant in Ohio. Reliability studies suggest that if X measures the life span (in years) of a randomly selected circuit from the California plant and Y measures the life span (in years) of a randomly selected circuit from the Ohio plant, then the joint probability density function for X and Y is

$$f(x, y) = \begin{cases} 0.4e^{-x}e^{-0.4y} & \text{if } x \geq 0 \text{ and } y \geq 0 \\ 0 & \text{otherwise} \end{cases}$$

If the smoke detector will operate as long as either of its circuits is operating, find the probability that a randomly selected smoke detector will fail within 1 year.

Solution

Since the smoke detector will operate as long as either of its circuits is operating, it will fail within 1 year if and only if *both* of its circuits fail within 1 year. Thus, we want the probability that both $0 \leq X \leq 1$ and $0 \leq Y \leq 1$. The points (x, y) for which both of these inequalities hold lie inside and on the boundary of the square R shown in Figure 11.10. The corresponding probability is the double integral of the density function f over this region R. That is,

$$P(0 \leq X \leq 1 \text{ and } 0 \leq Y \leq 1) = \int_0^1 \int_0^1 0.4e^{-x}e^{-0.4y}\, dx\, dy$$

$$= \int_0^1 0.4e^{-0.4y}\left(\frac{e^{-x}}{-1}\right)\Big|_{x=0}^{x=1} dy$$

$$= 0.4(-e^{-1} + 1)\int_0^1 e^{-0.4y}\, dy$$

$$= 0.4(1 - e^{-1})\left(\frac{e^{-0.4y}}{-0.4}\right)\Big|_{y=0}^{y=1}$$

$$= (1 - e^{-1})(-e^{-0.4} + 1)$$

$$\approx 0.2084$$

Thus, it is approximately 21% likely that a randomly selected smoke detector will fail within 1 year.

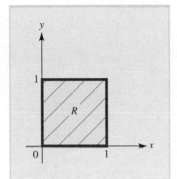

FIGURE 11.10 Square consisting of all points (x, y) for which $0 \leq x \leq 1$ and $0 \leq y \leq 1$.

EXAMPLE 11.2.6 Using a Joint Distribution to Study Waiting Time

Sanford plans to visit his dentist and then go to the bank. Suppose the random variable X measures the time (in minutes) that a randomly selected patient waits in the office of Sanford's dentist and Y measures the time (in minutes) that a person stands in line at Sanford's bank. Express the probability that Sanford's total waiting time will be no more than 20 minutes in terms of a double integral involving the joint probability density function $f(x, y)$ for X and Y.

Solution

The goal is to find the probability that $X + Y \leq 20$. The points (x, y) for which $x + y \leq 20$ lie on or below the line $x + y = 20$. Moreover, since X and Y stand for nonnegative quantities, only those points in the first quadrant are meaningful in this particular context. The problem, then, is to find the probability that the point (X, Y) lies in R, where R is the region in the first quadrant bounded by the line $x + y = 20$ and the coordinate axes (Figure 11.11). This probability is given by the double integral of the density function f over the region R. That is,

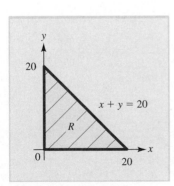

FIGURE 11.11 Triangle consisting of all points (x, y) for which $x + y \leq 20$.

$$P[(X, Y) \text{ in } R] = \iint_R f(x, y)\, dA = \int_0^{20}\int_0^{20-x} f(x, y)\, dy\, dx$$

EXERCISES ▪ 11.2

In Exercises 1 through 14, determine whether the given function is a probability density function.

1. $f(x) = \begin{cases} \dfrac{1}{10} & \text{for } 30 \le x \le 40 \\ 0 & \text{otherwise} \end{cases}$

2. $f(x) = \begin{cases} \dfrac{1}{e} & \text{for } 0 \le x \le 1 \\ 0 & \text{otherwise} \end{cases}$

3. $f(x) = \begin{cases} \dfrac{1}{4}(7 - 2x) & \text{for } 1 \le x \le 5 \\ 0 & \text{otherwise} \end{cases}$

4. $f(x) = \begin{cases} x - \dfrac{1}{2} & \text{for } 1 \le x \le 2 \\ 0 & \text{otherwise} \end{cases}$

5. $f(x) = \begin{cases} \dfrac{10}{(x + 10)^2} & \text{for } x \ge 0 \\ 0 & \text{for } x < 0 \end{cases}$

6. $f(x) = \begin{cases} \dfrac{1}{2}e^{-2x} & \text{for } x \ge 0 \\ 0 & \text{for } x < 0 \end{cases}$

7. $f(x) = \begin{cases} xe^{-x} & \text{for } x \ge 0 \\ 0 & \text{for } x < 0 \end{cases}$

8. $f(x) = \begin{cases} \dfrac{1}{9}\sqrt{x} & \text{for } 0 \le x \le 9 \\ 0 & \text{otherwise} \end{cases}$

9. $f(x) = \begin{cases} \dfrac{3}{2}x^2 + 2x & \text{for } -1 \le x \le 1 \\ 0 & \text{otherwise} \end{cases}$

10. $f(x) = \begin{cases} \dfrac{1}{2}x^2 + \dfrac{5}{3}x & \text{for } 0 \le x \le 1 \\ 0 & \text{otherwise} \end{cases}$

11. $f(x, y) = \begin{cases} 2e^{-(6y + x/3)} & \text{for } x \ge 0, y \ge 0 \\ 0 & \text{otherwise} \end{cases}$

12. $f(x, y) = \begin{cases} 4x^2 & \text{for } 0 \le y \le 1 \text{ and } y \le x \le 1 \\ 0 & \text{otherwise} \end{cases}$

13. $f(x, y) = \begin{cases} 3x^2 + 6xy & \text{for } 0 \le x \le 1 \text{ and } x \le y \le 1 \\ 0 & \text{otherwise} \end{cases}$

14. $f(x, y) = \begin{cases} 2e^{-x - y/2} & \text{for all } (x, y) \text{ with } x \ge y \ge 0 \\ 0 & \text{otherwise} \end{cases}$

In Exercises 15 through 24, either find a number k such that the given function is a probability density function or explain why no such number exists.

15. $f(x) = \begin{cases} k - 3x & \text{for } 0 \le x \le 1 \\ 0 & \text{otherwise} \end{cases}$

16. $f(x) = \begin{cases} 4 + 3kx & \text{for } 0 \le x \le 2 \\ 0 & \text{otherwise} \end{cases}$

17. $f(x) = \begin{cases} (kx - 1)(x - 2) & \text{for } 0 \le x \le 2 \\ 0 & \text{otherwise} \end{cases}$

18. $f(x) = \begin{cases} kx(3 - x) & \text{for } 0 \le x \le 3 \\ 0 & \text{otherwise} \end{cases}$

19. $f(x) = \begin{cases} x - kx^2 & \text{for } 0 \le x \le 1 \\ 0 & \text{otherwise} \end{cases}$

20. $f(x) = \begin{cases} x^3 + kx & \text{for } 0 \le x \le 2 \\ 0 & \text{otherwise} \end{cases}$

21. $f(x) = \begin{cases} xe^{-kx} & \text{for } x \ge 0 \\ 0 & \text{otherwise} \end{cases}$

22. $f(x) = \begin{cases} k(x + 2)^{-1} & \text{for } -1 \le x \le 2 \\ 0 & \text{otherwise} \end{cases}$

23. $f(x) = \begin{cases} \ln kx & \text{for } 1 \le x \le 2 \\ 0 & \text{otherwise} \end{cases}$

24. $f(x) = \begin{cases} \sqrt{kx} & \text{for } 0 \le x \le 2 \\ 0 & \text{otherwise} \end{cases}$

In Exercises 25 through 32, f(x) is a probability density function for a particular random variable X. Use integration to find the indicated probabilities.

25. $f(x) = \begin{cases} \dfrac{1}{3} & \text{if } 2 \le x \le 5 \\ 0 & \text{otherwise} \end{cases}$

  a. $P(2 \le X \le 5)$
  b. $P(3 \le X \le 4)$
  c. $P(X \ge 4)$

26. $f(x) = \begin{cases} \dfrac{x}{2} & \text{if } 0 \le x \le 2 \\ 0 & \text{otherwise} \end{cases}$

  a. $P(0 \le X \le 2)$
  b. $P(1 \le X \le 2)$
  c. $P(X \le 1)$

27. $f(x) = \begin{cases} \dfrac{1}{8}(4 - x) & \text{if } 0 \le x \le 4 \\ 0 & \text{otherwise} \end{cases}$

  a. $P(0 \le X \le 4)$
  b. $P(2 \le X \le 3)$
  c. $P(X \ge 1)$

28. $f(x) = \begin{cases} \dfrac{3}{32}(4x - x^2) & \text{if } 0 \le x \le 4 \\ 0 & \text{otherwise} \end{cases}$

  a. $P(0 \le X \le 4)$
  b. $P(1 \le X \le 2)$
  c. $P(X \le 1)$

29. $f(x) = \begin{cases} \dfrac{3}{x^4} & \text{if } x \ge 1 \\ 0 & \text{if } x < 1 \end{cases}$

  a. $P(1 \le X < +\infty)$
  b. $P(X \le 2)$
  c. $P(X \ge 2)$

30. $f(x) = \begin{cases} \dfrac{1}{10}e^{-x/10} & \text{if } x \ge 0 \\ 0 & \text{if } x < 0 \end{cases}$

  a. $P(0 \le X < +\infty)$
  b. $P(X \le 2)$
  c. $P(X \ge 5)$

31. $f(x) = \begin{cases} 2xe^{-x^2} & \text{if } x \ge 0 \\ 0 & \text{if } x < 0 \end{cases}$

  a. $P(X \ge 0)$
  b. $P(1 \le X \le 2)$
  c. $P(X \le 2)$

32. $f(x) = \begin{cases} \dfrac{1}{4}xe^{-x/2} & \text{if } x \ge 0 \\ 0 & \text{if } x < 0 \end{cases}$

  a. $P(0 \le X < +\infty)$
  b. $P(2 \le X \le 4)$
  c. $P(X \ge 6)$

In Exercises 33 through 36, find a number k so the given function f(x, y) will be a joint probability density function or explain why no such number exists.

33. $f(x, y) = \begin{cases} k\sqrt{x(1 - y)} & 0 \le x \le 1, 0 \le y \le 1 \\ 0 & \text{otherwise} \end{cases}$

34. $f(x, y) = \begin{cases} ky(2x - x^2) & 0 \le x \le 2, -1 \le y \le 5 \\ 0 & \text{otherwise} \end{cases}$

35. $f(x, y) = \begin{cases} ke^{-(x+3y)} & \text{for } x \ge 0, y \ge 0 \\ 0 & \text{otherwise} \end{cases}$

36. $f(x, y) = \begin{cases} 4e^{-(kx+y)} & \text{for } x \ge 0, y \ge 0 \\ 0 & \text{otherwise} \end{cases}$

In Exercises 37 through 40, f(x, y) is a joint probability density function for the random variables X and Y. Use integration to find the indicated probabilities.

37. $f(x, y) = \begin{cases} \dfrac{1}{16}(x^2 + 2y) & 0 \le x \le 1, 1 \le y \le 4 \\ 0 & \text{otherwise} \end{cases}$

  a. $P(0 \le X \le 1, 1 \le Y \le 2)$
  b. $P(0 \le X \le 1, 1 \le Y \le 2X + 1)$

38. $f(x, y) = \begin{cases} 2e^{-2x}e^{-y} & x \ge 0, y \ge 0 \\ 0 & \text{otherwise} \end{cases}$

  a. $P(0 \le X \le 1, 1 \le Y \le 2)$
  b. $P(2X + Y \le 1)$

39. $f(x, y) = \begin{cases} \dfrac{1}{6}e^{-x/2}e^{-y/3} & x \ge 0, y \ge 0 \\ 0 & \text{otherwise} \end{cases}$

  a. $P(1 \le X \le 2, 0 \le Y \le 2)$
  b. $P(X + Y \le 1)$

40. $f(x, y) = \begin{cases} xe^{-x-y} & x \ge 0, y \ge 0 \\ 0 & \text{otherwise} \end{cases}$

  a. $P(0 \le X \le 1, 0 \le Y \le 2)$
  b. $P(X + Y \le 7)$

BUSINESS AND ECONOMICS APPLIED PROBLEMS

41. USEFUL LIFE OF A MACHINE   The useful life X of a particular kind of machine is a random variable with density function

$$f(x) = \begin{cases} \dfrac{3}{28} + \dfrac{3}{x^2} & \text{if } 3 \le x \le 7 \\ 0 & \text{otherwise} \end{cases}$$

where x is the number of years a randomly selected machine stays in use.

  a. Find the probability that a randomly selected machine will be useful for more than 4 years.
  b. Find the probability that a randomly selected machine will be useful for less than 5 years.
  c. Find the probability that a randomly selected machine will be useful between 4 and 6 years.

42. USEFUL LIFE OF A MACHINE   Answer the questions in Exercise 41 if the useful life of the machine is a random variable X with density function

$$f(x) = \begin{cases} \dfrac{2}{15} + \dfrac{2}{x^2} & \text{if } 2 \le x \le 5 \\ 0 & \text{otherwise} \end{cases}$$

43. PRODUCT RELIABILITY   The life span of the lightbulbs manufactured by a certain company is measured by a random variable X with probability density function

$$f(x) = \begin{cases} 0.01e^{-0.01x} & \text{if } x \geq 0 \\ 0 & \text{if } x < 0 \end{cases}$$

where x denotes the life span (in hours) of a randomly selected bulb.
   a. What is the probability that the life span of a randomly selected bulb is between 50 and 60 hours?
   b. What is the probability that the life span of a randomly selected bulb is no greater than 60 hours?
   c. What is the probability that the life span of a randomly selected bulb is greater than 60 hours?

44. PRODUCT RELIABILITY   The useful life of a particular type of printer is measured by a random variable X with probability density function

$$f(x) = \begin{cases} 0.02e^{-0.02x} & \text{if } x \geq 0 \\ 0 & \text{if } x < 0 \end{cases}$$

where x denotes the number of months a randomly selected printer has been in use.
   a. What is the probability that a randomly selected printer will last between 10 and 15 months?
   b. What is the probability that a randomly selected printer will last less than 8 months?
   c. What is the probability that a randomly selected printer will last longer than 1 year?

45. CUSTOMER SERVICE   Suppose the time X a customer must spend waiting in line at a certain bank is a random variable that is exponentially distributed with probability density function

$$f(x) = \begin{cases} \frac{1}{4}e^{-x/4} & \text{if } x \geq 0 \\ 0 & \text{if } x < 0 \end{cases}$$

where x is the number of minutes a randomly selected customer spends waiting in line.
   a. Find the probability that a customer will have to stand in line at least 8 minutes.
   b. Find the probability that a customer will have to stand in line between 1 and 5 minutes.

46. WARRANTY PROTECTION   A certain appliance consisting of two independent electronic components will be usable as long as at least one of its components is still operating. The appliance carries a warranty from the manufacturer guaranteeing replacement if the appliance becomes unusable within 1 year of the date of purchase. Let the random variable X measure the life span (in years) of the first component and Y measure the life span

(also in years) of the second component, and suppose that the joint probability density function for X and Y is

$$f(x, y) = \begin{cases} \frac{1}{4}e^{-x/2}e^{-y/2} & \text{if } x \geq 0 \text{ and } y \geq 0 \\ 0 & \text{otherwise} \end{cases}$$

You purchase one of these appliances, selected at random from the manufacturer's stock. Find the probability that the warranty will expire before your appliance becomes unusable. [*Hint:* You want the probability that the appliance *does not* fail during the first year. How is this related to the probability that the appliance *does* fail during this period?]

47. TIME MANAGEMENT   Suppose the random variable X measures the time (in minutes) that a person spends in the waiting room of a doctor's office and Y measures the time required for a complete physical examination (also in minutes). You arrive at the doctor's office for a physical 50 minutes before you are due to leave for a meeting. If the joint probability distribution for X and Y is

$$f(x, y) = \begin{cases} \frac{1}{500}e^{-x/10}e^{-y/50} & \text{if } x \geq 0 \text{ and } y \geq 0 \\ 0 & \text{otherwise} \end{cases}$$

find the probability that you will be late leaving for your meeting. [*Hint:* The probability you want is $1 -$ probability of being on time.]

LIFE AND SOCIAL SCIENCE APPLIED PROBLEMS

48. TRAFFIC FLOW   A certain traffic light remains red for 45 seconds at a time. You arrive (at random) at the light and find it red. Use an appropriate uniform density function to find:
   a. The probability that the light will turn green within 15 seconds.
   b. The probability that the light will turn green between 5 and 10 seconds after you arrive.

49. COMMUTING   During the morning rush hour, commuter trains run every 20 minutes from the station near Jaime's home into the city center. Jaime arrives (at random) at the station and finds no train at the platform. Assuming that the trains are running on schedule, use an appropriate uniform density function to find:
   a. The probability Jaime will have to wait at least 8 minutes for a train.
   b. The probability Jaime will have to wait between 2 and 5 minutes for a train.

50. **EXPERIMENTAL PSYCHOLOGY** Suppose the length of time that it takes a laboratory rat to traverse a certain maze is measured by a random variable X that is exponentially distributed with probability density function

$$f(x) = \begin{cases} \dfrac{1}{3}e^{-x/3} & \text{if } x \geq 0 \\ 0 & \text{if } x < 0 \end{cases}$$

where x is the number of minutes a randomly selected rat spends in the maze.

a. Find the probability that a randomly selected rat will take more than 3 minutes to traverse the maze.

b. Find the probability that a randomly selected rat will take between 2 and 5 minutes to traverse the maze.

51. **TRANSPORTATION** Suppose that the monorail from Newark Liberty Airport to New York City leaves every hour. Let X be the random variable that represents the time a person arriving at the monorail terminal at a random time must wait until the next train leaves. Assume X has a uniform distribution.

a. Find the probability density function for X.

b. Find the probability that a person has to wait at least 45 minutes for the next train to New York.

c. Find the probability that a person has to wait between 5 and 15 minutes for the next train.

52. **ENTERTAINMENT** A 2-hour movie runs continuously at a local theater. Eileen leaves for the theater without first checking the show times. Use an appropriate uniform density function to find the probability that she will arrive at the theater within 10 minutes of the start of the film (before or after).

53. **TROPICAL ECOLOGY** Let X be the random variable that represents the time (in hours) between successive visits by hummingbirds to feed on the flowers of a particular tropical plant. Suppose X is distributed exponentially with parameter $k = 0.5$. Find the probability that the time between the arrival of successive hummingbirds is greater than 2 hours.

54. **MEDICINE** Let X be the random variable that represents the number of years a patient with a particular type of cancer survives after receiving radiation therapy. Suppose X has an exponential distribution with parameter k. Research indicates that a patient with this type of cancer has an 80% chance of surviving for at least 5 years after radiation therapy begins.

a. Use this information to determine k.

b. What is the probability of a cancer patient surviving at least 10 years after beginning radiation therapy?

55. **EXPERIMENTAL PSYCHOLOGY** Suppose the length of time that it takes a laboratory rat to traverse a certain maze is measured by a random variable X with probability density function

$$f(x) = \begin{cases} \dfrac{1}{16}xe^{-x/4} & \text{if } x \geq 0 \\ 0 & \text{if } x < 0 \end{cases}$$

where x is the number of minutes a randomly selected rat spends in the maze.

a. Find the probability that a randomly selected rat will take no more than 5 minutes to traverse the maze.

b. Find the probability that a randomly selected rat will take at least 10 minutes to traverse the maze.

56. **AIRPLANE ARRIVALS** The time interval between the arrivals of successive planes at a certain airport is measured by a random variable X with probability density function

$$f(x) = \begin{cases} 0.2e^{-0.2x} & \text{for } x \geq 0 \\ 0 & \text{for } x < 0 \end{cases}$$

where x is the time (in minutes) between the arrivals of a randomly selected pair of successive planes.

a. What is the probability that two successive planes selected at random will arrive within 5 minutes of one another?

b. What is the probability that two successive planes selected at random will arrive more than 6 minutes apart?

57. **BIOLOGY** Let X be a random variable that measures the age of a randomly selected cell in a particular population. Suppose X is distributed exponentially with a probability density function of the form

$$f(x) = \begin{cases} ke^{-kx} & \text{for } x \geq 0 \\ 0 & \text{otherwise} \end{cases}$$

where x is the age of a randomly selected cell (in days) and k is a positive constant. Experiments indicate that it is twice as likely for a cell to be less than 3 days old as it is for it to be more than 3 days old.

a. Use this information to determine k.

b. Find the probability that a randomly selected cell is at least 5 days old.

58. INSECT POPULATIONS Entomologists are studying two insect species that interact within the same ecosystem. Let X be the random variable that measures the population of species A and let Y measure the population of species B. Research indicates that the joint probability density function for X and Y is

$$f(x, y) = \begin{cases} xe^{-x-y} & \text{if } x \geq 0 \text{ and } y \geq 0 \\ 0 & \text{otherwise} \end{cases}$$

What is the probability that species A outnumbers species B?

59. HEALTH CARE Suppose the random variables X and Y measure the length of time (in days) that a patient stays in the hospital after abdominal and orthopedic surgery, respectively. On Monday, the patient in bed 107A undergoes an emergency appendectomy while her roommate in bed 107B undergoes orthopedic surgery for the repair of torn knee cartilage. Suppose the joint probability density function for X and Y is

$$f(x, y) = \begin{cases} \dfrac{1}{12}e^{-x/4}e^{-y/3} & \text{if } x \geq 0 \text{ and } y \geq 0 \\ 0 & \text{otherwise} \end{cases}$$

a. What is the probability that both patients will be discharged from the hospital within 3 days?

b. What is the probability that at least one of the patients is still in the hospital after 3 days?

60. SPY STORY The spy manages to escape from Coldfinger's "chat room" (Exercise 55, Section 10.1). He locates his equipment, but his weapons have all been removed except for a cleverly hidden pearl-handled pistol he received as a birthday gift from an old girlfriend exactly 1 year ago. He recalls that this particular kind of gun has a 1-year warranty and that its life span X is a random variable exponentially distributed with probability density function

$$f(t) = \begin{cases} 0.08e^{-0.08t} & \text{for } t \geq 0 \\ 0 & \text{for } t < 0 \end{cases}$$

where t is the number of months since the pistol was purchased. Assuming the pistol was new when he received it and that it had been chosen randomly from the factory stock, what is the probability that the pistol fails and the spy expires along with the warranty?

The **median** of a random variable X is the real number m such that $P(X \leq m) = \dfrac{1}{2}$. This term is used in Exercises 61 through 64.

61. Find the median time you have to wait for the light to turn green in Example 11.2.3.

62. Find the median length of a telephone call in Example 11.2.4.

63. Find the median of a uniformly distributed random variable over the interval $A \leq x \leq B$.

64. Find the median of a random variable that has an exponential distribution with parameter λ.

65. The *reliability function $r(x)$* of an electronic component is the probability that the component continues to work for more than x days. Suppose that the working life of a component is measured by a random variable X with a probability density function $f(x)$.
a. Find a formula for $r(x)$ in terms of $f(x)$.
b. Suppose the working life (days) of a memory chip has an exponential distribution with parameter $\lambda = 20$. Find $r(10)$, the probability that the memory chip continues to work for more than 10 days.

A continuous random variable X has a Pareto distribution if its probability density function has the form

$$f(x) = \begin{cases} \dfrac{a\lambda^a}{x^{a+1}} & \text{if } x \geq \lambda \\ 0 & \text{otherwise} \end{cases}$$

where a and λ are parameters (positive real numbers). The Pareto distribution is often used to model the distribution of incomes in a society. Exercises 66 and 67 involve the Pareto distribution.

66. Verify that for all positive numbers a and λ, the region under the graph of the Pareto probability density function $f(x)$ has area 1, as required of all probability density functions.

67. Suppose X is a random variable that has a Pareto distribution with parameter $a = 1$ and $\lambda = 1.2$.
a. Sketch the graph of the probability density function $f(x)$ for X.
b. Find the probability $P(10 \leq X \leq 20)$.
c. Read an article on the Pareto distribution, and write a paragraph on how it can be used.

SECTION 11.3 Expected Value and Variance of Continuous Random Variables

Learning Objectives

1. Compute and use expected value.
2. Interpret variance and standard deviation.
3. Find expected value for a joint probability density function.

Suppose it has been determined by medical researchers that when a new drug is used with patients suffering from a certain kind of cancer, the number of years X that a patient survives after receiving the drug is a continuous random variable with a particular probability distribution. On average, how long can a randomly selected patient expect to live after receiving the drug? Further, suppose that reliability studies of a certain manufactured product indicate that the life span X of the product is a continuous random variable with a known probability distribution. What is the average life span of a randomly selected unit, and to what extent are the experimental outcomes of the reliability testing spread out in relation to the average life span? In Section 11.1, we answered questions of this nature for discrete random variables by examining the expected value and variance of such variables, and in this section we will use integration to extend these concepts to random variables that are continuous.

Expected Value of a Continuous Random Variable

In general, if a discrete random variable X takes on values $x_1, x_2, \ldots, x_n$ with respective probabilities $p_1, p_2, \ldots p_n$, then the expected value (or mean) of X is given by the weighted sum

$$x_1p_1 + x_2p_2 + \cdots + x_np_n = \sum_{j=1}^{n} x_jp_j$$

Now consider a continuous random variable X with probability density function $f(x)$. For simplicity, restrict your attention to a bounded interval $A \leq x \leq B$. Divide this interval into n subintervals of width Δx, and let x_j be the left endpoint of the jth subinterval. Then

$$p_j = \begin{array}{c} \text{probability that } x \text{ is} \\ \text{in the } j\text{th subinterval} \end{array} = \begin{array}{c} \text{area under the graph} \\ \text{of } f \text{ from } x_j \text{ to } x_{j+1} \end{array} \approx f(x_j)\,\Delta x$$

where $f(x_j)\,\Delta x$ is the area of an approximating rectangle (Figure 11.12).

To approximate the expected value of X over the interval $A \leq x \leq B$, treat X as if it were a discrete random variable that takes on the values $x_1, x_2, \ldots, x_n$ with respective probabilities $p_1, p_2, \ldots p_n$. Then the expected value of X is approximated by the weighted sum

$$\sum_{j=1}^{n} x_jp_j \approx \sum_{j=1}^{n} x_jf(x_j)\,\Delta x$$

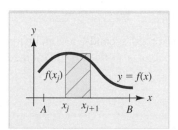

FIGURE 11.12 The jth subinterval with an approximating rectangle.

The actual expected value of X on the interval $A \leq x \leq B$ is then obtained by taking the limit of this approximation as n increases without bound; that is,

$$\begin{array}{c} \text{Expected value of } X \\ \text{on } A \leq x \leq B \end{array} = \lim_{n\to\infty} \sum_{j=1}^{n} x_jf(x_j)\,\Delta x = \int_{A}^{B} xf(x)\,dx$$

An extension of this argument over all real numbers shows that the expected value of X on $-\infty < x < \infty$ is

$$\int_{-\infty}^{\infty} xf(x)\,dx$$

To summarize:

Expected Value ■ If X is a continuous random variable with probability density function f, the expected value (or mean) of X is

$$E(X) = \int_{-\infty}^{\infty} xf(x)\,dx$$

The use of this integral formula to compute the expected value of a continuous random variable is illustrated in Examples 11.3.1 and 11.3.2.

EXAMPLE 11.3.1   Finding Expected Waiting Time

Continuing Example 11.2.3, a certain traffic light remains red for 40 seconds at a time. Brie arrives (at random) at the light and finds it red. Let X denote the random variable that measures the time (in seconds) that she must wait for the light to change, and assume that X is distributed uniformly with the probability density function

$$f(x) = \begin{cases} \dfrac{1}{40} & \text{if } 0 \le x \le 40 \\ 0 & \text{otherwise} \end{cases}$$

How long should Brie expect to wait for the light to turn green?

Solution

The expected value of X is

$$E(X) = \int_{-\infty}^{\infty} xf(x)\,dx = \int_{0}^{40} x\left(\frac{1}{40}\right) dx = \frac{x^2}{80}\Big|_{0}^{40} = \frac{1{,}600}{80} = 20$$

Therefore, Brie should expect to wait at the red light for 20 seconds, a conclusion that should come as no surprise since the random variable is uniformly distributed between 0 and 40.

EXAMPLE 11.3.2   Finding the Expected Length of a Phone Call

Continuing Example 11.2.4, let X be the random variable that measures the duration of phone calls in a certain city and assume that X is distributed exponentially with the probability density function

$$f(x) = \begin{cases} 0.5e^{-0.5x} & \text{if } x \ge 0 \\ 0 & \text{if } x < 0 \end{cases}$$

How long would you expect a randomly selected call to last?

EXPLORE!

Refer to Example 11.3.2. Store
0.5X ∗ e^(−0.5X)(X≥0) into Y1
of the equation editor of your
graphing calculator, and graph
using the window [0, 30]1 by
[−0.1, 0.4]0.1. Approximate
the value of the improper
integral from X = 0 to ∞,
setting the lower limit of
integration to 0 and the upper
limit to 10, 20, and 30,
successively. What value do
these results seem to be
approaching? What is the
significance of this value?

Solution

The expected value of X is

$$E(X) = \int_{-\infty}^{\infty} xf(x)\,dx = \int_{0}^{\infty} 0.5xe^{-0.5x}\,dx$$

$$= \lim_{N \to \infty} \int_{0}^{N} 0.5xe^{-0.5x}\,dx \qquad \text{integration by parts}$$

$$= \lim_{N \to \infty} \left(-xe^{-0.5x}\Big|_{0}^{N} + \int_{0}^{N} e^{-0.5x}\,dx \right)$$

$$= \lim_{N \to \infty} \left(-xe^{-0.5x} - 2e^{-0.5x} \right)\Big|_{0}^{N}$$

$$= \lim_{N \to \infty} \left(-Ne^{-0.5N} - 2e^{-0.5N} + 2 \right)$$

$$= 2$$

Thus, the expected duration of telephone calls in the city is 2 minutes.

Variance and Standard Deviation

Recall from Section 11.1 that the variance of a random variable provides a measure of the extent to which the distribution is spread out in relation to the expected value $E(X)$. The variance of a discrete random variable was given in terms of a sum, but for a continuous random variable, it is expressed as an integral.

> **Variance and Standard Deviation of a Continuous Random Variable**
>
> If X is a continuous random variable with probability density function f, then the **variance** of X is given by the integral
>
> $$\text{Var}(X) = \int_{-\infty}^{\infty} [x - E(X)]^2 f(x)\,dx$$
>
> and the **standard deviation** is
>
> $$\sigma(X) = \sqrt{\text{Var}(X)}$$

The definition of the variance is not as mysterious as it may seem at first glance. Notice that the formula for the variance is the same as that for the expected value, except that x has been replaced by the expression $[x - E(X)]^2$, which represents the square of the deviation of x from the mean $E(X)$. The variance, therefore, is the expected value of the squared deviations of the values of X from the mean. If the values of X tend to cluster about the mean as in Figure 11.13a, most of the deviations from the mean will be small and the variance, which is the mean of the squares of these deviations, will also be small. On the other hand, if the values of the random variable are widely scattered as in Figure 11.13b, there will be many large deviations from the mean, and the variance will be large. As with the discrete case, since the computation of variance involves the term $[x - E(X)]^2$, it is often more convenient to measure the spread of a distribution by the *square root* of the variance; that is, by the standard deviation $\sigma(X) = \sqrt{\text{Var}(X)}$.

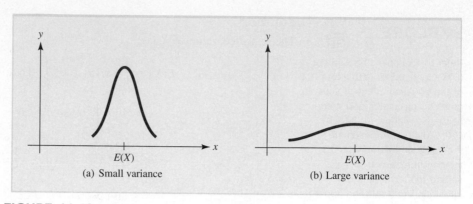

FIGURE 11.13 The variance of a continuous random variable as a measure of the spread of a distribution.

For all but the simplest density functions, the formula

$$\text{Var}(X) = \int_{-\infty}^{\infty} [x - E(X)]^2 f(x)\, dx$$

is cumbersome to use for the actual calculation of the variance because the term $x - E(X)$ must be squared and multiplied by $f(x)$ before the integration can be performed. Here is an equivalent formula for the variance that is easier to use in computations.

Alternative Formula for Variance

$$\text{Var}(X) = \int_{-\infty}^{\infty} x^2 f(x)\, dx - [E(X)]^2$$

The derivation of this formula is straightforward. It involves expanding the integrand in the definition of the variance and rearranging the resulting terms until the new formula emerges. As you read the steps, keep in mind that the expected value $E(X)$ is a constant and can be brought outside integrals by the constant multiple rule.

$$\text{Var}(X) = \int_{-\infty}^{\infty} [x - E(X)]^2 f(x)\, dx \qquad \text{expand integrand}$$

$$= \int_{-\infty}^{\infty} \{x^2 - 2xE(X) + [E(X)]^2\} f(x)\, dx$$

split into 3 integrals and factor out $E(X)$ and $[E(X)]^2$

$$= \int_{-\infty}^{\infty} x^2 f(x)\, dx - 2E(X) \int_{-\infty}^{\infty} x f(x)\, dx + [E(X)]^2 \int_{-\infty}^{\infty} f(x)\, dx$$

use the definition of $E(X)$ and the fact that $\int_{-\infty}^{\infty} f(x)\, dx = 1$

$$= \left[\int_{-\infty}^{\infty} x^2 f(x)\, dx \right] - 2E(X) \cdot E(X) + [E(X)]^2 (1)$$

$$= \int_{-\infty}^{\infty} x^2 f(x)\, dx - [E(X)]^2$$

This formula is used to compute variance in Examples 11.3.3 and 11.3.4.

EXAMPLE 11.3.3 Finding Variance and Standard Deviation

Find the variance and standard deviation of the uniformly distributed random variable X from Example 11.2.3 with probability density function

$$f(x) = \begin{cases} \dfrac{1}{40} & \text{if } 0 \le x \le 40 \\ 0 & \text{otherwise} \end{cases}$$

Solution

The first step is to compute $E(X)$. This was done in Example 11.3.1, where we found that $E(X) = 20$. Using this value in the variance formula, we get

$$\text{Var}(X) = \int_{-\infty}^{\infty} x^2 f(x)\, dx - [E(X)]^2$$

$$= \int_0^{40} x^2 \left(\frac{1}{40}\right) dx - 20^2 = \frac{x^3}{120}\Big|_0^{40} - 400$$

$$= \frac{64{,}000}{120} - 400 = \frac{16{,}000}{120} = \frac{400}{3}$$

So the standard deviation is

$$\sigma(X) = \sqrt{\text{Var}(X)} = \sqrt{\frac{400}{3}} = \frac{20\sqrt{3}}{3}$$

EXAMPLE 11.3.4 Finding Variance and Standard Deviation

Find the variance and standard deviation of the exponentially distributed random variable X from Example 11.2.3 with probability density function

$$f(x) = \begin{cases} 0.5e^{-0.5x} & \text{if } x \ge 0 \\ 0 & \text{if } x < 0 \end{cases}$$

Solution

Using the value $E(X) = 2$ obtained in Example 11.3.2 and integrating by parts twice, we get

$$\text{Var}(X) = \int_{-\infty}^{\infty} x^2 f(x)\, dx - [E(X)]^2$$

$$= \int_0^{\infty} 0.5x^2 e^{-0.5x}\, dx - 4$$

$$= \lim_{N \to \infty} \int_0^N 0.5x^2 e^{-0.5x}\, dx - 4$$

$$= \lim_{N \to \infty} \left(-x^2 e^{-0.5x}\Big|_0^N + 2\int_0^N xe^{-0.5x}\, dx \right) - 4$$

$$= \lim_{N \to \infty} \left[(-x^2 e^{-0.5x} - 4xe^{-0.5x})\Big|_0^N + 4\int_0^N e^{-0.5x}\, dx \right] - 4$$

$$= \lim_{N \to \infty} (-x^2 - 4x - 8)e^{-0.5x}\Big|_0^N - 4$$

$$= \lim_{N \to \infty} [(-N^2 - 4N - 8)e^{-0.5N} + 8] - 4$$

$$= 4$$

and the standard deviation is

$$\sigma(X) = \sqrt{\mathrm{Var}(X)} = \sqrt{4} = 2$$

Expected Value for a Joint Probability Distribution

If X and Y are continuous random variables with joint probability density function $f(x, y)$, then the expected values of X and Y are given by the integrals

$$E(X) = \int_{-\infty}^{+\infty} \int_{-\infty}^{+\infty} xf(x, y)\, dA \qquad \text{and} \qquad E(Y) = \int_{-\infty}^{+\infty} \int_{-\infty}^{+\infty} yf(x, y)\, dA$$

In Example 11.3.5, we compute the expected value of one of the random variables in Example 11.2.5.

EXAMPLE 11.3.5   Finding Expected Reliability for a Joint Distribution

Continuing Example 11.2.5, smoke detectors manufactured by a certain firm contain two independent circuits, one manufactured at the firm's California plant and the other at its plant in Ohio. Reliability studies suggest that if X measures the life span (in years) of a randomly selected circuit from the California plant and Y measures the life span of a randomly selected circuit from the Ohio plant, then the joint probability density function for X and Y is

$$f(x, y) = \begin{cases} 0.4e^{-x}e^{-0.4y} & \text{if } x \geq 0 \text{ and } y \geq 0 \\ 0 & \text{otherwise} \end{cases}$$

How long would you expect a randomly selected circuit from the Ohio plant to last?

Solution

You need to compute the expected value of the random variable Y; that is,

$$E(Y) = \int_{-\infty}^{+\infty} \int_{-\infty}^{+\infty} y(0.4e^{-x}e^{-0.4y})\, dx\, dy = \int_{0}^{+\infty} 0.4ye^{-0.4y}\left(\int_{0}^{+\infty} e^{-x}\, dx \right) dy$$

$$= \int_{0}^{+\infty} 0.4ye^{-0.4y}\left[\lim_{N \to +\infty} \left(\frac{e^{-x}}{-1} \right)\Big|_{x=0}^{x=N} \right] dy \qquad \qquad \lim_{N \to +\infty} -(e^{-N} - e^{0}) = 1$$

$$= \int_{0}^{+\infty} 0.4ye^{-0.4y}\,(1)\, dy \qquad\qquad \begin{array}{l} \text{integration by parts:} \\ u = 0.4y \quad dv = e^{-0.4y}\, dy \end{array}$$

$$= \lim_{N \to +\infty} \left(-ye^{-0.4y} + \frac{e^{-0.4y}}{-0.4} \right)\Big|_{y=0}^{y=N}$$

$$= \lim_{N \to +\infty} [(-Ne^{-0.4N} - 2.5e^{-0.4N}) - (0 - 2.5)] \qquad \lim_{N \to +\infty} (-N - 2.5)e^{-0.4N} = 0$$

$$\approx 2.5$$

Thus, you would expect a randomly selected circuit from the Ohio plant to last approximately 2.5 years.

EXERCISES ▪ 11.3

In Exercises 1 through 10, the probability density function for a continuous random variable X is given. In each case, find the expected value E(X) and the variance Var(X) of X.

1. $f(x) = \begin{cases} \dfrac{1}{3} & \text{if } 2 \le x \le 5 \\ 0 & \text{otherwise} \end{cases}$

2. $f(x) = \begin{cases} \dfrac{x}{2} & \text{if } 0 \le x \le 2 \\ 0 & \text{otherwise} \end{cases}$

3. $f(x) = \begin{cases} \dfrac{1}{8}(4 - x) & \text{if } 0 \le x \le 4 \\ 0 & \text{otherwise} \end{cases}$

4. $f(x) = \begin{cases} \dfrac{3}{32}(4x - x^2) & \text{if } 0 \le x \le 4 \\ 0 & \text{otherwise} \end{cases}$

5. $f(x) = \begin{cases} \dfrac{3}{x^4} & \text{if } 1 \le x < \infty \\ 0 & \text{if } x < 1 \end{cases}$

6. $f(x) = \begin{cases} \dfrac{1}{10}e^{-x/10} & \text{if } x \ge 0 \\ 0 & \text{otherwise} \end{cases}$

7. $f(x) = \begin{cases} 4e^{-4x} & \text{if } x \ge 0 \\ 0 & \text{otherwise} \end{cases}$

8. $f(x) = \begin{cases} x & \text{if } 0 \le x < 1 \\ 2 - x & \text{if } 1 \le x \le 2 \\ 0 & \text{otherwise} \end{cases}$

9. $f(x) = \begin{cases} 20(x^3 - x^4) & \text{if } 0 \le x \le 1 \\ 0 & \text{otherwise} \end{cases}$

10. $f(x) = \begin{cases} xe^{-x} & \text{if } x \ge 0 \\ 0 & \text{otherwise} \end{cases}$

In Exercises 11 through 14, the joint probability density function f(x, y) for two continuous random variables X and Y is given. In each case, find the expected values E(X) and E(Y).

11. $f(x, y) = \begin{cases} \dfrac{1}{2}e^{-x/2}e^{-y} & \text{if } x \ge 0 \text{ and } y \ge 0 \\ 0 & \text{otherwise} \end{cases}$

12. $f(x, y) = \begin{cases} \dfrac{1}{3}e^{-2x}e^{-y/6} & \text{if } x \ge 0 \text{ and } y \ge 0 \\ 0 & \text{otherwise} \end{cases}$

13. $f(x, y) = \begin{cases} ye^{-x-y} & \text{if } x \ge 0 \text{ and } y \ge 0 \\ 0 & \text{otherwise} \end{cases}$

14. $f(x, y) = \begin{cases} \dfrac{24}{5}x(1 - y) & \text{if } 0 \le y \le x \text{ and } 0 \le x \le 1 \\ 0 & \text{otherwise} \end{cases}$

BUSINESS AND ECONOMICS APPLIED PROBLEMS

15. PRODUCT RELIABILITY The useful life of a particular type of printer is measured by a random variable X with probability density function

$$f(x) = \begin{cases} 0.05e^{-0.05x} & \text{if } x \ge 0 \\ 0 & \text{if } x < 0 \end{cases}$$

where x denotes the number of months a randomly selected printer has been in use. What is the expected life of a randomly selected printer?

16. CUSTOMER SERVICE Suppose the time X a customer must spend waiting in line at a certain bank is a random variable that is exponentially distributed with density function

$$f(x) = \begin{cases} \dfrac{1}{4}e^{-x/4} & \text{if } x \ge 0 \\ 0 & \text{if } x < 0 \end{cases}$$

where x is the number of minutes a randomly selected customer spends waiting in line. Find the expected waiting time for customers at the bank.

17. USEFUL LIFE OF A MACHINE The useful life X of a particular kind of machine is a random variable with density function

$$f(x) = \begin{cases} \dfrac{3}{28} + \dfrac{3}{x^2} & \text{if } 3 \le x \le 7 \\ 0 & \text{otherwise} \end{cases}$$

where x is the number of years a randomly selected machine stays in use. What is the expected useful life of the machine?

18. WORKER EFFICIENCY Let X be the random variable that measures the amount of time a randomly selected worker takes to perform a particular task each day. Suppose X is distributed with probability density function

$$f(t) = \begin{cases} \dfrac{7}{288} \sqrt[3]{t}\,(8 - t) & \text{if } 0 \leq t \leq 8 \\ 0 & \text{otherwise} \end{cases}$$

where t is the number of hours the average worker works each day.

a. Sketch the graph of the density function $f(t)$.
b. What is the probability that a randomly selected worker takes at least 5 hours to complete the task?
c. On average, how long would you expect it to take workers to complete the task?

19. **COST ANALYSIS**   A product consists of two components, A and B, which are manufactured separately. To estimate the cost of labor for the process, the manufacturer assigns a random variable X to measure the number of worker-hours required during an 8-hour shift to produce each unit of component A and another random variable Y to measure the number of worker-hours required to produce each unit of component B. Analysis suggests assigning to X and Y the joint probability density function

$$f(x, y) = \begin{cases} \dfrac{1}{5,120}(x^2 + xy + 2y^2) & \text{for } 0 \leq x \leq 8, 0 \leq y \leq 8 \\ 0 & \text{otherwise} \end{cases}$$

a. What is the probability that the total number of worker-hours required to produce both components is less than 8?
b. How many worker-hours should the manufacturer expect to be expended on producing a unit of component A? A unit of component B?
c. Suppose the workers assigned to producing component A receive $18 per hour, while those producing component B receive $20 per hour. What is the expected cost of producing one complete (combined) unit of the product?

LIFE AND SOCIAL SCIENCE APPLIED PROBLEMS

20. **AIRPLANE ARRIVALS**   The time interval between the arrivals of successive planes at a certain airport is measured by a random variable X with probability density function

$$f(x) = \begin{cases} 0.2e^{-0.2x} & \text{for } x \geq 0 \\ 0 & \text{for } x < 0 \end{cases}$$

where x is the time (in minutes) between the arrivals of a randomly selected pair of successive planes.

If Sadie arrives at the airport just in time to see a plane landing, how long should she expect to wait for the next arriving flight?

21. **COMMUTING**   During the morning rush hour, commuter trains run every 20 minutes from the station near your home into the city center. You arrive (at random) at the station and find no train at the platform. Assuming that the trains are running on schedule, use an appropriate uniform density function to find how long you should expect to wait for a train.

22. **LEARNING**   Let X be the random variable that measures the amount of time a randomly selected calculus student at my college spends studying the subject each day. Suppose X is distributed with probability density function

$$f(x) = \begin{cases} \dfrac{5}{324} \sqrt{x}\,(9 - x) & \text{for } 0 \leq x \leq 9 \\ 0 & \text{otherwise} \end{cases}$$

where x is the number of hours the student studies each day.

a. Sketch the graph of the density function $f(x)$.
b. What is the probability that a randomly selected calculus student studies at least 2 hours per day?
c. On average, how long should I expect my students to study calculus each day?

23. **TROPICAL ECOLOGY**   Let X be the random variable that represents the time (in hours) between successive visits by hummingbirds to feed on the flowers of a particular tropical plant. Suppose X is distributed exponentially with parameter $\lambda = 0.5$. If a hummingbird has just left a flower you are observing, how long would you expect to wait for the next hummingbird to arrive? What is the variance of the random variable X?

24. **TRAFFIC FLOW**   A certain traffic light remains red for 45 seconds at a time. You arrive (at random) at the light and find it red. Use an appropriate uniform density function to find how long you should expect to wait for the light to turn green.

25. **MEDICAL RESEARCH**   A group of patients with a potentially fatal disease has been treated with an experimental drug. Assume the survival time X for a patient receiving the drug is a random variable exponentially distributed with probability density function

$$f(x) = \begin{cases} \lambda e^{-\lambda x} & \text{if } x \geq 0 \\ 0 & \text{if } x < 0 \end{cases}$$

where x is the number of years a patient survives after first receiving the drug.

a. Research indicates that the expected survival time for a patient receiving the drug is 5 years. Based on this information, what is λ?

b. Using the value of λ determined in part (a), what is the probability that a randomly selected patient survives for less than 2 years?

c. What is the probability that a randomly selected patient survives for more than 7 years?

26. **EXPERIMENTAL PSYCHOLOGY** Suppose the length of time that it takes a laboratory rat to traverse a certain maze is measured by a random variable X that is distributed with a probability density function of the form

$$f(x) = \begin{cases} axe^{-bx} & \text{if } x \geq 0 \\ 0 & \text{otherwise} \end{cases}$$

where x is the number of minutes a randomly selected rat spends in the maze and a and b are positive numbers. Experiments indicate that on average, a rat will take 6 minutes to traverse the maze.

a. Use the given information to find a and b. $\Big[$ *Hint*:

Why must it be true that $\dfrac{a}{b^2} = 1$ and $\dfrac{2a}{b^3} = 6?\Big]$

b. Is it more likely that the rat spends less than 5 minutes in the maze or more than 7 minutes?

27. **VIROLOGY** A research institute for public health has identified a new deadly virus. To study the virus, researchers consider a model in which it is assumed that the life span of an individual chosen from a sample population of the virus is a random variable X distributed with density function of the form

$$f(x) = \begin{cases} Ae^{-bx} & \text{for } x \geq 0 \\ 0 & \text{otherwise} \end{cases}$$

where x is time (in weeks) and A and b are positive constants. Experiments suggest that after 1 week, 90% of the population is still alive.

a. Use this information to find A and b.

b. What percentage of the sample population survives for 10 weeks?

c. How long does it take for 90% of the sample population to die off?

d. What is the expected life span of a virus selected randomly from the population?

e. What is the variance of X? Interpret this result.

28. **LEGISLATIVE TURNOVER** A mathematical model in political science* asserts that the length of time served by a legislator may be modeled as a random variable X that is exponentially distributed with probability density function

$$N(t) = \begin{cases} ce^{-ct} & \text{for } t \geq 0 \\ 0 & \text{otherwise} \end{cases}$$

where t is the number of years of continuous service and c is a positive constant that depends on the nature and character of the legislative body.

a. For the U.S. House of Representatives, it was found that $c = 0.0866$. Find the probability that a randomly selected House member will serve at least 6 years.

b. For the British House of Commons, it was found that $c = 0.135$. What is the probability that a member of this body will serve at least 6 years?

c. The module cited in the footnote to this exercise also shows how the given density function can be used to estimate how many members of the Soviet Central Committee may have been purged by Nikita Khruschev in the period 1956–1961. Read this module, and write a paragraph on whether or not you think the method of analysis is valid.

29. **HEALTH CARE** Suppose the random variables X and Y measure the length of time (in days) that a patient stays in the hospital after abdominal and orthopedic surgery, respectively. On Monday, the patient in bed 107A undergoes an emergency appendectomy while her roommate in bed 107B undergoes surgery for the repair of torn knee cartilage. Suppose the joint probability density function for X and Y is

$$f(x, y) = \begin{cases} \dfrac{1}{12}e^{-x/4}e^{-y/3} & \text{if } x \geq 0 \text{ and } y \geq 0 \\ 0 & \text{otherwise} \end{cases}$$

How long should each patient expect to stay in the hospital?

30. **TIME MANAGEMENT** Let the random variable X measure the time (in minutes) that a person spends in the waiting room of a doctor's office and let Y measure the time required for a complete physical examination (also in minutes).

*Thomas W. Casstevens, "Exponential Models for Legislative Turnover," *UMAP Modules 1978: Tools for Teaching,* Lexington, MA: Consortium for Mathematics and Its Applications, Inc., 1979.

Suppose the joint probability density function for X and Y is

$$f(x, y) = \begin{cases} \dfrac{1}{500}e^{-x/10}e^{-y/50} & \text{if } x \geq 0 \text{ and } y \geq 0 \\ 0 & \text{otherwise} \end{cases}$$

How much time would you expect to spend in the waiting room? Is this more or less than the time you would expect to spend during your examination?

MISCELLANEOUS PROBLEMS

31. Show that a uniformly distributed random variable X with probability density function

$$f(x) = \begin{cases} \dfrac{1}{b-a} & \text{if } a \leq x \leq b \\ 0 & \text{otherwise} \end{cases}$$

has expected value

$$E(X) = \frac{a+b}{2}$$

32. Show that the uniformly distributed random variable in Exercise 31 has variance

$$\text{Var}(X) = \frac{(b-a)^2}{12}$$

33. Show that an exponentially distributed random variable X with probability density function

$$f(x) = \begin{cases} \lambda e^{-\lambda x} & \text{if } x \geq 0 \\ 0 & \text{otherwise} \end{cases}$$

has expected value $E(X) = \dfrac{1}{\lambda}$.

34. Show that the exponentially distributed random variable in Exercise 33 has variance $\text{Var}(X) = \dfrac{1}{\lambda^2}$.

35. **THE INTERNET** Suppose that during evening hours, the time between successive hits on a popular Web page can be represented by an exponentially distributed random variable X whose expected value is $E(X) = 0.5$ seconds.
    a. Use the result of Exercise 33 to find the probability density function $f(x)$ for X.
    b. Find the probability that the time between successive hits on the Web page is more than 1 second.
    c. Find the probability that the time between successive hits on the Web page is between $\dfrac{1}{4}$ second and $\dfrac{1}{2}$ second.

36. **TELECOMMUNICATIONS** Suppose that during business hours, the time between successive wireless calls at a network switch can be represented by an exponentially distributed random variable X with expected value $E(X) = 0.01$ second.
    a. Use the result of Exercise 33 to find the probability density function for X.
    b. Find the probability that the time between the arrival of successive wireless calls at the switch is more than 0.05 seconds.
    c. Find the probability that the time between the arrival of successive wireless calls at the switch is between 0.05 and 0.08 seconds.

37. Recall from Section 11.2 that a continuous random variable X has a Pareto distribution if its probability density function has the form

$$f(x) = \begin{cases} \dfrac{a\lambda^a}{x^{a+1}} & \text{if } x \geq \lambda \\ 0 & \text{otherwise} \end{cases}$$

where a and λ are positive real numbers.
    a. Find the expected value $E(X)$ for $a > 1$.
    b. Find the variance $\text{Var}(X)$ for $a > 2$.

SECTION 11.4 Normal and Poisson Probability Distributions

Learning Objectives

1. Compute probability using a standard normal distribution table, and explore nonstandard normal distributions.
2. Use normal distributions to study quality control and other applications.
3. Examine and use the Poisson probability distribution.

In this section, we introduce two important families of probability distributions, one continuous and the other discrete. We begin by discussing normal distributions, which are continuous and are arguably the best known and most widely used of all probability distributions. Then we shall examine Poisson distributions, which are discrete and somewhat narrower in scope than normal distributions but are nonetheless extremely useful for modeling certain phenomena. We shall consider both computational issues and practical applications involving these two distribution classes.

Normal Distributions

A **normal density function** is one of the form

$$f(x) = \frac{1}{\sigma\sqrt{2\pi}} e^{-(x-\mu)^2/2\sigma^2}$$

where μ and σ are real numbers, with $\sigma > 0$. A random variable X with a density function of this form is called a **normal random variable,** and its values are said to be **normally distributed.** The graph of a normal density function is called a **normal curve** or less formally, a "bell curve" because of its distinctive shape (Figure 11.14).

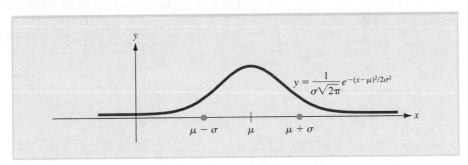

FIGURE 11.14   A bell curve, the graph of a normal density function.

Historically, the normal distribution was originally introduced in 1733 by the French mathematician Abraham de Moivre, who used it to study the results of certain games of chance. Other applications soon followed. For instance, the Belgian scientist Adolph Quetelet showed that height and chest measurements taken of French and Scottish soldiers were approximately normally distributed.

Indeed, a primary reason for the importance of the normal distribution is that many random variables of practical interest behave as if their distributions are either normal or essentially normal. Examples of quantities closely approximated by normal distributions include time until first failure of certain products, the height of people of a particular age and gender, the daily caloric intake of a grizzly bear, the number of shark attacks in a given time period, length of life of an organism, and the concentration of heavy metals in blood. Moreover, the normal distribution arises frequently in psychological and educational studies. Examples include variables such as the job satisfaction rating of a company's employees, the reading ability of children of a particular age, the number of crimes of a certain type committed annually, and the number of words a person can memorize during a given time period. Finally, other important distributions can be approximated in terms of normal distributions. In fact, a key result in probability theory, called the *central limit theorem*, shows that the sample mean based on a random sampling of a large number of independent observations drawn from a given distribution (e.g., exponential, geometric) is approximately normally distributed.

Our first goal is to examine some of the features and properties of the normal density function. Recall that any probability density function $f(x)$ must satisfy $\int_{-\infty}^{+\infty} f(x)\, dx = 1$. Using advanced integration techniques, it can be shown that

$$\int_{-\infty}^{+\infty} e^{-(x-\mu)^2/2\sigma^2}\, dx = \sigma\sqrt{2\pi}$$

which means that

$$\frac{1}{\sigma\sqrt{2\pi}} \int_{-\infty}^{+\infty} e^{-(x-\mu)^2/2\sigma^2}\, dx = 1$$

as required.

We have already commented on the bell shape of the graph of the normal density function. However, the "bell" can be flattened or peaked depending on the parameters μ and σ. The graph of any normal density function is symmetric about the vertical line $x = \mu$ and has a high point where $x = \mu$ and inflection points where $x = \mu + \sigma$ and $x = \mu - \sigma$ (see Exercise 60). Three different normal curves are shown in Figure 11.15a. The normal density curve with $\mu = 0$ and $\sigma = 1$, called the *standard normal curve*, is shown in Figure 11.15b. Recall that this curve was analyzed using derivative methods in Example 4.4.2.

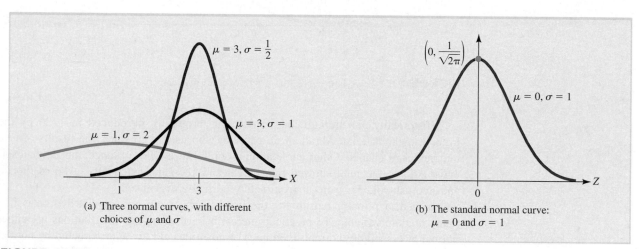

(a) Three normal curves, with different choices of μ and σ

(b) The standard normal curve: $\mu = 0$ and $\sigma = 1$

FIGURE 11.15 Graphs of several normal density functions (normal curves).

The parameters μ and σ that appear in the normal density function are closely related to the expected value, variance, and standard deviation of the associated normal random variable, as summarized in the following box.

Formulas for the Expected Value, Variance, and Standard Deviation of a Normal Random Variable ■ For a normal random variable X with density function

$$f(x) = \frac{1}{\sigma\sqrt{2\pi}}e^{-(x-\mu)^2/2\sigma^2}$$

we have

Expected value	$E(X) = \mu$	
Variance	$\mathrm{Var}(X) = \sigma^2$	
Standard deviation	$\sqrt{\mathrm{Var}(X)} = \sigma$	

Most of the area under a normal curve lies relatively close to the mean μ. In fact, it can be shown that roughly 68.3% of normally distributed data lies within one standard deviation of the mean μ and that almost all the data (99.7%) lies within three standard deviations of μ. These observations are part of the **empirical rule** summarized in the following box and displayed in Figure 11.16. The analysis in Example 11.4.1 is typical of how the empirical rule may be used to quickly draw important conclusions about normally distributed data.

Empirical Rule for Area Under a Normal Curve

For any normal curve:

1. Roughly 68.3% of the area under the curve lies between $\mu - \sigma$ and $\mu + \sigma$, that is, within one standard deviation of the mean.

2. Roughly 95.4% of the area under the curve lies between $\mu - 2\sigma$ and $\mu + 2\sigma$, that is, within two standard deviations of the mean.

3. Roughly 99.7% of the area under the curve lies between $\mu - 3\sigma$ and $\mu + 3\sigma$, that is, within three standard deviations of the mean.

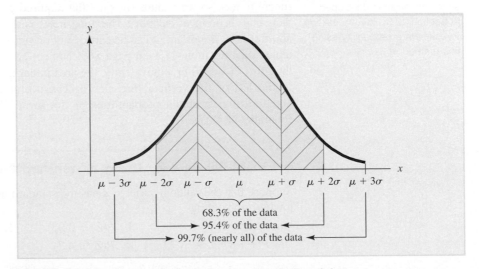

FIGURE 11.16 The empirical rule for normally distributed data.

EXAMPLE 11.4.1 Using the Empirical Rule to Study Height

The height of women between the ages of 18 and 24 in the United States is normally distributed with a mean of 65.5 inches and a standard deviation of 2.5 inches. Describe the distribution of these heights using the empirical rule.

Solution

Using the empirical rule, we see that 68.3% of all women in the United States have heights between $65.5 - 2.5 = 63.0$ inches and $65.5 + 2.5 = 68.0$ inches, 95.4% of all women in the United States have heights between $65.5 - 2(2.5) = 60.5$ inches and $65.5 + 2(2.5) = 70.5$ inches, and 99.7% of all women in the United States have heights between $65.5 - 3(2.5) = 58.0$ inches and $65.5 + 3(2.5) = 73.0$ inches.

Computations Involving Standard Normal Distributions

There are many normal density distributions, but fortunately, it is not necessary to study them individually. Indeed, we shall find that any normal distribution can be "standardized" by transforming its density function into standard normal form

$$S(x) = \frac{1}{\sqrt{2\pi}} e^{-x^2/2}$$

with mean $\mu = 0$ and standard deviation $\sigma = 1$. The graph of the standard normal density function is shown in Figure 11.15b. Notice that it is symmetric about the y axis, with its peak at $\left(0, \frac{1}{\sqrt{2\pi}}\right)$. We shall begin our study of normal distributions by focusing on the standard normal case before turning to more general distributions.

It is common practice to denote the standard normal random variable by Z. To compute the probability $P(a \le Z \le b)$ that the standard normal random variable Z lies between a and b, we need to evaluate the integral

Just-In-Time REVIEW

Example 10.3.7 illustrates how integrals like the one used to evaluate $P(a \le Z \le b)$ can be approximated using the Taylor series expansions developed in Section 10.3. You are asked to estimate a probability using this method in Exercise 61.

$$P(a \le Z \le b) = \int_a^b S(x)\, dx = \int_a^b \frac{1}{\sqrt{2\pi}} e^{-x^2/2}\, dx$$

The bad news is that there is no simple antiderivative for the standard normal density function $S(x)$, so we cannot compute the required probability directly using the fundamental theorem of calculus. The good news is that because computing such probabilities is so important, it can be carried out on most scientific calculators or by using tables such as Table 11.1 on page 869. Notice that Table 11.1 gives probabilities of the form $P(Z \le b)$ or, equivalently, the area under the standard normal density curve to the left of the vertical line $x = b$. These probabilities can then be combined algebraically to obtain probabilities of the form $P(a \le Z \le b)$, as illustrated in Example 11.4.2.

EXAMPLE 11.4.2 Using a Standard Normal Table

Suppose that the random variable Z has the standard normal distribution. Find the following probabilities.

a. $P(Z \le 1)$ **b.** $P(Z \ge 0.03)$ **c.** $P(Z \le -1.87)$

d. $P(-1.87 \le Z \le 1)$ **e.** $P(Z \le 4.7)$

TABLE 11.1 Areas Under the Standard Normal Curve to the Left of Positive Z Values: $P(Z \le b)$

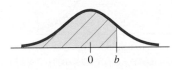

b	.00	.01	.02	.03	.04	.05	.06	.07	.08	.09
.0	.5000	.5040	.5080	.5120	.5160	.5199	.5239	.5279	.5319	.5359
.1	.5398	.5438	.5478	.5517	.5557	.5596	.5636	.5675	.5714	.5753
.2	.5793	.5832	.5871	.5910	.5948	.5987	.6026	.6064	.6103	.6141
.3	.6179	.6217	.6255	.6293	.6331	.6368	.6406	.6443	.6480	.6517
.4	.6554	.6591	.6628	.6664	.6700	.6736	.6772	.6808	.6844	.6879
.5	.6915	.6950	.6985	.7019	.7054	.7088	.7123	.7157	.7190	.7224
.6	.7257	.7291	.7324	.7357	.7389	.7422	.7454	.7486	.7517	.7549
.7	.7580	.7611	.7642	.7673	.7704	.7734	.7764	.7794	.7823	.7852
.8	.7881	.7910	.7939	.7967	.7995	.8023	.8051	.8078	.8106	.8133
.9	.8159	.8186	.8212	.8238	.8264	.8289	.8315	.8340	.8365	.8389
1.0	.8413	.8438	.8461	.8485	.8508	.8531	.8554	.8577	.8599	.8621
1.1	.8643	.8665	.8686	.8708	.8729	.8749	.8770	.8790	.8810	.8830
1.2	.8849	.8869	.8888	.8907	.8925	.8944	.8962	.8980	.8997	.9015
1.3	.9032	.9049	.9066	.9082	.9099	.9115	.9131	.9147	.9162	.9177
1.4	.9192	.9207	.9222	.9236	.9251	.9265	.9279	.9292	.9306	.9319
1.5	.9332	.9345	.9357	.9370	.9382	.9394	.9406	.9418	.9429	.9441
1.6	.9452	.9463	.9474	.9484	.9495	.9505	.9515	.9525	.9535	.9545
1.7	.9554	.9564	.9573	.9582	.9591	.9599	.9608	.9616	.9625	.9633
1.8	.9641	.9649	.9656	.9664	.9671	.9678	.9686	.9693	.9699	.9706
1.9	.9713	.9719	.9726	.9732	.9738	.9744	.9750	.9756	.9761	.9767
2.0	.9772	.9778	.9783	.9788	.9793	.9798	.9803	.9808	.9812	.9817
2.1	.9821	.9826	.9830	.9834	.9838	.9842	.9846	.9850	.9854	.9857
2.2	.9861	.9864	.9868	.9871	.9875	.9878	.9881	.9884	.9887	.9890
2.3	.9893	.9896	.9898	.9901	.9904	.9906	.9909	.9911	.9913	.9916
2.4	.9918	.9920	.9922	.9925	.9927	.9929	.9931	.9932	.9934	.9936
2.5	.9938	.9940	.9941	.9943	.9945	.9946	.9948	.9949	.9951	.9952
2.6	.9953	.9955	.9956	.9957	.9959	.9960	.9961	.9962	.9963	.9964
2.7	.9965	.9966	.9967	.9968	.9969	.9970	.9971	.9972	.9973	.9974
2.8	.9974	.9975	.9976	.9977	.9977	.9978	.9979	.9979	.9980	.9981
2.9	.9981	.9982	.9982	.9983	.9984	.9984	.9985	.9985	.9986	.9986
3.0	.9987	.9987	.9987	.9988	.9988	.9989	.9989	.9989	.9990	.9990
3.1	.9990	.9991	.9991	.9991	.9992	.9992	.9992	.9992	.9993	.9993
3.2	.9993	.9993	.9994	.9994	.9994	.9994	.9994	.9995	.9995	.9995
3.3	.9995	.9995	.9995	.9996	.9996	.9996	.9996	.9996	.9996	.9997
3.4	.9997	.9997	.9997	.9997	.9997	.9997	.9997	.9997	.9997	.9998

Source: From *Introduction to the Theory of Statistics*, 3d ed., by A. M. Mood, F. A. Graybill, and D. C. Boes. Copyright © 1974 by McGraw-Hill, Inc. Used with permission of McGraw-Hill Book Company.

Just-In-Time **REVIEW**

The standard normal curve is the graph of a probability density function, so the region under the entire curve must have area 1.

Solution

a. $P(Z \le 1)$ is the area under the normal curve to the left of $b = 1$ (Figure 11.17a). Looking up $b = 1$ in the table, we find that

$$P(Z \le 1) = 0.8413$$

b. $P(Z \ge 0.03)$ is the area under the curve to the right of 0.03 (Figure 11.17b). Since the table only gives areas to the *left* of values of Z, we cannot get this area directly from the table. Instead, we use the fact that

$$P(Z \ge 0.03) = 1 - P(Z \le 0.03)$$

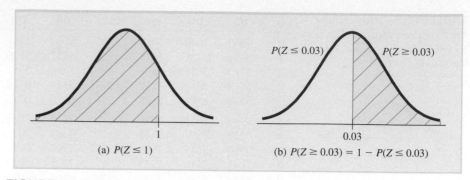

FIGURE 11.17   Probability as area under the standard normal density curve.

That is, the area to the right of 0.03 is 1 minus the area to the left of 0.03. From the table, we get

$$P(Z \le 0.03) = 0.5120$$

and so

$$P(Z \ge 0.03) = 1 - 0.5120 = 0.4880$$

c. $P(Z \le -1.87)$ is the area under the curve to the left of -1.87 (Figure 11.18a). Since the curve is symmetric, this is the same as the area to the right of 1.87 (Figure 11.18b). That is,

$$P(Z \le -1.87) = P(Z \ge 1.87)$$

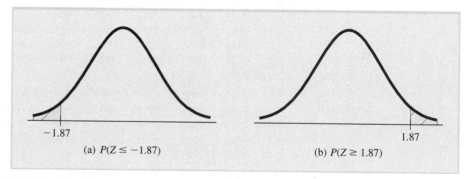

FIGURE 11.18 $P(Z \le -1.87) = P(Z \ge 1.87) = 1 - P(Z \le 1.87)$.

Proceeding as in part (b), we get

$$P(Z \ge 1.87) = 1 - P(Z \le 1.87) = 1 - 0.9693 = 0.0307$$

Hence, $P(Z \le -1.87) = 0.0307$.

d. $P(-1.87 \le Z \le 1)$ is the area under the curve between -1.87 and 1 (Figure 11.19c). This can be computed by subtracting the area to the left of -1.87 (Figure 11.19b) from the area to the left of 1 (Figure 11.19a). By using the results of parts (a) and (c), we find that

$$P(-1.87 \le Z \le 1) = P(Z \le 1) - P(Z \le -1.87)$$
$$= 0.8413 - 0.0307 = 0.8106$$

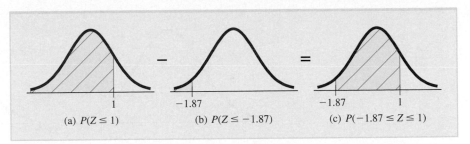

FIGURE 11.19 $P(-1.87 \leq Z \leq 1) = P(Z \leq 1) - P(Z \leq -1.87)$.

e. $P(Z \leq 4.7)$ is the area under the curve to the left of 4.7. However, the table only covers the values $-3.49 \leq Z \leq 3.49$. In fact, outside the interval $-3 \leq Z \leq 3$, the curve is so close to the x axis that there is virtually no area under it (Figure 11.20). Since almost all the area under the curve is to the left of 4.7, we have, approximately, $P(Z \leq 4.7) \approx 1$.

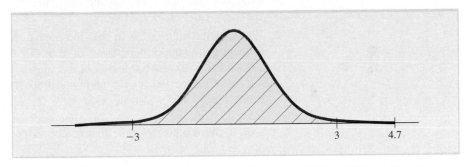

FIGURE 11.20 There is virtually no area to the right of 4.7, so $P(Z \leq 4.7) \approx 1$.

Example 11.4.3 illustrates how to use the table to find Z values corresponding to given probabilities.

EXAMPLE 11.4.3 Finding Z Values Using a Table

a. Find the value b so that $P(Z \leq b) = 0.9$.
b. Find the value b so that $P(Z \geq b) = 0.01$.
c. Find the value b so that $P(Z \leq b) = 0.1$.
d. Find the positive number b so that $P(-b \leq Z \leq b) = 0.95$.

Solution

a. We want the value of b with the property that the area to the left of b is 0.9 (Figure 11.21a). Looking in the area columns of the normal table, we find no entry equal to 0.9, so we choose the entry closest to it, 0.8997, which corresponds to a b value of 1.28. That is, rounded off to 2 decimal places, $P(Z \leq 1.28) = 0.9$, and so the required value is (approximately) $b = 1.28$.

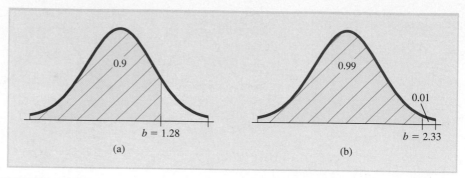

FIGURE 11.21   (a) $P(Z \le b) = 0.9$ when $b = 1.28$ and (b) $P(Z \ge b) = 0.01$ when $b = 2.33$.

b. We want the value of b with the property that the area to the right of it is 0.01, or equivalently, the area to the left of it is 0.99 (Figure 11.21b). Searching the table, we find that 0.9901 is the area closest to 0.99, and so we take the corresponding value, $b = 2.33$.

c. We want the value of b so that the area to the left of b is 0.1 (Figure 11.22). Since the area under the curve to the left of 0 is 0.5, it follows that the desired value of b is less than 0 and does not appear in the table. Proceeding indirectly, we first find the value of b with the property that the area to the right is 0.1. Equivalently, this is also the value of b for which the area to the left is 0.9, and from part (a), we know that this value is 1.28. By symmetry, the value of b we seek is the negative of this; that is, the value of b such that $P(Z \le b) = 0.1$ is $b = -1.28$.

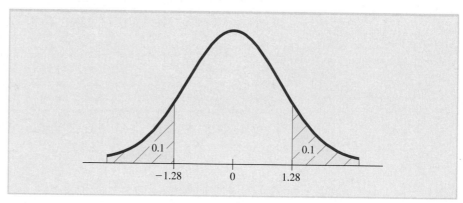

FIGURE 11.22 $P(Z \le b) = 0.1$ when $b = -1.28$.

d. We want the value b of Z so that the area to the left of b is

$$0.95 + \frac{1}{2}(1 - 0.95) = 0.95 + 0.025 = 0.975$$

(Figure 11.23). From Table 11.1, we find that $b = 1.96$.

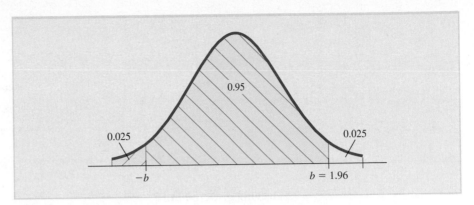

FIGURE 11.23 $P(-b \leq Z \leq b) = 0.95$ when $b = 1.96$.

Other Normal Distributions

You will not find separate tables of $P(X \leq b)$ for nonstandard normal variables. Instead, we compute $P(a \leq X \leq b)$ using a table for standard normal variables together with the transformations displayed in this box.

> **Transforming a Nonstandard Normal Distribution to Standard Form**
>
> If the random variable X has a normal distribution with mean μ and standard deviation σ, then the transformed variable $Z = \dfrac{X - \mu}{\sigma}$ has a standard normal distribution. Furthermore, we have
>
> $$P(a \leq X \leq b) = P\left(\frac{a - \mu}{\sigma} \leq Z \leq \frac{b - \mu}{\sigma}\right)$$

Geometrically, this result says that the area between a and b under the normal curve with mean μ and standard deviation σ is the same as the area under the standard normal curve between $z_1 = \dfrac{a - \mu}{\sigma}$ and $z_2 = \dfrac{b - \mu}{\sigma}$. This is illustrated in Figure 11.24.

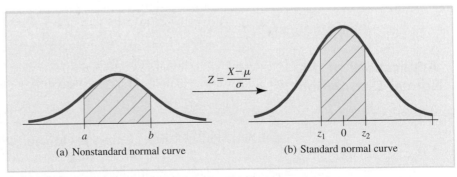

(a) Nonstandard normal curve   (b) Standard normal curve

FIGURE 11.24   The area under the nonstandard normal curve between a and b is the same as the area under the standard normal curve between z_1 and z_2.

To see why this is true, recall that

$$P(a \leq X \leq b) = \int_a^b \frac{1}{\sigma\sqrt{2\pi}} e^{-(x-\mu)^2/2\sigma^2} \, dx$$

and transform the integral on the right using this change of variable:

$$z = \frac{x - \mu}{\sigma}$$

limits of integration

when $x = a$, then $z = \dfrac{a - \mu}{\sigma}$

$$dz = \frac{1}{\sigma} \, dx$$

when $x = b$, then $z = \dfrac{b - \mu}{\sigma}$

Substituting, we find that

$$\int_a^b \frac{1}{\sigma\sqrt{2\pi}} e^{-[(x-\mu)^2/\sigma^2]/2} \, dx = \int_{(a-\mu)/\sigma}^{(b-\mu)/\sigma} \frac{1}{\sqrt{2\pi}} e^{-z^2/2} \, dz$$

$$= P\left(\frac{a - \mu}{\sigma} \leq Z \leq \frac{b - \mu}{\sigma}\right)$$

The use of this transformation is illustrated in Example 11.4.4.

EXAMPLE 11.4.4   Studying a Nonstandard Normal Distribution

Suppose X has a normal distribution with $\mu = 20$ and $\sigma = 4$. Find the following probabilities:

a. $P(X \leq 26)$ **b.** $P(X \geq 18)$ **c.** $P(15 \leq X \leq 21)$

Solution

In each case, we apply the formula $Z = \dfrac{X - \mu}{\sigma}$ and then use the values in Table 11.1.

a. $P(X \leq 26) = P\left[Z \leq \dfrac{26 - 20}{4}\right] = P(Z \leq 1.5) = 0.9332$

b. $P(x \geq 18) = P\left[Z \geq \dfrac{18 - 20}{4}\right] = P(Z \geq -0.5) = 0.6915$

c. $P(15 \leq X \leq 21) = P\left[\dfrac{15 - 20}{4} \leq Z \leq \dfrac{21 - 20}{4}\right]$

$$= P(-1.25 \leq Z \leq 0.25)$$

$$= 0.5987 - 0.1056 = 0.4931$$

Applications of the Normal Distribution

Next, we will examine several practical situations in which the normal distribution can be used. Example 11.4.5 deals with random errors made by measuring devices.

EXAMPLE 11.4.5   Using a Normal Distribution to Study Inventory

The number of loaves of bread that can be sold during a day by a certain supermarket is normally distributed with $\mu = 1,000$ loaves and $\sigma = 100$ loaves. If the market stocks 1,200 loaves on a given day, what is the probability that the loaves will be sold out before the day is over?

Solution

If X denotes the number of loaves that can be sold during a day, then X is normally distributed with $\mu = 1,000$ and $\sigma = 100$. We wish to find $P(X \geq 1,200)$. Transforming to Z and using the table, we obtain

$$P(X \geq 1,200) = P\left(Z \geq \frac{1,200 - 1,000}{100}\right) = P(Z \geq 2) = 0.0228$$

Thus, there is about a 2% chance that the market will run out of loaves before the day is over.

EXAMPLE 11.4.6 Using a Normal Distribution to Study Trout Length

The lengths of trout in a certain lake are normally distributed with a mean of 7 inches and a standard deviation of 2 inches. If the fish and game department would like anglers to keep only the largest 20% of the trout, what should the minimum size for "keepers" be?

Solution

Let X denote the length of a randomly selected trout and c the minimum size for a keeper. Since only the largest 20% are to be keepers, c must satisfy the equation $P(X > c) = 0.2$. Using $\mu = 7$ and $\sigma = 2$, we transform to Z and obtain

$$P\left(Z \geq \frac{c - 7}{2}\right) = 0.2 \quad \text{or} \quad P\left(Z \leq \frac{c - 7}{2}\right) = 0.8$$

From Table 11.1 we see that

$$P(Z \leq 0.84) = 0.8$$

Hence,

$$\frac{c - 7}{2} = 0.84 \quad \text{or} \quad c = 8.68$$

We conclude that the fish and game department should set 8.68 inches as the minimum size for keepers.

EXAMPLE 11.4.7 Using a Normal Distribution to Study Errors

On a particular scale, measurement errors are made that are normally distributed with $\mu = 0$ and $\sigma = 0.1$ ounce. If you weigh an object on this scale, what is the probability that the measurement error is no greater than 0.15 ounce?

Solution

Let X denote the measurement error made when an object is weighed on the scale. Our goal is to find $P(-0.15 \leq X \leq 0.15)$. Since $\mu = 0$ and $\sigma = 0.1$, we transform to

Z using the formula $Z = \dfrac{X - 0}{0.1}$ and find that

$$P(-0.15 \leq X \leq 0.15) = P\left(\frac{-0.15 - 0}{0.1} \leq Z \leq \frac{0.15 - 0}{0.1}\right)$$
$$= P(-1.5 \leq Z \leq 1.5) = 0.8664$$

That is, there is an 87% chance that the error in measurement is no greater than 0.15 ounce.

Poisson Distributions

Next, we examine the Poisson distribution, which is a discrete distribution named for the 19th-century French mathematician Siméon Denis Poisson. Although Poisson introduced this distribution in theoretical work done in 1837, its first practical application was made in 1898 by Ladislaus Josephowitsch Bortkiewicz, who used it to study the accidental deaths of Prussian soldiers from kicks by horses, a relatively rare event. Since then, Poisson distributions have become an important tool for modeling other rare events, such as the number of people struck by lightning during a year or the number of substitutions that have occurred in a particular amino acid sequence over a million-year period. More generally, Poisson distributions are used for modeling counts in time or space, such as the number of customers that arrive at a store during an hour, or the number of typographical errors on a page in a particular book, or the density of trees with tracts of a fixed size throughout a large forest, or the number of bacterial colonies in a Petri dish.

A discrete random variable X is said to be **Poisson distributed** with parameter λ ($\lambda > 0$) if X describes an experiment whose possible outcomes are the nonnegative integers $n = 0, 1, 2, \ldots$, and for each nonnegative integer n, the event $(X = n)$ is assigned the probability

$$P(X = n) = \frac{\lambda^n}{n!}e^{-\lambda}$$

To verify that this satisfies the requirements of a probability assignment, we must show that $P(X = n) \geq 0$ for $n = 0, 1, 2, \ldots$ and that $\displaystyle\sum_{n=0}^{\infty} P(X = n) = 1$. The condition $P(X = n) \geq 0$ for all n is satisfied since $\lambda > 0$ and $e^{-\lambda} > 0$. To show that the probability assignments of the distribution sum to 1, we require the Taylor series expansion $e^x = \displaystyle\sum_{n=0}^{\infty} \frac{x^n}{n!}$ developed in Section 10.3. Note that

$$\sum_{n=0}^{\infty} P(X = n) = P(X = 0) + P(X = 1) + P(X = 2) + \cdots$$

$$= \frac{\lambda^0}{0!}e^{-\lambda} + \frac{\lambda^1}{1!}e^{-\lambda} + \frac{\lambda^2}{2!}e^{-\lambda} + \cdots$$

$$= e^{-\lambda}\left(1 + \lambda + \frac{\lambda^2}{2!} + \cdots\right) = e^{-\lambda}\sum_{n=0}^{\infty} \frac{\lambda^n}{n!}$$

$$= e^{-\lambda}e^{\lambda} \quad \text{since } \sum_{n=0}^{\infty} \frac{\lambda^n}{n!} \text{ is the Taylor series for } e^{\lambda}$$

$$= 1$$

The use of the Poisson distribution is illustrated in Examples 11.4.8 and 11.4.9.

EXPLORE!

The Poisson distribution appears on several graphing calculator models as a built-in function. On a TI-84 Plus, this can be found on the **DISTR** key **(2nd VARS)** key. Use this feature to verify the results of Example 11.4.8.

EXAMPLE 11.4.8   Studying Plant Growth with a Poisson Distribution

A botanist researching the growth patterns of a particular kind of orchid models the number of orchid plants per square meter as a random variable X with a Poisson distribution. If the distribution has parameter $\lambda = 0.2$, find

a. The probability that there are no orchid plants located within a randomly selected square meter.

b. The probability that there is no more than one orchid plant within a randomly selected square meter.

Solution

a. The probability that there are no orchid plants is

$$P(X = 0) = \frac{(0.2)^0}{0!}e^{-0.2} = e^{-0.2} \quad \text{since } (0.2)^0 = 1 \text{ and } 0! = 1$$

$$\approx 0.82$$

b. The event that there is no more than one orchid plant is $(X \leq 1)$, and the probability of this event is

$$P(X \leq 1) = P(X = 0) + P(X = 1) = \frac{(0.2)^0}{0!}e^{-0.2} + \frac{(0.2)^1}{1!}e^{-0.2}$$

$$= (1 + 0.2)e^{-0.2}$$

$$\approx 0.98$$

Thus, a randomly selected square meter is not likely to contain many orchids.

EXAMPLE 11.4.9   Studying Industrial Safety with a Poisson Distribution

The number of accidents in a particular industry that occur monthly can be modeled as a random variable X with a Poisson distribution if the industrial accidents tend to occur independently of one another at an approximately constant rate. Suppose studies indicate that there is only a 10% likelihood of no accidents occurring during a randomly chosen month.

a. Determine the parameter λ for the distribution.

b. Find the probability that at least three accidents occur during a randomly chosen month.

Solution

a. Since there is a 10% likelihood that no accidents will occur, we have

$$P(X = 0) = \frac{\lambda^0}{0!}e^{-\lambda} = 0.10$$

$$e^{-\lambda} = 0.10 \quad \text{since } \lambda^0 = 1 \text{ and } 0! = 1$$

Taking logarithms on both sides of this equation, we get

$$\ln(e^{-\lambda}) = \ln(0.10)$$
$$-\lambda = -2.3$$

so the parameter is $\lambda = 2.3$.

b. The event that at least three industrial accidents occur is $(X \geq 3)$, and the probability of this event satisfies

$$P(X \geq 3) = 1 - P(X < 3)$$

where $(X < 3)$ is the complementary event that less than three accidents occur. We find that

$$P(X < 3) = P(X = 0) + P(X = 1) + P(X = 2)$$
$$= \frac{(2.3)^0}{0!}e^{-2.3} + \frac{(2.3)^1}{1!}e^{-2.3} + \frac{(2.3)^2}{2!}e^{-2.3}$$
$$\approx 0.6$$

so that

$$P(X \geq 3) \approx 1 - 0.6 = 0.4$$

Thus, there is approximately a 40% probability that at least three accidents will occur.

If the random variable X is Poisson distributed with parameter λ, then the expected value of X is $E(X) = \lambda$. To show that this is true, note that

$$E(X) = \sum_{n=0}^{\infty} nP(X = n) \qquad \text{definition of expected value}$$

$$= \sum_{n=0}^{\infty} n\left(\frac{\lambda^n e^{-\lambda}}{n!}\right) \qquad \text{definition of the Poisson distribution}$$

$$= 0 \cdot \left(\frac{\lambda^0 e^{-\lambda}}{0!}\right) + 1 \cdot \left(\frac{\lambda^1 e^{-\lambda}}{1!}\right) + 2 \cdot \left(\frac{\lambda^2 e^{-\lambda}}{2!}\right) + \cdots + n \cdot \left(\frac{\lambda^n e^{-\lambda}}{n!}\right) + \cdots$$

$$= \lambda e^{-\lambda} + \frac{\lambda^2}{1}e^{-\lambda} + \frac{\lambda^3}{2!}e^{-\lambda} + \cdots \quad \text{since } \frac{n}{n!} = \frac{1}{(n-1)!}$$

$$= \lambda e^{-\lambda}\left[1 + \lambda + \frac{\lambda^2}{2!} + \cdots\right] \qquad \text{factor } \lambda e^{-\lambda} \text{ from each term}$$

$$= \lambda e^{-\lambda}\sum_{n=0}^{\infty}\frac{\lambda^n}{n!}$$

$$= \lambda e^{-\lambda}e^{\lambda} \qquad \text{since } \sum_{n=0}^{\infty}\frac{\lambda^n}{n!} \text{ is the Taylor series for } e^{\lambda}$$

$$= \lambda \qquad \text{since } e^{-\lambda}e^{\lambda} = 1$$

We leave it as an exercise (Exercise 58) for you to show that the variance of X also equals λ. To summarize:

> **Expected Value, Variance, and Standard Deviation of a Random Variable with a Poisson Distribution** ■ If the random variable X has a Poisson distribution with parameter λ, then
>
> $$P(X = n) = \frac{\lambda^n}{n!} e^{-\lambda}$$
>
> and
>
> | | |
> |---|---|
> | Expected value | $E(X) = \lambda$ |
> | Variance | $\text{Var}(X) = \lambda$ |
> | Standard deviation | $\sigma(X) = \sqrt{\text{Var}(X)} = \sqrt{\lambda}$ |

EXAMPLE 11.4.10   Studying Telemarketing Calls with a Poisson Distribution

Suppose that the number of calls you receive from telemarketers during the dinner hour on a randomly selected day follows a Poisson distribution, and that on average, you expect to get three such calls.

a. What is the parameter λ of the distribution?

b. What is the probability that you will get exactly three calls?

c. What is the probability that you will receive more than three calls?

Solution

a. Let X be the random variable that counts the number of calls you receive during a randomly selected dinner hour. Interpret the statement that "on average, you expect to get three calls" as meaning that the expected value of X is 3. Thus, $E(X) = \lambda = 3$.

b. The probability that you will get exactly three calls is

$$P(X = 3) = \frac{3^3 e^{-3}}{3!} \approx 0.22$$

c. The event that you receive more than three calls is $(X > 3)$, and the probability of this event satisfies

$$P(X > 3) = 1 - P(X \le 3)$$

where $(X \le 3)$ is the complementary event that no more than three calls are received. We find that

$$P(X \le 3) = P(X = 0) + P(X = 1) + P(X = 2) + P(X = 3)$$
$$= \frac{3^0 e^{-3}}{0!} + \frac{3^1 e^{-3}}{1!} + \frac{3^2 e^{-3}}{2!} + \frac{3^3 e^{-3}}{3!}$$
$$\approx 0.647$$

so that

$$P(X > 3) = 1 - P(X \le 3) \approx 1 - 0.647$$
$$= 0.353$$

Thus, it is approximately 35% likely that you will receive more than three calls from telemarketers during a randomly selected dinner hour.

EXERCISES ■ 11.4

In Exercises 1 through 4, the density function f(x) of a normal random variable X is given. In each case, find the expected value E(X), the variance Var(X), and the standard deviation σ.

1. $f(x) = \dfrac{1}{2\sqrt{2\pi}} e^{-x^2/8}$

2. $f(x) = \dfrac{1}{7\sqrt{2\pi}} e^{-(x-3)^2/98}$

3. $f(x) = \dfrac{1}{\sqrt{6\pi}} e^{-(x+1)^2/6}$

4. $f(x) = \dfrac{1}{\sqrt{2\pi}} e^{-(x+6)^2/2}$

In Exercises 5 through 10, find the indicated probability assuming that the random variable Z has a standard normal distribution.

5. $P(Z \le 1.24)$

6. $P(Z \le -1.20)$

7. $P(Z \le 4.26)$

8. $P(Z \ge 0.19)$

9. $P(-1 \le Z \le 1)$

10. $P(-1.20 \le Z \le 4.26)$

In Exercises 11 through 14, find the appropriate value of b for the given probability, assuming that the random variable Z has a standard normal distribution.

11. $P(Z \le b) = 0.8413$

12. $P(Z \ge b) = 0.2266$

13. $P(Z \le b) = 0.8643$

14. $P(-b \le Z \le b) = 0.9544$

15. Suppose X has a normal distribution with $\mu = 60$ and $\sigma = 4$. Find the following probabilities:
    a. $P(X \ge 68)$ b. $P(X \ge 60)$
    c. $P(56 \le X \le 64)$ d. $P(X \le 40)$

16. Suppose X has a normal distribution with $\mu = 35$ and $\sigma = 3$. Find the following probabilities:
    a. $P(X \ge 38)$ b. $P(X \ge 35)$
    c. $P(32 \le X \le 38)$ d. $P(X \le 20)$

17. Suppose X has a normal distribution with $\mu = 4$ and $\sigma = 0.5$. For each of the following probabilities, find the appropriate value of x.
    a. $P(X \ge x) = 0.0228$
    b. $P(X \le x) = 0.1587$

18. Suppose X has a normal distribution with $\mu = 5$ and $\sigma = 0.6$. For each of the following probabilities, find the appropriate value of x.
    a. $P(5 - x \le X \le 5 + x) = 0.6826$
    b. $P(X \le x) = 0.6$

19. A random variable X has a normal distribution with $\mu = 15$ and $\sigma = 4$. Using the empirical rule, find:
    a. $P(11 \le X \le 19)$
    b. $P(7 \le X \le 23)$

20. A random variable X has a normal distribution with $\mu = 18$ and $\sigma = 3$. Using the empirical rule, find:
    a. $P(15 < X < 21)$
    b. $P(9 \le X \le 27)$

21. Suppose that X is a Poisson random variable with parameter $\lambda = 3$. Give the exact formula and the value to three decimal places for these probabilities.
    a. $P(X = 1)$ b. $P(X = 3)$
    c. $P(X = 4)$ d. $P(X = 8)$

22. Suppose that X is a Poisson random variable with parameter $\lambda = \dfrac{3}{2}$. Give the exact formula and the value to three decimal places for these probabilities.
    a. $P(X = 0)$ b. $P(X = 1)$
    c. $P(X = 3)$ d. $P(X = 5)$

23. Draw the histogram for the Poisson random variable with $\lambda = 2$. Show 5 rectangles.

24. Draw the histogram for the Poisson random variable with $\lambda = \dfrac{1}{2}$. Show 5 rectangles.

BUSINESS AND ECONOMICS APPLIED PROBLEMS

25. **EMPLOYEE COMPENSATION** The incomes of industrial workers in a certain region are normally distributed with a mean of $42,500 and a standard deviation of $3,000. Find the probability that a randomly selected worker has an income between $41,000 and $44,000.

26. **PRODUCT MANAGEMENT** A meat company produces hot dogs having lengths that are normally distributed with a mean length of 4 inches and standard deviation of 0.2 inch. Find the fraction of hot dogs that are less than 3.5 inches long.

27. QUALITY CONTROL A certain company produces glass jars. On the average, the jars hold 1 quart, but there is some variation among them. Assume the volumes are normally distributed with standard deviation $\sigma = 0.01$ quart.
   a. Find the probability that a randomly chosen jar will hold between 0.98 and 1.02 quarts.
   b. Find the fraction of jars that hold less than 0.97 quart.

28. QUALITY CONTROL The breaking strength of a certain type of rope is normally distributed. If the rope has an average breaking strength of 1,200 pounds with standard deviation 100 pounds, find the probability that a randomly selected piece of rope will break under a strain of less than 1,000 pounds.

29. FARMING A certain carrot farm is famous for growing carrots of almost identical length. If the length is normally distributed with mean 5 inches, what must the standard deviation be if 99% of the carrots are between 4.9 and 5.1 inches long?

30. AGRICULTURE Suppose that the weight of a Granny Smith apple from a particular orchard is normally distributed with a weight of 10 ounces and a standard deviation of 1 ounce.
   a. Find the probability that a Granny Smith apple from this orchard weighs at least 9 ounces.
   b. Find the probability that a Granny Smith apple from this orchard weighs between 8 and 12 ounces.
   c. Find the weight so that 99% of all Granny Smith apples from this orchard weigh more than this weight.

31. MANUFACTURING Suppose that the weight that a processing plant puts in each can of peas is normally distributed with a mean of 10 ounces and a standard deviation of 0.05 ounce. The quality control department of the factory has specified that a can must contain at least 9.9 ounces so that consumers do not feel cheated, but no more than 10.15 ounces to make sure the can is not overfilled. What is the probability that a randomly selected can from the production line passes these specifications?

32. MANUFACTURING When a certain type of engine is built, a metal rod is inserted into a 6.2-centimeter round hole. If the diameter of the rod is greater than 6.2 centimeters, the rod cannot be used. Find the percentage of rods that cannot be used in the assembly of the engine if the diameter of these rods follows a normal distribution with a mean of 6.05 centimeters and a standard deviation of 0.1 centimeter.

33. QUALITY CONTROL A manufacturer needs washers that are between 0.18 and 0.22 inch thick. Any other thickness will render a washer unusable. Machine shop A sells washers for $1.00 per thousand, and the thicknesses of its washers are modeled as a normal random variable with $\mu = 0.2$ inch and $\sigma = 0.010$ inch. Machine shop B sells washers for $0.90 per thousand, and the thicknesses of its washers are also modeled as a normal random variable, but with $\mu = 0.2$ inch and $\sigma = 0.011$ inch. Using the price per usable washer as the criterion, which shop offers the manufacturer the better deal?

34. QUALITY CONTROL Suppose in Exercise 33, machine shop C wants to compete for the washer contract. Its quality control model is based on the assumption that the thickness of its washers is normally distributed with mean $\mu = 0.195$ inch and standard deviation $\sigma = 0.009$ inch. What is the maximum price p machine shop C can charge per thousand washers so that its price per usable washer will be better than the deals offered by the competition, machine shops A and B?

35. WEBSITE HITS Suppose that the number of hits on a popular website in a 1-minute period follows a Poisson distribution. It is estimated that there is a 5% likelihood of no hits occurring during the period.
   a. Find the parameter λ for the distribution.
   b. Find the probability that during a randomly chosen 1-minute period, there are exactly five hits.
   c. Find the probability that the website receives no more than five hits during a 1-minute period.
   d. Find the probability that the website receives at least one hit during a 1-minute period.
   e. Determine the most likely number of hits the website receives during a 1-minute interval.

36. INDUSTRIAL SAFETY Sandra McAfee, the safety officer for a large firm, determines that the number of business-related accidents that occur at a certain plant follows a Poisson distribution. Records suggest that there is a 2% chance of an accident occurring on a randomly selected day. Sandra decides that additional safety measures will be instituted if there is a greater than 80% chance of at least one accident occurring during a randomly selected 6-day work week. Is this action necessary or not? Explain.

37. VETERINARY MEDICINE The weights of adult basset hounds are normally distributed with $\mu = 50$ pounds and standard deviation $\sigma = 5$ pounds. Use the empirical rule to describe this distribution of weights.

38. ACADEMIC TESTING The scores on a certain test are normally distributed with $\mu = 78$ and standard deviation $\sigma = 7$. Use the empirical rule to describe the distribution of these scores.

39. ACADEMIC TESTING Suppose all fourth-graders in a certain school are taught to read by the same teaching method and that at the end of the year they are tested for reading speed. Suppose reading speed is normally distributed and that average reading speed is 150 words per minute with standard deviation of 25 words per minute. Find the percentage of students who read more than 180 words per minute.

40. ACADEMIC TESTING The scores on a psychology exam are normally distributed with a mean of 75 and a standard deviation of 10.
  a. If a score of at least 60 is needed to pass, find the fraction of students who pass the exam.
  b. If the instructor wishes to pass 70% of those taking the test, what should the lowest passing score be?
  c. How high must someone score to be in the top 10%?

41. SOCIAL PATTERNS A certain club is having a meeting. There are 1,000 members who will attend, but only 200 seats. It is decided that the oldest members will get seats. Suppose the age of members is normally distributed. If the average age of members is 40 with a standard deviation of 5, find the age of the youngest person who will get a seat.

42. VETERINARY MEDICINE The weights of healthy 10-week-old domesticated kittens follow a normal distribution with a mean weight of 24.5 ounces and standard deviation of 5.25 ounces. Find the symmetric intervals around the mean of 24.5 ounces that include the weights of 68.3%, 95.4%, and 99.7% of 10-week-old domesticated kittens, respectively.

43. VETERINARY MEDICINE Use the information in Exercise 42 to find

  a. The percentage of healthy 10-week-old domesticated kittens that weigh more than 32 ounces.
  b. The weight that is exceeded by exactly 90% of 10-week-old domesticated kittens.

44. PEDIATRIC MEDICINE The cholesterol level in children follows a normal distribution with a mean level of 175 mg/dL and a standard deviation of 35 mg/dL. Find the symmetric intervals around the mean of 175 mg/dL that respectively include the cholesterol levels of 68.3%, 95.4%, and 99.7% of all children.

45. PEDIATRIC MEDICINE Using the information in Exercise 44, find
  a. The percentage of children with a cholesterol level above 225 mg/dL.
  b. The cholesterol level exceeded by 98% of children.

46. WASTE DISPOSAL Suppose that the amount of garbage a household in the United States discards in a month follows a normal distribution with an average of 225 pounds and a standard deviation of 19 pounds.
  a. Find the percentage of households that discard more than 250 pounds of garbage per month.
  b. Find the percentage of households that discard less than 100 pounds of garbage per month.
  c. Find the percentage of households that discard between 250 and 300 pounds of garbage per month.
  d. Find the weight of garbage discarded per month by the households that produce the highest 5% of garbage.
  e. Find the weight of garbage discarded per month by the households that produce the lowest 10% of garbage.

47. MEDICINE The gestation period of humans approximately follows a normal distribution with a mean of 266 days and a standard deviation of 16 days. Find the probability of each of the following:
  a. A gestation period that is greater than 290 days.
  b. A gestation period less than 260 days.
  c. A gestation period that is between 260 and 280 days.

48. MEDICINE Assume that the human gestation period follows a normal distribution with a mean of 266 days and a standard deviation of 16 days. Find the value of A in each of the following cases:

a. The probability is 0.33 that a gestation period is less than A days.

b. The probability is 0.74 that a gestation period exceeds A days.

c. The probability is 0.5 that a gestation period is between $266 - A$ and $266 + A$ days.

49. **WEATHER FORECASTING** It is very cold in Duluth, Minnesota, in the winter. Suppose that during January, the minimum daily temperature in Duluth has a normal distribution with a mean of $-1.6°F$ and a standard deviation of $8°F$.

a. Find the probability that during a day in January, the minimum daily temperature in Duluth lies between $0°$ and $10°$. What is the probability that the minimum daily temperature lies between $-5°$ and $5°$?

b. What is the probability that during a day in January, the minimum daily temperature in Duluth is greater than $5°$?

c. Find the probability that during a day in January, the minimum daily temperature in Duluth is less than $5°$.

50. **POLITICAL POLLS** A political pollster announces that Senator Able leads his challenger, Ms. Baker, by 51.2% to 48.8%, where the error in the poll results is normally distributed with a mean of 0 percentage points and a standard deviation of 3 percentage points.

a. Find the probability that Ms. Baker is actually leading. [*Hint:* How large must the error in Senator Able's share be for the lead to switch? Assume there are no undecided votes.]

b. Find the probability that Senator Able's lead is actually at least 3 percentage points instead of the indicated 2.4.

c. Research polling procedures, and write a paragraph on whether you believe political polls provide useful, reliable information.

51. **DRUG TESTING** The recovery time of patients with Goleta rash is normally distributed with a mean of 11 days and a standard deviation of 2 days.

a. Find the probability that a patient takes between 10 and 18 days to recover.

b. Acme Drugs claims that its treatment of Goleta rash accelerates recovery. A patient takes the Acme medication and recovers in 9 days. Are you impressed? [*Hint:* What is the probability that the patient would have recovered in 9 days or less without the medication?]

c. Another patient takes the Acme medication and recovers in 6 days. Now are you impressed?

d. Actual drug testing involves conducting carefully monitored experiments using large numbers of subjects, with a goal of achieving consistent results that are statistically "significant." Research the policies for drug certification required by the Food and Drug Administration (FDA), and write a paragraph on these policies. Do you think the FDA is justified in being so strict?

52. **SHARK ATTACKS** Suppose that the number of shark attacks in coastal waters around the United States follows a Poisson distribution with a mean of 0.3 shark attacks per day.

a. What is the parameter λ of the distribution?

b. Find the probability that on a randomly selected day, there will be no shark attacks.

c. Find the probability that on a randomly selected day, there will be at least two shark attacks.

53. **BACTERIOLOGY** Suppose that the number of *Escherichia coli* bacteria in a water sample drawn from a swimming beach in New Jersey follows a Poisson distribution with a mean of 6.1 bacteria per cubic millimeter of water.

a. What is the parameter λ of the distribution?

b. Find the probability that a 1 cubic millimeter random sample of water contains no *E. coli* bacteria.

c. Find the probability that a random sample of 1 cubic millimeter of water contains exactly three *E. coli* bacteria.

d. Find the probability that a random sample of 1 cubic millimeter of water contains less than six *E. coli* bacteria.

54. **DISEASE CONTROL** Assume that the number of people who die from hepatitis A each year follows a Poisson distribution and that an average of 3 people per 100,000 die from hepatitis A each year in the United States.

a. Find the probability that in an American city of 100,000, no people die during a randomly selected year from hepatitis A.

b. Find the probability that in an American city of 100,000, no more than four people die from hepatitis A during a randomly selected year.

c. Find the probability that in an American city of 100,000, at least five people die from hepatitis A during a randomly selected year.

55. DISEASE CONTROL Using the information from Exercise 54, in a randomly selected year, how much more likely is it for at least one person to die from hepatitis A in an American city of 1,000,000 people than in a city of 100,000 people?

56. NUTRITION Suppose a bakery makes a batch of 2,400 chocolate chip cookies using a total of 12,000 chocolate chips. The random variable X that represents the number of chocolate chips in a cookie can be modeled using a Poisson distribution with parameter

$$\lambda = \frac{12,000}{2,400} = 5$$

  a. Find the probability that a cookie has exactly five chocolate chips.

  b. Find the probability that a cookie has no chocolate chips.

  c. Find the probability that a cookie has three or fewer chocolate chips.

  d. Find the probability that a cookie has seven or more chocolate chips.

MISCELLANEOUS PROBLEMS

57. Decide whether the following statements are true or false. Explain your answer.

  a. If X has a normal distribution, then

$$P(X \geq 2.1) = P(X \leq -2.1)$$

  b. If Z has the standard normal distribution, then

$$P(Z \geq 2.1) = P(Z \leq -2.1)$$

58. Show that a Poisson random variable X with parameter λ has variance $\mathrm{Var}(X) = \lambda$.

59. Verify that normally distributed data are 95.4% likely to be within two standard deviations of the mean and 99.7% likely to be within three standard deviations of the mean.

60. Show that the probability density function

$$f(x) = \frac{1}{\sigma\sqrt{2\pi}}e^{-(x-\mu)^2/2\sigma^2}$$

has an absolute maximum at $x = \mu$ and inflection points at $x = \mu + \sigma$ and $x = \mu - \sigma$.

61. Refer to Example 10.3.7. Let Z be the standard normal density function, and estimate the probability

$$P(0 \leq Z \leq 1) = \int_0^1 \frac{1}{\sqrt{2\pi}}e^{-x^2/2}\,dx$$

by integrating the first five nonzero terms of the Taylor series expansion for $e^{-x^2/2}$ term by term. Compare your estimate with the probability found by using Table 11.1.

Important Terms, Symbols, and Formulas

Probability density function $f(x)$ for the continuous random variable X: (843)

$$f(x) \geq 0 \quad \text{for all } x$$

$$\int_{-\infty}^{\infty} f(x)\, dx = 1$$

$$P(a \leq X \leq b) = \int_{a}^{b} f(x)\, dx$$

Uniform density function: (845)

$$f(x) = \begin{cases} \dfrac{1}{b-a} & \text{if } a \leq x \leq b \\ 0 & \text{otherwise} \end{cases}$$

Exponential density function: (846)

$$f(x) = \begin{cases} \lambda e^{-\lambda x} & \text{if } x \geq 0 \\ 0 & \text{if } x < 0 \end{cases}$$

Joint probability density function $f(x, y)$, for random variables X and Y: (847)

$$f(x, y) \geq 0 \quad \text{for all } (x, y)$$

$$\int_{-\infty}^{\infty} \int_{-\infty}^{\infty} f(x, y)\, dA = 1$$

$$P[(X, Y) \text{ in } R] = \int\!\!\int_{R} f(x, y)\, dA$$

Features of a continuous random variable X:

Expected value

$$E(X) = \int_{-\infty}^{\infty} x f(x)\, dx \quad (856)$$

Variance

$$\text{Var}(X) = \int_{-\infty}^{\infty} [x - E(X)]^2 f(x)\, dx \quad (857)$$

Expected values for a joint probability density function $f(x, y)$:   (860)

$$E(X) = \int_{-\infty}^{\infty} \int_{-\infty}^{\infty} x f(x, y)\, dA$$

$$E(Y) = \int_{-\infty}^{\infty} \int_{-\infty}^{\infty} y f(x, y)\, dA$$

Features of a normal random variable X:

Normal probability density function

$$f(x) = \frac{1}{\sigma \sqrt{2\pi}} e^{-(x-\mu)^2/2\sigma^2} \quad (865)$$

Expected value $E(X) = \mu$ (867)
Variance $\text{Var}(X) = \sigma^2$ (867)
Standard variation $\sigma(X) = \sigma$ (867)

Empirical rule for area under a normal curve   (867)
Standard normal distribution ($\mu = 0$, $\sigma = 1$)   (868)
Transformation to standard normal form   (873)
Features of a Poisson distributed random variable X with parameter λ:
Probability

$$P(X = n) = \frac{\lambda^n}{n!} e^{-\lambda} \quad (879)$$

Expected value $E(X) = \lambda$ (879)
Variance $\text{Var}(X) = \lambda$ (879)
Standard deviation $\sigma(X) = \sqrt{\lambda}$ (879)

Checkup for Chapter 11

1. The outcomes and corresponding probability assignments for a discrete random variable X are listed in the table. Draw the histogram for X. Then find the expected value $E(X)$, the variance $\text{Var}(X)$, and the standard deviation $\sigma(X)$.

Outcomes for X	0	1	2	3	4
Probability	$\dfrac{1}{6}$	$\dfrac{1}{6}$	$\dfrac{1}{5}$	$\dfrac{2}{5}$	$\dfrac{1}{15}$

2. In each case, determine whether the given function is a probability density function.

   a. $f(x) = \begin{cases} \dfrac{4}{3} \sqrt[3]{x} & \text{if } 0 \leq x \leq 1 \\ 0 & \text{otherwise} \end{cases}$

   b. $f(x) = \begin{cases} 2e^{-x/2} & \text{if } x \geq 0 \\ 0 & \text{if } x < 0 \end{cases}$

3. The probability density function for a particular random variable X is

$$f(x) = \begin{cases} \dfrac{1}{3} e^{-x/3} & \text{if } x \geq 0 \\ 0 & \text{if } x < 0 \end{cases}$$

Find each of the following probabilities:
   a. $P(2 \leq X \leq 3)$
   b. $P(X \geq 3)$

4. The probability density function for a particular random variable X is

$$f(x) = \begin{cases} \dfrac{1}{36}(6x - x^2) & \text{if } 0 \le x \le 6 \\ 0 & \text{otherwise} \end{cases}$$

Find the expected value $E(X)$ and variance $\text{Var}(X)$ for X.

5. Suppose the random variable Z has a standard normal distribution. Find these probabilities:
   a. $P(Z \ge 0.85)$
   b. $P(-1.30 \le Z \le 3.25)$

6. A random variable X has a normal distribution with mean $\mu = 12$ and standard deviation $\sigma = 2$. Use the empirical rule to find
   a. $P(10 \le X \le 14)$
   b. $P(8 \le X \le 16)$

7. **MAINTENANCE**   The failure of a certain key component will shut down the operation of a system until a new component is installed. If installation time is distributed uniformly from 1 to 4 hours, what is the probability that the system will be inoperable for at least 2 hours?

8. **QUALITY CONTROL**   Suppose that 5% of the items produced by a certain company are defective. A quality control inspector tests each item before it leaves the factory. Find the probability that the inspector passes at least five items before detecting a defective item.

9. **MEDICINE**   Let X be the random variable that represents the number of years a person in a test group with a certain kind of virulent cancer lives after receiving an experimental drug. Suppose X has an exponential distribution with parameter $\lambda = 0.5$.

a. What is the probability of a randomly selected cancer victim in this group surviving at least 3 years after beginning therapy with the drug?
b. Find the expected value of X. Describe what $E(X)$ represents.

10. **ACADEMIC TESTING**   To qualify for the police academy, a candidate must score in the top 20% on a qualifying test. Assuming that the test scores are normally distributed and the average score on the test is 150 points with a standard deviation of 24 points, find the minimum qualifying score.

11. **ANIMAL CONTROL**   Assume that the number of people who die each year from dog bites follows a Poisson distribution and that on average, five people per million die in this fashion in the United States each year. Find the probability that at least five people will die from dog bites in an American city of population 1 million during a randomly selected year.

12. **MILITARY TRAINING**   As part of his training, a paratrooper is required to jump into a circular area marked out on the ground. A bulls-eye that occupies 10% of the circle is marked 10, and there are three other circles, numbered 5, 3, and 1 that respectively occupy 20%, 30%, and 40% of the total area of the target. Assuming the trainee lands at a random position inside the target, what score should he expect to receive?

13. Decide whether the following statements are true or false. Explain your answer.
    a. If X has a normal distribution with $\mu = 8$, then
    $$P(X \ge 8.21) = P(X \le 7.79)$$
    b. If X has a normal distribution, then the probability is about 0.87 that its value will be within 1.5 standard deviations of the mean.

Review Exercises

In Exercises 1 and 2, the outcomes and corresponding probability assignments for a discrete random variable X are listed. Draw the histogram for X. Then find the expected value E(X), the variance Var(X), and the standard deviation σ(X).

1.
Outcomes for X	1	2	3	4	5
Probability	$\dfrac{1}{9}$	$\dfrac{2}{9}$	$\dfrac{1}{3}$	$\dfrac{1}{9}$	$\dfrac{2}{9}$

2.
Outcomes for X	0	2	4	6	8
Probability	$\dfrac{1}{8}$	$\dfrac{3}{8}$	$\dfrac{1}{4}$	$\dfrac{1}{8}$	$\dfrac{1}{8}$

In each of Exercises 3 through 6, determine whether the given random variable X is discrete or continuous.

3. X counts the number of eggs laid by a randomly selected fruit fly.

4. X measures the annual distance flown by a randomly selected airplane from a particular airline.

5. X measures the annual salary of a randomly selected player on a particular major league baseball team.

6. X counts the number of books in the library of a randomly selected professor at your school.

In Exercises 7 through 10, $f(x)$ is a probability density function for a particular continuous random variable X. In each case, find the indicated probabilities.

7. $f(x) = \begin{cases} \dfrac{1}{5} & \text{if } 3 \le x \le 8 \\ 0 & \text{otherwise} \end{cases}$
$P(2 \le X \le 7)$ and $P(X \ge 5)$

8. $f(x) = \begin{cases} \dfrac{1}{x^2} & \text{if } x \ge 1 \\ 0 & \text{if } x < 1 \end{cases}$
$P(1 \le X \le 3)$ and $P(X \ge 2)$

9. $f(x) = \begin{cases} 0.75x(2x - x^2) & \text{if } 0 \le x \le 2 \\ 0 & \text{otherwise} \end{cases}$
$P(X \le 1)$ and $P(1 \le X \le 3)$

10. $f(x) = \begin{cases} 0.25xe^{-x/2} & \text{if } x \ge 0 \\ 0 & \text{otherwise} \end{cases}$
$P(X \ge 2)$ and $P(0 \le X \le 3)$

11. Find the expected value $E(X)$ and variance $\text{Var}(X)$ for the random variable X in Exercise 9.

12. Find the expected value $E(X)$ and variance $\text{Var}(X)$ for the random variable X in Exercise 10.

13. Find a number c so that the following function $f(x)$ is a probability density function:
$$f(x) = \begin{cases} cxe^{-x/4} & \text{if } x \ge 0 \\ 0 & \text{otherwise} \end{cases}$$

14. Find a number c so that the following function $f(x)$ is a probability density function:
$$f(x) = \begin{cases} \dfrac{c}{x^4} & \text{if } x \ge 1 \\ 0 & \text{otherwise} \end{cases}$$

15. If the random variable X is normally distributed with mean $\mu = 7$ and standard deviation $\sigma = 2$, what is $P(X \ge 9)$?

16. Find b if $P(Z \ge b) = 0.73$, where Z is a random variable with a standard normal distribution ($\mu = 0$, $\sigma = 1$).

17. **QUALITY CONTROL** A toy manufacturer makes hollow rubber balls. The thickness of the outer shell of such a ball is normally distributed with mean 0.03 millimeter and standard deviation 0.0015 millimeter. What is the probability that the outer shell of a randomly selected ball will be less than 0.025 millimeter thick?

18. **EFFECT OF AN EPIDEMIC** A study of the effect of an epidemic of mononucleosis (mono) on the students at a particular small private college determines that during the 30 days of November, there were:

    5 days when no students had mono
    7 days when exactly one student had mono
    4 days when exactly two students had mono
    9 days when exactly six students had mono
    3 days when exactly seven students had mono
    2 days when exactly eight students had mono

Let X be the random variable that measures the number of students with mononucleosis on a randomly selected day in November.

  a. Find the probability distribution for X. Then construct a histogram for this distribution.

  b. How many students would you expect to have mononucleosis on a randomly selected day in November?

19. **PORTFOLIO MANAGEMENT** Noreen estimates that it is 30% likely for a certain speculative stock to go up by $5 a share, while it is equally likely for the stock to go up by only $2 a share. However, it is also 40% likely that the stock will go down by $3 a share. How much should Noreen expect to gain or lose on each share of the stock she purchases?

20. **WHEEL OF FORTUNE** A wheel of fortune is divided into 20 circular sectors of equal area. The wheel is spun and a payoff is made according to the color of the region on which the indicator lands. One region is gold and pays $50 when hit; five regions are blue and pay $25; and four other regions are red and pay $10. The remaining 10 regions are black and pay nothing. Would you be willing to pay $10 to play this game?

21. **PEDIATRICS** There are 200 children in a certain school, and the weight of the children is a random variable X that is normally distributed with mean $\mu = 80$ pounds and standard deviation $\sigma = 7$ pounds.

a. How many children weigh more than 90 pounds?

b. How many children weigh less than 70 pounds?

c. How many children weigh exactly 80 pounds?

22. **PRODUCT RELIABILITY** Electronic components made by a certain process have a time to failure that is measured by a normal random variable. If the mean time to failure is 15,000 hours with a standard deviation of 800 hours, what is the probability that a randomly selected component will last no more than 10,000 hours?

23. **LOTTERY** The probability of winning $100 in a particular lottery is 0.08, the probability of winning $20 is 0.12, the probability of winning $5 is 0.2, and the probability of losing is 0.6. What is a fair price to pay for a lottery ticket?

24. **TIME MANAGEMENT** A bakery turns out a fresh batch of chocolate chip cookies every 45 minutes. Tina arrives (at random) at the bakery, hoping to buy a fresh cookie. Use an appropriate uniform density function to find the probability that Tina arrives within 5 minutes (before or after) the time that the cookies come out of the oven.

25. **DEMOGRAPHICS** A study recently commissioned by the mayor of a large city indicates that the number of years a current resident will continue to live in the city may be modeled as an exponential random variable with probability density function

$$f(t) = \begin{cases} 0.4e^{-0.4t} & \text{for } t \geq 0 \\ 0 & \text{otherwise} \end{cases}$$

a. Find the probability that a randomly selected resident will move within 10 years.

b. Find the probability that a randomly selected resident will remain in the city for more than 20 years.

c. How long should a randomly selected resident be expected to remain in town?

26. **TRAFFIC CONTROL** Suppose the time (in minutes) between the arrivals of successive cars at a toll booth is measured by the random variable X with probability density function

$$f(t) = \begin{cases} 0.5e^{-0.5t} & \text{if } t \geq 0 \\ 0 & \text{otherwise} \end{cases}$$

a. Find the probability that a randomly selected pair of successive cars will arrive at the toll booth at least 6 minutes apart.

b. Find the average time between the arrivals of successive cars at the toll booth.

27. **TRAFFIC MANAGEMENT** The distance (in feet) between successive cars on a freeway is modeled by the random variable X with probability density function

$$f(x) = \begin{cases} 0.25xe^{-x/2} & \text{if } x \geq 0 \\ 0 & \text{otherwise} \end{cases}$$

a. Find the probability that a randomly selected pair of cars will be less than 10 feet apart.

b. What is the average distance between successive cars on the freeway?

28. **TRAFFIC MANAGEMENT** Suppose the random variable X in Exercise 27 is normally distributed with mean $\mu = 12$ feet and standard deviation $\sigma = 4$ feet. Now what is the probability that a randomly selected pair of cars will be less than 10 feet apart?

29. **INSURANCE POLICY** An insurance company charges $10,000 for a policy insuring against a certain kind of accident and pays $100,000 if the accident occurs. Suppose it is estimated that the probability of the accident occurring is $p = 0.02$. Let X be the random variable that measures the insurance company's profit on each policy it sells.

a. What is the probability distribution for X?

b. What is the company's expected profit per policy sold?

c. What should the company charge per policy to double its expected profit per policy?

30. **PERSONAL HEALTH** Jules decides to go on a diet for 6 weeks, with a goal of losing between 10 and 15 pounds. Based on his body configuration and metabolism, his doctor determines that the amount of weight he will lose can be modeled by a continuous random variable X with probability density function $f(x)$ of the form

$$f(x) = \begin{cases} k(x - 10)^2 & \text{for } 10 \leq x \leq 15 \\ 0 & \text{otherwise} \end{cases}$$

If the doctor's model is valid, how much weight should Jules expect to lose? [*Hint:* First determine the value of the constant k.]

31. **FISHERY MANAGEMENT** Brooke, the manager of a fishery, determines that the age X (in weeks) at which a certain species of fish dies follows an

exponential distribution with probability density function

$$f(t) = \begin{cases} \lambda e^{-\lambda t} & \text{for } t \geq 0 \\ 0 & \text{otherwise} \end{cases}$$

Brooke observes that it is twice as likely for a randomly selected fish to die during the first 10-week period as during the next 10 weeks (from week 10 to week 20).

a. What is λ?

b. What is the probability that a randomly chosen fish will die within the first 5 weeks?

c. How long should Brooke expect a randomly selected fish to live?

32. **METALLURGY** The proportion of impurities by weight in samples of copper ore taken from a particular mine is measured by a random variable X with probability density function

$$f(x) = \begin{cases} 21x^2(1 - \sqrt{x}) & \text{for } 0 \leq x \leq 1 \\ 0 & \text{otherwise} \end{cases}$$

a. What is the probability that the proportion of impurities in a randomly selected sample will be less than 5%?

b. What is the probability that the proportion of impurities will be greater than 50%?

c. What proportion of impurities would you expect to find in a randomly selected sample?

33. **BEVERAGES** Suppose that the volume of soda in a bottle produced at a particular plant is normally distributed with a mean of 12 ounces and a standard deviation of 0.05 ounce.

a. Find the probability that a bottle filled at this plant contains at least 11.8 ounces.

b. Find the volume of soda so that 95% of all bottles filled at this plant contain less than this amount.

34. **QUALITY CONTROL** An automobile manufacturer claims that its new cars get an average of 30 miles per gallon in city driving. Assume the manufacturer's claim is correct and that gas mileage is normally distributed, with standard deviation of 2 miles per gallon.

a. Find the probability that a randomly selected car will get less than 25 miles per gallon.

b. If you test two cars, what is the probability that both get less than 25 miles per gallon?

35. **ACADEMIC TESTING** The results of a calculus exam are normally distributed with a mean of 72.3 and a standard deviation of 16.4. Find the

probability that a randomly chosen student's score is between 50 and 75. If there are 82 students in the class, about how many have scores between 50 and 75?

36. **MEDICINE** Suppose that the number of children who die each year from leukemia follows a Poisson distribution and that on average, 7.3 children per 100,000 die from leukemia. For a city with 100,000 children, find the probability of each of the following events:

a. Exactly seven children in the city die from leukemia each year.

b. Fewer than two children in the city die from leukemia each year.

c. More than five children in the city die from leukemia each year.

37. **ECOLOGY** The pH level of a liquid measures its acidity and is an important issue in studying the effects of acid rain. Suppose that a test is conducted under controlled conditions that allow the change in pH in a particular lake resulting from acid rain to be recorded. Let X be a random variable that measures the pH of a sample of water taken from the lake, and assume that X has the probability density function.

$$f(x) = \begin{cases} 0.75(x - 4)(6 - x) & \text{for } 4 \leq x \leq 6 \\ 0 & \text{otherwise} \end{cases}$$

a. Find the probability that the pH of a randomly selected sample will be at least 5.

b. Find the expected pH of a randomly selected sample.

38. **SPORTS MEDICINE** Suppose that the number of injuries a team suffers during a typical football game follows a Poisson distribution with an average of 2.5 injuries.

a. Find the probability that during a randomly chosen game, the team suffers exactly two injuries.

b. Find the probability that during a randomly chosen game, the team suffers no injuries.

c. Find the probability that during a randomly chosen game, the team suffers at least one injury.

39. **PACKAGE DELIVERY** Suppose that the number of overnight packages a business receives during a business day follows a Poisson distribution and that, on average, the company receives four overnight packages per day.

a. Find the probability that the company receives exactly four overnight packages on a randomly selected business day.

b. Find the probability that the company does not receive any overnight packages on a randomly selected business day.

c. Find the probability that the company receives fewer than four overnight packages on a randomly selected business day.

40. **JOURNALISM**   Suppose that the number of typographical errors on a page of a local newspaper follows a Poisson distribution with an average of 2.5 errors per page.

a. Find the probability that a randomly selected page is free from typographical errors.

b. Find the probability that a randomly selected page has at least one typographical error.

c. Find the probability that a randomly selected page has at least three typographical errors.

d. Find the probability that a randomly selected page has fewer than three typographical errors.

41. **HIGHWAY ACCIDENTS**   A report models the number of automobile accidents on a particular highway as a random variable with a Poisson distribution. Suppose it is found that on average, there is an accident every 10 hours.

a. Find the probability that there are no accidents on this highway during a randomly selected 24-hour period.

b. Find the probability that there is at least one accident on this highway during a randomly selected 12-hour period.

c. Find the probability that there are no accidents on this highway during a randomly selected hour.

42. **PUBLIC HEALTH**   As part of a campaign to combat a new strain of influenza, public health authorities are planning to inoculate 1 million people. It is estimated that the probability of an individual having a bad reaction to the vaccine is 0.0005. Suppose the number of people inoculated who have bad reactions to the vaccine is modeled by a random variable with a Poisson distribution.

a. What is λ for the distribution?

b. What is the probability that of the 1 million people inoculated, exactly five will have a bad reaction?

c. What is the probability that of the 1 million people inoculated more than 10 will have a bad reaction?

43. **LABOR EFFICIENCY**   A company wishes to examine the efficiency of two members of its senior staff, Jack and Jill, who work independently of one another. Let X and Y be random variables that measure the proportion of the work week that Jack and Jill, respectively, actually spend performing their duties. Assume that the joint probability density function for X and Y is

$$f(x, y) = \begin{cases} 0.4(2x + 3y) & \text{if } 0 \le x \le 1, 0 \le y \le 1 \\ 0 & \text{otherwise} \end{cases}$$

a. Verify that $f(x, y)$ satisfies the requirements for a joint probability density function.

b. Find the probability that Jack spends less than half his time working while Jill spends more than half her time working.

c. Find the probability that Jack and Jill each spend at least 80% of the work week performing their assigned tasks.

d. Find the probability that Jack and Jill combine for less than a full work week. [*Hint:* This is the event that $X + Y < 1$.]

44. **LABOR EFFICIENCY**   For the situation described in Exercise 43, find the expected values of X and Y. The company expects its senior staff members to be spending at least 75% of their time performing their duties. Are either Jack or Jill in danger of a reprimand?

45. **TIME MANAGEMENT**   Suppose the random variable X measures the time (in minutes) that a person stands in line at a certain bank and Y measures the duration (in minutes) of a routine transaction at the teller's window. Assume that the joint probability density function for X and Y is

$$f(x, y) = \begin{cases} 0.125e^{-x/4} e^{-y/2} & \text{if } x \ge 0 \text{ and } y \ge 0 \\ 0 & \text{otherwise} \end{cases}$$

a. What is the probability that neither activity takes more than 5 minutes?

b. What is the probability that you will complete your business at the bank (both activities) within 8 minutes?

46. **INSURANCE SALES**   Let X be a random variable that measures the time (in minutes) that a person spends with an agent choosing a life

insurance policy, and let Y measure the time (in minutes) the agent spends doing paperwork once the client has selected a policy. Suppose the joint probability density function for X and Y is

$$f(x, y) = \begin{cases} \dfrac{1}{300} e^{-x/30} e^{-y/10} & \text{for } x \geq 0, y \geq 0 \\ 0 & \text{otherwise} \end{cases}$$

a. Find the probability that choosing the policy takes more than 20 minutes.

b. Find the probability that the entire transaction (policy selection and paperwork) will take more than half an hour.

c. How much more time would you expect to spend selecting the policy than completing the paperwork?

47. **TIME MANAGEMENT**   A shuttle tram arrives at a tram stop at a randomly selected time X within a 1-hour period, and a tourist independently arrives at the same stop also at a randomly selected time Y within the same hour. The tourist has the patience to wait for the tram for up to 20 minutes before calling a taxi. The joint probability density function for X and Y is

$$f(x, y) = \begin{cases} 1 & \text{if } 0 \leq x \leq 1, 0 \leq y \leq 1 \\ 0 & \text{otherwise} \end{cases}$$

a. What is the probability that the tram takes longer than 20 minutes to arrive?

b. What is the probability that the tourist arrives after the tram?

c. What is the probability that the tourist connects with the tram? [*Hint:* The event of this occurring has the form

$$Y + a \leq X \leq Y + b$$

for suitable numbers a and b.]

48. **WARRANTY PROTECTION**   A major appliance contains two components that are vital for its operation in the sense that if either fails, the appliance is rendered useless. Let the random variable X measure the useful life (in years) of the first component, and let Y measure the useful life of the second component (also in years). Suppose the joint probability density function for X and Y is

$$f(x, y) = \begin{cases} 0.1e^{-x/2} e^{-y/5} & \text{if } x \geq 0 \text{ and } y \geq 0 \\ 0 & \text{otherwise} \end{cases}$$

a. Find the probability that the appliance fails within the first 5 years.

b. Which component of a randomly selected appliance would you expect to last longer? How much longer?

EXPLORE! UPDATE

Complete solutions for all EXPLORE! boxes throughout the text can be accessed at the book-specific website, www.mhhe.com/hoffmann.

Solution for Explore! on Page 830

Find the expected value of the random variable X in Example 11.1.5 using a graphing calculator. Clear lists L1 through L5, if necessary, and enter the numbers of ill students into list L1. Since there were no days when 3, 4, or 5 students were ill, you can omit these values as shown in the following left screen. Then enter the corresponding numbers of days in list L2. Put the probability of each number of ill students into list L3 by letting L3 = L2 ÷ 30. These values are shown in the following middle screen; these are the decimal forms of the probabilities found in Example 10.1.5.

Next multiply each number of ill students by its probability and put the result in list L4 by letting L4 = L1 × L3. These are the values being summed on page 830. To find the sum, let L5 = cumSum(L4) where **6:cumSum(** is found under the **OPS** menu of **LIST (2nd STAT).** The final value in L5 is the sum of all the values in L4 and is the expected value of the random variable X representing the number of students ill on a given day. This value is close to 5, as on page 830.

Solution for Explore! on Page 835

Find the probabilities in parts (a) and (b) of Example 11.1.10 using your graphing calculator. The appropriate functions are **E:geometpdf(** and **F:geometcdf(** found under **DISTR (2nd VARS)** as shown in the following left screen. To find the probability that we have to toss the coin exactly three times until heads comes up, use geometpdf(p, n) where p is the probability of success on any given trial and the first success occurs on the nth trial. The middle screen shows geometpdf(0.5, 3), where $p = 0.5$ is the probability of heads coming up on any given toss.

To find the probability that the coin comes up heads within the first three tosses, use geometcdf(p, n) where p is the probability of success on any given trial and the first success occurs on or before the nth trial. The right screen shows geometcdf(0.5, 3). This probability is the sum of the probabilities that heads comes up for the first time on the first toss, the second toss, and the third toss. These probabilities are the decimal forms of the probabilities found in parts (a) and (b) of Example 11.1.10.

Solution for Explore! on Page 845

Following Example 11.2.3, store $f(x)$ into Y1 as shown in the following left screen, using the window $[0, 47]5$ by $[-0.01, 0.05]0.01$. The area below this uniform density function from $x = 15$ to 40 is $0.625 = \dfrac{5}{8}$, as graphically displayed in the right screen.

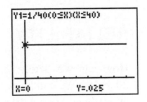
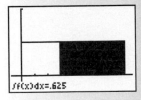

Solution for Explore! on Page 846

Following Example 11.2.4, set $Y1 = 0.5e^\wedge(-0.5X)(X \geq 0)$, using the window $[-2, 10]1$ by $[-0.1, 0.6]0.1$. The graph of Y1 is shown on the following left screen. Now write $Y2 = \text{fnInt}(Y1, X, 0, X)$ and observe the table of values for Y2, for X values starting at $X = 20$ in unit increments. Notice that Y2 is approximately 1 when $X = 25$. The numerical integration feature can be used to confirm that $P(2 \leq X \leq 3) = 0.1447$.

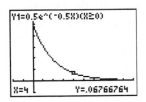
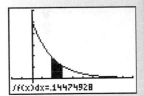
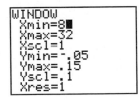

Solution for Explore! on Page 874

On a TI-84 Plus, find **1:ShadeNorm**(lower bound, upper bound, μ, σ) via the **DRAW** menu of the **DISTR** key (**2nd VARS**). First set the window $[8, 32]1$ by $[-0.05, 0.15]0.1$. For part (a) of Example 11.4.4, write $\text{ShadeNorm}(-100, 26, 20, 4)$, where $\mu = 20$, $\sigma = 4$, and the lower bound is -100, a sufficiently large negative number.

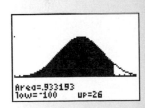

For part (b), write ShadeNorm(18, 100, 20, 4), and for part (c), write ShadeNorm(15, 21, 20, 4). The following screens display the results. It is important to clear the previous screen by pressing **1:ClrDraw** on the **DRAW** key (**2nd PRGM**), before doing the next calculation.

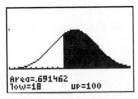
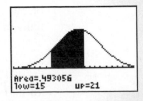

EXPLORE! UPDATE

THINK ABOUT IT

RELIABILITY

How can we determine the probability that an artificial knee will last more than 20 years? We can answer this question if we have a probability distribution for the time an artificial knee works before failing. But what type of distribution is appropriate for the life span of an artificial knee? In this essay, we introduce probability models that can be used to study the reliability, or life span, of systems. The study of reliability is important because all human-made systems, ranging from simple components to complex systems, ultimately fail. The same holds for biological systems, as living organisms have finite life spans. Many practical questions involve determining the reliability of a system, either human-made or biological.

Using what we have already learned about probability, we can begin studying reliability. Given a particular system, let the random variable X denote the time it operates properly before failing, assuming that it was put into service at time $t = 0$. We can answer many practical questions such as finding the average time a system operates until it fails and the probability a system operates for at least a specified length of time by studying properties of the random variable X. In reliability studies, we make use of several important functions related to X. First, we let $f(t)$ denote the probability density function of X. Then, the probability that the system fails before time t is

$$F(t) = \int_0^t f(x)\, dx$$

We call $F(t)$ the **cumulative distribution function** of X. The probability density function $f(t)$ is the derivative of the cumulative distribution $F(t)$ (this follows from one of the two parts of the fundamental theorem of calculus). That is,

$$\frac{dF(t)}{dt} = f(t)$$

To study questions involving reliability, we introduce the function $R(t)$, which is aptly called the **reliability function** $R(t)$. This is the probability that the system survives past time t; that is,

$$R(t) = P(X > t)$$

Note that $R(t) = 1 - P(X \le t) = 1 - F(t)$ and that if we know any of the three functions $f(t)$, $F(t)$, and $R(t)$, we can easily find the other two functions.

Given the reliability function associated with the random variable X, we would like to find the rate at which failures are occurring at a particular time t. This should be the limit as s approaches zero of the probability that the system fails during the s units of time following time t, divided by s. That is, the failure rate at time t is

$$\lim_{s \to 0} \frac{\text{probability of failure in the time interval from time } t \text{ to time } t + s}{s}$$

To find this failure rate, first observe that the probability that a system operating at time t continues to operate during the s units of time immediately following time t is the proportion of systems working at time t that are still working at time $t + s$. This proportion is

$$\frac{P(t < X \le t + s)}{P(t < X)} = \frac{F(t + s) - F(t)}{R(t)}$$

Consequently, the failure rate at time t is

$$\lim_{s \to 0} \frac{1}{s} \frac{F(t + s) - F(t)}{R(t)} = \frac{1}{R(t)} \lim_{s \to 0} \frac{F(t + s) - F(t)}{s} = \frac{F'(t)}{R(t)} = \frac{f(t)}{R(t)}$$

This leads us to define the **failure rate function** $r(t)$, which represents the failure rate at time t of the system being studied, by

$$r(t) = \frac{f(t)}{R(t)}$$

It follows that once we know one of the three functions $f(t)$, $F(t)$, and $R(t)$, we also can find $r(t)$. Furthermore, given the failure rate function $r(t)$, we can find the reliability function using the equation

$$R(t) = e^{-\int_0^t r(x)\, dx} \tag{1}$$

To see why Equation (1) is true, first note that because $R(t) = 1 - F(t)$, we have $R'(t) = -F'(t)$. Moreover, because $f(t) = F'(t)$, it follows that

$$r(t) = \frac{f(t)}{R(t)} = \frac{F'(t)}{R(t)} = \frac{-R'(t)}{R(t)}$$

This implies that

$$-\int_0^t r(x)\, dx = \int_0^t \left[-\frac{R'(x)}{R(x)} \right] dx = \ln R(t) - \ln R(0) = \ln R(t) - \ln 1 = \ln R(t)$$

where we have used the fact that $R(0) = P(X > 0) = 1$. We can now conclude that $e^{-\int_0^t r(x)\, dx} = e^{\ln R(t)} = R(t)$.

When modeling the reliability of a system, certain assumptions are made about its failure rate. The simplest model of the reliability of a system assumes that the system has a constant failure rate. We leave it as an exercise (Question 1) to use Equation (1) to show that if $r(t) = \lambda$ where $\lambda > 0$, then $f(t)$ has an exponential distribution. Assuming a constant failure rate produces a simple mathematical model using the exponential distribution. However, the failure rates of most, if not all, human-made products and biological systems are not constant. In practice, many systems have failure rates that increase with time. For example, artificial knees wear out at a more rapid rate as components wear out. Older cars suffer a higher failure rate than new cars as their engines, transmissions, and other systems age. Many electronic components that are put into service after an initial burn-in period exhibit low failure rates for a while but then degrade over time. Even computer software experiences increasing failure rates. For instance, a newly rebooted Windows PC is less likely to crash than one that has been left running for a day. Living organisms also have a higher failure (or mortality) rate as they get older. It is more likely that a 100-year-old person will die in the next year than for a 50-year-old person to die in this period. Also note that for some systems, failure rates decrease with time, such as some mechanical components whose failures are mostly caused by manufacturing defects. Such components fail at a faster rate initially, but once the components with flaws have failed, this rate decreases.

How can we model increasing failure rates, as well as failure rate decreases, with time? Finding adequate models challenged mathematicians and engineers for many years. After extensive study, it was found that an extremely wide class of applications can be accurately modeled using a failure rate function of the form

$$r(t) = \alpha\beta t^{\beta-1}$$

where α and β are positive constants. This function looks quite complicated, but by carefully choosing values of α and β, engineers and scientists have had success using failure rate functions of this form in many applications. Note that the function $r(t)$ is an increasing function when $\beta > 1$; values of β greater than 1 are used in models where the failure rate increases with time. The failure $r(t)$ is a decreasing function when $0 < \beta < 1$; values of β less than 1 are used in models where the failure rate decreases with time. When $\beta = 1$, there is a constant failure rate.

Using this function for $r(t)$ shows (see Question 4) that the reliability function $R(t)$ is given by

$$R(t) = e^{-\alpha t^{\beta}}$$

and that the cumulative distribution function is given by

$$F(t) = 1 - e^{-\alpha t^{\beta}}$$

for $t > 0$. By differentiating $F(t)$, we can find the probability density function. That is,

$$f(t) = F'(t) = \alpha\beta t^{\beta-1} e^{-\alpha t^{\beta}}$$

A random variable X with this probability density function is said to have a **Weibull distribution** with parameters α and β, where $\alpha > 0$ and $\beta > 0$. This distribution is widely used in reliability studies and in many other models. It is named for its inventor, Waloddi Weibull, who introduced the distribution that bears his name in 1939. Weibull was a Swedish scientist, inventor, and statistician. He also consulted extensively on projects for Saab and for the U.S. Air Force. We display graphs of $f(t)$ for three different Weibull distributions in Figure 1. In all three of these distributions, the parameter α equals 0.5, while the parameter β takes on the values 1, 2, and 3, respectively.

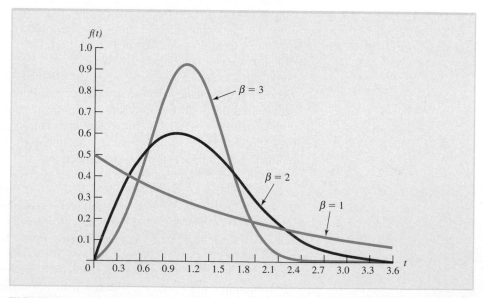

FIGURE 1   Three Weibull probability density function graphs, with parameters $\alpha = 0.5$ in each case and $\beta = 1$, 2, and 3.

Source: Steven Nahmias, *Production and Operations Analysis*, 4th ed., © 2001. The McGraw-Hill Companies, Inc., New York, NY.

We can now return to the question of modeling the life span of an artificial knee. Suppose it has been shown that the life span of an artificial knee after it is implanted can be modeled using a Weibull distribution with parameters $\alpha = 0.0000939$ (measured in years) and $\beta = 3$. This means that $R(t) = e^{-0.0000939t^3}$. We can now answer the question posed at the beginning of this essay using the formula we have found for the reliability function for a Weibull distribution. We find that the probability that the artificial knee lasts more than 20 years is $R(20) = e^{-0.0000939 \cdot 20^3} \approx 0.472$.

Weibull models have been used to solve a variety of problems from many different disciplines. For example, the Weibull distribution has been used to determine the expected costs of pumping sand to replenish beaches damaged by hurricanes, to determine how effective new cancer treatments are, to set premiums for insurance policies, to find the best mix of tree species that should be planted for paper production, to set up maintenance procedures for replacing street lights in a city, to estimate the risk of an offshore platform collapsing if it is designed to withstand waves up to a certain height, to determine the viability of windmills, to determine the best blasting strategy for ores, to manage inventories of spare parts, and to study conflicts in highway lane merging. They have also been used to study the yield strength and the fatigue life of steel, the tensile strength of optical fibers, the fineness of coal, the size of droplets in sprays, the pitting corrosion in pipes, the time an organism takes to die after exposure to carcinogens, fracturing in concrete, the pace of technology change, the fracture strength of glass, the size of icebergs, wave heights, flood and earthquake frequency, temperature fluctuations, precipitation amounts, the size of raindrops, inventory lead times, and wind speed distributions, among other things.

For many complex systems, the failure rate is neither an increasing nor a decreasing function. Instead, the failure rate resembles a bathtub, as shown in Figure 2. When the product is young, manufacturing defects cause a high initial failure rate in the infant mortality phase. After bad components have failed, the failure rate remains relatively constant until the system reaches a phase where it begins to wear out and the

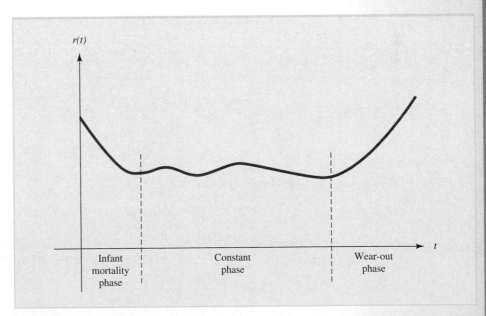

FIGURE 2   Graph of a typical bathtub failure rate function for a complex system.

Source: Steven Nahmias, *Production and Operations Analysis*, 4th ed., © 2001. The McGraw-Hill Companies, Inc., New York, NY.

failure rate increases. To build models where the failure rate resembles a bathtub we need more complicated functions than are introduced here. To learn more about such models, the reader is invited to consult books that focus on reliability theory.

Questions

1. Suppose the failure function is a positive constant; that is, $r(t) = \lambda$ for $\lambda > 0$. Show that the reliability function $R(t)$ satisfies the separable differential equation $R'(t) = -\lambda R(t)$. Solve this differential equation to find $R(t)$. What kind of probability density function $f(t)$ corresponds to $R(t)$ in this case?

2. Show that when the parameter β in a Weibull distribution satisfies $\beta = 1$, then the distribution is exponential.

3. Suppose the random variable X that represents the time to failure of a system (the time it works until it fails) has an exponential distribution.

   a. Show that the probability that a system working at time t will fail in the next s units of time does not depend on t. That is, the probability of the system failing in a particular time span does not depend on how long it has been working.

   b. The property in part (a) is described by saying that *an exponential random variable has no memory*. Show that a geometric random variable also has no memory.

4. Verify that the equation given for the reliability function $R(t)$ for a Weibull distribution with parameters α and β is correct.

5. Suppose that the reliability function for a computer model is $R(t) = e^{-0.008t^{1.98}}$. Find the failure rate function for these computers.

6. Suppose that the failure rate function for a copier model is $r(t) = 3.11t^{1.5}$, with time measured in months. Find the probability that one of these copiers continues to operate for at least 6 months.

7. Find the probability that the artificial knee discussed in the text lasts at least 30 years.

8. Let $\Gamma(\alpha) = \int_0^\infty x^{\alpha-1} e^{-x}\, dx$, where $\alpha > 0$. Show that if the random variable X has a Weibull distribution with parameters α and β, then

$$E(X) = \alpha^{-1/\beta}\Gamma\left(1 + \frac{1}{\beta}\right)$$

and

$$\text{Var}(X) = \alpha^{-2/\beta}\Gamma\left(1 + \frac{2}{\beta}\right) - [E(X)]^2$$

References

Steven Nahmias, *Production and Operations Analysis,* 4th ed., New York: McGraw-Hill, 2001.

Wallace R. Blischke and D. N. Prabhakar Murthy, *Reliability: Modeling, Prediction, and Optimization,* New York: John Wiley and Sons, 2000.

Appendix **A**

Algebra Review

SECTION A.1   A Brief Review of Algebra

There are many techniques from elementary algebra that are needed in calculus. This appendix contains a review of such topics, and we begin by examining numbering systems.

The Real Numbers

An **integer** is a "whole number," either positive or negative. For example, 1, 2, 875, -15, -83, and 0 are integers, while $\frac{2}{3}$, 8.71, and $\sqrt{2}$ are not.

A **rational number** is a number that can be expressed as the quotient $\frac{a}{b}$ of two integers, where $b \neq 0$. For example, $\frac{2}{3}$, $\frac{8}{5}$, and $-\frac{4}{7}$ are rational numbers, as are

$$-6\frac{1}{2} = -\frac{13}{2} \qquad \text{and} \qquad 0.25 = \frac{25}{100} = \frac{1}{4}$$

Every integer is a rational number since it can be expressed as itself divided by 1. When expressed in decimal form, rational numbers are either terminating or infinitely repeating decimals. For example,

$$\frac{5}{8} = 0.625 \qquad \frac{1}{3} = 0.33\ldots \qquad \text{and} \qquad \frac{13}{11} = 1.181818\ldots$$

A number that cannot be expressed as the quotient of two integers is called an **irrational number.** For example,

$$\sqrt{2} \approx 1.41421356 \qquad \text{and} \qquad \pi \approx 3.14159265$$

are irrational numbers.

The rational numbers and irrational numbers form the **real numbers** and can be visualized geometrically as points on a line, called the **real number line.** To construct such a representation, choose a point on a line as the location of the number 0. This is called the **origin.** Select a point to represent the number 1. This determines the scale of the number line, and each number is located an appropriate distance (multiple of 1) from the origin. If the line is horizontal, the positive numbers are located to the right of the origin and the negative numbers to the left, as indicated in Figure A.1. The **coordinate** of a particular point on the line is the number associated with it.

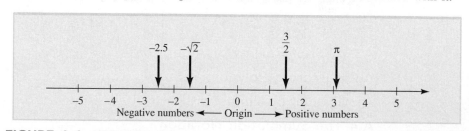

FIGURE A.1   The number line.

Inequalities

If a and b are real numbers and a is to the right of b on the number line, we say that **a is greater than b** and write $a > b$. If a is to the left of b, we say that **a is less than b** and write $a < b$ (Figure A.2). For example,

$$5 > 2 \qquad -12 < 0 \qquad \text{and} \qquad -8.2 < -2.4$$

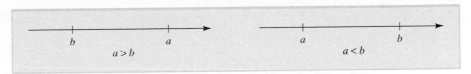

FIGURE A.2   Inequalities.

Moreover,

$$\frac{6}{7} < \frac{7}{8}$$

as you can see by noting that

$$\frac{6}{7} = \frac{48}{56} \quad \text{and} \quad \frac{7}{8} = \frac{49}{56}$$

A few basic properties of inequalities are presented in the following box. Note especially property 3, which states that the direction of an inequality is preserved if both sides are multiplied by a positive number, but is *reversed* if the multiplier is negative.

Properties of Inequalities

1. **Transitive property:** If $a > b$ and $b > c$, then $a > c$.
2. **Additive property:** If $a > b$ and $c \geq d$, then $a + c > b + d$.
3. **Multiplicative property:** If $a > b$ and $c > 0$, then $ac > bc$, but if $a > b$ and $c < 0$, then $ac < bc$.

For example, since $7 > 3$, we have $7 - 9 > 3 - 9$ or $-2 > -6$. Since $5 > 2$ and $3 > 0$, it follows that $5 \cdot 3 > 2 \cdot 3$, or $15 > 6$. Since $5 > 2$ and $-2 < 0$, we have $5(-2) < 2(-2)$, or $-10 < -4$.

The symbol $\geq$ stands for **greater than or equal to,** and the symbol $\leq$ stands for **less than or equal to.** Thus, for example,

$$-3 \geq -4 \qquad -3 \geq -3 \qquad -4 \leq -3 \qquad \text{and} \qquad -4 \leq -4$$

A real number is said to satisfy a particular inequality involving a variable if the inequality is satisfied when the number is substituted for the variable. The inequality is said to be **solved** when all numbers that satisfy it have been found. The set of all solutions is called the **solution set** of the inequality.

EXAMPLE A.1.1   Solving an Inequality

Solve the two-sided inequality $-5 < 2x - 3 \leq 1$.

Solution
Add 3 to both sides of the inequality (property 2) to obtain

$$-2 < 2x \leq 4$$

Then multiply each side of this new inequality by $\frac{1}{2}$:

$$-1 < x \leq 2$$

Thus the solution set is comprised of all real numbers between -1 and 2, including 2 (but not -1).

Intervals

A set of real numbers that can be represented on the number line by a line segment is called an **interval.** Inequalities can be used to describe intervals. For example, the interval $a \leq x < b$ consists of all real numbers x that are between a and b, including a but excluding b. This interval is shown in Figure A.3. The numbers a and b are known as the **endpoints** of the interval. The square bracket at a indicates that a is included in the interval, while the rounded bracket at b indicates that b is excluded.

Intervals may be finite or infinite in extent and may or may not contain either endpoint. The possibilities (including customary notation and terminology) are illustrated in Figure A.4.

FIGURE A.3   The interval $a \leq x < b$.

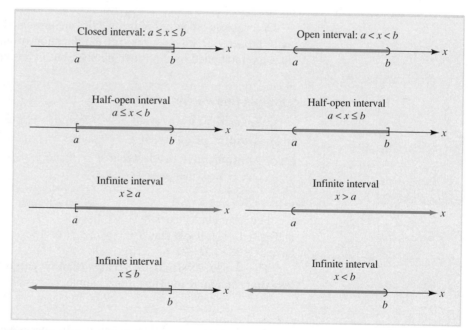

FIGURE A.4   Intervals of real numbers.

EXAMPLE A.1.2   **Describing Intervals with Inequalities**

Use inequalities to describe these intervals.

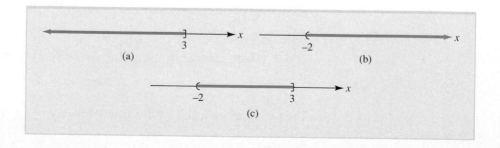

Solution

a. $x \leq 3$ **b.** $x > -2$ **c.** $-2 < x \leq 3$

▌ **EXAMPLE A.1.3**   Graphing an Inequality

Represent each of these intervals as a line segment on a number line.

    a. $x < -1$ **b.** $-1 \leq x \leq 2$ **c.** $x > 2$

Solution

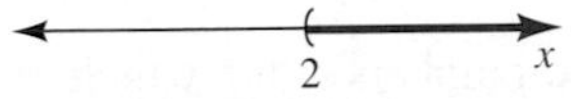

Absolute Value

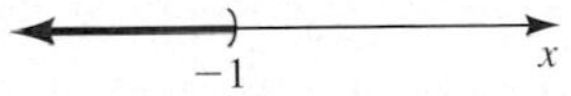

The **absolute value** of a real number x, denoted by $|x|$, is the distance from x to 0 on a number line. Since distance is always nonnegative, it follows that $|x| \geq 0$. For example,

$$|4| = 4 \qquad |-4| = 4 \qquad |0| = 0 \qquad |5 - 9| = 4 \qquad |\sqrt{3} - 3| = 3 - \sqrt{3}$$

Here is a general formula for absolute value.

> **Absolute Value** ▪ For any real number x, the absolute value of x is
> $$|x| = \begin{cases} x & \text{if } x \geq 0 \\ -x & \text{if } x < 0 \end{cases}$$

Notice that $|-a| = |a|$ for any real number a. This is one of several useful properties of absolute value listed in the following box.

> **Properties of Absolute Value**
> Let a and b be real numbers. Then
> 1. $|-a| = |a|$
> 2. $|ab| = |a|\,|b|$
> 3. $\left|\dfrac{a}{b}\right| = \dfrac{|a|}{|b|}$ if $b \neq 0$
> 4. $|a + b| \leq |a| + |b|$ (the triangle inequality)

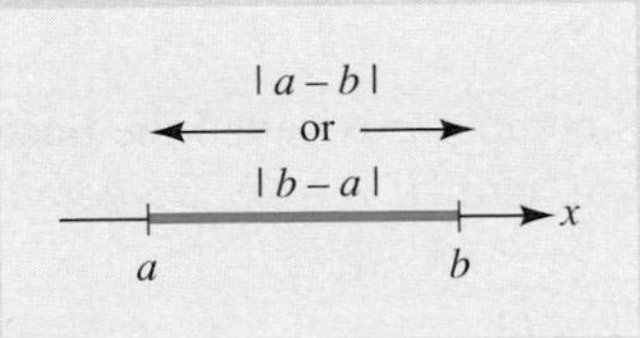

FIGURE A.5 The distance between a and $b = |a - b|$.

The distance on a number line between any two numbers a and b is the absolute value of their difference taken in either order ($a - b$ or $b - a$), as illustrated in Figure A.5. For instance, the distance between $a = -2$ and $b = 3$ is $|-2 - 3| = 5$ (Figure A.6).

FIGURE A.6 Distance between -2 and 3.

The solution set of an inequality of the form $|x| \leq c$ for $c > 0$ is the interval $-c \leq x \leq c$, that is, $[-c, c]$. This property is used in Example A.1.4.

EXAMPLE A.1.4 Solving an Absolute Value Inequality

Find the interval consisting of all real numbers x such that $|x - 1| \leq 3$.

Solution

In geometric terms, the numbers x for which $|x - 1| \leq 3$ are those whose distance from 1 is less than or equal to 3. As illustrated in Figure A.7, these are the numbers that satisfy $-2 \leq x \leq 4$.

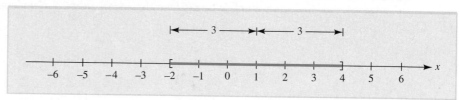

FIGURE A.7 The interval on which $|x - 1| \leq 3$ is $-2 \leq x \leq 4$.

To find this interval algebraically, without relying on the geometry, rewrite the inequality $|x - 1| \leq 3$ as

$$-3 \leq x - 1 \leq 3$$

and add 1 to each part to get

$$-3 + 1 \leq x - 1 + 1 \leq 3 + 1$$

or

$$-2 \leq x \leq 4$$

Exponents and Roots

If a is a real number and n is a positive integer, the expression

$$a^n = \underbrace{a \cdot a \cdot \cdots a}_{n \text{ terms}}$$

indicates that a is to be multiplied by itself n times. This number a is called the **base** of the **exponential expression** a^n, and n is called the **exponent.** If $a \neq 0$, we define

$$a^{-n} = \frac{1}{a^n} \quad \text{and} \quad a^0 = 1$$

Note that 0^0 is not defined.

If m is a positive integer, then $a^{1/m}$ denotes the number whose mth power is a. This is called the **mth root** of a and is also denoted by $\sqrt[m]{a}$; that is,

$$a^{1/m} = \sqrt[m]{a}$$

The mth root of a negative number is not defined when m is even. For example, $\sqrt[4]{-5}$ is not defined since there is no real number whose 4th power is -5.

By convention, if m is even, $a^{1/m}$ is taken to be positive even when there is a negative number whose mth power is a. For example, 2^4 and $(-2)^4$ both equal 16, but the 4th root of 16 is defined to be 2. Thus,

$$\sqrt[4]{16} = 16^{1/4} = 2$$

not ± 2.

Finally, we write $a^{n/m}$ to denote the nth power of the mth root of the real number a, which is the same as the mth root of the nth power. That is,

$$a^{n/m} = (a^{1/m})^n = (a^n)^{1/m}$$

For example,

$$8^{-2/3} = (8^{-2})^{1/3} = \left(\frac{1}{8^2}\right)^{1/3} = \left(\frac{1}{64}\right)^{1/3} = \frac{1}{4} \quad \text{Since } \left(\frac{1}{4}\right)^3 = 64$$

or, equivalently,

$$8^{-2/3} = (8^{1/3})^{-2} = 2^{-2} = \frac{1}{2^2} = \frac{1}{4}$$

Here is a summary of the exponential notation.

Exponential Notation ▪ Let a be a real number and m and n be positive integers. Then:

Integer powers: $\quad a^n = \underbrace{a \cdot a \cdot \cdots \cdot a}_{n \text{ terms}} \quad$ and $\quad a^0 = 1$

Negative integer powers: $\quad a^{-n} = \dfrac{1}{a^n}$

Reciprocal integer powers (roots): $\quad a^{1/m} = \sqrt[m]{a}$

Fractional exponents: $\quad a^{n/m} = (a^{1/m})^n = (a^n)^{1/m}$

EXAMPLE A.1.5 Evaluating Expressions with Exponents

Evaluate these expressions (without using your calculator).

a. $9^{1/2}$ **b.** $27^{2/3}$ **c.** $8^{-1/3}$ **d.** $\left(\dfrac{1}{100}\right)^{-3/2}$ **e.** 5^0

Solution

a. $9^{1/2} = \sqrt{9} = 3$

b. $27^{2/3} = (\sqrt[3]{27})^2 = 3^2 = 9$
$$= \sqrt[3]{(27)^2} = \sqrt[3]{729} = 9$$

c. $8^{-1/3} = \dfrac{1}{8^{1/3}} = \dfrac{1}{\sqrt[3]{8}} = \dfrac{1}{2}$

d. $\left(\dfrac{1}{100}\right)^{-3/2} = 100^{3/2} = (\sqrt{100})^3 = 10^3 = 1{,}000$

e. $5^0 = 1$

Exponents obey these useful laws.

Laws of Exponents ■ For real numbers a, b and integers m, n the following laws are valid whenever the quantities are defined.

Identity law: If $a^m = a^n$, then $m = n$.

Product law: $a^m \cdot a^n = a^{m+n}$

Quotient law: $\dfrac{a^m}{a^n} = a^{m-n}$ if $a \neq 0$

Power laws: $(a^m)^n = a^{mn}$ and $(ab)^n = a^n \cdot b^n$

The laws of exponents are illustrated in Examples A.1.6 through A.1.9.

EXAMPLE A.1.6   Evaluating Expressions with Exponents

Evaluate these expressions (without using a calculator).

a. $(2^{-2})^3$ b. $\dfrac{3^3}{3^{1/3}(3^{2/3})}$ c. $2^{7/4}(8^{-1/4})$

Solution

a. $(2^{-2})^3 = 2^{-6} = \dfrac{1}{2^6} = \dfrac{1}{64}$

b. $\dfrac{3^3}{3^{1/3}(3^{2/3})} = \dfrac{3^3}{3^{1/3+2/3}} = \dfrac{3^3}{3^1} = 3^2 = 9$

c. $2^{7/4}(8^{-1/4}) = 2^{7/4}(2^3)^{-1/4} = 2^{7/4}(2^{-3/4}) = 2^{7/4-3/4} = 2^1 = 2$

EXAMPLE A.1.7   Solving for an Exponent

Solve each of these equations for n.

a. $\dfrac{a^5}{a^2} = a^n$ b. $(a^n)^5 = a^{20}$

Solution

a. Since $\dfrac{a^5}{a^2} = a^{5-2} = a^3$, it follows that $n = 3$.

b. Since $(a^n)^5 = a^{5n}$, it follows that $5n = 20$ or $n = 4$.

EXAMPLE A.1.8 Simplifying Expressions with Exponents

Simplify the following expressions, and express each in terms of positive exponents.

a. $(x^3)^{-2}$ **b.** $(x^{-5})^{-2}$ **c.** $(x^{-2}y^{-3})^{-4}$

d. $\left(\dfrac{x^{-3}}{y^4}\right)^{-2}$ **e.** $\dfrac{4x^{-3}y^2}{2x^2y^{-5}}$

Solution

a. $(x^3)^{-2} = x^{3(-2)} = x^{-6} = \dfrac{1}{x^6}$

b. $(x^{-5})^{-2} = x^{(-5)(-2)} = x^{10}$

c. $(x^{-2}y^{-3})^{-4} = x^{(-2)(-4)}y^{(-3)(-4)} = x^8 y^{12}$

d. $\left(\dfrac{x^{-3}}{y^4}\right)^{-2} = (x^{-3}y^{-4})^{-2} = x^{(-3)(-2)}y^{(-4)(-2)} = x^6 y^8$

e. $\dfrac{4x^{-3}y^2}{2x^2y^{-5}} = \dfrac{4}{2}x^{-3-2}y^{2-(-5)} = 2x^{-5}y^7 = \dfrac{2y^7}{x^5}$

EXAMPLE A.1.9 Simplifying Root Expressions

Simplify each of the following root expressions.

a. $3\sqrt{64} + 5\sqrt{72} - 9\sqrt{50}$

b. $\sqrt{a^{-5}b^{-8}c^{10}}, \, a > 0, b \neq 0$

c. $\sqrt{\dfrac{36x^3}{y^3}} \sqrt{\dfrac{y^8}{25x}}, x > 0, y > 0$

Solution

a. $3\sqrt{64} + 5\sqrt{72} - 9\sqrt{50} = 3\sqrt{8^2} + 5\sqrt{6^2 \cdot 2} - 9\sqrt{5^2 \cdot 2}$

$= 3(8) + 5(6)\sqrt{2} - 9(5)\sqrt{2} = 24 - 15\sqrt{2}$

b. $\sqrt{a^{-5}b^{-8}c^{10}} = \sqrt{\dfrac{c^{10}}{a^5 b^8}} = \dfrac{c^5}{b^4\sqrt{a^4}\sqrt{a}} = \dfrac{c^5}{b^4 a^2 \sqrt{a}}$

c. $\sqrt{\dfrac{36x^3}{y^3}} \sqrt{\dfrac{y^8}{25x}} = \sqrt{\dfrac{36}{25}} \sqrt{\dfrac{x^3 y^8}{xy^3}} = \dfrac{6}{5}\sqrt{x^2 y^5} = \dfrac{6}{5}\sqrt{x^2(y^4 \cdot y)}$

$= \dfrac{6}{5}\sqrt{x^2}\sqrt{y^4}\sqrt{y} = \dfrac{6}{5}xy^2\sqrt{y}$

Example A.1.10 illustrates how factoring and division by a common factor are used to simplify a type of algebraic expression that is important in calculus.

EXAMPLE A.1.10 Simplifying an Algebraic Expression

Simplify the expression $\dfrac{4(x+3)^4(x-2)^2 - 6(x+3)^2(x-2)^3}{(x+3)(x-2)^3}$

Solution

First, factor the common factor $2(x+3)^3(x-2)^2$ out of the numerator to get

$$\frac{4(x+3)^4(x-2)^2 - 6(x+3)^2(x-2)^3}{(x+3)(x-2)^3} = \frac{2(x+3)^3(x-2)^2[2(x+3) - 3(x-2)]}{(x+3)(x-2)^3}$$

$$= \frac{2(x+3)^3(x-2)^2[2x + 6 - 3x + 6]}{(x+3)(x-2)^3}$$

$$= \frac{2(x+3)^3(x-2)^2[12 - x]}{(x+3)(x-2)^3}$$

Now we can divide both numerator and denominator by the common factor $(x+3)(x-2)^2$ to conclude that

$$\frac{4(x+3)^4(x-2)^2 - 6(x+3)^2(x-2)^3}{(x+3)(x-2)^3} = \frac{2(x+3)^2(12 - x)}{(x-2)}$$

Rationalizing Sometimes it is necessary, or at least desirable, to write a fraction so that either the numerator or denominator contains no roots. The algebraic procedure for achieving this is called **rationalizing.** Example A.1.11 illustrates how a root is removed from the denominator.

EXAMPLE A.1.11 Rationalizing a Denominator

Rationalize the denominator in the expression $\dfrac{5}{3\sqrt{x}}$.

Solution

Multiply both the numerator and denominator of the given expression by $\sqrt{x}$:

$$\frac{5}{3\sqrt{x}} = \frac{5(\sqrt{x})}{3\sqrt{x}(\sqrt{x})} = \frac{5(\sqrt{x})}{3(\sqrt{x})^2}$$

$$= \frac{5\sqrt{x}}{3x}$$

The algebraic identity

$$(x+y)(x-y) = x^2 - y^2$$

can be used to rationalize fractions when the numerator or denominator contains a factor of the form $a + \sqrt{b}$. The key lies in noting that the root can be removed from $a + \sqrt{b}$ by multiplying by the complementary expression $a - \sqrt{b}$ since

$$(a + \sqrt{b})(a - \sqrt{b}) = a^2 - (\sqrt{b})^2 = a^2 - b$$

An expression of the form $\sqrt{a} + b$ can be rationalized in a similar fashion by using its complement $\sqrt{a} - b$. This procedure is illustrated in Example A.1.12.

> **EXAMPLE A.1.12** Rationalizing a Numerator
>
> Rationalize the numerator in the expression $\dfrac{4 - \sqrt{3}}{7}$.
>
> **Solution**
>
> We multiply both the numerator and denominator by $4 + \sqrt{3}$ and obtain
>
> $$\frac{4 - \sqrt{3}}{7} = \frac{(4 - \sqrt{3})(4 + \sqrt{3})}{7} = \frac{4^2 - (\sqrt{3})^2}{7(4 + \sqrt{3})} = \frac{16 - 3}{7(4 + \sqrt{3})} = \frac{13}{7(4 + \sqrt{3})}$$

EXERCISES ▪ A.1

In Exercises 1 through 4, use inequalities to describe the indicated interval.

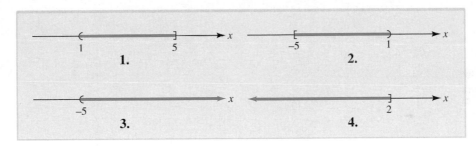

1. **2.** **3.** **4.**

In Exercises 5 through 8, represent the given interval as a line segment on a number line.

5. $x \geq 2$

6. $-6 \leq x < 4$

7. $-2 < x \leq 0$

8. $x > 3$

In Exercises 9 through 12, find the distance on the number line between the given pair of real numbers.

9. 0 and -4

10. 2 and 5

11. -2 and 3

12. -3 and -1

In Exercises 13 through 18, find the interval or intervals consisting of all real numbers x that satisfy the given inequality.

13. $|x| \leq 3$

14. $|x - 2| \leq 5$

15. $|x + 4| \leq 2$

16. $|1 - x| < 3$

17. $|x + 2| \geq 5$

18. $|x - 1| > 3$

In Exercises 19 through 26, evaluate the given expression without using a calculator.

19. 5^3

20. 2^{-3}

21. $16^{1/2}$

22. $36^{-1/2}$

23. $8^{2/3}$

24. $27^{-4/3}$

25. $\left(\dfrac{1}{4}\right)^{1/2}$

26. $\left(\dfrac{1}{4}\right)^{-3/2}$

In Exercises 27 through 34, evaluate the given expression without using a calculator.

27. $\dfrac{2^5(2^2)}{2^8}$

28. $\dfrac{3^4(3^3)}{(3^2)^3}$

29. $\dfrac{2^{4/3}(2^{5/3})}{2^5}$

30. $\dfrac{5^{-3}(5^2)}{(5^{-2})^3}$

31. $\dfrac{2(16^{3/4})}{2^3}$

32. $\dfrac{\sqrt{27}\,(\sqrt{3})^3}{9}$

33. $[\sqrt{8}\,(2^{5/2})]^{-1/2}$

34. $[\sqrt{27}\,(3^{5/2})]^{1/2}$

In Exercises 35 through 42, solve the given equation for n. (Assume $a > 0$ and $a \neq 1$.)

35. $a^3 a^7 = a^n$

36. $\dfrac{a^5}{a^2} = a^n$

37. $a^4 a^{-3} = a^n$

38. $a^2 a^n = \dfrac{1}{a}$

39. $(a^3)^n = a^{12}$

40. $(a^n)^5 = \dfrac{1}{a^{10}}$

41. $a^{3/5} a^{-n} = \dfrac{1}{a^2}$

42. $(a^n)^3 = \dfrac{1}{\sqrt{a}}$

In Exercises 43 through 76, simplify the given expression as much as possible. Assume a, b, and c are positive real numbers.

43. $(a^3b^2c^5)(a^2b^6c^3)$

44. $(a^5b^2c)^3$

45. $\left(\dfrac{a^2c^3}{b}\right)^4$

46. $\left(\dfrac{a^{-2}b}{c^{-3}}\right)^2$

47. $\left(\dfrac{a^2b^3c^{-3}}{a^{-3}b^4c^4}\right)^2$

48. $\left(\dfrac{a^{-3}b^{-2}c^{-4}}{a^4b^3c^5}\right)^{-3}$

49. $[(a^3b^2)^{-2}c^2]^{-3}$

50. $[a^3(b^3c^{-1})^{-3}]^{-2}$

51. $\dfrac{a^{-2}b^{-3}+a^{-3}b+bc^{-1}}{ab^2c^3}$

52. $\left(\dfrac{3a^{-3}}{c^2}\right)^{-1}\left(\dfrac{2c^{-2}}{a^3}\right)^2$

53. $\dfrac{a^{-3}+b^{-1}}{(ab)^{-2}}$

54. $(a^{-1}+b^{-2})^2$

55. $\sqrt[7]{128}+\sqrt[3]{-64}$

56. $\sqrt{18}+\sqrt[3]{-162(27)}$

57. $\sqrt[3]{6^5 5^8 3^6}$

58. $\sqrt[3]{(-2)^{15}(-3)^{18}}$

59. $2\sqrt{32}+5\sqrt{72}$

60. $3\sqrt{96}+\sqrt{294}$

61. $3\sqrt{24}-2\sqrt{54}+\sqrt{486}$

62. $3\sqrt[3]{15}-\sqrt[3]{120}+5\sqrt[3]{405}$

63. $\sqrt[5]{a^{15}b^{20}c^{35}}$

64. $\dfrac{\sqrt[3]{-64a^9b^{-6}}}{\sqrt{a^2b^4}}$

65. $\sqrt{\dfrac{25a^2}{b}}\sqrt{\dfrac{b^3}{49a^4}}$

66. $\sqrt[3]{\dfrac{a^6b^9}{64c^{15}}}$

67. $\sqrt[3]{\dfrac{a^5}{b^7c^9}}$

68. $\sqrt[5]{\dfrac{a^8b^{-16}}{c^7}}$

69. $(a^4b^2c^{12})^{-1/2}$

70. $\dfrac{a^2b}{(a^6b^4)^{-1/4}}$

71. $(a^{1/6}b^{-1/3}c^{1/4})^{12}$

72. $\dfrac{(a^{25}b^{35})^{-3/5}}{(a^{16}b^{12})^{-3/4}}$

73. $(a^{1/2}+b^{1/4})(a^{1/2}-b^{1/4})$

74. $(a^{2/3}+b^{2/3})(a^{2/3}-b^{2/3})$

75. $\sqrt[3]{\dfrac{a^{17}b^9}{c^{11}}}$

76. $\sqrt[5]{(a^{24}b^{-8}c^{11})^4}$

In Exercises 77–88, factor and simplify the given expression as much as possible.

77. x^5-4x^4

78. $3x^3-12x^4$

79. $100-25(x-3)$

80. $60-20(4-x)$

81. $8(x+1)^3(x-2)^2+6(x+1)^2(x-2)^3$

82. $12(x+3)^5(x-1)^3-8(x+3)^6(x-1)^2$

83. $x^{-1/2}(2x+1)+4x^{1/2}$

84. $x^{-1/4}(3x+5)+4x^{3/4}$

85. $\dfrac{(x+3)^3(x+1)-(x+3)^2(x+1)^2}{(x+3)(x+1)}$

86. $\dfrac{3(x-2)^2(x+1)^2-2(x-2)(x+1)^3}{(x-2)^4}$

87. $\dfrac{4(1-x)^2(x+3)^3+2(1-x)(x+3)^4}{(1-x)^4}$

88. $\dfrac{6(x+2)^2(1-x)^4-4(x+2)^6(1-x)^3}{(x+2)^8(1-x)^2}$

In Exercises 89 through 96, rationalize the root (or roots) in the given expression.

89. $\dfrac{\sqrt{3}-\sqrt{2}}{5}$

90. $\dfrac{\sqrt{7}+3}{2}$

91. $\dfrac{7}{3-\sqrt{3}}$

92. $\dfrac{5}{\sqrt{5}+\sqrt{2}}$

93. $\dfrac{\sqrt{5}+2}{3}$

94. $\dfrac{\sqrt{5}-\sqrt{11}}{4}$

95. $\dfrac{5}{\sqrt{5}+1}$

96. $\dfrac{3}{2-\sqrt{7}}$

97. Show that
$$\sqrt{x+h}-\sqrt{x}=\dfrac{h}{\sqrt{x+h}+\sqrt{x}}$$
where x and h are positive numbers.

98. Simplify the expression
$$\dfrac{1}{\sqrt{x+h}}-\dfrac{1}{\sqrt{x}}$$
where x and h are positive constants.

99. ECOLOGY   The atmosphere above each square centimeter of Earth's surface weighs 1 kilogram (kg).

a. Assuming Earth is a sphere of radius $R=6{,}440$ km, use the formula $S=4\pi R^2$ to calculate the surface area of Earth and then find the total mass of the atmosphere.

b. Oxygen occupies approximately 22% of the total mass of the atmosphere, and it is estimated that plant life produces approximately 0.9×10^{13} kg of oxygen per year. If none of this oxygen were used up by plants or animals (or combustion), how long would it take to build up the total mass of oxygen in the atmosphere [part (a)]?*

100. Show that $(\sqrt[n]{x})^m=\sqrt[n]{x^m}$ in the case where m is a negative integer.

---

*Adapted from a problem in E. Batschelet, *Introduction to Mathematics for Life Scientists,* 2nd ed., New York: Springer-Verlag, 1979, p. 31.

SECTION A.2   Factoring Polynomials and Solving Systems of Equations

A **polynomial** is an expression of the form

$$a_0 + a_1 x + a_2 x^2 + \cdots + a_n x^n$$

where n is a nonnegative integer and $a_n, a_{n-1}, \ldots, a_0$ are real numbers known as the **coefficients** of the polynomial. Polynomials appear in a variety of mathematical contexts, and the first goal of this section is to examine some important algebraic properties of polynomials.

If $a_n \neq 0$, n is said to be the **degree** of the polynomial. A nonzero constant is said to be a **polynomial of degree 0.** (Technically, the number 0 is also a polynomial, but it has no degree.) For example, $3x^5 - 7x + 12$ is a polynomial of degree 5, with terms $3x^5$, $-7x$, and 12. **Similar terms** in two polynomials in the variable x are terms with the same degree. Thus, in the fifth-degree polynomial $3x^5 - 5x^2 + 3$ and the third-degree polynomial $-2x^3 + 2x^2 + 7x - 9$, the terms $-5x^2$ and $2x^2$ are similar terms. Polynomials are multiplied by constants and added and subtracted by combining similar terms, as illustrated in Example A.2.1.

EXAMPLE A.2.1   Combining Polynomials

Let $p(x) = 3x^2 - 5x + 7$ and $q(x) = -4x^2 + 9$. Find the polynomials $2p(x)$ and $p(x) + q(x)$.

Solution
We have

$$2p(x) = 2(3)x^2 - 2(5)x + 2(7) = 6x^2 - 10x + 14$$

and

$$p(x) + q(x) = [3 + (-4)]x^2 + [-5 + 0]x + [7 + 9]$$
$$= -x^2 - 5x + 16$$

A convenient way to remember how to multiply two first-degree polynomials $p(x) = ax + b$ and $q(x) = cx + d$ is the "FOIL" method:

$$
\begin{array}{cccc}
\text{F} & \text{O} & \text{I} & \text{L} \\
\text{First} & \text{Outer} & \text{Inner} & \text{Last} \\
\text{product} & \text{product} & \text{product} & \text{product}
\end{array}
$$

$$(ax + b)(cx + d) = (ac)x^2 + (ad)x + (bc)x + (bd)$$

Example A.2.2 demonstrates this.

EXAMPLE A.2.2   Multiplying Using FOIL

Find $(3x + 5)(-2x + 7)$.

Solution

Applying the FOIL method, we get

$$
\begin{array}{cccc}
\text{F} & \text{O} & \text{I} & \text{L} \\
\text{First} & \text{Outer} & \text{Inner} & \text{Last} \\
\underline{\text{product}} & \underline{\text{product}} & \underline{\text{product}} & \underline{\text{product}}
\end{array}
$$

$$
\begin{aligned}
(3x + 5)(-2x + 7) &= (3)(-2)x^2 + (3)(7)x + (5)(-2)x + (5)(7) \\
&= -6x^2 + 11x + 35
\end{aligned}
$$

To multiply two polynomials that are not both of degree one, we use the distributive laws of real numbers, namely,

$$
a(b + c) = ab + ac \qquad \text{and} \qquad (a + b)c = ac + bc
$$

Example A.2.3 illustrates this procedure.

EXAMPLE A.2.3 Multiplying Polynomials

Find $(-x^2 + 3x + 5)(x^2 + 2x - 4)$.

Solution

To find the required product, we must multiply each term of $-x^2 + 3x + 5$ by each term of $x^2 + 2x - 4$, and then combine similar terms. We have

$$
\begin{aligned}
(-x^2 + 3x + 5)&(x^2 + 2x - 4) \\
&= -x^2(x^2 + 2x - 4) + 3x(x^2 + 2x - 4) + 5(x^2 + 2x - 4) \\
&= [-x^4 - 2x^3 + 4x^2] + [3x^3 + 6x^2 - 12x] + [5x^2 + 10x - 20] \\
&= -x^4 + (-2 + 3)x^3 + (4 + 6 + 5)x^2 + (-12 + 10)x - 20 \\
&= -x^4 + x^3 + 15x^2 - 2x - 20
\end{aligned}
$$

The computation can also be done "vertically":

$$
\begin{array}{r}
-x^2 + 3x + 5 \\
\underline{x^2 + 2x - 4} \\
4x^2 - 12x - 20 \\
-2x^3 + 6x^2 + 10x \\
\underline{-x^4 + 3x^3 + 5x^2 \qquad\qquad} \\
-x^4 + x^3 + 15x^2 - 2x - 20
\end{array}
$$

Factoring Polynomials with Integer Coefficients

Many of the polynomials that arise in practice have integer coefficients (or are closely related to polynomials that do). Techniques for factoring polynomials with integer coefficients are illustrated in Examples A.2.4 and A.2.5. In each, the goal is to rewrite the given polynomial as a product of polynomials of lower degree that also have integer coefficients.

EXAMPLE A.2.4 Factoring a Polynomial

Factor the polynomial $x^2 - 2x - 3$ using integer coefficients.

Solution

The goal is to write the polynomial as a product of the form

$$x^2 - 2x - 3 = (x + a)(x + b)$$

where a and b are integers. The distributive law gives us

$$(x + a)(x + b) = x^2 + (a + b)x + ab$$

Hence, we must find integers a and b such that

$$x^2 - 2x - 3 = x^2 + (a + b)x + ab$$

or, equivalently, such that

$$a + b = -2 \quad \text{and} \quad ab = -3$$

From the list

$$1, -3 \quad \text{and} \quad -1, 3$$

of pairs of integers whose product is -3, choose $a = -3$ and $b = 1$ as the only pair whose sum is -2. It follows that

$$x^2 - 2x - 3 = (x - 3)(x + 1)$$

which you should check by multiplying out the right-hand side.

EXAMPLE A.2.5 Factoring a Polynomial

Factor the polynomial $12x^2 - 11x - 15$ using integer coefficients.

Solution

We wish to write the polynomial as a product of the form

$$12x^2 - 11x - 15 = (ax + b)(cx + d)$$

Expanding the product on the left by the FOIL method, we get

$$12x^2 - 11x - 15 = (ac)x^2 + (bc + ad)x + bd$$

Our goal is to find integers a, b, c, d such that

$$ac = 12 \quad bc + ad = -11 \quad \text{and} \quad bd = -15$$

Since ac is to be positive, there is no harm in assuming that a and c are both positive. (What happens if both are negative?) The factors of the coefficients 12 and -15 are as follows:

12		-15	
a	*c*	*b*	*d*
12	1	15	-1
6	2	5	-3
4	3	3	-5
3	4	1	-15
2	6		
1	12		

We try each pair on the left with each pair on the right, with a goal of finding a combination that produces the middle term $bc + ad = -11$. By trial and error, we find that $a = 4$ and $c = 3$ matched with $b = 3$ and $d = -5$ gives the correct middle term. We obtain the following factorization:

$$12x^2 - 11x - 15 = (4x + 3)(3x - 5)$$

Certain polynomial types occur so often that it is useful to have the following formulas for factoring them:

Factorization Formulas

Square of sum: $A^2 + 2AB + B^2 = (A + B)^2$

Square of difference: $A^2 - 2AB + B^2 = (A - B)^2$

Difference of squares: $A^2 - B^2 = (A - B)(A + B)$

Difference of cubes: $A^3 - B^3 = (A - B)(A^2 + AB + B^2)$

Sum of cubes: $A^3 + B^3 = (A + B)(A^2 - AB + B^2)$

EXAMPLE A.2.6   Factoring a Difference of Cubes

Factor the polynomial $x^3 - 8$ using integer coefficients.

Solution

Since $8 = 2^3$, we can use the difference of cubes formula, with $A = x$ and $B = 2$, to obtain the factorization

$$x^3 - 8 = x^3 - 2^3 = (x - 2)(x^2 + 2x + 4)$$

Sometimes a polynomial can be factored by grouping terms strategically. Example A.2.7 illustrates this.

EXAMPLE A.2.7   Factoring Polynomials

Factor the following polynomials:
 a. $p(x) = 4(x - 2)^3 + 3(x - 2)^2$
 b. $q(x) = 9x^2 - 49$

Solution

 a. By factoring out the common term $(x - 2)^2$, we find that
$$4(x - 2)^3 + 3(x - 2)^2 = (x - 2)^2[4(x - 2) + 3]$$
$$= (x - 2)^2(4x - 5)$$

 b. The polynomial $q(x) = 9x^2 - 49$ can be written as a difference of squares $A^2 - B^2$, with $A = 3x$ and $B = 7$. Thus, we have
$$9x^2 - 49 = (3x)^2 - 7^2 = (3x - 7)(3x + 7)$$

Rational Expressions   The quotient of two polynomials is called a **rational expression.** For instance,

$$\frac{1}{x} \qquad \frac{4}{2x^2 + 3} \qquad \frac{-2x^3 + 7x - 1}{5x^2 + 3x + 9} \qquad \text{and} \qquad \frac{x^3 + x - 6}{2}$$

are all rational expressions. One of our goals in working with rational expressions is to reduce such an expression to *lowest terms,* that is, to eliminate all common factors from the numerator and denominator. The following properties of fractions will be useful in this process.

Properties of Fractions

1. **Sum rule:** $\dfrac{a}{b} + \dfrac{c}{d} = \dfrac{ad + bc}{bd}$

2. **Product rule:** $\left(\dfrac{a}{b}\right)\left(\dfrac{c}{d}\right) = \dfrac{ac}{bd}$

3. **Quotient rule:** $\dfrac{a/b}{c/d} = \dfrac{a}{b} \cdot \dfrac{d}{c} = \dfrac{ad}{bc}$

> **EXAMPLE A.2.8**   Simplifying Rational Expressions

Write each of the following as a rational expression in lowest terms:

a. $\dfrac{-2}{x^2 - 1} + \dfrac{x}{x - 1}$ **b.** $\left(\dfrac{x^3 - 7x^2 + 10x}{x^2 + 6x + 9}\right)\left(\dfrac{x + 3}{x - 5}\right)$

Solution

a. $\dfrac{-2}{x^2 - 1} + \dfrac{x}{x - 1} = \dfrac{-2}{x^2 - 1} + \dfrac{x}{x - 1}\dfrac{x + 1}{x + 1}$

$\qquad\qquad = \dfrac{-2}{x^2 - 1} + \dfrac{x^2 + x}{x^2 - 1} = \dfrac{x^2 + x - 2}{x^2 - 1}$

$\qquad\qquad = \dfrac{(x + 2)(x - 1)}{(x + 1)(x - 1)} = \dfrac{x + 2}{x + 1} \qquad \text{for } x \neq 1, -1$

b. $\left(\dfrac{x^3 - 7x^2 + 10x}{x^2 + 6x + 9}\right)\left(\dfrac{x + 3}{x - 5}\right)$

$\qquad = \dfrac{x(x^2 - 7x + 10)(x + 3)}{(x + 3)^2(x - 5)}$

$\qquad = \dfrac{x(x - 2)(x - 5)(x + 3)}{(x + 3)(x - 5)(x + 3)} = \dfrac{x^2 - 2x}{x + 3} \qquad \text{for } x \neq 5, -3$

A rational expression with fractions in both the numerator and the denominator is known as a **compound fraction.** It is often useful to represent a compound fraction as the quotient of two polynomials. This procedure is illustrated in Example A.2.9.

EXAMPLE A.2.9   Simplifying a Compound Fraction

Simplify the compound fraction

$$\frac{1 + 3/x - 4/x^2}{1 + 4/x - 5/x^2}$$

Solution

Writing both the numerator and the denominator as rational expressions and then simplifying, we obtain

$$\frac{1 + 3/x - 4/x^2}{1 + 4/x - 5/x^2} = \frac{\dfrac{x^2 + 3x - 4}{x^2}}{\dfrac{x^2 + 4x - 5}{x^2}}$$

$$= \frac{(x^2 + 3x - 4)x^2}{(x^2 + 4x - 5)x^2} \qquad \text{since } \frac{a/b}{c/d} = \frac{ad}{bc}$$

$$= \frac{(x + 4)(x - 1)x^2}{(x + 5)(x - 1)x^2}$$

$$= \frac{x + 4}{x + 5} \qquad \text{for } x \neq 0, 1, -5$$

Solving Equations by Factoring

The **solutions** of an equation are the values of the variable that make the equation true. For example, $x = 2$ is a solution of the equation

$$x^3 - 6x^2 + 12x - 8 = 0$$

because substitution of 2 for x gives

$$2^3 - 6(2^2) + 12(2) - 8 = 8 - 24 + 24 - 8 = 0$$

In Examples A.2.10 and A.2.11, you will see how factoring can be used to solve certain equations. The technique is based on the fact that if the product of two (or more) terms is equal to zero, then at least one of the terms must be equal to zero. For example, if $ab = 0$, then either $a = 0$ or $b = 0$ (or both).

EXAMPLE A.2.10   Solving an Equation Using Factoring

Solve the equation $x^2 - 3x = 10$.

Solution

First subtract 10 from both sides to get

$$x^2 - 3x - 10 = 0$$

and then factor the resulting polynomial on the left-hand side to get

$$(x - 5)(x + 2) = 0$$

Since the product $(x - 5)(x + 2)$ can be zero only if one (or both) of its factors is zero, it follows that the solutions are $x = 5$ (which makes the first factor zero) and $x = -2$ (which makes the second factor zero).

EXAMPLE A.2.11   Solving a Rational Equation

Solve the equation $1 - \dfrac{1}{x} - \dfrac{2}{x^2} = 0$.

Solution

Put the fractions on the left-hand side over the common denominator x^2 and add to get

$$\frac{x^2}{x^2} - \frac{x}{x^2} - \frac{2}{x^2} = 0$$

or

$$\frac{x^2 - x - 2}{x^2} = 0$$

Now factor the polynomial in the numerator to get

$$\frac{(x + 1)(x - 2)}{x^2} = 0$$

A quotient is zero only if its numerator is zero and its denominator is *not* zero, so it follows that $x = -1$ and $x = 2$ are the required solutions.

Completing the Square

An equation of the form

$$ax^2 + bx + c = 0 \qquad \text{for } a \neq 0$$

is called a **quadratic equation.** A quadratic equation can have at most two solutions. As you have seen, one way to find the solutions is to factor the equation. Another is by the algebraic procedure called **completing the square,** in which the equation is rewritten in the form

$$(x + r)^2 = s$$

for real numbers r and s. Here are the steps in the procedure.

Step 1. Divide both sides of the given equation

$$ax^2 + bx + c = 0$$

by a (remember, $a \neq 0$) to obtain

$$x^2 + \left(\frac{b}{a}\right)x + \left(\frac{c}{a}\right) = 0$$

Then subtract $\dfrac{c}{a}$ from both sides:

$$x^2 + \left(\frac{b}{a}\right)x = -\frac{c}{a}$$

Step 2. Add the square of $\dfrac{1}{2}\left(\dfrac{b}{a}\right)$ to both sides:

$$x^2 + \left(\frac{b}{a}\right)x + \left(\frac{b}{2a}\right)^2 = -\frac{c}{a} + \left(\frac{b}{2a}\right)^2$$

Step 3. Notice that the left side of the equation is $\left(x + \dfrac{b}{2a}\right)^2$. Thus, the equation can be written as

$$\left(x + \frac{b}{2a}\right)^2 = -\frac{c}{a} + \left(\frac{b}{2a}\right)^2$$

EXAMPLE A.2.12 Solving an Equation by Completing the Square

Solve the quadratic equation $x^2 + 5x + 4 = 0$ by completing the square.

Solution

$$x^2 + 5x + 4 = 0 \qquad \text{subtract 4 from both sides}$$

$$x^2 + 5x = -4 \qquad \begin{array}{l}\text{add the square of } \frac{1}{2}(5) \text{ to}\\ \text{both sides}\end{array}$$

$$x^2 + 5x + \left(\frac{5}{2}\right)^2 = -4 + \left(\frac{5}{2}\right)^2 \quad \begin{array}{l} x^2 + 5x + (5/2)^2\\ = (x + 5/2)^2\end{array}$$

$$\left(x + \frac{5}{2}\right)^2 = \frac{9}{4}$$

So

$$x + \frac{5}{2} = \sqrt{\frac{9}{4}} = \frac{3}{2} \quad \text{and} \quad x + \frac{5}{2} = -\sqrt{\frac{9}{4}} = -\frac{3}{2}$$

and the solutions are

$$x = \frac{3}{2} - \frac{5}{2} = -1 \quad \text{and} \quad x = -\frac{3}{2} - \frac{5}{2} = -4$$

EXAMPLE A.2.13 Solving an Equation by Completing the Square

Solve the quadratic equation $3x^2 + 5x + 7 = 0$ by completing the square.

Solution
We have

$$3x^2 + 5x + 7 = 0 \qquad \text{divide each term by 3}$$

$$x^2 + \left(\frac{5}{3}\right)x + \left(\frac{7}{3}\right) = 0 \qquad \text{subtract } \frac{7}{3} \text{ from each side}$$

$$x^2 + \left(\frac{5}{3}\right)x = -\frac{7}{3} \qquad \begin{array}{l}\text{add the square of}\\ \frac{1}{2}\left(\frac{5}{3}\right) \text{ to both sides}\end{array}$$

$$x^2 + \left(\frac{5}{3}\right)x + \left(\frac{5}{6}\right)^2 = -\frac{7}{3} + \left(\frac{5}{6}\right)^2$$

$$\left(x + \frac{5}{6}\right)^2 = -\frac{59}{36}$$

Since it is impossible for the square $\left(x + \dfrac{5}{6}\right)^2$ to equal the negative number $-\dfrac{59}{36}$, the given quadratic equation has no (real) solutions.

The Quadratic Formula

By completing the square in the general quadratic equation

$$ax^2 + bx + c = 0 \qquad \text{(for } a \neq 0\text{)}$$

we can obtain a general form for the solutions of the equation called the **quadratic formula.**

The Quadratic Formula ▪ The solutions of the quadratic equation

$$ax^2 + bx + c = 0 \qquad \text{(for } a \neq 0\text{)}$$

are given by the formula

$$x = \frac{-b \pm \sqrt{b^2 - 4ac}}{2a}$$

The term $b^2 - 4ac$ in the quadratic formula is called the **discriminant** of the quadratic equation. If the discriminant is positive, the equation has two solutions, one coming from the formula with the sign $\pm$ replaced by $+$ and the other with $\pm$ replaced by $-$. If the discriminant is zero, the equation has only one solution since the formula reduces to $x = \dfrac{-b}{2a}$. If the discriminant is negative, the equation has no real solutions since negative numbers do not have real square roots.

The use of the quadratic formula is illustrated in Examples A.2.14 through A.2.16.

EXAMPLE A.2.14   Using the Quadratic Formula

Solve the equation $x^2 + 3x + 1 = 0$.

Solution

This is a quadratic equation with $a = 1$, $b = 3$, and $c = 1$. Using the quadratic formula, you get

$$x = \frac{-3 + \sqrt{5}}{2} \approx -0.38 \qquad \text{and} \qquad x = \frac{-3 - \sqrt{5}}{2} \approx -2.62$$

EXAMPLE A.2.15   Using the Quadratic Formula

Solve the equation $x^2 + 18x + 81 = 0$.

Solution

This is a quadratic equation with $a = 1$, $b = 18$, and $c = 81$. Using the quadratic formula, you find that the discriminant is zero and that the formula for x gives

$$x = \frac{-18 \pm \sqrt{0}}{2} = -\frac{18}{2} = -9$$

EXAMPLE A.2.16 Using the Quadratic Formula

Solve the equation $x^2 + x + 1 = 0$.

Solution

This is a quadratic equation with $a = 1$, $b = 1$, and $c = 1$. Using the quadratic formula, you get

$$x = \frac{-1 \pm \sqrt{-3}}{2}$$

Since there is no real square root of -3, it follows that the equation has no real solution.

Systems of Equations

A collection of equations that are to be solved simultaneously is called a **system of equations.** Some of the calculus problems in Chapter 7 involve the solution of systems of two (or more) equations in two (or more) unknowns. For example, you may wish to find the real numbers x and y that satisfy the system

$$2x + 3y = 5$$
$$x + 2y = 4$$

The procedure for solving a system of two equations in two unknowns is to (temporarily) eliminate one of the variables, thereby reducing the problem to a single equation in one variable, which you then solve for its variable. Once you have found the value of one of the variables, you can substitute it into either of the original equations and solve to get the value of the other variable.

The most common techniques for the elimination of variables are illustrated in Examples A.2.17 and A.2.18.

EXAMPLE A.2.17 Solving a System of Equations

Solve the system

$$4x + 3y = 13$$
$$3x + 2y = 7$$

Solution

To eliminate y, multiply both sides of the first equation by 2 and both sides of the second equation by -3 so that the system becomes

$$8x + 6y = 26$$
$$-9x - 6y = -21$$

Then add the equations to get

$$-x + 0 = 5 \qquad \text{or} \qquad x = -5$$

To find y, you can substitute $x = -5$ into either of the original equations. If you choose the second equation, you find that

$$3(-5) + 2y = 7 \qquad 2y = 22 \qquad \text{or} \qquad y = 11$$

That is, the solution of the system is $x = -5$ and $y = 11$.

To check this answer, substitute $x = -5$ and $y = 11$ into each of the original equations. From the first equation you get

$$4(-5) + 3(11) = -20 + 33 = 13$$

and from the second equation you get

$$3(-5) + 2(11) = -15 + 22 = 7$$

as required.

EXAMPLE A.2.18 Solving a System of Equations

Solve the system

$$2y^2 - x^2 = 14$$
$$x - y = 1$$

Solution

Solve the second equation for x to get

$$x = y + 1$$

and substitute this into the first equation to eliminate x. This gives

$$2y^2 - (y + 1)^2 = 14$$
$$2y^2 - (y^2 + 2y + 1) = 14$$
$$2y^2 - y^2 - 2y - 1 = 14$$
$$y^2 - 2y - 15 = 0$$

or

$$(y + 3)(y - 5) = 0$$

from which it follows that

$$y = -3 \qquad \text{or} \qquad y = 5$$

If $y = -3$, the second equation gives

$$x - (-3) = 1 \qquad \text{or} \qquad x = -2$$

and if $y = 5$, the second equation gives

$$x - 5 = 1 \qquad \text{or} \qquad x = 6$$

Hence the system has two solutions,

$$x = 6, y = 5 \qquad \text{and} \qquad x = -2, y = -3$$

To check these answers, substitute each pair x, y into the first equation. If $x = 6$ and $y = 5$, you get

$$2(5^2) - 6^2 = 50 - 36 = 14$$

and if $x = -2$ and $y = -3$, you get

$$2(-3)^2 - (-2)^2 = 18 - 4 = 14$$

as required.

EXERCISES ■ A.2

In Exercises 1 through 10, find the indicated product.

1. $3x(x - 9)$

2. $-2x^2(3 - 4x)$

3. $(x - 7)(x + 2)$

4. $(x + 1)(x + 5)$

5. $(3x - 7)(4 - 2x)$

6. $(-x - 3)(5 - 3x)$

7. $(x - 1)(x^2 + 2x - 3)$ **8.** $(3x^2 - 5x + 4)(x + 2)$

9. $(x^3 - 3x + 4)(x^2 - 3x + 2)$

10. $(2x^3 + x^2 - 5)(x^2 - x - 3)$

In Exercises 11 through 28, simplify the given rational expression.

11. $\dfrac{x + 3}{x - 3} + \dfrac{x}{x + 3}$

12. $\dfrac{4}{x^2 + 5x + 6} + \dfrac{x - 2}{x + 3}$

13. $\dfrac{-5x - 6}{x^2 + 2x - 3} + \dfrac{x + 2}{x - 1}$

14. $\dfrac{x - 6}{x^2 + 3x - 10} - \dfrac{x + 3}{x - 5}$

15. $\dfrac{x - 2}{2x^2 - 7x - 15} - \dfrac{1}{2x + 3}$

16. $\left(\dfrac{x^3 - 8}{x}\right)\left(\dfrac{x^2 - 3x}{x - 2}\right)$

17. $\dfrac{4}{x + 2} - \dfrac{3}{x - 1} - \dfrac{2x}{x^2 + x - 2}$

18. $\dfrac{-2}{x - 4} + \dfrac{1}{x + 4} + \dfrac{1 - 2x}{x^2 - 16}$

19. $\dfrac{4}{x + 3} - \dfrac{2}{x + 4} - \dfrac{2x + 3}{x^2 + 7x + 12}$

20. $\dfrac{7}{x - 1} + \dfrac{5}{2x + 3} - \dfrac{x + 2}{2x^2 + x - 3}$

21. $\dfrac{1/x - 1/3}{1/x + 1/3}$

22. $\dfrac{1/x}{1 + (1/x)}$

23. $\dfrac{\dfrac{x - 3}{x + 3} - \dfrac{x + 3}{x - 3}}{\dfrac{x}{x - 3} - \dfrac{x}{x + 3}}$

24. $\dfrac{\dfrac{3x^2 + 5x - 8}{x^3 - 1}}{\dfrac{3x + 8}{x^2 + x + 1}}$

25. $1 - \dfrac{1}{1 + \dfrac{x}{2x - 1}}$

26. $3 + \dfrac{5}{1 - \dfrac{x - 1}{x + 1}}$

27. $\dfrac{\dfrac{1}{x} - 2 + \dfrac{x}{x + 1}}{\dfrac{3x - 1}{x^2 + x}}$

28. $\dfrac{\dfrac{x}{x^2 - 9} - \dfrac{1}{x + 3}}{\dfrac{3}{x - 3}}$

In Exercises 29 through 58, factor the given polynomial using integer coefficients.

29. $x^2 + x - 2$

30. $x^2 + 3x - 10$

31. $x^2 - 7x + 12$

32. $x^2 + 8x + 12$

33. $x^2 - 2x + 1$

34. $x^2 + 6x + 9$

35. $16x^2 - 25$

36. $3x^2 - x - 14$

37. $x^3 - 1$

38. $x^3 - 27$

39. $x^7 - x^5$

40. $x^3 + 2x^2 + x$

41. $2x^3 - 8x^2 - 10x$

42. $x^4 + 5x^3 - 14x^2$

43. $x^2 + x - 12$

44. $x^2 - 9x + 14$

45. $2x^2 - x - 15$

46. $3x^2 - 22x + 35$

47. $x^2 - 7x - 18$

48. $x^2 + 8x + 15$

49. $28x^2 + 2x - 6$

50. $12x^2 - x - 20$

51. $x^3 + 2x^2 - 15x$ **52.** $25x^3 - 16x$

53. $x^3 + 27$ **54.** $25x^2 - 81$

55. $x^5 + x^2$ **56.** $x^4 - 9x^2$

57. $3(x + 2)^3 - 5(x + 2)^2$

58. $5(x - 1)^4 + 3(x - 1)^2$

In Exercises 59 through 74, solve the given equation by factoring.

59. $x^2 - 2x - 8 = 0$ **60.** $x^2 - 4x + 3 = 0$

61. $x^2 + 10x + 25 = 0$ **62.** $x^2 + 8x + 16 = 0$

63. $x^2 - 16 = 0$ **64.** $x^2 - 25 = 0$

65. $2x^2 + 3x + 1 = 0$ **66.** $x^2 - 2x + 1 = 0$

67. $4x^2 + 12x + 9 = 0$ **68.** $6x^2 + 7x - 3 = 0$

69. $1 + \dfrac{4}{x} - \dfrac{5}{x^2} = 0$ **70.** $\dfrac{9}{x^2} - \dfrac{6}{x} + 1 = 0$

71. $2 + \dfrac{2}{x} - \dfrac{4}{x^2} = 0$ **72.** $\dfrac{3}{x^2} - \dfrac{5}{x} - 2 = 0$

73. $\dfrac{x}{x - 2} - \dfrac{4}{x + 3} - \dfrac{10}{x^2 + x - 6} = 0$

74. $\dfrac{x}{x + 1} + \dfrac{3}{2x + 3} - \dfrac{11x + 10}{2x^2 + 5x + 3} = 0$

In Exercises 75 through 82, solve the given quadratic equation by completing the square.

75. $x^2 + 2x - 3 = 0$ **76.** $2x^2 + 11x + 15 = 0$

77. $15x^2 - 14x + 3 = 0$ **78.** $21x^2 + 11x - 2 = 0$

79. $x^2 + 5x + 11 = 0$ **80.** $4x^2 + 3x + 1 = 0$

81. $6x^2 + 17x - 4 = 0$ **82.** $7x^2 + 12x - 5 = 0$

In Exercises 83 through 88, use the quadratic formula to solve the given equation.

83. $2x^2 + 3x + 1 = 0$ **84.** $-x^2 + 3x - 1 = 0$

85. $x^2 - 2x + 3 = 0$ **86.** $x^2 - 2x + 1 = 0$

87. $4x^2 + 12x + 9 = 0$ **88.** $x^2 + 12 = 0$

In Exercises 89 through 94, solve the given system of equations.

89. $\begin{aligned} x + 5y &= 13 \\ 3x - 10y &= -11 \end{aligned}$ **90.** $\begin{aligned} 2x - 3y &= 4 \\ 3x - 5y &= 2 \end{aligned}$

91. $\begin{aligned} 5x - 4y &= 12 \\ 2x - 3y &= 2 \end{aligned}$ **92.** $\begin{aligned} 3x^2 &- 9y = 0 \\ 3y^2 &- 9x = 0 \end{aligned}$

93. $\begin{aligned} 2y^2 - x^2 &= 1 \\ x - 2y &= 3 \end{aligned}$ **94.** $\begin{aligned} 2x^2 - y^2 &= -7 \\ 2x + y &= 1 \end{aligned}$

SECTION A.3   Evaluating Limits with L'Hôpital's Rule

L'Hôpital's Rule: $\dfrac{0}{0}$ and $\dfrac{\infty}{\infty}$ Forms

In curve sketching and other applications of calculus, it is often necessary to compute a limit of the form

$$\lim_{x \to c} \frac{f(x)}{g(x)}$$

where c is either a finite number or ∞. If $\lim\limits_{x \to c} g(x) \neq 0$, then the quotient rule for limits may be used, but if both $f(x)$ and $g(x)$ approach 0 as x approaches c, practically anything can happen. For example,

$$\lim_{x \to \infty} \frac{(1/x^3) - (1/x^2)}{1/x} \qquad \lim_{x \to 0} \frac{2x^3 + 3x^2}{x^5 + x^4} \qquad \text{and} \qquad \lim_{x \to 1} \frac{x - 1}{x^3 - 1}$$

all have this property, but the limit on the left is 0, the one in the center is ∞, and the one on the right is $\dfrac{1}{3}$.

Limits such as these are called $\dfrac{0}{0}$ **indeterminate forms.** Similarly, limits of quotients in which both the numerator and denominator increase or decrease without bound as $x \to c$ are called $\dfrac{\infty}{\infty}$ **indeterminate forms.**

There is a powerful technique, known as **L'Hôpital's rule,** which you can use to analyze indeterminate forms. The rule says, in effect, that if your attempt to find the limit of a quotient leads to either a $\frac{0}{0}$ or $\frac{\infty}{\infty}$ indeterminate form, then take derivatives of the numerator and the denominator and try again. Here is a more symbolic statement of the procedure.

L'Hôpital's Rule

If $\lim_{x\to c} f(x) = 0$ and $\lim_{x\to c} g(x) = 0$, then

$$\lim_{x\to c} \frac{f(x)}{g(x)} = \lim_{x\to c} \frac{f'(x)}{g'(x)}$$

If $\lim_{x\to c} f(x) = \infty$ and $\lim_{x\to c} g(x) = \infty$, then

$$\lim_{x\to c} \frac{f(x)}{g(x)} = \lim_{x\to c} \frac{f'(x)}{g'(x)}$$

The use of L'Hôpital's rule is illustrated in Examples A.3.1 through A.3.4. As you read through these examples, pay particular attention to the following two points:

1. L'Hôpital's rule involves differentiating the numerator and the denominator *separately*. A common mistake is to differentiate the entire quotient using the quotient rule.

2. L'Hôpital's rule applies only to quotients whose limits are indeterminate forms $\frac{0}{0}$ or $\frac{\infty}{\infty}$. Limits of the form $\frac{0}{\infty}$ or $\frac{\infty}{0}$ are *not* indeterminate (the first is 0, and the second is ∞).

EXAMPLE A.3.1 Using L'Hôpital's Rule

Use L'Hôpital's rule to compute the limit

$$\lim_{x\to\infty} \frac{x}{(x+1)^2}$$

Solution

This is a $\frac{\infty}{\infty}$ indeterminate form, so L'Hôpital's rule applies, and we get

$$\lim_{x\to\infty} \frac{x}{(x+1)^2} = \lim_{x\to\infty} \frac{(x)'}{[(x+1)^2]'} = \lim_{x\to\infty} \frac{1}{2(x+1)} = 0$$

EXAMPLE A.3.2   Using L'Hôpital's Rule

Use L'Hôpital's rule to compute the limit

$$\lim_{x \to 1} \frac{x^5 - 3x^4 + 5x - 3}{4x^5 + 2x^3 - 5x^2 - 1}$$

Solution

By substituting $x = 1$ into the numerator and denominator, we see that this is a $\frac{0}{0}$ indeterminate form. We could evaluate this limit by the factor method developed in Chapter 1, but notice how much easier it is to use L'Hôpital's rule:

$$\lim_{x \to 1} \frac{x^5 - 3x^4 + 5x - 3}{4x^5 + 2x^3 - 5x^2 - 1} = \lim_{x \to 1} \frac{(x^5 - 3x^4 + 5x - 3)'}{(4x^5 + 2x^3 - 5x^2 - 1)'}$$

$$= \lim_{x \to 1} \frac{5x^4 - 12x^3 + 5}{20x^4 + 6x^2 - 10x} = -\frac{2}{16} = -\frac{1}{8}$$

EXAMPLE A.3.3   Recognizing a Limit That Is Not Indeterminate

Evaluate $\lim_{x \to 2} \dfrac{2x + 5}{x^2 + 3x - 10}$.

Solution

If you blindly apply L'Hôpital's rule, you get

$$\lim_{x \to 2} \frac{2x + 5}{x^2 + 3x - 10} = \lim_{x \to 2} \frac{2}{2x + 3} = \frac{2}{7}$$

However, if you use your calculator to evaluate the given quotient at a number very close to 2 (say, at 2.0001), you find that the number you get is much larger than $\frac{2}{7}$. Why? The answer you got with L'Hôpital's rule was wrong because the given limit is not indeterminate. In fact, by simply substituting $x = 2$, you get

$$\lim_{x \to 2} \frac{2x + 5}{x^2 + 3x - 10} = \frac{9}{0}$$

This tells us that the limit is infinite.

EXAMPLE A.3.4   Using L'Hôpital's Rule

Find $\lim_{x \to \infty} \dfrac{3 - e^x}{x^2}$.

Solution

The limit is indeterminate of the form $\dfrac{\infty}{\infty}$. Applying L'Hôpital's rule, we get

$$\lim_{x \to \infty} \frac{3 - e^x}{x^2} = \lim_{x \to \infty} \frac{-e^x}{2x}$$

Since this new limit is also of the form $\dfrac{\infty}{\infty}$, we apply L'Hôpital's rule again to get

$$\lim_{x \to \infty} \frac{-e^x}{2x} = \lim_{x \to \infty} \frac{-e^x}{2} = -\infty$$

and we conclude that

$$\lim_{x \to \infty} \frac{3 - e^x}{x^2} = -\infty$$

Although L'Hôpital's rule only applies to $\dfrac{0}{0}$ and $\dfrac{\infty}{\infty}$ indeterminate forms, other kinds of indeterminate forms can often be computed by combining L'Hôpital's rule with a little algebra. This procedure is illustrated in Examples A.3.5 and A.3.6.

EXAMPLE A.3.5 Applying L'Hôpital's Rule to the Form $0 \cdot \infty$

Find $\lim\limits_{x \to \infty} e^{-x} \ln x$.

Solution

This limit is of the indeterminate form $0 \cdot \infty$ and can be rewritten as

$$\lim_{x \to \infty} \frac{e^{-x}}{1/\ln x} \qquad \left(\text{of the form } \frac{0}{0} \right)$$

or as

$$\lim_{x \to \infty} \frac{\ln x}{e^x} \qquad \left(\text{of the form } \frac{\infty}{\infty} \right)$$

Applying L'Hôpital's rule to the simpler second quotient, you get

$$\lim_{x \to \infty} e^{-x} \ln x = \lim_{x \to \infty} \frac{\ln x}{e^x} = \lim_{x \to \infty} \frac{1/x}{e^x} = 0$$

As a final illustration of this technique, here is the limit that was used in Section 4.1 to define the number e.

EXAMPLE A.3.6 Applying L'Hôpital's Rule to the Form 1^{∞}

Find $\lim\limits_{x \to \infty} \left(1 + \dfrac{1}{x} \right)^x$.

Solution

This limit is of the indeterminate form 1^∞. To simplify the problem let

$$y = \left(1 + \frac{1}{x}\right)^x$$

Then $\ln y = x \ln\left(1 + \frac{1}{x}\right)$

$$\lim_{x \to \infty} \ln y = \lim_{x \to \infty} x \ln\left(1 + \frac{1}{x}\right) \qquad (\infty \cdot 0)$$

$$\lim_{x \to \infty} \ln y = \lim_{x \to \infty} \frac{\ln(1 + 1/x)}{1/x} \qquad \left(\frac{0}{0}\right) \qquad \text{L'Hôpital's rule}$$

$$= \lim_{x \to \infty} \frac{\dfrac{d}{dx}[\ln(1 + 1/x)]}{\dfrac{d}{dx}[1/x]} = \lim_{x \to \infty} \frac{\dfrac{(-1/x^2)}{(1 + 1/x)}}{-1/x^2} \qquad \text{algebraic simplification}$$

$$= \lim_{x \to \infty} \frac{1}{1 + 1/x}$$

$$= 1$$

Since $\ln y \to 1$, it follows that $y \to e^1 = e$. That is,

$$\lim_{x \to \infty} \left(1 + \frac{1}{x}\right)^x = e$$

EXERCISES ▪ A.3

In Exercises 1 through 16, use L'Hôpital's rule to evaluate the given limit if the limit is an indeterminate form.

1. $\displaystyle\lim_{x \to 0} \frac{x^3 - 3x^2}{3x^4 + 2x}$

2. $\displaystyle\lim_{x \to 0} \frac{x^2(x - 1)}{3x^3 + 2x - 5}$

3. $\displaystyle\lim_{x \to \infty} \frac{x^2 - 2x + 3}{2x^2 + 5x + 1}$

4. $\displaystyle\lim_{x \to \infty} \frac{x^2 + x - 5}{1 - 2x - x^3}$

5. $\displaystyle\lim_{x \to \infty} \frac{(1/x) - (2/x^2)}{(1/x^3) + (2/x^2) - (3/x)}$

6. $\displaystyle\lim_{x \to 3} \frac{x^2 + 2x - 15}{x^3 - 19x + 3}$

7. $\displaystyle\lim_{x \to -1} \frac{x^3 + 3x^2 + 3x + 1}{2x^3 + 3x^2 - 1}$

[*Hint:* Use L'Hôpital's rule twice.]

8. $\displaystyle\lim_{x \to 1/2} \frac{-8x^3 + 2x^2 + 3x - 1}{(2x - 1)^3}$

9. $\displaystyle\lim_{x \to \infty} \frac{e^{-x}}{1 + e^{-2x}}$

10. $\displaystyle\lim_{x \to \infty} x^2 e^{-x}$

11. $\displaystyle\lim_{t \to 0} \frac{\sqrt{t}}{e^t}$

12. $\displaystyle\lim_{t \to \infty} \frac{\ln \sqrt{t}}{t}$

13. $\displaystyle\lim_{x \to \infty} \frac{(\ln x)^2}{x}$

14. $\displaystyle\lim_{x \to \infty} x^{1/x}$

15. $\displaystyle\lim_{x \to 0} (1 + 2x)^{1/x}$

16. $\displaystyle\lim_{x \to \infty} \left(1 + \frac{1}{x}\right)^{x^2}$

SECTION A.4   The Summation Notation

Sums of the form $a_1 + a_2 + \cdots + a_n$ appear so often in mathematics that a special notation has been developed to handle them. To describe such a sum, it suffices to characterize the general term a_j and to indicate that n terms of this form are to be added, starting with the term where $j = 1$ and ending with the term where $j = n$. It is customary to use the Greek uppercase letter Σ (sigma) to denote summation and to express the sum compactly as follows.

> **Summation Notation** ■ The sum of the numbers $a_1, \ldots, a_n$ is given by
>
> $$a_1 + a_2 + \cdots + a_n = \sum_{j=1}^{n} a_j$$

The use of summation notation is illustrated in Examples A.4.1 and A.4.2.

EXAMPLE A.4.1   Evaluating a Sum

Evaluate these sums.

a. $\displaystyle\sum_{j=1}^{4}(j^2 + 1)$

b. $\displaystyle\sum_{j=1}^{3}(-2)^j$

Solution

a. $\displaystyle\sum_{j=1}^{4}(j^2 + 1) = (1^2 + 1) + (2^2 + 1) + (3^2 + 1) + (4^2 + 1)$

$$= 2 + 5 + 10 + 17 = 34$$

b. $\displaystyle\sum_{j=1}^{3}(-2)^j = (-2)^1 + (-2)^2 + (-2)^3 = -2 + 4 - 8 = -6$

EXAMPLE A.4.2   Using Summation Notation

Use summation notation to represent these sums.

a. $1 + 4 + 9 + 16 + 25 + 36 + 49 + 64$

b. $(1 - x_1)^2\Delta x + (1 - x_2)^2\Delta x + \cdots + (1 - x_{15})^2\Delta x$

Solution

a. This is a sum of 8 terms of the form j^2, starting with $j = 1$ and ending with $j = 8$. Hence,

$$1 + 4 + 9 + 16 + 25 + 36 + 49 + 64 = \sum_{j=1}^{8} j^2$$

b. The jth term of this sum is $(1 - x_j)^2\Delta x$. Hence,

$$(1 - x_1)^2\Delta x + (1 - x_2)^2\Delta x + \cdots + (1 - x_{15})^2\Delta x = \sum_{j=1}^{15}(1 - x_j)^2\Delta x$$

EXERCISES ▪ A.4

In Exercises 1 through 4, evaluate the given sum.

1. $\displaystyle\sum_{j=1}^{4}(3j+1)$

2. $\displaystyle\sum_{j=1}^{5}j^2$

3. $\displaystyle\sum_{j=1}^{10}(-1)^j$

4. $\displaystyle\sum_{j=1}^{5}2^j$

In Exercises 5 through 10, use summation notation to represent the given sum.

5. $1+\dfrac{1}{2}+\dfrac{1}{3}+\dfrac{1}{4}+\dfrac{1}{5}+\dfrac{1}{6}$

6. $3+6+9+12+15+18+21+24+27+30$

7. $2x_1+2x_2+2x_3+2x_4+2x_5+2x_6$

8. $1-1+1-1+1-1$

9. $1-2+3-4+5-6+7-8$

10. $x-x^2+x^3-x^4+x^5$

Important Terms, Symbols, and Formulas

Integer   (900)
Rational number   (900)
Irrational number   (900)
Real numbers   (900)
Number line   (900)
Inequality   (901)
Interval   (902)
Absolute value: (903)

$$|x| = \begin{cases} x & \text{if } x \geq 0 \\ -x & \text{if } x < 0 \end{cases}$$

Distance on a number line   (904)
Exponential notation:

$$a^n = \underbrace{a \cdot a \cdots a}_{n \text{ terms}}$$

$$a^{-n} = \frac{1}{a^n} \quad \text{(negative power)} \quad (904)$$

$$a^{n/m} = (\sqrt[m]{a})^n = \sqrt[m]{a^n} \quad \text{(fractional powers)} \quad (905)$$

Laws of exponents: (906)

$$a^r a^s = a^{r+s} \quad \text{(product law)}$$

$$\frac{a^r}{a^s} = a^{r-s} \quad \text{(quotient law)}$$

$$(a^r)^s = a^{rs} \quad \text{(power law)}$$

Rationalizing   (908)
Polynomial:
  Coefficients of a polynomial   (911)
  Degree of a polynomial   (911)
Factoring   (912)
Laws of numbers used in factoring:
  Distributive law $ab + ac = a(b+c)$ (911)
  Difference of two squares   (914)

$$a^2 - b^2 = (a+b)(a-b)$$

Rational expressions   (915)
Properties of fractions   (915)

$$\frac{a}{b} + \frac{c}{d} = \frac{ad+bc}{bd}$$

$$\left(\frac{a}{b}\right)\left(\frac{c}{d}\right) = \frac{ac}{bd}$$

$$\frac{a/b}{c/d} = \frac{ad}{bc}$$

Solving an equation by factoring   (916)
Completing the square   (917)
The quadratic formula: (919)
  The solutions of $ax^2 + bx + c = 0$ for $a \neq 0$ are

$$x = \frac{-b \pm \sqrt{b^2 - 4ac}}{2a}$$

Discriminant of the quadratic equation
  $ax^2 + bx + c = 0$ is $b^2 - 4ac$ (919)
System of equations   (920)
Solving a system of equations by elimination   (921)
L'Hôpital's rule: (924)

$\left(\dfrac{0}{0} \text{ form}\right)$ If $\displaystyle\lim_{x\to c} f(x) = 0$ and $\displaystyle\lim_{x\to c} g(x) = 0$, then

$$\lim_{x\to c}\frac{f(x)}{g(x)} = \lim_{x\to c}\frac{f'(x)}{g'(x)}$$

$\left(\dfrac{\infty}{\infty} \text{ form}\right)$ If $\displaystyle\lim_{x\to c} f(x) = \infty$ and $\displaystyle\lim_{x\to c} g(x) = \infty$, then

$$\lim_{x\to c}\frac{f(x)}{g(x)} = \lim_{x\to c}\frac{f'(x)}{g'(x)}$$

The summation notation: (928)

$$\sum_{j=1}^{n} a_j = a_1 + a_2 + \cdots + a_n$$

Review Exercises

In Exercises 1 and 2, use inequalities to describe the given interval.

1.

2.

In Exercises 3 through 6, represent the given interval as a line segment on a number line.

3. $-3 \le x < 2$

4. $-1 < x < 5$

5. $x \ge 1$

6. $2 \le x < 7$

In Exercises 7 and 8, find the distance on the number line between the given pair of real numbers.

7. 0 and 3

8. -5 and -2

In Exercises 9 and 10, find the interval or intervals consisting of all real numbers x that satisfy the given inequality.

9. $|x - 3| \le 1$

10. $|2x + 1| > 3$

In Exercises 11 through 20, evaluate the given expression without using a calculator.

11. 3^5

12. 4^{-2}

13. $8^{2/3}$

14. $49^{-3/2}$

15. $\dfrac{4(32)^{3/4}}{(\sqrt{2})^3}$

16. $\left(\dfrac{1}{9}\right)^{-5/2}$

17. $16^{3/2} + 27^{2/3}$

18. $\dfrac{2^{3/2}(4^{5/2})}{8^{2/3}}$

19. $\dfrac{\sqrt[3]{54}\,\sqrt[6]{2}}{\sqrt{8}}$

20. $\dfrac{\sqrt[3]{81}(6^{2/3})}{2^{4/3}}$

In Exercises 21 through 24, solve the given equation for n (assume $a > 0$, $a \ne 1$).

21. $a^{2/3}a^{1/2} = a^{3n}$

22. $\dfrac{a^3}{(\sqrt{a})^5} = a^{2n}$

23. $a^2 a^{-5} = (a^n)^3$

24. $a^{2n}a^3 = a^{-7}$

In Exercises 25 through 28, evaluate the given sum.

25. $\displaystyle\sum_{k=1}^{3} (2k + 3)$

26. $\displaystyle\sum_{k=1}^{4} (k + 1)^2$

27. $\displaystyle\sum_{k=1}^{5} (2k^2 - k)$

28. $\displaystyle\sum_{k=1}^{4} \left[\dfrac{k - 1}{k + 3}\right]^2$

In Exercises 29 and 30, express the given sum in terms of the summation notation.

29. $1 + \dfrac{1}{2} - \dfrac{1}{3} + \dfrac{1}{4} - \dfrac{1}{5} + \dfrac{1}{6} - \dfrac{1}{7}$

30. $3 + 12 + 27 + 48 + 75$

In Exercises 31 through 34, factor the given expression.

31. $x^4 - 9x^2$

32. $x^3 + 3(x - 12)$

33. $x^{16} - (2x)^4$

34. $2(x - 3)^2(x + 1) - 5(x - 3)^3(2x)$

In Exercises 35 and 36, simplify the given quotient as much as possible.

35. $\dfrac{x^2(x - 1)^3 - 2x(x - 1)^2}{x^2 - x - 2}$

36. $\dfrac{x(x + 2)^4 - x^3(x + 2)^2}{x^2 + 3x + 2}$

In Exercises 37 through 42, factor the given polynomial using integral coefficients.

37. $x^2 + 2x - 15$

38. $2x^2 + 5x - 3$

39. $4x^2 + 12x + 9$

40. $12x^2 + 5x - 3$

41. $x^3 + 3x^2 - x - 3$

42. $x^4 - 5x^2 + 4$

In Exercises 43 through 48, solve the given equation by factoring.

43. $x^2 + 3x - 4 = 0$

44. $2x^2 - 3x - 2 = 0$

45. $x^2 + 14x + 49 = 0$

46. $x^2 - 64 = 0$

47. $1 - \dfrac{1}{x} - \dfrac{2}{x^2} = 0$

48. $4 + \dfrac{9}{x^2} = \dfrac{12}{x}$

In Exercises 49 through 54, use the quadratic formula to find all real numbers x that satisfy the given equation.

49. $14x^2 - x - 3 = 0$

50. $24x^2 + x - 10 = 0$

51. $x^2 - 3x + 5 = 0$

52. $7x^2 + 3x - 2 = 0$

53. $3x^2 + 5x - 2 = 0$

54. $2x^2 + 12x + 11 = 0$

In Exercises 55 through 58, solve the given system of equations.

55. $3x + 5y = -1$
$2x + 7y = 3$

56. $2x + y = 7$
$-x + 4y = 1$

57. $3x^2 - y^2 = -1$
$2x + y = 4$

58. $5x^2 - 2y^2 = 2$
$5x - 2y = 4$

In Exercises 59 through 64, use L'Hôpital's rule to evaluate the given limit if the limit is an indeterminate form.

59. $\displaystyle \lim_{x \to 1} \frac{x^2 - 1}{2x^3 + x - 3}$

60. $\displaystyle \lim_{x \to -2} \frac{x^3 + 8}{3x^3 - 7x + 10}$

61. $\displaystyle \lim_{x \to \infty} \frac{e^{-2x}}{3 + 2e^{-2x}}$

62. $\displaystyle \lim_{x \to \infty} \sqrt{x}\, e^{-x}$

63. $\displaystyle \lim_{x \to \infty} x(e^{1/x} - 1)$

64. $\displaystyle \lim_{x \to \infty} \left(1 - \frac{3}{x}\right)^{2x}$

LIVING SPACE

PROBLEM

For comfortable modern living, it is estimated that each person needs roughly 60 m^2 for housing, 40 m^2 for his or her job, 50 m^2 for public buildings and recreation facilities, 90 m^2 for transportation (e.g., highways), and 4,000 m^2 for the production of food.

Questions

1. Switzerland has approximately 11,000 km^2 of livable space (arable and habitable land). How many people can comfortably live in Switzerland? Look up the actual population of Switzerland. Based on the figures given here, is Switzerland overcrowded or is there still room for comfortable growth?

2. Use an almanac, an encyclopedia, or some other source to obtain the population of India. How much livable space would there have to be in India to accommodate its current population without overcrowding? Now look up the total area of India. Even if all of India is comprised of livable space, is there enough room for the population to live comfortably?

3. You probably knew that India is overcrowded. Pick another country where the answer is not so obvious (Bolivia? Zimbabwe? San Marino?). Describe the "comfort of living" in that country.

Source: Adapted from a problem in E. Batschelet, *Introduction to Mathematics for Life Scientists,* 2nd ed., New York: Springer-Verlag, 1979, p. 31. You may find it interesting to examine several of the applied life science problems on pages 31–33 of the Batschelet text.

TABLES

TABLE I Powers of e

x	e^x	e^{-x}	x	e^x	e^{-x}	x	e^x	e^{-x}
0.00	1.0000	1.00000	0.50	1.6487	.60653	1.00	2.7183	.36788
0.01	1.0101	0.99005	0.51	1.6653	.60050	1.10	3.0042	.33287
0.02	1.0202	.98020	0.52	1.6820	.59452	1.20	3.3201	.30119
0.03	1.0305	.97045	0.53	1.6989	.58860	1.30	3.6693	.27253
0.04	1.0408	.96079	0.54	1.7160	.58275	1.40	4.0552	.24660
0.05	1.0513	.95123	0.55	1.7333	.57695	1.50	4.4817	.22313
0.06	1.0618	.94176	0.56	1.7507	.57121	1.60	4.9530	.20190
0.07	1.0725	.93239	0.57	1.7683	.56553	1.70	5.4739	.18268
0.08	1.0833	.92312	0.58	1.7860	.55990	1.80	6.0496	.16530
0.09	1.0942	.91393	0.59	1.8040	.55433	1.90	6.6859	.14957
0.10	1.1052	.90484	0.60	1.8221	.54881	2.00	7.3891	.13534
0.11	1.1163	.89583	0.61	1.8404	.54335	3.00	20.086	.04979
0.12	1.1275	.88692	0.62	1.8589	.53794	4.00	54.598	.01832
0.13	1.1388	.87809	0.63	1.8776	.53259	5.00	148.41	.00674
0.14	1.1503	.86936	0.64	1.8965	.52729	6.00	403.43	.00248
0.15	1.1618	.86071	0.65	1.9155	.52205	7.00	1096.6	.00091
0.16	1.1735	.85214	0.66	1.9348	.51685	8.00	2981.0	.00034
0.17	1.1853	.84366	0.67	1.9542	.51171	9.00	8103.1	.00012
0.18	1.1972	.83527	0.68	1.9739	.50662	10.00	22026.5	.00005
0.19	1.2092	.82696	0.69	1.9937	.50158			
0.20	1.2214	.81873	0.70	2.0138	.49659			
0.21	1.2337	.81058	0.71	2.0340	.49164			
0.22	1.2461	.80252	0.72	2.0544	.48675			
0.23	1.2586	.79453	0.73	2.0751	.48191			
0.24	1.2712	.78663	0.74	2.0959	.47711			
0.25	1.2840	77880	0.75	2.1170	.47237			
0.26	1.2969	.77105	0.76	2.1383	.46767			
0.27	1.3100	.76338	0.77	2.1598	.46301			
0.28	1.3231	.75578	0.78	2.1815	.45841			
0.29	1.3364	.74826	0.79	2.2034	.45384			
0.30	1.3499	.74082	0.80	2.2255	.44933			
0.31	1.3634	.73345	0.81	2.2479	.44486			
0.32	1.3771	.72615	0.82	2.2705	.44043			
0.33	1.3910	.71892	0.83	2.2933	.43605			
0.34	1.4049	.71177	0.84	2.3164	.43171			
0.35	1.4191	.70469	0.85	2.3396	.42741			
0.36	1.4333	.69768	0.86	2.3632	.42316			
0.37	1.4477	.69073	0.87	2.3869	.41895			
0.38	1.4623	.68386	0.88	2.4109	.41478			
0.39	1.4770	.67706	0.89	2.4351	.41066			
0.40	1.4918	.67032	0.90	2.4596	.40657			
0.41	1.5068	.66365	0.91	2.4843	.40252			
0.42	1.5220	.65705	0.92	2.5093	.39852			
0.43	1.5373	.65051	0.93	2.5345	.39455			
0.44	1.5527	.64404	0.94	2.5600	.39063			
0.45	1.5683	.63763	0.95	2.5857	.38674			
0.46	1.5841	.63128	0.96	2.6117	.38289			
0.47	1.6000	.62500	0.97	2.6379	.37908			
0.48	1.6161	.61878	0.98	2.6645	.37531			
0.49	1.6323	.61263	0.99	2.6912	.37158			

Excerpted from R. S. Burington, *Handbook of Mathematical Tables and Formulas*, 5th ed. Copyright © 1973 by McGraw-Hill, Inc. Used with permission of McGraw-Hill Book Company.

TABLE II The Natural Logarithm (Base e)

x	$\ln x$	x	$\ln x$	x	$\ln x$	x	$\ln x$
.01	−4.60517	0.50	−0.69315	1.00	0.00000	1.5	0.40547
.02	−3.91202	.51	.67334	1.01	.00995	1.6	7000
.03	.50656	.52	.65393	1.02	.01980	1.7	0.53063
.04	.21888	.53	.63488	1.03	.02956	1.8	8779
		.54	.61619	1.04	.03922	1.9	0.64185
.05	−2.99573	.55	.59784	1.05	.04879	2.0	9315
.06	.81341	.56	.57982	1.06	.05827	2.1	0.74194
.07	.65926	.57	.56212	1.07	.06766	2.2	8846
.08	.52573	.58	.54473	1.08	.07696	2.3	0.83291
.09	.40795	.59	.52763	1.09	.08618	2.4	7547
0.10	−2.30259	0.60	−0.51083	1.10	.09531	2.5	0.91629
.11	.20727	.61	.49430	1.11	.10436	2.6	5551
.12	.12026	.62	.47804	1.12	.11333	2.7	9325
.13	.04022	.63	.46204	1.13	.12222	2.8	1.02962
.14	−1.96611	.64	.44629	1.14	.13103	2.9	6471
.15	.89712	.65	.43078	1.15	.13976	3.0	9861
.16	.83258	.66	.41552	1.16	.14842	4.0	1.38629
.17	.77196	.67	.40048	1.17	.15700	5.0	1.60944
.18	.71480	.68	.38566	1.18	.16551	10.0	2.30258
.19	.66073	.69	.37106	1.19	.17395		
0.20	−1.60944	0.70	−0.35667	1.20	.18232		
.21	.56065	.71	.34249	1.21	.19062		
.22	.51413	.72	.32850	1.22	.19885		
.23	.46968	.73	.31471	1.23	.20701		
.24	.42712	.74	.30111	1.24	.21511		
.25	.38629	.75	.28768	1.25	.22314		
.26	.34707	.76	.27444	1.26	.23111		
.27	.30933	.77	.26136	1.27	.23902		
.28	.27297	.78	.24846	1.28	.24686		
.29	.23787	.79	.23572	1.29	.25464		
0.30	−1.20397	0.80	−0.22314	1.30	.26236		
.31	.17118	.81	.21072	1.31	.27003		
.32	.13943	.82	.19845	1.32	.27763		
.33	.10866	.83	.18633	1.33	.28518		
.34	.07881	.84	.17435	1.34	.29267		
.35	−1.04982	.85	−0.16252	1.35	.30010		
.36	.02165	.86	.15032	1.36	.30748		
.37	−0.99425	.87	.13926	1.37	.31481		
.38	.96758	.88	.12783	1.38	.32208		
.39	.94161	.89	.11653	1.39	.32930		
0.40	−0.91629	0.90	−0.10536	1.40	.33647		
.41	.89160	.91	.09431	1.41	.34359		
.42	.86750	.92	.08338	1.42	.35066		
.43	.84397	.93	.07257	1.43	.35767		
.44	.82098	.94	.06188	1.44	.36464		
.45	.79851	.95	.05129	1.45	.37156		
.46	.77653	.96	.04082	1.46	.37844		
.47	.75502	.97	.03046	1.47	.38526		
.48	.73397	.98	.02020	1.48	.39204		
.49	.71335	.99	.01005	1.49	.39878		

From S. K. Stein, *Calculus and Analytic Geometry.* Copyright © 1973 by McGraw-Hill, Inc. Used with permission of McGraw-Hill Book Company.

TABLE III   Trigonometric Functions

Degrees	Radians	Sin	Cos	Tan	Degrees	Radians	Sin	Cos	Tan
0	0.0000	0.0000	1.000	0.0000	45	0.7854	0.7071	0.7071	1.000
1	0.01745	0.01745	0.9998	0.01746	46	0.8028	0.7193	0.6947	1.036
2	0.03491	0.03490	0.9994	0.03492	47	0.8203	0.7314	0.6820	1.072
3	0.05236	0.05234	0.9986	0.05241	48	0.8378	0.7431	0.6691	1.111
4	0.06981	0.06976	0.9976	0.06993	49	0.8552	0.7547	0.6561	1.150
5	0.08727	0.08716	0.9962	0.08749	50	0.8727	0.7660	0.6428	1.192
6	0.1047	0.1045	0.9945	0.1051	51	0.8901	0.7772	0.6293	1.235
7	0.1222	0.1219	0.9926	0.1228	52	0.9076	0.7880	0.6157	1.280
8	0.1396	0.1392	0.9903	0.1405	53	0.9250	0.7986	0.6018	1.327
9	0.1571	0.1564	0.9877	0.1584	54	0.9425	0.8090	0.5878	1.376
10	0.1745	0.1736	0.9848	0.1763	55	0.9599	0.8192	0.5736	1.428
11	0.1920	0.1908	0.9816	0.1944	56	0.9774	0.8290	0.5592	1.483
12	0.2094	0.2079	0.9782	0.2126	57	0.9948	0.8387	0.5446	1.540
13	0.2269	0.2250	0.9744	0.2309	58	1.012	0.8480	0.5299	1.600
14	0.2444	0.2419	0.9703	0.2493	59	1.030	0.8572	0.5150	1.664
15	0.2618	0.2588	0.9659	0.2680	60	1.047	0.8660	0.5000	1.732
16	0.2792	0.2756	0.9613	0.2868	61	1.065	0.8746	0.4848	1.804
17	0.2967	0.2924	0.9563	0.3057	62	1.082	0.8830	0.4695	1.881
18	0.3142	0.3090	0.9511	0.3249	63	1.100	0.8910	0.4540	1.963
19	0.3316	0.3256	0.9455	0.3443	64	1.117	0.8988	0.4384	2.050
20	0.3491	0.3420	0.9397	0.3640	65	1.134	0.9063	0.4226	2.144
21	0.3665	0.3584	0.9336	0.3839	66	1.152	0.9136	0.4067	2.246
22	0.3840	0.3746	0.9272	0.4040	67	1.169	0.9205	0.3907	2.356
23	0.4014	0.3907	0.9205	0.4245	68	1.187	0.9272	0.3746	2.475
24	0.4189	0.4067	0.9136	0.4452	69	1.204	0.9336	0.3584	2.605
25	0.4363	0.4226	0.9063	0.4663	70	1.222	0.9397	0.3420	2.748
26	0.4538	0.4384	0.8988	0.4877	71	1.239	0.9455	0.3256	2.904
27	0.4712	0.4540	0.8910	0.5095	72	1.257	0.9511	0.3090	3.078
28	0.4887	0.4695	0.8830	0.5317	73	1.274	0.9563	0.2924	3.271
29	0.5062	0.4848	0.8746	0.5543	74	1.292	0.9613	0.2756	3.487
30	0.5236	0.5000	0.8660	0.5774	75	1.309	0.9659	0.2588	3.732
31	0.5410	0.5150	0.8572	0.6009	76	1.326	0.9703	0.2419	4.011
32	0.5585	0.5299	0.8480	0.6249	77	1.344	0.9744	0.2250	4.332
33	0.5760	0.5446	0.8387	0.6494	78	1.361	0.9782	0.2079	4.705
34	0.5934	0.5592	0.8290	0.6745	79	1.379	0.9816	0.1908	5.145
35	0.6109	0.5736	0.8192	0.7002	80	1.396	0.9848	0.1736	5.671
36	0.6283	0.5878	0.8090	0.7265	81	1.414	0.9877	0.1564	6.314
37	0.6458	0.6018	0.7986	0.7536	82	1.431	0.9903	0.1392	7.115
38	0.6632	0.6157	0.7880	0.7813	83	1.449	0.9926	0.1219	8.144
39	0.6807	0.6293	0.7772	0.8098	84	1.466	0.9945	0.1045	9.514
40	0.6981	0.6428	0.7660	0.8391	85	1.484	0.9962	0.08716	11.43
41	0.7156	0.6561	0.7547	0.8693	86	1.501	0.9976	0.06976	14.30
42	0.7330	0.6691	0.7431	0.9004	87	1.518	0.9986	0.05234	19.08
43	0.7505	0.6820	0.7314	0.9325	88	1.536	0.9994	0.03490	28.64
44	0.7679	0.6947	0.7193	0.9657	89	1.553	0.9998	0.01745	57.29
45	0.7854	0.7071	0.7071	1.000	90	1.571	1.000	0.0000	——

Answers to Odd-Numbered Exercises, Checkup Exercises, and Review Exercises

CHAPTER 1   Section 1

1. $f(0) = 5; f(-1) = 2; f(2) = 11$

3. $f(0) = -2; f(-2) = 0; f(1) = 6$

5. $g(-1) = -2; g(1) = 2; g(2) = \dfrac{5}{2}$

7. $h(2) = 2\sqrt{3}; h(0) = 2; h(-4) = 2\sqrt{3}$

9. $f(1) = 1; f(5) = \dfrac{1}{27}; f(13) = \dfrac{1}{125}$

11. $f(1) = 0; f(2) = 2; f(3) = 2$

13. $h(3) = 10; h(1) = 2; h(0) = 4; h(-3) = 10$

15. Yes

17. No, $f(t)$ is not defined for $t > 1$

19. All real numbers x except $x = -2$

21. All real numbers x for which $x \geq -3$

23. All real numbers t for which $-3 < t < 3$

25. $f(g(x)) = 3x^2 + 14x + 10$

27. $f(g(x)) = x^3 + 2x^2 + 4x + 2$

29. $f(g(x)) = \dfrac{1}{(x-1)^2}$

31. $f(g(x)) = |x|$

33. $\dfrac{f(x+h) - f(x)}{h} = -5$

35. $\dfrac{f(x+h) - f(x)}{h} = 4 - 2x - h$

37. $\dfrac{f(x+h) - f(x)}{h} = \dfrac{1}{(x+1)(x+h+1)}$

39. $f(g(x)) = \sqrt{1 - 3x}; g(f(x)) = 1 - 3\sqrt{x};$
$f(g(x)) = g(f(x))$ if $x = 0$

41. $f(g(x)) = x; g(f(x)) = x; f(g(x)) = g(f(x))$ for all
real numbers except $x = 1$ and $x = 2$

43. $f(x - 2) = 2x^2 - 11x + 15$

45. $f(x - 1) = x^5 - 3x^2 + 6x - 3$

47. $f(x^2 + 3x - 1) = \sqrt{x^2 + 3x - 1}$

49. $f(x + 1) = \dfrac{x}{x + 1}$

Note: In answers 51 to 55, responses may vary.

51. $h(x) = x - 1; g(u) = u^2 + 2u + 3$

53. $h(x) = x^2 + 1; g(u) = \dfrac{1}{u}$

55. $h(x) = 2 - x; g(u) = \sqrt[3]{u} + \dfrac{4}{u}$

57. a. $C = \$12{,}000; AC = \$1{,}200/\text{unit}$
b. $C(10) - C(9) = \$1{,}090$

59. a. $R(x) = -0.02x^2 + 29x;$
$P(x) = -1.45x^2 + 10.7x - 15.6$
b. $P(x) > 0$ if $2 < x < 5.38$

61. a. $R(x) = -0.5x^2 + 39x;$
$P(x) = -2x^2 + 29.8x - 67$
b. $P(x) > 0$ if $2.76 < x < 12.14$

63. a. All real numbers x except $x = 300$
b. All real numbers x for which $0 \leq x \leq 100$
c. $W(50) = 120$ hours
d. $W(100) = 300$ hours
e. $W(x) = 150$ hours implies that $x = 60\%$

65. a. \$8.70 in 1990; \$47.20 in 2006
b. Early in 2008
c. \$812.80

67. a. $Q(p(t)) = \dfrac{4{,}374}{(0.04t^2 + 0.2t + 12)^2}$
b. $Q(p(10)) = 13.5$ kilograms/week
c. $t = 0$

69. a. All real numbers x except $x = 200$
b. All real numbers x for which $0 \leq x \leq 100$
c. $C(50) = \$50$ million
d. $C(100) - C(50) = \$100$ million
e. $C(x) = 37.5$ million implies that $x = 40\%$

71. a. $P(9) = \dfrac{97}{5}$; 19,400 people
b. $P(9) - P(8) = \dfrac{1}{15}$; 67 people
c. $P(t)$ approaches 20 (20,000 people)

73. a. $S(0) = 25.344$ cm/sec
b. $S(6 \times 10^{-3}) = 19.008$ cm/sec

75. a. $s(8) = 5.8 \approx 6$ species
b. $s_2 = \sqrt[3]{2s_1}$
c. Approximately 41,000 square miles

77.
  a. $H(2) = 192$ feet

  b. $H(3) - H(2) = -80$, it falls 80 feet

  c. $H(0) = 256$ feet

  d. $H(t) = 0$ at $t = 4$ seconds

79. All real numbers x except $x = 1$ and $x = -1.5$

81. $f(g(2.3)) = 6.31$

CHAPTER 1   Section 2

1.

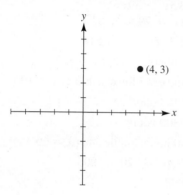

3.

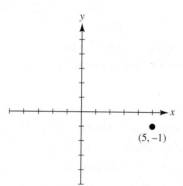

5.

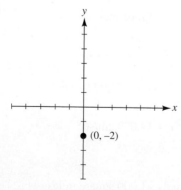

7. $D = 2\sqrt{5}$

9. $D = 2\sqrt{10}$

11.
  a. Power function

  b. Polynomial

  c. Polynomial

  d. Rational function

13. $f(x) = x$

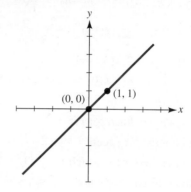

15. $f(x) = \sqrt{x}$

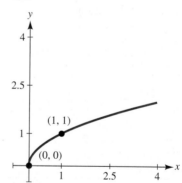

17. $f(x) = 2x - 1$

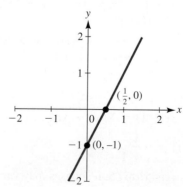

19. $f(x) = x(2x + 5)$

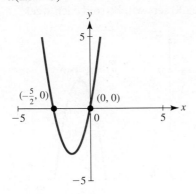

$\left(-\frac{5}{2}, 0\right)$ $(0, 0)$

21. $f(x) = -x^2 - 2x + 15$

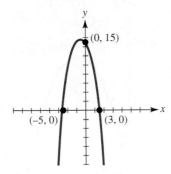

$(0, 15)$

$(-5, 0)$ $(3, 0)$

23. $f(x) = x^3$

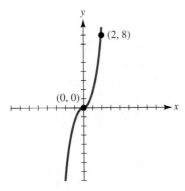

$(2, 8)$

$(0, 0)$

25. $f(x) = \begin{cases} x - 1 & \text{if } x \le 0 \\ x + 1 & \text{if } x > 0 \end{cases}$

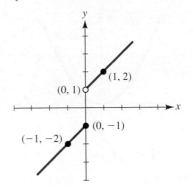

$(1, 2)$

$(0, 1)$

$(0, -1)$

$(-1, -2)$

27. $f(x) = \begin{cases} x^2 + x - 3 & \text{if } x < 1 \\ 1 - 2x & \text{if } x \ge 1 \end{cases}$

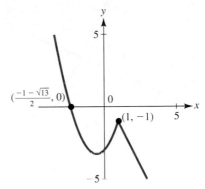

$\left(\dfrac{-1 - \sqrt{13}}{2}, 0\right)$

$(1, -1)$

29. $y = 3x + 5$ and $y = -x + 3$; $\left(-\dfrac{1}{2}, \dfrac{7}{2}\right)$

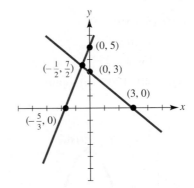

$(0, 5)$

$\left(-\dfrac{1}{2}, \dfrac{7}{2}\right)$ $(0, 3)$

$(3, 0)$

$\left(-\dfrac{5}{3}, 0\right)$

ANSWERS

31. $y = x^2$ and $y = 3x - 2$; (2, 4) and (1, 1)

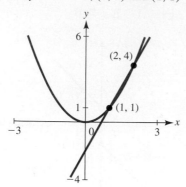

33. $3y - 2x = 5$ and $y + 3x = 9$; (2, 3)

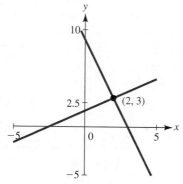

35. **a.** $(0, -1)$
 b. $(1, 0)$
 c. $f(x) = 3$ when $x = 4$
 d. $f(x) = -3$ when $x = -2$

37. **a.** $(0, 2)$
 b. $(-1, 0), (3.5, 0)$
 c. $f(x) = 3$ when $x = 2$
 d. $f(x) = -3$ when $x = 4$

39. $P(p) = (p - 40)(120 - p)$; optimal price is $80 per recorder.

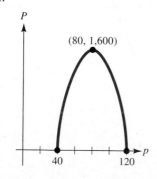

41. $P(x) = (27 - x)(5x - 75)$; optimal price is $21 per game; 30 sets will be sold each week.

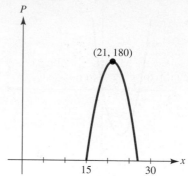

43. **a.** $E(p) = -200p^2 + 12,000p$

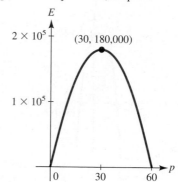

 b. The p intercepts represent prices at which consumers spend no money on the commodity.
 c. $30 per unit

45. **a.** $P(x) = -0.07x^2 + 35x - 574.77$

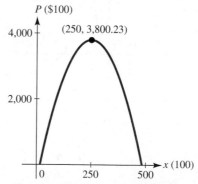

 b. When $P = 37$, $x = 20$, and $AP(20) = $4.86/unit
 c. 25,000 units; $25.50

47. a.

Days of Training	Mowers per Day
2	6
3	7.23
5	8.15
10	8.69
50	8.96

b. The number of mowers per day approaches 9.

c.

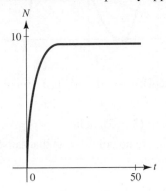

49. a.

$$C(m) = \begin{cases} 19 & \text{if } 0 \le m \le 200 \\ 19 + 0.04 - (m - 200) & \text{if } 200 < m \le 1,000 \end{cases}$$

b.

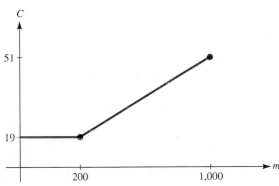

51. a. $R(p) = -0.05p^2 + 210p$

b.

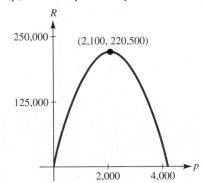

c. $2,100 per month; $220,500

53. a.

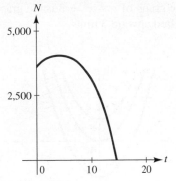

b. 3,967 thousand tons

c. 4.27 years after 1990, or in March 1994.

d. No, the formula predicts negative emissions after December 2004.

55. $D(v) = 0.065v^2 + 0.148v$

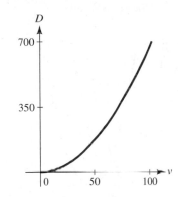

57. a. $H(t) = -16t^2 + 160t$

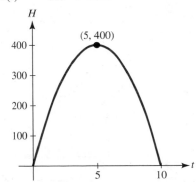

b. After 10 seconds

c. 400 ft

59. Function

61. Not a function

63. Answers will vary.

ANSWERS

65. a. The graph of $y = x^2 + 3$ is the graph of $y = x^2$ shifted upward 3 units.

b.

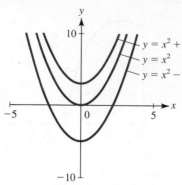

c. The graph of $g(x)$ is the graph of $f(x)$ shifted $|c|$ units upward if $c > 0$ or downward if $c < 0$.

67. a. The graph of $y = (x - 2)^2$ is the graph of $y = x^2$ shifted to the right 2 units.

b.

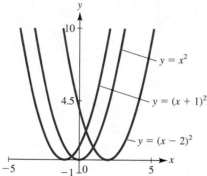

c. The graph of $g(x)$ is the graph of $f(x)$ shifted $|c|$ units to the right if $c > 0$ and to the left if $c < 0$.

69.

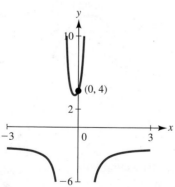

$f(x)$ is defined for $x \neq \dfrac{-1 \pm \sqrt{17}}{8}$

71.

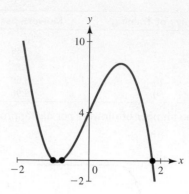

x-intercepts: $x = -1, -0.76, 1.76$

73. a. $(x - 2)^2 + (y + 3)^2 = 16$

b. Center: $(2, -3)$; radius: $2\sqrt{6}$

c. There are no points (x, y) that satisfy the equation.

CHAPTER 1   Section 3

1. $m = -\dfrac{7}{2}$

3. $m = -1$

5. m is undefined.

7. $m = 0$

9. Slope: 2; intercepts: $(0, 0)$; $y = 2x$

11. Slope: $-\dfrac{5}{3}$; intercepts: $(0, 5)$, $(3, 0)$;

$y = -\dfrac{5}{3}x + 5$

13. Slope: undefined; intercepts: $(3, 0)$

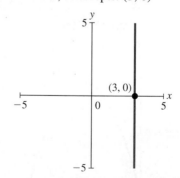

15. Slope: 3; intercepts: $(0, 0)$

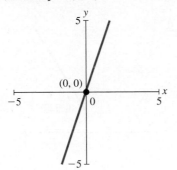

17. Slope: $-\dfrac{3}{2}$; intercepts: $(2, 0)$, $(0, 3)$

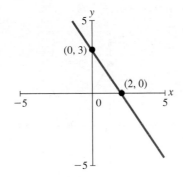

19. Slope: $-\dfrac{5}{2}$; intercepts: $(2, 0)$, $(0, 5)$

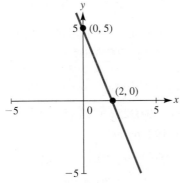

21. $y = x - 2$

23. $y = -\dfrac{1}{2}x + \dfrac{1}{2}$

25. $y = 5$

27. $y = -x + 1$

29. $y = -\dfrac{45}{52}x + \dfrac{43}{52}$

31. $y = 5$

33. $y = -2x + 9$

35. $y = x + 2$

37. **a.** $y = C(x) = 60x + 5{,}000$

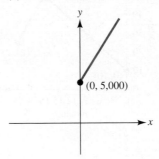

    b. $AC(x) = \dfrac{5{,}000}{x} + 60$; $AC(20) = \$310/\text{unit}$

39. **a.** $y = D(t) = 254.8t + 7{,}853$

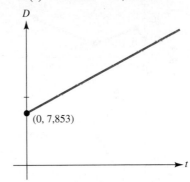

    b. $\$10{,}401$

    c. In about 2036

41. $f(t) = -150t + 1{,}500$

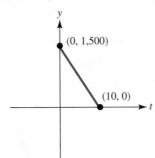

43. **a.** $B = \left(\dfrac{S - V}{N}\right)t + V$ **b.** $\$30{,}800$

45. a.

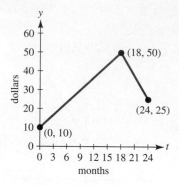

b.

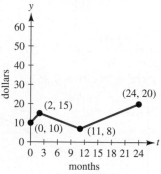

c.

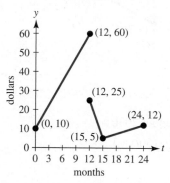

47. a. $v(1930) = \$800$; $v(2000) = \$102,400$; $v(2020) = \$409,600$

b. No, it is not linear.

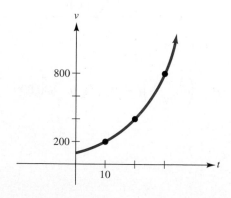

49. a. $y = f(t) = 35t + 220$ **b.** 325   **c.** 220

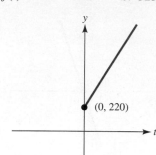

51. a. 95.5 cm   **b.** 15.4 years old
c. 50 cm; yes   **d.** 180 cm; yes

53. a. $y = f(t) = -4t + 248$

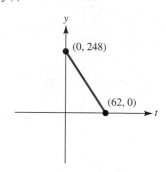

b. $f(8) = 216$ million gallons

55. a. $F = \dfrac{9}{5}C + 32$ **b.** 59°F   **c.** 20°C

d. $-40°$ Celsius $= -40°$ Fahrenheit

57. a. $y = -6(t - 2005) + 575$
b. 515
c. 2013

59. a. $N(x) = 0.0325x + 93.75$
b. 104; 192 mg/m^3
c. Answers may vary.

61.

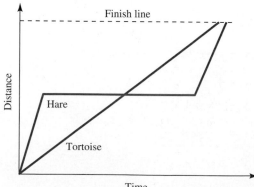

63. The two lines are not parallel; they do not have the same slope.

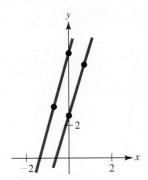

65. Lines L_1 and L_2 are given as perpendicular, and the slope of L_1 is $m_1 = b/a$ while that of L_2 is $m_2 = c/a$. In right triangle OAB, the hypotenuse has length $|AB| = |b - c|$ while the two legs have lengths $|OA| = \sqrt{a^2 + b^2}$ and $|OB| = \sqrt{a^2 + c^2}$. Thus, by the Pythagorean theorem

$$(a^2 + b^2) + (a^2 + c^2) = (b - c)^2$$
$$a^2 + b^2 + a^2 + c^2 = b^2 - 2bc + c^2$$
$$2a^2 = -2bc$$
$$\frac{bc}{a^2} = -1$$
$$\left(\frac{b}{a}\right)\left(\frac{c}{a}\right) = -1$$
$$m_1 m_2 = -1$$
so
$$m_2 = \frac{-1}{m_1}$$

CHAPTER 1 Section 4

1. $S = x + \dfrac{318}{x}$

3. $R = kP$; R = rate of population growth; P = size of population

5. $A = 2w(500 - w)$

7. $A = x(160 - x)$; 80 m by 80 m

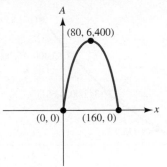

9. $V = x\left(1,000 - \dfrac{x^2}{2}\right)$

11. $R = k(T_0 - T_e)$; R = rate of temperature change; T_0 = temperature of object; T_e = temperature of surrounding medium.

13. $R = kP(T - P)$; R = rate of implication; P = number of people already implicated; T = total number of people involved.

15. $C = \dfrac{k_1}{R} + k_2 R$; R = speed of truck

17. **a.** $P(x) = 3x - 17,000$
  b. $P(5,000) = -\$2,000$ (loss); $P(20,000) = \$43,000$ profitable for $x \geq 5,667$
  c. $AP(x) = 3 - \dfrac{17,000}{x}$; $AP(10,000) = \$1.30$/unit

19. $P(p) = (53,000 - 1,000p)(p - 29)$

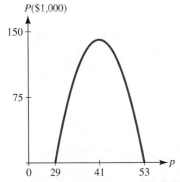

Optimal price is $41.

21.
  a. $f(x) = \begin{cases} 0.1x & \text{if } 0 \leq x \leq 8,375 \\ 0.15x - 419 & \text{if } 8,375 < x \leq 34,000 \\ 0.25x - 3,819 & \text{if } 34,000 < x \leq 82,400 \\ 0.28x - 6,291 & \text{if } 82,400 < x \leq 171,850 \end{cases}$

b.

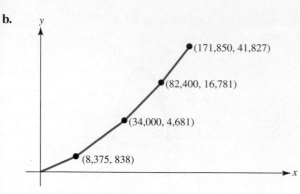

The slopes of the segments are 0.1, 0.15, 0.25, and 0.28. The slopes increase with the taxable income. The more taxable income, the greater the portion of income that will be taxed.

23. $R(x) = (8 - 0.05x)(140 + x)$, where x = days from July 1

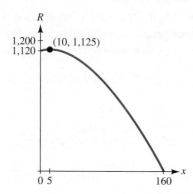

Farmer should harvest when $x = 10$ or July 11.

25. $R(x) = \begin{cases} 2{,}400 & \text{if } 1 \le x \le 40 \\ x\left(80 - \dfrac{1}{2}x\right) & \text{if } 40 < x < 80 \\ 40x & \text{if } x \ge 80 \end{cases}$

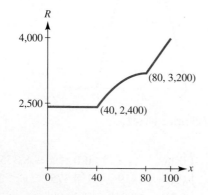

27. If $N \le 6{,}000$ or $N \ge 126{,}000$, choose publisher A, otherwise publisher B.

29. $C(x) = 20x + \dfrac{5{,}120}{x}$

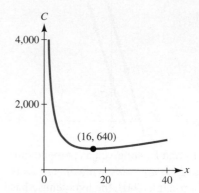

The number of machines is 16.

31. a. $x_e = 25; p_e = 225$

b.

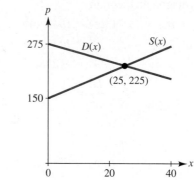

c. $0 < x < 25; x > 25$

33. a. $x_e = 9; p_e = 25.43$

b.

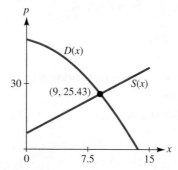

c. $0 < x < 9; x > 9$

35. a. $x_e = 10$; $p_e = 35$

b.

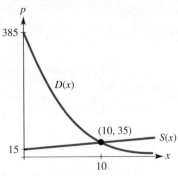

c. $S(0) = 15$; No units will be produced until the price is at least \$15.

37. a. $R(x) = 110x$; $C(x) = 7{,}500 + 60x$; $P(x) = 50x - 7{,}500$

b. 150

c. −\$2,500 (loss)

d. 175

39. a. $V(x) = 20x(160 - x)$

b. Height: 20 m; sides: 80 m by 80 m

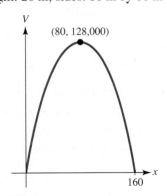

c. \$9,880,000

41. a. $a > 0$; $b > 0$; $c < 0$; $d > 0$

b. $q_e = \dfrac{d - b}{a - c}$; $p_e = \dfrac{ad - bc}{a - c}$

c. As a increases, q_e decreases; as d increases, q_e increases.

43. $I = k\pi r^2$

45. a. $B(t) = 0.31t + 46$; $E(t) = 0.07t + 76$

b. A is 84.75 years.

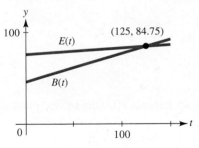

47.

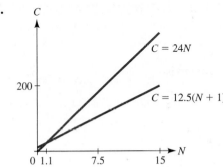

49. a. 95.2 mg

b. Since $0.0072(2W)^{0.425}(2H)^{0.725}$
$= 0.0072(2)^{0.425}(W)^{0.425}(2)^{0.725}(H)^{0.725}$
$= (2)^{0.425}(2)^{0.725}0.0072(W)^{0.425}(H)^{0.725}$
$\approx 2.22[0.0072(W)^{0.425}(H)^{0.725}]$

the larger child has 2.22 times the surface area of the smaller. If S is multiplied by 2.22 in the formula $C = \dfrac{SA}{1.7}$, then C grows by the same factor.

51. $y = \dfrac{K_m}{R_m}x + \dfrac{1}{R_m}$

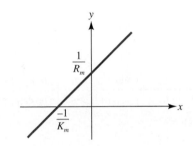

53. $V = \pi r(60 - r^2)$

55. $C = 0.08\pi\left(r^2 + \dfrac{2}{r}\right)$

57. $A(x) = 8x + \dfrac{100}{x} + 57$

59. $V(x) = 4x - \dfrac{x^3}{3}$

61. 2 hours, 45 minutes after the second plane leaves

CHAPTER 1 Section 5

1. $\lim\limits_{x \to a} f(x) = b$

3. $\lim\limits_{x \to a} f(x) = b$

5. Limit does not exist.

7. 4

9. 7

11. 16

13. $\dfrac{4}{7}$

15. Limit does not exist.

17. 2

19. 7

21. $\dfrac{5}{3}$

23. 5

25. $\dfrac{1}{4}$

27. $+\infty; -\infty$

29. $-\infty; -\infty$

31. $\dfrac{1}{2}; \dfrac{1}{2}$

33. $0; 0$

35. $+\infty; -\infty$

37. $1; -1$

39.

x	1.9	1.99	1.999	2	2.001	2.01	2.1
$f(x)$	1.71	1.9701	1.997001		2.003001	2.0301	2.31

$\lim\limits_{x \to 2} f(x) = 2$

41.

x	0.9	0.99	0.999	1	1.001	1.01	1.1
$f(x)$	-17.29	-197.0299	$-1{,}997.002999$		2,003.003001	203.0301	23.31

$\lim\limits_{x \to 1} f(x)$ does not exist.

43. $\lim\limits_{x \to c} [2\,f(x) - 3g(x)] = 2(5) - 3(-2) = 16$

45. $\lim\limits_{x \to c} \sqrt{f(x) + g(x)} = \sqrt{5 + (-2)} = \sqrt{3}$

47. $\lim\limits_{x \to c} \dfrac{f(x)}{g(x)} = -\dfrac{5}{2}$

49. $\lim\limits_{x \to \infty} \dfrac{2f(x) + g(x)}{x + f(x)} = \lim\limits_{x \to \infty} \dfrac{\dfrac{2f(x)}{x} + \dfrac{g(x)}{x}}{1 + \dfrac{f(x)}{x}}$

$= \dfrac{0 + 0}{1 + 0} = 0$

51. a. $P(t) = \dfrac{\sqrt{9t^2 + 0.5t + 179}}{0.2t + 1{,}500}$ thousand dollars per person

b. $\lim\limits_{t \to \infty} P(t) = \$15{,}000$ per person

53. \$7.50; As the number of units produced increases, the contribution of fixed costs to the average cost decreases to 0.

55. \$700 $(C = 7)$

57. \$1,051.27

59. a. $C(0) = 0.413$ mg/ml

b. $C(5) - C(4) \approx -0.013$

The concentration decreases by about 0.013 mg/ml.

c. $\lim\limits_{t \to \infty} C(t) = 0.013$

The residual concentration is 0.013 mg/ml.

61. a. Species I: 10,000; species II: 16,000

b. As t increases, $P(t)$ approaches 0; as t increases to approach 4, $Q(t)$ approaches ∞.

c.

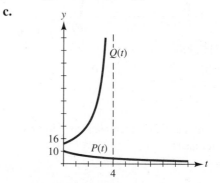

d. Answers will vary.

63. a. $\lim\limits_{S \to +\infty} I(S) = a$. No matter how large the bite size, the required vigilance limits the amount of food intake.

b. Answers will vary.

65. a. 0

b. $\dfrac{a_n}{b_m}$

c. $+\infty$ if a_n and b_m have the same sign, $-\infty$ if a_n and b_m have different signs

67. 1.8 in.

CHAPTER 1   Section 6

1. -2; 1; does not exist

3. 2; 2; exists and equals 2

5. 39

7. 0

9. $\dfrac{5}{4}$

11. 0

13. $\dfrac{1}{4}$

15. 15; 0

17. Yes

19. Yes

21. No

23. No

25. No

27. Yes

29. $f(x)$ is continuous for all real numbers x.

31. $f(x)$ is continuous for all $x \ne 2$.

33. $f(x)$ is continuous for all $x \ne -1$.

35. $f(x)$ is continuous for all $x \ne -3, 6$.

37. $f(x)$ is continuous for all $x \ne 0, 1$.

39. $f(x)$ is continuous for all real numbers x.

41. $f(x)$ is continuous for all $x \ne 0$.

43. a. $C(0) \approx 0.333$; $C(100) \approx 7.179$

b. $C(x)$ is not continuous on the interval $0 \le x \le 100$ because $C(80)$ is not defined.

45. a. \$4,000; \$12,000

b.

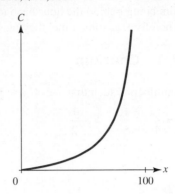

c. $\lim\limits_{x \to 100} C(x) = \infty$; it is not possible to remove all the pollution.

47. The graph is discontinuous at $t = 10$ and $t = 25$. The driver stops at a gas station and purchases some amount of gas at these times.

49. $\lim\limits_{t \to \infty} p(t) = 20$ and $c(20) = 8.4$

51. a. $80°$; $86°$

b. 70%

c. Continuous at 40 and at 80

53. $p(x)$ is discontinuous at $x = 1$ and 2.

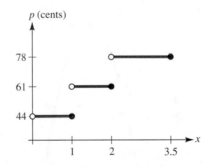

55. $f(x)$ is continuous on the open interval $0 < x < 1$. $f(x)$ is continuous at every point of the closed interval $0 \le x \le 1$ except $x = 0$.

57. $A = 6$

59. Let $f(x) = \sqrt[3]{x - 8} + 9x^{2/3} - 29$. Then $f(x)$ is continuous on $0 \le x \le 8$, $f(0) = -31 < 0$ and $f(8) = 7 > 0$. By the IVT, $f(x) = 0$ for some $0 < x < 8$.

61. a. $\lim\limits_{x \to 2} f(x)$ exists at $x = 2$, but $f(x)$ is not continuous there.

b. $\lim\limits_{x \to -2} f(x)$ does not exist, so $f(x)$ is not continuous at $x = -2$.

ANSWERS

63. During each hour, the minute hand moves continuously from being behind the hour hand to being ahead. Therefore, at some time, they must coincide.

CHAPTER 1   Checkup

1. All real numbers x such that $-2 < x < 2$

2. $g(h(x)) = \dfrac{2x + 1}{4x + 5}; x \neq -\dfrac{1}{2}$

3. **a.** $y = -\dfrac{1}{2}x + \dfrac{3}{2}$

   b. $y = 2x - 3$

4. **a.**

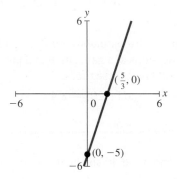

   b.

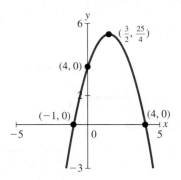

5. **a.** 2

   b. 4

   c. 1

   d. $-\infty$

6. $f(x)$ is not continuous at $x = 1$.

7. **a.** $p(t) = 0.02t + 2.7$

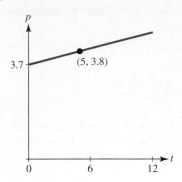

   b. $3.70 per gallon

   c. $3.88 per gallon

8. $D(t) = 30\sqrt{5t^2 - 20t + 100}$

9. **a.** $A = 3, B = -1$; $52

   b.

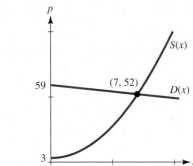

   c. $-$26$; $54

10. **a.** $t = 9$

   b. The function $g(t) = f(t) - 10$ is continuous and $g(1) = -2$ and $g(7) = 6$, so by the IVT $g(t) = 0$ or $f(t) = 10$ sometime between $t = 1$ and $t = 7$.

11. $M = 2.5D + 0.2$; 0.2%

CHAPTER 1   Review Exercises

1. **a.** All real numbers x

   b. All real numbers x except $x = 1$ and $x = -2$

   c. All real numbers x for which $|x| \geq 3$

3. **a.** $g(h(x)) = x^2 - 4x + 4$

   b. $g(h(x)) = \dfrac{1}{2x + 5}$

5. **a.** $f(3 - x) = -x^2 + 7x - 8$

   b. $f(x^2 - 3) = x^2 - 4$

   c. $f(x + 1) - f(x) = \dfrac{-1}{x(x - 1)}$

7. *Note:* Answers will vary.

   a. $g(u) = u^5; h(x) = x^2 + 3x + 4$

   b. $g(u) = (u - 1)^2 + \dfrac{5}{2u^3}; h(x) = 3x + 2$

9. $f(x) = x^2 + 2x - 8$

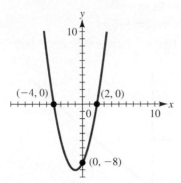

11. **a.** $m = 3, b = 2$

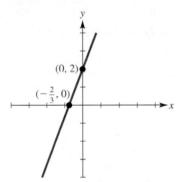

   b. $m = \dfrac{5}{4}, b = -5$

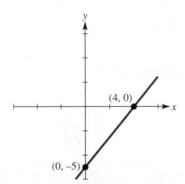

13. **a.** $y = 5x - 4$

   b. $y = -2x + 5$

   c. $2x + y = 14$

15. **a.** $(3, -4)$

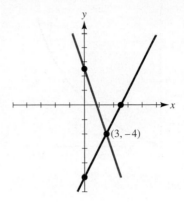

   b. No intersection

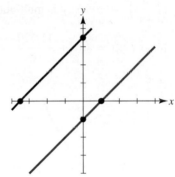

17. $c = -4$

19. $\dfrac{3}{2}$

21. -12

23. Limit does not exist.

25. 0

27. $-\infty$

29. 0

31. 0

33. Not continuous for $x = -3$

35. Continuous for all real numbers x

37. **a.** $P(5) = \$45$

   b. $P(5) - P(4) = -1$

      (a \$1 drop)

   c. In 9 months

   d. The price approaches \$40.

ANSWERS

39. a.

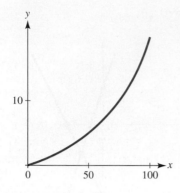

b. 5 weeks

c. 20 weeks

41. $V = \dfrac{4\pi}{3}\left(\sqrt{\dfrac{S}{4\pi}}\right)^3 = \dfrac{S^{3/2}}{6\sqrt{\pi}}$; V is multiplied by $2^{3/2}$.

43. For x machines $C(x) = 80x + \dfrac{11{,}520}{x}$

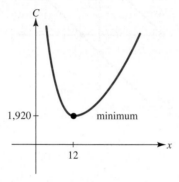

Minimum cost when $x = 12$

45. $P(p) = 2(360 - p)(p - 150)$; optimal price $p = \$255$

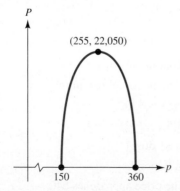

47. Choose Proposition A if $V < 30{,}000$, otherwise choose Proposition B.

49. a. 150 units

b. \$1,500 profit

c. 180 units

51. $y = k(N - x)$, where y is the recall rate, x is the number of facts that have been recalled, and N is the total number of facts.

53. $C = 60x + (2\pi - 6)x^2$

55. $C(x) = 1{,}500 + 2x$, for $0 \le x \le 5{,}000$

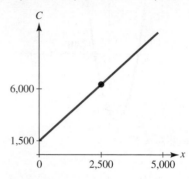

$C(x)$ is continuous for $0 \le x \le 5{,}000$.

57. $A = \dfrac{B}{(4{,}000)^3}$

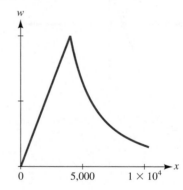

59. The limit exists and is 0.

61.

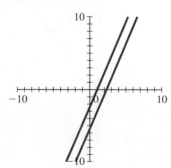

No, because $\dfrac{21}{9} \ne \dfrac{654}{279}$.

63. The function is discontinuous at $x = 1$.

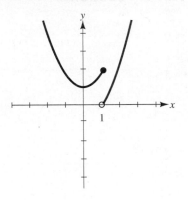

CHAPTER 2   Section 1

1. $f'(x) = 0; m = 0$

3. $f'(x) = 5; m = 5$

5. $f'(x) = 4x - 3; m = -3$

7. $f'(x) = 3x^2; m = 12$

9. $g'(t) = -\dfrac{2}{t^2}; m = -8$

11. $H'(u) = -\dfrac{1}{2u\sqrt{u}}; m = -\dfrac{1}{16}$

13. $f'(x) = 0; y = 2$

15. $f'(x) = -2; y = -2x + 7$

17. $f'(x) = 2x; y = 2x - 1$

19. $f'(x) = \dfrac{2}{x^2}; y = 2x + 4$

21. $f'(x) = \dfrac{1}{\sqrt{x}}; y = \dfrac{1}{2}x + 2$

23. $f'(x) = -\dfrac{3}{x^4}; y = -3x + 4$

25. $\dfrac{dy}{dx} = 0$

27. $\dfrac{dy}{dx} = 3$

29. $\dfrac{dy}{dx} = 3$

31. $\dfrac{dy}{dx} = 2$

33. **a.** $m_{\text{sec}} = -3.9$

   b. $m_{\text{tan}} = -4$

35. **a.** $m_{\text{sec}} = 3.31$

   b. $m_{\text{tan}} = 3$

37. **a.** $\text{rate}_{\text{ave}} = -\dfrac{13}{16}$

   b. $\text{rate}_{\text{ins}} = -1$

39. **a.** $\text{rate}_{\text{ave}} = 4$

   b. $\text{rate}_{\text{ins}} = 8$

41. **a.** The average rate of temperature change between t_0 and $t_0 + h$ hours after midnight. The instantaneous rate of temperature change t_0 hours after midnight.

   b. The average rate of change in blood alcohol level between t_0 and $t_0 + h$ hours after consumption. The instantaneous rate of change in blood alcohol level t_0 hours after consumption.

   c. The average rate of change of the 30-year fixed mortgage rate between t_0 and $t_0 + h$ years after 2005. The instantaneous rate of change of the 30-year fixed mortgage rate t_0 years after 2005.

43. **a.** $P'(x) = 4{,}000(17 - 2x)$

   b. $P'(x) = 0$ when $x = \dfrac{17}{2}$ or 850 units. At this level of production, profits are neither increasing nor decreasing.

45 **a.** 2.94 per unit ($2,940)

   b. $C'(10) = 2.9$ per unit ($2,900); increasing

49. $V'(30) \approx \dfrac{65 - 50}{50 - 30} = \dfrac{3}{4}$; decreases to 0

51. Approx. $-0.01°$C/meter; approx. $0°$C/meter

53. **a.** $H'(t) = 4.4 - 9.8t$; $H(t)$ is decreasing at the rate of -5.4 m/sec when t is 1 sec.

   b. $H'(t) = 0$ when $t = 0.449$ sec; this is the highest point of the jump.

   c. Lands when $t = 0.898$ sec; -4.4 m/sec; decreasing

55. **a.** 0.0211 mm per mm of mercury

   b. 0.022 mm per mm of mercury; increasing

   c. 72.22 mm of mercury. At this pressure the aortic diameter is neither increasing nor decreasing.

57. **a.** $v_{\text{ins}} = \dfrac{2}{\sqrt{t + 1}}$

   b. 2 m/sec

   c. 4 m; 1 m/sec

59. a. The graph of $y = x^2 - 3$ is the same as the graph of $y = x^2$ shifted down by 3 units. Thus, both curves have the same slope for each x, and their derivatives are the same, both equal to $y' = 2x$.

b. $y' = 2x$

61. a. $\dfrac{dy}{dx} = 2x; \dfrac{dy}{dx} = 3x^2$

b. $\dfrac{dy}{dx} = 4x^3; \dfrac{dy}{dx} = 27x^{26}$

63. For $x > 0$, we have $f(x) = x$ and

$$f'(x) = \lim_{h \to 0} \frac{(x+h) - (x)}{h} = 1$$

and for $x < 0$, we have $f(x) = -x$ so that

$$f'(x) = \lim_{h \to 0} \frac{-(x+h) - (-x)}{h} = -1$$

However, the derivative for $x = 0$ would be

$$f'(0) = \lim_{h \to 0} \frac{|0+h| - 0}{h} = \lim_{h \to 0} \frac{|h|}{h}$$

which does not exist since the two one-sided limits at $x = 0$ are not the same (limit from the left is -1 and from the right is $+1$).

65. $f(x)$ is not continuous at $x = 1$, so it cannot be differentiable there.

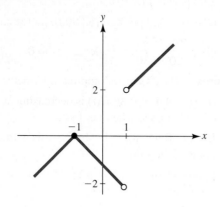

CHAPTER 2   Section 2

1. $\dfrac{dy}{dx} = 0$

3. $\dfrac{dy}{dx} = 5$

5. $\dfrac{dy}{dx} = -4x^{-5}$

7. $\dfrac{dy}{dx} = 3.7x^{2.7}$

9. $\dfrac{dy}{dr} = 2\pi r$

11. $\dfrac{dy}{dx} = \dfrac{\sqrt{2}}{2\sqrt{x}}$

13. $\dfrac{dy}{dt} = \dfrac{-9}{2\sqrt{t^3}}$

15. $\dfrac{dy}{dx} = 2x + 2$

17. $f'(x) = 9x^8 - 40x^7 + 1$

19. $f'(x) = -0.06x^2 + 0.3$

21. $\dfrac{dy}{dt} = -\dfrac{1}{t^2} - \dfrac{2}{t^3} + \dfrac{1}{2\sqrt{t^3}}$

23. $f'(x) = \dfrac{3}{2}\sqrt{x} - \dfrac{3}{2\sqrt{x^5}}$

25. $\dfrac{dy}{dx} = -\dfrac{x}{8} - \dfrac{2}{x^2} - \dfrac{3}{2}x^{1/2} - \dfrac{2}{3x^3} + \dfrac{1}{3}$

27. $\dfrac{dy}{dx} = 2x + \dfrac{4}{x^2}$

29. $y = 10x + 2$

31. $y = -\dfrac{1}{16}x + 2$

33. $y = x + 3$

35. $y = -4x - 1$

37. $y = 3x - 3$

67.

h	-0.02	-0.01	-0.001	0	0.001	0.01	0.02
$x + h$	3.83	3.84	3.849	3.85	3.851	3.86	3.87
$f(x)$	4.37310	4.37310	4.37310	4.37310	4.37310	4.37310	4.37310
$f(x + h)$	4.35192	4.36251	4.37204	4.37310	4.37415	4.38368	4.39426
$\dfrac{f(x+h) - f(x)}{h}$	1.05880	1.05870	1.05860	undefined	1.05858	1.05849	1.05838

39. $y = -3x + \dfrac{22}{3}$

41. $f'(-1) = -5$

43. $f'(1) = -\dfrac{3}{2}$

45. $f'(1) = \dfrac{1}{2}$

47. $\dfrac{f'(x)}{f(x)} = \dfrac{6x^2 - 10x}{2x^3 - 5x^2 + 4}; \dfrac{f'(1)}{f(1)} = -4$

49. $\dfrac{f'(x)}{f(x)} = \dfrac{4x + 3\sqrt{x}}{2(x\sqrt{x} + x^2)}; \dfrac{f'(4)}{f(4)} = \dfrac{11}{24}$

51. **a.** $10,800 per year

  b. 17.53%

53. **a.** $T'(0) = $40 per year

  b. $480

55. **a.** $C(x) = 4x + \dfrac{9,800}{x}$ dollars

  b. $C'(40) = -2.125$ dollars per mile per hour; decreasing

57. **a.** $f(t) = \dfrac{100(2,000)}{45,000 + 2,000t} = \dfrac{200}{45 + 2t}$

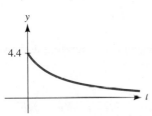

  b. 4.26%

  c. The percentage rate of change approaches zero.

59. **a.** $f'(x) = -6$ points per year

  b. The change in the average SAT score is the same each year. The average SAT score decreases each year.

61. $P'(x) = 2 + 6x^{1/2}$

  a. $P'(9) = 20$ persons per month

  b. 0.39%

63. Approximately 2,435 people per day.

65. **a.** $T'(t) = -204.21t^2 + 61.96t + 12.52$

  b. $T'(0) = 12.52$, increasing; $T'(0.713) = -47.12$, decreasing

  c. $t = 0.442$ days or 10.61 hours; $T(0.442) = 42.8°C$, which is the maximum temperature during the period.

67. **a.** 0.2 parts per million per year

  b. 0.15 parts per million

  c. 0.4 parts per million

69. Mars

71. **a.** $v(t) = 6t + 2; a(t) = 6$

  b. Not stationary

73. **a.** $v(t) = 4t^3 - 12t^2 + 8; a(t) = 12t^2 - 24t$

  b. $t = 1, 1 + \sqrt{3}$

75. **a.** 32 ft/sec

  b. 128 ft

  c. -32 ft/sec

  d. -96 ft/sec

77. $a = -1, b = 5, c = 0$

79. $(f + g)'(x)$

$$= \lim_{h \to 0} \dfrac{(f + g)(x + h) - (f + g)(x)}{h}$$

$$= \lim_{h \to 0} \dfrac{[f(x + h) + g(x + h)] - [f(x) + g(x)]}{h}$$

$$= \lim_{h \to 0} \dfrac{[f(x + h) - f(x)] + [g(x + h) - g(x)]}{h}$$

$$= \lim_{h \to 0} \left[\dfrac{f(x + h) - f(x)}{h} + \dfrac{g(x + h) - g(x)}{h} \right]$$

$$= \lim_{h \to 0} \dfrac{f(x + h) - f(x)}{h} + \lim_{h \to 0} \dfrac{g(x + h) - g(x)}{h}$$

$$= f'(x) + g'(x)$$

CHAPTER 2   Section 3

1. $f'(x) = 12x - 1$

3. $\dfrac{dy}{du} = -300u - 20$

5. $f'(x) = \dfrac{1}{3}\left(6x^5 - 12x^3 + 4x + 1 + \dfrac{1}{x^2} \right)$

7. $\dfrac{dy}{dx} = \dfrac{-3}{(x - 2)^2}$

9. $f'(t) = \dfrac{-(t^2 + 2)}{(t^2 - 2)^2}$

11. $\dfrac{dy}{dx} = \dfrac{-3}{(x + 5)^2}$

13. $f'(x) = \dfrac{11x^2 - 10x - 7}{(2x^2 + 5x - 1)^2}$

15. $f'(x) = 10(2 + 5x)$

ANSWERS

17. $g'(t) = \dfrac{4\sqrt{t^5} + 20\sqrt{t^3} - 2t + 5}{2\sqrt{t}\,(2t + 5)^2}$

19. $y = 17x - 4$

21. $y = 3x + 2$

23. $y = -\dfrac{11}{2}x + \dfrac{19}{2}$

25. $(1, -4), (-1, 0)$

27. $(0, 1), \left(-2, -\dfrac{1}{3}\right)$

29. -18

31. 4

33. $y = \dfrac{2}{5}x + \dfrac{3}{5}$

35. $y = \dfrac{1}{31}(-x - 371)$

37. 483

39. $-\dfrac{13}{64}$

41. a.–d. $y' = \dfrac{9 - 4x}{x^4}$

43. $f''(x) = 8x^3 - 24x + 18$

45. $\dfrac{d^2y}{dx^2} = \dfrac{4}{3x^3} + \dfrac{\sqrt{2}}{4x^{3/2}} - \dfrac{1}{8x^{5/2}}$

47. $\dfrac{d^2y}{dx^2} = 36x^2 + 30x + 12$

49. a. $S'(2) = \$378{,}072$ per year
b. Sales approach a limit of $\$6{,}666{,}666.67$.

51. a. $P'(5) = \dfrac{1{,}900}{441} \approx 4.3\%$ increase per week
b. $P(t)$ approaches 100% in the long run; the rate of change approaches 0.

53. a. $R(x) = 17.5x - 0.0125x^2$;
$R'(x) = 17.5 - 0.025x$ dollars per unit
$R'(1000) = -7.5$ dollars per unit; decreasing
b. The average revenue is $\dfrac{R(x)}{x} = 17.5 - 0.0125x$ dollars per unit.
The average revenue is changing at the constant negative rate of -0.0125 dollars per unit per unit, and is therefore decreasing.

55. a. $P'(16) \approx -0.631\%$ per $\$1$ million; decreasing
b. Increasing for $0 < x < 10$; decreasing for $x > 10$

57. a. $S = \dfrac{2}{3}KM - M^2$
b. $\dfrac{dS}{dM} = \dfrac{2}{3}K - 2M$ and represents the rate of change of sensitivity with respect to the amount of medicine absorbed.

59. a. $P'(t) = \dfrac{6}{(t + 1)^2}$
b. 1,500 people per year
c. 1,000 people
d. 60 people per year
e. Approaches 0

61. a. $v(t) = 15t^4 - 15t^2$; $a(t) = 60t^3 - 30t$
b. $a(t) = 0$ when $t = 0, \dfrac{\sqrt{2}}{2}$

63. a. $v(t) = -3t^2 + 14t + 1$
$a(t) = -6t + 14$
b. $a(t) = 0$ when $t = \dfrac{7}{3}$

65. a. 9.8 meters per minute
b. 9.83 meters

67. a. $a(t) = -32$
b. $a(t)$ is constant.
c. The object is accelerating downward.

69. $\dfrac{3}{8x^{5/2}} + \dfrac{3}{x^4}$

71. a. $\dfrac{d}{dx}\left(\dfrac{fg}{h}\right) = \dfrac{hfg' + hf'g - fgh'}{h^2}$
b. $\dfrac{dy}{dx} = \dfrac{12x^3 + 51x^2 + 70x - 33}{(3x + 5)^2}$

75.

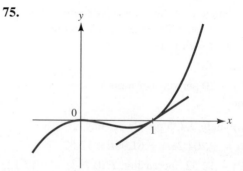

$f'(x) = 0$ where $x = 0$ and $x = \dfrac{2}{3}$.

77.

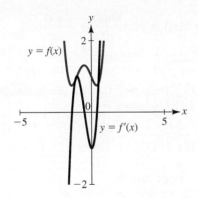

The x intercepts of $f'(x)$ occur at $x = -0.5$,
$x = -\dfrac{1}{2} + \dfrac{\sqrt{3}}{2} \approx 0.366$, and
$x = -\dfrac{1}{2} - \dfrac{\sqrt{3}}{2} \approx -1.366$. The x intercepts are
those points where $f'(x) = 0$, which are the points
where the tangent to the graph of $f(x)$ is horizontal,
that is, the maxima and minima of $f(x)$.

CHAPTER 2 Section 4

1. $\dfrac{dy}{dx} = 6(3x - 2)$

3. $\dfrac{dy}{dx} = \dfrac{x + 1}{\sqrt{x^2 + 2x - 3}}$

5. $\dfrac{dy}{dx} = \dfrac{-4x}{(x^2 + 1)^3}$

7. $\dfrac{dy}{dx} = \dfrac{-2x}{(x^2 - 1)^2}$

9. $\dfrac{dy}{dx} = 1 + \dfrac{1}{\sqrt{x}}$

11. $\dfrac{dy}{dx} = -\dfrac{2 + x}{x^3}$

13. 20

15. -160

17. $\dfrac{2}{3}$

19. -16

21. $f'(x) = 2.8(2x + 3)^{0.4}$

23. $f'(x) = 8(2x + 1)^3$

25. $f'(x) = 8x^2(x^5 - 4x^3 - 7)^7(5x^2 - 12)$

27. $f'(t) = \dfrac{-2(5t - 3)}{(5t^2 - 6t + 2)^2}$

29. $g'(x) = \dfrac{-4x}{(4x^2 + 1)^{3/2}}$

31. $f'(x) = \dfrac{24x}{(1 - x^2)^5}$

33. $h'(s) = \dfrac{15(1 + \sqrt{3s})^4}{2\sqrt{3s}}$

35. $f'(x) = (x + 2)^2(2x - 1)^4(16x + 17)$

37. $f'(x) = \dfrac{(x + 1)^4(9 - x)}{(1 - x)^5}$

39. $y = \dfrac{3}{4}x + 2$

41. $y = -48x - 32$

43. $y = -12x + 13$

45. $y = \dfrac{2}{3}x - \dfrac{1}{3}$

47. $x = 0; x = -1; x = -\dfrac{1}{2}$

49. $x = -\dfrac{2}{3}$

51. $x = 2$

53. $f'(x) = 6(3x + 5)$

55. $f''(x) = 180(3x + 1)^3$

57. $h''(t) = 80(t^2 + 5)^6(3t^2 + 1)$

59. $f''(x) = (1 + x^2)^{-3/2}$

61. $\dfrac{1}{2}$

63. 6

65. **a.** \$2,295 per year
    b. 10.4% per year

67. **a.** -12 pounds per dollar
    b. $(-12)(0.5) = -6$ pounds per week; decreasing

69. 875 units per month; increasing

71. **a.** \$64,000; 8,000 units
    b. 6,501 units per month; increasing

73. **a.** $A'(r) = 1,000\left(1 + \dfrac{0.01r}{12}\right)^{119}$

    $A'(5) = \$1,640.18/\%$

    b. $A(6) - A(5) = \$1,723.87$

75. a. 0.4625 ppm per thousand people

b. 0.308 ppm per year; increasing

77. a. $L'(w) = 0.65w^{1.6}$; $L'(60) \approx 455$ mm/kg

b. A 100-day-old tiger weighs $w(100) = 24$ kg and is $L(24) \approx 969$ mm long. By the chain rule,
$$L'(A) = L'(w)w'(A) = (0.65w^{1.6})(0.21)$$
so that when $A = 100$, $w = 24$, we have
$$L'(100) = (0.65)(0.21)(24)^{1.6} \approx 22.1$$
That is, the tiger's length is increasing at the rate of about 22.1 mm per day.

79. a. Decreasing at about 0.2254% per day

b. Increasing

c. Eventually the oxygen proportion returns to its typical level.

81. a. $V'(T) = 0.41(-0.02T + 0.4)$

b. $m'(V) = \dfrac{0.39}{(1 + 0.09V)^2}$

c. $V(10) = 2.6732$ cm^3; 0.02078 gm/°C

83. a. $v(t) = \dfrac{3}{2}(1 - 2t)(3 + t - t^2)^{1/2}$

$a(t) = \dfrac{24t^2 - 24t - 33}{4(3 + t - t^2)^{1/2}}$

b. $t = \dfrac{1}{2}$; $s\left(\dfrac{1}{2}\right) = \dfrac{13\sqrt{13}}{8} \approx 5.86$

$a\left(\dfrac{1}{2}\right) = \dfrac{-3\sqrt{13}}{2} \approx -5.41$

c. $a(t) = 0$ when $t = \dfrac{2 + \sqrt{26}}{4} \approx 1.775$

$s(1.775) \approx 2.07$; $v(1.775) \approx -4.875$

d.

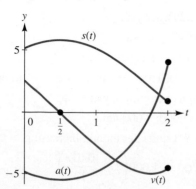

e. The object is slowing down for $0 \le t < 0.5$ and $1.775 < t \le 2$.

85. $\dfrac{dy}{dx} = \dfrac{d}{dx}[h(x)h(x)] = h(x)\dfrac{dh(x)}{dx} + \dfrac{dh(x)}{dx}h(x)$

$= 2h(x)\dfrac{dh(x)}{dx}$

87. $f'(1) \approx 0.2593$, $f'(-3) \approx -0.4740$; one horizontal tangent, $y = 2.687$, where $x = 0$.

89. $g'(x) = \dfrac{2}{1 + (2x + 1)^2}$

CHAPTER 2   Section 5

1. a. $C'(x) = \dfrac{2}{5}x + 4$; $R'(x) = 12 - \dfrac{x}{2}$

b. $C'(20) = \$12$; $C(21) - C(20) = \$12.20$

c. $R'(20) = \$2$; $R(21) - R(20) = \$1.75$

3. a. $C'(x) = \dfrac{2}{3}x + 2$; $R'(x) = -3x^2 - 20x + 4{,}000$

b. $C'(20) = \$15.33$; $C(21) - C(20) = \$15.67$

c. $R'(20) = \$2{,}400$; $R(21) - R(20) = \$2{,}329$

5. a. $C'(x) = \dfrac{x}{2}$; $R'(x) = \dfrac{2x^2 + 4x + 3}{(1 + x)^2}$

b. $C'(20) = \$10.00$; $C(21) - C(20) = \$10.25$

c. $R'(20) = \$2.00$; $R(21) - R(20) = \$2.00$

7. 2.1

9. $\dfrac{100f'(4)}{f(4)}(0.3) \approx 0.20$; 20%

11. a. \$232; yes

b. $R(81) - R(80) = \$231.95$

13. a. \$241

b. \$244

15. Revenue will decrease by approximately \$150.80.

17. Daily output will increase by approximately 8 units.

19. Daily output will increase by approximately 825 units.

21. 0.2 units

23. 200

25. a. $R(t) = -3t^2 + 18t + 48$

b. $R'(t) = -6t + 18$

c. $\dfrac{1}{12}R'(3) = 0$ change; $R\left(3 + \dfrac{1}{12}\right) - R(3) = -0.021$ or 21 people per year less

27. $\dfrac{3c}{|c - b|}$

29. 46.67%

31. 3.85 cm

33. The tangent line at $x = x_0$ is

$y = f'(x_0)(x - x_0) + f(x_0)$.

Setting $y = 0$ and solving for $x = x_1$ gives

$0 = f'(x_0)(x_1 - x_0) + f(x_0) \Rightarrow x_1 - x_0 = \dfrac{-f(x_0)}{f'(x_0)} \Rightarrow$

$x_1 = x_0 - \dfrac{f(x_0)}{f'(x_0)}$. Repeating the process shows

$x_n = x_{n-1} - \dfrac{f(x_{n-1})}{f'(x_{n-1})}$.

35. 3.82070437, 1.61179338

37. a. $x_{n+1} = x_n - \dfrac{f(x_n)}{f'(x_n)} = x_n - \dfrac{x_n^{1/3}}{\frac{1}{3}x_n^{-2/3}}$

$= x_n - 3x_n = -2x_n$

b. For any $x_0 \neq 0$ the sequence $x_0, -2x_0, 4x_0, \dots$ grows larger in magnitude and alternates sign, so it cannot approach a limit.

CHAPTER 2 Section 6

1. $\dfrac{dy}{dx} = -\dfrac{2}{3}$

3. a. $\dfrac{dy}{dx} = \dfrac{3x^2}{2y} = \pm\dfrac{3x^2}{2\sqrt{x^3 - 5}}$

b. $\dfrac{dy}{dx} = \pm\dfrac{3x^2}{2\sqrt{x^3 - 5}}$

5. a. $\dfrac{dy}{dx} = -\dfrac{y}{x} = \dfrac{-(4/x)}{x}$

b. $\dfrac{dy}{dx} = -\dfrac{4}{x^2}$

7. a. $\dfrac{dy}{dx} = \dfrac{-y}{x + 2} = \dfrac{-[3/(x + 2)]}{x + 2}$

b. $\dfrac{dy}{dx} = \dfrac{-3}{(x + 2)^2}$

9. $\dfrac{dy}{dx} = -\dfrac{x}{y}$

11. $\dfrac{dy}{dx} = \dfrac{y - 3x^2}{3y^2 - x}$

13. $\dfrac{dy}{dx} = \dfrac{3 - 2y^2}{2y(1 + 2x)}$

15. $\dfrac{dy}{dx} = -\dfrac{\sqrt{y}}{\sqrt{x}}$

17. $\dfrac{dy}{dx} = \dfrac{y - 1}{1 - x}$

19. $\dfrac{dy}{dx} = \dfrac{1}{3(2x + y)^2} - 2$

21. $\dfrac{dy}{dx} = \dfrac{y - 5x(x^2 + 3y^2)^4}{15y(x^2 + 3y^2)^4 - x}$

23. $y = \dfrac{1}{3}x + \dfrac{4}{3}$

25. $y = -\dfrac{1}{2}x + 2$

27. $y = \dfrac{5}{8}x - \dfrac{9}{4}$

29. $y = \dfrac{13}{12}x + \dfrac{11}{12}$

31. a. None

b. $(9, 0)$

33. a. None

b. $(0, 0)$ and $(64, 2)$

35. a. $(1, -2), (-1, 2)$

b. $(-2, 1), (2, -1)$

37. $\dfrac{d^2y}{dx^2} = \dfrac{-3y^2 - x^2}{9y^3} = \dfrac{-5}{9y^3}$

39. $\Delta y \approx -1.704$ hours

41. $\dfrac{dx}{dt} = 1.74$ or increasing by 174 units per month

43. $\dfrac{dD}{dt} = -2$ toasters per month; decreasing

45. $\dfrac{dK}{dt} = -\dfrac{2}{5}(\$1,000) = -\$400$ per week

47. $\dfrac{dR}{dt} = 20$ mm/min

49. 0.476 cm^3 per month

51. a. 14.04 kcal per day per day

b. -9.87 kcal per day per day

53. Since $v = \dfrac{KR^2}{L} = $ constant at the center of the vessel ($r = 0$), we have

$$0 = v'(t) = K\left[\frac{2R}{L}R' - \frac{R^2}{L^2}L'\right]$$

Thus,

$$\frac{2R}{L}R' = \frac{R^2}{L^2}L'$$

$$\frac{L'}{L} = 2\frac{R'}{R}$$

so the relative rate of change of L is twice that of R.

55. a. $\dfrac{dF}{dC} = -\dfrac{kD^2}{2\sqrt{A - C}}$; increases

b. $\dfrac{50}{A - C}\%$

57. 4 feet per second

59. $\dfrac{dP}{dt} = -21$ lb/in.^{2} per second; decreasing

61. $\dfrac{x^2}{a^2} + \dfrac{y^2}{b^2} = 1; \dfrac{2x}{a^2} + \dfrac{2yy'}{b^2} = 0; 2b^2x + 2a^2yy' = 0,$

$$y' = -\frac{2b^2x}{2a^2y} = -\frac{b^2x}{a^2y}. \text{ At } P(x_0, y_0), m = -\frac{b^2x_0}{a^2y_0}$$

so the equation of the tangent line is

$$y - y_0 = -\frac{b^2x_0}{a^2y_0}(x - x_0)$$

$$a^2yy_0 - a^2y_0^2 = -b^2xx_0 + b^2x_0^2$$

$$b^2xx_0 + a^2yy_0 = b^2x_0^2 + a^2y_0^2$$

$$\frac{x_0x}{a^2} + \frac{y_0y}{b^2} = \frac{x_0^2}{a^2} + \frac{y_0^2}{b^2} = 1$$

since $P(x_0, y_0)$ lies on the curve and thus satisfies the equation of the curve.

63. Let $y = x^{r/s}$, then $y^s = x^r$ and $sy^{s-1}\dfrac{dy}{dx} = rx^{r-1}$,

$$\frac{dy}{dx} = \frac{rx^{r-1}}{sy^{s-1}}. \text{ But } y^{s-1} = \frac{y^s}{x^{r/s}} = \frac{x^r}{x^{r/s}}, \text{ so}$$

$$\frac{dy}{dx} = \frac{r}{s} \cdot x^{r-1} \cdot \frac{x^{r/s}}{x^r}$$

$$= \frac{r}{s} \cdot x^{r-1+r/s-r}$$

$$= \frac{r}{s} \cdot x^{r/s-1}$$

65. Horizontal tangent lines are $y = 1.24$ and $y = -1.24$.

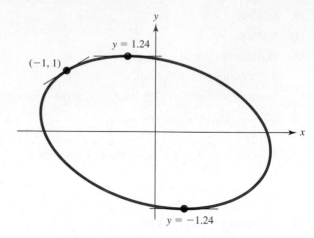

67. Horizontal tangent lines are $y = 1.23$ and $y = -1.23$.

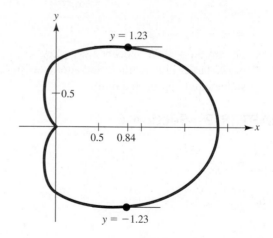

CHAPTER 2   Checkup

1. a. $\dfrac{dy}{dx} = 12x^3 - \dfrac{2}{\sqrt{x}} - \dfrac{10}{x^3}$

b. $\dfrac{dy}{dx} = -15x^4 + 39x^2 - 2x - 4$

c. $\dfrac{dy}{dx} = \dfrac{-10x^2 + 10x + 1}{(1 - 2x)^2}$

d. $\dfrac{dy}{dx} = (9x - 6)(3 - 4x + 3x^2)^{1/2}$

2. $f''(t) = 24t + 8$

3. $y = -4x$

4. $\dfrac{3}{8}$

5. a. 58 dollars per year

b. 2.98% per year

6. a. $v(t) = 6t^2 - 6t;\ a(t) = 12t - 6$

b. $t = 0, 1$; retreating $0 < t < 1$; advancing $1 < t < 2$.

c. 6

7. $C'(4.09)(0.01) = 5.3272(0.01) = 0.053272$ thousand dollars

$C(4.1) - C(4.09) = 0.053276$ thousand dollars

8. Output increases by approximately 10,714 units.

9. 0.001586 m^2 per week

10. a. 2.25π cm^2 per cm

b. $\dfrac{8}{3}\%$

CHAPTER 2 Review Exercises

1. $f'(x) = 2x - 3$

3. $f'(x) = 24x^3 - 21x^2 + 2$

5. $\dfrac{dy}{dx} = \dfrac{-14x}{(3x^2 + 1)^2}$

7. $f'(x) = 10(20x^3 - 6x + 2)(5x^4 - 3x^2 + 2x + 1)^9$

9. $\dfrac{dy}{dx} = 2\left(x + \dfrac{1}{x}\right)\left(1 - \dfrac{1}{x^2}\right) + \dfrac{5}{2\sqrt{3x^3}}$

11. $f'(x) = 3\sqrt{6x + 5} + \dfrac{9x + 3}{\sqrt{6x + 5}}$

13. $\dfrac{dy}{dx} = \dfrac{-7}{2(3x + 2)^2}\sqrt{\dfrac{3x + 2}{1 - 2x}}$

15. $y = -x - 1$

17. $y = -\dfrac{2}{3}x + \dfrac{5}{3}$

19. a. $f'(0) = 0$

b. $f'(1) = -\dfrac{1}{4}$

21. a. -400%

b. -100%

23. a. $\dfrac{dy}{dx} = -2(2 - x)$

b. $\dfrac{dy}{dx} = -\dfrac{1}{(2x + 1)^{3/2}}$

25. a. 2

b. $\dfrac{3}{2}$

27. a. $f''(x) = 24x$

b. $f''(x) = 24(x + 4)(x + 2)$

c. $f''(x) = \dfrac{2(x - 5)}{(x + 1)^4}$

29. a. $\dfrac{dy}{dx} = -\dfrac{2y}{x}$

b. $\dfrac{dy}{dx} = -\left[\dfrac{1 + 10y^3(1 - 2xy^3)^4}{4 + 30xy^2(1 - 2xy^3)^4}\right]$

31. a. $m = -\dfrac{5}{9}$

b. $m = -1$

33. $\dfrac{d^2y}{dx^2} = \dfrac{6y^2 - 9x^2}{4y^3} = \dfrac{-9}{2y^3}$

35. a. 8,000 per year

b. $-18,000\ \dfrac{\text{per year}}{\text{year}}$

37. a. $v(t) = \dfrac{-2(t + 4)(t - 3)}{(t^2 + 12)^2}$,

$a(t) = \dfrac{2(2t^3 + 3t^2 - 72t - 12)}{(t^2 + 12)^3}$. The object is advancing for $0 < t < 3$, retreating for $3 < t < 4$. It is always decelerating for $0 < t < 4$.

b. $\dfrac{11}{42}$

39. a. Output will increase by approximately 12,000 units.

b. Output will increase by 12,050 units.

41. Output will decrease by approximately 5,000 units per day.

43. Pollution will increase by approximately 10%.

45. a. 0.2837 individuals per square kilometer

b. 2.61 million

c. 55 years; 60.67 animals per year

47. 1.5%

49. $425.25 \le A \le 479.53$; accurate to 6%

51. $100\dfrac{\Delta Q}{Q} \approx 0.67\%$

53. $17.01 \le S \le 19.18$; accurate to 6%

55. $\dfrac{dx}{dt} = 0.15419$ or increasing by 15.419 units per month

57. 10.7%

ANSWERS

59. 5.5 seconds; 242 feet

61. **a.** $195 per unit per month

  b. −$16 per unit per month per month

  c. −$8 per unit per month

  d. −$8.75 per unit per month

63. −$99 per month

65. 28.37 cubic inches

67. 3 ft/sec

69. 2.25 ft/sec

71. −3.29 ft/sec

73. $\dfrac{8}{5}$ ft/sec^2

75. The percentage rate of change approaches 0 since,

  if $y = mx + b$, $\dfrac{100y'}{y} = \dfrac{100m}{mx + b}$, which approaches

  0 as x approaches ∞.

77.

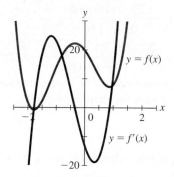

  $f'(x) = 0$ when $x \approx -1.78, -0.35, 0.88$

79. **a.**

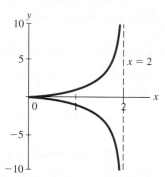

  b. At $(1, 1)$, the tangent line is $y = 2x - 1$ and at $(1, -1)$, it is $y = -2x + 1$.

  c. As $x \to 2^-$, the top branch of the curve rises indefinitely $(y \to +\infty)$, while the bottom branch falls indefinitely $(y \to -\infty)$.

  d. The x axis $(y = 0)$ is a double tangent line to the curve at the origin.

CHAPTER 3   Section 1

1. $f'(x) > 0$ for $-2 < x < 2$; $f'(x) < 0$ for $x < -2$ and $x > 2$

3. $f'(x) > 0$ for $x < -4$ and $0 < x < 2$; $f'(x) < 0$ for $-4 < x < -2$, $-2 < x < 0$, and $x > 2$

5. B

7. D

9. $f(x)$ is increasing for $x > 2$; $f(x)$ is decreasing for $x < 2$.

11. $f(x)$ is increasing for $x < -1$ and $x > 1$; $f(x)$ is decreasing for $-1 < x < 1$.

13. $g(t)$ is increasing for $t < 0$ and $t > 4$; $g(t)$ is decreasing for $0 < t < 4$.

15. $f(t)$ is increasing for $0 < t < 2$ and $t > 2$; $f(t)$ is decreasing for $t < -2$ and $-2 < t < 0$.

17. $h(u)$ is increasing for $-3 < u < 0$; $h(u)$ is decreasing for $0 < u < 3$.

19. $F(x)$ is increasing for $x < -3$ and $x > 3$; $F(x)$ is decreasing for $-3 < x < 0$ and $0 < x < 3$.

21. $f(x)$ is increasing for $x > 1$; $f(x)$ is decreasing for $0 < x < 1$.

23. $x = 0, 1$; $(0, 2)$ relative minimum; $(1, 3)$ neither

25. $x = -1$; $(-1, 3)$ neither

27. $x = 1$; $(1, 0)$ neither

29. $t = -\sqrt{3}, \sqrt{3}$; $\left(\sqrt{3}, \dfrac{\sqrt{3}}{6}\right)$ relative maximum; $\left(-\sqrt{3}, -\dfrac{\sqrt{3}}{6}\right)$ relative minimum

31. $t = -2, 0, 1, 4$; $(0, 0)$ relative maximum; $\left(4, \dfrac{8}{9}\right)$ relative minimum

33. $t = 0, -1, 1$; $(0, 1)$ relative maximum; $(-1, 0)$ and $(1, 0)$ relative minima

35.

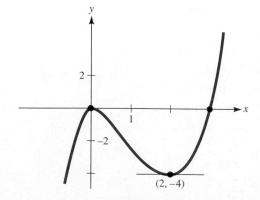

37.

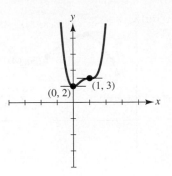

(0, 2) (1, 3)

39.

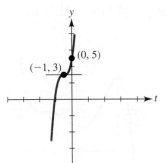

(0, 5) (−1, 3)

41.

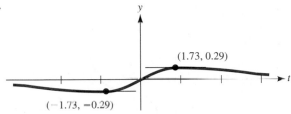

(1.73, 0.29)

(−1.73, −0.29)

43.

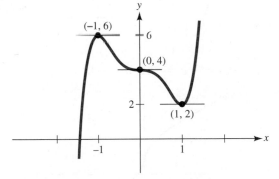

(−1, 6) (0, 4)

2

(1, 2)

−1   1

45.

Critical Numbers	Classification
−2	Relative minimum
0	Neither
2	Relative maximum

47.

Critical Numbers	Classification
−1	Neither
$\frac{4}{3}$	Relative maximum

49. One possibility:

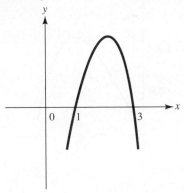

0   1   3

51. One possibility:

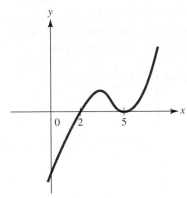

0   2   5

53. a. $A'(x) = 2x - 20 - \dfrac{242}{x^2}$

    b. Increasing: $x > 11$; decreasing: $0 \leq x < 11$

    c. Average cost is minimized when $x = 11$; minimum average cost: \$102,000/unit.

55. $R(x) = x(10 - 3x)^2; \dfrac{dR}{dx} = (10 - 3x)(10 - 9x)$;

    Revenue is maximized when $x = \dfrac{10}{9}$ units.

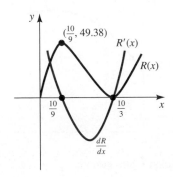

$\left(\frac{10}{9}, 49.38\right)$ $R'(x)$

$R(x)$

$\frac{10}{9}$ $\frac{10}{3}$

$\dfrac{dR}{dx}$

ANSWERS

57. a.

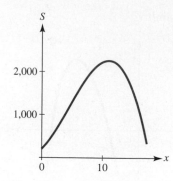

b. 207

c. $11,000; 2,264 units

59. a. $1 \leq r \leq 5.495$

b. 5.495%; 1,137 mortgages

61. a. 1971, 1976, 1980, 1983, 1988, 1996

b. 1973, 1979, 1981, 1985, 1989

c. Approximately $\frac{1}{2}$% per year

d. Approximately $\frac{1}{2}$% per year

63. Maximum concentration occurs when $t = 0.9$ hours.

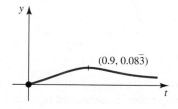

65. a. $Y(t) = \dfrac{9,300}{31 + t}(3 + t - 0.05t^2)$

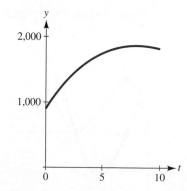

b. 8 weeks; 1,860 pounds

67.

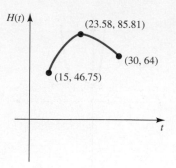

Maximum of 85.81% at 23.58°C.

69.

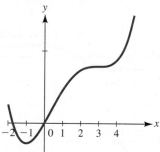

71.

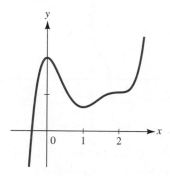

73. $a = 2; b = 3; c = -12; d = -12$

75.

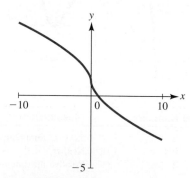

77. By the product rule, $\dfrac{dy}{dx} = (x - p)(1) + (1)(x - q)$

$= 2x - p - q$. Solving $\dfrac{dy}{dx} = 0$ yields $x = \dfrac{p + q}{2}$, the point midway between the x intercepts p and q.

79. $f'(x) = 0$ at $x = -3; -2.529, -1.618, -0.346,$ 0.618

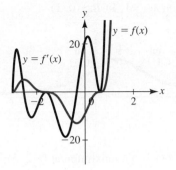

81. $f'(x)$ is never 0.

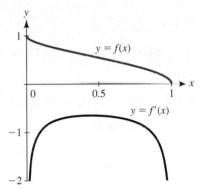

83. The top half of a circle.

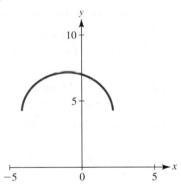

CHAPTER 3 Section 2

1. $f''(x) > 0$ for $x > 2; f''(x) < 0$ for $x < 2$

3. $f''(x) > 0$ for $x < -1$ and $x > 1; f''(x) < 0$ for $-1 < x < 1$

5. Concave upward for $x > -1$; concave downward for $x < -1$; inflection at $(-1, 2)$

7. Concave upward for $x > -\dfrac{1}{3}$; concave downward for $x < -\dfrac{1}{3}$; inflection at $\left(-\dfrac{1}{3}, -\dfrac{1}{27}\right)$

9. Concave upward for $t < 0$ and $t > 1$; concave downward for $0 < t < 1$; inflection at $(1, 0)$

11. Concave upward for $x < 0$ and $x > 3$; concave downward for $0 < x < 3$; inflection at $(0, -5)$ and $(3, -65)$

13. Increasing for $x < -3$ and $x > 3$; decreasing for $-3 < x < 3$; concave upward for $x > 0$; concave downward for $x < 0$; maximum at $(-3, 20)$; minimum at $(3, -16)$; inflection at $(0, 2)$

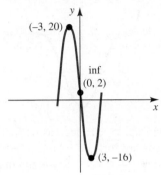

15. Increasing for $x > 3$; decreasing for $x < 3$; concave upward for $x < 0$ and $x > 2$; concave downward for $0 < x < 2$; minimum at $(3, -17)$; inflection at $(0, 10)$ and $(2, -6)$

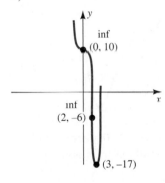

17. Increasing for all x; concave upward for $x > 2$; concave downward for $x < 2$; inflection at $(2, 0)$

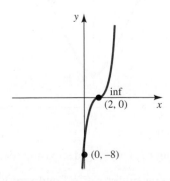

ANSWERS

19. Increasing for $x > 0$; decreasing for $x < 0$; concave upward for $x < -\sqrt{5}$, $-1 < x < 1$, $x > \sqrt{5}$; concave downward for $-\sqrt{5} < x < -1$ and $1 < x < \sqrt{5}$; minimum at $(0, -125)$; inflection points at $(-\sqrt{5}, 0)$, $(\sqrt{5}, 0)$, $(-1, -64)$, and $(1, -64)$

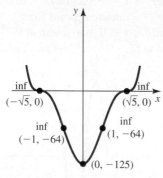

21. Increasing for $s > -1$; decreasing for $s < -1$; concave upward for $s < -4$ and $s > -2$; concave downward at $-4 < s < -2$; minimum at $(-1, -54)$; inflection at $(-4, 0)$ and $(-2, -32)$

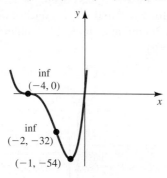

23. Increasing for $x > 0$; decreasing for $x < 0$; concave upward for all real x; minimum at $(0, 1)$

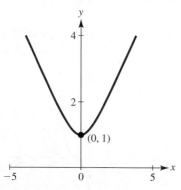

25. Increasing for $x < -\dfrac{1}{2}$; decreasing for $x > -\dfrac{1}{2}$; concave upward for $x < -1$ and $x > 0$; concave

downward for $-1 < x < 0$; maximum at $\left(-\dfrac{1}{2}, \dfrac{4}{3}\right)$; inflection at $(-1, 1)$ and $(0, 1)$

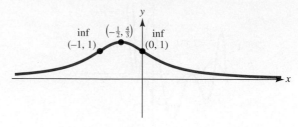

27. $f''(x) = 6(x + 1)$; maximum at $(-2, 5)$; minimum at $(0, 1)$

29. $f''(x) = 12(x^2 - 3)$; maximum at $(0, 81)$; minimum at $(3, 0)$ and $(-3, 0)$

31. $f''(x) = \dfrac{36}{x^3}$; maximum at $(-3, -11)$; minimum at $(3, 13)$

33. $f''(x) = 12x^2 - 60x + 50$; maximum at $\left(\dfrac{5}{2}, \dfrac{625}{16}\right)$; minimum at $(0, 0)$ and $(5, 0)$

35. $f''(t) = \dfrac{4(3t^2 - 1)}{(1 + t^2)^3}$; maximum at $(0, 2)$

37. $f''(x) = \dfrac{24(x - 2)}{x^4}$; maximum at $(-4, -13.5)$. Test fails for $x = 2$ [there is an inflection point at $(2, 0)$].

39. Concave upward for $x < 0$, for $0 < x < 1$, and for $x > 3$; concave downward for $1 < x < 3$; inflection points at $x = 1, 3$

41. Concave upward for $x > 1$; concave downward for $x < 1$; inflection point at $x = 1$

43. **a.** Increasing for $x < 0$ and $x > 4$; decreasing for $0 < x < 4$

     b. Concave upward for $x > 2$ and concave downward for $x < 2$

     c. Relative minimum at $x = 4$, relative maximum at $x = 0$; inflection point at $x = 2$

     d.

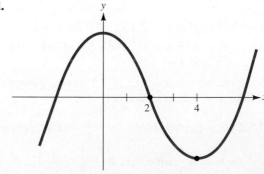

45. a. Increasing for $-\sqrt{5} < x < \sqrt{5}$; decreasing for $x > \sqrt{5}$ and $x < -\sqrt{5}$

b. Concave upward for $x < 0$ and concave downward for $x > 0$

c. Relative maximum at $x = \sqrt{5}$ and relative minimum at $x = -\sqrt{5}$; inflection point at $x = 0$

d.

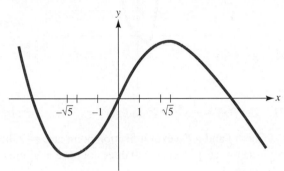

47. A typical graph is shown.

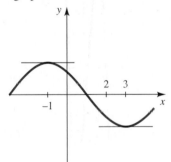

49. $f(x)$ is increasing for $x > 2$.
$f(x)$ is decreasing for $x < 2$.
$f(x)$ is concave upward for all real x.
$f(x)$ has a relative minimum at $x = 2$.
$f(x)$ has no inflection points.

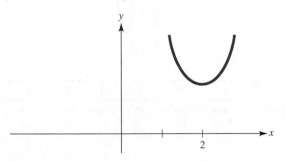

51. $f(x)$ is increasing for $x > 2$.
$f(x)$ is decreasing for $x < -3$ and $-3 < x < 2$.
$f(x)$ is concave upward for $x < -3$ and $x > -1$.
$f(x)$ is concave downward for $-3 < x < -1$.
$f(x)$ has a relative minimum at $x = 2$.
$f(x)$ has inflection points at $x = -3$ and $x = -1$.

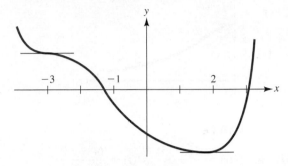

53. $C(x) = 0.3x^3 - 5x^2 + 28x + 200$

a. $C'(x) = 0.9x^2 - 10x + 28$

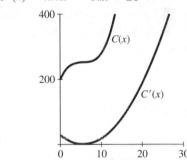

b. Only inflection number of $C(x)$ is $x = 5.56$. It corresponds to a minimum on the graph of $C'(x)$.

55. a. 1,000 units will be sold.

b. Inflection point when $x = 11$. Sales are increasing at the largest rate when $11,000 is spent on marketing.

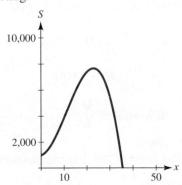

57. a. When $t = \dfrac{3}{2}$ (8:30 A.M.)

b. When $t = 5$ (noon)

59. a. $M'(r) = \dfrac{0.02 - 0.018r - 0.00018r^2}{(1 + 0.009r^2)^2}$

$M''(r) = \dfrac{-0.018 - 0.00108r + 0.000486r^2 + 0.00000324r^3}{(1 + 0.009r^2)^3}$

b.

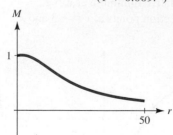

c. 7.10%

61. a. $S'(t) = \dfrac{18 - 3t}{(t + 2)^3}$; $S''(t) = \dfrac{6t - 60}{(t + 2)^4}$

b. After 6 months, the sales reach a maximum level of 519 pairs.

c. After 10 months, the sales rate is minimized; the sales level is 517 pairs, and the sales rate is -0.7 pairs per month.

63. a. $N'(t) = \dfrac{60 - 5t^2}{(12 + t^2)^2}$;

$N''(t) = \dfrac{10t^3 - 360t}{(12 + t^2)^3}$

b. 3.5 weeks after the outbreak; 722 new cases

c. 6 weeks after the outbreak; 62.5 new cases

65. Rate of growth is $P'(t) = -3t^2 + 18t + 48$.

a. Rate is largest when $t = 3$ years.

b. Rate is smallest when $t = 0$ years.

c. Rate of growth of $P'(t)$ is $P''(t) = -6t + 18$, which is largest when $t = 0$.

67. a. $R'(t) = A''(t) = \dfrac{d}{dt}[(k\sqrt{A(t)})(M - A(t))]$

$= k\left(\dfrac{1}{2}\right)\dfrac{A'(t)}{\sqrt{A(t)}}[M - A(t)] + k\sqrt{A(t)}(-A'(t))$

$= \dfrac{kA'(t)}{2\sqrt{A(t)}}[M - 3A(t)]$

$= 0$ when $A = \dfrac{M}{3}$.

b. Greatest

c. The graph of $A(t)$ has an inflection point where $A(t) = \dfrac{M}{3}$.

69. $f'(x) = 4x^3 + 1; f''(x) = 12x^2$. Although $f'(0) = f''(0) = 0$, neither $f'(x)$ nor $f''(x)$ change sign at $x = 0$, so the graph of f has neither a relative extremum nor an inflection point where $x = 0$.

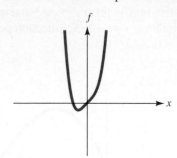

71. Let $f(x) = \dfrac{1}{6}x^3 - x^2$ and $g(x) = -\dfrac{1}{6}x^3 + x^2$; then both f and g have an inflection point at $x = 2$, but $h(x) = f(x) + g(x) = 0$ does not have any inflection points.

73. a.

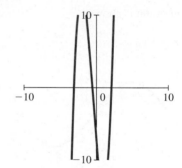

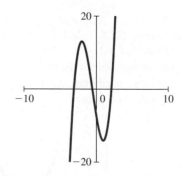

b.

x	-4	-2	-1	0	1	2
$f(x)$	-39	13	6	-7	-14	-3
$f'(x)$	60	0	-12	-12	0	24
$f''(x)$	-42	-18	-6	6	18	30

c. $(-3.08, 0)$, $(-0.54, 0)$, $(2.12, 0)$; $(0, -7)$

d. Relative maximum at $(-2, 13)$; relative minimum at $(1, -14)$

e. $x < -2$ and $x > 1$

f. $-2 < x < 1$

g. $\left(-\dfrac{1}{2}, -\dfrac{1}{2}\right)$

h. $x > -\dfrac{1}{2}$

i. $x < -\dfrac{1}{2}$

j. Answers will vary.

k. 13; -39

CHAPTER 3   Section 3

1. Vertical asymptote, $x = 0$; horizontal asymptote, $y = 0$

3. No vertical asymptotes; horizontal asymptote at $y = 0$

5. Vertical asymptotes, $x = -2$, $x = 2$; horizontal asymptotes, $y = 2$ and $y = 0$ (x axis)

7. Vertical asymptote, $x = 2$; horizontal asymptote, $y = 0$

9. Vertical asymptote, $x = -2$; horizontal asymptote, $y = 3$

11. No vertical asymptotes; horizontal asymptote, $y = 1$

13. Vertical asymptotes, $t = 2$, $t = 3$; horizontal asymptote, $y = 1$

15. Vertical asymptotes, $x = 0$, $x = 1$; horizontal asymptote, $y = 0$

17.

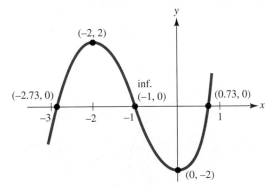

19.

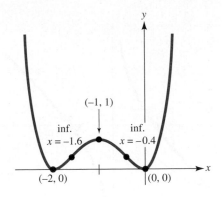

21.

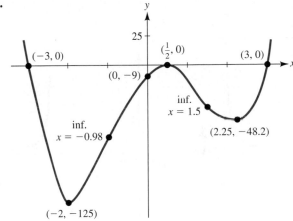

23.

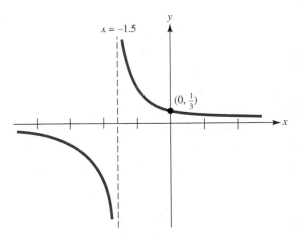

25.

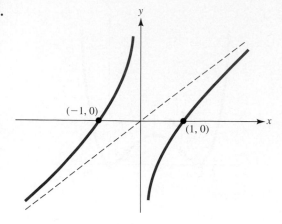

27.

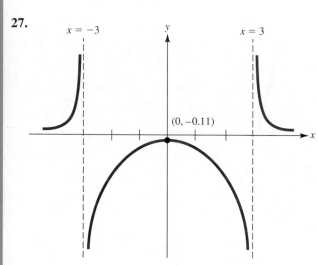

29.

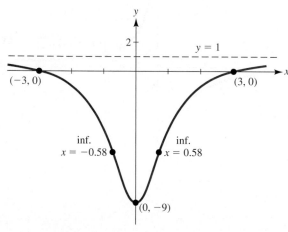

31.

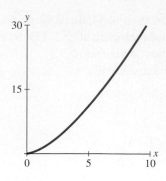

33. Answers may vary.

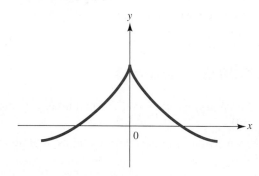

35. Answers may vary.

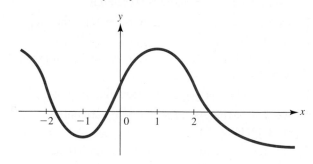

37. Answers may vary.

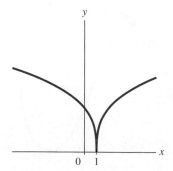

39. a. $f(x)$ is increasing ($f'(x) > 0$) for $0 < x < 2$ or $x > 2$; $f(x)$ is decreasing ($f'(x) < 0$) for $x < 0$.

  b. Relative minimum at $x = 0$

  c. $f''(x) = x^2(x - 2)(5x - 6)$; $f(x)$ is concave up for $x < 0, 0 < x < \dfrac{6}{5}$, and $x > 2$; concave down for $\dfrac{6}{5} < x < 2$.

  d. $x = \dfrac{6}{5}$ and $x = 2$

41. a. $f(x)$ is increasing ($f'(x) > 0$) for $-3 < x < 2$ or $x > 2$; $f(x)$ is decreasing ($f'(x) < 0$) for $x < -3$.

  b. Relative minimum at $x = -3$

  c. $f''(x) = \dfrac{-x - 8}{(x - 2)^3}$; $f(x)$ is concave up for $-8 < x < 2$; concave down for $x < -8$ and $x > 2$.

  d. $x = -8$

43. $B = -\dfrac{5}{2}; A = -10$

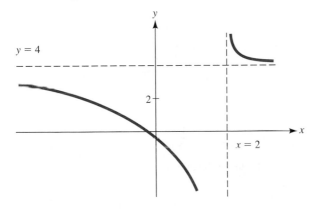

45. a. Vertical asymptote, $x = 0$; no horizontal asymptotes

  b. The average cost curve $A(x)$ approaches the line $y = 3x + 1$ as x gets larger.

  c.

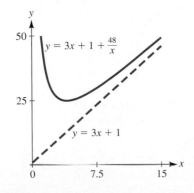

47. a.

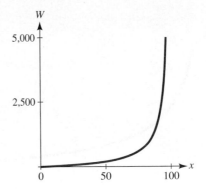

  b. 11.8%

49. a.

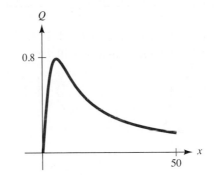

  b. $5,196; 674 units

  c. $9,000

51. a. $C(x) = \dfrac{7,880}{x} + 4.25x$

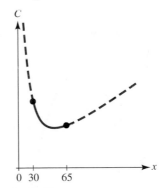

  b. 43 mph; $366

53. Answers may vary.

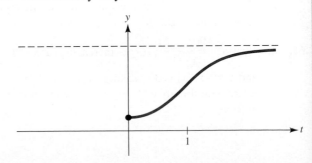

55. a.

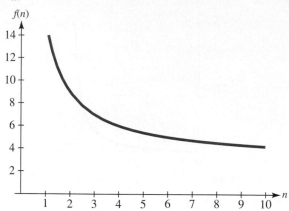

b. Points corresponding to $n = 1, 2, 3, \ldots$
c. Writing exercise; answers will vary.

57. a.

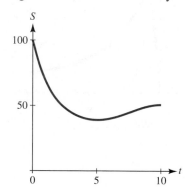

b. $t = 5$; 41.2%
c. Positive; decreasing

59. a.

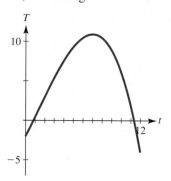

b. The temperature is greatest at 1:00 P.M. The high temperature is 10.9°C.

61. a. $f'(x) = \dfrac{10(x - 1)}{3x^{1/3}}$; $f(x)$ is increasing for $x < 0$ and $x > 1$; $f(x)$ is decreasing for $0 < x < 1$; relative minimum at $(1, -3)$; relative maximum at $(0, 0)$.

b. $f''(x) = \dfrac{10(2x + 1)}{9x^{4/3}}$; $f(x)$ is concave upward for $x > -\dfrac{1}{2}$; $f(x)$ is concave downward for $x < -\dfrac{1}{2}$; inflection point at $\left(-\dfrac{1}{2}, -3\sqrt[3]{2}\right)$.

c. $(0, 0), \left(\dfrac{5}{2}, 0\right)$; no asymptotes

d.

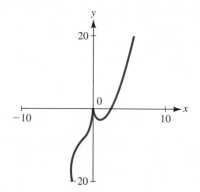

63. a.

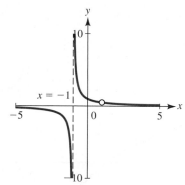

The graph has a hole at $x = 1$.

b.

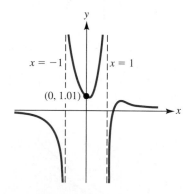

Vertical asymptotes at $x = 1, x = -1$; horizontal asymptote $y = 0$.

CHAPTER 3   Section 4

1. Absolute maximum at $(1, 10)$; absolute minimum at $(-2, 1)$

3. Absolute maximum at $(0, 2)$; absolute minimum at $\left(2, -\dfrac{40}{3}\right)$

5. Absolute maximum at $(-1, 2)$; absolute minimum at $(-2, -56)$

7. Absolute maximum at $(-3, 3{,}125)$; absolute minimum at $(0, -1{,}024)$

9. Absolute maximum at $\left(3, \dfrac{10}{3}\right)$; absolute minimum at $(1, 2)$

11. Absolute minimum at $(1, 2)$; no absolute maximum

13. $f(x)$ has no absolute maximum or minimum for $x > 0$.

15. Absolute maximum at $(0, 1)$; no absolute minimum

17. **a.** $R(q) = 49q - q^2$; $R'(q) = 49 - 2q$;

$C'(q) = \dfrac{1}{4}q + 4$;

$P(q) = -\dfrac{9}{8}q^2 + 45q - 200$;

maximum when $q = 20$

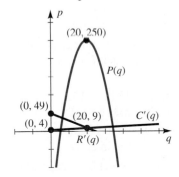

b. $A(q) = \dfrac{1}{8}q + 4 + \dfrac{200}{q}$; $A(q)$ is minimized at $q = 40$.

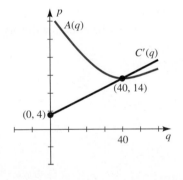

19. **a.** $R(q) = 180q - 2q^2$; $R'(q) = 180 - 4q$;
$C'(q) = 3q^2 + 5$;
$P(q) = -q^3 - 2q^2 + 175q - 162$;
$P(q)$ is maximized at $q = 7$.

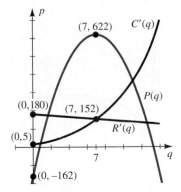

b. $A(q) = q^2 + 5 + \dfrac{162}{q}$; $A(q)$ is minimized at $q = 4.327$.

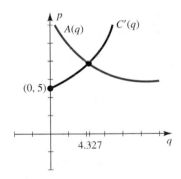

21. **a.** $R(q) = 1.0625q - 0.0025q^2$;
$R'(q) = 1.0625 - 0.005q$;
$C'(q) = \dfrac{q^2 + 6q - 1}{(q + 3)^2}$;
$P(q) = \dfrac{-0.0025q^3 + 0.055q^2 + 3.1875q - 1}{q + 3}$;
Profit $P(q)$ is maximized when $P'(q) = 0$; when $q = 17.3$ units.

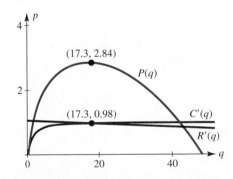

b. $A(q) = \dfrac{q^2 + 1}{q(q + 3)}$ is minimized when

$q = \dfrac{1 + \sqrt{10}}{3} \approx 1.3874.$

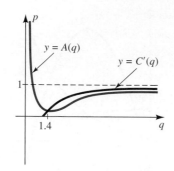

23. $E(p) = \dfrac{1.3p}{-1.3p + 10}; E(4) = \dfrac{13}{12}$, elastic

25. $E(p) = \dfrac{-2p^2}{p^2 - 200}; E(10) = 2$, elastic

27. $E(p) = \dfrac{-30}{p - 30}; E(10) = \dfrac{3}{2}$, elastic

29. The slope $f'(x) = 4x - x^2$ has its largest absolute value when $x = -1$. The graph is steepest at $\left(-1, \dfrac{7}{3}\right)$. The slope of the tangent is -5.

31. **a.** $P'(q) = -4q + 68;$

$A(q) = \dfrac{P(q)}{q} = -2q + 68 - \dfrac{128}{q}$

b. $P'(q) = A(q)$ when $q = 8$

c. A is increasing ($A' > 0$) if $0 < q < 8$ and decreasing ($A' < 0$) if $q > 8$

d.

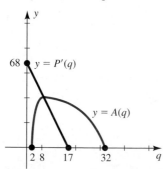

33. **a.** $E = \dfrac{3p}{2q + 3p}$

b. When $p = 3$, $q = 2$, and $E = \dfrac{9}{13}$, demand is inelastic.

35. **a.** $225 \le p \le 250$

b. $E(p) = \dfrac{p}{250 - p}$

Demand is elastic when $p > 125$, inelastic when $p < 125$, and of unit elasticity when $p = 125$.

c. Total revenue is increasing for $p < 125$ and decreasing for $p > 125$.

d. If any number of prints are available, then $p = 125$ maximizes total revenue. If only 50 prints are available, then $p = 225$ maximizes total revenue.

37. 400 Floppsies and 700 Moppsies

39. 11:00 A.M.

41. **a.** $E(p) = \dfrac{ap}{b - ap}$

b. $E(p) = 1 \Rightarrow ap = (1)(b - ap)$

$\Rightarrow 2ap = b \Rightarrow p = \dfrac{b}{2a}$

c. Elastic for $\dfrac{b}{2a} < p \le \dfrac{b}{a}$, inelastic for $0 \le p < \dfrac{b}{2a}$

43. $E(p) = -\dfrac{p}{q}\dfrac{dq}{dp} = -\dfrac{p}{\left(\dfrac{a}{p^m}\right)}\left(\dfrac{-ma}{p^{m+1}}\right)$

$= -\left(\dfrac{p^{m+1}}{a}\right)\left(\dfrac{-ma}{p^{m+1}}\right) = m$

If $m = 1$, demand is of unit elasticity. If $m > 1$, demand is elastic, and if $0 < m < 1$, demand is inelastic.

45. **a.** Largest in 2013 ($x = 15$); smallest in 2009 ($x = 11$)

b. Largest: 58,500 in 2013; smallest: 12,100 in 2009

47. The speed of the blood is greatest when $r = 0$, that is, at the central axis.

49. $p = \dfrac{n}{m}$

51. **a.** $v = 39$ km/hr

b. Writing exercise; responses will vary.

53. **a.** $D = \dfrac{C}{2}; \dfrac{C^2}{4}$

b. $\dfrac{C^3}{12}$

55. a. $R(x) = \dfrac{AB + A(1 - m)x^m}{(B + x^m)^2}$; $x = \left(\dfrac{B}{m - 1}\right)^{1/m}$

   b. $R'(x) = \dfrac{-Amx^{m-1}[(1 - m)x^m + (1 + m)B]}{(B + x^m)^3}$;

   $= 0$ when $x = 0$ and $x = \left[\dfrac{B(m + 1)}{m - 1}\right]^{1/m}$

   c. Relative maximum; use the first derivative test.

57. $R = r$

59. $v = \sqrt[4]{\dfrac{B}{A}}$

   F is decreasing ($F' < 0$) if $0 < v < \sqrt[4]{\dfrac{B}{A}}$ and

   increasing ($F' > 0$) if $v > \sqrt[4]{\dfrac{B}{A}}$, so there is a

   minimum at $v = \sqrt[4]{\dfrac{B}{A}}$.

61. Use $R'(p) = q[-E(p) + 1]$ with $q > 0$.
   Elastic demand $E(p) > 1$; $-E(p) + 1 < 0$,
       so $R'(p) < 0$ and R is decreasing.
   Inelastic demand $E(p) < 1$; $-E(p) + 1 > 0$,
       so $R'(p) > 0$ and R is increasing.
   Unitary demand $E(p) = 1$; $-E(p) + 1 = 0$,
       so $R'(p) = 0$ and R is maximized.

CHAPTER 3   Section 5

1. $\dfrac{1}{2}$

3. $x = 25$, $y = 25$

5. Make the playground square with side $S = 60$ m.

7. Let x be the length of the rectangle and y the width,
   and let p be the fixed value of the perimeter, so that

   $p = 2(x + y)$ and $y = \dfrac{1}{2}(p - 2x)$. The area is

   $$A = xy = x\left[\dfrac{1}{2}(p - 2x)\right] = -x^2 + \dfrac{1}{2}px$$

   Differentiating, we find that

   $$A' = -2x + \dfrac{1}{2}p = 0$$

   when $x = \dfrac{p}{4}$. Since $A'' = -2 < 0$, the maximum

   area occurs when $x = \dfrac{p}{4}$ and

   $$y = \dfrac{1}{2}\left[p - 2\left(\dfrac{p}{4}\right)\right] = \dfrac{p}{4}$$

   that is, when the rectangle is a square.

9. 6 by 2.5

11. $40.83 \approx $41.00

13. 80 trees

15. $p = 8.12 (or $8.13) per card

17. Entirely under water

19. 5 years from now

21. 17 floors

23. a. 200 bottles
   b. every 3 months

25. a. 10 machines
   b. $400
   c. $200

27. Suppose the setup cost and operating cost are aN

   and $\dfrac{b}{N}$, respectively, for positive constants a and b.

   Total cost is then $C = aN + \dfrac{b}{N}$. The minimum cost

   occurs when

   $$C' = a - \dfrac{b}{N^2} = 0$$

   $$aN = \dfrac{b}{N}$$

   that is, when setup cost aN equals operating

   cost $\dfrac{b}{N}$.

29. a. $P(x) = x\left(15 - \dfrac{3}{8}x\right) - \dfrac{7}{8}x^2 - 5x - 100 - tx$;

   Thus, $P'(x) = -\dfrac{5}{2}x + 10 - t = 0$

   when $x = \dfrac{2}{5}(10 - t)$

   b. $t = 5$
   c. The monopolist will absorb $4.25 of the $5 tax
       per unit. $0.75 will be passed on to the consumer.
   d. Writing exercise; responses will vary.

31. 4.5 miles from plant A

33. The point P should be $\dfrac{5\sqrt{3}}{3} \approx 2.9$ miles from A.

35. a.

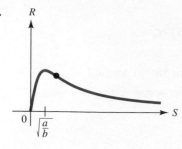

The graph appears to have a highest point
$\left(\text{at } x = \sqrt{\dfrac{a}{b}}\right)$, a lowest point (at $x = 0$), and
a point of inflection. The growth rate seems to
level off (toward 0) as S grows larger and larger.

b. Writing exercise; responses will vary.

37. a. $E'(v) = \dfrac{Cv^{k-1}[(v - v_w)k - v]}{(v - v_w)^2} = 0$ when the

numerator is zero, or when $v = \dfrac{v_w k}{k - 1}$.

It corresponds to a relative minimum.

b. $F(k) = \dfrac{v_w k}{k - 1}, k > 2$

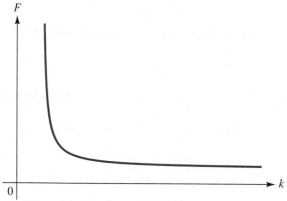

When k is very large, $F(k)$ is approximately v_w.

39. Width: 22 cm; length: 44 cm

41. $r = 1.51$ inches; $h = 3.02$ inches

43. $r = \dfrac{2}{3}h$

45. 2 by 2 by $\dfrac{4}{3}$ meters

47. Frank is right; in Example 3.5.5 replace 3,000 with
any fixed distance $D \geq 1,200$. The result is the same
since D drops out when $C'(x)$ is computed.

49. Yes, he has 5 minutes, 17 seconds to spare.

51. $S = Kwh^3 = Kh^3\sqrt{225 - h^2};$

$S'(h) = \dfrac{675h^2 - 4h^4}{\sqrt{225 - h^2}} = 0$ when $h \approx 13$ in.;

$w \approx 7.5$ in.

53. $x = 18$; $y = 36$; $V = 11,664$ in.^{3}

CHAPTER 3   Checkup

1. (a) is the graph of $f(x)$ and (b) is the graph of $f'(x)$.
Answers will vary. One reason is the x intercepts of
the graph in (b) correspond to the high and low
points in (a).

2. a. Increasing for $x < 0$ and $0 < x < 3$; decreasing
for $x > 3$

Critical Numbers	Classification
0	Neither
3	Relative maximum

b. Increasing for $t < 1$ and $t > 2$; decreasing for
$1 < t < 2$

Critical Numbers	Classification
1	Relative maximum
2	Relative minimum

c. Increasing for $-3 < t < 3$; decreasing for
$t < -3$ and $t > 3$

Critical Numbers	Classification
-3	Relative minimum
3	Relative maximum

d. Increasing for $x < -1$ and $x > 9$; decreasing for
$-1 < x < 9$

Critical Numbers	Classification
-1	Relative maximum
9	Relative minimum

3. a. Concave upward for $x > 2$; concave downward
for $x < 0$ and $0 < x < 2$; inflection at $x = 2$

b. Concave upward for $-5 < x < 0$ and $x > 1$;
concave downward for $x < -5$ and $0 < x < 1$;
inflection at $x = -5$, $x = 0$, and $x = 1$

c. Concave upward for $t > 1$; concave downward
for $t < 1$; no inflection points

d. Concave upward for $-1 < t < 1$; concave
downward for $t < -1$ and $t > 1$; inflection
at $t = -1$ and $t = 1$

4. a. Vertical asymptote, $x = -3$; horizontal asymptote, $y = 2$

b. Vertical asymptotes, $x = -1$, $x = 1$; horizontal asymptote, $y = 0$

c. Vertical asymptotes, $x = -\dfrac{3}{2}$, $x = 1$; horizontal asymptote, $y = \dfrac{1}{2}$

d. Vertical asymptote, $x = 0$; horizontal asymptote, $y = 0$

5. a. No asymptotes; intercepts at $(0, 0)$ and $\left(\dfrac{4}{3}, 0\right)$; relative minimum at $(1, -1)$; inflection points at $(0, 0)$ and $\left(\dfrac{2}{3}, -\dfrac{16}{27}\right)$

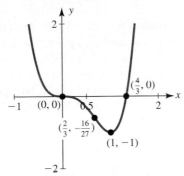

b. No asymptotes; y intercept at $(0, 1)$; relative minimum at $(0, 1)$; inflection points at $(1, 2)$ and $\left(\dfrac{1}{2}, \dfrac{23}{16}\right)$

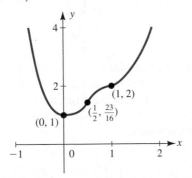

c. Vertical asymptote, $x = 0$; horizontal asymptote, $y = 1$; x intercept at $(-1, 0)$; relative minimum at $(-1, 0)$; inflection point at $\left(-\dfrac{3}{2}, \dfrac{1}{9}\right)$

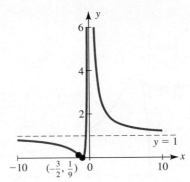

d. Vertical asymptote, $x = 1$; horizontal asymptote, $y = 0$; x intercept at $\left(\dfrac{1}{2}, 0\right)$; y intercept at $(0, 1)$; relative maximum at $(0, 1)$; inflection point at $\left(-\dfrac{1}{2}, \dfrac{8}{9}\right)$

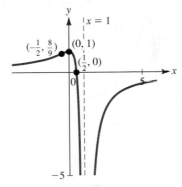

6. Answers will vary.

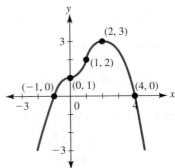

7. a. Absolute maximum value of 6 where $x = -1$; absolute minimum value of -26 where $x = 3$

b. Absolute maximum value of 23 where $t = 2$; absolute minimum value of -69 where $t = 4$

c. Absolute maximum value of 19 where $u = 16$; absolute minimum value of 3 where $u = 0$

8. $f''(t) = 0$ when $t = \dfrac{7}{3}$; 8:20 A.M.

ANSWERS

9. $135

10. a.

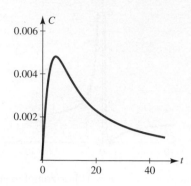

b. $t = 9$

c. The concentration tends to 0.

11. a. 1.667 million

b. $t = 2$ hours; 5 million

c.

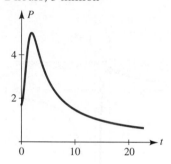

The population dies off.

CHAPTER 3   Review

1. $f(x)$ is increasing for $-1 < x < 2$; decreasing for $x < -1$ and $x > 2$; concave up for $x < \frac{1}{2}$; concave down for $x > \frac{1}{2}$; relative maximum $(2, 15)$; relative minimum $(-1, -12)$; inflection point $\left(\frac{1}{2}, \frac{3}{2}\right)$

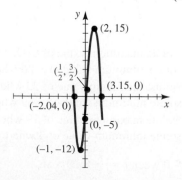

3. $f(x)$ is increasing for $x < -0.79$ and $1.68 < x$; $f(x)$ is decreasing for $-0.79 < x < 1.68$; $f(x)$ is concave downward for $x < \frac{4}{9}$; $f(x)$ is concave upward for $x > \frac{4}{9}$. There is a relative maximum at $(-0.79, 22.51)$, a relative minimum at $(1.68, -0.23)$. There is one inflection point, at $(0.44, 11.14)$.

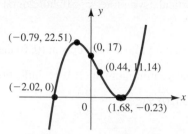

5. $f(t)$ is increasing for $t < -2$ and $t > 2$; decreasing for $-2 < t < 2$; concave up for $-\sqrt{2} < t < 0$ and for $t > \sqrt{2}$; concave down for $t < -\sqrt{2}$ and for $0 < t < \sqrt{2}$. Relative maximum at $(-2, 64)$; relative minimum at $(2, 64)$; inflection points $(-\sqrt{2}, 39.6)$ and $(\sqrt{2}, -39.6)$

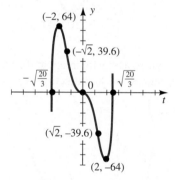

7. $g(t)$ is increasing for $t < -2$ and for $t > 0$; decreasing for $-2 < t < -1$ and for $-1 < t < 0$; concave up for $t > -1$; concave down for $t < -1$. Relative maximum $(-2, -4)$; relative minimum $(0, 0)$; no inflection points.

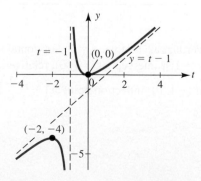

9. $F(t)$ is increasing for $x < -2$ and for $x > 2$; decreasing for $-2 < x < 0$ and for $0 < x < 2$; concave up for $x > 0$; concave down for $x < 0$. Relative maximum $(-2, -6)$; relative minimum $(2, 10)$; no inflection points.

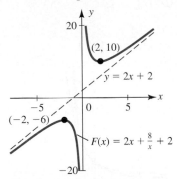

11. The graph of $f(x)$ is (b) and (a) is the graph of $f'(x)$. Answers will vary. One reason is that the graph in (b) is always increasing and the graph in (a) is always positive.

13.

Critical Numbers	Classification
-1	Relative minimum
0	Relative maximum
3/2	Neither
7	Relative minimum

15.

Critical Numbers	Classification
0	Relative minimum
2	Neither

17. Here is one possible graph.

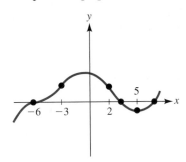

19. Here is one possible graph.

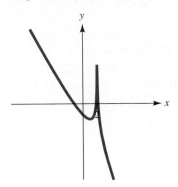

21. Relative maximum at $(2, 15)$; relative minimum at $(-1, -12)$

23. Relative maximum at $(-2, -4)$; relative minimum at $(0, 0)$

25. Absolute maximum value of 40 where $x = -3$; absolute minimum value of -12 where $x = -1$

27. Absolute maximum value of $\frac{1}{2}$ where $s = -\frac{1}{2}$ or $s = 1$; absolute minimum value of 0 where $s = 0$

29. **a.** $f(x)$ is increasing for $0 < x < 1$ and $x > 1$; it is decreasing for $x < 0$.

b. $f(x)$ is concave upward for $x < \frac{1}{3}$ and $x > 1$; it is concave downward for $\frac{1}{3} < x < 1$.

c. $f(x)$ has a relative minimum at $x = 0$ and inflection points at $x = \frac{1}{3}$ and $x = 1$.

d.

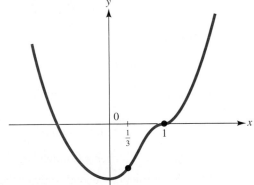

31. $12.50

33. $r = \dfrac{2}{3}h$

35. a. An 80 ft by 80 ft square

    b. 160 ft parallel to the wall, 80 ft wide

37. Row all the way to town.

39. 12 machines

41. a. $E = \dfrac{2p^2}{100 - p^2}$

    b. At $p = 6$, $E = \dfrac{9}{8}$. Since $E > 1$, demand is elastic (i.e., as price increases, revenue decreases).

    c. $5.77

43. a. $E(p) = \dfrac{1.4p^2}{300 - 0.7p^2}$

    b. $E(8) = 0.351$; raise the price

45. Rectangle: 3.9 feet by 4.2 feet; side of triangle: 3.9 feet

47. 4,000 maps per batch

49. *Hint:* If x units are ordered, $C = k_1 x + \dfrac{k_2}{x}$.

51. a. $x = \sqrt{\dfrac{pq}{ns}}$

    b. Let $x = \sqrt{\dfrac{pq}{ns}}$. Then $C_s = sx = s\sqrt{\dfrac{pq}{ns}}$ and

$$C_o = pt = p\left(\dfrac{q}{nx}\right) = \dfrac{pq}{n\sqrt{\dfrac{pq}{ns}}} = s\sqrt{\dfrac{pq}{ns}}$$

    so that $C_s = C_o$.

53. a. Relative minimum at $x = \dfrac{1}{c}$

    b. Maximum of $\dfrac{\pi}{3}(5 - 3\sqrt{2})$; minimum of $\dfrac{\pi}{6}$

    c. Maximum of $\dfrac{\sqrt{3}\pi}{16}$; minimum of $\dfrac{\sqrt{3}\pi}{4(2 + \sqrt{2})^2}$

    d. $\lim\limits_{x \to \infty} f(x) = Kc^2$, when r is much larger than R, the packing fraction depends only on the cell structure in the lattice.

CHAPTER 4   Section 1

1. $e^2 \approx 7.389$, $e^{-2} \approx 0.135$, $e^{0.05} \approx 1.051$, $e^{-0.05} \approx 0.951$, $e^0 = 1$, $e \approx 2.718$, $\sqrt{e} \approx 1.649$, $\dfrac{1}{\sqrt{e}} \approx 0.607$

3.

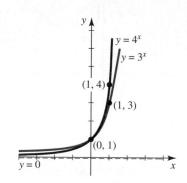

5. a. 9

    b. $\dfrac{1}{27}$

7. a. 12

    b. $\dfrac{189}{1,331} \sqrt{7}$

9. a. 3

    b. 4

11. a. 243

    b. $e^{14/3}$

13. a. $9x^4$

    b. $2x^{2/3}y$

15. a. $\dfrac{1}{x^{1/3}y^{1/2}}$

    b. $x^{1.1}y^2$

17. a. $\dfrac{1}{t}$

    b. t

19. $\dfrac{3}{2}$

21. 1

23. 1

25. $-2, 2$

27. $0, \dfrac{3}{2}$

29.

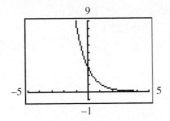

31.

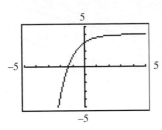

33. $b = 2, C = 3$

35.
  a. $1,967.15
  b. $2,001.60
  c. $2,009.66
  d. $2,013.75

37.
  a. $7,129.86
  b. $7,068.25
  c. $7,047.12
  d. $7,046.88

39. $r_e \approx 6.14\%$

41. $r_e \approx 5.13\%$

43. $4,093.65

45.
  a. $40.60
  b. $4,060
  c. Revenue is less by $1,458 when 100 units are produced.

47. d, c, b, a

49. $608.33

51.
  a. 0.5488
  b. 0.1813
  c. 0.1215

53. The sellers would have gotten the better deal by 2.7676 trillion dollars.

55.
  a. GDP $= 500(1.027)^t$ billion dollars
  b. $652.6 billion

57. $1,206.93

59.
  a. No. A fair monthly payment is $166.07.
  b. Writing exercise; responses will vary.

61.
  a. 12,000 people per square mile
  b. 5,959 people per square mile

63.
  a. 50,000,000
  b. 91,105,940

65.
  a. 3 mg/ml; 1.78 mg/ml
  b. -0.72 mg/ml per hour

67.
  a. $A = \dfrac{10,000}{2^{0.01}} \approx 9931$
  b. 9931; 10,070; 10,353
  c. 440 bacteria per hour or 7.33 bacteria per minute

69.
  a. 0.13 g/cm^3
  b. 0.1044 g/cm^3; 0.0795 g/cm^3
  c. -0.0016 g/cm^3 per minute
  d. As $t \to \infty$, $C(t) \to 0.065$ g/cm^3
  e.

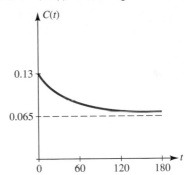

71. $\dfrac{1}{\sqrt[3]{10}} I_0 \approx 0.46 I_0$

73. 329.7 g

75.

x	-2.2	-1.5	0	1.5	2.3
$f(x)$	10.5561	4	0.5	0.0625	0.0206

77. As $n \to -\infty$, $\left(1 + \dfrac{1}{n}\right)^n \to e \approx 2.71828$

79. $\displaystyle\lim_{n \to +\infty} \left(2 - \dfrac{5}{2n}\right)^{n/3} = +\infty$

CHAPTER 4   Section 2

1. $\ln 1 = 0$, $\ln 2 \approx 0.693$, $\ln e = 1$, $\ln 5 \approx 1.609$, $\ln\dfrac{1}{5} \approx -1.609$, $\ln e^2 = 2$, $\ln 0$ and $\ln -2$ are undefined; e^x cannot be negative or equal to zero.

3. 3

5. 5

7. $\dfrac{8}{25}$

9. $3 + \log_3 2 + \log_3 5$

11. $2 \log_3 2 + 2 \log_3 5$

13. $4 \log_2 x + 3 \log_2 y$

15. $\dfrac{1}{3}[\ln x + \ln(x - 1)]$

17. $2 \ln x + \dfrac{2}{3}\ln(3 - x) - \dfrac{1}{2}\ln(x^2 + x + 1)$

19. $3 \ln x - x^2$

21. $\dfrac{\ln 53}{\ln 4} \approx 2.864$

23. 5

25. $\dfrac{\ln 2}{0.06} \approx 11.552$

27. $\dfrac{\ln 5}{4} \approx 0.402$

29. $e^{-C-t/50}$

31. 4

33. $\dfrac{2}{\ln 3} \approx 1.820$

35. $10 \ln 2 \approx 6.931$

37. $5 \ln 2 \approx 3.4657$

39. $7 \ln 5 - \ln 2 \approx 10.5729$

41. -5.5

43. $\dfrac{\ln 2}{0.06} \approx 11.55$ years

45. $\dfrac{\ln 2}{13} \approx 5.33\%$

47. $\dfrac{12 \ln 3}{\ln 2} \approx 19.02$ years

49. $\ln 1.06 \approx 5.83\%$

51. $Q(t) = 500 - 200e^{-0.1331t}$; 459.5 units

53. a. $10 - \ln 11 \approx \$7.60$
 b. $\ln 102 \approx \$4.62$
 c. $x_e = \dfrac{-3 + \sqrt{1 + 4e^{10}}}{2} \approx 147$ units
  $P_e \approx \$5$

55. a. $300 \ln 3 \approx \$329.6$ million
 b. 6 years
 c. $e^{10/3} - 3 \approx 25.03$ years

57. a. 0.765 g/cm^3; 0.784 g/cm^3
 b. $-50 \ln\left(\dfrac{0.125}{0.13}\right) \approx 1.96$ sec

59. a.

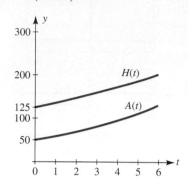

 b. $A = \dfrac{2H^2}{625}$

61. 10,523 years

63. 24.84 years ago; 95.6%

65. a. $51 + 100 \ln 3 \approx 161$ thousand
 b. $e^{271/100} - 3 \approx 12$ years
 c. $10 \ln\dfrac{13}{3} \approx 14{,}700$ people/year

67. a. 45%
 b. 2.34%

69. a. 0.89
 b. $\dfrac{\ln 0.557}{\ln 0.85} \approx 3.6$ sec
 c.

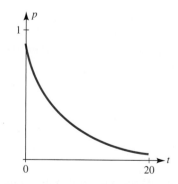

71. 5,614 years

73. $f(t) = 70 + 142e^{-kt}$; ideal temperature is 179.95°F.

75. Scélérat; Wednesday morning at 1:27 A.M.

77. a. 8.25
 b. $10^{17.9}$ joules

79. The line $y = x$ has slope 1, so it is perpendicular to the line through $A(a, b)$ and $B(b, a)$, which has slope $m = \dfrac{a - b}{b - a} = -1$. If O is the origin and M is the point where $y = x$ intersects the line segment AB, then right triangles OMA and OMB are similar since they share side OM and

$$|OA| = \sqrt{a^2 + b^2} = \sqrt{b^2 + a^2} = |OB|$$

It follows that $|AM| = |BM|$, which means that A and B are reflections of one another in $y = x$.

81. For $y = Cx^k$, let $Y = \ln y$ and $X = \ln x$. Then $Y = mX + b$ where $m = k$ and $b = \ln C$.

83. $x \approx -17.4213$

85. $x \approx 1.1697$ (using a graphing utility)

87. **a.** $(\log_a b)(\log_b a) = \left(\dfrac{\ln b}{\ln a}\right)\left(\dfrac{\ln a}{\ln b}\right) = 1$

b. $\log_a x = \dfrac{\ln x}{\ln a}$

$= \dfrac{(\ln x)(\ln b)}{(\ln b)(\ln a)} = (\log_b x)(\log_a b)$

$= \dfrac{\log_b x}{\log_b a}$ using part (a)

CHAPTER 4   Section 3

1. $f'(x) = 5e^{5x}$

3. $f'(x) = xe^x + e^x$

5. $f'(x) = -0.5e^{-0.05x}$

7. $f'(x) = (6x^2 + 20x + 33)e^{6x}$

9. $f'(x) = -6e^x(1 - 3e^x)$

11. $f'(x) = \dfrac{3}{2\sqrt{3x}}e^{\sqrt{3x}}$

13. $f'(x) = \dfrac{3}{x}$

15. $f'(x) = 2x \ln x + x$

17. $f'(x) = \dfrac{2}{3}e^{2x/3}$

19. $f'(x) = \dfrac{-2}{(x + 1)(x - 1)}$

21. $f'(x) = -2e^{-2x} + 3x^2$

23. $g'(s) = (e^s + 1)(2e^{-s} + s) +$
    $(e^s + s + 1)(-2e^{-s} + 1)$
    $= 1 + 2s + e^s + se^s - 2se^{-s}$

25. $h'(t) = \dfrac{te^t \ln t + t \ln t - e^t - t}{t (\ln t)^2}$

27. $f'(x) = \dfrac{e^x - e^{-x}}{2}$

29. $f'(t) = \dfrac{t + 1}{2t\sqrt{\ln t + t}}$

31. $f'(x) = \dfrac{1 - e^{-x}}{x + e^{-x}}$

33. $g'(u) = \dfrac{1}{\sqrt{u^2 + 1}}$

35. $f'(x) = \dfrac{2^x(x \ln 2 - 1)}{x^2}$

37. $\dfrac{1 + \ln x}{\ln 10}$

39. e; 1

41. $3e^{-4/3}$; -1

43. $\dfrac{3\sqrt{3}}{8}e^{-3/2}$; 0

45. $\dfrac{1}{e}$; 0

47. $y = x$

49. $y = e^2$

51. $y = \dfrac{1}{2}x - \dfrac{1}{2}$

53. $f''(x) = 4e^{2x} + 2e^{-x}$

55. $f''(t) = 2 \ln t + 3$

57. $f'(x) = f(x)\left[\dfrac{4}{2x + 3} + \dfrac{1 - 10x}{2(x - 5x^2)}\right]$

59. $f'(x) = f(x)\left[\dfrac{5}{x + 2} - \dfrac{1}{2(3x - 5)}\right]$

61. $f'(x) = f(x)\left[\dfrac{3}{x + 1} - \dfrac{2}{6 - x} + \dfrac{2}{3(2x + 1)}\right]$

63. $f'(x) = x(2 \ln 5)5^{x^2}$

65. **a.** $E(p) = 0.04p$; elastic for $p > 25$, inelastic for $p < 25$, of unit elasticity for $p = 25$

b. Demand will decrease by approximately 1.2%.

c. $R(p) = 3{,}000pe^{-0.04p}$; $p = 25$

67. **a.** $E(p) = \dfrac{p^2 + p}{10(p + 11)}$; elastic when $p > 15.91$, inelastic when $p < 15.91$, of unit elasticity when $p = 15.91$

b. Demand will decrease by approximately 1.85%.

c. $R(p) = 5{,}000\, p(p + 11)e^{-0.1p}$; $p = 15.91$

69. a. $C'(x) = 0.2e^{0.2x}$

b. 5 units

71. a. $C'(x) = \dfrac{6e^{x/10}}{\sqrt{x}}\left(1 + \dfrac{x}{5}\right)$

b. 5 units

73. a. Value is decreasing at the rate of $1,082.68 per year.

b. Constant rate of -40% per year

75. a. Approximately 406 copies

b. 368 copies

77. $\dfrac{R'(t_0)}{R(t_0)} = \dfrac{0.09(11) - 0.02(8)}{19} \approx 0.0437;\ 4.37\%$

79. $\dfrac{-(\ln q)'}{(\ln p)'} = \dfrac{-\dfrac{1}{q}\dfrac{dq}{dp}}{\dfrac{1}{p}\dfrac{dp}{dp}} = -\dfrac{p}{q}\dfrac{dq}{dp} = E(p)$

81. a. $F'(t) = -k(1 - B)e^{-kt}$ is the rate at which you are forgetting material.

b. $F'(t) = -k(F(t) - B)$, which says that the rate you forget is proportional to the fraction you have left to forget.

c.

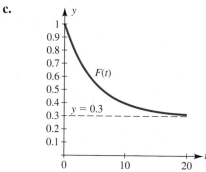

83. a. Population is increasing at the rate of 1.22 million people per year.

b. Constant rate of 2% per year

85. a. $N'(t) = \dfrac{36e^{-0.02t}}{(1 + 3e^{-0.02t})^2}$; the population is increasing at all times t.

b. Increasing for $t < 50 \ln 3$, decreasing for $t > 50 \ln 3$

c. $N(t)$ approaches 600.

87. a. $P_1'(10) \approx 1.556$ cm/day

$P_2''(10) = -0.257$ cm/day per day; decreasing

b. Plants have the same height, approximately 20 cm, after 20.71 days; $P_1'(20.71) \approx 0.286$, $P_2'(20.71) \approx 0.001$, so the first plant is growing more rapidly.

89. a.

$$b^x = e^{x \ln b}$$

$$\dfrac{d}{dx}(b^x) = e^{x \ln b}[\ln b]$$

$$= (\ln b)\, b^x$$

b.

$$y = b^x$$

$$\ln y = x \ln b$$

$$\dfrac{1}{y}\dfrac{dy}{dx} = \ln b$$

$$\dfrac{dy}{dx} = (\ln b)\, y = (\ln b)\, b^x$$

91. $f(x) = \dfrac{1}{3}\ln(x + 1) - 4\ln(1 + 3x)$

$f'(x) = \dfrac{1}{3(x + 1)} - \dfrac{12}{1 + 3x};\ f'(0.65) \approx -3.87$

tangent line at $(0.65, -4.16)$ is $y = -3.87x - 1.65$.

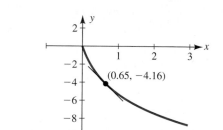

93. $100\left[\dfrac{tk - 1}{t}\right]$

CHAPTER 4   Section 4

1. $f_5(x)$

3. $f_3(x)$

5. $f(t)$ is increasing for all real t; concave upward for all real t. There is a horizontal asymptote at $y = 2$.

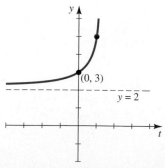

7. $g(x)$ is decreasing for all real x; concave downward for all real x; $y = 2$ is a horizontal asymptote.

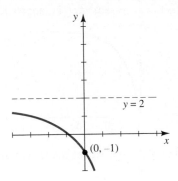

9. $f(x)$ is increasing for all real x; concave upward for $x < 0.549$; concave downward for $x > 0.549$. Inflection point is $(0.549, 1)$, and $y = 2$ and $y = 0$ are horizontal asymptotes.

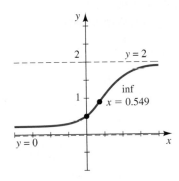

11. $f(x)$ is increasing for $x > -1$; decreasing for $x < -1$; concave upward for $x > -2$; concave downward for $x < -2$. Relative minimum is $\left(-1, -\dfrac{1}{e}\right)$ and inflection point is $\left(-2, -\dfrac{2}{e^2}\right)$. The x axis ($y = 0$) is a horizontal asymptote.

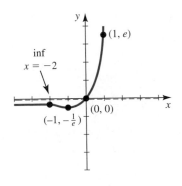

13. $f(x)$ is increasing for $x < 1$; decreasing for $x > 1$; concave upward for $x > 2$; concave downward for $x < 2$. Relative maximum is $(1, e)$, inflection point is $(2, 2)$. The x axis ($y = 0$) is a horizontal asymptote.

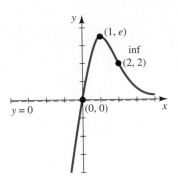

15. $f(x)$ is increasing for $0 < x < 2$; decreasing for $x < 0$ and $x > 2$; concave upward for $x < 0.6$ and $x > 3.4$; concave downward for $0.6 < x < 3.4$. Relative minimum is $(0, 0)$; relative maximum is $\left(2, \dfrac{4}{e^2}\right)$; inflection points are $(0.6, 0.2)$ and $(3.4, 0.4)$. The x axis is a horizontal asymptote.

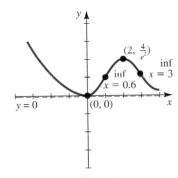

17. $f(x)$ is increasing for all real x; concave upward for $x < 0$; concave downward for $x > 0$. Inflection point is $(0, 3)$. The x axis ($y = 0$) and $y = 6$ are horizontal asymptotes.

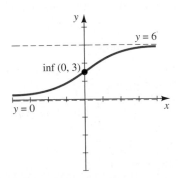

19. $f(x)$ is increasing for $x > 1$; decreasing for $x < 1$; concave upward for $x < e$; concave downward for $x > e$. Relative minimum is $(1, 0)$; inflection point is $(e, 1)$. The y axis $(x = 0)$ is a vertical asymptote.

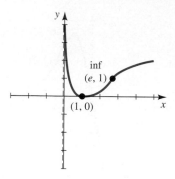

21. 36 billion hamburgers

23. **a.** As $t \to \infty, f(t) \to 1$

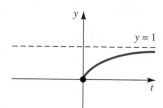

   b. 0.741
   c. 0.089

25. 37.5 units per day

27. **a.** Approximately 403 copies
   b. 348 copies

29. $C = \dfrac{1}{199}; k = 0.1745$

Rate of change of proportion of day-trading is

$$p'(t) = \frac{199ke^{kt}}{(199 + e^{kt})^2}$$

which is maximized when

$$p''(t) = 0; t \approx 30.33 \text{ weeks}$$
$$p(30.33) \approx 0.5$$

31. 69.44 years from now

33. 6.5 years

35. **a.** $V(5) = \$207.64$;

$$V'(t) = V_0\left(1 - \frac{2}{L}\right)^t \ln\left(1 - \frac{2}{L}\right)$$

$$V'(5) \approx -\$60/\text{year}$$

   b. $100 \ln\left(1 - \dfrac{2}{L}\right)$

37.

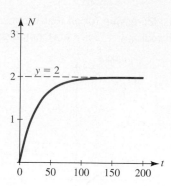

The value of N approaches the maximum of 2 million people.

39. 202.5 million

41. **a.** $e - 1 \approx 1.7$ years
   b. The learning rate $L'(t)$ is largest when $t = 0$ (at birth).

43. **a.** 0.15% per year
   b. 70.24 years; 0.15% per year

45. **a.** $C = 9, k = \dfrac{1}{2} \ln 3$
   b. 4 hours
   c. 4 hours

47. **a.**

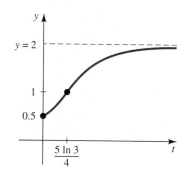

   b. 500
   c. 1,572
   d. 2,000

49. 0.45 years

51. **a.** $C = \dfrac{b}{aR}; E''\left(\dfrac{b}{aR}\right) = \dfrac{2a^3R^3}{b} > 0$
   b. $k = \dfrac{b}{aR}; m = \dfrac{4}{e}a^2R^2$

53. **a.** $Q(t) = 1{,}139e^{0.06t}, Q(7) = 1{,}734$ staff members
   b. $t \approx 11.6$ years

55. a. $E(t) = 1{,}000w(t)p(t)$

b. $t = 82$ days; 3,527 pounds

c.

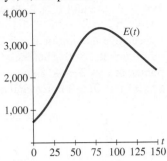

57. a.

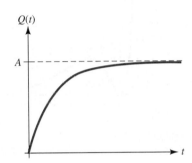

b. Given enough time, a person can recall all relevant facts in his or her memory.

59. $f''(t) = 0$ when $t = \dfrac{\ln C}{k}$ and $f\left(\dfrac{\ln C}{k}\right) = \dfrac{A}{2}$ or half the number of susceptible residents.

61. a. $t_{\max} - \dfrac{1}{b-a}\ln\left(\dfrac{b}{a}\right)$

In the long run, the concentration approaches 0.

b.

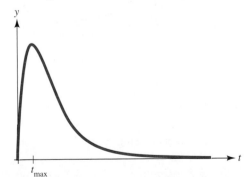

63. a. $N(0) = 15$ employees; $N(5) = 482$ employees; 2.10 years; 500 employees

b.

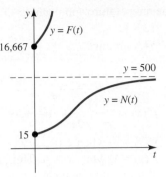

65. a. $x = r;\ p''(r) < 0$

b. The inflection points occur at $x = \dfrac{r(s - \sqrt{s})}{s}$ and $x = \dfrac{r(s + \sqrt{s})}{s}$. Both are positive since $s > 1$. The rate of blood cell production is maximized and minimized at the two inflection points, respectively.

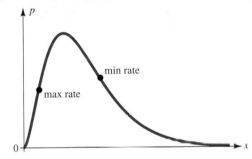

c.

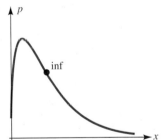

There is only one inflection point.

67. a. $A = 85,\ k = \dfrac{1}{20}\ln\dfrac{17}{6}$

b.

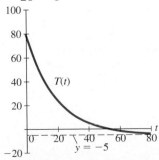

The temperature approaches $-5°C$.

c. 12.8°C

d. 54.4 minutes

69. a. $f'(x) = -\dfrac{1}{\sigma^3\sqrt{2\pi}}(x-\mu)e^{-(x-\mu)^2/2\sigma^2}$

$f''(x) = \dfrac{1}{\sigma^5\sqrt{2\pi}}[-\sigma^2 + (x-\mu)^2]e^{-(x-\mu)^2/2\sigma^2}$

$f'(x) = 0$ when $x = \mu$; $f'(x) > 0$ when $x < \mu$; $f'(x) < 0$ when $x > \mu$. Thus, $f(x)$ has one absolute maximum at $x = \mu$.
$[(\mu \pm \sigma) - \mu]^2 = \sigma^2$ so $f''(x) = 0$ when $x = \mu + \sigma$ and $x = \mu - \sigma$. Thus, $f(x)$ has inflection points at $x = \mu + \sigma$ and $x = \mu - \sigma$.

b. $[(\mu \pm c) - \mu]^2 = c^2$ so $f(\mu + c) = f(\mu - c)$ for every number c. Thus, the graph of $f(x)$ is symmetric about the line $x = \mu$.

CHAPTER 4   Checkup

1. a. 1

    b. $\dfrac{10}{3}$

    c. 0

    d. $\dfrac{16}{81}$

2. a. $27x^6y^3$

    b. $\dfrac{1}{\sqrt{3}xy^{2/3}}$

    c. $\dfrac{y^{7/6}}{x^{1/6}}$

    d. $\dfrac{1}{x^{6.5}y^8}$

3. a. $x = 3, x = -1$

    b. $x = \dfrac{1}{\ln 4}$

    c. $x = -4, x = 4$

    d. $t = -2\ln\dfrac{11}{3}$

4. a. $\dfrac{dy}{dx} = \dfrac{e^x(x^2 - 5x + 3)}{(x^2 - 3x)^2}$

    b. $\dfrac{dy}{dx} = \dfrac{3x^2 + 4x - 3}{x^3 + 2x^2 - 3x}$

    c. $\dfrac{dy}{dx} = x^2(1 + 3\ln x)$

    d.

5. a. $f(x)$ is increasing for $0 < x < 2$; decreasing for $x < 0$ and $x > 2$; concave upward for $x < 2 - \sqrt{2}$ and $x > 2 + \sqrt{2}$; concave downward for $2 - \sqrt{2} < x < 2 + \sqrt{2}$. There is a relative maximum at $x = 2$ and a relative minimum at $x = 0$. There are two inflection points, at $x = 2 - \sqrt{2}$ and $x = 2 + \sqrt{2}$. The x axis ($y = 0$) is a horizontal asymptote.

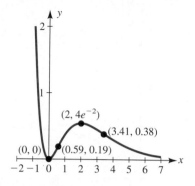

b. $f(x)$ is increasing for $0 < x < \sqrt{e}$; decreasing for $x > \sqrt{e}$; concave upward for $x > e^{5/6}$; concave downward for $x < e^{5/6}$. There is a relative maximum at $x = \sqrt{e}$. There is an inflection point at $(2.30, 0.08)$, where $x = e^{5/6}$. The y axis ($x = 0$) is a vertical asymptote, and the x axis ($y = 0$) is a horizontal asymptote.

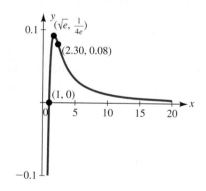

c. $f(x)$ is increasing for $0 < x < \dfrac{1}{4}$ and $x > 1$; decreasing for $\dfrac{1}{4} < x < 1$; concave downward for $x \neq 1$. There is a relative maximum at $x = \dfrac{1}{4}$. There are no inflection points, and $x = 0$ and $x = 1$ are vertical asymptotes.

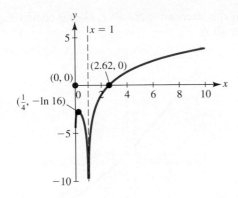

d. $f(x)$ is increasing for all x; concave upward for $x < 0$; concave downward for $x > 0$. There is an inflection point at $x = 0$, and $y = 0$ and $y = 4$ are horizontal asymptotes.

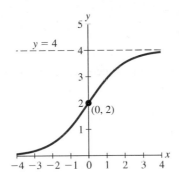

6. $2,323.67; 8.1 years

7. **a.** $4,323.25
   b. $4,282.09

8. **a.** Increasing for $0 \le t < e - 1$; decreasing for $t > e - 1$
   b. $t = e^{3/2} - 1$
   c. The price approaches $500.

9. **a.** $q'(p) = -1{,}000e^{-p}(p + 1) < 0$ for $p \ge 0$
   b. $p = \sqrt{2}$ or $141.42; $117,387.14

10. 6,601 years old

11. **a.** 80,000
   b. 2 hours; 81,873
   c. The population dies off completely.

CHAPTER 4   Review Exercises

1.

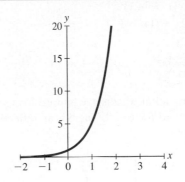

3.

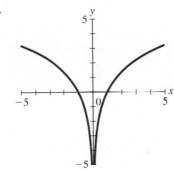

5. **a.** $f(4) = \dfrac{3{,}125}{8}$

   b. $f(3) = \dfrac{100}{3}$

   c. $f(9) = \dfrac{65}{2}$

   d. $f(10) - \dfrac{6}{5}$

7. $x = 25 \ln 4$

9. $x = e^2$

11. $x = \dfrac{41}{2}$

13. $x = 0$

15. $\dfrac{dy}{dx} = xe^{-x}(2 - x)$

17. $\dfrac{dy}{dx} = 2 \ln x + 2$

19. $\dfrac{dy}{dx} = \dfrac{2}{x \ln 3}$

21. $\dfrac{dy}{dx} = e^x$ (Note that $y = e^x$.)

23. $\dfrac{dy}{dx} = \dfrac{-(1 + 2e^{-x})}{1 + e^{-x}}$

ANSWERS

25. $\dfrac{dy}{dx} = \dfrac{-e^{-x}(x^2 + x + 1 + x \ln x)}{x(x + \ln x)^2}$

27. $\dfrac{dy}{dx} = \dfrac{ye^{x-x^2}(2x - 1) + 1}{e^{x-x^2} - 1}$

29. $\dfrac{dy}{dx} = 2y\left[\dfrac{3x + 3e^{2x}}{x^2 + e^{2x}} - 1 - \dfrac{1 - 2x}{3(1 + x - x^2)}\right]$

31. $f(x)$ is increasing for all x; concave upward for $x > 0$; concave downward for $x < 0$. There is an inflection point at $x = 0$.

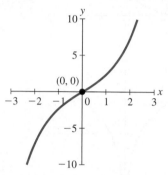

33. $f(t)$ is increasing for $t > 0$; decreasing for $t < 0$; concave upward for all t. There is a minimum at $t = 0$. There are no inflection points.

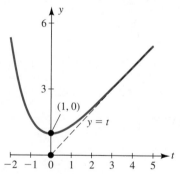

35. $F(u)$ is increasing for $-2 < u < -1$ and $u > -1$; concave upward for $u > -1$; concave downward for $-2 < u < -1$. There is an inflection point at $u = -1$, and $u = -2$ is a vertical asymptote.

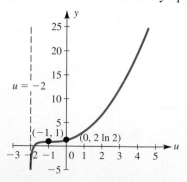

37. $G(x)$ is decreasing for all x. $G(x)$ is concave upward for all x.

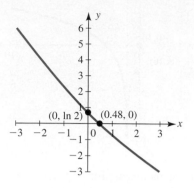

39. $\ln 4$; $\ln 3$

41. $\left(e + \dfrac{1}{e}\right)^5$; 32

43. $y = 2x - 2$

45. $y = 4x$

47. 8

49. The original investment will have quadrupled.

51. 204.8 grams

53. 20,480 bacteria

55. a.

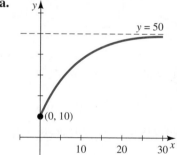

b. 10,000

c. 32,027

d. 9.81 thousand dollars ($9,808.29)

e. Just under 50,000 units

57. a. The third quarter of the 12th year

b. 11.45 years

59. a. $4,975.96

b. $5,488.12

61. 8.20% per year compounded continuously

63. a.

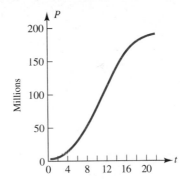

b. 10 million

c. 17.28 million (17,283,507)

d. The population will approach 30 million.

65. a. 0.13 parts per million per year

b. Constant rate of 3%

67. After 200 years. Realistically, Beth should keep the investment as long as possible.

69. a. Since $\lambda = \dfrac{\ln 2}{k}$, we have $k = \dfrac{\ln 2}{\lambda}$ and
$Q(t) = Q_0 e^{-(\ln 2/\lambda)t}$.

b. $Q_0 (0.5)^{kt} = Q_0 e^{-(\ln 2/\lambda)t}$

$kt \ln 0.5 = -\left(\dfrac{\ln 2}{\lambda}\right)t$

So, $k = \dfrac{1}{\lambda}$

71. Bronze age began about 5,000 years ago (3,000 B.C.); maximum percentage is 55%.

73. 0.8110 minutes = 48.66 seconds; $-8.64°C$ per minute

75. a. $A = 5e, k = \dfrac{1}{2}$

b. $t = 9.78$ hours

77. $C = \dfrac{37}{3}, k = 0.021,$

$P(50) \approx 7.525$ billion

79. $10^{-1.6} \approx 0.0251$

81. a. $D(10) = 0.00195; D(25) = 0.000591$

b.

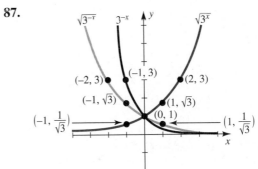

83. a. 2.31×10^{-198} percent left, too little for proper measurement

b. Writing exercise

85. a.

1790	3,867,087
1800	5,256,550
1830	12,956,719
1860	30,207,500
1880	50,071,364
1900	77,142,427
1920	108,425,601
1940	138,370,607
1960	162,289,822
1980	178,782,499
1990	184,566,652
2000	189,034,385

b. The model predicts that the population was increasing most rapidly around 1915.

c. Writing exercise; responses will vary.

87.

89. $x = 1.066$

91.

n	$(\sqrt{n})^{\sqrt{n+1}}$	$(\sqrt{n+1})^{\sqrt{n}}$
8	22.63	22.36
9	32.27	31.62
12	88.21	85.00
20	957.27	904.84
25	3,665	3,447
31	16,528	15,494
37	68,159	63,786
38	85,679	80,166
43	261,578	244,579
50	1,165,565	1,089,362
100	1.12×10^{10}	1.05×10^{10}
1,000	2.87×10^{47}	2.76×10^{47}
	$(\sqrt{n})^{\sqrt{n+1}}$ > $(\sqrt{n+1})^{\sqrt{n}}$	

CHAPTER 5   Section 1

1. $-3x + C$

3. $\dfrac{x^6}{6} + C$

5. $-\dfrac{1}{x} + C$

7. $4\sqrt{t} + C$

9. $\dfrac{5}{3}u^{3/5} + C$

11. $t^3 - \dfrac{2\sqrt{5}}{3}t^{3/2} + 2t + C$

13. $2y^{3/2} + y^{-2} + C$

15. $\dfrac{e^x}{2} + \dfrac{2}{5}x^{5/2} + C$

17. $\dfrac{u^{1.1}}{3.3} - \dfrac{u^{2.1}}{2.1} + C$

19. $x + \ln x^2 - \dfrac{1}{x} + C$

21. $-\dfrac{5}{4}x^4 + \dfrac{11}{3}x^3 - x^2 + C$

23. $\dfrac{2}{7}t^{7/2} - \dfrac{2}{3}t^{3/2} + C$

25. $\dfrac{1}{2}e^{2t} + 2e^t + t + C$

27. $\dfrac{1}{3}\ln|y| - 10\sqrt{y} - 2e^{-y/2} + C$

29. $\dfrac{2}{5}t^{5/2} - \dfrac{2}{3}t^{3/2} + 4t^{1/2} + C$

31. $y = \dfrac{3}{2}x^2 - 2x - \dfrac{3}{2}$

33. $y = \ln x^2 + \dfrac{1}{x} - 2$

35. $f(x) = 2x^2 + x - 1$

37. $f(x) = -\dfrac{1}{3}x^3 - \dfrac{1}{2}x^2 + \dfrac{31}{6}$

39. $f(x) = \dfrac{x^4}{4} + \dfrac{2}{x} + 2x - \dfrac{5}{4}$

41. $f(x) = -e^{-x} + \dfrac{x^3}{3} + 5$

43. $y = 3e^{-2x}$

45. $e^{-y} = 2 - e^x$

47. $22,360$

49. 646.20

51. 3,253

53. **a.** $P(q) = 100q - q^2 - 200$
 b. $q = 50; \$2,300$

55. $c(x) = 0.9x + 0.2x^{3/2} + 10$

57. $986,880$

59. 10,128 people

61. 206,152

63. **a.** $f'(x)$ is maximized when $x = 10$; 7 items per minute
 b. $f(x) = x + 0.6x^2 - 0.02x^3$
 c. $f'(20.8) = 0; f(20.8) \approx 100$ items

65. **a.** $18\dfrac{1}{3}$ (18 items)

 b. $48\dfrac{1}{3}$ (48 items)

67. **a.** $V(t) = 0.15t - 15e^{0.006t} + 45$
 b. 32.5 cm^3; 32.2 cm^3
 c. No, since $V(90) = 32.8$ cm^3

69. $v(r) = \dfrac{1}{2}a(R^2 - r^2)$

71. **a.** $T(t) = 16 - 20e^{-0.35t}$
 b. 6.1°C
 c. 3.44 hours

73. The car travels 199.89 feet before stopping, so the camel gets nudged.

75. a. $v(t) = -23t + 67$; $s(t) = -\dfrac{23}{2}t^2 + 67t$

b.

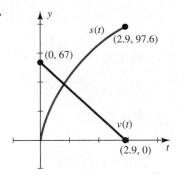

c. $v(t) = 0$ when $t = 2.9$ sec and $s(2.9) = 97.6$ ft; $s(t) = 45$ ft when $t \approx 0.77$ sec or 5.05 sec and $v(0.78) \approx 49.30$ ft/sec while $v(5.05) \approx -49.15$.

77. $\displaystyle\int b^x \, dx = \int e^{x \ln b} \, dx = \dfrac{e^{x \ln b}}{\ln b} + C = \dfrac{b^x}{\ln b} + C$

CHAPTER 5   Section 2

1. a. $u = 3x + 4$

   b. $u = 3 - x$

   c. $u = 2 - t^2$

   d. $u = 2 + t^2$

3. $\dfrac{1}{12}(2x + 6)^6 + C$

5. $\dfrac{1}{6}(4x - 1)^{3/2} + C$

7. $-e^{1-x} + C$

9. $\dfrac{1}{2}e^{x^2} + C$

11. $\dfrac{1}{12}(t^2 + 1)^6 + C$

13. $\dfrac{4}{21}(x^3 + 1)^{7/4} + C$

15. $\dfrac{2}{5}\ln |y^5 + 1| + C$

17. $\dfrac{1}{26}(x^2 + 2x + 5)^{13} + C$

19. $\dfrac{3}{5}\ln |x^5 + 5x^4 + 10x + 12| + C$

21. $-\dfrac{3}{2}\left(\dfrac{1}{u^2 - 2u + 6}\right) + C$

23. $\dfrac{1}{2}(\ln 5x)^2 + C$

25. $\dfrac{-1}{\ln x} + C$

27. $\dfrac{1}{2}[\ln (x^2 + 1)]^2 + C$

29. $\ln |e^x - e^{-x}| + C$

31. $\dfrac{1}{2}x - \dfrac{1}{4}\ln |2x + 1| + C$

33. $\dfrac{1}{10}(2x + 1)^{5/2} - \dfrac{1}{6}(2x + 1)^{3/2} + C$

35. $2 \ln (\sqrt{x} + 1) + C$

37. $y = -\dfrac{1}{6}(3 - 2x)^3 + \dfrac{9}{2}$

39. $y = \ln |x + 1| + 1$

41. $y = \dfrac{1}{2}\ln |x^2 + 4x + 5| - \dfrac{1}{2}\ln 2 + 3$

43. $f(x) = \dfrac{1}{5} - \dfrac{1}{5}(1 - 2x)^{5/2}$

45. $f(x) = \dfrac{3}{2} - \dfrac{1}{2}e^{4-x^2}$

47. $y = 2 + Ce^{1/(x+1)}$

49. $y^2 = 2 + Cx^{-2}$

51. a. $C(q) = (q - 4)^3 + 64 + k$, where k is the overhead

   b. \$1,500

53. a. $R(x) = 50x - 175e^{-0.01x^2} + 175$

   b. \$50,175

55. \$7,120

57. a. $p(x) = \dfrac{300}{\sqrt{x^2 + 9}} + 15$

   b. \$66.45; \$115

   c. 265

59. a. $p(x) = \ln (x + 3) + \dfrac{3}{x + 3} - 0.25$

   b. \$2.55

ANSWERS

61. a. The rate relationship can be expressed as

$$\frac{dV}{dt} = \begin{bmatrix} \text{rate money is} \\ \text{added to account} \end{bmatrix} - \begin{bmatrix} \text{rate money} \\ \text{is withdrawn} \end{bmatrix}$$
$$= rV - W$$

Separating the variables, we find that

$$\frac{dV}{rV - W} = dt$$

The particular solution with $V(0) = S$ is

$$V = \frac{W}{r} + \left(S - \frac{W}{r}\right)e^{rt}$$

b. $175,639

c. $25,000

d. 7.5 years

63. a. $p(t) = -0.01t^2 + 0.06t + 1$

b. $p(4) = 1.08$

c. $p \to -\infty$; The price decreases without bound.

65. a. $p(t) = 2 - e^{-2t/25}$

b. $p(4) = 1.27$

c. $p \to 2$

67. 2.3 meters

69. a. $C(t) = \dfrac{1}{e^{0.01t} + 1}$

b. 0.3543 mg/cm^3; 0.1419 mg/cm^3

c. 294 minutes

71. a. $L(t) = 0.03\sqrt{-t^2 + 16t + 36} + 0.07$; at $t = 8$ (3:00 P.M.); 0.37 parts per million

b. The ozone level at 11:00 A.M. ($t = 4$) is $L(4) = 0.345$. The same level occurs at $t = 12$ (7:00 P.M.).

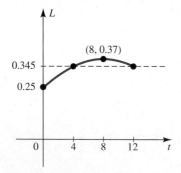

73. a. $x(t) = -\dfrac{4}{9}(3t + 1)^{3/2} + \dfrac{40}{9}$

b. $x(4) = -16.4$

c. $t = 0.4$

75. a. $x(t) = \sqrt{2t + 1} - 1$

b. $x(4) = 2$

c. $t = \dfrac{15}{2}$

77. $e^x + 1 - \ln(e^x + 1) + C$

79. $\dfrac{3}{7}(x^{2/3} + 1)^{7/2} - \dfrac{3}{5}(x^{2/3} + 1)^{5/2} + C$

81. $\displaystyle\int \frac{dx}{1 + e^x} = \int \frac{e^{-x}\,dx}{e^{-x} + 1}$

$$= \int \frac{-1}{u}\,du = -\ln(e^{-x} + 1) + C$$

(Substitute $u = e^{-x} + 1$ and $du = -e^{-x}\,dx$)

CHAPTER 5   Section 3

1. 4

3. 9; 8

5. 2.148; 2.333

7. 0.725; 0.693

9. 7; 8

11. 21.500; 25.333

13. 1.429; 1.609

15. 15

17. $\dfrac{95}{2}$

19. $\dfrac{6}{5}$

21. $-\dfrac{6}{5}$

23. $3 - \dfrac{4}{e}$

25. 1.95

27. 144

29. $\dfrac{8}{3} + \ln 3 \approx 3.7653$

31. $\dfrac{2}{9}$

33. 3.2

35. $\dfrac{4}{3}$

37. $\dfrac{7}{6}$

39. e

41. $\dfrac{8}{3}$

43. $e^3 - e^2$

45. -20

47. 0

49. 3

51. $\dfrac{33}{5}$

53. $\dfrac{112}{9}$

55. 4

57. $\dfrac{3}{2}\ln 3 \approx 1.6479$

59. $V(5) - V(0)$

61. \$480

63. \$75

65. A decrease of \$1,870

67. $1{,}500\left(\dfrac{3}{2} + \dfrac{5}{4}\ln\dfrac{11}{9}\right) \approx 2{,}626$ telephones

69. **a.** $-\$48{,}036.33$

     b. \$28,546.52

71. 0.75 ppm

73. About 98 people

75. $2\ln 2 \approx 1.386$ grams

77. $8\sqrt{11} - 8\sqrt{6} \approx 7$ facts

79. The concentration decreases by 0.8283 mg/cm³.

81. 96 ft

83. **a.** $\dfrac{\pi}{4}$

     b. $\dfrac{\pi}{4}$; part of the area inside the circle
     $(x - 1)^2 + y^2 = 1$

CHAPTER 5   Section 4

1. $\dfrac{5}{12}$

3. $2\ln 2 - \dfrac{1}{2}$

5. Area = 1

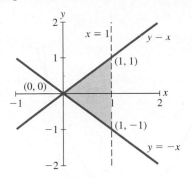

7. Area = $\dfrac{4}{3}$

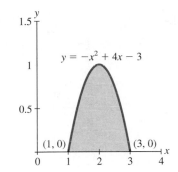

9. Area = $\dfrac{4}{3}$

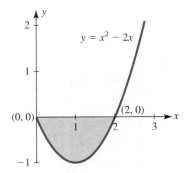

11. Area = 9

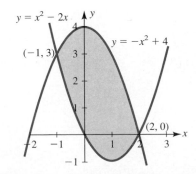

ANSWERS

13. Area $= \dfrac{443}{6}$

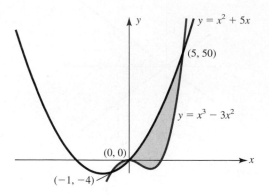

15. Area $= 18$

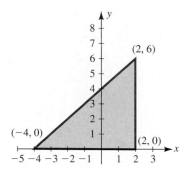

17. Area $= 14$

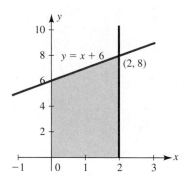

19. -2

21. $\dfrac{3}{2}\left(e - \dfrac{1}{e}\right)$

23. $\dfrac{\ln 5 - \ln 3}{\ln 3}$

25. Average value $= \dfrac{2}{3}$

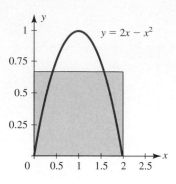

27. Average value $= \dfrac{\ln 2}{2}$

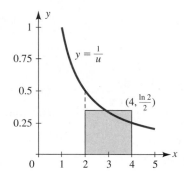

29. 0.5

31. 0.1833

33. 0.383

35. 2,400 units

37. 30,000 kg

39. **a.** $11,361.02

   b. Writing exercise; response will vary.

41. **a.** 16 years

   b. $209,067

   c.

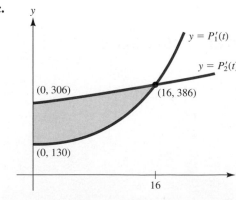

43. a. 14.7 years

b. \$582,221

c.

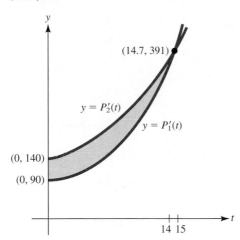

45. a. 65,244 units

b. 1,491 worker-hours

47. \$88,480

49. \$241,223.76

51. Baseball: $\frac{1}{3}$; football: $\frac{5}{18}$; basketball: $\frac{9}{25}$. Football is the most equitable sport, basketball the least.

53. 2,272.2

55. 5,710 people

57. a. $M_0 + 20.833$

b. $t = \sqrt{5}$, $M(\sqrt{5}) = M_0 + 50\sqrt{5}/e$

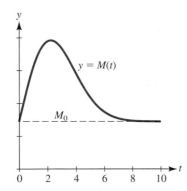

59. $\frac{1}{40}$ mg/cm^3

61. a. $S' = F''(M) = \frac{1}{3}(2k - 6M) = 0$ for $M = \frac{k}{3}$. Maximum since $S'' = -2 < 0$.

b. $\frac{k^3}{108}$

63. a. 0°C

b. 8 A.M. and 2 P.M.

65.

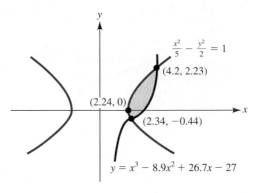

$$A = \int_{\sqrt{5}}^{2.34}\left[\sqrt{\frac{2x^2}{5} - 2} - \left(-\sqrt{\frac{2x^2}{5} - 2}\right)\right]dx$$
$$+ \int_{2.34}^{4.2}\left[\sqrt{\frac{2x^2}{5} - 2} - (x^3 - 8.9x^2 + 26.7x - 27)\right]dx$$
$$\approx 2.097$$

67. Writing exercise; response will vary.

CHAPTER 5   Section 5

1. a. \$624

b.

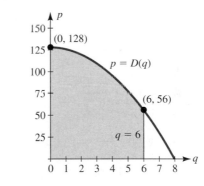

3. a. $1,600 \ln 2 \approx \$1,109.04$

b.

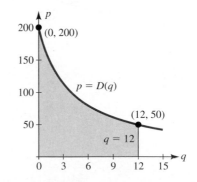

5. a. $800\left(1 - \dfrac{1}{\sqrt{e}}\right) \approx \$314.78$

b.

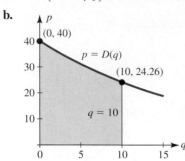

7. $p_0 = \$110;\ CS = \$36$

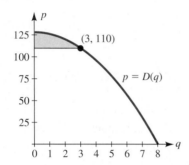

9. $p_0 = \$31.15;\ CS = \$21.21$

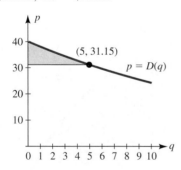

11. $p_0 = \$34.80;\ PS = \$12.80$

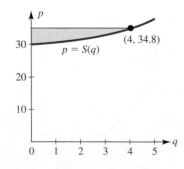

13. $p_0 = \$26.41;\ PS = \$2.14$

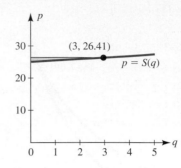

15. a. $104

   b. $CS = \$162,\ PS = \$324$

17. a. $40

   b. $CS = \$200,\ PS = \$116.67$

19. a. $1

   b. $CS = \$3.09,\ PS = \$0.67$

21. a. 11 years

   b. $26,620

   c.

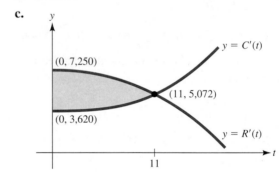

23. a. 8

   b. $15,069

   c. Net earnings are represented as the area between the curve $R'(t) = 6{,}537e^{-0.3t}$ and the horizontal line $y = 593$.

25. $17,182.82

27. $237,730; $319,453

29. $5,308.78

31. The $50,000 plan is better, producing net income of $37,465 versus $22,479 for the $30,000 plan over 5 years.

33. a. $P(q) = -q^3 + 24q^2 + 108q - 3{,}000$

   b. 18 units

   c. $162

35. **a.** $P(t) = 32.5e^{0.04t} - 32.5$; 4.14 billion barrels; 4.67 billion barrels

    b. 12 years

    c. 1,646.44 billion dollars

    d. Answers will vary.

37. **a.** $P(t) = 60e^{0.02t} - 60$; 3.71 billion barrels; 3.94 billion barrels

    b. 9.12 years

    c. 1,218 billion dollars

    d. Answers will vary.

39. $1,929,148

41. **a.** $137,334.29

    b. $44,585,04

43. **a.** $1,287,360

    b. Answers will vary.

45. **a.** $\text{FV} = \int_0^T f(t)e^{r(T-t)}\,dt$

$$= \int_0^T Me^{r(T-t)}\,dt$$

$$= Me^{rT}\int_0^T e^{-rt}\,dt$$

$$= Me^{rT}\left(\frac{1}{r} - \frac{e^{-rT}}{r}\right)$$

$$= \frac{M}{r}(e^{rT} - 1)$$

    b. $\text{PV} = \int_0^T f(t)e^{-rt}\,dt$

$$= \int_0^T Me^{-rt}\,dt$$

$$= M\int_0^T e^{-rt}\,dt$$

$$= M\left(\frac{1}{r} - \frac{e^{-rT}}{r}\right)$$

$$= \frac{M}{r}(1 - e^{-rT})$$

CHAPTER 5   Section 6

1. 30,484

3. 468,130

5. 451,404

7. $7\pi \approx 21.99$ cubic units

9. $\dfrac{1,532}{15}\pi \approx 320.86$ cubic units

11. $\dfrac{32}{3}\pi \approx 33.51$ cubic units

13. $2\pi \approx 6.28$ cubic units

15. 61,070,138

17. About 80 members

19. 4,097.62 (4,098 people)

21. 515.48 billion barrels

23. 4,207 members

25. As computed in the example in the text, the total quantity of blood flowing through the artery per second is

$$\int_0^R 2\pi kr(R^2 - r^2)\,dr = \frac{\pi kR^4}{2}$$

The area of the artery is πR^2, and the average velocity of the blood through the artery is

$$V_{\text{ave}} = \frac{\pi kR^4/2}{\pi R^2} = \frac{kR^2}{2}$$

The maximum speed for the blood occurs at $r = 0$, so $S(0) = kR^2$. Thus the average velocity is one-half the maximum speed.

27. Approximately 208,128 people

29. **a.** The LDL level decreases 6.16 units.

    b. $L(t) = \dfrac{3}{28}(49 - t^2)^{1.4} + 120 - \dfrac{21}{4}(49)^{0.4}$

    c. 5.8 days

31. 1,565.83 (1,566 animals)

33. 10,125 people

35. $\int_0^{12}[W'(t) - D'(t)]\,dt = 0.363$; about 36 people; 18.1%

37. **a.** 55 years; 74.7 years

    b. 70.78 years

    c. 86.36 years; they have already exceeded their life expectancy.

    d. 71.69 years

39. **a.** 2.37 sec

    b. 0.905 L

    c. 0.382 L/sec

41. a.

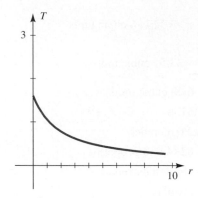

b. $r(T) = \dfrac{3}{T} - 2$

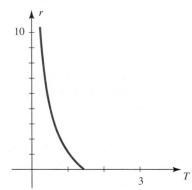

c. $\pi\left[12\left(\ln\dfrac{1}{3} - \ln\dfrac{3}{2}\right) + \dfrac{77}{3}\right] \approx 23.93 \text{ ft}^3$

43. a. $100\pi \ln \dfrac{23}{5} \approx 479.42$ units

b. $L = \dfrac{3\sqrt{10}}{2} \approx 4.74$ miles; $100\pi \ln 10 \approx 723.38$ units

45. $V = \displaystyle\int_0^3 \pi[\sqrt{x}(3-x)]^2\, dx$

≈ 21.21 cubic centimeters

47. After T years, the first population is

$$P_1(T) = \left(100{,}000 - \dfrac{50}{0.011}\right)e^{-0.011T} + \dfrac{50}{0.011}$$

Then $P_1(T) < P(T)$ for $T = 50$ and $T = 100$, but $P_1(300) > P(300)$.

49. The hypotenuse of the triangle has equation $y = \dfrac{r}{h}x$, and the volume is

$$V = \pi\int_0^h \left(\dfrac{r}{h}x\right)^2 dx = \pi\int_0^h \dfrac{r^2}{h^2}x^2\, dx$$

$$= \dfrac{\pi r^2}{h^2}\left(\dfrac{1}{3}x^3\right)\Bigg|_0^h = \dfrac{\pi r^2}{3h^2}(h^3 - 0)$$

$$= \dfrac{1}{3}\pi r^2 h$$

CHAPTER 5 Checkup

1. a. $\dfrac{x^4}{4} - \dfrac{2\sqrt{3}}{3}x^{3/2} - \dfrac{5}{2}e^{-2x} + C$

b. $\dfrac{x^2}{2} - 2x + 4\ln|x| + C$

c. $\dfrac{2}{7}x^{7/2} - 2x^{1/2} + C$

d. $\dfrac{-1}{2\sqrt{3 + 2x^2}} + C$

e. $\dfrac{1}{4}(\ln x)^2 + C$

f. $\dfrac{1}{2}e^{1+x^2} + C$

2. a. $\dfrac{62}{5} + 4\ln 2$

b. $e^3 - 1$

c. $1 - \ln 2$

d. $\sqrt{31} - 2$

3. a. $\dfrac{73}{6}$

b. 36

4. $1 - 2\ln 2$

5. \$10,333.33

6. 71.14 billion dollars; increase

7. \$4,266.67

8. \$16,183.42

9. 45,055

10. 0.1 mg/cm^3

CHAPTER 5 Review Exercises

1. $\dfrac{1}{4}x^4 + \dfrac{2}{3}x^{3/2} - 9x + C$

3. $\dfrac{x^5}{5} + \dfrac{5}{2}e^{-2x} + C$

5. $\dfrac{5}{3}x^3 - 3\ln|x| + C$

7. $\dfrac{1}{6}t^6 - t^3 - \dfrac{1}{t} + C$

9. $\dfrac{2}{9}(3x + 1)^{3/2} + C$

11. $\dfrac{1}{12}(x^2 + 4x + 2)^6 + C$

13. $\dfrac{-3}{4(2x^2 + 8x + 3)} + C$

15. $\dfrac{1}{14}(v - 5)^{14} + \dfrac{5}{13}(v - 5)^{13} + C$

17. $-\dfrac{5}{2}e^{-x^2} + C$

19. $\dfrac{2}{3}(\ln x)^{3/2} + C$

21. 0

23. $\dfrac{1}{2}(e^2 + 5)$

25. $1{,}710$

27. $1 - \dfrac{1}{e}$

29. $e - 2$

31. Area $= \dfrac{101}{6}$

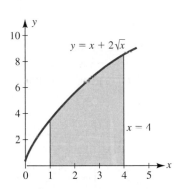

33. Area $= \ln 2 + \dfrac{7}{3}$

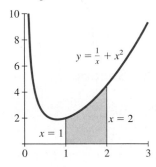

35. Area $= \dfrac{15}{2} - 8 \ln 2$

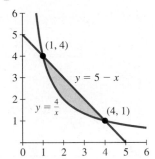

37. Area $= \dfrac{9}{2}$

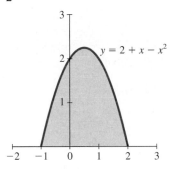

39. $\dfrac{11{,}407}{84} - \dfrac{2\sqrt{2}}{21} \approx 135.7$

41. $\dfrac{1}{4}\left(1 - \dfrac{1}{e^4}\right) \approx 0.245$

43. \$128; \$21.33

45. \$6.70; \$6.16

47. GI $= \dfrac{1}{5}$

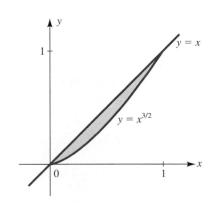

49. $GI = \dfrac{1}{10}$

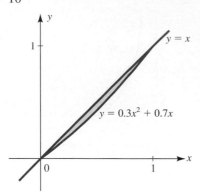

b. $p_2(x) = 0.3x + 0.001x^3 + 250$; $p_2(10) = \$2.54$ per dozen

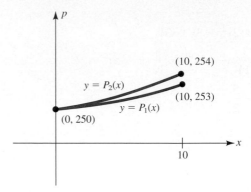

51. 43,984

53. 14,308

55. $\dfrac{78}{5}\pi \approx 49.01$ cubic units

57. $\pi \ln 3 \approx 3.45$ cubic units

59. $y = 2x + 10$

61. $x = \dfrac{9}{2} - \dfrac{1}{2}e^{-2t}$

63. $y = \dfrac{1}{2}\ln(x^2 + 1) + 5 - \dfrac{1}{2}\ln 2$

65. $87.57

67. 1,220 people

69. 11,250 commuters

71. In 2013 (0.2554 billion barrels versus 0.1682 billion barrels in 2014)

73. $7,377.37

75. 61.65 (about 62 homes)

77. 14,860 pounds

79. $3,447,360

81. $6.32 per pound

83. Temperature decreases by 2.88°C

85. **a.** $p_1(x) = 0.2x + 0.001x^3 + 250$; $p_1(10) = \$2.53$ per dozen

87. 30 meters

89. Physical therapists

91. 2,255 trout

93. **a.** $\dfrac{1}{N}\displaystyle\int_0^N S(t)\,dt$

b. $\displaystyle\int_0^N S(t)\,dt$

c. The average speed is equal to the total distance divided by the total number of hours.

95. The region bounded by the curves is between $x = -4.66$ and $x = -1.82$; the curves also intersect at $x = 4.98$. The area is approximately 3.

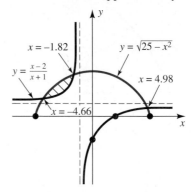

CHAPTER 6   Section 1

1. $-(x + 1)e^{-x} + C$

3. $(2 - x)e^x + C$

5. $\dfrac{1}{2}t^2\left(\ln 2t - \dfrac{1}{2}\right) + C$

7. $-5(v + 5)e^{-v/5} + C$

9. $\dfrac{2}{3}x(x-6)^{3/2} - \dfrac{4}{15}(x-6)^{5/2} + C$

11. $\dfrac{1}{9}x(x+1)^9 - \dfrac{1}{90}(x+1)^{10} + C$

13. $2x\sqrt{x+2} - \dfrac{4}{3}(x+2)^{3/2} + C$

15. $\dfrac{8}{3}$

17. $\dfrac{1}{4}(1-3e^{-2})$

19. $\dfrac{1}{12}(3e^4+1)$

21. $\dfrac{1}{16}(e^2+1)$

23. $-\dfrac{1}{x}(\ln x + 1) + C$

25. $\dfrac{1}{2}e^{x^2}(x^2-1) + C$

27. $\dfrac{1}{25}(3 - 5x - 3\ln|3-5x|) + C$

29. $\dfrac{-\sqrt{4x^2-9}}{x} + 2\ln|2x + \sqrt{4x^2-9}| + C$

31. $\dfrac{1}{2}\ln\left|\dfrac{x}{2+3x}\right| + C$

33. $\dfrac{\sqrt{3}}{24}\ln\left|\dfrac{4+\sqrt{3}u}{4-\sqrt{3}u}\right| + C$

35. $x(\ln x)^3 - 3x(\ln x)^2 + 6x\ln x - 6x + C$

37. $-\dfrac{1}{25}\left[\dfrac{5+4x}{x(5+2x)} + \dfrac{4}{5}\ln\left|\dfrac{x}{5+2x}\right|\right] + C$

39. $y = \left(-\dfrac{x}{2} - \dfrac{1}{4}\right)e^{-2x} + \dfrac{1}{4}$

41. $\ln y = \dfrac{2}{3}(x+1)^{3/2} - 2(x+1)^{1/2} + \dfrac{4}{3}$

43. $f(x) = 5 + \dfrac{3}{e} - \dfrac{x+2}{e^x}$

45. 176 units

47. $62{,}000e^{1/2} - 63{,}000 \approx \$39{,}220.72$

49. \$11,417

51. **a.** \$4.47 per unit
   b. \$14,284.57

53. $1 - \dfrac{2}{e} \approx 0.2642$

55. \$34,555

57. 2,008,876

59. $\dfrac{40}{27}(5e^2 - 14e^{1/5}) \approx 29.4$ mg/mL

61. 4,367

63. $N(t) = \dfrac{60}{1 + 29e^{-0.42t}}$

$N(t) = 20$ when $t = 6.37 < 7$. He's a dead duck before finishing his mission.

65. $(0.244, 0.353)$

67. **a.** Locate the kiosk at the center $(\bar{x}, \bar{y})$ of the parking lot. The curved boundary of the lot has the equation $2x^2 - y^2 = 1$ or, equivalently, $y = \sqrt{2x^2 - 1}$, so the lot has area

$$A = \int_1^5 \sqrt{2x^2 - 1}\, dx = 16.38$$

(Use integral formula 18 in Table 6.1.) Therefore, we have

$$\bar{x} = \dfrac{1}{A}\int_1^5 x\sqrt{2x^2-1}\, dx = \dfrac{57}{16.38} = 3.48$$

and

$$\bar{y} = \dfrac{1}{2A}\int_1^5 (\sqrt{2x^2-1})^2\, dx = \dfrac{236/3}{2(16.38)} = 2.40$$

The kiosk should be located at $(3.48, 2.40)$.

   b. Writing exercise.

69. Let $\quad U = u^n \qquad\qquad dV = e^{au}\, du$

$\qquad\qquad dU = nu^{n-1}\, du \qquad V = \dfrac{1}{a}e^{au}$

Then

$$\int u^n e^{au}\, du$$

$$= u^n\left(\dfrac{1}{a}e^{au}\right) - \int \dfrac{1}{a}e^{au}(nu^{n-1}\, du)$$

$$= \dfrac{1}{a}u^n e^{au} - \dfrac{n}{a}\int u^{n-1}e^{au}\, du$$

71. Integration by parts:

$u = e^{kx}, dv = x^{-n}\, dx$

$$\int x^{-n}e^{kx}\, dx = \dfrac{1}{n-1}\left[-x^{-n+1}e^{kx} + k\int x^{-n+1}e^{kx}\, dx\right]$$

73. Area ≈ 0.75834

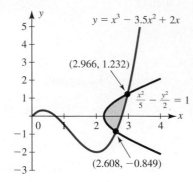

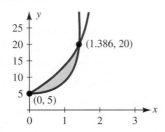

75. Area ≈ 1.95482

77. 4.2265

79. 0.4509

CHAPTER 6   Section 2

1. a. 2.343750
 b. $\dfrac{7}{3}$

3. a. 0.782794
 b. 0.785392

5. a. 1.151479
 b. 1.147782

7. a. 0.742984
 b. 0.746855

9. a. 1.930756
 b. 1.922752

11. a. 1.096997
 b. 1.094800

13. a. 0.849195
 b. 0.836203

15. a. $0.508994; |E_n| \le 0.031250$
 b. $0.500418; |E_n| \le 0.002604$

17. a. $2.796731; |E_n| \le 0.001667$
 b. $2.797432; |E_n| \le 0.000017$

19. a. $1.490679; |E_n| \le 0.084946$
 b. $1.463711; |E_n| \le 0.004483$

21. a. 164
 b. 18

23. a. 36
 b. 6

25. a. 179
 b. 8

27. a. 3.0898
 b. 3.1212

29. 0.138569

31. 0.358531 cubic units

33. $26,072.45

35. $586,653

37. $5,950

39. $34,200

41. GI ≈ 0.39425

43. 3,496 people

45. 230 ft^2

47. 475,197

49. 320 ft^2

CHAPTER 6   Section 3

1. $\dfrac{1}{2}$

3. Diverges

5. Diverges

7. $\dfrac{1}{10}$

9. $\dfrac{5}{2}$

11. $\dfrac{1}{9}$

13. Diverges

15. $\dfrac{2}{e} = 2e^{-1}$

17. $\dfrac{2}{9}$

19. Diverges

21. Diverges

23. 2

25. $\dfrac{3}{4}$

27. 1

29. Diverges

31. $60,000

33. $2,360,000

35. a. Cost of M_1 = $28,518.50
   Cost of M_2 = $20,222.20
   The company should purchase M_2.
   b. Writing assignment

37. $\text{PV} = \int_0^\infty Qe^{-rt}\, dt$

$$= \lim_{R\to\infty}\int_0^R Qe^{-rt}\, dt$$

$$= \lim_{R\to\infty}\left[-\frac{Q}{r}e^{-rt}\right]_0^R$$

$$-\lim_{R\to\infty}-\frac{Q}{r}\left[e^{-rR}-1\right]$$

$$=\frac{Q}{r}$$

39. 200 patients

41. 50 units

43. $C = \dfrac{ab}{b-a}$

45. 7,853,982 people

CHAPTER 6   Checkup

1. a. $\dfrac{4\sqrt{2}}{9}x^{3/2}(-2 + 3\ln|x|) + C$

   b. $25 - 20e^{1/5}$

   c. $-\dfrac{298}{15}$

   d. $-xe^{-x} + C$

2. a. 10

   b. $\dfrac{3}{4}e^{-2}$

   c. Diverges

$$\lim_{N\to\infty}\int_1^N \frac{x}{(x+1)^2}\, dx$$

$$= \lim_{N\to+\infty}\left(\frac{1}{x+1} + \ln|x+1|\right)\Big|_1^N$$

$$= \lim_{N\to+\infty}\left(\frac{1}{N+1} + \ln|N+1| - \frac{1}{2} - \ln 2\right) = \infty$$

   d. 0

3. a. $\dfrac{x}{4}\left[(\ln|3x|)^2 - 2\ln|3x| + 2\right] + C$

   b. $-\dfrac{1}{2}\ln\left|\dfrac{\sqrt{4+x^2}+2}{x}\right| + C$

   c. $\dfrac{\sqrt{x^2-9}}{9x} + C$

   d. $-\dfrac{1}{4}\ln\left|\dfrac{x}{3x-4}\right| + C$

4. The waste will increase without bound.

5. Approximately $1,666,666.67

6. 3.5 mg

7. 16,000 units

8. By the trapezoidal rule, $\displaystyle\int_3^4 \frac{\sqrt{25-x^2}}{x}\, dx \approx 1.0276$.

   The exact answer is $-1 + 5\ln\dfrac{3}{2} \approx 1.0273$.

   The error of the approximation is ≈ 0.0003.

9. a. $N(t) = \dfrac{2{,}000e^{0.4184t}}{39 + e^{0.4184t}}$

   b. 648 people

   c. 8.76 days $\approx$ 9 days

CHAPTER 6   Review Exercises

1. $-(1+t)e^{1-t} + C$

3. $\dfrac{x}{3}(2x+3)^{3/2} - \dfrac{1}{15}(2x+3)^{5/2} + C$

5. $-2 + 4\ln 2$

7. $\dfrac{74}{7}$

9. $\dfrac{x^2}{9}(3x^2+2)^{3/2} - \dfrac{2}{135}(3x^2+2)^{5/2} + C$

11. $\dfrac{5}{8}\ln\left|\dfrac{2+x}{2-x}\right| + C$

13. $-3(18 + 6w + w^2)e^{-w/3} + C$

15. $x[-6 + 6\ln 2x - 3(\ln 2x)^2 + (\ln 2x)^3] + C$

17. $y = \dfrac{1}{2}x^2\ln\sqrt{x} - \dfrac{1}{8}x^2 + \dfrac{25}{8}$

19. $(y+1)e^{-y} = 1 - \ln|x|$

21. Diverges

$$\lim_{N\to\infty}\int_0^N \frac{1}{\sqrt[3]{1+2x}}\, dx = \lim_{N\to\infty}\left[-\frac{3}{4} + \frac{3(1+2N)^{2/3}}{4}\right]$$

$$= \infty$$

23. Diverges

$$\lim_{N\to\infty} \int_0^N \frac{3t}{t^2 + 1}\, dt = \lim_{N\to\infty} \frac{3}{2} \ln (1 + N^2) = \infty$$

25. $\dfrac{1}{4}$

27. $\dfrac{1}{4}$

29. Diverges

$$\lim_{N\to\infty} \int_1^N \frac{\ln x}{\sqrt{x}}\, dx = \lim_{N\to\infty} 2(2 - 2\sqrt{N} + \sqrt{N} \ln N) = \infty$$

31. -2

33. 0

35. \$320,000

37. The population increases without bound.

39. 15,000 lb

41. **a.** 1.1016   with $|E| \le \dfrac{1}{75}$

   b. 1.0987   with $|E| \le \dfrac{4}{9{,}375}$

43. **a.** 3.0607   with $|E| \le \dfrac{e}{1{,}200}$

   b. 3.0591   with $|E| \le \dfrac{e}{200{,}000}$

45. **a.** 58

   b. 8

47. **a.** $\displaystyle\int_0^8 \sqrt{q}\, e^{0.01q}\, dq$

   b. 15.6405 (\$15.64)

49. **a.**

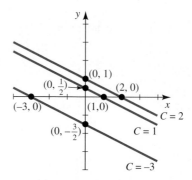

$y = x \ln x$

$(1, 0)$

$y = -x^3 - 2x^2 + 5x - 2$

$(0.406, -0.366)$

   b. 0.1692

51. $\dfrac{2}{3} \ln 2 \approx 0.4621$

53. $I(1) \approx 0.4214$
   $I(10) \approx 0.5$
   $I(50) \approx 0.5$

   $I(N)$ converges to $\dfrac{1}{2}$ as $N \to \infty$.

CHAPTER 7   Section 1

1. $f(-1, 2) = 1; f(3, 0) = 15$

3. $g(1, 1) = 0; g(-1, 4) = -5$

5. $f(2, -1) = -3, f(1, 2) = 16$

7. $g(4, 5) = 3, g(-1, 2) = \sqrt{3} \approx 1.7321$

9. $f(e^2, 3) = \dfrac{3}{2}; f(\ln 9, e^3) = 25.515$

11. $g(1, 2) = 2.5; g(2, -3) = -\dfrac{13}{6} \approx -2.167$

13. $f(1, 2, 3) = 6; f(3, 2, 1) = 6$

15. $F(1, 1, 1) = \dfrac{\ln 2}{3} \approx 0.2310; F(0, e^2, 3e^2) \approx 0.1048$

17. All ordered pairs (x, y) of real numbers for which
   $y \ne \dfrac{-4}{3} x$

19. All ordered pairs (x, y) of real numbers for which
   $y \le x^2$

21. All ordered pairs (x, y) of real numbers for which
   $x > 4 - y$

23.

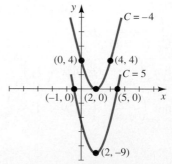

$(0, 1)$
$\left(0, \frac{1}{2}\right)$
$(2, 0)$
$(-3, 0)$
$(1, 0)$
$C = 2$
$\left(0, -\frac{3}{2}\right)$
$C = 1$
$C = -3$

25.

$C = -4$
$(0, 4)$
$(4, 4)$
$C = 5$
$(-1, 0)$
$(2, 0)$
$(5, 0)$
$(2, -9)$

27.

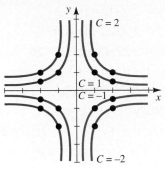

29.

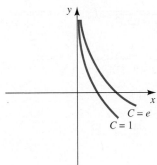

31. a. 160,000 units
   b. Production will increase by 16,400 units.
   c. Production will increase by 4,000 units.
   d. Production will increase by 20,810 units.

33. a. $R(x_1, x_2) - 200x_1 - 10x_1^2 + 25x_1x_2 + 100x_2 - 10x_2^2$
   b. \$7,230

35. a. If $a + b > 1$, production is more than doubled.
   b. If $a + b < 1$, production is increased (but not doubled).
   c. If $a + b = 1$, production is doubled.

37. $R(x, y) = 60x - \dfrac{x^2}{5} + \dfrac{xy}{10} + 50y - \dfrac{y^2}{10}$

39. a. 70 units

   b. $y = -\dfrac{3}{2}x + 35$

   c.

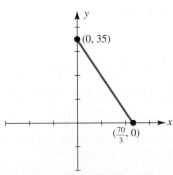

   d. Unskilled labor should be decreased by three workers.

41. 260

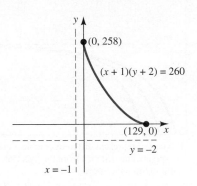

43. a. \$2,003.13; \$110,563.40
   b. \$1,435.20; \$266,672

45. For $Q(K, L) = A[\alpha K^{-\beta} + (1 - \alpha)L^{-\beta}]^{-1/\beta}$,
$$\begin{aligned}
Q(sK, sL) &= A[\alpha(sK)^{-\beta} + (1 - \alpha)(sL)^{-\beta}]^{-1/\beta} \\
&= A[\alpha s^{-\beta}K^{-\beta} + (1 - \alpha)s^{-\beta}L^{-\beta}]^{-1/\beta} \\
&= A[s^{-\beta}\{\alpha K^{-\beta} + (1 - \alpha)L^{-\beta}\}]^{-1/\beta} \\
&= A[s^{-\beta}]^{-1/\beta}[\alpha K^{-\beta} + (1 - \alpha)L^{-\beta}]^{-1/\beta} \\
&= sA[\alpha K^{-\beta} + (1 - \alpha)L^{-\beta}]^{-1/\beta} \\
&= sQ(K, L)
\end{aligned}$$

47. a. $S(15.83, 87.11) = 0.5938$

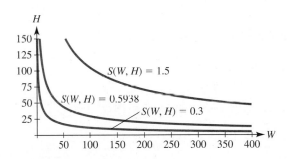

The curves represent different combinations of height and weight that result in the same surface area.
   b. Height = 90.05 cm
   c. 254%

49. a. 2,105.03 kilocalories
   b. 1,428.84 kilocalories
   c. Approximately 27 years
   d. Approximately 24.4 years

51. a. 0.866 cm/sec

b.

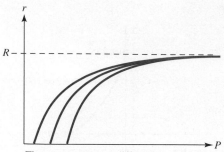

The curves represent different combinations of pressure and distance from the axis that result in the same speed.

53. 23.54

55. a.

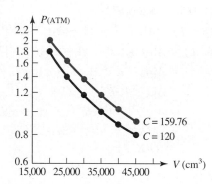

b. 159.76 °C

CHAPTER 7   Section 2

1. $f_x = 7; f_y = -3$

3. $f_x = 12x^2 - 6xy + 5; f_y = -3x^2$

5. $f_x = 2y^5 + 6xy + 2x; f_y = 10xy^4 + 3x^2$

7. $\dfrac{\partial z}{\partial x} = 15(3x + 2y)^4; \dfrac{\partial z}{\partial y} = 10(3x + 2y)^4$

9. $f_s = -\dfrac{3t}{2s^2}; f_t = \dfrac{3}{2s}$

11. $\dfrac{\partial z}{\partial x} = (xy + 1)e^{xy}; \dfrac{\partial z}{\partial y} = x^2 e^{xy}$

13. $f_x = \dfrac{-e^{2-x}}{y^2}; f_y = \dfrac{-2e^{2-x}}{y^3}$

15. $f_x = \dfrac{5y}{(y - x)^2}; f_y = \dfrac{-5x}{(y - x)^2}$

17. $\dfrac{\partial z}{\partial u} = \ln v; \dfrac{\partial z}{\partial v} = \dfrac{u}{v}$

19. $f_x = \dfrac{1}{y^2(x + 2y)}; f_y = \dfrac{2[y - (x + 2y)\ln(x + 2y)]}{y^3(x + 2y)}$

21. $f_x(1, -1) = 2; f_y(1, -1) = 3$

23. $f_x(0, -1) = 2; f_y(0, -1) = 0$

25. $f_x(-2, 1) = -22; f_y(-2, 1) = 26$

27. $f_x(0, 0) = 1; f_y(0, 0) = 1$

29. $f_{xx} = 60x^2y^3; f_{xy} = 2(30x^3y^2 + 1);$
$f_{yx} = 2(30x^3y^2 + 1); f_{yy} = 30x^4y$

31. $f_{xx} = 2y(2x^2y + 1)e^{x^2y}; f_{xy} = 2x(x^2y + 1)e^{x^2y};$
$f_{yx} = 2x(x^2y + 1)e^{x^2y}; f_{yy} = x^4 e^{x^2y}$

33. $f_{ss} = \dfrac{t^2}{\sqrt{(s^2 + t^2)^3}}; f_{st} = \dfrac{-st}{\sqrt{(s^2 + t^2)^3}};$
$f_{ts} = \dfrac{-st}{\sqrt{(s^2 + t^2)^3}}; f_{tt} = \dfrac{s^2}{\sqrt{(s^2 + t^2)^3}}$

35. $\dfrac{dz}{dt} = 4t + 15$

37. $\dfrac{dz}{dt} = \dfrac{3}{y} - \dfrac{6xt}{y^2} = \dfrac{-3}{t^2}$

39. $\dfrac{dz}{dt} = 2ye^{2t} - 3xe^{-3t} = -e^{-t}$

41. Substitute

43. Neither

45. Substitute

47. Daily output will increase by approximately 10 units.

49. a. $Q_K = 60K^{-3/2}[0.4K^{-1/2} + 0.6L^{-1/2}]^{-3}$
$Q_L = 90L^{-3/2}[0.4K^{-1/2} + 0.6L^{-1/2}]^{-3}$

   b. If $K = 5,041$ and $L = 4,900$,
   $Q_K \approx 58.48$ and $Q_L \approx 91.54$.

   c. Labor

51. The monthly demand for bicycles decreases by approximately 4 (actually, 3.84).

53. a. An increase in x will decrease the demand $D(x, y)$ for the first brand of mower. An increase in y will increase the demand $D(x, y)$ for the first brand of mower.

   b. $\dfrac{\partial D}{\partial x} < 0, \dfrac{\partial D}{\partial y} > 0$

   c. $b < 0, c > 0$

55. a. $\dfrac{\partial^2 Q}{\partial L^2} < 0$; for a fixed level of capital investment,

   the effect on output of the addition of 1 worker-hour is greater when the workforce is small than when it is large.

b. $\dfrac{\partial^2 Q}{\partial K^2} < 0$; for a fixed workforce, the effect on output of the addition of \$1,000 in capital investment is greater when the capital investment is small than when it is large.

57. Substitute commodities; pencils

59. **a.** $Q(37, 71) = 304{,}691$; $Q(38, 71) = 317{,}310$;
$Q(37, 72) = 309{,}031$

    b. $Q_x(37, 71) = 12{,}534$ units;
$Q(38, 71) - Q(37, 71) = 12{,}619$ units

    c. $Q_y(37, 71) = 4{,}344$ units;
$Q(37, 72) - Q(37, 71) = 4{,}340$ units

61. The number of units produced decreases by about 55.

63. **a.** 424 units/month

    b. 16.31

65. The number of units produced increases by about 61.6 units per day.

67. **a.** $P_x = -8x + 10y - 10$
$P_y = -14y + 10x + 185$

    b. For $x = 70$ and $y = 73$,
$P_x = 160$ and $P_y = -137$

    c. Decreases profit by 114 cents.

    d. Increases profit by 457 cents.

69. **a.** $F(3.17, 0.085) = 60{,}727.24k$;

$\dfrac{\partial F}{\partial L} = \dfrac{k}{r^4} = 19{,}156.86k$;

$\dfrac{\partial F}{\partial r} = -\dfrac{4kL}{r^5} - -2{,}857{,}752.58k$

    b. $F(1.2L, 0.8r) = \dfrac{k(1.2L)}{(0.8r)^4} = 2.93F(L, r)$;

$\dfrac{\partial F}{\partial L}(1.2L, 0.8r) = 2.44\dfrac{\partial F}{\partial L}(L, r)$;

$\dfrac{\partial F}{\partial r}(1.2L, 0.8r) = 3.66\dfrac{\partial F}{\partial r}(L, r)$

71. $\dfrac{\partial F}{\partial z} = \dfrac{-c\pi x^2}{8\sqrt{y - z}}$; decreasing since $F_{zz} < 0$

73. Yes

75. No

77. The volume is increased by 72π cm^3.

79. $\dfrac{\partial V}{\partial T}\dfrac{\partial T}{\partial P}\dfrac{\partial P}{\partial V} = \dfrac{nR}{P} \cdot \dfrac{V}{nR} \cdot \left(-\dfrac{nRT}{V^2}\right) = -\dfrac{nRT}{PV} = -1$

81. **a.** $C(R, H) = 0.001\pi(R^2 + RH + R^2 H)$

    b. The cost increases by about 0.08 cents per can.

83. Approximately 1062 ft^3

85. $\dfrac{dy}{dx} = \dfrac{1}{2}$; $x - 2y = -3$

CHAPTER 7 Section 3

	Relative maximum	Relative minimum	Saddle point
1.	(0, 0)	None	None
3.	None	None	(0, 0)
5.	None	None	(2, −1)
7.	(−2, −1)	(1, 1)	(−2, 1); (1, −1)
9.	None	$\left(4, \dfrac{19}{2}\right)$	$\left(2, \dfrac{7}{2}\right)$
11.	(0, 0)	None	(3, 6); (3, −6)
13.	(0, 1); (0, −1)	(0, 0)	(1, 0); (−1, 0)
15.	None	$\left(\dfrac{4}{3}, \dfrac{4}{3}\right)$	(0, 0)
17.	(1, 1); (−1, −3)	None	(0, −1)
19.	$\left(-\dfrac{3}{2}, 1\right)$	None	None
21.	(e, 1); (e, −1)	None	None

23. Points (3, 1), (2, 2); max 5 at (5, 5); min −5 at (5, 0)

25. Points (0, 0), (−1, 2), (−1, −2), (−5, 0), and (5, 0); max 198 at (5, 5) and (5, −5); min −52 at (−5, 5) and (−5, −5)

27. Points (0, 0), (1, 0), (−1, 0), $\left(\dfrac{\sqrt{3}}{2}, \dfrac{1}{2}\right)$, and $\left(\dfrac{-\sqrt{3}}{2}, \dfrac{1}{2}\right)$; max 1 at (1, 0) and (−1, 0); min 0 at (0, 0)

29. Duncan shirts $x = \$2.70$; James shirts $y = \$2.50$

31. $x = \$20$, $y = \$20$

33. $x = 200$; $y = 300$

35. \$4,000 on development, \$9,000 on promotion

37. **a.** The quadrilateral with vertices (0, 0), (300, 0), (200, 200), and (0, 400)

    b. Minimum cost of \$4,300 when $x = \$140$ and $y = \$190$

39. **a.** You need $x \geq 2$ and $y \geq 2$ to ensure nonnegative profits per item and $40 - 50x + 40y \geq 0$ and $20 + 60x - 70y \geq 0$ to ensure nonnegative

ANSWERS

quantities sold. The triangular region has

vertices $(2, 2)$, $\left(\dfrac{12}{5}, 2\right)$, and $\left(\dfrac{36}{11}, \dfrac{34}{11}\right)$.

b. The solution does not change. The smallest profit is obtained at any of the vertices.

41. $x = \dfrac{\sqrt{2}}{2}; y = \dfrac{\sqrt{2}}{2}$

43. $S\left(\dfrac{5}{4}, -\dfrac{1}{4}\right)$

45. Maximum of $P = \dfrac{2}{3}$, when $p = q = r = \dfrac{1}{3}$

47. Problem can be stated as:

Maximize $V = 6\left[x - 2\left(\dfrac{6}{\sqrt{3}}\right)\right]y$

subject to $2xy + 2\left(\dfrac{1}{2}\dfrac{\sqrt{3}}{2}x^2\right) = 500$

Solution is $x \approx 13.86$ ft; $y \approx 12.02$ ft.

49. a. Time is minimized when $x = 0.424$ and $y = 2.236$. Dick should wait at the point $(0.424, 1.2)$, and Mary should wait at $(2.66, 3.7)$.

b. Tom, Dick, and Mary will win by 0.2080 hours (12.5 minutes).

c. Writing exercise; responses will vary.

51. The base of the box is a 2 ft by 2 ft square. The height is 8 ft.

53. $\dfrac{\partial f}{\partial x} = 2x - 4y; \dfrac{\partial f}{\partial y} = 2y - 4x$. Thus, $(0, 0)$ is a

critical point. Since $\dfrac{\partial^2 f}{\partial^2 x} = 2 > 0$, the second

derivative test tells us there is a minimum in the x

direction. Likewise, $\dfrac{\partial^2 f}{\partial^2 y} = 2 > 0$ implies a

minimum in the y direction. However, along the curve determined by $y = x$, we have $f = -2x^2$, which has a relative maximum at $(0, 0)$.

55. $f_x = \dfrac{x^2 - 7y^2}{x^2 \ln y}$;

$f_y = \dfrac{y(x + 14y) \ln y - (x^2 + xy + 7y^2)}{xy\, (\ln y)^2}$

Critical points: $(\sqrt{7}e, e), (-\sqrt{7}e, e)$

57. $f_x = 8x^3 - 22xy + 36x; f_y = 4y^3 - 11x^2$

Critical point: $(0, 0)$

CHAPTER 7   Section 4

1. $y = \dfrac{1}{4}x + \dfrac{3}{2}$

3. $y = 3$

5. $y = \dfrac{7}{9}x + \dfrac{19}{18}$

7. $y = -\dfrac{1}{2}x + 4$

9. $y = 1.018x + 0.802$

11. $y = -0.915x + 1.683$

13. $y = 15.018e^{0.04x}$

15. $y = 20.03e^{-0.201x}$

17. a.

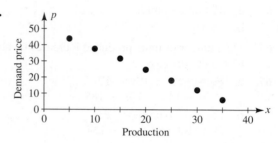

b. $p = -1.29x + 50.71$

c. The predicted price is negative. All 4,000 units cannot be sold at any price.

19. a. $V = 53.90e^{0.041t}$; 4.09%

b. \$122,380

c. Approximately 42 years

d. Frank's formula is $V = 54.52e^{0.044t}$. It is easier to find but does not take into account the trend in account value.

21. a.

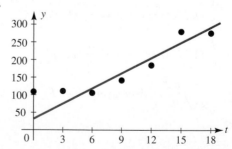

b. $y = 10.96t + 73.61$; reasonably good fit.

c. \$3.25 per gallon

23. a. $y = 3{,}828.5t + 15{,}120.4$
  b. 76,376.4 billion yuan

25. a.

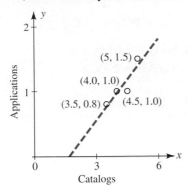

  b. $y = 0.42x - 0.71$
  c. When 4,800 catalogs ($x = 4.8$) are requested,
    $y = 0.42(4.8) - 0.71 = 1.306$ or 1,306
    applications are predicted to be received.

27. a. Approximately 12.5% per decade
  b. 308.4 million; 328.4 million

29. a.

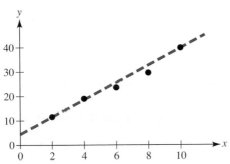

  b. $y = 3.05x + 6.10$
  c. When the polls close at 8:00 P.M., $x = 12$ and so
    $y = 3.05(12) + 6.1 = 42.7$, which means that
    approximately 42.7% of the registered voters
    can be expected to vote.

31. a.

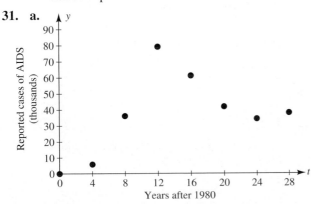

  b. $y = 1{,}255.9t + 20{,}040.2$
  c. 60,229
  d. No, the line has positive slope, but the number of
    cases appears to be decreasing in recent years.

33. a.

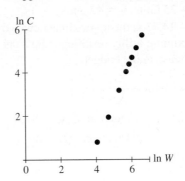

  b. $y = 1.631x - 4.975$
  c. $C = 0.0069W^{1.631}$

CHAPTER 7 Section 5

1. $f\left(\dfrac{1}{2}, \dfrac{1}{2}\right) = \dfrac{1}{4}$

3. $f(1, 1) = f(-1, -1) = 2$

5. $f(0, 2) = f(0, -2) = -4$

7. $f\left(\dfrac{\sqrt{3}}{2}, -\dfrac{1}{2}\right) = f\left(-\dfrac{\sqrt{3}}{2}, -\dfrac{1}{2}\right) = \dfrac{3}{2}$ (max);
  $f(0, 1) = -3$ (min)

9. $f(8, 7) = -18$

11. $f(\sqrt{2}, \sqrt{2}) = f(-\sqrt{2}, -\sqrt{2}) = e^2$ (max);
  $f(\sqrt{2}, -\sqrt{2}) = f(-\sqrt{2}, \sqrt{2}) = e^{-2}$ (min)

13. $f\left(8, 4, \dfrac{8}{3}\right) = \dfrac{256}{3}$ (max)

15. $f\left(\dfrac{4}{\sqrt{14}}, \dfrac{8}{\sqrt{14}}, \dfrac{12}{\sqrt{14}}\right) = \dfrac{56}{\sqrt{14}}$ (max);
  $f\left(-\dfrac{4}{\sqrt{14}}, -\dfrac{8}{\sqrt{14}}, -\dfrac{12}{\sqrt{14}}\right) = \dfrac{-56}{\sqrt{14}}$ (min)

17. 12,500 Deluxe and 17,500 Standard

19. a. $36,000 on development and $24,000 on
    promotion; 103,680 copies
  b. Approximately 4,320 more books will be sold.

21. a. $40,000 on labor and $80,000 on equipment
  b. Approximately 31.75 more units will be
    produced.

23. $\lambda = 306.12$, which gives the approximate change per \$1,000. Since the difference is only \$100, the maximum profit is increased by approximately $0.1(\$306.12) = \$30.61$.

25. **a.** $x = 35$ units, $y = 42$ units
   b. $\lambda = 14.33$ is the approximate change in the maximum utility resulting from a one-unit increase in the budget.

27. Increases by $\lambda = \left(\dfrac{\alpha}{a}\right)^{\alpha}\left(\dfrac{\beta}{b}\right)^{\beta}$

29. Let $Q(x, y) = $ production; $C(x, y) = px + qy = k$. $C_x = p, C_y = q$. Therefore, $Q_x = \lambda p; Q_y = \lambda q$; $\dfrac{Q_x}{p} = \dfrac{Q_y}{q}$.

31. $Q(40, 14) \approx 1{,}398$

33. The Lagrange equations are
$$A\alpha K^{-\beta-1}[\alpha K^{-\beta} + (1 - \alpha)L^{-\beta}]^{-1/\beta-1} = c_1\lambda$$
$$A(1 - \alpha)L^{-\beta-1}[\alpha K^{-\beta} + (1 - \alpha)L^{-\beta}]^{-1/\beta-1} = c_2\lambda$$
$$c_1K + c_2L = B$$
Solve the first two equations for λ and simplify to get
$$c_2\alpha K^{-\beta-1} = c_1(1 - \alpha)L^{-\beta-1}$$

35. $f(-0.49, -2.91) \approx 4.18$ mi

37. $H = 2R$

39. 4,060 (3,500 rabbits, 560 foxes)

41. 40 meters by 80 meters

43. 11,664 cubic inches, when $x = 18, y = 36$

45. $r = 1.51$ inches; $h = 3.02$ inches

47. $s_{\max} = 4L$

49. $x = 8.93$ cm, $y = 10.04$ cm

51. $x = y = z = \sqrt[3]{V_0}$

53. Front length 11.5 ft; side length 15.4 ft; height 7.2 ft

55. $x = 0, y = -2$. The critical point $(0, -2)$ is an inflection point, not a relative extremum.

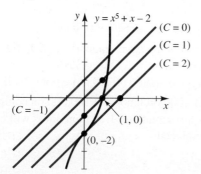

57. $\dfrac{dy}{dx} = \dfrac{\dfrac{y}{x^2} - (1 + xy^2)e^{xy^2} - \ln(x + y) - \dfrac{x}{x + y}}{2x^2ye^{xy^2} + \dfrac{1}{x} + \dfrac{x}{x + y}}$

59. $f(2.1623, 1.5811) = 1.6723$

61. $f(0.9729, -0.1635) = 2.9522$

CHAPTER 7 Section 6

1. $\dfrac{7}{6}$

3. -1

5. $4 \ln 2 = \ln 16$

7. 0

9. $\ln 3$

11. 32

13. $\dfrac{1}{3}$

15. $\dfrac{32}{9}$

17. $\dfrac{e^2 - 1}{8}$

19. Vertical cross sections: $0 \le x \le 3$
   $$x^2 \le y \le 3x$$
   Horizontal cross sections: $0 \le y \le 9$
   $$\dfrac{y}{3} \le x \le \sqrt{y}$$

21. Vertical cross sections: $-1 \le x \le 2$
   $$1 \le y \le 2$$
   Horizontal cross sections: $1 \le y \le 2$
   $$-1 \le x \le 2$$

23. Vertical cross sections: $1 \le x \le e$
   $$0 \le y \le \ln x$$
   Horizontal cross sections: $0 \le y \le 1$
   $$e^y \le x \le e$$

25. $\dfrac{3}{2}$

27. $\dfrac{1}{2}$

29. $\dfrac{44}{15}$

31. 1

33. $\dfrac{3}{2}\ln 5$

35. $2(e-2)$

37.

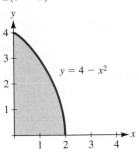

$$\int_{y=0}^{y=4}\int_{x=0}^{x=\sqrt{4-y}} f(x,y)\,dx\,dy$$

39.

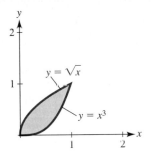

$$\int_{y=0}^{y=1}\int_{x=y^2}^{x=y^{1/3}} f(x,y)\,dx\,dy$$

41.

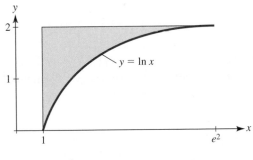

$$\int_{y=0}^{y=2}\int_{x=1}^{x=e^y} f(x,y)\,dx\,dy$$

43.

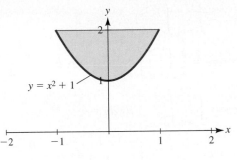

$$\int_{y=1}^{y=2}\int_{x=-\sqrt{y-1}}^{x=\sqrt{y-1}} f(x,y)\,dx\,dy$$

45. $\displaystyle\int_{x=-4}^{x=2}\int_{y=0}^{y=x+4} 1\,dy\,dx = 18$

47. $\displaystyle\int_{x=0}^{x=4}\int_{y=\frac{1}{2}x^2}^{y=2x} 1\,dy\,dx = \dfrac{16}{3}$

49. $\displaystyle\int_{x=1}^{x=3}\int_{y=x^2-4x+3}^{0} 1\,dy\,dx = \dfrac{4}{3}$

51. $\displaystyle\int_{x=1}^{x=e}\int_{y=0}^{y=\ln x} 1\,dy\,dx = 1$

53. $\displaystyle\int_{y=0}^{y=3}\int_{x=y/3}^{x=\sqrt{4-y}} 1\,dx\,dy = \dfrac{19}{6}$

55. $\displaystyle\int_{y=0}^{y=2}\int_{x=0}^{x=1} (6-2x-2y)\,dx\,dy = 6$

57. $\displaystyle\int_{x=1}^{x=2}\int_{y=1}^{y=3} \dfrac{1}{xy}\,dy\,dx = (\ln 3)(\ln 2)$

59. $\displaystyle\int_{x=0}^{x=1}\int_{y=0}^{y=2} xe^{-y}\,dy\,dx = \dfrac{1}{2}\left(1-\dfrac{1}{e^2}\right)$

61. $\displaystyle\int_{y=0}^{y=1}\int_{x=y}^{x=2-y} (2x+y)\,dx\,dy = \dfrac{7}{3}$

63. $\displaystyle\int_{x=-2}^{x=2}\int_{y=x^2}^{y=8-x^2} (x+1)\,dy\,dx = \dfrac{64}{3}$

65. Area $= \displaystyle\int_{x=-2}^{x=3}\int_{y=-1}^{y=2} 1\,dy\,dx = 15$

Average $= \dfrac{1}{15}\displaystyle\int_{x=-2}^{x=3}\int_{y=-1}^{y=2} xy(x-2y)\,dy\,dx = \dfrac{1}{6}$

ANSWERS

67. Area $= \displaystyle\int_{x=0}^{x=1} \int_{y=0}^{y=2} 1 \, dy \, dx = 2$

Average $= \dfrac{1}{2} \displaystyle\int_{y=0}^{y=2} \int_{x=0}^{x=1} xye^{x^2y} \, dx \, dy = \dfrac{e^2 - 3}{4}$

69. Area $= \displaystyle\int_{x=0}^{x=3} \int_{y=x/3}^{y=1} 1 \, dy \, dx = \dfrac{3}{2}$

Average $= \dfrac{2}{3} \displaystyle\int_{x=0}^{x=3} \int_{y=x/3}^{y=1} 6 \, xy \, dy \, dx = \dfrac{9}{2}$

71. Area $= \displaystyle\int_{x=-2}^{x=2} \int_{y=0}^{y=4-x^2} 1 \, dy \, dx = \dfrac{32}{3}$

Average $= \dfrac{3}{32} \displaystyle\int_{x=-2}^{x=2} \int_{y=0}^{y=4-x^2} x \, dy \, dx = 0$

73. $\displaystyle\int_{x=1}^{x=3} \int_{y=2}^{y=5} \dfrac{\ln(xy)}{y} \, dy \, dx$

$= (3 \ln 3 - 2) \ln \dfrac{5}{2} + (\ln 5)^2 - (\ln 2)^2$

75. $\displaystyle\int_{x=0}^{x=1} \int_{y=0}^{y=1} x^3 e^{x^2y} \, dy \, dx = \dfrac{e - 2}{2}$

77. Area $= \displaystyle\int_{x=0}^{x=5} \int_{y=0}^{y=7} 1 \, dy \, dx = 35$

Average $= \dfrac{1}{35} \displaystyle\int_{x=0}^{x=5} \int_{y=0}^{y=7} (2x^3 + 3x^2y + y^3) \, dy \, dx$

$= \dfrac{943}{4}$

79. Average $= \dfrac{1}{25 \cdot 19} \displaystyle\int_{x=100}^{x=125} \int_{y=70}^{y=89} [(x - 30)(70 + 5x - 4y)$

$+ (y - 40)(80 - 6x + 7y)] \, dy \, dx$

$= 24,896.5 \; (\$2,489,650)$

81. Value $= \displaystyle\int_{x=-1}^{x=1} \int_{y=-1}^{y=1} (300 + x + y)e^{-0.01x} \, dy \, dx$

$= 79,800e^{0.01} - 80,200e^{-0.01} \approx 1,200$

That is, $\$1,200,000$.

83. $\dfrac{9}{2}(5 \ln 5 - 4) \approx 18.212 \; (18,212 \text{ people})$

85. Average $= \dfrac{1}{12} \displaystyle\int_{x=0}^{x=4} \int_{y=0}^{y=3} 90(2x + y^2) \, dy \, dx$

$= 630 \text{ feet}$

87. $\dfrac{17,408}{105} \approx 166 \text{ m}^3$

89. $\dfrac{304}{27} \approx 11.26$

91. 62,949 people

93. 64 (64,000 people)

95. **a.** 0.991 square meters

b. No, it can only be considered the average surface area from birth until the time at which the person reached adulthood.

97. $\dfrac{7e^{-6}}{9} + \dfrac{17}{9} \approx 1.891$ cubic units

CHAPTER 7   Checkup

1. **a.** Domain: all ordered pairs (x, y) of real numbers

$$f_x = 3x^2 + 2y^2$$
$$f_y = 4xy - 12y^3$$
$$f_{xx} = 6x$$
$$f_{yx} = 4y$$

b. Domain: all ordered pairs (x, y) of real numbers for which $x \neq y$

$$f_x = \dfrac{-3y}{(x - y)^2}$$
$$f_y = \dfrac{3x}{(x - y)^2}$$
$$f_{xx} = \dfrac{6y}{(x - y)^3}$$
$$f_{yx} = \dfrac{-3(x + y)}{(x - y)^3}$$

c. Domain: all ordered pairs (x, y) of real numbers for which $y^2 > 2x$

$$f_x = 2e^{2x-y} - \dfrac{2}{y^2 - 2x}$$
$$f_y = -e^{2x-y} + \dfrac{2y}{y^2 - 2x}$$
$$f_{xx} = 4e^{2x-y} - \dfrac{4}{(y^2 - 2x)^2}$$
$$f_{yx} = -2e^{2x-y} + \dfrac{4y}{(y^2 - 2x)^2}$$

2. a. Circles centered at the origin and the single point $(0, 0)$

b. Parabolas with vertices on the x axis and opening to the left

3. a. Relative maximum: $(0, 0)$; relative minimum: $(1, 4)$; saddle points: $(1, 0)$, $(0, 4)$

b. Saddle point: $(-1, 0)$

c. Relative minimum: $(-1, -1)$

4. a. $\dfrac{16}{5}$ at $\left(\dfrac{4}{5}, \dfrac{8}{5}\right)$

b. Maximum value of 4 at $(1, 2)$ or $(1, -2)$; minimum value of -4 at $(-1, 2)$ or $(-1, -2)$

5. a. 16

b. $\dfrac{1}{4}(e^2 + 3e^{-2})$

c. $2 \ln 2 - \dfrac{3}{4}$

d. $1 - \dfrac{1}{e^2}$

6. $Q_K - 180; Q_L = 3.75$

7. 20 DVDs and 2 video games

8. 30 units of drug A and 25 units of drug B, which results in an equivalent dosage of $E(30, 25) = 83.75$ units. Since the total number of units is 55, which is less than 60, there is no risk of side effects, and since $E(30, 25) > 70$, the combination is effective.

9. $\dfrac{5}{2}(1 + e^{-2}) \approx 2.84°C$

10. a.

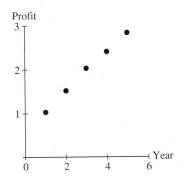

b. $y = 0.45x + 0.61$

c. 3.31 million dollars

CHAPTER 7 Review Exercises

1. $f_x = 6x^2y + 3y^2 - \dfrac{y}{x^2}; f_y = 2x^3 + 6xy + \dfrac{1}{x}$

3. $f_x = \dfrac{3x - y^2}{2\sqrt{x}}; f_y = -2y\sqrt{x}$

5. $f_x = \dfrac{1}{2\sqrt{xy}} - \dfrac{\sqrt{y}}{2x^{3/2}}; f_y = \dfrac{1}{2\sqrt{xy}} - \dfrac{\sqrt{x}}{2y^{3/2}}$

7. $f_x = \dfrac{2x^3 + 3x^2y - y^2}{(x + y)^2}; f_y = \dfrac{-x^2(x + 1)}{(x + y)^2}$

9. $f_x = \dfrac{2(x^2 + xy + y^2)}{(2x + y)^2}; f_y = \dfrac{-x^2 - 4xy - y^2}{(2x + y)^2}$

11. $f_{xx} = (4x^2 + 2)e^{x^2+y^2}; f_{yy} = (4y^2 + 2)e^{x^2+y^2};$
$f_{xy} = 4xy\, e^{x^2+y^2}; f_{yx} = 4xy\, e^{x^2+y^2}$

13. $f_{xx} = 0; f_{yy} = -\dfrac{x}{y^2}; f_{xy} = \dfrac{1}{y}; f_{yx} - \dfrac{1}{y}$

15. a.

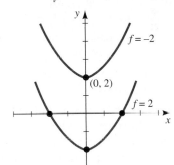

b.

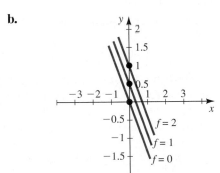

$f_y = \dfrac{x}{y(x + 3y)} = \dfrac{1}{y} - \dfrac{3}{x + 3y}$

17. Saddle point at $(6, -6)$

19. Relative maximum at $(-2, 0)$; relative minimum at $(0, 2)$; saddle points at $(0, 0)$ and $(-2, 2)$

21. Relative minimum at $\left(-\dfrac{23}{2}, 5\right)$; saddle point at $\left(\dfrac{1}{2}, 1\right)$

23. Saddle points at $\left(\dfrac{2}{3}, -\dfrac{5}{6}\right)$ and $\left(-\dfrac{2}{3}, \dfrac{5}{6}\right)$

25. Points $(-1, 2)$, $(-1, 0)$, $\left(-\dfrac{3}{2}, \dfrac{5}{2}\right)$, $\left(\dfrac{7}{17}, \dfrac{40}{17}\right)$; max of 20 at $(-4, 0)$; min of 7 at $(-1, 2)$

27. Points $(2, 4)$, $\left(\dfrac{2\sqrt{3}}{3}, 2\right)$, $\left(\dfrac{4\sqrt{3}}{3}, 5\right)$; max of 52 at $(4, 2)$; min of 0 at $(2, 4)$

29. Point $(-2, 0)$; max of 1 at any point on the boundary; min of $e^{-4} \approx 0.018$ at $(-2, 0)$

31. Maximum value of 12 at $(1, \pm\sqrt{3})$; minimum value of 3 at $(-2, 0)$

33. Maximum value of 17 at $(1, 8)$; minimum value of -17 at $(-1, -8)$

35. Daily output will increase by approximately 16 units.

37. The level of unskilled labor should be decreased by approximately 2 workers.

39. Maximize area $A = xy$ subject to fixed perimeter $P = 2x + 2y$. Lagrange conditions are $y = \lambda(x)$, $x = \lambda(y)$, and $2x + 2y = C$. We must have $\lambda > 0$ since x and y are positive, so $x = y$ and the optimum rectangle is actually a square.

41. Development $x = \$4{,}000$; promotion $y = \$7{,}000$

43. We have $f_x = y - \dfrac{12}{x^2}$ and $f_y = x - \dfrac{18}{y^2}$, so $f_x = f_y = 0$ when $x = 2$ and $y = 3$. Since $f(x, y)$ is large when either x or y is large or small, a relative minimum is indicated at $(2, 3)$, and the minimum value is $f(2, 3) = 18$. To verify this claim, note that

$$D = \left(\dfrac{24}{x^3}\right)\left(\dfrac{24}{y^3}\right) - 1 \quad \text{and} \quad f_{xx} = \dfrac{24}{x^3}$$

so that $D(2, 3) > 0$ and $f_{xx}(2, 3) > 0$.

45. $\dfrac{e^3 - e^2 - e + 1}{e^3} \approx 0.5466$

47. $\dfrac{1}{4}(e^2 - e^{-2})$

49. $2e - 2$

51. $\dfrac{3}{2}(e - 1)$

53. 2

55. $\dfrac{3}{2}(e^{-2} - e^{-3})$ cubic units

57. $x = y = z = \dfrac{20}{3}$

59. $\sqrt{10}$; at $(0, \pm\sqrt{10}, 0)$

61. **a.**

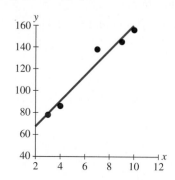

b. $y = 11.54x + 44.45$

c. Approximately $\$102{,}150$

63. 5.94; demand is increasing at the rate of about 6 quarts per month.

65. -3; demand is decreasing at the rate of 3 pies per week.

67. The amount of pollution is decreasing by about 113 units per day.

69. About 7.056 units

71. $Q(x, y) = x^a y^b$
$Q_x = ax^{a-1}y^b$; $Q_y = bx^a y^{b-1}$
$xQ_x + yQ_y = x(ax^{a-1}y^b) + y(bx^a y^{b-1})$
$= (a + b)x^a y^b = (a + b)Q$
If $a + b = 1$, then $xQ_x + yQ_y = Q$.

CHAPTER 8   Section 1

1. **a.** $30°$
  b. $60°$

3. **a.** $-120°$
  b. $390°$

5. **a.** $\dfrac{3\pi}{4}$
  b. $\dfrac{\pi}{3}$

7. **a.** $-\dfrac{\pi}{3}$
  b. $\dfrac{5\pi}{6}$

9. a. $60°$

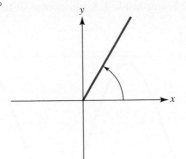

b. $120°$

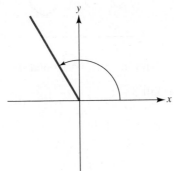

11. a. $45°$

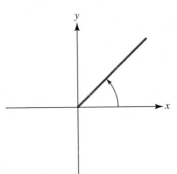

b. $-150°$

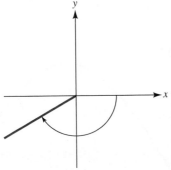

13. a. $\dfrac{3\pi}{4}$

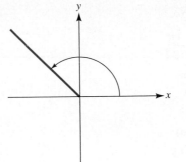

b. $-\dfrac{2\pi}{3}$

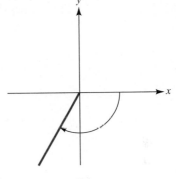

15. a. $-\dfrac{5\pi}{3}$

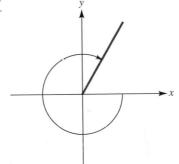

b. $\dfrac{3\pi}{2}$

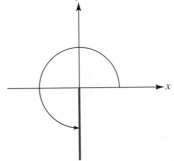

17. a. $\dfrac{\pi}{12}$

  b. $-\dfrac{4\pi}{3}$

19. a. $\dfrac{3\pi}{4}$

  b. 3π

21. a. $150°$

  b. $-15°$

23. a. $540°$

  b. $\dfrac{180°}{\pi}$

25. 0

27. 1

29. 0

31. 1

33. $-\dfrac{1}{2}$

35. $\dfrac{1}{2}$

37. $\dfrac{\sqrt{3}}{3}$

39. -1

41. $\dfrac{2\sqrt{3}}{3}$

43. $-\sqrt{2}$

45. $\cos\dfrac{\pi}{3} = \dfrac{1}{2}, \sin\dfrac{\pi}{3} = \dfrac{\sqrt{3}}{2}$

47.

θ	$\dfrac{7\pi}{6}$	$\dfrac{5\pi}{4}$	$\dfrac{4\pi}{3}$	$\dfrac{3\pi}{2}$	$\dfrac{5\pi}{3}$	$\dfrac{7\pi}{4}$	$\dfrac{11\pi}{6}$	2π
$\sin\theta$	$-\dfrac{1}{2}$	$-\dfrac{\sqrt{2}}{2}$	$-\dfrac{\sqrt{3}}{2}$	-1	$-\dfrac{\sqrt{3}}{2}$	$-\dfrac{\sqrt{2}}{2}$	$-\dfrac{1}{2}$	0
$\cos\theta$	$-\dfrac{\sqrt{3}}{2}$	$-\dfrac{\sqrt{2}}{2}$	$-\dfrac{1}{2}$	0	$\dfrac{1}{2}$	$\dfrac{\sqrt{2}}{2}$	$\dfrac{\sqrt{3}}{2}$	1

49. $\dfrac{3}{4}$

51. $\dfrac{4}{3}$

53. $\dfrac{5}{4}$

55. Period 4, amplitude 1.5, horizontal shift 1, vertical shift 2

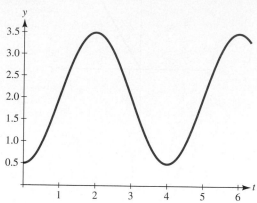

57. Period π, amplitude 1, horizontal shift $\dfrac{\pi}{2}$, vertical shift 3

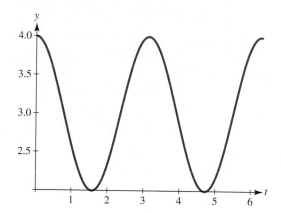

59. $\theta = \dfrac{\pi}{6}, \dfrac{5\pi}{6}$

61. $\theta = \dfrac{\pi}{3}, \pi$

63. a.

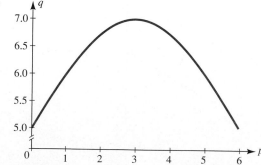

  b. Smallest = 5,000 units, largest = 7,000 units

  c. $p = \$1/\text{unit}$ or $\$5/\text{unit}$

65. a. Week 1; 181 pairs of skis

b.

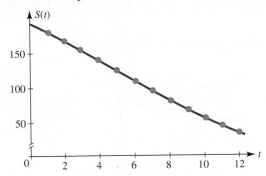

67. a. Since the sine and cosine functions can never be less than -1, $S(t)$ can never be less than $12 - 3 - 2 = 7$. The upper bound is $12 + 3 + 2 = 17$.

b. The ideal minimum level \$7 is never attained (the lowest price is about \$8.18). The ideal maximum level \$17 is attained when $t = 12$ weeks.

69. a. Maximum volume $= 0.9$ liter at times $t = (2n + 1)(1.75)$ where n is an integer

b. Minimum volume $= 0.1$ liter at times $t = (2n)(1.75)$ where n is an integer

c.

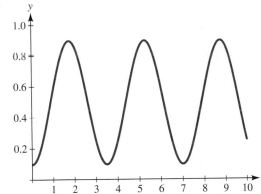

71. a. $E_0 \sin\left(\dfrac{2\pi}{28} \cdot 21 \cdot 365\right) = -E_0$

b. $P_0 \sin\left(\dfrac{2\pi}{23} t\right)$

$P_0 \sin\left(\dfrac{2\pi}{23} \cdot 21 \cdot 365\right) = P_0 \sin\left(\dfrac{15{,}330}{23} \pi\right)$

$\approx 0.998 P_0$

c. The E_0 cycle has a maximum when $t = 7(4n + 1)$ days and n is an integer. The P_0 cycle has a maximum when $t = \dfrac{23}{4}(4m + 1)$ days and m is

an integer. There are no integers n and m that give the same t. Therefore, Jake's 2 cycles will never both be at their peak levels simultaneously.

73. a. $A = 46$; $B = 24$; $C = 431$; $D = 204$; the cycle for both species is 24 years.

b. 431 rabbits; 431 rabbits

c. 46 foxes; 46 foxes

d.

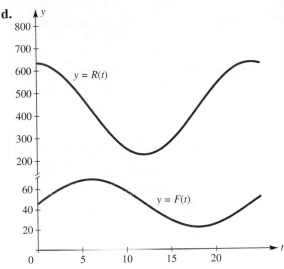

75. $\sin\left(\dfrac{\pi}{2} - \theta\right) = \sin\left(\dfrac{\pi}{2}\right)\cos(-\theta) + \cos\left(\dfrac{\pi}{2}\right)\sin(-\theta)$

$= \cos(-\theta)$

$= \cos\theta$

77. a. $f(t + p) = a + b\sin\left[\dfrac{2\pi}{p}(t + p - d)\right]$

$= a + b\sin\left[\dfrac{2\pi}{p}(t - d) + 2\pi\right]$

$= a + b\left[\sin\left(\dfrac{2\pi}{p}(t - d)\right)\cos(2\pi)\right.$

$\left. + \sin(2\pi)\cos\left(\dfrac{2\pi}{p}(t - d)\right)\right]$

$= a + b\sin\left[\dfrac{2\pi}{p}(t - d)\right]$

$= f(t)$

b. $f(t + p) = a + b\cos\left[\dfrac{2\pi}{p}(t + p - d)\right]$

$= a + b\cos\left[\dfrac{2\pi}{p}(t - d) + 2\pi\right]$

$$= a + b\left[\cos\left(\frac{2\pi}{p}(t-d)\right)\cos(2\pi) \right.$$
$$\left. - \sin\left(\frac{2\pi}{p}(t-d)\right)\sin(2\pi) \right]$$
$$= a + b\cos\left[\frac{2\pi}{p}(t-d)\right]$$
$$= f(t)$$

79. An angle θ corresponds to a point on the unit circle with coordinates $(x, y) = (\cos\theta, \sin\theta)$. The correspondence is one-to-one for $0 \le \theta < 2\pi$. From the following graphs we see that for $0 \le \theta \le 2\pi$, if (x_0, y_0) corresponds to angle θ, then (y_0, x_0) corresponds to angle $\frac{\pi}{2} - \theta$. This gives us

$$(y_0, x_0) = \left(\cos\left(\frac{\pi}{2} - \theta\right), \sin\left(\frac{\pi}{2} - \theta\right) \right)$$

If $(x_0, y_0) = (\cos\theta, \sin\theta)$, then $(y_0, x_0) = (\sin\theta, \cos\theta)$. Combining this with the earlier result, we have

$$(y_0, x_0) = \left(\cos\left(\frac{\pi}{2} - \theta\right), \sin\left(\frac{\pi}{2} - \theta\right) \right) =$$
$$(\sin\theta, \cos\theta)$$

1st quadrant

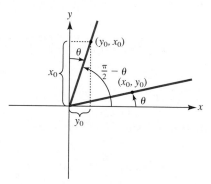

2nd quadrant

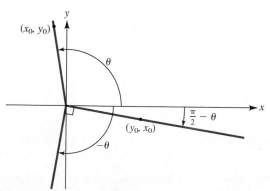

3rd quadrant

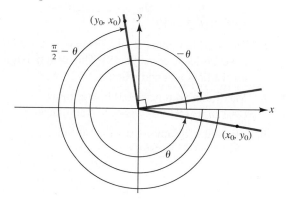

4th quadrant

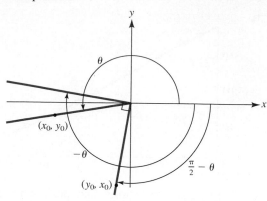

CHAPTER 8   Section 2

1. $3\cos 3t$

3. $-2\cos(1 - 2t)$

5. $-3t^2 \sin(t^3 + 1)$

7. $2\cos\left(\frac{\pi}{2} - t\right)\sin\left(\frac{\pi}{2} - t\right)$

9. $-6(1 + 3x)\sin(1 + 3x)^2$

11. $e^{-u/2}\left(-\frac{1}{2}\cos 2\pi u - 2\pi \sin 2\pi u \right)$

13. $\dfrac{\cos t}{(1 + \sin t)^2}$

15. $-3t^2 \sec^2(1 - t^3)$

17. $-2\pi \sec\left(\frac{\pi}{2} - 2\pi t\right)\tan\left(\frac{\pi}{2} - 2\pi t\right)$

19. $2\cot t$

21.

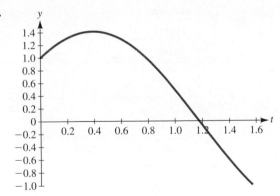

Maximum at $t = \dfrac{\pi}{8} \approx 0.39$,

minimum at $t = \dfrac{\pi}{2} \approx 1.57$

23.

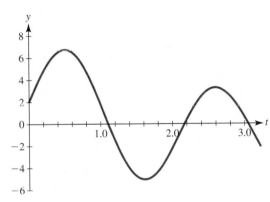

Maximum at $t \approx 0.502$,
minimum at $t \approx 1.615$

25.

Maximum at $t = \dfrac{\pi}{2}$,

minimum at $t = \pi$

27. Maximum value $= f\left(\dfrac{\pi}{4}\right) = -1$,

no minimum value, $\lim\limits_{t \to \pi} f(t) = -\infty$

29. $\approx -5.69 \dfrac{\text{thousand units}}{\text{month}}$, decreasing

31. **a.** $W(2) \approx 50$ units

 b. $W'(4) \approx 18$ units/hr; increasing

 c. $W''(t) = 0$ when $t = \dfrac{3\pi}{4} \approx 2.35$ hr; most
 efficient rate is about 48 units/hr.

33. **a.** $P'(t) = 0$ when $t = 3, 7,$ or 11. The largest
 percentage is 95%, when $t = 3$ or 11. The
 smallest percentage is 43%, when $t = 7$.

 b. $P''(t) = 0$ when $t = 1, 5,$ or 9; 69%.

35. **a.** Minimum of ≈ 97.87 at time $t = 0$ (midnight),
 maximum of 98.7 at time $t = 15$ (3:00 P.M.)

 b. $\approx 0.03°$F/hr

 c. At $t = 3$ (3:00 A.M.)

37. ≈ -3.69 deg/hr, cooler

39. **a.** $P'(t) = 0$ when $t = 3, 10$; largest percentage is
 35% when $t = 3$; smallest percentage is 25%
 when $t = 10$.

 b. $P''(t) = 0$ when $t = 6.5$, but $P(t)$ is *decreasing*
 most rapidly when $t = 6.5$; it is *increasing* most
 rapidly when $t = 0$.

41. **a.** 460, 2,483, 13,211, respectively

 b. 156, 834, 4,398, respectively

 c.

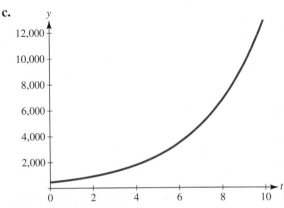

 d. Writing exercise; responses will vary.

43. $\sin \theta = \dfrac{E_L}{E_W}$

ANSWERS

45. $F'(\theta) = kh \csc \theta \left(\dfrac{\csc \theta}{R^4} - \dfrac{\cot \theta}{r^4} \right), 0 < \theta < \dfrac{\pi}{2}$

Critical point at θ_0 where $\cos \theta_0 = \dfrac{r^4}{R^4}$.

$\csc \theta > 0$ for $0 < \theta < \dfrac{\pi}{2}$ and $k > 0, h > 0$

$\Rightarrow$ sign of $F'(\theta) = $ sign of $\left(\dfrac{\csc \theta}{R^4} - \dfrac{\cot \theta}{r^4} \right)$

$\dfrac{\csc \theta}{R^4} - \dfrac{\cot \theta}{r^4} = \dfrac{1}{R^4 \sin \theta} - \dfrac{\cos \theta}{r^4 \sin \theta}$

$= \dfrac{1}{\sin \theta} \left(\dfrac{1}{R^4} - \dfrac{\cos \theta}{r^4} \right)$

$\sin \theta > 0$ for $0 < \theta < \dfrac{\pi}{2}$, so, consider

$\left(\dfrac{1}{R^4} - \dfrac{\cos \theta}{r^4} \right)$.

$\cos \theta$ is strictly decreasing on $\left(0, \dfrac{\pi}{2} \right)$

$\Rightarrow$ if $\theta < \theta_0$ then $\cos \theta > \cos \theta_0$

$\Rightarrow \left(\dfrac{1}{R^4} - \dfrac{\cos \theta}{r^4} \right) < \left(\dfrac{1}{R^4} - \dfrac{\cos \theta_0}{r^4} \right) = 0$

$\Rightarrow F'(\theta) < 0.$

Likewise, if $\theta > \theta_0$, then $\cos \theta < \cos \theta_0$

$\Rightarrow \left(\dfrac{1}{R^4} - \dfrac{\cos \theta}{r^4} \right) > \left(\dfrac{1}{R^4} - \dfrac{\cos \theta_0}{r^4} \right) = 0$

$\Rightarrow F'(\theta) > 0.$

There is a minimum at θ_0 since
$F(\theta)$ is decreasing ($F'(\theta) < 0$) for $\theta < \theta_0$
and increasing ($F'(\theta) > 0$) for $\theta > \theta_0$.

47. a.

$v(t) = -\pi \cos \pi t + \dfrac{1}{2} t^{-1/2}; a(t) = \pi^2 \sin \pi t - \dfrac{1}{4} t^{-3/2}$

b. $x(1) = 1; v(1) = \pi + \dfrac{1}{2}; a(1) = -\dfrac{1}{4}$

49. $\dfrac{d\theta}{dt} = \dfrac{4}{15}$ rad/sec

51. 60 rad/hr

53. $\theta = \dfrac{\pi}{2}$

55.

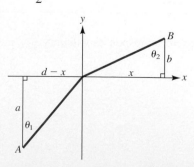

Time $= T = \dfrac{\text{distance}}{\text{speed}}$

$T = \dfrac{\sqrt{a^2 + (d - x)^2}}{v_1} + \dfrac{\sqrt{b^2 + x^2}}{v_2}$

$\dfrac{dT}{dx} = \left(\dfrac{1}{2} \right) \dfrac{2(d - x)(-1)}{v_1 \sqrt{a^2 + (d - x)^2}} + \left(\dfrac{1}{2} \right) \dfrac{2x}{v_2 \sqrt{b^2 + x^2}}$

$= \dfrac{-(d - x)}{v_1 \sqrt{a^2 + (d - x)^2}} + \dfrac{x}{v_2 \sqrt{b^2 + x^2}}$

$= \dfrac{-\sin \theta_1}{v_1} + \dfrac{\sin \theta_2}{v_2}$

$\dfrac{dT}{dx} = 0 \Rightarrow \dfrac{\sin \theta_1}{\sin \theta_2} = \dfrac{v_1}{v_2}$

57. $x = \dfrac{v_0}{\sqrt{k}} \sin (\sqrt{k}\, t)$

59.

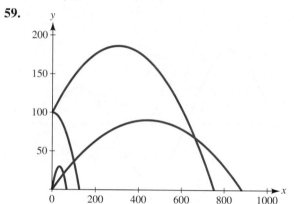

a. Parabola

b.

Highest point	100	187.89	29.30	91.53
Point of impact	125.00	749.62	67.66	883.88

61. a. 29.93 feet

b. 48.18 ft/sec

63. $2\sqrt{2} \approx 2.83$ ft

65. $y = 3.332x - 2.921$

67. $\lim\limits_{x \to 0} \dfrac{\cos x - 1}{x} = 0$

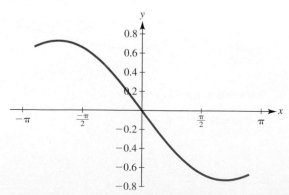

69. $\dfrac{d}{dt}(\cos t) = \lim\limits_{h\to 0} \dfrac{\cos(t+h) - \cos(t)}{h}$

$= \lim\limits_{h\to 0} \dfrac{\cos(t)\cos(h) - \sin(t)\sin(h) - \cos(t)}{h}$

$= \lim\limits_{h\to 0} \dfrac{\cos(t)(\cos(h) - 1) - \sin(t)\sin(h)}{h}$

$\lim\limits_{h\to 0} \dfrac{\cos(h) - 1}{h} = 0,$

and $\lim\limits_{h\to 0} \dfrac{\sin(h)}{h} = 1$, so

$\dfrac{d}{dt}(\cos t) = -\sin t.$

71. a. $\dfrac{d}{dt}\sec t = \dfrac{d}{dt}\left(\dfrac{1}{\cos t}\right)$

$= -\dfrac{1}{\cos^2 t}(-\sin t)$

$= \left(\dfrac{1}{\cos t}\right)\left(\dfrac{\sin t}{\cos t}\right)$

$= \sec t \tan t$

b. $\dfrac{d}{dt}\cot t = \dfrac{d}{dt}\left(\dfrac{\cos t}{\sin t}\right)$

$= \dfrac{\sin t(-\sin t) - \cos t(\cos t)}{\sin^2 t}$

$= \dfrac{-\sin^2 t - \cos^2 t}{\sin^2 t}$

$= -\dfrac{1}{\sin^2 t}$

$= -\csc^2 t$

c. $\dfrac{d}{dt}\csc t = \dfrac{d}{dt}\left(\dfrac{1}{\sin t}\right)$

$= -\dfrac{1}{\sin^2 t}(\cos t)$

$= -\left(\dfrac{1}{\sin t}\right)\left(\dfrac{\cos t}{\sin t}\right)$

$= -\csc t \cot t$

CHAPTER 8   Section 3

1. $-2\cos\left(\dfrac{t}{2}\right) + C$

3. $-\dfrac{1}{6}\sin(1 - 3x^2) + C$

5. $\dfrac{1}{4}\cos(2x) + \dfrac{x}{2}\sin(2x) + C$

7. $\dfrac{1}{2}\sin^2 x + C$

9. $\sin x - x\cos x + C$

11. 0

13. π

15. $\dfrac{3\sqrt{3} - 4}{2\pi}$

17. $\dfrac{2}{\pi} + \dfrac{1}{6}$

19. 2π

21. $2\sqrt{3}\pi$

23. $y = \cos x + \sin x + 2$

25. $x = 1 - 2\cos\sqrt{t}$

27. $x = \dfrac{1}{1 - \sin t}$

29. About \$10.91

31. \$10

33. \$18

35. About \$21.86

37. FV $\approx$ 9,401; PV $\approx$ 7,697

39. FV $\approx$ 2,000; PV $\approx$ 1,828

41. a. $p_e = 6$; $q_e = \dfrac{10}{3}$ (hundred units)

b. CS = \$48,800; PS = \$16,300

43. \$14,895.74

45. a. \$2,546

b. $t = 13.628$, or 42.371 weeks

47. $B_0 + 32\sqrt{6}$

49. $\displaystyle\int \dfrac{dP}{P} = \int (b + a\cos 2\pi t)\,dt$

$\ln|P| = bt + \dfrac{a}{2\pi}\sin 2\pi t + C_1$

$P(t) = C_2 e^{bt + \frac{a}{2\pi}\sin 2\pi t}$, where $C_2 = e^{C_1}$

$P_0 = P(0) = C_2$

51. a. $t = 35$ (February 5); 47

b. $t = 217.5$ (August 6); 3

c. January 1: ≈ 0.2146; increasing

April 1: ≈ -0.3111; decreasing

June 1: ≈ -0.3421; decreasing

ANSWERS

d.

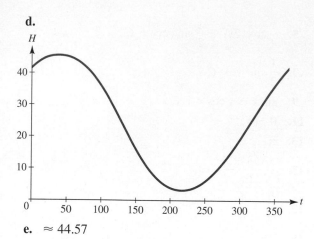

e. ≈ 44.57

53. a. 609 lbs

b. 0

55. a. $L = \dfrac{2\pi}{0.65} \approx 9.67$ sec

b. 2.68 liters; yes

c. $I = 9.67$ sec; inhale 6.47 liters; no

57. a. $T'(t) = -\dfrac{0.46\,\pi}{12} \sin\left[\dfrac{\pi(t - 16.37)}{12}\right]$

b. Maximum at $t = 16.37$

Maximum temperature = 37.75°C

c. Minimum temperature = 36.83°C at time $t = 4.37$

d. The average over period from 7 A.M. to 10 P.M. = 37.48°C.

59. a.

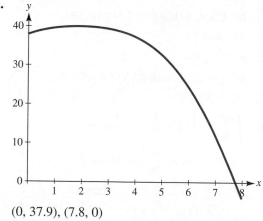

$(0, 37.9)$, $(7.8, 0)$

b. ≈ 244.73

61. a. $v(t) = \dfrac{1}{2}(3 - 3\cos 2t)$

b. $x(t) = -\dfrac{3}{4}\sin 2t + \dfrac{3}{2}t - 2$

c. $AV = \dfrac{1}{4}(6 - \sin 6) \approx 1.57$

63. a. 0.828155

b. 0.828116

65. a. $-\ln|\cos x| + C$

b. $\ln|\sin x| + C$

67. a. $\sec x$

b. $\ln|\sec x + \tan x| + C$

c. Writing exercise, responses will vary.

CHAPTER 8   Checkup

1. a. $\dfrac{1}{2}$

b. $-\dfrac{1}{2}$

c. -1

d. $-\dfrac{2\sqrt{3}}{3}$

2. $\cos\theta = \dfrac{3\sqrt{13}}{13}$

3. a. $2\cos(2x)$

b. $\cos x - x\sin x$

c. $\dfrac{x\sec^2 x - \tan x}{x^2}$

d. $\sin(x^2) + 2x^2\cos(x^2)$

4. a. $-\dfrac{1}{2}\cos(2x) + C$

b. $\dfrac{1}{2}\sin(x^2) + C$

c. ∞

d. $\dfrac{1}{24}$

5. $\theta = \dfrac{\pi}{6}, \dfrac{\pi}{2}, \dfrac{5\pi}{6}$

6. $\sqrt{2} - 1$

7. $\sin^2 x + \cos^2 x = 1$

$\left(\dfrac{1}{\sin^2 x}\right)(\sin^2 x + \cos^2 x) = \left(\dfrac{1}{\sin^2 x}\right)$

$1 + \cot^2 x = \csc^2 x$

8. a. $\approx -4.92\,\dfrac{\text{thousand units}}{\text{month}}$; decreasing

b. Largest sales: 25.9 thousand units; February and May

Smallest sales: 6.5 thousand units; October

9. **a.** $3,750
  b. $1,500 per unit; $\approx$ $1,871 per unit per unit
10. **a.** $\approx -0.74°F/day$
  b. $\approx 98.28°F$
11. **a.** $k \approx 8.38$
  b.

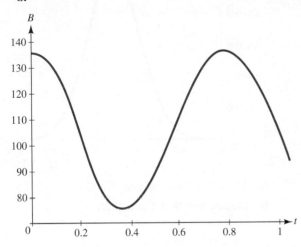

  c. Systolic: 136; diastolic: 74
  d. 105

CHAPTER 8   Review Exercises

1. **a.** $120° = \dfrac{2\pi}{3}$ radians

  b. $-225° = -\dfrac{5\pi}{4}$ radians

3. **a.** $\dfrac{45°}{\pi} \approx 14.32°$

  b. $\dfrac{180°}{\pi} \approx 57.30°$

  c. $-\dfrac{270°}{\pi} \approx 85.94°$

5. $\dfrac{4}{3}$

7. $\cos\left(\dfrac{\pi}{2} + \theta\right) = \cos\left(\dfrac{\pi}{2}\right)\cos\theta - \sin\left(\dfrac{\pi}{2}\right)\sin\theta$
  $= -\sin\theta$
  $\sin\left(\dfrac{\pi}{2} + \theta\right) = \sin\left(\dfrac{\pi}{2}\right)\cos\theta + \sin\theta\cos\left(\dfrac{\pi}{2}\right)$
  $= \cos(\theta)$

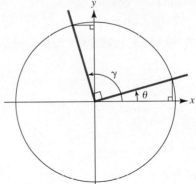

Adding 90° to θ effectively interchanges the x and y axes, so that if $\gamma = \theta + \dfrac{\pi}{2}$, then

$|\cos\gamma| = |\sin\theta|$ and $|\sin\gamma| = |\cos\theta|$.
By quadrant,

$\sin > 0$	$\sin > 0$
$\cos < 0$	$\cos > 0$
$\sin < 0$	$\sin < 0$
$\cos < 0$	$\cos > 0$

Adding 90° to θ moves the angle to the next quadrant. We see that $\cos\left(\dfrac{\pi}{2} + \theta\right) = -\sin\theta$ and $\sin\left(\dfrac{\pi}{2} + \theta\right) = \cos\theta$.

9. **a.** $\cos 2\theta = \cos^2\theta - \sin^2\theta$
  $= \cos^2\theta - (1 - \cos^2\theta)$
  $= 2\cos^2\theta - 1$
  so $\cos^2\theta = \dfrac{1}{2}(1 + \cos 2\theta)$
  Similarly, $\cos 2\theta = 1 - 2\sin^2\theta$
  so $\sin^2\theta = \dfrac{1}{2}(1 - \cos 2\theta)$

  b. $\displaystyle\int \cos^2 x\,dx = \int \dfrac{1}{2}(1 + \cos 2x)\,dx$
  $= \dfrac{1}{2}x + \dfrac{1}{4}\sin 2x + C$
  $\displaystyle\int \sin^2 x\,dx = \int \dfrac{1}{2}(1 - \cos 2x)\,dx$
  $= \dfrac{1}{2}x - \dfrac{1}{4}\sin 2x + C$

  c. AV = 3.5 m/sec

ANSWERS

11. $f'(x) = 5 \sin(1 - 5x)$

13. $f'(x) = -2 \cos x \sin x$

15. $f'(x) = 6 \tan(3x + 1) \sec^2(3x + 1)$

17. $f'(x) = -\dfrac{2 \sin x}{\cos x}$

19. $-\dfrac{1}{2} \cos 2t + \dfrac{1}{2} \sin 2t + C$

21. $\dfrac{1}{2} \sin^2 x + C$

23. $\ln|\tan t| + C$

25. $\dfrac{3\sqrt{3}}{2}$

27. Angle $A = -52° \, 14' \, 20''$, $\theta = 232° \, 14' \, 20''$

29. $x = 0.436, 1.468$

31. $\dfrac{2}{3}\pi^2 - \dfrac{3\pi}{4}$

33. a. $p = 25$
$b = 27$
$d = 11$
$a = 33$

b.

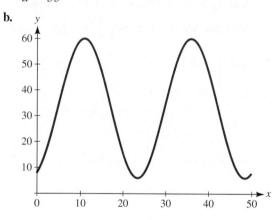

35. $C = (3{,}000 - 900 \cot \theta)\,4 + (900 \csc \theta)\,5$
$C' = -900(-\csc^2 \theta)\,4 + 900(-\csc \theta \cot \theta)\,5$
$\quad = 900 \csc^2 \theta\,(4 - 5 \cos \theta)$

$C' = 0 \Rightarrow \cos \theta = \dfrac{4}{5}$ or $\csc \theta = 0$

but $\csc \theta \neq 0$

37. a.

t	0.125	0.25	0.325	0.5
$p(t)$	70,570	548,317	255,673	500

b.

t	0.125	0.25	0.325	0.5
$p'(t)$	0.2195×10^7	0	-0.5105×10^7	$-21{,}991.15$

c.

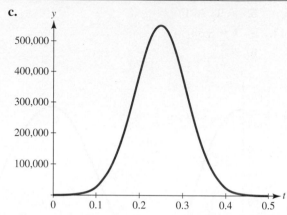

39. a. 1st year 3,300; 2nd year 3,300

b. Largest month is December; smallest month is January.

41. a.

Jan. 1	Mar. 15	June 21
9.11	11.17	15.21

b. Maximum on 185th day, July 4; minimum on 3rd day, January 3

c. 12.2 hours

43. 44.94%

45. $-\dfrac{3}{250}$ radians/sec

47. a. $L = \dfrac{4.25}{\cos \theta \, \sin^2 \theta}$

b. L minimized $\theta = 0.9553$; minimum length $= 11.04$ inches

c. Area $= \dfrac{1}{2} L \sin \theta = \dfrac{1}{8} 8.5^2 \sec \theta \, (\csc \theta)^3$; area is minimized when $\theta = \dfrac{\pi}{3}$; minimum area is 27.81 square inches.

49. $\sin y = -\cos x + 1 + \sin 1$

51. $\dfrac{3\sqrt{3}}{8} \approx 0.65$

53. a. $\dfrac{du}{dt} = (-c^2k^2e^{-c^2k^2t})\sin kx = -c^2k^2e^{-c^2k^2t}\sin kx$

$\dfrac{du}{dx} = e^{-c^2k^2t}(k\cos kx)$

$\dfrac{d^2u}{dx^2} = e^{-c^2k^2t}(-k^2\sin kx) = -k^2e^{-c^2k^2t}\sin kx$

$\Rightarrow u_t = c^2u_{xx}$

b. Writing exercise; responses will vary.

55. 1.848; at $\left(-\dfrac{\pi}{8}, \dfrac{\pi}{8}\right)$

CHAPTER 9   Section 1

1. $y = x^3 + \dfrac{5}{2}x^2 - 6x + C$

3. $y = Ce^{3x}$

5. $y = -\ln(C - x)$

7. $y = \pm\sqrt{x^2 + C}$

9. $y = \left(\dfrac{1}{3}x^{3/2} + C\right)^2$

11. $y = C|x - 1|$

13. $y = -3 + Ce^{[-(2x-5)^{-5}]/10}$

15. $x = \dfrac{Ce^{t/2}}{(2t+1)^{1/4}}$

17. $y = \ln(xe^x - e^x + C)$

19. $y = Ce^{t(\ln t - 1)/2}$

21. $y = \dfrac{1}{5}(e^{5x} + 4)$

23. $y = \left(\dfrac{3}{2}x^2 + 21\right)^{1/3}$

25. $y = \dfrac{6}{4(4-x)^{3/2} + 3}$

27. $y - 2\ln|y + 1| = \ln|t| + 2(1 - \ln 3)$

29. $\dfrac{dQ}{dt} = 0.07Q$ where $Q(t) =$ amount of investment at time t

31. $\dfrac{dQ}{dt} = kQ$ where $Q(t) =$ number of bacteria at time t

$k =$ constant of proportionality

33. $\dfrac{dP}{dt} = 500$ where $P(t) =$ number of people at time t

35. $\dfrac{dT}{dt} = k(T - T_m)$ where $T(t) =$ temperature of the object at time t

$T_m =$ temperature of the medium

$k =$ constant of proportionality

37. $\dfrac{dR}{dt} = k(F - R)$ where $R(t) =$ number of facts recalled at time t

$F =$ total number of facts

$k =$ constant of proportionality

39. $\dfrac{dP}{dt} = kP(C - P)$ where $P(t) =$ number of people implicated at time t

$C =$ number of people involved

$k =$ constant of proportionality

41. $\dfrac{dy}{dx} = \dfrac{d}{dx}(Ce^{kx}) = kCe^{kx} = ky$

43. $\dfrac{dy}{dx} = C_1 e^x + C_2(xe^x + e^x)$

$\dfrac{d^2y}{dx^2} = C_1 e^x + C_2(xe^x + 2e^x)$

$\dfrac{d^2y}{dx^2} - 2\dfrac{dy}{dx} + y$

$= [C_1 e^x + C_2(xe^x + 2e^x)] - 2[C_1 e^x + C_2(xe^x + e^x)]$
$\quad + [C_1 e^x + C_2 xe^x] = 0$

45. 8.35 days

47. 14 months

49. About \$22,857

51. $\dfrac{dp}{dt} = k(1 - p), \quad p(8) = 0.05$

$p(t) = 1 - (0.95)^{t/8}$

53. About \$1,252,608

55. With S in thousands of units

a. $S' = k(10 - S)$ and $S(4) = 1$, $S(6) = 2$,
$S(p) = 10 - 9e^{0.0589(4-p)}$

b. 2,889 units

c. \$14

57. $\dfrac{dP}{dt} = kP(C - P)$

$\dfrac{d^2P}{dt^2} = k\dfrac{dP}{dt}(C - P) - kP\dfrac{dP}{dt}$

$\quad = k(C - 2P)\dfrac{dP}{dt}$

ANSWERS

$$\frac{d^2P}{dt^2} = 0 \Leftrightarrow \frac{dP}{dt} = 0 \quad \text{or} \quad P = \frac{C}{2}$$

We want the maximum of $\frac{dP}{dt}$, so we eliminate

$\frac{dP}{dt} = 0$ and check $P = \frac{C}{2}$.

$\frac{d^2P}{dt^2} = k(kP(C - P))(C - 2P)$, where $k > 0$, $P \geq 0$

Assume $P < \frac{C}{2}$; then $\frac{d^2P}{dt^2} > 0$.

Assume $P > \frac{C}{2}$ and $P < C$; then $\frac{d^2P}{dt^2} < 0$.

Thus, $\frac{dP}{dt}$ has a local maximum at $P = \frac{C}{2}$.

59. 45

61. $P(t) = \dfrac{10}{1 + 1.5e^{-0.363t}}$

$P(10) = 9.6$ million

63. a. Equation 2

b. $k = \dfrac{1}{5,000}$

c. $P(t) = \dfrac{5,000}{50 - t}$

d. $P(12) = 132$; $P(48) = 2,500$

e.

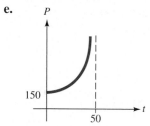

$P(t) \to \infty$ as $t \to 50$

65. a. $H(t) = 300e^{(12/\pi)\sin(\pi t/12)}$

b. 300 grams

67. a. $V(t) = \dfrac{1}{125}[(180^{1/3} - 6)t + 30]^3$

b. Approximately 85 weeks

69. Let $F = \dfrac{dQ}{dt} = kQ(B - Q) = k(QB - Q^2)$.

This function has a critical point at

$\frac{dF}{dQ} = k(B - 2Q) = 0 \quad \text{or} \quad Q = \frac{B}{2}.$

The rate of increase of the dependent variable grows until halfway toward the asymptote. Note that

$\frac{d^2F}{dQ^2} < 0$. The rate of increase is greatest at $Q = \dfrac{B}{2}$.

CHAPTER 9   Section 2

1. $y = \dfrac{x^2}{5} + \dfrac{C}{x^3}$

3. $y = \dfrac{1}{\sqrt{x}}(xe^x - e^x + C)$

5. $y = \dfrac{2}{x}\ln x + \dfrac{C}{x}$

7. $y = \dfrac{x}{2e^{2x}} + \dfrac{C}{xe^{2x}}$

9. $y = \dfrac{1}{1 + x}(Ce^x - 1)$

11. $y = \dfrac{1}{1 + t}\left(\dfrac{t^2}{2} + \dfrac{t^3}{3} + C\right)$

13. $y = (t + 1)(1 + Ce^{-t})$

15. $y = \dfrac{C - \cos x}{x}$

17. $y = \cos x[C - \ln(\cos x)]$

19. $y = 2x^3$

21. $y = 1 + (x - 2)e^{-x^2/2}$

23. $y = \dfrac{1}{x}(\ln x - 2)$

25. $y = \dfrac{5}{3} + Ce^{-3x}$

27. $y = 2 - \dfrac{C}{x + 1}$

29. $p(t) = -15e^{-t/5} - t + 17$

$p(4) = \$6.26$

31. a. $N' = 5 - 0.1N$; $N(0) = 60$

b. $N(t) = 10e^{-t/10} + 50$

c. $N(30) \approx 50.5$ units; no, $N(t)$ decreases from 60.

33. About \$260,578

35. a. The change in the value of the account is due to accrued interest, rV, plus deposits, D

$V(t) = \dfrac{D}{r}(e^{rt} - 1).$

b. About \$245,108.19

c. About \$11,488.68

37. About $3,860.03

39. About $39,346.93

41. With A in millions:

   a. $A' = 18 - \dfrac{18A}{5,000}$ and $A(0) = 0$

   b. $A(t) = 5,000(1 - e^{-9t/2,500})$

   c. $A(t) = 0.9(5,000)$ when $t \approx 640$ days

43. $p' = 9 - p - \sin t$ and $p(0) = 2$

$$p(t) = \frac{1}{2}(\cos t - \sin t + 18) - \frac{15}{2}e^{-t}$$

$$p(4) = \$8.91$$

45. **a.** $\dfrac{dA}{dt} = 3 - kA; A = 0$ when $t = 0$

$$A(t) = \frac{3}{k}(1 - e^{-kt})$$

   b. $A(8) \approx 5.32$ mg

     $\lim\limits_{t \to \infty} A(t) \approx 5.39$ mg

   c. The rate should be about 5.013 mg/hr.

47. **a.** $D(t) = D_0\, e^{-7t/50}$

   b. 2,961 mg

49. **a.** $P' = 1,200 - 0.02P$ and $P(0) = 5,000$

   b. $P(t) = 60,000 - 55,000e^{-t/50}$

   c. $P(21) \approx 23,862$ is less than half (48%) the voter pool, so she loses.

   d. $P' = 125t - 0.02P$ and $P(0) = 5,000$

     $P(t) = 317,500e^{-t/50} + 6,250t - 312,500$

     $P(21) \approx 27,362$ is more than half (55%) the voter pool, so she wins.

51. **a.** $V' = 100 - 0.005V$ and $V(0) = 588,792$

     $V(t) = 568,792e^{-t/200} + 20,000$

   b. $V(t) = \dfrac{1}{2}V(0)$ when $t \approx 146$ years

53. **a.** $\dfrac{dS}{dt} = \dfrac{-3}{40 - 2t}S, S(0) = 5$

$$S(t) = \frac{1}{8\sqrt{5}}(20 - t)^{3/2} \quad \text{for } 0 \le t \le 20$$

   b. ≈ 2.76 min

   c. 0 lbs

55. $y = -x - 1 + 2e^{x+1}$

57. **a.** $P' = 10 - 0.2P$ and $P(0) = 10$

   b. $P(t) = 50 - 40e^{-t/5}$

     $P(15) \approx 48$

   c. $10(0.8)^{15} + \displaystyle\int_0^{15} 10(0.8)^{15-t}\, dt \approx 44$

CHAPTER 9   Section 3

1. $xy^2 = C$, orthogonal family $y^2 - 2x^2 = C$

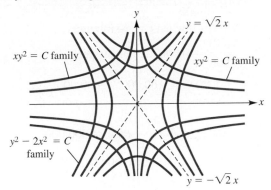

3. $y = Cx^2$, orthogonal family $2y^2 + x^2 = C$

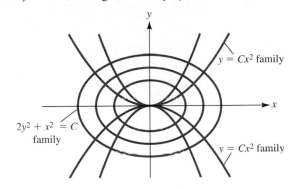

5. Money market: about $9,499.10
   Stock fund: about $30,274.65

7. $\dfrac{dP}{dx} - \dfrac{kP}{x}$

$$P(x) = Cx^k$$

9. **a.** $p(t) = \dfrac{49}{5}e^{5t/3} - \dfrac{9}{5}$

   b. Unstable

11. **a.** $k = \dfrac{1}{30}\ln 10$

   b. $D(t) = 50e^{-\frac{\ln 10}{30}t}$

     $S(t) = 5e^{\frac{\ln 10}{15}t}$

   c. About 23.2 units

13. $\dfrac{dA}{dt} = 6(0.16) - 6\left(\dfrac{A}{200}\right)$

$$A(t) = 32(1 - e^{-0.03t})$$

15. Time of death $\approx$ 12:40 P.M.

ANSWERS

17. a. $\dfrac{dN}{dt} = 0.2te^{-t/40}\dfrac{75}{3,000} - N\left(\dfrac{75}{3,000}\right)$

$N(t) = \dfrac{1}{400}t^2e^{-t/40}$

b. $N'(t) = 0$ when $t = 80$ minutes

$\dfrac{N(80)}{3,000} = 0.000722$ ounces per gallon

19. $P(t) = \dfrac{1,000}{3}t + \dfrac{100,000}{9} + \dfrac{800,000}{9}e^{0.03t}$

21. $\dfrac{dP}{dt} = 0.02P + 100e^{-t}$; $P(0) = 300,000$

$P(t) \approx -98.04e^{-t} + 300,098.04e^{0.02t}$

23. $y = x^{1/3}$

25. a. With populations in thousands,

$N(t) = \dfrac{2.5(250 - N_0)e^{-9.9t} + 250(N_0 - 2.5)}{(250 - N_0)e^{-9.9t} + (N_0 - 2.5)}$

b. If $N_0 = 5(5,000$ people), then as $t \to \infty$

$N(t) = \dfrac{2.5(250 - 5)e^{-9.9t} + 250(5 - 2.5)}{(250 - 5)e^{-9.9t} + (5 - 2.5)} \to 250$

or 250,000 people

c. If $N_0 = 1(1,000$ people), then $N(t) = 0$ when the numerator satisfies

$2.5(250 - 1)e^{-9.9t} + 250(1 - 2.5) = 0$

$t = 0.051$ days

d. In general, if $N_0 > m$, then $N(t) \to N_s$ as $t \to \infty$. However, if $N_0 < m$, then $N(t) = 0$ when

$t = \dfrac{1}{(N_s - m)k}\ln\left(\dfrac{m(N_s - N_0)}{N_s(m - N_0)}\right)$

27. a. $\dfrac{dP_1}{dt} = -\dfrac{15P_1}{7,000}$; $P_1(0) = 2,000$

$P_1(t) = 2,000e^{-3t/1,400}$

b. $\dfrac{dP_2}{dt} = \dfrac{15P_1}{7,000} - \dfrac{15P_2}{4,000}$; $P_2(0) = 0$

$a = \dfrac{15}{7,000}$ and $b = \dfrac{15}{4,000}$

$P_2(t) = \dfrac{8,000}{3}(e^{-3t/1,400} - e^{-3t/800})$

c. $P'_2(t) = 0$ when $t = 348.2$ hours.

$P_2(348.2) \approx 542$ pounds of pollutant, the maximum

29. a. $R_e = 26$; $S_e = 21$

b. $26\ln R - R = 63\ln S - 3S - 43.3$

31. a. If $A(t)$ is the amount of drug at time t, then

$\dfrac{dA}{dt} = 5 - \dfrac{4A}{800 + 2t}$ where $A(0) = 50$

$A(t) = \dfrac{5}{6}(800 + 2t) - \dfrac{296,000,000}{3(t + 400)^2}$

$C(t) = \dfrac{A(t)}{800 + 2t} = \dfrac{5}{6} - \dfrac{148,000,000}{3(t + 400)^3}$

b. About 12.41 hours

33. 6 minutes

35. About 97.27°C

37. a. $V = (9)^2h = 81h$ so

$81\dfrac{dh}{dt} = -k\sqrt{h}$ with $h'(0) = -\dfrac{1}{10}$

$81\left(\dfrac{1}{10}\right) = -k\sqrt{9}$ means that $k = \dfrac{27}{10}$

$h(t) = \left(3 - \dfrac{t}{60}\right)^2$

Drains when $h(t) = 0$; $t = 180$ minutes

b. Answers will vary.

39. a. Answers will vary.

b. Solve $x' = k(x - a)^2$

$x = \dfrac{a^2kt}{akt + 1}$

c. $x = \dfrac{ab[e^{akt} - e^{bkt}]}{ae^{akt} - be^{bkt}}$

CHAPTER 9   Section 4

1.

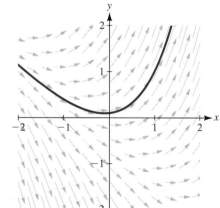

3.

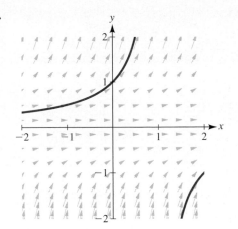

5.

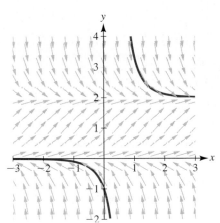

7.

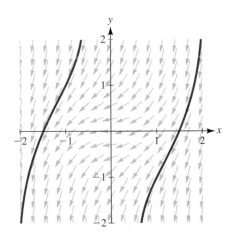

9.

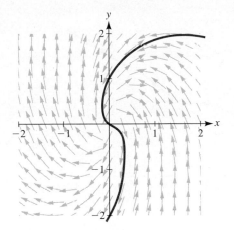

11.

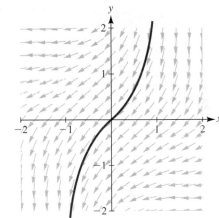

13.

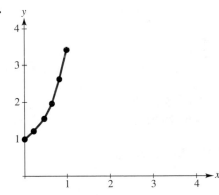

15.

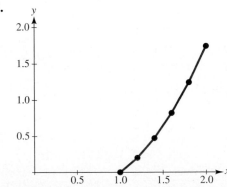

ANSWERS

17.

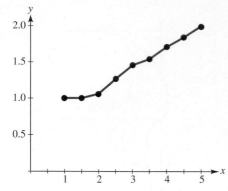

19.

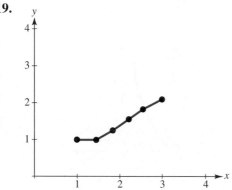

21. $y(1)$ is approximately 5.9728.

23. $y(1)$ is approximately 0.666546.

25. $y(2)$ is approximately 2.77973.

27. $L(4)$ is approximately 3.4526.
$n = 6, h = 0.5$

29. About 870 units will be produced.

31. **a.** At time $t = 2$, the fish will be about 4.9415 inches long.
     b. $L(t) = 5.3 - 5.1e^{-1.03t}$
        $L(2) = 4.65$ inches

CHAPTER 9   Section 5

1. $y_0 = 1$
   $y_1 = 3$
   $y_2 = 9$
   $y_3 = 27$
   $y_4 = 81$

3. $y_0 = y_1 = y_2 = y_3 = y_4 = 1$

5. $y_0 = 1$
   $y_1 = 1$
   $y_2 = 2$

   $y_3 = 3$
   $y_4 = 5$

7. $y_0 = 1$
   $y_1 = \dfrac{1}{3}$
   $y_2 = \dfrac{1}{7}$
   $y_3 = \dfrac{1}{15}$
   $y_4 = \dfrac{1}{31}$

9. $A = 1, B = -1$

11. $y_n = (-1)^n$

13. $y_n = 2 - \dfrac{1}{2^{n-1}}$

15. $y_n = 2\left(\dfrac{9}{8}\right)^n - 2$

17. $y_n = \dfrac{11^n}{10^{n-1}}$

19. $y_n = (1.05)y_{n-1} + 50, y_0 = 1{,}000$
    $y_n = 1{,}000[2(1.05)^n - 1]$

21. **a.** $C_n = 0.75\,C_{n-1} + 50, C_0 = 300$
       $C_n = (0.75)^n\,100 + 200$
     b. 205; 200

23. $M = \dfrac{iD}{1 - (1 + i)^{-tk}}$, with $i = \dfrac{r}{k}$

25. \$446.08

27. $A = \$180.63$

29. **a.** $C_n = C_{n-1} + 18 - \left(\dfrac{C_{n-1}}{5{,}000}\right)18, C_0 = 0$
       $C_n = 5{,}000\left(1 - \left(\dfrac{4{,}982}{5{,}000}\right)^n\right)$
     b. About 638.46 years
     c. $t \approx 639.61$ years; very similar times
     d. Writing exercise; responses will vary

31. **a.** $A_n = 1.07A_{n-1} + 25{,}000$
       $A_n = \dfrac{25{,}000}{0.07}(1.07^{n+1} - 1)$
     b. At age 40 (after 19 years)
     c. \$283,389.88

33. **a.** $p_e = 0.825$
     b. $p_n = -3\,p_{n-1} + 3.3; p_0 = 1$
       $p_n = 0.175(-3)^n + 0.825$

c.

n	p_n	S_n	D_n
1	0.3	1	1
2	2.4	−1.1	−1.1
3	−3.9	5.2	5.2

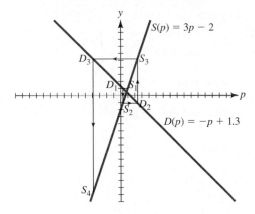

35. a. $p_e = 2$

b. $p_n = -\dfrac{1}{2}p_{n-1} + 3$

$p_n = (-0.5)^n + 2$

c.

n	p_n	S_n	D_n
1	1.5	4	4
2	2.25	2.5	2.5
3	1.875	3.25	3.25

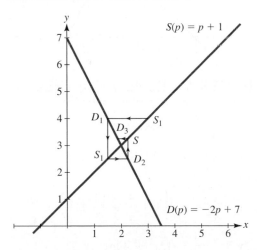

37. $I_n = \left(I_0 - \dfrac{h}{c}\right)(1 + c)^n + \dfrac{h}{c}$

39. a. $S_{n+1} - S_n = k(P - S_n) - 0.2S_n$

where $P = 10{,}000,\ S_0 = 500,\ S_1 = 1{,}100,\ k = \dfrac{7}{95}$

b. $S_n = \left(0.8 - \dfrac{7}{95}\right)^n 500$

$\qquad + \left[\dfrac{1 - (0.8 - \frac{7}{95})^n}{1 - (0.8 - \frac{7}{95})}\right]\dfrac{7}{95}(10{,}000)$

$\lim\limits_{n\to\infty} S_n = 2{,}692.3$

41. a. $25{,}000(1.08^n + 1)$

b. $h = 4{,}000$

43. a. $D_n = 0.94D_{n-1}$

$D_n = (0.94)^n D_0$

b. Approximately 71.4 mg

45. a. Suppose $P_m = P_{m+1} = L$ for some m.

Then $P_{m+2} = P_{m+1} + kP_{m+1}\left(1 - \dfrac{1}{K}P_{m+1}\right)$

$\qquad = P_m + kP_m\left(1 - \dfrac{1}{K}P_m\right)$

$\qquad = L$

Thus, once two consecutive values are the same, all the rest of the values will be the same.

b. $K = 105,\ k = \dfrac{21}{110},\ L = 105$

c. $K = 110,\ k = \ln\dfrac{5}{6},\ L = K = 110$

The values are similar and $L = K$ in both cases.

47. P_n in thousands

a. $P_n = \dfrac{1{,}000}{n}\left[\dfrac{1 - 1.05^{-n}}{0.05}\right]$

b. $A = \$30{,}242.59$

CHAPTER 9   Checkup

1. a. $y = C\sqrt{x^2 + 1}$

b. $y = \dfrac{1}{2}x^4 + Cx^2$

c. $y = -\ln(|x + 1 - Ce^x|) + x$

2. a. $y = \sqrt{\dfrac{4 + 5x}{x}}$

b. $y = \dfrac{5}{2}\left(x - \dfrac{1}{x}\right)$

c. $y = \dfrac{1}{3}\left[x^2 + 1 - \dfrac{1}{\sqrt{x^2 + 1}}\right]$

ANSWERS

3. **a.** $y_n = -\left(-\dfrac{1}{2}\right)^n$

    b. $y_n = 2\left(\dfrac{4}{3}\right)^n$

    c. $y_n = 2\left(1 - \left(\dfrac{5}{4}\right)^n\right)$

4. **a.**

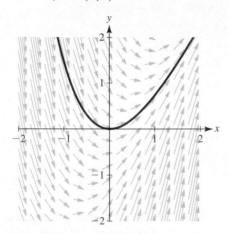

    b.

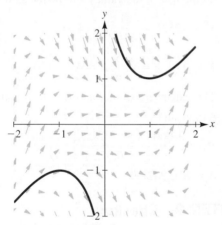

    c.

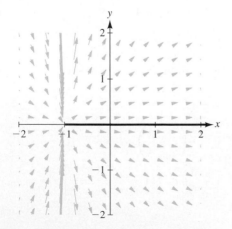

5. $\dfrac{dP}{dt} = -0.11p$

6. $\dfrac{dP}{dt} = k[S(p) - D(p)]$

7. $A_0 \approx 2{,}546$ mg

8. $t = 16$ years

9. $\dfrac{dA}{dt} = 8 - \dfrac{A(t)}{200 + 3t}$

$$S(t) = 400 + 6t - 325\left[\frac{200}{200 + 3t}\right]^{1/3}$$

There are about 760.54 pounds of salt when the tank is full.

10. **a.** $p_e = 10$

    b. $p_n = -\dfrac{1}{2}p_{n-1} + 15$

$$p_n = 10\left(1 + \left(-\frac{1}{2}\right)^n\right)$$

    c.

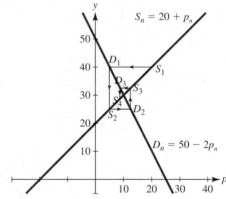

CHAPTER 9   Review Exercises

1. $y = \dfrac{1}{4}x^4 - x^3 + 5x + C$

3. $y = 80 + Ce^{-kx}$

5. $y = -\ln(-e^x - C)$

7. $y = \dfrac{C}{x^4} - e^{-x}\left[1 + \dfrac{4}{x} + \dfrac{12}{x^2} + \dfrac{24}{x^3} + \dfrac{24}{x^4}\right]$

9. $y = \dfrac{1}{(x+1)^2}[2x^2 + 4\ln(x+2) + C]$

11. $y = \pm\sqrt{(\ln x)^2 + C}$

13. $y = x^5 - x^3 - 2x + 6$

15. $y = 100e^{0.06x}$

17. $y = -\dfrac{1}{2}x^3 - \dfrac{7}{2}x^5$

19. $y = (4 + x)e^{x^2/2}$

21. $y = -4x^{x/2}e^{(1-x)/2}$

23. $y = 2e^{1 - \sqrt{1-x^2}}$

25. $y_n = 1 + 2n$

27. $y_n = 1 + 3n$

29. $y_n = \dfrac{1}{2}(-1 + 5^{1+n})$

31.

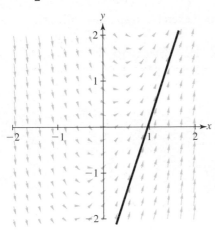

33.

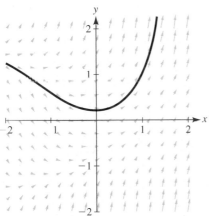

35.

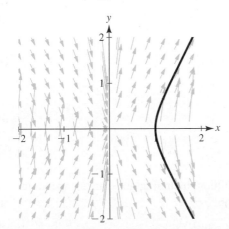

37. $x = 0.1 \quad y = 0$
$x = 0.2 \quad y = 0.03$
$x = 0.3 \quad y = 0.087$
$x = 0.4 \quad y = 0.1683$
$x = 0.5 \quad y = 0.27147$

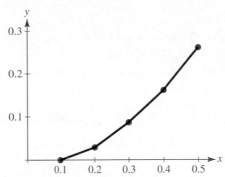

39. $x = 1.2 \quad y = 1.4$
$x - 1.4 \quad y = 1.91333$
$x = 1.6 \quad y = 2.56933$
$x = 1.8 \quad y = 3.40437$

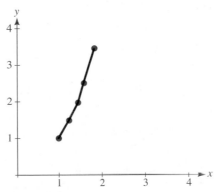

41. $y(1)$ is approximately 19.7642.

43. $y(2)$ is approximately 678.6282.

45. $\dfrac{dN}{dt} = k(A - N), N(0) = 0$
$N(t) = A(1 - e^{-kt})$

47. About 81.20 lb

49. $-\dfrac{V}{A} \ln\left(1 - \dfrac{L}{c_0}\right)$ seconds

51. a. $\dfrac{dC}{dt} = R - kC, k > 0$

$C(t) = \dfrac{1}{k}(R - (R - C_0 k)e^{-kt})$

b. $\lim\limits_{t \to \infty} C(t) = \dfrac{R}{k}$

c. $T = -\dfrac{1}{k} \ln\left(\dfrac{1}{2}\left(1 + \dfrac{R}{R - C_0 k}\right)\right)$

53. $E_n = 1.05E_{n-1}, E_0 = 100$

$E_n = (1.05)^n 100$

55. a. $p' = 0.02[(3 + 7e^{-t}) - (2 + p)]$ and $p(0) = 4$

$p(t) = \dfrac{22}{7}e^{-t/50} - \dfrac{1}{7}e^{-t} + 1$

b. $3.79

c. $p'(t) = 0$ when $t = 0.838$

$p_{max} = p(0.838) = 4.03$;

$S(0.838) = 6.03$ and $D(0.838) = 6.03$

d.

As $t \to \infty, p(t) = \dfrac{22}{7}e^{-t/50} - \dfrac{1}{7}e^{-t} + 1 \to 0 - 0 + 1 = 1$;

$D(t) = 3 + 7e^{-t} \to 3$ and $S(t) = 2 + p \to 2 + 1 = 3$

57. a. $p_n = 0.3p_{n-1}$

$p_n = (0.3)^n p_0$

b. The likelihood that a defective unit obtains its defect at some time during production equals 1.

$p_n = 0.7(0.3)^n$

$p_3 = 0.0189$

59. a. Since $P_0 > 0, a > 0, b > 0$,

then $P_1 = \dfrac{bP_0}{a + P_0} > 0$

likewise $P_2 > 0, P_3 > 0, \ldots, P_{n-1} > 0, \ldots$

Since $P_{n-1} > 0$,

then $a + P_{n-1} > a$

Then $P_n = \dfrac{bP_{n-1}}{a + P_{n-1}} < \dfrac{b}{a}P_{n-1}$

$< \left(\dfrac{b}{a}\right)^2 P_{n-2}$

$< \ldots < \left(\dfrac{b}{a}\right)^n P_0$

b. Assume $a > b$. Then

$\lim_{n \to \infty} P_n < \lim_{n \to \infty} \left[\left(\dfrac{b}{a}\right)^n P_0\right] = 0$

Hence, $\lim_{n \to \infty} P_n = 0$.

c. By assumption

$x = \lim_{n \to \infty} P_n = \lim_{n \to \infty} P_{n-1} < \infty$

Then x satisfies $x = \dfrac{bx}{a + x}$

or $x = 0$ (which is discarded)

or $x = b - a$. So, $\lim_{n \to \infty} P_n = b - a$.

61. a. Each month you have all the pairs of rabbits from the previous month, R_{n-1}, plus a pair of rabbits for each pair of rabbits that is at least 2 months old, so $R_n = R_{n-1} + R_{n-2}$.

b. $\dfrac{1 + \sqrt{5}}{2}, \dfrac{1 - \sqrt{5}}{2}$

c. $A = \dfrac{1}{\sqrt{5}}$ $B = -\dfrac{1}{\sqrt{5}}$

d. Writing exercise; responses will vary.

CHAPTER 10   Section 1

1. $\displaystyle\sum_{n=1}^{\infty} \dfrac{1}{3^n}$

3. $\displaystyle\sum_{n=2}^{\infty} \dfrac{n-1}{n}$

5. $\displaystyle\sum_{n=1}^{\infty} (-1)^{n+1} \dfrac{n^2}{n+1}$

7. $\dfrac{15}{16}$

9. $-\dfrac{7}{12}$

11. 5

13. 3

15. Diverges

17. $\dfrac{3}{20}$

19. 45

21. $\dfrac{3}{16}$

23. 100

25. 10

27. 4

29. 1

31. $\dfrac{1}{2}$

33. $\dfrac{4}{9}$

35. $\dfrac{25}{99}$

37. $625 billion, including government spending

39. $25,503

41. $\dfrac{A}{1+r} + \dfrac{A}{(1+r)^2} + \cdots = \displaystyle\sum_{n=1}^{\infty} \dfrac{A}{(1+r)^n} = \dfrac{A}{r}$

43. 2,037

45. $\dfrac{d(1-q)}{q}$

47. $\dfrac{20}{\sqrt{e}-1}$ units

49. **a.** $1 + 2\sqrt{0.36} + 2(0.36) + 2(0.36)^{3/2} + \cdots = 4$

  b. $\dfrac{\sqrt{H}}{4}(1 + 2\sqrt{r} + 2r + 2r^{3/2} + \cdots)$

  $= \dfrac{\sqrt{H}}{4}\left(1 + 2\displaystyle\sum_{n=1}^{\infty} r^{n/2}\right) = \dfrac{\sqrt{H}}{4}\left(\dfrac{1+\sqrt{r}}{1-\sqrt{r}}\right)$

  c. 15 feet

51. 856.52125

53. **a.** When $r = 1$, $S_n = na$. Hence $\lim\limits_{n\to\infty} S_n$ is not finite if $a \neq 0$.

  b. When $r = -1$, S_n, odd $= a$ and S_n, even $= 0$. Hence $\lim\limits_{n\to\infty} S_n$ does not exist.

55. 8 hours

CHAPTER 10 Section 2

1. $\lim\limits_{n\to\infty} a_n = 2 \neq 0$

3. $\lim\limits_{n\to\infty} a_n$ does not exist

5. Diverges

7. Diverges

9. Diverges

11. Converges

13. Converges

15. Diverges

17. Diverges

19. Converges

21. Diverges

23. Diverges

25. Converges

27. Diverges

29. Converges

31. Converges

33. Converges

35. Diverges

37. $\displaystyle\sum_{k=1}^{\infty} \dfrac{1}{2^k}, \sum_{k=1}^{\infty} \dfrac{1}{k^2}$

39. If $p > 1$, then $\displaystyle\int_1^{\infty} \dfrac{dx}{(1+2x)^p} = \dfrac{3^{1-p}}{2(p-1)d}$.

  If $p \leq 1$, then $\displaystyle\int_1^{\infty} \dfrac{dx}{(1+2x)^p}$ does not converge.

41. $c < 1$

43. $\displaystyle\sum_{n=0}^{\infty} \dfrac{1}{3n+1} > \sum_{n=0}^{\infty} \dfrac{1}{3n+3}$

  $= \dfrac{1}{3}\displaystyle\sum_{n=0}^{\infty} \dfrac{1}{n+1} = \dfrac{1}{3}\sum_{m=1}^{\infty} \dfrac{1}{m}$

  and this last series is the divergent harmonic series.

45. As in Example 10.2.2, determining the breaking strength of the first beam will set a record minimum breaking strain and also break the beam. The next beam will, with probability $\dfrac{1}{2}$, set a new record minimum breaking strain and also break the beam. The total number of broken beams is the number of new record minimum breaking strains, or

  $1 + \dfrac{1}{2} + \dfrac{1}{3} + \cdots + \dfrac{1}{100} \approx 5.19$

47. An infinite amount of time

49. Compare to $\displaystyle\sum_{k=1}^{\infty} \dfrac{1}{k^{2/3}}$; therefore it diverges.

51. Compare to $\displaystyle\sum_{k=1}^{\infty} \left(\dfrac{2}{5}\right)^k$; therefore it converges.

53. Diverges, because $\displaystyle\sum_{n=3}^{\infty} \dfrac{1}{(\ln n)^2} > \sum_{n=3}^{\infty} \dfrac{1}{n \ln n}$, which diverges by the integral test.

CHAPTER 10 Section 3

1. $R = \dfrac{1}{5}$; the interval of absolute convergence is $-\dfrac{1}{5} < x < \dfrac{1}{5}$.

3. $R = 1$; the interval of absolute convergence is $-1 < x < 1$.

5. $R = \dfrac{1}{4}$; the interval of absolute convergence is $-\dfrac{1}{4} < x < \dfrac{1}{4}$.

7. $R = \infty$; the interval of absolute convergence is all real x.

9. $f(x) = \sum_{n=1}^{\infty} (-1)^{n+1} x^n$; the interval of absolute convergence is $-1 < x < 1$.

11. $f(x) = \sum_{n=1}^{\infty} x^{2n}$; the interval of absolute convergence is $-1 < x < 1$.

13. $f(x) = \ln 2 + \sum_{n=1}^{\infty} \frac{(-1)^{n+1}}{n\,2^n} x^n$; the interval of absolute convergence is $-2 < x < 2$.

15. $f(x) = \sum_{n=0}^{\infty} \frac{3^n}{n!} x^n$

17. $f(x) = \sum_{n=0}^{\infty} \frac{1}{(2n)!} x^{2n}$

19. $\sum_{n=0}^{\infty} \frac{(-1)^n e^2 (x+1)^n}{n!}$

21. a. $f(x) = 3x^2 \sum_{n=0}^{\infty} \frac{x^{3n}}{n!} = \sum_{n=0}^{\infty} \frac{3}{n!} x^{3n+2}$

b. $f(x) = \frac{d}{dx}\left(\sum_{n=0}^{\infty} \frac{x^{3n}}{n!} \right) = \sum_{n=1}^{\infty} \frac{3x^{3n-1}}{(n-1)!}$

$= \sum_{n=0}^{\infty} \frac{3}{n!} x^{3n+2}$

23. $f(x) = \sum_{n=1}^{\infty} \frac{n(n+1)}{2} x^{n-1}$

25. $\sum_{n=0}^{\infty} \frac{(-1)^n}{(2n+3)n!} x^{2n+3}$

27. 1.0955

29. -0.3565

31. 0.6065

33. 0.2970

35. 0.4854

37.

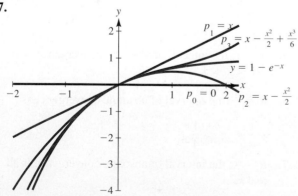

39.

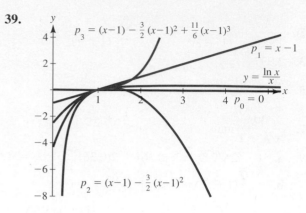

41. Approximately \$1,862

43. 13,548

45. a. $c = 0$: $P_{ave} = 2$

$c = 5$: $P_{ave} \approx 0.98280$

$c = 10$: $P_{ave} \approx 0.99981$

b. Using $c = 5$ yielded the best approximation.

47. a. $f(x) = \cos x$

$f^{(2n)}(x) = (-1)^n \cos x \qquad f^{(2n)}(0) = (-1)^n$

$f^{(2n+1)}(x) = (-1)^n \sin x \qquad f^{(2n+1)}(0) = 0$

so

$$\cos x = 1 - \frac{x^2}{2!} + \frac{x^4}{4!} - \cdots = \sum_{n=0}^{\infty} \frac{(-1)^n x^{2n}}{(2n)!}$$

b. $\cos x = \dfrac{d}{dx}\left[\sum_{n=0}^{\infty} \frac{(-1)^n x^{2n+1}}{(2n+1)!} \right]$

$= \sum_{n=0}^{\infty} \frac{(-1)^n (2n+1) x^{2n}}{(2n+1)!}$

$= \sum_{n=0}^{\infty} \frac{(-1)^n x^{2n}}{(2n)!}$

49. a. $\cos(x^2) = \sum_{n=0}^{\infty} \frac{(-1)^n (x^2)^{2n}}{(2n)!}$

$= \sum_{n=0}^{\infty} \frac{(-1)^n x^{4n}}{(2n)!}$

b. $\displaystyle\int_0^1 \cos(x^2)\,dx = \sum_{n=0}^{\infty} \frac{(-1)^n}{(2n)!} \int_0^1 x^{4n}\,dx$

$= \sum_{n=0}^{\infty} \frac{(-1)^n}{(2n)!} \frac{1}{4n+1} \approx 0.9045$

CHAPTER 10 Checkup

1. a. Converges to $\dfrac{1}{8}$

 b. Diverges

2. a. p-series test with $p = \dfrac{3}{2}$

 b. Ratio test $L = \lim\limits_{n\to\infty}\left|\dfrac{a_{n+1}}{a_n}\right| = \lim\limits_{n\to\infty}\dfrac{3}{n+1} = 0$

3. a. Divergence test: $\lim\limits_{n\to\infty}\dfrac{n(n^2+1)}{100n^3+9} = \dfrac{1}{100} \neq 0$

 Therefore, diverges.

 b. $\sum\limits_{n=3}^{\infty}\dfrac{\ln n}{\sqrt{n}} > \sum\limits_{n=3}^{\infty}\dfrac{1}{\sqrt{n}}$ which diverges by p-series test. Therefore, diverges.

4. a. Diverges **b.** Converges
 c. Converges **d.** Converges

5. a. $-1 < x < 1$

 b. All real x

6. a. $\sum\limits_{n=0}^{\infty}(-1)^n x^{2n+1}$

 b. $\sum\limits_{n=0}^{\infty}\dfrac{5}{2}\left(-\dfrac{3}{2}\right)^n x^n$

7. a. $\sum\limits_{n=0}^{\infty}\dfrac{2x^{2n}}{(2n)!}$

 b. $\sum\limits_{n=1}^{\infty}\dfrac{(-1)^{n+1}x^{2n}}{n}$

8. $\dfrac{25}{e^{1/3}-1} \approx 63.2$

9. \$122,517

10. Using $n = 10,000$, the estimate is 0.7854.

CHAPTER 10 Review Exercises

1. $\dfrac{1}{7}$

3. $\dfrac{65}{6}$

5. $\dfrac{\sqrt{e}}{\sqrt{e}-1}$

7. Diverges

9. Diverges

11. Diverges

13. Diverges

15. Converges

17. Converges

19. $-\dfrac{\sqrt{3}}{2} < x < \dfrac{\sqrt{3}}{2}$

21. $-\dfrac{1}{e} < x < \dfrac{1}{e}$

23. $\sum\limits_{n=0}^{\infty}\dfrac{2}{(2n+1)!}x^{2n+1}$

25. $\sum\limits_{n=1}^{\infty}\dfrac{(-1)^n(2^n-1)x^n}{n}$

27. $\sum\limits_{n=0}^{\infty}(-1)^n(n+1)(x-2)^n$

29. $x - 1 + \sum\limits_{n=2}^{\infty}\dfrac{(-1)^n}{n(n-1)}(x-1)^n$

31. 0.9487

33. a.

 b. The integral is approximately $\dfrac{25}{48}$.

35. $\dfrac{25,417}{495}$

37. 10 feet

39. \$102,521

41. $k = \ln\dfrac{19}{14} \approx 0.305$

43. 4,014 in 20 years, 4,633 in the long run

45. a. $\sum\limits_{k=1}^{\infty}\dfrac{T}{k(k+1)}$ is the number of different words; it should sum to T.

 b. $\sum\limits_{k=1}^{\infty}\dfrac{T}{k(k+1)} = T$

 c. Writing exercise; responses will vary.

ANSWERS

47. a. $t = \dfrac{\lambda}{\ln 2}\left[\sqrt{2\dfrac{S(t) - S(0)}{R(t)} + 1} - 1\right]$

b. $t = 5.587 \times 10^8$ years

c. Writing exercise; responses will vary.

49. a. The total time to collision is

$$\frac{1,000}{90} + \frac{1,000}{90}\left(1 - \frac{2}{3}\right) + \frac{1,000}{90}\left(1 - \frac{2}{3}\right)^2 + \cdots$$

$$= \frac{50}{3}\text{ seconds.}$$

The distance the bee travels is

$$\frac{50}{3} \cdot 60 = 1,000\text{ feet.}$$

b. The trains will collide in

$$\frac{1,000\text{ ft}}{60\text{ ft/sec}} = \frac{50}{3}\text{ seconds.}$$

The distance the bee travels is

$$\frac{50}{3} \cdot 60 = 1,000\text{ feet.}$$

c. The trains travel for $\dfrac{1,000 - 180}{60} = \dfrac{820}{60}$

seconds before starting to decelerate. They then travel for another 6 seconds before stopping with the bee cradled gently between their headlights. Thus, the total travel time for the bee is $\dfrac{820}{60} + 6 = \dfrac{1,180}{60}$ seconds, and in that time, it travels $\left(\dfrac{1,180}{60}\right)(60) = 1,180$ feet.

CHAPTER 11   Section 1

1. Discrete

3. Continuous

5. Continuous

7. Discrete

9.

X	P(x)
0	$\frac{1}{8}$
1	$\frac{3}{8}$
2	$\frac{3}{8}$
3	$\frac{1}{8}$

11. {0, 1, 2}

13. {0, 1, 2}

15. a. {11, 12, 13, 14, 15, 16, 21, 22, 23, 24, 25, 26, 31, 32, 33, 34, 35, 36, 41, 42, 43, 44, 45, 46} (*Note:* The number on the ticket is listed first.)

b. 11, 13, 15, 22, 24, 26, 31, 33, 35, 42, 44, 46

c. 11, 22, 33, 44

17. a.

H H H H, $X = 0$	H T T H, $X = 2$
H H H T, $X = 1$	T H T H, $X = 2$
H H T H, $X = 1$	T T H H, $X = 2$
H T H H, $X = 1$	H T T T, $X = 3$
T H H H, $X = 1$	T H T T, $X = 3$
H H T T, $X = 2$	T T H T, $X = 3$
H T H T, $X = 2$	T T T H, $X = 3$
T H H T, $X = 2$	T T T T, $X = 4$

b. H T T T, T H T T, T T H T, T T T H

c. All outcomes

d. $P(X = 0) = \dfrac{1}{16}$, $P(X \le 3) = \dfrac{15}{16}$,

$P(X > 3) = \dfrac{1}{16}$

19. $E(X) = \dfrac{21}{8}$, $\text{Var}(X) = \dfrac{143}{64}$, $\sigma(X) = \dfrac{\sqrt{143}}{8}$

21. $E(X) = \dfrac{20}{7}$, $\text{Var}(X) = \dfrac{160}{49}$, $\sigma(X) = \dfrac{4}{7}\sqrt{10}$

23.

$E(X) = 2$

$\text{Var}(X) = 1$

$\sigma(X) = 1$

25. Up 16%

27. a. About 0.00050

b. Page 2,000

c. About 0.98757

29. a. 0.05236

  b. 0.30431

  c. 0.74805

  d. About 14.3 chips

31. a.

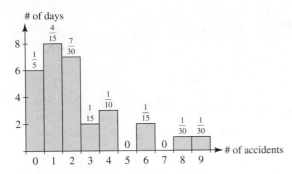

  b. $E(X) = \dfrac{23}{10}$

   We expect about two accidents to occur on any given day.

  c. $\text{Var}(X) = \dfrac{1{,}603}{300}$ $\sigma(X) = \dfrac{\sqrt{4{,}809}}{30}$

33. $E(X) = 0.096$

35. a. About 0.03397

  b. About 0.18463

  c. About 0.61271

  d. 25 students

37. About 0.13533

39. $20

41. $-\$0.027$ (loss of 2.7 cents)

43. a. $E(X) = pw + (1 - p)(-1)$
         $= pw + p - 1$

  b. The game is fair when $E(X) = 0$.
   $pw + p - 1 = 0$
   $pw = 1 - p$
   $w = \dfrac{1 - p}{p}$
   Fair odds are $\dfrac{1 - p}{p}$ to 1.

45. $\dfrac{19}{18}$ to 1

47. $P(X = 1) = 0.8$, $P(X = 2) = 0.16$,
   $P(X = 3) = 0.032$, $P(X = 4) = 0.0064$,
   $P(X = 5) = 0.00128$

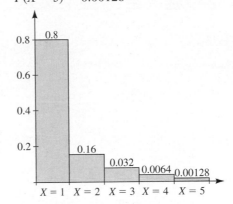

49. $E(X) = \displaystyle\sum_{n=1}^{\infty} np(1 - p)^{n-1}$

   $= p \displaystyle\sum_{n=1}^{\infty} n(1 - p)^{n-1}$

   $= p \dfrac{1}{(1 - (1 - p))^2}$

   $= \dfrac{1}{p}$

51. $\displaystyle\sum_{n=1}^{\infty} P(X = n) - \sum_{n=1}^{\infty} \dfrac{a}{n^b} = a \sum_{n=1}^{\infty} \dfrac{1}{n^b} = 1$

   Therefore, $a = \dfrac{1}{\zeta(b)}$ where $\zeta(b) = \displaystyle\sum_{n=1}^{\infty} \dfrac{1}{n^b}$.

CHAPTER 11   Section 2

1. Yes

3. No

5. Yes

7. Yes

9. No

11. Yes

13. Yes

15. No such number exists; $\displaystyle\int_{-\infty}^{+\infty} f(x)\, dx = 1$ for $k = \dfrac{5}{2}$,

   but then $f(1) = -\dfrac{1}{2}$.

17. No such number exists; $\displaystyle\int_{-\infty}^{+\infty} f(x)\, dx = 1$ for $k = \dfrac{3}{4}$,

   but then $f(x) < 0$ for $\dfrac{4}{3} < x < 2$.

ANSWERS

19. $k = -\dfrac{3}{2}$

21. $k = 1$

23. $\dfrac{e^2}{4}$

25. a. 1

   b. $\dfrac{1}{3}$

   c. $\dfrac{1}{3}$

27. a. 1

   b. $\dfrac{3}{16}$

   c. $\dfrac{9}{16}$

29. a. 1

   b. $\dfrac{7}{8}$

   c. $\dfrac{1}{8}$

31. a. 1

   b. $\dfrac{1}{e} - \dfrac{1}{e^4}$

   c. $1 - \dfrac{1}{e^4}$

33. $k = \dfrac{9}{4}$

35. $k = 3$

37. a. $\dfrac{5}{24}$

   b. $\dfrac{23}{96}$

39. a. About 0.11612

   b. About 0.06347

41. a. $\dfrac{9}{14}$

   b. $\dfrac{43}{70}$

   c. $\dfrac{13}{28}$

43. a. About 0.05772

   b. About 0.45119

   c. About 0.54881

45. a. $\dfrac{1}{e^2}$

   b. $\dfrac{1}{e^{1/4}} - \dfrac{1}{e^{5/4}} \approx 0.49230$

47. About 0.45816

49. a. $\dfrac{3}{5}$

   b. $\dfrac{3}{20}$

51. a. $f(t) = \begin{cases} \dfrac{1}{60} & \text{if } 0 \le t \le 60 \\ 0 & \text{otherwise} \end{cases}$

   b. $\dfrac{1}{4}$

   c. $\dfrac{1}{6}$

53. $\dfrac{1}{e} \approx 0.36788$

55. a. $1 - \dfrac{9}{4}\left(\dfrac{1}{e^{5/4}}\right) \approx 0.35536$

   b. $\dfrac{7}{2}\left(\dfrac{1}{e^{5/2}}\right) \approx 0.28730$

57. a. $k \approx 0.3662$

   b. $P(X \ge 5) \approx 0.1602$

59. a. About 0.33353

   b. About 0.66647

61. 20 seconds

63. $\dfrac{A + B}{2}$

65. a. $r(x) = \displaystyle\int_x^\infty f(x)\,dx = 1 - \int_0^x f(x)\,dx$

   b. e^{-200}

67. a.

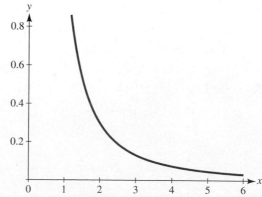

b. 0.06

c. Writing exercise; responses will vary.

CHAPTER 11   Section 3

1. $E(X) = \dfrac{7}{2}$, $\mathrm{Var}(X) = \dfrac{3}{4}$

3. $E(X) = \dfrac{4}{3}$, $\mathrm{Var}(X) = \dfrac{8}{9}$

5. $E(X) = \dfrac{3}{2}$, $\mathrm{Var}(X) = \dfrac{3}{4}$

7. $E(X) = \dfrac{1}{4}$, $\mathrm{Var}(X) = \dfrac{1}{16}$

9. $E(X) = \dfrac{2}{3}$, $\mathrm{Var}(X) = \dfrac{2}{63}$

11. $E(X) = 2$, $E(Y) = 1$

13. $E(X) = 1$, $E(Y) = 2$

15. $E(X) = 20$ months

17. $E(X) \approx 4.685$ years

19. a. $\dfrac{7}{30}$

   b. A: $\dfrac{24}{5}$ worker-hours, B: $\dfrac{16}{3}$ worker-hours

   c. About $193.07

21. $E(X) = 10$ minutes

23. 2 hours, $\mathrm{Var}(X) = 4$

25. a. $\lambda = \dfrac{1}{5}$

   b. About 0.32968

   c. About 0.24660

27. a. $A = b = -\ln 0.9$

   b. About 34.9%

   c. About 21.9 weeks

   d. About 9.5 weeks

   e. About 90.1 (weeks)^2; the variability is large compared to the expected value.

29. Patient in 107A: 4 days
   Patient in 107B: 3 days

31. $E(X) = \displaystyle\int_a^b \left(\dfrac{x}{b-a}\right) dx = \dfrac{x^2}{2(b-a)} \bigg|_{x=a}^{x=b}$

$= \dfrac{(b^2 - a^2)}{2(b-a)} = \dfrac{b+a}{2}$

33. $E(X) = \displaystyle\int_0^\infty \lambda x e^{-\lambda x}\, dx = \lim_{N\to\infty} \int_0^N \lambda x e^{-\lambda x}\, dx$

$= \lim_{N\to\infty}\left[-xe^{-\lambda x}\Big|_0^N + \int_0^N e^{-\lambda x}\, dx \right]$

$= \lim_{N\to\infty}\left(-xe^{-\lambda x} - \dfrac{1}{\lambda}e^{-\lambda x} \right)\bigg|_0^N = \dfrac{1}{\lambda}$

35. a. $f(x) = \begin{cases} 2e^{-2x} & \text{if } x \ge 0 \\ 0 & \text{otherwise} \end{cases}$

   b. $e^{-2} \approx 0.13534$

   c. $e^{-1/2} - e^{-1} \approx 0.23865$

37. a. $\dfrac{a\lambda}{a-1}$

   b. $\dfrac{a\lambda^2}{a-2} - \dfrac{a^2\lambda^2}{(a-1)^2}$

CHAPTER 11   Section 4

1. $E(X) = 0$, $\mathrm{Var}(X) = 4$, $\sigma = 2$

3. $E(X) = -1$, $\mathrm{Var}(X) = 3$, $\sigma = \sqrt{3}$

5. 0.8925

7. 1.0000

9. 0.6827

11. 1.00

13. 1.10

15. a. 0.0228

   b. 0.5

   c. 0.6826

   d. 0.0000

17. a. 5

   b. 3.5

19. a. 0.683 or 68.3%

   b. 0.954 or 95.4%

21. a. $3e^{-3} \approx 0.149$

   b. $\dfrac{9}{2}e^{-3} \approx 0.224$

   c. $\dfrac{27}{8}e^{-3} \approx 0.168$

   d. $\dfrac{3^8}{8!}e^{-3} \approx 0.008$

23.

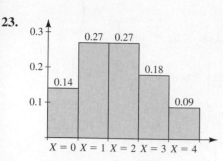

25. 0.383

27. **a.** 0.9544
b. 0.0013

29. About 0.0388 inch

31. 0.9759

33. Shop B

35. **a.** $\lambda = \ln 20 \approx 3$
b. 0.1005
c. ≈ 0.9165
d. 0.95
e. 3 hits

37. **a.** Roughly 68.3% of adult basset hounds have weight w in the range $45 \le w \le 55$ lb.
b. Roughly 95.4% of adult basset hounds have weight w in the range $40 \le w \le 60$ lb.
c. Roughly 99.7% of adult basset hounds have weight w in the range $35 \le w \le 65$ lb.

39. 11.51%

41. 44.2 years old

43. **a.** 7.64%
b. 17.78 oz

45. **a.** 7.64%
b. 103.25 mg/dL

47. **a.** 0.0668
b. 0.3538
c. 0.4554

49. **a.** 0.3472, 0.4599
b. 0.2047
c. 0.7953

51. **a.** 0.6912
b. Yes; there's less than a 16% chance of healing in 9 days or less, 0.1587
c. Yes, 0.0062
d. Writing exercise; responses will vary.

53. **a.** $\lambda = 6.1$
b. 0.0022

c. 0.0848
d. 0.4298

55. About 5% more likely

57. **a.** False; only true if mean $= 0$.
b. True; it is symmetric about 0.

59. **a.** $\int_{x=\mu-2\sigma}^{x=\mu+2\sigma} \frac{1}{\sigma\sqrt{2\pi}} e^{\frac{-(x-\mu)^2}{2\sigma^2}} \, dx$
$= \int_{-2}^{2} \frac{1}{\sqrt{2\pi}} e^{-x^2/2} \, dx \approx 0.954$
b. $\int_{x=\mu-3\sigma}^{x=\mu+3\sigma} \frac{1}{\sigma\sqrt{2\pi}} e^{-\frac{(x-\mu)^2}{2\sigma^2}} \, dx$
$= \int_{-3}^{3} \frac{1}{\sqrt{2\pi}} e^{-x^2/2} \, dx \approx 0.997$

61. 0.3414 by integration, table answer is $0.8413 - 0.5 = 0.3413$.

CHAPTER 11 Checkup

1.

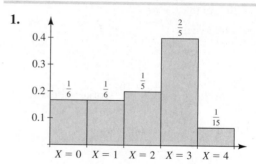

$E(X) = \frac{61}{30} \approx 2.0333$

$\text{Var}(X) = \frac{1{,}349}{900} \approx 1.4989$

$\sigma(X) = \frac{\sqrt{1{,}349}}{30} \approx 1.2243$

2. **a.** Yes
b. No

3. **a.** $e^{-2/3} - e^{-1} \approx 0.1455$
b. $e^{-1} \approx 0.3679$

4. $E(X) = 3$, $\text{Var}(X) = \frac{9}{5}$

5. **a.** 0.1977
b. 0.9026

6. **a.** 0.683
b. 0.954

7. $\frac{2}{3}$

8. 0.7738

9. a. $e^{-1.5} \approx 0.2231$

   b. 2 years; the life expectancy of those on the drug

10. 170 points

11. 0.5595

12. 3.3

13. a. True

   b. True

CHAPTER 11 Review Exercises

1.

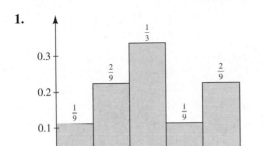

$$E(X) = \frac{28}{9} \approx 3.1111$$

$$\text{Var}(X) = \frac{134}{81} \approx 1.6543$$

$$\sigma(X) = \frac{\sqrt{134}}{9} \approx 1.2862$$

3. Discrete

5. Discrete

7. $P(2 \leq X \leq 7) = 0.8$
   $P(X \geq 5) = 0.6$

9. $P(X \leq 1) = 0.3125$
   $P(1 \leq X \leq 3) = 0.6875$

11. $E(X) = 1.2$, $\text{Var}(X) = 0.16$

13. $c = \frac{1}{16}$

15. 0.1587

17. 0.0004

19. Gain 90 cents per share

21. a. 15 children

   b. 15 children

   c. 0 children

23. $11.40

25. a. $1 - e^{-4} \approx 0.9817$

   b. $e^{-8} \approx 0.0003$

   c. 2.5 years

27. a. $1 - 6e^{-5} \approx 0.9596$

   b. 4 feet

29. a. $P(X = 10{,}000) = 0.98$
        $P(X = -90{,}000) = 0.02$

   b. $8,000

   c. $18,000

31. a. $\lambda = \dfrac{\ln 2}{10}$

   b. $1 - \dfrac{\sqrt{2}}{2}$

   c. $\dfrac{10}{\ln 2} \approx 14.2$ weeks

33. a. 1.0000

   b. 12.08 ounces

35. 0.4784; 39 students

37. a. 0.5

   b. 5

39. a. $\dfrac{32}{3}e^{-4} \approx 0.1954$

   b. $e^{-4} \approx 0.0183$

   c. $\dfrac{71}{3}e^{-4} \approx 0.4335$

41. a. $e^{-2.4} \approx 0.0907$

   b. $1 - e^{-1.2} \approx 0.6988$

   c. $e^{-0.1} \approx 0.9048$

43. a. $\displaystyle\int_{y=0}^{y=1}\int_{x=0}^{x=1} 0.4(2x + 3y)\, dx\, dy = 1$
        and for $0 \leq x \leq 1,\, 0 \leq y \leq 1,\, f(x, y) \geq 0$

   b. 0.275

   c. 0.072

   d. $\dfrac{1}{3} \approx 0.333$

45. a. $(e^{-5/4} - 1)(e^{-5/2} - 1) \approx 0.6549$

   b. $1 + e^{-4} - 2e^{-2} \approx 0.7476$

47. a. $\dfrac{2}{3}$

   b. $\dfrac{1}{2}$

   c. $\dfrac{5}{18}$

Appendix Section A.1

1. $1 < x \le 5$

3. $x > -5$

5.

7.

9. 4

11. 5

13. $-3 \le x \le 3$

15. $-6 \le x \le -2$

17. $x \le -7$ or $x \ge 3$

19. 125

21. 4

23. 4

25. $\dfrac{1}{2}$

27. $\dfrac{1}{2}$

29. $\dfrac{1}{4}$

31. 2

33. $\dfrac{1}{4}$

35. $n = 10$

37. $n = 1$

39. $n = 4$

41. $n = \dfrac{13}{5}$

43. $a^5 b^8 c^8$

45. $\dfrac{a^8 c^{12}}{b^4}$

47. $\dfrac{a^{10}}{b^2 c^{14}}$

49. $\dfrac{a^{18} b^{12}}{c^6}$

51. $\dfrac{1}{a^3 b^5 c^3} + \dfrac{1}{a^4 b c^3} + \dfrac{1}{a b c^4}$

53. $a^{-1} b^2 + a^2 b$

55. -2

57. $1{,}350 \sqrt[3]{900}$

59. $38\sqrt{2}$

61. $9\sqrt{6}$

63. $a^3 b^4 c^7$

65. $\dfrac{5b}{7a}$

67. $\dfrac{a\sqrt[3]{a^2 b^2}}{b^3 c^3}$

69. $\dfrac{1}{a^2 b c^6}$

71. $\dfrac{a^2 c^3}{b^4}$

73. $a - \sqrt{b}$

75. $\dfrac{a^5 b^3 \sqrt[3]{a^2 c}}{c^4}$

77. $x^4 (x - 4)$

79. $-25(x - 7)$

81. $2(x + 1)^2 (x - 2)^2 (7x - 2)$

83. $x^{-1/2}(6x + 1)$

85. $2(x + 3)$

87. $\dfrac{2(x + 3)^3 (5 - x)}{(1 - x)^3}$

89. $\dfrac{1}{5(\sqrt{3} + \sqrt{2})}$

91. $\dfrac{7(3 + \sqrt{3})}{6}$

93. $\dfrac{1}{3(\sqrt{5} - 2)}$

95. $\dfrac{5(\sqrt{5} - 1)}{4}$

97. $\sqrt{x + h} - \sqrt{x} = \dfrac{(\sqrt{x + h} - \sqrt{x})(\sqrt{x + h} + \sqrt{x})}{\sqrt{x + h} + \sqrt{x}}$

$$= \dfrac{x + h - x}{\sqrt{x + h} + \sqrt{x}}$$

$$= \dfrac{h}{\sqrt{x + h} + \sqrt{x}}$$

99. **a.** Surface area is approximately 5.212×10^8 km^2; mass of the atmosphere is 5.212×10^{18} kg.

b. 127,400 years

Appendix Section A.2

1. $3x^2 - 27x$
3. $x^2 - 5x - 14$
5. $-6x^2 + 26x - 28$
7. $x^3 + x^2 - 5x + 3$
9. $x^5 - 3x^4 - x^3 + 13x^2 - 18x + 8$
11. $\dfrac{2x^2 + 3x + 9}{x^2 - 9}$
13. $\dfrac{x^2}{x^2 + 2x - 3}$
15. $\dfrac{3}{2x^2 - 7x - 15}$
17. $-\dfrac{x + 10}{x^2 + x - 2}$
19. $\dfrac{7}{x^2 + 7x + 12}$
21. $\dfrac{-x + 3}{x + 3}$
23. -2
25. $\dfrac{x}{3x - 1}$
27. $-\dfrac{x^2 + x - 1}{3x - 1}$
29. $(x + 2)(x - 1)$
31. $(x - 3)(x - 4)$
33. $(x - 1)^2$
35. $(4x + 5)(4x - 5)$
37. $(x - 1)(x^2 + x + 1)$
39. $x^5(x + 1)(x - 1)$
41. $2x(x - 5)(x + 1)$
43. $(x + 4)(x - 3)$
45. $(2x + 5)(x - 3)$
47. $(x + 2)(x - 9)$
49. $2(2x + 1)(7x - 3)$
51. $x(x + 5)(x - 3)$
53. $(x + 3)(x^2 - 3x + 9)$
55. $x^2(x + 1)(x^2 - x + 1)$
57. $(3x + 1)(x + 2)^2$
59. $x = 4; x = -2$
61. $x = -5$
63. $x = 4; x = -4$
65. $x = -\dfrac{1}{2}; x = -1$
67. $x = -\dfrac{3}{2}$
69. $x = 1; x = -5$
71. $x = 1; x = -2$
73. $x = -1$
75. $x = 1; x = -3$
77. $x = \dfrac{1}{3}; x = \dfrac{3}{5}$
79. No real solutions
81. $x = \dfrac{-17 + \sqrt{385}}{12}; x = \dfrac{-17 - \sqrt{385}}{12}$
83. $x = -\dfrac{1}{2}; x = -1$
85. No real solutions
87. $x = -\dfrac{3}{2}$
89. $x = 3; y = 2$
91. $x = 4, y = 2$
93. $x = -7; y = -5$ and $x = 1, y = -1$

Appendix Section A.3

1. 0
3. $\dfrac{1}{2}$
5. $-\dfrac{1}{3}$
7. 0
9. 0
11. 0
13. 0
15. e^2

Appendix Section A.4

1. 34
3. 0
5. $\displaystyle\sum_{j=1}^{6} \dfrac{1}{j}$

ANSWERS

7. $\displaystyle\sum_{j=1}^{6} 2x_j$

9. $\displaystyle\sum_{j=1}^{8} (-1)^{j+1} j$

Appendix Review Exercises

1. $-2 \le x < 3$

3.

5.

7. 3

9. $2 \le x \le 4$

11. 243

13. 4

15. $16\sqrt[4]{2}$

17. 73

19. $\dfrac{3}{2}$

21. $n = \dfrac{7}{18}$

23. $n = -1$

25. 21

27. 95

29. $1 + \displaystyle\sum_{k=2}^{7} \dfrac{(-1)^k}{k}$

31. $x^2(x + 3)(x - 3)$

33. $x^4(x^6 + 4)(x^3 + 2)(x^3 - 2)$

35. $x(x - 1)^2$

37. $(x + 5)(x - 3)$

39. $(2x + 3)^2$

41. $(x + 1)(x - 1)(x + 3)$

43. $x = -4; x = 1$

45. $x = -7$

47. $x = -1; x = 2$

49. $x = -\dfrac{3}{7}; x = \dfrac{1}{2}$

51. No real solutions

53. $x = -2; x = \dfrac{1}{3}$

55. $x = -2; y = 1$

57. $x = 1, y = 2$ and $x = 15, y = -26$

59. $\dfrac{2}{7}$

61. 0

63. 1

INDEX

Index of Applications (continued)

Social Sciences Problems